D1070927

ADVANCES IN BUILDING TECHNOLOGY

Proceedings of the
International Conference on Advances in Building Technology

4-6 December 2002, Hong Kong, China

Volume I

Elsevier Science Internet Homepage - http://www.elsevier.com

Consult the Elsevier homepage for full catalogue information on all books, journals and electronic products and services.

Elsevier Titles of Related Interest

CHAN & TENG
ICASS '99, Advances in Steel Structures.
(2 Volume Set). *ISBN: 008-043015-5*

FRANGOPOL, COROTIS & RACKWITZ
Reliability and Optimization of Structural Systems. *ISBN: 008-042826-6*

FUKUMOTO
Structural Stability Design.
ISBN: 008-042263-2

HOLLAWAY & HEAD
Advanced Polymer Composites and Polymers in the Civil Infrastructure.
ISBN: 008-043661-7

KELLY & ZWEBEN
Comprehensive Composite Materials.
ISBN: 008-042993-9

KO & XU
Advances in Structural Dynamics.
(2 Volume Set). *ISBN: 008-043792-3*

LUNDQUIST, LETERRIER, SUNDERLAND & MÅNSON
Life Cycle Engineering of Plastics: Technology, Economy and the Environment
ISBN: 008-043886-5

MAKELAINEN
ICSAS'99, Int. Conf. on Light-Weight Steel and Aluminium Structures.
ISBN: 008-043014-7

USAMI & ITOH
Stability and Ductility of Steel Structures.
ISBN: 008-043320-0

VASILIEV & MOROZOV
Mechanics and Analysis of Composite Materials. *ISBN: 008-042702-2*

WANG, REDDY & LEE
Shear Deformable Beams and Plates.
ISBN: 008-043784-2

Related Journals

Free specimen copy gladly sent on request. Elsevier Science Ltd, The Boulevard, Langford Lane, Kidlington, Oxford, OX5 1GB, UK

Advances in Engineering Software
CAD
Composites Part A: Applied Science and Manufacturing
Composites Part B: Engineering
Composite Structures
Computer Methods in Applied Mechanics and Engineering
Computers and Structures
Computer Science and Technology
Construction and Building Materials
Engineering Failure Analysis
Engineering Fracture Mechanics
Engineering Structures
International Journal of Fatigue
International Journal of Mechanical Sciences
International Journal of Solids and Structures
Journal of Constructional Steel Research
Mechanics of Materials
Mechanics Research Communications
Structural Safety
Thin-Walled Structures

To Contact the Publisher

Elsevier Science welcomes enquiries concerning publishing proposals: books, journal special issues, conference proceedings, etc. All formats and media can be considered. Should you have a publishing proposal you wish to discuss, please contact, without obligation, the publisher responsible for Elsevier's civil and structural engineering publishing programme:

Keith Lambert
Senior Publishing Editor
Elsevier Science Ltd
The Boulevard, Langford Lane
Kidlington, Oxford, OX5 1GB, UK

Phone: +44 1865 843411
Fax: +44 1865 843931
E.mail: k.lambert@elsevier.com

General enquiries, including placing orders, should be directed to Elsevier's Regional Sales Offices – please access the Elsevier homepage for full contact details (homepage details at the top of this page).

ADVANCES IN BUILDING TECHNOLOGY

Proceedings of the
International Conference on Advances in Building Technology

4-6 December 2002, Hong Kong, China

Volume I

Edited by
M. Anson
J.M. Ko
E.S.S. Lam
The Hong Kong Polytechnic University

Organized by
The Faculty of Construction and Land Use,
The Hong Kong Polytechnic University

Sponsored by
Hong Kong Institute of Construction Managers
The Association of Consulting Engineers of Hong Kong
The Hong Kong Construction Association Ltd.
The Hong Kong Institute of Architects
The Hong Kong Institution of Engineers

2002

ELSEVIER
Amsterdam – Boston – London – New York – Oxford – Paris
San Diego – San Francisco – Singapore – Sydney – Tokyo

ELSEVIER SCIENCE Ltd
The Boulevard, Langford Lane
Kidlington, Oxford OX5 1GB, UK

First edition 2002

Library of Congress Cataloging in Publication Data: A catalog record from the Library of Congress has been applied for.
British Library Cataloguing in Publication Data: A catalogue record from the British Library has been applied for.

ISBN: 0-08-044100-9 (2 volume set)

∞ The paper used in this publication meets the requirements of ANSI/NISO Z39.48-1992 (Permanence of Paper). Printed in the Netherlands.
The papers presented in these proceedings have been reproduced directly from the authors' 'camera ready' manuscripts. As such, the presentation and reproduction quality may vary from paper to paper.

PREFACE

The International Conference on Advances in Building Technology has been organized by the Faculty of Construction and Land Use, The Hong Kong Polytechnic University, and held in Hong Kong from December 4 to December 6, 2002. The Conference aims to provide an international forum for scientists, researchers, engineers, and other professionals to present and discuss recent research and advances in the design, construction and operation of our buildings.

The Conference also represents an initiative of a specially promoted Hong Kong Polytechnic University programme in Advanced Buildings Technology, a three-year programme which began in mid 1999. Although some hundreds of papers and reports have already been published by the Structural, Building Performance and Fire Safety Engineering teams, this Conference provides a technology transfer outlet in a more concentrated fashion for the programme as a whole.

These two volumes of proceedings contain 9 invited keynote papers, 72 papers delivered by 11 teams, and 132 contributed papers from over 20 countries around the world. These papers cover a wide spectrum of topics across the three technology sub-themes of structures and construction, environment and information technology. The variety within these categories is also wide and the Conference, therefore, provides the delegates with a good general oversight of building research advances, as opposed to the benefit, different in nature, to be gained from a specialist topic conference.

The organization of the Conference would not have been possible without the support and contributions of many individuals and organizations. We sincerely appreciate the support from the Hong Kong Institute of Construction Managers, The Association of Consulting Engineers of Hong Kong, The Hong Kong Construction Association Ltd., The Hong Kong Institute of Architects and The Hong Kong Institution of Engineers. We would like to thank the members of the International Advisory Committee for their valuable assistance, the Executive Committee, the Conference Organizing Committee, and all those who generously acted as referees. We also wish to express our gratitude to all the contributors for their careful preparation of the manuscripts. Thanks are also due to those who devoted time and effort to the organization of the Conference and publication of these Proceedings, including the secretarial staff and research students of the Faculty.

Finally, we gratefully acknowledge our pleasant cooperation with Mr. Ian Salusbury, Mr. Keith Lambert, Mrs. Lorna Canderton and Ms. Noël Blatchford of Elsevier Science Ltd. in the U.K.

M. Anson
J. M. Ko
E. S. S. Lam

INTERNATIONAL ADVISORY COMMITTEE

ORGANIZING COMMITTEE

Chairman :	**Prof. J. M. KO** The Hong Kong Polytechnic University
Secretary :	**Dr. E. S. S. LAM** The Hong Kong Polytechnic University

Members :

Mr. F. S. Y. BONG	Maunsell Consultants Asia Ltd.
Mr. P. W. T. CHAN	The Hong Kong Construction Association Ltd.
Mrs. E. CHENG	Architectural Services Department, HKSAR Government
Mr. K. M. CHEUNG	Buildings Department, HKSAR Government
Dr. C. GIBBONS	Ove Arup & Partners Hong Kong Ltd.
Mr. M. HADAWAY	Gammon Skanska Ltd.
Mr. R. S. H. LAI	Electrical and Mechanical Services Department, HKSAR Government
Mr. D. H. M. LEE	Housing Department, HKSAR Government
Mr. P. K. W. MOK	Henderson Land Development Co. Ltd.
Mr. F. S. H. NG	Paul Y. – ITC Construction Holdings Ltd.
Mr. C. K. WONG	Meinhardt (Hong Kong) Ltd.
Mr. C. T. C. WONG	Yau Lee Construction Co. Ltd.
Mr. P. K. K. LEE	The University of Hong Kong
Prof. A. Y. T. LEUNG	The City University of Hong Kong
Prof. W. H. TANG	The Hong Kong University of Science and Technology
Prof. M. ANSON	The Hong Kong Polytechnic University
Prof. J. BURNETT	The Hong Kong Polytechnic University
Prof. Y. Q. CHEN	The Hong Kong Polytechnic University
Prof. Y. S. LI	The Hong Kong Polytechnic University
Prof. W. SEABROOKE	The Hong Kong Polytechnic University
Dr. A. K. D. WONG	The Hong Kong Polytechnic University
Prof. F. W. H. YIK	The Hong Kong Polytechnic University

EXECUTIVE COMMITTEE

CONTENTS

Volume I

Earthquake Engineering

FRP Composite, Retrofit and Repair Technologies Engineering

High Performance Concrete/Composites

Volume II

CONSTRUCTION TECHNOLOGIES (CONTINUED)

Monitoring Technologies, Damage Detection and Vibration Control

Structural Systems and Design

Fire Safety Engineering

Indoor Environmental Control

Recycling Technologies

KEYNOTE PAPERS

Advances in Building Technology, Volume 1
M. Anson, J.M. Ko and E.S.S. Lam (Eds.)

ADVANCES IN MATERIALS AND MECHANICS

K. P. Chong[1-2], M. R.VanLandingham[2], and L. Sung[2]

[1]Directorate for Engineering, National Science Foundation
Arlington, VA 22230, U.S.A.
[2]National Institute of Standards and Technology,
Gaithersburg, MD 20899, U.S.A.

ABSTRACT

Mechanics and materials are essential elements in all of the transcendent technologies that are the primary drivers of the twenty first century and in the new economy. The transcendent technologies include nanotechnology, microelectronics, information technology and biotechnology as well as the enabling and supporting civil infrastructure systems and materials. Research opportunities and challenges in mechanics and materials, including nanomechanics, microscopy, radiation scattering and other nanoscale metrologies, as well as improved engineering and design of materials in the exciting information age are presented and discussed.

KEYWORDS

Atomic force microscope, biotechnology, designer materials, durability, information technology, nanomechanics, nanotechnology, microelectronics, confocal microscopy, multi-scale modeling, scattering metrology, simulations.

INTRODUCTION

The National Science Foundation (NSF) has supported basic research in engineering and the sciences in the United States for a half century and it is expected to continue this mandate through the next century. As a consequence, the United States is likely to continue to dominate vital markets, because diligent funding of basic research does confer a preferential economic advantage (Wong, 1996). Concurrently over this past half century, technologies have been the major drivers of the U. S. economy, and as well, NSF has been a major supporter of these technological developments. According to the former NSF Director for Engineering, Eugene Wong, there are three *transcendental* technologies:

- Microelectronics – Moore's Law: doubling the capabilities every two years for the last 30 years; unlimited scalability; nanotechnology is essential to continue the miniaturization process and efficiency.
- Information Technology [IT] – NSF and DARPA started the Internet revolution about three decades ago; the confluence of computing and communications.
- Biotechnology – unlocking the molecular secrets of life with advanced computational tools as well as advances in biological engineering, biology, chemistry, physics, and engineering including mechanics and materials.

Efficient civil infrastructure systems as well as high performance materials are essential for these technologies. By promoting research and development at critical points where these technological areas intersect, NSF can foster major developments in engineering. The solid mechanics and materials engineering (M&M) communities will be well served if some specific linkages or alignments are made toward these technologies. Some thoughtful examples for the M&M communities are:

- Bio-mechanics/materials
- Thin-film mechanics/materials
- Wave Propagation/NDT
- Nano-mechanics/materials
- Simulations/modeling
- Micro-electro-mechanical systems (MEMS)
- Smart materials/structures
- Designer materials

Considerable NSF resources and funding will be available to support basic research related to these technologies. These opportunities will be available for the individual investigator, teams, small groups and larger interdisciplinary groups of investigators. Nevertheless, most of the funding at NSF will continue to support unsolicited proposals from individual investigator on innovative "blue sky" ideas.

The National Institute of Standards and Technology, or NIST, is a non-regulatory federal agency within the U. S. Commerce Department's Technology Administration. NIST's mission is to develop and promote measurements, standards, and technology to enhance productivity, facilitate trade, and improve the quality of life. NIST carries out its mission in four cooperative programs, the NIST Laboratories, the Baldrige National Quality Program, the Manufacturing Extension Partnership (MEP), and the Advanced Technology Program (ATP). The Baldrige National Quality Program promotes performance excellence among U. S. manufacturers, service companies, educational institutions, and health care providers by conducting outreach programs and managing the annual Malcolm Baldrige National Quality Award, which recognizes performance excellence and quality achievement. The MEP is a nationwide network of local centers offering technical and business assistance to small manufacturers, and the ATP provides co-funding of research and development partnerships with the private sector to accelerate the development of innovative technologies for broad national benefit.

Within the seven NIST Laboratories, research is conducted in support of U. S. industry that advances the nation's technology infrastructure and enables U. S. industry to continually improve products and services. Recent strategic planning efforts have lead to the development of four programmatic focus areas, which include nanotechnology, health care, information and knowledge management, and homeland security – very much along the transcendental technologies mentioned above. NIST will be working to provide distinct contributions to each of these focus areas within the framework of its mission. New funding will continue to be requested from the U. S. Congress in support of these focus areas. For the measurement laboratories, research will continue to be conducted in the development of measurement science and technology both in support of and outside of these focus areas.

NIST plays a significant role in enabling technological advances. Current project areas at NIST of particular interest to the materials and mechanics community include:

- Biomaterials
- Thin films
- Surface and Interface Characterization
- High-throughput/Combinatorial methods
- Nanoparticles and Nanocomposites
- Fire Retardant Materials and Structures
- Materials for electronics, photonics, and magnetics
- Sensor materials and systems
- Cementitious and other construction materials
- Tissue Engineering

The Center for High Resolution Neutron Scattering, or CHRNS, is a national user facility that is jointly funded by NSF and the NIST Center for Neutron Research (NCNR). The CHRNS develops and operates state-of-the-art neutron scattering instrumentation with broad applications in materials research for use by the general scientific community. When used in combination, CHRNS instruments can provide structural information on length scales from 1 nm to approximately 10 μm, and dynamic information on energy scales from approximately 30 neV (nano electron volt) to 100 meV (micro electron volt). These ranges are the widest accessible at any neutron research center in North America. The instruments are used by university, government and industrial researchers in materials science, chemistry, biology and condensed matter physics to investigate materials such as polymers, metals, ceramics, magnetic materials, porous media, fluids and gels, and biological molecules. Proposals for use of the CHRNS facilities are considered on the basis of scientific merit or technological importance. For additional information, please visit http://rrdjazz.nist.gov/programs/CHRNS/.

INFORMATION TECHNOLOGY

In addition to NSF, there is also a sense that U. S. Federal agency research support is increasingly being driven by broad systemic initiatives. One of these initiatives has been *Information Technology* (IT), listed previously as one of the *transcendental* technologies. The former President's Information Technology Advisory Committee (PITAC) advised that an "immediate and vigorous information technology research and development (R&D) effort in Information Technology be initiated. IT is essential for the United States to have economic growth and prosperity in the 21st Century." PITAC (www.ccic.gov) concluded that current U. S. Federal support for research in IT is inadequate and these current efforts also take "a short-term focus for immediate returns." PITAC recommended IT R&D with long term priorities focusing on "software development that is far more usable, reliable, and powerful; scalable information infrastructures that satisfy the demands of large numbers of users; high-end computing systems with both rapid calculation and rapid data movement; and IT education and training for the citizenry." Achieving these ends requires diversified modes of research support to foster projects of broader scope and longer duration with emphasis on projects involving multiple investigators over several years. Of most importance is the use of these new information technologies to advance critical application domains to benefit our nation and the world.

NANOTECHNOLOGY

Nanomechanics Workshop

Initiated by the senior author [KPC], with the organization and help of researchers from Brown [K.

S. Kim, et al], Stanford, Princeton and other universities, a NSF Workshop on Nano- and Micro-Mechanics of Solids for Emerging Science and Technology was held at Stanford in October 1999. The following is extracted from the Workshop Executive Summary. Recent developments in science have advanced capabilities to fabricate and control material systems on the scale of nanometers, bringing problems of material behavior on the nanometer scale into the domain of engineering. Immediate applications of nanostructures and nano-devices include quantum electronic devices, bio-surgical instruments, micro-electrical sensors, functionally graded materials, and many others with great promise for commercialization. The branch of mechanics research in this emerging field can be termed nano- and micro-mechanics of materials, highly cross-disciplinary in character. A subset of these, which is both scientifically rich and technologically significant, has mechanics of solids as a distinct and unifying theme. The presentations at the workshop and the open discussion precipitated by them revealed the emergence of a range of interesting lines of investigation built around mechanics concepts that have potential relevance to microelectronics, information technology, bio-technology and other branches of nanotechnology. It was also revealed, however, that the study of complex behavior of materials on the nanometer scale is in its infancy. More basic research that is well coordinated and that capitalizes on progress being made in other disciplines is needed if this potential for impact is to be realized.

Recognizing that this area of nanotechnology is in its infancy, substantial basic research is needed to establish an engineering science base. Such a commitment to nano- and micro-mechanics will lead to a strong foundation of understanding and confidence underlying this technology based on capabilities in modeling and experiment embodying a high degree of rigor. The instruments and techniques available for experimental micro- and nano-mechanics are depicted in Fig. 1, courtesy of K. S. Kim of Brown University.

Field of View (Gage Length in m)

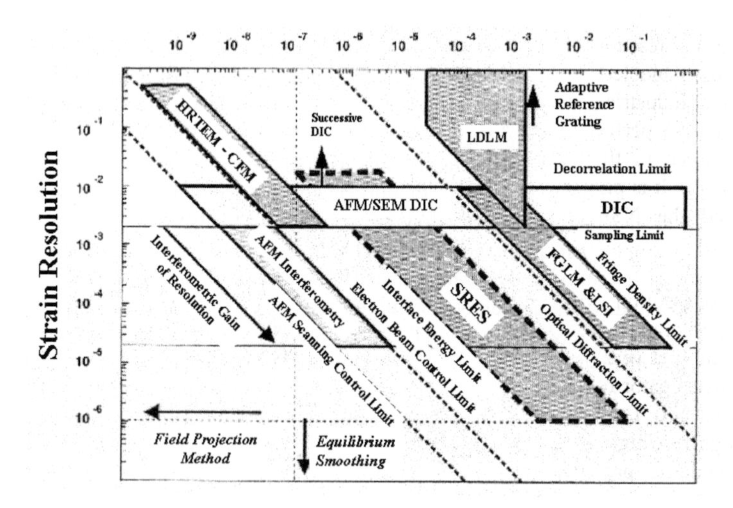

Figure 1. Instruments and techniques for experimental micro and nano mechanics [courtesy of K. S. Kim of Brown University] where

HRTEM High Resolution Transmission Electron Microscopy
SRES Surface Roughness Evolution Spectroscopy
CFTM Computational Fourier Transform Moire
FGLM Fine Grating Laser Moiré
AFM Atomic Force Microscopy
LSI Laser Speckle Interferometry
SEM Scanning Electron Microscopy
DIC Digital Image Correlation
LDLM Large Deformation Laser Moire

The potential of various concepts in nanotechnology will be enhanced, in particular, by exploring the nano- and micro-mechanics of coupled phenomena and of multi-scale phenomena. Examples of coupled phenomena discussed in this workshop include modification of quantum states of materials caused by mechanical strains, ferroelectric transformations induced by electric field and mechanical stresses, chemical reaction processes biased by mechanical stresses, and changes of bio-molecular conformality of proteins caused by environmental mechanical strain rates. Multi-scale phenomena arise in situations where properties of materials to be exploited in applications at a certain size scale are controlled by physical processes occurring on a size scale that is orders of magnitude smaller. Important problems of this kind arise, for example, in thermo-mechanical behavior of thin-film nanostructures, evolution of surface and bulk nanostructures caused by various material defects, nanoindentation, nanotribological responses of solids, and failure processes of MEMS structures. Details of this workshop report can be found by visiting http://en732c.engin.brown.edu/nsfreport.html.

Nanoscale Science and Engineering Initiatives

Coordinated by M. Roco (IWGN, 2000), NSF recently announced a second year program [NSF 01-157; see: www.nsf.gov] on collaborative research in the area of nanoscale science and engineering (NSE). This program is aimed at supporting high risk/high reward, long-term nanoscale science and engineering research leading to potential breakthroughs in areas such as materials and manufacturing, nanoelectronics, medicine and healthcare, environment and energy, chemical and pharmaceutical industries, biotechnology and agriculture, computation and information technology, improving human performance, and national security. It also addresses the development of a skilled workforce in this area as well as the ethical, legal and social implications of future nanotechnology. It is part of the interagency National Nanotechnology Initiative (NNI). Details of the NNI and the NSE initiative are available on the web at http://www.nsf.gov/nano or http://nano.gov.

The NSE competition will support Nanoscale Interdisciplinary Research Teams (NIRT) and Nanoscale Exploratory Research (NER). Nanoscale Science and Engineering Centers (NSEC) awarded in the first year (FY 2001) competition will be funded on a continuing basis. In addition, individual investigator research in nanoscale science and engineering will continue to be supported in the relevant NSF Programs and Divisions outside of this initiative. This NSE initiative focuses on seven high risk/high reward research areas, where special opportunities exist for fundamental studies in synthesis, processing, and utilization of nanoscale science and engineering. The seven areas are:

- Biosystems at the nanoscale
- Nanoscale structures, novel phenomena, and quantum control
- Device and system architecture
- Nanoscale processes in the environment
- Multi-scale, multi-phenomena theory, modeling and simulation at the nanoscale
- Manufacturing processes at the nanoscale

- Societal and educational implications of scientific and technological advances on the nanoscale

The National Nanotechnology Initiative started in 2000 ensures that investments in this area are made in a coordinated and timely manner (including participating federal agencies – NSF, DOD, DOE, DOC [including NIST], NIH, DOS, DOT, NASA, EPA and others) and will accelerate the pace of revolutionary discoveries now occurring. Current request of Federal agencies on NNI is $519 million. The NSF share of the budget is $199 million [on NSE, part of NNI].

Microscopy

In atomic force microscopy (AFM), a probe consisting of a sharp tip (nominal tip radius on the order of 10 nm) located near the end of a cantilever beam is raster scanned across the sample surface using piezoelectric scanners. Changes in the tip-sample interaction are often monitored using an optical lever detection system, in which a laser beam is reflected off of the cantilever and onto a position-sensitive photodiode. During scanning, a particular operating parameter is maintained at a constant level, and images are generated through a feedback loop between the optical detection system and the piezoelectric scanners. A schematic illustration of a scanning stylus atomic force microscope is shown in Figure 2a. In this design, the probe tip is scanned above a stationary sample, while in a scanning sample design, the sample is scanned below a fixed probe tip. Operationally, little difference exists between these two designs, because the relative motion of the tip to the sample is used to generate topographic images. Applications of AFM and other types of scanning probe microscopy continue to grow rapidly in number and include biological materials (e.g., studying DNA structure), polymeric materials (e.g., studying morphology, mechanical response, and thermal transitions), and semiconductors (e.g., detecting defects). In particular, AFM can be utilized to evaluate the surface quality of products such as paint and coating systems, contact lenses, optical components (mirrors, beamsplitters, etc.), and semiconductor wafers after various cleaning, etching, or other manufacturing processes.

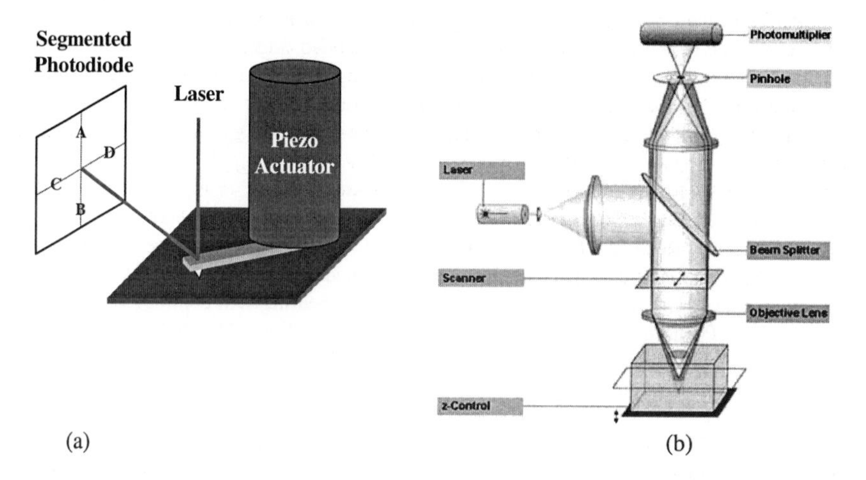

Figure 2. Schematic illustrations of (a) a scanning-probe atomic force microscope (AFM) and (b) a laser-scanning confocal microscope (LSCM).

Laser scanning confocal microscopy (LSCM) is a non-destructive, powerful tool for characterizing the microstructure of polymeric and biological materials. In Fig. 2b, the confocal principle is illustrated. A LSCM uses coherent incident light and collects reflected or scattered light exclusively from a single plane, rejecting light out of the focal plane. The laser scanning confocal microscope scans the sample sequentially point by point and line by line and assembles the pixel information into one image. Optical slices of the specimen are thus created with high contrast and high resolution in x, y and z directions. By moving the focal plane systematically, a series of optical slices are created that can be used to construct a three-dimensional image stack that can be digitally processed. The wavelength, numerical aperture (N.A.) of the objective, and the size of the collecting pinhole in front of the detector determine the resolution in the thickness or axial direction (Corle et al., 1996). Microstructure information obtained from LSCM results can be used, for example, to model scattering properties of coating materials, which can then be compared to scattering measurements using light scattering (Sung et al., 2002).

Nanomechanics

Measurements of mechanical behavior of very small volumes of material have been enabled by the development of instruments that can sense and apply very small forces and displacements, including AFMs and nanoindenters. These devices are often capable of producing contact areas and penetration depths characterized by nanometer dimensions while also providing lateral motion capabilities for studying tribological behavior. One objective of using these devices is to provide methods for characterizing mechanical response of material systems with nanoscale spatial resolution. Such measurements can be a key to understanding technologically important material systems, including those used in magnetic storage, microelectronic and telecommunication devices. Further, these types of measurements are important in support of nanotechnology developments, e.g., nanoelectronic devices and nanostructured materials such as ultra-thin films and nanocomposites.

Nanoindentation via the AFM is performed in a non-scanning mode, termed force mode, in which the AFM probe is moved toward the sample surface, pushed into the surface, and then lifted back off of the surface (see Fig. 2a). A force-displacement curve is produced from which a force-penetration curve related to the nanoindentation process can be determined. The force applied is related to the deflection (bending) of the cantilever beam through the cantilever spring constant, and the penetration into the material is the difference between the overall vertical displacement of the cantilever and the probe tip displacement related to cantilever bending, which must be calibrated. For nanoindentation devices, shown schematically in Fig. 3a, the application of force and the measurement of displacement are often done using electromagnetic or electrostatic transducers. System compliance must be calibrated to eliminate displacement of the load frame, such that force-penetration curves are produced (see Fig. 3b). Typically, the unloading data is assumed to be primarily elastic recovery of the material and is often analyzed using a power law curve fit of the form:

$$P = \alpha\left(h - h_f\right)^m$$

where P is force, h is penetration, h_f is the penetration after unloading, and α and m are curve fitting parameters. The slope, dP/dh, taken at the point of maximum load, P_{max}, and maximum displacement, h_{max}, is the contact stiffness, S, which is used to first determine the contact depth, h_c,

$$h_c = h_{\max} - \frac{\varepsilon P_{\max}}{S}$$

here ε is a parameter related to the contact geometry, and subsequently to calculate elastic modulus,

$$E_r = \frac{\sqrt{\pi}}{2\beta}\frac{S}{\sqrt{A}}$$

where A is the contact area, which is calculated from knowledge of the tip shape and h_c, β is a parameter related to tip geometry, and E_r is the reduced or effective modulus, which is related to values of modulus and Poisson's ratio for the sample (E_s and ν_s) and indentation tip (E_i and ν_i):

$$\frac{1}{E_r} = \frac{\left(1-\nu_s^2\right)}{E_s} + \frac{\left(1-\nu_i^2\right)}{E_i}$$

Indentation hardness, H, can also be calculated as the maximum load, P_{max}, divided by the contact area, A.

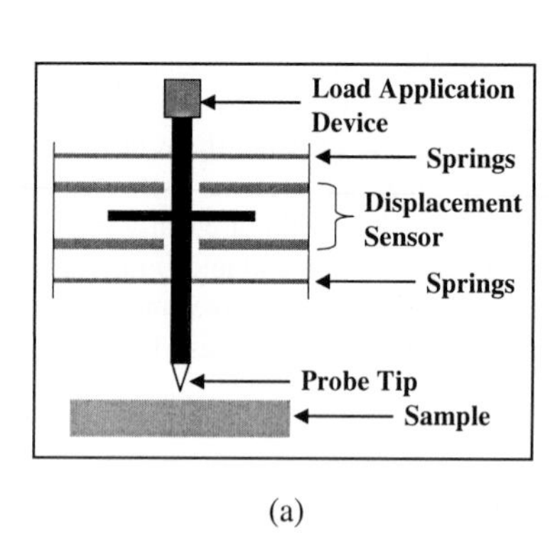

(a)

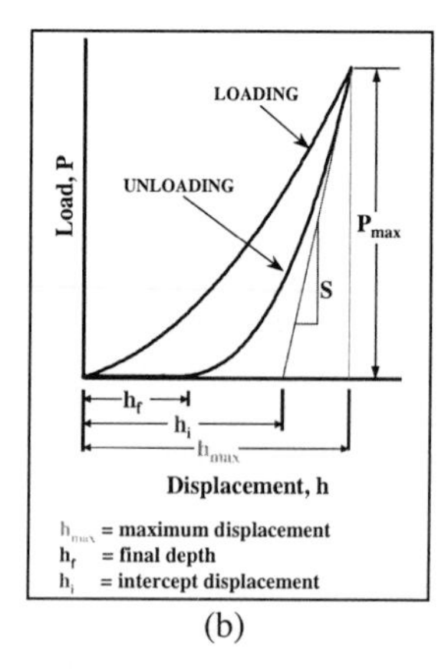

(b)

Figure 3. Schematic illustrations of (a) a nanoindentation system and (b) load-displacement or load-penetration data produced from a nanoindentation measurement, including key characteristics.

Recent advances in instrumentation have focused on improvements in sensitivity, the addition of testing capabilities, and the integration of force measurements and imaging. For nanoindentation devices, improvements in sensitivity have been made by adding dynamic capabilities to these systems, which previously have operated only in a quasi-static mode. Often, a dynamic signal with a displacement amplitude on the order of 1 nm is superimposed over a quasi-static loading history. This type of technique allows hundreds of measurements of elastic modulus and hardness to be calculated as opposed to a single measurement from a quasi-static indentation test. Also, dynamic behavior of the materials, e.g., storage and loss characteristics, can be studied, often over a range of frequencies, similar to dynamic mechanical analysis. Other additional capabilities include lateral motion and lateral force measurement capabilities for tribological studies, including scratch testing, and improvements to the control systems that allow for automated testing using a variety of user-specified loading histories.

Integration of force measurements and imaging capabilities has been made in both AFM systems and in hybrid AFM-nanoindentation devices. AFM images are now produced routinely using particular aspects of the tip-sample force interactions. Each pixel in the image represents a position on the sample at which a force-distance curve was measured. The image can then be set up to indicate changes in local tip-sample adhesion or sample stiffness, for example. However, mechanical property measurements associated with SPM systems have significant uncertainties in the probe spring constant and tip shape, which typically render them useful only as qualitative information (VanLandingham et al., 2001). More quantitative measures of tip-sample forces can be achieved for systems that employ force transducers in combination with scanning capabilities similar to AFM. Examples of such systems include the interfacial force microscope or IFM developed at Sandia National Laboratories (Joyce & Houston, 1991) and a commercial force transducer that interfaces with many commercial AFM systems (Bhushan et al., 1996; Asif et al., 2001).

While advances in instrumentation continue, questions remain regarding uncertainties associated with measurement and calibration techniques. Current research at NIST is focused on measurement systems for calibrating micro-Newton forces. Questions regarding the use of reference materials for calibrating load-frame compliance and tip shape as well as other analysis and procedural issues persist with regard to nanoindentation measurements that must be addressed. For example, recent studies, including an interlaboratory comparison (Jennett & Meneve, 1998), have shown the calibration results to have poor reproducibility and large uncertainties.

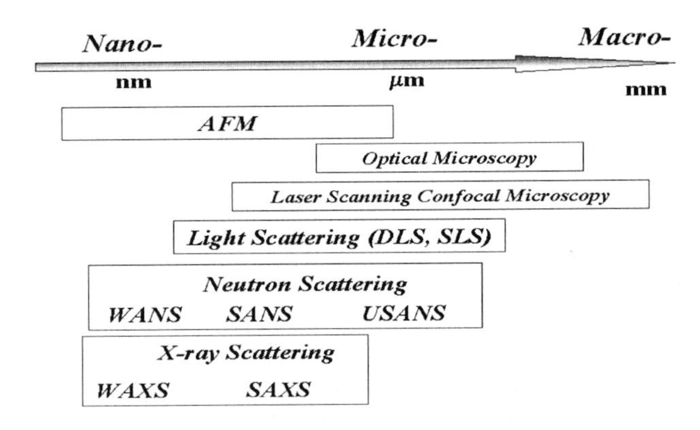

Figure 4. Capabilities of different microscopy and scattering techniques.

Scattering Metrology

Scattering metrology is widely used for characterizing nano-/ micro-domains and nano-structural features of polymers and other materials, and is a complim entary technique to microscopy tools. Many forms of radiation can be used for scattering purposes: X-rays, neutron, laser light, electrons, etc. Each has different characteristics and is used for different purposes. In general, the principle and operation of measurements is similar. However, the microstructural information generated could be different due to the different interactions between the scattering source and the materials. For example, neutron scattering is sensitive to inhomogeneities in density of nuclei in the materials, while light or X-ray scattering is sensitive to inhomogeneities in the refractive index or electron density in the materials. In this paper, we will only describe the neutron scattering method.

In Fig. 4 the relative measurable length scale using different scattering metrology and microscopy techniques are shown. The small-angle neutron and X-ray scattering methods (SANS, SAXS) are useful for polymer research (or condense matter research in general) because they probe size scales from angstrom-level to micron-level. With recently developed Ultra-SANS (USANS) instruments, the length scales using neutron scattering method can be extended to 10 μm (Drews et al., 1998).

Static light scattering (SLS) complements these techniques by focusing on the micron length scale. Dynamic light scattering (DLS) can be used for measuring diffusive motion and particle sizing in complex fluid systems from 1 nm to 5 μm. Other methods such as wide-angle neutron and X-ray scattering (WANS, WAXS) probe very local (atomic) structure on the order of a few nanometers.

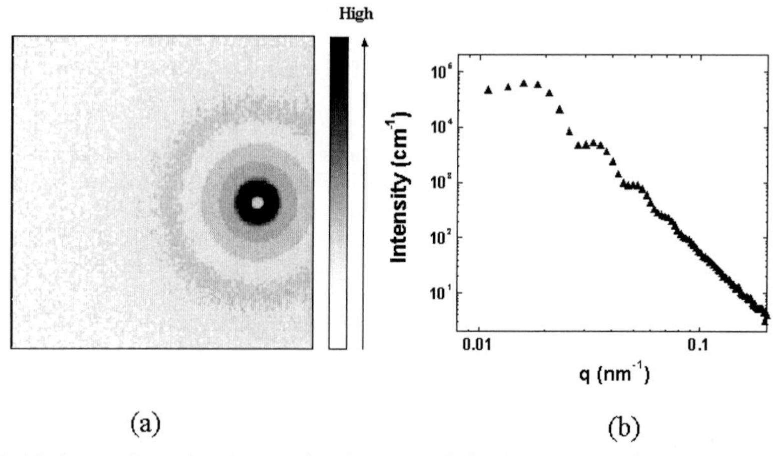

Figure 5. (a) A two-dimensional scattering image, and (b) the corresponding absolute scattering intensity curve as a function of the scattering wave vector q, for a polymeric system with a well-ordered structure.

The NIST Center for Neutron Research (NCNR), which maintains the leading neutron scattering facility in the U. S., offers advanced measurement capabilities for use by all qualified applicants. Information on instrumental description may be obtained at the NCNR website [http://www.ncnr.nist.gov/]. Many of its instruments rely on intense neutron beams produced by a liquid hydrogen cold neutron source (Cappelletti et al., 2001). The small angle neutron scattering (SANS) instrument is a powerful tool for characterizing nano-/ micro-domains and nanostructures of various polymeric systems as well as particles or fillers in solution or in a polymer matrix. In Fig. 5, SANS results obtained from nano-size fillers (for example: inorganic pigment) in a polymer matrix are shown. After standard calibrations and taking into account the sample transmission and film thickness, two-dimensional scattering images (see Fig. 5a) are averaged azimuthally to produce a one-dimensional absolute scattering intensity curve (see Fig. 5b) as a function of the scattering wave vector q. ($q = 4\pi\sin(\theta/2)/\lambda$, where θ is the scattering angle and λ is the wavelength). In this case, the filler has a higher neutron scattering coefficient than the polymer matrix, and the neutron scattering intensity is proportional, to the first order, to differences in the local filler concentrations within the sample. Typically the scattering intensity, I(q), can be modeled as following:

$$I(q) = I_o\, P(q)\, S(q),$$

where $P(q)$ is the form factor, and $S(q)$ is the structure factor (Higgins & Benoit, 1994). The wavelength independent part of the equation: I_o is proportional to the scattering amplitude (scattering contrast between the fillers and the polymer matrix), the number density, and the volume of the fillers. The form factor $P(q)$ comes from the intramolecular interferences and is characteristic of the size and shape of the particles. For a particle with well-defined shape, such as a spherical particle with uniform scattering density (see example of Fig. 5b) or a spherical particle with a core-shell structure, the form factor $P(q)$ can be calculated, and the size of the particles can be determined. The structure factor S(q) is related to the correlation function describing the radial distribution function between particles or micro-domains. A peak in the scattering profile at a scattering wave vector q indicates the existence of micro-domains with a characteristic average length $d = 2\pi/q$. The microstructure/morphology information, such as the dimension of ordered domains or the correlation length of concentration fluctuations, can be determined from the intensity profile.

MODELING AND SIMULATION

The initiative *Engineering Sciences for Modeling and Simulation-Based Life-Cycle Engineering* (Program Announcement NSF 99-56) is a three-year collaborative research program by NSF and the Sandia National Laboratories (Sandia) focusing on advancing the fundamental knowledge needed to support advanced computer simulations. This collaborative initiative capitalizes on the missions of both organizations. NSF's mission is to advance the fundamental science and engineering base of the United States. Sandia has the responsibility to provide solutions to a wide range of engineering problems pertinent to national security and other national issues. It is moving toward engineering processes in which decisions are based heavily on computational simulations including meshless methods in moving boundary problems [see e.g. Boresi, Chong and Saigal, 2002]; thus, capitalizing on the available high performance computing platforms. This initiative has sought modeling and simulation advances in key engineering focus areas such as thermal sciences, mechanics and design.

The NSF Civil and Mechanical Systems (CMS) Division developed an initiative on *Model-based Simulation* (MBS), see NSF 00-26. Model-based simulation is a process that integrates physical test equipment with system simulation software in a virtual test environment aimed at dramatically reducing product development time and cost. This initiative will impact many civil/mechanical areas: "structural, geotechnicial, materials, mechanics, surface science, and natural hazards (e.g., earthquake, wind, tsunami, flooding and land-slides)." MBS would involve "combining numerical methods such as finite element and finite difference methods, together with statistical methods and reliability, heuristics, stochastic processes, etc., all combined using super-computer systems to enable simulations, visualizations, and virtual testing." Expected results could be less physical testing, or at best, better strategically planned physical testing in the conduct of R&D. Examples of the use of MBS in research, design and development exist in the atmospheric sciences, biological sciences, and the aerospace, automotive and defense industries. The manufacturing of the prototype Boeing 777 aircraft, for example, was based on computer-aided design and simulation, cutting costs and shortening production time significantly.

The NIST Center for Theoretical and Computational Materials Science (CTCMS) was established to investigate important problems in materials theory and modeling with novel computational approaches. The Center creates opportunities for collaboration where CTCMS can make a positive difference by virtue of its structure, focus, and people. It develops powerful new tools for materials theory and modeling and accelerates their integration into industrial research. The current technical focus areas include: modeling microstructure and mechanical response using object-oriented finite element analysis, micromagnetic modeling, nanofilled polymer melts, deformation of metals...etc. Detailed information can be found on the center web site at http://www.ctcms.nist.gov/programs. In addition to the current projects, the CTCMS is continuously soliciting proposals for workshops on materials theory and modeling, short-term and long-term visiting fellowship, guest researcher, and NRC postdoctoral positions, and creative and ambitious new projects in materials theory and modeling.

In the future one should expect the continued introduction of bold innovative research initiatives related to important national agenda issues such as the environment, civil and mechanical infrastructure, the service industry and the business enterprises.

CHALLENGES

The challenge to the mechanics and materials research communities is: How can we contribute to these broad-base and diverse research agendas? Although the mainstay of research funding will support the traditional programs for the foreseeable future, considerable research funding will be directed toward addressing these research initiatives of national focus. At the request of the author [KPC], a NSF research workshop has been organized by F. Moon of Cornell University to look into the research needs and challenges facing the mechanics communities. A website with recommendations of research needs and challenges will soon be available.

Mechanics and materials engineering are really two sides of a coin, closely integrated and related. For the last decade this cooperative effort of the M&M Program has resulted in better understanding and design of materials and structures across all physical scales, even though the seamless and realistic modeling of different scales from the nano-level to the system integration-level (Fig. 6) is not yet attainable. In the past, engineers and material scientists have been involved extensively with the characterization of given materials. With the availability of advanced computing and new developments in material science, researchers can now characterize processes and design and manufacture materials with desirable performance and properties. One of the challenges is to model short-term micro-scale material behavior, through meso-scale and macro-scale behavior into long-term structural systems performance. Accelerated tests to simulate various environmental forces and impacts are needed (NSF, 1998). Supercomputers and/or workstations used in parallel are useful tools to solve this scaling problem by taking into account the large number of variables and unknowns to project micro-behavior into infrastructure systems performance, and to model or extrapolate short term test results into long term life-cycle behavior (NSF, 1998; Chong, 1999). Twenty-four awards were made totaling $7 million. A grantees' workshop was held recently in Berkeley and a book of proceedings has been published (Monteiro et al., 2001).

MATERIALS		STRUCTURES		INFRASTRUCTURE
nano-level (10^{-9})	micro-level (10^{-6})	meso-level (10^{-3})	macro-level (10^{+0})	systems-level (10^{+3}) m
Molecular Scale	*Microns*		*Meters*	*Up to Km Scale*
*nano-mechanics	*micro-mechanics	*meso-mechanics	*beams	*bridge systems
*self-assembly	*micro-structures	*interfacial-structures	*columns	* lifelines
*nanofabrication	*smart materials	*composites	*plates	*airplanes

Figure 6. Physical scales in materials and structural systems (Boresi & Chong, 2000)

Many current NIST projects are focused on multidisciplinary and multi-scale problems, including in the areas of materials and mechanics. One successful program in the NIST Building and Fire Research Laboratory is the Service Life Prediction or SLP program, which includes partnerships with industry and other U. S. Federal Agencies. The mission of the SLP program is to greatly reduce the time-to-market for new products through the development and implementation of advanced methods and metrologies for quantitative and reliable prediction of the service life of existing and new products. Competition within the construction materials community has drastically increased with industrial globalization. As has been the case in the high technology industries, globalization has greatly increased the competitive pressures for reducing the time-to-market for new construction materials and products. Currently, time-to-market is controlled by the time required to generate a performance history for a product, which is established through a series of time-consuming field and laboratory exposure studies, typically taking between 5 and 15 years to complete.

The methodology that is being used to reduce the time to generate these durability performance histories to months instead of years is called reliability theory and life testing analysis. This methodology has had a long and successful history of application in predicting the service life of electronic, aerospace, nuclear, and medical products, and is currently being applied in five different, but highly overlapping, project areas including: 1) appearance (the primary measure of durability for a coating system), 2) coatings and pigments, 3) composites, 4) polymeric interphases, and 5) sealants. This research effort includes implementing and developing use-inspired scientific and technical advances in methodologies and metrologies for improving and advancing the appearance, service life, and performance of polymeric construction materials. Included in this effort is an integration of materials science, metrology from nanoscale though macroscale including nanomechanics, high-throughput testing and analysis, and information management.

ACKNOWLEDGMENTS AND DISCLAIMER

The authors would like to thank their colleagues and many members of the research communities for their comments and inputs during the writing of this paper. Information on NSF initiatives, announcements and awards can be found in the NSF website: www.nsf.gov. Information pertaining to NIST can be found on its website: www.nist.gov. The opinions expressed in this article are the authors' only, not necessarily those of the National Science Foundation or the National Institute of Standards and Technology.

REFERENCES

Asif, S. A. S., Wahl, K. J., Colton, R. J., and Warren, O. L. (2001). Quantitative Imaging of Nanoscale Mechanical Properties Using Hybrid Nanoindentation and Force Modulation. *J. Appl. Phys.* **90**, 5838-5838.

Boresi, A. P. and Chong, K. P. (2000). *Elasticity in Engineering Mechanics*, John Wiley, New York.

Boresi, A. P., Chong, K. P. and Saigal, S. (2002). *Approximate Solution Methods in Engineering Mechanics*, John Wiley, New York.

Bhushan, B., Kulkarni, A. V., Banin, W., and Wyrobek, J. T. (1996). Nanoindentation and Picoindentation Measurements Using a Capacitive Transducer System in Atomic Force Microscopy. *Phil. Mag. A* **74**, 1117-1128.

Cappelletti, R.L., Glinka, C.J., Krueger, S., Lindstrom, R.A., Lynn, J.W., Prask, H.A., Prince, E., Rush, J. J., Rowe, J.M., Satija, S.K. Toby, B.H., Tsai, A., and Udovic, T.J. (2001). Materials Research With Neutrons at NIST. *J. Res. Natl. Inst. Stand. Technol.* **106**,187-230.

Chong, K. P. (1998, 1999). Smart Structures Research in the U.S. *Keynote paper, Proc. NATO Adv. Res. Workshop on Smart Structures*, held in Pultusk, Poland, 6/98, *Smart Structures,* Kluwer Academic Publ. 37-44 (1999).

Corle, T.R., and Kino, G.S. (1996). *Confocal Scanning Optical Microscopy and Related Imaging Systems*, Academic Press, San Diego.

Drews, A.R., Barker, J.G., Glinka, C.J., Agamalian, M. (1998). Development of a thermal-neutron double-crystal diffractometer for USANS at NIST. *Physica B* **241-243**, 189-191.

Higgins, J.S. and Benoit, H.C. (1994). *Polymers and Neutron Scattering*. Clarendon Press, Oxford, UK.

Interagency Working Group on Nano Science, Engineering and Technology (IWGN). (2000). *Nanotechnology Research Directions*. Kluwer Academic Publ. 37-44.

Jennett, N. M. and Meneve, J. (1998). Depth Sensing Indentation of Thin Hard Films: A Study of Modulus Measurement Sensitivity to Indentation Parameters. *Fundamentals of Nanoindentation and Nanotribology* **522**, Materials Research Society, Pittsburgh, PA, 239-244.

Joyce, S. A. and Houston, J. E. (1991). A New Force Sensor Incorporating Force-Feedback Control For Interfacial Force Microscopy. *Rev. Sci. Instrum.* **62**, 710-715.

Monteiro, P. J. M., Chong, K. P., Larsen-Basse, J. and Komvopoulos, K., eds.(2001). *Long-term Durability of Structural Materials*, Elsevier, Oxford, UK.

NSF. (1998). Long Term Durability of Materials and Structures: Modeling and Accelerated Techniques. *NSF 98-42,* National Science Foundation, Arlington, VA.

NSF. (2001). Nanoscale Science and Engineering. *NSF 01-157*, National Science Foundation, Arlington, VA.

Sung, L, Nadal, M. E., McKnight, M.E. Marx, E., and Laurenti, B. (2002). Optical Reflectance of Metallic Coatings: Effect of Aluminum Flake Orientation. *J. Coating Technology, in press.*

VanLandingham, M. R., Villarrubia, J. S., Guthrie, W. F., and Meyers, G. F. (2001). Nanoindentation of Polymers: An Overview. *Macromolecular Symposia* **167**: *Advances in Scanning Probe Microscopy of Polymers*, V. V. Tsukruk and N. D. Spencer, eds., 15-44.

Wong, E. (1996). An Economic Case for Basic Research. *Nature* **381,** 187-188.

Advances in Building Technology, Volume I
M. Anson, J.M. Ko and E.S.S. Lam (Eds.)

ADVANCES IN CONCRETE TECHNOLOGY

M.F. Cyr and S.P. Shah

Center for Advanced Cement Based Materials,
Department of Civil Engineering, Northwestern University, Evanston, IL
60208 USA

ABSTRACT

A survey of recent advances in concrete technology, with a focus on research performed at the Center for Advanced Cement Based Materials (ACBM Center) at Northwestern University, is presented. Ultra-high-strength concrete (UHSC), with compressive strength of 200MPa, has been developed. The properties and applications of reactive powder concrete, one type of UHSC, are discussed. Fiber reinforcement is used to overcome the inherent brittleness and increase the tensile strength of concrete, especially high- and ultra-high-strength concrete. Fiber-reinforced cementitious composites can be designed for specific applications with the use of special processing techniques, such as extrusion, and hybrid fiber reinforcement. Significant reductions in drying shrinkage are achieved with a newly developed shrinkage reducing admixture. Construction costs can be reduced with the use of self-compacting concrete (SCC), which does not require vibration at placement. The design of SCC is facilitated with a newly developed rheological model. A nondestructive evaluation technique has been developed to monitor the hardening process of fresh concrete.

KEYWORDS

High-performance concrete, ultra-high-strength concrete, fibers, extrusion, shrinkage, self-compacting concrete, nondestructive evaluation

INTRODUCTION

As concrete technology developed, an initial goal was to increase its strength. High-strength concrete (HSC) columns were first used in the construction of high-rise buildings in the 1970s. The same changes that increased the strength also improved the durability and other aspects of concrete performance. The term high-performance concrete (HPC) began to be used. Today, HPC refers to concrete with many different attributes. It is produced with specifically designed matrices, often containing special chemical and mineral admixtures and fiber reinforcement. HPC performance criteria include high strength and elastic modulus, improved toughness and impact resistance, high early-age strength, high durability—including low permeability, resistance to chemical attack and free-thaw damage—and ease of placement and compaction without segregation.

A selection of current research in HPC, with an emphasis on work performed at the Center for Advanced Cement Based Materials (ACBM) at Northwestern University, is presented. Ultra-high-strength concrete, with a compressive strength of 200 MPa, has been produced for specialized applications. The ductility of concrete, especially high- and ultra-high-strength concrete, has been enhanced with fiber reinforcement. Hybrid fiber reinforcement and special processing techniques, such as extrusion, have enabled the optimization of composite performance for specific applications. A shrinkage reducing admixture has been developed to improve shrinkage cracking resistance. A model to facilitate the design of self-compacting concrete has been developed. In addition to improving concrete performance, new nondestructive techniques have been employed to monitor the setting of fresh concrete and to assess early deterioration in hardened concrete.

ULTRA-HIGH-STRENGTH CONCRETE

The strength of brittle materials, such as concrete, is related to the porosity of the material. As the porosity decreases, the strength exponentially increases (Mindess and Young 1981). Powers and Brownyard (1948) showed that decreasing the water-to-cement (w/c) ratio reduced the porosity of the concrete, increasing the strength. This reduction in porosity also makes the concrete more durable. In addition to having a lower water-to-binder (w/b) ratio, HSC usually contains superplasticizer, and mineral admixtures, such as silica fume or fly ash. Its compressive strength is around 100 MPa compared to compressive strengths of 20-40 MPa for normal-strength concrete (Kosmatka *et al.* 2001).

Recently, special processing techniques have been used to produce concrete with even higher compressive strengths. Ultra-high-strength concrete (UHSC) can reach compressive strengths of 200 MPa. Two commonly produced UHSCs are macro-defect-free (MDF) cement and reactive powder concrete (RPC). Macro-defect free cement is a mixture of cement and a water-soluble polymer. High shear mixing causes a mechano-chemical reaction between the cement and the polymer resulting in tensile strengths of up to 200 MPa (Shah and Weiss 1998a).

Reactive powder concrete typically has a compressive strength of 200 MPa, although strengths as high as 810 MPa have been recorded (Semioli 2001). Its high strength and low porosity are obtained by optimizing particle packing and reducing water content. The mixture contains no coarse aggregates. Instead fine powders, such as sand, crushed quartz, and silica fume, with particle sizes ranging from 0.02 to 300 μm are used. The grain size distribution is optimized to increase the matrix density. Superplasticizer is used to reduce the w/b to 0.2 as compared with w/b of 0.4-0.5 for typical normal-strength concrete. Steel and synthetic fibers are typically added to improve the ductility, and a post-set heat treatment is applied to improve the microstructure.

RPC is produced commercially by two French construction companies (Semioli 2001). Béton Spécial Industriel (BSI) is produced by the Eiffage Group (EGI) in conjunction with Sika Corp., and Ductal® is made by Bouygues Construction in partnership with Lafarge Corp. Ductal® is reinforced with high-strength steel microfibers to improve ductility. It has a compressive strength of 200 MPa, a tensile strength of 8 MPa and a flexural strength of 40-50 MPa. Ductal® is 100 times more resistant to water diffusion than normal-strength concrete and 10 times more resistant than typical French HPC. It has zero shrinkage after setting and 85-95% less creep than conventional concrete. The relatively low tensile strength requires the use of prestressing in more severe applications. With Ductal®, very strong, lightweight and durable thin sections can be produced. Currently, there are plans to use Ductal® in a 120-m long, slender arch pedestrian bridge near Seoul, Korea. The bridge will consist of six post-tensioned segments with a 30-mm thick walkway. In the United States, the Federal Highway Administration (FHWA) is currently testing Ductal® in prestressed concrete girders (FHWA 2002).

BSI has been used in French nuclear power plants. The producers of RPC also see potential for its use in pipes, tunnel and canal linings, paving, floors, liquid storage structures, nuclear waste containment, and long-spanning, slender, self-supporting structures such as stadium domes.

The high strength of UHSC is due to the homogeneity and low porosity of the matrix. These same characteristics also cause the material to be extremely brittle and increase the likelihood of shrinkage cracking. Microfiber reinforcement can increase the ductility of the concrete and improve shrinkage cracking resistance. The use of shrinkage reducing admixture also improves shrinkage performance.

FIBER-REINFORCED CONCRETE

Microfiber reinforcement

Microfiber reinforcement reduces the inherent brittleness of concrete, especially UHSC. Fibers spaced at the micron scale can interact with microcracks, delaying localization and increasing the tensile strength of the matrix (Shah 1991), in ways that steel reinforcing bars with spacing at the millimeter scale cannot. In addition, microfiber reinforcement delays the age of the first visible crack and reduces the crack width in restrained shrinkage tests (Grysbowski and Shah 1990).

The performance of fiber-reinforced concrete (FRC) is governed by the ratio of the elastic modulus of the fiber to the elastic modulus of the matrix, the strength of the fiber-matrix bond, the aspect ratio (fiber length/fiber diameter) of the fiber, and the material properties of the fiber (Bentur and Mindess 1991). Different fibers types and geometries yield different composite performance. Only relatively small amounts, usually less than 1% by volume, of fiber reinforcement can be added to conventional concrete because the fibers significantly reduce the workability of the fresh concrete. At this dosage, fibers can improve shrinkage cracking resistance and slightly enhance ductility. To maximize the benefits of fibers, including increasing the tensile strength and ductility of the composite and producing a strain-hardening response, larger doses must be added to the matrix. In cement-based materials, this requires special processing techniques. These composites are referred to as high-performance, fiber-reinforced cementitious composites (HPFRCC). The relative performance of plain, normal-strength concrete, conventional FRC, and HPFRCC is shown in Figure 1.

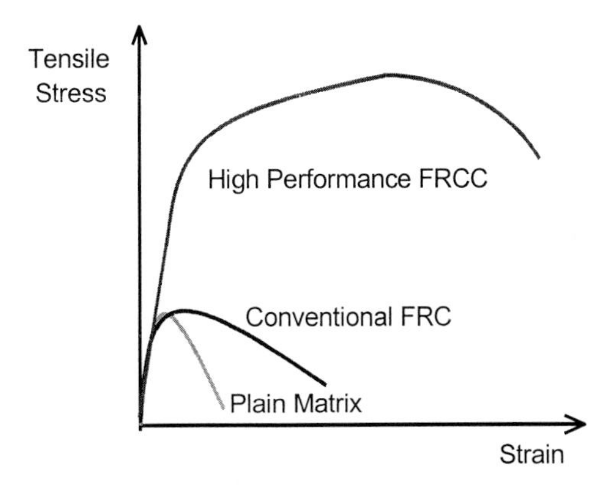

Figure 1: Tensile response of concrete, FRC, and HPFRCC.

Extruded fiber-reinforced cementitious composites

Extrusion, a common processing technique for polymers and ceramics, was recently adapted for the production of HPFRCC at ACBM (Shao and Shah 1997 and Shao *et al.* 1995). A highly viscous, dough-like mixture of cement paste and fibers (2%-10% by volume) is forced through a die to produce an element of desired cross-section. Production is continuous, and a variety of shapes, such as thin sheets for siding and roofing tiles, pipe, and cellular sections can be extruded, as shown in Figure 2. The high compressive and shear forces required to extrude the composite result in a dense matrix, a strong fiber-matrix bond and alignment of fibers in the direction of extrusion. The extruded composites are stronger and tougher than a cast composite of the same material, as shown in Figure 3 (Shah *et al.* 1998b).

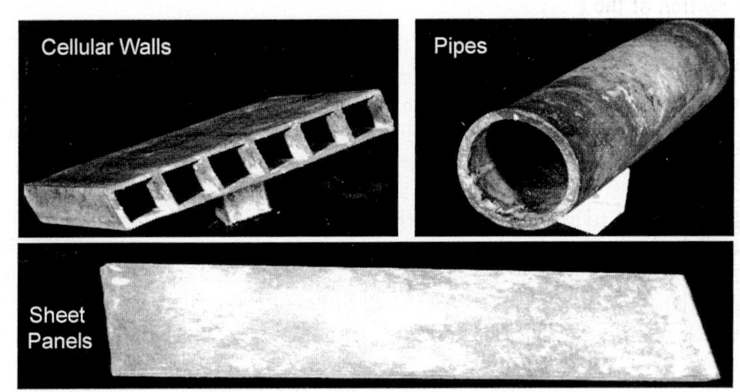

Figure 2: Extruded cross-sections.

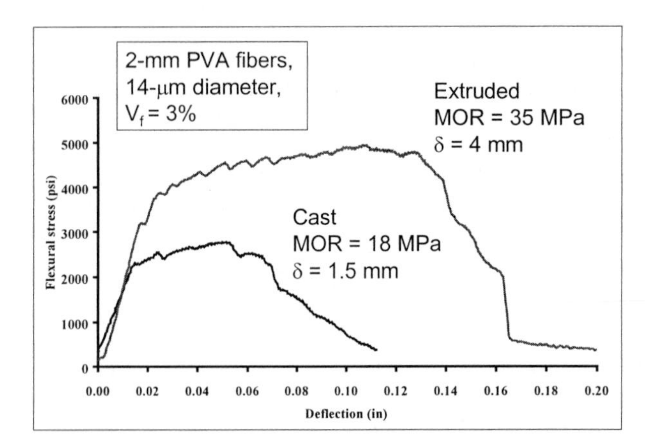

Figure 3: Flexural response of extruded and cast composites (Shah *et al.* 1998b).

Successful extrusion requires the mixture to be soft enough to flow through the extruder but stiff enough to maintain its shape after exiting the die. The highly specialized cementitious matrix has a low w/b (w/b ~ 0.25) with admixtures, and mineral additives. The rheology of the extrudate has been improved with the replacement of a portion of the cement with Class F fly ash (Peled *et al.* 2000a). The spherical fly ash particles make the paste easier to extrude. Fly ash is also less expensive and more environmentally friendly than cement. It improves the flexural performance of fiber-reinforced extruded composites by increasing the likelihood of fiber pullout instead of fiber fracture at failure. Fly ash also improves the durability of glass-fiber-reinforced extruded composites by reducing the alkalinity of the cement matrix (Cyr *et al.* 2001)

Hybrid fiber reinforcement

Material performance can be optimized for given applications by combining different types of fiber reinforcement in hybrid fiber composites. For example, glass fibers tend to be strong but relatively brittle and form a strong fiber-matrix bond. Polyvinyl alcohol (PVA) fibers are weaker, but more ductile than glass fibers. These fibers were combined in extruded composites to produce composites that were both strong and tough with strain hardening behavior (Peled *et al.* 2000b). In addition, it was shown that a portion of the PVA fibers can be replaced with less expensive polypropylene (PP) fibers without any significant reduction in performance. The performance of extruded single-fiber composites and extruded hybrid-fiber composites is shown in Figure 4.

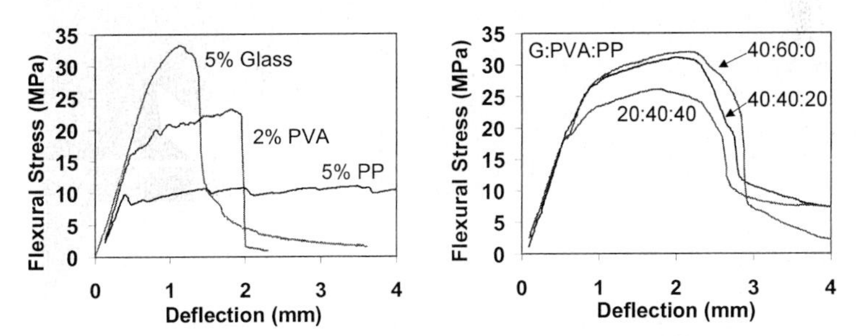

Figure 4: Flexural response of single-fiber and hybrid-fiber reinforced extruded composites. Total $V_f = 5\%$ for hybrids (Peled *et al.* 2000b).

Different size fibers can also be combined to enhance performance. Lawler (2001) found that combining 0.5% steel macrofibers (500 μm diameter, 30 mm length) with 0.5% PVA microfibers (14 μm diameter, 12 mm length) significantly improved both the pre- and post-peak performance of mortar. The microfibers bridge microcracks as they form, preventing them from coalescing and increasing the tensile strength of the composites. As the cracks grow, the steel macrofibers bridge the larger cracks, increasing the ductility of the composite. This hybrid combination also significantly reduces the permeability of cracked concrete.

SHRINKAGE REDUCING ADMIXTURES

It is well known that concrete shrinks as it dries. If the concrete is restrained, tensile stresses will develop and cracking can occur. This is of particular concern in pavements, bridge decks, and industrial floors where the volume-to-surface ratio is low. The likelihood of shrinkage cracking depends on the free shrinkage, the creep relaxation, age-dependent material properties, such as tensile strength, and the degree of restraint of the concrete. Shrinkage cracking is of greater concern in HSC where increased early-age free shrinkage, reduced creep and increased brittleness result in earlier cracking (Wiegrink *et al.* 1996). Attempts to reduce shrinkage cracking have included the use of secondary reinforcement to keep cracks from widening, the use of fiber reinforcement to prevent microcracks from coalescing, the use of expansive cement—a cement that expands during hydration creating a compressive prestress that counteracts the tensile stresses that develop under restrained shrinkage—and reducing the w/c. Research at the ACBM Center has included the development of experimental techniques and theoretical models to assess the cracking potential due to shrinkage (Grysbowski and Shah 1990, Weiss *et al.* 1998, Shah *et al.* 1998c). In addition, a shrinkage reducing admixture (SRA) has been developed to improve the shrinkage cracking resistance of concrete (Shah *et al.* 1992).

The SRA is a propylene glycol derivative sold by Grace Construction Products as Eclipse™ that reduces free shrinkage by reducing the surface tension of water. One cause of drying shrinkage in concrete is the surface tension that develops in small pores as water evaporates (Balogh 1996). As cement reacts with water, calcium silicate hydrate (CSH) forms in water-filled spaces. These spaces are not completely filled by the CSH so a capillary pore network develops. As the water evaporates, a meniscus forms in the pores. The surface tension of the water pulls the pore walls inward causing the concrete to shrink. This phenomenon occurs pores with radii from 2.5 nm to 50 nm. The reduction in the surface tension of the water by the SRA reduces the capillary pore forces that cause shrinkage, thus reducing the drying shrinkage of the concrete.

The shrinkage reducing admixture was tested in both normal- and high-strength concrete at 1% and 2% by weight of cement (Weiss *et al.* 1998). In general, 2% SRA showed much greater improvements over mixtures without SRA than 1% SRA did. With 2% SRA, the free shrinkage at 49 days was reduced by 42% in both normal- and high-strength concrete. Because the SRA greatly reduced the free shrinkage, the age of cracking of restrained ring specimens containing SRA was increased. Rings of normal-strength concrete cracked 10 days after casting, on average. With 2% SRA, only one of the rings cracked before the end of the tests, 50 days after casting. For high-strength concrete, the mixture without SRA cracked at 3.2 days, while 2% SRA delayed cracking until 11.6 days. The rings that cracked at later ages also had much smaller cracks.

SELF-COMPACTING CONCRETE

Self-compacting concrete (SCC) is concrete that is designed to flow under its own weight. This eliminates the need for vibration, making it easy to place in dense reinforcement and complicated formwork and reducing construction time and costs. SCC must be fluid enough to fill a mold without vibration but not segregate. Viscosity agents, such as superplasticizer, and fine mineral admixtures are commonly used.

SCC is characterized by deformability, segregation resistance, and passing ability. Deformability, represented by the flow or fluidity of the SCC, is a measure of yield stress (Ozawa *et al.* 1992). It is quantified as the slump flow diameter, which is obtained from a modified slump test (Takada 2000). The segregation resistance is sufficient if the aggregates are uniformly distributed throughout the cement paste. It is evaluated using a penetration apparatus (Bui 2000). The segregation of aggregates has been modeled using Stoke's Law at the ACBM Center (Saak *et al.* 2000). The passing ability indicates how well the fresh SCC can flow through the spaces between rebar. It is measured using an L-box apparatus (Tangtermsirikul and Khayat 2000). These properties of SCC are affected by the rheology of the cement paste and the average diameter and spacing (D_{ss}) of aggregates.

To facilitate the design of SCC, a paste rheology model was developed at the ACBM Center (Bui *et al.* 2001). The goal was to determine the rheology of cement paste required to obtain SCC with sufficient deformability and segregation resistance for given aggregate properties. The paste rheology is characterized by yield stress, measured as the paste flow diameter, and the viscosity, measured with a standard rheometer. The model establishes, for given aggregate properties and doses, the minimum apparent paste viscosity, the minimum paste flow and the optimum flow-viscosity ratio required to achieve SCC with acceptable segregation resistance and deformability.

In a recent study at the ACBM Center, the parameters for the paste rheology were examined by varying the total aggregate ratio, the paste volume, the w/b, and the cement, fly ash and superplasticizer contents (Shah *et al.* 2002). These mixes had different degrees of deformability and segregation resistance. Plots of D_{ss} vs. flow diameter and D_{ss} vs. apparent viscosity were obtained for a constant average aggregate diameter (Shah *et al.* 2002). For a constant aggregate diameter and a given D_{ss}, there exists a minimum paste flow diameter below which the SCC exhibits poor deformability and a minimum viscosity below which the SCC segregates. Fresh SCC with a larger D_{ss} requires a smaller paste flow diameter and higher viscosity to achieve acceptable performance. To increase the aggregate spacing (D_{ss}), the paste volume of the mixture is increased, or the aggregate volume is decreased. This results in reduced friction between aggregates (Bui and Montgomery 1999). A higher viscosity is required to hold the aggregates together, and a smaller paste flow diameter is necessary to achieve good deformability.

NONDESTRUCTIVE EVALUATION

Nondestructive evaluation (NDE) uses stress waves to determine mechanical properties, the presence, location, and extent of damage, or the degree of hydration of concrete structures. Stress pulses are applied to the structure, and the transmission or reflection of the resulting waves or the vibration response of the structure is measured. Early NDE techniques relied on the transmission of a wave through a structure, which required access to both sides of the structure, making them inappropriate for concrete pavements or slabs. They were also unable to detect early stages of deterioration, i.e., the presence of microcracks, resulting from damage due to freeze-thaw cycling, sulfate attack, or rebar corrosion.

Recently, several new NDE techniques have been developed at the ACBM Center (Shah *et al.* 2000). These techniques are sensitive to the early stages of damage. A self-calibrating, one-sided technique, suitable for use on concrete pavements, was developed and shown to be sensitive to the presence of cracks (Popovics *et al.* 1998). The structural vibration frequency response can be tracked during test loading (Subramaniam *et al.* 1998) and was used to predict the remaining life of a specimen subjected to fatigue loading. Another newly developed technique, discussed in detail here, uses the reflection of wave energy at a steel-concrete interface to monitor the setting of fresh concrete.

A one-sided, ultrasonic technique has been developed to monitor the hardening and setting of fresh concrete (Öztürk *et al.* 1999, Rapoport *et al.* 2000, and Akkaya *et al.* 2001). The change in ultrasonic shear wave reflections over time between a steel plate and hardening concrete are measured to monitor the setting process. High-frequency, ultrasonic shear waves are transmitted through the steel plate into the fresh concrete. As the wave reaches the steel-concrete interface, a portion of it is transmitted through the concrete and the rest is reflected back to the transducer. The wave energy reflected at the steel-concrete interface is called the wave reflection factor (WRF). It depends on the differences in the acoustic impedance, the product of material density and wave velocity, of the steel and the hardening concrete. Initially, the value of the WRF is unity because the concrete is in a liquid state, which cannot transmit shear waves, so all of the wave energy is reflected at the interface. As the concrete stiffens, more of the wave energy is transmitted into the concrete and the WRF decreases. A typical WRF vs. time plot is shown in Figure 5. In the initial stage, the WRF equals one. At the end of the induction period, there is a sharp drop in the WRF as the concrete begins to harden. The WRF eventually approaches a final asymptote. The significance of this asymptote is not yet known. The WRF test setup is shown in Figure 6.

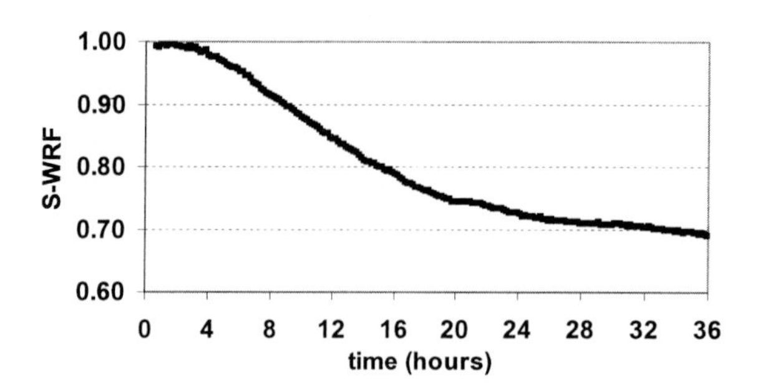

Figure 5: A typical WRF curve (Rapoport *et al.* 2000).

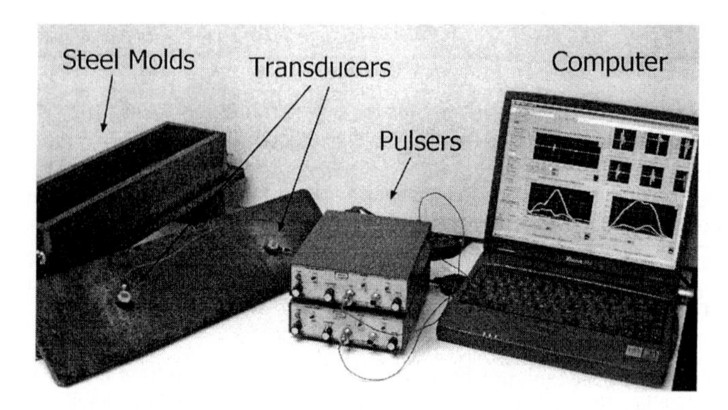

Figure 6: The WRF test apparatus.

Initial tests demonstrated the sensitivity of the WRF method to the presence of different admixtures, including accelerators, retarders, superplasticizer, and silica fume (Rapoport *et al.* 2000). WRF, pin-penetration measurements and dynamic modulus tests were performed simultaneously. The results were correlated and critical points on the WRF curve were shown to correspond to critical points in set time, temperature and dynamic modulus curves (Rapoport *et al.* 2000).

Additional work has further demonstrated the sensitivity of the technique to changes in mixture design and curing conditions and correlated the wave energy attenuation, or the inverse of the WRF, with early-age strength gain. Akkaya *et al.* (2001) measured wave energy attenuation for two different mixtures at three different curing temperatures. It is well understood that temperature affects hydration (Mindess and Young 1981). This trend is evidenced in WRF attenuation measurements. Akkaya *et al.* (2001) also compared the wave energy attenuation in mortar and concrete. Three batches of each mixture were tested. The mortar specimens showed good repeatability, while the concrete did not. The attenuation curves for mortar had the same shape and approached the same final asymptote. The shape of the attenuation curves for concrete was consistent—different stages of hydration, as indicated by distinct points on the curve, were reproduced—but each sample approached a different asymptote. This might be due to local differences in the concrete, and suggests that the homogeneity of the mixture and the final attenuation value are related.

Compressive strength tests and ultrasonic measurements were performed simultaneously to correlate wave energy attenuation with early-age strength gain (Akkaya *et al.* 2001). The results were correlated to predict strength evolution from the change in wave energy attenuation. The relationship between strength and attenuation is linear up to three days. The linearity is not affected by changes in curing temperature or mixture design. Several tests performed outdoors demonstrated that fluctuating ambient temperatures also did not influence the linearity of the relationship between strength evolution and change in attenuation. This procedure requires the determination of one or two compressive strength values at the beginning of strength evolution—within the first day—to calibrate the strength-change-in-attenuation relationship.

CONCLUSIONS

Some of the most recent developments in concrete technology were discussed. Reactive powder concrete, one form of UHSC, has a compressive strength of 200 MPa. It is currently used in long, slender pedestrian bridges and nuclear power plants and has potential for use in pipes, tunnel linings, and nuclear waste containment. Microfiber reinforcement is used to overcome the inherent brittleness of concrete. Fiber-reinforced cementitious composites for specialized applications are produced with special processing techniques, such as extrusion, and hybrid-fiber reinforcement. Significant reductions in drying shrinkage, and thus the potential for shrinkage cracking, are achieved with the use of a newly designed shrinkage reducing admixture. The design of SCC is facilitated with a recently developed rheological model. Finally, a new nondestructive technique is used to monitor the setting and predict the strength gain of fresh concrete.

ACKNOWLEDGEMENTS

Much of the work presented here was funded by the Center for Advanced Cement Based Materials at Northwestern University. In addition, the research on extrusion is currently being funded by a grant from the National Science Foundation in support of the Partnership for Advancing Technology in Housing. The assistance of these organizations is gratefully acknowledged.

REFERENCES

Akkaya Y., Voigt, T., Kolluru, S. and Shah, S.P. (2001). Non-destructive measurement of concrete strength by an ultrasonic wave reflection method. Submitted to *RILEM Journal.*

Balogh, A. (1996). New admixture combats concrete shrinkage. *Concrete Construction.* 546-551.

Bentur, A. and Mindess, S. (1991), *Fibre Reinforced Cementitious Composites,* Elsevier Applied Science, London, UK.

Bui, V.K., Montgomery, D. (1999). Mixture proportioning method for self-compacting concrete with minimum paste volume. *Proceedings of the first RILEM Symposium on Self-Compacting Concrete.* Stockholm, Sweden. 373-384.

Bui, V.K. (2000). Penetration test of resistance to segregation. *Self-Compacting Concrete – State of the Art Report of RILEM Technical Committee 174-SCC.* ed. A. Skarendahl and O. Petersson. RILEM Report 23, 136-138.

Bui, V.K., Akkaya, Y. and Shah, S.P. (2001). A rheological model for self-consolidating concrete. Submitted to *ACI Materials Journal.*

Cyr, M.F., Peled, A. and Shah, S.P. (2001). Improving the performance of glass fiber reinforced extruded cementitious composites. *Proceedings of the 12th International Congress of the International Glassfibre Reinforced Concrete Association. GRC 2001.* ed. N. Clarke, R. Ferry. Dublin, Ireland. 163-172.

Federal Highway Administration (FHWA). (2002). A new and improved high-performance concrete. *Focus.* http://www.tfhrc.gov/focus/feb02/highperformance.htm.

Grysbowski, M. and Shah, S.P. (1990). Shrinkage cracking of fiber reinforced concrete. *ACI Materials Journal.* **87:2**, 138-148.

Kosmatka, S.H., Kerkhoff, B., Panarese, W.C. (2001), *Design and Control of Concrete Mixtures,* 14th ed., Portland Cement Association, Skokie, IL.

Lawler, J.S. (2001), *Hybrid Fiber-Reinforcement in Mortar and Concrete*, Ph.D. Thesis, submitted to Northwestern University, Evanston, IL.

Mindess, S., and Young, J.F. (1981), *Concrete,* Prentice Hall, Englewood Cliffs, New Jersey.

Ozawa, K., Maekawa, K. and Okamura, H. (1992). Development of high performance concrete. *Journal of the Faculty of Engineering. The University of Tokyo.* **3**, 381-439.

Öztürk, T., Rapoport, J., Popovics, J.S. and Shah, S.P. (1999). Monitoring the setting and hardening of cement-based materials with ultrasound. *Concrete Science and Engineering* **1**, 83-91.

Peled, A., Cyr, M.F. and Shah, S.P. (2000a). High content of fly ash (Class F) in extruded cementitious composites. *ACI Materials Journal.* **97:5**, 509-517.

Peled, A., Cyr, M. and Shah, S.P. (2000b). Hybrid fibers in high performance extruded cement composites. *Proceedings of International RILEM Symposium BEFIB'2000.* Lyon, France.

Popovics, J.S., Song, W. and Achenbach, J.D. (1998). A study of surface wave attenuation measurement for application to pavement characterization. *Structural Materials Technology III: An NDT Conference.* ed. R.D. Medlock and D.C. Laffey. Vol. 3400. International Society for Optical Engineering. Bellingham, WA. 300-308.

Powers, T.C., and Brownyard, T.L. (1948). Studies on the physical properties of hardened portland cement paste. *Bulletin 22 of the Portland Cement Association.* Chicago, IL.

Rapoport, J., Popovics, J.S., Subramaniam, K.V. and Shah, S.P. (2000). The use of ultrasound to monitor the stiffening process of portland cement concrete with admixtures. *ACI Materials Journal* **97:6**, 675-683.

Saak, A.W., Jennings, H.M. and Shah, S.P. (2001). New methodology for designing self-compacting concrete. *ACI Materials Journal.* **98:6**, 429-438.

Semioli, W.J. (2001). The new concrete technology. *Concrete International.* **23:11**, 75-79.

Shah, S.P. (1991). Do fibers increase the tensile strength of cement-based matrices? *ACI Materials Journal.* **88:6**, 595-602.

Shah S.P., Karaguler, M.E. and Sarigaphuti M. (1992). Effects of shrinkage reducing admixture on restrained shrinkage cracking of concrete. *ACI Materials Journal*. **89:3**, 289-295.

Shah, S.P. and Weiss, W.J. (1998a). Ultra high performance concrete: a look to the future. *Presented at Zia Symposium at ACI Spring Convention*. Houston, TX.

Shah, S.P., Peled, A., DeFord, D., Akkaya, Y., and Srinivasan, R. (1998b). Extrusion technology for the production of fiber-cement composites. *Sixth International Inorganic-Bonded Wood and Fiber Composite Materials Conference*. Sun Valley, Idaho.

Shah, S.P., Ouyang, C., Marikunte, S., Yang, W. and Becq-Giraudon, E. (1998c). A fracture mechanics model for shrinkage cracking of restrained concrete rings. *ACI Materials Journal*. **95:4**, 339-346.

Shah, S.P., Popovics, J.S., Subramaniam, K.V. and Aldea, C.M. (2000). New directions in concrete-health monitoring technology. *Journal of Engineering Mechanics*. **126:7**, 754-760.

Shah, S.P., Akkaya, Y. and Bui, V.K. (2002). Innovations in microstructure, processing, and properties. Accepted for publication in *Proceedings of the International Congress – Challenges of Concrete Construction*. September 2002, Dundee, Scotland.

Shao, Y., Marikunte, S. and Shah, S.P. (1995). Extruded fiber-reinforced composites. *Concrete International*. **17:4**, 48-52.

Shao, Y. and Shah, S.P. (1997). Mechanical properties of PVA fibers reinforced cement composites fabricated by extrusion processing. *ACI Materials Journal*. **94:6**, 555-564.

Subramaniam, K.V., Popovics, J.S. and Shah, S.P. (1998), Monitoring fatigue damage in concrete. *Materials Research Society Symposium Proceedings*, Materials Research Society, Warrendale, Pa. **503**, 151-158.

Takada, K. (2000). Slump flow test. *Self-Compacting Concrete – State of the Art Report of RILEM Technical Committee 174-SCC*. ed. A. Skarendahl and O. Petersson. RILEM Report 23, 117-119.

Tangtermsirikul, S. and Khayat, K. (2000). Fresh concrete properties. *Self-Compacting Concrete – State of the Art Report of RILEM Technical Committee 174-SCC*. ed. A. Skarendahl and O. Petersson. RILEM Report 23, 17-22.

Wiegrink, K., Marikunte, S. and Shah, S.P. (1996). Shrinkage cracking of high-strength concrete. *ACI Materials Journal*. **93:5**, 409-415.

Weiss, W.J., Yang, W. and Shah, S.P. (1998). Shrinkage cracking of restrained concrete slabs. *Journal of Engineering Mechanics*. **124:7**, 765-774.

Advances in Building Technology, Volume I
M. Anson, J.M. Ko and E.S.S. Lam (Eds.)

HUMAN REQUIREMENTS IN FUTURE AIR-CONDITIONED ENVIRONMENTS

P. Ole Fanger, Professor, D.Sc., Hon.D.Sc.

Director, International Centre for Indoor Environment and Energy,
Technical University of Denmark, DK-2800 Lyngby, Denmark
www.ie.dtu.dk

ABSTRACT

Air-conditioning of buildings has played a very positive role for economic development in warm climates. Still its image is globally mixed. Field studies demonstrate that there are substantial numbers of dissatisfied people in many buildings, among them those suffering from SBS symptoms, even though existing standards and guidelines are met. A paradigm shift from rather mediocre to excellent indoor environments is foreseen in buildings in the 21st century. Based on existing information and on new research results, five principles are suggested as elements behind a new philosophy of excellence in the built environment: better indoor air quality increases productivity and decreases SBS symptoms; unnecessary indoor pollution sources should be avoided; the air should be served cool and dry to the occupants; personalized ventilation, i.e. small amounts of clean air, should be provided gently, close to the breathing zone of each individual; individual control of the airflow and or the thermal environment should be provided. These principles of excellence should be combined with energy efficiency and sustainability of future buildings.

KEYWORDS

Indoor air quality; Thermal comfort; Productivity; Personalized ventilation; Ventilation

INTRODUCTION

Air-conditioning of buildings has been essential for economic development in areas with warm climates or warm summers. There are numerous examples of the positive impact of air-conditioning, e.g. in eastern Asia, where over the last 30 years a remarkably strong economic growth rate has been experienced which would hardly have been possible without the widespread use of air-conditioning.

Today air-conditioning is used in many parts of the world, often in combination with heating and ventilation in HVAC systems. The image of such systems, however, is not always positive. The purpose of most systems is to provide thermal comfort and an acceptable indoor air quality for human occupants. But numerous field studies (Fisk et al., 1993; Mendell, 1993; Bluyssen et al., 1996; Lee et al., 1999; Sekhar, S.C. et al., 2000) have documented substantial rates of dissatisfaction with the indoor environment in many buildings. One of the main reasons is that the requirements of existing ventilation standards and guidelines (ASHRAE, 1984; CEN, 1998) are quite low. The philosophy of these documents has been to establish an indoor air quality where less than a certain percentage (e.g. 15%, 20% or 30%) of people are dissatisfied with the indoor air quality, while the rest may find the IAQ barely acceptable. A similar thinking applies to the thermal environment. This philosophy behind the design of HVAC systems has led in practice to quite a number of dissatisfied persons (as predicted), while few seem to be ready to characterize the indoor environment as outstanding. At the same time numerous negative effects on human health are reported: many persons suffer from SBS (Sick Building Syndrome) symptoms (Fisk et al., 1993; Mendell, 1993, Bluyssen et al., 1996) and a dramatic increase in cases of allergy and asthma have been related to poor IAQ, especially in dwellings (Bornehag et al., 2001).

I think it is fair to say that the indoor air quality is quite mediocre in many air-conditioned or mechanically ventilated buildings, even though existing standards may be met (Lee et al, 1999; Sekhar et al., 2000). Two factors have contributed to a lowering of indoor air quality during recent decades: a general decrement in ventilation rates to save energy; and an increment in indoor air pollution sources from the many new materials and electronic processes introduced during this period. We need a paradigm shift in the new century to search for excellence in the indoor environment. Our aim should be to provide indoor air that is perceived as fresh, pleasant and stimulating, with no negative effects on health, and a thermal environment perceived as comfortable by almost all occupants. In achieving this aim, due consideration must be given to energy efficiency and sustainability. Do we have the necessary information to implement this in practice? Yes, on thermal comfort we do have a comprehensive database, while our knowledge on indoor air quality is still rather incomplete. This reflects the complexity of the interaction between indoor air quality and human comfort and health. But we do have some information on IAQ, however, as well as important new research results that will have a significant impact on the design of future air-conditioned or ventilated spaces for human occupants.

This article will discuss some principles and new research results believed to be essential for providing excellence in future indoor environments.

PRODUCTIVITY AND INDOOR AIR QUALITY

Three recent independent studies document that the quality of indoor air has a significant and positive influence on the productivity of office workers. In one study, a well-controlled normal office (field lab) was used in which two different air qualities were established by including or excluding an extra pollution source, invisible to the occupants (Wargocki et al., 1999). The two cases corresponded to a low-polluting and a non-low-polluting building as specified in the new European guidelines for the design of indoor environments (CEN, 1998).The same subjects worked for 4-1/2 hours on simulated office work in each of the two air qualities. The ventilation rate and all other environmental factors were the same under the two conditions. The productivity of the subjects was found to be 6.5% higher ($P<0.003$) in good air quality (Figure 1) and they also made fewer errors and experienced fewer SBS symptoms. This study performed in Denmark has later been repeated in Sweden with similar results (Lagercrantz et al., 2000ree different ventilation rates: 3, 10 and 30 l/s·person (Wargocki et al., 2000a). The productivity increased significantly with increased ventilation (Figure 2). The three studies involving seven experimental conditions and 90 subjects have been analysed as a whole, relating productivity to perceived air quality (Wargocki et al., 2000b). The results are presented in Figure 3 and show a significant influence of perceived air quality on productivity in offices. An improvement of perceived air quality by 1 decipol increased productivity by 0.5%. The results of three blind studies document that improved air quality increases productivity significantly.

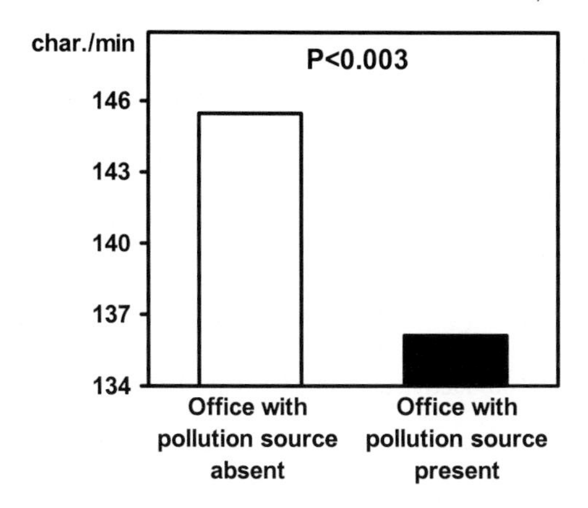

Figure 1. Impact of indoor air pollution on productivity, i.e. number of characters typed on a PC (Wargocki et al., 1999)

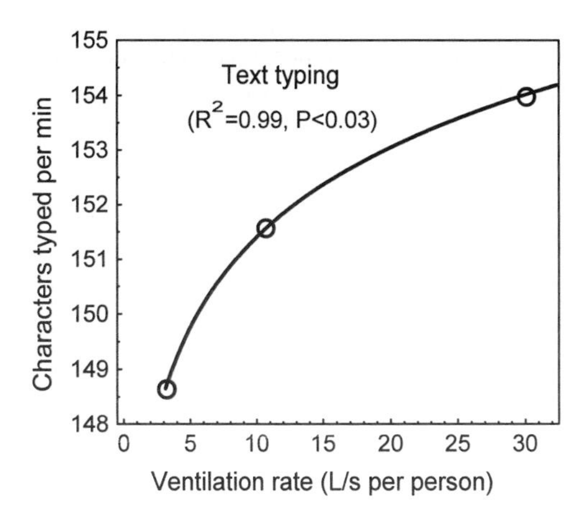

Figure 2. Impact of ventilation rate on productivity (Wargocki et al., 2000a)

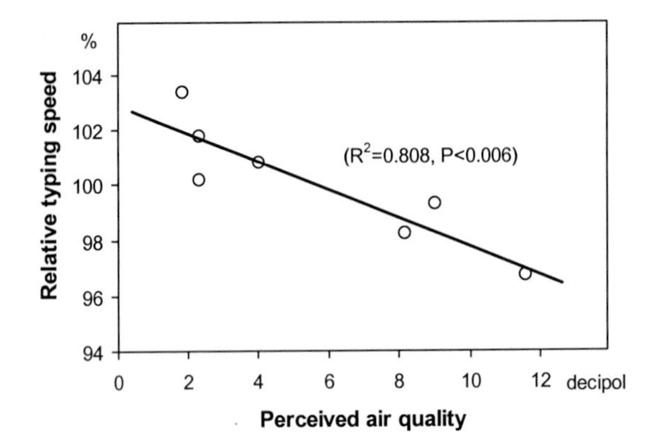

Figure 3. Relation between perceived air quality and productivity (Wargocki et al., 2000b)

POLLUTION SOURCE CONTROL AND VENTILATION

Avoiding unnecessary indoor air pollution sources is the most obvious way of improving the IAQ. Its effect on productivity and SBS symptoms has been demonstrated in the studies discussed above (Wargocki et al., 1999; Lagercrantz et al., 2000). Source control has also been used with great success outdoors and is the reason why the outdoor air quality in many cities in the developed world is much better today than it was 20 or 50 years ago.

In the new European guidelines for the indoor environment (CEN CR 1752) (CEN, 1998), there is strong encouragement to design low-polluting buildings and recommendations on low-polluting building materials are given. The same applies to ASHRAE Standard 62 (1989). Systematic selection of materials planned to be used in a building to avoid the well-known cases of SBS caused by polluting materials is common practice in several countries, e.g. in Scandinavia. Other pollution sources are electronic equipment in offices. A recent study (Bakó-Biró, Z. et al., 2002) has thus documented that personal computers can be a serious indoor pollution source. Air pollution from PCs of a common brand used for three months were found to decrease the perceived air quality and productivity significantly. Each computer had a sensory pollution source strength of 3 olfs, i.e. polluted equally as much as 3 persons. PCs and other electronic equipment provide in many cases an important pollution load requiring much extra ventilation. An obvious solution is that the PCs be modified so that their pollution load is negligible in the future. Pollution sources in the HVAC system are a serious fault, degrading the quality of the air even before it is supplied to the conditioned space. Used particle filters have been shown to degrade the perceived air quality and to contribute to SBS symptoms (Clausen, 2002). The selection of materials, components and processes, as well as maintenance of the HVAC system, should be given high priority in future.

Source control is the obvious way to provide good indoor air quality with a simultaneous decrease in the consumption of energy. But increased ventilation also improves the IAQ and decreases SBS symptoms as demonstrated by Sundell's classic field studies in 160 office buildings studies (Sundell, 1994) (Fig. 4). A multidisciplinary group pf scientists in Europe (EUROVEN) recently reviewed the peer-reviewed literature concerning the relation between health and ventilation. The group concluded that decreasing the outdoor air supply rates below 25 L/s per person increased the risk of SBS symptoms, increases short-term sick leave and decreases procuctivity among occupants of office buildings (Wargocki et al., 2002).

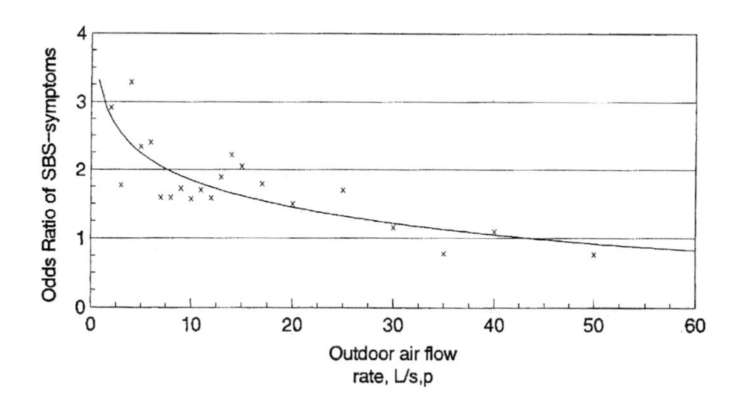

Fig. 4. SBS symptoms as a function of outdoor ventilation rate in 160 office buildings in Sweden (Sundell, 1994)

SERVE THE AIR COOL AND DRY

In ventilation standards, indoor air humidity has for decades been overlooked. It has been generally accepted that the relative humidity was rather unimportant for human beings as long as it was kept between approximately 30% and 70% (CEN, 1998; ASHRAE, 1989). This consensus stems from the fact that the humidity in the comfort range of temperatures has a minor impact on the thermal sensation of the entire human body (ASHRAE, 1992, ISO, 1993).

All existing ventilation standards and guidelines are based on the following thinking: there are certain pollution sources in a space and ventilation is required in order to dilute the chemical pollutants to a level where they are perceived as acceptable to humans. The thinking is that air is perceived exclusively by the olfactory and chemical senses and that the perception depends only on the chemical composition of the air. The implied conclusion is that the required ventilation is independent of temperature and humidity. However, a paper by Berglund and Cain indicated in 1989 that temperature and humidity have an impact on the perception of clean air in a climate chamber (Berglund and Cain, 1989).

New comprehensive studies at the International Centre for Indoor Environment and Energy have confirmed that perceived air quality is strongly influenced by the humidity and temperature of the air we inhale. People prefer rather dry and cool air. The strong effect of humidity and temperature on perceived air quality was proven in experiments where 36 subjects judged the acceptability of air polluted by different typical building materials in a climate chamber (Fang et al., 1997).

It is the effect of humidity and temperature combined in the enthalpy of the air that is essential for the perceived air quality as shown in Figure 5. The enthalpy was changed in the room while the chemical composition of the air was constant and the thermal sensation for the entire body was kept neutral by modification of the subjects' clothing. The acceptability did not change with time, i.e. no adaptation took place.

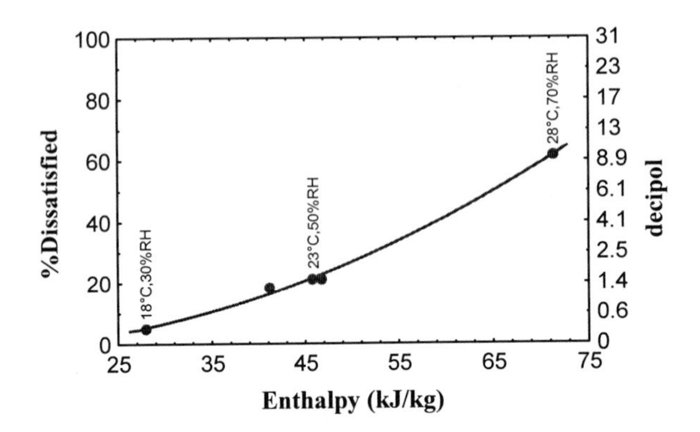

Fig. 5. Perception of clean air during whole-body exposure of persons to different levels of indoor air enthalpy (Fang et al., 1997).

The impact of enthalpy on acceptability or on perceived air quality expressed in percent dissatisfied or decipol is strong. Two other independent studies at the International Centre for Indoor Environment and Energy where approximately 70 subjects were exposed to numerous combinations of humidity and temperature on the face also showed an excellent correlation between enthalpy and acceptability (Fang et al., 1998; Toftum et al., 1998) with an even stronger impact of enthalpy.

Humans obviously like a sensation of cooling of the respiratory tract each time air is inhaled. This causes a sensation of freshness which is felt pleasant. If proper cooling does not occur, the air may be felt stale, stuffy and unacceptable. A high enthalpy means a low cooling power of the inhaled air and therefore an insufficient convective and evaporative cooling of the respiratory tract, and in particular the nose. This lack of proper cooling is closely related to poorly perceived air quality. The phenomenon is analogous to the well-known strong impact of temperature on perceived quality during intake of drinks, e.g. water or wine.

Heat loss through respiration is only around 10% of the total heat loss from the body and humidity and temperature of the inhaled air have therefore only a small impact on the thermal sensation for the human body as a whole. This is presumably why humidity has previously been overlooked. The new studies show that the local effect of temperature and humidity on the respiratory tract and therefore on perceived air quality is one order of magnitude higher than for whole-body thermal sensations. This new evidence has quite dramatic practical consequences. It is obvious that the enthalpy has a strong impact on ventilation requirements and therefore on energy consumption. Fang et al. (1999) showed thus in the most recent study that people perceive the IAQ better at 20°C and 40% RH and a small ventilation rate of 3.5 l/s·person than at 23°C and 50% RH at a ventilation rate of 10 l/s·person.

It is advantageous to maintain a moderately low humidity and a temperature that is at the lower end of the range required for thermal neutrality for the body as a whole. This will improve the perceived air quality and decrease the required ventilation. Field studies (Andersson et al., 1995; Krogstad et al., 1991) show that moderate air temperatures and humidities also decrease SBS symptoms. Moderate temperatures and humidities may furthermore result in potential energy savings, during both winter and summer. The advice is therefore: serve the air cool and dry.

SERVE THE AIR WHERE IT IS CONSUMED

In many ventilated rooms the outdoor air supplied is of the order of magnitude of 10 l/s·person. Of this air, only 0.1 l/s·person, or 1%, is inhaled. The rest, i.e. 99% of the supplied air, is not used. What a huge waste! And the 1% of the ventilation air being inhaled is not even clean. It is polluted by bioeffluents, emissions from building materials and sometimes even by ETS before it is inhaled.

According to normal engineering practice, full mixing of clean air and pollutants seems to be an ideal. With displacement ventilation systems, one may be proud of reaching a ventilation effectiveness of 1.2. What I foresee in the future are systems that supply rather small quantities of clean air close to the breathing zone of each individual. The idea is to serve to each occupant, clean air that is as unpolluted as possible by the pollution sources in the space. We would hesitate to drink water from a swimming pool polluted by human bioeffluents. Still we accept consuming indoor air polluted by human bioeffluents and other contaminants. Why not serve small quantities of high-quality air direct to each individual rather than plenty of mediocre air throughout the space? Such air may be provided by a personalized ventilation system with a terminal device close to the breathing zone (Fig. 6). The person inhales clean, cool and dry air from the core of the jet where the air is unmixed with polluted room air and has a low velocity and turbulence which do not cause draught. The position of the air terminal devide should be easily movable by each individual, who also should be able to control the airflow, its direction and its temperature. Studies on personalized ventilation have recently been published by Kaczmarczyk et al. (2002) and Cermak (2002).

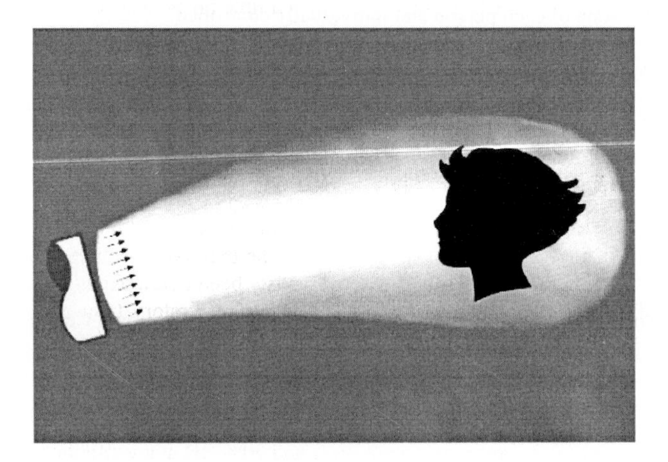

Fig. 6. The principle of personalized ventilation: small amounts of cool, dry, and clean air supplied directly and gently to a person's breathing zone

INDIVIDUAL THERMAL CONTROL

In buildings where many people occupy the same space it is difficult to provide thermal comfort for everyone at the same time .A compromise is therefore usually aimed at which satisfies most people. The PMV model (Fanger, 1970) included in ISO, ASHRAE and European standards and guidelines (ASHRAE, 1989; ISO, 1993) specifies such compromises. Due to well-known differences in preferred temperature between people, there will still be persons who are dissatisfied with the thermal environment. To accommodate these differences between people, individual thermal control is an obvious solution. This may in practice be realized by slight local heating or cooling of each individual by conduction (through seat and chair back), radiation or convection.

CONCLUSIONS

Air-conditioning of buildings has played an important role for the economic development in areas of the world with warm climates or warm summers. Still, the indoor environment in many air-conditioned buildings is rather mediocre and gives rise to frequent complaints, even though present standards are met. A paradigm shift is foreseen with a search for excellence of the indoor environment rather than the present effort to reduce dissatisfaction and complaints.

The following principles may be useful steps in realizing such a new philosophy of excellence.

- Better indoor air quality pays as it results in higher productivity and causes fewer SBS symptoms.
- Unnecessary pollution sources frequently occurring in present buildings should be avoided.
- The air should be served cool and dry for people.
- Small amounts of clean air should be served where it is consumed, i.e. by personalized ventilation close to the breathing zone of each person and individually controlled.
- Individual thermal control should be established to handle personal differences in thermal preference.

The challenge for the engineering community in the future is to combine the above principles of excellence for humans with energy efficiency and sustainability.

ACKNOWLEDGEMENT

The Danish Technical Research Council is acknowledged for its support to the International Centre for Indoor Environment and Energy.

REFERENCES

Andersson, N.H., Frisk, P., Löfstedt, B., Wyon, D.P,. 1975, *Human Responses to Dry, Humidified and Intermittently Humidified Air in Large Office Buildings*. Gävle, Swedish Building Research (D11).

ASHRAE Standard 62-1989, 1989, *Ventilation for Acceptable Indoor Air Quality*, Atlanta, USA, American Society of Heating, Refrigerating and Air-Conditioning Engineers, Inc.

ASHRAE, 2002, *Thermal Environmental Conditions for Human Occupancy*. Atlanta, GA, American Society of Heating, Refrigerating and Air-Conditioning Engineers (ANSI/ASHRAE Standard 55.2002).

Bakó-Biró, Z.m Wargocki, P., Weschler, C., Fanger, P.O., 2002, Personal comuters pollute indoor air: effects on perceived air quality, SBS symptoms and productivity in offices. In: Proceedings of Indoor Air 2002, Monterey.

Berglund, L., Cain, W.S., 1989. Perceived air quality and the thermal environment, *Proc. of IAQ'89: The Human Equation: Health and Comfort*, Atlanta, GA, American Society of Heating, Ventilating and Air-Conditioning Engineers, pp. 93-99.

Bluyssen, P.M., de Oliveira Fernandes, E., Groes, L., Clausen, G., Fanger, P.O., Valbjørn, O., Bernhard, C.A., Roulet C.A., 1996, European indoor air quality audit project in 56 office buildings, *Indoor Air*, Vol. 6, pp. 221-238.

Bornehag, C.-G. et al., 2001, Dampness in buildings and health, *Indoor Air*, **11**(2), 72-86.

CEN, 1998, *Ventilation for Buildings: Design Criteria for the Indoor Environment*, Brussels, European Committee for Standardization (CR 1752).

Cermak, R., Majer, M., Melikov, A., 2002, Measurements and prediction of inhaled air quality with personalized ventilation. In: Proceedings of Indoor Air 2002, Monterey.

Clausen, G., Alm, O., Fanger, P.O., 2002, The impact of air pollution from used ventilation filters on human comfort and health. In: Proceedings of Indoor Air 2002, Monterey.

Fang, L, Clausen, G., Fanger, P.O., 1997, Impact of temperature and humidity on perception of indoor air quality during immediate and longer whole-body exposure, *Indoor Air*, Vol. 8, No. 4, pp. 276-284.

Fang, L., Clausen, G., Fanger, P.O., 1998, Impact of temperature and humidity on the perception of indoor air quality, *Indoor Air*, Vol. 8, No. 2, pp. 80-90.

Fang, L., Wargocki, P., Witterseh, T., Clausen, G., Fanger, P.O., 1999, Field study on the impact of temperature, humidity and ventilation on perceived air quality, *Proc. of Indoor Air '99*, Vol. 2, pp. 107-112.

Fanger, P.O., 1970, *Thermal comfort – analysis and applications in environmental engineering*. Copenhagen, Danish Technical Press.

Fisk, W.J., Mendell, M.J., Daisey, J.M., Faulkner, D., Hodgson, A.T., Macher, J.M. , 1993, The California healthy building study, Phase 1: a summary, *Proc. of Indoor Air '93*, Vol. 1, pp. 279-284.

Fisk, W.J. and Rosenfeld, A.H., 1997, Estimates of improved productivity and health from better indoor environments, *Indoor Air*, Vol. 7, pp. 158-172 [Errata in *Indoor Air*, 1998, Vol. 8, p. 301].

International Organization for Standardization. 1993. *Moderate Thermal Environments - Determination of the PMV and PPD Indices and Specification of the Conditions for Thermal Comfort* (EN ISO Standard 7730).

Kaczmarczyk, J., Zeng, Q., Melikov, A., Fanger, P.O. 2002, The effect of a personalized ventilation system on perceived air quality and SBS symptoms. In: Proceedings of Indoor Air 2002, Monterey.

Krogstad, A.L., Swanbeck, G., Barregård, L. et al., 1991, *Besvär vid kontorsarbete med olika temperaturer i arbetslokalen - en prospektiv undersökning* [A prospective study of indoor climate problems at different temperatures in offices], Göteborg, Volvo Truck Corp.

Lagercrantz, L., Wistrand, M., Willén, U., Wargocki, P., Witterseh, T., Sundell, J., 2000, Negative impact of air pollution on productivity: previous Danish findings repeated in new Swedish test room, Paper submitted to *Healthy Buildings 2000.*
Lee, S.C., Chan, L.Y., Chiu, M.Y., 1999, Indsoor and outdoor air quality investigation at 14 public places in Hong Kon,. *Environment International*, **25**, 443-450.
Mendell, M.J., 1993, Non-specific symptoms in office workers: a review and summary of the epidemiologic literature, *Indoor Air*, Vol. 3, pp. 227-236.
Sekhar, S.C., Tham, K.W., Cheong, K.W., Wong, N.H., 2000, A study of indoor pollutant standard index (IPSI) and building symptom index (BSI). In: Proceedings of Healthy Buildings, **1**, 145-150.
Seppänen, O., 1999, Estimated cost of indoor climate in Finnish buildings. *Proc. of Indoor Air '99*, Edinburgh, Vol. 4, pp. 13-18.
Sundell, J., 1994, On the association between building ventilation characteristics, some indoor environmental exposures, some allergic manifestations and subjective symptom reports, *Indoor Air*, Supplement No. 2.
Toftum, J., Jørgensen, A.S., Fanger, P.O., 1998, Upper limits for air humidity to prevent warm respiratory discomfort, *Energy and Buildings*, Vol. 28, No. 1, pp. 15-23.
Wargocki, P., Wyon, D.P., Baik, Y.K., Clausen, G., Fanger, P.O. 1999, Perceived air quality, sick building syndrome (SBS) symptoms and productivity in an office with two different pollution loads, *Indoor Air*, Vol. 9, No. 3, pp. 165-179.
Wargocki, P., Fanger, P.O., 1999, Impact of ventilation rates on SBS symptoms and productivity in offices, *Proc. of DKV-Jahrestagung*, Berlin, Vol. IV.
Wargocki, P., Wyon, D.P., Sundell, J., Clausen, G., Fanger, P.O., 2000, The effects of outdoor air supply rate in an office on perceived air quality, Sick Building Syndrome (SBS) symptoms and productivity, *Indoor Air*, Vol. 10, No. 4, pp. 222-236.
Wargocki, P., Wyon, D.P., Fanger, P.O., 2000, Productivity is affected by the air quality in offices, Proc. of Healthy Buildings 2000, Vol. 1, pp. 635-640.
Wargocki, P., Sundell, J., Bischof, W., Brundrett, G., Fanger, P.O., Gyntelberg, F., Hanssen, S.O., Harrison, P., Pickering, A., Seppänen, O., Wouters, P., 2002, Ventilation and health in nonindustrial indoor environments: report from a European Multidisciplinary Scientific Consensus Meeting, *Indoor Air*, **12**(2).
Wyon, D.P., 1996, Indoor environmental effects on productivity, *Proc. of IAQ '96*, Atlanta, GA, American Society for Heating, Refrigerating and Air-Conditioning Engineers, pp. 5-15.

Advances in Building Technology, Volume 1
M. Anson, J.M. Ko and E.S.S. Lam (Eds.)

CREATING COMPETITIVE ADVANTAGE AND PROFITS WITH TECHNOLOGY IN THE CONSTRUCTION SECTOR

Roger Flanagan
School of Construction Management and Engineering
The University of Reading, Reading, RG6 6AW UK

ABSTRACT

The construction sector is often accused of a lack of investment in R&D. However, this does not reflect the huge investment made by design organisations, contractors, plant manufacturers and materials and component manufacturers in the development of their services and products. A number of countries have selected 'critical technologies' for investment over the medium to long term; these include smart materials, biomimetics, and nanotechnology. The construction sector has not been good at applying technologies from other industries, and frequently duplication of effort occurs. There must be a balance between technology push and market pull, which is reflected in enterprises' profits. New technology does not guarantee profits, but it can offer competitive advantage. The development of new products and processes cannot be divorced from the development of codes and standards.

Keywords: technology, competitive advantage, construction sector.

INTRODUCTION

This paper investigates five facets of how enterprises can create competitive advantage from technology.

1) Global drivers of change - how technology can be used to respond.
2) A technology map - how technology is shaping the future for construction enterprises in the construction sector
3) A review of construction technology strategies in selected countries.
4) New technology - some examples and how the construction sector can learn from other industries.
5) Profiting from technology.

FACET 1 GLOBAL DRIVERS OF CHANGE

Globalisation has created an interconnected and interdependent world but also an uneven one. We live in a world with a growing gulf between the haves and have-nots, the digitally rich and the digitally deprived. Economies, culture, organisations and interest groups are becoming linked together across national boundaries. Sources of saving and investment, capital markets, that were once distinct, have now become coupled together with the consequence of near-universal standards of performance of capital assets. Investors want the best they can find anywhere in the world. There is a shift from public to private sector financing of infrastructure and the utilities.

Construction sectors around the world cannot be immune from the global drivers that have an impact upon the economy and quality of life. The drivers of change for the beginning of the 21st century involve social, technological, economic, environmental and political trends; they are the factors that cannot be changed and are an inevitable result of development in the broadest sense. Their effects can be global or national.

- **Technology, knowledge and skills**

  The growing importance of technology, knowledge, and skills are now crucial with the move towards knowledge-based business with knowledge technology becoming central to all business. The level of performance of the world's best is constantly being raised as a result of innovation in communications, technology and learning. The centre of gravity in business success is shifting from the exploitation of physical assets to the realisation of the creativity and learning potential of people.

- **Globalisation - economic liberalisation and barriers to trade falling**

  Globalisation in construction can be seen with the growth in imported construction components and materials, and the increase in foreign companies working overseas. The fastest growing element in world trade is information, increasingly sent by satellite or down telephone wires. The implication of globalisation is that instead of the world being made up of a series of different economies that determine their own business conditions, the economic conditions in most parts of the world will be determined by the world economic position.

- **Demographic trends are impacting the world**

  Age distribution has significant effects on economic growth. The world's human population currently numbers about 5.8 billion people, and the figure grows by more than 80 million people each year, or around 220,000 each day. The industrialised countries face the problems of an ageing population and the implications of pensions provision, an ageing workforce, availability of human resources, and the provision of more facilities for an ageing population. In the developed world technology will provide the answer to improving productivity and wealth creation with a smaller workforce supporting more people. The developing countries give a different picture, unsustainable population growth is more of an issue. In the developing world technology will provide the means to alleviate poverty and increase competitiveness.

- **Pace of economic and technological change is accelerating**

  During the industrial revolution it took 60 years for productivity per person to double, it recently took China 10 years to achieve the same improvement. 70% of all the scientists who have ever lived are alive today. Companies must focus on service and productivity - increasing fast, flexible, customised, networked and global with clear specialisation. Rapid improvement based on continuous learning or a step-change in performance has become a critical success factor.

- **More competition**

  This will come from the developing world with the spread of literacy and the growth in the birth rate. In the next 15 years, technology will improve the literacy and skills of people in the developing world.

- **New employment patterns and organisational structures**

  New employment patterns with the old idea of the "employer and employee" becoming obsolete. No one can feel secure in the sense of lifetime employment. Only those who learn new skills will achieve long-term employability. Work is becoming less full-time; in 1994 less than 60% of employment was full time with a growth in temporary and part time workers. Service providers are growing in importance with outsourcing to specialist providers.

- **Changing client expectations**

  Customers are becoming more demanding and seeking reliability, consistency, and value for money. Ethical issues are high on customer's priorities. There is increasing need for companies to maintain public confidence in the legitimacy of their operations and business conduct - a licence to operate. Growing importance of developing community relationships by increasing participation in national and local community partnerships.

- **Growing importance of sustainability and environmental issues**

  Issues with sustainability and the greening of construction are high on the agenda of global summits. Water and other natural resources, including energy, will become more scarce and more valuable. Over 1 billion people in the world are without safe drinking water. Almost 3 billion people (roughly half the world's population) are without adequate sanitation in developing countries. Technology has the solutions to provide safe drinking water, but the issue is at what cost.

- **New patterns of transportation**
  Traffic congestion on the roads is becoming a major issue in all big cities. Car manufacturers produce cars, irrespective of the available space on the roads. As a result, the demand for road and rail transportation in the future will be for the infrastructure to be built underground. Therefore, new technology for tunnels and ventilation systems will be required that will make the process more cost effective. Advanced Transport Telematics (ATT), is specifically concerned with improving safety and efficiency in all forms of transport and reducing damage to the environment.
- **Growth of privatisation of infrastructure and built facilities**
  In the developed world there is a shift towards privatisation of new infrastructure for roads, rail, telecommunications, water, sewerage, and power supplies.

Drivers cannot be controlled; they need to be addressed. The construction sector may be tempted to say that these will not impact their business: they would be wrong.

FACET 2 TECHNOLOGY MAP

The word 'technology' comes from the Greek 'technologia', meaning systematic treatment and describes the use of scientific knowledge for practical purposes. 'Technology' can be understood as the systematic use of knowledge usually seen in actual machines or in industry. It can also be defined as the application of the existing body of knowledge (science) to the production of goods and services; it embraces equipment, tools, techniques, materials, systems, processes, information, etc., (*Ofori, 1994*). Technology takes different forms: 'hardware', such as machinery and equipment; 'software', such as drawings and process specification; and the 'service' of technicians and professionals for tasks such as quality improvement, management and marketing know-how, process and product design. The pervasive nature of the concept of technology raises several problems in relating technology to growth (*Jeremy, 1994*).

Figure 1 shows a simplistic overview of the process and impact of technology for construction.

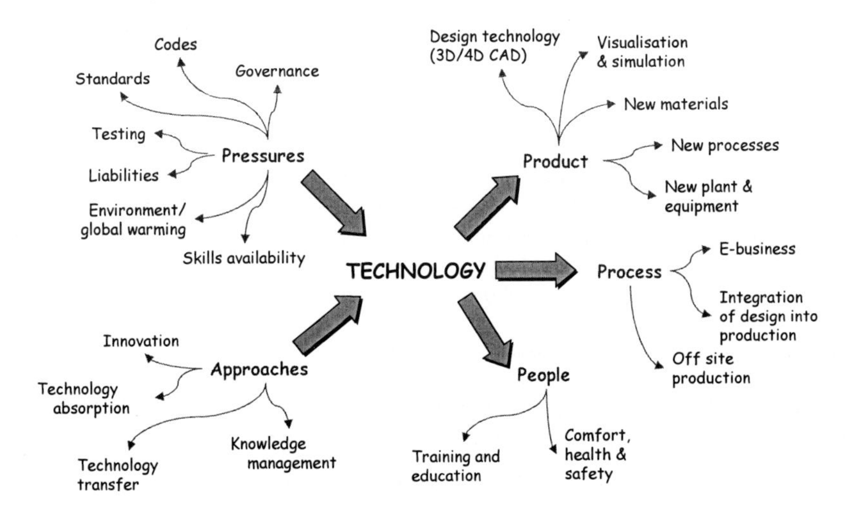

Figure 1 Technology map

- **Pressures** come from such issues as the requirement of codes and standards for new materials and products, both nationally and internationally. These should not be underestimated in importance, particularly with the liability issues and risk with new products. Similarly, governments, and even global regimes, will impose standards that must be met. New emission regulations are a good example of this.
- **Approaches** are the way in which technology is absorbed, transferred and applied, and how knowledge is 'managed'.
- The **Product** heading relates to new materials, plant and processes, such as new tunnel equipment, soil remediation, fire retardation processes and concrete pumping innovation.
- **Process** refers to the way in which business can increase competitiveness by use of information and communication technologies, off-site prefabrication and customisation of standard components.
- **People** relates to how technology can make sites safer, such as using virtual reality to simulate accidents on site; and its role in novel training and education.

FACET 3 TECHNOLOGY STRATEGIES AROUND THE WORLD

Technological developments are irreversible. The biggest question is how fast technology will change between now and say 2020, and how that change can be properly managed. Technology enables almost anything to be done; deciding what to do becomes the critical skill. Our ability to perceive and interact with complicated, remote and, either huge or tiny, abstract or concrete things will be unprecedented. Technology gives us more and more access to information, so life gets more and more chaotic. 'Information chaos' prevails and we need to help people find the information that they want, when they want it.

Many countries in the developed world are looking to the future and developing national strategies. Japan was one of the first countries to take a futures perspective, a process that is often called 'Foresight'. There are three ways to look at the future: forecasting, assessment, and foresight. Forecasting makes probabilistic predictions of future developments; assessment anticipates future societal impacts of known new and existing technologies; and foresight identifies present science and technology priorities in the light of hypothetical projections of future economic and societal developments. They each have different functions and objectives (*Gavigan and Cahill, 1997*).

Our research showed that there was a lot of duplication of research and development effort (*Flanagan et al, 2001*). There seems to be a reluctance to learn from each other. It appears to be more attractive to start new research programmes in a country rather than learning from overseas or, at industry level, learning from other industries. Figure 2 shows the potential of learning from other industries.

Table 1 shows the issues in a number of technology strategies for construction sectors in a number of countries. Our review of these strategies revealed a number of common threads. Most governments are trying to make their construction sector more competitive by leveraging competitive advantage internationally through the use of technology, becoming more efficient, and constructing to a lower unit cost. The review revealed that many of the countries have developed very similar strategies.

FACET 4 NEW TECHNOLOGIES

A number of countries have selected 'critical technologies' for investment over the medium to long term; these include smart materials, biomimetics, and nanotechnology. Embedded systems and e-business are also areas that have been selected; they are evolving very quickly and having a huge impact on the construction sector.

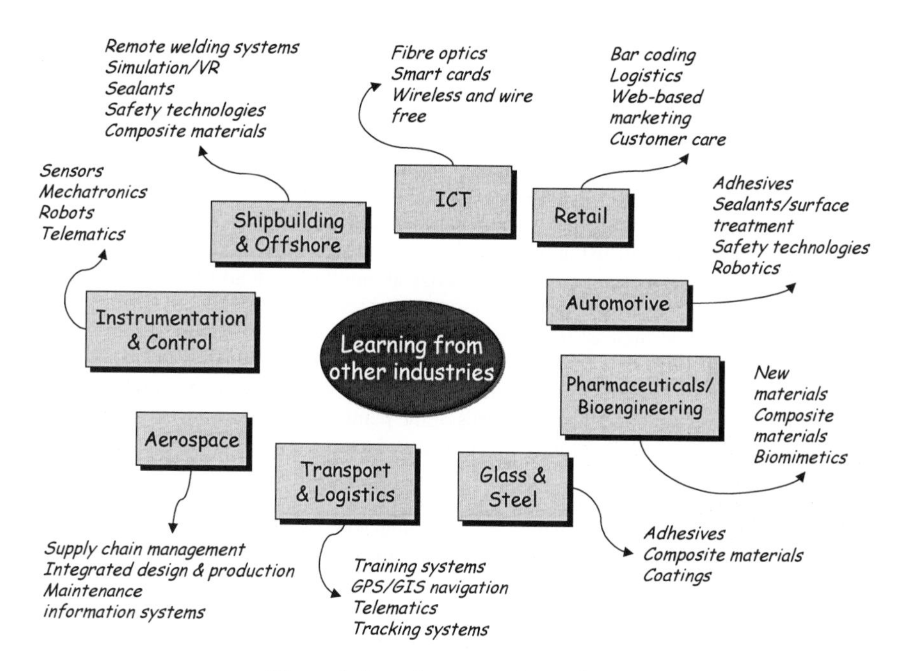

Figure 2: Learning from other industries

Table 1
CONSTRUCTION STRATEGY ISSUES AROUND THE WORLD

	Australia	Canada	Germany	Japan	Sweden	UK	USA
Intensive use of information and communication technologies	✔	✔		✔	✔	✔	✔
Development of products against market needs			✔				
Becoming a service rather than product provider						✔	✔
Process and system -orientated planning and realisation concepts			✔				
Governance, standards, and legislation	✔	✔					
Environment - embracing sustainability					✔	✔	✔
Integration and networking of all participants into all phases (feedback). Seamless industry	✔		✔				
Focus on people issues	✔			✔	✔	✔	
New construction methods				✔	✔		✔
Smart buildings and infrastructure				✔	✔	✔	
Integration of new technologies in planning, construction, and use				✔	✔	✔	
Blue sky and emerging technologies	✔	✔	✔		✔	✔	✔
Impact of new plant and equipment			✔	✔			✔
Creating competitive advantage	✔						✔

Source: Construct New South Wales, National Research Council, Bauen 21, RICE, CMB, UK Foresight reports ,CERF.

The technology of intelligent or **smart materials** uses the knowledge of a number of different technologies such as materials science, biotechnology, biomimetics, nanotechnology, molecular electronics, neural networks and artificial intelligence. These technologies are inter-related. Smart materials can be further defined as:

- Materials functioning as both sensing and actuating.
- Materials that have multiple responses to one stimulus in a co-ordinated fashion.
- Passively smart materials self-repairing or stand-by characteristics to withstand sudden changes.
- Actively smart materials utilising feedback.
- Smart materials and systems reproducing biological functions in load-bearing structural systems.

Biomimetics has been defined as 'the abstraction of good design from nature'. From the beginnings of humanity, man has been intent upon using (or abusing), and taming nature. There is a wide range of possible applications of biomimetics in building. Nature provides many examples of structures built to respond to the environment and meet the needs of the plant or animal, for example, termite towers. The fur or feathers of birds and animals are perfectly adapted for their environment and way of life. Biomimetic engineering could provide clothing that is light, responsive and strong to be used in harsh conditions on site. Mimicking nature could produce new designs in civil engineering that are lighter, stronger and adaptable to a changing environment. New adhesives, based on those produced in nature, could revolutionise the building process. Buildings could be 'glued' together, giving stronger, faster and cleaner construction techniques.

The essence of **nanotechnology** is the ability to create large structures from the bottom up, that is by starting with materials at a molecular level an building them up. The structures created – 'nanostructures' are the smallest human-made objects whose building blocks are understood from first principles in terms of their biological, chemical and physical properties. This comprehensive understanding of the make-up of nanostructures has allowed scientists to create very strong materials at a fraction of the size and weight of traditional materials. Carbon nanotubes can be constructed which are ten times the strength of steel but a sixth of its weight. Scientists can make single molecules that can behave like transistors - which makes computers the size of a sugar cube a possibility! Nanoparticles can target and kill cancer cells.

As with biomimetics and smart materials and structures, nanotechnology embraces and requires cross-disciplinary work involving physics, chemistry, biology, medicine, engineering and computer simulation. The understanding and determination of properties and processes at a molecular scale, means that nanotechnology could impact on virtually every human-made object, from cars to computer circuits, and from tissue replacements to bridges. Nanostructured ceramics, polymers and metals will have vastly improved mechanical properties. In particular, nanostructured silicates and polymers will act as contaminant scavengers to create a cleaner environment.

A number of industrial applications are beginning to emerge that exploit the newly emerging Internet capabilities of **embedded systems**. Embedded systems differ markedly from desktop systems, being fitted with just enough functionality to handle a specific application, enabling them to be produced at low-cost. Such systems have a more limited processing speed, CPU power, display capability and persistent storage capability. The challenge for developers is to produce embedded systems that are able to provide network functionality within these constraints. The future is where all electronic devices are ubiquitous and so are networked and every object, whether it is physical or electronic, is electronically tagged with information pertinent to that object. The use of physical tags will allow remote, contactless interrogation of their contents; thus, enabling all physical objects to act as nodes in a networked physical world. This technology will benefit supply chain management and inventory control, product tracking and location identification, and human-computer and human-object interfaces. In the construction sector, auto-ID technologies would have a huge impact on the supply chain, the design process, the construction process and facilities management.

FACET 5 PROFITING FROM TECHNOLOGY

The construction sector is often accused of a lack of investment in R&D: This statement does not reflect the huge investment made by design organisations, plant manufacturers and materials and component manufacturers in the development of their services and products. Whilst research and development is a good long-term investment; enterprises can only invest sums that can be justified to the shareholders. The challenge is how to measure the returns from research investment in developing technology. For example, in the manufacturing, chemical, automotive, aerospace, pharmaceutical and electronic industries, research is often laboratory-based, and results in new products or materials that can be patented. More importantly, in a less fragmented industry, such as aerospace, there are a few large clients and a small number of suppliers – BAE accounts for 60% of the UK supplier output. The aerospace industry also benefits from fixed locations, high barriers to entry and long time frames compared to the construction sector. This provides an environment in which research can be fostered. In the UK aerospace industry there are 420 SMEs, in construction there are 123,000 (DTI, UK).

The fragmentation in the construction sector does not preclude research, but makes it more difficult to quantify, justify, and to retain for competitive advantage. Therefore, construction enterprises need to balance new technology against their prime aim of maintaining sensible profit margins that show a reasonable return on the capital employed in the business. Any technology that involves high levels of risk must correspondingly carry better margins. Owners and shareholders want faster returns; construction enterprises aim to improve productivity, and correspondingly aim to increase margins, and; enterprises must balance risk and reward. With products, plant and equipment, it is easier to obtain a patent and to earn money from the application of new technology. For example, Pilkington have developed Activ™ self-cleaning glass. "Its unique dual-action uses the forces of nature to help keep the glass free from organic dirt"; this means less cleaning and better-looking windows. Hilti have developed a 'smart' drill with embedded intelligence that removes the guesswork out of drilling into different materials.

CONCLUSIONS

Technology is making a difference in construction, it is possible to leverage profit from innovation and new technology. Global drivers will put pressure on enterprises to find solutions using technology. For example, in the developed world changes in demographics will mean fewer workers leading to a skill shortage. Ultimately, there will need to be more off-site prefabrication and more mechanisation, which can be achieved through new technologies such as advances in materials science an information and communication technologies.

The importance of codes and standards should not be underestimated in the development of new technology. A new high-strength concrete will not be used unless it complies with the standard developed to cope with it, and that has been fully tested. The differences between market pull and technology push should be recognised.

Knowledge sitting on the bookshelf is of absolutely no value; this is often called 'knowing at rest'. There is a need to connect knowledge and knowing with connections between the way that people work, the way they are organised into their work, the way they get information and the way they use it.

REFERENCES

Aichholzer, G. (2000) **Delphi Austria: an example of tailoring Foresight to the needs of a small country**. Institute of Technology Assessment (ITA), Austrian Academy of Sciences on United Nations Industrial Development Organisation (UNIDO) web site http://www.unido.org/userfiles/kaufmanC/AichholzerPaper.pdf

Blind, Dr K. (2001) **Delphi As A Technology Foresight Methodology: Experiences From Germany**. Fraunhofer Institute for Systems and Innovation Research on United Nations Industrial Development Organisation (UNIDO) web site http://www.unido.org/userfiles/kaufmanC/BlindPaper.pdf

Flanagan, R. (1999) **Lessons for UK Foresight from around the world**. Construction Associate Programme, Office of Science and Technology, London.

Flanagan, R., Jewell, C.A., Larsson, B. and Sfeir, C. (2001) **Vision 2020 – Building Sweden's future**. CMB, Gothenburg, Sweden. ISBN 0 7049 12066.

Foresight (2001) **Constructing the Future**. Construction Associate programme, Built Environment and Transport Panel, Foresight, Office of Science and Technology, UK.DTI Pub 5567 3k/06/01/NP. URN 01/884

Gavigan, J.P. and Cahill E.A. (1997) **Overview of recent European and non-European National Technology Foresight studies**, Technical report TR97/02 European Commission JRC Institute for Prospective Technological Studies, Seville

Jeremy, D.J. (1994) **Technology transfer and business enterprise**. Edward Elga Publishing, UK.

Klusacek, K. (2000) **Technology Foresight In The Czech Republic**, Technology Centre, Academy of Sciences of the Czech Republic on United Nations Industrial Development Organisation (UNIDO) web site http://www.unido.org/userfiles/kaufmanC/KlusacekPaper.pdf

Kozlowski, J. (2001) **Adaptation of foresight exercises in Central and Eastern European countries** on United Nations Industrial Development Organisation (UNIDO) web site http://www.unido.org/userfiles/kaufmanC/Koslowskipaper.pdf

Ksenofontov, M. (2000) **Technology foresight in the Russian Federation: Background, modern social, economic and political context and agenda for the future**. Institute of Economic Forecasting, Russian Academy of Sciences on United Nations Industrial Development Organisation (UNIDO) web site http://www.unido.org/userfiles/kaufmanC/KsenofontovPaper.pdf

Lübeck, L. (2000) **The Swedish Technology Foresight Project: Tecknisk Framsym** on United Nations Industrial Development Organisation (UNIDO) web site http://www.unido.org/

Marsh, L., Flanagan R. and Finch, E. (1997) **Enabling technologies: a primer on bar coding for construction**. The Chartered Institute of Building, ISBN 1 85380 081 3.

Martin, B. (2001) **Technology Foresight in a rapidly globalizing economy**. Vienna, Austria: Regional Conference on Technology Foresight for CEE and NIS countries, 4-5 April 2001.

Ofori, G. (1994) Construction industry development: role of technology transfer. *Construction Management & Economics*, Spon. 379-391.

Shakeri, C. Noori, M.N. and Hou, Z. (1996) Smart materials and structures: a review. *Proceedings of the Fourth Materials Engineering Conference*, Washington DC, November 10-14, 1996. **Vol 2**, pp863-876.

United Nations Industrial Development Organisation (UNIDO) web site http://www.unido.org/

Zgurovsky, M. (2000) **Technological Foresight In Ukraine**, National Technical University of Ukraine "Kiev Polytechnic Institute" on United Nations Industrial Development Organisation (UNIDO) web site http://www.unido.org/userfiles/kaufmanC/ZurovskyPaper.pdf

Advances in Building Technology, Volume 1
M. Anson, J.M. Ko and E.S.S. Lam (Eds.)

ADVANCES IN IT FOR BUILDING DESIGN

J. S. Gero

Key Centre of Design Computing and Cognition, University of Sydney,
NSW, 2006, Australia

ABSTRACT

Computers have been used building design since the 1950s. Their first use was in structural analysis and construction planning. The use of computers in building design analysis has included extensive developments in structural analysis as well as programs for the analysis of the HVAC and environmental performance of buildings. Recently, more sophisticated analyses of environmental behaviour and the behaviour of building users have been developed and implemented. Computer graphics was developed initially in the 1960s and formed the basis of computer-aided drafting systems – called CAD systems. These early CAD systems were used during the documentation phase of building design. These CAD systems developed beyond simply drafting to modelling the geometry of the building. Today's commercial CAD systems are used at various stages in the building design process and are integrated with analysis tools. This paper briefly traces these developments before introducing current research on IT for building design that has the potential to impact the way buildings will be designed in the future. Three strands of research are presented. The first strand deals with virtual environments for designing and designing virual environments. In virtual environments the focus moves away from documents to models and from a virtual building model on a designer's computer that is then sent to other members of the design team to a building model in a virtual environment that is accessible by any authorised member of the design team anywhere at any time. The issue of designing within virtual environments is raised. The second strand deals with new ways of carrying out simulations of the behaviour of the building users rather than the behaviour of the building. The third strand deals with novel computational agent technologies and examines the potential of their use in building design.

KEYWORDS

CAD, building design, virtual environments, virtual models, simulations, design agents, computational agents

BACKGROUND

The invention of the computer goes back to the 1940s, however, their industrial use commenced only a decade later. The University of Sydney built and installed its first computer in 1953 for research purposes.

The possible use of computers in the building industry commenced with research into automating structural analysis through the development of the matrix method of frame analysis. Thus, by the late 1950s/early 1960s exceptionally unusual buildings (such as the Sydney Opera House) were having their structures analysed by computer. Standardised structural analysis programs appeared in the late 1960s and the early 1970s. Today the finite element method has the capacity to analyse virtually any building structure to a sufficient degree of accuracy. Further, the cost of analysis has dropped dramatically in the intervening 30 years. This cost reduction has been brought about by three factors: the reduction in the cost of computation, the reduction in the cost of describing the structure and the reduction in the cost of presenting the results in a coherent graphical form. The effect of this increase in ability to analyse building structures and the reduction in the cost of analysis has been to allow designers to explore a much wider range of structural forms and shapes than would have been feasible earlier.

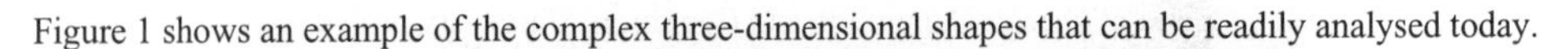

Figure 1 shows an example of the complex three-dimensional shapes that can be readily analysed today.

Figure 1: The Guggenheim Museum in Bilbao, Spain, designed by Frank Gehry
(model by David Ju, University of Sydney)

Following on from the developments in structural analysis were the developments in environmental analysis of buildings. These commenced with energy analysis and expanded to include a variety of other environmental factors as well as the influence of building services on the environmental performance of buildings. Today with computational fluid dynamics it is possible to model of environmental behaviour including smoke spread and fire spread in buildings.

The development by the US Navy of the PERT (Program Evaluation and Review Technique) and the later development of the CPM (Critical Path Method) based on the mathematical technique of dynamic programming provided a formalisable means construction project management that was readily computable. As a consequence this was also amongst the early applications of computers in the building industry.

In 1963 Ivan Sutherland submitted his PhD thesis titled *Sketchpad: A Man-machine Graphical*

Communications System to MIT and in it he invented interactive computer graphics. *Sketchpad* (Sutherland 1963) pioneered the concepts of interactive graphical computing, including memory structures to store objects, rubber-banding of lines, the ability to zoom in and out on the display, and the ability to make perfect lines, corners, and joints. This was the first GUI (Graphical User Interface) long before the term was coined. It also had parameterised objects. Figure 2 shows Sutherland using the system.

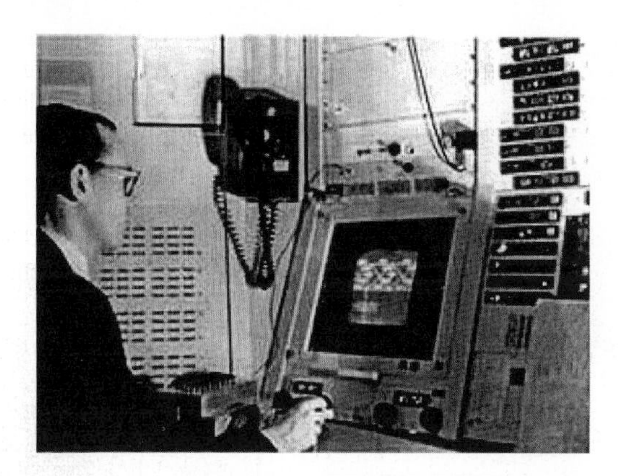

Figure 2: Ivan Sutherland with the *Sketchpad* system at MIT in 1963

The concepts in *Sketchpad* laid the foundation for computer-aided drafting systems for years to come. These concepts along with research on object modelling continues to provide the bases of current commercial CAD systems. More recently commercial CAD systems have begun to implement object-oriented technology that allows users a greater flexibility in their use. Most current CAD systems are world-wide web enabled, meaning that users can access them across the web.

VIRTUAL ENVIRONMENTS FOR DESIGNING

Virtual environments are computational representations of objects and spaces in "cyberspace". They fall into two distinct classes. In the first are virtual environments that are CAD models generated using the standard CAD packages. Here the focus is entirely on the CAD model that happens to be accessed in a virtual space. There are advantages in placing CAD models into virtual environments since this allows designers to collaborate asynchronously, without being co-located. One of the issues in this approach is the size of files that need to be transferred from one designer to another is generally increasingly large. This has performance effects.

The second approach is to design within a virtual environment where the model of the building is developed as it is being designed. It can then be converted into a CAD model. The advantage of this approach is the ease with which members of the design team can gain access to the design and the means by which they can communicate with each other. In general the technology that supports this is different, often having come from the distributed computer games world, and as a consequence the files that are moved around are very small and hence performance is high. Each member of the design team can be connected to the same virtual environment at the same time and can have a different view of it than any other member of the team (Maher and Gu 2001).

Figure 3 show a virtual design environment made up of objects. The virtual environment is being used by a number of designers concurrently, designers who both design within it and communicate with each other in it.

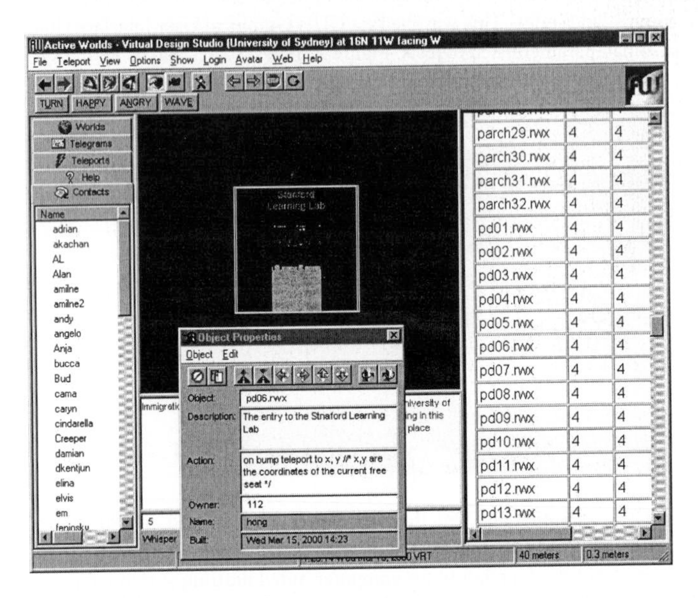

Figure 3: A virtual design environment showing objects whose information can be directly accessed

Virtual design environments offer functionality beyond CAD modelling in that all documents are treated as objects as well. So an office with all its attendant data and information content can become part of the environment (Maher and Gu 2001). Figure 4 shows a virtual environment where the office and design information is part of the environment. This opens up a variety of novel collaboration opportunities. Virtual design teams, within the overall team, can form and carry out sub-designs. Further each of the different design disciplines (architect, structural engineer, façade engineer, etc) can work on the current version of the design.

SIMULATING THE BEHAVIOUR OF BUILDING USERS

The user of buildings has been notoriously absent from most simulations of the behaviour of buildings. The behaviour of the building has been assumed to be largely independent of the users. In recent research into simulations of groups of animals Reynolds (1987) demonstrated that realistic results could be produced using simple computational agents. The "social force model" that is the result is a model of pedestrian behaviour, similar to animal flocking, used to model self-organising phenomena in crowds (Helbing and Molnár, 1995). Computer simulations have shown that the social force model is capable of realistically describing several interesting aspects of collective pedestrian behaviours (Helbing and Molnár, 1997).

A simple crowd management problem is used to illustrate the behaviour pedestrians. The problem is to design a doorway to facilitate the efficient and comfortable movement of pedestrians travelling in opposite

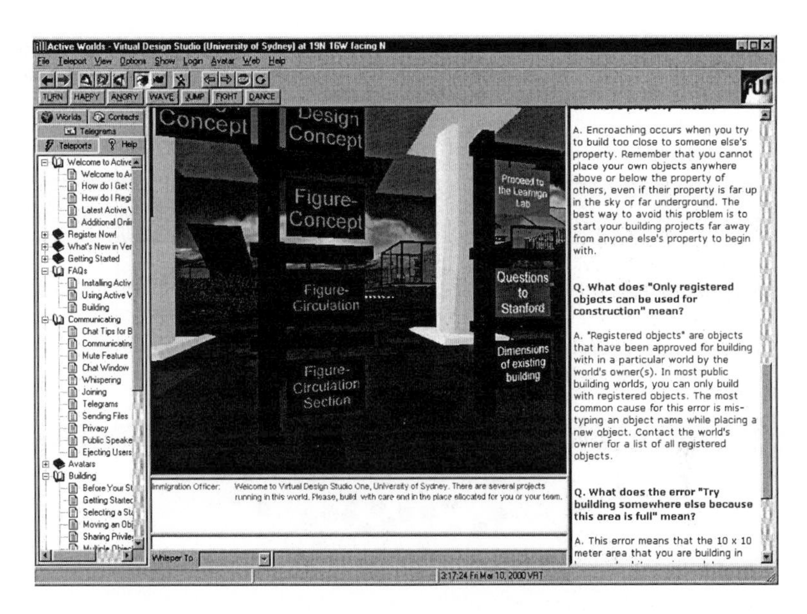

Figure 4: A virtual environment where both the design and the information about the design office are part of a single environment

directions. A pedestrian simulator was developed to evaluate doorway designs. Pedestrian movement is simulated using a microscopic model of crowd behaviour developed to account for empirically observed self-organizing phenomena. The social forces modelled in the simulations of pedestrian crowds are listed in TABL, detailed mathematical descriptions of these forces can be found in Helbing and Molnár (1995).

TABLE 1

THE SOCIAL FORCES MODELLED IN THE SIMULATIONS OF PEDESTRIAN CROWDS

	Description of social force
1.	Pedestrians are motivated to move as efficiently as possible to a destination.
2.	Pedestrians wish to maintain a comfortable distance from other pedestrians.
3.	Pedestrians wish to maintain a comfortable distance from obstacles like walls.
4.	Pedestrians may be attracted to other pedestrians (e.g. family) or objects (e.g. posters).

This model is used to simulate the behaviour of pedestrians using three different designs for the doors. The three designs are: narrow door, wide door and two separated doors. The behaviour of the pedestrians using a narrow door can be seen in Figure 5, where pedestrians moving in opposite direction block each other's paths.

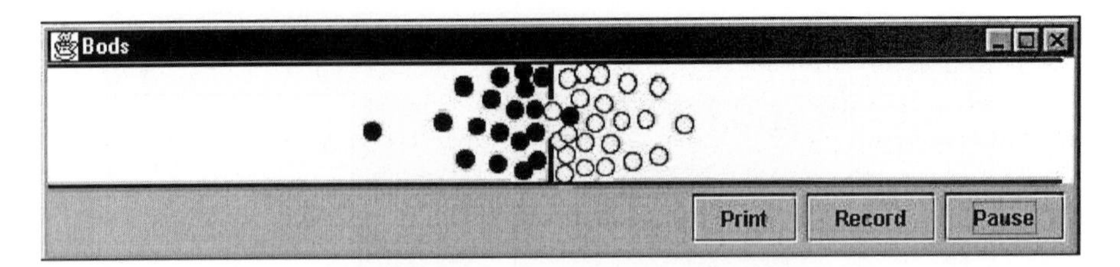

Figure 5: Simulating the behaviour of pedestrians using a narrow door (Saunders and Gero 2001)

Figure 6 shows the same pedestrians using a wide door. Here the pedestrians generate a behaviour that produces two passing streams, indicating that the size of the door is satisfactory.

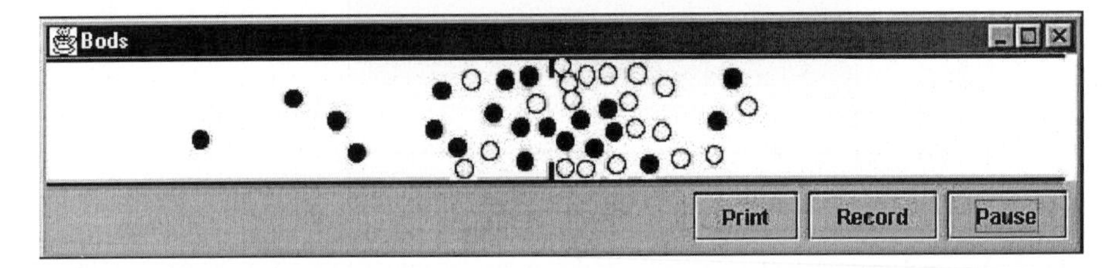

Figure 6: Simulating the behaviour of pedestrians using a wide door (Saunders and Gero 2001)

Figure 7 shows the same pedestrians using two separated doors. Here the pedestrians form two streams, one for each door. These streams are not programmed inot the system but emerge from the behaviour of each individual.

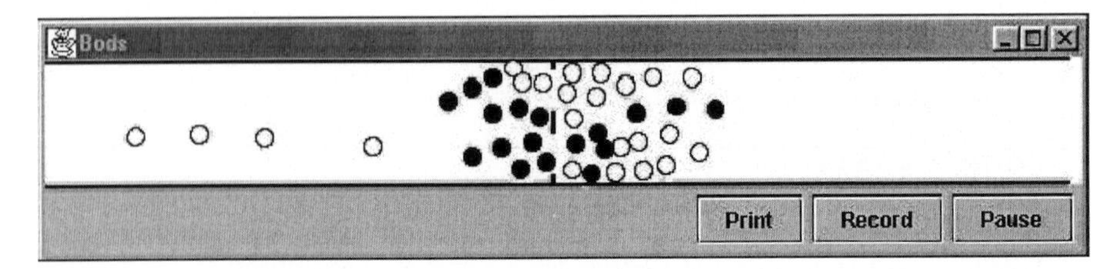

Figure 7: Simulating the behaviour of pedestrians using two doors (Saunders and Gero 2001)

DESIGNING WITH COMPUTATIONAL AGENTS

Computational agents are computer programs that are designed to exhibit a degree of autonomy and as a consequence are able to influence and change the environment within which they sit. Already many people who use the world-wide web unwittingly make use of agents during their web searches. Designing with computational agents is at the cutting edge of current design computing research. There are proposals to develop agent-based design systems (Maher and Gero 2002). These use a multi-agent system as the core of a 3D multi-user virtual world. Each object in the world is an agent in a multi-agent system. The agent model provides a common vocabulary for describing, representing, and implementing agent knowledge and communication. The agent can sense its own environment and can generate or modify the spatial infrastructure needed for a specific collaborative or communication need of the users of the world. Each agent has five kinds of reasoning: sensation, perception, conception, hypothesizer, and action (Gero and Fujii 2000). This agent approach is derived from recent developments in cognitively-based design agents, where design is considered as a situated act (Gero 1998). The agents are developed to interact with the design and the design knowledge (Smith and Gero 2001; Saunders and Gero 2001). The agent approach to virtual worlds provides for new kinds of interaction among the elements of the virtual world representation and between individuals and project teams with the components of the virtual world that makes both the virtual environment and interactions with it dynamic.

Agents can function in three modes based on their internal processes: reflexive, reactive, and reflective (Maher and Gero 2002). Each mode requires increasingly sophisticated reasoning, where reflexive is the simplest.

Reflexive mode: here the agent responds to sense data from the environment with a preprogrammed response – a reflex without any reasoning. In this mode the agent behaves as if it embodies no intelligence. Only preprogrammed inputs can be responded to directly. Actions are a direct consequence of sense data. This mode is equivalent to the kinds of behaviors that are available in current virtual worlds.

Reactive mode: here the agent exhibits the capacity to carry out reasoning that involves both the sense data, the perception processes that manipulate and operate on that sense data and knowledge about processes. In this mode the agent behaves as if it embodies a limited form of intelligence. Such agent behavior manifests itself as reasoning carried out within a fixed set of goals. It allows an agent to change the world to work towards achieving those goals once a change in the world is sensed. Actions are a consequence not only of sense data but also how that data is perceived by the agent. The agent's perception will vary as a consequence of its experience.

Reflective mode: here the agent partially controls its sensors to determine its sense data depending on its current goals and beliefs. The agent also partially controls its perception processes depending on its current goals and beliefs and its concepts may change as a consequence of its experiences. The concepts it has form the basis of its capacity to "reflect", ie not simply to react but to hypothesize possible desired external states and propose alternate actions that will achieve those desired states through its effectors. The reflective mode allows an agent to re-orient the direction of interest by using different goals at different times in different situations (Gero and Kannengiesser 2002).

We can expect such systems to change the way we design and potentially what we design.

References

Gero J.S. (1998). Towards a model of designing which includes its situatedness, *in* H. Grabowski, S. Rude and G. Grein (eds), *Universal Design Theory,* Shaker Verlag, Aachen, 47-55.

Gero J. S. and Fujii H. (2000). A computational framework for concept formation in a situated design agent, *Knowledge-Based Systems* **13:6,** 361-368.

Gero J.S. and Kannengiesser U. (2002). The situated Function-Behavior-Structureframework, *in* JS Gero (ed.), *Artificial Intelligence in Design'02,* Kluwer, Dordrecht,. 89-104.

Helbing D. and Molnár P. (1995). Social force model for pedestrian dynamics.*Physical Review E* **51,** 4282-4286.

Helbing D. and Molnár P (1997). Self-organization phenomena in pedestrian crowds, *in* F. Schweitzer (ed.), *Self-Organization of Complex Structures: From Individual to Collective Dynamics,* London: Gordon

and Breach, London, 569 – 577.
Maher M.L. and Gero J.S. (2002). Agent models of 3D virtual worlds, *ACADIA* (to appear)
Maher, M. L. and Gu, N. (2001). 3D virtual world. *in* M. Engeli and P. Carrard (eds), *ETH World: Virtual and Physical Presence,* Karl Schwegler AG, Zurich, 146-147.
Reynolds, C. W. (1987). Flocks, herds, and schools: A distributed behavioural model, *Computer Graphics* **21:4,** 25-34.
Saunders R. and Gero J.S. (2001). Designing for interest and novelty: Motivating design agents. In *CAADFutures 2001,* B de Vries, J van Leeuwen and H Achten (eds), Dordrecht: Kluwer, 725 - 738.
Saunders R. and Gero J.S. (2001). A curious design agent, *in* JS Gero, S Chase and M Rosenman (eds), *CAADRIA'01,* Key Centre of Design Computing and Cognition, University of Sydney, 345-350.
Smith G. and Gero J.S. (2001). Situated design interpretation using a configuration of actor capabilities, *in* Gero, JS, Chase, S and Rosenman, MA (eds) (2001) *CAADRIA2001,* Key Centre of Design Computing and Cognition, University of Sydney, Sydney, 15-24.
Sutherland I.E. (1963). Sketchpad – A Man-Machine Graphical Communication System. *Proceedings of the Spring Joint Computer Conference*, Detroit, Michigan, May 1963, and *Technical Report #296,* MIT Lincoln Laboratory

Advances in Building Technology, Volume 1
M. Anson, J.M. Ko and E.S.S. Lam (Eds.)

INFORMATION TECHNOLOGY SUPPORT TO IMPROVED CONSTRUCTION PROCESSES: INTER-DISCIPLINARITY IN RESEARCH AND PRACTICE

Amanda Marshall-Ponting, Ange Lee, Martin Betts,
Ghassan Aouad, Rachel Cooper[1], Martin Sexton

School of Construction & Property Management,
University of Salford, Salford, M7 9NU, UK.
[1] School of Art and Design,
University of Salford, Salford, M7 9NU, UK.

ABSTRACT

Research and innovation in the built environment is increasingly taking on an inter-disciplinary nature. The built environment industry and professional practice have long adopted multi and inter-disciplinary practices. The application of IT in Construction is moving beyond the automation and replication of discrete mono and multi-disciplinary tasks to replicate and model the improved inter-disciplinary processes of modern design and construction practice. A major long-term research project underway at the University of Salford seeks to develop IT modelling capability to support the design of buildings and facilities that are buildable, maintainable, operable, sustainable, accessible, and have properties of acoustic, thermal and business support performance that are of a high standard. Such an IT modelling tool has been the dream of the research community for a long time. Recent advances in technology are beginning to make such a modelling tool feasible.

Some of the key problems with its further research and development, and with its ultimate implementation, will be the challenges of multiple research and built environment stakeholders sharing a common vision, language and sense of trust. This paper explores these challenges as a set of research issues that underpin the development of appropriate technology to support realisable advances in construction process improvements.

KEYWORDS

Construction process improvement; inter-disciplinarity; multi-disciplinarity; construction IT; nD modelling; trust; multi-criteria decision making.

INTRODUCTION

The building design process is complex encapsulating a great number and variety of factors - which can be social, economic and legislative - in order to satisfy the client's requirements. In fact, it is rarely the case that there will be one homogeneous client, but a number and variety of stakeholders who will be the end users of the design and its resultant product. These can include, not only the organisations and individuals who will occupy the building on a daily basis, but also those who provide it, manage it, own it, maintain it and be affected by it by its existence in their environment.

Project Context

There are increasing demands being made by multiple stakeholders for inclusion of design features such as buildability (cost and time), maintainability, sustainability, accessibility, crime deterrence and acoustic and energy performance. This diversity in stakeholder aspiration arises from increasing numbers of construction specialists who may now be involved in design. Each design parameter that these stakeholders seek to consider will have a host of social, economic and legislative constraints which often are in conflict with one another. The criteria for successful design therefore will include a measure of the extent to which all these factors can be co-ordinated and mutually satisfied to meet the expectations of all the parties involved. In this era of increasing global competition, being able to deliver value to clients - and integrated design, if this is what the market dictates - is what could set one company ahead of the competition.

The volume of information required by so many experts make it difficult for the client to visualise design changes and their impacts upon the time and cost of the project. Design adaptations which aid a client's decision-making ability in this way are also costly, time consuming and laborious. A multi-dimensional computer model, which allows the entire design and construction process to be portrayed visually, would enhance both the decision-making and construction processes, by enabling true and holistic 'what-if' analyses to be performed. This is the ultimate aim of the 3D to nD modelling research project underway at the University of Salford. Conceptually, this will involve taking 3-dimensional modelling in the built environment to an *n* number of dimensions.

The research seeks to develop the infrastructure, methodologies and technologies that will facilitate the integration of time, cost, buildability, accessibility, sustainability, maintainability, acoustics, crime, lighting and thermal requirements. It aims to assemble and combine the leading advances that have been made in discrete information communication technologies (ICTs) and process improvement to produce an integrated prototyping platform for the construction and engineering industries. This output will allow seamless communication, simulation and visualisation, and intelligent and dynamic interaction of emerging building design prototypes, so that their fitness for purpose for economic, environmental, building performance, and human usability will be considered in an integrated manner. It is essential to ensure that the proposed system would be durable and be compatible with new developments in IT, a far as these could be anticipated.

Project Aims and Structure

The Engineering and Physical Sciences Research Council (EPSRC) in the UK fund the 3D2nD project as a platform grant. Platform grants are a recent innovation in UK government research funding and are intended to allow longer time-scale and more fundamental basic research within leading national centres of excellence of an internationally-leading standing, into strategic research of global importance. Salford University has secured one of the first of these grants within the Built Environment on the back of its well-established research in IT, management and environmental and design systems. The project has a large pool of senior academics from multiple disciplines within its project team. It was agreed by the project members that the nD tool should encompass people, technology, process (organisation of work), organisational relevance (business strategy/market positioning) and the business model (cost effective, capable of being developed) issues. This has led to the emergence of 5 work packages, whose interrelationships are illustrated by figure 1.

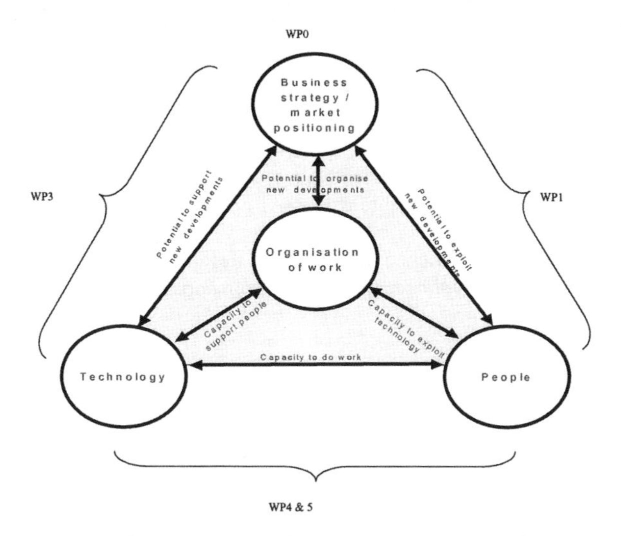

Figure 1: Work packages of the 3D2nD project

The Focus of this Paper

In view of the number and diversity of project researchers representing the requirements of construction industry stakeholders and professionals, it can be expected that gaining consensus on the characteristics of a vision for the project will be problematic and require management. Each work package inherently embodies different epistemological approaches, for example WP3's model testing will be objective and likely empirical in nature, whereas WP4's technology transfer will be more qualitative in nature and concerned with the organisational and social features and processes which will aid or inhibit the technology implementation. Should consensus be achieved and if lacking, how should it be handled so as to not jeopardise the projects' overall deliverables? This has been a key challenge for the research team in the project and is indicative of the issues associated with resolving conflicting perspectives and requirements of multiple stakeholders that the model itself will be called on to perform. As a means of exploring these trade-offs and multi-criteria maximisation issues, this early paper from the research explores how research teams might resolve these differences in approach within their own research methodology and management. Two earlier examples of multi-disciplinary research projects are examined as case studies to surface the associated risks and issues and provide some guidelines for effective inter-disciplinary research working on which this project can draw. It is argued that such inter-disciplinary approaches will increasingly become key to our broader advances in building and construction technology research and innovation.

CASE STUDY 1: AN INTERDISCIPLINARY EU-FUNDED RESEARCH NETWORK

Cooper (2002) uses the BEQUEST network, an EU concerted action with 14 partners from 6 countries, as an example of a major research exercise that illustrates the problems of multi-disciplinary working. The network's aim was to bridge discipline-based differences held by those working in the urban sustainability field to develop a shared platform for assessing sustainable urban development (SUD). This objective necessitated consensus building between the project's immediate partners, but also within the wider built environment community through its Extranet. It is of relevance to the nD modelling project because of its varied stakeholders working in the built environment field. Cooper argues (p117) that BEQUEST faced a triple challenge, because:

(1) There is no consensus of the meaning of SUD (Palmer et al 1997) or how it should be implemented;
(2) There is no agreement and little advice (Epton et al 1983) on how interdisciplinary research should be conducted, despite expressed needs to support this activity in the UK (e.g. HEFCE 2001);
(3) There is little shared experience on how to design and run effective virtual organisations, despite the rapid recognition of their appearance after the rise of the Internet

Multi- vs. Inter-disciplinary working

Cooper distinguishes between 'multi-disciplinary and 'inter-disciplinary' working (p118). The former term applies when two or more disciplines work together without stepping outside their own traditional discipline boundaries. In contrast, the latter emphasises the development of a shared perspective transcending traditional boundaries. Newell and Swan (2000) use the analogies of jigsaws and kaleidoscopes: in multi-disciplinary work, the individual pieces of the jigsaw do not change their identities as a result of being combined with other pieces, whilst with the inter-disciplinary kaleidoscope, it is not possible to establish characteristics of each individual from the whole. BEQUEST is an example of the latter as the emergent conceptual frameworks the members created transcended those owned by any one discipline making up that group (p121).

Examining and building consensus

Cooper draws attention to the 'wheel of cognate disciplines' (Eclipse Research Consultants 1997) which suggests that the closer disciplines are located to each other, the more likely they are to share a common theoretical parentage and therefore the more open their boundaries are to each other. In BEQUEST's case, the most common partner disciplines - engineering, planning and architecture - formed an arc and so it could be expected that collaboration would be fairly straightforward. However, this was not the case and Cooper argues this might be partially explained by the situated nature of professional learning and expertise: "'Legitimate peripheral participation' provides a way to speak about the relations between newcomers and old-timers and concerns the process by which newcomers become part of a community of practice" (Lave and Wenger, 1991). In BEQUEST's case, there was no newcomer-old-timer network and the collectively created shared conceptual space was still being developed.

Eclipse Research Consultants (1996) developed a self-assessment technique, PICABUE, which would gauge potential partners' individual and collective commitment to the principles underpinning sustainability at the start of the project (Curwell et al, 1998). Eclipse embedded the principles into a simple mapping technique for ease of comparison which showed that not only were the different disciplines committed to different aspects and to greatly varying degrees, but that the partner averages increased and became more uniform between the aspects over time. This technique also allowed the 'common platform' to be established, which is a value representing the minimum amount of agreement between members. Represented as a percentage, it increased from 25% in 1996 to 45% in 2001.

New vs. Old Knowledge Production

One of the BEQUEST project deliverables was "an effective, multi-professional, international interactive networked community" to be mediated electronically over the Internet (BEQUEST 2001 p5). Pursued in this way, Cooper argues (2002 p117) that: "...the concerted action displayed some of the characteristics of a new approach to conducting global collaborative research that, in the UK at least, has very recently been given the very grand title of 'e-science'". Two defining characteristics of 'e-science' are the spatial distribution of those participating and the infrastructure, usually in the form of ICTs, which enable it. Its purpose is to achieve "world beating science through the effective use of the latest information technologies" by conducting cross-disciplinary research "at the intersection of many scientific disciplines", and that the resulting 'e-science' "will change the dynamic of the way science is undertaken" (Boyd, 2001, cited in Cooper 2002 p120).

E-science is one example of the pervasive trend towards the 'new' production of knowledge. Gibbons et al (1994) argue that the *way* in which knowledge is being produced is changing. They describe this knowledge (p 3-8) as being produced through 'mode 2 research' and some defining attributes include:

- Knowledge produced in the context of application (as opposed to problem solving using the codes of practice or a particular discipline);
- Trans-disciplinarity;
- Heterogeneity and organisational diversity (in terms of the skills and experiences brought to it);
- Social accountability and reflexivity;
- Quality control.

'New' knowledge may not show all of these characteristics, and it may not even be new: the description by Gibbons *et al* argues that innovation can occur through the reconfiguration of existing knowledge, so that it can be used in new contexts or by new users.

The expansion of the number, nature and range of communicative interactions between the different sites of knowledge production leads not only to more knowledge being produced but also to more knowledge of different kinds; not only of sharing of resources, but to their continuous configuration. Each new configuration becomes itself a potential source of new knowledge which in turn is transformed into the site of further possible configurations. The multiplication of the numbers and kinds of configurations are at the core of the diffusion process resulting from increasing density of communication...This process has been greatly aided by information technologies which not only speed up the rate of communication, but also create more new linkages. (ibid, p35). The development of communication linkages has played an important part in allowing this to occur, and this underlines the importance of the growth of the Internet and other computer-mediated communication and collaboration tools as enablers.

One of the most important conclusions Cooper (2002 p126) draws is that the BEQUEST concerted action has demonstrated that it is not necessary for all members to agree absolutely and at all times, within a research project team, for true progress to be made. What is important though, is that they are willing to negotiate openly about what they are trying to jointly achieve. The test of the validity of any group agreement will be in the reaction from other, wider stakeholders. With so little guidance on how to work effectively in inter-disciplinary groups, Cooper suggests (2002 p126) making explicit the amount of time and effort required to develop a shared perspective and that this should necessarily include trust and consensus building techniques and methods for identifying and resolving conflicts.

CASE STUDY 2: A MULTI-DISCIPLINARY UK UNIVERSITY RESEARCH PROJECT

Newell and Swan's (2000) case study is concerned with the development and maintenance of trust within a multi-disciplinary research team and the implications that this can have for inter-organisational working. The team was comprised of 4 project leaders and 4 researchers based at 3 geographically dispersed universities in the UK.

The Nature of Trust

It is generally agreed that trust is a necessary condition for co-operative behaviour among individuals, groups or organisations (Jones and George 1998) and that it facilitates the levels of communication necessary to generate learning and innovation in networked organisations (Dodgson 1993). Despite different definitions, the two central issues are that trust is about dealing with risk and uncertainty and that it is about accepting vulnerability. From their literature review, Newell and Swan (2000 p1295) conclude that a 3-fold trust typology exists:

1. Companion trust – based on goodwill judgements or personal friendships, it has an expectation of openness and honesty and although slow to develop, it is also the most resilient.
2. Competent trust – based on the perceptions of others' abilities to carry out necessary tasks, it can be driven by contextual cues – such as institution reputation. It develops more quickly, but is more fragile.
3. Commitment trust – stemming from contractual agreements between parties, the continuing existence of the contract in the absence of other trust forms makes it more resilient than competence trust.

Observer Findings

Through their in-depth analyses as invited outsider observers and team building facilitators, Newell and Swan found that the team started to break down very quickly after the project funding was received. Companion trust between two of the principal investigators (PIs) and perceptions of being able to enjoy working together led to the formation of the PI team. Temporary 'swift trust' (Meyerson et al 1996) developed based on competence cues taken from involvement on an expert panel and the quality and reputations of the universities to which each PI belonged. Crucially, in the rush to meet the proposal deadline the negotiation of the deliverables was omitted and in the light of little or no experience of working together and combined with epistemological differences of their disciplines, despite attempts to reach consensus, once work started, conflicts ensued.

The solution was to divide the project into semi-autonomous pieces which, through avoidance of conflict allowed their friendship to remain intact, but created centrifugal tensions which were later difficult to correct: this approach was agreed by everyone involved and it was often the case that the decisions made by the sub-groups would then be rejected by the other members. Newell and Swan concluded that had proper negotiations been held at the proposal stage, the group would have found that the idea would have shown itself to be unworkable. In the final phase, the research officers (ROs) got on board, but few attempts were made by the PIs to develop the informal, social coordination mechanisms and repeated changes to the project structure served to increase ambiguity and actually led to the PIs becoming less involved in the project. Disciplinary differences eroded competence trust between two of the RO s and this stared to undermine their companion trust and this eventually led to one of them leaving the project.

A Model of Trust

Newell and Swan argue that this work demonstrates how the three forms of trust interrelate and that 'it is the way in which they worked together which provides the greatest understanding of the development of trust'. The PIs and ROs differed in terms of the commitment trust demonstrated and this appeared to affect the development of companion and competence trust. As a result of their observations, they make a number of propositions which could form the basis of a model:

1. *In a low commitment trust situation, competence and companion trust are reciprocally dependent, one reinforcing the other in the same direction*
   The spiral that results can be either positive or negative: Risk among the ROs was low if the network failed to deliver its original objectives and none would be held accountable - commitment trust was low. There is little to hold the individuals together and the downward spiral develops its own momentum, which can finish with total trust and maybe group breakdown. This occurred in the RO group - 2/4 left within the first 8 months.

2. *In a high commitment trust situation, competence and companion trust are held in mutual tension such that, as one decreases, the other increases to compensate*
   Commitment trust had been established via the funding body's allocation of money which meant that, although there was high competence distrust, the same downward spiral didn't occur in the way that it had for the ROs. Also, concerted efforts were made to remain friends and confrontation was avoided - through the development of federated sub-projects. Unlike the RO situation where competence distrust reinforced companion distrust, it acted to strengthen companion trust and, consequently, led to compromises and detracted from the benefits of interdependence. Newell and Swan (p1319) argue that this demonstrates why it is important to differentiate between different types of trust as they were affected differentially by the interactions which occurred.

3. *Putting in place formal integration mechanisms will not guarantee the development of the more informal integration mechanisms which underpin the emergence of at least companion and competence trust.*
   Here, the research team did not spend enough time on social integration. Newell and Swan go further by questioning the extent to which, if at all, multi-disciplinary teams actually produce more innovative and useful knowledge, especially given their epistemological and ontological differences. Knights and Wilmott (1997) argue that the most common response to the pressures of inter-disciplinary research is 'mechanistic pooling' - each member takes a different 'slice' of the project and works with a minimum of communication with other members. This, they argue,

is the most common response because "..no time is 'wasted' in confronting differences in theoretical perspective.." and "..each member is able to maximise publication output in journals that cater for their particular specialism.." (cited in Newell and Swan 2000 p1321). Here, the research team acted in a multi-disciplinary way.

4. *The greater the variety in epistemological and ontological perspectives within a research network, the more likely the output is to be a compromise, which is actually less creative and innovative than would have been achieved within a single discipline.*
   They conclude that any examples of really creative and interdisciplinary research are likely to be the exception rather than the rule and those multi-disciplinary approaches may inhibit creation and diffusion of innovative ideas.

INTER-DISCIPLINARY APPROACHES IN 3D2ND

Taking note of the above discussion, an attempt has been made to develop consensus amongst members of the project research team on the Salford 3D2nD modelling project. A workshop was conducted in February 2002 with the whole research team to start defining the theoretical and ontological approaches for a vision for the nD tool. An electronic voting tool was used during the workshop to ascertain any participant consensus. In order to define a vision, the workshop participants explored several existing visions governing 'the future of construction IT'.

Although aspects of the existing visions can be strategically placed within the context of this project, it was clear from the results that the majority of workshop participants – 88% - wanted to develop a new collective vision of their own. The preference for an approach based and focussed both on implementation and on a blue sky approach is the crux of the project as it will dictate the type of technology that would be used and the subsequent implementation issues of the chosen technology. The majority voted for a mixture of both aspects and the voting on the ratios of this showed that, of the two most popular responses, 28% wanted a 50:50 split, but that 27% favoured 20:80. The implementation timeframe for industry was most popularly agreed at 25 years, but the question provoked the least consensus of all, with a quarter of the group thinking that some sort of implementation should be possible within 3 years. It was generally concluded from the workshop that more work would need to be done with the group to develop consensus and a shared view around the vision, but that due to the nature of the grant it would be possible to develop a range of scenarios to accommodate all of these factors, each being comprised of varying degrees of applied and blue-sky/innovative technology. The scenarios would include technology that could be utilised readily and technology/visions that could be utilised in the future, encompassing multiple timescales and a vision that would be continuously and iteratively developed so that it would harness the future direction and application of technology.

DISCUSSION

The current situation in the 3D2nD project demonstrates a relatively low level of agreement on a number of the issues voted upon in the workshop. This would be expected given the multi-disciplinary nature of the group make-up and the lack of a unanimously agreed vision at this point in time. On the 'wheel of cognate disciplines' (Eclipse Research Consultants 1997), the nD group members' disciplines are represented at almost all stages around the circle, meaning that the groups furthest from each other may never agree and that, in fact, group decisions in general would be expected to be more polarized than was the case for BEQUEST. It would be quite fair to say that in the absence of a fully shared vision, it is human nature for individuals to try to steer a project towards their own goals and areas of expertise. However, it is also likely that with more formalised and specific goals in place, the research team will 'rally around' these. This was demonstrated by Cooper's (2002) case study of BEQUEST and the greater level of consensus demonstrated by the PICABUE assessment technique over the project's duration. It could be interesting and relevant to replicate this assessment exercise with the 3D2nD group, with some alteration to the concepts for consideration.

Newell and Swan (2000) highlighted a number of pitfalls which could be avoided through management, not only of the goals and tangible aspects, but also of the wider context and the inter-relationships of all those involved. Effective teamwork can only occur if there is a sense of ownership towards the project and its objectives by all the members of the research team. This does not mean that consensus on all aspects, by all members and at all times should be the only aim; rather than the process of striving to achieve these should help facilitate the development of factors which, in absence could bring down the project. The dominant theme in a set of case studies presented at the ISI seminar (Newell 2002) was the existence of trust, and although this may be the hardest aspect to control favourably, in the absence of consensus, if enough trust exists between members, collective goals and relationships are less likely to be lost and eroded.

Following from Coopers (2002 p118) distinction between inter- and multi-disciplinary work, it is difficult at this stage to determine which the nD project is. The project has not yet required any member to think and work beyond their discipline boundary; indeed the tool is currently envisioned as being a 'pool' for all information relating to building design and construction. The tool may take on an inter-disciplinary nature when applied to real world problems – if this is the remit – as it will be capable of more than the initial individual's inputs and should be able to cross the traditionally held discipline boundaries.

With the scarcity of previous research on how to best carry out multi-disciplinary work, an important by-product would be a paper on the management of the 3D2nD project and it mechanisms and contingencies for success and the processes by which these were achieved (or not). The encouragement of research councils to partake in research of this kind, but the ironic lack of knowledge on how to do this, highlights an area worthy of further consideration.

REFERENCES

BEQUEST (2001) Final Report 2000-1, Contract No. ENV 4 CT/97-607, EC Environment and Climate Programme (1994-1998), *Research Theme 4, Human Dimensions of Environmental Change*, University of Salford, UK.

Cooper, I. (2002) Transgressing discipline boundaries: Is BEQUEST an example of the 'new production of knowledge'? *Building Research & Information*, **30:2,** 116-129.

Curwell, S., Hamilton, A. & Cooper, I. (1998) The BEQUEST network: towards sustainable development. *Building Research and Information*, **26:1,** 56-65.

Eclipse Research Consultants (1996) *PICABUE: Mapping Commitment to Sustainable Development - A Self-assessment Technique for Policy-Makers and Practitioners*, ERC, Cambridge.

Eclipse Research Consultants (1997) *Support for Inter-disciplinary Research, workshop prepared for the Research and Graduate College, University of Salford, and for the Committee for Inter-disciplinary Environmental Studies*, University of Cambridge, ERC, Cambridge.

Egan, J., (1998) *Rethinking construction : the report of the Construction Task Force to the Deputy Prime Minister, John Prescott, on the scope for improving the quality and efficiency of UK construction.* Department of the Environment, Transport and the Regions, London.

Epton, S., Payne, R. & Pearson, A. (eds.) (1983) *Managing Interdisciplinary Research*, John Wiley & Sons, Chichester.

Gibbons, M., Limoges, C., Nowotny, H., Schwartzman, S., Scott, P. & Trow, M. (1994) *The new production of knowledge.* Sage, London.

HEFCE (2001) *Interdisciplinary Research and the RAE (Research Assessment Exercise)*, Briefing Note 14, Higher Education Funding Council (England), Bristol.

Lave, J. & Wenger, E. (1991) *Situated Learning: Legitimate Peripheral Learning*, Cambridge University Press, Cambridge.

Newell, S. (2002) Managing knowledge in project teams - some lessons from some failed projects. Seminar held at Information Systems Institute, Salford University, April 24th 2002.

Newell, S. & Swan, J. (2000) Trust and inter-organizational networking. *Human Relations*, 53 (10) pp 1287-1328.

Palmer, J., Cooper, I. & van der Vorst, R. (1997) Mapping out fuzzy buzz words – who sits where on sustainability and sustainable development. *Sustainable Development,* **5:2,** 87-93.

Swan, J., Newell, S., Scarbrough, H. & Hislop, D. (1999) Knowledge management and innovation: networks and networking, *Journal of Knowledge Management*, **3:4,** 262-275.

Advances in Building Technology, Volume 1
M. Anson, J.M. Ko and E.S.S. Lam (Eds.)

RECENT TOPICS IN WIND ENGINEERING FOCUSING ON MONITORING TECHNIQUES

Y. Tamura and M. Matsui

Department of Architecture, Tokyo Institute of Polytechnics,
Atsugi, Kanagawa, JAPAN, 243-0297

ABSTRACT

This paper presents recent results relating to wind speed monitoring, wind pressure monitoring and wind response monitoring techniques and methods. For wind speed monitoring, LDV, PIV, PTV and so on were used in laboratory tests, and Doppler sodars and GPS drop-sonde were used in atmospheric boundary layer wind measurements. For wind pressure monitoring, various applications of simultaneous multi-channel pressure measuring systems are presented, and their efficiency in research, education and practical design are demonstrated. For response monitoring, the efficiency of RTK-GPS for wind-induced responses is mainly demonstrated.

KEYWORDS
LDV, PIV, Doppler Sodar, RTK-GPS, GPS Sonde, Simultaneous Multi-channel Pressure Measuring System, POD, Extreme Pressure Distribution, Full-scale Measurement

INTRODUCTION

This paper introduces recent topics focusing on monitoring techniques in wind engineering. Techniques for monitoring wind speeds, wind pressures and building responses are introduced, such as Doppler Sodars, the Simultaneous Multi-channel Pressure Measuring System, and the Global Positioning System. Their efficient contributions and potential ability in wind engineering are demonstrated, with reference to the author's group and other researchers' recent results. It should be noted that this paper does intend to review all the relevant works and literature, so many interesting results may have been omitted.

MONITORING OF WINDS

Wind Speed Monitoring for Wind Tunnel Testing

In wind tunnel experiments, wind speed is usually measured by Pitot static tubes and hot-wire anemometers. However, optical techniques have recently been playing important roles in comprehending complex flow fields. There are two main optical techniques. One uses the optical

Doppler shift and the interference effect, and to the other obtains the loci of particles from two successive images. Both methods require seeding of tracer particles. They have several advantages, such as non-contact measurement without any disturbance and applicability even in wake regions where there are complex reverse flows. The principle of Laser Doppler Velocimetry (LDV) was first introduced by Cummins et al. (1964). This method is widely used in fluid mechanics. It is very reliable, but it provides point-wise measurement, making it difficult to obtain simultaneous spatial data. Its application to wind engineering is seen in Akins & Reinhold (1998), Havel et al. (2001), and Becker et al. (2002). Figure 1 is an example of successfully illustrated streamlines in the wake region of bluff bodies captured by LDV (Havel et al., 2001). Particle Image Velocimetry (PIV) and Particle Tracking Velocimetry (PTV) are image processing techniques that solve a spatial velocity field by carrying out synchronous photography of particles in a plane recognized by a laser light sheet that emits light for a very short duration. Particle images are acquired by Charge-Coupled Device (CCD) cameras to detect particle movements. A sophisticated method has also been used to obtain 3D flow velocity vectors, using spatial information obtained by multiple CCD cameras from different positions. However, measurement is limited to the range of the laser light sheet. Some trials for obtaining flow velocity fields around buildings with complex geometry have been carried out using a transparent material for the building models. The flow field in the space surrounded by buildings becomes very complicated, and it is difficult to apply hot-wire anemometers. PIV has advantages in measuring in such a limited space, although there are some problems to be solved, such as too high a concentration of tracer particles, failure of the laser light sheet reaching the space of interest and the laser light sheet reflecting from the model surface. Figure 2 shows an instantaneous wind velocity field inside a dome that has wide openings at the roof and side walls (Kondo et al., 2001).

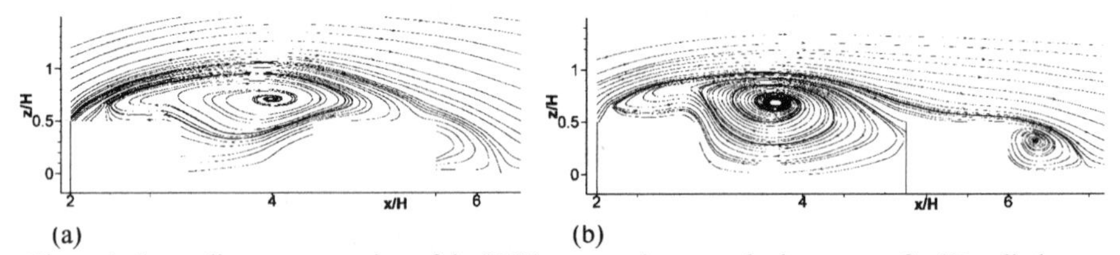

Figure 1: Streamline representation of the LDV measured mean velocity vectors for 2D cylinders. (Due to symmetry only half the plane is shown. Havel et al. 2001)

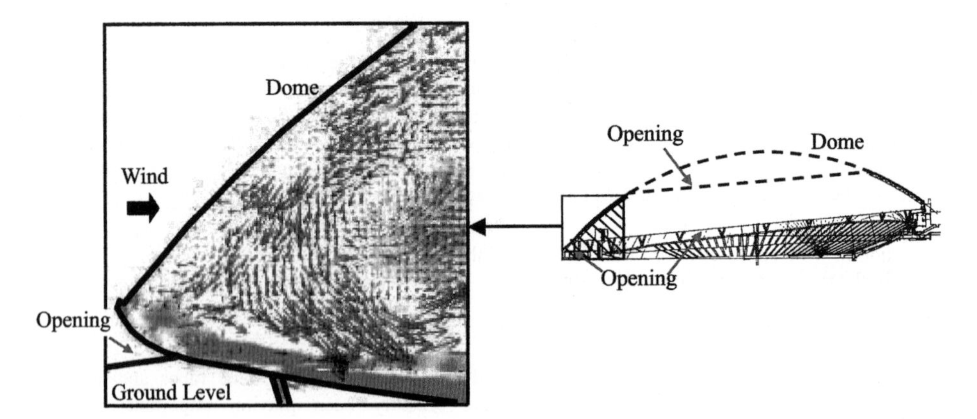

Figure 2: Instantaneous wind velocity fields in a dome measured by PIV (Kondo et al. 2001)

Wind Speed Monitoring in Atmospheric Boundary Layer

It is very important in wind-resistant structural design to accurately estimate the design wind speeds. Two topics relating to the monitoring technique are presented here: vertical wind speed profile in the atmospheric boundary layer during tropical storms such as hurricanes, cyclones and typhoons; and estimation of terrain roughness. Because of difficulties in observing extremely rare phenomena and measuring atmospheric boundary layer wind speeds at high altitudes, wind speed profiles for various terrain conditions are not well known, although almost all wind load codes specify them. Doppler radar (Hayashida et al., 1996), GPS drop-sondes (Powell et al., 1999) and Doppler sodars (Tamura et al., 1999a, 2001, Amano et al., 1999) are powerful devices for measuring atmospheric boundary layer wind profiles. Powel et al. (1999) showed wind speed profiles inside hurricanes using GPS drop-sondes as shown in Fig.3(a). Powel (2002) also obtained the normalized wind speed profiles for 14 hurricanes of 330 profiles near the eye walls. These profiles revealed some characteristics of friction at the sea surface. It is expected that a lot of useful information will be obtained from this challenging research in the near future.

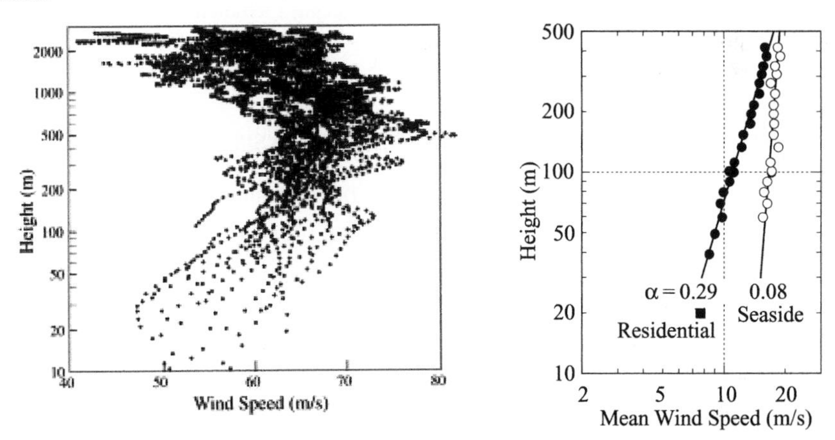

(a) 10 eye-wall profiles from Hurricane Mitch (1998) (b) Seaside → 23km (Residential)→ Residential
Figure 3: Wind speed profiles measured by (a) GPS drop-sonde (Powel et al., 1999 and Powel, 2002) and (b) Doppler sodars (Tamura et al., 2001a)

To assess the design wind speed at a particular site, it is necessary to determine the variation in wind speed with terrain roughness. Wind observations at a single site provide some information on the wind speed profile, but they give no information on variations in wind speed with terrain roughness. Tamura et al. (1999a, 2001a) conducted simultaneous wind observations over two sites with different roughnesses using Doppler sodars to find a reasonable method for estimating design wind speed for a given terrain roughness. They described the results of boundary layer wind observations at seaside sites and inland sites of the main island of Japan, and discussed the change in the horizontal wind speed profiles and the reduction in wind speed at lower altitudes due to inland roughness. Mono-static Doppler sodars were used for the measurements. Figure 3(b) compares the 10min mean wind speed profiles at two sites 23km apart for sea winds during the same storms to assess the effects of terrain roughness between them. One is located in a seaside area and another is in a residential area near Tokyo. The profiles were examined as ensemble averages of several 10min mean profiles. It can be seen that, due to so-called friction, the mean wind speeds decreased at lower altitudes after the wind blew over the inland terrain roughness, and consequently the power law indices α increased considerably.

Systematic multiple-site observations were carried out by Maeda et al. (1995) to capture the wind speed distribution over a wide area. They introduced a network of 121 anemometers installed on a transmission tower network in Kyushu, Japan. This network provides useful information on horizontal wind structures.

Terrain Roughness Monitoring with Airborne Sensors

Although the height and land use pattern of the ground surface are digitized and have been utilized as the Geological Information System (GIS), the aerodynamic effects of roughness including buildings and trees are most important in wind engineering. It has been thought to be difficult to directly measure ground roughness. Recently, a laser profiler mounted on an aircraft with a GPS and an Inertial Navigation System has been developed. This technique enables monitoring of the displacement between the ground surfaces including obstacles, and then evaluation of the terrain roughness (Kraus and Pfeifer, 1998). The measured results are obtained in the form that can be handled by a computer as a Digital Terrain Model (DTM), and is utilized for the shadow evaluation of a building condensed urban area (Yoshida et al. 2001). In wind engineering, these data can be utilized in two patterns. One is direct use of the displacement data to generate a numerical grid point to making a terrain model in boundary layer wind tunnels. The other is aerodynamic modeling of the terrain roughness (Maruyama, 1999). These are on-going studies, and results are expected in the near future.

WIND PRESSURE MONITORING AND SPATIAL DATA ANALYSIS

Multi-channel Wind Pressure Measuring System for Wind Tunnel Testing

The Simultaneous Multi-channel Pressure Measuring System (SMPMS) was first proposed by Fujii et al. (1986) and Ueda et al. (1994), to measure unsteady pressure fields on building models. Although its original version used 512 electronically scanned pressure sensors consisting of 32/16-port modules, every pressure tap has its own transducer without any scanning system in the current version. Animation showing the fluctuating pressure field is thought to be CFD's specialty. However, the SMPMS can provide beautiful animations showing the temporal and spatial fluctuations of the pressure fields as for the CFD results. The evidences of forming and drifting down of vortices of various sizes and shapes near the surfaces and their dynamic structures can be observed precisely. The most important advantage is that the phenomena observed by the SMPMS are the actual physical phenomena. Although SMPMS is not a recent technology, a lot of information has been obtained from the SMPMS. Here, some interesting SMPMS results are introduced.

Instantaneous pressure distributions causing extreme wind load effects

Wind pressure distributions specified for frame analyses in various wind loading codes are basically set proportional to the mean pressure distributions. However, these are quite different from the actual extreme wind pressures causing maximum load effects as in Holmes (1988, 1992), Kasperski (1992), Tamura et al.(2000). Kasperski & Niemann (1988) and Kasperski (1992) proposed the Load Response Correlation (LRC) formula to estimate the extreme pressure distributions, and Holmes (1992) and Tamura et al. (2001b) proved its validity. Figures 4(a) and 4(b) show examples of instantaneous pressure distributions on the surface of high-rise and low-rise building models.

These images give very useful information on the precise structures of fluctuating surface pressures. Based on the SMPMS data, Tamura et al. (2001c) pointed out a particular case in which the LRC formula has a limitation in its applicability. Tamura et al. (2002a) also discussed wind load combinations and extreme pressure distributions causing maximum wind forces, and clearly demonstrated the importance of considering wind force combinations even for low- or middle-rise buildings.

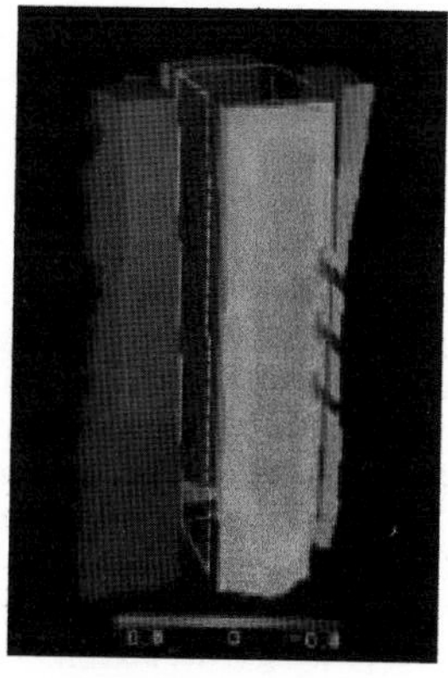

(a) High-rise model

(b) Low-rise model

Figure 4: Instantaneous pressure distributions acting on a low-rise and a high-rise building model (Kikuchi & Hibi, 2002)

Combinations of SMPMS and POD

Proper Orthogonal Decomposition (POD) is a kind of Karhunen-Loève decomposition, which is a probabilistic expression of a method called Factor Analysis (Armit, 1968). POD is a useful tool for capturing hidden deterministic structures in random fields and for reducing information necessary to reproduce the phenomena, and it has also been used in wind engineering, e.g. Lee (1975), Best & Holmes (1983), and Kareem & Cermak (1984). When the POD technique is combined with SMPMS, its efficiency significantly increases as demonstrated by Bienkiewicz et al. (1995), Kikuchi et al. (1997), and Tamura et al. (1999b). Tamura et al. (1999b) summarized the points to note for POD analyses using SMPMS data. Their results for high-rise building models showed that the 1st and 5th modes closely relate to the vortex shedding and greatly contribute to the across-wind force, and the 2nd, 3rd and 4th principal coordinates contribute to along-wind force. Only these few of 500 modes in total can predict the wind-induced response of a high-rise building model within an error of 6%, showing an amazing reduction in necessary amount of information.

Calculations of internal forces based on SMPMS and practical applications

Information including animation based on SMPMS data showing complex temporal and spatial wind pressure fluctuations clearly demonstrates the limitation of simple wind load provisions specified in current wind load codes and the necessity of using 3D precise pressure fluctuation data for the practical design of important buildings such as long-span stadium roofs, domes or high-rise buildings. Whalen et al. (1998) proposed a wind resistant design system using electronic databases containing authorized aerodynamic data sets for various types of buildings, because of the discrepancy between the maximum load effects calculated by some major wind load codes and actual fluctuating pressures by SMPMS. SMPMS can be an efficient tool in conjunction with data processing and computer graphic techniques not only for basic academic research or educational purposes, but also for practical applications, for example Yasui et al. (1999). In practical applications, SMPMS can provide direct information on fluctuating pressures acting on building models with a sufficiently fine spacing. Internal forces or stresses in members of frames and displacements of particular points can be directly obtained by dynamic analysis of 3D frame models. Thus, errors in modeling of wind forces or lack of wind load combinations can be minimized. Taniguchi et al. (2001) directly compared the internal stresses and displacements caused by strong winds and earthquakes. Incidentally, Zhou et al. (2001) also proposed an interactive aerodynamic database based on High Frequency Force Balance data. These trends to open web sites providing an electronic aerodynamic database may accelerate the use of direct information on pressure records obtained by SMPMS for more precise information on wind loads to realize more accurate and economic wind resistant design.

Full-scale Measurements of Fluctuating External Wind Pressures

Full-scale measurements on fluctuating pressures have been conducted since the '60s, and some sophisticated experimental studies have been conducted, such as Aylesbury Experimental House (Eaton & Mayne, 1975), Texas Tech Building (Ng & Mehta, 1990, Levitan et al., 1991), Silsoe Structures Building (Richradson et al., 1990). Some investigations have used flow visualization techniques, e.g. Banks et al. (2000). Animation of pressure fluctuation has also been carried out to visually observe actual phenomena, e.g. Hibi et al. (1994), which measured fluctuating pressures acting on a 30.4 m-high building using one hundred and fourteen pressure transducers on the walls. Figures 5(a) - 5(e) show examples of instantaneous pressure distributions acting on a full-scale building (Hibi et al., 1994).

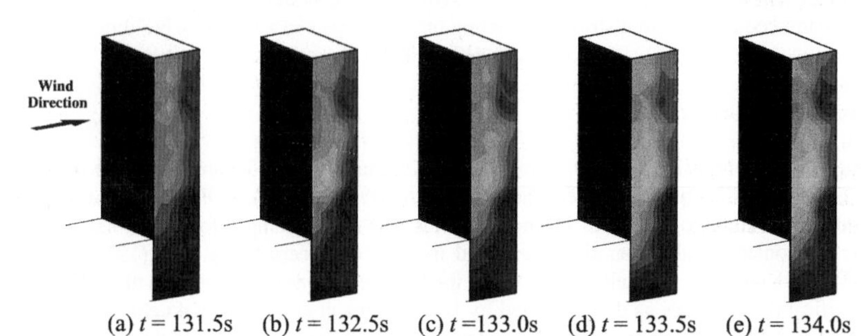

(a) t = 131.5s (b) t = 132.5s (c) t =133.0s (d) t = 133.5s (e) t = 134.0s
Figure 5: Instantaneous pressure distributions on a 30.4m-high building (Full-scale data) (Hibi et al., 1994)

Full-scale Measurements of Internal Wind Pressures

A recent remarkable full-scale result is the measurement of internal pressures on a 120m-high office building by Kato et al. (1997). They used Highly Precise Absolute Pressure Transducers, which had a measurement range of 800-1100hPa and a resolution of 5ppm. They clearly demonstrated the variation of internal pressures with height, wind direction and so on.

RESPONSE MONITORING

Monitoring of Displacement by RTK-GPS

Accelerometers have been commonly used for field measurements of building responses, and occasionally displacement transducers have been used. These transducers can be very useful for identifying dynamic characteristics of buildings. Tamura et al. (2002) identified the dynamic properties of a 15-story building up to the 9th mode very accurately by using the Frequency Domain Decomposition technique for ambient vibrations. Xu & Zhan (2001) measured wind-induced responses of Di Wang Tower with accelerometers and displacement transducers during a strong typhoon, and determined its dynamic components. As is well known, wind-induced responses consist of a static component, i.e. a mean value; a quasi-static component; and a resonant component. The static and quasi-static components are difficult to measure with general accelerometers. In the past few decades, various attempts have been made to measure wind-induced responses including these components by using optical sensors or strain gauges. For example, Kobayashi (1962) used a He-Ne gas laser transmitter and a photo-cell to measure the wind-induced tip displacement of a 170m-high building. Maeda et al. (1999) utilized a CCD video camera and a target mark to measure the displacement of a power transmission tower.

Çelebi (1998) proposed using a RTK-GPS (Real-Time Kinematic Global Positioning System) for building response measurements. RTK-GPS (Leica MC1000) has a nominal accuracy of 1cm +1ppm for horizontal displacements and 2cm +2ppm for vertical displacements with a sampling rate of 10Hz (Celebi 1998, Tamura 2001d). Tamura et al. (2001d) reported that RTK-GPS could measure displacements when the vibration frequency was less than 2Hz and the amplitude was more than 2cm. Considering the static component and the first mode predominance of wind-induced responses, GPS can be a powerful tool for monitoring wind-induced responses. For this reason, many projects on monitoring of wind-induced responses of buildings and structures have been carried out, e.g. Toriumi et al. (2000), Ding et al. (2002), Kijewski & Kareem (2002), and Breuer & Konopka (2002).

Figure 6 shows an example of the temporal variation of response of a 108m-high steel tower. The RTK-GPS data consists of a static displacement of about 4cm, a slowly fluctuating quasi-static component with a period of 20s - 30s, and a resonant component with the first mode natural period of the tower: 1.8s (0.57Hz). The acceleration record seems to correspond closely to the fluctuating component of the displacement by RTK-GPS, and their power spectrum densities show very good agreement (Tamura et al., 2001d). Since the RTK-GPS measurement system can measure the static displacement, the tower deformation caused by heat stress can also be measured.

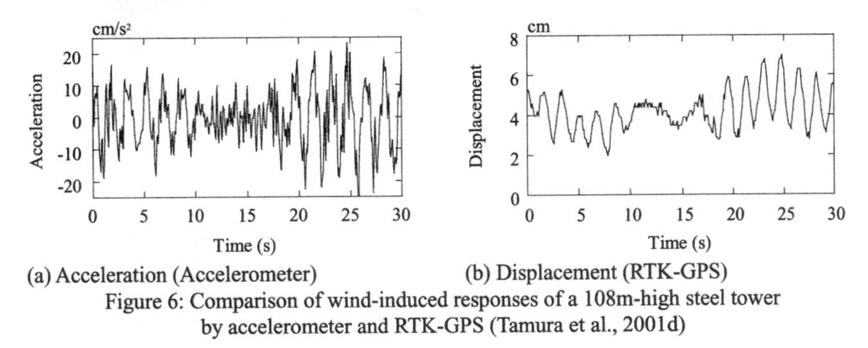

(a) Acceleration (Accelerometer) (b) Displacement (RTK-GPS)

Figure 6: Comparison of wind-induced responses of a 108m-high steel tower by accelerometer and RTK-GPS (Tamura et al., 2001d)

Stress Monitoring by Hybrid Use of FEM and RTK-GPS

Tamura et al. (2002c) developed a hybrid method for evaluating structural integrity. The tip displacement observed with the GPS was converted into structural member stresses. The conversion was based on the stiffness matrix of a finite element modeling. This concept can be developed for a future disaster prevention system consisting of an RTK-GPS monitoring system of a group of high-rise buildings in a city and a database containing analytical virtual building models having structural reality constructed in a computer.

CONCLUDING REMARKS

Recent developments and applications of various monitoring devices and techniques for detecting wind speeds, wind pressures and wind-induced responses of buildings and structures have been briefly overviewed. Developments that we know about have mainly resulted from developments of monitoring systems and the visualization techniques. In order to understand the phenomena and their natures, a sufficient amount of accurate information is necessary. These are only supported by high quality monitoring techniques. Many of the techniques have just been introduced into the wind engineering field or are still under development, and new information will be provided in the future to deepen and widen our knowledge.

References

Akins R.E. and Reinhold T.A. (1998). Laser Doppler velocimeter measurements of separated shear layers on bluff bodies, Journal of Wind Engineering and Industrial Aerodynamics 74-76, 455-461

Amano T., Fukushima H., Ohkuma T. and Kawaguchi A. (1999). The observation of typhoon winds in Okinawa by Doppler sodar., *Journal of Wind Engineering and Industrial Aerodynamics* **83**, 11-20

Armitt J. (1968). Eigenvector analysis of pressure fluctuations on the West Burton instrumented cooling tower, *Central Electricity Research Laboratories (U.K.) Internal Report* RD/L/N114/68

Banks D., Meroney R.N., Sarkar P.P., Zhao Z. and Wu F. (2000). Flow visualization of conical vortices on flat roofs with simultaneous surface pressure measurement, *Journal of Wind Engineering and Industrial Aerodynamics* **84:1**, 65-85

Becker S., Lienhart H. and Durst F. (2002). Flow around three-dimensional obstacles in boundary layers, *Journal of Wind Engineering and Industrial Aerodynamics* **90:4-5**, 265-279

Best R.J. and Holmes J.D. (1983). Use of eigenvalues in the covariance integration method for determination of wind load effects, *Journal of Wind Engineering and Industrial Aerodynamics* **13**, 359-370

Bienkiewicz B., Tamura Y., Ham H.J., Ueda H. and Hibi K. (1995). Proper orthogonal decomposition and reconstruction of multi-channel roof pressure, *Journal of Wind Engineering and Industrial Aerodynamics* **54-55**, 369-381

Breuer P., Chmielewski T., Gorski P. and Konopka E. (2002). Application of GPS technology to measurements of displacements of high-rise structures due to weak winds, *Journal of Wind Engineering and Industrial Aerodynamics* **90**, 223-230

Celebi M. (1998). GPS and/or strong and weak motion structural response measurements -Case studies, *Structural Engineering World Congress '98, San Francisco, Conference Proceedings on CD-ROM*, T193-1

Cummins H.Z., Knable N. and Yeh Y. (1964). Observation of Diffusion Broadening of Rayleigh Scattered Light, *Phyics Review Letter* **12**, 150

Ding X.L., Chen Y.Q., Huang D.F, Xu Y.L. and Ko J.M. (2002). Measurement of Vibrations of Tall Buildings with GPS, *Structure Congress and Exposition, ASCE, Performance of Structure from Research to Design, Denver, Colorado*, 201-202

Eaton K.J. and Mayne J.R. (1975). The measurement of wind pressures on two-story houses at Aylesbury, *Journal of Industrial Aerodynamics* **1**, 67-109

Fujii K., Hibi K. and Ueda H. (1986). A new measuring system of unsteady pressures using electronically scanned pressure sensors (ESP) and its applications to building models, *Proceedings of the 9th National Symposium on Wind Engineering*, 313-318 (in Japanese)

Gerdes F. and Olivari D. (1999). Analysis of pollutant dispersion in an urban street canyon, *Journal of Wind Engineering and Industrial Aerodynamics* **82:1-3**, 105-124.

Hayashida H., Fukao S., Kobayashi T, NIrasawa H., Mataki Y., Ohtake K. and Kikuchi M. (1996) Remote sensing of wind velocity by boundary layer lader, *Journal of Wind Engineering* **67**, 39-42, (in Japanese)

Havel B., Hangan H. and Martinuzzi R. (2001). Buffeting for 2D and 3D sharp-edged bluff bodies, *Journal of Wind Engineering and Industrial Aerodynamics* **89:14-15**, 1369-1381

Hibi K., Kikuchi H. and Ueda H. (1994). Visualization of fluctuating surface pressure distribution on a full-scale building surface, *Proceedings of Third Asian Symposium on Visualization*, 862-867

Holmes J.D. (1988). Distribution of peak wind loads on a low-rise building, *Journal of Wind Engineering and Industrial Aerodynamics* **29**, 59-67

Holmes J.D. (1992). Optimized peak load distributions, *Journal of Wind Engineering and Industrial Aerodynamics* **41-44**, 267-276

Kareem A. and Cermak J.E. (1984). Pressure fluctuations on a square building model in boundary-layer flows, *Journal of Wind Engineering and Industrial Aerodynamics* **16**, 17-41

Kasperski M. (1992). Extreme wind load distributions for linear and nonlinear design, *Engineering Structures* **14-1**, 27-34

Kasperski M. and Niemann H.J. (1988). On the correlation of dynamic wind loads and structural response of natural-draught cooling towers, *Journal of Wind Engineering and Industrial Aerodynamics* **30**, 67-75

Kato N., Niihori Y., Kurita T. and Ohkuma T. (1997). Full-scale measurement of wind-induced internal pressures in a high-rise building, *Journal of Wind Engineering and Industrial Aerodynamics* **69-71**, 619-630

Kikuchi H., Tamura Y., Ueda H. and Hibi K. (1997). Dynamic wind pressures acting on a tall building model - proper orthogonal decomposition, *Journal of Wind Engineering and Industrial Aerodynamics* **69-71**, 631-646

Kikuchi H. and Hibi K. (2002). Personal communication

Kijewski T. and Kareem A. (2002). GPS for monitoring the dynamic response of tall buildings, *Structure Congress and Exposition, ASCE, Performance of Structure from Research to Design, Denver, Colorado*, 197-198

Kobayashi S. (1962). Sway observation of tall building in strong wind (Keio Plaza Hotel Building 170m high), *Transactions of The Architectural Institute of Japan*, **192**, 33-39

Kondo K., Hongo T., Suzuki M., Tsuchiya M., Nakano R. and Hayashida K. (2001). Flow Visualization Techniques in Architectural and Civil Engineering Fields, *Journal of the Visualization Society of Japan* **21:1**, (in Japanese)

Kraus K. and Pfeifer N. (1998). Determination of terrain models in wooded areas with airborne laser scanner data, *ISPRS Journal of Photogrammetry and Remote Sensing* **53**, 193-203

Lee B.E. (1975). The effects of turbulence on the surface pressure field of a square prism, *Journal of Fluid Mechanics* **69**, 263-282

Lee S. and Kim H. (1999). Laboratory measurements of velocity and turbulence field behind porous fences, *Journal of Wind Engineering and Industrial Aerodynamics* **80:3**, 311-326

Levitan M.L., Holmes J.D., Mehta K.C. and Vann W.P. (1991). Field measurements of pressures on the Texas Tech building, *Journal of Wind Engineering and Industrial Aerodynamics* **38**, 227-234

Maeda J., Tomonobu N. and Miyazaki F. (1995) Wide area network system of wind measurement utilizing power transmission line systems, *International Symposium on cable dynamics*,

Maeda J., Imamura Y. and Morimoto Y. (1999). Wind response behaviour of a power transmission tower using new displacement measurement, *Proceedings of the 10th International Conference on Wind Engineering* **1**, 481-486

Maruyama T. (1999). Surface and inlet boundary conditions for the simulation of turbulent boundary layer over complex rough surfaces, *Journal of Wind Engineering and Industrial Aerodynamics* **81:1-3**, 311-322

Naito G., Ito Y. and Hayashi T. (1999). Strong wind features of the lower planetary boundary layer on a complex terrain, *Wind Engineering into the 21st Century, Larsen, Larose and Livesey(eds)* 281-286

Ng H.T. and Mehta K.C. (1990). Pressure measuring system for wind-induced pressures on building surfaces, *Journal of Wind Engineering and Industrial Aerodynamics* **36**, 351-360

Richardson G.M., Robertson A.P., Hoxey R.P. and Surry D. (1990). Full-scale and model investigations of pressures on an industrial/agricultural building, *Journal of Wind Engineering and Industrial Aerodynamics* **36**, 1053-1062

Powell M.D., Reinhold T.A. and Marshall R.D. (1999). GPS sonde insights on boundary layer wind structure in hurricanes, *Wind Engineering into the 21st Century, Lasen, Larose and Livesey (eds),* 307-314

Powel M.D. (2002). (Personal communication)

Tamura Y., Suda K., Sasaki A., Iwatani Y., Fujii K., Hibi K. and Ishibashi R.(1999a). Wind speed profiles measured over ground using Doppler sodas, *Journal of Wind Engineering and Industrial Aerodynamics* **83**, 83-93

Tamura Y., Suganuma S., Kikuchi H. and Hibi K. (1999b). Proper orthogonal decomposition of random wind pressure field, *Journal of Fluids and Structures* **13**, 1069-1095

Tamura Y., Kikuchi H. and Hibi K. (2000). Wind load combinations and extreme pressure distributions on low-rose buildings, *Wind & Structures, International Journal* **3:4**, 279-289

Tamura Y., Suda K., Sasaki A., Iwatani Y., Fujii K., Ishibashi R. and Hibi K. (2001a). Simultaneous measurement of wind speed profiles at two sites using Doppler sodas, *Journal of Wind Engineering and Industrial Aerodynamics* **89**, 325-335

Tamura Y., Kikuchi H. and Hibi K. (2001b). Extreme wind pressure distributions on low-rise building models, *Journal of Wind Engineering and Industrial Aerodynamics* **89**, 1635-1646

Tamura Y., Kikuchi H. and Hibi K. (2001c). Actual pressure distributions and LRC formula, *The Fifth Asia-Pacific Conference on Wind Engineering, Kyoto, Journal of Wind Engineering* **89**, 589-592

Tamura Y., Matsui M., Pagnini L., Ishibashi R. and Yoshida A. (2001d). Measurement of wind-induced response of buildings using RTK-GPS, *Proceedings of The Fifth Asia-Pacific Conference on Wind Engineering, Kyoto, Japan*, 93-97

Tamura Y., Kikuchi H. and Hibi k. (2002a). Quasi-static wind load combinations for low- and middle-rise buildings, *Preprints of Engineering Symposium to Honor Alan G. Davenport, The University of Western Ontario*, C3-1-C3-12

Tamura Y., Zhang L., Yoshida A., Nakata S. and Itoh T. (2002b). Ambient vibration tests and modal identification of structures by FDD and 2DOF-RD Technique, *Proceedings of Structural Engineers World Congress 2002, Yokohama, Japan* (to be published)

Tamura Y., Yoshida A., Ishibashi R., Matsui M. and Pagnini L.C. (2002c). Measurement of wind-induced response of buildings using RTK-GPS and integrity monitoring, *The Second International Symposium on Advances in Wind and Structures (AWAS'02)* (to be published)

Taniguchi H., Hirose K., Watanabe Y., Gouda H. and Saitoh M. (2001). On design and construction of the roof structure of the swimming stadium for national athletic meet in Kochi, *IASS International Symposium on Theory, Design and Realization of Shell and Spatial Structures, October, Nagoya, Japan* TP172

Toriumi R., Katsuchi H. and Furuya N. (2000). A study on spatial correlation of natural wind, *Journal of Wind Engineering and Industrial Aerodynamics* **87**, 203-216

Ueda H., Hibi K., Tamura Y. and Fujii K. (1994). Multi-channel simultaneous fluctuating pressure measurement system and its applications, *Journal of Wind Engineering and Industrial Aerodynamics* **51**, 93-104

Whalen T., Simiu E., Harris G., Lin J. and Surry D. (1998). The use of aerodynamic databases for the effective estimation of wind effects in main wind-force resisting systems: application to low buildings, *Journal of Wind Engineering and Industrial Aerodynamics* **77-78** ,685-693

Xu Y.L. and Zhan S. (2001). Field measurements of Di Wang Tower during Typhoon York, *Journal of Wind Engineering and Industrial Aerodynamics* **89:1**, 73-93

Yasui H., Marukawa H., Katagiri J., Katsumura A., Tamura Y. and Watanabe K. (1999). *Journal of Wind Engineering and Industrial Aerodynamics* **83**, 277-288

Yoshida H., Ishibashi K. and Kurisaki N. (2001). A simulation of complex insolation in CBD by using airborne LIDAR data set, *Papers and Proceedings of the Geographic Information Systems, 2001* 315-319 (in Japanese)

Zhou Y., Kijewski T. and Kareem A. (2001). Aerodynamic Loads Database, *http://www.nd.edu/~nathaz/database/index.html*

Advances in Building Technology, Volume I
M. Anson, J.M. Ko and E.S.S. Lam (Eds.)

A CONCEPTION OF CASUALTY CONTROL BASED SEISMIC DESIGN FOR BUILDINGS

Li-Li Xie

College of Civil Engineering, Harbin Institute of Technology &.
Institute of Engineering Mechanics, China Seismological Bureau
Harbin, 150080, China

ABSTRACT

Earthquake disaster is still the number one among all other natural disasters, particularly, in terms of destructive power in causing deaths. Can earthquake engineers control seismic casualties through the seismic design of buildings? For this purpose, a conception of casualty control based seismic design is presented and a "two-step decision-making" method is proposed for determining the optimum seismic design intensity (or ground-motion) for controlling both seismic death and economic losses. The key problems in establishing the model are to determine the appropriate socially acceptable level of earthquake mortality and establish the corresponding objective function and /or constraint conditions for optimum seismic design intensity. Ten different grades of socially acceptable mortality are initially suggested and the final socially seismic acceptable mortality level was proposed based on a questionnaire that was carried out nationwide in China. Finally, the effect of various grades of acceptable earthquake mortality on seismic design intensity is analyzed.

KEYWORDS

Seismic Death control, Performance-based design, Seismic design criterion, Seismic vulnerability analysis, Earthquake economic loss estimation, Acceptable level for earthquake human mortality, Optimum seismic design load, Decision-making method

INTRODUCTION

Recently, with the development of Performance-Based Earthquake Engineering, it seems that control of economic losses to a reasonable level has become the main objective in determining seismic design loads (Cornell C. Allin and Helmut Krawinkler. (2000)). This may be acceptable for a developed country, but might not be appropriate for a developing country. Table 1 lists the number of human deaths and economic losses in several recent earthquakes (China Seismological Bureau (1996), Fu Zheng-xiang and Li Ge-ping (1993)). Table 1 indicates that the number of deaths is higher, but economic losses are lower in developing countries than in developed countries. For example, an earthquake of magnitude 7.1 caused deaths in Iran about 24 times those in the U.S.A, but economic loss in Iran is only about 1/3 that in the U.S.A.

TABLE 1
LIFE AND ECONOMIC LOSSES IN MAJOR EARTHQUAKE DISASTERS

Earthquake	Country	Year	Magnitude	Number of deaths	Economic loss
Haicheng	China	1975	7.3	1328	13 billions
Tangshan	China	1976	7.8	242,769	100 billions
North Iran	Iran	1997	7.1	1573	$5 billion
North Philippines	Philippines	1990	8.0	2,600	$5.3 billion
Armenia	Former USSR	1988	6.8	25,000	
Mexico	Mexico	1985	8.1	10,000	
Loma Prieta	U.S.A	1989	7.1	65	$15 billions
Northridge	U.S.A	1994	6.7	57	$151 billions
Kobe	Japan	1995	7.2	5,500	$1000 billions

It is generally accepted that the main reason causing death and injury during earthquakes is collapse of and damage to structures. How can seismic mortality be reduced at an effective cost is a central problem of seismic design of buildings (Xie Li-li (1996))? How can we control the seismic mortality through determining the seismic design load in seismic design? At present, neither seismic hazard analyses nor seismic design codes have mentioned that earthquake casualties can be quantitatively controlled by selecting seismic design load at an effective cost. The available seismic design can answer neither how great the potential mortality in a future earthquake will be, nor whether the casualty level is acceptable (Xie Li-li (1996)). In order to consider the casualty factor in determining seismic design load so as to quantitatively control casualties by adjusting design intensity, this paper specifically includes the concept of a socially acceptable level for earthquake human fatalities. For a city or a region, socially acceptable mortality is defined as a ratio of possible maximum death toll to the total population of the concerned region in a future earthquake, which can be accepted by society (Ma Yu-hong (1996)). Based on statistics of the deaths caused from historical natural disasters and devastating earthquakes (particularly the recent destructive earthquakes occurring in modern cities, such as the 1975 Haicheng, and 1976 Tangshan earthquakes, the 1988 Armenia earthquake, the 1994 Northridge earthquake, the 1995 Kobe earthquake and the 1999 Izmit earthquake in Turkey), ten grades of "acceptable" mortality are proposed in this study. After an analysis of response to a questionnaire distributed to the experts in the field of earthquake disaster prevention and officers from civil defense departments, a realistic compromise for a socially acceptable mortality level is recommended. Furthermore, a "two-step decision-making" method is developed for determining the optimum seismic design intensity conditional on reducing the earthquake fatalities. Finally, a method for determining seismic design load based on various grades of socially acceptable earthquake mortality is presented.

MODEL OF DECISION-MAKING ANALYSIS FOR SEISMIC DESIGN INTENSITY IN CONTROLLING SEISMIC FATALITIES

In the models of decision-making analysis for seismic design intensity, a reasonable objective function keeps a balance between extra cost of a structure to ensure earthquake resistance and possible earthquake losses. In quantitative form, the sum of the extra cost of construction for withstanding an earthquake and the earthquake economic losses should be at a minimum. This model of decision-making for seismic design intensity is depicted in Fig 1.

Find I to minimize

$$S(I) = C(I) + LP(I) \tag{1}$$

with constraint condition:

Numbers of fatalities $\leqslant$ socially acceptable level of mortality (2)

Figure 1. Mathematical model of optimum decision-making for seismic design

In the figure, $S(I)$ is the objective function, $C(I)$ is the extra cost of structures to resist the earthquake, $LP(I)$ is the expected earthquake economic losses of structures designed and constructed, and I is the adopted seismic design intensity for the region. The expected earthquake economic loss $LP(I)$ is a function of the design intensity I and the intensities i that are likely to happen in the region during future earthquakes. Therefore, $LP(I)$ can be expressed as:

$$LP(I) = \tau \cdot (L_1(I) + L_2(I)) \tag{3}$$

$$L_1(I) = \sum_{i=6^\circ}^{10^\circ} \sum_{j=1}^{5} P(D_j \mid I, i) \cdot l_1(D_j) \cdot W \cdot P(i) \tag{4}$$

$$L_2(I) = \sum_{i=6^\circ}^{10^\circ} \sum_{j=1}^{5} P(D_j \mid I, i) \cdot l_2(D_j) \cdot Y \cdot P(i) \tag{5}$$

Where, $L_1(I)$ is the direct economic loss due to structural damage, $L_2(I)$ is the economic loss due to the damage of contents, τ is a modified coefficient obtained from a statistical analysis and varies from 1.4 to 3.0, accounting for the indirect economic losses; for example, τ takes 1.4 for residential buildings and 3.0 for commercial and industrial buildings (Ma Yu-hong (1996)), $P(D_j \mid I, i)$ is the vulnerability matrix of buildings (Ma Yu-hong (1996)), W is the construction cost, Y is the total value of building contents, $l_1(D_j)$ and $l_2(D_j)$ are the ratio of direct economic losses of structures to W and losses of contents to Y respectively, and $P(i)$ is occurrence probability of earthquake intensity i worked out from seismic hazard analysis.

In case only objective function equation (1) is minimized, we can obtain the optimal design intensity (or **ground motion**) for assuring minimal economic losses and it can be named the **optimum economic design intensities (ground motion)**. It is apparent that the optimum economic design intensities (ground motion) can only make economic losses minimal and do nothing directly to reduce seismic mortality. However, the concept of the **optimum safe design intensity (ground motion)** is defined as the seismic design intensity or (ground motion) that can be used not only to minimize the economic losses but also to control the seismic mortality below the acceptable level. For this purpose, a "two-step decision-making" method is presented.

"TWO-STEP DECISION-MAKING" METHOD FOR DETERMINING THE OPTIMUM SAFE SEISMIC DESIGN INTENSITY TO CONTROL SEISMIC MORTALITY POTENTIAL

It has been mentioned that the optimum safe design intensity (ground motion) should be optimal in economic senses and also assure seismic mortality below the acceptable levels. In the optimum decision-making model, the equation (1) is the objective function and the controlled mortality (equation 2) is regarded as the constraint condition. The optimum safe seismic design intensity (ground motion) can be determined when both objective function and constraint condition are satisfied. It is assumed that the optimum safe seismic intensity should ensure earthquake mortality below the socially acceptable levels for each of the "frequently occurring earthquake", "occasionally occurring earthquake" and "rarely occurring earthquake"(see Table 2). Therefore, the constraint conditions of socially acceptable mortality can be expressed as:

$$\begin{cases} RD_{I,I_{63}} \leq RD_{acc,I_{63}} \\ RD_{I,I_{10}} \leq RD_{acc,I_{10}} \\ RD_{I,I_{5}} \leq RD_{acc,I_{5}} \end{cases} \tag{6}$$

where, I_{63} stands for intensity of a frequently occurring earthquake with exceedance probability of 63% in 50 years, I_{10} is the intensity of an occasionally occurring earthquake with exceedance probability of 10% in 50 years, I_5 is the intensity of a rarely occurring earthquake with exceedance probability of 5% in 50 years, RD_{I,I_J} is the estimated mortality caused during the earthquake of intensity I_J in a region where the buildings were designed against the earthquake of intensity I, RD_{acc,I_J} is socially acceptable mortality for the concerned region, I_J is earthquake intensity that might be considered as one of I_{63}, I_{10} and I_5.

TABLE 2.
DEFINITIONS OF THREE SEISMIC DESIGN LEVELS

Seismic design levels I_J	Probability of exceedance in 50 years	Estimated earthquake mortality caused during earthquake I_J	Acceptable mortality
Frequently occurring earthquake intensity I_{63}	63%	$RD_{I,I_{63}}$	$RD_{acc,I_{63}}$
Occasionally occurring earthquake intensity I_{10}	10%	$RD_{I,I_{10}}$	$RD_{acc,I_{10}}$
Rarely occurring earthquake intensity I_5	5%	RD_{I,I_5}	RD_{acc,I_5}

The first step in determining seismic design loads is to calculate the optimum economic intensity by minimizing the objective function, and then, substitute the obtained intensity into the inequality (4). If the inequality is satisfied, this optimum economic intensity will be the final optimum safe design intensity, otherwise, the intensity must be recalculated by finding the next minimum of the objective function and then substituting into the inequality (4); the remaining steps are the same as above. Finally, the optimum safe design intensity can be determined for a case when both objective function and constraint condition are satisfied. This method is named here as "two-step decision-making" method and is illustrated in Figure 2.

The expression of estimated mortality RD_{I,I_J} under the earthquake of intensity I_J is given as:

$$RD_{I,I_J} = f_t \cdot f_\rho \cdot \sum_{J=1}^{5} P(D_j \mid I, I_J) \cdot l_5(D_j) \cdot P(I_J) \qquad (7)$$

where, I is the design intensity, $l_5(D_j)$ is the mortality ratio corresponding to various categories D_j

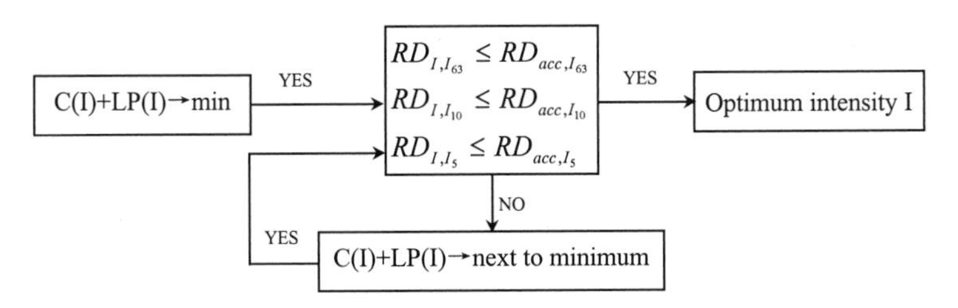

Figure 2. Two-step decision-making method for determining the optimum seismic safe intensity (ground motion)

of damage to the buildings (see Table 3), obtained from statistical data at home and abroad (Ma Yu-hong (1996)), $P(I_J)$ is the occurrence probability of intensity I_J, mainly depending on the results of seismic hazard analysis, $P(D_j \mid I, I_J)$ is the vulnerability matrix, namely the conditional probability of damage level D_j over the intensity I_J for the buildings with earthquake resistance design against intensity I, and f_ρ and f_t are the modified coefficients obtained from statistical analysis in accounting for the effects of different population density (see Table 4) and different occurrence time of earthquake on casualties (see Table 5) respectively. This is because of the fact that seismic death tolls are usually quite different with earthquake occurring in a densely populated area or in a less populated area, and whether occurring in the day time or the night. (Ma Yu-hong (1996))

TABLE 3
MORTARITY CORRESPONDING TO VARIOUS DAMAGE CATEGORIES (%)

Damage level D_j	Intact	Slight	Moderate	Extensive	Complete collapse
Mortality	0	0	0~0.1	0.01~1	2~30

TABLE 4
MODIFIED COEFFICIENT CORRESPONDING WITH POPULATION DENSITY

Population density ρ (persons/km^2)	< 50	50-200	200-500	> 500
Modified coefficient f_ρ	0.8	1.0	1.1	1.2

TABLE 5
MODIFIED COEFFICIENT CORRESPONDING TO OCCURRENCE TIME OF EARTHQUAKE

Intensity	VI	VII	VIII	IX	X
Modified coefficient f_t (night)	17	8	4	2	1.5

INFLUENCE OF SOCIALLY ACCEPTABLE EARTHQUAKE MORTALITY GRADE ON SEISMIC DESIGN INTENSITY

Different communities may make different choices in fixing an adequate level of socially acceptable earthquake mortality, according to their local social and economic situations. Therefore, the level should be flexible for various choices. In order to analyze the influence of different choices of acceptable seismic mortality on optimal safe design intensities, ten grades of socially acceptable earthquake mortality are given in Table 6. The Grade ① corresponds to the highest requirement, or the minimum level of acceptable seismic mortality and Grade ⑩ to the lowest requirement or the maximum acceptable seismic mortality .

TABLE 6
GRADE OF SOCIALLY ACCEPTABLE EARTHQUAKE MOTARITY

Intensity	VI	VII	VIII	IX	X	Intensity	VI	VII	VIII	IX	X
Grade①	1×10^{-8}	2×10^{-8}	1×10^{-7}	5×10^{-7}	2×10^{-6}	Grade⑥	2×10^{-5}	5×10^{-5}	2×10^{-4}	1×10^{-3}	5×10^{-3}
Grade②	2×10^{-8}	5×10^{-8}	2×10^{-7}	1×10^{-6}	5×10^{-6}	Grade⑦	1×10^{-4}	2×10^{-4}	1×10^{-3}	5×10^{-3}	2×10^{-2}
Grade③	4×10^{-8}	1×10^{-7}	4×10^{-7}	2×10^{-6}	1×10^{-5}	Grade⑧	2×10^{-4}	5×10^{-4}	2×10^{-3}	1×10^{-2}	5×10^{-2}
Grade④	2×10^{-7}	5×10^{-7}	2×10^{-6}	1×10^{-5}	5×10^{-5}	Grade⑨	1×10^{-3}	2×10^{-3}	1×10^{-2}	5×10^{-2}	2×10^{-1}
Grade⑤	6×10^{-6}	1×10^{-5}	2×10^{-5}	4×10^{-4}	3×10^{-3}	Grade⑩	2×10^{-3}	5×10^{-3}	2×10^{-2}	1×10^{-1}	5×10^{-1}

With different acceptable mortality levels, the seismic design intensities determined by "two-step decision-making method" should be different. As a numerical example, eight cities in China are selected for determining optimum safe design intensities. For these eight selected cities, optimum safe seismic design intensities determined by the "two-step decision-making method" are shown in Table 7. From Table 7, it can be found:

1. Generally, the optimum safe seismic design intensity is increased with decrease of socially acceptable mortality grade. The lower the acceptable mortality which is adopted the higher the optimum safe design intensity required, and the more seismically resistant the designed structures will be. With increase of the grade from ① to ⑩, the corresponding optimum safe seismic design intensity will be decreased. Once an acceptable mortality ratio is chosen, the corresponding optimum safe seismic design intensity can be determined.

2. Optimum safe design intensity corresponding to different acceptable mortality levels sometimes is different from optimum economic intensity, which is conditioned on minimum economic losses only. This result means that if structures were designed to withstand optimum economic seismic intensity, constraint condition (4) can only be satisfied at rather high mortality levels.

3. The term "Basic Intensity" means the intensity or peak ground acceleration with 10% probability of exceedance in 50 years given by the existing seismic zoning map. It is widely adopted as the design intensity in the current seismic design code; however, this is with little consideration of controlling the death toll. Table 7 demonstrates that the optimum economic design intensity is generally in accordance with the Basic Intensity, but lower than the optimum safe seismic design intensity, and especially lower than the intensity required for the lower socially acceptable mortality grade. If a structure is designed to withstand only the Basic Intensity, therefore the safety of occupants cannot be guaranteed.

TABLE 7
OPTIMUM SAFE DESIGN INTENSITY DETERMINED
BY "TWO STEP DECISION-MAKING METHOD"

	City	Harbin	Hangzhou	Nanjing	Dalian	Shanghai	Beijing	Xian	Xicang
Basic Intensity		6	6	7	7	7	8	8	9
Economic optimum intensity		5.5	5.5	7	7	7	8	8	9
Optimum safe design intensity	Grade ①	8	8	9	9	9	9	9	10
	Grade ②	8	8	8	8	8	9	9	9
	Grade ③	8	8	8	8	8	9	9	9
	Grade ④	8	7	8	8	8	8	8	9
	Grade ⑤	7	7	7	7	7	8	8	9
	Grade ⑥	6.5	6.5	7	7	7	8	8	9
	Grade ⑦	6.5	6.5	7	7	7	8	8	9
	Grade ⑧	5.5	5.5	7	7	7	8	8	9
	Grade ⑨	5.5	5.5	7	7	7	8	8	9
	Grade ⑩	5.5	5.5	7	7	7	8	8	9

The analysis illustrates that the influence of various grades of socially acceptable earthquake mortality on optimum safe seismic design intensity is significant. Human casualty control must be accounted for in determining seismic design load. The method presented in this paper stresses this point. It should take into account not only the need to minimize the possible economic losses but also to reduce the potential seismic mortality when determining seismic design load (or intensity) and emphasizes that the decision-makers should select carefully the socially acceptable mortality grade in determining the appropriate seismic design load.

CONCLUSIONS AND DISCUSSION

In this paper, a "two-step decision-making method" is presented for determining the optimum safe seismic design intensity in consideration of both optimal investment and controlling of seismic fatalities. The method as well as the ten recommended grades of socially acceptable seismic mortality presented in this paper intends to provide a tool by which both economic losses and life losses during earthquakes can be controlled through decision-making in determining the seismic design intensities or seismic design ground motion. It may provide a basis for government as well as building owners to formulate seismic design load (or intensity) and make an optimum balance between the economic investment and life-safe requirements.

It should be pointed out that the conception and the method developed in this paper provide only a framework for seismic design to control both economic losses and human deaths in future earthquakes. More detail of the parameters, such as socially acceptable mortality grade and several coefficients accounting for the effects of damage levels, occurrence time of earthquake and population density level on the mortality can differ from place to place and also needs to be further improved.

ACKNOWLEDGMENTS

This work was jointly funded by the Chinese National Natural Science Foundation with the Grant No.59895410 and the China Basic Research and Development Project: "the Mechanism and Prediction of the Strong Earthquake of the Continental" under the Grant No.95130603. The authors are grateful to Professor Bruce Bolt of University of California, Berkeley, for his valuable comments.

REFERENCES

China Seismological Bureau. (1996). A compilation for earthquake losses evaluation in Chinese Mainland (in Chinese). Earthquake Publishing House. Beijing.

Cornell C. Allin and Helmut Krawinkler. (2000). Progress and Challenges in Seismic Performance Assessment. PEER CENTER NEWS. **Vol.3, No.2**.

Fu Zheng-xiang and Li Ge-ping. (1993). Studies on life losses from earthquake [M]. Earthquake Publishing House. (in Chinese). Beijing.

Ma Yu-hong. (1996). Research on performance-based seismic design load [D], Institute of Engineering Mechanics, China Seismological Bureau. (in Chinese with English summary).

Xie Li-li (1996). On the design earthquake level for earthquake resistant works [C], Proceedings of the PRC-USA Bilateral Workshop on Seismic codes. December 3-7, 1996, Guangzhou, China.

models, since it is an inherent part of the physics which would otherwise distort the fire-safety design results, if the radiation effects were not properly included. The purpose of this paper then is to promote the adequate inclusion of radiation calculations in fire models by first briefly discussing the physical role and effects of thermal radiation in room fires, followed by reviewing the theoretical basis for radiation calculations and how they are formally incorporated into the fire model. Also discussed will be the various approximation methods of incorporating simplified radiation calculations in fire models within the spirit of performance-based engineering practices. Explicit references will be cited to give an indication as to how the existing models including radiation perform as compared to available experimental data in room fires.

ROLE AND EFFECTS OF THERMAL RADIATION IN ROOM FIRES

As briefly mentioned previously, the room-fire phenomena are essentially governed by the coupled interactions among combustion, buoyancy, turbulence, and radiation. Consequently, it is difficult to discuss radiation without accounting for combustion, turbulence, and other physical effects (Yang, 1994). In room fires, the combustion phenomenon is really that of turbulent combustion, in that the flow turbulence induces mixing, which brings the reactants (fuel and oxygen molecules) together so that reactions can take place and generate heat and produce species as products of combustion. The heated gas species such as water vapor, carbon mono- and dioxides and carbon soot, which are all radiatively participating, then emit, absorb, and scatter radiation, and reach all far corners of the room. The net result is that the radiation heats up all room surfaces, producing heat sources in the gas medium, and provides a mechanism for changing local gas densities, thus resulting in additional buoyancy, which in turn further affects the behaviors of the convective flow and turbulence fields. It then enhances the turbulent combustion process and brings the entire interactions to a full circle. It is here seen that radiation acts as an intermediary for maintaining and enhancing the fire growth in the room. Therefore, despite the complexity of the role of radiation in the room-fire phenomena, no fire model can be considered truly adequate without accounting for thermal radiation and its interaction with other effects such as combustion, buoyancy, and turbulence.

From a quantitative theoretical framework, the role and effects of thermal radiation in the room-fire phenomena are known to be accommodated in a single source term in the energy equation (Siegel and Howell, 2001; Yang, 1994), which is written as the negative of the divergence of the radiative flux vector. It is also known that this energy equation is part of a set of coupled differential equations governing the dynamics and interactions in the fire phenomena, including the mass continuity equation, the flow equations, the energy equation, and the species-conservation equations (Yang, 1994). In fact, this set of governing equations forms the basis for all the fire models known as the fire field models, which, along with initial and boundary conditions and appropriate submodels, are usually solved by computational fluid dynamics (CFD) methodologies (Minkowycz et al., 1988). Examples of fire field models are UNDSAFE (Yang et al., 1994, Raycraft et al., 1990), and KAMELEON (Holen et al., 1990). Unfortunately, this deceptively simple source term to account for all radiation effects belies the fact that radiative fluxes are extremely difficult to compute, involving not only the entire physical domain of the room, but also the spatial variations of all radiation-participating species and the temperature field at every time instant into the fire. Further complications are uncertainties in the determination of radiative properties of the gas and soot mixtures and the approximate computations associated with the geometrical distribution of the radiation rays traversing the entire room. Moreover, the entire computational process for the radative fluxes must be repeated many times as the fire phenomenon evolves in time and space. Since the overall fire problem without the radiation component is already very computationally intensive for its solution, even with the available CFD solvers, there is still a need to make appropriate simplifications for the determination of the radiative fluxes. These issues will be addressed in the following sections.

THEORETICAL BASIS AND NUMERICAL METHODS FOR DETERMINING RADIATIVE FLUXES FOR ROOM FIRES

For room fires, the domain for radiation heat transfer covers basically the wall, ceiling, and floor surfaces, the surfaces of all material contents, and the volume of gas and particulate medium bounded by the solid surfaces. Also, doorways and windows, either closed or open, all play a role in the radiation heat transfer process. Once ignited, the evolving fire becomes a part of the medium in the room, where the temperature rises due to the combustion process, and heat is imparted to the rest of the room by direct radiation exchange and natural convection. In room fires, radiation exchange occurs between surface and surface, gas (including flame) and surface, and gas (including flame) and gas, and is governed by different independent physics known as radiative transfer. The easiest way to consider the propagation of radiation in a fire room is to simply look along a specific ray of radiation in a specific direction that traverses the room medium from one surface to another. At the starting point on a surface, the radiation leaving is that due to emission there and reflection of all radiation arriving at the starting surface, all in the same direction. As the radiation travels along the ray in the medium, it encounters additional emission due to the local medium temperature, absorption by the medium, and radiation scattering into and out of the ray due to soot particles from the combustion process. As a result, the original radiation intensity continues to change until it reaches the target surface, where the arriving radiation is absorbed and reflected. One inherent complexity here is that both emission and absorption are present for radiation-participating gases such as water vapor and carbon mono- and dioxides in most room fires, as they possess discrete absorption bands in the wave lengths, and such emission and absorption are also functions of local concentrations of the participating gases, in additional to the local temperatures. Furthermore, the radiation scattering is dependent on the size of the particulates in the medium. Therefore, in the calculation of variations of the radiation intensities, even along a single ray requires information on the absorption and scattering coefficients, which are the medium radiation properties. This single-ray calculation at a given solid angle characterizing the ray direction will have to be repeated for all rays spanning all the boundary surfaces in the room. The resulting accumulated local radiation intensities in all directions and the corresponding fluxes can then be resolved into the radiation fluxes in the principal directions of the chosen coordinate system. The local negative divergence of these radiation vectors then becomes the radiative heat source term in the energy equation, which is to be solved, together with other coupled equations, thus resulting in the data for the evolution of the fire dynamics phenomena of the room fires.

The radiation submodel in the fire field model is just the theoretical quantification of the determination of the radiative flux vectors and the corresponding radiative heat source based on the ray-tracing process just mentioned. Before some details are shown, it is pertinent and desirable to mention here a companion analysis needed to determine the relative geometrical positions of the chosen two surfaces which bound a particular group of rays of radiation in the fire room. In order to resolve the specific direction of a giving ray, the surfaces involved are taken to be sufficiently small so that each surface has a uniform temperature. Under this condition, it is convenient to introduce the well-known view factor or configuration factor to characterize the relative positions of any pair of surfaces in a given room. Since this factor is well discussed and tabulated in the literature (Siegel and Howell, 2001; Howell, 1982) and also in many undergraduate texts on Heat Transfer, its definition and derivation thus will not be repeated here. However, it is important to note that these view factors can be used to determine the radiation heat transfer between any pair of the surfaces with known temperatures in a black enclosure (room) with nonparticipating medium, in terms of the difference between the black-body emissions at the two surfaces at the respective temperatures in accordance with the Pranck's Law (Siegel and Howell, 2001; Modest, 1993). Furthermore, this enclosure theory for a black enclosure can be readily extended to that of a gray enclosure with gray surfaces and nonparticipating medium with prescribed emissivities in accordance with the radiosity method (Siegel and Howell, 2001). A similar role is also played by these same view factors in rooms with radiation participating medium to characterize the geometry of radiation rays for the purpose of determining the corresponding radiation intensities and eventually the radiation flux vectors.

As mentioned previously, in the product of combustion in room fires, the water vapor, carbon mono- and dioxides, in view of their respective absorption bands, and soot particles because of their scattering properties, collectively represent a participating medium that emits, absorbs, and scatters radiation. These characteristics are determined by the equation of radiative transfer and their respective radiation properties (Siegel and Howell, 2001; Modest, 1993; Brewster, 1992). As also mentioned, the interaction of radiation with other physical effects in the room-fire dynamics manifests itself through the presence of distributed radiation heat sources in the energy equation, which can be written as

$$\rho c \frac{DT}{Dt} = -\nabla \cdot \dot{\mathbf{q}}_c - \nabla \cdot \dot{\mathbf{q}}_r \tag{1}$$

where D/Dt is the substantial derivative, - $\dot{\mathbf{q}}_c$ is the conductive flux vector, $\dot{\mathbf{q}}_r$ is the radiation flux vector, and $\nabla\cdot$ represents the scalar divergence of a vector. In the general case, the equation of radiative transfer is written as (Siegel and Howell, 2001)

$$\nabla \cdot (\mathbf{s} I_\lambda) = -(a_\lambda + \sigma_\lambda) I_\lambda + S_\lambda \tag{2}$$

where I_λ is the spectral (wave-length dependent) radiative intensity, generally expressed in terms of W/m^2 µm sr, a_λ and σ_λ are the spectral absorption and scattering coefficient, respectively, and S_λ is a source function given by

$$S_\lambda = a_\lambda I_{b\lambda}(T) + \frac{\sigma_\lambda}{4\pi} \int_0^{4\pi} I_\lambda(\mathbf{r}, \mathbf{s}') \Phi_\lambda(\mathbf{s}' \rightarrow \mathbf{s})\, d\Omega' \tag{3}$$

where $I_{b\lambda}$ is the spectral Planck function of blackbody emission, Φ_λ (s'—>s) is the phase function for scattering from **s'** into **s**, Ω' is the solid angle, and **r** is the position vector. Here the prime refers to the incoming radiation, and its relation to the ray direction **s** is shown in Fig. 1. By introducing an optical thickness τ_λ based on the extinction coefficient $\gamma_\lambda = a_\lambda + \sigma_\lambda$ and defined by

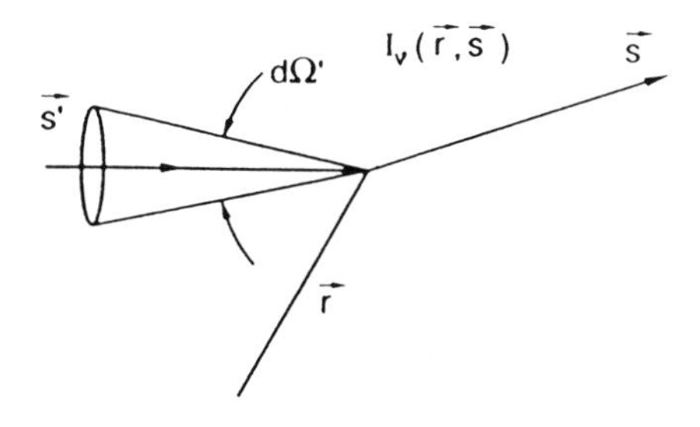

Figure 1: Geometry of radiative transfer

$$\tau_\lambda = \int_0^s \gamma_\lambda \, ds' \tag{4}$$

Eqn. 2 can be rewritten as

$$\frac{dI_\lambda}{d\tau_\lambda} + I_\lambda = (1-\omega) I_{b\lambda} + \frac{\omega}{4\pi} \int_0^{4\pi} I_\lambda(\tau_\lambda, \Omega') \Phi_\lambda(\Omega' \to \Omega) \, d\Omega' \tag{5}$$

where ω is the single-scattering albedo defined by $\sigma_\lambda / (\kappa_\lambda + \sigma_\lambda)$. Eqn. 5 can be integrated to yield the intensity

$$I_\lambda(\tau_\lambda) = I_\lambda(0) e^{-\tau_\lambda} + \int_0^{\tau_\lambda} \left[(1-\omega) I_{b\lambda}(\tau'_\lambda) + \frac{\omega}{4\pi} \int_0^{4\pi} I_\lambda(\tau'_\lambda, \Omega') \Phi_\lambda(\Omega' \to \Omega) \, d\Omega' \right] e^{-(\tau_\lambda - \tau'_\lambda)} \, d\tau'_\lambda \tag{6}$$

where $I_\lambda(0)$ is the radiation intensity leaving the boundary at s = 0 due to both emission and reflection. This equation can now be used to determine the divergence of the radiative flux vector, which is equivalent to the net out flow of radiant energy per unit volume, by integrating Eqn. 2 over the solid angle and wave lengths, resulting in

$$\nabla \cdot \dot{\mathbf{q}}_r = 4 \int_0^\infty \left[a_\lambda e_{b\lambda} - \pi \gamma_\lambda \bar{I}_\lambda + \frac{\sigma_\lambda}{4} \int_0^{4\pi} I_\lambda(\Omega') \bar{\Phi}_\lambda(\Omega') \, d\Omega' \right] d\lambda \tag{7}$$

where $e_{b\lambda}$ is the hemispherical blackbody emissive power and the barred quantities are the average values, as given by

$$\bar{I}_\lambda = \frac{1}{4\pi} \int_0^{4\pi} I_\lambda(\Omega) \, d\Omega \qquad \bar{\Phi}_\lambda = \frac{1}{4\pi} \int_0^{4\pi} \Phi_\lambda(\Omega, \Omega') \, d\Omega \tag{8}$$

For isotropic scattering $\Phi_\lambda = \bar{\Phi}_\lambda = 1$, and Eqn. 7 takes on a simpler form

$$\nabla \cdot \dot{\mathbf{q}}_r = 4 \int_0^\infty a_\lambda (e_{b\lambda} - \pi \bar{I}_\lambda) \, d\lambda \tag{9}$$

which can also be used for anisotropic scattering, provided that the phase function Φ_λ is independent of Ω'. Eqn. 7 is deceptively simple and in general is complicated by the fact that much additional information is needed form for its evaluation. To implement the solutions to the equation of radiative transfer, two more steps are yet to be taken. One is the evaluation of the needed gas and soot radiation properties, and the second is the numerical solution methodology for the three-dimensional geometry in the fire room under consideration. For radiation properties of the medium in room fires (Yang, 1986; Modest,1993; Brewster, 1992), it is sufficient to deal with only those of water vapor, carbon dioxide, and soot. For these gases, the exponential wide-band model, along with provisions for overlapping bands and nonhomogeneous mixtures is of sufficient accuracy, and soot can certainly be considered as a gray medium in view of the extremely small smoke particles involved (Yuen and Tien, 1977). However, simpler gray medium models and nongray-gas models such as the weighted-sum-of-gray-gases model (Modest, 1993), have also been used. A comparison of the various nongray models for combustion products has been given by Song (1993). The second step that needs to be considered is in the calculation of radiative transfer in multidimensional spaces, primarily in accommodating the geometric complexities in evaluating the integral with respect to the solid angle in Eqn. 7. Approximate, but accurate methods are known, but are in general still rather cumbersome to use. Many versions of the classical methods such as Monte Carlo, zonal, ray-tracing, P-N, integral-evaluation and discrete ordinate methods are now available, as reviewed by Viskanta (1984) and Yang (1986), and are still being updated to broaden their utility and capability (Yuen and Takara, 1997; Fiveland, 1988; Jamaluddin and Smith, 1988). Additional new methodologies are continually being developed to improve accuracy and calculation efficiencies. Good examples are the discrete transfer method of Lockwood and Shah (1981), the radial-flux method of Chang et al. (1983), and the finite-volume method of Raithby and Chui (1990). At the present time, the common practice is simply the preference of the user, even though the multidimensional discrete-ordinate method (Fiveland, 1988) seems to be the preferred by many practitioners. Since ultimately any of these methods must be used simultaneously with the solver of the convection field in a given room fire problem, how well the radiative transfer calculations can be integrated into the convection solver has become an important issue in the selection of the method for radiation calculations. On the other hand, as the computing resources become increasingly more plentiful in both data storage and processing speed, the choice of the radiation calculation methodology may become less critical in time. For instance, the radiation source term in the unsteady energy equation can be treated as an inhomogeneous term evaluated at the last time instant where the temperatures are already known by using any of the radiative transfer methodology, and the energy equation is then solved for the unknown temperatures and species distributions at the current time instant.

So far the discussion on the theoretical basis and the numerical methodologies for including radiation calculations in the interaction phenomena in room fires has been centered on the fire field model which is more natural in assessing the significance of radiation, because of the first-principle basis in its formulation. This discussion is perhaps not complete without mentioning the zone fire models, which have been found increasing use for room and building fires for fire-risk analysis and the development of performance-based fire-safety codes. A zone model is one in which the individual flow regions with unique and distinguishable characteristics such as the flame, fire plume, ceiling layer, floor and wall layers, and regions next to windows and doorways are treated as zones. Popular zone models are, for example, CFAST (Peacock et al., 1993), FIRECAM (Yung et al., 1997), and FRESCO (Mathews et al., 1997) which have been utilized to deal with realistic room fire situations. Unfortunately, the current status of incorporating radiation effects in fire zone models is very limited, and in fact most zone models do not include radiation at all. Among some zone-model developers, there is a consensus that the critical radiation effect in room fires is in the flashover phenomenon which projects radiation heat transfer through open doorways next to the fire room and thus provides a mechanism for fire spread into the rest of a building. The critieria for flashover to occur are very *ad hoc*, and are still debated in terms of whether such criteria have the correct physical basis in radiation heat transfer.

RELEVANT APPROXIMATIONS TO ACCOUNT FOR RADIATION HEAT TRANSFER IN ROOM FIRES

While it is now recognized that there is a need to develop the best optimum method to deal with radiation in room-fire analysis in terms of accuracy and computing efficiency. One should also recognize that in real physical applications not every room-fire problem requires the full application of a complete optimum solution, even when one does exist. Simplifications in the radiation calculations in many fire-safety applications are indeed possible, and in fact are more preferable, depending on what specific information is needed. This becomes clear when one realizes that normally one is not particularly interested in actually predicting the physical fire phenomena, but will be happy to obtain information with reasonable accuracy to assess parametric effects of physical variables in specific applications for design purposes. One significant example in this regard lies in the field of fire modeling for fire-safety engineering for rooms and buildings, which are playing an ever increasingly important role in the practice and development of Performance-Based Fire Codes advocated by the majority of developed and developing nations of the world (Yang, 1999). Several such relevant and useful approximations for accommodating radiation in room-fire analysis are described:

1. As already mentioned previously, one expedient way to deal with the radiative transfer calculations for fire field-model analysis of an evolving room fire is to treat the radiation source term in the energy equation as an inhomogeneous term based on the known field variables for the previous time instant, and then to solve the coupled equations for the field variables for the current time instant. This numerical method of solution, because of its explicit nature, does require small time steps, though it is simple to use. In this method, there is no need to compute the radiation source term at every single time step. As long as the field variables do not change greatly, subject to certain prescribed variability constraint, the same source term can be used for a relatively large number of time steps without introducing any undue noticeable errors. As a result, it only needs to be updated several times throughout the fire-growth period, thus greatly reducing the radiation calculations. An example of this practice is given in the study of Chang et al. (1983) with rather good results.

2. There is a very practical and useful approximation in dealing with all the complexities for nonuniform temperatures and species distributions in the fire-room space, and it is to ignore all the effects of participating medium (Raycraft et al., 1990; Yang et al., 1994). This approximation simply reduces the full problem to one of evaluating the radiation exchange in a gray enclosure with a nonparticipating medium. In this case, the fire room consists only of boundary surfaces such as walls, ceiling and floor with prescribed emissivities, reflectivities and surface temperatures, and calculated radiation exchanges are those between surface-to-surface, and flame-to-surface pairs, taking into account the effects of shading. The flame is represented by a conveniently-chosen volume with its bounding surfaces at the flame temperature. This approxinmation, which greatly simplifies the radiation calculations, generally overpredicts the temperatures in the hot gas region in the fire room and underpredicts the temperatures in regions of low temperatures such as the floor region of the room, due to the diffusive nature of the radiation effects of the participating medium. In addition, this approximation also simplifies the combustion model, which would normally be needed for determining the heat release rates, in that it can be replaced by a prescribed heat release rate, which in turn becomes a simulation parameter for the room fire under consideration for fire-risk assessment purposes (Babrauskas and Peacock, 1992). The results from using this approximation are not strictly accurate. However, they do lead to predictable behavior as compared to that when the participating medium effect is included in the simulations. In fact, it has been found that such a simplification still leads to reasonably good results when compared to the experimental full-scale fire data even in rather complex compartments (Raycraft et al., 1990, Yang et al., 1994).

3. Another possible simplification of radiation calculations in room fires is to take advantage of the high degree of stratification of the smoke and hot-gas layers just below the ceiling, evidently because

of the strong buoyancy resulted from the fire. These layers generally are of uniform thickness throughout the room and also essentially at a uniform temperature, except possibly in a very thin region close to the ceiling in which the gas temperature drops substantially due to ceiling heat losses. Under these physical conditions, the radiation within the hot ceiling layer can be well approximated by one-dimensional radiation behaviors of a participating gas (Siegel and Howell, 2001). Further simplification is still possible if the medium that consists of hot gas mixture and soot (smoke) is taken to be a gray medium. Just as an example, if the hot smoke layer is thick, implying an optically thick medium, the radiation behavior is simply given by the optically-thick or diffusion approximation, which replaces the radiative transfer equation by a conduction-type of diffusion equation that can be be readily solved (Modest, 1993). An example of this approximation is in the study of Lloyd et al. (1979).

4. There is another approximate treatment of radiation heat transfer particularly useful in the analysis of fire phenomena by fire zone models. In the case of two-layer zone models for a room with a hot ceiling layer at a uniform temperature and an air layer below, but without a hot fire plume, radiation heat loss in the hot-layer zone can be accommodated by energy losses to the smoke-air interface, according to one-dimensional radiative heat flux calculations, which may be considered in the either optically thick or optically thin limits, depending on the instantaneous thickness of the smoke layer, and with absorption coefficients calculated from reasonably-prescribed uniform compositions of the medium mixture. The radiation heat transfer loss can then be added to the heat balance of the hot-layer zone in the energy equation of the zone. A similar adjustment can also be made for the zones that are in contact with the hot ceiling-layer zone. This procedure for incorporating radiation contribution is also applicable to a fire room with a hot fire plume zone above the fire, in which case the hot ceiling layer is treated as two separate zones, the ceiling layer away from the plume which penetrates into the ceiling layer and the plume zone itself.

There are also other radiation approximations for estimation purposes only, and are certainly more *ad hoc* than having any real rational basis. One such practice is to boast the fire heat release rates in the fire by a certain percentages to account for radiation contributions. This would inevitably increase the maximum flame temperature for a given fire (Satoh and Yang, 2000). This practice, however, is certainly not recommended unless it can be justified physically in a given situation.

CONCLUDING REMARKS

In this paper, the role and effects of thermal radiation in room fires have been described on the physical basis of the interaction of thermal radiation with strong buoyancy, turbulence, and combustion in the room-fire phenomena, and it is this interaction that governs the dynamics of fire growth and spread. Also discussed are the theoretical basis of thermal radiation itself and how it is determined quantitatively. This then becomes the formal framework to develop a radiation submodel to join the other submodels of buoyancy, turbulance, and combustion for incorporation into a fire field model based on conservation laws. The framework of the radiation submodel is extremely complex because of the complexity in obtaining the solutions to the equation of radiative transfer in the three-dimensional room space with nonuniform temperature and participating-gas and soot distributions as functions of both space and time as the fire evolves. While optimum methods for the general radiation calculations are still being sought, promising approximate calculations suitable for use in fire-safety engineering analyses can be delineated with reasonable accuracy. As more approximate methods become known, more pertinent experimental data in real fires will be needed to validate such methods so that they can be used in practice with some confidence. It is ,however, realized that large room-fire tests are extremely costly to carry out, and only slow progress in this direction is expected now and in the future.

REFERENCES

Babrauskas, V. B. and Peacock, R. D. (1992). Heat Release Rate: The Single Most Important Variable in a Fire Hazard, *Fire Safety Journal*, **18**, 255-272.

Brewster, M. Q. (1992). *Thermal Radiation Transfer and Properties*, John Wiley, New York, USA.

Chang, L. C., Yang, K. T. and Lloyd, J. R. (1983). Radiation-Natural Convection Interaction in Two-Dimensional Complex Enclosures, *Journal of Heat Transfer*, **105**, 89-95.

Fiveland, W. A. (1988). Three Dimensional radiative Heat-Transfer Solutions by the Discrete-Ordinates Method, *Journal of Thermophysics and Heat Transfer*, **2:4**, 309-316.

Holen, J., Brostrom, M. and Magnussen, B. F. (1990). Finite Difference Calculation of Pool Fires, *Proceedings of the 23rd Symposium (International) on Combustion*, The Combustion Institute, 1677-1683.

Howell, J. R. (1982). *A Catalog of Radiation Configuration Factors*, McGraw Hill, New York, USA.

Jamaluddin, A. S. and Smith, P. J. (1988). Discrete-Ordinates Solution of Radiative Transfer Equation in Non-Axisymmetric Cylindrical Enclosures, *Proceedings of the 1988 National Heat Transfer Conference*, ASME HTD **96**, 227-232.

Lockwood, F. C. and Shah, N. G. (1981). A New Radiation Solution Method for Incorporation in General Combustion Prediction Procedures, *Proceedings of the 18th Symposium (Internatioanl) on Combustion*, 1405-1414.

Mathews, M. K., Karydas, D. M. and Delichatsios, M. A. (1997). A Performance-Based Approach for Fire Safety Engineering : a Comprehensive Engineering Risk Analysis Methodology, a Computer Model, and a Case Study, *Proceedings of the 5th Symposium on Fire Safety Science*, 595-606.

Minkowycz, W. J., Sparrow, E. M., Schneider, G. E. and Pletcher, R. H. (eds.) (1988). *Handbook of Numerical Heat Transfer*, John Wiley, New York, USA.

Modest, M. F. (1993). *Radiative Heat Transfer*, McGraw Hill, New York, USA.

Peacock, R. D., Forney, G. P., Renecke, P., Portier, R. and Jones, W. W. (1993). CFAST, The Consolidated Model of Fire Growth and Smoke Transport, *Technical Note 1299*, NIST, U. S. Department of Comerce, USA.

Raithby, G. D. and Chui, E. H. (1990). A Finite-Volume Method for Predicting Radiant Heat Transfer in Enclosures with Participating Media, *Journal of Heat Transfer*, **112**, 415-423.

Raycraft, J., Kelleher, M. D., Yang, H. Q. and Yang, K. T. (1990). Fire Spread in a Three-Dimensional Pressure Vessel with Radiation Exchange and Wall Heat Losses, *Mathematical and Computer Modeling*, **14**, 795-800.

Song, T. H. (1993). Comparison of Engineering Models of Nongray Behavior of Combustion Products, *International Journal of Heat and Mass Transfer*, **16**, 3975-3984.

Siegel, R. and Howell, J. R. (2001). *Thermal Radiation Heat Transfer*, 4th edition, Taylor & Francis, Washington, D. C., USA.

Viskanta, R. (1984). Radiative Heat Transfer, *Fortschreft der Verfahrenstechnik*, **22A**, 51-81.

Yang, K. T. (1986). Numerical Modeling of Natural Convection-Radiation Interactions in Enclosures, *Proceedings of the 8th International Heat Transfer Conference*, **1**, 131-140.

Yang, K. T. (1994). Recent Development in Field Modeling of Compartment Fires, *JSME International Journal*, Ser. B, **37:4**, 702-717.

Yang, K. T. (1999). Role of Fire Field Models as a Design Tool for Performance-Based Fire-Code Implementation, *International Journal on Engineering Performance-Based Fire Codes*, **1:1**, 11-17.

Yang, K. T., Nicolette, V. F. and Huang, H. J. (1994). Field Model Simulation of Full-Scale Forced Ventilation Room-Fire Test in the HDR Facility in Germany, *Heat Transfer in Fire and Combustion Systems*, ASME HTD **272**, 13-20.

Yuen, W. W. and Takara, E. E. (1997). The Zonal Method: A Practical Solution Method for Radiative Transfer in Nonisothmal Inhomogeneous Media, *Annual Review of Heat Transfer*, **VIII**, 153-215.

Yuen, W. W. and Tien, C. L. (1977). A Simple Calculation Scheme for the Luminous-Flame Emissivity, *Proceedings of the 16th Symposium (International) on Combustion*, 1481-1487.

Yung, D., Hadjisophocleous, G. V. and Proulx, G. (1997). Modelling Concepts for the Risk-Cost Assessment Model FIRECAM and its Application to a Canadian Government Office Building, *Proceedings of the 5th Symposium on Fire Safety Science*, 619-630.

TEAM PRESENTATION:

ADVANCES IN PANEL SYSTEMS

Advances in Building Technology, Volume I
M. Anson, J.M. Ko and E.S.S. Lam (Eds.)

WOOD FRAME SHEAR WALLS WITH METAL PLATE CONNECTED FRAMEWORK

Robert N. Emerson

School of Civil and Environmental Engineering, Oklahoma State University,
Stillwater, OK 74078-5033, USA

ABSTRACT

The vast majority of residential and light commercial structures in the United States are constructed of wood. These structures are susceptible to damage during lateral loading events. An improved shear wall system is under development that employs toothed metal plates to connect the framework of shear walls. These connectors provide moment resistant connections that transform the framework into integral moment resistant frames that assist the sheathing in resisting lateral loads. Preliminary investigations show great promise.

KEYWORDS

Engineered wood construction, wood frame shear walls, prefabricated wall panels, metal plate connections, lateral load resistance, seismic, wind

INTRODUCTION

The vast majority of homes and light commercial structures in the United States are constructed of wood. Ninety percent of new houses built in America are of woodframe construction [1]. Many of these structures reside in locations exposed to lateral forces resulting from seismic activity or high winds due to hurricanes or tornadoes. These lateral forces can cause significant damage with devastating results. While many factors lead to damage during extreme lateral loading events, inadequate shear walls result in some of the most devastating damage. Another major cost is the repair and/or replacement of non-serviceable structures associated with damage produced during lesser lateral loading events.

It is evident that the performance of wood frame shear walls directly affects the performance of wood frame structures during lateral loading events. Wood frame shear walls with increased resistance to extreme lateral loads as well as improved durability to multiple median lateral loads will save many lives and billions of dollars in repair and replacement costs. Previously proposed improvements have not been widely accepted by the building industry due to unacceptable increased costs associated with the improvements. A strong motivation for the current research is the desire and need for an improved wood frame shear wall system that has increased performance and decreased or similar costs relative to conventional shear wall construction.

BACKGROUND

Conventional light-frame wood construction employs horizontal diaphragms and shear walls to resist lateral forces. Horizontal diaphragms are designed as horizontal beams that transfer lateral forces to the supporting shear walls. Shear walls are designed as vertical deep cantilever beams supported by the structure's foundation [2,3,4]. A typical shear wall is composed of dimension lumber framing overlaid with sheathing on one or both sides. Nails connect the framework together. The top plate is connected to the studs by end nailing. The studs are connected to the bottom plate by either toenailing or end nailing. The end studs are anchored to the foundation to resist uplift forces resulting from applied moments. The bottom plate is anchored to the foundation to resist base shear forces. The sheathing is attached to the dimension lumber framework with nails. The nails are spaced closely around the edges and spaced farther apart in the interior of the individual sheathing panels.

Conventional Wood Frame Shear Wall Behavior

Substantial research has been performed regarding the behavior, analysis, modeling, and design of conventional wood frame shear walls. The dimension lumber chords and sheathing are designed to resist applied moment and shear, respectively. However, the behavior of the system is governed by connections between the top and bottom plates and studs and connections between the sheathing and framework. Under lateral loads nailed connections between the plates and studs provide little rigidity. As a result, the framework distorts as a parallelogram. The sheathing remains rectangular and rotates as a rigid body. Relative displacement of the framework and sheathing due to racking forces is depicted in figure 1. The connections between the sheathing and framework then become the controlling factor in racking behavior. The nails connecting the sheathing to the framing resist the lateral load. Typically, failure occurs when the nail heads pull through the sheathing panel or withdraw from the framework due to large relative displacements between the sheathing and the framework at the corners of sheathing panels.

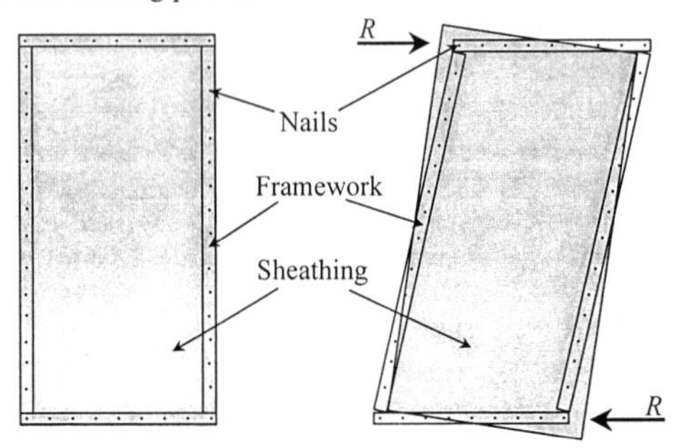

Figure 1. Conventional wood frame shear wall panel behavior due to racking forces

Tuomi and McCutcheon developed a method for calculating the racking strength of frame panels and then compared the method with experimental tests [5]. Racking strength was calculated by accounting for panel geometry, number and spacing of nails, and the lateral resistance of a single nail. The method also accounted for minimal frame resistance. Frame resistance was assumed to contribute less than ten percent of the overall racking strength. Calculated racking strength compared favorably to results from experimental testing.

Gupta and Kuo developed a model for shear wall behavior that was governed by nonlinear nail deformation but also accounted for bending and shear stiffness of the studs and sheathing, respectively [6]. They found that assuming the studs were infinitely rigid in bending gave comparable results to assuming the studs held normal bending stiffness. This confirms that the nailed connections between the plates and studs are too flexible to develop bending within the studs.

The NAHB Research Center investigated the performance of perforated shear walls [7]. Variables included in their study were narrow wall segments, reduced base restraint, and alternative framing methods. The investigation of alternative framing methods provided an interesting result. Two nearly identical perforated shear walls were tested. Both walls had a sheathing area ratio equal to 0.57 as determined from Sugiyama's empirical equation [8,9]. The control wall consisted of conventionally connected framing with moment anchors at both ends of the wall. The experimental wall with alternative framing consisted of framing connected with metal truss plate connectors at the corners and around openings. This wall also had no moment anchors to resist uplift. Even with the omission of moment anchors, the experimental wall outperformed the control wall dramatically. The experimental wall resisted 40% more lateral load and dissipated 68% more energy than the conventionally connected perforated shear wall.

Metal Plate Connected Truss Technology

Prefabricated light frame wood trusses are commonly connected with toothed metal connector plates. The trusses are designed and fabricated according to ANSI/TPI 1-1995, National Design Standard for Metal Plate Connected Wood Truss Construction [10]. Wood members are designed according to the National Design Specification for Wood Construction [11]. Guidelines for metal plate connector design regarding tooth holding, net section steel tensile capacity, net section shear capacity, and combined shear and tension capacity are presented in ANSI/TPI 1-1995.

In the past, metal plate connected trusses were modeled and designed assuming pinned connections. However, toothed metal plate connectors provide semi-rigid connections that will develop moment capacity. Noguchi investigated five models describing the moment resistance of metal plate connections [12]. These models were (i) completely elastic behavior; (ii) elastic behavior of the plate with compressive yielding of the wood member; (iii) tensile yielding of the plate with elastic behavior of the wood; (iv) plastic behavior of both the plate and wood; and (v) Edlund's model for bending moment [13]. The models were fit to experimental data and the elastic model was found to be the most conservative while the plastic model was recommended for design purposes. Kevarinmäki used both elastic and plastic models to investigate the tooth withdrawal capacity of metal plate connected truss joints subjected to bending moments [14]. O'Regan, Woeste, and Lewis developed a design method for the steel net-section of truss joints subjected to tension and moment [15]. Their design method was based on Noguchi's model where the steel is fully plastic in tension and the wood is linearly elastic in compression. O'Regan, Woeste, and Brakeman added to the design method by developing design methodology that forces steel yielding as the failure mechanism [16].

The ability to withstand repeated dynamic cyclic loading is a concern for connections in lateral load resisting systems. Emerson and Fridley investigated the tooth holding performance of toothed metal plate truss connections subjected to repeated dynamic cyclic loading [18,19]. A variety of joint configurations were tested to evaluate the effect of tooth, load, and grain orientation. Connections were subjected to a loading regime of 20, 40, and 60% of the mean maximum static load cycled at 10 Hz for 200 cycles. Dynamic stiffness degradation and residual static strength were evaluated. The connections behaved elastically when cycled at 20% of the maximum static load and experienced stiffness degradation during the 40 and 60% of maximum static load cycles. The connections maintained 100, 75, and 56% of their initial dynamic stiffness after 200 load cycles at the 20, 40, and 60% load levels, respectively. Most of the stiffness degradation occurred within the first 50 cycles and then tapered off. After being subjected to the cyclic dynamic loading regime the connections maintained at least 87% of their mean maximum static strength.

Conclusions from Background

Conventional wood frame shear walls are designed as vertical deep cantilever beams. The sheathing provides the shear resistance. The end studs, acting as tension and compression chords, provide the bending resistance. The interior wall studs are only employed to support the sheathing. Nailed connections between the top and bottom plates and wall studs are quite flexible. This connection flexibility prevents any moment to be developed in the studs. As a result, conventional shear wall resistance is governed by the nailed connections between the sheathing and dimension lumber framework. Failure of the shear wall occurs when the nailed connections between the sheathing and framework fail due to large relative displacements between the sheathing and framework at the corners of sheathing panels. This failure mechanism prevents efficient utilization of the dimension lumber framework and dissipates little energy before significant damage to the system occurs.

Toothed metal plate connectors provide semi-rigid connections that are able to resist moment. This moment resistance has been investigated and accounted for in truss design and may be incorporated into shear wall design.

PRELIMINARY STUDIES

The potential behavior and benefits of using toothed metal plate connectors to transform underlying wood framework into moment resistant frames that are integral to the lateral force resistant system were investigated in preliminary analytical and experimental studies.

Analytical Study

Preliminary analytical studies of the proposed wall system have been performed. An 8 ft by 20 ft shear wall composed of conventional construction was employed as a model for this study. Vertical studs were positioned at 16 in. on center with double end studs, a double top plate, and a single bottom plate. Design racking forces were determined for shear walls sheathed with $^{15}/_{32}$, $^{7}/_{16}$, and $^{3}/_{8}$ in. thick structural sheathing with various nailing patterns. Design racking forces were determined using both Uniform Building Code design tables [17] and the Tuomi McCutcheon method [7]. Subsequent to calculating design racking forces it was determined how much racking force needed to be carried by the framework to reduce the sheathing thickness while maintaining overall strength. Two methods employing matrix structural analysis were used to establish reasonable limits for the design racking resistance of moment resistant frames composed of wood framework and toothed metal connector plates. First, the connections were assumed to be infinitely rigid and strong. This assumption forced failure of the wood members (end studs) and was used as an extreme upper limit to the frame's design racking strength. The design racking strength was determined assuming connection failure. Connections were designed according to ANSI/TPI 1-1995 guidelines [10] incorporating methods recommended by O'Regan, Woeste, and Lewis [15] and Kevarinmäki [14].

While differences in stiffness between the sheathing and framing were neglected, results of the preliminary analytical study are promising. An 8 ft by 20 ft shear wall with $^{15}/_{32}$, $^{7}/_{16}$, or $^{3}/_{8}$ in. thick sheathing and nails spaced at 6 in. along the panel edges has an allowable lateral load capacity of 5200, 4800, and 4400 lbs, respectively. With infinitely rigid and strong connections (unrealistic assumption), Southern Pine 2 by 4 framework has a design lateral load capacity of 640 to 2400 lbs for stud and dense select structural framing members, respectively. With connections designed according to TPI and recommendations from previous research the design lateral load capacity of the framework was 220 lbs. If moment resistant frame design capacity is assumed directly additive to shear wall design capacity based on sheathing performance, then the *combined system* will have 4 to 55% more design lateral capacity than conventional shear walls. This range envelops the 40% increase seen in the NAHB test on one shear wall with openings reinforced with toothed metal connector plates.

Experimental Study

The relative behavior between conventional construction and the proposed system was investigated in a small-scale pilot study. A total of eight frame configurations were tested. The basic frame consisted of kiln-dried Douglas fir No.2 2x4 dimension lumber connected together with 8d common nails or 3"x3" toothed metal connector plates. Four specimens were tested for each basic frame configuration. These included bare frames and frames sheathed with 19/32" structural sheathing connected to one side of the frame with 8d common nails spaced at 4" along the perimeter. Each of theses frames was tested with and without tension anchorage between the tension chord and the support. Conventional wood frame construction was represented by an end-nailed frame with sheathing and tension anchorage. The basic test configurations are depicted in figure 2.

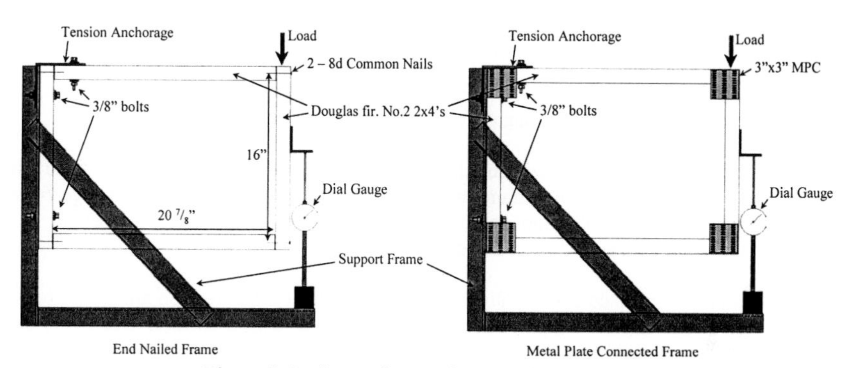

Figure 2. Basic test frames shown without sheathing

Each of the specimens was tested seven days after fabrication. Load was recorded at 0.05" displacement intervals. The test results are presented in table 1. Load-displacement plots are presented in figure 3. Elastic stiffness was calculated as the slope of the line that passes through the origin and a point representing 40% of the maximum load. The yield load was determined at the point where the load-displacement plot diverges from the line representing elastic stiffness

Table 1. Test Results of MPC Frame Pilot Study

	Elastic Stiffness (lb./in.)	Maximum Load (lbs.)	Displacement at Max Load (in.)
MPC Frame with 3x3 Plates, 19/32" Sheathing, and Tension Tie	4070	2435	1.35
MPC Frame with 3x3 Plates and 19/32" Sheathing	5770	2030	1.05
End Nailed Frame with 19/32" Sheathing and Tension Tie	2530	1850	1.1
End Nailed Frame with 19/32" Sheathing	1730	632	0.75
MPC Frame with 3x3 Plates and Tension Tie	440	392	2.05
MPC Frame with 3x3 Plates	560	368	1.85
End Nailed Frame with Tension Tie	120	145	2.35
End Nailed Frame	70	80	1.23

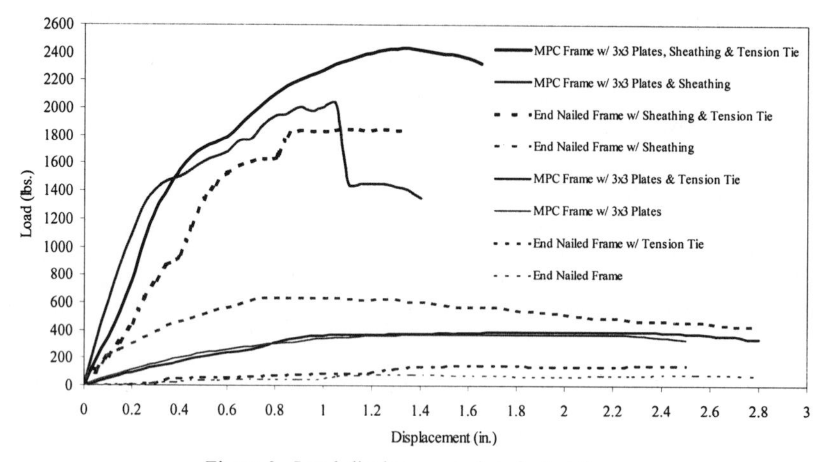

Figure 3. Load-displacement plots for test frames

The following results were derived from the pilot study tests.

1. End-nailed frames without sheathing developed virtually no lateral load resistance.

2. Metal plate connected frames without sheathing developed approximately 60% of the lateral load resistance of the end nailed frame with sheathing but without tension anchorage between the tension chord and the supports. The tension anchorage provided no benefit to the unsheathed systems because they behaved as true frames.

3. The metal plate connected frame with sheathing but without direct tension anchorage between the tension chord and the support developed 228% of the elastic stiffness and 110% of the lateral load resistance of the end-nailed frame with sheathing and tension anchorage between the tension chord and the support. The MPCs transferred tension from the tension chord into the bottom plate. The bottom plate transferred tension forces from the MPCs to the anchor bolt and support frame through cross-grain bending. This specimen displayed a sudden drop in load as a result of local tension perpendicular to the grain failure of the bottom plate between the MPCs and the support anchor.

4. The metal plate connected frame with sheathing and direct tension anchorage between the tension chord and the support developed 161% of the elastic stiffness and 132% of the lateral load resistance of the end nailed frame with sheathing and tension anchorage between the tension chord and the support.

Although the pilot study was far from definitive, it does suggest probable behavior and benefits of the proposed system. An initial concern was that stiffness of the sheathing would prevent any benefits of a more flexible underlying moment resistant frame. This concern was alleviated when both of the MPC frames with sheathing developed substantially more elastic stiffness and ultimate strength than the conventional end-nailed frames with sheathing and tension anchorage. The increased strength and stiffness provided by the metal plate connected framework should provide the following benefits to wood frame structures: (a) improved elastic and ultimate resistance to lateral loads, (b) improved utilization of the wood framework, (c) decreased reliance on the sheathing effect, (d) improved tension anchorage to the foundation, and (e) reduced reliance on expensive tension anchorage hardware.

ONGOING RESEARCH

Current research activity is focused on preparing the foundation for the development of an engineered wood frame wall panel system integrating prefabricated metal plate connected truss technology. Design methodology for moment resistant frames integrating TPI guidelines and previous metal plate connector research recommendations are being developed and refined. The ability of the wood framework to perform as a moment resistant frame is under investigation. The behavior of the moment resistant frames covered with sheathing is being assessed. Finally, design procedures combining the lateral load resistance provided by both moment resistant frames and sheathing will be developed.

The design methodology for moment resistant frames connected with toothed metal plate connectors is a direct adaptation of procedures specified in ANSI/TPI 1-1995 [10] and recommendations of Kevarinmäki [14] and O'Regan, Woeste, and Lewis [15]. The first attempt at a design procedure for combining lateral load resistance of a moment resistant frame with that of the sheathing is an additive process. This design methodology will be refined to better represent the actual behavior of moment resistant frames and the combined systems as well.

A variety of parameters expected to affect the behavior of the wall panel system are under investigation. These include framing member stiffness, connection strength and stiffness, sheathing thickness, and wall height to width ratio. The conventional shear wall system is being tested in conjunction with the proposed wall panel system to quantify performance benefits of the experimental system.

REFERENCES CITED

[1] Portland Cement Association. 1997. "Home Builder Report of 1997."

[2] Breyer, D.E., Fridley, K.J., and Cobeen, K.E. 1998. *Design of Wood Structures, ASD.* McGraw-Hill.

[3] Stalnaker, J.J. and Harris, E.C. 1997. *Structural Design in Wood, 2nd Edition.* Kluwer Academic Publishers.

[4] Faherty, K.F. and Williamson, T.G. 1997. *Wood Engineering and Construction Handbook.* McGraw-Hill.

[5] Tuomi, R.L. and McCutcheon, W.J. 1978. "Racking Strength of Light-Frame Nailed Walls." *Journal of the Structural Division.* American Society of Civil Engineers. Vol. 104. No. ST7. pp. 1131-1140.

[6] Gupta, A.K. and Kuo, G.P. 1985. "Behavior of Wood-Framed Shear Walls." *Journal of Structural Engineering.* American Society of Civil Engineers. Vol. 111. No.8. pp. 1722-1733.

[7] McKee, S., Crandell, J., Hicks, R., and Marchman, L. (National Association of Home Builders Research Center). 1998. "The Performance of Perforated Shear Walls with Narrow Wall Segments, Reduced Base Restraint, and Alternative Framing Methods." Prepared for: The U.S. Department of Housing and Urban Development.

[8] Sugiyama, H. and Matsumoto, T. 1993. "A Simplified Method of Calculation the Shear Strength of a Plywood-Sheathed Wall With Openings II. Analysis of the Shear Resistance and Deformation of a Shear Wall with Openings." Mokuzai Gakkaishi, 39(8) pp. 924-929.

[9] Sugiyama, H. and Matsumoto, T. 1994. "Empirical Equations for the Estimation of Racking Strength of a Plywood-Sheathed Shear Wall with Openings." Summaries of Technical Papers of Annual Meeting, *Trans of A.I.J.*

[10] ANSI/TPI 1-1995. National Design Standard for Metal Plate Connected Wood Truss Construction. 1995. Truss Plate Institute. Madison, WI.

[11] ANSI/NfoPA NDS-1991 National Design Specification for Wood Construction. 1991. American Forest and Paper Association. Washington, D.C.

[12] Noguchi, M. 1980. "Ultimate Resisting Moment of Butt Joints With Plate Connectors Stressed in Pure Bending." Wood Science. Vol.12 No.3 168-175.

[13] Edlund, G. 1971. "Längdskarving av trabälkar med spikplatsforband. Byggforskningen Rapport R40. Stockholm, Sweden.

[14] Kevarinmäki, A. 1996. "Moment Capacity and Stiffness of Punched Metal Plate Fastener Joints." *In*: Proceedings of the International Wood Engineering Conference. 1:385-392.

[15] O'Regan, P.J., Woeste, F.E., and Lewis, S.L. 1998. "Design Procedure for the Steel Net-Section of Tension Splice Joints in MPC Wood Trusses." Forest Products Journal. Vol.48, No.5, 35-42.

[16] O'Regan, P.J., Woeste, F.E., and Brakeman, D.B. 1998. "Design Procedure for the Lateral Resistance of Tension Splice Joints in MPC Wood Trusses." Forest Products Journal. Vol.48, No.6, 66-69.

[17] *Uniform Building Code*. 1997. International Conference of Building Officials. Whittier, California.

[18] Emerson, Robert N. and Fridley, Kenneth J., "Resistance of Metal Plate Connections to Dynamic Loading", *Forest Products Journal*, Vol. 46, No. 5, 83-90.

[19] Emerson, Robert N. and Fridley, Kenneth J., "Probabilistic Modeling of Metal Plate Connections," Proceedings of the 7th ASCE Specialty Conference on Probabilistic Mechanics and Structural Reliability.

Advances in Building Technology, Volume 1
M. Anson, J.M. Ko and E.S.S. Lam (Eds.)

Toward Engineering Design of Gypsum Board Fire Barriers

O. M. Friday[1] and S. M. Cramer[1]

[1]Department of Civil and Environmental Engineering,
University of Wisconsin-Madison, Madison, WI, 53706, USA

ABSTRACT

Gypsum board sheathing panels are widely used throughout different building types to provide passive fire protection, but in residential construction it is the primary structural fire protection. Most fires and most fire deaths occur in residential construction so maintaining the integrity of the gypsum board fire barrier during a fire is critical to the longevity and the fire safety of the structure. Reported here is progress in a study examining the engineering properties of Type X gypsum board. Little is published concerning the engineering properties of this common construction material. The objective of this research is to measure these properties and to organize them into a rational failure theory that will allow engineering design of gypsum board fire barriers. Test were undertaken with a small oven within a test machine measuring mass loss, shrinkage, bending strength and modulus of elasticity at elevated temperatures. The research shows that near-complete calcination of the gypsum board occurs only after 60 minutes of exposure at 400°C. Only after complete calcination, where all moisture is driven off, does the bending strength approach 0. Strength appeared to decrease linearly with increasing temperature. Mass loss and modulus of elasticity exhibited similar trends with the greatest decreases occurring from 100°C to 200°C followed by more gradual decrease to 400°C. The reinforcing glass fibers play a significant stiffening and strengthening role in the long direction of sheathing panels up to 60 minutes of exposure at 300°C.

KEYWORDS

fire, gypsum, light-frame construction, wood, material properties, ASTM E119

INTRODUCTION

Problem Statement

The majority of residential construction in the United States consists of 1- and 2-family dwellings and 1 to 4 story apartment complexes, and is considered light-frame construction. This typically consists of sawn timber, engineered wood products or light gage steel as the floor and wall structural members. Engineered wood products include I-joists and trusses for floor and roof systems. Gypsum board is widely used for wall and ceiling linings and in addition to providing an enclosure surface, acoustic and fire protection are also provided. Most fire deaths in the United States each year occur in the home, with the majority of them occurring in 1- and 2- family dwellings.

Although the number of fire deaths per capita in the United States is decreasing, it is well above other industrialized nations. For example, the United States deaths per capita from fire are 20% above the European average. Fires in the United States kill many more people each year than floods, hurricanes, earthquakes, tornadoes and other natural disasters combined (United States Fire Authority 1999). Fires in homes caused 3,420 deaths, 16,975 injuries, and $5.5 billion dollars of property damage in 2000 (NFPA website). This is the result of 368,000 home fires, which represents 73% of all structural fires. Nearly 85% of the fire deaths in 2000 occurred in the home and 72% of all fire deaths occurred in 1- and 2- family dwellings. These values are representative of the previous twelve years (United States Fire Authority 1999).

The primary fire resistant strategy for residential construction is based on the use of gypsum board in certain locations to provide a passive protective membrane over wood or steel framing members. Active fire resistance systems are not customary in 1- and 2- family dwellings in the United States, so all of the fire resistance relies on the combined resistance of the structural members and the gypsum board. The gypsum board and the attachment to the framing are typically prescribed in building codes based on fire ratings achieved directly or indirectly from standard tests such as that specified in ASTM E-119.

Previous ASTM E119 tests indicate that floor/ceiling structural assemblies typically fail within minutes after the gypsum board pulls away from the ceiling structural members, allowing hot gases to flow into the plenum and ignite the wood members. It is clear from test observations that the longer the gypsum board remains intact and attached to the wall or ceiling framing members, the longer the structure will withstand the fire to which it is exposed.

It is hypothesized that if the material properties of gypsum board are known, the time and temperature conditions associated with gypsum board failure in an ASTM E-119 test or other fire condition can be computed. The significance of this research is that once the response of the gypsum board can be categorized and predicted, innovative designs that provide equivalent or superior fire safety can be designed. Costly fire testing can be used to verify proposed designs rather than serving as a trial-and-error design tool.

The objective of the research reported here was to establish the engineering properties of typical Type X gypsum board. These properties are not well established and their absence makes it impossible to conduct an engineered structural design of walls or ceilings that incorporate the structural performance of the gypsum board for either standard load or fire conditions. The research project is in-progress and our final outcomes have not yet been achieved.

Little information exists on the mechanical properties of gypsum. A series of studies have established some of the thermal properties of gypsum (Harmathy 1995 and 1983, Lawson 1977). Fuller (1990) was one investigator who examined both thermal and mechanical properties of gypsum. He established bending strength and bending stiffness of regular 13-mm (1/2-in) gypsum board at temperatures up to 140°C.

Gypsum Board Chemistry

Gypsum board generally consists of hydrous calcium sulfate with paper on either side (Lawson 1977). Different additives may be included to improve the fire resistance of the board, such as glass fibers or vermiculite. In the United States, Type X board is a fire-rated board containing glass fibers, and other additives are possible. Type X board is a generic label not specific to any particular company and is defined based only on performance as a sheathing that will provide a 60-minute load-bearing fire resistance when one 15.9 mm board is attached to each side of a wood or steel stud wall assembly (Buchanan 2001). Gypsum board undergoes calcination when exposed to temperatures above 100°C. Chemically combined water is expelled by the gypsum board between 100°C and 125°C, reducing the gypsum to plaster of paris as shown chemically in Eq. 1 (Harmathy 1995).

$$CaSO_4 \bullet 2H_2O \rightarrow CaSO_4 \bullet \tfrac{1}{2}H_2O + 1\tfrac{1}{2}H_2O \qquad (1)$$

There is no information available on the rate at which this reaction occurs. With fire exposed to one side of the board, the calcination process proceeds gradually through the board thickness consuming thermal energy and limiting temperatures on the backside of the board to about 120°C. Board density and initial moisture content are important factors that influence the fire resistant capabilities of the gypsum board. The glass fibers in Type X gypsum also help retard shrinkage cracks and vermiculite also aids in this effort. Shrinkage cracks could open up and allow hot gases to flow into the cavity of a wall or floor system and ignite the structural system.

Measuring Material Properties

To predict the time to failure of gypsum board in a fire, the material properties at elevated temperatures are needed. Obviously, when complete calcination occurs the remaining plaster of paris is in powder form and no usable strength or stiffness remains. But in a fire this change does not occur immediately and the gypsum board retains structural integrity for considerable time. The properties of interest include dimensional and mass changes, and strength and stiffness losses of gypsum board at elevated temperatures. These properties were measured with an oven configured within a small-scale test machine. Through a series of bending and load-free tests, properties were measured at ambient, 100°C, 200°C, 300°C and 400°C. A high temperature linear differential induction transducer was used to measure displacements and dimensional changes of the specimen within the oven. Load tests were conducted at elevated temperatures after soak times of 15 minutes and 60 minutes at the target temperature where no load was applied. This paper is limited to presentation of the 60-minute results. A total of nine tests were conducted for each condition with at least 4 specimens obtained from cuts along the length of the panel and at least 4 specimens cut from the width of the panel. These specimens were cut from a minimum of 3 different panels. A single source of 15.9 mm thick Type X gypsum board was used for all tests. The specimens were conditioned at 21°C and 50% relative humidity for at least 30 hours prior to testing.

The test setup was configured to provide repeatable measures of material properties at different temperature exposures and not to precisely replicate conditions during a fire. Two significant differences between test conditions and fire conditions exist. First, in the tests both sides of the gypsum board specimen were exposed to the same temperature. Second, test temperatures were limited to 400°C even though much higher temperatures will likely occur in a fire. Applications of these research results must consider these factors.

TEST RESULTS

Mass and Dimensional Change Tests

According to Harmathy (1995), gypsum will expand dimensionally 0.1% from 0 to about 100°C or 200°C, and then remain somewhat constant until about 300°C. Between 300°C and 400°C it will shrink 1.3% followed by slowed shrinkage to 700°C. Our test data showed similar trends. We observed 0.1% shrinkage at 200°C and 300°C with the greatest shrinkage observed at 400°C with a value of 0.65%. Details of Harmathy's tests were not available so the differences may be due to specimen size or the heating regime used. The specimen size in our tests was 203 mm long by 51 mm wide.

Gypsum board contains approximately 21% wat er by weight and thus as calcination occurs the mass of the board is reduced. We found that after a 60-minute soak period at 100°C that only 3% of the original board weight was lost. Harmathy (1995) found the mass at 100°C reduced by about 6% of original room temperature mass, with no significant changes from 200°C through 600°C. At 200°C, we observed a 16% loss in mass that increased to 22% at 400°C (Figure 1). This observation is consistent with complete calcination and loss of all moisture.

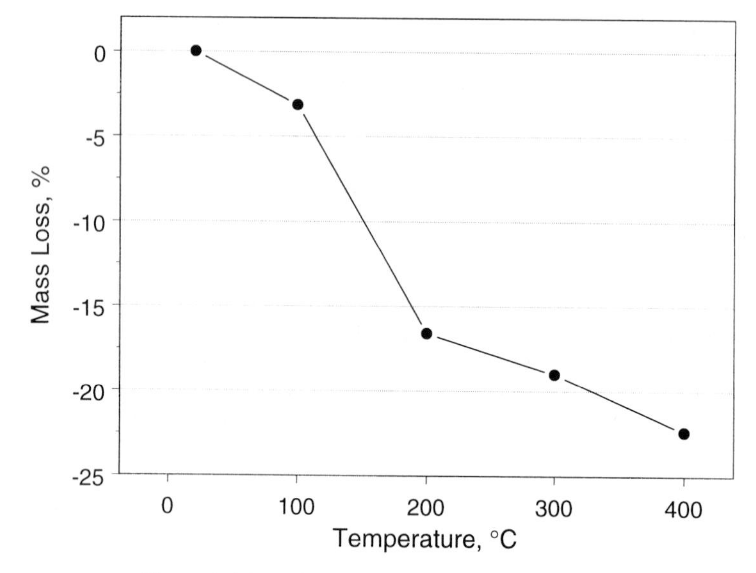

Figure 1: Mass change after 60 minutes of exposure

Tension and Bending Tests

Loads and deflections from three-point bending tests were used to evaluate modulus of elasticity and bending strength. The specimen size, 177.8 mm by 50.8 mm, was selected to meet the constraints of the test chamber, resulting in a short beam where shear deformations were significant. Shear effects were considered by assuming a fixed relationship between modulus of elasticity and shear modulus. The brittle nature and low tensile strength of the gypsum board specimens rendered tension tests impractical. All tests were conducted with the paper intact on both faces of the gypsum board as it is used in service.

The bending strength of the gypsum board specimens was defined as the maximum load achieved in the initial linear portion of the load-displacement plot. Sometimes additional load was resisted after the first crack formed but these were not considered in the initial data analysis presented here.

The strengths and bending stiffnesses of the specimens depended on the orientation of the specimen in the original gypsum board panel. When the length of the specimen coincided with the long direction, or machine direction, of the panel, the strength and stiffness were significantly higher than they were when the length of the specimen was oriented with the panel width. This distinction was reduced at higher temperatures. The glass fibers were primarily oriented in the machine direction of the panel to provide additional tensile and bending strength in the long direction.

Bending strengths at ambient conditions were in the range of 2 to 4 MPa depending on specimen direction and other factors. Figure 2 shows the average degrade in bending strength that occurs after 60 minutes of exposure at temperatures up to 400°C. As indicated by the mass loss in Figure 1, calcination is complete after 60 minutes of exposure at 400°C as all available water has been driven off. Similarly the bending strength decreases to a value close to zero after 60 minutes at 400°C. Figure 2 shows that the bending strength decreases in a near linear manner. With specimens cut from the long direction of the panel, the role of the glass reinforcing fibers is clearly revealed in Figure 2. At 300°C the strengthening influence of the glass reinforcing fibers has all but disappeared.

An apparent modulus of elasticity was computed that depended on the gross cross section. The presence of the surface paper on each face of the gypsum board results in a composite section. The separate layers most likely degrade differently as temperature exposure is increased and eventually the paper burns away completely. The gross section was used in this analysis and was assumed to be homogeneous.

The degradation of apparent modulus of elasticity relative to the ambient value to is shown in Figure 3. Ambient modulus of elasticity is in the range of 1700 to 2200 MPa. The trend of degradation is similar to that of mass loss. After 60 minutes of exposure at 200°C, the loss in modulus of elasticity in the gypsum matrix is essential complete. Sufficient adhesion to the reinforcing glass fibers continues however and it is not until exposure of 400°C that the stiffening effect of the glass fibers is completely lost.

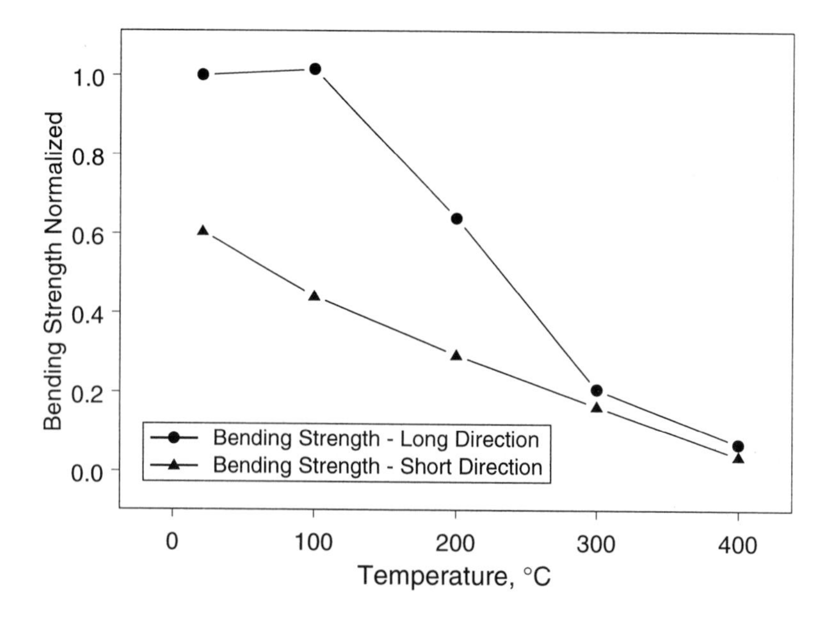

Figure 2: Degrade in bending strength after 60 minutes of exposure

Application of Test Results

To foresee the application of this research, consider a gypsum ceiling membrane attached to wood joist ceiling members. Under ambient nonfire conditions, the gypsum is held in place with screws embedded into the wood members or into a resilient channel. The gypsum experiences internal stresses that are induced from its self-weight. In addition, as the structural components of the ceiling are loaded, more stress is transferred into the gypsum board as a result of partial composite action. Different deflections of the individual ceiling joists may also induce additional stresses in the gypsum board panels that are spanning a series of joists in a continuous fashion. As heat from fire is added to this scenario, the gypsum board panel begins to shrink and the restraint of the attachments to the ceiling joists results in additional stresses. As the temperatures increase and calcination occurs, the gypsum board panel will lose stiffness and bending strength as discussed. When critical stresses are reached the gypsum board will fracture, allowing hot gases and flames to reach the wood members.

Although the described scenario is complex, it is not intractable and not unlike other structural analysis problems. With the baseline of data established from this research, structural analysis of assemblies as described above will be possible and in turn, engineering design of fire resistant barriers will also be possible.

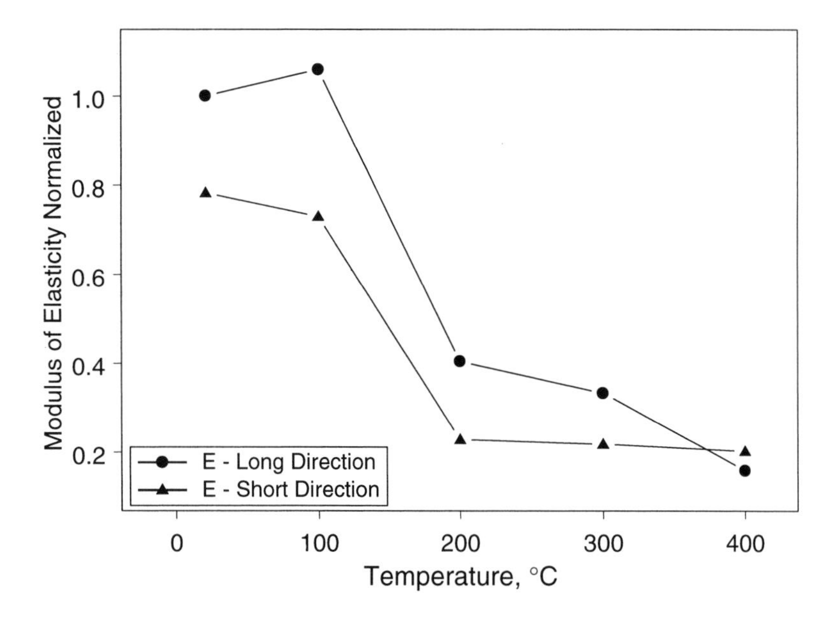

Figure 3: Degrade in modulus of elasticity after 60 minutes of exposure

CONCLUSIONS

The fire resistance of residential construction is primarily achieved through the use of gypsum board panels. The longevity of light-frame wood and steel construction in a fire depends on the maintaining the gypsum board lining over framing members. The complete loss of moisture (calcination) of the gypsum board occurred after 60 minutes of exposure at 400°C. The largest decreases in mass and shrinkage occurred between 100°C and 200°C. Bending strength, less the influence of glass reinforcing fibers, decreased linearly as temperature increased and approached 0 at 400°C. Modulus of elasticity based on an assumed homogenous section decreased the most from 100°C to 200°C. The glass reinforcing fibers played a significant stiffening role through exposure at 300°C. This research reveals the engineering properties of gypsum board such that gypsum board fire barriers can advance from simple prescriptive construction requirements to engineering design.

ACKNOWLEDGEMENTS

The authors gratefully acknowledge the support of the National Science Foundation through grant CMS-0080293 as part of the Partnership for Advanced Technologies for Housing. The assistance of US Gypsum and the technical advice of Dr. Robert White of the USDA Forest Products Lab are appreciated.

REFERENCES

Buchanan, A.H. (2001). *Structural Design for Fire Safety*, John Wiley & Sons, Ltd. Chichester, England.

Fuller, J. (1990). *Predicting the Thermo-Mechanical behavior of a Gypsum-To-Wood Nailed Connection.* Master of Science Thesis, Department of Forest Products, Oregon State University, Corvallis, OR.

Harmathy, T. (1983). Properties of building materials at elevated temperatures, *DBR Paper No. 1080 of the Division of Building Research*, National Research Council of Canada.

Harmathy, T. (1995). *The SFPE Handbook of Fire Protection, Second Edition.* National Fire Protection Association, Quincy, MA.

Lawson, J.R. (1977). An Evaluation of Fire Properties of Generic Gypsum Board Products, NBSIR 77-1265, Center for Fire Research, Institute for Applied Technology, National Bureau of Standards, Washington D.C.

National Fire Protection Association (2001). NFPA Fact Sheet: Home Fire Statistics, NFPA website, www.nfpa.org/Research/NFPAFactSheets/NFPAHomeFireStats/NFPAHomeFireStats.asp

United States Fire Authority (1999). Fire in the United States 1989-1998 (Twelfth Edition), United States Fire Authority website, www.usfa.fema.gov

Advances in Building Technology, Volume 1
M. Anson, J.M. Ko and E.S.S. Lam (Eds.)

BARRIER MATERIALS TO REDUCE CONTAMINANT EMISSIONS FROM STRUCTURAL INSULATED PANELS

J. C. Little[1], D. Kumar[1], S. S. Cox[1], and A. T. Hodgson[2]

[1]Civil & Environmental Engineering Dept., Virginia Tech, Blacksburg, VA, USA
[2]Indoor Environment Dept., Lawrence Berkeley National Laboratory, Berkeley, CA, USA

ABSTRACT

The use of Structural Insulated Panels (SIPs) to create very tight building envelopes will help reduce the environmental impact and energy use of new housing. Typically, SIPs are constructed from oriented strand board (OSB) and rigid foam in multi-layered sandwich-like structures. Although environmental and energy advantages make panelized systems very attractive, the tighter building envelopes may result in degraded indoor air quality. The potential release of volatile contaminants from SIPs must also be considered. A physically based diffusion model that predicts emissions from a single layer of vinyl flooring has recently been developed and successfully validated. A logical and promising extension of this approach is to apply the model to predict emissions from multi-layer systems such as SIPs. As a first step towards this goal, a double-layer model is developed to predict the rate of mass transfer from double-layered building material to indoor air. It is assumed that the two layers are flat homogeneous slabs, that internal mass transfer is governed by diffusion, and that the indoor air is well mixed. An analytical solution to the double-layer model is presented and used to demonstrate the potential for a thin surface barrier layer to reduce contaminant emission rates from building materials.

KEYWORDS

Air, barrier, building, contaminant, diffusion, indoor, layer, material, model, source, VOC

INTRODUCTION

One of the primary goals of the Partnership for Advancing Technology in Housing (PATH) is to cut the environmental impact and energy use of new housing by 50 percent. The use of Structural Insulated Panels (SIPs) in new construction and major renovation to create very tight building envelopes will help realize this goal. The environmental advantages of SIPs include:

- less job-site waste;
- lower energy consumption;
- greater use of fast-growth harvested farm trees rather than old-growth forests;
- reduced consumption of dimensional lumber.

These advantages make panelized systems very attractive from both environmental impact and energy use perspectives. However, degradation of indoor air quality is an important and well-documented result of tighter building envelopes and the use of engineered wood products as construction materials (Hodgson et al., 2001).

Indoor sources of volatile organic compounds (VOCs) are one determinant of residential indoor air quality (Hodgson et al., 2000). Many materials used to construct and finish the interiors of new houses emit VOCs such as formaldehyde. These emissions are a probable cause of acute health effects and discomfort among occupants (Andersson et al., 1997). Ventilation is another determinant of indoor air quality in houses (Hodgson et al., 2000). Ventilation serves as the primary mechanism for removal of gaseous contaminants generated indoors. The trend in new construction is to make house envelopes tighter. This practice improves energy efficiency by decreasing the infiltration of unconditioned outdoor air. Consequently, natural ventilation rates in new houses without supplemental forced ventilation can be relatively low with a related potential for degraded indoor air quality (Hodgson et al., 2000). In many cases, these ventilation rates may be below recommended guidelines (ASHRAE, 1989).

There have been very few investigations of VOC contamination in new houses. In one recent study (Hodgson et al., 2000), the concentrations and emission rates of VOCs were shown to be similar among 11 new manufactured and site-built houses in four different locations. This was attributed to strong similarities in construction materials and building practices. Formaldehyde, other aldehydes, and terpene hydrocarbons (HCs) were the predominant compounds. Formaldehyde concentrations had a geometric mean value for all houses of 40 ppb. Exposures to formaldehyde are of concern because formaldehyde is a potent sensory irritant and is classified as a probable human carcinogen (Lui et al., 1991; U.S. EPA, 1994). The State of California has set an allowable daily formaldehyde exposure limit of 40 μg, which equates to an indoor air concentration of just 1.6 ppb (Kelly et al., 1999). Higher molecular weight aldehydes can produce objectionable odors at low concentrations. The odor thresholds for hexanal and other aldehydes are often exceeded in new houses and may remain elevated for months after construction (Lindstrom et al., 1995; Hodgson et al., 2000). Terpene HCs are of potential concern because they react with ozone to produce ultrafine particles (Weschler and Shields, 1997). Animal studies also indicate that strong sensory irritants are formed by terpene-ozone reactions (Wolkoff et al., 2000). Wood and engineered wood products (e.g., particleboard, medium density fiberboard, plywood and OSB) are the likely major sources of aldehydes and terpene HCs in new houses (Hodgson et al., 2001).

There is a potential for houses constructed with SIPs to have degraded air quality relative to conventionally constructed houses that use fewer engineered wood products. Some SIP systems are assembled with an OSB layer on the interior facing side. OSB is a source of formaldehyde emissions approximately equivalent to phenol-formaldehyde plywoods (Kelly et al., 1999). OSB also emits pentanal and hexanal, two odorous aldehydes (Barry and Corneau, 1999). These contaminants originate in the wood drying process through the breakdown of wood tissue

(Otwell et al., 2000) and are, thus, inherent to most engineered wood products. In SIPs applications, the OSB is typically finished with a layer of gypsum board. However, gypsum board and other interior finishes may not provide an effective barrier for volatile contaminants released by OSB. The large surface area of installed SIP systems, combined with the resulting decrease in ventilation rate due to very low infiltration exacerbates the potential problem.

The ability to predict and consequently minimize the potential impact of panel systems on indoor concentrations of contaminants of concern would be extremely useful for housing technologies. Recently, a physically-based model that predicts emissions from vinyl flooring, a single-layer, diffusion-controlled VOC source, has been developed (Little et al., 1994; Little and Hodgson, 1996) and successfully validated (Cox et al., 2001a; Cox et al., 2001b; Cox et al., 2002). The model validation process achieved a much higher degree of scientific integrity than previous diffusion-controlled emission models because the three model parameters were determined in a completely independent fashion. A logical and promising extension of this approach is to apply the model to predict emissions from multi-layer systems such as SIPs. The multi-layer model will be valuable when designing new panel systems, because the emission characteristics can be predicted in advance, based on a few simple and direct measurements of the key model parameters. In addition, the model can be used to develop strategies to design panel systems in such a way as to reduce or eliminate emissions. In particular, the model could be used to optimize the required properties of suitable diffusion barriers. In this paper, the recently validated single-layer emissions model will be extended to a double-layer system. The double-layer model will then be used to investigate required barrier layer properties.

VALIDATION OF SINGLE-LAYER MODEL

In prior research, a fundamental model was developed for diffusion-controlled building material, as well as rapid procedures to measure model parameters.

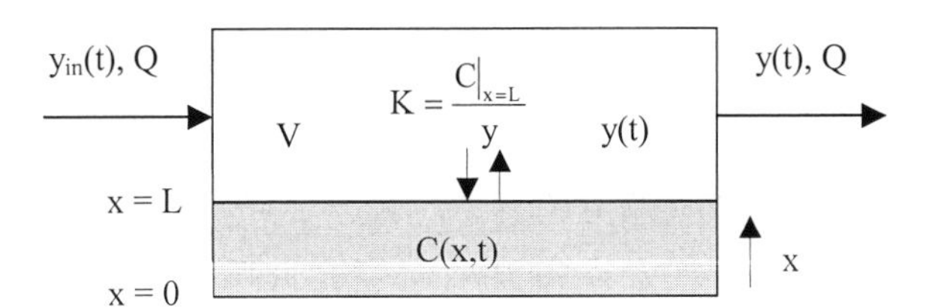

Figure 1: Schematic representation of a homogenous source in a room or chamber

The model for predicting the emission rate of VOCs from diffusion-controlled building materials is summarized in Figure 1 (Cox et al., 2002). The resulting equation for predicting the concentration of VOC in the source material is

$$C(x,t) = 2C_0 \sum_{n=1}^{\infty} \left[\frac{\exp(-Dq_n^2 t)(h - kq_n^2)\cos(q_n x)}{\left[L(h - kq_n^2)^2 + q_n^2(L+k) + h\right]\cos(q_n L)} \right] \tag{1}$$

where $h = \frac{Q}{ADK}$, $k = \frac{V}{AK}$, and the q_ns are the positive roots of $q_n \tan(q_n L) = h - kq_n^2$.

Referring to Equation 1 and Figure 1, C(x, t) is the material-phase VOC concentration, t is time, x is distance from the base of the material, y(t) is the gas-phase VOC concentration in the well-mixed chamber air, D is the concentration independent material-phase diffusion coefficient, K is the concentration independent material/air partition coefficient, Q is the volumetric flow rate of air through the chamber, V is the chamber volume, A is the exposed surface area of the material, and L is the material thickness. The initial concentration in the material (C_0) is assumed to be uniform and $y_{in}(t)$ is assumed to be zero for the entire time period.

The model was successfully validated for three VOCs found in VF (Cox et al., 2002). Model predictions were compared to gas-phase concentrations obtained during small-scale chamber tests, and to material-phase concentrations measured in the VF at the conclusion of the chamber tests. The model parameters (the initial VOC concentration in the material-phase (C_0), the material/air partition coefficient (K), and the material-phase diffusion coefficient (D)) were measured in completely independent tests (Cox et al., 2001a; Cox et al., 2001b).

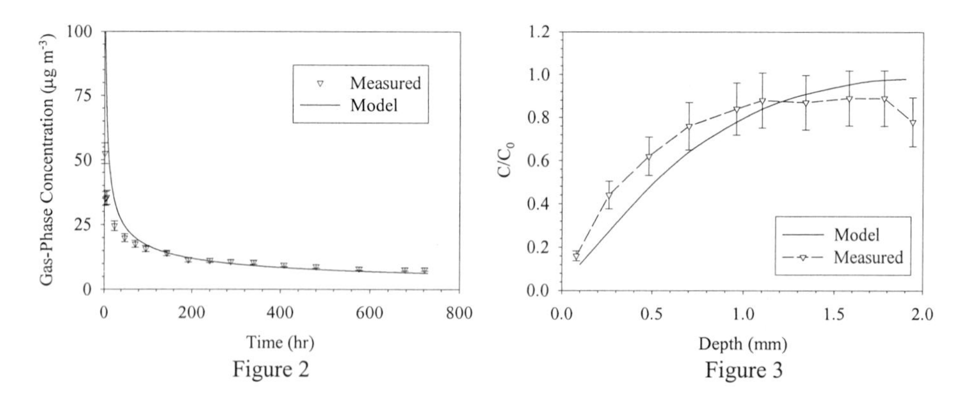

Figure 2

Figure 3

Figure 2 shows predicted gas-phase concentrations compared to concentrations of n-tetradecane measured in chamber experiments. Figure 3 shows the measured material-phase contaminant concentrations of n-tetradecane after 722 hours of exposure in the chamber. The model predicts the measured concentrations very well in both cases.

DEVELOPMENT OF DOUBLE-LAYER MODEL

The model for the double-layer system depicted in Figure 4 is a logical extension of the single-layer model. The diffusion equation applies to each of the layers, or

$$\frac{\partial^2 C_i}{\partial x^2} = \frac{1}{D_i}\frac{\partial C_i}{\partial t} \tag{2}$$

where i represents the specific layer, C is the material-phase concentration, D is the diffusion coefficient, t is time and x is distance as shown in Figure 4.

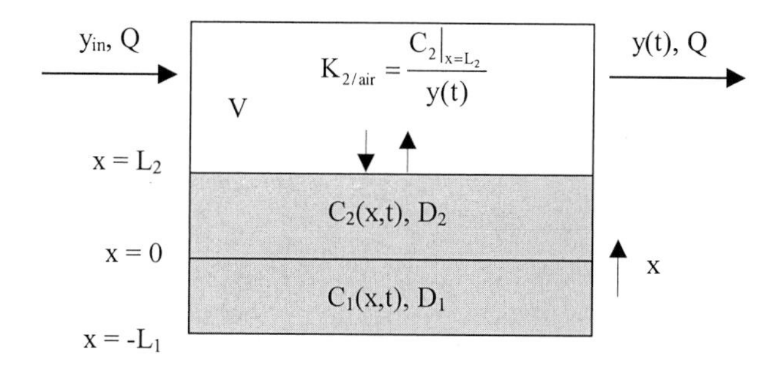

Figure 4: Schematic representation of a double-layer source in a room or chamber

The initial conditions for uniform material-phase concentrations are

$$C_1 = C_{10} \quad \text{for} \quad -L_1 \le x \le 0 \quad \text{and} \quad C_2 = C_{20} \quad \text{for} \quad 0 \le x \le L_2 \tag{3}$$

The first boundary condition is based on a mass balance about the chamber. It is assumed that the air within the chamber is in instantaneous equilibrium (partition coefficient $K_{2/air}$) with the upper surface of the top layer ($K_{2/air}\, y(t) = C_2$ at $x = L_2$). Thus,

$$V\frac{\partial y(t)}{\partial t} = -AD_2\left.\frac{\partial C_2}{\partial x}\right|_{x=L_2} - Qy(t) \tag{4}$$

where V is the volume of the chamber and Q is flow rate though the chamber. The second boundary condition assumes that the flux through the base of lower layer is zero, or

$$\left.\frac{\partial C_1}{\partial x}\right|_{x_1=-L_1} = 0 \tag{5}$$

The other two boundary conditions are obtained using the assumptions that the two layers are in interfacial equilibrium (with partition coefficient $K_{1/2} = K_{1/air}/K_{2/air}$) and that there is no loss of mass at the interface, or

$$C_1\big|_{x=0} = K_{1/2}C_2\big|_{x=0} \tag{6}$$

$$D_1\left.\frac{\partial C_1}{\partial x}\right|_{x=0} = D_2\left.\frac{\partial C_2}{\partial x}\right|_{x=0} \tag{7}$$

Using Laplace transforms, a solution is found and the inverse is obtained by the method of residues:

$$C_1 = 2\alpha\sum_{n=1}^{\infty}\left[\frac{J(q_n)\{\cos(\alpha q_n x)\cos(\alpha q_n L_1) - \sin(\alpha q_n x)\sin(\alpha q_n L_1)\}\exp(-D_2 q_n^2 t)}{q_n G(q_n)}\right] \quad (8)$$

$$C_2 = 2\sum_{n=1}^{\infty}\left[\frac{J(q_n)\{\alpha\cos(q_n x)\cos(\alpha q_n L_1) - K_{1/2}\sin(q_n x)\sin(\alpha q_n L_1)\}\exp(-D_2 q_n^2 t)}{K_{1/2} q_n G(q_n)}\right] \quad (9)$$

where

$$J(q_n) = C_{20}K_{1/2}\{h_2 - k_2 q_n^2(1-\beta)\} - (K_{1/2}C_{20} - C_{10})\{(h_2 - k_2 q_n^2)\cos(q_n L_2) - q_n\sin(q_n L_2)\} \quad (10)$$

$$\begin{aligned} G(q_n) = {} & q_n\{K_{1/2}(L_2 + 2k_2) + \alpha^2 L_1\}\sin(q_n L_2)\sin(\alpha q_n L_1) \\ & -\{K_{1/2} + (K_{1/2}L_2 + \alpha^2 L_1)(h_2 - k_2 q_n^2)\}\cos(q_n L_2)\sin(\alpha q_n L_1) \\ & -\alpha q_n\{2k_2 + L_2 + L_1 K_{1/2}\}\cos(q_n L_2)\cos(\alpha q_n L_1) \\ & -\alpha\{1 + (L_1 K_{1/2} + L_2)(h_2 - k_2 q_n^2)\}\sin(q_n L_2)\cos(\alpha q_n L_1) \end{aligned} \quad (11)$$

$$\alpha = \sqrt{\frac{D_2}{D_1}} \; ; \; \beta = \frac{C_{L_2}}{C_{20}} \; ; \; h_2 = \frac{Q}{AK_{2/air}D_2} \; ; \; k_2 = \frac{V}{AK_{2/air}} \quad (12)$$

and the q_ns are the roots of

$$(h_2 - k_2 q_n^2)\{K_{1/2}\tan(\alpha q_n L_1)\tan(q_n L_2) - \alpha\} + q_n\{\alpha\tan(q_n L_2) + K_{1/2}\tan(\alpha q_2 L_1)\} = 0 \quad (13)$$

Although the periodicity of the roots is complex, they can be obtained using software such as Solver in Microsoft Excel.

IMPACT OF BARRIER LAYER

The model developed can now be used to study the required properties of barrier materials used to reduce the rate of contaminant emissions from building materials. The input parameters for the model are listed in Table 1. Phenol is used as the example contaminant and the conditions employed approximate laboratory chamber experiments conducted using a sample of vinyl flooring (Cox et al., 2002).

TABLE 1
PARAMETER VALUES USED FOR CALCULATIONS

Parameter	Value
Volume of chamber, V	0.0105 m^3
Flow rate through chamber, Q	1.7×10^{-5} m^3 s^{-1}
Partition coefficient, $K_{1/air}$	120,000
Diffusion coefficient, D_1	1.2×10^{-13} m^2 s^{-1}
Initial concentration, C_{10}	2×10^8 μg m^{-3}
Surface area of VF exposed to air, A	0.0195 m^2
Thickness of VF, L_1	0.0017 m
Thickness of barrier layer, L_2	0.0002 m

To simulate a baseline scenario for the single-layer experimental VF system, the properties of layer 2 (D_2, $K_{2/air}$, and C_{20}) are set equal to those of layer 1 (D_1, $K_{1/air}$, and C_{10}). In a second scenario the surface layer has the exact same D and K values as the lower layer, but has an initial concentration of zero. In this case, the barrier layer acts as a buffer to reduce the initial peak in the chamber concentration. Finally, to simulate the properties of a diffusion barrier, D_2 is decreased by a factor of 10. The series of results is shown in Figure 5. Decreasing D of the barrier layer results in a significant drop in gas-phase concentration.

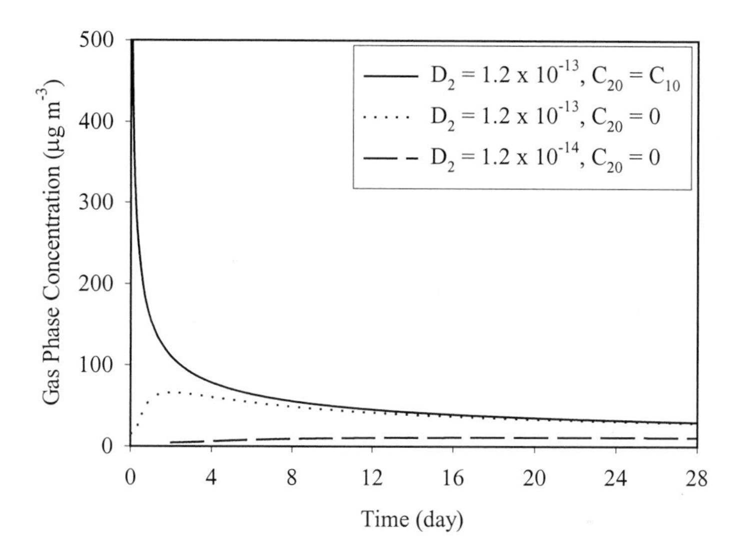

Figure 5: Effect on gas-phase concentration of decreasing D_2 of barrier layer

CONCLUSIONS

The results suggest that thin barrier layers can be used to reduce the rate of emissions from diffusion-controlled materials, including SIPs. The model provides a means to predict the combined effect that diffusion barriers and material transport properties have on overall emission rates. Note that while the emission rate is reduced, the same amount of mass is ultimately emitted, unless the initial material-phase contaminant concentration could itself be decreased. The double-layer model should prove useful in predicting the source/sink behavior of other layered building materials. It will also be used to validate the more mathematically complex multi-layer model that is currently being developed to predict contaminant transport through SIPs. Although the single-layer version of this model has been validated, experimental work has yet to be done to formally validate the double-layer model.

ACKNOWLEDGEMENTS

Financial support was provided by the National Science Foundation (NSF) through an NSF CAREER Award (Grant No. 9624488) and an NSF PATH Award (Grant No. 0122165).

REFERENCES

Andersson, K., Bakke, J.V., Bjorseth, O., Bornehag, C.-G., Clausen, G., Hongslo, J.K., Kjellman, M., Kjaergaarrd, S., Levy, F., Molhave, L., Skerfving, S. and Sundell, J. (1997). TVOC and Health in Non-industrial Indoor Environments *Indoor Air* **7**, 78-91.

ASHRAE. (1989). Standard 62, Ventilation for Acceptable Indoor Air Quality, American Society of Heating, Refrigeration and Air-Conditioning Engineers.

Barry, A. and Corneau, D. (1999). The Impact of Wood Composite Panel Products such as OSB, Particleboard, MDF and Plywood on Indoor Air Quality *Indoor Air 99, Proceedings of the 8th International Conference on Indoor Air Quality and Climate* **5**, 129-134.

Cox, S.S., Hodgson, A.T. and Little, J.C. (2001a). Measuring Concentrations of Volatile Organic Compounds in Vinyl Flooring *Journal AWMA* **51**, 1195-1201.

Cox, S.S., Little, J.C. and Hodgson, A.T. (2002). Predicting the Emission Rate of Volatile Organic Compounds from Vinyl Flooring *Environmental Science & Technology* **36**, 709-714.

Cox, S.S., Zhao, D.Y. and Little, J.C. (2001b). Measuring Partition and Diffusion Coefficients of Volatile Organic Compounds in Vinyl Flooring *Atmospheric Environment* **35**, 3823-3830.

Hodgson, A.T., Rudd, A.F., Beal, D. and Chandra, S. (2000). Volatile Organic Compounds Concentrations and Emission Rates in New Manufactured and Site-built Houses *Indoor Air* **10**, 178-192.

Hodgson, A.T., Beal, D. and McIlvaine, J. (2001). Sources of Formaldehyde, other Aldehydes and Terpenes in a New Manufactured House *Indoor Air* – In press.

Kelly, T.J., Smith, D.L. and Satola, J. (1999). Emission Rates of Formaldehyde from Materials and Consumer Products found in California Homes *Environmental Science and Technology* **33**, 81-88.

Lindstrom, A.B., Proffitt, D. and Fortune, C.R. (1995). Effects of Modified Residential Construction on Indoor Air Quality *Indoor Air* **5**, 258-269.

Little, J. C. and Hodgson, A. T. (1996). A Strategy for Characterizing Homogeneous, Diffusion-controlled, Indoor Sources and Sinks *Standard Technical Publication 1287, American Society for Testing and Materials* 294-304.

Little, J. C., Hodgson, A. T., and Gadgil, A. J. (1994). Modeling Emissions of Volatile Organic Compounds from New Carpets *Atmospheric Environment* **28**, 227-234.

Lui, K.-S., Huang, F.-Y., Hayward, S.B., Wesolowski, J. and Sexton, K. (1991). Irritant Effects of Formaldehyde Exposure in Mobile Homes *Environmental Health Perspectives* **94**, 91-94.

Otwell, L.P, Hittmeier, M.E., Hooda, U., Yan, H., Wei, S. and Banerjee, S. (2000). HAPs Release from Wood Drying *Environmental Science and Technology* **34**, 2280-2283.

U.S. EPA. (1994). Review Draft of the Health Effects Notebook for Hazardous Air Pollutants. Environmental Protection Agency, Air Risk Information Support Center, (Contract N. 68-D2-0065).

Weschler, C.J. Shields, H.C. (1997). Potential Reactions Among Indoor Pollutants *Atmospheric Environment* **31**, 3487-3495.

Wolkoff, P., Clausen, P.A., Wilkins, C.K. and Nielsen, G.D. (2000). Formation of Strong Airway Irritants in Terpene/Ozone Mixtures *Indoor Air* **10**, 82-91.

Advances in Building Technology, Volume I
M. Anson, J.M. Ko and E.S.S. Lam (Eds.)

NOVEL MICROCELLULAR PLASTICS FOR LIGHTWEIGHT AND ENERGY EFFICIENT BUILDING APPLICATIONS

Krishna Nadella, Vipin Kumar[1], and Wei Li

Department of Mechanical Engineering
Box 352600
University of Washington
Seattle, WA 98195-2600

ABSTRACT

This paper presents thick microcellular plastics for lightweight, energy efficient building applications. The technology is capable of reducing polymer density and thermal conductivity, while increasing its specific strength. The resulted plastic materials can significantly contribute to reducing the house construction time, energy consumption, maintenance cost, and risk of injury during construction. The proposed thick microcellular plastics are produced in a batch process. In this research, a systematic study of various process variables is conducted using a two-stage, sliding-level design of experiment approach. The preliminary results have shown that the foamed thick ABS sheets have clearly defined skins and foamed core structures. 70% relative density reduction has been achieved.

KEYWORDS

Microcellular plastics, polymer processing, building material, plastic foam, design of experiment

INTRODUCTION

Microcellular Plastics refer to closed-cell thermoplastic foams with a very large number of very small bubbles. Typically, the cells are of order 10 micrometers in diameter, and there are 10^8 or more cells per cubic centimeter (cm^3) of the foam. The idea to introduce such small bubbles in plastics was first introduced by Suh *et al* (1982, 1984) as a means to reduce the density of solid plastics, and thus to save on material costs in applications where the full mechanical properties of solid plastics were not needed. Being a foam material, the microcellular polymers have inherited beneficial characteristics such as lightweight and low thermal conductivity. The bubbles inside the polymers are in very small scales and therefore the materials maintain reasonably high tensile and shear strength. These

[1] Corresponding author: vkumar@u.washington.edu

characteristics offer an excellent solution to many of the problems encounter in the current construction practice. Particularly, thick microcellular polymer sheet can be used in advanced panel systems to reduce house construction time and maintenance cost, cut energy consumption, reduce the risk of injury during construction, and improve the durability of the houses.

A house consists of many panels such as walls, roofs, floors, doors, and windows. The manufacture and thermal properties of these panels directly affect the construction time, energy consumption, and thus the cost of new houses. In order to reduce construction time and achieve better thermal properties, various advanced panel systems have been developed. These include precast concrete, tilt-up concrete panel, structural insulated panel system (SIPS), and insulated concrete foam (ICF). These systems often consist of a sandwich type of structure in which a foamed core is used with a layer of other materials on each side. Fig. 1 shows an example SIP system where the sheathing material used is compressed wood.

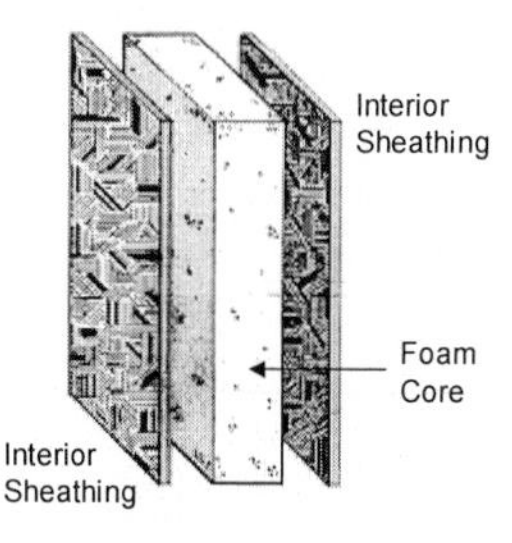

Figure 1: A schematic of an advanced panel system.

Although the concept of advanced panel systems has the potential to reduce the house construction time and thus the cost of new houses, currently available panel systems have not been widely accepted. In order to meet the requirements of home construction, the materials used for advanced panel systems have to be lightweight, reasonably strong, and easy to manufacture. Another important characteristic of such panel system is high surface quality. Good surface quality of panels can reduce large amount of interior work, which accounts for a major portion of the selling price of an average new house.

Foam materials are lightweight and low thermal conductive. However, currently available construction foam materials usually have low strengths. The sandwich type of panel systems helps increase the overall strength of the structures. But the sheathing materials used usually have poor surface quality. More often, the interior side of the walls need large amount of finish work. Recently, carbon fiber reinforced precasts and composite panels have been developed. Such panels are low cost, lightweight, and have the potential to be used as a "stand-alone" construction component. However, this type of panels has to be used together with insulation and still needs interior finish work.

The long-range objective of this research is to develop an industrially viable technology of manufacturing thick microcellular polymer sheets (6 mm–10 mm) that can be used in advanced panel systems. The proposed microcellular technology is capable of reducing polymer density and thermal conductivity while increasing its specific strength. The resulted polymer materials can significantly contribute to reducing the house construction time and maintenance cost, cutting energy consumption, reducing the risk of injury during construction, and improving the durability of the houses.

This paper presents the proposed thick microcellular sheet manufacturing process. A two-stage, sliding-level design of experiment approach is used to study the effects of various process variables on the microstructure of the polymer sheet. Preliminary results have shown that thick ABS sheets can be foamed and the structure has clearly defined skins and a foamed core. 70% relative density reduction has been achieved in this preliminary study.

MICROCELLULAR PLASTICS

The basic solid-state microcellular process is a two-stage batch process. In the first stage, the polymer is placed in a pressure vessel with a high-pressure and non-reacting gas. This step is usually conducted at room temperature. Over time, the gas diffuses into the polymer, and attains a uniform concentration throughout the polymer specimen. When this specimen is removed from the pressure vessel and brought to the atmospheric pressure, a "supersaturated" specimen that is thermodynamically unstable due to the excessive gas dissolved into the polymer is produced. In the second stage, the specimen is heated to what is termed the foaming temperature. This step is typically carried out in a heated bath with temperature control. The dissolved gas lowers the glass transition temperature of the polymer (Zhang & Handa, 1998) and the foaming temperature needs only to be above the glass transition temperature of the gas-polymer system in order for the bubbles to nucleate and grow. Since the polymer is still in a solid state, the foams thus produced are called "solid-state foams" to distinguish them from the conventional foams that are produced in an extruder from a polymer melt. Details of the batch process can be found in Kumar & Weller (1994), where the polycarbonate-CO_2 system is described.

Fig. 2 (a) shows a CPET microcellular foam with a 26% reduction of density. The size of the cells is of the order of 1 micrometer and the cell density is estimated to be 10^{13} cells per cm^3. Fig. 2 (b) shows a PVC microcellular foam produced by the solid-state process. Here we find a bimodal distribution of the bubble size, with the bubble diameters clustered around 2 and 10 micrometers. The density of the PVC specimen was 40% of the solid PVC. A unique feature of the microcellular process is that a solid, unfoamed skin can be created as an integrated part of the process (Kumar &Weller, 1994). A brief overview of microcellular polymers can be found in Kumar (1993).

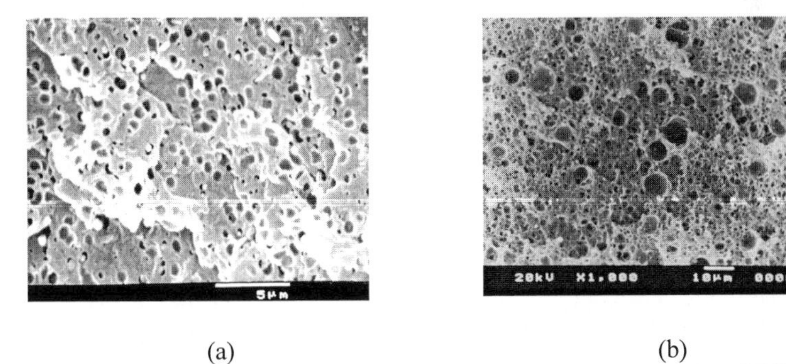

(a) (b)

Figure 2: (a) Microcellular CPET (originally 1 mm thick) with a relative density of 74% (or a density reduction of 26%). The cell density is estimated to be 10^{13} per cm^3, while the average cell diameter is approximately one micrometer. (b) The cells in this 40% Relative Density PVC example (originally 1.5 mm thick) are bi-modal in diameter distribution; some are about 10 micrometer and most about two micrometer.

It was recognized that although the batch process could be used to study new systems and characterize mechanical properties, the potential of these novel materials could not be realized unless the process could be scaled up for mass-production. An extrusion process to produce microcellular filaments was developed by Park and Suh (1992). At the University of Washington, a semi-continuous process to produce rolls of microcellular foams from rolls of polymer film was developed by Kumar and Schirmer (1995,1997) while Seeler et al (1996) demonstrated the feasibility of producing a net-shape part in a sintering process based on saturating and foaming polycarbonate powder. Meanwhile, further innovations in extrusion based microcellular processing have been reported from Japan (Shimbo et al, 1998) and Germany (Seibig et al, 2000). Thus, the race to develop industrial processes and reap the potential benefits of microcellular foams was on in the U.S., Europe, Canada, and Japan.

The end product of this research is envisioned to be an industrial-scale commercial "batch" process, which will produce, for example, 1.2 m × 2.4 m (about 4 ft × 8 ft) microcellular sheets of desired thickness and density. These sheets will have a microstructure that is tailored to the strength and stiffness requirements. The microcellular process has the capability to control material density distribution across the thickness and to create a solid skin of desired thickness that is integral to the foam itself. These materials will be the new, lightweight building blocks for future house construction.

EXPERIMENT DESIGN

In this research, a preliminary experimental study is being conducted to study the feasibility of producing thick microcellular sheets and the effects of various process variables. The ABS-CO_2 gas polymer system is chosen for this study. The batch process described above is used to create flat sheets that are 12.5 mm thick. The primary characteristics of interest include the skin thickness and the average density of the core. In additions, it is also of interest to control the bubble size distribution through the control of the gas concentration profile and/or the temperature profile. The equipment used in the experiment includes a hot press and a high-pressure vessel.

The process variables are examined in order to determine a set of potentially significant ones for further study. High and low priorities are assigned to the variables. Table 1 summarizes the results. The first four variables in the table have a high priority and will be systematically studied.

TABLE 1
VARIABLES THAT COULD AFFECT THE QUALITY
OF THICK MICROCELLULAR SHEET FOAMING PROCESS

No	Parameters	Priority
1	Saturation pressure	High
2	Temperature of platens (Foaming temperature)	High
3	Desorption time	High
4	Thickness of starting material	High
5	Force exerted by the platens	Low
6	Foaming time	Low
7	Shim thickness	Low
8	Environmental temperature during desorption	Low
9	Environmental pressure during desorption	Low
10	Composition of raw material	Low
11	Saturation temperature	Low
12	Humidity of the environment	Low

Hot press foaming is a complicated process. There is interdependency among the process variables to achieve desirable results. For example, to obtain a foamed structure, the foaming temperature is dependent on the saturation pressure. This interdependency can be observed in Fig. 3. In general, if the saturation pressure is low, the foaming temperature has to be high. Without a foamed structure, the experiment does not provide useful information on relative density and skin thickness. Therefore, conventional design of experiment, in which the settings of process variables are independently chosen, does not provide an efficient experiment plan, because they result in data points outside of the interest area or considerably smaller response area to be explored. In this study, a two-stage, sliding-level design of experiment approach (Li, et al, 2001) is used to resolve the parameter interdependency problem. In this experiment, the foaming temperature is chosen as a "slid factor" whose settings will be determined based on those of other variables. Table 2 shows the settings the process variables. The saturation pressure and desorption time are chosen as three-level variables, and the starting thickness of the unfoamed sheet as two-level variable. The foaming temperature will be determined for each of the combinations of the factor levels.

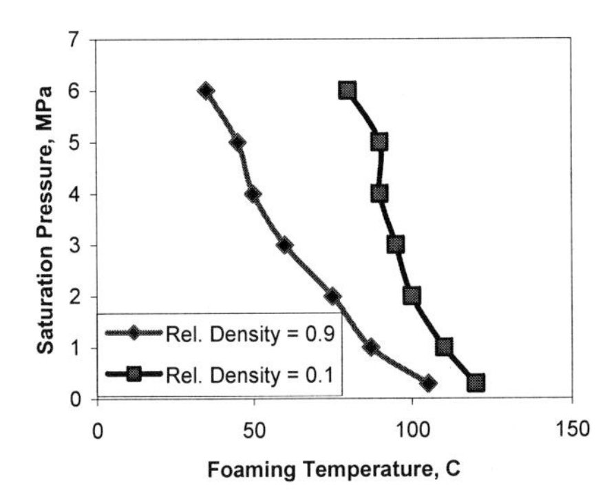

Figure 3: The interdependency between foaming temperature and saturation pressure.

TABLE 2
FACTORS AND THEIR LEVELS IN THE DESIGN OF EXPERIMENT

Factors	Levels		
	1	2	3
Saturation Pressure	1 MPa	3 MPa	5 MPa
Desorption Time	0	1	2
Starting Thickness of Unfoamed Sheet	3/16"	7/16"	--
Foaming Temperature (Temperature of Platens)	Five levels of foaming temperature at each combination of factor levels in the matrix for the above three factors		

As shown in in Table 3, given any setting of the other three variables, the maximum and minimum foaming temperatures have to be found first. These two temperatures will yield the maximum and minimum possible relative density for different pressures. For the ABS material, the minimum foaming temperature for a given saturation pressure can be estimated by using Chow's equation. The maximum foaming temperature, on the other hand, is not easy to find. Fortunately, prior experience can help with the search. For example, the maximum foaming temperature is known to be close to the glass transition temperature T_g of unsaturated ABS. It decreases as the saturation pressure increases. Once the maximum foaming temperature is determined, another three levels of temperature will be chosen equally spaced within the two extreme temperatures. Since there will be five temperature data points for each setting of other variables, the maximum and minimum temperatures do not have to be exact. Based on the above design, the total number of experiments to be conducted is 3*3*2*5 = 90.

TABLE 3
DESIGN MATRIX OF THE EXPERIMENT

Saturation Pressure (p)	Desorption Time (t)	Thickness (h)	Foaming Temperature (T)	Relative Density (ρ)
1	1	1	T_{11} T_{12} T_{13} T_{14} T_{15}	Min Foaming Temperature; Lowest relative density will be explored
1	2	1	T_{21} T_{22} T_{23} T_{24} T_{25}	
1	3	1	T_{31} T_{32} T_{33} T_{34} T_{35}	
1	1	2	T_{41} T_{42} T_{43} T_{44} T_{45}	
...	...	...	...	

PRELIMINARY RESULTS

Preliminary experiments are conducted with 0.25" thick ABS sheets using the hot press foaming process. The resulting foamed boards have the desired structure with a relative density of 30% (thus a 70% density reduction) and a final thickness of 0.385". The surfaces of the sheet are unfoamed skins and the core is uniform microcellular foam. Figures 4, 5, 6 and 7 show micrographs of the cross-section of 30% relative density ABS sheet at different magnifications. Fig. 4 is a low magnification image showing the unfoamed skin, the skin-core transition area, and the foamed core on the ABS sample. Fig. 5 shows a close up view of the skin, which is about 400 μm thick. The image of the skin-core transition area in Fig. 6 shows a gradual change from solid material near the surface on the left to the microcellular bubbles towards the core on the right. Fig. 7 shows the uniformly nucleated microcellular bubbles in the core. The size of the microcellular bubbles is in the range of 2 – 10 μm. The result of the preliminary experiments is very encouraging. It demonstrates that it is possible to not only create an ideal thick microcellular polymer sheet with desired density reduction and foam structure, but also maintain the flatness and smoothness on the surface of the board to make them viable for construction panel applications.

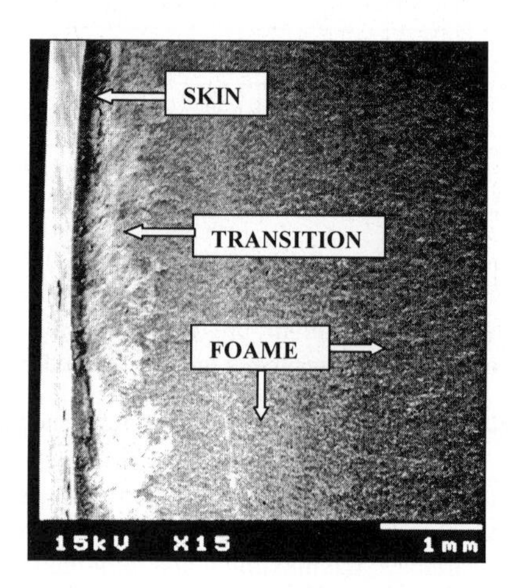

Figure 4: 30% relative density ABS sample showing skin, transition area and core.

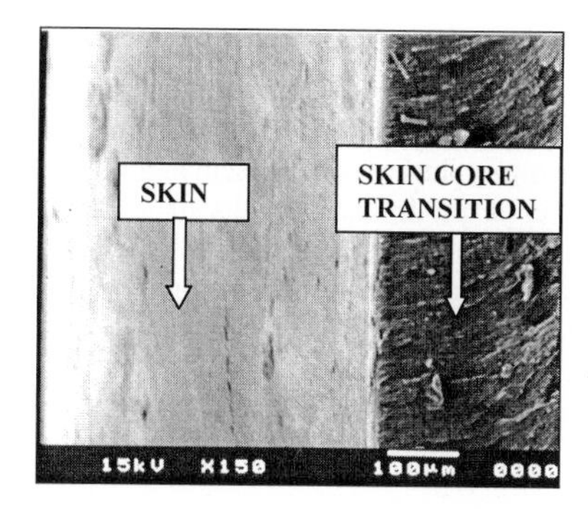

Figure 5: 30% relative density ABS sample showing close up of skin and skin-core transition area.

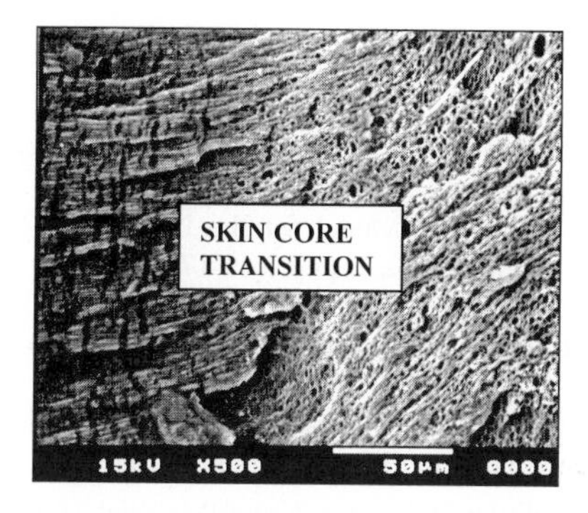

Figure 6: 30% relative density ABS sample micrograph showing close up of skin-core transition area.

Figure 7: 30% relative density ABS sample micrograph showing foamed core.

SUMMARY

This paper presents thick ABS microcellular sheets technology for lightweight, energy efficient building applications. A systematic study of various process variables is being conducted using a two-stage, sliding-level design of experiment approach. The preliminary results have shown that the foamed ABS sheet has clearly defined skins and foamed core structures. 70% relative density reduction is achieved. In general, the proposed technology is expected to be capable of reducing polymer density and thermal conductivity while increasing its specific strength. The resulted polymer materials can significantly contribute to reducing the house construction time and maintenance cost, energy consumption, and risk of injury during construction and improving the durability of the houses. The work presented in this paper is preliminary results of a long-range research in developing an industrially viable industrial process. Future work includes the study of gas polymer interaction to reduce the gas saturation time, investigation on other gas-polymer systems, and optimization of the processes.

Acknowledgement

The authors wish to acknowledge the United States National Science Foundation for the financial support of this study under Grant No. CMS-0122055.

References

1. Kumar V. and Schirmer H.G. (1997). A Semi-Continuous Process to Produce Microcellular Foams. **US patent 5,684,055**, issued November 4, 1997.
2. Kumar V. and Weller J.E. (1994). A Model for the Un-foamed Skin on Microcellular Foams. **Polymer Engineering and Science: Vol. 34,** pp. 169-173.
3. Kumar V. and Weller J.E. (1994). Production of Microcellular Polycarbonate using Carbon Dioxide for Bubble Nucleation. **ASME Journal of Engineering for Industry: Vol. 116,** pp. 413-420.
4. Kumar V., and Schirmer H. G. (1995). Semi-Continuous Production of Solid State PET Foams. **Society of Plastics Engineers Technical Papers: Vol. XLI**, pp. 2189-2192.
5. Kumar, V., (1993). Microcellular Polymers: Novel Materials for the 21st Century. **Progress in Rubber and Plastics Technology: Vol. 9,** No. 1, pp. 54-70.
6. Li W., Hu S.J., and Ni J., (2001). On-line Quality Estimation in Resistance Spot Welding. **ASME Journal of Manufacturing Science and Engineering: Vol. 122,** pp. 511-512.
7. Martini, J. Waldman F.A., and Suh N.P. (1982). The Production and Analysis of Microcellular Foam. **SPE Tech. Papers: XXVIII,** pp. 674-676
8. Martini-Vvedensky J.E., Suh N.P., and Waldman F.A. (1984). U.S. Patent # 4,473,665.
9. Park C. B. and Suh N. P., (1992). Extrusion of Microcellular Filament: A case Study of Axiomatic Design. **ASME Winter Annual Meeting, Anaheim, CA: MD-Vol. 38**, pp. 69-91.
10. Seeler K.A., Billington S.A., Drake B.D. and Kumar V. (1996). Net-Shape Forming of Sintered Microcellular Foam Parts. **1996 ASME Cellular and Microcellular Materials: MD-Vol. 76**, pp. 65-70.
11. Seibig B., Huang Q., and Paul D. (2000). Design of a Novel Extrusion System for Manufacturing Microcellular Polymer. **Cellular Polymers: Vol. 19, No. 2,** pp. 93-102.
12. Shimbo M., Nishid, K., Nishikawa S., Sueda T., and Eriguti M. (1998). Foam Extrusion Technology for Microcellular Foams. **Porous, Cellular and Microcellular Materials: MD-Vol. 82, V. Kumar, Ed.,** pp. 93-98.
13. Zhang Z. and Handa Y. P. (1998). An In-Situ Study of Plasticization of Polymers by High-Pressure Gases. **Journal of Polymer Science: Part B: Polymer Physics: Vol. 36,** pp. 977-982.

Advances in Building Technology, Volume 1
M. Anson, J.M. Ko and E.S.S. Lam (Eds.)

LONG-TERM CREEP RESPONSE OF BORATE-MODIFIED ORIENTED STRANDBOARD

Qinglin Wu and Jong N. Lee

School of Renewable Natural Resources
Louisiana State University, Baton Rouge, LA 70803, USA

ABSTRACT

Creep performance of zinc and calcium borate-modified OSB was assessed under both constant and varying moisture conditions. The influence of initial borate content, wood species, and stress level on the creep deformation was studied. Under the constant moisture condition, there was practically no difference in creep for boards at various borate levels for both types of borate. The creep data were fitted well with a spring-dashpot model. Predicted fractional creep validated the current adjustment factor up to a 30-year duration under a constant moisture content level. Under the varying moisture condition, large creep deflection developed due to the mechano-sorptive effect. The effect of borate on wood deformation became significant for both zinc and calcium borate OSB, indicating a reduced load carrying capacity of the OSB at higher borate levels. This result indicates the need for studying long-term duration of load properties of the modified OSB under combined mechanical and moisture loadings.

KEYWORDS

Borate, creep, mechano-sorptive, modeling, OSB, stress level, termite, wood composites

INTRODUCTION

Borate-modified oriented strandboard (OSB) has been developed to combat Formosan Subterranean Termite (FST) infestation on wood buildings in the southern United States (Laks and Manning 1995, Wu et al. 2002). One large uncertainty on the use of the product is its long-term structural performance under sustained loading conditions. It has been shown that boards bonded with both phenolic and isocyanate adhesives displayed a reduction in bending strength upon the incorporation of borate. Thus, durability issues of borate-modified OSB will arise both in load-bearing (e.g., OSB shear wall, roof, and I-beams) and non-load bearing (e.g., OSB siding and sheathing) situations. Because the chemical is an inorganic salt, it diffuses throughout the wood with moisture movement (LeVan and Winandy 1990). In some situations, such as roofs, elevated temperatures and humidity changes cause shifts in the equilibrium moisture content (MC) of the wood. As the moisture moves, so does the inorganic salt. This cycling could cause migration of the salt within the wood. At each new site, the acidic salt can cause wood degradation. It is well-known that cyclic MC changes accelerate creep due to mechano-sorptive (MS) effect. In the presence of borate chemicals, creep may be further accelerated and degradation of other mechanical properties becomes more pronounced.

Creep studies on untreated structural wood composite panels were well documented (Laufenberg 1988, McNatt and Laufenberg 1991, and Fridley et al. 1992). Various analytical models have also been proposed to describe the observed behavior under combined mechanical and moisture loading (Ranta-Maunus 1975, Fridley et al. 1992). However, creep data on treated OSB are completely lacking and the applicability of the creep models to describe its time-dependent behavior has yet to be evaluated. This work forms part of a larger project on investigating long-term durability of borate-modified OSB from southern wood species. The specific objective of this study was to investigate the creep response of the modified OSB as influenced by wood species, borate type, borate content level, moisture cycling, and other panel processing variables.

MATERIAL AND METHODS

Panel production and sample preparation

The details of panel manufacturing are presented in Wu et al. (2002). In brief, green boards from eight southern wood species were selected for the study. These species included ash (Fraxinus spp.), cottonwood (Populus spp.), cypress (Taxodium distichum L.), elm (Ulmus americana L.), locust (R. pseudoacacia L.), pecan (Carya spp.), red oak (Quercus spp.), and southern pine (Pinus taeda L.). The boards were cross-cut into 152-mm sections, which were flaked to produce 76-mm long flakes using a disc flaker. The flakes were dried to 2-3% MC and screened to eliminate fines prior to panel manufacturing. The wood flakes were used to manufacture single species of OSB for zinc borate (ZB) and mixed hardwood and southern pine OSB for calcium borate (CB).

During the panel manufacturing, certain amounts of dry wood flakes, liquid phenol formaldehyde resin, wax, and borate were blended together. The loading rates for resin and wax were 4 and 1% based on the oven-dry wood weight. There were one type of ZB and two types of CB with two different particle sizes. The borate was loaded at several levels to achieve target BAE (Boric Acid Equivalent) levels (Wu at al. 2002). The formed mats were hot pressed in a single opening press with a regulated platen temperature of 200°C for 5 minutes. The target thickness and panel density were 12.7 mm and 0.75 g/cm^3, respectively. Tests were done to determine basic panel properties as shown in Wu et al. (2002). Creep samples were cut from the experimental panels. The samples were conditioned. Their weight and dimension were measured for the determination of sample density prior to the creep measurements. Four samples from each board type were randomly selected to test their static bending modulus of rupture (MOR). The mean MOR value from each group was used to determine actual stress levels (SLs) for the creep test.

Creep testing

A twenty-four position loading frame for constant load creep tests was constructed (Figure 1). The frame has two levels with twelve samples at each level arranged side by side. Each sample was supported and dead weight (lead) was applied to the sample at the central point of the span with a pneumatic loading device for loading and unloading. One linear variable differential transducer (LVDT) was used to measure sample deflections at each level. During measurements, each LVDT was traveling to individual samples with a miniature carriage on two precision stainless steel rods. One reference plate was set on each level for positioning the LVDT before each measurement. The deflection data (D in Figure 1) were collected through a Strawberry-Tree DataShuttle card controlled by a specially designed Visual Basic program. The program enables a three-way communication with the WorkBench program (data acquisition) and Microsoft Excel (data processing) using dynamic data exchange tools (Figure 1). During testing, the loading frame was placed in a climate-controlled laboratory. Stress levels equivalent to 15%, 25%, and/or 40% of the MOR of each board type were applied to the specimens. Several unloaded samples from each board type were used to monitor sample MC and thickness change.

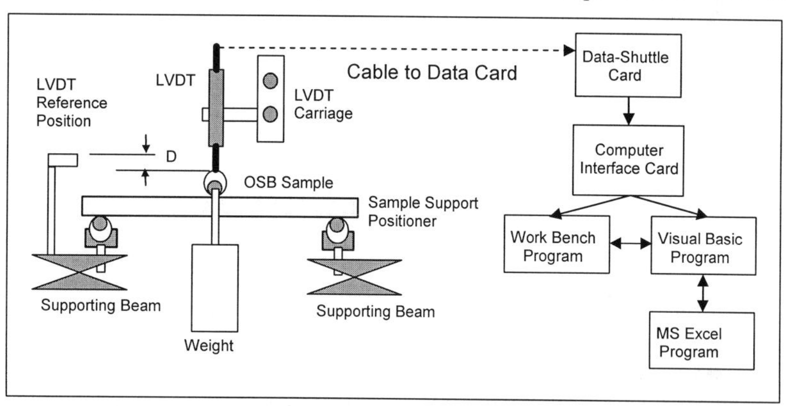

Figure 1. A Schematic of the Loading Frame and Data Acquisition System for Creep Testing

Data reduction and modeling

The measured deflection data were converted to strain using the geometric relationship for three point bending. The effect of sample thickness swelling (TS) on measured deflection during moisture cycling was accounted in the strain calculation. A four-element spring and dashpot model was used to fit the measured creep data under constant MC condition (Pierce and Dinwoodie 1977). The model, consisting of elastic (e), viscoelastic (k), and viscoplastic (v) deformations, has the following form:

$$\varepsilon_{Creep}(t) = \varepsilon_e + \varepsilon_K + \varepsilon_V = \frac{\sigma}{K_e} + \frac{\sigma}{K_k}\left\{1 - \exp\left[-\frac{K_k t}{\eta_k}\right]\right\} + \frac{\sigma t}{\eta_V} \quad (1)$$

where, ε represents deformation, t is time, σ is the applied stress, K_e is the elastic spring constant, K_k and η_k are the spring constant and viscosity for the viscoelastic deformation, and η_f

is the viscosity of dashpot for the viscoplastic deformation. A nonlinear curve fitting technique was employed to determine the model parameters (Pierce and Dinwoodie 1977). The model parameters were then used to predict creep, recovery strain, and fractional creep over an extended time period (i.e., 30 years).

To model the mechano-sorptive effect under changing MC condition, an additional element was added to the creep model (Equation 1) following the approach proposed by Fridley (1990). The MS element has a form (Fridley et al 1992):

$$\varepsilon_{ms}(t_\omega) = \frac{\sigma}{\mu'_{ms}} |\Delta\omega| \, [1 - \exp(-B_\omega t_\omega)] \tag{2}$$

where, $\Delta\omega = \omega_e - \omega_i$ with ω_e as the final equilibrium MC and ω_i as initial MC, μ'_{ms} is viscosity of the MS element, σ is applied stress, t_ω is elapsed time for MC change, and B_ω is a constant related to the time required for reaching moisture equilibrium. Equation 1 can be rewritten with the MS element as:

$$\varepsilon_{Total}(t) = \frac{\sigma}{K_e} + \frac{\sigma}{K_k}\left\{1 - \exp\left[-\frac{K_k t}{\eta_k}\right]\right\} + \frac{\sigma t}{\eta_V} + \frac{\sigma}{\mu'_{ms}}|\Delta\omega|[1 - \exp(-B_\omega t_\omega)] \tag{3}$$

to describe the total strain under combined mechanical and moisture loading. The MS component was actually modeled by first determining the MC factor B_w in Equation 2. This was done by fitting the measured MC data as a function of time with the exponential function. The MS strain was obtained by subtracting the predicted creep component from the total strain. Assuming a piece-wise linear relationship between MS strain and MC change in a given time step, the coefficient, μ'_{ms}, was determined through a linear regression analysis.

RESULTS AND DISCUSSION

Creep strain

Typical creep curves at the 40% stress level are shown in Figure 2 for zinc borate OSB. The data represent an average of four replicate samples for each board type. The sample MC remained about 6% throughout the test period. For both southern pine and mixed hardwood OSBs, the observed creep was generally higher at the higher borate loading levels. This is very obvious for zinc borate modified hardwood OSB shown in Figure 2B. However, a statistical analysis on instantaneous, total, and permanent plastic strains from various species indicated an insignificant effect of the borate level on the creep deflection. The effect of load level on the observed creep strain is shown in Figure 3 for single species OSB from southern pine (A) and red oak (B). Compared with data at the 15% SL, the increase in the strain at the 40% SL was mainly due to the initial elastic strain for both species. Thus, the creep behavior at the SLs used was basically linear. This is true regardless the borate type and its content level. The effect of wood species on creep under the influence of load level, borate type and content was also inconsistent among the species. OSB made of mixed hardwoods and calcium borate showed smaller strain than those of southern pine. This was mainly due to the difference in the initial moduli of the panels, which were influenced largely by panel density and other processing variables. Therefore, the creep behavior of borate-treated OSB under constant MC conditions was not practically different from those of untreated controls.

The four-element creep model (Equation 1) fitted the individual creep data set well (Lines in Figures 2 and 3). The model parameters varied with borate type, borate content level, and wood species. The model parameters allow predicting creep performance of the OSB under similar loading and environmental conditions.

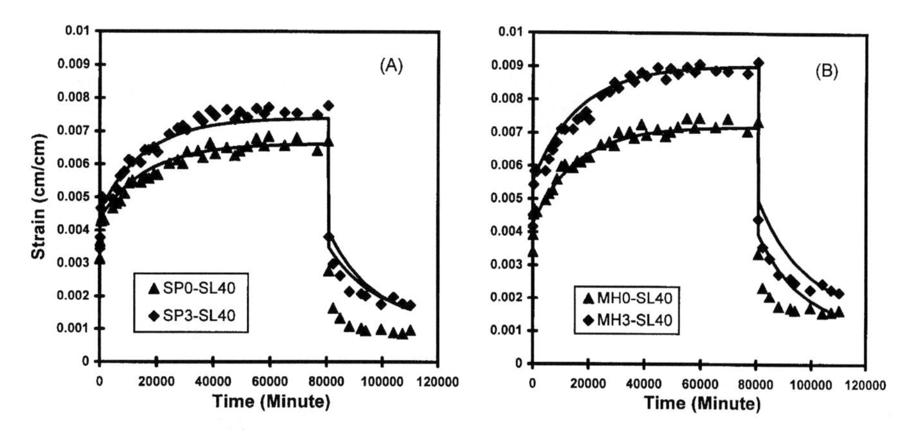

Figure 2. Creep Response of Zinc-Borate Modified OSB under the 40% Stress Level. A) Southern Pine with BAE= 1.78% for SP3 and B) Mixed Hardwoods with BAE=1.72% for MH3. The Lines Show The Predicted Values By The Model (Equation 1).

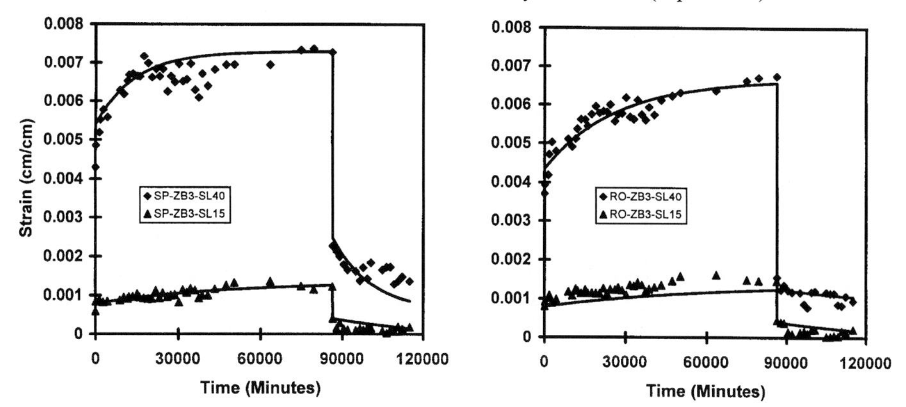

Figure 3. Effect of Load Level on Creep Response of Zinc Borate Modified OSB from Southern Pine (Left) and Red Oak (Right) under the 40% and 15% Stress Levels. BAE=1.95% for SP and 1.265% for RO. The lines show the predicted values by the creep model (Equation 1).

Figure 4 shows predicted fractional creep for southern pine and red oak OSB at various borate levels. The predicted fractional creep under the specific test condition meets the recommended level in the national design standards for wood construction (NFPA 1991). Thus, the current design practice can be applied to borate-modified OSB up to a 30-year service period. After 30 years, a rapid increase of the fractional creep was predicted at both of the stress levels.

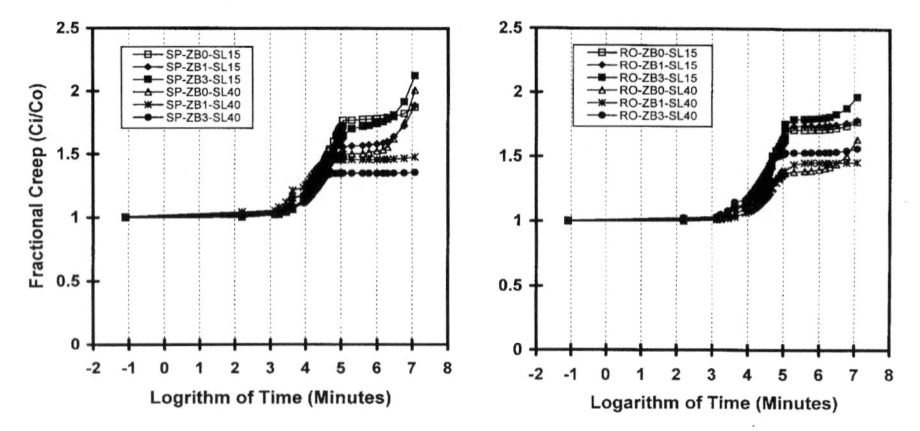

Figure 4. Predicted Fractional Creep for Single-Species OSB from Southern Pine (Left) and Red Oak (Right) under the 15% and 40% Stress Levels. BAE= 1.95% SP-ZB3, 0.89% SP-ZB1, 1.265% for RO-ZB3 and 0.39% for RO-ZB1.

Mechano-sorptive Strain

Typical plots showing time-dependent fractional deflection under varying MC conditions are shown in Figure 5. Tests consisted of an 18-day creep under a constant MC (8%), a 27-day wetting (from 8% to 23% MC), and a 12-day drying (from 23% to 12.5% MC) under load. The mean maximum MC reached during the wetting cycle was somewhat lower from the mixed hardwood group compared with the southern pine group.

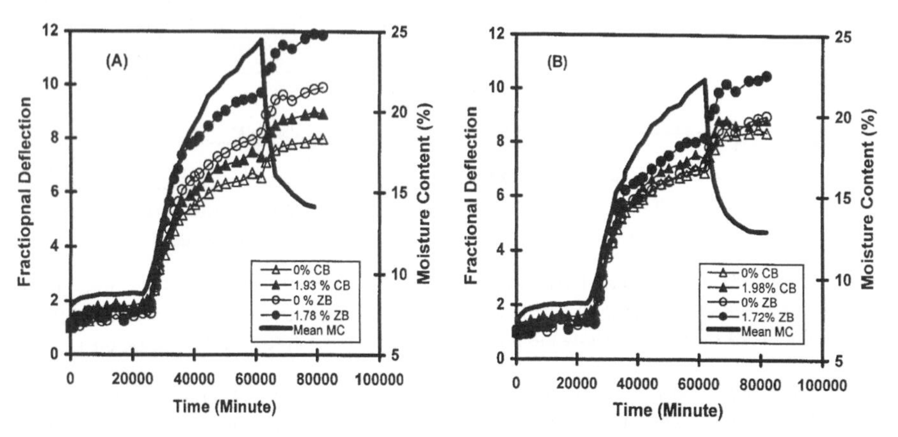

Figure 5. Mechano-sorptive Response (Fractional Deflection) of Zinc and Calcium Borate Modified OSB from Southern Pine (A) and Mixed Hardwoods (B) under the 25% Stress Level.

Creep at 8% MC developed in a regular manner. MC increases and decreases under load led to large deflections in all groups (i.e., MS effect). The effect of borate on the fractional deflection is obvious, considering difference between treated and untreated groups for either ZB or CB OSB. The control groups for ZB and CB OSB showed a significant difference in creep deflection. This was due to difference of initial panel modulus between the groups as those boards were made separately.

Significant TS occurred during testing especially for CB treated OSB. The TS had a significant negative effect on the deformation. Figure 6 shows predicted strain using Equation 3 based on model parameters from each group (ZB or CB) under the same stress level. For both zinc and calcium borates, all treated groups showed a significant larger strain compared with the control group. CB showed a larger negative effect on the deformation than ZB due to its effect on swelling. The increased creep deformation will definitely affect long-term load carrying capacity of the modified product under varying moisture conditions. Thus, a proper account of the MS effect on long-term duration of load behavior of treated OSB has to be made. Further testing and modeling on the MS deformation of the OSB are still on-going.

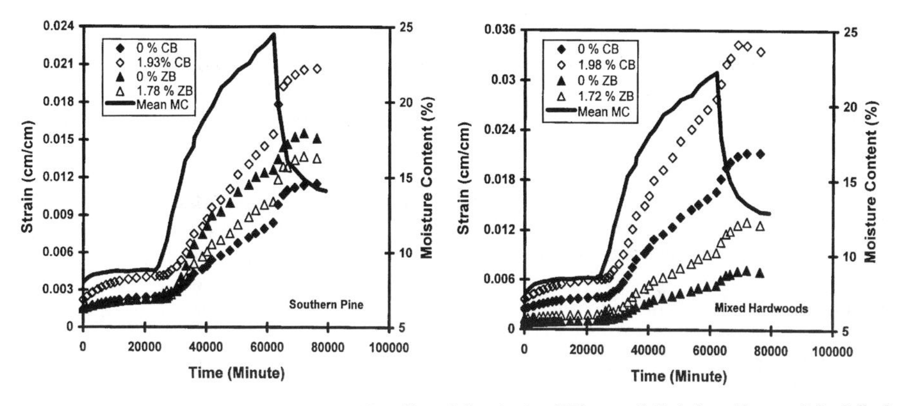

Figure 6. Mechano-Sorptive Response (Predicted Strains) of Zinc and Calcium Borate Modified OSB from Southern Pine (A) and Mixed Hardwoods (B) under the 25% Stress Level.

CONCLUSIONS

Creep behavior of borate-modified OSB was investigated under controlled environments. The experimental work was carried out with OSB made of zinc and calcium borate. A four-element spring-dashpot model was used to fit the creep data under constant moisture condition. The model was combined with an MS element to predict measured strain under varying MCs. There was practically no difference in creep for boards at various borate levels under the constant moisture condition. Predicted fractional creep validated the current adjustment factor up to 30-year duration under a constant moisture content level. However, the added borate led to a significant larger deformation under varying moisture condition due to the mechano-sorptive effect and the deteriorating effect of borate on the thickness swelling properties of the OSB. This indicates the need for studying long-term duration of load properties of treated OSB under combined mechanical and moisture loadings.

ACKNOWLEDGMENTS

This material is based upon work supported by the National Science Foundation under Grant No. 0080248. Any opinions, findings, and conclusions or recommendations expressed in this material are those of the authors and do not necessarily reflect the view of the National Science Foundation.

REFERENCES

Fridley, K. J., R.C. Tang, and L.A. Soltis. 1992. Creep behavior model for structural lumber. J. of Structural Engineering 118(8):2261-2277.

Fridley, K.J. 1990. Load-duration behavior of structural lumber: effect of mechanical and environmental load histories. Ph.D. Dissertation, Auburn University. 316 pp.

Laks, P.E., and M.J. Manning. 1995. Preservation of wood composites with zinc borate. IRG/WP 95-30074. Intl. Res. Group on Wood Pres., Stockholm, Sweden.

Laufenberg, T. L. 1988. Composite products rupture under long-term loads: a technology assessment. In Proc. 22nd International Particleboard /Composite Materials Symposium, T. M. Maloney Ed. Washington State University, Pullman, WA. pp. 247-256.

LeVan, S.L. and J.E. Winandy. 1990. Effects of fire retardant treatments on wood strength: a review. Wood Fiber Sci. 22(1):113-131.

McNatt, J.D., and T. L. Laufenberg. 1991. Creep and creep-rupture of plywood and oriented strandboard. In Proc. of International Timber Engineering Conference, London, England. pp. 3457-3464.

National Forest Products Association (NFPA). 1991. National design specifications for wood construction. AFPA. Washington D.C.

Pierce, C.B., and J.M. Dinwoodie. 1977. Creep in chipboard. Part 1. Fitting 3- and 4-element response curves to creep data. J. Materials Sci. 12:1955-1960.

Ranta-Maunus. 1975. The viscoelasticity of wood at varying moisture content. Wood Sci. Technol. (9):189-205.

Wu, Q., S. Lee, and J. N. Lee. 2002. Mechanical, physical, and biological properties of borate-modified oriented strandboard. In Proc. International Conference on Advances in Building Technology. Hong Kong. December 4-6, 2002.

Advances in Building Technology, Volume 1
M. Anson, J.M. Ko and E.S.S. Lam (Eds.)

MECHANICAL, PHYSICAL, AND BIOLOGICAL PROPERTIES OF BORATE-MODIFIED ORIENTED STRANDBOARD

Qinglin Wu, Sunyoung Lee, and Jong N. Lee

School of Renewable Natural Resources
Louisiana State University, Baton Rouge, LA 70803, USA

ABSTRACT

Borate-modified oriented strandboard (OSB) was manufactured using zinc and calcium borate to provide resistance to Formosan Subterranean Termites (FSTs). Mechanical, physical, and biological properties of the product were evaluated. It was shown that static bending properties of the OSB were affected little at the room condition by the added borate. The internal bond strength was, however, generally smaller at the higher borate loading levels. Thickness swelling (TS) under the 24-hour water soaking increased with increased borate content, especially for calcium borate. Part of the borate leached out under the water soaking condition due to the glueline washing and decomposition of the borate to form less water-soluble boric acid. The use of borate with a smaller particle size helped reduce TS and leaching rate significantly. Laboratory no-choice termite tests indicated that both zinc and calcium borate modified OSB resisted FSTs well. Thus, termite resistant OSB with required mechanical and physical performance could be successfully developed using southern wood species. This technology will allow more OSB producers to manufacture chemically modified OSB to meet increasing market demands.

KEYWORDS

Borate, leaching, treated wood composites, stiffness, strength, swelling, termites

INTRODUCTION

The Formosan Subterranean Termites (FSTs) pose a growing threat to all structural wood materials in residential construction. The species is one of the most aggressive and voracious insects in the world, eating cellulose - a main component of wood material. FSTs cause over $1 billion structural damage per year in the United States, with a large portion of that in Louisiana (Henderson and Dunaway 1999, Ring1999.) Chemical treatments of wood members with chromated copper arsenate (CCA) and borate have been shown to be effective against the termites (Laks 1988, Barnes et al. 1989). Structural lumber and plywood can be successfully treated after manufacturing (e.g., pressure treatments with CCA). Structural composite panels such as oriented strandboard (OSB), however, cannot be pressure-treated once it is made into panel form due to its large swelling characteristics. The product is made of wood flakes glued with a thermal-setting resin. It is widely used as sheathing, flooring, and I-joist materials in light-frame wood construction, replacing more traditional plywood. Thus, alternative techniques for protecting OSB against FSTs have to be developed.

Work has been done to combine powder borate with wood flakes during the manufacturing of OSB to provide termite resistance of the finished products (Laks et al. 1988; Laks et al. 1991; Sean et al. 1999). However, information on the effect of wood species, borate type, borate particle size, initial borate content level, and other panel processing variables on long-term durability of borate-treated OSB is still missing. The information is highly desired for developing durable structural panels for residential construction using southern wood species.

This project was conducted to incorporate powder borate into OSB furnish in order to provide required biological performance characteristics of OSB. The objectives of this work were to investigate the effects of wood species, borate type and content on short-term loading stiffness and strength, swelling, water leaching, and biological performance of the product. Long-term structural performance of the modified OSB under sustained loading conditions is discussed in other related publications.

MATERIAL AND METHODS

Panel Manufacturing and Testing

Green boards from eight southern wood species were selected from a local sawmill in south Louisiana. These species included ash (Fraxinus spp.), cottonwood (Populus spp.), cypress (Taxodium distichum L.), elm (Ulmus americana L.), locust (R. pseudoacacia L.), pecan (Carya spp.), red oak (Quercus spp.), and southern pine (Pinus taeda L.). The boards were cross-cut into 152-mm sections, which were flaked to produce 76-mm long flakes using a disc flaker. The flakes were dried to 2-3% moisture content. They were screened to eliminate fines and stored in polyethylene bags until needed. The wood flakes were used to manufacture single species of OSB for zinc borate (ZB) and mixed hardwood and southern pine OSB for calcium borate (CB).

During the panel manufacturing, certain amounts of dry wood flakes, liquid phenol formaldehyde resin, wax, and borate were blended together. The loading rates for resin and wax were 4 and 1% based on the oven-dry wood weight. There were one type of zinc borate ($2ZnO3B_2O_33.5H_2O$) and two types of calcium borate ($Ca_2B_6O_{II}5H_2O$) with two different particle sizes. Both types of borate are considered as water insoluble at room temperature. The loading rates for zinc borate were 0 (control), 0.5, 1, and 3% based on dry flake weight in the panel. The loading rates for calcium borate were 0 (control), 0.75, 1.5, 3, and 4.5%. Several single species panels with zinc borate were made with addition of PEG, $H(OCH_2CH_2)_{n>4}OH$, to study its effect on panel properties. The formed mats were hot pressed in a single opening press with the regulated platen temperature of 200°C for 5 minutes. The target thickness and panel density were 12.7 mm and 0.75 g/cm^3, respectively. Two replicate panels at each borate level were made. Samples were taken from each panel to test its chemical content. Each sample was ground to pass through a 20-mesh screen with a Wiley mill. Approximately 5 g oven-dry wood meal was extracted under acid condition for 2 hours at boiling temperature. After extraction, the wood meal was filtered out and the liquid filtrate was analyzed with an Inductively Coupled Plasma (ICP) spectrograph. The percent of boron in the sample was calculated based on the oven-dry weight of wood meal. The result was expressed as boric acid equivalent (BAE)

Tests were conducted to determine the panel's static bending stiffness and strength, internal bond (IB) strength, and thickness swelling (TS) according to the American Society for Testing and Materials standard D1037-96 (ASTM 1998). Water leaching experiments were done to evaluate the leachability of the modified products. Laboratory no-choice tests were conducted according to the American Wood Preservation Association (AWPA) Standard E1-97, Standard Method for Laboratory Evaluation to Determine Resistance to Subterranean Termites (AWPA 1997) to evaluate termite resistance of the products.

RESULTS AND DISCUSSION

Mechanical Properties

The specific modulus of elasticity (static bending modulus/sample specific gravity) was affected little at room condition (i.e., 5% moisture content at 70°F temperature) by borate up to the 3.5% BAE level (Figure 1A for zinc borate and Figure 2A for calcium borate).

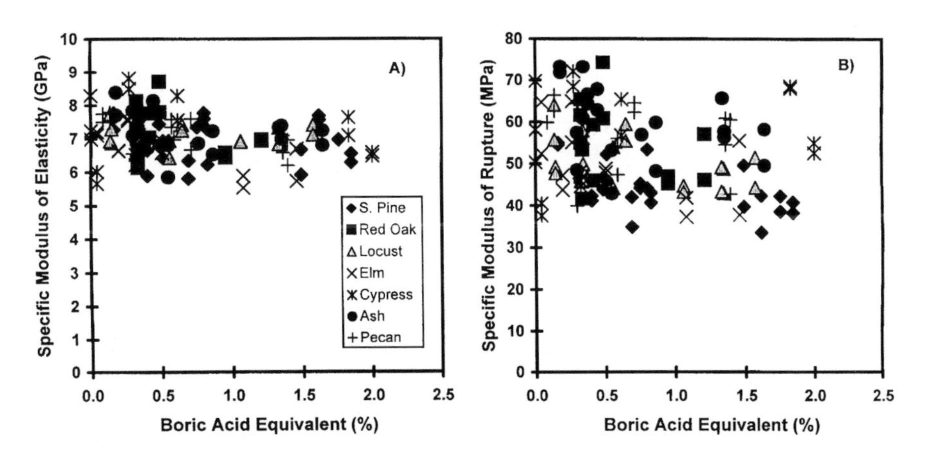

Figure 1. Bending Properties of Zinc Borate OSB (A: Bending Modulus of Elasticity/Specific Gravity and B: Bend Modulus of Rupture/Specific Gravity)

There was some reduction for the specific modulus of rupture (i.e., bending strength/sample specific gravity) at higher borate loading levels (Figure 1B for zinc borate and Figure 2B for calcium borate), indicating a negative effect of borate on panel strength. Wood species and borate type had an insignificant influence on both properties as shown in the graphs.

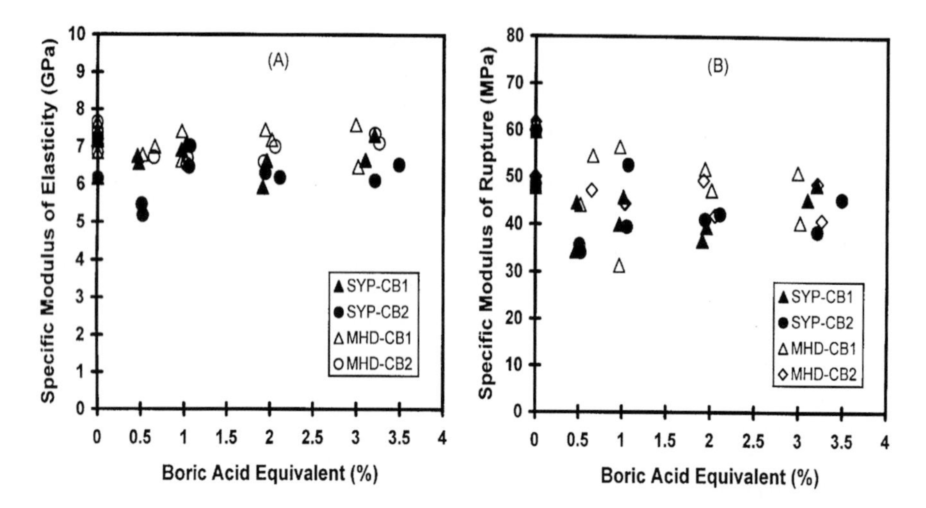

Figure 2. Bending Properties of Calcium Borate OSB (A: Bending Modulus of Elasticity/Specific Gravity and B: Bend Modulus of Rupture/Specific Gravity)

The effect of borate on the specific IB strength varied with borate type and wood species. Zinc borate generally showed less negative effects on the IB values (Figure 3A), compared with calcium borate (Figure 3B) for hardwood OSB. Southern pine OSB showed an IB reduction at higher BAE levels for both zinc and calcium borates. For calcium borate, the particle size had some influence on the IB strength. At a comparable BAE level, CB2, which had a smaller particle size, led to a higher IB strength compared with boards made of CB1, which had a larger particle size. For both types of borate, acceptable bending stiffness, strength, and IB values (based on the industry standard) can be achieved.

Physical Properties

Thickness swelling from the 24-hour water soaking (Figure 4) generally increased with borate content in the panel for both ZB and CB OSB, indicating a negative effect of borate on the panel properties under high moisture content levels. For the given exposure condition, ZB-modified OSB had a smaller TS than CB OSB at a comparable BAE level. Borate particle size had a significant influence on the swelling properties. Calcium borate with a larger particle size (CB1) had large TS, especially at the higher BAE levels. However, reducing the particle size of the chemical (CB2) helped bring the thickness welling to a stable and acceptable level (Figure 4B). Wood species had a large influence on TS for single species OSB with ZB. The single species hardwood OSB had relatively smaller TS.

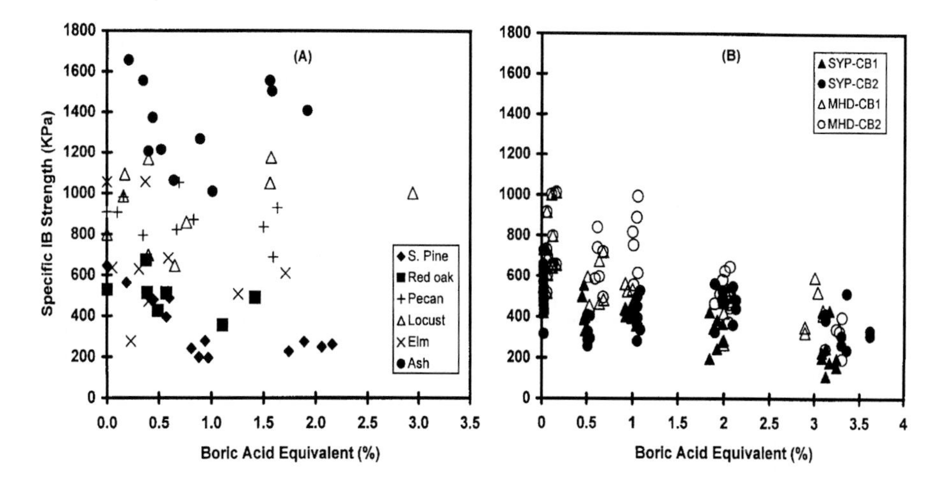

Figure 3. Internal Bond Strength of Zinc (A) and Calcium (B) Borate OSB. The Value Shown Represents A Ratio of Internal Bond Strength and Sample Specific Gravity.

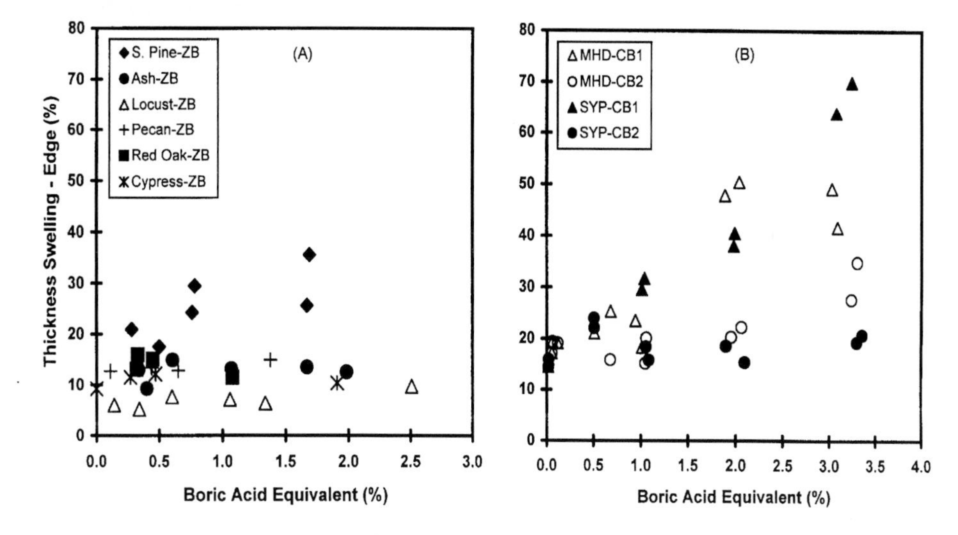

Figure 4. A Comparison of the 24-hour Water Soaking Thickness Swelling Properties of Zinc (A) and Calcium (B) Borate OSB.

A significant amount of borate leached out under the water soaking condition (Figure 5). The leaching of the OSB occurred upon the initial water exposure, and the leaching rate decreased as the leaching time increased. Wood species, borate type, initial BAE level, and sample TS significantly influenced the leaching rate. This was due to a combined effect of water washing on the glueline under the swelling condition and decomposition of the borate to form a more water-soluble chemical (i.e., boric acid). Calcium borate with a smaller particle size (CB2) helped reduce sample thickness swelling and leaching rate. There was no consistent effect of polyethylene glycol (PEG) on zinc borate leaching. Thus, boron fixation with other chemical agents is necessary for borate-modified OSB under the extreme water exposure condition. Protection of the panel from direct water exposure can help reduce this problem significantly. A unified method specifying sample size and exposure condition for testing the leachability of treated wood composite materials is highly needed.

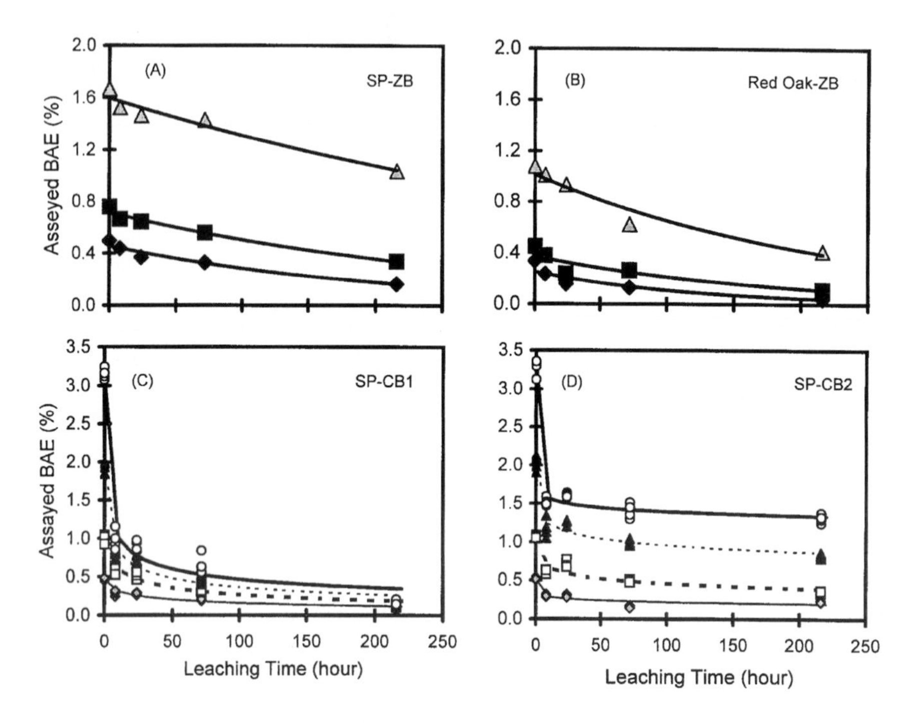

Figure 5. Leaching Properties of Zinc (A and B) and Calcium (C and D) Borate OSB.

Biological Performance

Laboratory no-choice termite tests indicated that both zinc and calcium borate modified OSB resisted FSTs well (Figure 6 and Table 1 for ZB OSB). As the borate loading increased, termite mortality rate (Figure 6A) increased and wood sample weight loss (Figure 6B) decreased. At the higher borate levels, there was little damage on wood samples, showing a significant deterring effect of boron on termites. The observed weight loss in these samples was due to loss of volatile materials in the samples. A similar behavior was seen for calcium borate OSB. Wood species showed an insignificant effect. There were strong correlations among the visual damage rating as stipulated in the AWPA standards (AWPA 1997), wood sample weight loss, and termite mortality. The established correlations among the three variables allow predicting FST damage and mortality based on visual rating for treated OSB at various borate levels.

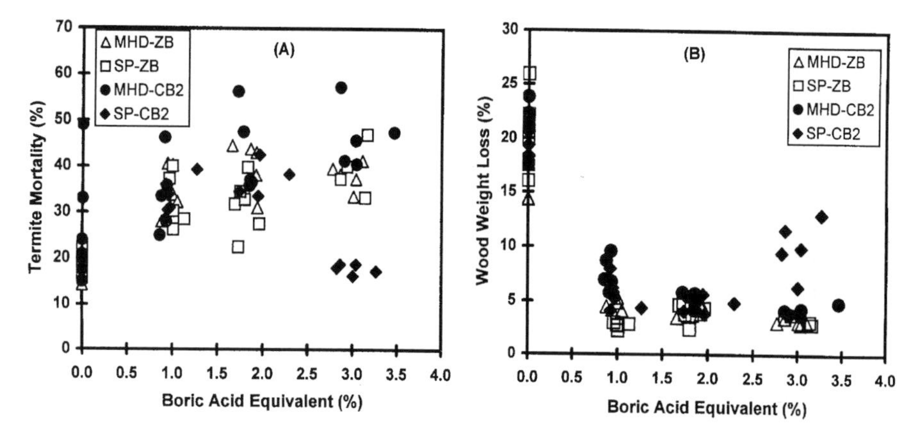

Figure 6. Termite Resistance Properties of Zinc and Calcium Borate OSB. A) Termite Mortality and B) Wood Sample Weight Loss.

Table 1. Results of Laboratory Termite Tests for Zinc-Borate Modified OSB from Southern Pine and Mixed Hardwoods.

Wood Species	Zinc Borate Weight [1] (%)	Sample Density (g/cm^3)	Visual Damage Rating [2,3] (1-10)	Wood Weight Loss (%)	Termite Mortality (%)
Mixed Hardwood	0	0.74	2.96c	16.48a	17.50b
	1.5	0.69	8.54b	4.58b	34.95a
	3.0	0.59	8.52b	4.17b	40.05a
	4.5	0.71	9.54a	3.08b	37.95a
Southern Pine	0	0.74	2.36c	21.02a	19.50b
	1.5	0.72	9.80a	2.70b	32.45a
	3.0	0.70	8.86b	3.84b	32.25a
	4.5	0.67	9.18ab	3.51b	37.00a

[1] Target weight percentage based on dry flake weight used for manufacturing the board.
[2] 1 showing the most amount of damage and 10 showing the least amount of damage.
[3] Each mean represents five replicates. Means within each column followed by the same letter are not significantly different based on the Tukey's multiple range tests at the 5% significance level.

CONCLUSIONS

Borate-modified OSB using southern wood species was manufactured and their performance characteristics were evaluated according to the industry test standards. The results of this study indicated that termite-resistant structural OSB with desired mechanical and physical performance could be successfully developed with a right combination of wood species, borate type and content, and other processing variables. The results provide comparative properties between zinc and calcium borate modified OSB and an alternative treating method for structural OSB (i.e., using calcium borate). This technology will allow more OSB producers to manufacture chemically modified OSB to meet increasing market demands.

ACKNOWLEDGMENTS

This material is based upon work supported by the National Science Foundation under Grant No. 0080248. Any opinions, findings, and conclusions or recommendations expressed in this material are those of the authors and do not necessarily reflect the view of the National Science Foundation. The authors also wish to thank several wood products, resin, and chemical companies for their material and technical support.

REFERENCES

American Society for Testing Materials (ASTM). 1998. Standard test methods for evaluating properties of wood-based fiber and particle panel materials, D 1037-96a. Philadelphia, Pa.

American Wood Preservation Association (AWPA). 1997. Standard Method for Laboratory Evaluation to Determine Resistance to Subterranean Termites, E1-97. AWPA, Woodstock, MD.

Barnes, H.M., T.L. Amburgey, L.H. Williams, and J.J. Morrell. 1989. Borates as wood preserving compounds: The status of research in the United States. IRG/WP/3542. International Research Group on Wood Preservation, Stockholm.

Henderson, G., and C. Dunaway. 1999. Keeping Formosan termites away from underground telephone lines. Louisiana Agriculture 42:5-7.

Laks, P.E. 1988. Wood preservations – looking ahead. The Construction Specifier: 10:61-69.

Laks, P.E., B.A. Haataja, R.D. Palardy, and R.J. Bianchini. 1988. Evaluation of adhesives for bonding borate-treated flakeboards. Forest Prod. J. 38(11/12):23-24.

Laks, P.E., X. Quan, and R.D. Palardy.1991. The effects of sodium octaborate and zinc borate on the properties of isocyanate-bonded waferboard. In Proc. Adhesives and Bonded Wood Symposium (C.Y. Hse and B. Tomita, eds.) FPRS. PP. 44-57.

Ring, D. 1999. Need for integrated pest management of the Formosan Subterranean Termite. Technical Note, LSU Agricultural Center. 3 pp.

Sean, T., G. Brunnett, and F. Cote. 1999. Protection of oriented strandboard with borate. Forest Prod. J. 49(6):47-51.

TEAM PRESENTATION:

COMMISSIONING AND ENERGY MANAGEMENT

Advances in Building Technology, Volume 1
M. Anson, J.M. Ko and E.S.S. Lam (Eds.)

Re-commissioning of an Air Handling Unit

Cristian CUEVAS*, Jean LEBRUN*, Patrick LACÔTE, Philippe ANDRE**

*University of Liège – Laboratory of Thermodynamics
**Fondation Universitaire Luxembourgeoise – Department of Environmental Monitoring

1. Presentation of the building and HVAC system

The building, the Administrative Center of the Walloon Ministry of Equipments and Transport (CA-MET), is designed for a one thousand occupants and is made of 13 modules representing together 15 000 m^2 of gross floor area. Each module can be subdivided in three sections (figure 1, 2) :

1. The southern building
2. The atrium (or the interior street)
3. The northern building

Figure 1 : General view of the CA-MET in Namur, Belgium

The division appears different for the central module, which constitutes a "welcome area" for the whole building. Most of the useful area of the building consists in offices distributed on four rows (two in each building, separated by a corridor) and three to five levels.

The CA-MET building is characterized by a complex HVAC plant made of the following components:

- Centralized production of heating and cooling using :
- 3 boilers (operating in cascade)
- 2 chillers (reciprocating compressors with air condensers) situated at both ends of the building

- Heating and cooling power is distributed through collectors to 14 substations (figure 4). Most of the substations, divided in northern and southern parts are feeding two building modules. In each substation, distribution of heat is organized from local collectors who are feeding the different entities connected thereto.

- In each substation, the AHU's feeding the offices and the atrium are connected to each other in that a fraction of the air extracted from the offices is ventilated in the atrium. The part that is not injected in the atrium AHU's is re-circulated in the offices or extracted through the toilets. Ventilation in the atrium is happening at constant flow rate (CAV system) while ventilation in the offices is made through a VAV system. A constant fresh air flow rate is furthermore provided in the office. Figure 2 shows a schematic diagram of a coupled offices-atrium AHU. Some rooms (meeting-rooms) are provided with fan-coils which ventilate air, pre-heated at 20° C.

- The Air Handling Units are made of the following components: mixing box, filter, heating coil, cooling coil (not present in the fan coils AHU's), humidifier, fan

- Energy is distributed in the room by means of radiators (heating) and VAV boxes (cooling and ventilation). Thermostatic valves or VAV terminals provide a local control.

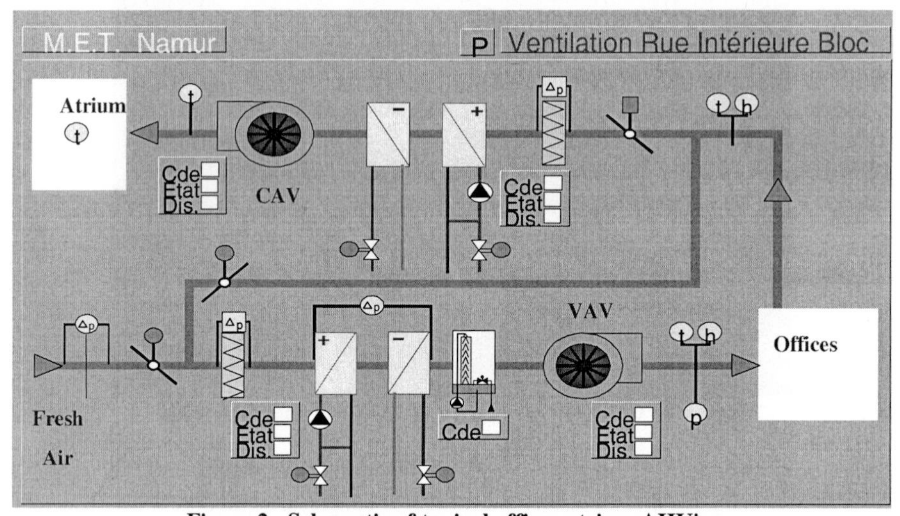

Figure 2 : Schematic of typical offices-atrium AHU's.

The Building Energy Management system is controlling both the Air Handling Units and the radiators heating circuits. For AHU's, the strategy consists in controlling the valve of the heating or cooling coil in order to modify the supply air temperature according to the return air and the outside air temperature conditions.
The control is thus performed following two steps :

- Return temperature set point calculation (including summer weather compensation)

- Supply air temperature set point calculation from the difference between return temperature measurement and its set point (including winter weather compensation).
- Valves signal control (by a classical PI algorithm).

In order to assume a minimum air flow rate into the offices whatever the opening of the VAV terminals, speed of the fan is regulated in order to keep a constant static pressure at the its outlet.

For radiators heating circuits, control is changing the water flow rate depending upon both the external conditions (temperature) and the internal climate (reflected by a room temperature sensor).

2. MEASUREMENTS

The AHU studied was the GP Bs1, which supplies air to the modules B and C. Figure 2 shows a schematic of the AHU, which supplies an airflow rate varying between 8600 and 18900 m^3/h.

A constant part of this air flow rate (3400 m^3/h) is fresh air; the (variable) remaining part is recirculation air. The non-recirculated part of the air, extracted from the offices, is used to feed the atrium AHU GP Bs2.

During the tests presented here, the AHU GPBSs2 was cut off, in order to make easier the study only the performance of GPBs1.

In air side, the following variables were measured:

ΔP_{hx} : Pressure drop through the heat exchangers, Pa

$P_{st,AHU}$: AHU exhaust air static pressure, Pa

$T_{exhaust\ AHU}$: AHU exhaust air temperature, °C

Movable pressure transducers were installed in parallel with existing ones. Their locations are indicated in Figure 3 . It can also be seen in that figure that the static pressure used by the fan control ($P_{st,AHU}$) is located just after an elbow.

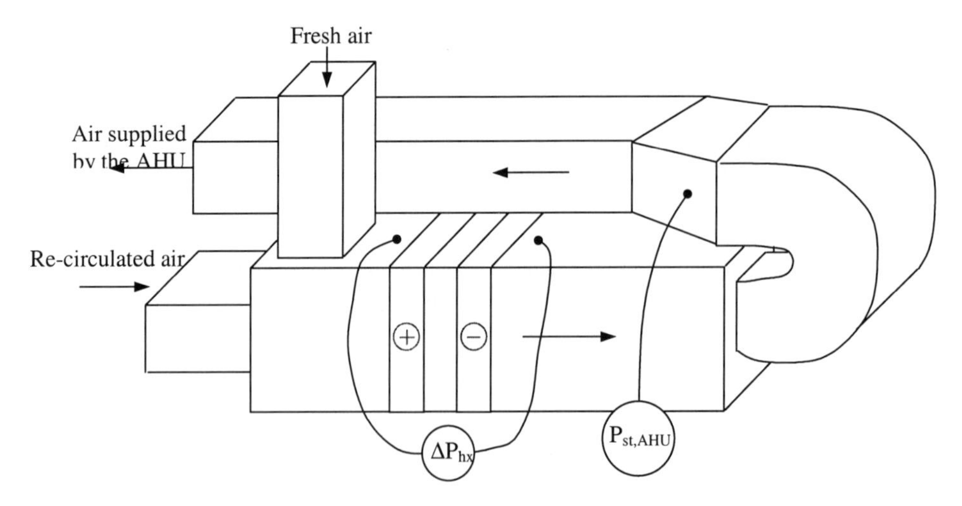

Figure 3: Location of the pressure sensors

In the water side, the movable measuring points were:

$T_{w,su,cc,GPBs1}$: Cooling coil supply temperature GPBs1, °C

$T_{w,ex,cc,GPBs1}$: Cooling coil exhaust temperature GPBs1, °C

$M_{w,cc,GPBs1}$: Cooling coil water mass flow rate, kg/s

The cooling coil water flow rate was measured using transit time flow meter, which was located just after the derivation of the main water network to the cooling coil.

3.RESULTS AND ANALYSIS OF THE MEASUREMENTS

3.1.Air handling unit fan control performance

The air handling unit fan control regulates the AHU exhaust pressure. It must stay constant at the set point fixed by the monitoring system, which ensures a good performance of the terminal units. In this case, it was fixed to 400 Pa.

Figure 4 and 5 show an example of test made in order to check the fan control system.

At the beginning of the test the AHU return temperatures is 21 °C.

At 14:13 h the radiators are turned on in order to heat the offices of the modules B and C. After 2 hour 40 min. the return temperature has only increased of 3 K, which is not yet enough for the test considered.

In order to speed-up the temperature increase, the fresh air valve is closed at 15:50 h and then the heating coil of the AHU is opened to 100 % at 16:35 h.

Due to the temperature increase, all VAV boxes are opening in such a way to increase the airflow rate, which makes that the fan speed has also to increase, in order to cope with the exhaust pressure set point, as it can be seen in Figure 4. According to the results shown in Figure 5, the fan control is able to keep the value fixed by the monitoring system.

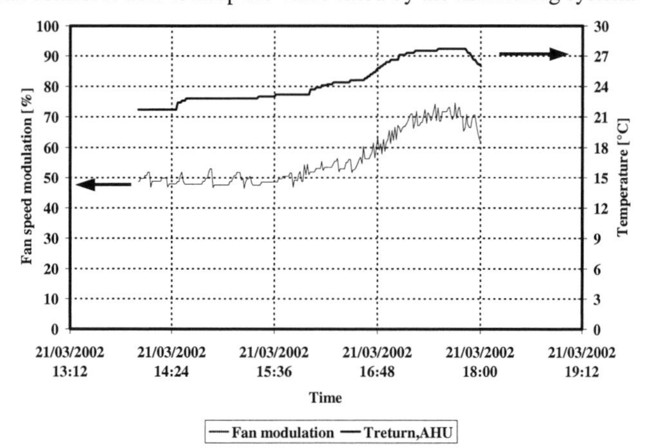

Figure 4: AHU fan control performance

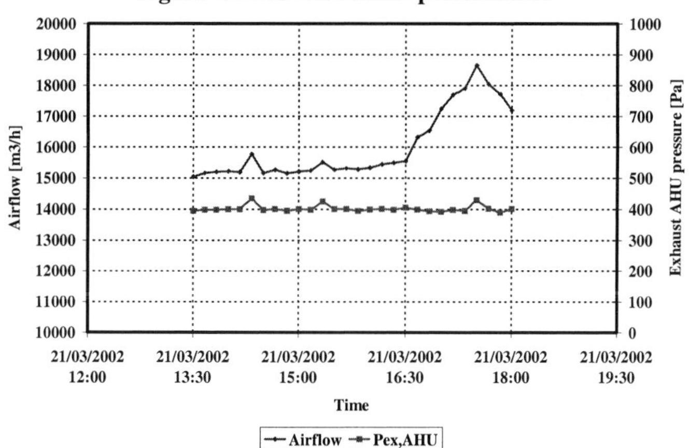

Figure 5: Fan speed modulation

3.2 Air handling unit cooling coil performance

The cooling coil imposes the exhaust AHU temperature. This heat exchanger works with a variable water flow, controlled by a solenoid valve. It regulates the water flow rate to keep the AHU exhaust temperature at the value fixed by the monitoring system. In this case, the law imposed by the monitoring system is calculated according to the AHU return temperature, as shown in Figure 6. In the actual case, the control is working without seasonal compensation because of the external temperature is lower than 20°C (Andre et al. 2000).

According to the measurements, when the AHU return temperature is higher than 22 °C, the set point of the AHU exhaust temperature is 14°C. For AHU return temperatures between 20 and 22°C, there is a linear relationship between the control temperature and the return temperature. At return temperatures lower than 20°C the control temperature is fixed at 35 °C.

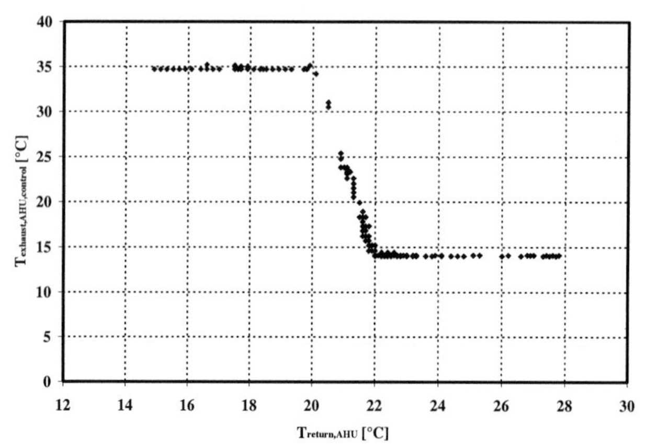

Figure 6: Control law of the AHU exhaust temperature

Time variations of temperatures and water flow rate are presented in Figure 7.

The solenoid valve modulates the water flow rate between 0 and 5.4 l/s (nominal value).

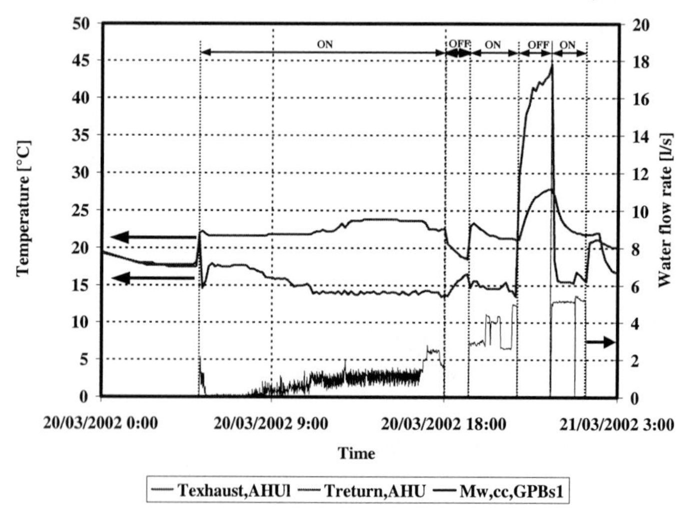

Figure 7: Modulation of the water mass flow in the cooling coil

The cooling coil model proposed by Ding et al. (1991) was used to verify the performance of the heat exchanger. The model has two parameters to be identified: the water and the air side heat transfer coefficients respectively. The identification was done in this case on the basis of cooling coil capacities *given by the manufacturer*. A total of 60 data points were used in this identification. All the values given by the manufacturer correspond to wet regime. The water flow rates and air flow rates covered by these points are shown in Figure 8. The two dry steady state regimes (point 1 and 2) selected from the recordings are also indicated.

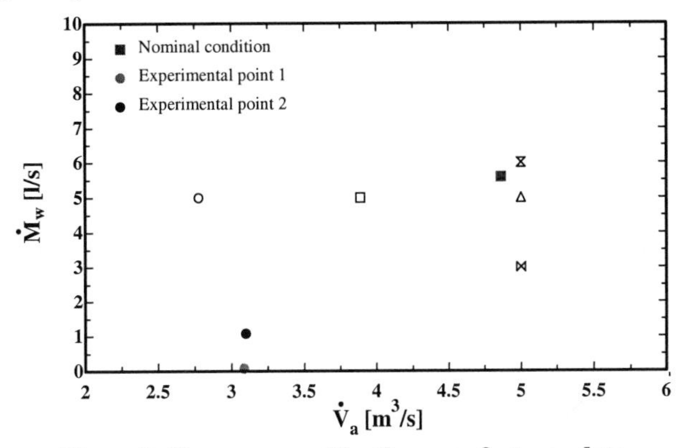

Figure 8: Range covered by the manufacturer data

The results of this modeling are presented in Figure 9: most of the values are predicted with an accuracy of ±7 %, except for two points, where the error reaches -12 %. One of these two points was not included in the regression analysis: it is the nominal regime.

No further parametric adjustment was done; the model was then supposed to underestimate the nominal cooling.

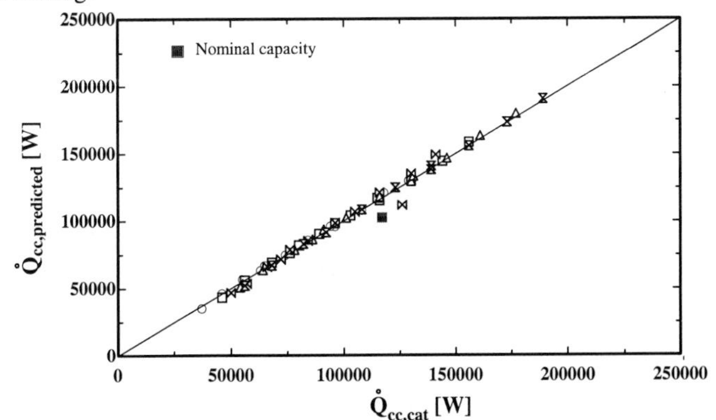

Figure 9: Predicted against measured cooling coil capacity

Table 1 gives the results of the modeling for the two steady state periods already selected. Cooling powers are predicted (in dry regime), with errors of -0.2 and +7 %, respectively. This agreement is very good. But it has to be so for the first regime: when one of the capacity flow rate is very low, the heat exchanger effectiveness has to stay very near to 100 % and there is no more room for any simulation mistake!

The second point is still satisfactory, but an error analysis should still be made in order to estimate the corresponding heat transfer coefficient uncertainty.

We may conclude that the cooling coil performances are in the expected range, but other tests at higher power would be welcome in order to confirm this diagnosis.

Table 1 Cooling coil capacity

Point	$\dot{Q}_{cc,measured}$ [W]	$\dot{Q}_{cc,model}$ [W]
1	3677	3671
2	23890	25603

4. CONCLUSIONS

The commissioning of the AHU permits us to conclude that the summertime problem of high temperatures is probably not due to bad performances of the AHU.

This was verified in different ways:

- The air supplied for the AHU corresponds to the design value and the fan control performance is good;
- The AHU exhaust temperature was always kept to the set point value, which confirms the good performance of the cooling coil, at least in the conditions considered.

Other tests should still be done at higher cooling powers to confirm these results.

In the mean time, the commissioning of the central control system has shown that the air pressure set point was fixed to a much too low value, making impossible a satisfactory behavior of the VAV boxes, but that is another story…

5. REFERENCES

Andre, Ph.; Lacôte, P.; Lebrun, J.; Dewitte, J.; Georges, B. "QG-MET building in Namur: Identification of typical errors, evaluation of their influence on energy consumption using simulation and proposals for automatic detection". IEA Annex 34 working document, Liège meeting, April 2000.

Ding X., Eppe J. P., Lebrun J. and Wasacz M. 1991. "Cooling Coil Models to Be Used in Transient and/or Wet Regimes. Theoretical Analysis and Experimental Validation". System Simulation in Building, December 3-5, pp. 405-441.

Advances in Building Technology, Volume 1
M. Anson, J.M. Ko and E.S.S. Lam (Eds.)

AUTOMATED COMMISSIONING OF AIR-HANDING UNITS USING FUZZY AIR TEMPERATURE SENSORS

P.S. Lee and A.L. Dexter

Department of Engineering Science, University of Oxford, Oxford, OX1 3PJ, UK

ABSTRACT

Measurement errors can result in faults being incorrectly diagnosed during commissioning. One of the main causes of errors is the use of a single or multi-point sensor to measure the average value of a spatially distributed variable whose distribution changes at different operating conditions. In such cases, the problem is not the accuracy of the sensor itself, but estimation bias. The paper describes a new approach to automating the commissioning of air-handling units based on the use of fuzzy air-temperature sensors whose outputs indicate the measurement uncertainties. The fuzzy sensors are developed from data obtained by simulating the air temperature and flow around the sensing element(s) using a computational fluid dynamics package. The data obtained during commissioning are analysed automatically using generic fuzzy reference models, which describe the behaviour of the air-handling unit when it is operating correctly and when particular faults are present. To assess the benefits of using the fuzzy sensors, automated commissioning tests are performed on the cooling coil subsystem of an air-handling unit. Experimental results are presented, which show that the use of a fuzzy mixed air temperature sensor can reduce the number of false alarms and lead to a more precise diagnosis of any faults.

KEYWORDS

Bias errors, fuzzy sensors, fault diagnosis, automated commissioning, air-conditioning systems

INTRODUCTION

For some time, there has been interest in automating the process of commissioning HVAC control systems, Dexter et al. (1993). Early work, Haves et al. (1996), demonstrated that data collected during manually controlled commissioning tests could be analysed using expert rules. An automatic commissioning tool, which analyses the test data using a model-based fault diagnosis scheme, has also been developed, Ngo & Dexter (1998).

Schemes for automating the commissioning of air-handling units (AHUs) are usually based on measurements provided by the temperature and flow sensors, which are connected to the building energy management and control system. The accurate measurement of the average temperature of the air flowing down a large duct is extremely difficult as there can be significant variations in the temperature and velocity over the cross-section of the duct, Carling & Isakson (1999). Errors arising from the spatial variations are difficult to avoid, particularly if the spatial distribution changes with operating conditions, because the problem is not the accuracy of the sensor itself but estimation bias. Such errors can result in either low sensitivity to faults, imprecise diagnosis or unacceptably high false alarm rates. The presence of these measurement errors is one of the main barriers to the practical application of schemes for automatically commissioning air-conditioning systems, Dexter & Pakanen (2001).

A measurement of the mixed air temperature is often used when commissioning the sub-systems of AHUs. This measurement is particularly problematic because space restrictions often result in the mixed air temperature sensor being located very close to the outlet of the mixing-box where there can be significant stratification, Robinson (1999). Even commercial averaging sensors can produce large errors if they are used downstream of a mixing-box, Lee & Dexter (2001b).

A number of ways of dealing with sensor bias have been proposed. The sensor bias can be estimated and eliminated, Wang & Wang (1999). This approach requires a large number of temperature and flow sensors to be installed, assumes that the bias is constant or slowly varying and is computationally demanding. Alternatively, the detection threshold can be adjusted to take sensor bias into account. The main difficulty of this approach is the selection of an appropriate value for the threshold, Carling (2002). Too high a value reduces the fault sensitivity too much; too low a value causes too many false alarms. The effects of sensor bias can be incorporated into the reference models used by the diagnosis scheme, Lee & Dexter, (2001a). This approach avoids false alarms but can produce highly ambiguous results, especially if a constant, worst-case, bias is assumed, Ngo & Dexter (1999). Another approach is to use other available measurements to estimate the measured variable, Lee et al. (1997). The main weakness of this approach is that estimates based on the outputs of other sensors may exhibit even greater bias because the estimation will be sensitive to modelling errors as well as the biases on the other sensors.

There is widespread interest in the development of smart sensors, which make better use of other sources of information and provide an indication of their accuracy and reliability. Previous work has shown that the fusion of measurements and *a priori* information can reduce measurement errors, Filippidis (1996). Sensor fusion is an attractive option in this area of application because many other sources of information are already available through the building energy management system. Fuzzy approaches to sensor fusion, Mauris et al. (1996), have been shown to have particular strengths in complex and highly uncertain systems, Perrot et al. (1996).

The paper describes a new approach to automating the commissioning of air-handling units based on the use of a fuzzy mixed air temperature sensor. The fuzzy sensor, Schodel (1994), is developed using a fuzzy identification scheme and training data obtained by simulating the air temperature and flow around the sensing element(s) with a computational fluid dynamics package. The fuzzy sensor removes the bias errors and provides an indication of the uncertainty associated with the air temperature measurement at the current operating conditions. Sensor fusion is used to reduce the measurement uncertainty. Fuzzy reference models, which take account of sensor bias, are identified using the output from the fuzzy sensor together with training data generated by both CFD and conventional lumped-parameter simulations. The reference models, which provide a semi-qualitative description of the plant behaviour when it is fault-free and when particular faults are present, are incorporated into a fuzzy model-based fault diagnosis scheme, Dexter & Ngo (2001). The results of automated commissioning tests performed on a laboratory test rig are presented.

DEVELOPMENT OF A FUZZY MIXED-AIR TEMPERATURE SENSOR

A CFD package is used to simulate, in two dimensions, the mixing of two air streams in a mixing-box for a range of outside and return air temperatures, positions of the sensor and positions of the dampers in the mixing-box. A fuzzy model, which relates the true mixed air temperature to the output from a six-point averaging sensor mounted downstream of the mixing-box, is identified using a fuzzy identification scheme, which takes "frequency of occurrence" into account, Ridley et al. (1988), and simulation data obtained at a wide range of the operating conditions. Figures 1 to 2 show the output generated by the fuzzy sensor at two different measured temperatures (287K and 292K).

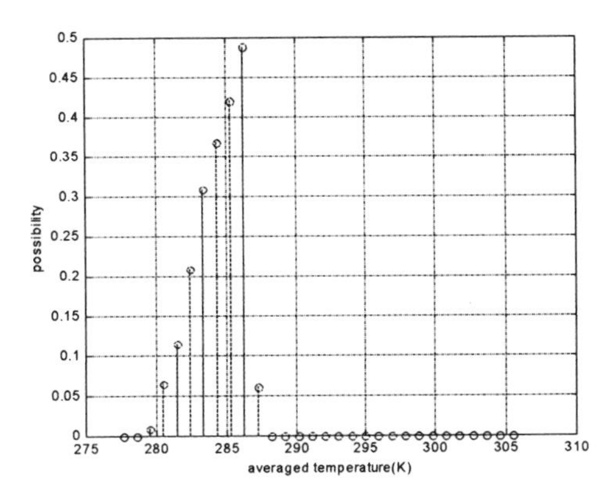

Figure 1: Fuzzy sensor output when the measured mixed air temperature is 287K

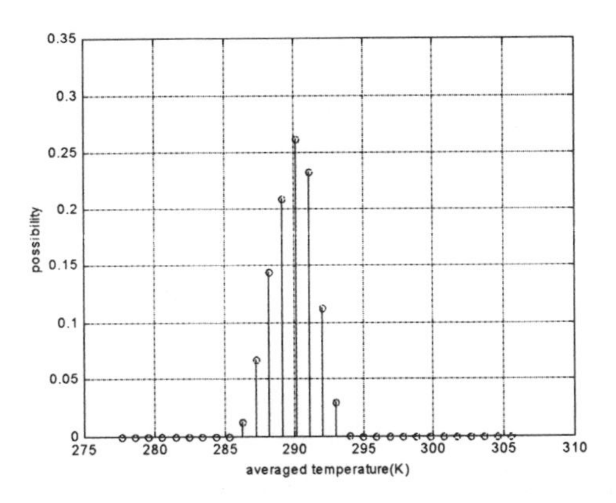

Figure 2: Fuzzy sensor output when the measured mixed air temperature is 292K

The results show that there is significant uncertainty associated with the estimate of the average temperature in both cases and that the averaging sensor generates a biased estimate of the mixed air temperature.

A version of the fuzzy sensor, which uses the return damper position as an auxiliary input, has also been developed to reduce the levels of uncertainty associated with estimating the average mixed air temperature. The results of an earlier study showed that the size of the bias error is related to the position of the dampers, Lee & Dexter (2001a). Figures 3 to 5 show the output generated by the fuzzy sensor for three different positions of the return dampers when the measured temperature is 287K.

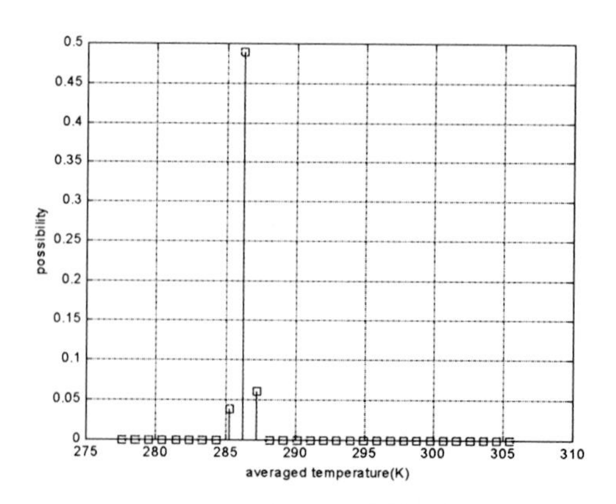

Figure 3: Fuzzy sensor output when the measured mixed air temperature is 287K and the return damper angle is 20 degrees

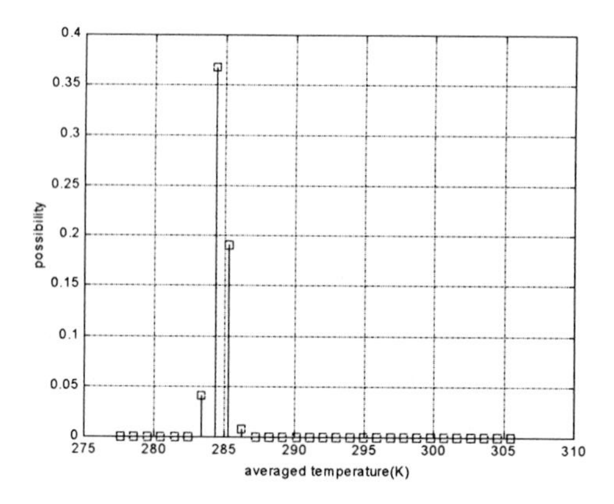

Figure 4: Fuzzy sensor output when the measured mixed air temperature is 287K and the return damper angle is 45 degrees

The results demonstrate that knowledge of the damper angle reduces the uncertainty associated with estimating the average mixed air temperature, when the return damper is less than half open. The results also show that the uncertainty and the bias vary with damper angle.

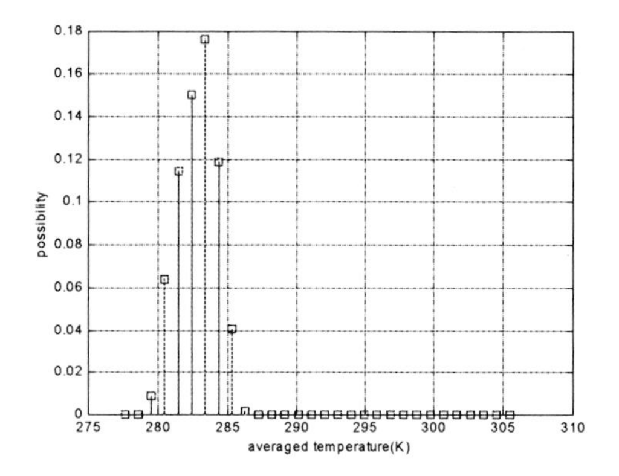

Figure 5: Fuzzy sensor output when the measured mixed air temperature is 287K and the return damper angle is 70 degrees

GENERATION OF THE REFERENCE MODELS

The test data collected during commissioning are analysed by comparing the observed behaviour to the behaviour predicted by a set of fuzzy reference models. The fuzzy reference models relate the temperature difference across the coil to the position of the control valve, the air flow rate, and the on-coil air temperature and humidity. The mixed air temperature sensor is used as a proxy for the on-coil air temperature.

The reference models are identified using training data obtained from a lumped-parameter model of the chilled water, cooling coil subsystem and from the output of the fuzzy mixed air temperature sensor. To obtain generic reference models, which can be used for commissioning a class of chilled-water cooling coils, nine different designs of cooling coil subsystems are simulated over a range of operating conditions, when there are no faults present and when various faults have occurred, Dexter & Ngo (2001). The output of the fuzzy mixed air temperature sensor is combined with the data obtained from the cooling coil simulations using the following procedure. Data from the CFD simulation of the mixing-box is used to identify those values of the mixed air temperature measured by the averaging sensor that correspond to the particular value of the average mixed air temperature used in the lumped-parameter simulation. The outputs of the fuzzy sensor, which correspond to the largest and smallest values measured by the averaging sensor, are then included in the training data.

COMMISSIONING THE COOLING COIL SUBSYTEM IN AN AIR-HANDLING UNIT

The commissioning tests are performed on the cooling coil subsystem of the laboratory air-handling unit shown in Figure 6. During the commissioning test, the cooling coil valve is automatically stepped through a pre-specified series of values from its fully closed to its fully open position. The commissioning software checks whether steady-state operating conditions have been achieved before it moves on to the next step in the sequence, Ngo & Dexter (1998). The test is performed for two different positions of the mixing-box dampers: when the recirculation damper is nearly closed (return damper angle = 18°) and when it is half open (return damper angle = 45°). The AHU operates at its full design air flow rate throughout the tests. The mixed air temperature, which is used as a proxy for the on-coil temperature, is measured using either a commercial averaging sensor or the fuzzy sensor with the auxiliary input (the return damper angle). A commercial averaging sensor is used to measure the supply air temperature, which is used as a proxy for the off-coil air temperature. The on-coil relative humidity is assumed to be 50% as no measurement was available. Six fuzzy reference models are used for the diagnosis: correct operation (no faults), leaky valve, fouled coil, valve stuck closed, valve stuck midway and valve stuck open. Tables 1 to 3 show the results obtained from the automated commissioning tests for two different positions of the mixing-box dampers. The symbol " c " indicates correct operation; " f " indicates fouled coil; " l " indicates leaky valve; " cf " indicates either correct operation or fouled coil; " cl " indicates correct operation or leaky valve; and "cfl" indicates either correct operation or fouled coil or a leaky valve.

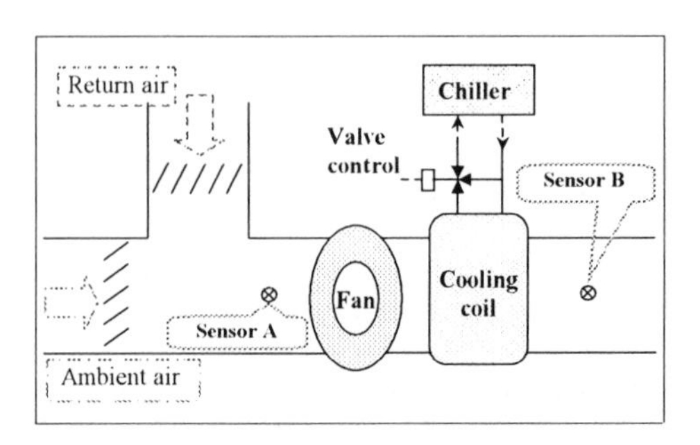

Figure 6: The air-handling unit.

The results of the commissioning test using the fuzzy sensor to measure the mixed-air temperature are analysed using the generic reference models of the cooling coil subsystem described in the previous section (reference models A). The reference models do not take the damper angle into account. When the commercial averaging sensor is used to measure the mixed-air temperature, the reference models are trained using data obtained only from the lumped-parameter simulation. It is assumed either that there are no bias errors (reference models B) or that the bias errors are less than $\pm 2^{o}C$ (reference models C).

TABLE 1
RESULTS OF AN AUTOMATED COMMISSIONING TEST IN AN OCCUPIED BUILDING WHEN NO FAULTS WERE PRESENT

Mixed Air Temperature Measurement	Reference Models	Return Damper Angle	
		18°	45°
		Largest Least Ambiguous Belief	
Fuzzy Sensor	A	cf 2.6%	cf 3.9%
Averaging Sensor	B	cf 0.4%	cl 2.6%
Averaging Sensor	C	cfl 63.3%	cfl 71.0%

Table 1 shows the results obtained when no faults were present in the cooling coil subsystem. When the return damper is nearly closed, the use of the fuzzy sensor produces slightly higher levels of belief in correct operation or fouled coil compared to diagnosis based on the averaging sensor and the type B reference models. The results are less ambiguous than those produced by diagnosis based on the averaging sensor and the type C reference models at both damper angles.

Table 2 shows the results obtained when the cooling coil control valve was leaking (20% of the valve control signal). At both positions of the return damper, the use of the fuzzy sensor results in lower values of belief in a leaky valve than are obtained using the averaging sensor and the references models that assume there is no sensor bias (type B reference models). However, when the damper is nearly closed, the fuzzy sensor does produce a far less ambiguous result than is obtained with the averaging sensor and type C reference models.

TABLE 2
RESULTS OF AN AUTOMATED COMMISSIONING TEST IN AN OCCUPIED BUILDING WHEN THE COOLING COIL CONTROL VALVE WAS LEAKING

Mixed Air Temperature Measurement	Reference Models	Return Damper Angle	
		18°	45°
		Largest Least Ambiguous Belief	
Fuzzy Sensor	A	l 2.1%	l 5.7%
Averaging Sensor	B	l 2.6%	l 22.1%
Averaging Sensor	C	cfl 16.8%	l 5.6%

The size of the bias errors depends on the difference between the return air temperature and the ambient air temperature. If commissioning takes place when the building is unoccupied and the cooling load is negligible, the return air temperature will be lower, and the effect of the bias errors on the results will be greater. Table 3 shows the results of an automated commissioning test when no faults were introduced into the cooling coil subsystem and the building was unoccupied. The use of the fuzzy sensor avoids the false alarm, which is generated when the return damper is nearly closed and the averaging sensor and the type B reference models are used. The results of the diagnosis using the fuzzy sensor are less ambiguous than those, which are obtained with the averaging sensor and the type C reference models, when the return damper is half open.

TABLE 3
RESULTS OF AN AUTOMATED COMMISSIONING TEST IN AN UNOCCUPIED BUILDING WHEN NO FAULTS WERE PRESENT

Mixed Air Temperature Measurement	Reference Models	Return Damper Angle	
		18°	45°
		Largest Least Ambiguous Belief	
Fuzzy Sensor	A	cf 8.5%	c 0.3%
Averaging Sensor	B	f 0.4%	cl 0.9%
Averaging Sensor	C	cf 9.8%	cfl 36.2%

CONCLUSIONS

Fuzzy sensors can eliminate bias errors and provide an indication of the uncertainty associated with air temperature measurements in ducts. Automated commissioning of AHUs based on fuzzy air temperature sensors can reduce the number of false alarms or result in a more precise diagnosis. However, bias errors can sometimes increase the belief in a fault. For example, when the measured mixed air temperature is larger than the actual mixed air temperature, and the valve is leaking, the apparent temperature difference across the coil will be larger than expected and the evidence of a leaky valve will be greater.

Acknowledgements

Miss P S Lee would like to acknowledge the financial support provided by the Rhodes Trust.

References

Carling P. and Isakson P. (1999). Temperature Measurement Accuracy in an Air-handling Unit Mixing Box. Trans. 3rd International Symposium on HVAC, ISHVAC '99. Shenzhen, PRC, 922-928.

Carling P. (2002). Comparison of Three Fault Detection Methods based on Field Data of an Air-Handling Unit. *Trans. ASHRAE* 108:1.

Dexter A.L, Haves P. and Jorgensen D.R. (1993). Automatic Commissioning of HVAC Control Systems. CD ROM of technical papers presented at Clima 2000 World Congress, London, UK.

Dexter A.L. and Ngo D. (2001). Fault Diagnosis in HVAC Systems: a Multi-step Fuzzy Model-based Approach. *Int. J. of HVAC&R Research* 7:1, 83-102.

Dexter A.L. and Pakanen J. (2002). Demonstrating Automated Fault Detection and Diagnosis Methods in Real Buildings. *VTT Building Technology*, Finland.

Filippidis A. (1996). Data Fusion using Sensor Data and A Priori Information.*Control Engineering Practice* 4:1, 43-53.

Haves P., Jorgensen R., Salsbury T.I. and Dexter A.L. (1996). Development and Testing of a Prototype Tool for HVAC Control System Commissioning. *Trans. ASHRAE* 102:1, 483-491.

Lee P.S. and Dexter A.L. (2001a). Generating Fuzzy Reference Models for Fault Diagnosis when Sensor Bias is the Main Uncertainty. Proc. UK Workshop on Computational Intelligence, Edinburgh, 28-34.

Lee P.S. and Dexter A.L. (2001b). A Fuzzy Approach to Fault Diagnosis in the Presence of Sensor Bias. Proc. IEEE Conference FUZZ'01, Melbourne, Australia.

Lee W-Y, Park C., House J.M and Shin D.R. (1997). Fault Diagnosis and Temperature Sensor Recovery for an Air-handling Unit. *Trans. ASHRAE* 103:1.

Mauris G., Benoit E. and Foulloy L. (1996). The Aggregation of Complementary Information via Fuzzy Sensors. *Measurement* 17:4, 235-249.

Ngo D. and Dexter A.L. (1998). Automatic Commissioning of Air-conditioning Plant, Proc. Control '98, IEE Conf. Pub. 455:2, 1694-1699.

Ngo D. and Dexter A.L. (1999). A Robust Model-based Approach to Diagnosing Faults in Air-handling Units. *Trans ASHRAE* 105:1.

Perrot N., Trystram G., Le Guennec D. and Guely F. (1996). Sensor Fusion for Real-time Quality Evaluation of Biscuit during Baking. Comparison between Bayesian and Fuzzy Approaches. J. *of Food Engineering* 29, 301-315.

Ridley J.N., Shaw I.S. and Kruger J.J. (1988). Probabalistic Fuzzy Models for Dynamic Systems. *Electronic Letters* 24:14, 890-892.

Robinson K.D. (1999). Mixing Effectiveness of AHU Combination Mixing/Filter Box with and without Filters. *Trans. ASHRAE* 105:1.

Schodel H. (1994). Utilization of Fuzzy Techniques in Intelligent Sensors. *Fuzzy Sets and Systems* 63, 271-292.

Wang S.W. and Wang J.B. (1999) Law-Based Sensor Fault Diagnosis and Validation for Building Air-conditioning Systems, *Int. J. of HVAC&R Research* 5:4, 353-378.

Advances in Building Technology, Volume 1
M. Anson, J.M. Ko and E.S.S. Lam (Eds.)

MODELING OF PACKAGE AIR-CONDITIONING EQUIPMENT

M.Reichler[1] J.W. Mitchell[2] W. A. Beckman[2]
[1] Graduate student, Solar Energy Laboratory
[2] Professor, Mechanical Engineering Department
University of Wisconsin, Madison, Wisconsin, 53717, USA

ABSTRACT

A model for package air-conditioning units that can accurately replicate equipment performance data is described in this paper. The model is suitable for use in building energy simulation programs. The performance of each component is described with mechanistic models using the Engineering Equation Solver (EES) software package (Klein and Alvarado (1999). Generally available manufacturer catalog data are employed to determine the model parameters. Using characteristic parameters based on manufacturer's performance data allows extrapolation of performance data to different operating conditions. The method has been successfully used to predict the full load performance of package air-conditioning systems over the range of operating conditions.

KEYWORDS: Air-conditioning, modeling, performance, prediction, simulation

INTRODUCTION

In the design of an air-conditioning system it is important to evaluate the energy performance of different alternatives. Computer simulations are a convenient method, but models that predict the performance under all operating conditions are required. The goal of this paper is to describe a model of a package air-conditioning system that is suitable for use in such simulations. The model uses available manufacturer catalog data to allow extrapolation over a wide range of operating conditions.

Package air-conditioning systems and central air-conditioning systems are both used in buildings. Typically, central air-conditioning systems are used for large buildings and package units for small buildings or zones. Package air conditioners, which are self-contained units that provide the cooling, heating, and circulation of the conditioned air, are the subject of this work. The cooling of the airflow is provided by a vapor compression cycle using either a reciprocating or scroll compressor. The cooling coils are direct expansion coils in which the refrigerant directly cools the circulating air. The condenser is air-cooled. The components in package units from different manufacturers are basically the same, which allows a general model to be developed. The available cooling capacity of common package units ranges from about 10 to 1000 kW and the circulation airflow rate covers a range from 400 to 40,000 l/s. Package units are equipped with a supply fan and, if needed, an additional return air fan. Thus, package units cover a wide range of applications.

Manufacturers of package air-conditioning units supply catalogs that contain detailed performance information. The total cooling capacity, the sensible cooling capacity and the power draw are given as functions of ambient temperature and return air temperature and humidity, and volumetric flow rate of the supply air. The model that is developed uses this generally available catalog information together with separate catalog information for the compressor that includes the refrigerant flow rate and power draw as a function of inlet and outlet pressure.

In developing the system model, the different components of a standard air-conditioning system (compressor, evaporator, condenser, and expansion valve) were modeled. One modeling approach considered was the use of polynomial curve fits. Although the characteristics of a component are readily represented with this method, the resulting equations have no physical meaning and prediction outside the range of the fitted data may be inaccurate. Engineering models that describe the behavior of a component were also considered. However, these require detailed knowledge on many parameters such as the geometric specifications, fin efficiencies, and other quantities that are not generally given in manufacturer catalogs. A semi-mechanistic modeling technique was therefore chosen in order to avoid the problems of uncertain extrapolation and to utilize the available information. Models for the components were developed that contain the basic engineering relations underlying the performance, with values of the parameters obtained by fitting the model to the catalog data.

COMPONENT MODELS

A single stage vapor compression cycle, consisting of a compressor, condenser, expansion valve, and evaporator, was used as the basis for the model. It was assumed that the refrigerant leaving the evaporator was saturated vapor and the refrigerant leaving the condenser was saturated liquid. With these assumptions the thermodynamic state points were calculated using basic engineering principles.

Compressor Model

The power and mass flow rate through the compressor were predicted using the clearance volumetric efficiency model of McQuiston and Parker (1988). The clearance volumetric efficiency is defined as the ratio of the actual mass flow rate of the refrigerant through the compressor to that theoretically possible. The theoretically possible mass flow rate assumes that the displacement volume is fully filled with refrigerant vapor at the suction inlet state. The clearance volumetric efficiency was calculated as:

$$\eta_V = \frac{\dot{m}_{r,act}}{\dot{m}_{r,th}} = \left(1 + C - C \cdot \left(\frac{p_{dis}}{p_{suc}}\right)^{1/n}\right) \cdot \frac{v_{suc}}{v_a} \cdot \frac{RPM}{RPM} \tag{1}$$

and the corresponding compressor power draw was calculated as:

$$P_{th} = \dot{m}_{r,act} \cdot \left(\frac{n}{n-1}\right) \cdot p_{suc} \cdot v_{suc} \cdot \left[\left(\frac{p_{dis}}{p_{suc}}\right)^{n-1/n} - 1\right] \tag{2}$$

Jaehnig (1999) proposed introducing a pressure drop parameter and a combined compressor efficiency to account for motor and mechanical inefficiencies and for heat transfer in the compressor. The suction pressure and combined efficiency were determined as:

$$p_{suc} = p_{evap} \cdot (1 - \Delta p) \tag{3}$$

$$\eta_{comb} = \frac{P_{th}}{P_{act}} = a + \frac{b}{p_{suc}} \tag{4}$$

Using this formulation, four model parameters are required to model a compressor: the clearance factor C, a pressure drop parameter Δp, and parameters a and b for the compressor efficiency. These four parameters are obtained by fitting the compressor model to compressor performance data.

The compressor model was validated using data from two different manufacturers over a capacity range from 4 to 52 kW. The minimum information required is five data points chosen as combinations of low and high values for the operating variables (condensing and evaporating temperature) and one other data point within this range. The effect of using fourteen data points in the fitting process was also evaluated.

Figure 1 shows the fit of power data to the compressor model. Using five data points, the mean RMS-error for the power draw was 3.9 %, and that for the mass flow rate was 2.6 %. With fourteen data points spread over the whole range, the RMS-error for the power draw was decreased to 2.9 % and that for the mass flow rate to 2.0 %. Increasing the number of points increased the accuracy only slightly.

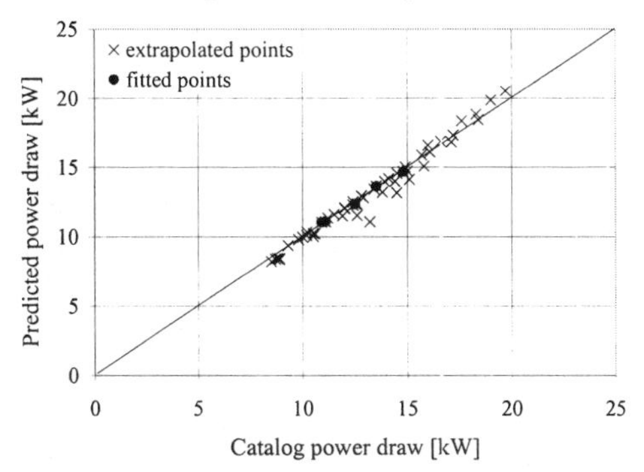

Figure 1: Predicted compressor power draw compared to catalog values

Cooling Coil Model

The modeling technique of Rabehl (1997) was used for the cooling coil (evaporator) and the condenser. In this approach, the overall heat transfer coefficient is related to the mass flow rates using fundamental heat and mass transfer correlations with heat exchanger specific parameters. These parameters depend on heat exchanger geometry and flow arrangement and are determined by fitting the system model to catalog data. With the overall heat transfer coefficient determined, the effectiveness-NTU approach was used to predict performance. The prediction is good and extrapolation is possible since generally valid fundamental heat and mass transfer relations are used (Rabehl (1997)). The condenser model uses the same approach as the cooling coil, but with different relations, and is described separately.

For the direct expansion coil the heat exchanger analogy method (Braun et al. (1989)) was employed. In this approach, a wet coil is analyzed using the effectiveness-Ntu heat exchanger equations for a sensible heat exchanger with the refrigerant flow inside the tubes represented by an equivalent flow of saturated air, and with enthalpy used as the driving force instead of temperature. Heat transfer calculations are made for both totally dry and totally wet operation. If the air entering dew point temperature is higher than the tube surface temperature at the entrance, the coil is totally wet. Otherwise, the coil is partially wet and the performance is approximated depending on which calculation yields the higher heat transfer rate. The error in using the maximum heat transfer rate of the two conditions is usually less than 5 % (Braun et al. (1989)).

The overall heat transfer coefficient for the cooling coil was calculated as the resistance between the refrigerant flow and the inside of the tube wall in series with that between the finned tube surface on the outside and the air stream. The resistance due to conduction through the tube was neglected. For the heat transfer coefficient between the outside surface and the air stream, the correlation of Zhukauskas (Incropera and DeWitt (1996)) was used:

$$\mathrm{Nu_D} = \mathrm{C_0} \cdot \mathrm{Re_D^m} \cdot \mathrm{Pr}^{0.36} \cdot \left(\mathrm{Pr} / \mathrm{Pr_s} \right)^{0.25} \tag{5}$$

where C_0 and m are specific values that depend on the heat exchanger configuration. Rearranging the equations and introducing a convection coefficient correction factor C_f for wet coil operation (Brandemuehl (1992)) allowed the following equation for the overall conductance-area product to be obtained:

$$(\eta_0 hA)_a = C_f \cdot C_1 \cdot k_a \cdot \left(\frac{\dot{m}_a}{\mu_a} \right)^{C_2} \cdot \mathrm{Pr}_a^{0.36} \cdot \left(\mathrm{Pr}_a / \mathrm{Pr}_{a,s} \right)^{0.25} \tag{6}$$

where η_o is the overall efficiency of the surface and the correction factor C_f is determined by

$$C_f = 1.07 \cdot u_a^{0.101} \quad \text{with } u_a \text{ in [m/s]} \tag{7}$$

A single correlation that describes the convection coefficient for boiling inside tubes over the complete range of quality was not available. Rabehl (1997) combined an empirical heat transfer correlation with that developed by Pierre (1955) (ASHRAE Fundamentals (1997).

$$(hA)_r = C_3 \cdot k_r \cdot \left[\left(\frac{\dot{m}_r}{\mu_r} \right) \cdot \Delta h_r \right]^{0.45} \tag{8}$$

The direct expansion coil model incorporates the three model parameters C_1, C_2, and C_3. The range of C_2 is between 0.4 and 0.8 (Incropera and de Witt (1998)). Setting the parameter C_2 to a fixed value of 0.5 allowed an accurate prediction of performance. Figure 2 compares the predicted and catalog heat transfer rate for a number of different cooling coils.

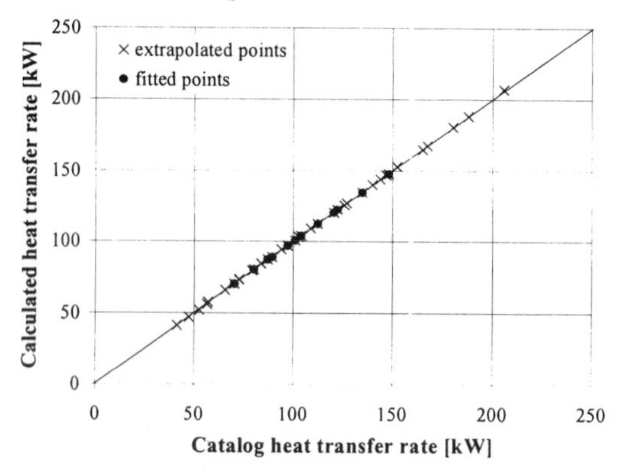

Figure 2: Predicted evaporator heat transfer rate compared to catalog values

Condenser Model

The condenser coil was modeled using the effectiveness-NTU method for sensible heat exchangers. In a condenser, the superheated refrigerant vapor discharged from the compressor is first cooled down, condenses when it reaches the dew point temperature, and then is subcooled before it exits. A detailed condenser model would require correlations for these three different regions. It was assumed that the refrigerant leaving the condensing coil was saturated liquid, and therefore a heat transfer correlation was not required for the subcooling region. The sensible heat transfer in the superheated region is usually small and was taken into account by including it in the overall heat transfer area coefficient product. The refrigerant flow was thus treated to be an isothermal phase change.

As with the direct-expansion coil, the condenser was modeled with two convective resistances, one on the inside of the tubes where condensation of the refrigerant occurs and the other one on the outside exposed to the air stream. The same heat transfer correlation as for the direct expansion coil, but with different coefficients, was used to determine the airside heat transfer area coefficient product.

$$(\eta_0 hA)_a = C_4 \cdot k_a \cdot \left(\frac{\dot{m}_a}{\mu_a}\right)^{C_5} \cdot \mathrm{Pr}_a^{\ 0.36} \cdot \left(\mathrm{Pr}_a / \mathrm{Pr}_{a,s}\right)^{0.25} \tag{9}$$

Inside the tubes, a correlation for condensation in horizontal tubes was found to work best (Rohsenow (1998)):

$$h_c = F \cdot \left[\frac{g \cdot \rho_f \cdot (\rho_f - \rho_g) \cdot k_f^3 \cdot h_{fg}}{D \cdot \mu_f \cdot (T_{sat} - T_s)}\right]^{0.25} \tag{10}$$

where F is a coefficient that is a function of the tube circumference that is covered by refrigerant. The coefficient F was assumed constant. An additional parameter C_6 was introduced that contains all of the constants in equation 10.

$$(\mathrm{hA})_r = C_6 \cdot \left[\frac{\rho_{f,r} \cdot (\rho_{f,r} - \rho_{g,r}) \cdot k_{f,r}^3 \cdot h_{fg,r}}{\mu_{f,r} \cdot (T_{sat} - T_s)}\right]^{0.25} \tag{11}$$

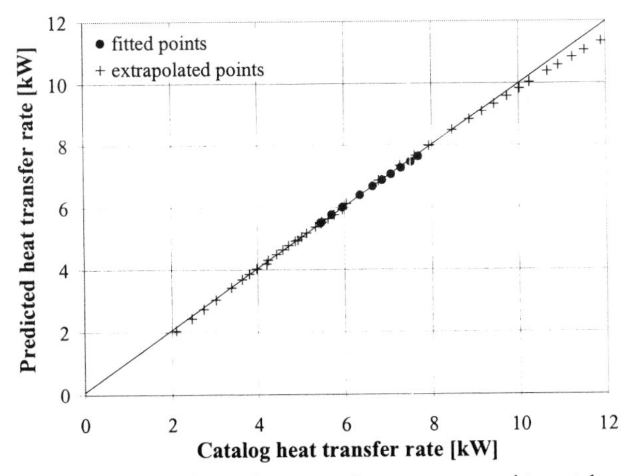

Figure 3: Predicted condenser heat transfer rate compared to catalog values

The model for the condensing coil was validated using condensing coil catalog performance data for a number of coils. Figure 3 shows the predicted coil performance versus the catalog coil performance.

System Model

The compressor, cooling coil, and condenser models described in the preceding sections were coupled together and modeled using EES. In fitting the system model to data, the compressor parameters (C, Δp, a, and b) need to be determined first from compressor performance data. This allows the thermodynamic state points inside the vapor compression cycle to be calculated. Catalog information available for package units generally consists of the capacity (sensible and total) and the compressor power draw as functions of the air temperature entering the condenser, the dry and wet bulb air temperature entering the evaporator, and the air flow rate through the evaporator. Because performance data for the condenser and evaporator separately are not given, system data are used to determine the heat exchanger parameters.

The air conditioner model was validated for units from three manufacturers that used the refrigerant R-22. The ability of different data sets to extrapolate performance was evaluated. It was found that data points that were combinations of high and low values for each of the operating point parameters yielded the best fit.

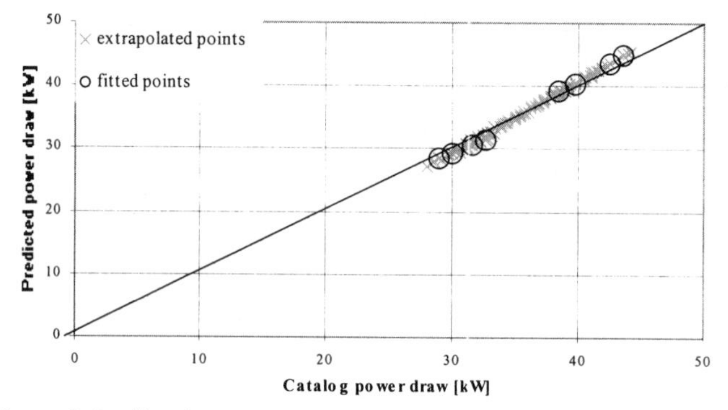

Figure 4: Predicted compressor power draw compared to catalog values

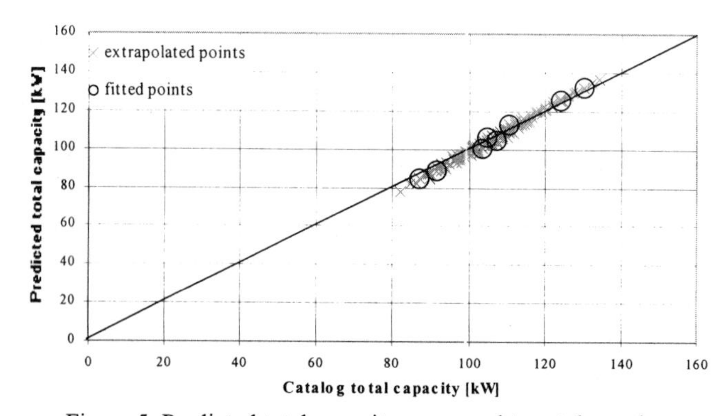

Figure 5: Predicted total capacity compared to catalog values

For the four operating parameters (circulation flow rate, ambient temperature, and supply air dry and wet bulb temperature), 16 data points chosen as indicated were found to be satisfactory. The improvement obtained using data sets with more points within the range of low and high values was not significant. Figures 4, 5, and 6 show the performance prediction for the power draw, the total capacity, and the sensible capacity, respectively, for a units with a nominal 105 kW capacity. The error was less than 3.3 % for the power prediction, less than 2.3 % for the total capacity, and less than 3.2 % for the sensible capacity.

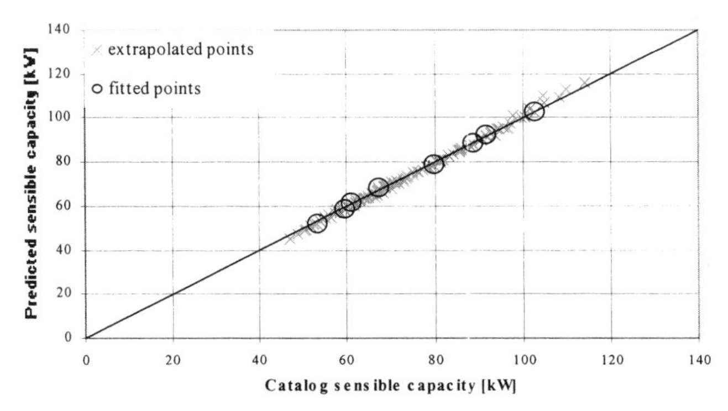

Figure 6: Predicted sensible capacity compared to catalog values

CONCLUSIONS

A model for package air-conditioning equipment was developed that predicts the full load performance (power draw, total and sensible capacity) for different operating conditions. The model is based on semi-mechanistic component models that are interconnected to form the system model.

The semi-empirical model for the compressor contains four parameters that are obtained in a fitting process to performance data. The model is applicable to both reciprocating and scroll compressors. Compressor performance was predicted to within 5 % of manufacturer's catalog data. The condenser and evaporator are modeled using the effectiveness-NTU method. Each heat exchanger model contains three characteristic parameters obtained by fitting the model to system data. The condenser model predicted the performance within 3.9 % and that for the evaporator model was within 1.8 %.

Fitting the characteristic parameters with only a portion of the catalog data set allowed an accurate prediction over the entire range of catalog data. Data points that were combinations of high and low values for each of the operating point parameters (e.g. flow rate and temperatures) produced a good fit. With four operating parameters, 16 data points were found to be satisfactory. Using more data points within the range of low and high values did not significantly improve the performance prediction.

The ability of the model to predict performance was evaluated using data for three package units. The overall prediction of performance was within 3.3 % over the entire operating range. The model allows simulation of the power and capacity of package units outside the range of the fitted data.

NOMENCLATURE

a, b	parameter for compressor efficiency
C	clearance factor
C_0	coefficient in heat transfer correlation
$C_{1..9}$	heat exchanger parameters
C_f	coefficient in heat transfer correlation
D	tube diameter
Δhr	enthalpy difference in condensation
Δp	pressure drop parameter
F	coefficient in condensation correlation
η_0	fin efficiency
η_{comb}	combined efficiency
η_v	volumetric efficiency
g	gravitational constant
h_c	convection coefficient
h_{fg}	enthalpy of vaporization
hA	overall conductance area product
k	thermal conductivity
m	exponent in heat transfer correlation
$\dot{m}$	mass flow rate
μ	dynamic viscosity
n	polytropic exponent
Nu_D	Nusselt number
P	power
p	pressure
Pr	Prandtl number
Re_D	Reynolds number
ρ	density
RPM	motor speed
T	Temperature
u	velocity
v	specific volume
va	specific volume before compression

Subscripts

a	air
act	actual
dis	discharge
evap	evaporator
f	liquid
g	vapor
r	refrigerant, at refrigerant temperature
s	at surface temperature
sat	saturated
suc	suction
th	theoretical

REFERENCES

American Society of Heating, Refrigeration and Air-Conditioning Engineers (ASHRAE), *1997 ASHRAE Handbook Fundamentals*, American Society of Heating Refrigeration and Air-Conditioning Engineers, Atlanta, GA.

Brandemuehl, M.J. et al. (1992), *HVAC2 Toolkit Algorithms and Subroutines for Secondary HVAC Systems Energy Calculations,* American Society of Heating Refrigeration and Air-Conditioning Engineers, Atlanta, GA

Braun, J.E., Klein S.A., and Mitchell, J.W. (1989), "Effectiveness Models for Cooling Towers and Cooling Coils", ***ASHRAE Transactions***, **Vol 95, Pt 2**

Incropera, F.P., and de Witt, D.P. (1998), *Introduction to Heat Transfer*, John Wiley & Sons, New York, NY

Jaehnig, D. (1999), *A Semi-Empirical Method for Modeling Reciprocating Compressors in Residential Refrigerators and Freezers*, MS Thesis, University of Wisconsin

Klein, S.A. and Alvarado, F.L. (1999), *EES – Engineering Equation Solver*, F-Chart Software, Middleton, WI

McQuiston, F.C., Parker J. D. (1988), *Heating, Ventilating and Air Conditioning Analysis and Design*, 3rd edition, John Wiley & Sons, New York

Rabehl, R.J., Mitchell, J. W., and Beckman, W. A. (1999), *Parameter Estimation and the Use of Catalog Data in Modeling Thermal System Components*, **Intl J of HVAC&R Research,** Vol 5, No. 1

Rohsenow, R.M., Hartnett, J.P., and Young, I. C. (1998), *Handbook of Heat Transfer*, 3rd edition, McGraw Hill, New York

Advances in Building Technology, Volume 1
M. Anson, J.M. Ko and E.S.S. Lam (Eds.)

AUTOMATIC DIAGNOSIS AND COMMISSIONING OF CENTRAL CHILLING SYSTEMS

Shengwei Wang

Department of Building Services Engineering
Faculty of Construction and Land Use, The Hong Kong Polytechnic University, Hong Kong

ABSTRACT

A strategy is developed to automatically diagnose and evaluate the BMS sensors and building refrigeration systems during commissioning or periodical check (re-commissioning). The strategy evaluates soft sensor faults (biases) by examining and minimizing the weighted sum of the squares of the concerned mass and/or steady state energy balance residuals represented by the corrected measurements over a period, on the basis of the measurements downloaded from BMS. A Genetic Algorithm is employed to determine the global minimal solution to the multimodal objective function. This paper presents the strategy and examples of application.

KEYWORDS

Chilling system, Building Management System, Sensor fault, Commissioning, Fault diagnosis

INTRODUCTION

Proper commissioning of BMS is a prerequisite for the successful application of the Building Management Systems (BMS) [1]. In engineering practice, all sensors (measurements) should be checked against calibrated instruments at the normal operation conditions manually on site [1,2]. They also need to be checked periodically during operation. This involves large amount of labor and time costs, and may encounter difficulties in practice [3,4,5]. On site checking of some sensors can be difficult and even impossible.

The accuracy of the on site manual checking is limited also. Besides, sensor faults, particularly the soft ones (drifts or biases), might be undetected due to ignorance or other reasons, such as the inadequate thermal contact between thermistor and medium to be monitored. It is, therefore, highly desirable to develop some convenient methods for assessing the health states of the monitoring sensors. Application of such FDD methods would help to ease the burdens and difficulties in on site manual sensor check during the commissioning or re-commissioning of BMS.

On sensor faults in HVAC systems, Usoro et al. [6] studied the detection and diagnosis of an abrupt bias in a room temperature sensor using a model-based method. Stylianou and Nikanpour [7] also used a model-based method to detect soft sensor fault aiming at making sure that the measurements were reliable when monitoring the performance of a laboratory chiller. Lee et al. [8] investigated the detection and automatic recovery of a faulty supply air temperature sensor in AHU. Recently, the authors of this paper developed a "law-based" strategy for the fault detection, diagnosis and evaluation (FDD&E) of the temperature sensors and flow meters in building refrigeration plants [9,10].

The law-based sensor FDD&E strategy is based on the fundamental mass and (steady state) energy conservation (balance) relationships in statistics. These relationships are easy to build and their validity is absolute and independent of plant performance degradations and change of working conditions. Sensor bias values are estimated basically by minimizing the weighted sum of the squares of the corrected residuals of each of the involved balances. On this basis, a software package in prototype is developed to evaluate the BMS sensors automatically on a personal computer using the measurements recorded in a period, downloaded from BMS, during BMS sensor commissioning or periodical check. This paper presents the basic principle and approach of the sensor FDD&E method, the structure of the software, and the examples of the application of the software in central chilling systems.

OUTLINE OF STRATEGY

Basic Principle of Strategy

Figure 1 shows the schematic of a typical central chilling system and its monitoring instruments commonly used in large HVAC systems. In principle, heat balance and mass balance exist in each control volume. Using the measurements from the sensors shown in Figure 1, the residuals (unbalances) of the mass and heat balances for the control volumes can be presented in Equation (1)-(4). Equation (1)-(4) refer to the mass balance and the heat balance of the control volume B, the heat balance of the control volume C (physical redundancy) and the heat balance of the control volume A, respectively. Where, *Ix* or *I(x)* equals to 1 if its corresponding sensor is in use, otherwise it equals to zero.

$$\hat{r}_M^{[i]} = \hat{M}_b^{[i]} + \hat{M}_{bp1}^{[i]} I_{bp1}^{[i]} + \hat{M}_{bp2}^{[i]} I_{bp2}^{[i]} - \sum_{j=1}^{N} \left[\hat{M}^{[i]}(j) I^{[i]}(j) \right] \tag{1}$$

$$\hat{r}_B^{[i]} = \rho c_{pw} \sum_{j=1}^{N} \left[I_{bp1}^{[i]} I^{[i]}(j) \hat{M}^{[i]}(j) \left(\hat{T}_s^{[i]}(j) - \hat{T}_{sb}^{[i]} \right) \right] \tag{2}$$

$$\hat{r}_C^{[i]} = \left(\hat{T}_{rb}^{[i]} - \hat{T}_{rch}^{[i]} \right) I_{bp2}^{[i]} \tag{3}$$

$$\hat{r}_A^{[i]} = \rho c_{pw} \left\{ \hat{M}_b^{[i]} \left(\hat{T}_{rb}^{[i]} - \hat{T}_{sb}^{[i]} \right) - \sum_{j=1}^{N} \left[\hat{M}^{[i]}(j) I^{[i]}(j) \left(\hat{T}_{rch}^{[i]} - \hat{T}_s^{[i]}(j) \right) \right] \right\} \tag{4}$$

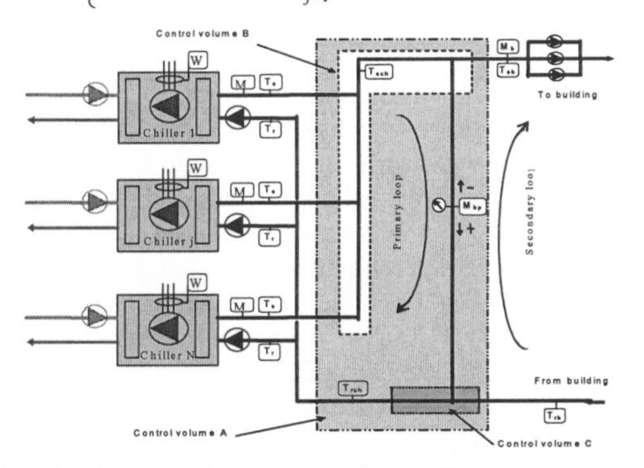

Figure 1: Schematic of primary-secondary refrigeration system

In ideal case, these residuals should be equal to zero when there are no heat losses, thermal storage and water leakage associated with the control volumes. Unfortunately, perfect balance can never be achieved using the realistic measurement data for any system due to various errors in measurements, such as biases, drifts, noises and failures. Any sampled measurement ($\hat{x}^i$) consists of three components: true value of the process variable (x^i), sensor bias (δx) and random noise ($\nu_x^{\ i}$), as shown by the measurement model, Equation (5). When the raw measurements in the equations (1)-(4) are corrected by eliminating the sensor biases from the raw measurements as represented by equations (6) and (7), the balance residual should be reduced significantly if not to zero.

The basic idea of the strategy is that the best estimates of the sensor biases are these values of δx, that minimize the sum of the squares of the balance residuals over certain period when the raw measurements are corrected using the values (δx) as presented by Equation (8).

$$\hat{x}^i = x^i + \delta x + \nu_x^{\ i} \tag{5}$$

$$x^i \cong \hat{x}^i - \delta x \tag{6}$$

$$r^{[i]} = r(\hat{x}^{[i]} - \delta_x) \tag{7}$$

$$\text{minimize}\Big|_{\delta_x} \sum_i \left(r^{[i]}\right)^2 \; ; \; [r^{[i]} = r_M^{[i]}, r_C^{[i]}, r_B^{[i]}, r_A^{[i]}] \tag{8}$$

Basic and Robust FDD&E Schemes

Based on the concept presented by equation (8), four estimator are developed to evaluate the biases of chilled water flow meters, biases of relative supply water temperatures, bias of building return temperature sensor and biases of building supply flow meter and temperature sensors respectively. The sequence of their application is illustrated in Figure 2 (as basic scheme).

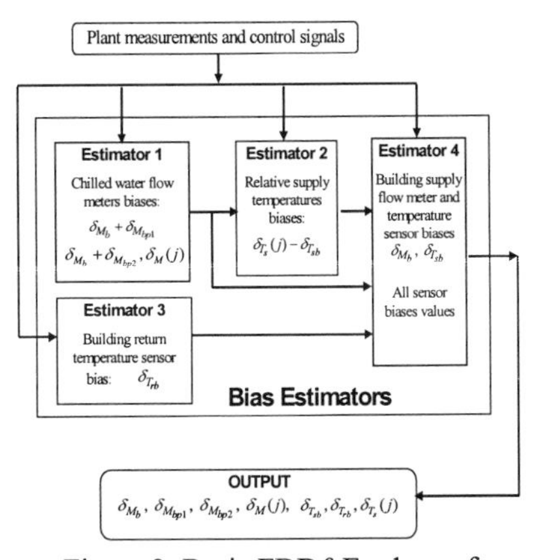

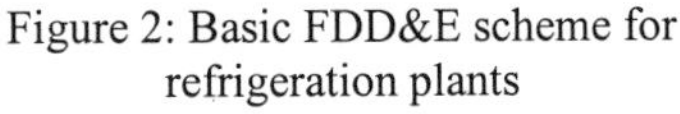

Figure 2: Basic FDD&E scheme for refrigeration plants

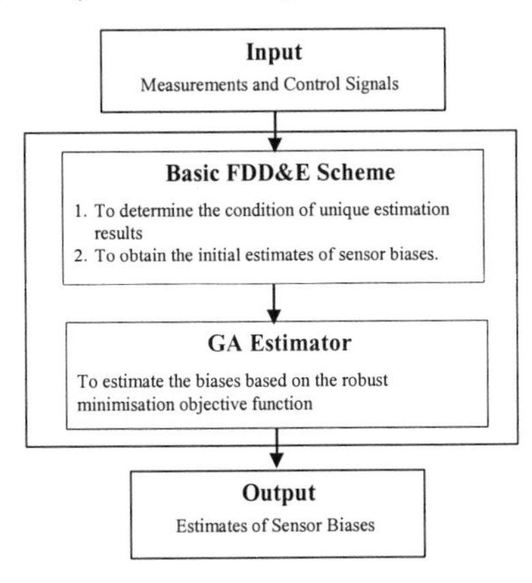

Figure 3: Robust scheme for sensor fault detection, diagnosis and estimation

As the mass and heat balance residuals for the control volumes are minimized individually in the basic scheme, and outputs of estimators are used as the known parameters of the other estimators. The uncertainty of the estimation might be accumulated. The estimation errors of an estimator used earlier might be amplified by the other estimators used later. Tests using data from existing refrigeration systems and simulation show that the robustness of the basic scheme might not be satisfactory in some cases. To overcome this problem, a robust scheme is developed, which minimizes systematically the heat balance residuals of the control volume A and B. A Genetic Algorithm (GA) is employed to cope with the minimization problem of multiple variables.

Figure 3 illustrates the overall structure of the robust FDD&E scheme. It employs the basic FDD&E scheme and a robust GA Estimator. The GA Estimator is designed to estimate the biases of the building supply flow meter and temperature sensor (δ_{M_b}, $\delta_{T_{sb}}$), and the chiller supply temperature sensors ($\delta_{T_s}(j)$). The biases of the chiller flow meters and the building return temperature sensor ($\delta_M(j)$, δ_{Trb},) are estimated by the basic scheme. The GA Estimator estimates δ_{M_b}, $\delta_{T_{sb}}$ and $\delta_{T_s}(j)$ by minimizing the sum of the mean squares of the normalized corrected residuals of the control volume A and control volume B heat balances, as shown in Equation (14). Where, $\dot{r}_A^{[i]}$, $\dot{r}_B^{[i]}$ are the normalized corrected residuals at the instant *[i]*, which are normalized against the number of the chillers operating.

$$\text{minimise}\Big|_{\delta_x}\left(\frac{\dot{S}r_{sq.A}}{n_A}+\frac{\dot{S}r_{sq.B}}{n_B}\right);\quad [\delta_x=\delta_{M_b},\ \delta_{T_{sb}},\ \delta_{T_s}(j),\ j=1,\ldots,N] \tag{9}$$

where,

$$\dot{S}r_{sq.A}=\sum_i(\dot{r}_A^{[i]})^2 \tag{10}$$

$$\dot{S}r_{sq.B}=\sum_i(\dot{r}_B^{[i]})^2 \tag{11}$$

In the robust scheme, the basic scheme is employed not only to estimate the biases of the individual chiller and bypass flow meters and the building return temperature sensor, but also to assist the GA Estimator in two ways. First, the basic scheme is used to check whether a unique set of estimates can be obtained based on the collected measurement data. Secondly, the basic scheme is used to produce a set of initial estimates, which are then used to determine and narrow the search space of the GA Estimator. A fitness function (*f*), Equation (17), is defined, which is the reciprocal of the objective function of the original minimization problem (Equation (14))

$$f=f\left(\delta_{M_b},\delta_{T_{sb}},\delta_{T_s}(j)\right)=1\Bigg/\left(\frac{\dot{S}r_{sq.A}}{n_A}+\frac{\dot{S}r_{sq.B}}{n_B}\right) \tag{12}$$

STRUCTURE OF SOFTWARE

Overview of Software

The software consists of three modes: *Preparation*, *FDD&E programs*, and *Presentation*. The *Preparation* mode is designed for users to input necessary information for configuring and running the FDD&E programs. The FDD&E programs are a series of sensor bias estimators, the corresponding confidence estimators, and the routines for generating data for presenting the results. Execution (*Running* mode) of those programs is the core of the package. The *Presentation* mode allows users to review the FDD&E results. The results include the estimates of the sensor biases, the confidence intervals for the estimates, and the statistics of the balance residuals based on the raw and corrected measurement data.

Preparation Mode

The *Preparation* mode is divided into two stages, while two different kinds of information are requested. Stage I concerns the refrigeration plant configuration and sensor installation condition on which the selection and configuration of the appropriate FDD&E programs are based. Stage II asks for the information regarding the measurement data collected for examination.

The program enters Stage I first if the relevant system and sensor information has not been specified before, or, if re-specifying is necessary. On completion of this stage, the program enters Stage II. If the information required in Stage I has been available in the software from previous runs, the user can enter Stage II directly.

The required specifications regarding the refrigeration plant include the types of the chilled water distribution system, the number of chillers in the plant, the number of chiller types, the specifications of each chiller type, etc. This information is used for selecting the FDD&E programs. Sensor installation conditions are required also for selecting and configuring appropriate FDD&E programs. Figure 4 shows an example of the windows designed as the interface for the users to select the sensors installed in the chilling systems.

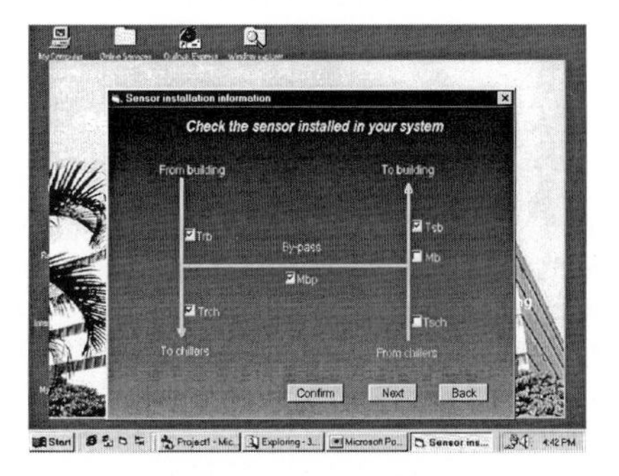

Figure 4: A window for specifying the installation condition of the sensors

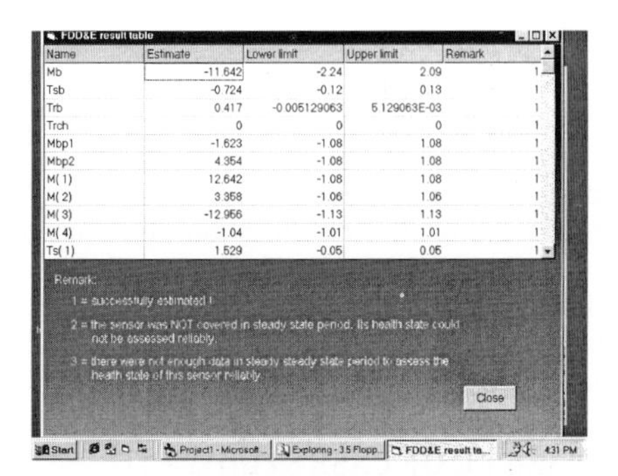

Name	Estimate	Lower limit	Upper limit	Remark
Mb	-11.642	-2.24	2.09	1
Tsb	-0.724	-0.12	0.13	1
Trb	0.417	-0.005129063	5.129063E-03	1
Trch	0	0	0	1
Mbp1	-1.623	-1.08	1.08	1
Mbp2	4.354	-1.08	1.08	1
M(1)	12.642	-1.08	1.08	1
M(2)	3.358	-1.06	1.06	1
M(3)	-12.956	-1.13	1.13	1
M(4)	-1.04	-1.01	1.01	1
Ts(1)	1.529	-0.05	0.05	1

Remark:
1 = successfully estimated !
2 = the sensor was NOT covered in steady state period. Its health state could not be assessed reliably.
3 = there were not enough data in steady steady state period to assess the health state of this sensor reliably.

Figure 5: A window presenting sensor bias estimates and the confidence intervals

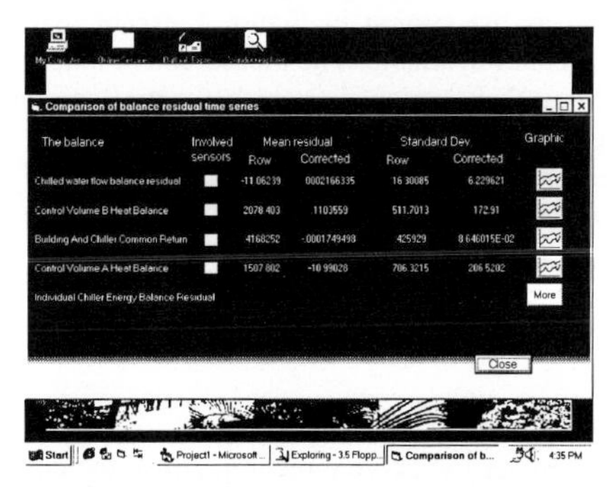

Figure 6: A window showing the statistics of the balance residuals

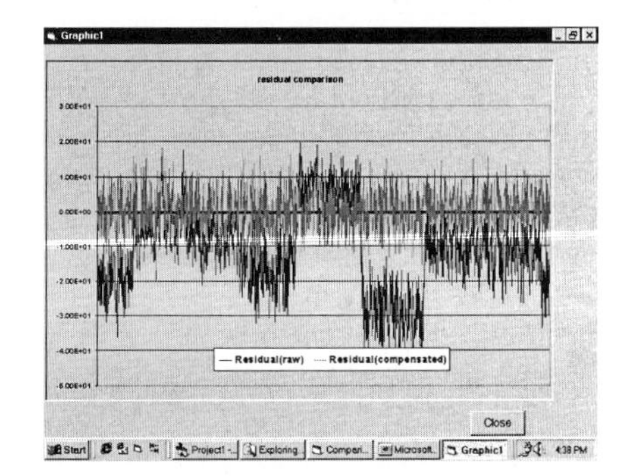

Figure 7: A window showing the time series of the raw and corrected flow balance

After the information about a refrigeration system and the sensor installation conditions have been specified, a program makes decisions accordingly on what FDD&E programs (estimators) should be used, and configures the selected programs into a proper execution sequence. A bank of 12 sensor bias estimators and the GA estimator, the corresponding confidence estimators and post data processing routines have been developed for all the possible sensor installation conditions.

The second preparation stage also asks for the information regarding the collected measurement data. It is required that the collected data have been prepared in a data file with given format. The software will check if the data file is in the format required.

Running Mode

On entering this mode, the software checks the availability of the files that record the required information regarding the system and sensor conditions and the properly manipulated measurement data. If the preparation is complete, the three batch files created in the preparation stage are activated to run. Otherwise, the user will be prompt to re-enter the preparation mode.

At the end of the running mode, three types of files are generated after running estimation programs. The first type is the one that records the estimated biases of the examined sensors. The second type is the one that records the confidence intervals for all the estimated sensor biases. The third type is the one that records the statistical quantities and the time series of the raw and the corrected balance residuals.

Presentation Mode

In the *Presentation* mode, the user can view the FDD&E results through graphic interfaces. The three types of files resulting from the Running mode are further processed. The final results are presented to user in two tables. One table (see Figure 5) shows the values of the estimated sensor biases and confidence intervals. The second table (see Figure 6) shows the statistics of the raw and corrected balance residuals. Time series of the raw and corrected residuals of individuals can also be viewed graphically as shown in Figure 7, when it is selected on the second table.

EXAMPLE OF CASE STUDY

An example of applying the software to an existing building central chilling system of five chillers is presented. The system has the same configuration as the system in Figure 1 except that the common return water temperature measurement (T_{rch}) is not available. The measurement data from the monitoring sensors are recorded in BMS, which are then retrieved from the central computer station. Data of several intermittent days within one month is collected. Sensor faults (biases) are introduced to three of the chilled water temperature sensors (i.e., biases introduced into T_{sb}, $T_s(2)$ and $T_s(3)$ are 1.5°C, -1.0°C and 1.5°C respectively). Prior to introducing these faults, check and calibration of the temperature sensors in the refrigeration plant are conducted. Table 1 presents the output of the robust FDD&E scheme based on the collected measurement data. The biases introduced to the three temperature sensors are successfully diagnosed. The largest error of the three estimates is 0.25°C ($\delta_{T_{sb}}$).

The estimated biases of the bypass chilled water flow meter (negative direction, δ_{Mbp2}), the chilled water flow meters of chiller 2, 3, and 5 ($\delta_M(2)$, $\delta_M(3)$, $\delta_M(5)$) turn out to be significant. For chiller 2, 3, and 5, the biases are more than ten percent of the measured values. The remaining four chilled water flow meter biases are negligible, considering the measurement uncertainties of the flow meters. Since there is no other simple method to compare accurately the flow meter bias estimation results with the actual unknown condition, the chilled flow rate measurements before and after correction are compared with each other and with the historical commissioning data.

TABLE 1 Bias estimates of the sensor in an existing refrigeration plant

Sensor	$M(1)$	$M(2)$	$M(3)$	$M(4)$	$M(5)$	M_b	M_{bp1}	M_{bp2}
Bias Estimate (L/s)	3.2	17.9	17.7	6.8	17.0	-4.9	-2.2	16.8
Sensor	$T_s(1)$	$T_s(2)$	$T_s(3)$	$T_s(4)$	$T_s(5)$	T_{sb}		
Bias Estimate (°C)	-0.08	1.10	-1.47	0.14	0.27	1.75		
Sensor	$T_r(1)$	$T_r(2)$	$T_r(3)$	$T_r(4)$	$T_r(5)$			
Bias Estimate (°C)	-0.24	-0.11	0.04	0.24	0.08			

The balance residuals are the effective indicators for validating the results of bias estimates. Two examples of the balance residuals using the raw measurements and corrected measurements are presented for comparison. The raw flow balance residual indicates the existence of the flow meters biases as shown in Figure 8. The raw residual deviates apparently from zero. This indicates that some, if not all, of the flow meters are biased.

It has been carefully checked if chilled water flowed through any of the evaporators of the idle chillers, by tracing the individual supply and return temperatures. The rising rate of each of those temperature measurements was found normal after the chillers were shut down. This indicates that there was no bypass flow through the evaporators. There cannot be any other reason why some of the chilled water disappeared or was generated. Therefore, the violation of the law of mass conservation by the raw flow measurements presents the clear evidence that the flow meters must have problems. On the other hand, the corrected flow balance residual varies randomly around zero approximately. Of the raw residual, the mean and the standard deviation are -9.01 and 21.4 litres per second, respectively. Of the corrected residuals, they are 0.04 and 7.45 litres per second.

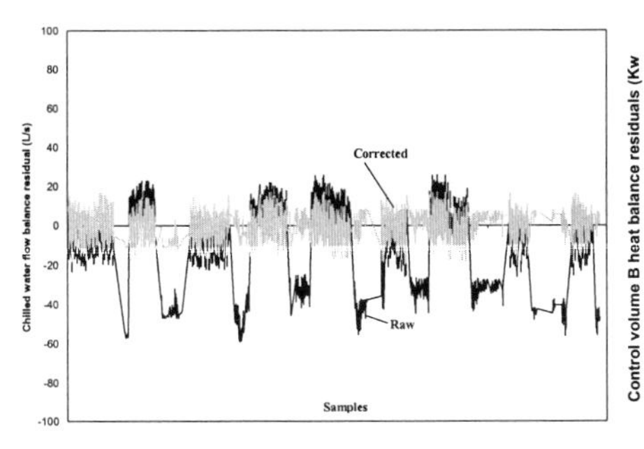

Figure 8: Raw and corrected chilled water flow balance residuals

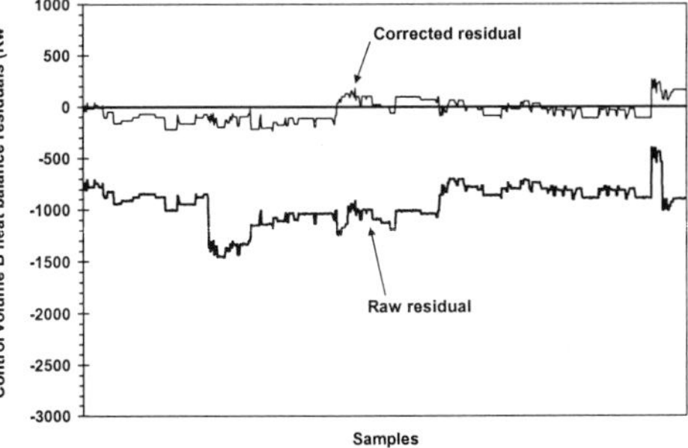

Figure 9: Comparison of the control volume B heat balance residuals (normalized)

Similar to the flow balance residual, the heat balance residuals also indicate the existence of the biases of the associated sensors. The raw and the corrected heat balance residuals of the control volume B are shown in Figure 9. It can be seen that the raw heat balance residuals of the control volume deviate from zero severely. Obviously, the main reason was that three large temperature sensor biases were introduced. After the raw measurements have been corrected with the obtained bias estimates, the large biases in the balance residuals diminished.

CONCLUSIONS

The balance residuals are sensitive indicators of the existence of flow meter and temperature sensor biases. Analysis of the residuals under various operating conditions of the chilling systems and minimization of the sum of the squares of the corrected balance residuals make it possible to locate biased sensors and to estimate the magnitudes of the biases.

In Building Management Systems, there are enormous data stored. It is very convenient for the users to download some data and evaluate the BMS measurements in a computer in commissioning stage or any time during normal operation. When no significant error is detected by the software, the engineer may not need to spend too much effort to check or calibrate the sensors frequently because the measurement quality is ensured. When there is significant error reported by the software, the engineers can focus their effort to check the faulty sensors reported. As remote monitoring via Internet is becoming a regular function of BMS nowadays, the automatic commissioning software provides a means for remote check and diagnosis of the BMS sensors.

ACKNOWLEDGEMENT

The research work of presented in the paper is supported by a grant from the Research Grants Council of the Hong Kong SAR.

REFERENCES

1. Pike P. G. and Pennycook K. 1992. Commissioning of BEMS - A code of practice. AH 2/92. BSRIA.
2. Pike P.G. (1994) BEMS Performance Testing, BSRIA
3. Bilas F., Bourdouxhe J.P., et al. 1997 "Commissioning Of A Centralized Cooling Plant", CLIMA 2000, Brussels. P292.
4. Haves P. Salsbury T.I, Jorgensen D.R. and Dexter A.L. 1996. Development and Testing of A Prototype Tool for HVAC Control System Commissioning, ASHRAE Transactions, 102(1), 467-475
5. Ngo D. and Dexter A.L. 1998. Automatic Commissioning of Air-conditioning Plant, Proceedings of The 1988 International Conference on Control, Pt2, 1690-1699
6. Usoro, P.B., L.C. Schick and S. Negahdaripour. 1985, An Innovation-Based Methodology for HVAC System Fault Detection. *Journal of Dynamics systems, Measurement, and Control. Transactions of the ASME*, 107: 284-285.
7. Stylianou, M. and D. Nikanour. 1996. Performance Monitoring, Fault Detection, and Diagnosis of Reciprocating Chillers. *ASHRAE Transactions,* 102(1): 615-627.
8. Lee, W.Y., J.M. House and D.R. Shin. 1997. Fault Diagnosis and Temperature Sensor Recovery for An Air-Handling Unit. *ASHRAE Transactions,* 103(1): 621-633.
9. Wang S. and J.B. Wang 1999. Law-Based Sensor Fault Diagnosis and Validation for Building Air-conditioning Systems. *Int. J. of HVAC&R Research*, 5(4): 353-380.
10. Wang S.W. and Wang J.B. 2002. Automatic Sensor Evaluation in BMS Commissioning of Building Refrigeration Systems, Automation in Construction, V11(1), pp59-73.

TEAM PRESENTATION:

COMPOSITE CONSTRUCTION

Advances in Building Technology, Volume I
M. Anson, J.M. Ko and E.S.S. Lam (Eds.)

DESIGN OF SIMPLY-SUPPORTED AND CONTINUOUS BEAMS IN STEEL-CONCRETE COMPOSITE CONSTRUCTION

P.A. Berry and M. Patrick

Centre for Construction Technology & Research,
College of Science, Technology & Environment, University of Western Sydney, Locked Bag 1797, Penrith South DC, NSW 1797, Australia

ABSTRACT

Steel-concrete composite beams in Australia are predominantly designed as simply-supported with web-side-plate steel connections, which has some structural limitations. The inevitable slope discontinuity at internal supports can cause wide cracks, whilst its remedy, sufficient tension reinforcement to control cracking, can overload the cleat of the web-side-plate connection, which is in corresponding compression.

Continuous composite beams offer better crack control and lower deflections, even when the depth of the steel section is less than that required for the equivalent simply-supported beam. Innovative design methods for continuous and semi-continuous composite beams have recently been developed which eliminate the need for iterative approaches and allow accurate preliminary sizing of members. This paper reviews the design of simply-supported and continuous composite beams in Australia, highlighting the benefits of continuous construction with two typical design examples.

KEYWORDS

Composite beam, continuous composite beam, semi-continuous composite beam, composite connection, web-side-plate connection, end plate connection.

INTRODUCTION

Current Australian practice is to support composite beams with web-side-plate connections, which are assumed to behave as nominally pinned joints. If the connections do behave as pins, the resulting high end rotations can cause unacceptable cracking of the concrete surface and there are a number of composite carparks in Australia that are excessively cracked due to inadequate reinforcement over the connection (Patrick *et al.* 2002). Conversely, typical levels of crack control reinforcement may invalidate the pinned joint assumption and can lead to compressive forces in the cleat that exceed the current Australian design limit for web-side-plate connections (Berry *et al.* 2000).

Recent design guidelines (Berry *et al.* 2001a), which supplement AS 2327.1 (Standards Australia 1996), have demonstrated the benefits of continuous composite beams. Continuous composite beams may be defined in the broadest sense to refer to any composite member (beam or cantilever) subject to negative curvature at one or more of its supports. Continuity in a composite beam may be achieved either with internal supports, or by the use of suitable connections within a frame. In the latter case, this paper is limited to rigid connections in braced frames, and in order to limit the rotation demand on the connection, the design moment capacity at the critical positive moment cross-section is limited to $0.85\phi M_{bv}^{+}$ (Steel Construction Institute 1998).

COMPOSITE CONSTRUCTION

Classification of Connections

A composite connection must have adequate strength, ductility, and stiffness at the strength limit state. These three attributes form the basis of a useful classification system for both the bare steel and composite states, as shown in Figure 1.

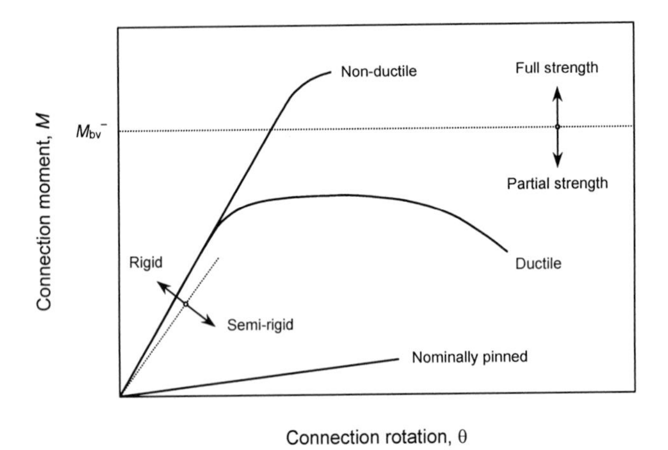

Figure 1: Classification of Connections (adapted from Steel Construction Institute 1998)

Types of Framing

Three different types of framing can be defined by the requirements of global analysis (British Standards Institution 1994) and the corresponding connections that are applicable.

1. *Simple*. Only equilibrium need be considered in the global analysis. Simple framing uses only nominally pinned connections.
2. *Continuous*. Both equilibrium and the structural properties of the member need to be considered in the global analysis. Any connections that are not nominally pinned must be full-strength rigid connections, so that the connection properties do not affect the member behaviour.
3. *Semi-continuous*. The structural properties of the connections also need to be considered in the global analysis. For the purposes of this paper, which excludes semi-rigid connections, semi-continuous framing may use partial-strength rigid connections in addition to those permitted for continuous framing.

DESIGN FACTORS

Simply-supported beams

The bending moment diagram for a simply-supported beam depends solely on satisfying equilibrium for the given applied loads and is unaffected by the variation of other design parameters. The designer has little scope for creative input into the design process. Given the steel beam and concrete slab details, it is possible to determine directly the minimum degree of shear connection at each potentially critical cross-section that is required to satisfy the strength limit state (Patrick & Dayawansa 1998).

Simply-supported beams, although very efficient at satisfying strength requirements, do not perform well under serviceability conditions. The end rotations are high, which can cause unacceptable cracking over internal supports, and the deflections frequently govern the design, negating the benefits of efficiency at the strength limit state. It is not uncommon for only 70% of the strength capacity to be utilised under ultimate loads.

Continuous beams

For continuous and semi-continuous beams, the bending moment diagram must satisfy compatibility as well as equilibrium. Varying the relative stiffness and strength of each cross-section can therefore be used in conjunction with moment redistribution to produce different design options. Innovative design charts have recently been developed that allow designers to obtain accurate preliminary designs for a range of parameters in order to find the optimum solution (Berry *et al.* 2001b).

The degree of shear connection at the peak positive moment cross-section, β_{m}^{+}, may be varied within the range $0.5 \le \beta_{\mathrm{m}}^{+} \le 1$. It is most economical to use $\beta_{\mathrm{m}}^{+} = 0.5$: increasing the degree of shear connection above this level leads to decreasing returns in composite strength. Based on the values given in Table A4 of the Composite Beam Design Handbook (Australian Institute of Steel Construction and Standards Australia 1997), for typical universal beams (310 UB – 610 UB), 50% shear connection ($\beta_{\mathrm{m}}^{+} = 0.5$) produces 70% of the maximum composite benefit ($\phi M_{\mathrm{bc}}^{+} - \phi M_{\mathrm{s}}^{+}$).

The designer may also choose the desired level of reinforcement: the suggested practical range that satisfies minimum crack control requirements is $0.75\% \le p_{\mathrm{r}} \le 1.5\%$. In general, it is most economical to use $p_{\mathrm{r}} = 0.75\%$, but there may be situations in which the overall structural depth is the governing factor. Based on the values given in Charts G2.1 – G2.4 of the Composite Structures Design Booklet DB2.1 (Berry *et al.* 2001a), increasing the reinforcement from the minimum value of $p_{\mathrm{r}} = 0.75\%$ to the suggested maximum value of $p_{\mathrm{r}} = 1.5\%$ increases the load-carrying capacity on average by 20% for a semi-continuous composite beam and by 10% for a continuous beam.

The design of continuous and semi-continuous composite beams is rarely governed by deflections, which means they perform well under serviceability conditions. By the very nature of continuity, the end rotations are low, which minimises any cracking over internal supports. Economics and ease of construction are often raised as factors in support of simply-supported composite beams, but this perception may not be soundly based. A partial depth end plate connection requires no more fabrication in terms of drilling and welding than a web-side-plate connection and, since the one set of bolts connect adjacent beams, the total number of bolts required during erection may be halved. A rigorous cost-benefit analysis, including the cladding savings of reduced floor-to-floor heights, may show that continuous and semi-continuous composite beams should be the preferred construction option.

DESIGN PROCEDURE FOR CONTINUOUS BEAMS

Geometry

It is assumed that the general layout is known, including details of the:

- *beam* (length and spacing to adjacent beams);
- *concrete* (grade and depth of slab); and,
- *profiled steel sheeting* (type and orientation of ribs).

Loads

It is assumed that all design loads, together with their stage of application, are known. The one exception at the outset of design is the beam self-weight, which can typically be approximated by a uniformly distributed load in the range 0.5 – 1 kN/m.

Supports in the Composite State

The supports in the composite state must be either pinned or rigid, since semi-rigid connections are beyond the scope of this paper. Rigid composite supports rely on symmetry for properly anchored negative moment reinforcement, and therefore should normally only be used at internal supports.

Rigid supports may be either partial-strength or full-strength, which correspond to *semi-continuous* and *continuous* framing respectively. Semi-continuous framing relies on high levels of moment redistribution and requires a 0.85 positive moment capacity reduction factor to be applied in design, to minimise the rotation demand at the supports (Steel Construction Institute 1998).

Degree of Positive Moment Shear Connection

Partial shear connection is permitted for critical cross-sections in positive moment regions, so the designer may choose the desired degree of shear connection at the peak positive moment cross-section, β_m^+, such that

$$0.5 \le \beta_m^+ \le 1 \tag{1}$$

Level of Negative Moment Reinforcement

The designer may choose the desired level of reinforcement, within certain limits. One of the requirements for crack control in AS 3600 (Standards Australia 2001) equates to $p_r = 0.75\%$ for N12 reinforcing bars, which represents a suitable minimum, and 1.5% has been chosen as a convenient maximum for the purposes of preliminary design. Therefore,

$$0.75\% \le p_r \le 1.5\% \tag{2}$$

If a beam is relatively short and heavily loaded, levels of reinforcement towards the higher end of this range will be required to satisfy the stiffness requirements for a rigid connection.

Initial Member Sizing

Semi-continuous and continuous beams are more likely to be governed by strength than deflection, which means that the strength charts can provide accurate member sizing. The charts are based on fully built-in beams with complete shear connection at the peak positive moment cross-section. In other cases, approximate solutions can be obtained by making the following adjustments before using the charts (Figure 2):

- for propped cantilevers, increase the beam span by 15%; and,
- for partial shear connection ($\beta_m^+ = 0.5$), increase the loads by 10%.

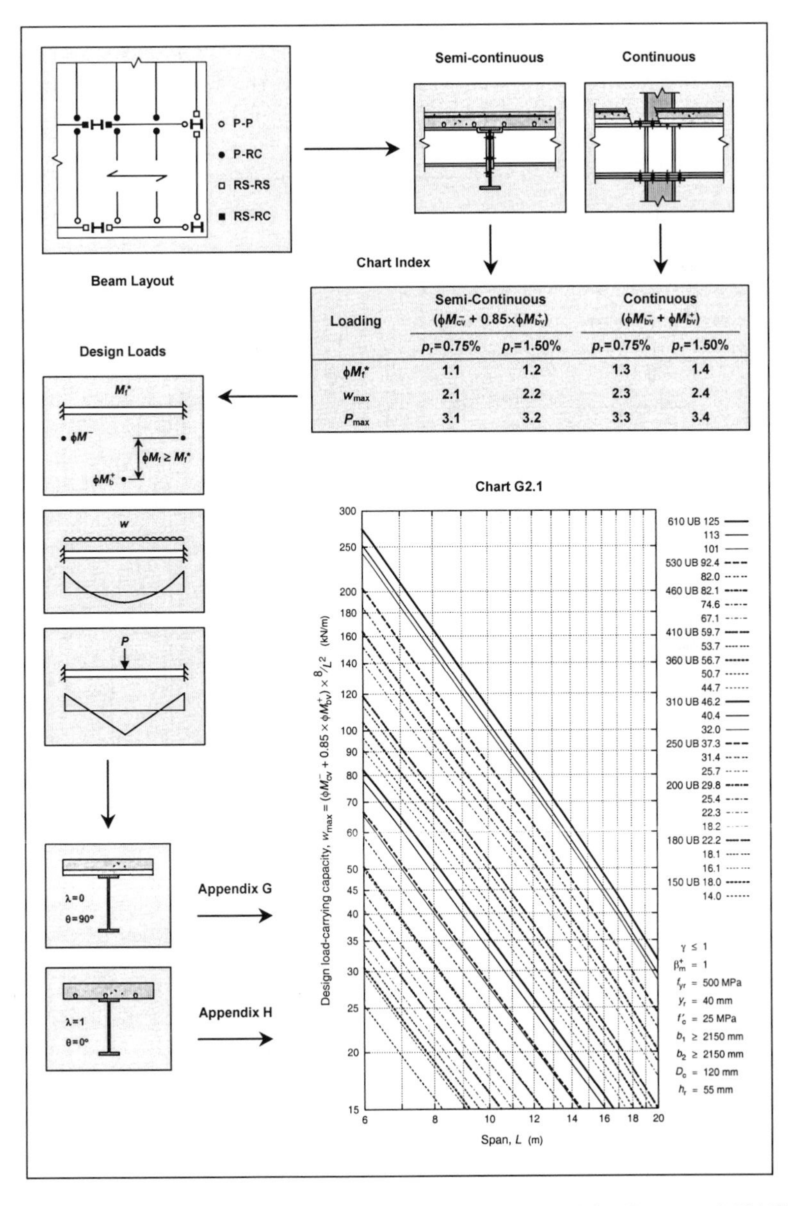

Figure 2: Design Flowchart for Continuous Composite Beams (after Berry *et al.* 2001b)

DESIGN EXAMPLES

Secondary Beam in a Carpark

Two semi-continuous beam options have been added to the simply-supported secondary beam 'B1' from scheme 1C of the BHP Design Guide for Economical Carparks (Watson *et al.* 1998), as shown in Figure 3. Not only has the beam size been reduced from a 460 UB 74.6 to a 360 UB 50.7 in each case, but the incremental deflection, which governed the simply-supported design, has also reduced by 40% and 25% for the unpropped and propped options respectively. The detailed design calculations are given in Berry *et al.* (2001a).

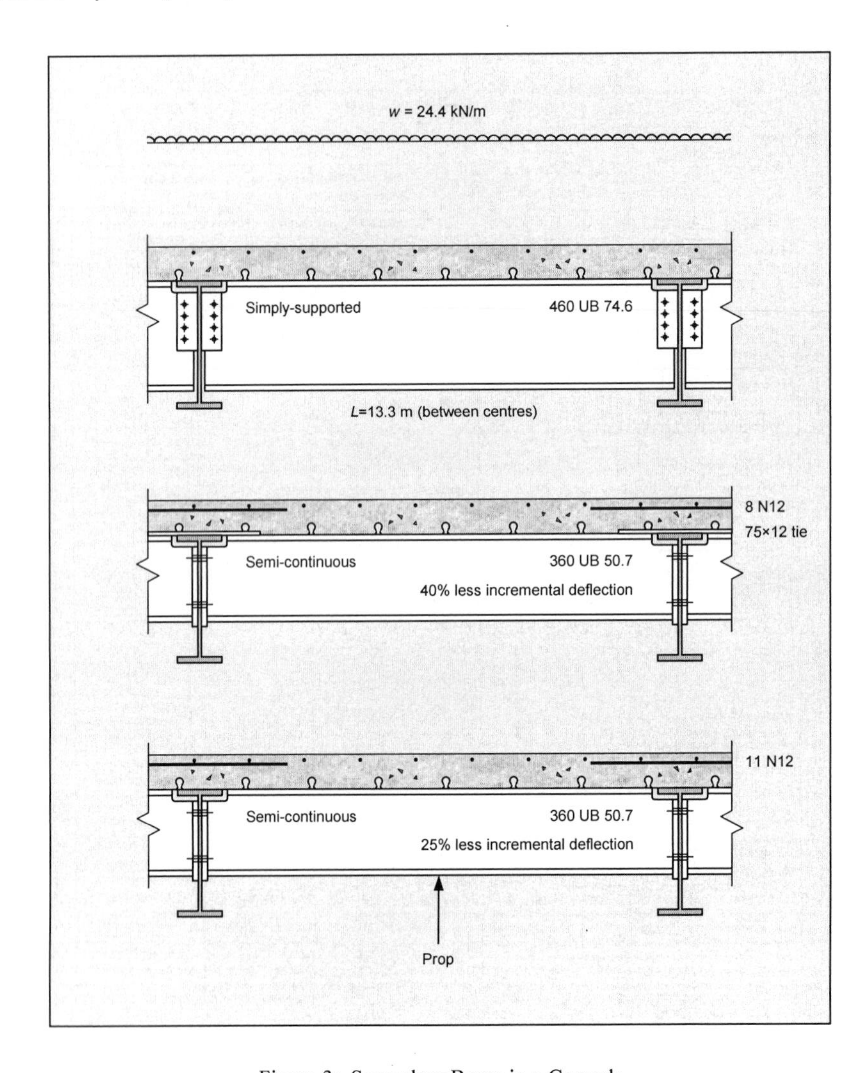

Figure 3: Secondary Beam in a Carpark

Primary Beam in a Carpark

A semi-continuous and a continuous beam option have been added to the simply-supported primary beam 'PB1' from scheme 1C of the BHP Design Guide for Economical Carparks (Watson *et al.* 1998), as shown in Figure 4. Again, continuous construction has resulted in significantly smaller beam sizes, while the total deflection, which was the governing factor in the simply-supported design, still reduced by 50% and 25% for the semi-continuous and continuous options respectively.

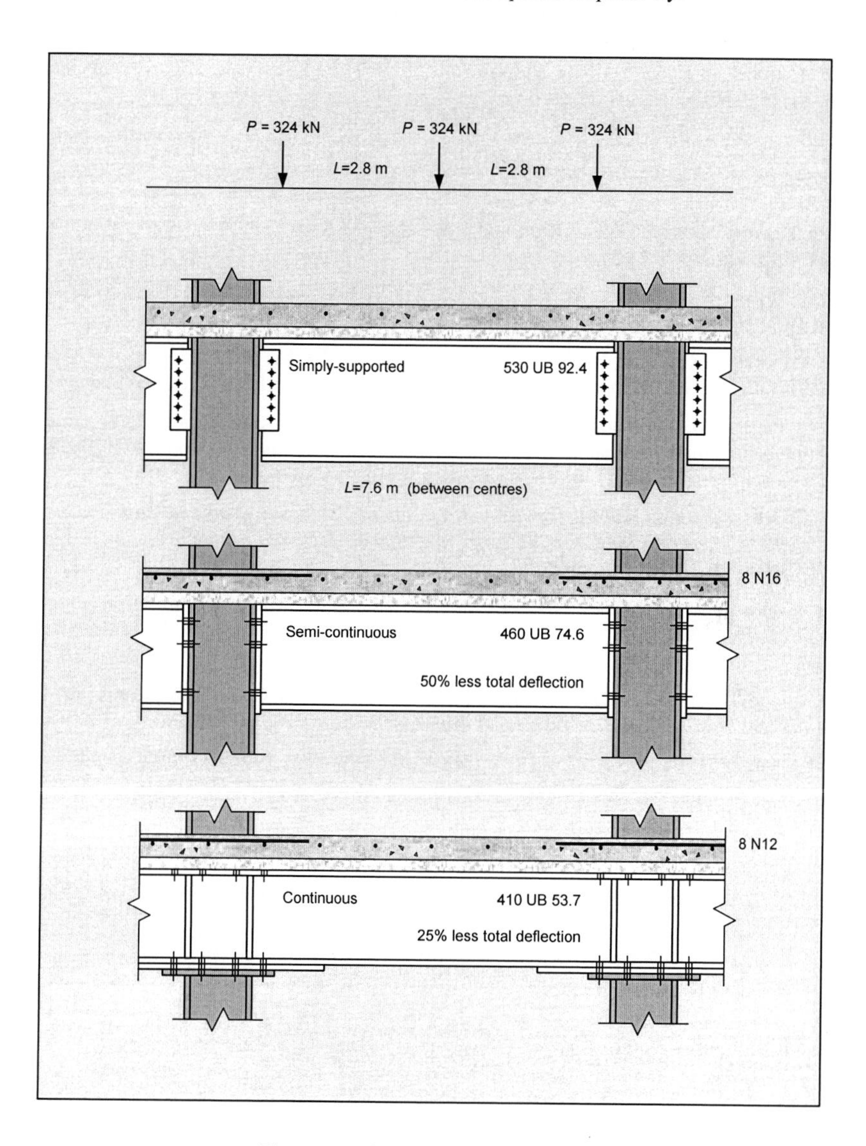

Figure 4: Primary Beam in a Carpark

CONCLUSIONS

Continuous composite beams have significant advantages compared to simply-supported composite beams, particularly with regard to serviceability performance. Continuous composite beams offer better crack control and lower deflections, even when the depth of the steel section is less than that required for the equivalent simply-supported beam. Continuous beams provide the designer with greater scope to vary design parameters, such as positive moment shear connection and negative moment reinforcement levels, to find the optimum solution for each situation. The perception that simply-supported composite beams have economic and construction benefits may be overemphasized. A rigorous cost-benefit analysis may show that continuous and semi-continuous composite beams should be the preferred construction option.

REFERENCES

Australian Institute of Steel Construction and Standards Australia (1997). *Composite Beam Design Handbook (SAA HB–1997)*, Sydney.

Berry, P.A., Bridge, R.Q., Patrick, M. and Wheeler, A.T. (2000). *Design Booklet DB5.1, Design of the Web-Side-Plate Steel Connection*, Ed. 1, OneSteel Market Mills, Sydney.

Berry, P.A., Bridge, R.Q. and Patrick, M. (2001a). *Design Booklet DB2.1, Design of Continuous Composite Beams with Rigid Connections for Strength*, Ed. 1, OneSteel Market Mills, Sydney.

Berry, P.A , Patrick, M. and Bridge, R.Q. (2001b). Development of Design Charts for Continuous and Semi-Continuous Composite Beams, *Proceedings, The First International Conference on Steel and Composite Structures (ICSCS'01)*, 14-16 June, Pusan, Korea, 369-376.

British Standards Institution (1994). *Eurocode 4: Design of Composite Steel and Concrete Structures, Part 1.1. General Rules and Rules for Buildings (together with United Kingdom National Application Document)*, DD ENV 1994-1-1, London.

Patrick, M., Berry, P.A. and Wheeler, A.T. (2002). Web-Side-Plate Framing Connections in both Steel and Composite Construction. *Proceedings, Advances in Building Technology*, 4-6 December, Hong Kong.

Patrick, M. and Dayawansa, P.H. (1998). Design of Continuous Composite Beams for Bending Strength. *Proceedings, Australasian Structural Engineering Conference*, Auckland.

Standards Australia (1996). *AS 2327.1–1996: Composite structures, Part 1: Simply supported beams*, Sydney.

Standards Australia (2001). *AS 3600–2001: Concrete Structures*, Sydney.

Steel Construction Institute (1998). *Joints in Steel Construction: Composite Connections*, SCI Publication 213, The Steel Construction Institute (in association with The British Constructional Steelwork Association), Ascot, United Kingdom.

Watson, K., Cottam, J. and Dallas, S. (1998). *Economical Carparks - A Design Guide*, Ed. 1., BHP Integrated Steel, Sydney.

Advances in Building Technology, Volume 1
M. Anson, J.M. Ko and E.S.S. Lam (Eds.)

INNOVATIONS IN COMPOSITE SLABS INCORPORATING PROFILED STEEL SHEETING

R.Q. Bridge and M. Patrick

Centre for Construction Technology & Research, College of Science, Technology & Environment, University of Western Sydney, Kingswood, NSW 2747, Australia,

ABSTRACT

The use of structural steel decking as a dual formwork/reinforcement system in Australian building practice has developed steadily since the 1960's. The pressure to improve conventional construction practices and to reduce construction time and overall cost coupled with competition from a number of manufacturers has led to new technological developments, the emergence of new applications, the development of new design methods and the improved education of design engineers.

KEYWORDS

Composite slab; design; friction; mechanical resistance; shear; standards; steel decking; tests.

INTRODUCTION

Non-composite, cellular steel decking originated in the USA in the early 1930's for constructing light-weight floors with in-floor services in steel-frame buildings, which led to the development of structural steel decking as we know it today (Dallaire, 1971). As part of this development, a less steel-intensive deck which acted compositely with the hardened concrete on account of steel reinforcing mesh welded to the decking, first appeared in 1950. Even more economical embossed decks were first developed in the 1960's (Viest, 1997) and composite slabs are now the preferred flooring alternative for steel-frame buildings in most developed countries. Despite its success in steel-frame buildings, steel decking has long been perceived as a "short-span (7-15 ft) construction, unsuitable to the longer spans of a concrete frame" (Dallaire, 1971). Several of the new developments described herein will eventually allow the Australian decks to be used structurally in concrete-frame

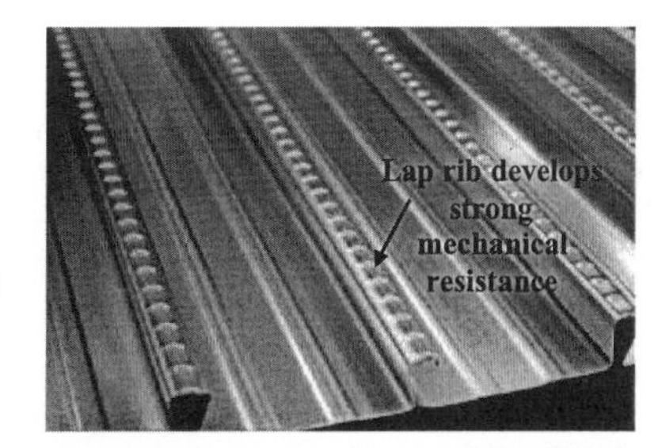

Fig. 1 An Australian Structural Steel Deck

buildings without any span restrictions applying (Patrick, 1998).

Composite slab construction has a long history of successful use in Australia, dating back to the mid-1960's. At that time there were no embossments on the dovetail ribs of the steel decking, Bondek, and the bond between the galvanised steel and concrete was principally achieved by chemical adhesion of the cement paste to the zinc coating. The major Australian decks now all include an asymmetric lap rib like that shown in Fig. 1 for Bondek II. It is now known that this type of rib develops strong mechanical resistance with hardened concrete, even if the rib is unembossed, although embossments are still very beneficial. An innovative small-scale test, developed by the authors, shows this. This and other innovations concerning composite slab behaviour and design are described herein.

STRUCTURAL DECKING - COMBINED FORMWORK/REINFORCEMENT SYSTEM

As steel decking can serve dual roles as both the formwork and as the main longitudinal tensile reinforcement in the bottom face of the slab, economic advantages should result compared with using removable formwork systems. In the case of short-span composite slabs typical of steel-frame buildings constructed unpropped, the large cross-sectional area of the decking, or more specifically its large tensile capacity, cannot normally be utilised efficiently in design for the composite stage. In this situation, the economic viability of the system is often governed by the formwork stage. Steel decking in its conventional trapezoidal form can also significantly reduce the strength of the shear connection between the steel beam and the slab, thereby reducing the efficiency of the composite beam.

For structural steel decking to be an advanced, efficient system, it must at least be possible for the decking to also be utilised efficiently during the composite stage. Neither should the decking impinge unduly on the performance of the composite member. Decks that develop strong mechanical resistance allow its tensile capacity to be well utilised in long-spanning or heavily-loaded situations. Otherwise, end anchorage is required, which may not be a practical or an economic proposition.

COMPOSITE SLAB CONSTRUCTION

Structural steel decking is produced with a wide variety of cross-section profiles. In the USA, there is a move to treat decking generically during the composite stage, in a bid to eliminate the need for standard slab testing. However, the amount of mechanical resistance they each develop with the concrete, which largely determines how effectively a deck can be used as longitudinal reinforcement in a slab, varies enormously depending on the decking profile and its features (see Fig. 2). The major American, British and European Standards for composite slabs, i.e. ANSI/ASCE 3-91, BS 5950: Part 4: 1994 and Eurocode 4: Part 1.1, permit any deck to be used compositely, irrespective of the amount of mechanical resistance it develops. However, the failure mode of a slab can be significantly affected by the strength (and ductility) of the shear connection (Patrick, 1989). This is at least recognised in Eurocode 4.

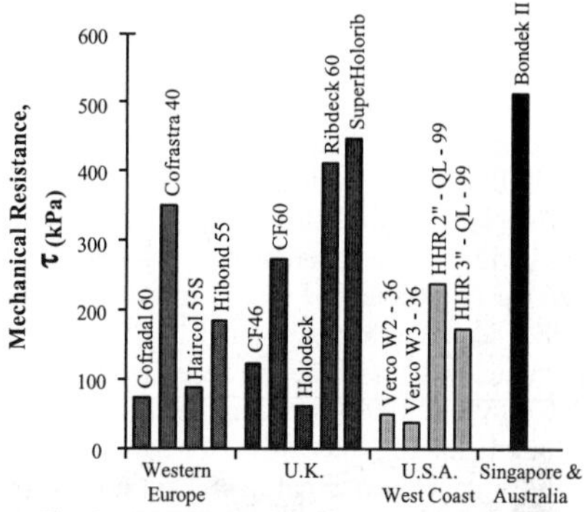

Fig. 2 Mechanical resistance developed by different steel decks (Patrick, 1990a)

The decks with very low mechanical resistance cannot be used in long spans. They can still be used without penalty in short spans and designed using the empirical shear-bond method (ASCE 3-91, BS5950: Part 4: 1994) even though their behaviour may be brittle. If the behaviour is ductile, i.e. if the mechanical resistance is sufficiently high, then the more advanced partial shear connection strength theory (Patrick, 1990b, Bode and Sauerborn, 1991, Patrick and Bridge, 1994) may be used as an alternative design method. Partial shear connection strength theory has many significant advantages over the empirical shear-bond theory used in BS 5950: Part 4. The strength and ductility of the mechanical resistance that the steel decking develops in the hardened concrete is known which improves the understanding of the design engineer; the loading arrangement and termination of the decking ends can be directly included; and small-scale tests can be used to derive design values for the major parameters used in the strength model.

COMPOSITE SLABS INCORPORATING STRUCTURAL STEEL DECKING

Although designers of composite slabs often focus on the determination of shear-bond capacity, for the Australian decks which can develop strong, ductile mechanical resistance, there are many other issues that should be of equal importance to the designer. Some of these are touched upon in this section.

Structural Testing

Detailed reports have been prepared which explain the tests that were conducted on nine composite slabs incorporating Bondek II structural steel decking. The basis for the tests, performed on slabs and Slip-Block™ Test specimens, has been fully documented after a detailed review of BS 5950: Part 4: 1982 (Patrick, 1991). The test results have been examined in detail (Patrick, 1994), and are briefly discussed in the analysis of test results that follows. All of these tests were concerned with the strength limit states of shear-bond or longitudinal slip, and flexure. A separate experimental investigation has been made into the effects of vertical shear, another important strength limit state (Patrick, 1993). This investigation showed that the design provisions for vertical shear in BS 5950: Part 4 and Eurocode 4: Part 1.1 are incorrect (Johnson, 1994).

Six slabs were tested in strict accordance with BS 5950: Part 4: 1982 to develop the shear-bond relationship (m and k values) to be used in design for a particular bare metal thickness t_b = 0.75 mm. Therefore, these slabs were designed to fail by longitudinal slip. Partial shear connection strength theory, and prior knowledge of the shear connection performance of the decking obtained from the Slip-Block™ Test, made this possible. It was necessary to limit the span of each of these test slabs so that they would not fail by flexure. The Slip-Block™ Test set-up is shown in Fig. 3(a), while a group of typical test specimens is shown prior to being poured in Fig. 3(b), and the test procedure is described elsewhere (Patrick, 1990b, 1994; Standards Australia, 1997).

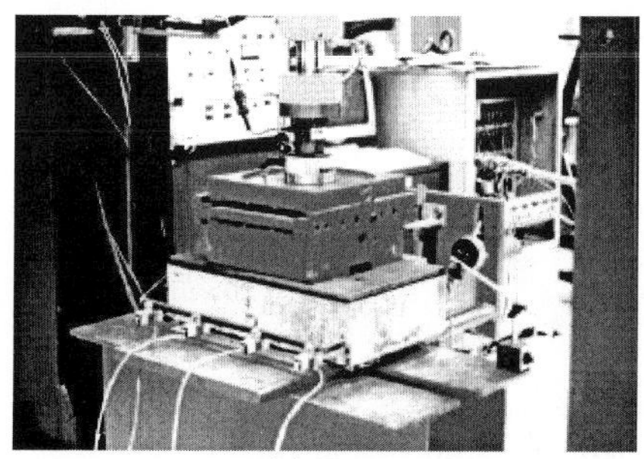

(a) Test set-up

(b) Test specimens as companions to slabs

Fig. 3 The Slip-Block™ Test

Fourteen simply-supported composite slabs were tested for vertical shear (Patrick, 1993). These are the first-known tests of their kind. The slabs were designed such that they could not fail by longitudinal slip if loaded at any location between the supports. Diagonal splitting failure was observed in some of the tests when loaded very close to the supports (see Fig. 4(a)). Flexural failure occurred if they were not loaded close to the supports. A slab could experience very large deflections in excess of span/30 in order to fracture the steel decking (see Fig. 4(b)).

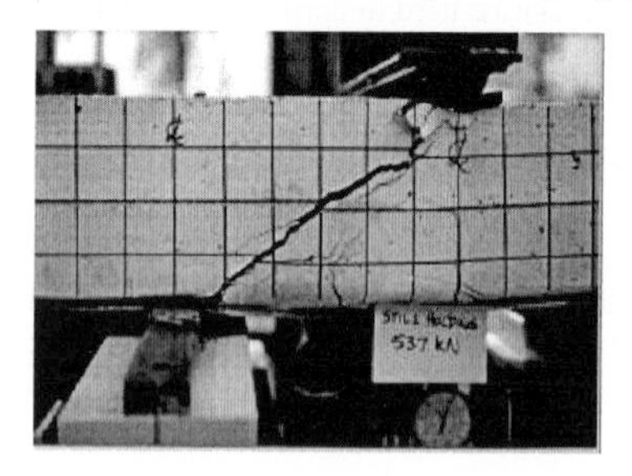

(a) Diagonal splitting failure

(b) Flexural failure

Fig. 4 Vertical Shear and Flexural Slab Tests

Analysis of Test Results for Design

The results from testing over eighty Slip-Block™ Test specimens have been used to derive a design relationship between mechanical resistance τ (kPa), bare metal thickness t_b (mm) and concrete compressive strength f'_c (MPa), viz. $\tau = 88\sqrt{(t_b f'_c)}$. The strength model assumed using partial shear connection strength theory which includes support friction is explained elsewhere (Patrick, 1989). The results of the Slip-Block™ Test gave accurate predictions of the slab strengths gained in the tests. Accurate prediction of the ultimate strength of slabs that failed by either longitudinal slip or flexure is illustrated by Fig. 5(a). The results in Fig. 5(b) illustrate the importance of including frictional resistance developed at the support in the strength prediction.

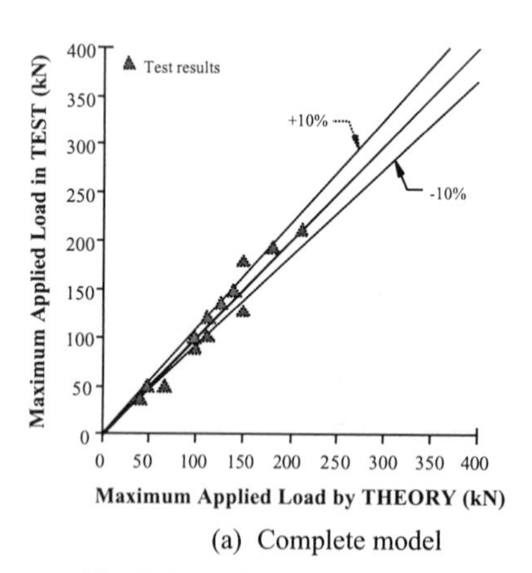

(a) Complete model

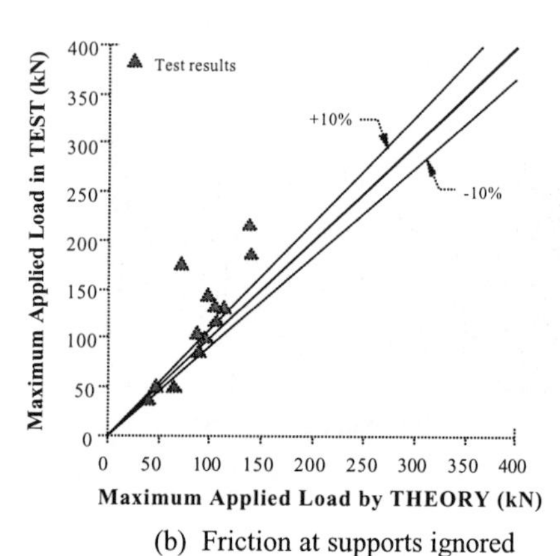

(b) Friction at supports ignored

Fig. 5 Strength Predictions Using Partial Shear Connection Strength Theory (Patrick, 1994)

It can be seen from this latter figure that the strength of some of the slabs cannot be explained if friction is ignored. Support friction can have a significant effect on the strength of composite slabs when the decking passes over supports, which is the normal case in steel-frame building construction (Johnson, 1994; Johnson and Anderson, 1993; Bode and Minas, 1997).

STRUCTURAL STEEL DECKING AS PERMANENT FORMWORK

Structural Testing

Extensive structural testing of the Bondek steel decking in single and multiple span arrangements has been performed in order to model the behaviour of the decking during the formwork stage. Simple design rules have been developed for calculating design action effects, design ultimate moments of resistance, design end shear resistance, effective bending stiffness and minimum bearing lengths under uniformly-distributed loading conditions. Some tests were also performed using concentrated loads, while all the tests were conducted in the absence of any wet concrete. Thus the decking can be designed using a consistent approach through the whole construction process, ignoring any increase in strength due to the stabilising effect of wet concrete.

The tests were performed in accordance with British Standard BS 5950: Part 4: 1982, which has subsequently been superseded by BS 5950: Part 4: 1994. The 1982 edition of BS 5950: Part 4 allowed the strength of the decking to be determined by testing. This option was taken with regard to both strength and deflection since it had been determined that the analytical methods in BS 5950: Part 4: 1982 were not appropriate for Bondek and resulted in unconservative designs. This is on account of the complex, undesirable behaviour of the lap joint which causes the decking to exhibit such effects as flange curling (with associated amplified lateral flange displacements and pan lifting), warping of lap joint flanges (consistent with shear lag effects), distortional buckling, moment softening, etc.

Reference is made in BS 5950: Part 4: 1994 to Section 7 of BS 5950: Part 6: 1995 to determine the strength and stiffness of the steel decking during the formwork stage. These testing provisions are essentially the same as those referred to in BS 5950: Part 4: 1982. The tests comprised: mid-span bending under distributed loading (21 tests); bending at a support under distributed loading (13 two-span and 10 three-span tests); shear at an end support (12 tests); and point-loading tests (20 tests).

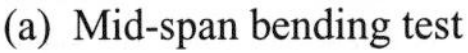

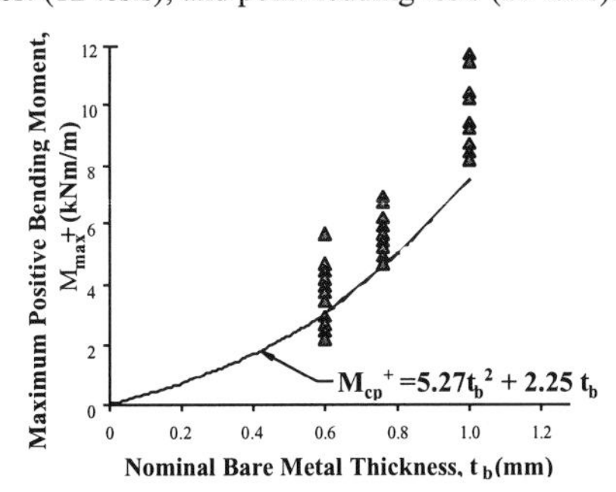

(a) Mid-span bending test (b) Design relationship and test results

Fig. 6 Design Ultimate Positive Moment of Resistance

Analysis of Test Results for Design

Forty-four test results were available to determine the design ultimate positive moment of resistance M_{cp}^+. Some of these came from mid-span bending tests (see Fig. 6(a)). The design relationship between M_{cp}^+ and bare metal thickness t_b is shown in Fig. 6(b) along with the test results. The maximum positive bending moments obtained in the tests were used. This was also the case for the multi-span tests (see Fig. 7(a)) in which moment-softening was exhibited at internal supports (see Fig. 7(b)).

The tests for bending at a support under distributed loading (see Fig. 7(a)) revealed that the continuous spans effectively became simply-supported due to substantial moment-softening of the support regions prior to the maximum positive bending moment being reached in the mid-span regions (see Fig. 7(b)). The sheets tended to shed load suddenly when the maximum positive moment was reached. Therefore, the design ultimate negative moment of resistance is assumed to equal zero, i.e. $M_{cp}^- = 0$.

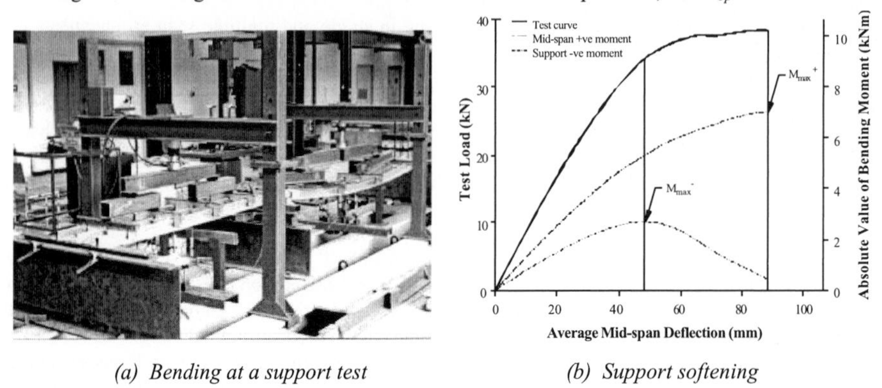

(a) Bending at a support test *(b) Support softening*

Fig. 7 Design Ultimate Negative Moment of Resistance

Estimates for the effective bending stiffness $E_s I_{ef}$ were determined from the slope of the linear-elastic portion of the load-deflection curves for the forty-four tests mentioned above (see Fig. 8(a)). The assumptions made were that the decking had a constant section, response was linear-elastic up to a mid-span deflection of span(L)/130, and $E_s = 210$ GPa. Linear regression analysis of the test data gave the relationships shown in Fig. 8(b) for single- and multiple-span situations. It can be noticed that the minimum and maximum spans tested were 1000 and 4000 mm, respectively.

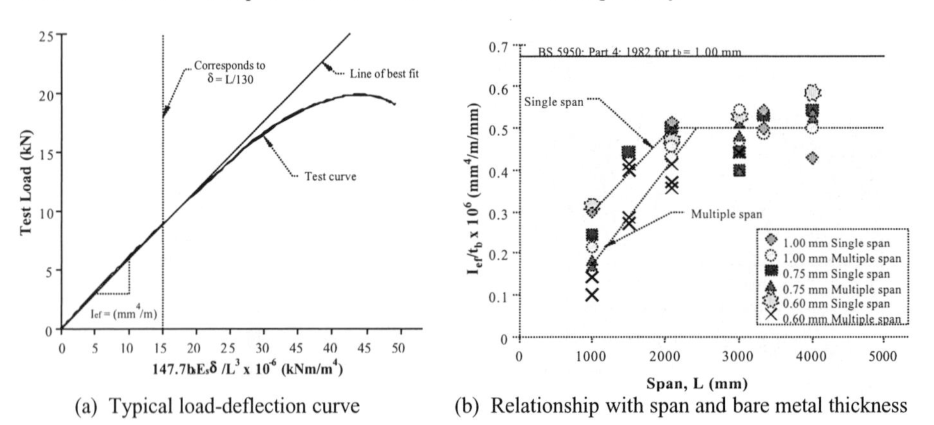

(a) Typical load-deflection curve (b) Relationship with span and bare metal thickness

Fig. 8 Effective Bending Stiffness

The relationship between design end shear capacity V_c and bare metal thickness t_b (see Fig. 9(b)) was determined from the end shear tests (see Fig. 9(a)). Two end bearing lengths of 25 and 50 mm were investigated. It was found that the bearing length had only a small effect on the shear capacity at a simple support.

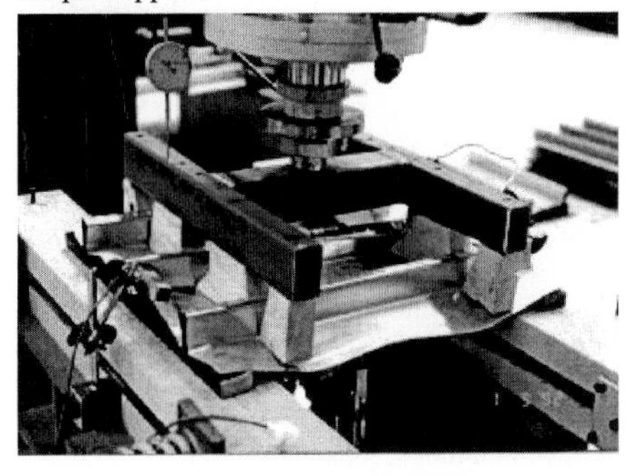

(a) End shear test

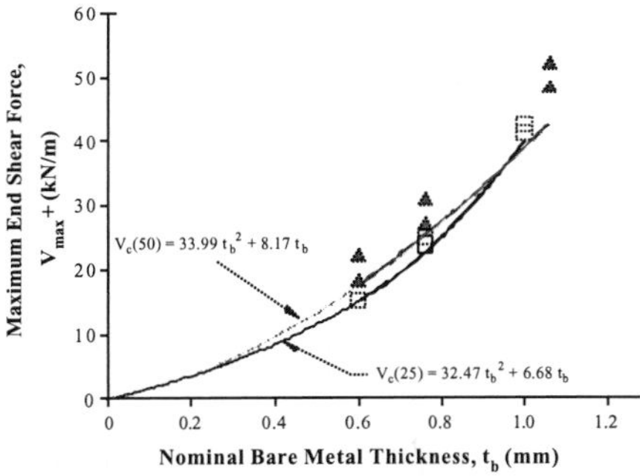

(b) Relationships with bare metal thickness

Fig. 9 Design End Shear Resistance

Relationships were also derived for the design ultimate point load (P_u) that could be applied in pans away from the decking edge. Span (L) and bare metal thickness (t_b) had a significant effect on the strength of the decking when supporting point loads. The load must be applied over an area of at least 100×100 mm^2, and at least one screw fastener must be placed in the lap joint at each mid-span location. Maximum spans were derived so that a point load of 2.3 kN can be supported with a load factor of 1.5, although for t_b =1.0 mm, L was limited to the maximum span tested (3250 mm).

NEW DEVELOPMENTS

The Centre for Construction Technology and Research at the University of Western Sydney has been developing a new long-spanning composite steel-concrete floor system called TRUSSDEK™ (OneSteel Reinforcing, 2001) in conjunction with a leading Australian company OneSteel Reinforcing. TRUSSDEK™ is a patented, ultra-lightweight, permanent combined formwork and reinforcement system incorporating light steel trusses with a structural deck that offers unrivalled spanning capabilities compared with other steel decking systems in Australia and is the winner of the prestigious 2001 Australian Design Award™. The product and its use on a project are shown in Fig. 10.

Fig. 10 TRUSSDEK™ long-spanning floor construction system

CONCLUSIONS

Structural steel decking has advanced as a dual formwork/reinforcement system in Australian building practice. Decks with strong mechanical resistance and ductile behaviour have been developed. Some innovative concepts and testing that have allowed these developments to occur have been described.

REFERENCES

ASCE 3-91 (1994). *Standard for the Structural Design of Composite Slabs*. American Society of Civil Engineers, New York.

BHP Structural Steel (1998). *Composite Structures Design Manual - Design of Composite Slabs for Strength.* Design Booklet DB3.1, May.

Bode, H. and Minas, F. (1997) Composite Slabs with and without End Anchorage under Static and Dynamic Loading. *Conference Report*, International Conf. on Composite Construction - Conventional and Innovative, IABSE, Innsbruck, Sept.

Bode, H. and Sauerborn, I. (1991). Partial Shear Connection Design of Composite Slabs. *Proceedings*, 3rd Int. Conf. Steel-Concrete Composite Structures, ICCS-3, Fukuoka, Japan, 467- 472.

BS 5950: Part 4: 1982. *Structural Use of Steelwork in Building, Part 4. Code of Practice for Design of Floors with Profiled Steel Sheeting*. British Standards Institution, London.

BS 5950: Part 4: 1994. *Structural Use of Steelwork in Building, Part 4. Code of Practice for Design of Design of Composite Slabs with Profiled Steel Sheeting*. British Standards Institution, London,

BS 5950: Part 6: 1995. *Structural Use of Steelwork in Building, Part 6. Code of Practice for Design of Light Gauge Profiled Steel Sheeting*. British Standards Institution, London.

Dallaire, E.E. (1971). Cellular Steel Floors Mature. *Civil Engineering*, ASCE, July, 70-74.

Eurocode 4: (1994). *Design of Composite Steel and Concrete Structures, Part 1.1. General Rules and Rules for Buildings (United Kingdom National Application Document).* DD ENV 1994-1-1: 1994, British Standards Institution, London.

Johnson, R.P. (1994). *Composite Structures of Steel and Concrete, Volume 1: Beams, Slabs, Columns, and Frames for Buildings*. 2nd ed., Blackwall Scientific Publ. London.

Johnson R.P. and Anderson, D. (1993). *Designers' Handbook to Eurocode 4, Part 1.1: Design of Composite Steel and Concrete Structures*. Thomas Telford, London.

OneSteel Reinforcing (2001). *Trussdek Span Charts.* OneSteel Reinforcing, Sydney, Australia

Patrick, M., (1989). Design of Continuous Composite Slabs - The Issue of Ductility. *AISC Steel Construction Journal*, **23:3**, 2-10.

Patrick, M. (1990a). The Slip Block Test — Experience with Some Overseas Profiles (Part A). *BHP Melb. Res. Labs Report*, MRL/PS64/90/021, June.

Patrick, M. (1990b). A New Partial Shear Connection Strength Model for Composite Slabs. *AISC Steel Construction Journal*, **24:3**, 2-17.

Patrick, M. (1991). Proposed Bondek II Composite Slab Tests to British Standard BS 5950: Part 4: 1982. *BHP Melb. Res. Labs Report*, BHPRML/PS64/91/029, Jan., 1991.

Patrick, M. (1993). Testing and Design of Bondek II Composite Slabs for Vertical Shear. *AISC Steel Construction Journal*, **27:2**, 2-26.

Patrick, M. (1994). *Shear Connection Performance of Profiled Steel Sheeting in Composite Slabs.* PhD Thesis, School of Civil Engineering, Univ. of Sydney.

Patrick, M., (1998). The Application of Structural Steel Decking to Commercial and Residential Buildings, *Malaysian Structural Steel Association Convention*, Dec.

Patrick, M. and Bridge, R.Q. (1994) Partial Shear Connection Design of Composite Slabs. *Engineering Structures Journal*, **16:4**, 348-362.

Standards Australia (1997). *Methods of Test for Elements of Composite Construction, Method 1: Slip-Block Test*. Proposed Standard, BD/32-0090-203, Oct.

Viest, I.M. (1997). Studies of Composite Construction at Illinois and Lehigh, 1940-1978. *Proceedings, Composite Construction in Steel and Concrete III*, ASCE, Irsee, 1-14.

Advances in Building Technology, Volume 1
M. Anson, J.M. Ko and E.S.S. Lam (Eds.)

WEB-SIDE-PLATE FRAMING CONNECTIONS IN BOTH STEEL AND COMPOSITE CONSTRUCTION

M. Patrick, P.A. Berry and A.T. Wheeler

Centre for Construction Technology & Research, College of Science, Technology & Environment, University of Western Sydney, Kingswood, NSW 2747, Australia,

ABSTRACT

The web-side-plate connection is by far the most common connection used in Australia to support simply-supported steel or composite beams in steel building frames. It is also used extensively in other developed countries, although it is known by different names and its details may vary to some degree.

An up-to-date, detailed review was made of Australian and overseas research and design practice for the web-side-plate steel connection. This led to an extensive series of new Australian tests being undertaken on the connection using a novel test rig, and an improved design model being developed for the connection in the bare steel state. This allowed some significant restrictions to be removed that applied at the time to the Australian limit-state design method for the steel connection.

In composite construction, longitudinal reinforcement located in the concrete slab over the connection, which is needed for crack control, can cause significant compressive force to develop in the cleat. The effect of this force has normally been ignored in design, irrespective of its potential magnitude. This can cause problems in practice under both serviceability and ultimate load conditions, which can only be avoided if the behaviour of the web-side-plate composite connection is better understood and appropriate design rules are developed. Preliminary results from several types of tests undertaken for this purpose are presented and briefly discussed. Future work is also described.

KEYWORDS

Web-side-plate connection; single-plate connection; fin plate; steel connection; composite connection.

INTRODUCTION

The web-side-plate steel connection is used to connect steel or composite beams to a supporting member or element such as a steel beam, a steel column or a concrete core wall. The type of construction envisaged is shown in Figure 1, noting that the uncoped primary and coped secondary steel beams shown could equally be elements of composite beams if shear connectors and a concrete slab were located on top of the steel beams. It has been assumed to qualify as being suitable for simple construction according to AS 4100 (Standards Australia 1998).

When the web-side-plate steel connection is used with composite beams, the longitudinal reinforcement in the slab can lead to continuity, thus developing a resultant compressive force in the cleat. In this instance, the connection is referred to as a web-side-plate composite connection. The compressive force in the cleat can affect the behaviour of the connection at ultimate load. Design rules developed by Berry et al. (2001) place an upper limit on the compressive force developed in the cleat at ultimate load, and provided that this condition is met, the connection may be treated as a web-side-plate steel connection. Design rules that allow for composite action are currently under development.

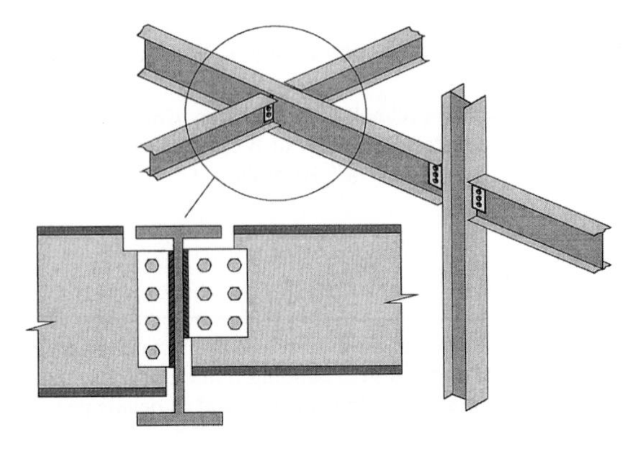

Figure 1: Typical construction using the web-side-plate steel connection

WEB-SIDE-PLATE STEEL CONNECTION

Review of Research and Design Practice

An up-to-date, detailed review has been made of Australian and overseas research and design practice for the web-side-plate steel connection (Berry, et al. 2000b). This led to an extensive series of new Australian tests being undertaken on the connection using a novel test rig, and an improved design model for the bare steel connection being developed. Moreover, the rotation capacity or ductility of the connection was found to be much greater than previously believed. This allowed some significant restrictions to be removed that applied at the time to the limit-state Australian design method for the connection (Hogan & Thomas 1994). Some significant advantages that followed, which lead to economic improvements, are that much more flexible beams may be supported by the connection (allowing much longer spans), much smaller welds are required, and fewer bolts and/or thinner cleats may be used in equivalent situations. Also, thicker cleats than were previously allowed may be specified, whereby the large additional strength of double-column bolt groups over single-column groups can be utilised leading to shallower connections with improved rotation capacity. Australian design engineers now use the improved design method (Berry et al. 2000a) and associated software.

Test Program

A highly innovative program of laboratory testing was undertaken to study the moment-rotation behaviour of the web-side-plate steel connection, and to gain values for key parameters used in the design model when the connection is attached to a rigid support (Berry et al. 2000a). The test rig that was used is unique, and it allowed over 50 tests to be performed to study the effect of major parameters such as plate thickness, steel grade and different bolt group arrangements. Moreover, the rig was also designed to simulate the conditions experienced for the web-side-plate composite connection - see Figure 4. It is planned in the future that other innovative types of tests will be performed to greatly reduce the need for testing full-scale composite connections, the results from which other researchers have previously depended too heavily upon.

In previous work in Australia, the web-side-plate steel connection has been tested in the rigid support condition using a relatively complex procedure of lowering the free end of a long steel beam attached by the connection to a column (Pham & Mansell 1982, Patrick et al. 1986). Xiao et al. (1994) and other overseas researchers have performed tests on the web-side-plate composite connection, although only small numbers of tests have been performed, typically in a conventional cruciform arrangement involving twin cantilever steel beams connected to a continuous composite slab. Patrick et al. (1986) have explained some of the technical limitations with tests of this type. The economic disadvantage is obvious, and largely explains why so few tests have been performed; a problem it seems is common to connection testing. Also, the compressive force in the steel part of the connection has not been measured, and nor can it necessarily be determined accurately from the test results, even though this is a critical consideration during the modelling and design of the connection components. Ren (1995) claims to have performed the first tests of their type on several other types of bare steel connections (e.g. double angle cleats) to investigate the effect of axial compressive force on moment-rotation response. Ren attempted to simulate the conditions experienced by a steel connection at a support in the hogging moment region of a continuous composite beam. A test rig was devised whereby a vertical force was applied at the end of a steel beam which could be rotated and fixed in one of four possible positions, ranging from a cantilever (no axial force with the beam horizontal) through to a column (no bending with the beam vertical). Otherwise, the load was applied at either 30 or 60 degrees to the longitudinal axis of the steel beam. A disadvantage with this system is that the ratio of axial force to shear force effectively remains constant during a test, and is not affected by the behaviour of the connection, which it would be in a beam in a real structure.

The special test rig shown in Figure 2(a) was developed which uses small plate specimens like that shown in Figure 2(b) to simulate a web-side-plate connection subjected to rotation in either the bare steel or composite states. Shear force is assumed to not significantly affect the response and is not applied during the test. The test specimens do not incorporate a steel beam or a concrete slab, whereby the cost and time associated with their construction and testing is very considerably reduced. Therefore, it is possible to conduct a comparatively large number of tests, and thus systematically examine the effect that different variables have on moment-rotation response.

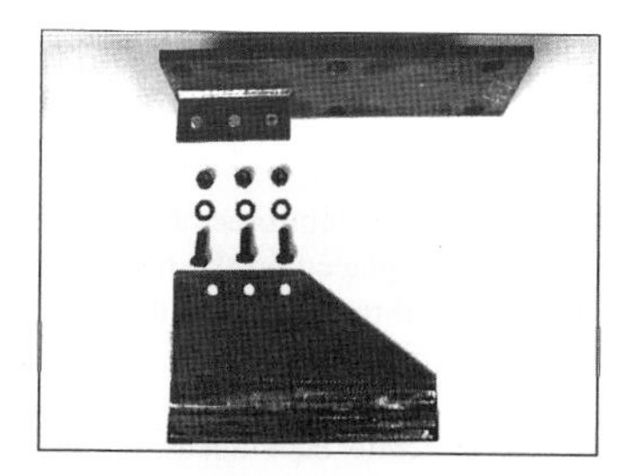

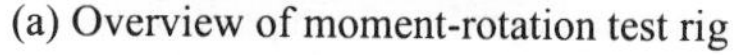

(a) Overview of moment-rotation test rig (b) Typical test specimen components

Figure 2: Special moment-rotation test rig and typical unassembled test specimen

WEB-SIDE-PLATE COMPOSITE CONNECTION

Review of Research and Design Practice

Overseas researchers are currently focussing their attention on the development of composite connections or joints that can be used with semi-continuous or continuous composite beams (Anderson 2000, Kemp & Nethercot 2000, Crisinel & Kattner 2000). The behaviour of the web-side-plate composite connection has previously been investigated in full-scale tests (Xiao et al. 1994), and a design method has been proposed for predicting moment capacity using the results of six tests (Ahmed et al. 1997). However, like some other types of connections that do not permit the compressive force in the bottom flange of the steel beam to be transmitted directly through the connection, its moment capacity and rotational stiffness are low. Some modifications have been made to the web-side-plate composite connection, which involve the use of a contact plate, or an angle, placed between the bottom flange and the support, in order to improve the strength and stiffness (Ahmed et al. 1997, Bode & Kronenberger 1996, Robertson 1997). However, the primary objective of this investigation will be to study the web-side-plate composite connection as it is constructed in practice in many composite buildings in Australia. Modifications to the connection to improve performance for use with continuous composite beams are beyond the scope of the investigation.

Limited studies have been made into the behaviour of the web-side-plate composite connection, and adequate design models do not exist, the connection normally being designed ignoring composite action and treating it as an ordinary web-side-plate steel connection. Design models are needed to address serviceability design issues, particularly crack control, as well as strength.

Australian Construction Practice

Extensive use is made of composite construction in multi-storey buildings in developed countries. Australian Standard AS 2327.1 addresses the design of simply-supported composite beams, and most composite beams in Australian buildings take this form. This is because the design calculations are simple to perform, and the web-side-plate steel connection can be used in the steel frame, which is simple to fabricate and erection of the structural steel is rapid.

An important restriction has been placed on the use of the new, improved design method for the web-side-plate steel connection (Berry et al. 2000b), viz.: the compressive force that develops in the cleat at the strength limit state must not exceed 15 percent of its nominal squash load, i.e. $0.15f_{yi}t_id_i$, where these variables are the yield stress, thickness and overall depth of the cleat, respectively. Larger compressive forces may adversely affect the moment-rotation response of the connection, which is critical in design. No guidance is given to designers using the improved method if the compressive force in the cleat exceeds this limit. Then they do not have a design model to use.

In a floor of a building comprising simply-supported composite beams, compressive force develops in the cleats wherever the concrete slab extends between adjacent ends of the composite beams, and continuity of longitudinal reinforcement in the slab causes the reinforcement to be in tension over the connections. The limit placed on the maximum compressive force that may act on the cleat of a web-side-plate connection directly affects the maximum amount of longitudinal reinforcing steel that may be placed in the concrete slab.

Cracking of the concrete slab needs to be controlled in composite buildings where durability is important such as in multi-storey carparks. There are cases in Australia of composite carparks that are excessively cracked due to an inadequate amount of reinforcement in the slab over the web-side-plate connections. Some of them are in disrepair despite still being used (see Figure 3). Cracking can also cause concern in buildings with brittle floor finishes, or where the cracks can form directly along lines of shear connectors raising questions about the integrity of the shear connection. Again, this is a common occurrence in Australian composite buildings incorporating web-side-plate steel connections. Temporary propping of the supporting steelwork during construction greatly increases the chance of excessive cracking occurring, and normally the concrete cracks immediately the props are removed. If the quantity of reinforcing steel is small and it yields at the same time the props are removed, then the crack widths become uncontrollable. Even in unpropped construction, normal drying shrinkage of the concrete during the early life of a floor in a building can cause excessively wide cracks to form over web-side-plate connections at the beam ends. In order to control cracking, there needs to be a sufficient quantity of longitudinal reinforcement over the connections.

(a) Open-deck layout

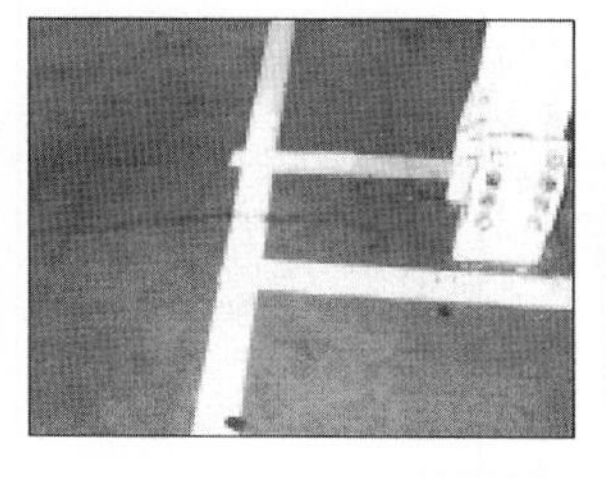

(b) Excessively wide cracks

(c) Corrosion of steel

Figure 3: Australian carpark constructed with web-side-plate connections

The latest Australian design guidance on controlling cracking in concrete slabs of continuous composite beams (Berry et al. 2001) indicates that, typically, the tensile capacity of the reinforcement will be greatly in excess of 15 percent of the cleat squash load. Clear examples of this situation can readily be taken from an Australian design guide for carparks (Watson et al. 1998), that requires additional reinforcement to be placed over the web-side-plate connections to provide crack control. Typically, the tensile capacity of the additional bars alone is more than four times the limit placed on the maximum cleat compressive force. Moreover, in designs involving the largest web-side-plate connections, the compressive force in the cleat could be in excess of 100 tonnes. Therefore, new design rules are needed to cater for greater compressive force in the cleat if cracking is to be controlled. The problem cannot be simply overcome by increasing the thickness and/or yield stress of the cleat, and therefore its squash load, to enable the connection to be designed ignoring composite action. This is because the beam web or the bolt group would then become the component that controls the moment-rotation response of the connection rather than the cleat. An upper limit on the compressive force is necessary using the current design model.

It follows from above that design rules for the web-side-plate composite connection are required with some urgency, viz.: (a) design engineers who choose to comply with the limit on cleat compressive force and use light amounts of reinforcement will continue to experience excessively wide cracks, and therefore problems associated with poor durability and aesthetics; while (b) those who ignore the limit by using large amounts of reinforcement needed for crack control, may be well outside the scope of the existing design model, and risk premature or sudden structural failure of the connections and therefore the beams.

New Australian design rules for more substantial types of composite connections, including partial strength and full strength versions, have recently been developed (Berry et al. 2001). However, many situations will still require the web-side-plate composite connection to be widely used in various parts of a composite floor. For example, it is likely to continue to be the preferred alternative for composite secondary beams, or where the beam ends frame into building edges and logically remain simple supports.

Test Program

The test rig shown in Figure 2(a) has been designed to also simulate the behaviour of the web-side-plate composite connection by allowing a steel pin to be inserted into the rig as explained by Figure 4.

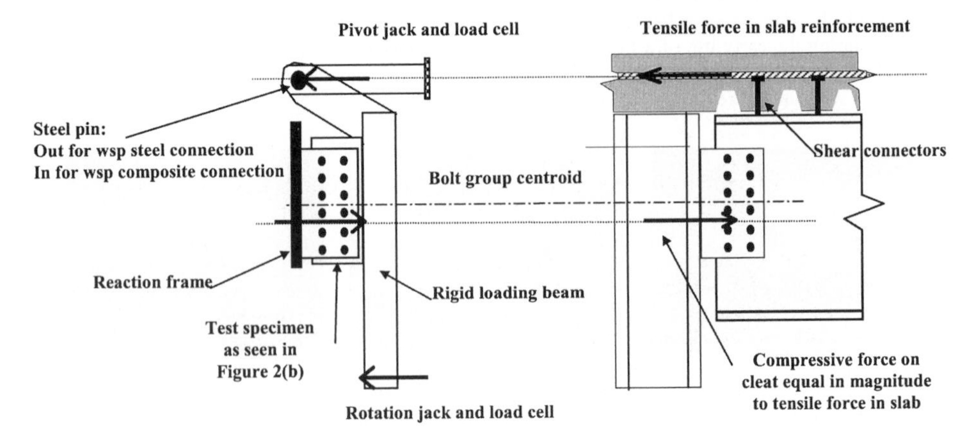

Figure 4: Testing of the web-side-plate composite connection

When large compressive force acts with the steel pin engaged, the connection could buckle as seen in Figure 5(a). Other small-scale tests like that shown in Figure 5(b) have also been performed to investigate the complex interaction between the cleat, beam web and bolt group.

(a) Buckled specimen in moment-rotation test rig (b) Buckled bolted web and cleat strip

Figure 5: Tests involving buckling of web-side-plate connection

Some preliminary test results have already been obtained using the test rig shown in Figures 2(a) and 4, with the pin engaged to simulate the situation for a web-side-plate composite connection. Referring to Figures 6(a) and (b), which came from tests on web-side-plate steel and composite connections, respectively, with a 6×2 bolt group, the 300 MPa grade cleat and web plates were both 10 mm thick. Part of the same bare steel curve is shown dashed in Figure 6(b). It can be seen that both the secant stiffness and moment capacity were significantly increased in the composite connection, while rotation capacity was greatly reduced. Also in this test on the composite connection, the compressive force on the cleat reached 988 kN, which explains the change in failure mode from a cleat tear to a cleat/web buckle, cf. Figure 6(a).

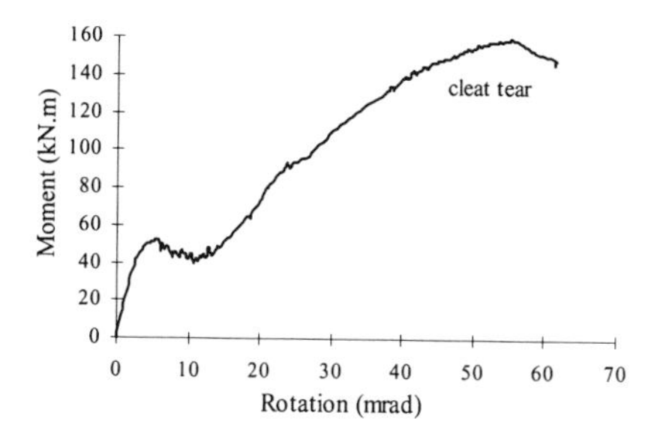

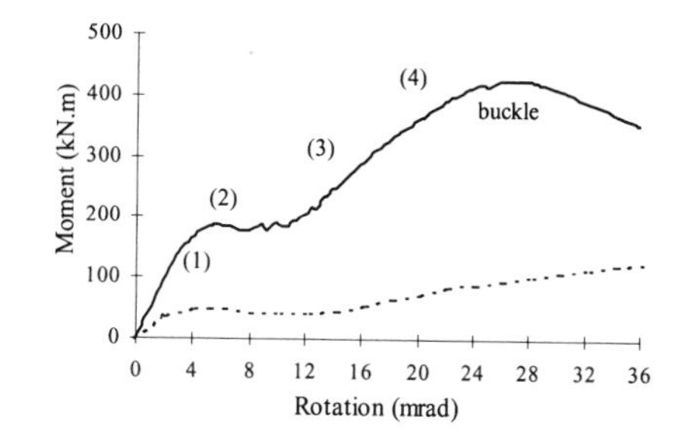

(a) Web-side-plate steel connection (6×2) (b) Web-side-plate composite connection (6×2)

Figure 6: Comparison of results from steel and composite web-side-side plate tests

Another special test rig shown in Figure 7(a) has also already been built to investigate the development of cracking in concrete or composite slabs under direct tension (Adams & Patrick 1998). For example, the test specimens (see Figure 7(b)) can represent a composite slab incorporating profiled steel sheeting situated in the vicinity of a secondary-to-primary beam connection. In this region, tension is created from secondary-beam end rotation. The sheeting spans perpendicular to the secondary beams, and therefore to the direction of the tensile force. Control of cracking in this region is of particular concern due to the discontinuity in the concrete created by the sheeting ribs.

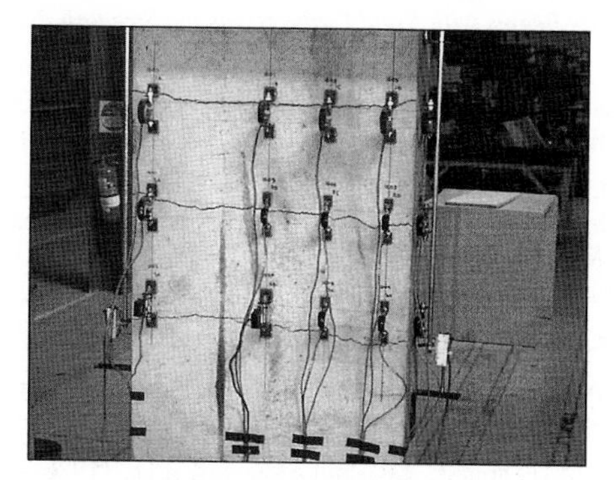

(a) Overview of test rig (b) Typical test specimen

Figure 7: Direct tension tests on reinforced composite slab

CONCLUSIONS

New Australian research has led to improvements in the design of the web-side-plate steel connection, and as a consequence, also of the simply-supported beams they support.

A number of composite buildings in Australia that have been constructed using the web-side-plate connection have had excessively wide cracks develop in the concrete slabs. This is because the beams have been designed as simply-supported without any consideration being given to this serviceability issue. The compressive force that develops on the connection when the slab is continuous, which can reduce its shear capacity and ductility, has also been ignored. Further research, supported by the results of a carefully-planned experimental program using the types of tests described in the paper, is required before the web-side-plate composite connection can also be designed with sufficient confidence to avoid problems in practice.

REFERENCES

Adams, J.C. and Patrick, M. (1998). *Testing of the Web-Side-Plate Semi-Rigid Composite Connection*, Proc. Australasian Structural Engineering Conference (ASEC-98), Auckland, 133-140.

Ahmed, B., Li, T.Q. and Nethercot, D.A. (1997). Design of Composite Finplate and Angle Cleated Connections, *J. Construct. Steel Research,* **41**, 1-29.

Anderson, D. (2000). *European Recommendations for the Design of Composite Joints*, Theorie und Praxis im Konstruktiven Ingenieurbau (Festschrift zu Ehren von Prof. Dr.-Ing. Helmut Bode), Ibidem-Verlag, Stuttgart, 457-464.

Berry, P.A., Bridge, R.Q., Patrick, M. and Wheeler, A.T. (2000a). *Design Booklet DB5.1, Design of the Web-Side-Plate Steel Connection*, Ed. 1, OneSteel Market Mills.

Berry, P.A., Bridge, R.Q., Patrick, M. and Wheeler, A.T. (2000b). *Design Booklet SDB5.1, Supplement - Web-Side-Plate Steel Connection Improved Design Method*, Ed. 1, OneSteel Market Mills.

Berry, P.A., Bridge, R.Q. and Patrick, M. (2001). *Design Booklet DB2.1, Design of Continuous Composite Beams with Rigid Connections for Strength*, Ed. 1, OneSteel Market Mills.

Bode, H. and Kronenberger, H.J. (1996). *Behaviour of Composite Joints and their Influence on Semi-Continuous Composite Beams*, Composite Construction in Steel and Concrete III, Proc. Engineering Foundation Conference, Irsee, Germany, ASCE, 766-779.

Crisinel, M. and Kattner, M. (2000). *A Design Method for Braced Steel Frames Comprising Semi-Continuous Composite Joints*, Proc. Engineering Foundation Conf., Composite Construction IV, Vol. 1, M1: Composite Systems I.

Hogan, T. J. and Thomas, I. R. (1994). *Design of Structural Connections*, 4th Edition, Australian Institute of Steel Construction, Sydney.

Kemp, A.R. and Nethercot, D.A., *Satisfying Ductility Criteria in Continuous Composite Beams*, Proc. Engineering Foundation Conf., Composite Construction IV, Vol. 1, T1: Composite Beams, May, 2000.

Patrick, M., Thomas, I.R. and Bennetts, I.D. (1986). Testing of the web-side-plate connection, *Australian Welding Research,* AWRA Report P6-1-87.

Pham, L. and Mansell, D.S. (1982). Testing of standardised connections, *Australian Welding Research,* AWRA Report P6-22-81.

Ren, P. (1995). *Numerical modelling and experimental analysis of steel beam-to-column connections allowing for the influence of reinforced-concrete slabs*, Swiss Federal Institute of Technology, Lausanne, Thesis No. 1369.

Robertson, B.W. (1997). *Semi-continuous Composite Beams*, BE Thesis, Dept. Civil, Mining and Environmental Engineering, University of Wollongong.

Standards Australia (1998), *AS 4100-2000: Steel Structures*, Sydney.

Watson, K.B., et al. (1998). *Economical Carparks - A Design Guide*, 1st Ed., BHP Integrated Steel.

Xiao, Y., Choo, B.S. and Nethercot, D.A. (1994). Composite connections in steel and concrete. I. Experimental behaviour of composite beam-to-column connections, *J. Construct. Steel Research,* **31**, 3-30.

Advances in Building Technology, Volume 1
M. Anson, J.M. Ko and E.S.S. Lam (Eds.)

NOVEL NEW REINFORCING COMPONENTS FOR COMPOSITE BEAMS

M. Patrick and R.Q. Bridge

Centre for Construction Technology & Research, College of Science, Technology & Environment, University of Western Sydney, Kingswood, NSW 2747, Australia

ABSTRACT

Several novel new reinforcing components have been developed in Australia to improve detailing of the shear connection in composite beams. By using these components: unsatisfactory failure modes are suppressed; the amount of reinforcement required for longitudinal shear resistance is significantly reduced compared with conventional detailing; design is simplified; and installation of reinforcement on site is much easier and faster. Rules developed for design engineers to detail beams incorporating the new products are described in detail in the paper.

One type of component has been developed to prevent a special type of longitudinal shear failure from occurring in the concrete flange of secondary composite edge beams when profiled steel sheeting is laid transverse to the longitudinal axis of the steel beam. It is now mandatory in Australian Standard AS 2327.1 "Simply-Supported Composite Beams" to provide reinforcement for this purpose. A method for designing this reinforcement is proposed for the first time. Another type of component was developed to prevent a similar type of longitudinal shear failure in primary composite beams with widely-spaced shear connectors. A further two types of components have been developed for use in any primary and secondary composite internal or edge beams where the other types of components are not required. All the components can be used in either continuous or simply-supported composite beams, as described in the paper.

KEYWORDS

Composite beam; shear connector; longitudinal shear; reinforcing steel.

INTRODUCTION

The shear connection of a composite beam comprises the five components shown in Figure 1. They can all influence the behaviour of the shear connection, and can be considered to include the shear connectors, profiled steel sheeting, slab reinforcement (including longitudinal bars if beams are continuous), concrete slab and steel beam top flange.

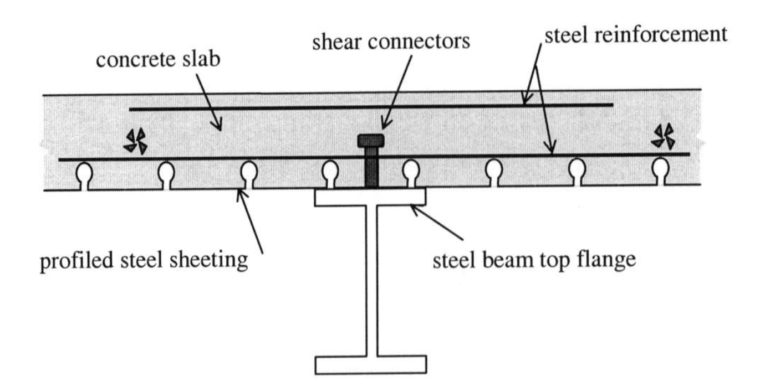

Figure 1: Components of the shear connection of a composite beam

Three new, patented reinforcing components have been developed in Australia to ensure that design engineers can detail economical longitudinal shear reinforcement in the presence of profiled steel sheeting. Otherwise, it is possible that non-composite beams, or even other forms of construction, will be more economical. The new components are described, and were designed to be used with automatically welded, headed studs, which are the only type of shear connector referred to in AS 2327.1 (Standards Australia 1996) that may be attached directly through profiled steel sheeting.

The strength design method given in AS 2327.1 requires that the shear connection is "ductile", which can be assessed by knowing the shape of the load-slip curves from push-out tests, and also the magnitude of the slips required along typical composite beams at ultimate load. No specific guidance is given to designers on either of these matters in AS 2327.1. However, Eurocode 4, Part 1.1 (British Standards Institution 1994) requires ductile shear connectors to have a slip capacity of at least 6 mm. Ductile and brittle behaviour of the shear connection, as well as the assumed model for design in AS 2327.1, are shown diagrammatically in Figure 2. A feature of this model for the Australian shear connection covered by AS 2327.1 is that it reaches its design shear capacity, f_{ds}, with very little slip (in practice, typically 2 to 4 mm) and maintains it indefinitely (typically 8 to 10 mm).

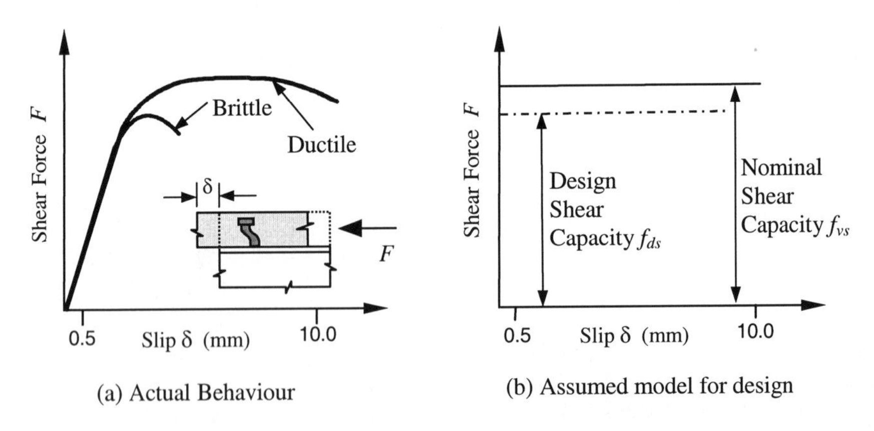

Figure 2: Load-slip behaviour of the shear connection

WAVE-FORM REINFORCEMENT – RIBS PERPENDICULAR

Rib Shearing Failure Mode

A patented wave-form reinforcing component, DECKMESH™ (OneSteel Reinforcing 2001), has been developed to prevent rib shearing failure in composite edge beams when profiled steel sheeting is laid transverse to the longitudinal axis of the steel beam, and welded-stud shear connectors are fastened directly through the sheeting. This brittle failure mode involves horizontal cracks that form in the concrete ribs between the tops of the steel sheeting ribs in pans where there are shear studs, while the failure surface locally avoids the studs by passing over their heads. It is now mandatory in AS 2327.1 to reinforce against rib shearing failure, in certain circumstances, which is referred to therein as Type 4 longitudinal shear failure. The product has to date been developed for use with BONDEK II and CONDECK HP (see Figure 3), but can potentially be used with many other profiles including those with wide steel ribs, e.g. Patrick (2001).

(a) DECKMESH™ -B for BONDEK II

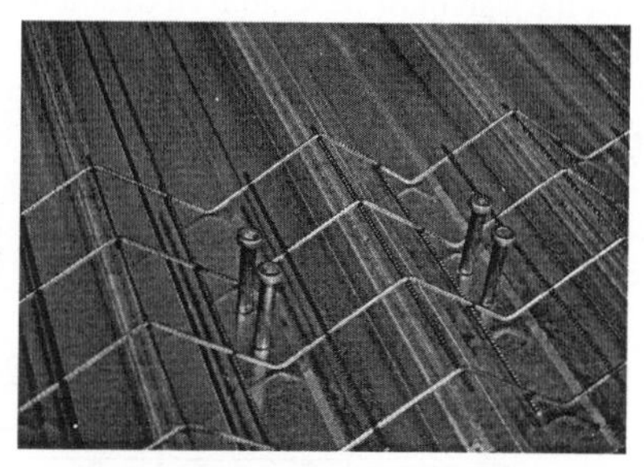

(b) DECKMESH™-C for CONDECK

Figure 3: DECKMESH™ longitudinal shear reinforcement

A more detailed account of the development of DECKMESH™ to reinforce against rib shearing failure is given by Patrick (2000). The severe delamination of the cover slab of the composite beam occurred as a result of a rib shearing failure over each pan along the beam which were all fitted with shear connectors. At the end of the test, the maximum slip had only reached 2.5 mm. It is worth noting that the horizontal cracks did not penetrate to the far side of the slab. The beam was detailed in accordance with BS 5950: Part 3.1 (British Standards Institution 1990) and Eurocode 4, Part 1.1 as seen in Figure 4(b). While the heavy, 16 mm diameter U-bars successfully controlled longitudinal splitting, they could not prevent rib shearing failure, and in fact possibly weakened the failure surface.

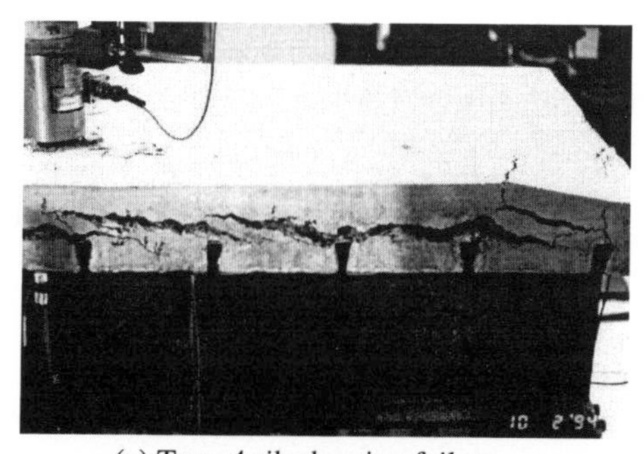

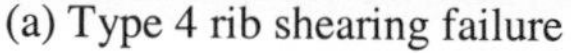

(a) Type 4 rib shearing failure

(b) U-bars required by some Standards

Figure 4: Rib shearing failure in edge beams detailed using BS 5950 and Eurocode 4

The situations when Type 4 shear surfaces must be reinforced are described in Clause 9.8.1 of AS 2327.1. The product DECKMESH™, shown in Figure 3, can be used to reinforce a Type 4 shear surface in a beam incorporating either BONDEK II or CONDECK HP without requiring design calculations. DECKMESH™ should only be used when the sheeting ribs are deemed perpendicular to the longitudinal axis of the steel beam. Some out-of-alignment of the sheeting ribs (up to about 15 degrees) can be accommodated by either twisting the DECKMESH™ panels around slightly (CONDECK HP option) or off-setting them by up to 150 mm (BONDEK II option).

Designing Against Rib Shearing Failure

Tests have shown that rib shearing failure is caused by a flexural crack that initiates in the concrete at the top of the sheeting rib on the opposite side to where the compressive force in the slab thrusts on the shear connectors (point A in Figure 5). The crack propagates horizontally across the top of the concrete rib, locally passing over the top of the welded studs, as shown in Figure 5. Tests have also shown that the slab must be reinforced over a width of at least 400 mm in order to suppress the failure mode completely. Clause 9.8 of AS 2327.1 requires four reinforcing wires or bars to be placed on each side of the concrete rib, with a maximum centre-to-centre spacing of 150 mm measured in the transverse direction. Correspondingly, it will be assumed that the effective width of the slab, b_{eff}, with respect to resisting rib shearing failure, normally equals 600 mm in edge beams, and is centred about the steel beam, except when the concrete flange outstand on one side of the beam is less than 300 mm measured from the vertical centreline of the steel beam.

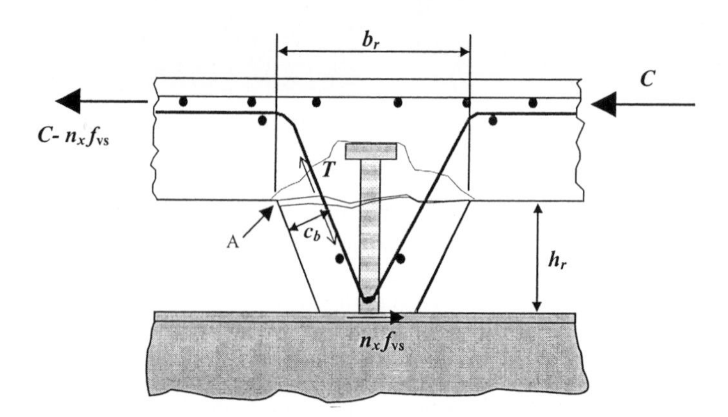

Figure 5: Design of wave-form reinforcement to prevent rib shearing failure

It is proposed that reinforcement like that shown in Figure 5 is not required to control rib shearing failure provided the cracking moment of the concrete rib is not exceeded, viz.:

$$n_x f_{vs} h_r \leq f'_{cf} b_{eff} b_r^2 / 6 \qquad (1)$$

where –

n_x = number of welded studs per pan;

f_{vs} = nominal shear capacity of a welded stud in accordance with AS 2327.1;

h_r = overall height of profiled steel sheeting ribs;

f'_{cf} = flexural tensile strength of concrete determined in accordance with AS 3600 (Standards Australia 2001);

b_{eff} = effective width of concrete rib resisting rib shearing failure (≤ 600 mm); and

b_r = width across top of concrete rib.

It follows that when Eqn. 1 is not satisfied, rib shearing reinforcement is required. With four bars placed on each side of a concrete rib containing shear connectors, and ignoring any inclination of the reinforcement to the horizontal plane, it can be shown that the diameter of each wire or bar, d_b, must be such that:

$$d_b \geq \sqrt{\frac{n_x f_{vs} h_r}{2.8 \phi f_{sy} (b_r - c_b) k_b}} \quad \textbf{(2)}$$

where –

f_{sy} = design yield stress of steel reinforcement (=500 MPa in AS 3600);

ϕ = capacity factor for bending of concrete rib (=0.8 in AS 3600);

c_b = concrete cover measured from nearest edge of sheeting rib to centre of bar (see Figure 5); and

k_b = bar anchorage factor, which equals 1.0 if the bars can develop f_{sy} across the horizontal failure surface, but is otherwise less than 1.0.

When the design of the product is based on the results of structural tests involving the actual profiled steel sheeting to be used, such as has been the case for DECKMESH™ with BONDEK II and CONDECK HP, then bars of a lesser diameter than given by Eqn. 2 may be used.

WAVE-FORM REINFORCEMENT – RIBS PARALLEL

As recognised in AS 2327.1, a Type 3 longitudinal shear surface may form instead of a Type 2 surface if the steel sheeting ribs on at least one side of the steel beam run parallel to the steel beam. If the nearest rib is sufficiently close to the shear connectors, the failure surface first passes to the top of the rib instead of to the base of the slab, i.e. the failure surface will form along the path of least resistance.

In Figures 6(a) and (b) the different features of a potential Type 3 failure surface are shown. The failure surface in the region of a shear connector is shown in Figure 6(a) passing over the top of the connector, then down to the top of each adjacent sheeting rib. If the shear connectors are well spaced apart along the steel beam (which is parallel to the sheeting ribs), then it is possible that the shear surface will become approximately horizontal in the regions between the connectors, passing directly between the tops of the sheeting ribs as shown in Figure 6(b). The narrower the width of the concrete rib and the greater the tension in the shear connectors, then the more likely that this will occur. A horizontal shear surface like that in Figure 6(b) has been observed in tests when the shear connectors were so short that they did not even protrude above the tops of the sheeting ribs (Veldanda & Hosain 1992).

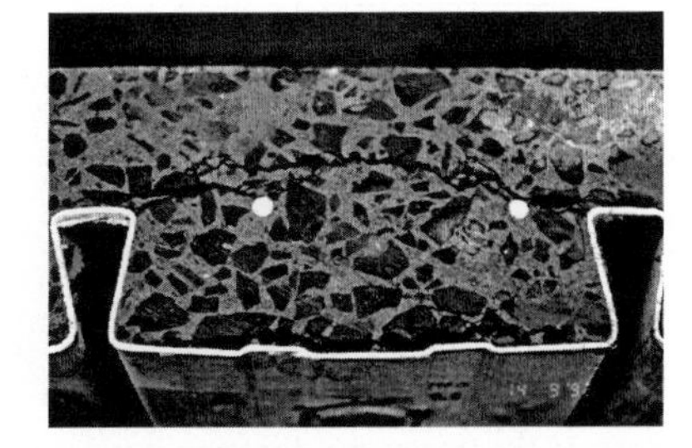

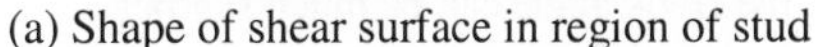

(a) Shape of shear surface in region of stud (b) Shape of shear surface away from stud

Figure 6: Possible Type 3 longitudinal shear surfaces (Patrick 2001)

It follows that various parameters can affect the shape of a Type 3 longitudinal shear surface. Limited account has been taken of this in AS 2327.1 to date. Moreover, if the shear surface takes the form shown in Figure 6(a), then horizontal reinforcing bars placed transverse to the steel beam will pass through the shear surface and therefore be effective in resisting longitudinal shear. However, in regions where the shear surface is effectively horizontal (Figure 6(b)), then wave-form reinforcement like that shown in Figure 7 is required.

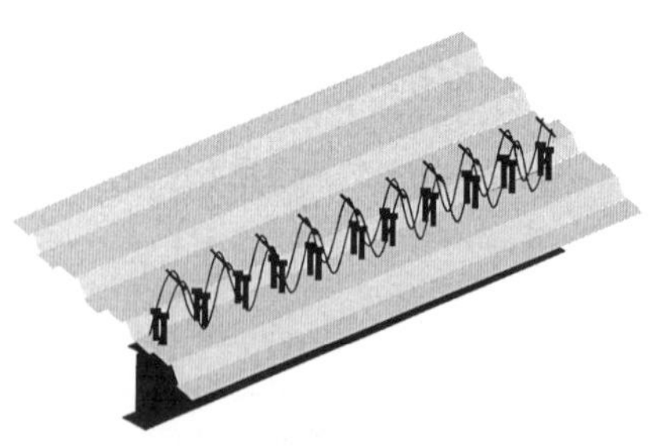

(a) Wave-form reinforcement shown along beam at shear connectors

Potential longitudinal shear surface crossed by "vertical" legs only

(b) Wave-form reinforcement crossing potential shear failure surface

Figure 7: Possible Type 3 longitudinal shear reinforcement for well-spaced shear connectors and narrow concrete ribs with sheeting ribs parallel to steel beam

In view of the discussion above, it is proposed that an additional requirement must be satisfied when determining the maximum longitudinal spacing of shear connectors in accordance with Clause 8.4.1(b) of AS 2327.1. This requirement applies if the profiled steel sheeting on at least one side of the steel beam is deemed parallel to the steel beam, and therefore a Type 3 longitudinal shear failure is possible. The requirement is explained diagrammatically by Figure 8, and is particularly relevant to primary beams. The objective of this rule is to ensure that only conventional Type 1 (vertical) or Type 3 (see Figure 6(a)) shear surfaces can form. Therefore, horizontal bars placed transverse to the steel beam will be effective as longitudinal shear reinforcement. Otherwise, waveform reinforcement like that shown in Figure 7 is required, in order to cross the horizontal shear surface that can form in the region between shear connectors. This rule may sometimes necessitate either using more-closely-spaced single shear connectors rather than pairs of connectors, or increasing the overall height of the connectors.

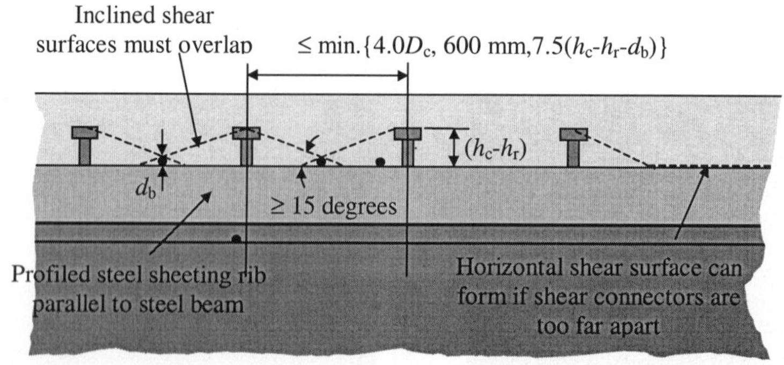

Note: In the transverse direction it is assumed that the crack angle can be much sharper so that the width of the concrete rib does not control the longitudinal stud spacing.

Figure 8: Maximum longitudinal spacing of shear connector groups when formation of a Type 3 shear surface is possible

TRANSVERSE REINFORCEMENT

Improved Detailing

A new reinforcing component STUDMESH® (OneSteel Reinforcing 2001) has been specially developed to overcome problems with designing and detailing horizontal Type 2 and 3 longitudinal shear reinforcement (see Figure 9). The practice of tying individual transverse bars to other reinforcement in order to reinforce these shear surfaces, even if done reasonably simply, is costly and labour intensive and should be avoided whenever possible. STUDMESH® is produced in N10 (i.e. Normal ductility, 10 mm diameter), 500 MPa reinforcing bar because it is more economical to anchor small diameter bar rather than using larger N12 or N16 bar. Also, the light 6 mm longitudinal wires to which the N10 bars are attached assist in the anchorage of the transverse bars, further reducing the total mass of the product. Cutting panels of mesh across their shorter dimension into 2.4 metre long strips comprising four transverse bars to form fully-anchored Type 2 or 3 reinforcement is an improvement to using ordinary bars. However, there is wastage of the steel when it is cut, and also, mesh is normally only available with wires in standard diameters, which further reduces its efficiency in many cases. Using STUDMESH® can significantly reduce the mass of the reinforcement, and also the labour required during construction (since it doesn't have to be cut into strips) compared with using mesh.

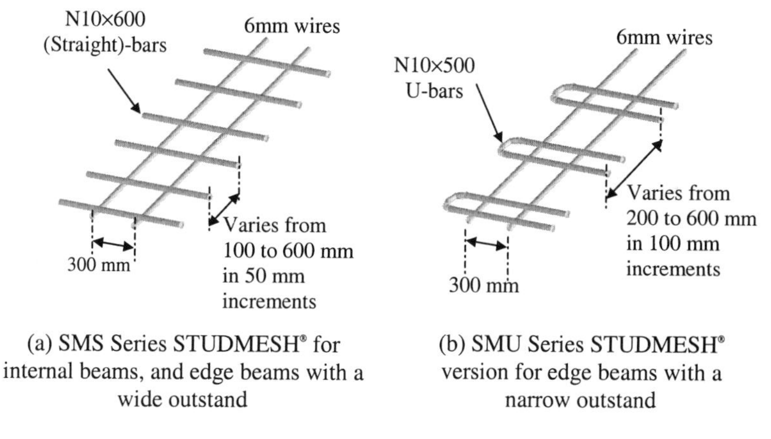

(a) SMS Series STUDMESH® for internal beams, and edge beams with a wide outstand

(b) SMU Series STUDMESH® version for edge beams with a narrow outstand

Figure 9: STUDMESH® as Type 2 or 3 Reinforcement

Reinforcement that crosses a shear surface must be fully anchored on both sides of the crack for the shear-friction model to be valid. Rules for the anchorage of straight deformed bars, and welded-wire mesh with ribbed wires, are given in Section 13 of Australian Standard AS 3600. However, recourse has also been made to Eurocode 2, Part 1 (British Standards Institution 1992) to determine suitable anchorage details for STUDMESH®. A provisional patent has been lodged for the product.

Standard Solutions

Tables of standard solutions for STUDMESH® have been published by Patrick (2001). Solutions are given for 19 mm diameter welded studs spaced at 100 to 600 mm centres. Significant savings of over 30 percent can be made in the mass of the reinforcing steel by choosing STUDMESH® instead of mesh, but the market price of the different products should also be considered before making a final decision. The solutions for primary beams, for which the sheeting ribs are deemed parallel to the steel beam, cannot be used unless the stud height, h_c, is increased enough to satisfy the requirements in Figure 8. Otherwise, wave-form reinforcement is required in accordance with Figure 7.

CONCLUSIONS

The shear connector types permitted in AS 2327.1 all exhibit ductile load-slip behaviour reaching magnitudes of slip required along typical composite beams at ultimate load. When used in conjunction with the current types of Australian profiled steel sheeting with narrow re-entrant steel ribs, the strength of shear studs is the same as the strength in a solid slab provided the shear connection is detailed in accordance with AS 2327.1.

A patented wave-form reinforcing component DECKMESH™ has been developed and is now in use in Australia. It is designed to prevent rib shearing failure from occurring in composite edge beams when profiled steel sheeting is laid transverse to the longitudinal axis of the steel beam. While designed specifically for the Australian types of profiled steel sheeting, tests have shown that its use in slabs incorporating trapezoidal profiled steel sheeting prevents rib shearing failure and results in a ductile load-slip performance rather than the brittle performance observed in tests without this wave-form reinforcing component. A simple method for designing the wave-form reinforcement has been presented for the first time in the paper. The method can be applied to internal beams, as well as edge beams, if the concrete ribs are narrow. Another patent-pending type of waveform reinforcement may be required in primary composite beams if the shear connectors are widely spaced along the steel beam. A new patent-pending reinforcing component STUDMESH® has been specially developed to overcome problems with designing and detailing Type 2 and 3 shear surfaces. The practice of tying individual transverse bars to other reinforcement in order to reinforce these shear surfaces, even if done reasonably simply, is costly and labour intensive. This new component is an improvement on using ordinary reinforcing bars or pieces of cut mesh. It comes in a range of standard sizes to cover most applications and the anchorage length requirements are automatically satisfied.

REFERENCES

British Standards Institution (1990). *BS 5950: Part 3: Section 3.1: Code of Practice for Design of Simple and Continuous Composite Beams*, London.

British Standards Institution (1992). *Eurocode 2: Design of Concrete Structures, Part 1 – General Rules and Rules for Buildings*, DD ENV-1992-1-1.

British Standards Institution (1994). *Eurocode 4: Design of Composite Steel and Concrete Structures, Part 1.1. General Rules and Rules for Buildings (together with UK National Application Document)*, DD ENV 1994-1-1: 1994, London.

OneSteel Reinforcing (2001). *DECKMESH™ & STUDMESH® – Composite Slab Shear Reinforcement Solutions.*

Patrick, M. (2000). Experimental Investigation and Design of Longitudinal Shear Reinforcement in Composite Edge Beams, *Journal of Progress in Structural Engineering and Materials*, **2:2**, April-June, 196-217.

Patrick, M. (2001). *Design of the Shear Connection of Simply-Supported Composite Beams, Composite Structures Design Manual - Design Booklet DB2.1*, OneSteel Market Mills.

Standards Australia (1996). *AS 2327.1-1996: Composite Structures - Simply Supported Beams*, Sydney.

Standards Australia (2001), *AS 3600-2001: Concrete Structures*, Sydney.

Veldanda, M.R. and Hosain, M.U. (1992). Behaviour of Perfobond Rib Shear Connectors: Push-out Tests, *Canadian Journal of Civil Engineering*, **19:1**, 1-10.

Advances in Building Technology, Volume 1
M. Anson, J.M. Ko and E.S.S. Lam (Eds.)

THE INFLUENCE OF MATERIAL PROPERTIES ON THE LOCAL BUCKLING OF CONCRETE FILLED STEEL TUBES

A. Wheeler and M. Pircher

Centre for Construction Technology and Research, College of Science, Technology and Environment, University of Western Sydney, NSW, 2747 Australia

ABSTRACT

Extensive studies have been conducted on thin walled concrete filled steel tubes, subjected to predominantly axial loading. However, limited documented research has been carried out into the behaviour of these sections subjected to pure flexure. The University of Western Sydney is currently carrying out an extensive investigation which includes both numerical and experimental studies into flexural behaviour of concrete filled tubes.

In this paper the effect of material properties of the circular hollow sections, with an emphasis on the effect of the yield stress on the on the local buckling of the tube when subjected to flexural actions, is investigated numerically. The study looks at both the bare steel tube and the concrete filled tube to determine the effect of the concrete infill on the flexural behaviour of the tube. The investigation is conducted using the finite element method, with a numerical model developed to simulate experimental work being carried out at the University of Western Sydney.

KEYWORDS

Buckling Capacity, Tubes, Concrete Filled, Flexural Loading

INTRODUCTION

The use of concrete filled steel tubes in the construction industry continues to increase as an economical alternative to traditional construction methods. The system utilises the compressive strength of the concrete, while the steel tube provides tensile capacity under flexural loading. The tube also provides the necessary formwork for construction. Current practice employs concrete filled tubes in members subjected to predominantly axial loads (columns). Consequently, extensive research into the axial behaviour of concrete filled tubes exists (O'Shea & Bridge, 2000). However, for tubes subjected to flexural loading, limited documented research exists (Shilling, 1965 and Sherman, 1976). An investigation into the flexural behaviour of concrete filled tubes is currently being carried out at the Centre for Construction Technology and Research, University of Western Sydney, in which concrete filled tubes subjected to pure bending are being investigated.

The current work takes into account imperfections (Wheeler & Pircher, 2002) and focuses on the behaviour of the composite system as well as on the behaviour of the individual components. The concrete infill in a steel tube limits the deformations of the steel tube in bending such as ovalisation and inward buckling and thus increases the capacity of such members. This paper will report on the studies performed to determine the influence of the yield stress of the steel tube on the buckling behaviour. Tubes with D/t–ratios (diameter to thickness ratios) in the range from 40 to 110 have been investigated numerically and the buckling behaviour of these tubes with and without concrete infill has been recorded. While the results presented in this paper are generated using a numerical model, the paper will also demonstrate that the model accurately predicts results obtained in an experimental programme carried out at the University of Western Sydney.

FINITE ELEMENT ANALYSIS

The finite element analysis carried out in this paper was conducted using the ABAQUS finite element analysis package (HKS, 1998). To accurately predict the behaviour of the tube, a model was developed that represented a section of tube subjected to four-point bending. This method enabled studies to be carried out on a section of tube subjected to constant moment with no shear. The model also replicated experimental tests being carried out at the University of Western Sydney.

In this study both bare steel tubes and concrete filled steel tubes were considered. The model of the tube is shown in Figure 1. Symmetries at mid-span and in the longitudinal direction reduce the model to one quarter of the beam. The elements used for the analysis were four-noded shell elements with nine integration points through the thickness. To negate local effects at the support and loading points, resulting from loading, a pair of stiff diaphragms were placed at the loading points. The accuracy of the model, in particular for modelling of the local buckling effects, was further improved by refining the mesh in the compression region of constant moment zone. To instigate the buckling of the tube, imperfections based on eigen-modes were super-imposed on the perfect geometry, with a magnitude of one percent of the wall thickness.

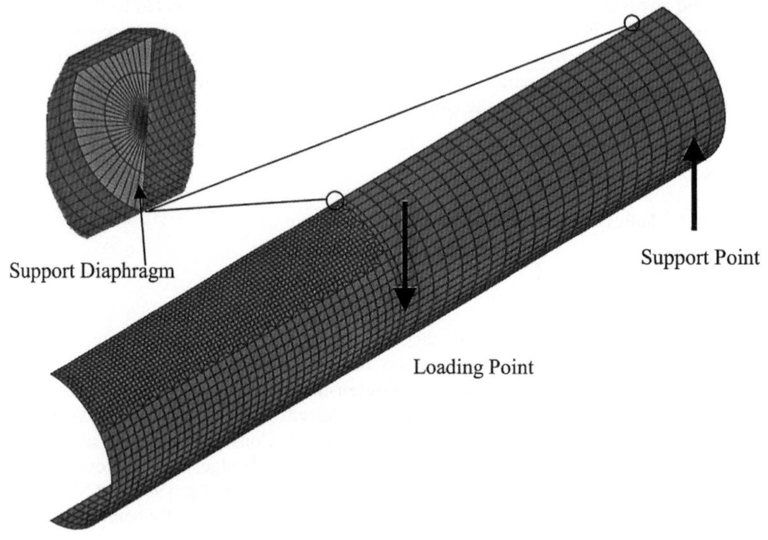

FIGURE 1 FINITE ELEMENT MODEL

The behaviour of the bare steel tube subjected to flexural loading has been studied by Wheeler *et al* (1999). This study looked primarily at the effect of imperfections on the buckling behaviour of the tubes. However, it was observed that prior to buckling considerable ovalisation occurred within the initially circular cross-section. Consequently one of the major effects of the concrete infill was assumed to be to prevent any ovalisation. The infill was assumed to also prevent any inward buckles forming along the tube. To model the effect of the concrete infill on the behaviour of the tube incompressible springs placed with in the section, allowing outward deformations but limiting any inward deformations. The element type used for the springs were two dimensional with a high stiffness in compression and zero stiffness in tension.

The accuracy of the model is demonstrated in Figure 2, were the experimental results for a 406 mm tube (solid line) is compared to the results for the same tube generated using the finite element model (dashed line). Also shown in the figure are the load levels at which yielding commences, the limit imposed by AS 4100 and the theoretical load when the section is fully plastic.

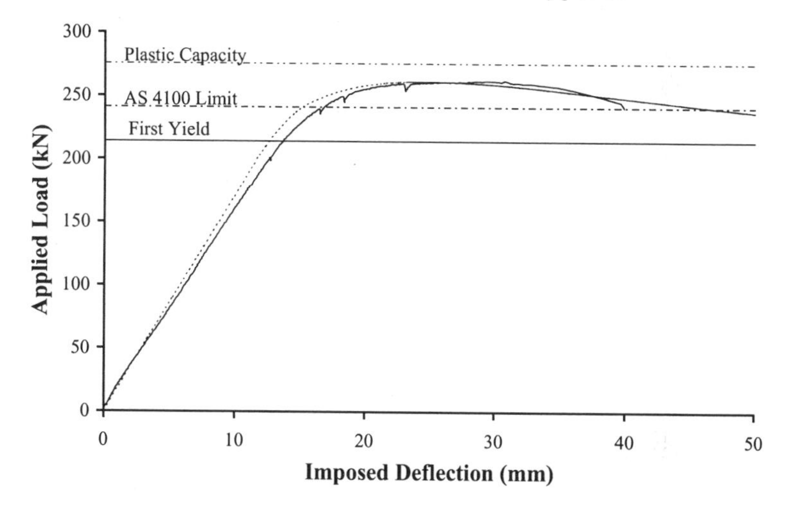

FIGURE 2 ABAQUS MODEL VS EXPERIMENTAL RESULTS

The match between the numerical and experimental results is close, with the initial stiffness and the peak loads demonstrating a good correlation. The variation in stiffness as the load increases is explained by the fact that the numerical model uses a bi-linear elastic plastic material relationship, while the true material properties are generally non-linear.

GEOMETRY AND MATERIAL PROPERTIES

The geometry of the tubes selected for this investigation was limited to D/t ratios that are generally used in the structural building industry. All tubes in this study had an equal wall thickness of $t = 6.4$ mm, with the span of the load introduction points adjusted to maintain a constant span to depth ratio. The diameter (D) and the D/t–ratios for the tubes investigated are given in Table 1.

The material properties of commercially available steel tubes vary significantly, depending on both the geometry and the process used in manufacturing the tube. The stress-strain behaviour for typical cold-formed tubes are generally highly non-linear with no clear yield point and yield plateau. However, this paper focuses on the effect of the yield stress on the local buckling of tubes. Thus, a perfectly elastic-plastic stress strain curve will be used for the steel tubes.

TABLE 1
TUBE DIAMETERS

D	D/t
219	34
457	71
710	111

BARE STEEL TUBE

Initial analyses were carried out to determine the behaviour of the tube without the presence of the concrete infill. Firstly the classic elastic buckling load (M_{Buckle}) for each tube was determined using an eigenmode-extraction and perfect tube geometry. Secondly, seven analyses were performed on the same tube with varying material properties to determine the influence of the yield-stress. In these analyses the yield stress of the tube material was varied between 200 MPa to 800 MPa with geometric non-linearity being considered. The moment-deflection behaviour for these tubes are presented in Figure 2, Figure 3 and Figure 4 for the 219 mm, 457 mm and 710 mm tubes respectively. The moment has been normalised against the elastic buckling moment (M_{Buckle}), and the deflection at mid span (Δ) has been normalised using the wall thickness of the tube.

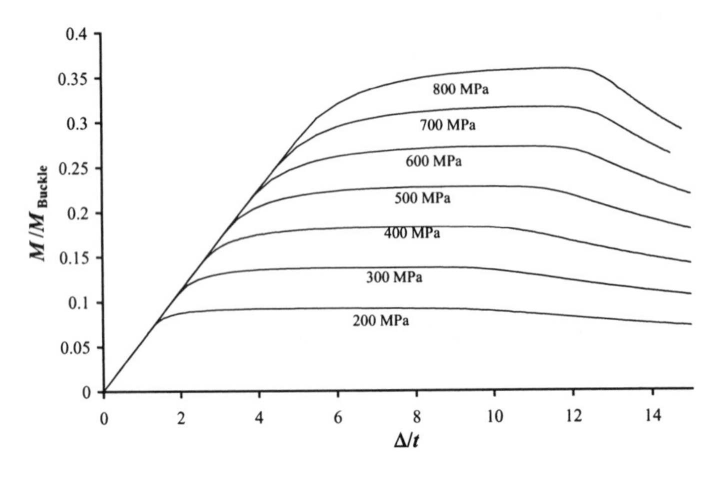

FIGURE 3 BARE STEEL D=219 MM

Figure 3 depicts the results for a 219 mm tube, in this case the D/t is considered to be low at a value of approximately 34. The behaviour of the tube for all yield stresses selected behaves as would be expected for a compact section, with an increase in moment capacity until plastic section capacity is approached. The moment then remains stable until the onset of local plastic buckling at large strains, resulting in a reduction of the moment capacity.

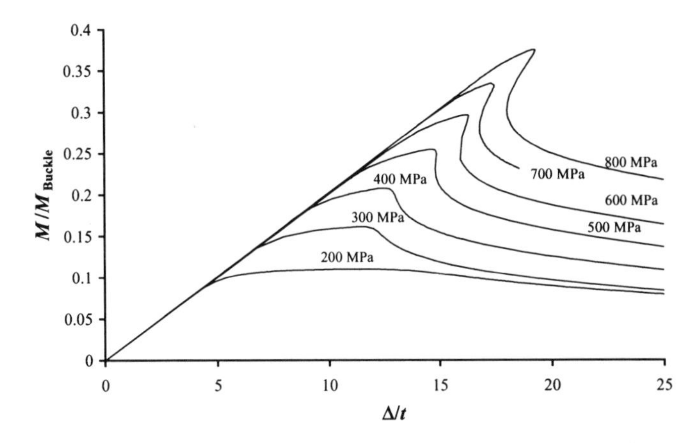

FIGURE 4 BARE STEEL D = 457 mm

As the *D/t* ratio is increased the tube behaviour varies significantly depending on the yield stress as shown in Figure 4. In this case the load-deflection curves for a 457 mm tube is presented with a *D/t* ratio of approximately 71. For low yield stresses (≤ 200MPa), the behaviour is similar to the compact sections with a long yield plateau followed by a gradual reduction in moment capacity. However, as the yield stress increases the tube behaviour becomes more dependent on local buckling effects with significant buckling occurring. As the yield stress increase the effect of the local buckling on the post buckling behaviour increases with significant snap back being observed for the higher yield stresses.

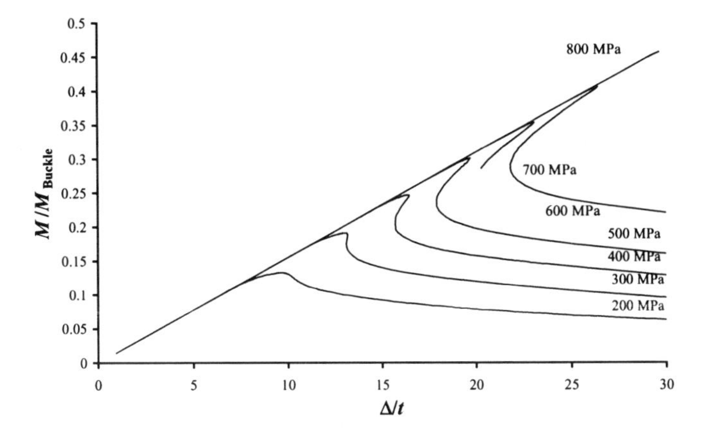

FIGURE 5 BARE STEEL D = 710 mm

Figure 5 presents the moment-displacement behaviour for a 710 mm-diameter tube with a *D/t* ratio of approximately 111. In this case the tube would be considered to be either non-compact or slender (AS4100, 1998), with the formation of local buckles having a distinct effect on the post buckling behaviour of the tubes. It is shown that even for low yield stresses (300 MPa) significant losses in the moment capacity can be observed with the onset of local buckling.

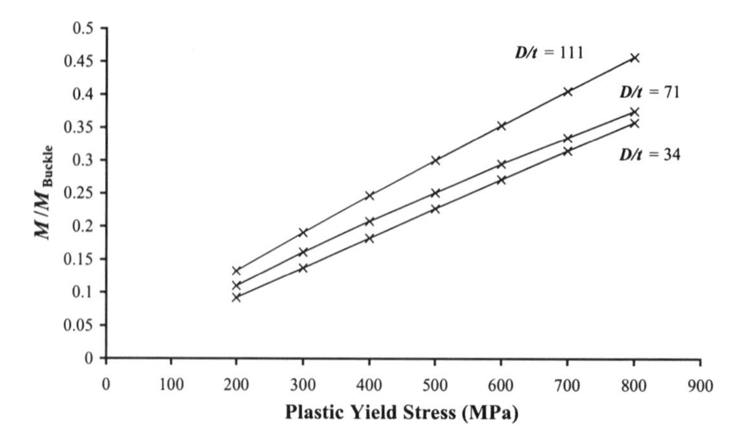

FIGURE 6 PEAK MOMENT TO YIELD STRESS – TUBE ONLY

In Figure 6 the buckling moments versus the yield stresses of the three considered tubes are shown. In each case the relationship between the buckling moment and the yield stress is very close to linear. However, the slopes vary significantly between the three tubes. This figure supports the assumption made in most standards that the relationship between the buckling moment and the nominal yield stress is linear. For example, the definition for section slenderness in the Australian standard AS 4100 (SA, 1998) is defined as

$$\lambda_s = \left(\frac{d_o}{t}\right)\left(\frac{f_y}{250}\right) \tag{1}$$

In the Australian standard a tube is considered to be compact if is λ_s is less that 50, and slender if λ_s is greater than 120. While this formula may be sufficient for most materials, at the extremes it may be un-conservative. For example, the slenderness λ_s for the tube with d = 219mm equals λ_s =109 which classifies this tube as non-compact. However, the analysis results shown in Figure 3 suggest that the section is in fact compact.

CONCRETE FILLED TUBE

In studying the effect of the concrete infill on the steel tube, it was assumed that the concrete within the tube has cracked and that there is no longitudinal interaction such as friction or mechanical interlocking. Consequently, the concrete infill makes no direct contribution to the flexural strength. It is recognised that frictional forces and mechanical interlocking will increase the flexural capacity of the concrete filled tube. However, this will be the subject of further studies.

Subjecting the bare steel tube to flexural action results in the cross-sectional deformations, with a decrease in depth and a corresponding increase in the width of the section. Such a deformation is referred to as "*ovalisation*". Wheeler et al. (1999) demonstrated that this ovalisation decreases the buckling capacity of the tube. By including the concrete infill in the tube, the cross-sectional shape of the tube is retained, thus preventing any significant ovalisation. The concrete infill also prevents any inward buckles from forming, causing the buckles to form in the shape of a buckling mode with higher associated buckling moments thus increasing the bending capacity of the tube.

Analyses corresponding to those carried out on the bare steel tube were repeated with the simulated concrete infill. In all cases a significant increase in the flexural capacity of the tube was observed.

Typical results are shown in Figure 7, where the dashed line is the response of the bare steel tube while the solid line represented the concrete filled tube. Beside the increase in critical moment a significant change in the post-buckling behaviour was observed. The change became more pronounced for higher yield stresses of the considered tubes and can be characterised as an increase in ductility of the system due to the concrete infill.

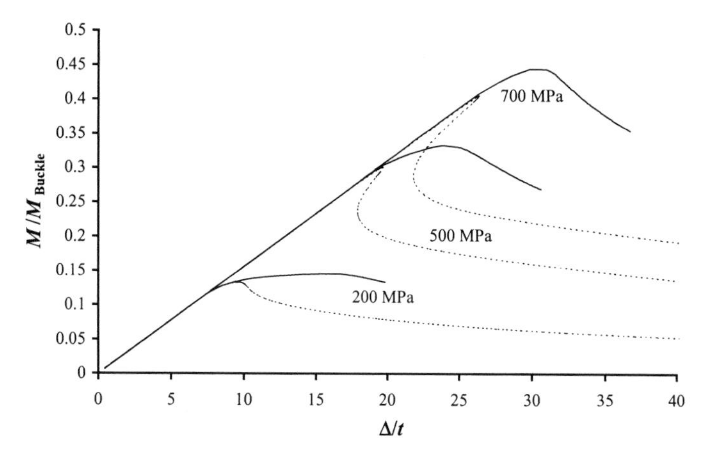

FIGURE 7 EFFECT OF CONCRETE INFILL ON FLEXURAL BEHAVIOUR

The overall effect of the concrete infill on the tube behaviour is shown in Figure 8, with the ratio of buckling moment to elastic buckling moment is plotted against the yield stress of the tubes. For the bare steel tubes as well as for the concrete-filled tubes, the relationships between the yield stress and the buckling moment is very close to linear. From this figure it is also evident that the effect of the concrete infill is related to the D/t ratio, with larger increases in moment capacity between the bare and concrete filled tubes observed for the tubes with higher D/t ratios. For the D/t of 71 at 200 MPa the capacities differ by one percent while for 800 MPa the difference increases to 10 percent.

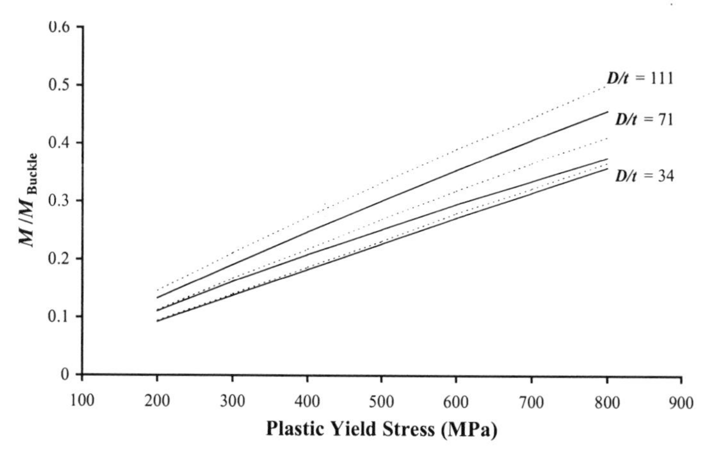

FIGURE 8 PEAK MOMENT TO YEILD STRESS - CONCRETE FILLED TUBE

CONCLUSION

The accuracy of the numerical model used in this paper was demonstrated with excellent correlation between the experimental and numerical results. This accuracy of the finite element model suggests that parametric studies may be used in lieu of expensive experimental tests to predict the effect of material properties on the buckling behaviour of thin wall tubes

In this paper the influence of the yield stress in concrete filled steel tubes on the buckling behaviour in pure bending was investigated. It was found that there are significant increases in the flexural capacity of steel tubes when filled with concrete, these increase include both the stiffening of the section and the realisation of higher buckling loads. The gain in strength is directly related to the yield stress. However, the *D/t* ratio also influences this relationship, with higher stresses experienced in the tubes with lower *D/t* ratio's and higher increases in capacity due to concrete infill observed for the more thin-walled members investigated for this paper.

REFERENCES

HKS (1998). *ABAQUS/Standard Users Manual, Version 5.8.* Hibbit, Karlsson and Sorensen , Inc.

O'Shea M.D. and Bridge R.Q. (2000). Design of Thin-walled Concrete Filled Tubes. *Journal of Structural Engineering, ASCE,* **126**(11), 1295-1303.

SA (1998), *AS 4100-1998: Steel Structures*, Standards Australia, Sydney.

Sherman D.R. (1976) Tests of Circular Steel Tubes in Bending. *Journal of the Structural Division, ASCE,* **102(**11), 2181-2195.

Schilling, C.G., "Buckling Strength of Circular Tubes", *Journal of the Structural Division*, American Society of Civil Engineers, Vol 91, No. 5, 1965, pp. 325-348.

Wheeler A.T., Pircher M., Bridge R.Q. (1999). Modelling of Thin-walled Tubular Sections Subjected to Pure Bending. *Proceeding, The Thirteenth Australian Compumod Users' Conference*, Melbourne Australia, 33.1-33.11.

Wheeler A. and Bridge R.Q. (2000). Thin-walled Steel Tubes Filled with High Strength Concrete in Bending. *Proceedings, Engineering Foundation Conference, Composite Construction IV*, Banff, Canada.

Wheeler A. and Pircher M. (2002). Measured Imperfections in Six Cold-Formed Thin-Walled Steel Tubes. *Research Report CCTR: 001*, Centre of Construction Technology & Research, University of Western Sydney

TEAM PRESENTATION:

COMPOSITES FOR HOUSING

LOW-COST, BIO-BASED COMPOSITE MATERIALS FOR HOUSING APPLICATIONS

L.T. Drzal, A. K. Mohanty, G. Mehta, M. Misra

Composite Materials and Structures Center
Dept of Chemical Engineering and Materials Science
Michigan State University, East Lansing, Michigan, USA , 48824

ABSTRACT

In an on-going project, sponsored by the National Science Foundation, to develop bio-based composite materials for the next generation of American housing panel applications, a group of researchers at Michigan State University are striving to generate eco-friendly greener composite materials for structural applications. The objective of this research is to determine if bio-composites designed and engineered from natural/bio-fibers and blends of polyester resin and derivitized soybean oil can replace existing glass fiber-polyester composites for use in housing structures. Bio-composites provide environmental gains, reduced energy consumption, lighter weight, insulation and sound absorption properties, elimination of health hazards and reduce dependence on petroleum based and forest product based materials. A three cornered approach including the use of engineered natural fibers, polymer resin modification and development of a new high volume continuous manufacturing processes(Bio-Composite Sheet Molding Compound Panel-BCSMCP) is required to achieve the objective of producing an affordable alternative construction material for the housing industry of the 21st century .

KEYWORDS
Housing Panels, Biocomposites, Unsaturated Polyester Resins, Derivitized soybean oil

INTRODUCTION

Contemporary housing needs are challenging designers and builders to utilize more inventive materials in order to provide housing that are environmentally benign and at the same time provides for the requisite occupant environment and operating efficiency. Increasing attention is turning to bio-based materials from renewable resources. The future for bio-based building materials is bright (Figure 1, ref.1) and several new agricultural natural fiber-based building materials are already making

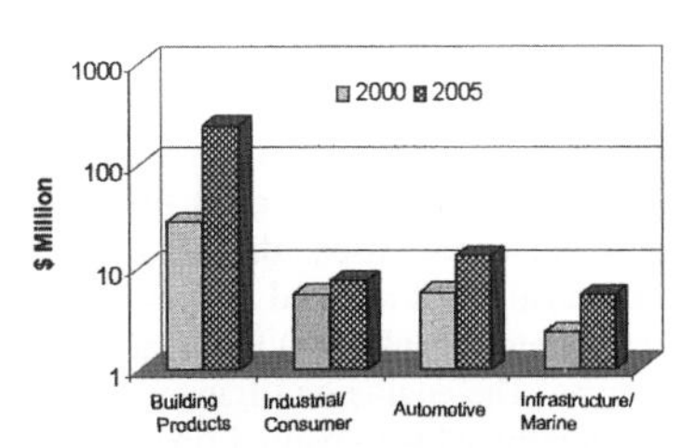

Figure 1: Growth Outlook For Bio-based Composites by Application in United States, 2000-2005

their mark in the building industry. Natural fiber composites (bio-based composite materials) are economical commodity composites that have useable structural properties at relatively low cost. The manufacture, use and replacement of conventional building materials (e.g. petroleum-based plastics, glass fibers, etc.) are becoming priorities because of the growing environmental consciousness[2]. There is a growing interest in the use of natural/bio-fibers as reinforcing components for thermoplastics and thermosets. Advantages of natural fibers over traditional reinforcing fibers such as glass and carbon are: low cost, low density, acceptable specific properties, ease of separation, enhanced energy recovery, CO_2 sequesterization and biodegradability. Although thermoplastics have the added advantage of recyclability, thermosets have the necessary mechanical properties for use as structural bio-composites. Bio-composites derived from natural fibers and traditional thermoplastics or thermosets are not fully environmentally friendly because matrix resins are non-biodegradable. However these bio-composites do now maintain a balance between economics and environment allowing them to be considered for applications in the automotive, building, furniture and packaging industries.

The use of reinforced thermoset composites by automakers has nearly doubled in the last decade, and is expected to increase 47 percent during the next five years[3] through 2004. The majority of resins

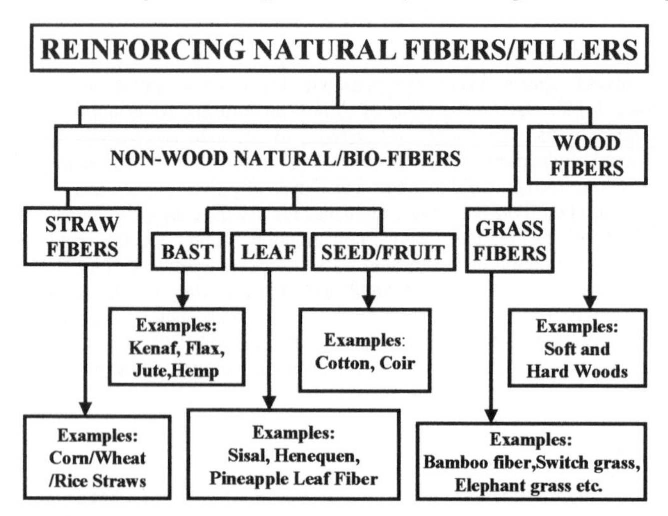

Figure 2: Broad classification of Natural/Bio-fibers: Strong potential reinforcing fibers in designing bio-composites for Building Products Industry

used in the composite industry are thermosets[4]. About 65% of all composites produced currently for various applications; use glass fiber and polyester or vinyl ester resins. Unsaturated polyester (USP) resins are widely used, thanks to a relatively low price, ease of handling and a good balance of mechanical, electrical and chemical properties. Natural fibers have recently made a comeback in FRP (fiber reinforced plastics) to replace glass fiber in many applications such as automotive and building products. The engine and transmission covers of *Mercedes-Benz* transit buses now contain polyester resin reinforced with natural fiber[5]. Natural fiber reinforced polyester composites have received much commercial success in the infrastructure area primarily for low cost housing applications.

There is, however, a major drawback associated with the application of bio-fibers for reinforcements of organic matrix resins. Due to the presence of hydroxyl and other polar groups on the surface and throughout bio-fibers, moisture absorption can be high which leads to poor wettability by the matrix resin and weak interfacial bonding between fibers and hydrophobic matrices. In order to develop composites with better mechanical properties, it is necessary to impart hydrophobicity to biofibers by suitable chemical treatments[6].

The mechanical behavior of fiber, the types of fiber (non-woven, woven, short/long fibers), the nature of matrix, and fiber-matrix adhesion play vital role in controlling the properties of composites. This paper gives an overview of natural fiber reinforced polyester composites highlighting the future aspects of such bio-based composite materials in building applications.

REINFORCING NATURAL/BIO-FIBERS

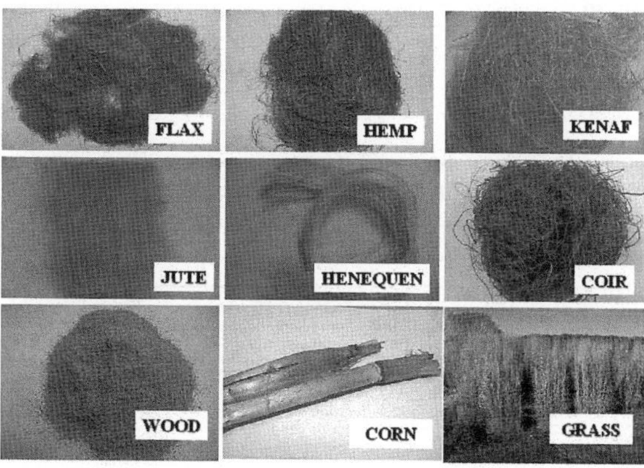

Figure 3. Various Sources for BioFibers.

Plastics, by themselves are not fit for load bearing application due to their lack of strength, stiffness and dimensional stability. However fibers possess high strength and sufficiently stiffness but cannot be used for load bearing applications because of their fibrous structure. In fiber-reinforced composites, the fibers serve as reinforcement by giving the strength and stiffness to the structure. Bio-fibers may be classified in two broad categories: Non-wood fibers and Wood fibers (Fig.2). Interest in the use of non-wood cellulose fibers in plastic composite structures has increased rather dramatically in recent years. The non-wood bast (from the stem part of the plant) fibers are poised to be utilized to a greater extent than wood fibers in bio-composite housing structures. The best-known examples are: (i) bast fibers: flax, ramie, kenaf/mesta, hemp and jute (ii) Leaf fibers: sisal, pineapple leaf fiber (PALF), and henequen (iii) seed fibers: cotton; fruit fibers: coconut fiber i.e. coir. (Fig.3) All natural fibers (wood and non-wood) are lingo-cellulosic in nature with the basic components being cellulose and lignin (Fig.4). The density (g/cm^3) of natural fibers (varies from ~1.2-1.5) is much less than that of E-glass (2.55); the specific strength of natural fibers is quite comparable to glass fibers. The elastic modulus and specific modulus[7] of natural fibers are comparable or even superior to E-glass fibers (Table 1 and Fig.5).

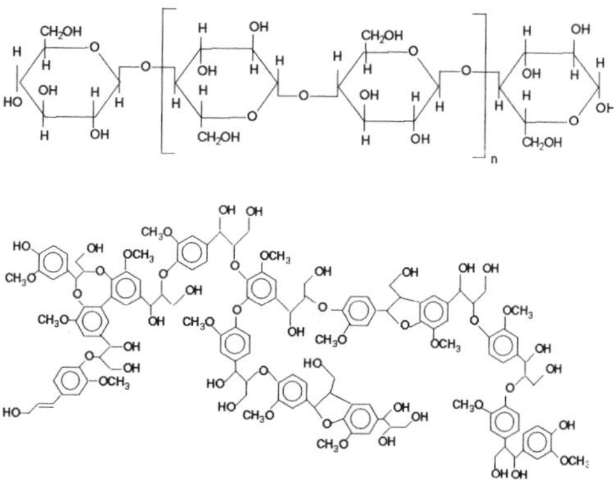

Figure 4: Structure of Cellulose (shown top) and lignin (shown bottom): basic constituents of natural fibers

POLYESTER RESIN MATRIX POLYMERS FOR BIO-COMPOSITES

The polyester resins, because of their versatility and low cost, are widely used in polymer composites. Polyester resins are classified (Scheme 1) as: (i) ortho resins, (ii) isoresins, (iii) bisphenol-A-fumarates, (iv) chlorendics, and (v) vinyl ester. Ortho-resins, known as general-purpose polyester resin, are based on phthalic anhydride, maleic anhydride and glycols (Scheme 2). Ortho-resin is the least expensive among all the polyester resins. The solutions of unsaturated polyesters and styrene vinyl monomers (reactive diluents) are known as unsaturated polyester (UP) resins. Considerable work has been reported on the synthesis, characterization, and curing behavior of UP resins[8]. The curing reaction of UP is a free-radical chain growth polymerization between reactive diluents styrene and UP resin. A wide range of peroxides, azo and azine compounds can be used as initiators, depending on the curing temperature. For room temperature curing as in the case of hand-lay-up structures, methyl ethyl ketone peroxide (MEKP) is used; for moderate temperature curing benzyl peroxide is used. For high temperature processing, di-t-butyl peroxide or t-butyl perbenzoate is used. A mixture of initiators is used when a large temperature increase is expected. To accelerate the decomposition of peroxides, some metal compounds, tertiary amines, and mercaptans can be used. Cobalt naphthenate (CoNp) and cobalt octanoate (CoOc) are the most widely used accelerators.

Table 1: Modulus comparison of E-glass and some important Natural Fibers (after ref.6 and partially modified)

Fiber Type	Density (g/cm^3)	Elastic Modulus (GPa)	Specific modulus
E-glass	**2.55**	**73**	**29**
Hemp	**1.48**	**70**	**47**
Flax	**1.4**	**60-80**	**43-57**
Jute	**1.46**	**10-30**	**7-21**
Sisal	**1.33**	**38**	**29**
Coir	**1.25**	**6**	**5**
Cotton	**1.51**	**12**	**8**

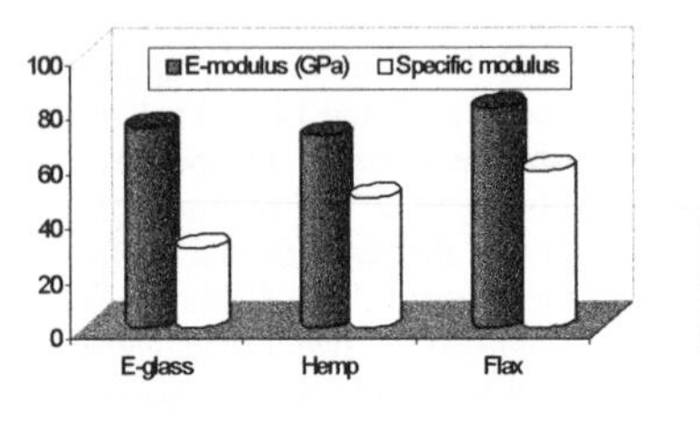

Figure 5: Modulus and Specific modulus: E-glass vs. some Natural Fibers

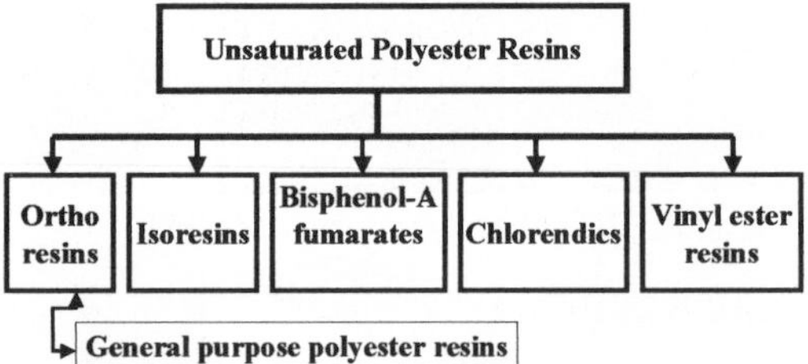

Scheme 1: Unsaturated Polyester resins--Classifications

Phthalic Anhydride + Maleic Anhydride + Propylene glycol → [polyester] + H_2O

Preparation of Polyester Resin

Scheme 2

OUR APPROACHES TO STRUCTURAL BIO-BASED COMPOSITE MATERIALS FOR HOUSING APPLICATIONS

Natural fiber composites are emerging as a realistic alternative to glass-reinforced composites. While they can deliver the same performance for lower weight, they can also be 25-30 percent stronger for the same weight. Research success from our lab.-scale experiments of our group[9-11] has shown that it is possible to produce bio-composites with properties that can compete with glass fiber reinforced composites. Natural fiber (NF) unsaturated polyester (UP) composites show lower density, equal flexural modulus, comparable flexural strength but relatively poor impact strength as compared to a glass fiber (GF) composites[12]. Our three-pronged approach to designing superior strength (Fig.6) bio-composites for housing structures includes:

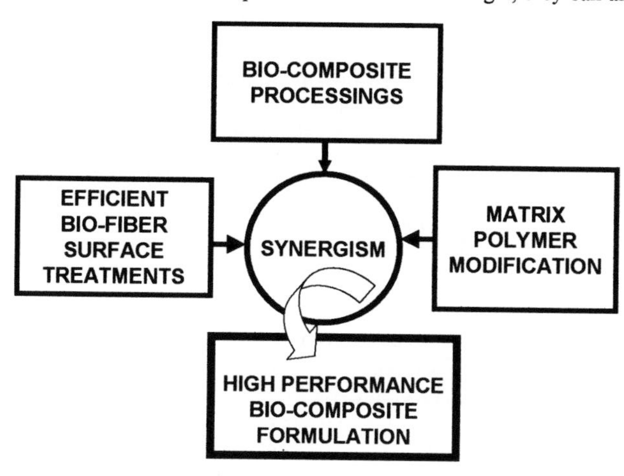

Figure 6: Design of superior strength Bio-Composite-Approaches

- An efficient (low cost but effective) bio-fiber surface treatment
- Matrix modification (Blending with functionalized vegetable oil)
- New, property-enhancing bio-composite process - - Sheet Molding Manufacturing.

Bio-fiber treatments and Design of "Engineered Natural/Bio-fibers"

Design of "Engineered Natural/Bio-fiber" is schematically represented in Fig.7. Engineered bio-fibers" are defined as a blend of surface treated bast (e.g. Kenaf, Hemp) and a leaf fiber (e.g. Pineapple leaf fiber, PALF). Selection of blends of bio-fibers is also based on the fact that the correct blend achieves an optimum balance in mechanical properties. The Kenaf and/or Hemp based composites exhibit excellent tensile and flexural properties, while leaf fiber (PALF) composites give the best impact properties of the composites. A blended composition of two bast and one leaf fiber is selected to achieve a balance of flexural and impact properties of the targeted bio-composites. Alkali treatment (AT) of natural fibers appears to be a very promising treatment, having the right combination of surface chemical and structural benefits along with low cost. The alkali treatment enhances the biofiber surface roughness, causes surface fibrillation and thus drastically improves

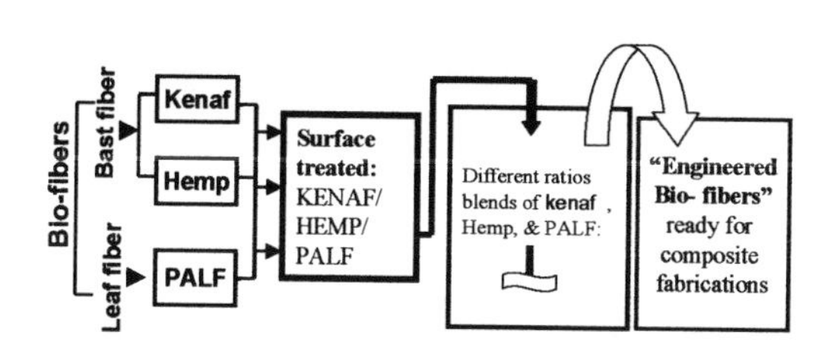

Figure 7: Design of Engineered Natural/Bio-fiber

fiber-matrix adhesion. An increase in strength for bio-composites can also be achieved through the use of silane treatment. However, the choice of the correct silane coupling agent and treatment conditions all play a vital role in achieving superior strength composites. A general compatibilization mechanism for silane treated natural cellulose fiber and polymer resin composites is represented in Scheme 3.

Matrix modification (Blend of polyester resin and derivitized vegetable oil)

Unsaturated polyester resin is petroleum-based and is not eco-friendly. Attempts are in progress in our group to replace polyester resin with vegetable oil based resins. Through successful blending of petroleum-based polyester resin and derivitized soybean oil we are developing a number of bio-based

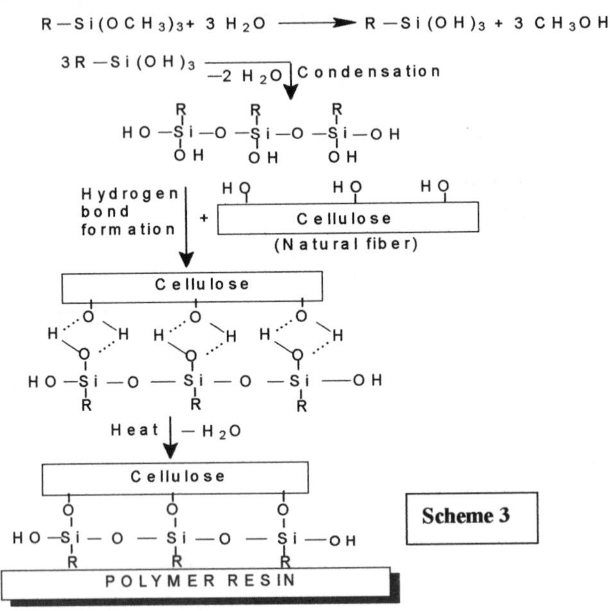

Scheme 3

resins for producing bio-composites. Combining a suitable reinforcement with a properly modified resin blend of polyester and vegetable oil will result in a novel bio-composite material. The incorporation of derivitized vegetable oil into polyester resins can improve the toughness of the resin and the resulting bio-composites. The content of derivitized vegetable oil and varying curing agent composition will affect the overall performance of the materials. Optimally maximizing the concentration of derivitized soybean oil under accurate curing conditions and incorporating suitably surface treated natural fibers will launch a new generation of eco-friendly bio-composites for housing applications.

Bio-Composite Processing

Bio-composites can now compete with glass-polyester composites both on a cost and performance basis. Most of the existing results on bio-composites are based on hand-lay-up labscale fabrication techniques. However the success of a high volume processing technique will be necessary to produce bio-composites in composite panels for housing applications economically. We have created a unique new process for fabrication of bio-composites. This process is similar to the Sheet Molding Compound (SMC) process but is a new and necessary approach for bio-composites.

SMC type of materials are often used in semi-structural components. The short production cycle times and the excellent surface appearance of some grades make it attractive as a panel material. Stronger SMC type material can be developed, mainly by incorporating high fiber contents. A schematic representation of a high volume bio-composite processing in making bio-composites from chopped

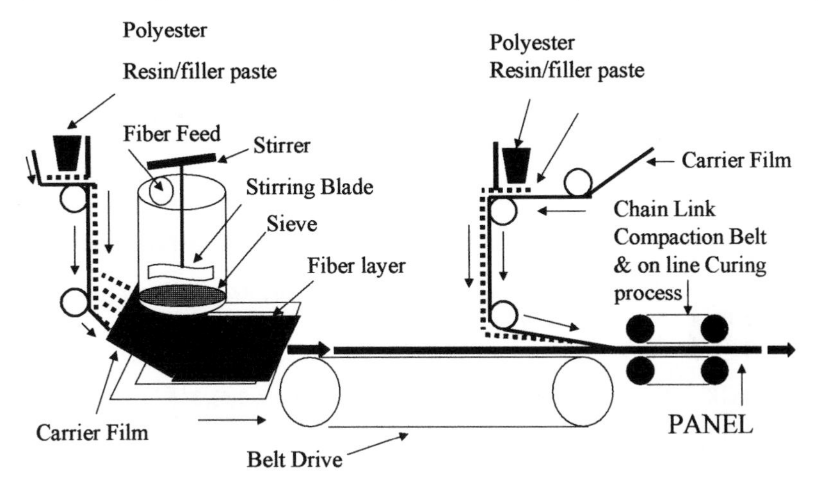

Figure 8: Schematic of Continuous Bio-Composite Sheet Molding Compound Panel (BCSMCP) Manufacturing Process being developed at Michigan State University

natural/bio-fiber and polyester resin or resin/filler paste is shown in Fig.8. Polyester resin in combination of $CaCO_3$ filler is required to be blended with biofibers to produce structural quality low cost housing panels.

CONCLUSIONS

New environmental regulations and societal concerns have triggered the search for new products and processes that are compatible with the environment. The incorporation of bio-resources into composite materials can reduce dependency on petroleum reserves. American market studies clearly identify the potential impact and opportunities for natural fiber composites. Natural fiber composites have the potential for major applications in building products. The majority of such applications include natural fiber-thermoplastic composites for applications like decking, window/siding/railing, flooring, furniture and automotive interiors. The vast majority of this market volume deals with wood fiber and remainder is comprised of non-wood agricultural natural fibers like kenaf, hemp and flax. Wood-fibers cannot impart the required strength for paneling applications, so our efforts target to use non-wood bast and leaf-based natural fibers. Natural fiber – thermoset composites show improved mechanical properties over thermoplastic-based bio-composites and such properties are required for paneling applications. In the U.S., current housing panels made from glass fiber – polyester composite contains ~10-20% glass fiber. Since natural fibers are less expensive (~ 25-30 cents/lb. in North America) than both glass fiber and PE resin, incorporation of the maximum permissible amount of inexpensive natural fibers into the composite structure would not only provide immense economical benefit, but also would result in housing structures with maximum permissible content of renewable resource based natural fibers. The impact/stiffness balance of bio-composites needs to be solved through the combination of bio-fibers (hybrid of bast and leaf fibers) and matrix modification through blending with derivitized vegetable oil. The future for bio-composites ("GREENER" than exiting glass composites) is bright because of the shortage of landfill availability, difficulty in recyclability of fiber glass composites, and the uncertainty of petroleum price/availability. Achieving this unique balance of mechanical properties and environmental benefits will create new markets for bio-composites for 21st century building product applications.

ACKNOWLEDGEMENTS

Financial support was [provided from the National Science Foundation (NSF) under Partnership for Advancing Technologies in Housing (PATH), 2001 Award # 0122108.

REFERENCES

1. Harris, T. (2001). Trends In New Filler And Reinforcement Technologies For Plastic Composites, *Conference on Functional Fillers*, San Antonio, TX.
2. Mohanty, A. K., Misra, M.. and Hinrichsen, G. (2000) *Macromol. Mater. Sci.* **276/277, 1**.
3. *Composite Week Newsletter*, (2000) **10(2)**.
4. *Composites Technology*, (2000).
5. *DaimlerChrysler HighTech Report* (1999), **88**.
6. Mohanty, A. K. , Misra, M. and Drzal, L. T., (2001) *Polym. Interf.* **8(5), 313**.
7. Brouwer, W. D., (2000). Natural Fiber Composites: Where Can Flax Compete with Glass ?, *SAMPE Journal*, **36(6), 18.**
8. M. Malik, V. Choudhary, I. K. Varma, J. Macromol. Sci. – Rev. Macromol. Chem. Phys; C40 (2&3), 139 (2000).
9. Mohanty, A. K. , Misra, M. and Drzal, L. T., (2001). Surface Modification of Natural Fibers to Improve Adhesion As Reinforcements for Thermoset Composites, *Proceedings of the 24th Annual Meeting of the Adhesion Society – Adhesion Science for the 21st Century"* **418-420**.
10. Mohanty, A. K. , Misra, M. and Drzal, L. T., (2001). Environmentally Benign Powder Impregnation Processing And Role Of Novel Water-Based Coupling Agents In Natural Fiber-Reinforced Thermoplastic Composites, *Polymer Preprint – Polymer Chemistry Division*, ACS, **42(2), 31**.
11. Belchler, L., Drzal, L. T., Misra, M. and Mohanty, A. K. (2001) Natural Fiber Reinforced Thermoset Composites: Studies on Fiber-Matrix Adhesion of Aligned Henequen Fiber – Epoxy Composites *Polymer Preprint – Polymer Chemistry Division*, ACS, **42(2), 73.**
12. Schloesser, T., (2000). Natural Fiber Composites in Automotive Applications, *3rd International Symposium- Bioresource Hemp & other fiber crops,* Wolfsburg, Germany.

Advances in Building Technology, Volume 1
M. Anson, J.M. Ko and E.S.S. Lam (Eds.)

DURABILITY OF PULP FIBER-CEMENT COMPOSITES TO WET/DRY CYCLING

N.H. El-Ashkar[1], B. Mohr[1], H. Nanko[2], and K.E. Kurtis[1]

[1]School of Civil and Environmental Engineering, Georgia Institute of Technology, Atlanta, GA, 30332-0355, USA
[2]Institute of Paper Science and Technology, Atlanta, GA, 30318-5794, USA

ABSTRACT

To ensure satisfactory long-term performance, focused research is required to more completely understand the moisture-related durability of natural fiber-reinforced cement-based materials. The results presented here are part of a comprehensive investigation to examine both the initial state of the fiber-cement matrix bond and the changes in the bond, and resulting mechanical performance of the composite, that occur due to moisture fluctuations. In this early phase of an extensive and systematic investigation, the appropriateness of a proposed environmental exposure test program is examined. Results show that the initial curing period should be extended beyond eight days for both the control samples and those samples subjected to wet/dry cycling to minimize the influence of continued cement hydration during environmental exposure.

KEYWORDS

cement, composite, concrete, durability, fiber, housing, paper, pulp

1.0 INTRODUCTION

Depending on their application, fiber-cement materials can offer a variety of advantages over traditional construction materials:

- As compared to wood, fiber-cement products offer improved dimensional stability, moisture resistance, decay resistance, and fire resistance.
- As compared to masonry, fiber-cement products enable faster, lower cost, lightweight construction.
- As compared to cement-based materials without fibers, fiber-cement products may offer improved toughness, ductility, and flexural capacity, as well as crack resistance and "nailability".

An alternative to more expensive steel fiber-, glass fiber-, and synthetic polymer fiber-reinforced cement products, wood fibers are well-suited for reinforcing cement-based materials because of their high strength-to-cost ratio, availability, renewability and recyclability, and non-hazardous nature [MacVicar *et al.*, 1999]. Increasing regulations surrounding the use of asbestos fiber reinforced materials has also prompted the growing interest in wood fiber-cement composites for construction. The growing worldwide acceptance and use of fiber-cement products in residential construction and the continuing development of new products have caused the industry to describe fiber-cement materials as "tomorrow's growth product" [Kurpiel, 2000].

Yet, despite continuing interest in the development of load-bearing or structural fiber-cement composites, the primary applications of this class of material remain non-structural. This is in part due to concerns about the degradation of fiber-cement composites. One issue of continuing concern is the effect of swelling of the fibers in the presence of moisture, and the effects of such swelling on composite performance over time. American Concrete Institute (ACI) Committee 544 "Fiber Reinforced Concrete" has identified moisture-related durability of natural fiber-reinforced cement-based materials as an area requiring focused research to ensure satisfactory long-term performance of such materials.

A comprehensive research plan is underway to improve understanding of the implications of fiber swelling/shrinking on the composite system – the changes that occur in the fiber, surrounding brittle matrix, and the performance of the composite as a whole. The objectives of this research are:

1) to assess the effects of repeated wetting and drying (rain and heat) cycles on dimensional changes of wood fibers, physical bonding at the fiber/cement paste interface, and bond strength between the fiber and matrix, and
2) to assess if improved bonding can be produced through tailoring the fiber processing for this application. (Specifically, the processes to be examined, individually and in combination, include fibrillation, bleaching, and pre-drying.)

Research to achieve these objectives is ongoing and includes both microstructural characterization by scanning electron microscopy and mechanical testing using both traditional flexural testing as well as specialized samples prepared assess fiber/matrix bonding. Here, the results of some initial work in this investigation to assess the proposed environmental exposure regimen are presented.

2. EXPERIMENT

2.1 Materials and Sample Preparation

Pulp fiber reinforced mortar bar samples, 1x1x4in. (2.54x2.54x10.16cm), were prepared with water-to-cement ratio of 0.60 and sand-to-cement ratio of 1. The cement used was commercially available Type I cement. The sand used was a natural siliceous sand (FM=1.80), obtained from Brown Brothers Quarry in Junction City, Georgia USA. Unbleached softwood (Slash Pine) kraft pulp was used without beating as fiber reinforcement at 1% fiber volume fraction in the mortar. Mortars were mixed in a 1.5L-capcity Hobart mixer at 120 rpm.

2.2 Exposure

All of the samples were cured in a saturated lime solution at room temperature for four days. After this initial curing period, the samples were divided into three groups and subjected to three different aging/exposure conditions. One group of samples was allowed to air cure at room temperature and humidity (RH = $60 \pm 5\%$) for one more day and was tested at 5 days of age. A second group was allowed to air cure, after the initial 4-day moist curing period, and was subsequently tested at 8 days of age. A third group was exposed to 10 cycles of wetting and drying in an environmental chamber (Thermotron Model SE-1200-3-3) over a 74-hour period. These samples were tested just after their environmental exposure, at an age of 8 days.

In the environmental chamber, the samples were kept for 2h 55min at 60° F (16°C) and 85% RH. Over a 30-minute period, the temperature was increased at a rate of 1.33°F/min (0.74°C/min) to 100°F (33°C) and the humidity was decreased at a rate of 2.17%/min to 20%; these conditions were held for 2h

55min. Over a 60-minute period, the environmental conditions were cycled to 60° F and 85% RH at rates of 0.67°F/min (0.37°C/min) and 1.08%RH/min, respectively, completing one wet/dry cycle. The selection of the "hold" period of 2h 55 min is based upon ASTM C 1185-99 and the selection of the rates is based upon the capabilities of the environmental chamber. A representation of these exposure conditions is shown in Figure 1.

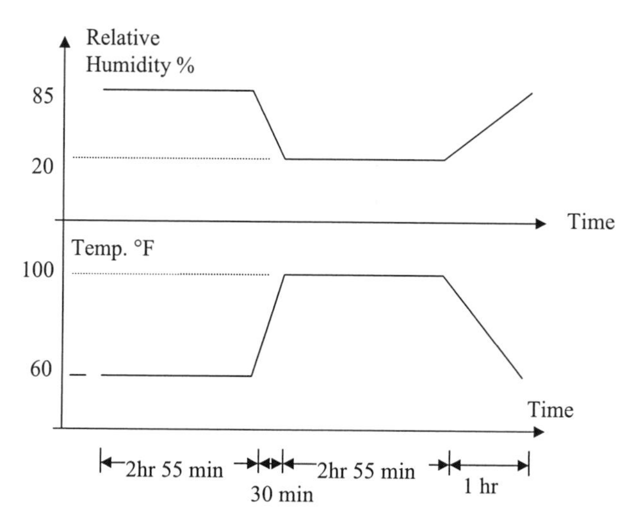

Figure 1: Schematic representing a single wet/dry cycle.

2.3 Test Method

Center-point flexure tests were performed on the mortar beams using a span of 3 in. (7.62 cm). The span-to-depth ratio of 3 was chosen to comply with ASTM C 348-97 and C 293-94. Tests were performed in triplicate using a 22-kip screw driven test frame (Satec model 22EMF). The deflection was captured using an electronic deflectometer, placed under the center point of the beam. The test was controlled using the deflection control system to better capture the post-peak behavior of the beams.

3.0 RESULTS AND DISCUSSION

Load-deflection curves for all three groups of samples are shown in Figure 2. Average behavior for each sample type is shown in Table 1. The results show that the average peak load is moderately increased due to age in the sample tested at 8 days without environmental exposure as compared to the sample tested at 5 days, also without environmental exposure. The average peak loads carried by these samples were 864 and 813 psi (5.96 and 5.61 MPa), respectively. A larger (46 %) increase in peak strength, as compared to the samples tested at 5 days, is seen in the data for the samples tested after 8 days and after being subjected to wet/dry cycling. This result was not expected, but it is proposed that this increased strength is due to accelerated cement hydration in a warm moist environment. Thus, this peak strength data can be seen to represent a change in the matrix with environmental exposure and time.

While peak load capacity may have increased with wet-dry cycling, toughness was found to decrease. The average post-peak toughness for the samples tested at 5 days, 8 days, and 8 days with wet/dry cycling were 1.176, 1.175, and 1.104 lb.in. (132.9, 132.8, and 124.7 N·mm), respectively. This 6% decrease in toughness observed between the samples tested at 8 days with and without environmental exposure may be due to increased matrix strength and perhaps increased matrix/fiber bond strength, which in turn changes the behavior of the composite in flexure. It is proposed that the increased strength of the fiber-cement bond causes the failure mode to result, to a greater extent, from fiber breakage rather than fiber pullout. These competing effects (i.e., increasing bond strength with continued cement hydration during aging and possible decreasing bond strength during wet/dry cycling) can make data interpretation challenging.

These results show that to isolate the effect of wet/dry cycling, the curing period and environmental exposure should be altered. First, the control samples should be continuously moist cured until testing. Second, samples should be cured for a longer period prior to wet/dry cycling, to minimize the effect of continued cement hydration during the environmental exposure.

TABLE 1
SUMMARY OF RESULTS FROM FLEXURAL TESTING

Specimen Type	Curing Time (days)	Age at Testing (days)	Number of Wet/dry Cycles	Average Flexural Strength (psi)	Post-peak Toughness (lb.in.)
1	4	5	0	813	1.176
2	4	8	0	864	1.175
3	4	8	10	1188	1.104

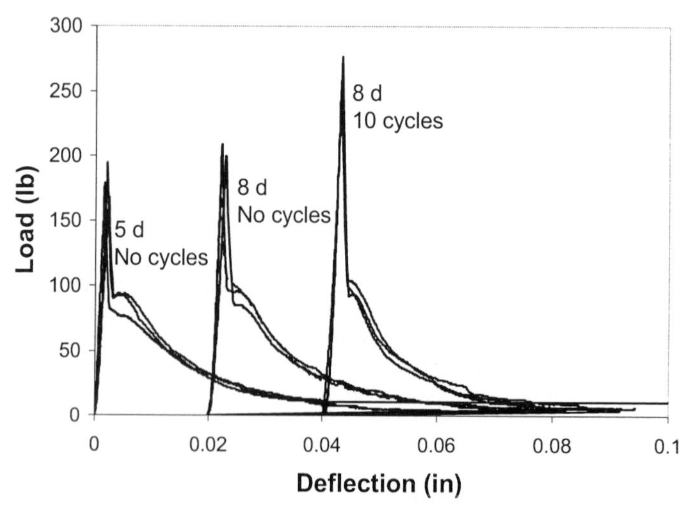

Figure 2 Load-deflection curves for samples cured for 5 and 8 days without wet/dry cycling and those cured for 4 days and subsequently subjected to 10 wet/dry cycles over a 4-day period.

4.0 CONCLUSIONS

This investigation shows that the wet/dry cycling employed here can result in improved flexural strength, which is likely due to the increase in the degree of hydration of the cement-based matrix during the environmental exposure. However, post-peak toughness is decreased as a result of the wet/dry cycling employed and this is believed to reflect changes occurring at the fiber/matrix interface. Therefore, to assess the effect of environmental exposure cycles on composite performance, data analysis should focus on the post-peak region of the load-deflection curve.

More than one factor may contribute to the observed decrease in toughness with environmental exposure. The loss in toughness may be due to fiber/matrix debonding during the wet/dry cycling and also due to the increased fiber/matrix bond strength due to continued cement hydration. Thus, it is recommended that control samples be moist cured until the day of testing and that the samples subjected to wet/dry cycling be moist cured for a more extended period prior to environmental exposure. Both of these measures are intended to reduce the effect of the bonding strength. Research is ongoing to identify appropriate conditions (i.e., curing times, exposure periods, temperature, relative humidity) for this environmental exposure.

ACKNOWLEDGEMENTS

The authors gratefully acknowledge support from U.S. National Science Foundation award CMS-0122068, which is part of an initiative co-sponsored by the Partnership for Advancing Technologies in Housing (PATH) as well as support from the Institute for Paper Science and Technology seed grant program and PATHWAYS program.

REFERENCES

American Concrete Institute Committee 544 (1996) "State-of-the-Art Report on Fiber Reinforced Concrete", Manual of Concrete Practice, ACI 544-1R96, Farmington Hills, MI.

American Society for Testing and Materials (1994) "Flexure Strength of Concrete (Using Simple Beam with Center-Point Loading)," *ASTM Spec. Tech. Publ. C 293.*

American Society for Testing and Materials (1997) "Flexure Strength of Hydraulic Cement Mortars," *ASTM Spec. Tech. Publ. C348.*

American Society for Testing and Materials (1999) "Standard Test Methods for Sampling and Testing Non-Asbestos Fiber-Cement Flat Sheet, Roofing and Siding Shingles, and Clapboards," *ASTM Spec. Tech. Publ. C 1185.*

Kurpiel, F. (2000) "Fiber-cement siding is tomorrow's growth product," *Wood Technology*, November.

MacVicar, R.; Matuana, L.M.; and Balatinecz, J.J. (1999) "Aging Mechanisms in cellulose fiber reinforced cement composites", *Cement and Concrete Composites*, **V.21**, 189-96.

Advances in Building Technology, Volume 1
M. Anson, J.M. Ko and E.S.S. Lam (Eds.)

Extruded Fiber-Reinforced Composite

B. Mu, M. F. Cyr and S. P. Shah

Center for Advanced Cement Based Materials, Northwestern University,
2145 Sheridan Road, Evanston, IL 60208, USA

ABSTRACT

In this study, high performance fiber-reinforced cementitious composites (HPFRCC) were produced using the extrusion technique. The reinforcing effects of three types of short fibers, glass, polyvinyl alcohol (PVA) and polypropylene (PP), and combinations of these fibers were tested. The mechanical properties and durability of extruded composites with high percentage of class F fly ash were also studied. It was found that the flexural performance of extruded composites can be controlled and optimized using hybrid fibers. The durability of glass-fiber composites was greatly improved by the addition of fly ash. A widely applicable rheological constitutive model was selected based on the experimental observations.

KEYWORDS

Fiber-reinforced concrete, Extrusion, Rheology, Fibers, Fly ash, Viscosity, ACBM

EXTRUSION

High performance fiber-reinforced cementitious composites (HPFRCC) exhibit increased tensile strength, enhanced toughness, and sometimes, a strain-hardening response. The amount of fiber reinforcement needed to achieve this behavior is higher than what can typically be added to conventional cast-in-place concrete. Special processing techniques are required to produce HPFRCC. Recently, extrusion technology has been adapted for the production of HPFRCC at the Center for Advanced Cement Based Materials (ACBM Center) at Northwestern University.

Extruded elements are formed by forcing a highly viscous, dough-like mixture of cement paste and fibers through a die of desired cross-section. The composites are formed under high shear and high compressive forces, producing a compact matrix with a good fiber-matrix bond and fiber alignment in the direction of extrusion. The ACBM center research group at Northwestern University started the trials to produce extruded HPFRCC (Shao *et al.* 1995). They have successfully extruded a variety of products including sheets, tubes and cellular shapes. Their research shows that extruded HPFRCC are much stronger and tougher than that made by traditional casting methods. This improvement can be attributed to the achievement of low porosity and good interfacial bond in HPFRCC under the high shear and compressive stresses during extrusion process (Shao and Shah 1996). Their findings and methodology have been adopted and verified by many other researchers (Stang and Pedersen 1996; Li and Mu 1998).

The relation of the modulus of elasticity of the fibers to the modulus of elasticity of the cement matrix influences the mechanical performance of the composite (Bentur and Mindess 1990). High-modulus; high-strength fibers that have a strong bond with the cement matrix can be used to increase composite strength, and low-modulus fibers that have a weaker bond with the cement matrix can be used to enhance composite ductility. Hybrid composites containing two or more fiber types can be considered to control the properties of HPFRCC by taking advantage of both high-modulus; high-strength and low-modulus fibers. The combination of the advantages of hybridization and the extrusion process enables the production of a large range of composite performance and the design of a specific composite performance.

Properly designed extrudate rheology, which controls both internal and external flow properties, is essential for successful extrusion. If the rheology is not correct, defects like lamination and surface tearing can occur during extrusion. Such extrusion defects could be reduced or eliminated with the use of a rheological model in the design of a HPFRCC mixture. The model could predict if the rheology of the matrix is appropriate for extrusion.

The mechanical properties of extruded HPFRCC containing several fiber types, along and in hybrid combinations were studied. The fibers included low elastic modulus PP fibers, as well as strong and stiff glass and PVA fibers. The flexural performance of composites containing large amounts of fly ash was examined. The durability of glass-fiber composites with and without fly ash was also examined. Based on the experimental observations of the fresh extrudate, a differential viscoelastic constitutive model was selected for the future description of the extrusion process.

PROPERTIES OF EXTRUDATES

Special processing techniques, such as extrusion, can enhance the benefits of fiber reinforcement and matrix. Short reinforcing fibers and mineral admixtures have been shown to increase strength and toughness of traditional cement-based composites. Figure 1 (Peled *et al.* 1999) shows that an extruded fiber-reinforced composite is stronger and more ductile than an identical cast fiber-reinforced composite. The effects of different reinforcing fibers and fly ash on the mechanical properties and durability of the extruded HPFRCC were investigated. The mechanical properties were characterized by flexural performance. Specimens were subjected to accelerated aging to assess durability.

The basic mixture proportions, by volume, of the composites were 43% Type I OPC, 12% silica fume, 1% superplasticiser, and 1% rheological enhancer, with a water/cement ratio of 0.24. When fly ash was added to the matrix, it replaced 70%, by volume, of the cement. The basic mixture proportions for composites containing fly ash were 30% Class F fly ash, and 13% cement, with remaining ingredients at the same proportions.

The cementitious matrices were reinforced with glass, PP and PVA fibers, individually and in hybrid combinations. The glass, PVA, and polypropylene fibers have strengths of 3500, 1900, and 700 MPa, respectively, and elastic moduli of 71, 41, and 5 GPa, respectively. The glass-fiber and polypropylene-fiber composites were prepared with a fiber volume fraction (V_f) of 5%. The PVA-fiber composite was prepared with a V_f of 3% because it was difficult to mix and achieve good dispersion of PVA fibers at higher volumes. The hybrid fiber composites were examined by preparing various combinations of glass, PP and PVA fibers (glass/PP/PVA): 20:80:0, 20:40:40, 40:20:40, 40:0:60. The total V_f in the hybrid combinations was 5%. The influence of fly ash on durability was studied in 5% glass-fiber reinforced composites only.

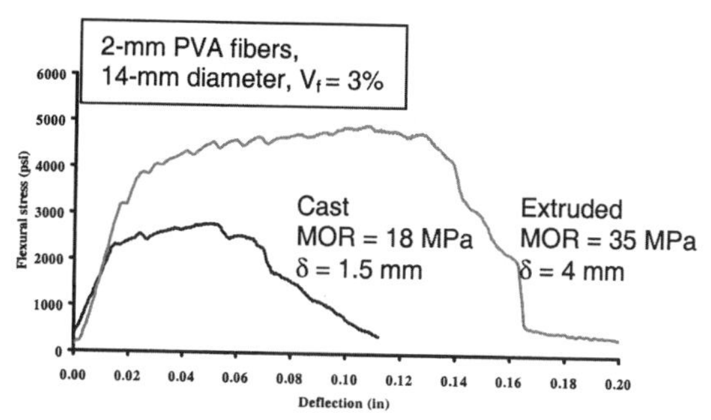

Figure 1: Flexural behavior of extruded HPFRCC and traditional cast FRC (Peled *et. al.* 1999).

Thin sheet specimens, 4mm x 25mm, were extruded using a laboratory-scale ram extruder. After extrusion, the specimens were covered with plastic to set for three days. Then, they were moist cured for 28 days at room temperature and 100% relatively humidity (RH). To evaluate long term durability of glass-fiber composites one batch of specimens was placed in an 80°C water batch for six week after a 28-day moist cure to accelerate aging. Prior to testing, the specimens were removed from the curing or aging environment and stored at 22°C, 50% RH for 24 hours to allow surface moisture to evaporate.

The three-point bend flexural tests were performed on an MTS machine with a 445 N load cell. The span of the extruded specimen was 101.6 mm. The test was conducted under stroke control at the rate of 0.0114 mm/sec. The modulus of rupture and area under the stress-deflection curves, referred to here as the toughness, were calculated. The toughness was calculated up to 20% of the maximum stress in the post-peak region, except from composites containing polypropylene, where the toughness was calculated to a deflection of 10 mm. The test results presented are an average of at least 5 specimens. Typical stress-deflection curves representing the flexural behavior of individual composites are chosen for comparison.

Effect of Hybridization

Hybrid composites containing two or more fiber types can be considered to optimize the desired properties of composites by taking advantage of the properties of each fiber. Figure 2 shows the flexural response of 100% glass, 100% PVA and 100% PP fiber composites with V_fs of 5%, 2%, and 5%, by volume, respectively. The relatively brittle behavior of the glass-fiber composite is clear in this figure.

PP fibers have a lower elastic modulus than the cementitious matrix and a relatively poor fiber-matrix bond, while glass fibers have a higher elastic modulus than the cementitious matrix and a relatively strong fiber-matrix bond. Figure 3 presents the flexural response of the 100% PP composite and the hybrid composite with a 20:80 glass/PP ratio, both with a V_f of 5%. When only 20% of the PP fibers are replaced with glass fibers, the strength of the composite increases by about 60%. However, the increase in toughness is quite small. It has been shown that further replacement of PP fibers with glass fibers will lead to a decrease in toughness but an increase in strength (Peled *et al.* 2000).

The performance of hybrid composites of various combinations of glass, PP, and PVA fibers with a total V_f of 5% is shown in Figure 4. Replacing half of the PP fibers in the 20:80 glass/PP hybrid with PVA fibers (glass/PP/PVA 20:40:40) significantly increases both the strength and toughness of the composite, producing a strain hardening response. Further replacement of some of the remaining PP fibers with glass fibers, 40:20:40 glass/PP/PVA composite leads to a greater improvement in flexural performance of the composite. Moreover, the 40:20:40 and 40:0:60 glass/PP/PVA hybrid combinations exhibit flexural strengths similar to the 100% glass fiber composite but show a significant improvement in toughness. It can be concluded that performance of extruded composites can be controlled and optimized with different hybrid combinations.

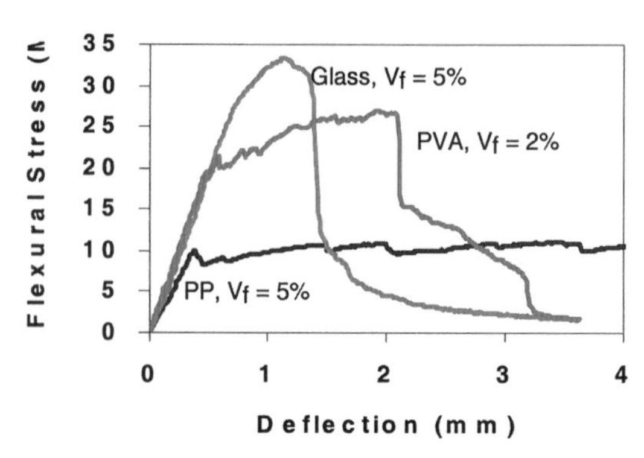

Figure 2: Flexural response of composites with different fiber types (Peled *et al.* 2000).

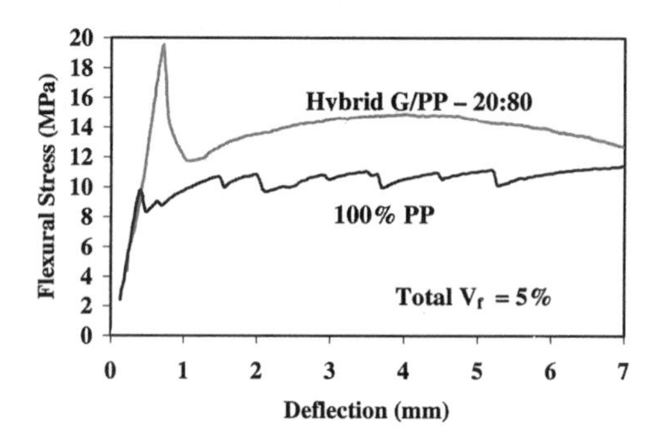

Figure 3: Flexural response of 100% PP fiber compared with glass/PP (20:80) hybrid composite (Peled *et al.* 2000).

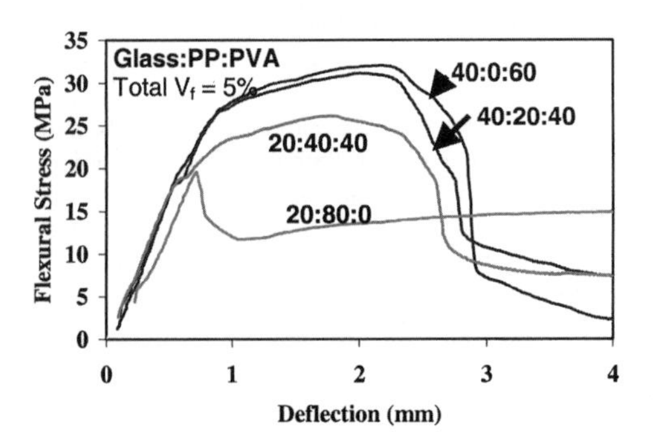

Figure 4: Flexural response of hybrid composite (Peled *et al.* 2000).

Effect of Fly Ash

Class F fly ash can replace cement in extruded composites. Its round particle morphology serves as a rheological aid. It is less expensive and more environmentally sound than cement. Fly ash also improves the mechanical performance of extruded composites, especially those containing glass fibers (Peled *et al.* 2000b). The replacement of 70% of the cement, by volume, with fly ash in a glass-fiber extruded composite increases the strength by 10% and the toughness by 50%, as shown in Figure 5 (Cyr *et al.* 2001). This improvement in performance might be a result of an increase in porosity with the addition of fly ash, which weakens the bond strength enough to increase fiber pullout at failure (Peled *et al.* 2000b).

Fly ash also improves the durability of extruded glass-fiber composites. Glass fibers, even those designed to be alkali resistant, may degrade in the alkaline environment of a cementitious matrix. Fly ash lowers the alkalinity of the matrix (Peled *et al.* 2000b) reducing the possible fiber degradation. Strength is maintained with aging and the reduction in toughness is only 30% when fly ash is added to the matrix, ash shown in Figure 6.

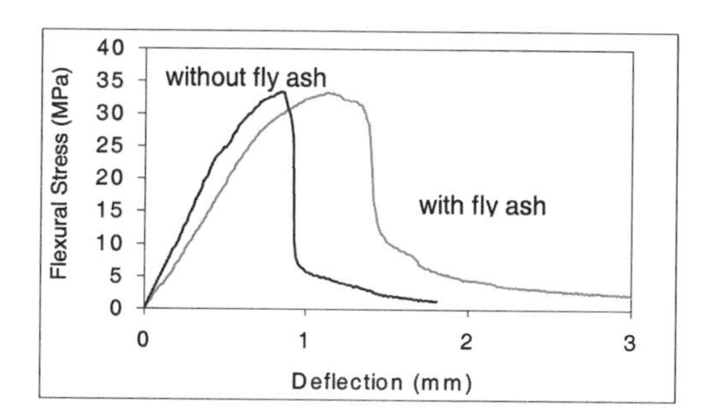

gure 5: Flexural performance of glass-fiber composites with and
ithout fly ash after 28-day moist cure. V_f=5% (Cyr *et al.* 2001)

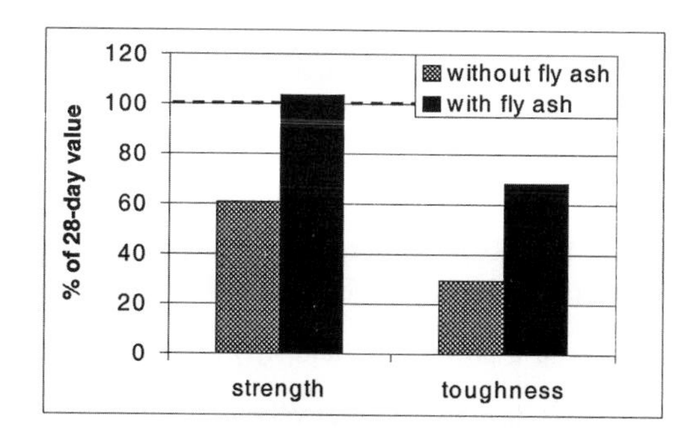

Figure 6: Change in strength and toughness of glass-fiber composites with and without fly ash after accelerated aging. V_f=5% (Cyr *et al.* 2001)

RHEOLOGY OF FRESH HPFRCC IN AN EXTRUDER

The rheology of the fresh HPFRCC is essential to the success of extrusion. If the rheology of the fresh HPFRCC is not ideal, defects will form during extrusion. To facilitate the extrusion process of the

HPFRCC, a study of the rheological behavior of the extrudate is required. Considering the deformation rate is small (described as a small Deborah number: <1), Li et al. (1999) employed the retarded-motion expansion constitutive model for analysis and obtained some preliminary results. However, the retarded-motion expansion was restricted to a small deformation rate. A more widely applicable nonlinear viscoelastic constitutive model is desired. The selected model should be able to describe the normal stress differences of the extrudates observed in the extrusion process.

To develop a suitable constitutive model, both the viscosity and elastic properties of the fresh HPFRCC were studied using two self-designed rheometers. One was a coaxial cylinder rheometer based on a screw extruder (Rheocord 9000, HAAKE), and the other was a cone-and-plate rheometer based on a torsion-compression Material Testing Machine (MTS 858 Table Top System). For the coaxial cylinder rheometer, two cylinders were installed at the head of the extruder. The extrudate was packed between the two coaxial cylinders. The extruder recorded the rotation speeds and the corresponding torques. From these data, the viscosities of the extrudate were calculated. The cone-and-plate rheometer was used to study the elastic properties of the extrudate. During the test, the cone was rotated with the MTS fixture and the plate was kept stationary. Under displacement control the MTS provided a compressive force to keep the cone and the plate in contact. From this compressive force, the elastic material parameter was obtained.

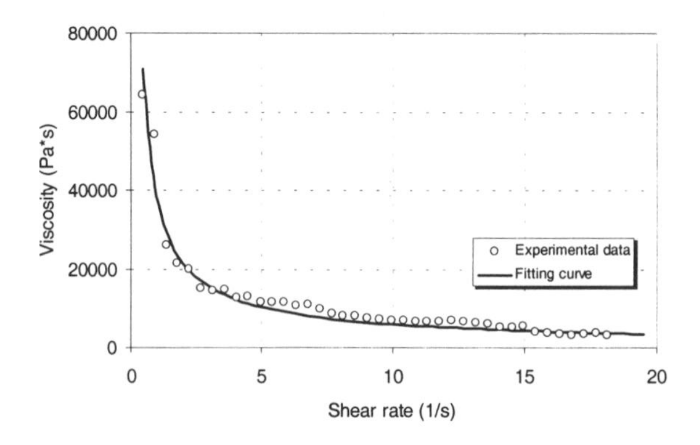

Figure 7: Viscosity and shear rate relationship of the fresh HPFRCC

The experimental results from the coaxial cylinder rheometer and the cone-and-plate rheometer are shown in Figure 7 and 8, respectively. The mixture proportions of the extrudate were cement:slag:sand-1:sand-2:water = 1:1:0.39:0.26:0.60, where sand-1 and sand-2 are silica sand with diameters of 600~300 μm and 150~90 μm, respectively. The mixture was reinforced with 2% PVA fibers. Figure 7 shows that this cementitious flow exhibits a shear thinning response. The curve fitting of the experimental data results in a power law fluid, $\eta(\dot{\gamma}) = m\dot{\gamma}^{n-1}$, with m = 37500 and n = 0.2, where $\eta(\dot{\gamma})$ is viscosity and $\dot{\gamma}$ is shear rate, respectively. The compressive force increases with the shear rate, as shown in Figure 8. This implies that the elastic material parameter is not only a function of the compressive force but also a function of the shear-thinning viscosity.

One model that explains the above experimental observations well is the White-Metzner model (White and Metzner 1963) given by

$$\boldsymbol{\tau} = -\eta(\dot{\gamma})(\boldsymbol{\gamma}_{(1)} + \frac{1}{G}\boldsymbol{\tau}_{(1)}) \tag{1}$$

where G is a constant modulus and $\eta(\dot{\gamma})$ is an arbitrary function to describe the shear rate dependent viscosity. $\boldsymbol{\tau}$ and $\boldsymbol{\gamma}_{(1)}$ are stress tensor and rate of strain tensor, respectively. $\boldsymbol{\tau}_{(1)}$ is the convected time derivative of the stress tensor. In the cone-and-plate rheological experiment (simple shear), Eq. (1) can be expressed as

$$G = \frac{\pi R_0^2 m^2 \dot{\gamma}^{2n}}{F} \tag{2}$$

where R_0 is the diameter of the cone and F is the compressive force.

The selected constitutive model gives a reasonable description of the observed behavior in the simple steady shear case. Since it is a differential constitutive model, it is not limited to small deformation gradients as are the linear viscoelastic models, or to small slowly changing deformation rates, as is the retarded-motion expansion model. There are only three unknown material parameters, m, n, and G, all of which can be obtained with standard rheometers. So, the White-Metzner model is applicable for simulating the extrusion of HPFRCC.

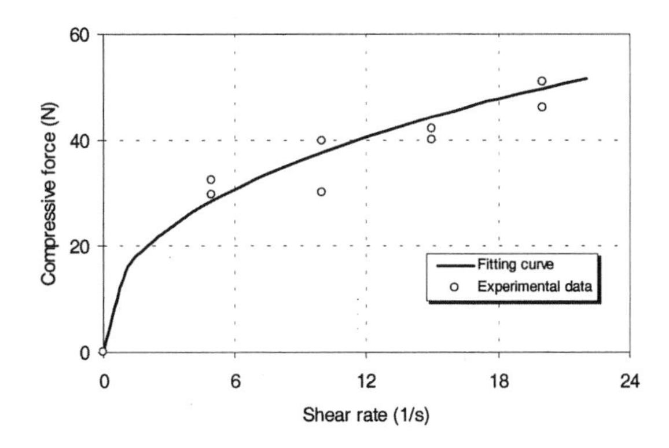

Figure 8: Compressive force and shear rate relationship of the fresh HPFRCC

FUTURE WORK

Future research will be the development of extruded HPFRCC suitable for residential construction applications, particularly advanced panel systems. Future work will focus on batch composition, development of cross-sections, and material testing.

New composites will be reinforced with different types of fibers, including glass, polypropylene, PVA and cellulose, and hybrid fiber combination, to tailor performance for specific applications. These composites will be developed with a high fly ash content to reduce cost, to improve ductility, and to make production more ecological. The effect of fiber movement and particle sizes on extrusion will be studied through a rheological analysis of the composites. Theoretical modeling will be used to guide the manufacture of extrudates with different cross-sections, including cellular and corrugated sheets.

Material testing will focus on determining extruded composite characteristics important to residential construction applications. Impact test will be highly emphasized. Its results will be employed to describe nailability of the extrudates. Impact wave propagation and fracture resistance of the extrudats will be modeled.

ACKNOWLEDGEMENT

This work is being funded by a grant from the National Science Foundation in support of the Partnership for Advancing Technologies in Housing (PATH), a presidential initiative to accelerate the creation and use of advanced technologies to improve the quality, durability, environmental performance, energy efficiency and affordability of housing in the US. Prof. Peter C. Chang, Program Director at Structural Systems and Hazards Mitigation at Division of Civil and Mechanical Systems at National Science Foundation is gratefully acknowledged.

References

Balaguru P. and Shah S. P. (1992). *Fiber Reinforced Cement Composites,* McGraw-Hill, New York.

Bentur A. and Mindess S. (1990). *Fiber Reinforced Cementitious Composites*, Elsevier Applied Science, UK.

Cry M. F., Peled A., and Shah S. P. (2001) Improving Performance of Glass Fiber-Reinfroced Extruded Composites. Proceedings of the 12th International Glassfiber Reinforced Concrete Congress GRC 2001 eds. N. Clarke and R. Ferry. Dublin, Ireland 143-172.

Li Z., and Mu B. (1998). Application of Extrusion for Manufacture of Short Fiber Reinforced Cementitious Composite, *Journal of Materials in Civil Engineering, ASCE,* **10**, 2-4.

Li Z., Mu B., and Chui S. (1999). Rheology Behavior of Short Fiber-Reinforced Cement-Based Extrudate, *Journal of Engineering Mechanics, ASCE,* **125:5**, 530-536.

Peled A., Cyr M., and Shah S. P. (2000). Hybrid Fibers in High Performance Extruded Cement Composites. *Proceedings of International RILEM Symposium BEFIB'2000.* Lyon, France.

Peled A., Cyr M., and Shah S. P. (2000b). High Content of Fly Ash (Class F) in Extruded Cementitious Composites. *ACI Materials Journal*, **97:5**, 509-517.

Peled A., Akkaya Y., and Shah S. P. (1999). Extruded Fiber-reinforced Cement Composites Containing Fly Ash. *Proceedings: 13th International Symposium on the Use and Management of Coal Combustion Products (CCPs).*ed. S.S. Tyson, B.R. Stewart, G.J. Deinhart. American Coal Ash Association, Orlando, FL. 32-1 – 32-16.

Stang H. and Pederson C. (1996). HPFRCC-extruded Pipes, Materials for the New Millennium, K. P. Chong, ed., *Proceedings of the 4th Materials Eng. Conference*, Washington, D. C., **2**, 261-270.

Shao Y., Marikunte S. and Shah S. P. (1995). Extruded Fiber-Reinforced Composites, *Concrete International* **17:4**, 48-52.

Shao Y., and Shah S. P. (1996). High Performance Fiber-Cement Composites by Extrusion Processing, Materials for the New Millennium, K. P. Chong, ed., *Proceedings of the 4th Materials Eng. Conference*, Washington, D. C., **2**, 251-260.

White J. L. and Metzner A. B. (1963). *J. Appl. Polym. Sci.*, **7**, 1867-1889.

Advances in Building Technology, Volume I
M. Anson, J.M. Ko and E.S.S. Lam (Eds.)

REVERSED IN-PLANE CYCLIC BEHAVIOR OF POST-TENSIONED CLAY BRICK MASONRY WALLS FOR HIGH PERFORMANCE MODULAR HOUSING

Owen A. Rosenboom, Graduate Research Assistant

Mervyn J. Kowalsky, Assistant Professor

Department of Civil Engineering, North Carolina State University, Raleigh, NC, 27695, USA

ABSTRACT

Initial experimental results of a research program aimed at investigation of post-tensioned clay masonry walls for high performance modular housing are presented. A series of five large scale reversed cyclic tests on double-wythe clay brick masonry walls are discussed. Each of the five walls contain different details with the objective being to identify the characteristics of each such that choices depending on required performance can be made. The basic behavioral mechanism that is a characteristic of unbonded post-tensioned masonry walls is discussed, followed by details of the performance of each test. Results indicate that walls with unbonded post-tensioning steel performed best in terms of deformation capacity, and strength degradation. Particularly good performance was achieved by confining the compression toe regions of unbonded walls.

KEYWORDS

Post-tensioned masonry, confined masonry, seismic design, cyclic wall response, in-plane wall response.

RESEARCH OBJECTIVE, MOTIVATION AND METHODS

The primary objective of the research described in this paper is to conduct fundamental research into the behavior of pre-stressed clay masonry walls that will then lead into a new modular housing technique that will advance the goals of the Partnership for Advancing Technology in Housing (PATH) program. The PATH program (http://www.pathnet.org) represents a partnership between industry, government, and researchers aimed at achieving the four goals listed below. Of the four PATH program goals, the research conducted at NC State specifically addresses goals one and two. Although it is difficult to quantify, it is also envisioned that strides in the third and fourth goals shown below would be made, particularly over the long term.

1. Reduce the risk of life, injury, and property destruction from natural hazards by at least 10% and reduce residential construction work illness and injuries by at least 20%.
2. Improve durability and reduce maintenance costs by 50%
3. Cut the environmental impact and energy use of new housing by 50%.

4. Reduce the monthly cost of new housing by 20% or more.

Specifically discussed in this paper is the behavior of post-tensioned clay brick masonry walls under seismic conditions. As discussed in the introduction, the motivation for the use of post-tensioning in clay brick masonry walls revolves around enhanced construction efficiency, cost, and most importantly, lateral force performance. Although the emphasis here is on seismic behavior, the details used could be readily employed in regions of low to moderate seismicity, as well advancing the behavior due to other lateral loads such as wind.

The methods used in this research center around five components that together form the first phase of a larger research project on the topic. A project website, http://www4.ncsu.edu/~kowalsky, may be accessed by the reader for complete details on all aspects of the research. The five components in this phase one research are as follows: (1) Literature review and evaluation of the most promising details to employ in the design of post-tensioned masonry walls and assessment of current design approaches. (2) Development of an analysis and design algorithm for the displacement-based design of post-tensioned clay masonry walls. (3) Reversed cyclic testing of a series of five large scale post-tensioned clay brick masonry walls. (4) Evaluation of displacement-based design parameters and revision of analysis and design algorithm. (5) Identification of Phase 2 research needs. In this paper, components 3, and 5 are discussed.

EXPERIMENTAL PROGRAM

Test Specimen Details

As part of the study described in this paper, a series of five large scale cantilever walls were tested in the Constructed Facilities Laboratory (CFL) at North Carolina State University (NCSU). All five walls were double wythe walls 1.22m long, 305mm thick and 2.44m tall to the center of seismic force as shown in Fig. 1. The walls were constructed from standard cored clay brick masonry units measuring 57mm high, 92mm wide, and 194mm long. All of the walls were post-tensioned with 3 equally spaced 25mm diameter dywidag bars as shown in Fig. 1. Test unit one represented the 'control' specimen and contained a fully grouted wall cavity. The tendons were unbonded, and no additional reinforcement was placed in the wall. Test unit two was the same as test unit one with the exception that confinement plates were placed in the compression toe regions of the wall to enhance the compression strain capacity. Test unit three was also the same as test unit one, with the exception that 4#4 grade 60 ASTM A706 mild reinforcing bars were placed near the end regions of the wall to provide some additional energy dissipation capacity. These mild reinforcing bars extended 914mm above the footing level and were unbonded over a length of 152mm below the footing interface and 457mm above the footing interface. Special considerations were needed in the design of this wall as too little reinforcing steel would not provide sufficient benefit, and too much would result in a large amount of residual deformation after cyclic loading. Test unit four was also the same as test unit one with the exception that the post-tensioning bars were bonded to the wall. The final test specimen was different from the first in that the wall was ungrouted, and as a result, the post-tensioning bars not laterally restrained. All five walls were post-tensioned with a target total force of 1000kN. The test matrix shown in Table 1 provides details of all 5 test units.

Design Approach

The pre-stress force for the walls was determined using a displacement-based design approach. It was stipulated that the pre-stress force in the walls should provide sufficient moment capacity to limit the wall drift ratio to 2% under the design level earthquake, which was obtained from the IBC for Charleston, SC (ICBO, 2000). For details regarding the displacement-based design approach, the reader is referred to Kowalsky et al (2003).

Test Setup, Instrumentation, and Load History

The test setup utilized is shown in Fig. 1. The walls were tested in single bending via a 980kN MTS hydraulic actuator mounted on one side to the strong wall, and at the test specimen via a cast in place concrete loading stub. A guidance frame (not shown in Fig. 1 for clarity) was used to provide out of plane lateral restraint. Instrumentation consisted of load cells to measure pre-stress force and actuator force, a string potentiometer to measure wall displacement, linear potentiometers to measure the profile of the crack at the wall base as well as wall curvature, and electrical resistance strain gages on test unit 3 to measure strain in the supplemental mild steel. The loading history for all five test specimens was identical and followed a pattern of reversed cyclic loading at increasing levels of drift ratio. At each drift ratio, the wall was cycled three times in each direction. The applied load history was based on the ACI recommendations for frame testing (ACI, 2002).

Test One Results - (Grouted, Unbonded, Unconfined)

Test unit one represents the 'control' specimen. During the course of this test, it was observed that the behavioral mechanism by which such a wall deforms is one of 'rocking', which is consistent with observations made by Laursen and Ingham (2000a,b) on post-tensioned concrete masonry walls. At a drift ratio of 0.2%, a crack formed at the base of the wall. Once this cracked formed, no further flexural cracking was observed, which is as expected given the absence of bond between masonry and steel post-tensioning steel. Cycling continued without any further observation of damage until a drift ratio of 0.5% when initial signs of crushing of masonry on the compression toe region was noted. Once the masonry started to crush, the extent of damage extended into the core region of the wall as the masonry was unconfined. At a drift ratio of 2.75% the maximum lateral force capacity of 327kN was achieved. The wall was cycled to a drift ratio of 6.5% with some gradual strength degradation. A photo of the test unit at 6.5% drift is shown in Fig. 2a, while the force-displacement hysteretic response is shown in Fig. 3a. From Fig. 2a, note the rather extensive damage to the compression toe regions, however, the remainder of the wall above this region remained largely crack free. The integrity of the masonry above the plastic hinge region is very important from the perspective of achieving a stable compression strut as will be noted in the discussion of tests four and five

From the force-displacement hysteretic response of Fig. 3a, there are two characteristics that must be noted. First, the unbonded post-tensioned system is largely self correcting as residual displacement is close to zero for the majority of the load history. This observation is consistent with the observations made for post-tensioned concrete (Priestley et al., 1999) and post-tensioned concrete masonry (Laursen and Ingham, 2000a,b). Second, the amount of energy dissipation is rather limited, however, this is not a considered to be a fatal flaw as the lateral strength is generally dependable without strength degradation until a drift ratio of 3.75%. Furthermore, test unit three will investigate the use of mild steel as a means for supplemental energy dissipation.

Test Two Results - (Grouted, Unbonded, Confined)

Test unit two was identical to test unit one, with the exception of the addition of galvanized steel plates to the mortar bed joints in the bottom five courses of masonry. Application of confinement plates to masonry was first proposed by Priestley and Bridgeman (1974). The initial behavior of the test specimens was identical to that of test unit one. Initial crushing was noted at the same drift ratio of 0.5%. Beyond initial crushing, however, the behavior of the wall changed dramatically. Whereas unit one started significant strength degradation after 3.75% drift, the confinement of unit two stabilized the compression toe region allowing the wall to achieve a drift ratio of 10% without significant strength degradation. By stabilizing the compression toe region, the diagonal compression strut was able to sustain much higher levels of lateral deformation until finally, the compression toe region degraded, forcing the compression strut resultant outside of the confined region at a drift ratio of 11.3%. Fig. 2b represents the test specimen at a drift ratio of 6.75%. Compared to test unit one, very little damage is noted in the compression toe regions. The test specimen at 11.3% drift ratio is shown in Fig. 2c. Note the diagonal shear crack that formed parallel to the compression strut once the compression strut resultant migrated outside of the confined region. It is likely that the wall could have sustained even higher deformations had the confinement plates extended further into wall core; however, the need to achieve a drift ratio higher than 11.3% is unlikely. The force-displacement hysteretic response is shown in Fig. 3b. Note the very stable loops with little residual deformation or loss of strength. The overall energy dissipation is rather low, consistent with the observation of test unit one.

Test Three Results - (Grouted, Unbonded, Unconfined, Supplemental Mild Steel)

The purpose of test unit three was to investigate the possibility of increasing the energy dissipation while retaining the desirable characteristics of little residual deformation of walls one and two. This test unit was identical to test unit one with the exception of the presence of mild steel to act as supplemental energy dissipation. Such an approach has been utilized for post-tensioned concrete structures (Priestley et al., 1999) and post-tensioned steel structures (Christopoulos et al., 2002), (Ricles et al., 2001) with some success. In order to determine the amount of mild steel to be placed in the wall, the following process was utilized: (1) Size mild steel such that applied PT force will be sufficient to return the mild steel strain to zero at zero column drift. This step ensures that the residual deformation of the wall will be kept to a minimum. It is important to note that the total PT force will vary as a function of applied lateral deformation, and that at high lateral drift, there is the potential for significant loss of pre-stress force. As a result, the initial pre-stress force should not be used to determine the compressive clamping force to return the mild steel to zero strain, but rather, the reduced PT force that can be expected at higher levels of response. (2) Determine the required length to debond the mild steel based on a pre-determined limit for mild steel tension strain. Due to the single large base crack, if the mild steel is not debonded, it is expected that the bars will rupture rather early on due to the high stress concentration at the base. Based on the kinematic relationship between wall drift ratio and base crack size, it is straightforward to determine the amount of debonded length that is required to keep the strain in the mild steel below an acceptable level. In the case of this test specimen, at a drift ratio of 6%, a base crack of 51mm is expected based on a neutral axis depth of 150mm. If it is desired to keep the strain in the mild steel below 8.5%, then the required debonded length is 610mm (51mm/0.085)

Just as the previous two tests, test three started in a similar manner. The base crack was observed to open at a drift ratio of 0.2%. Initial crushing was noted at a drift ratio of 0.35%. Fig. 2d represents a photo of the test specimen at a drift ratio of 6.5% which compares favorably to Fig. 2a for the control wall. The force-displacement hysteretic response is shown in Fig .3c. When compared to the control wall, it is noted that significant strength degradation did not occur until a drift ratio of 5.0%. It is also noted that the energy dissipated in the walls is greater than that from the control specimen. Since the wall was unconfined, once the masonry starts to crush, there is a rather rapid degradation of strength. Application of confinement would certainly enhance the performance of this configuration, however, the added benefit may not out way the added effort for designing and constructing this configuration.

Test Four Results - (Grouted, Bonded, Unconfined)

The goal of test four was to most closely mimic the behavior of a conventionally reinforced wall. In doing so, the PT bars were fully bonded along the length of the wall after applying the PT force. Although it was expected that this wall would be damaged more than the unbonded walls, it was still expected to achieve similar deformation capacity. Unfortunately, that was not the case. Unlike the previous three walls, in addition to the base crack, minor flexural cracks due to the bond stresses between PT steel and masonry occurred at a drift ratio of 0.2%. These cracks were spaced at 250mm. Further loading resulted in masonry crushing at 1.0% drift. Soon after, at a drift ratio of 1.75%, vertical cracks were noted at the positions of the PT bars. These cracks occurred due to exceedance of the bond stress capacity between PT duct and surrounding grout. As the bond stresses increased resulting in vertical cracks, the behavioral mechanism started to shift away from flexural deformation back to the rocking mechanism observed in the previous three tests. Significant strength degradation was noted after 3% drift ratio, and at a drift ratio of 3.75% (Fig. 2e) damage was extensive in the wall. Due to this rather extensive damage, the diagonal compression strut was unable to stabilize, resulting in wall failure. The force-displacement hysteretic response for the wall is shown in Fig. 3d, and although there is higher energy dissipation than the control wall, the level of damage, reduced deformation capacity, and quick loss of strength rendered the performance of this wall poor.

Test Five Results - (Ungrouted, Unbonded, Unconfined)

The final test of the series was the simplest to construct, and the motivation for this test stemmed from simplicity of the configuration. Unfortunately, this turned out to be the poorest performing configuration. Although it behaved initially just as tests one through three, this configuration suffered a diagonal shear failure on a plane parallel to diagonal compression strut at drift ratio of only 1% (Fig. 2f). Due to the lack of a grouted core, it became impossible for a diagonal compression strut to form in the wall, and as a result, the rocking deformation was not achievable. Laursen and Ingham observed similar behavior in partially grouted concrete masonry walls (Laursen and Ingham, 2000a,b). The force displacement hysteretic response is shown in Fig. 3e. It is clear that such a configuration is not for anything but the lowest seismic regions.

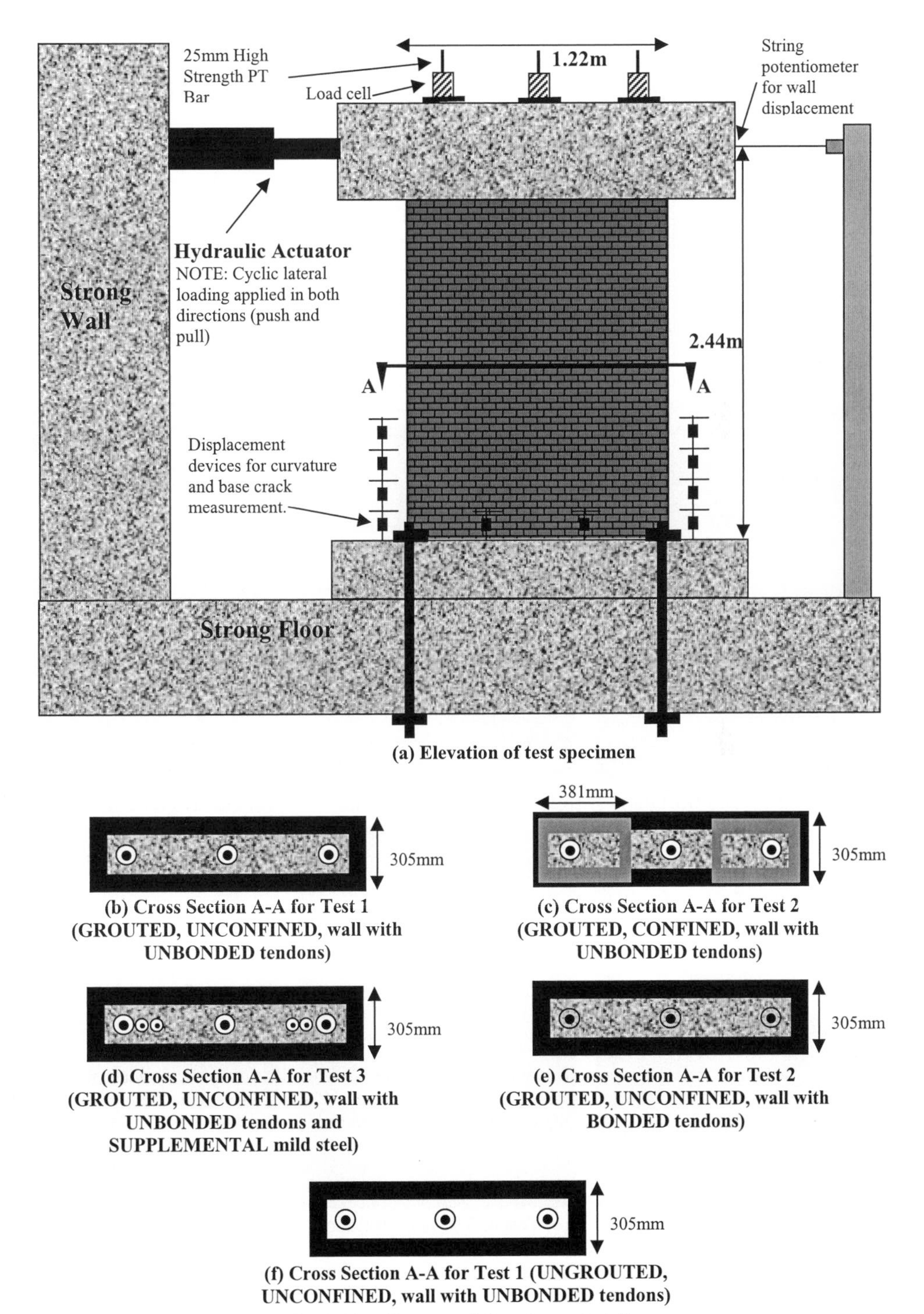

(a) Elevation of test specimen

(b) Cross Section A-A for Test 1 (GROUTED, UNCONFINED, wall with UNBONDED tendons)

(c) Cross Section A-A for Test 2 (GROUTED, CONFINED, wall with UNBONDED tendons)

(d) Cross Section A-A for Test 3 (GROUTED, UNCONFINED, wall with UNBONDED tendons and SUPPLEMENTAL mild steel)

(e) Cross Section A-A for Test 2 (GROUTED, UNCONFINED, wall with BONDED tendons)

(f) Cross Section A-A for Test 1 (UNGROUTED, UNCONFINED, wall with UNBONDED tendons)

Fig. 1 Test Setup and Specimen Details

Table 1 Test matrix and specimen details

Test #	Initial PT Force (kN)	Mild Steel	Confinement	Grouted	Bonded	*f'm* (Mpa)
1	983	No	No	Yes	No	22.2
2	962	No	3mm thick plates @ 67mm vertical spacing. f_y=269MPa.	Yes	No	25.5
3	944	4#4 Grade 60 ASTM A706, f_y = 408MPa	No	Yes	No	25.5
4	840	No	No	Yes	Yes	25.5
5	986	No	No	No	No	25.5

(a) Test 1: 6.5% Drift Ratio (b) Test 2: 6.5% Drift Ratio (c) Test 2: 11% Drift Ratio

(d) Test 3: 6.5% Drift Ratio (e) Test 4: 3.75% Drift Ratio (f) Test 5: 1% Drift Ratio

Fig. 2 Photos of test specimens during testing at various drift ratios

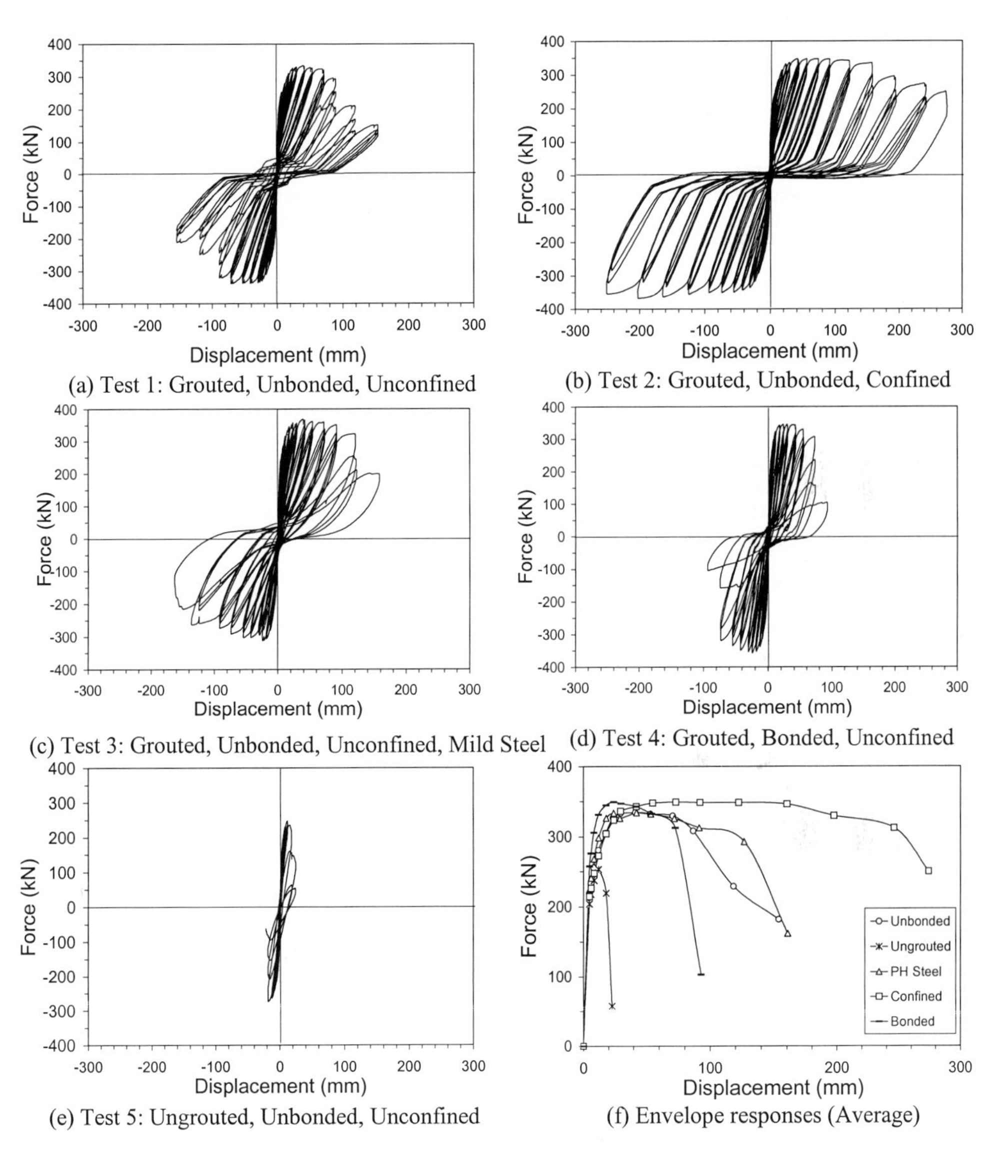

(a) Test 1: Grouted, Unbonded, Unconfined
(b) Test 2: Grouted, Unbonded, Confined
(c) Test 3: Grouted, Unbonded, Unconfined, Mild Steel
(d) Test 4: Grouted, Bonded, Unconfined
(e) Test 5: Ungrouted, Unbonded, Unconfined
(f) Envelope responses (Average)

Fig. 3 Force-displacement responses

PLANS FOR FUTURE RESEARCH

For a complete discussion of Phase I research, the reader is referred to (Rosenboom and Kowalsky, 2002), (Kowalsky et al., 2003). The second phase of the program is currently underway and involves the following components: (1) Shake-table testing of the most promising configuration from Phase I, namely, grouted, confined, unbonded walls. (2) Large-scale (cyclic and dynamic) testing and analysis of post-tensioned clay masonry walls with window openings. (3) Large-scale cyclic testing and analysis of "L" and "T" cross-section walls. (4) Development of cyclic section analysis software for analysis and design of post-tensioned walls. The third phase of the research will concentrate on practical applications and design details, namely, floor and roof diaphragm to wall connections and integration of non-structural components.

CONCLUSIONS

Based on the initial results, the following tentative conclusions are advanced. (1) Application of post-tensioning to clay masonry walls provides benefits similar to what has been observed for pre-cast concrete. (2) Walls with unbonded tendons performed best in terms of strength, deformation capacity, and strength degradation. (3) Application of galvanized steel confinement plates results in a marked effect on wall deformation capacity due to the ability of the plates to confine the compression toe region and stabilize the diagonal compression strut required for the wall rocking mechanism to develop.

ACKNOWLEDGEMENTS

The authors would like to acknowledge the support of the National Science Foundation (NSF) through grant #0080210. In addition, the support of North Carolina State University, Department of Civil Engineering is appreciated. Several local industry members are also thanked: Eric Johnson of the Carolinas Brick Association, Cody Cress of General Shale for material donations, Danks Burton of Pinnacle Masonry for mason's labor support, and John Little and Gary Rhodes of AmeriSteel. We would also like to thank Jerry Atkinson, the technician at the Constructed Facilities Laboratory at NC State without whom this work would not have been possible.

REFERENCES

American Concrete Institute (1999) "Acceptance Criteria for Moment Frames Based on Structural Testing" *ACI Provisional Standard.*

Christopoulos C., Filiatrault A., Uang C-M (2002). "Post-tensioned energy dissipating connections for steel frames". *Journal of Structural Engineering,* ASCE 2002.

International Code Council (2000) "International Building Code"

Kowalsky, M.J., Ewing, B.D. and Rosenboom, O.A. (2003) "Displacement-Based Design of Post-Tensioned Masonry Wall Buildings", ASCE Structures Congress, June, 2003.

Laursen, P., and Ingham J.M. (2000). "Cyclic In-Plane Structural Testing of Prestressed Concrete Masonry Walls, Volume A: Evaluation of Wall Structural Performance." *School of Engineering Report No. 599*, Department of Civil and Resource Engineering, University of Auckland, New Zealand.

Laursen, P., and Ingham J.M. (2000). "Cyclic In-Plane Structural Testing of Prestressed Concrete Masonry Walls, Volume B: Data Resource." *School of Engineering Report No. 600*, Department of Civil and Resource Engineering, University of Auckland, New Zealand.

Priestley and Bridgeman (1974) "Seismic Resistance of Brick Masonry Walls" *Bulletin of the New Zealand National Society for Earthquake Engineering*, Vol. 7, No. 4, December, pp167-187.

Priestley M.J.N, Sritharan S., Conley J.R., Pampanin S. (1999). "Preliminary Results and Conclusions from the PRESSS Five-Story Precast Concrete Test Building", *PCI Journal*, Vol. 44, No. 6, pp42-67

Ricles J.M., Sause R., Garlock M, Zhao C. (2001). "Post-tensioned seismic-resistant connections for steel frames", *Journal of Structural Engineering*, ASCE Vol. 127, No.2, pp113-121.

Rosenboom, O.A. and Kowalsky, M.J. (2002) "Reversed In-Plane Cyclic Behavior of Post-Tensioned Clay Brick Masonry Walls" Submitted to the *ASCE Journal of Structural Engineering.*

Advances in Building Technology, Volume 1
M. Anson, J.M. Ko and E.S.S. Lam (Eds.)

An All-Natural Composite Material Roof System for Residential Construction

H. W. Shenton III[1], R. P. Wool[2], B. Hu[1], A. O'Donnell[2], L. Bonnaillie[2], E. Can[2], R. Chapas[2] and C. Hong[2]

1-Department of Civil and Environmental Engineering, University of Delaware, Newark, Delaware, USA 19716
2-Department of Chemical Engineering and the Center for Composite Materials, University of Delaware, Newark, Delaware, USA 19716

ABSTRACT

In a collaborative effort between researchers from the Departments of Chemical Engineering and Civil and Environmental Engineering at the University of Delaware, work in underway to develop an all-natural composite material roof system for residential construction. The roof will be fabricated using innovative composite materials made from soy-oil based resins and all-natural fibers. The all-natural composites are significantly less expensive than traditional polymer-based composites and are environmentally friendly. The composite resin is derived from soy-oil, a plentiful, renewable crop, and the fibers are by-products or waste from other seasonal crops. The research will involve design and optimization of the composite material for the roof, development of an integral weather protection layer for the roof, and structural design, analysis and testing of the roof system. The preliminary design of the roof is a sandwich structure configuration, with top and bottom face sheets, periodically spaced webs, and a foam core. Tests are currently underway of the first unit beam section of the roof.

KEYWORDS

Composite material, fiber, natural, resin, roof, sandwich structure, seismic, soy oil, testing, wind

INTRODUCTION

Traditional composite materials are made from a petroleum-based matrix resin such as epoxy, vinyl-ester, or polyester, and a reinforcement of synthetic fibers such as glass, carbon, or aramid. These traditional composites can be quite expensive, are non-biodegradable, and depend upon our rapidly depleting petroleum reserves. In the

Affordable Composites from Renewable Resources (ACRES) program, headed by Professor Richard P. Wool at the University of Delaware, genetically engineered soybean oil and other plant oils have been made amenable to polymerization using a broad range of chemical routes (Williams and Wool, 2000). Using soy-based resins and natural fibers, the ACRES program has produced all-natural composite materials that have mechanical properties that are comparable to certain polymer-based composites and also some natural woods.

All-natural composites are environmentally friendly for a number of reasons. The composite resins are derived from soy beans, a seasonal crop that is already grown in large quantities in the United States. Inexpensive natural fibers such as hemp, flax, jute, and kenaf, are also seasonal crops grown specifically for the fiber, or are byproducts of other crops. These materials are renewable, biodegradable, and recyclable. When used as a substitute for conventional wood in low-rise commercial or residential construction, the composite system will reduce the demand on natural wood and its resulting deforestation. Finally, at the end of their useful life cycle, the composite materials can be ground up and used as low-cost filler for other thermosetting composites.

Recently, in a collaborative effort between the ACRES program and the Department of Civil and Environmental Engineering at the University of Delaware, researchers are exploring the use of all-natural composites in residential construction. With funding from the National Science Foundation, Partnership for Advanced Technology in Housing (PATH) program, the interdisciplinary research team is working to develop an advanced all-natural composite material roof system for residential construction. The composite roof fulfills the four main goals of the PATH program by, (1) reducing the monthly cost of housing through the roof's inherent insulation capability, (2) improving durability and reducing maintenance costs with the development of an integral weather protection layer (thus eliminating the need for shingles), (3) reducing environmental impact with the use of the all-natural composite materials, and (4) reducing the risk of life, injury and property destruction through the monolithic, ground based fabrication process.

Presented in the paper is a brief overview of the ongoing research effort. The all-natural composites are first described. Next, the concept for the composite roof is introduced, along with an outline of the work completed to date. Finally, a general discussion of the potential benefits of the all-natural composites is presented, with particular emphasis on the advantages for seismic and hurricane resistant design.

ALL-NATURAL COMPOSITES

Soybean oil is a naturally occurring triglyceride. Triglycerides are composed of three fatty acids connected by glycerol through ester linkages. The fatty acids in soyoil have zero to three double bonds. These double bonds are not suitable for polymerization but can be activated by different reactions. In the ACRES program, the chemically modified oils have been converted into a range of high-performance materials including plastics, rubbers, foams, and composite liquid molding resins. Several patent disclosures have been filed on these novel materials. Of the several chemical routes investigated by the ACRES group, at least ten have yielded materials that can be successfully used as matrix materials for polymer composites (Wool et al., 1998). These resins have been combined with synthetic fibers like glass and natural fibers like hemp and flax to make composites. In August 1999, the John Deere Company, an international supplier of farm equipment, announced their first fleet of harvesters that include parts made with ACRES soy-based resins.

The properties of composites made using ACRES resins have been found to be equivalent to those of synthetic-resin-based composites with similar reinforcement. Further development and commercialization of the ACRES resins and composites will serve three important purposes: (1) help the agricultural community through the use of a crop that is produced on a large scale in the United States; (2) help reduce the depletion of petroleum feedstock; and (3) in housing applications, help reduce reliance on natural wood and the resulting deforestation. Because of the very low cost of the soyoil from which the resins are derived, projections for the ACRES resins suggest a price half that of comparable vinyl-ester resins. This will represent a significant cost saving that comes with no sacrifice in properties and performance.

Combining the soy-based resins with natural fibers yields an all-natural composite material. The choice of natural fibers is numerous and includes hemp, jute, flax, kenaf and even chicken feathers. Many of these fibers are waste or by-products of seasonal crops, which normally have little practical use and in many cases are simply destroyed. These fibers are renewable, biodegradable and recyclable.

Preliminary studies have already been conducted by the ACRES group on selected all-natural composite materials (Williams and Wool, 2000). Presented in Table 1 are the tensile and flexural properties of two different all-natural composites, one a random, chopped flax fiber in the soy-based resin and the other, an oriented hemp fiber in the soy-based resin. Also shown in Table 1 are typical properties for two common softwoods, Douglas Fir and Cedar. Note that the all-natural composites have a tensile modulus lower than that of Douglas Fir, but about equal to the modulus of certain types of Cedar. Tensile strengths are lower than that of Cedar, but only by 20 to 30%. The flexural modulus of the all-natural composites are again lower than that of Douglas Fir, but very comparable to that of Cedar. The flexural strength, on the other hand, is about equal to the average strength of Douglas Fir.

Table 1: Comparison of Tensile and Flexural Properties of Various Materials

Material	Specific Gravity	Tensile Modulus[a] (ksi)	Tensile Strength[a] (ksi)	Flexural Modulus[a] (ksi)	Flexural Strength[a] (ksi)
Flax/soy oil matrix[b]	~1	680-870	4,4	610-650	8.8-9.3
Hemp/soy oil matrix[c]	~1	640	5.1	390	5.1-7.4
Douglas Fir[d]	0.48	1160-1940	15.5	1160-1940	6.8-13.1
Cedar[d]	0.50	640-1700	6.7-11.5	640-1700	4.2-12.8

[a]tensile properties for composites determined in accordance with ASTM D3039; flexural properties in accordance with ASTM D790; [b]random chopped fiber; 30-40% fiber content; [c]two layers placed at 0/90 degree to each other, 20% fiber content; [d]from Wood (1999)

It should be noted that while the results presented in Table 1 are encouraging, they do not in anyway depict the full potential of the all-natural composite material, since the materials tested for that study were not designed or optimized with the specific objectives for the roof system in mind. Nevertheless, the data does suggest that stiffness and not strength will govern the design of an all-natural composite structure. This has been found to be the case in other applications of composite materials to the civil infrastructure (e.g., in bridges).

COMPOSITE ROOF DESIGN

The objective of the NSF funded research is to develop a monolithic composite roof system for residential construction that is fabricated from low-cost, all-natural composite materials. The all-natural composite roof system will be designed to carry normal and in-plane loads, have a durable weather protection surface, and offer intrinsic insulation. The specific objectives are to (1) design and optimize the soy-based resin, all-natural composite material for roof applications; (2) develop and test a weather protection layer that is integral to the composite roof; and (3) design, analyze, and test the roof structural system.

The roof will be fabricated with a foam core, all-natural composite top and bottom face sheets, and all-natural composite ribs/webs. Two different proposed designs are shown in Figure 1. On the left in Figure 1 is the "rafter-core" design, in which foam cores run longitudinally from the roof ridge to the eave; on the right in Figure 1 is the "waffle-core" design, in which the foam cores form a honeycomb structure. The exterior face sheet will be fabricated with an integral weather protection layer, thus eliminating the need for shingles, tiles, or other conventional roofing material. In many ways, the roof is a hybrid system, similar to a conventional stressed-skin panel and a structural insulated panel (SIP), combined.

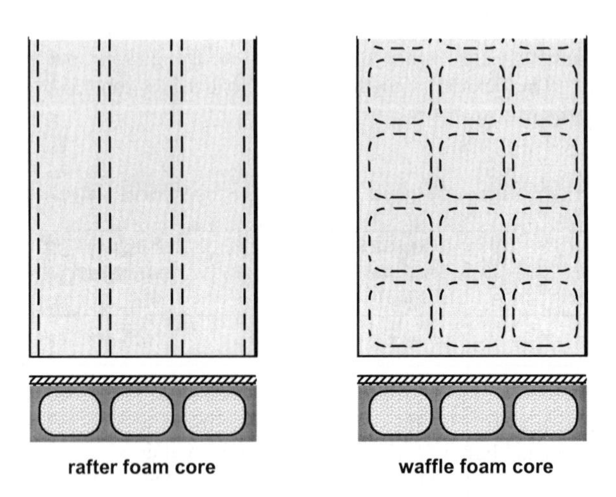

Figure 1: Two design concepts for the roof system - rafter foam core(left) and waffle foam core(right).

At least initially, the foam cores will be made of commercially available synthetic foam; however, the ACRES program is also working to develop an all-natural foam for use in the roof system.

Manufacturing

Minimizing the cost of fabrication of the roof and simplifying the manufacturing process are both very important goals in developing the roof system. It is envisioned that the roof will be fabricated as one single unit, or in very large components that will be bolted together in the field. Components will be fabricated using the Vacuum Assisted Resin Transfer Molding (VARTM) process. VARTM is a very flexible and inexpensive manufacturing process that has been used successfully in the past to create very large composite parts. The process does not require expensive one-of-a-kind molds, in fact, it is expected that the formwork for the roof will be constructed of plywood and framing materials. The nature of the composites is such that the roof will be air cured. As envisioned, the components could be fabricated in a plant and shipped by truck to the construction site, or, for larger new housing developments, fabricated on site in a temporary manufacturing facility.

Preliminary Design and Tests

The initial phase of the research has involved design of the composite material, and design and fabrication of a reduced-scale beam section of the "rafter-core" design. The current composite design uses Flax-Mat 85/15-20, produced by Cargill Limited (85% flax by volume, chopped, random orientation, 20 oz/yd^2 density and 0.25" thick) and Acrylated Epoxidized Soybean Oil (AESO) resin (Knot, et al, 2001). This is not an optimum design and work will be continuing to tailor the resin and fiber system for the roof application.

Presented in Figure 2 is a schematic of the reduced-scale beam section from the rafter-core design. The beam has a length of 3.5 ft, depth of 3.5 inches and width of 8 inches. The cross-section of the beam is a unit section of the rafter design roof. Based on the

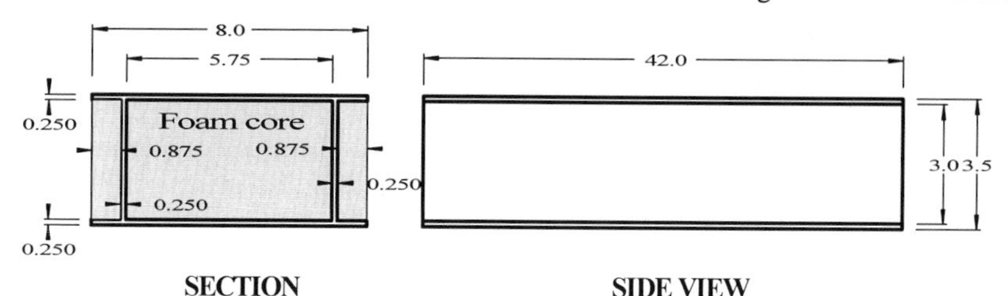

Figure 2. Reduced-scale unit beam of rafter-core roof design (units in inches).

material properties, the theoretical ultimate moment capacity of the beam is 6×10^4 in-lb. The beam will be tested under four point bending. Strains will be measured at several locations on the top and bottom surface of the beam, as will deflections. The tests will be used to measure the elastic flexural stiffness of the section, the yield and ultimate strength of the beam, the post-yield behavior and the failure mode. The data from the test will be analyzed and used to design the next, larger beam specimen, and also to calibrate a finite

(a.) Unit beam during fabrication

(b.) Completed beam

Figure 3. First unit beam

element model of the beam. Presented in Figures 3(a) and 3(b) are photographs of the beam during fabrication and the completed beam.

Tests will be conducted on smaller unit beam sections of the roof, of increasing size and length, as the composite material and section details are optimized, and the manufacturing process is refined. Reduced-scale tests will also be conducted of the "waffle-core" roof design. Ultimately, a large full panel section of the roof will be fabricated and tested to failure. Tests will be conducted for normal and in-plane loads acting on the roof.

SEISMIC AND HURRICANE RESISTANT DESIGN

The use of low-cost, all-natural composite materials should open the door for alternatives in design, fabrication and manufacturing in residential construction, that may lead to improved performance during earthquakes and hurricanes.

Traditional polymer-based composite materials typically have a high strength-to-weight ratio when compared to conventional materials (e.g., steel, concrete, and wood). Referring to Table 1, the non-optimal all-natural composites developed to date have a lower strength-to-weight ratio than typical woods; however, the real benefit of composites comes from being able to optimize the section details and to provide reinforcement where it is most needed. All-natural composite materials hold the potential for being able to create new structural members that are lighter, stronger and more durable than conventional wood members, which should translate into improved seismic and wind resistance of the structure. Another benefit of the randomly oriented fiber composite is the quasi-isotropic nature of the material: the complexity and design constraints that arise with grain direction in conventional woods would be eliminated with the composite material.

Flexibility in manufacturing should also lead to improved wind and seismic response of structures built from all-natural composites. Relatively large, monolithic components (e.g., walls, floors, rooms or entire homes) can be shop fabricated using VARTM and bolted or bonded together in the field. This should reduce the number of critical connections in the structure, a particular area of vulnerability for any structure. In addition, the manufacturing process can also open the door for more free-form design of the structure, for improved seismic and hurricane performance.

CONCLUSIONS

In a collaborative effort between researchers from the Departments of Chemical Engineering and Civil and Environmental Engineering of the University of Delaware, work is underway to develop an all-natural composite material roof for residential construction. The roof will be fabricated from innovative fiber reinforced composite materials made from soy-oil based resins and natural fibers. These low-cost, environmentally friendly materials offer great potential for their use in low-rise commercial and residential construction. The current project is focusing on the roof structure; however, other components such as walls, floors or entire homes could also be fabricated using the new composites. Work will be continuing to optimize the composite material for the roof system, to develop an integral weather protection layer for the roof, and to design, analyze and test sections of the roof and full-scale roof panels.

ACKNOWLEDGEMENTS

Funding for this work has been provided by the National Science Foundation under grant CMS-0122076. The authors would like to thank the NSF for their support.

REFERENCES

Khot, S. N., Lascala, J.J., Can, E., Morye, S.S., Williams, G.I., Palmese, G.R., Kusefoglu, S.H., and Wool, R.P. (2001), "Development and Application of Triglyceride-Based Polymers and Composites," Journal of Applied Polymer Science, Vol. 82, 703-723.

Williams, G.I. and Wool, R.P. (2000) "Composites from Natural Fibers and Soy Oil Resins," *Applied Composite Materials*, Vol 7, pp 421-432.

Wood Handbook, Forest Products Laboratory Technical Report FPL-GTR-113, Madison, WI, 1999.

Wool, R. P. S. Kusefoglu, R. Zhao, G. R. Palmese, and S. Khot (1998). “High Modulus Polymers and Composites from Plant Oils” (filed April 28,1998). Patent Application 09/067,743, PCT filed October 20, 1998.

Advances in Building Technology, Volume 1
M. Anson, J.M. Ko and E.S.S. Lam (Eds.)

COMPARISON OF OPTIMAL DESIGNS FOR STICK-FRAME WALL ASSEMBLIES OF DIFFERENT BAY SIZES

S. Van Dessel[1], A. Ismail-Yahaya[2], and A Messac[3]

[1] School of Architecture, Rensselaer Polytechnic Institute, Troy, NY 12180, USA
[2] Department of Mechanical, Aerospace, and Nuclear Engineering, Rensselaer Polytechnic Institute, Troy, NY 12180, USA
[3] Department of Mechanical, Aerospace, and Nuclear Engineering, Rensselaer Polytechnic Institute, Troy, NY 12180, USA

ABSTRACT

This paper aims to demonstrate a typical optimization process that is applicable to the design of stick-frame structures. A 10-meter load-bearing wall is optimized considering different bay sizes with the objective of minimizing the use of wood. The results of our study indicate that the amount of wood needed to build a single story wall increases dramatically at bay sizes larger than about 90 cm. This seems to correlate well with common notions that framing systems that use smaller lumber dimensions spaced at shorter distances are more efficient than heavy timber systems. The efficiency of such light framing systems however does not seem to be affected significantly when decreasing bay size to very small dimensions. This may suggest that more efficient manufacturing procedures can be more instrumental if we wish to decrease the cost of similar wall assemblies.

KEYWORDS

Design, Optimization, Frame, Wall, Wood, Bay

INTRODUCTION

Worldwide, an estimated fifty-five percent of trees cut for non-fuel uses are being used for making buildings [Roodman D. and Lenssen N., 1995]. While buildings can be constructed in various ways, stick-frame technology currently remains the most common way for making structures from wood. Being able to determine relative efficiencies for various types of framing systems may therefore have important economic and environmental consequences. In light of the above, this paper aims to demonstrate a typical optimization process that is applicable to the design of stick-frame structures. A single story 10-meter long wall composed of typical wood members and exposed to both vertical and lateral forces is being optimized. Optimal dimensional parameters were calculated for bay sizes ranging from 10m (1bay) to 0.20m (50 bays). Our interest here was to determine how efficiency changes with decreasing bay size for stick-frame wall assemblies. The total volume of wood needed to build the various walls was adopted as measure of efficiency. The total volume represents the sum of the volume of vertical, horizontal, and the infill members as expressed in Equation 1 (figure 1). Three cases were considered: In Case a, it is assumed that both vertical and horizontal wood members have the same dimensions and are placed in one plane (Figure 2a). This configuration conforms to current practice in the US. In Case b it is assumed that the vertical and horizontal members remain of the same dimensions, however the top member is rotated 90° accommodating a more optimum structural placement for that member (Figure 2b). In Case c both vertical and horizontal members were allowed to evolve independent from each other (Figure 2c). All three designs were assumed to be loaded in the same way. A uniformly distributed load was applied to the top member. This load was assumed to come from a flat roof 6 meters deep and 10 meters in length, half of this load was applied to the wall being studied. A lateral wind load was further applied to the wall enclosure. Roof and wind loads were calculated according to the international building code. The deflection for each member was constrained to *l*/360 of the member's length, this upper limit was used for all three cases. The upper bound for lumber size was set at 30 cm, while the lower bound was set at 1mm. The material was assumed to be solid wood. All joints within the frame were considered to be pinned connections. A more detailed description of the optimization follows.

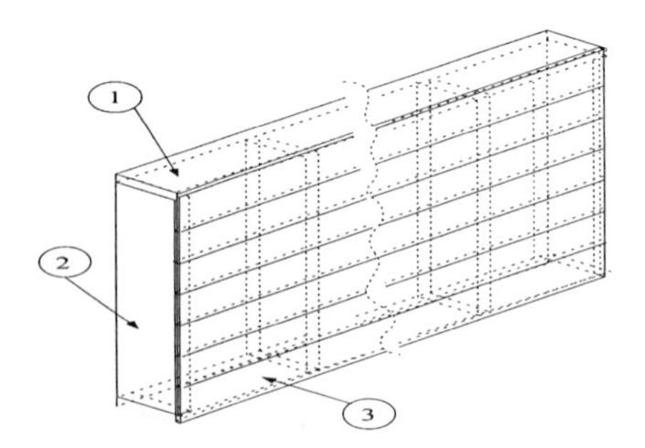

Figure 1: Stick-frame

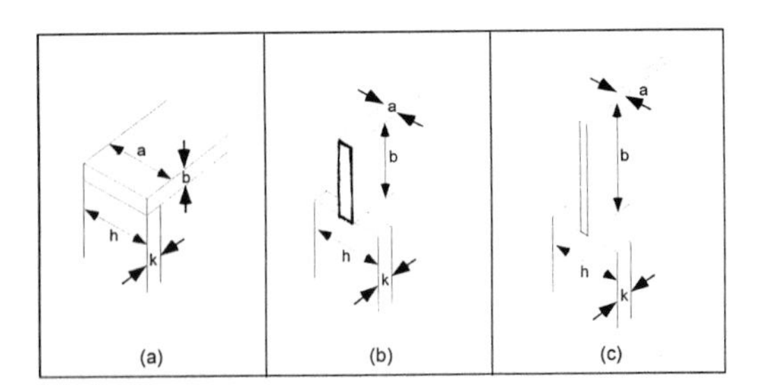

Figure 2: Three member-geometry cases

Nomenclature

Parameter	Value	Units
V	Volume	m^3
a,b,c,d,h,k	Wood dimensions	m
δ_{max}	Maximum deflection	m
n	Number of bays	
$m=20$	Number of in-fills	
L	Total length of the building	m
H	Total height of the building	m
D	Total width/span of the building	m
E = 14 GPa	Modulus of elasticity of the wood	Pa
σ_b =180MPa	Allowable stress wood	Pa
I	Moment of Inertia	m^4
S	Section Modulus	m^3
DL	Dead load	N/m^2
RL	Roof load	N/m^2
L_r	Live load	N/m^2
P_f	Snow load	N/m^2
W	Wind load	N/m^2
g =9.81	Gravitational force	m/s^2

**Note: Values are omitted if the parameters are design variables.

Design Objectives

An optimal design is defined as the structure having the least volume, while satisfying strength and geometrical requirements. The volume, *V*, of the structure is computed as follow:

$$V = 2abL + (n+1)(H - 2b)hk + HdL \tag{1}$$

Design Constraints

1. Geometry constrain

For the design to stay connected, the following equations are applied.

$$nl = L \tag{2}$$
$$mc = H \tag{3}$$

The length of k must be less than the length of h.

$$k \leq h \tag{4}$$

2. Deflection

a) Deflection of Component 1

$$\text{Maximum deflection, } (\delta_{max})_1 = \frac{5q_1 l^4}{384EI_1} \text{ (m)} \tag{5}$$

The deflection is limited to: *span length* /360

$$(\delta_{\max})_1 \le \frac{l}{360} \tag{6}$$

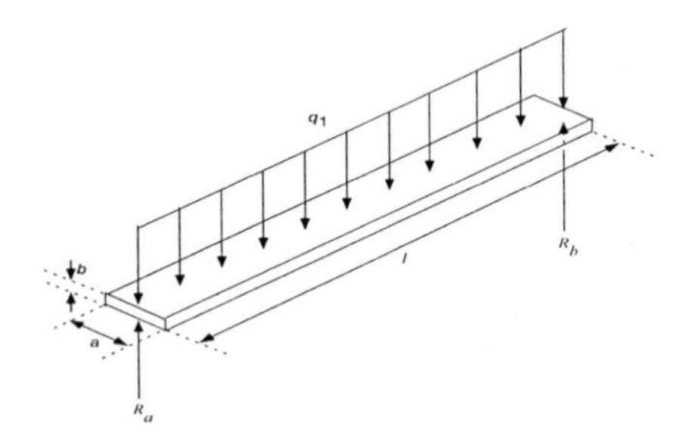

Figure 3: Component 1

b)Deflection of Component 2

$$\text{Maximum deflection, } (\delta_{\max})_2 = \frac{5q_2(H-2b)^4}{384EI_2} \text{ (m)} \tag{7}$$

The deflection is limited to: *span length*/360

$$(\delta_{\max})_2 \le \frac{H-2b}{360} \tag{8}$$

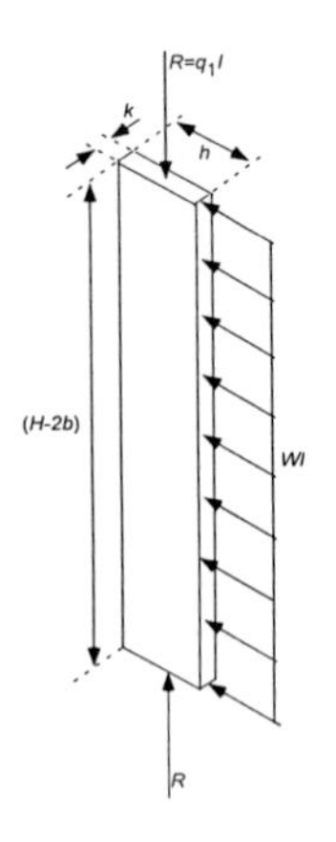

Figure 4: Component 2

c) Deflection of Component 3

The maximum deflection of the infill is computed by modeling the infill as a beam. The force over the area is modeled as a distributed force over the length by multiplying the distributed force by the infill width.

$$\text{Maximum deflection, } (\delta_{\max})_3 = \frac{5q_3 l^4}{EI_3} \text{ (m)} \tag{9}$$

The deflection is limited to: *span length*/360

$$(\delta_{\max})_3 \leq \frac{l}{360} \tag{10}$$

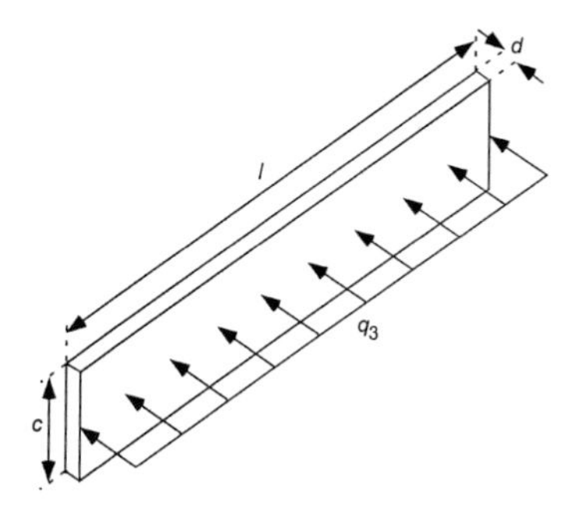

Figure 5: Component 3

3. Euler buckling

To prevent buckling, the reaction force, *R*, needs to be less than the critical force for buckling, P_{cr}, which can be computed using AFPA formulas

$$P_{cr} = \sigma_{allow} A = \sigma_{allow} hk \tag{11}$$

$$\sigma_{allow} = \begin{cases} F_c; \ 0 \leq (H-2b)/k \leq 11 \\ F_c\left[1 - \frac{1}{3}\left(\frac{(H-2b)/k}{K_c}\right)^4\right]; \ 11 \leq (H-2b)/k \leq K_c \\ \frac{0.3E}{((H-2b)/k)^2}; \ K_c \leq (H-2b)/k \leq 50 \end{cases} \tag{12}$$

Design constraint to be satisfied:

$$R \leq P_{cr} \tag{13}$$

4. Side Constraints

Component 1:

$$1mm(0.0394") \le a, b \le 30.48cm(12") \tag{14}$$

Component 2:

$$1mm(0.0394") \le h, k \le 30.48cm(12") \tag{15}$$

Component 3:

$$c = \frac{H}{m} = 0.12m \tag{16}$$

$$1mm(0.0394") \le d \le 30.48cm(12") \tag{17}$$

Optimization Problem Statement:

$$\min\ V = 2abL + (n+1)(H-2b)hk + HdL \tag{18a}$$

Subject to:

$$nl = L \tag{18b}$$
$$mc = H \tag{18c}$$
$$k \le h \tag{18d}$$
$$(\delta_{\max})_1 \le \frac{l}{360} \tag{18e}$$
$$(\delta_{\max})_2 \le \frac{H-2b}{360} \tag{18f}$$
$$(\delta_{\max})_3 \le \frac{l}{360} \tag{18g}$$
$$R \le P_{cr} \tag{18h}$$
$$1mm(0.0394") \le a, b \le 30.48cm(12") \tag{18i}$$
$$1mm(0.0394") \le d \le 30.48cm(12") \tag{18j}$$
$$1mm(0.0394") \le h, k \le 30.48cm(12") \tag{18k}$$

The following additional case specific constraints are applied:

Case a

$$a = h \tag{19a}$$
$$b = k \tag{19b}$$

Case b

$$a = k \tag{20a}$$
$$b = h \tag{20b}$$

There are no additional constraints for Case c.

RESULTS and DISCUSSION

Table 1 provides the optimal results for the three different scenarios. Figures 6, 7, and 8 represent changes in wood volumes with increasing numbers of bays for Cases a, b, and c, respectively. As might be expected, the result for Case c indicates that the design with more independent wood sizing yields the most optimal solution. Our results show that the volume of wood needed in Case c is approximately 10 % less than for Case a, and the results in Case b used about 5% less wood relative to Case a.

TABLE 1
OPTIMUM RESULTS FOR STICK-FRAME WALL

	n	a(m)	b(m)	h(m)	k(m)	l(m)	d(m)	Vol.(m^3)
Case a	30.0000	0.0877	0.0462	0.0877	0.0462	0.3333	0.0066	0.5280
Case b	30.0000	0.0446	0.0857	0.0857	0.0446	0.3333	0.0066	0.4977
Case c	28.0000	0.0215	0.0431	0.0899	0.0463	0.3571	0.0070	0.4464

Traditional wood light framed buildings in the US use dimension lumber of approximately 2x4 inches or 5 by 10 cm [Allen E., 1999]. The maximum spacing distance for these is typically 40 to 60 cm (16' to 24") depending on loading conditions and height of the studs [Ching F., 1991]. From Table 1 we can see that the design of Case a yields almost the actual dimension of such traditional design for both vertical and horizontal members of the frame. The common spacing distances of 40 to 60 cm (16 to 25 bays for our 10-meter wall) also correlates well with the results presented in Figure 6a. In general, the results for all 3 cases indicate that the use of wood starts to increases dramatically at n-values between 12 and 14, this represents a bay size of approximately 90 cm. For n-values starting at about 14 and higher, results further indicate that the optimal designs for each of these different n-values are very close to each other. Any solution within this range is relatively close to the optimum solution.

CONCLUSIONS

In this initial study we have developed a means to design and evaluate optimal dimensional parameters for load bearing single story timber walls utilizing optimization approaches. The results of our study indicate that the amount of wood needed to build a single story wall increases dramatically at bay sizes larger than about 90 cm. This seems to correlate well with common notions that framing systems that use smaller lumber dimensions spaced at shorter distances are more efficient than heavy timber systems. The efficiency of such light framing systems however does not seem to be affected significantly when decreasing bay size to very small dimensions. This may suggest that more efficient manufacturing procedures can be more instrumental, if we wish to decrease cost of similar wall assemblies.

ACKNOWLEDGEMENTS

Financial support from the National Science Foundation (Award No. CMS-0122022, Directorate for Engineering, Division of Civil and Mechanical Systems) and the US Department for Housing and Urban Development is much appreciated.

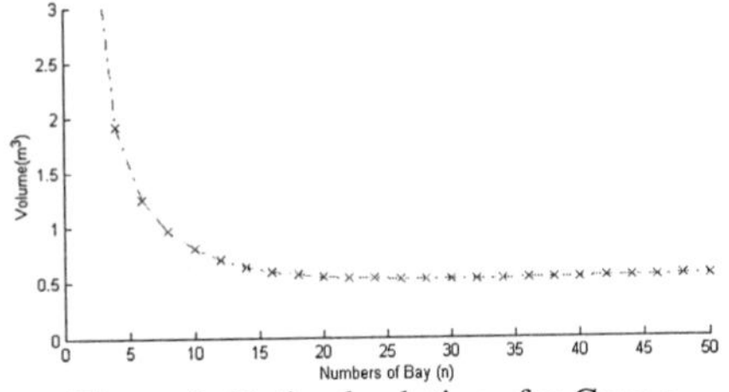

Figure 6: Optimal solutions for Case a

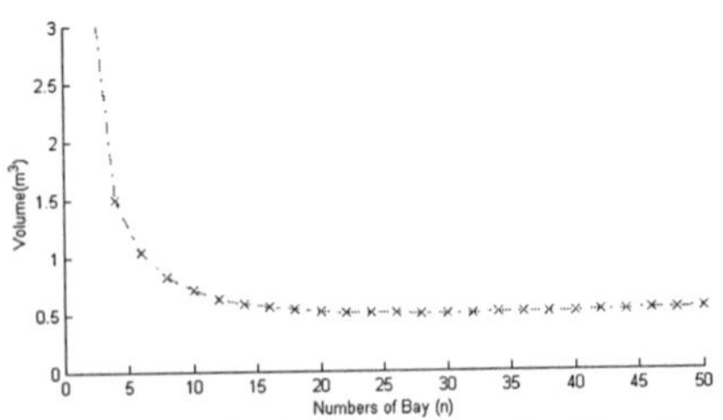

Figure 7: Optimal solutions for Case b

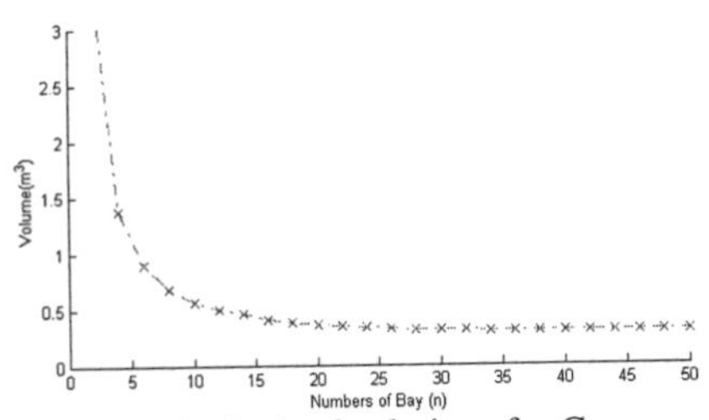

Figure 8: Optimal solutions for Case c

REFERENCES

Allen E. (1999), Fundamentals of Building Construction, Materials and Methods. third edition. John Wiley and Sons, Inc.

Ching F. (1991), Building Construction Illustrated, second edition, Van Nostrand Reinhold

Roodman D. and Lenssen N. (1995), A Building Revolution: How Ecology and Health Concerns Are Transforming Construction, *Worldwatch Paper 124*

TEAM PRESENTATION:

EARTHQUAKE RESISTANT THEORY AND APPLICATIONS

Advances in Building Technology, Volume 1
M. Anson, J.M. Ko and E.S.S. Lam (Eds.)

A MACRO-MODEL OF RC SHEAR WALL FOR PUSH-OVER ANALYSIS*

Chen qin Qian jiaru

Department of Civil Engineering, Tsinghua University,
Beijing, 100084, CHINA

ABSTRACT

Reinforced concrete shear wall is a kind of main vertical structural element of high-rise buildings to resist earthquakes. In order to perform push over analysis in the displacement-based seismic design of buildings, a simple, rational macro-model of shear wall is developed in this paper. The macro-model is composed of two kinds of element: concrete column element and concrete membrane element. In push-over analysis the shear wall is divided into layers along its height, and the wall panel is divided into columns. Two adjacent layers are connected by a rigid beam. There are only three degrees of freedom for one layer. A method called single degree of freedom compensation is adopted to solve the ill-conditioned equations, by which a complete capacity curve can be obtained. Experimental study of four shear wall specimens was carried out and the specimens were analyzed by the developed program. The analytical results have good agreement with the test results.

KEYWORDS

Shear wall, Macro-model, Push-over analysis, Column element, Membrane element, MCFT, Ill-conditioned equations

INTRODUCTION

In order to perform the displacement-based seismic design and push-over analysis of building structures, a simple and rational analytical model of shear wall should be developed. Various shear wall models have been proposed such as: beam model (Park Y.K 1987), spring macro-model (Linde P. 1994), braced wide-column analogy (Koumousis V.K. 1992). A macro-model which simulates the continuity model is developed in this paper. The degree of freedom of the model is condensed, and constitutive relation of cracked concrete is interpreted in the model, these make the calculation process much easy. The complete capacity curve can be obtained by using a method called single degree of freedom compensation to solve ill-conditioned equations. Comparison of the experimental data and the push-over analysis results verifies the accuracy of the proposed macro-model and the reliability of

*Supported by the National Natural Science Foundation of China (Grant No. 59895410)

d_0, $-d_0$ or 0), and the modified equilibrium equations: $P=(K_g+C)\Delta$ can be solved stably. The true solution can be obtained after iterative calculation of the modified equilibrium equations. Forced iteration approach is adopted to resolve the equilibrium equations after the stiffness matrix reaches negative condition.
So complete capacity curve of RC shear wall can be obtained by push-over analysis.

EXPERIMENT VERIFICATION

Specimens

Four isolated RC shear wall specimens (SW-1~SW-4) were tested to verify the shear wall model developed in this study. SW-1 and SW-2 were rectangular section, SW-3 was I section and SW-4 was T shape section. The details of specimens are shown in Fig.5.

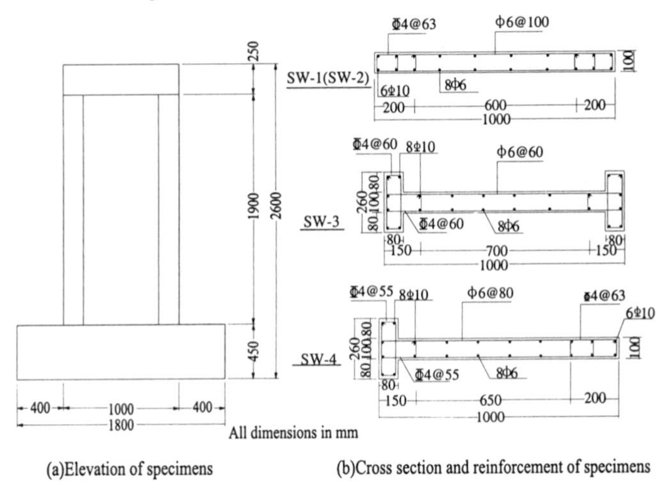

(a)Elevation of specimens (b)Cross section and reinforcement of specimens

Fig. 4 Details of specimens

The concrete cubic strength of specimens f_{cu} was 25.2Mpa, 22.9Mpa, 34.1Mpa and 34.4Mpa respectively. The yield strength f_y of reinforcement with diameter of 4mm, 6mm and 10mm was 631.7Mpa, 451.7Mpa and 395Mpa respectively.
The boundary zone of specimens was confined with stirrups and the wall panel was reinforced with distributed reinforcements. The degree of confinement of stirrups is measured with the stirrup characteristic-value λ_v, which is:

$$\lambda_v = \rho_v \cdot f_y / f_c \tag{16}$$

where ρ_v is the volumetric ratio of stirrup, f_c is the concrete compressive strength and f_c=0.67f_{cu}. For SW-1~SW-3, λ_v was 0.26, 0.28 and 0.25 respectively. For SW-4 λ_v of rectangular boundary zone and flanged boundary zone was 0.19 and 0.25 respectively.
Strains of vertical reinforcements in boundary zones were measured with strain gauges. Lateral displacement along the wall height was measured. Test data were recorded with the computer-based data acquisition system.
The specimens were subjected to a constant vertical load and cyclic lateral load at the top. The axial

the analysis methodology.

MODEL DESCRIPTION

The macro-model developed in this study is composed of two kinds of element: concrete membrane element which models the plane stress state of wall panel, and concrete column element which analogies tension or compression state of boundary zone respectively. Fig. 1 illustrates the modeling of shear wall in which the shear wall is divided into 5 layers and the wall panel is divided into 4 columns. Two column elements represent a boundary member to fit the different shape of the boundary member, such as rectangular section, flanged section or barbell section. In total, the wall shown in Fig.1 is modeled with 20 concrete column elements and 20 concrete membrane elements.
Two basic assumptions are adopted in the model: the cross section of shear wall remains plane section after deformation, and the stress in horizontal direction is equal to 0, i.e., $\sigma_x = 0$. By the plane section assumption, it is considered that there is a rigid beam between two adjacent layers. For one rigid beam there are three degrees of freedom at the center point: $\{u^g \quad v^g \quad w^g\}$, i.e., horizontal displacement, vertical displacement and rotation respectively. According to the assumption $\sigma_x = 0$, the strain in horizontal direction of any point in a membrane element is the same: $\varepsilon_x = \alpha$. If the wall panel is divided into k columns, the unknown numbers of horizontal strain for one layer of membrane element is $\{\alpha_1, \ \alpha_2 \cdots \alpha_k\}$. Then the total unknown numbers of one layer of membrane element is 3+k.

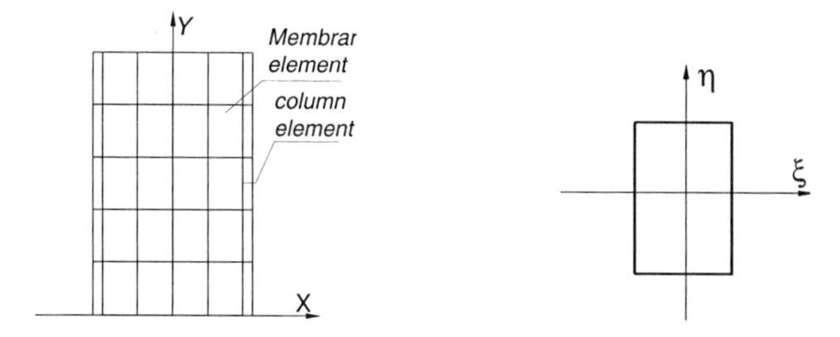

Fig.1 Macro-model of RC shear wall   Fig. 2 Coordinate system of membrane element

STIFFNESS MATRIX OF MODEL

Membrane Element

Fig.2 shows the local coordinate system of the membrane element. The displacement pattern of any point in membrane element is the following:

$$u = a_0 + a_1\eta + a_2\eta^2 + a_3\eta^3 + a_4\xi \tag{1a}$$

$$v = b_0 + b_1\eta + b_2\xi + b_3\eta\xi \tag{1b}$$

in which u is the horizontal displacement, v is the vertical displacement, ξ and η is the horizontal and vertical normal coordinate value of this point respectively; $a_0, a_1, a_2, a_3, a_4, b_0, b_1, b_2, b_3$ are the parameters to be determined. By derivating the displacement pattern and transferring the normal coordinate system (ξ, η) to global coordinate system (X, Y) the strains of membrane element of any point can be obtained:

$$\varepsilon = [B]u \tag{2}$$

in which the matrix $[B]$ is the shape function matrix. The stiffness matrix K^e of membrane element can be got by virtual work theory:

$$K^e = \int_e B^T DBtdxdy \tag{3}$$

in which $[D]$ is the stress-strain relation matrix of membrane element, t is the thickness of membrane element, i.e., the thickness of wall panel.

Layer of Membrane Element

The relationships of displacement increment between layer membrane element and membrane element can be obtained by their geometrical relations (Fig. 3):

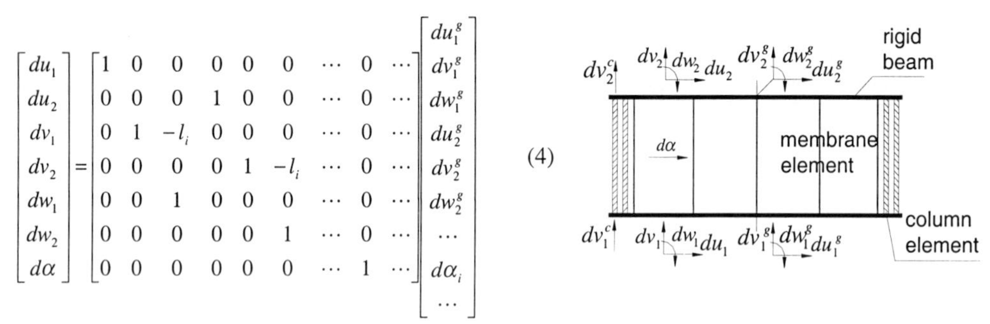

$$\begin{bmatrix} du_1 \\ du_2 \\ dv_1 \\ dv_2 \\ dw_1 \\ dw_2 \\ d\alpha \end{bmatrix} = \begin{bmatrix} 1 & 0 & 0 & 0 & 0 & 0 & \cdots & 0 & \cdots \\ 0 & 0 & 0 & 1 & 0 & 0 & \cdots & 0 & \cdots \\ 0 & 1 & -l_i & 0 & 0 & 0 & \cdots & 0 & \cdots \\ 0 & 0 & 0 & 0 & 1 & -l_i & \cdots & 0 & \cdots \\ 0 & 0 & 1 & 0 & 0 & 0 & \cdots & 0 & \cdots \\ 0 & 0 & 0 & 0 & 0 & 1 & \cdots & 0 & \cdots \\ 0 & 0 & 0 & 0 & 0 & 0 & \cdots & 1 & \cdots \end{bmatrix} \begin{bmatrix} du_1^g \\ dv_1^g \\ dw_1^g \\ du_2^g \\ dv_2^g \\ dw_2^g \\ \cdots \\ d\alpha_i \\ \cdots \end{bmatrix} \tag{4}$$

Fig.3 Displacement sketch of a layer

in which l_i is the distance between the center point of the membrane element and the center point of the rigid beam. Eqn. 4 can be marked as:

$$[du] = [B^g][du^g] \tag{5}$$

in which matrix $[du]$ is displacement increment of membrane element, matrix $[du^g]$ is the displacement increment of layer of membrane element, $[B^g]$ is the transfer matrix between the two kinds of displacement increment.
The relationships between the force increment and the displacement increment of the layer membrane element in the global coordinate system are:

$$[dX^g] = \left(\sum_k B^{g^T} K^e B^g \right) [du^g] \tag{6}$$

In which $dX^g = \{dX_1^g \quad dY_1^g \quad dM_1^g \quad dX_2^g \quad dY_2^g \quad dM_2^g \quad d\overline{X}_1 \quad d\overline{X}_2 \quad \cdots \quad d\overline{X}_n\}$, the parameter $d\overline{X}_i$ is the horizontal force of the membrane element corresponding to ε_x. By the assumption σ_x=0, $d\overline{X}_i$=0. The stiffness matrix of the layer of membrane element K_l can be obtained:

$$K_l = \sum_k B^{g^T} K^e B^g \tag{7}$$

Eqn. 6 can be expressed by block matrices:

$$\begin{Bmatrix} dX \\ 0 \end{Bmatrix} = \left[\begin{array}{c|c} K_{11} & K_{12} \\ \hline K_{21} & K_{22} \end{array}\right] \begin{Bmatrix} du \\ d\alpha \end{Bmatrix} \tag{8}$$

Resolve Eqn. 8 can get the strain increments $d\alpha$ of this layer：

$$d\alpha = -K_{22}^{-1} K_{21} du \tag{9}$$

From Eqn. 8 and Eqn. 9:

$$dX = K_m du \tag{10}$$

in which： $K_m = K_{11} - K_{12}K_{22}^{-1}K_{21}$, $dX = \{dX_1^g \;\; dY_1^g \;\; dM_1^g \;\; dX_2^g \;\; dY_2^g \;\; dM_2^g\}^T$, $du = \{du_1^g \;\; dv_1^g \;\; dw_1^g \;\; du_2^g \;\; dv_2^g \;\; dw_2^g\}$. The stiffness matrix of layer of membrane element K_m has been developed. The number of DOF of layer membrane element has been condensed from 3+k to 3.

Column Element

Two unknown numbers of displacement in the vertical direction at the two ends of column element are the only two degrees of freedom of the element. From the plane section assumption, the displacement of column element can be determined by the displacement of the rigid beam. The geometrical relationship of the two kinds of displacement is (Fig. 3):

$$du_c = \begin{bmatrix} 0 & 1 & -l_c & 0 & 0 & 0 \\ 0 & 0 & 0 & 0 & 1 & -l_c \end{bmatrix} du \tag{11}$$

Eqn. 11 can be marked as: $du_c = B_c du$, in which B_c is the shape function matrix of column element; du has the same meaning as Eqn. 10, l_c is the distance between the center of column element and the center point of rigid beam. The element stiff matrix of column element is:

$$K_c = B_c^T \begin{bmatrix} s & -s \\ -s & s \end{bmatrix} B_c \tag{12}$$

in which s is the tensile or compressive stiffness parameter, which can be determined by the constitutive relation.

Stiffness Matrix of Layer

The stiffness matrix of layer of the model is the sum of the stiffness matrices of membrane elements and column elements:

$$K_{layer} = K_m + \sum_4 K_c \tag{13}$$

in which 4 means there are 4 column elements in one layer.

CONSTITUTIVE RELATION OF MODEL

Constitutive Relation of Membrane Element

For the vertical and horizontal reinforcements are uniformly distributed in the wall panel, concrete membrane element is considered as a kind of continuous and uniform material composed of steel and concrete. Stress-strain relation matrix of this material is:

$$[D]=[D_c]+\sum_{2}[D_s] \tag{14}$$

in which $[D_c]$ is the stress-strain relation matrix of concrete; $[D_s]$is the stress-strain relation matrix of steel in one direction; $\sum_{2}[D_s]$means the sum of the stress-strain relation matrices of steel in horizontal and vertical directions.

Before cracking, concrete can be considered as an isotropic linear material. The modified compression field theory (MCFT) (Vecchio F.J. 1986, 1989, 1993) is adopted to determine the cracked concrete constitutive relations. The theory treats cracked reinforced concrete in terms of average stress and average strain, with the directions of principal stress and principal strain, and reflects the compression softening effects and tension stiffening effects. The relations of MCFT are formulized in complete curves.

The perfect elastic-plastic constitutive relation is used for the stress-strain relation of reinforcement embedded in concrete before concrete cracking. The experimental results (Abdeldjelil B. 1994) indicated that the constitutive relation of reinforcement embedded in concrete was different from that of bare reinforcement after concrete cracking, and the effect of crack should be considered. Based on the analysis of experiment data the linear strengthen constitutive relation for reinforcement in cracked concrete is adopted in this paper.

Constitutive Relation of Column Element

There are two kinds of boundary zone of shear wall: confined and unconfined. The stress-strain curve equations for confined concrete and for plain concrete suggested by Guo Z.H.(1999) were adopted. The longitudinal reinforcement is considered as perfect elastic-plastic material.

CALCULATION PROCESS

The equilibrium equations of push-over analysis of a structure are:

$$P = K_g \Delta \tag{15}$$

in which K_g is the global stiffness matrix, P and Δ is load and displacement field respectively. Before the peak of $P-\Delta$ relation curve, the stiffness matrix K_g is positive. After the peak, it is negative. The equilibrium equations are easy to be solved if the stiffness matrix is positive. When the $P-\Delta$ curve approaches its peak, the equations are ill-conditioned. A method called single degree of freedom compensation (SDFC) proposed by Zou J.L.(2001) is approached to solve ill-conditioned linear equations. The global stiffness matrix K_g should be decomposed into LDL^T style before the equilibrium equations are resolved by SDFC. The main concept of SDFC is to introduce a compensation value d_0 to keep stability of the resolution, so every diagonal element of matrix D is big enough. Then matrix K_g is modified to $K_g + C$ (C is a diagonal matrix whose elements are load ratio of specimens was 0.44, 0.21, 0.31 and 0.31 respectively. All specimens were designed with strong shear and weak bending and failed in flexure.

Comparison of Analytical and Experimental Results

To perform push-over analysis, each specimen was divided into 11 layers along the height, its wall panel was divided into 8 columns. In total there were 88 concrete membrane elements and 44 concrete column elements and 33 unknown numbers.

Complete process push-over analysis program of RC shear wall developed by this study was used to perform the analysis.

Lateral force-top displacement curves, i.e., capacity curves, of specimens obtained by experiment and push-over analysis are given in Fig. 5. The experimental results are skeleton curves of lateral force-top displacement hysteresis loops of specimens.

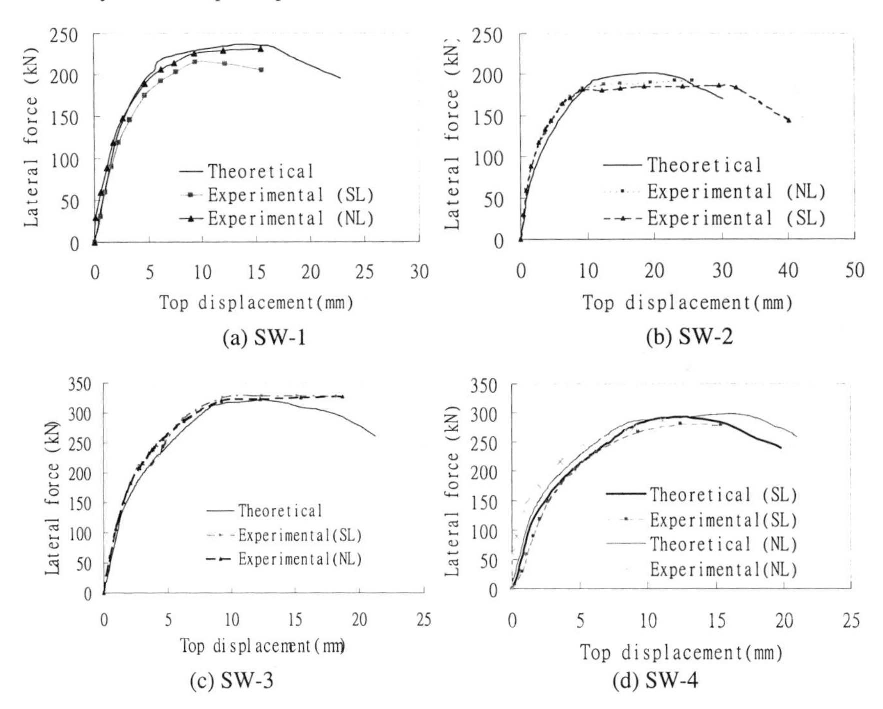

Fig. 5 Analytical and experimental lateral force-top displacement curves of specimens
(Note: the SL and NL mean the south loading and north loading respectively)

Table 1 lists the analytical and experimental maximum horizontal loads, i.e., the ultimate strength of specimens. The difference of two results is less than 10%. Table 1 also shows the displacement ductility ratios and the ultimate rotations of specimens. The displacement ductility ratio μ and the ultimate rotation θ are defined as:

$$\mu = \Delta u / \Delta y \tag{17a}$$

$$\theta = \Delta u / H \tag{17b}$$

in which, Δy and Δu are the yield and the ultimate lateral displacement at the top of specimen respectively, H is the displacement measure height.

From the comparison, the push-over analysis by using the macro-model provides a good estimate of the capacity curve of shear wall.

TABLE 1
COMPARISON OF ANALYTICAL AND EXPERIMENTAL ANALYSIS

Specimens		Ultimate lateral force (kN)		μ		θ	
		Exp	Ana	Exp	Ana	Exp	Ana
Sw-1	S	215.8	237.2	3.21	4.13	1/118	1/95
	N	232.1		3.68		1/119	
Sw-2	S	185.6	201.8	8.6	6.45	1/49	1/56
	N	192.9		5.58		1/72	
Sw-3	S	328.2	321.4	4.23	4.48	1/101	1/87
	N	326.1		4.35		1/99	
Sw-4	S	282.0	293.2	3.48	3.96	1/120	1/94
	N	286.4	297.9	2.92	3.43	1/166	1/110

(Note: S and N mean the loading direction of south and north respectively, Exp and Ana means the experimental and analytical results respectively.)

CONCLUSIONS

Membrane element and column element are used to simulate the wall panel and the boundary zone of RC shear wall respectively in the proposed macro-model. The model has advantages of finite element but the degree of freedom is greatly condensed. At the same time the constitutive relation derived from MCFT makes the analysis of cracked concrete very easy. The complete capacity curve of RC shear wall can be obtained by adopting single degree of freedom compensation approach. The macro-model of shear wall developed in the study is proved by comparison of experiment results and analytical results.

REFERENCES

Abdeldjelil B. and Hsu T.T.C. (1994). Constitutive Laws of Concrete in Tension and Reinforcing Bars Stiffened by Concrete. *ACI Structural Journal* 91:4, 465-474

Guo Z.H. (1999). *Theory of Reinforced Concrete*, Tsinghua University, Beijing, PRC.

Koumousis V.K. and Peppas G.A. (1992). Stiffness Matrices for Simple Analogous Frames for Shear Walls Analysis. *Computers and Structures* 43:4, 613-633

Linde P. and Bachmann H. (1994) Dynamic Modeling and Design of earthquake-resistant Walls. *Earthquake Engineering and Structural Dynamics* 23:12, 1331-1350

Park Y.H. and Reinhorm A.M. (1987). Inelastic Damage Analysis of RC Frame-shearwalls Structures. In: *National Center for Earthquake Engineering Research*, State University of New York at Baffalo, Baffalo, USA

Vecchio F.J. and Collins M.P. (1986) The Modified Compression-Field Theory for Reinforced Concrete Elements Subjected to Shear. *ACI Structural Journal* 83:2, 219-231

Vecchio F.J. (1989) Nonlinear finite Element Analysis of Reinforced Concrete Membranes. *ACI Structural Journal* 86:1, 26~35

Vecchio F.J. and Collins M.P. (1993). Compression Response of Cracked Reinforced Concrete. *Journal of Structural Engineering* 119:12 , 3590~3610

Zou JL. (2001). *3D RC frame structure full-range pushover analysis*. Ph.D. thesis, Tsinghua University, Beijing, P.R.C.

Advances in Building Technology, Volume 1
M. Anson, J.M. Ko and E.S.S. Lam (Eds.)

TORSIONAL EFFECT OF ASYMMETRIC R/C BUILDING STRUCTURE

Junwu DAI[1] Yuk-Lung WONG[2] and Minzheng ZHANG[1]

[1]Institute of Engineering Mechanics, China Seismological Bureau, China
[2]Department of CSE, The Hong Kong Polytechnic University, Hong Kong

ABSTRACT

This paper is composed by the following investigations: 1) Using the time-history response of the base shear-torque of 3 asymmetric RC single-story-building systems under a set of earthquake simulation tests on shake table, both of the advantages and shortcomings of the base shear-torque (BST) surface (J.C. De La Llera and A.K. Chopra) established on the assumption of idealized force-displacement relationship are checked. 2) Based on the experimental results and the assumption of the bi-linear force-displacement relationship, a more reasonable and effective analytical model, the state space of base shear-torque is developed for estimation of the seismic performance particularly the torsional effects of the RC asymmetric structures.

KEY WORDS

earthquake simulation test, torsional, asymmetric, seismic response, inelastic, shear-torque relationship

BACKGROUND AND OBJECTIVES

It has been observed repeatedly in strong earthquakes that the presence of asymmetry in the plan of a structure makes it more vulnerable to seismic damages. There are reports of extensive damages to buildings that are attributed to excessive torsional responses caused by asymmetry in earthquakes such as the 1972 Managua earthquake, the 1989 Loma Prieta earthquake, the 1995 Kobe earthquake and the 1999 Chi-Chi earthquake. Although torsion has been recognized as a major reason for poor seismic performance of asymmetric buildings and many studies have been done on this topic, the analytical and experimental studies on the inelastic seismic response of asymmetric buildings do not have a long history. Recently, De La Llera and Chopra (1995) developed a simple model for analysis and design of asymmetric buildings based on the assumption of idealized elasto-plastic force-displacement relationship. Each story of the building is represented by a single super-element in the simplified model. In this method, the story shear and torque interaction surface (Kan and Chopra 1981, Palazzo and Fraternali 1988, De La Llera and Chopra 1995) is used as an important component. The story shear and torque (SST) surface is basically the yield surface of the story due to the interaction between story shear and torque. Each point inside the surface represents a combination of story shear and torque that the story remains elastic. On the other hand, each point on the surface represent a combination of shear and torque that leads to the yielding of the story. How realistic is such a model to represent the behavior of ductile frame buildings in seismic regions is a subject that requires further investigation.

The objective of this paper presented here are: 1) to check the limitations of the story shear-torque surface based on the assumption of the idealized force-displacement relationship with the experimental results earthquake simulation tests on 3 asymmetric single-story structures and 2) to develop it to the more realistic three state spaces based on the bi-linear force-displacement relationship.

SHEAR-TORQUE RELATIONSHIP

There are three reinforced concrete single-story-building structures are designed to cater for the research objectives. All experiments are carried out on a single-axial shake table installed by MTS in HK Polytechnic University. Design detailing, instrumentations and ground excitations used in experiments have been described in reference [2].

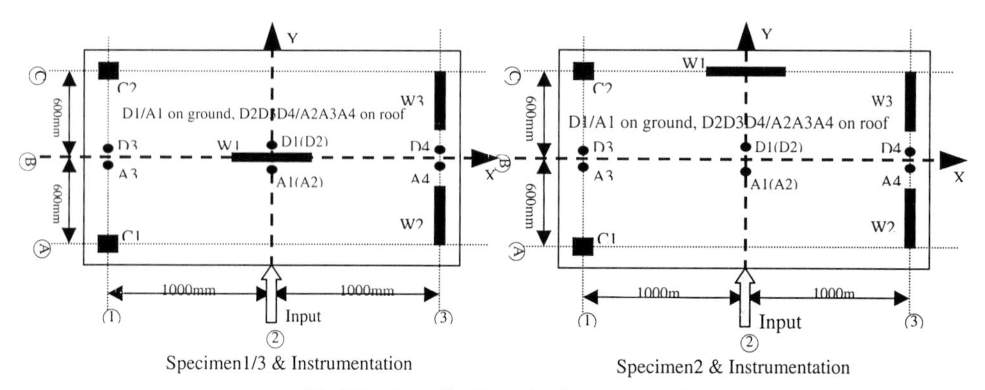

Fig.1 Specimen Configuration & Instrumentation

Verification of the shear-torque relationship

Based on the assumption of idealized force-displacement relationship of structural members, J.C. De Le Llera and A.K. Chopra (1995) developed a useful and effective method for understanding of fundamental issues on seismic torsional phenomena of asymmetric structures. The most important component of this approach is the use of the idealized yield base shear-torque (BST) surface of the studied system. As examples, abovementioned 3 model structures studied in earthquake simulation tests are used to examine both of the validity and the shortage of the idealized yield BST surface here for understanding of the real seismic behavior of asymmetric systems correctly.

According to the method of reference [3] developed, the BST surfaces of the 3 asymmetric model structures are founded respectively based on the assumption of idealized force-displacement relationship while only single-directional ground excitation is considered, shown in fig. 8 for convenience of comparison. The displacement-based approach that T. Paulay (2000) established is used in the determination of the stiffness of structural members. From fig.2, obviously, it can be found that the BST surface of specimen2 expanded uniformly on the basis of specimen1 while for that of specimen3, it is skewed and extended to the 1st and 3rd quadrant relatively. In fact, due to the strength of concrete used in the fabrication of specimen2 is higher than that of in specimen1, the strength of structural members in specimen2 is uniformly increased. Together with the wall w1 being arranged at the outer edge of specimen2 comparing with specimen1, the entire shear-torque resistant capacity is correspondingly increased in specimen2 than that of specimen1. The result is the space covered by the BST surface of

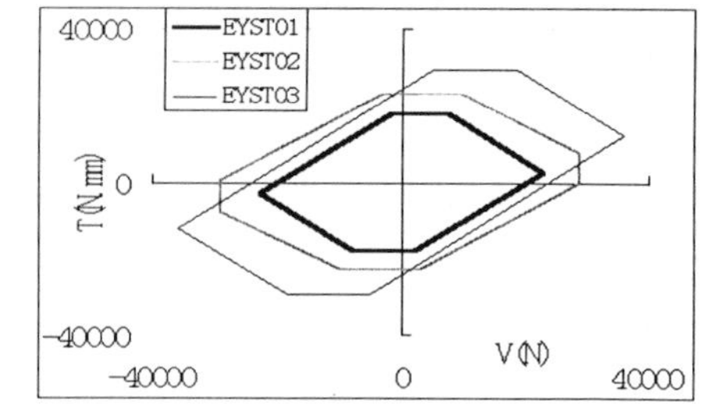

Fig.2 Comparison of BST surfaces of 3 model structures

specimen2 is larger than that of specimen1. Similarly, the higher strength of concrete and the reinforcement ratio used in specimen3 particularly in walls than that of in specimen1, intensified the strength eccentricity and simultaneously increased the entire shear-torque resistant capacity of specimen3. The associated consequence is that the BST surface of specimen3 is skewer and larger than that of specimen1. This figure illustrates the different seismic capacity among different model structures directly. From the area covered by the BST surface of each specimen, it can roughly predict that the seismic behavior of each specimen will be changed differently in a certain way.

In fact, there is no available method currently can be used to record the base shear and torque response directly in shake table tests but it is an easy thing to measure both of the acceleration and displacement response. In addition, because the roof plate of the 3 single-story model structures all performed as rigid body in experiments. Therefore, the rotational acceleration and the twist angle of the floor used to describe the torsional response can be determined directly from the lateral acceleration and displacement response based on the kinetic theory about rigid body. Further, the base torque response can be easily obtained. Associated base shear can be determined directly from the inertia force related lateral acceleration response.

The comparison between abovementioned theoretical results (solid line signifies the idealized BST surface of each specimens) and the experimental response (dotted area) in terms of the time history of the base shear-torque of each specimen is shown in fig.3~fig.5. Each sub-figure dictates one case of ground excitation. For example, A0.14g, E0.30g, L0.37g and N0.91g signify the time history response of base shear-torque in the case of ground excitation with intensity of artificial wave 0.14g, El Centro-NS component 0.30g, Loma Prieta record 0.37g and North Ridge record 0.91g, respectively.

From fig.3~fig.5, the following conclusions can be inferred:

1) For most cases of the intensity of ground motions are not higher than about 0.47g, the time history response of the base shear-torque does not or only lightly exceed the area the BST surface encircled. But it exceeds the boundary of the BST surface greatly in cases of the intensity of ground motions higher than about 0.47g for all 3 model structures. These facts tell the truth that the assumption of the idealized force-displacement relationship is only roughly reasonable for evaluation of the most common concrete structures in region with potential lower to moderate seismic intensity. However, for other relatively more important structures particularly those located in region with higher potential seismic intensity, more reasonable models such as bi- or tri-linear force-displacement relationship based strength/stiffness degradation/harden model should be used case to case. Correspondingly, the idealized BST surface should be developed further to be more realized bi- or tri-state spaces model to estimate the seismic performance (torsional effects) of asymmetric

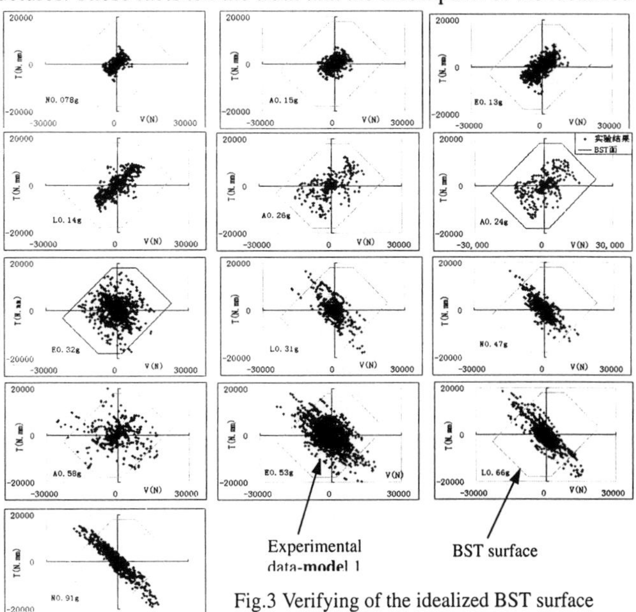

Fig.3 Verifying of the idealized BST surface

structures in a more general sense.

2) With the increase of the intensity level of the ground excitations, the time history response of the base shear-torque tends to be developed to the longest inclined side of the BST surface. From the definition of the BST surface according to reference [3], these two longest inclined sides signify the mechanism of other lateral resistant planes (the left column even the central wall planes) are damaged when the strongest right wall plane remains linear elastic. The experimental observation proved the fact that the damage sequence of all 3 specimens are started from the right column plane to the right wall plane at last although there are obvious difference among the seismic damage type of them. These experimental results testified that the theoretical results are reasonable. Therefore, the idealized BST surface can be used to estimate the seismic damage sequence of asymmetric structures roughly.

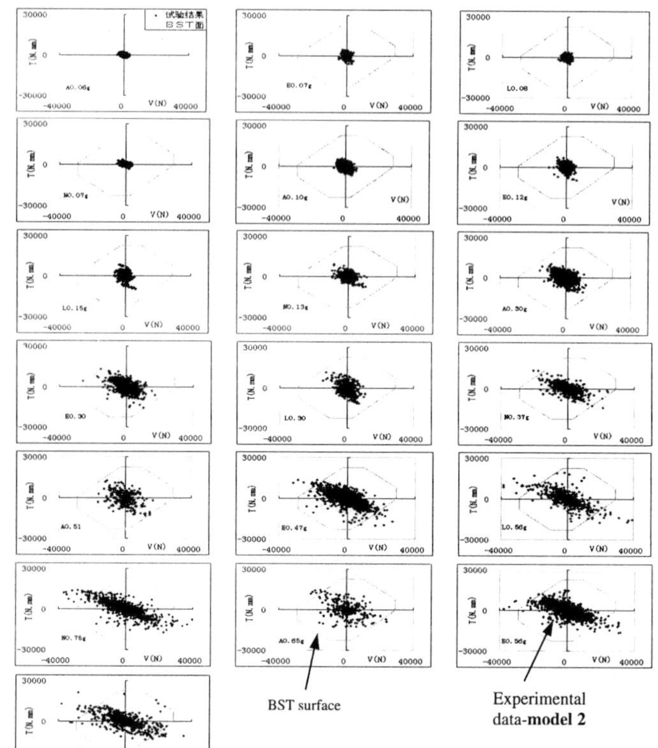

Fig.4 Verifying of the idealized BST surface

3) On the other hand, the trends of the time history of the base shear-torque lean to the longest sides of the BST surface illustrate effects of the change of the stiffness eccentricity to the seismic behavior of the model structure. That is, at the onset of the damage of the left column and central wall plane, the stronger right wall plane still remains linear elastic. Inevitably, the stiffness center shifts to the stronger right wall plane, intensified the stiffness eccentricity of the model structure. This phenomenon explains that although the stiffness eccentricity does not affect the shape of the BST surface, it really influences the developing trends of the time history response of the base shear-torque. In other words, the larger of the stiffness eccentricity is, the smaller of the possibility that the stronger members near the stiffness center are damaged will be and, the larger of the possibility that the weaker members far away from the stiffness center are damaged will be. The consequence is the larger of the possibility of the serious unbalanced damage occurring will be.

4) In most experimental cases, there is no pure torsional damage observed from the responses of the base shear-torque, i.e., only in few cases of model1 and model3, the torque resistance are exceeded lightly. In model2, such phenomena are never found. This fact illustrates, on one hand, the possibility of the occurring of the pure twist damage will be very small even for structures with larger eccentricity such as the model structures here studied. On the other hand, it shows that the location change of the central wall w1 improves the entire torque resistance directly and makes the coupling response more uniform and simultaneously, it tells the fact of the effects of the structural members in the orthogonal direction is

un-negligible when the seismic resistant capacity in one principal direction of the structure is studied and vice versa.

State space of base shear-torque relationship

In fact, the degradation type of bi- and tri-linear force-displacement relationship should be more suitable than the idealized elasto-plastic one for most cases of reinforced concrete building structures in seismic ductility design. This conclusion can be verified directly from the comparison of the abovementioned experimental response and the idealized BST surface shown in fig.3~5. Then, correspondingly, the single space surrounded by the BST surface can be developed to 2 or 3 state spaces signifying respectively the elastic, inelastic even collapsed state of the related structural system. In other words, for evaluating the seismic performance of asymmetric RC structures realistically, the ultimate BST surface in the idealized model should be developed to nonlinear space(s) of the base shear-torque responses. The first space signifies the elastic response while the second space represents the inelastic characteristics when yield damage occurred in structure, and the third space means that the shear-torque response exceeds the ultimate capacity of the system. The boundary between the first and the second space describes a kind of critical state that the yielding damage initiates in system. Whilst, the boundary between the second and the third space should theoretically signify that the system begins to collapse.

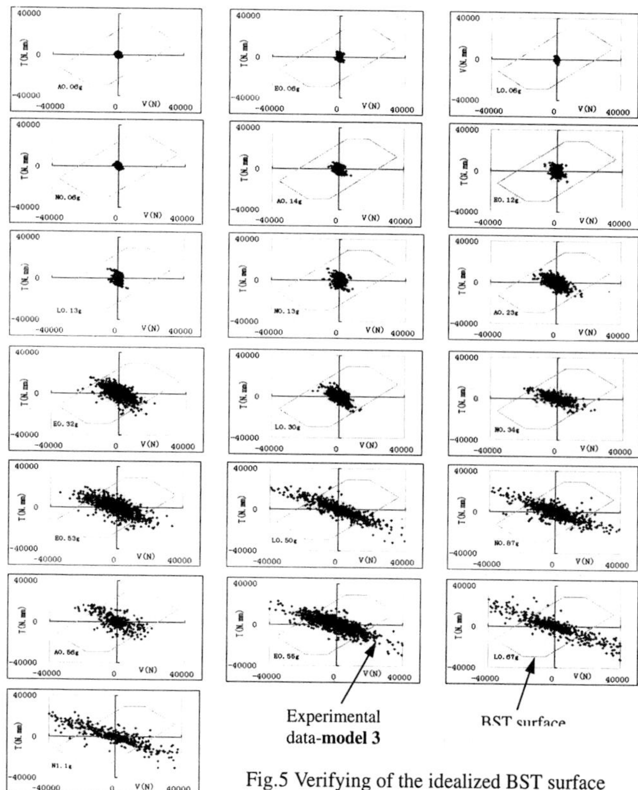

Fig.5 Verifying of the idealized BST surface

As example, abovementioned specimen2 (bi-directional asymmetry with 3 lateral resistant planes along the y-direction of ground excitations) here is used to show how the 3 state-spaces model for evaluation of the seismic response of the base shear-torque is established. From the experimental data for the yielding strength f_y of the reinforcement used in the model system shown in table 1 respectively, the yield boundary dividing the elastic and the inelastic response spaces can be determined according to the method that De La Llera and A.K. Chopra established, i.e., its' location is completely same to that of the BST surface in the idealized model but they have quiet different meaning. Using the similar way and the ultimate strength f_u shown in table 1, the eight vertices of the collapse boundary can be easily determined based on the following formulas[3]:

$$x_1 = V^*_{y0} \qquad y_1 = V^*_{y0}x^*_p + T^*_{\perp}(1-\hat{V}^*_x) \qquad x_2 = V^*_{yu} + V^*_{yc} \qquad y_2 = T^*_0 - T^*_{\perp}\hat{V}^*_x$$

$$x_3 = V^*_{yu} - V^*_{yc} \qquad y_3 = T^*_0 - T^*_{\perp}\hat{V}^*_x \qquad x_4 = -V^*_{y0} \qquad y_4 = -V^*_{y0}x^*_p + T^*_{\perp}(1-\hat{V}^*_x)$$

$$x_5=-x_1,\ x_6=-x_2,\ x_7=-x_3,\ x_8=-x_4$$
$$y_5=-y_1,\ y_6=-y_2,\ y_7=-y_3,\ y_8=-y_4 \quad (1)$$

where, in a more general sense, the symbol "*" signifies two kinds of critical state: the onset of yielding and reaching at the ultimate strength. If the structure system just starts to be damaged (reaches its yield strength), parameters in the above formulas have the same meanings with those of in the idealized BST surface. If the structural response reaches the maximum capacity of the system, these parameters can be determined based on the ultimate strength of structural members. As shown in fig.6, structural parameters controlling the response state space of the studied system are explained as,

1). $\hat{V}_x^* = V_x / V_{x0}^*$ is the normalized base shear along x-direction, while $V_{x0}^y = \sum_{i=1}^M f_{yxi}$ and $V_{x0}^u = \sum_{i=1}^M f_{uxi}$ are the yielding and ultimate lateral resistant capacities along the x-direction of the studied system respectively, in which, f_{yxi} and f_{uxi} are the corresponding yield and ultimate strength of the ith lateral resistant plane, and M is the number of the lateral resistant planes along x-direction. If the earthquake actions along the x-direction (besides slong the y-direction) should be taken into account, i.e., $V_x \neq 0$, the shape of the state space would be changed with the change of the value of V_x. Otherwise, if only single-directional earthquake excitation is studied, i.e., $V_x=0$, the effects of this item should be ignored.

2). $V_{y0}^y = \sum_{i=1}^N f_{yyi}$ and $V_{y0}^u = \sum_{i=1}^N f_{uyi}$ are the lateral yield and ultimate capacity of the system along y-direction, f_{yyi} and f_{uyi} are the lateral yield and ultimate strength of the lateral resistant planes, respectively. N is the number of the lateral resistant planes along y-direction. The value of V_{y0}^y and V_{y0}^u determine the span of the state space of base shear-toque on the shear axial. $V_{y0}^u - V_{y0}^y$ directly gives out the improvable space of the shear capacity of the structural system after yielding.

3). V_{yc}^y and V_{yc}^u are the lateral yield and ultimate capacity of the resistant plane passing through the center of mass. Generally, this parameter represents the sum of the resistant capacity of all lateral planes closing to the center of mass. From fig.6, it is obvious that the increase of the value $V_{yc}^u - V_{yc}^y$ means the increase of the possibility that the pure torsional mechanism will occur after yielding.

4). $T_0^y = \sum_{i=1}^N |f_{yyi} x_i| + \sum_{i=1}^M |f_{yxi} y_i|$ and $T_0^u = \sum_{i=1}^N |f_{uyi} x_i| + \sum_{i=1}^M |f_{uxi} y_i|$, respectively, define the torsional capacity of the system when pure twist yield and ultimate damage will be occurring. These two parameters control the maximum and minimum coordinates of the two critical boundaries along the torque axial. Moreover, the improvement of the torsional capacity of the system after yielding is also controlled by the value $T_0^u - T_0^y$.

5). $T_\perp^y = \sum_{i=1}^M |f_{yxi} y_i|$ and $T_\perp^u = \sum_{i=1}^M |f_{uxi} y_i|$ are

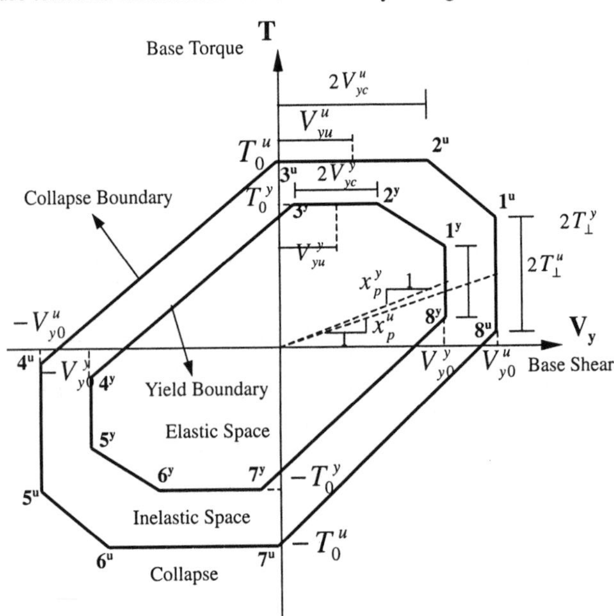

Fig.6 Structural parameters controlling the state space of base shear-torque

contribution of the torsional capacity provided by the orthogonal lateral planes at the level of initiating yield and collapse respectively. The possibility of the pure shear mechanism should be increased with the increase of the value $T_{\perp}^{u} - T_{\perp}^{y}$ particularly during the stage of inelastic response.

6). $x_p^y = \sum_{i=1}^{N} f_{yyi} x_i / V_{y0}^y$ and $x_p^u = \sum_{i=1}^{N} f_{uyi} x_i / V_{y0}^u$ determine the strength eccentricity of the system at two damage levels: before yielding and at the initiating of collapse, respectively. They signify independently the degree of the strength eccentricity on the two critical levels. Here, x_p^y is considered to be a constant during the elastic stage while x_p^u is also defined as a critical value at the instant of initiating collapse although the strength eccentricity is varied from x_p^y to its' maximum value, and then decreased to x_p^u with the increase of the intensity of the ground excitations gradually during the inelastic stage.

7). $V_{yu}^y = \sum_{i=1, i\neq 2}^{N} (f_{yyi} x_i / |x_i|)$ and $V_{yu}^u = \sum_{i=1, i\neq 2}^{N} (f_{uyi} x_i / |x_i|)$, at two levels: before yielding and at the initiating of collapse, respectively, define the degree of the unbalanced strength distribution at the two sides of the center of mass. The variation of the value $V_{yu}^u - V_{yu}^y$ shows the change of the possibility of the occurring of the pure shear mechanism. A maximum value of V_{yu} between V_{yu}^y and V_{yu}^u is also expected. Corresponding with the change of the strength eccentricity, this peak value should represent the state of some weaker structural members being damaged on one side of the center of mass while remaining elastic on the other side of the CM. Fig.6 does not show the peak value directly.

From the determination of the two critical boundary of the torsional-lateral coupling system, it can be inferred that the seismic response of the base shear-torque is naturally divided into 3 regions that represent respectively the linear elastic state, nonlinear state and the destroyed collapse state. The two critical boundaries, which signify the initiating yield and collapse, can be established independently based on the yield strength and the ultimate strength of structural members. This model can be directly used to estimate both of the overall seismic capacity and the developing trends of the time history response of the system base shear-torque to strong earthquakes even without complicated nonlinear analysis.

Fig.7 shows the comparison of the theoretical state space of the base shear-torque with the experimental results. The time-history shear-torque response of the bi-directional asymmetric single-story RC system in

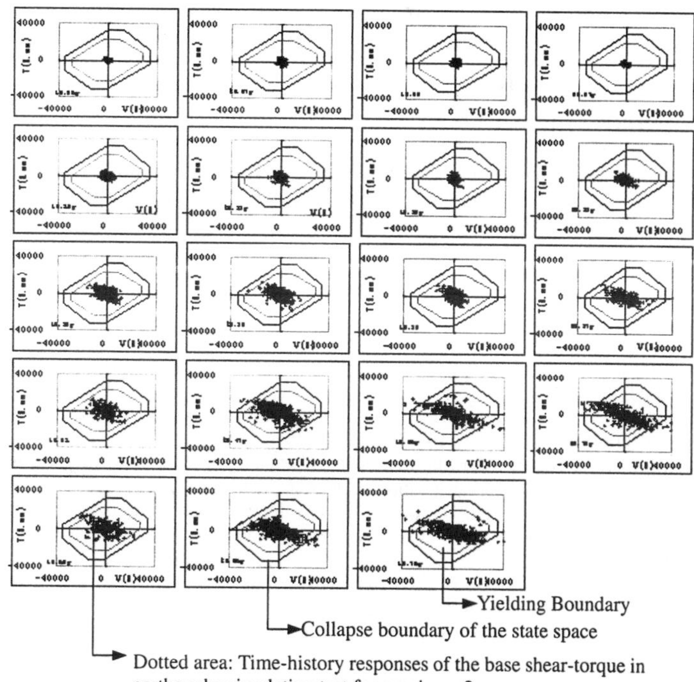

Fig.7 comparison of the theoretically established state space of the base shear-torque with experimental results

19 cases of ground excitations illustrates highly consistency with the state space established based on the assumption of the bi-linear force-displacement relationship. Such consistency is unreachable only by using the BST surface based on the idealized elasto-plastic assumption.

CONCLUDING REMARKS

In summary, the shake table tests provided useful data references for theoretical studies on seismic torsional effects of asymmetric building structures. Through the comparison between the experimental and the theoretical results, the following conclusions can be obtained:

1) Although the BST (base shear-torque) surface established on the assumption of idealized force-displacement relationship can be roughly used to estimate the seismic performance of asymmetric systems, it will bring obvious error on the prediction of the nonlinear response when the studied structure suffered strong earthquakes.
2) Therefore, based on the assumption of the bi-linear force-displacement relationship and the displacement-based method to determine the stiffness, the state space signifying the base shear-torque relationship is developed to describe the entire seismic capacity of the asymmetric systems. It shows high consistency with the experimental results. Therefore, it is no doubt a reliable model that can be directly used to estimate not only the entire seismic capacity but also the potential seismic damage may occur in the studied systems, even without relying on any complicated numerical analysis. It can be seen, this model will be very convenient and useful in evaluation of the lateral-torsional coupling effects of asymmetric structures particularly in the preliminary stage of design, retrofitting and research.

It should be noted that abovementioned conclusions are obtained on the basis of the experimental results of single story building structure. Their reliability and universality should be checked further with real seismic records or earthquake simulation tests for multi-story and high-rise building.

References

1. Dai Junwu, Yuklung Wong and Minzheng Zhang, *Torsion Response of Reinforced Concrete Building under Shaking Table Test*. Advances In Structural Dynamics, 2000, Vol. II, pp859-865.
2. Dai Junwu, *Research on nonlinear seismic performance of asymmetric RC building structure*, Ph.D. thesis, Institute of Engineering Mechanics, China Seismological Bureau, 2002,
3. De la Llera, J.C., Chopra, A. K., *Understanding the Inelastic Seismic Behaviour of Asymmetric Plan Buildings*, J. Earthquake Engineering & Structural Dynamics, 1995, v.24, pp.549-572.
4. De la Llera, J.C., Chopra, A. K., *Inelastic Behaviour of Asymmetric Multistory Buildings*, J. Structural Engineering, 1996, v.122, no.6, pp.597-606.
5. Guo Xun, Yuk-lung Wong and Yuan Yifan, *Investigation of Seismic Response of Soil site in HK*, HKIE Transactions, 2000, Vol. 7, No. 3.
6. Kan, C.L. and Chopra, A.K., *Torsional coupling and earthquake response of simple elastic and inelastic systems*, J. structural Division, ASCE, 1981, v.107, pp.1569-1588.
7. Minzheng Zhang, *Problem study on the application of the similitude law in earthquake simulation test*, Earthquake Engineering and Engineering Vibration, 1997.6, Vol.17, No.2.
8. Palazzo, B. and Fraternali, F., *Seismic ductility demand in buildings irregular in plan: A new single story nonlinear model*, Proceedings of the 9th World Conference on Earthquake Engineering, Tokyo-Kyoto, Japan, 1988, Vol.V, V-43 to V-48.
9. Paulay, T., *Displacement-based design approach to earthquake-induced torsion in ductile buildings*, Engineering Structures, 1997, Vol.19, No.9, pp.699-707.
11. Paulay, T., *Torsional mechanisms in ductile building systems*, Earthquake Engineering and Structural Dynamics, 1998, Vol.27, pp.1101-1121.
12. Paulay, T., *Understanding torsional phenomena in ductile systems*, Bulletin of the NZ Society for Earthquake Engineering, 2000, Vol.33, No.4, pp.403-420.

Advances in Building Technology, Volume 1
M. Anson, J.M. Ko and E.S.S. Lam (Eds.)

EXPERIMENTAL INVESTIGATION ON AXIAL COMPRESSIVE BEHAVIOR OF FRP CONFINED CONCRETE COLUMNS

Li Jing [1], Qian Jiaru [1] and Jiang Jianbiao [2]

[1] Dept. of Civil Engineering, Tsinghua University, Beijing
[2] Beijing Texida Technical Limited Company, Beijing

ABSTRACT

In this paper axial compression test results of 29 Fiber Reinforced Polymer (FRP) confined concrete columns and 2 reference plain concrete columns are introduced. Most of column specimens were confined with Carbon Fiber Reinforced Polymer (CFRP) sheet and some of specimens were hybrid confined with both CFRP sheet and Glass Fiber Reinforced Polymer (GFRP) sheet. The influences of FRP sheet layer, section height to width ratio, initial compressive ratio and FRP type on the axial compressive behavior of FRP confined concrete columns are investigated. Based on the experimental results equations of complete stress-strain curve and the parameters of equations are proposed for CFRP sheet confined concrete.

KEYWORDS

FRP confined concrete column, axial compressive behavior, complete stress-strain curve equation

PREFACE

One of the effective approaches to upgrade strength and ductility of concrete columns is to confine the column with FRP sheet, see Karbhari & Gao (1997), Yu (2001), Lam & Teng (2001), Zhao, Xie and Dai (2001). The method of confinement is to wrap concrete columns with FRP sheet, use epoxy resin to bond layers of FRP and to bond FRP sheet with the surface of column. Thus FRP confined column is formed as shown in Fig.1.

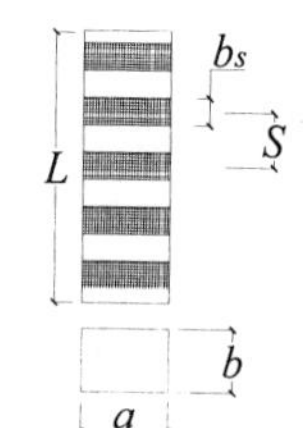

Figure 1: FRP confined column

The objectives of this paper were to study the compressive behavior of FRP confined concrete columns subjected to concentric axial compressive loads and to develop complete stress-strain curve equations for CFRP confined concrete.

TEST PROGRAM

Specimen Details

Thirty-one specimens including 29 FRP confined columns and 2 reference plain concrete columns were tested. All specimens were cast in 6 groups. Four groups were square cross section columns and two groups were rectangular cross section columns. Four parameters were varied: 1) the degree of confinement provided by FRP, 2) cross section dimensions, 3) initial axial compressive force before specimens were wrapped with FRP sheet, and 4) confinement material. All specimens were plain concrete column to neglect the influence of longitudinal reinforcements and stirrups on confinement. Two types of FRP sheet were used: CFRP and GFRP. The material properties of CFRP and GFRP are listed in Table 1.

A parameter called FRP characteristic value λ_{FS} is adopted to measure the degree of confinement provided by FRP. The FRP characteristic value accounts for the effect of FRP layer-wrapped, tension strength and thickness of FRP sheet, dimensions of cross section of specimen, axial compressive strength of plain concrete. It can be computed as:

$$\lambda_{FS} = \frac{V_{FS}}{V_C}\frac{f_{FS}}{f_c} = \frac{2(a+b)n_S t_S}{ab}\cdot\frac{f_{FS}}{f_c} \tag{1}$$

Where: V_{FS} and V_C are the volume of FRP sheet and the volume of concrete column specimen respectively, n_s and t_s are the layer-wrapped and thickness of FRP sheet, a and b are the length and width of cross section of the specimen, f_{FS} is the tension strength of CFRP or GFRP and f_c is the axial compression strength of concrete. For specimens hybrid confined with both CFRP and GFRP, the value of λ_{FS} is the sum of λ_{CFS} and λ_{GFS} calculated with Eqn.1. The layer-wrapped is defined as the ratio of the area of FRP sheet to the vertical surface area of specimen. With the definition, the layer-wrapped of the specimen shown in Fig.1 is 0.5. The 1-layer-wrapped means that allover the vertical surface of the specimen is wrapped with 1-layer FRP sheet.

Some specimens were loaded with an initial compressive force before they were wrapped with FRP. The initial compressive ratio n_p is adopted to measure the magnitude of initial force, which is defined as the ratio of the initial axial compressive force N to the produce of column cross section area ($a \times b$) and f_c. Specimens of group 2,3 and 4 were loaded with the initial compressive ratio n_p of 0.22, 0.46 and 0.3 respectively. The initial force was applied by prestressing steel cables. To avoid the local destroy and bending failure, the steel cable were symmetrically prestressed. The real force applied to the specimen was measured with force sensors. The prestressing facilities are schematically shown in Fig.2.

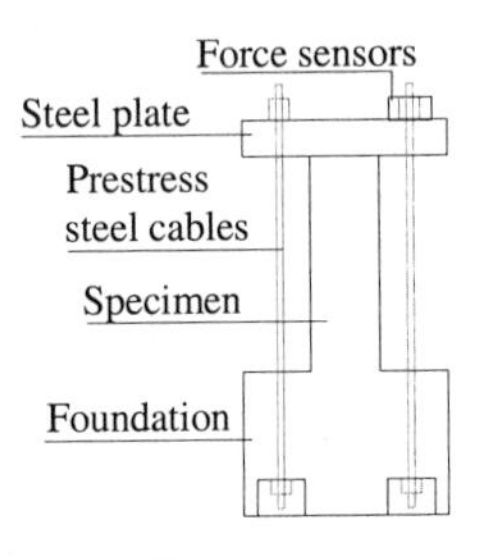

Figure 2: Prestressing facilities

The details of each specimen are listed in Table 2.

TABLE 1
MATERIAL PROPERTIES OF FRP SHEET

FRP sheet type	Tension strength f_{FS} (MPa)	Tension module E (GPa)	Thickness t_s (mm)	Ultimate tension strain
CFRP	4013	241	0.111	16650×10^{-6}
GFRP	2777.1	97	0.169	28630×10^{-6}

TABLE 2
DETAILS OF SPECIMENS

Group No	Specimen No	Dimension (mm)	f_c	n_p	FRP sheet type	Layer -wrapped	λ_{FS}
1	Z1-0	200×200×600	25.4	0	CFRP	-	0
	Z1-1					0.5	0.18
	Z1-2					1	0.37
	Z1-3					1.5	0.55
	Z1-4					2	0.74
	Z1-6					3	1.11
	Z1-4a		29.3			2	0.63
2	Z2-0	200×200×600	29.3	0.22	CFRP	-	0
	Z2-1					0.5	0.16
	Z2-2					1	0.31
	Z2-4					2	0.63
	Z2-6					3	0.94
3	Z3-1	200×200×600	26.6	0.46	CFRP	0.5	0.19
	Z3-2					1	0.38
	Z3-3					1.5	0.58
	Z3-4					2	0.77
	Z3-6					3	1.15
4	Z4-1	300×200×600	31.7	0.3	CFRP	0.5	0.17
	Z4-2					1	0.34
	Z4-3					1.5	0.50
	Z4-4					2	0.67
	Z4-5					3	1.01
	Z4-6	300×200×600	27.9	0.3	CFRP, GFRP	1, 1	0.55
	Z4-7					2, 2	1.09
5	Z5-1	400×200×900	26.7	0	CFRP	4/3	0.32
	Z5-2					8/3	0.64
	Z5-3				CFRP, GFRP	1, 1/3	0.33
6	Z6-1	300×300×900	28.3	0	CFRP	13/9	0.33
	Z6-2					3	0.68
	Z6-3				CFRP, GFRP	1, 4/9	0.33
	Z6-4					2, 1	0.69

Test Instrumentation

The axial deformation was measured with two couples of linear variable displacement transducer located on the symmetrical surfaces of specimen. One couple of transducer recorded the relative movement between the upper and lower load plates. One couple of transducer recorded the relative movement between two sections with 400mm spacing. The horizontal strains of FRP were measured with strain gauges. The gauges were located at the mid-height section of the specimen. The distributions of the gauges are shown in Fig.3. The specimens were tested by monotonically increasing concentric axial compressive load in a 5000KN capacity compressive testing machine. Fig.4 shows the photo of test setup. All of data were collected using a computer-based data acquisition system.

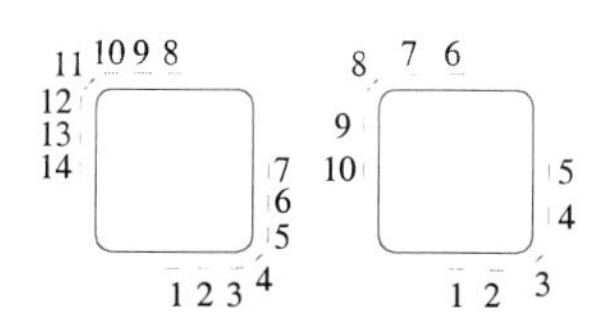

Figure 3: Location of strain gauges

Figure 4: Photo of test setup

EXPERIMENTAL RESULTS AND ANALYSES

Process of Destroy and Failure Modes

The measured axial compressive force-displacement curve was linear when the axial force was less than about half of the strength of specimen. When the curve reached its peak or started turn to level, concrete crack voice could be heard. Finally the rupture of FRP sheet characterized the failure of confined specimens.

Failure modes of FRP confined concrete column specimens can be classified into two types: diagonal shear failure and compression failure. If λ_{FS} was small, the specimen failed in shear as shown in Fig. 5(a). On the contrary, the specimen failed in compression and the failure surface was similar to a pyramid as shown in Fig.5(b). The force-displacement curve corresponding to compression failure is plump near the turning corner, and the descending branch was higher than that corresponding to diagonal shear failure. The FRP sheet of most specimens was broken at the corner of specimen. The rupture mode of FRP sheet was related to λ_{FS} and n_p. If λ_{FS} and n_p were smaller, the rupture was fluffy and unorderly by tearing out, on the contrary, it was straight like cut by scissors, as shown in Fig.5(c) and 5(d).

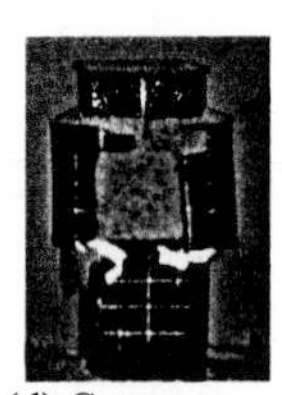

(a) Diagonal shear failure (b) Compression failure (c) Tearing rupture (d) Cut rupture

Figure 5: Failure modes of concrete and FRP sheet of specimen

Test Results

Fig.6 shows $\sigma/\sigma_0 - \varepsilon/\varepsilon_0$ curves of all specimens, where σ and ε are measured axial stress and strain respectively, σ_0 and ε_0 are peak stress and peak strain of plain concrete respectively. For specimens of group1 and 2, the maximum average stress (i.e., the maximum load divided by section area) of the reference column specimen is taken as the peak stress (σ_0) respectively. For other specimens the average value of σ_0/f_c of two reference columns is taken to calculate the peak stress (σ_0) of plain concrete. For all specimens, the average peak strain (0.002) of two reference column specimens is taken as the peak strain (ε_0) of plain concrete.

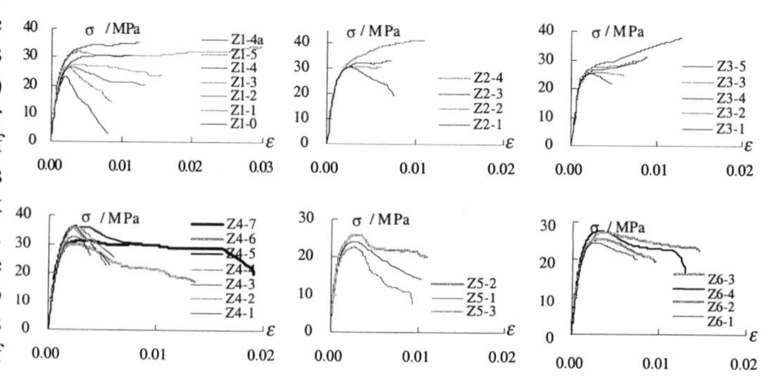

Figure 6: $y(\sigma/\sigma_0) - x(\varepsilon/\varepsilon_0)$ curves of specimens

Three types of curve shape can

be distinguished. (1) The curve has an ascending branch (before the peak of curve) and a descending branch (after the peak of curve). (2) The curve has an ascending branch and a horizontal branch. (3) The curve has two ascending branches, the slope of the latter branch is much less. Most of curves are type 1.

The main test results are listed in Table 3. In the table, f_{cc} and ε_{cc} are peak stress and peak strain. The peak point refers to the turning point for type 2 curves and type 3 curves. f_{ccu} is the ultimate stress corresponding to the end point for type 3. ε_{ccu} is the ultimate compressive strain which is defined as the strain corresponding to the residual stress of $0.6f_{cc}$ or the failure strain for type 2 and type 3. The average ultimate strain (0.004) of two reference columns is taken as the ultimate strain (ε_{cu}) of plain concrete for all specimens.

TABLE 3
TEST RESULTS

Group No	Specimen No	f_{cc} (MPa)	f_{cc}/σ_0	ε_{cc} ($\times10^{-6}$)	$\varepsilon_{cc}/\varepsilon_0$	f_{ccu} (MPa)	f_{ccu}/σ_0	ε_{ccu} ($\times10^{-6}$)	$\varepsilon_{ccu}/\varepsilon_{cu}$
1	Z1-0	23.02	1.0	2005	1.0			4501	1.13
	Z1-1	26.57	1.15	2296	1.13			8317	2.08
	Z1-2	26.54	1.15	3041	1.52			13388	3.35
	Z1-3	27.33	1.18	3458	1.74			15571	3.89
	Z1-4	30.38	1.31	6424	3.21	30.56	1.32	11765	2.94
	Z1-6	31.87	1.38	3745	1.87	33.97	1.46	29897	7.47
	Z1-4a	34.30	1.26	6532	3.27	34.44	1. 30	12417	3.10
2	Z2-0	27.20	1.0	2000	1.0			3600	0.90
	Z2-1	31.66	1.16	2628	1.31			7599	1.90
	Z2-2	32.40	1.19	5050	2.52	32.40	1.19	6070	1.52
	Z2-4	34.13	1.25	7026	3.51	34.13	1.26	8687	2.17
	Z2-6	40.14	1.48	7802	3.90	41.35	1.52	10905	2.73
3	Z3-1	25.33	1.14	2254	1.13			4769	1.19
	Z3-2	25.80	1.16	2572	1.29			6186	1.55
	Z3-3	27.64	1.24	3687	1.84	29.87	1.34	7622	1.91
	Z3-4	26.69	1.20	3312	1.66	31.09	1.40	8870	2.22
	Z3-6	29.24	1.31	4618	2.31	37.66	1.69	12985	3.25
4	Z4-1	33.24	1.17	1963	0.98			3095	0.77
	Z4-2	35.56	1.26	2148	1.07			4466	1.12
	Z4-3	35.75	1.26	2191	1.10			5292	1.32
	Z4-4	36.13	1.29	2416	1.21			5746	1.44
	Z4-5	36.03	1.28	2445	1.22			10090	2.53
	Z4-6	30	1.12	3588	1.22			11019	2.75
	Z4-7	31.24	1.16	2447	1.58			21443	5.36
5	Z5-1	24.05	0.90	2555	1.28			10439	2.61
	Z5-2	25.91	0.97	3215	1.61			11515	2.88
	Z5-3	22.60	0.85	2286	1.16			9812	2.45
6	Z6-1	24.53	0.98	2246	1.12			6134	1.54
	Z6-2	27.27	1.09	3588	1.79			14625	3.66
	Z6-3	25.49	1.02	3273	1.64			9553	2.39
	Z6-4	27.53	1.10	3334	1.67			12980	3.25

Influence Factors

FRP characteristic value λ_{FS}

The effect of λ_{FS} on the $\sigma/\sigma_0-\varepsilon/\varepsilon_0$ relationships of FRP confined concrete is obvious, as seen from Table 3, Fig.6 and Fig.8. The following conclusions can be obtained under the condition of $0 \le \lambda_{FS} < 1.1$. (1) The larger λ_{FS} is, the larger f_{cc}, f_{ccu}, ε_{cc} and ε_{ccu} are. (2) The initial stiffness of FRP confined concrete columns is larger than that of plain concrete columns. The larger λ_{FS} is, the larger initial stiffness is. (3) The descending branch of $\sigma/\sigma_0-\varepsilon/\varepsilon_0$ curve becomes more and more flat with the increasing of λ_{FS}. When λ_{FS} is large enough, the curve does not have descending branch.

Initial axial compressive ratio n_p

Initial axial compressive ratios n_p of group 1, 2 and 3 specimens are 0, 0.22 and 0.46 respectively. Comparing $\sigma/\sigma_0-\varepsilon/\varepsilon_0$ curves of group 1, 2 and 3 with about the same λ_{FS}, it is found that the larger n_p is, the flatter the descending branch is. Most specimens of group 2 and 3 do not have descending branch. For they had had initial lateral deformation before they were wrapped, the FRP sheet had more capacity to deform (i.e., FRP sheet provided more lateral confinement) after the turning point of $\sigma/\sigma_0-\varepsilon/\varepsilon_0$ curve than those with $n_p=0$.

Section length to width ratio λ_{ab}

FRP characteristic value λ_{FS} doesn't include the influence of section length to width ratio. Comparing $\sigma/\sigma_0-\varepsilon/\varepsilon_0$ curves of group 1, 4 and 5 specimens with λ_{ab} of 1.0, 1.5 and 2.0 with about the same λ_{FS}, the larger λ_{ab} is, the less f_{cc} and ε_{ccu} are.

FRP type

Comparing σ/σ_0 -$\varepsilon/\varepsilon_0$ curves of Z4-4 with Z4-6, Z4-5 with Z4-7 as shown in Fig.6, the second branch of the curve for specimens hybrid confined with both CFRP and GFRP is relatively higher, and the ultimate strain is larger. The results show that combining with GFRP to CFRP, especially allover the vertical surface of column was wrapped with GFRP, benefits deformation and ductility capacity, for the tension module of GFRP is less than that of CFRP and the ultimate tension strain is larger than that of CFRP.

EQUATION OF COMPLETE STRESS-STRAIN CURVE FOR CFRP CONFINED CONCRETE

Shape of Curves

As mentioned before, there are three types of stress-strain curve shape for CFRP confined concrete as shown in Fig.7. The coordinates in Fig.7 are defined as:

$$x=\frac{\varepsilon}{\varepsilon_{cc}},\quad y=\frac{\sigma}{f_{cc}} \tag{2}$$

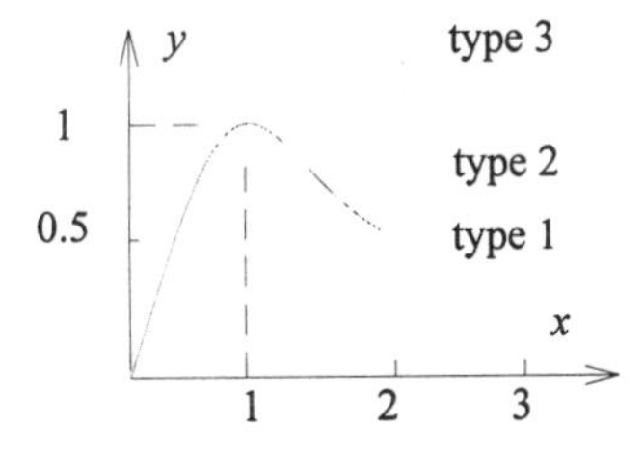

Figure 7: Three types of complete stress strain curve for CFRP confined concrete

All curves have two branches. The first branch is the ascending branch, the shape of which can be described with a single equation that will be presented below. The shape of the second branch is related to the value of λ_{FS} and the critical value λ_{FS0}. If $\lambda_{FS} < \lambda_{FS0}$, the curve shape is type 1. If $\lambda_{FS} = \lambda_{FS0}$, the curve shape is type 2. If $\lambda_{FS} > \lambda_{FS0}$, the curve shape is type 3. From experimental data, the critical value λ_{FS0} depends on n_p and λ_{ab}, i.e.:

$$\lambda_{FS0} = \frac{0.8(1-n_p)}{2-\lambda_{ab}} \tag{3}$$

Suggested Equation of Stress-strain Curve

The form of equation for plain concrete suggested by Sargin (1971) is adopted for the ascending branch of stress-strain curve of CFRP confined concrete, and a linear equation is adopted for the second branch, i.e.:

$$x \le 1: \quad y = \frac{a'x}{1+(a'-2)x+x^2} \tag{4}$$

$$x > 1: \quad y = \beta'(x-1)+1 \tag{5}$$

Five parameters are needed to define an equation of complete stress-strain curve of FRP confined concrete: peak stress f_{cc}, peak strain ε_{cc}, ultimate stress f_{ccu}, the ascending branch parameter a' and the second branch parameter β'. According to the experimental data under the condition of $0 \le \lambda_{FS} < 1.1$, $0 \le n_p < 0.5$ and $1 \le \lambda_{ab} < 2$, Eqn.(6)~Eqn.(9) are proposed to calculate f_{cc}, f_{ccu}, ε_{cc}, ε_{ccu}.

$$f_{cc} / f_c = 0.38\lambda_{FS} + 1 \tag{6}$$

$$f_{ccu} / f_c = 0.51\lambda_{FS} + 1 \tag{7}$$

$$\varepsilon_{cc} / \varepsilon_0 = 1.75\lambda_{FS} + 1 \tag{8}$$

$$\varepsilon_{ccu} / \varepsilon_{cu} = \begin{cases} 4.76\lambda_{FS} + 1, & n_p \le 0.2 \\ 1.81\lambda_{FS} + 1, & n_p > 0.2 \end{cases} \tag{9}$$

Eqn.(6) and Eqn.(7) are compared with tests in Fig.8(a), and Eqn.(8) and Eqn.(9) are compared with tests in Fig.8(b).

a' and β' are the parameters used to adjust the shape of the ascending branch and the second branch of stress-strain curve respectively. The FRP characteristic value λ_{FS} is selected to identify the curve shape so that the parameters a' and β' are recommended in Eqn.(10) and Eqn.(11).

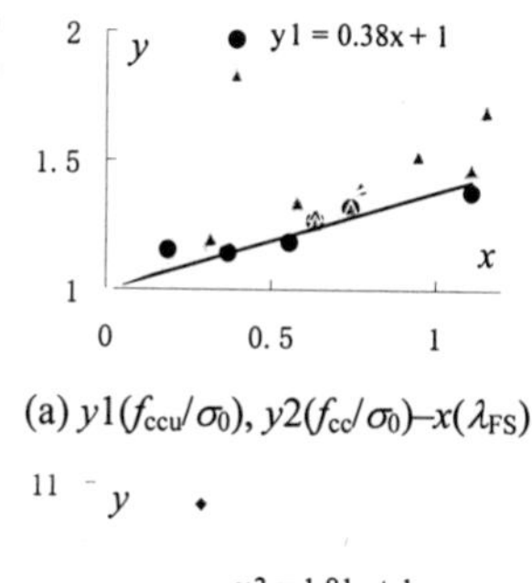

(a) $y1(f_{ccu}/\sigma_0)$, $y2(f_{cc}/\sigma_0)$–$x(\lambda_{FS})$

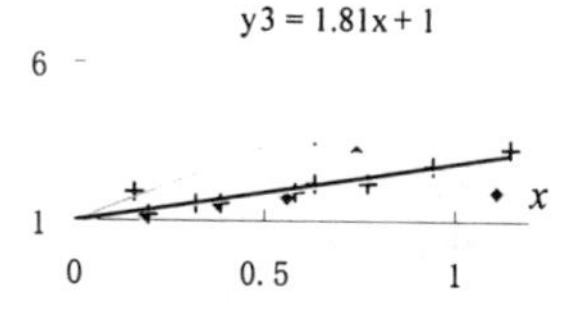

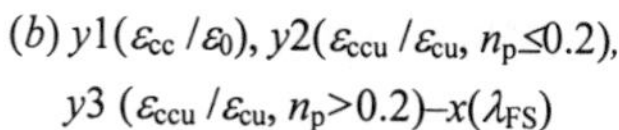

(b) $y1(\varepsilon_{cc}/\varepsilon_0)$, $y2(\varepsilon_{ccu}/\varepsilon_{cu}, n_p \le 0.2)$, $y3$ $(\varepsilon_{ccu}/\varepsilon_{cu}, n_p > 0.2)$–$x(\lambda_{FS})$

Figure 8: The influence of λ_{FS} on f_{cc}, f_{ccu}, ε_{cc} and ε_{ccu}

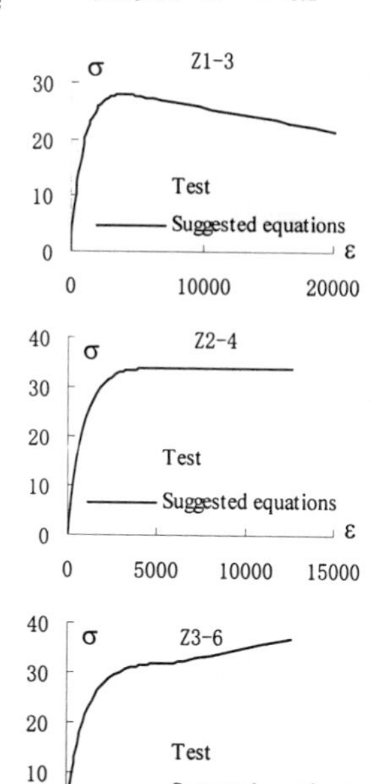

Figure 9: Comparison of σ(Mpa)-ε($\times 10^{-6}$) curves charted by test and by suggested equations

$$a' = 2(2.56\lambda_{FS} + 1) \tag{10}$$

$$\beta' = \begin{cases} \text{type 1}: -0.25(\lambda_{FS} - \lambda_{FS0})^2 + 0.17(\lambda_{FS} - \lambda_{FS0}) \\ \text{type 2}: 0 \\ \text{type 3}: 0.20(\lambda_{FS} - \lambda_{FS0}) \end{cases} \tag{11}$$

Using the above equations, the complete stress-strain curve of CFRP confined concrete can be described. The stress-strain curves of Z1-3 (type 1), Z2-4 (type 2) and Z3-6 (type 3) suggested by Eqn.(4) ~Eqn.(11) are charted in Fig.9, and the curves obtained from the test data are also drawn. It can be seen from Fig.9 that the relative curves have a good agreement. The equations recommended in this study to describe the behavior of CFRP confined concrete are suitable.

CONCLUSIONS

Concrete confined with FRP sheet is advantage to its strength, deformation and ductility. The strength and the ultimate strain of FRP confined concrete columns can be 1.6 and 7 times of those of plain concrete columns.

The effect of FRP on the strength and deformation of concrete columns is related to the degree of confinement, cross section dimensions, initial axial compressive ratio, and confinement material.

Two type failure modes of concentric axial loaded FRP confined concrete columns are diagonal shear failure and compression failure. The failure mode mostly depends on the characteristic value λ_{FS} of the column.

The stress-strain curve of CFRP confined concrete is classified into three types. A series of equations are obtained to describe the three type curves. Parameters of the equations are suggested by experimental data of this study.

REFERENCE

Karbhari V.M. and Gao Y. (1997). Composite Jacketed Concrete under Uniaxial Compression-Verification of Simple Design Equation. *Journal of Materials in Civil Engineering, ASCE* **9:4**, 185-193.

Sargin M. (1971). *Stress-Strain Relationships for Concrete and the Analysis of Structural Concrete Sections*. Canada

Lam L. and Teng J.G. (2001). A New Stress-strain Model for FRP-confined Concrete. *FRP Composites in Civil Engineering*, **Vol.I, J.-G. Teng (Ed.)**, 283-291.

Yu Qing. (2001). Investigation on Stress-strain Relationship of FRP Confined Concrete under Axial Compression. *Magazine of Industry Structure,* **31:4**, 18-25.

Zhao Tong, Xie Jian and Dai Ziqiang. (2001). Experimental Investigation on Full Stress-strain Curve of CFRP Confined Concrete. *Magazine of Structure Construction,* **30:7**, 32-38.

Advances in Building Technology, Volume 1
M. Anson, J.M. Ko and E.S.S. Lam (Eds.)

DYNAMIC PROPERTIES AND SEISMIC RESPONSE OF RC STRUCTURES CHARACTERIZED BY ABRUPT CHANGE IN HORIZONTAL CONNECTION STIFFNESS

Liu Jingbo[1] and Wang Zhenyu[1]

[1]Department of Civil Engineering, Tsinghua University, Beijing, 100084, China

ABSTRACT

More and more concrete structures characterized by abrupt change in stiffness are built because of requirements of building appearance and functions. Obviously, dynamic properties and seismic response of this kind of structure are very different from common structures. It is very interesting to discuss the design concept, analysis methods and other details of this kind of structure. Dynamic properties and seismic response of concrete structure characterized by abrupt change in horizontal connection stiffness are studied systematically in this paper. General mechanic concepts and analysis methods of this kind of structure are introduced firstly. Then for Putian Center of Radio, Film and TV example, complicated mode shapes and stress concentration resulted from abrupt change in horizontal connection stiffness are revealed. Moreover, effects of different connective plans located in place of abrupt change in horizontal connection stiffness on structure dynamic properties and seismic response are discussed. These effects are analyzed by FE numerical experiments. Finally, some useful suggestions are proposed which may be used as a reference in engineering.

KEYWORDS

abrupt change, concrete structures, dynamic properties, seismic response, connection stiffness, stress concentration

INTRODUCTION

It is very important that building structural systems have reasonable stiffness distribution. Strong stress and plastic deformation concentration because of partly weakening or abrupt change in stiffness should be avoided. However, abrupt change in horizontal or vertical stiffness usually exists because of the demand of architectural function and style in practice. In general, we must pay more attention to not only structural member's stress and deformation located in place of abrupt change in stiffness, but also the change of dynamic properties caused by abrupt change in stiffness. The design method of transfer beams or stories located in place of abrupt change in vertical stiffness has been studied by many authors. But few studies on abrupt change in horizontal stiffness have been performed. In fact, abrupt change in horizontal stiffness will lead to marked change of dynamic properties and acute stress and deformation concentration of structure members. In this paper, general problems of such structure dynamic analysis are discussed firstly. Then effects on structure dynamic properties and seismic responses are studied which caused by different connective plans located in place of abrupt change in horizontal stiffness. The effects are analyzed by finite element numerical experiments based on project of Putian Center of Radio, Film and TV, China. Methods and conclusions in this paper are referenced to such structure dynamic response analysis and optimization of seismic design.

GENERAL PROBLEMS OF DYNAMIC RESPONSE ANALYSIS

Two problems must be considered when we study seismic analysis of RC structure characterized by abrupt change in horizontal connection stiffness. The two problems consist of characteristics and selection of mode shapes, and acute stress concentration located in place of abrupt change in horizontal connection stiffness.

Mode superposition time history analysis and response spectra analysis are usually to be used in structural aseismic analysis. Enough number of mode shapes should be selected in order to reflect structure major dynamic properties. Few mode shapes are needed for structures characterized by reasonable stiffness distribution. However, only few mode shapes used in calculation will result in large extent of error for structures characterized by abrupt change in horizontal connection stiffness. The major reason lies in differences between mode shapes' curve. The structure characterized by abrupt change in horizontal connection stiffness has marked characteristics of mode shapes, such as many hybrid, interlocked and torsional modes. An uniform formula which can be used to determine number of mode shapes used in calculation is not given in this paper. But two principles are suggested for determining number of mode shapes used in calculation. One is that entirely deformation should be reflected in selected mode shapes, the other is that complicated mode shapes including hybrid, interlocked and torsional modes should be used in calculation. The last principle is specified for structures characterized by abrupt change in horizontal connection stiffness.

Stress concentration located in place of abrupt change in horizontal connection stiffness under horizontal earthquake motion can be given by finite element method. But small size element distribution in place of abrupt change in horizontal connection stiffness must be used in order to get precision results. We also pay attention to determination of member's stiffness located in place of abrupt change in connection stiffness. Small stiffness demand is given base on earthquake energy dissipation and slight stress concentration, which will result in large structure horizontal displacement. On the other hand, large stiffness demand is given based on small horizontal displacement, which will result in low energy dissipation capability and acute stress concentration. The above problems will be discussed in detail in following paper by numerical experiments based on project of Putian Center of Radio, Film and TV, China.

NUMERICAL CALCULATION

Structure and Finite Element Model

Putian Center of Radio, Film and TV, China is a multifunctional high-rise building. The total building area of the project is about 13560 square meters. It consists of main building and skirt building. Standard floor of the building is shown in figure 1. The building's plane layout is two connected polygonal tubes. The main building is composed of two stories underground and fifteen stories aground, which has frame-shear-wall structural system. The big tube of the main building is 59.4 meter high, while the small tube of the main building is 69.3 high including a turret. The first story aground is 4.8 meter high and the standard story height is 3.6 meter.

Characteristics of abrupt change in horizontal connection stiffness are marked because of connection of the big tube and the small tube of the main building, as shown in figure 1. Plate and beam members are used where the big tube is transferred to the small tube. Obviously, abrupt change in horizontal connection stiffness has considerable effects on structure dynamic properties, and most of important is acute stress concentration located in place of the connection of the big tube and the small tube under earthquake motion. In order to research this problem, the structure is simulated by means of discrete finite elements. Floor slabs and shear walls are simulated by discrete plate elements, while beams and columns are simulated by beam elements. Then we get the calculating model of finite element shown in figure 2.

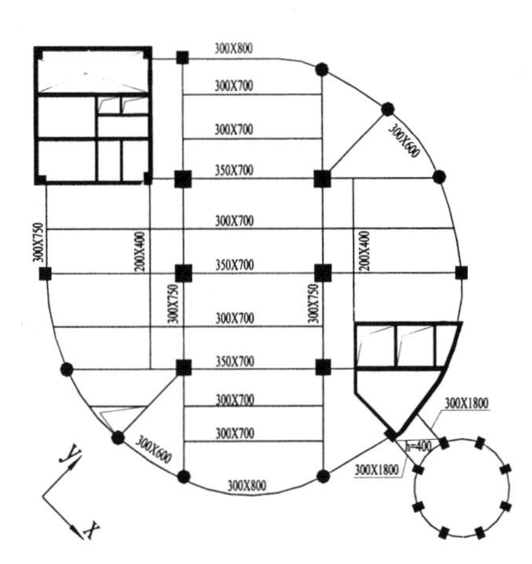

Figure 1: standard story planar graph

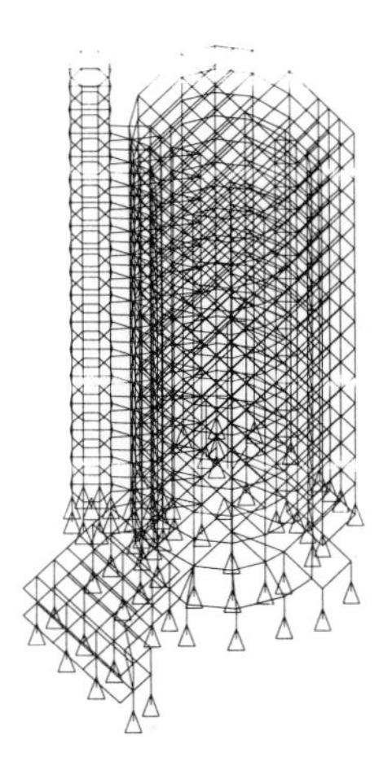

Figure 2: structural finite element modal

Structure Dynamic Properties

In order to research effects caused by abrupt change in horizontal connection stiffness on structural dynamic properties, mode shapes calculation of the structure is performed firstly. The first tenth natural frequencies and periods are given in table 1. We can discriminate natural period in coordinate direction and natural period of torsional vibration. Fundamental natural period in x direction is 1.339 second, and 1.482 second in y direction. Similarly, fundamental natural period of torsional and vertical vibration are respectively 0.854 and 0.226 second.

TABLE 1
THE FIRST TENTH NATURAL FREQUENCIES AND PERIODS OF THE STRUCTURE

Mode shape	1	2	3	4	5	6	7	8	9	10
Frequency(Hz)	0.675	0.747	1.172	2.452	2.904	3.505	3.803	4.015	4.430	4.659
Period(s)	1.482	1.339	0.854	0.408	0.344	0.285	0.263	0.249	0.226	0.215

We find many kinds of complex and partly mode shapes of the structure which include relative movement of the big tube and the small tube, partly torsion and so on. These mode shapes are caused by abrupt change in horizontal connection stiffness. Some mode shape graphs are shown in figure 3, which clearly illustrate relative movement, partly torsion and translation.

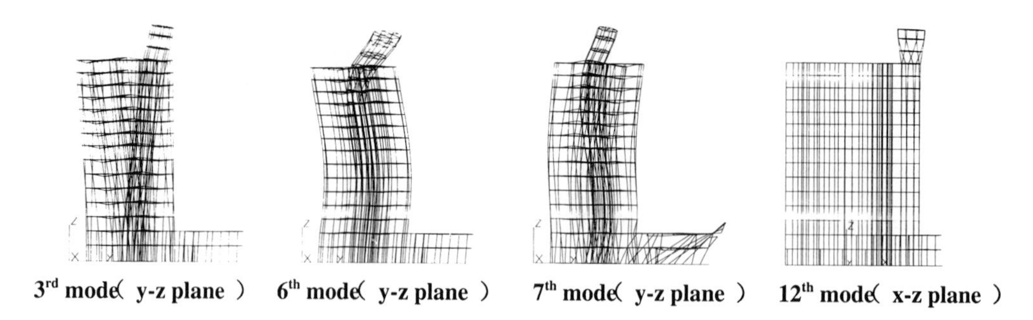

Figure 3: Some mode shapes of the structure

Horizontal Seismic Response Analysis

Response spectra method is used to calculate horizontal seismic response of the structure in this paper. The first 30th mode shapes including the complex mode shapes caused by abrupt change in horizontal connection stiffness are considered to ensure level of accuracy. Stiffness in y direction is lower than that in x direction, so only seismic response of the structure in y direction is discussed in this paper.

Calculating results illustrate that horizontal displacement is major deformation and torsional deformation is very small under earthquake motion in y direction. The top floor's horizontal displacement is about 0.0244 meter and the maximum displacement between adjacent floors is 0.00196 meter which located in the 17th floor of the small tube. The magnitude contours of stress in plane of floor slabs are shown in figure 4. Distributions of magnitude contour clearly illustrate acute stress concentration located in the connection of the big tube and the small tube. The maximum shearing stress and the positive stress in x and y directions are respectively 1177.00 KN/m^2、1838.75 KN/m^2 and 2509.46KN/m^2. Most of important, places which turn up these maximum values are all located in connection of the big tube and the small tube.

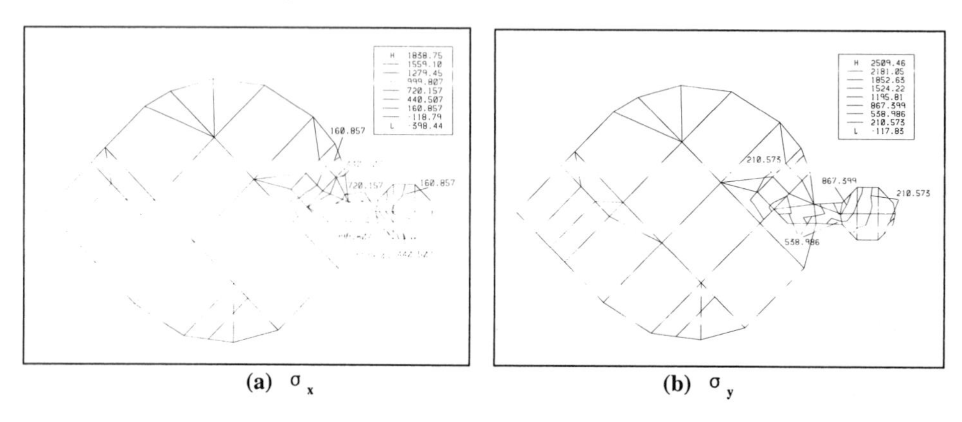

(a) σ_x (b) σ_y

Figure 4: Magnitude contour distribution of positive stress of 15th floor slabs

Acute stress concentration located in place of abrupt change in horizontal connection stiffness is revealed in extent of elasticity. It can be inferred that members located in place of abrupt change in horizontal connection stiffness will possibly become key members of energy dissipation, when the structure go through nonlinear range under strong earthquake motion. Therefore a key problem is how to handle the connection of members in the place. Effects of different connective members on structure seismic response will be studied by finite element numerical experiments in following paper. Obviously, the big tube isn't connected to the small tube when stiffness of the connection members is infinitely small, which becomes another structure plan.

COMPARISONS OF DIFFERENT CONNECTIVE PLANS

Structure Plan of Separated Tube

Because of acute stress concentration caused by abrupt change in horizontal connection stiffness, the possibility of removing the connective members is easily considered. The calculating results illustrate that structure plan of two independent tubes without connectivity improves dynamic properties and reduce dynamic response of the big tube. The small tube's fundamental natural period and maximum displacement under horizontal earthquake motion and wind loading are given in table 2. At the same time, the above values of the connected tubes are also given in table 2.

TABLE 2
COMPARISON OF DIFFERENT STRUCTURE PLANS

Name	Fundamental natural period(s)	Maximum displacement under earthquake motion(m)	Maximum displacement under wind loading(m)
The small tube	3.179	0.0585	0.1196
The connected tubes	1.482	0.0244	0.0222

The independent small tube has smaller stiffness and longer natural period than those of the connected tubes, as shown in table 2. Moreover, maximum displacement of the small tube under horizontal wind loading is unaccepted. In other words, the small tube must not be built independently and the connective members can't be removed.

Comparisons of Different Stiffness of the Connective Members

We have demonstrated that the connective members can't be removed, but there is a key problem about how to choose stiffness of the connective members. In general, stiffness variety of the connective members has considerable effects on natural period, horizontal displacement and stress concentration of the structure. Three kinds of stiffness of the connective members are used to calculate the structure seismic response in y direction in this paper, as shown in table 3.

TABLE 3
COMPARISONS OF DIFFERENT STIFFNESS OF THE CONNECTIVE MEMBERS

Stiffness of the connective members		Enlarging stiffness by 100 times	Design stiffness	Neglecting stiffness of the connective floor slabs
Fundamental Natural period (s)		1.476	1.482	1.491
Maximum displacement under earthquake motion (m)		0.0240	0.0244	0.0257
Maximum stress in plane of the connective floor slabs (KN/m^2)	σ_x	2548.02	1838.75	——
	σ_y	2845.44	2509.46	——
	τ_{xy}	1076.83	1177.00	——
Maximum stress of the connective beams (KN/m^2)		250.46	1949.63	7543.52

Results given in table 3 show that there isn't marked influence on structure fundamental natural period and maximum horizontal displacement if stiffness of the connective members is limited to certain range. The above results are given based on the fact that stiffness of the big tube is much larger than that of the small tube. By comparison, stiffness variety of the connective members has important influence on inner forces of the connective members. The increment of the connective members' stiffness maybe gives rise to acute stress concentration because of the increment of allocated earthquake loading. In regard to neglecting stiffness of the connective floor slabs, the maximum stress has great increase because the connective beams must resistant earthquake loading independently.

CONCLUSIONS

Dynamic properties and seismic response of RC structure characterized by abrupt change in horizontal connection stiffness are studied in this paper. Different structure plans and stiffness of the connective members are discussed based on finite element numerical experiments. Finally, we can get some conclusions as followed.

The structure characterized by uniform stiffness distribution has advantages of clear and definite transmission channel of earthquake loading. By contrast, the structure characterized by abrupt change in horizontal connection stiffness has disadvantages of complex mode shapes and acute stress concentration under earthquake motion.

The number of mode shapes in calculation must be selected carefully in seismic response analysis by method of response spectra analysis. Complex and partly mode shapes caused by abrupt change in horizontal connection stiffness must be contained in calculation. In addition, denser discrete elements located in place of abrupt change in horizontal connection stiffness should be used to ensure accuracy of stress field.

Stiffness of the connective members should be determined carefully in design. In general, stiffness of the connective members should be limited to a certain range. It will lead to acute stress concentration if stiffness of the connective members increases blindly. Moreover, there isn't obvious influence on structure horizontal displacements and fundamental natural period under earthquake motion.

This paper is limited to discuss linear elastic behaviors of RC structure characterized by abrupt change in horizontal connection stiffness. Nonlinear behaviors of the structure under strong earthquake motion will be studied in other papers.

References

Jingbo Liu and Yong Wang.(2000). Dynamic properties and seismic response of high-rise building with podium. *China Journal of Building Structures* **21:2,** 36-43

Jingbo Liu, etal.(2001). Preliminary study on seismic responses of large-span brick and concrete mixed structures. *China Earthquake Engineering and Engineering Vibration* **21:4,** 67-72

Jumin Shen, etal.(2000), *Aseismic Engineering*, China Architecture and Building Press

Kunyao Huang, etal.(2001). Influence of the connection stiffness on seismic response of the double-tower connected tall building. *China Journal of Building Structures* **22:3,** 22-26

Advances in Building Technology, Volume 1
M. Anson, J.M. Ko and E.S.S. Lam (Eds.)

A GIS SYSTEM INTEGRATED WITH VULNERABILITY FOR SEISMIC DAMAGE AND LOSS ESTIMATION OF HONG KONG

H.S. Lu[1], Z.P. Wen[1], K.T. Chau[2] and Y.X. Hu[1]

[1]Institute of Geophysics, China Seismological Bureau, Beijing, China
[2]Department of Civil and Structural Engineering, The Hong Kong Polytechnic University, Kowloon, Hong Kong, China

ABSTRACT

Estimation of damage and loss of a region due to earthquakes is of primary importance for disaster planning, mitigation and rehabilitation purposes. For regions of moderate seismicity, like Hong Kong, the social awareness of the potential risk of infrequent but high consequence events is very low. In this paper, a GIS is integrated with a simple vlnerability for reinforced concrete structures for estimating the seismic damage and loss for Hong Kong both in terms of deterministic scenario earthquake and probabilistic earthquake risk. This system is the first of this kind for Hong Kong. Two urban districts in Hong Kong, Tsim Sha Tsui East and Wanchai, are chosen for our preliminary analysis. Digital map developed by the Lands Development Department is adopted and modified, while the building inventory was initially supplied by Centaline Property Agency Limited which was subsequently updated and expanded for our purposes to include, building usage, number of story, and damage probability matrix. The evaluation of seismic hazard in the areas has been performed using the scenario earthquake and probabilistic approach. Earthquake intensity and response spectrum are used as input. The vulnerability of buildings is estimated by means of theoretical approach based on the modal response analysis. Seismic risk and loss are evaluated corresponding to 63%, 10% and 2% probability of exceedence in 50 years. Distribution of earthquake damage and loss under different ground motions are presented and managed by relational database management system in GIS. In addition, empirical formulas have been adopted to yield direct economic losses for both the deterministic and the probabilistic approach.

KEYWORDS

GIS, Vulnerability, Scenario earthquake, Risk, Hong Kong, Seismic damages, Loss, Digital map

INTRODUCTION

In the last few decades, civil engineering structures in many big cities become denser than ever due to the rapid population and economic growth. If earthquake hits such a metropolitan area, it may lead to catastrophe. The aftermath of the 1995 Kobe earthquake provides a vivid example on the problems associated with a densely populated city under the attack of a strong earthquake. For big cities like Kobe and Hong Kong, the seismic risk assessment and management is only feasible by using GIS based system, which should incorporate all inventory of buildings and lifelines. Such system can be used for prediction of future earthquake damage and loss, therefore it can support the seismic risk management by government agencies and by insurance companies Tasumi et al. (1992).

Using GIS, a program called RADIUS has been resulted from an initiative of the United Nation's International Decade for Natural Disaster Reduction (IDNDR) with financial assistance from the Japanese Government. This program has been carried out for seismic risk assessment of nine different cities worldwide (Ababa in Ethiopia; Antofagasta in Chile; Bandung in Indonesia; Guayaquil in Ecuador; Izmir in Turkey; Tashkent in Uzbekistan; Skopje in Macedonia; Tijuna in Mexico; and Zigong in China). In USA, the most popular GIS based system is the HAZUS99, which was originally proposed by the Federal Emergency Management Agency (FEMA, 1999). Such system has also been found useful in assessing seismic risk in other countries, such as Turkey. The evaluation of seismic damages in HAZUS99 bases primarily on ACT-13 (ATC, 1985) through the use of damage probability matrix.

Although HAZUS99 has been found useful in USA and other some countries, it cannot be used directly for the case of Hong Kong. One main problem is that the available vulnerability matrices in the HAZUS99 are not suitable for Hong Kong. In particular, there are for 16 different building types and, according to different story number, these are further subdivided into 36 building categories. According to this system 1-3 story buildings are identified as "low-rise", 4-7 story buildings are identified as "mid-rise", and 8 story or up buildings are identified as "high-rise". In Hong Kong, over 90% of the buildings are more than 8 stories. Probably, buildings over 26 stories in Hong Kong will be considered as "high-rise". Even so, many of these so-called buildings are of different structural forms and different design philosophy. Another issue of using HAZUS99 GIS system is that it is location dependent. That is, all local digital maps and building inventory are unique for different locations. Therefore, there is no such thing as simply modify the existing HAZUS99 system from USA to Hong Kong. All GIS programming has to be started from scratch.

In view of this, there is no advantage of using HAZUS99 system over any other system. More importantly, the vulnerability matrices of building available from HAZUS99 cannot be used. Thus, in this study a new GIS system (HKEHAS) is developed for Hong Kong. Limited by time and resources constraint, the present study only cover two districts in Hong Kong, namely, Wanchai and Tsim Sha Tsui (TST) East. These two areas are chosen because there are many different variety of buildings within these two areas. They include old pre Second World War buildings, medium age and mid-rise residential buildings (from 10 to 20 stories), mid-rise to high-rise new commercial buildings (including the tallest building in Hong Kong—The Central Plaza), sport stadium, temples and churches, markets, museums and schools.

Regarding the calculation of the seismic vulnerability of existing buildings in Hong Kong, there is a high uncertainty involved. All Hong Kong buildings are designed according to wind loads, while no special attention paid to reinforcement detailing. Typical new commercial and residential buildings are highly non-symmetric in both vertical stiffness distribution and structural forms, due to architectural requirement and the use of transfer-plate system to open up lower floor space for shopping mall or parking. The demand and capacity curve concept used in HAZUS99 assumes all buildings as single-degree-of-freedom (SDOF) oscillators, which is clearly too restrictive for most Hong Kong buildings. At this moment, it seems that no reliable and mature mathematical models available for estimating the earthquake resistance of buildings built according to wind code with no seismic provisions. Clearly, much works need to be done in this area.

In this study, we do not attempt to solve this complicated problem completely. Instead, a very simple approach developed by Yin (1995) will be adopted. The main feature of such approach is that the vulnerability of the structures depends on the total sectional areas of load bearing columns and shear wall. For the case of Hong Kong, the stiffness and ductility of wind-designed buildings can be taken into account appropriately. Another beauty of the present approach is that the local "high-rise" buildings are modeled more realistically as multi-degree-of-freedom (MDOF) oscillators, instead SDOF oscillators. A n-story building will be modeled as a n-DOF oscillator. The maximum seismic shear demand at each story level can be estimated and checked against the story shear capacity, estimated from the total load-bearing sections of columns and walls. For simplicity, only a small number of building classifications have been used here.

Nevertheless, the present study provides the first preliminary seismic risk and loss analyses using a GIS based system for Hong Kong. In addition, the present approach is quite compatible that applied to the seismic damage-loss estimation and disaster mitigation of many mainland Chinese cities, such as Urumqi (the capital of the Xinjiang Uygur Autonomous Region). The only differences arise from the vulnerability calculation and building categorization. One of the beauties of the GIS system is that the building inventory can be changed with time, the property price as well as the population can be revised, and, more importantly, a more accurate prediction for the damage probability matrix can be incorporated at any time. It is hoped that the present study can provide a framework for future seismic risk and loss assessment and management of Hong Kong.

FRAMEWORK OF SEISMIC DAMAGE AND LOSS ESTIMATION

Seismic hazard and risk are related but distinct concepts that need explanation. In earthquake engineering, hazard is related the probability of exceeding certain level of ground motion at a site, while risk is the probability of damage of a building exposed to and subjected to such hazard. If a building is more vulnerable to earthquake attack, it is more likely to suffer damages during earthquake. In mathematical terms, we have (Hu, 1998; Hu et al., 1996):

$$SR = SH(P/I) * \sum\left[VUL(D/I) * (DL(D) + SL(D)) * (1 + ID) * EXP\right] \tag{1}$$

where

EXP = exposure or everything subject to the attack of the earthquake (e.g. all items in natural, man-made, engineering and social environments);
SR= seismic risk in terms of damage-loss of the exposure;
SH(P/I) = seismic hazard on the exposure, in terms of probability P of the earthquake intensity I;
VUL(D/I) = vulnerability of the exposure, in terms of probability of damage D for intensity I;
DL(D) = direct loss-damage transfer matrix, as a function of damage D;
SL(D) = secondary or induced loss of the exposure item, as a function of damage D;
ID = indirect loss due to DL and SL (e.g. number of injury, number of death and monetary loss);
* = convolution operator or matrix multiplication;
Σ = summation over all items of exposure in an area.

However, the risk analysis conducted in the present study concerns only buildings and considers only direct economic loss due to building damage. The comprehensive frame of damage and loss estimation methodology can be outlined in Figure 1. As shown in Fig.1, the framework of risk analysis includes three components: seismic hazard (including the site condition), exposure inventory (or buildings), and vulnerability analysis (structural behavior). The final damage stages of large amount of buildings of different categories at different site conditions can be presented and managed by a GIS, which will be discussed in later section.

First, the building inventory is classified into different categories. Once the ground motion parameters are chosen, the corresponding vulnerability matrices will be assigned to each building category according to the risk level specified, far or near field earthquake, ground parameters, and site condition. For seismic load input, we can use either probabilistic or deterministic approaches. Each of these will be discussed briefly next.

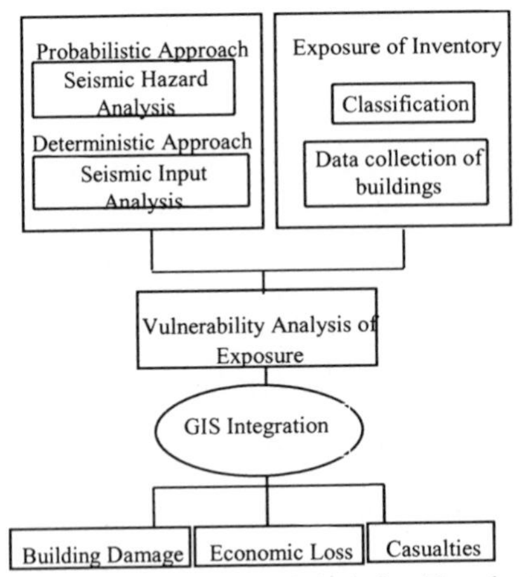

Figure 1: Seismic loss Calculation Flowchart

SEISMIC RISK BY DETERMINISTIC SCENARIO EARTHQUAKES

For scenario earthquake input, one has to assume the location, magnitude, and focal mechanism of an earthquake. The attenuation relation for either intensity or ground motions should also be input, together with the local site characteristics period and soil conditions. A response spectrum associated with the estimated macroseismic intensity or magnitude of shaking should also be included. The probability distribution of building damage at different levels can be determined by considering the structural response modeled by MDOF oscillators. In the present study, we follow the most commonly adopted 5-lvevl of damages stages. These damage stages are shown in Fig. 2.

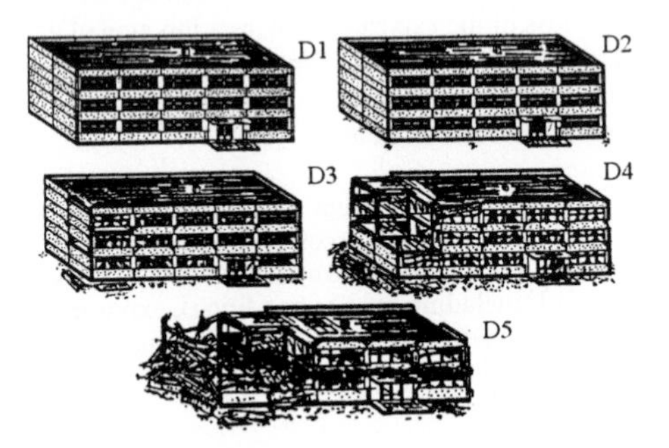

Figure 2: Five levels of damage stages, defined as: D1=none; D2=slight; D3=moderate; D4=extensive; and D5=complete.

The user can either create an event based on a historical earthquake epicenter, create an event from a map of potential source zones around Hong Kong, or create a scenario event by selecting a completely arbitrary epicenter. To help the user, we have input the potential seismic sources around Hong Kong in our GIS, together with the maximum creditable earthquake magnitude (see Fig. 3). Then, the ground-shaking hazard is displayed as an intensity map over the studied region for a prescribed scenario earthquake. For a given event, attenuation relation by Wang et al. (1988) is used to calculate ground shaking macroseismic intensity.

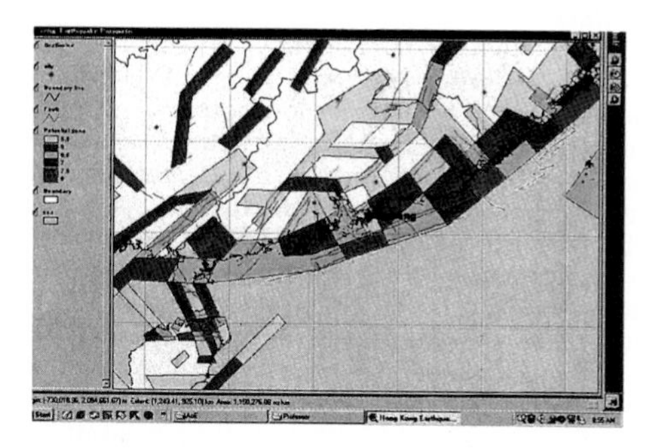

Figure 3: Potential seismic sources displayed in our GIS system HKEHAS.

Seismic vulnerability analysis

The reliability of a seismic risk analysis depends mainly on the vulnerability analysis for buildings. In this study, we modify Yin's (1995) method to obtain the damage probability matrices $P(D_j|I)$ in terms of intensity I, allowing for variations of local site conditions. The site-specific design response spectrum and inelastic building capacity was developed to describe the building damage (Wen et al, 2002). In particular, with the estimated ground shaking or macroseismic intensity and given soil condition, a simple model is used to estimate distribution of the ductility factor μ, which is estimated from the ratio of story yield shear to inter-story seismic shear. Note that $\mu > 1$ for inelastic deformation of buildings. The full details of this method are referred to Wen et al. (2002). These predicted nonlinear responses of building are used to obtain probability of the extent of damage to buildings. These damage estimates are expressed in terms of the probability $P(D_j|I,SC)$ that building is in one of the five damage states: none, slight, moderate, extensive, and complete damage. Sometimes, it is more information to express the probability of damages in a cumulative from:

$$P^*(D \geq D_j) = 1 - \sum_{i=1}^{j-1} P(D_i / I, SC) \tag{2}$$

The plots of $P^*(D \geq D_j)$ are normally referred as fragility curves.

Estimation of economic losses method

After completing the vulnerability analysis by combining the inventory and hazard, direct loss due to building damage, economic loss and casualties can be evaluated. In particular, the preliminary economic losses from the earthquake can be calculated by simple summing all building loss in the region. The losses are the results of combining the PDM (probability damage matrices), damage ratio, and values of the buildings. The expected economic loss of the buildings is obtained by

$$EL = \sum_{I} \sum_{D_j} P[D_j|I, SC] \cdot b_s(D_j) \cdot B_s \tag{3}$$

where $P[D_j|I,SC]$ is probability of building experiencing damage state D_j for intensity I and on the site condition SC; b_s is mean damage factor (percent) for building experiencing the damage state D_j. In the present study, the mean damage factors for none, slight damage, moderate, extensive and complete damages are set as 0%, 7.5%, 25%, 55%, 100% respectively, following Ying (1995). The value of the building B_s depends on the market price of the building inventory, which can be adjusted easily in our GIS system HKEHAS.

Estimation of human casualties

The estimation of human casualties in seismic events is one of the main objectives of seismic risk studies. Deaths and injuries are perhaps the most important to government. Protection of life is a primary function of a government and a prime incentive for undertaking hazard reduction. For most civilized societies, it is normally unacceptable to tolerate a very high fatality rate due to any earthquake event. Unfortunately, the ability to predict casualties is not as good as in the case of damage and direct economic loss. Data on which rational, systematic estimates can be made are very sparse (NRC, 1989). There are various ways to predict the number of casualties. One approach is that casualty rates are tied empirically to the damage states of buildings. This methodology can be used on the condition that there are detailed population distributions in the buildings available. Other approach to predict the casualty is to use historical rates of casualties per unit of population for wood-frame dwellings and extrapolated such data for other types of construction. In this way, a city-wide casualty rate can be established from previous earthquake data (Fu and Li, 1993). In the present work, since no detailed population distributions within each category of building are available for Wanchai and TST East, the estimation of fatality is carried out by means of methodologies developed Fu and Li (1993) based on the total population in the studied region. It is assumed that there are only two conditions of casualty from earthquake disaster, i.e., $m_j = 1,2$ corresponding to the conditions of the injuries and deaths respectively. The expected earthquake casualty $R(m_j)$ can be calculated form casualty probability matrix by:

$$R(m_j) = \sum_I P[m_j|I] \cdot B_p \tag{4}$$

where $P[m_j|I]$ is casualty probability matrix and B_p is total population in the region of study. Based on Yin (1995), the following values have been assumed for the injured at various seismic intensity: 5.4×10^{-4}(VI), 53×10^{-4}(VII), 460×10^{-4}(VIII), 4000×10^{-4}(IX); for death rate at various seismic intensity: 0.14×10^{-4}(VI), 3.1×10^{-4}(VII), 48×10^{-4}(VIII), 680×10^{-4}(IX).

SEISMIC RISK BY PROBABILISTIC ANALYSIS

The second type of risk analysis uses a probabilistic approach. The probabilities for various levels of loss being exceeded are estimated. It is important to realize that a particular loss data may result from earthquakes of different characteristics, for example, from a nearby earthquake of moderate size or from a distant earthquake of larger magnitude. Except for the initial phase of analysis, the loss produced by probabilistic seismic hazard is estimated using a procedure similar to scenario earthquake estimate, and therefore the details will not be repeated here. Only main differences will be discussed.

Probabilistic seismic hazard analysis provides a framework in which the uncertainties in size, location, and rate of recurrence of earthquakes can be avoided and results presented in a more rational manner to provide a more complete picture of the seismic hazard (Kramer, 1996). In the probabilistic approach, ground shaking may be characterized by response spectrum. The methodology includes ground shaking at three different hazard levels: 63%, 10% and 2% probability of being exceeded in 50 years. These probabilities of exceedance correspond to a return period of 50, 475 and 2475 years respectively.

In the present study, the probability distribution of building damage among different damage states for reinforced concrete structures are given as a function of response spectrum. The equivalent static force method is used to estimate the ductility factor distribution of different types of buildings at a specified probability. Similar to the scenario case, the distributions of ductility factor for none, slight, moderate, extensive and complete damages to building are denoted by $P[D_j|S_a(T)]$, with the only difference being using response spectrum as input. Similarly, the probable damage to buildings can be expressed as exceedance probability $P*(D \geq D_j)$. The estimation of economic loss is again very similar and can be given as

$$EL = \sum_{D_j} P[D_j|S_a(T)] \cdot b_s(D_j) \cdot B_s \tag{5}$$

where $P[D_j|S_a(T)]$ is probability of building experiencing damage state D_j in case of ground shaking $S_a(T)$; and all other parameters are the same as those given for equation (3).

Again, the data can be used for predicting deaths and injuries are very sparse, and considerable judgment is necessary in organizing available information to estimate casualties. There is no reliable and well accepted relation between casualty rate and probabilistic response spectrum which specifies the level of seismic hazard since the casualty rate must also depend on the structural response. In this study, we will simply correlate the designed peak ground acceleration to macroseismic intensity by using empirical relation adopted by CSB. It should be pointed that the conversion is very crude, and it is intended to yield some ideas only.

HKEHAS—A GIS SYSTEM FOR RISK ASSESSMENT

For hazard and disaster mitigation and management, there is a need to manage all data of potential damages in a graphical and user-friendly way for planning. Geographical Information System (GIS) has been a popular choice for seismic risk management, as it has been used for HAZUS99 and other systems. In the present study, ArcView GIS system of the ESRI was used since the digital map of the Hong Kong is available under software. All programming is made primarily using "Avenue" of ArcView. We will not, however, discuss the architecture as well as the development of our system HKEHAS (stands for Hong Kong Earthquake Hazard Analysis System).

A total of 1431 buildings have been included in our database system within Wanchai and TST East. All building data include the story number, the year of building, the usage of the building, and the site condition. Buildings are classified as residential, commercial, school, hotel, and museum. For residential buildings, 3 sub-categories are 1-7, 8-16 and 17-50 story; for commercial buildings, 3 sub-categories are 1-16, 17-40 and 41-80 story; for hotels, 2 sub-categories are 4-16 and 17-50. Therefore, a total of ten categories of buildings are used. More idealized categorization should consider the structural responses of buildings under seismic loads. With the limited knowledge on how Hong Kong buildings may perform under seismic attack, we cannot classify our buildings according to structural performance or damage potentials. Further works are required in this direction.

More detailed parametrical studies have been done, but we also aware of the fact that there is large uncertainty regarding the vulnerability of existing buildings in Hong Kong. Thus, these analyses will not be discussed here. In fact, the most important contribution of the present work lies on the introduction of a new GIS system for Hong Kong. Once a more reliable theoretical model is developed for estimating the vulnerability matrix of the existing Hong Kong buildings, the system can be updated by simply replacing the old vulnerability matrix with a more reliable one.

CONCLUSIONS

This paper has briefly described a procedure for calculating regional damage, loss and casualty, integrating information on regional earthquake hazard, site condition, building inventories, population inventories, and building damage functions by a new GIS system under the ArcView platform. Application of the procedure to Wanchi and TST East in Hong Kong has provided preliminary results that show the potential of the present GIS system—HKEHAS. This system provides the first ever GIS based system for seismic risk management for Hong Kong, and which will be extremely useful for policy maker of government and insurance companies in the future.

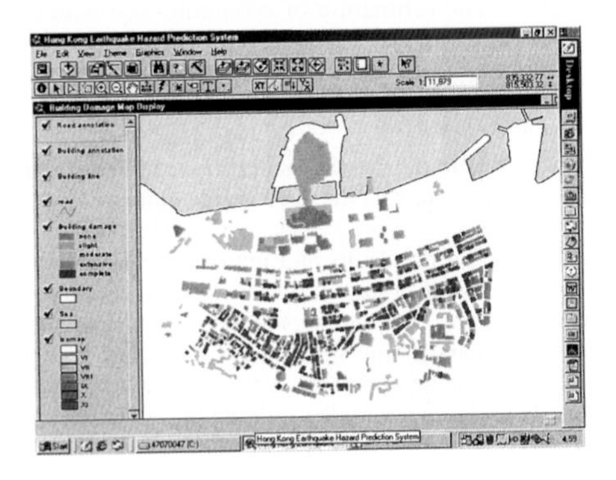

Figure 4: A typical display of the probable damage distribution for Wanchai

ACKNOWLEDGEMENTS

The research was supported by ASD funding of the Hong Kong Polytechnic University through Project A214—Earthquake Risk Assessment and Reliability Analysis Using GIS. We are grateful to Centaline Property Agency limited for providing a preliminary database for Hong Kong building inventory.

REFERENCES

ATC (Applied Technology Council) (1985). Earthquake Damage Evaluation data for California (ATC-13) Redwood City, CA.

FEMA (Federal Emergency Management Agency) (1999). Earthquake Loss Estimation Methodology, HAZUS 99: Technical Manual. Washington, D.C.

Fu Z. and Li G. (1993). *Report on Casualties Due to Earthquakes.* Seismological Press.

Hu, Y. (1998). *Earthquake Disaster Mitigation & Engineering Seismology.* IUGG/SSB International raining Course on Continental Earthquakes and Seismic Hazard, 70pp., Beijing, China.

Hu Y., Liu S.C. and Dong W.M. (1996). *Earthquake Engineering,* E & FN SPON, Chapman & Hall.

Kramer S.L. (1996). *Geotechnical Earthquake Engineering.* New Jersey: Prentice-Hall

NRC (National Research Council) (1989). Estimating losses from further earthquakes. (Washington, DC: National Academy Press)

Tasumi Y., Sugimoto M. and Seya. H. (1992) A seismic damage evaluation system for buildings. *Proc. 10th World conference of Earthquake Engineering.*

Wang S.Y., Liu H.Y., Wang Y.C. and Wu C.Y. (1988). The attenuation relationship for ground shaking parameters in the region of southern China. *Publication as Theme of Earthquake Research for Northern Hainan Island.* Seismological Press (in Chinese), 1988, 284-293.

Wen Z.P., Hu Y.X. and Chau K.T. (2002). Site effect on vulnerability of high-rise shear wall-buildings under near and far field earthquakes. Submitted to *Soil dynamics and Earthquake Engineering.*

Yin Z. (1995). *The Method of Seismic Damage and Loss Prediction.* Beijing: Seismological Press (in Chinese).

Advances in Building Technology, Volume 1
M. Anson, J.M. Ko and E.S.S. Lam (Eds.)

RULES OF DRIFT DECOMPOSITON AND DRIFT-BASED DESIGN OF RC FRAMES *

Luo Wenbin   Qian Jiaru

Department of Civil Engineering, Tsinghua University
Beijing, 100084, CHINA

ABSTRACT

In this paper, a drift-based design approach for RC frames is presented. Introducing the collapse mechanism factor χ as a pivotal indicator of the desirable collapse mechanism, the rules of drift decomposition are developed. By decomposing the target drift of the structure, the deformability demands of the columns and beams are determined. For detailing the columns and beams to meet the deformation demands, the $\theta_c^u - n - \lambda_{sv}$ diagram and $\mu_{b\delta} - \varepsilon_{cu} - \lambda_{sv}$ relations are proposed.

KEYWORDS

Drift decomposition; Drift-based design; Deformability; RC frame

INTRODUCTION

As is known, the displacement-based design approach begins with establishing a target displacement which is consistent with the selected performance objectives of the structure subjected to the expected intensity level of earthquake, and the capacity of the structure should conform to the anticipated target displacement or drift. It is now believed that the inter-storey drift ratio is a preferable performance criterion of RC frames. However, in realistic practice of the displacement-based seismic design of RC frames, three key issues must be solved. One is to predict the target drift ratio of the frame. The second

---

* Supported by the National Natural Science Foundation of China (No. 59895410)

is to determine the relationship between the drift ratio and the deformation of each member. The third is to ensure the deformability of members to meet the deformability demand when the frame reaches the target drift ratio.

The objectives of this study are to establish the drift decomposition rules for RC frame, and to develop the deformability design method for columns and beams. Then a drift-based design procedure is proposed.

TARGET DRIFT OF FRAME

For multi-storey frames, assuming that the anti-flexural point of beam or column is located at its mid-span or its mid-height, the inter-storey drift can be represented by the deformation of a typical beam/column sub-assemblage. As shown in Figure 1, the drift ratio θ of the frame is defined as the ratio of the inter-storey drift to the height of the storey, as follows:

$$\theta = \delta / h \tag{1}$$

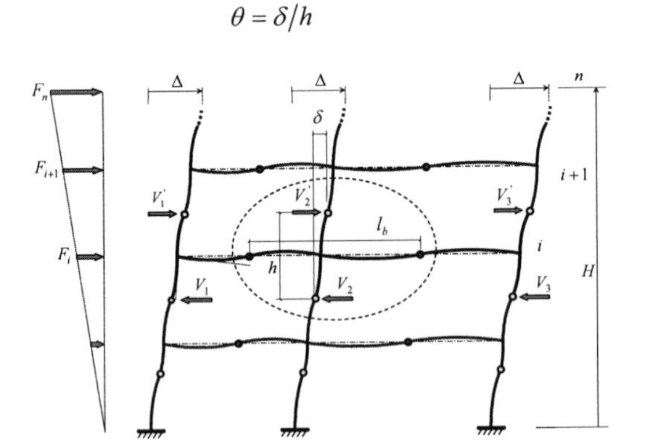

Figure 1: drift of frame and sub-assemblage

Based on the analysis of extensive experimental database of beam/column sub-assemblage, Priestley (1998) suggested an expression for the yield drift ratio θ_y in the form:

$$\theta_y = 0.5 \cdot \varepsilon_y \cdot (l_b / h_b) \tag{2}$$

in which, ε_y is the yield strength of longitudinal reinforcement of beam, l_b and h_b is the span length and the section depth of beam respectively. From Eqn.(2), the yield drift δ_y can be obtained.

Then the target drift ductility ratio μ_δ can be estimated as:

$$\mu_\delta = \delta_u / \delta_y \tag{3}$$

where δ_u is the target drift corresponding to the selected performance objectives. Many researches have been focused on the determination of the target drift or the target displacement, more detailed discussion can be found in Whittaker (1998).

RULES OF DRIFT DECOMPOSITION

Components of drift

Figure 2 shows that the drift of RC frames is the sum of drift due to the deformation of related beams, columns and joint, and the drift is composed of elastic drift and inelastic drift. Each of them could be expressed as:

$$\delta_y = \delta_b^y + \delta_c^y + \delta_{jV} \tag{4}$$

$$\delta_p = \delta_b^p + \delta_c^p + \delta_{js} \tag{5}$$

where δ_y and δ_p are yield drift and inelastic drift respectively, δ_b^y and δ_c^y are the contribution of elastic deformation of beam and column to δ_y, δ_b^p and δ_c^p are the contribution of inelastic deformation of beam and column to δ_p, δ_{jV} and δ_{js} are the contribution of shear deformation of joint core and bond slip of longitudinal reinforcements in core region to δ_y and δ_p respectively. The contribution of each component depends on the relative stiffness, relative strength and deformability of the beam, column and joint core.

Target Ductility Ratio of Members

According to the definition, the contribution of column and beam deformation to the target drift ductility ratio of the frame could be expressed as:

$$\mu_{c\delta} = \frac{\delta_c^y + \delta_c^p}{\delta_c^y} \Rightarrow \delta_c^p = (\mu_{c\delta} - 1)\delta_c^y \tag{6}$$

$$\mu_{b\delta} = \frac{\delta_b^y + \delta_b^p}{\delta_b^y} \Rightarrow \delta_b^p = (\mu_{b\delta} - 1)\delta_b^y \tag{7}$$

Given

$$\alpha_b \equiv \frac{\delta_b^y}{\delta_y}, \alpha_c \equiv \frac{\delta_c^y}{\delta_y}, \alpha_{js} \equiv \frac{\delta_{js}}{\delta_p} \tag{8}$$

Substituting Eqns. (6) and (7) into Eqn. (5), gives:

$$\delta_p = \frac{(\mu_{b\delta} - 1)\alpha_b + (\mu_{c\delta} - 1)\alpha_c}{(1 - \alpha_{js})}\delta_y \tag{9}$$

The drift ductility ratio of the frame can be written as:

$$\mu_\delta = \frac{\delta_y + \delta_p}{\delta_y} = 1 + \frac{(\mu_{b\delta} - 1)\alpha_b + (\mu_{c\delta} - 1)\alpha_c}{(1 - \alpha_{js})} \tag{10}$$

Defining the collapse mechanism factor χ to indicate the contribution ratio of column plastic deformation to beam plastic deformation:

$$\chi \equiv \frac{\delta_{cp}}{\delta_{bp}} = \frac{(\mu_{c\delta} - 1)\alpha_c}{(\mu_{b\delta} - 1)\alpha_b} \tag{11}$$

For RC frames designed according to the requirements of "Code for seismic design of buildings" (GBJ50011), with column overstrength factor η_c of 1.4, analytical results reveal that the collapse mechanism factor χ is about 0.24 when drift ductility ratio reaches 4. Based on a series of analysis, the relations between the collapse mechanism factor and column overstrength factor at different ductility levels are shown in Figure 3.

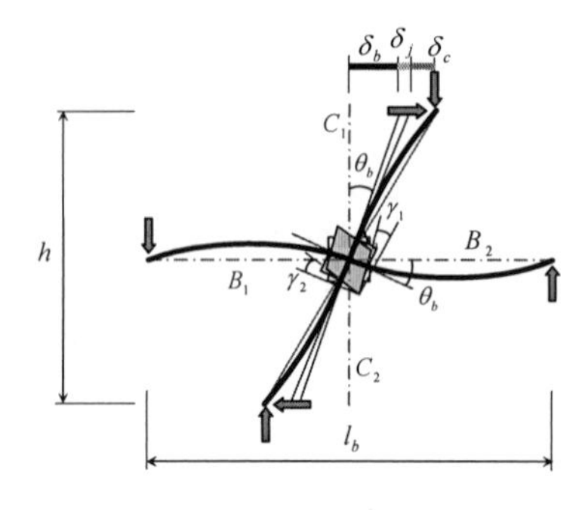

Figure 2: Drift decomposition of beam/column assemblage

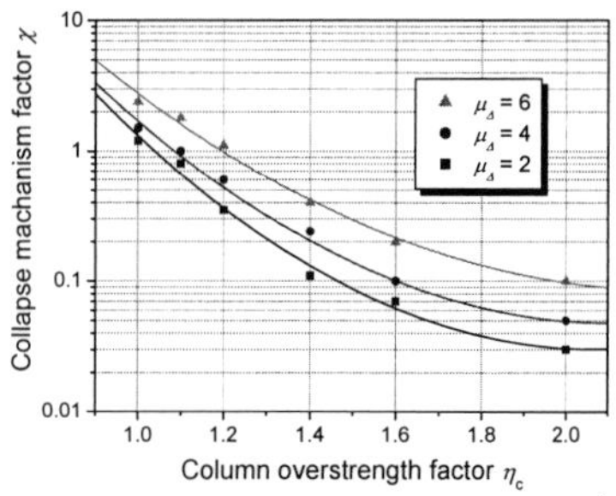

Figure 3: Relationship of $\chi - \mu - \eta_c$

The contribution of column and beam to the drift ductility ratio is obtained by substituting Eqn. (11) into Eqn. (10):

$$\mu_{c\delta} = 1 + \frac{(1-\alpha_{js})}{(1+1/\chi)\alpha_c}(\mu_\delta - 1) \tag{12}$$

$$\mu_{b\delta} = 1 + \frac{(1-\alpha_{js})}{(1+\chi)\alpha_b}(\mu_\delta - 1) \tag{13}$$

Clearly, Eqns. (12) and (13) show that the relationship between the contribution of members and the drift ductility ratio of frame is related to $\alpha_b, \alpha_c, \alpha_{js}$ and η. For a code based beam/column assemblage, the test results indicated that $\alpha_b = 56\%$, $\alpha_c = 19\%$, $\alpha_{js} = 13\%$. In this case, for the above mentioned frame with $\chi = 0.24$, the ductility ratio demand of beams would be:

$$\mu_{b\delta} = 1 + (1-0.13)(\mu_\delta - 1)/0.69 = 1.26\mu_\delta - 0.26 \tag{14}$$

It is found that the collapse mechanism factor χ, which depends on the column overstrength factor and drift ductility of the frame, is a very important parameter for determining the ductility demand and target deformability. Practically, coefficients α_b and α_c can be calculated by the elastic analysis, and α_{js} could approximately be taken 10% for normally proportioned beam-column joint.

DEFORMABILITY DESIGN OF MEMBERS

After the collapse mechanism factor χ is selected according to the expected collapse mechanism, with the target drift ductility ratio of the assemblage, the ductility demand of columns and beams can be evaluated by Eqns. (12) and (13).

Deformability of column and $\theta_c^u - n - \lambda_{sv}$ equations

According to the ductility demand of column $\mu_{c\delta}$ and the yield drift of column δ_c^y in the form of

$$\delta_c^y = \phi_{cy} h_c^2 / 3 \tag{15}$$

the target drift ratio of column θ_c^u is given by:

$$\theta_c^u = \frac{\mu_{c\delta}\delta_c^y}{h_c} = \frac{\mu_{c\delta}\phi_{cy}h_c}{3} \tag{16}$$

where ϕ_{cy} is the yield curvature of column section, h_c is the clear height of column.

The deformability of RC columns mainly depends on the axial compression ratio and transverse reinforcement, which is expressed by $\theta_c^u - n - \lambda_{sv}$ relations in this study. As shown in Eqn. (17), Luo & Qian (2001a) suggested an empirical expression based on the statistic analysis of the column test database, which collected hundreds of RC column test results worldwide. For columns in the database, the axial compression ratio ranged from 0 to 0.9 and the characteristic value of transverse reinforcement was 0.05~0.35. To establish Eqn.(17), the ultimate drift of column was taken as the drift corresponding to the point of strength decreasing 15% from peak strength.

$$\theta_c^u = \left(200 - 100\lambda_{sv}\frac{40}{1+10n}\sqrt{\frac{A_{co}}{A_g}}\right)^{-1} \tag{17}$$

where n is the axial compression ratio, defined as $n = N_c / f_c bh$, N_c is the axial compression force, f_c is the axial compression strength of concrete, b and h are the width and depth of column section respectively. λ_{sv} is the characteristic value of transverse reinforcement, $\lambda_{sv} = \rho_{sv} f_{yh} / f_c$. ρ_{sv} is the volumetric ratio of transverse reinforcement. A_g and A_{co} are the column gross section area and the confined core area respectively.

From Eqn.(17), the characteristic value of transverse reinforcement is obtained:

$$\lambda_{sv} = \left(\frac{200 - 1/\theta_c^u}{100}\right)\left(\frac{1+10n}{40}\right)\sqrt{\frac{A_g}{A_{co}}} \tag{18}$$

Figure 4 illustrates the above equation by $\theta_c^u - n - \lambda_{sv}$ diagram, by which the required characteristic value of transverse reinforcement of column with different axial compression ratio and with different drift ratio can be determined. For instance, a column with cross section dimensions of 800mm×800 mm, cover thickness of 25 mm, C40 concrete and transverse reinforcement yield strength of 310Mpa, the required λ_{sv} values are listed in table 1.

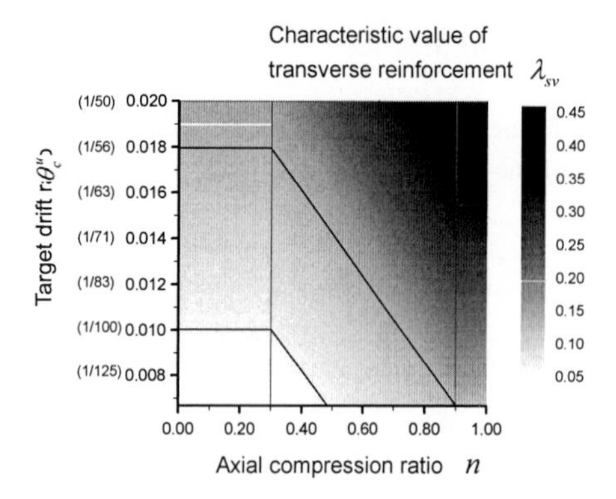

Figure 4: $\theta_c^u - n - \lambda_{sv}$ diagram for RC column

TABLE 1 REQUIREMENTS OF TRANSVERSE CONFINEMENT λ_{sv}

Target drift ratio of columns θ_c^u	Axial compression ratio n					
	0.4	0.5	0.6	0.7	0.8	0.9
1/100	0.15	0.18	0.21	0.24	0.27	0.30
1/120	0.13	0.15	0.17	0.19	0.22	0.24
1/150	0.08	0.09	0.11	0.12	0.14	0.15

Deformability of beam and $\mu_{b\delta} - \varepsilon_{cu} - \lambda_{sv}$ ***relations***

The deformation capacity of RC beams directly depends on the ultimate compression strain of concrete, while the ultimate compression strain mainly depends on the characteristic value of transverse reinforcement.

To assure deformation capacity of beams, there is a need to determine the required ultimate compression strain of confined concrete and then to numerically calculate the transverse reinforcement requirement. The former could be obtained as shown in Eqn. (19). More detailed sources could refer to Luo and Qian (2001).

$$\varepsilon_{cu} = \frac{1.7\varepsilon_{sy}(A_s - A_s')f_y}{k_1 f_c b h_b}\left[1 + (\mu_{b\delta} - 1)\frac{l_b^2}{3l_p(l_b - l_p/2)}\right] \tag{19}$$

In which ε_{sy} is the yield strain of the longitudinal reinforcement of the beam. l_p is the equivalent plastic hinge length. The coefficient k_1, as mentioned by Yie (2000), has the commendatory value listed in Table 2.

TABLE 2 THE VALUES OF COEFFICIENT k_1

Concrete strength	≤C50	C60	C70	C80
k_1	0.797	0.774	0.746	0.713

Applying relations between ultimate compression strain of confined concrete and characteristic value of transverse reinforcement proposed by Qian and et al. (2001), and with a little transfer, the requirement of transverse reinforcement is:

$$\lambda_{sv} = (\varepsilon_{cu} \times 10^3 - 4.2)/28.6 \tag{20}$$

DRIFT BASED DESIGN PROCEDURE FOR RC FRAMES

The procedure of drift based design for RC frames consists of the following steps:

(1) Designate the inter-storey drift demand δ_u or target drift ratio θ_u for the frame.

(2) Calculate the yield drift θ_y, and determine the required drift ductility ratio μ_δ.

(3) Determine the coefficient α_b and α_c by means of elastic analysis. Evaluate the collapse mechanism factor χ consistent with the expected yield and collapse mechanism. Estimate the required drift ductility ratio of column $\mu_{c\delta}$ and beam $\mu_{b\delta}$ using Eqns. (12) and (13).

(4) Perform the deformability design of columns by application of $\theta_c^u - n - \lambda_{sv}$ diagram.

(5) Detailing design of beams using $\mu_{b\delta} - \varepsilon_{cu}$ equations and $\varepsilon_{cu} - \lambda_{sv}$ relations.

DISCUSSION AND CONCLUSIONS

The drift based design approach developed in this study provides the designers a practical method for proportioning the RC frame which has been designed to resist minor earthquake without structural element and non-structural element damage. As an initial indicator in this proposed approach, the target drift ratio of the structure is designated according to the performance objects of the structure subjected to moderate or severe earthquakes.

In this approach, the collapse mechanism factor χ is a pivotal indicator of the desirable collapse mechanism of the frame and it manages the distribution of plastic deformation among the components. The factor χ can be straightforward treated as a function of column overstrength factor and drift ductility ratio. For a RC frame, using the rules of drift decomposition to determine the ductility demand of components according to the target drift ductility ratio of the frame, then the $\theta_c^u - n - \lambda_{sv}$ diagram and $\mu_{b\delta} - \varepsilon_{cu} - \lambda_{sv}$ relation can be respectively employed in proportioning the columns and beams to meet the drift and ductility demand.

REFERENCES

Priestley M.J.N. (1998). Displacement based approaches to rational limit states design of new structures, *Proceedings of the XI ECEE - Invited lectures*, Paris, 317-335.

Qian J.R. and Luo W.B. (2001). Displacement Based Seismic Design of Building Structures, *Building Structures* **31:4,** 3-6.

Whittaker A, Constantinou M and Tsopelas P. (2001). Displacement estimates for performance -based seismic design, *Journal of Structural Engineering* ASCE **124:8,** 905-912.

Luo W.B. and Qian J.R. (2001). Drift Based Design of RC Frame Sub-assemblages, *Reports of Research Advance in Displacement Based Seismic Design*, Department of Civil Engineering at Tsinghua University, Beijing.

Lu X.L., Guo Z.X. and Wang Y.Y. (2001). Experimental Study on Seismic Behaviour of Beam -column Sub-assemblages in RC Frame, *Journal of Building Structures* **22:1,** 2-7.

Yie L.P. (2000). *Concrete Structures and Masonry Structures*, Tsinghua University, Beijing.

Qian J.R., Zhou D.L. and Cheng L.R. (2001). Axial compression behaviour of concrete confined with general stirrups, Accepted by *Journal of Tsinghua University*.

Advances in Building Technology, Volume I
M. Anson, J.M. Ko and E.S.S. Lam (Eds.)

SEISMIC FRAGILITY CURVES FOR WIND-DESIGNED-BUILDINGS IN HONG KONG

Z.P. Wen [1], K.T. Chau [2], and Y.X. Hu [1]

[1]Institute of Geophysics, China Seismological Bureau, Beijing, China
[2]Department of Civil and Structural Engineering, The Hong Kong Polytechnic University, Kowloon, Hong Kong, China

ABSTRACT

This paper presents a simple analytical method to generate seismic fragility curves for low-rise to high-rise buildings in Hong Kong. In particular, the seismic vulnerability is assumed proportional to the sectional areas of the column and shear walls of the structure. Three wind-code-designed buildings were selected as examples for the present analysis, namely a 6-story school building by the Architectural Services Department, and a 21-story residential building at Mei Foo Sun Chuen and a 40-story residential building of the Hong Kong Housing Authority (HKHA). Instead of using intensity, seismic hazard is prescribed in terms of peak ground acceleration (PGA) in the present study. The adopted response spectra are site-dependent and depend on whether the design earthquakes are near field or far field. The seismic responses of buildings are analyzed using the equivalent static force approach for multi-degree-of-freedom-oscillators, expressed in terms of the fundamental site period and epicentral distances. The ductility distribution of buildings is then formulated as a log-normal distribution. Finally, the vulnerability of the structure is given as fragility curves. The results indicate that damage probability is highly sensitive to the number of story or building height, structural scheming, and functional usage of the buildings; other parameters such as site condition and epicentral distance are equally crucial to the structural seismic vulnerability. The conclusion should also be expected for other wind-designed buildings in Hong Kong and other parts of the world.

KEYWORDS

Fragility curves, Tall-buildings, Epicentral distance, Site condition, Near field and far field earthquakes, PGA input

INTRODUCTION

As reviewed by Docle et al. (1994), the most commonly adopted seismic input for vulnerability analysis is the macroseismic intensity (e.g. the Modified Mercalli Intensity or MMI). Less often, the seismic input is expressed in terms of ground motion parameters such as peak ground acceleration (PGA) or velocity (PGV). In the loss estimation methodologies, such as ATC-13 (ATC=Applied Technology Council) and FEMA-177 (FEMA=Federal Emergency Management Agency) used in the USA, intensity instead of PGA is still being used as the earthquake hazard. However, there is a recent trend of adopting PGA instead of intensity to quantify seismic hazard. In the last decade, many seismic codes in the world have been changed from specifying hazard in terms of intensity to specifying hazard in terms of PGA. For example, the newest Chinese seismic code GB 50011-2001 has adopted PGA contour map as hazard input comparing to the old code GBJ11-89 using intensity contour map as hazard input. In the newest seismic hazard evaluation methodology HAZUS99 (HAZUS=Hazards U.S.), PGA has been adopted as the hazard input. However, the most commonly used measure of the site hazard remains the macroseismic intensity in the vulnerability analysis of buildings, due to the lack of reliable PGA hazard map in many parts of the world.

For historical and pragmatic reasons, MMI has been used as ground-shaking measure in most earthquake loss studies conducted in the past. This procedure was popular in early loss studies because multiple, instrumental records of ground shaking were not available to correlate motion levels to damage. Even today, strong motion records at a particular site of a severely damaged building are rare. Thus, it remains difficult to statistically correlate damage distribution of buildings to PGA, PGV or response spectrum. Another reason is that intensity levels are partly defined by means of building damage data, it seems straightforward conceptually to predict future damage levels on the basis future expected intensity levels.

Despite all these obvious reasons, there are, however, conceptual and practical difficulties with the use of intensity as the principal hazard parameter in earthquake risk assessment. Intensity scales are essentially non-quantitative scales, using only a discrete scale system. Thus, intensity does not provide a satisfactory basis for the estimation of future losses (Coburn and Spence, 1992). The input parameters for the vulnerability analysis should be compatible with those from earthquake hazard analysis, which is typically in terms of PGA. By considering the frequency characteristics of ground motions, this paper further analyzes the effects of site condition and epicentral distance on fragility curves of low-rise to high-rise reinforced concrete (RC) wind-designed buildings in terms of PGA.

Past earthquake records clearly demonstrate that the effect of site conditions and source-to-site distance on damage probability distribution of buildings is very significant. Even in the 1920s, people realized from the damage data of buildings in the 1923 Kanto earthquake that rigid structures suffered heavier damage on the rigid ground and flexible structures suffered heavier damage on flexible ground. Similar observations on selective damages include those induced damages during the 1952 Kern earthquake of USA, the 1977 Vrancea earthquake of Romania, the 1985 Mexico earthquake, and the 1989 Loma Prieta earthqauke of USA (Hu, 1988; Hu et al., 1996). In general, short stiff structures suffer heavier damage under near field moderate earthquake; while, on the contrary, tall flexible structures suffer more damage under far field large earthquakes. Thus, the damages are highly selective in terms of both the fundamental frequency of structures and the frequency content of ground shaking. So both the frequency and amplitude characteristics of ground motions should be considered when the vulnerability analysis and loss estimation are conducted. This is precisely the aim of the present study.

FRAGILITY CURVES FOR VARIOUS SITES AND EPICENTRAL DISTANCES

The overall framework for evaluating the fragility curves of the building consists of three stages: input ground motion, seismic response analysis and probability distribution of building damage among different damage states. The PGA is resulted from seismic hazard assessment. The effect of soil condition and distance on ground motions is described by characterized period of ground motion. The structural responses are calculated by equivalent lateral force method. The probability distributions of damage states of building are predicted by damage functions, through the estimated ductility.

Input ground motions

It is known that local soil condition and seismic wave propagation can alter response spectra significantly. Therefore, an accurate modeling of ground motion should consider the frequency characteristics of ground motions. The dependence of response spectra on ground conditions is recognized in many earthquake-resistant design codes throughout the world (Paz, 1994; IAEE, 1996). For many years, the shape dependence of response spectrum on soil condition, at least approximately, was accounted for by using peak ground acceleration to scale down or up the design spectra in different frequency range (Newmark and Hall, 1982). Examination of the records from many earthquakes reveals that the distance of the recording station from seismic source can cause big difference in the spectral shapes at the rock stratum (Mohraz, 1992). The effects of both site condition and distance on shapes of response spectra are considered by in ATC-3, Romanian, Canadian and the building code for China when defining design spectra (site-dependent and earthquake-dependent spectra). In this study, the normalized response spectra S_a given in GBJ 11-89 are used (5% damping):

$$S_a(T)=\begin{cases} 2.\dot{2}a_{\max}(5.5T+0.45) & \text{for } T\leq 0.1\,\text{sec} \\ 2.\dot{2}a_{\max} & \text{for } 01.<T\leq T_g \\ 2.\dot{2}a_{\max}\left(\dfrac{T_g}{T}\right)^{0.9} & \text{for} T_g<\text{T}\leq 3.0\ \text{sec} \end{cases} \tag{1}$$

where T_g is given in GBJ 11-89 for various site categories and epicentral distances (i.e. far or near field earthquake); and a_{max} is the median of mean PGA expected at a site from seismic hazard analysis. To account for uncertainty, log-normal distribution of PGA is usually assumed.

Seismic shear of story

Applying the equivalent lateral force method, the shear force Q_x at the x-story is given by summing all lateral seismic forces above that story as

$$Q_x=\sum_{i=x}^{n}\frac{G_iH_i}{\sum_{j=1}^{n}G_jH_j}S_a(T)G_{eq}(1-\delta_n)+S_a(T)G_{eq}\delta_n \tag{2}$$

where $S_a(T)$ is the site-dependent response spectrum give in (2), which is derived from seismic hazard specified in terms of PGA $a_{\max}$; G_{eq} is the total equivalent weight of a structure; H_i and G_i are the height and weight at level i respectively; n is the total number of stories of the building; and δ_n is the additional seismic action coefficient which is intended to account approximately for the higher mode contributions and can be found from GBJ 11-89. In using the response spectrum at a site, the fundamental period T of the building is needed, and for the present case of RC frame/shear wall building we can estimate T (in s) by the following empirical formula (Guo, 1990):

$$T=0.33+0.00069\frac{H^2}{\sqrt[3]{B}} \tag{3}$$

where H and B are the height and length of the building (in m)along the shaking direction.

Distributions of ductility factor in terms of PGA

The capacity of each story is evaluated in terms of the yield shear force. For shear wall structures and frame structures with shear wall, the yield story shears are, respectively (GBJ 11-89)

$$Q_{yx} = 0.2F_C A_{wx} \text{ and } Q_{yx} = 0.25F_C A_{wx} \tag{4}$$

where F_C is compressive strength of concrete and A_{wx} is sectional area of shear walls which are parallel to the earthquake action in the x-story. Then, the yield shear coefficient of each story is calculated as:

$$R_x = \frac{Q_{yx}}{Q_x} \tag{5}$$

where Q_x and Q_{yx} are defined in (2) and (4) respectively. Many studies show that, in the case of multi-story frame structure with shear walls, nonlinear deformation will concentrate at the weakest stories (Newmark and Roseblueth, 1971; Yin et al., 1981), which corresponds to the minimum R_x among all stories. Note that the yield shear coefficient relates not only to the strength of the structure, but also to the characteristics of the seismic input. The next step to estimate the story ductility factor in terms of the minimum R_x, and subsequently to building damages. The story with minimum yield shear coefficient experiences the maximum deformation and the maximum ductility factor. By using the equivalent energy criterion (Veletsos and Newmark, 1960), the structural ductility may be evaluated from the following formula:

$$\mu_0 = \begin{cases} \dfrac{1+R^2}{2R^2} & R \leq 1 \\ \dfrac{1}{R} & R > 1 \end{cases} \tag{6}$$

where R is the minimum yield story shear coefficient calculated from (5). This empirical formula can further be refined by adding correction factors C_i (i=1,2,...,5) (Yin, 1995):

$$\bar{\mu} = \mu_0 \left(1 + \sum C_i\right) \tag{7}$$

How to apply these correction factors are discussed in Wen et al. (2002) and will not be reported here.

Naturally, a higher maximum mean ductility factor $\bar{\mu}$ also implies a more severely damaged building. But, due to the uncertainties involved in the estimation of seismic hazard as well as in the analysis of structural response, the ductility as well as the damage state is better represented in terms of probability distribution function. In particular, both the peak ground acceleration and seismic capacity of structures are often found satisfying a log-normal probability distribution versus the input ground parameters (Shibata, 1980; Kircher et al., 1997; Yamaguchi and Yamazaki, 2001; Yin et al., 1985). By analyzing 3120 cases of elastic-plastic seismic responses, Yin et al. (1985) suggested the following log-normal distribution for μ:

$$f(\mu)=\frac{1}{\sqrt{2\pi}\xi\mu}\exp\left[-\frac{(\ln\mu-\lambda)^2}{2\xi^2}\right] \tag{8}$$

where

$$\lambda=\ln\bar{\mu}-\frac{1}{2}\xi^2 \text{ and } \xi^2=\ln\left(1+\frac{\sigma^2}{\bar{\mu}^2}\right) \tag{9}$$

In these equations, $\bar{\mu}$ and σ are respectively the maximum mean value estimated from (7) and the standard deviation of the ductility factor of the stories. In this study, we assume that the uncertainty of the ductility factor mainly comes from the seismic hazard. Thus, the value of $\sigma/\bar{\mu}$ is taken from that of the attenuation relationship on peak ground acceleration. Note that this ductility distribution depends on the structural characteristics as well as the input ground motions through the calculation of $\bar{\mu}$ and variance. Using the threshold values of the ductility factor as integration limits, the probability of various damage states for a given ground motion can be integrated. For frame structures with shear walls, the threshold ductility factors for the onset of slightly damage, moderate damage, extensively damage and completely damage states are taken as 1.0,1.5,3.0 and 5.0 respectively (Yin, 1995).

Fragility curves

Identifying the relationship between ground shaking and distribution of damage is essential to vulnerability analyses (Hu, 1998). There are several ways in which this relationship about distribution of damage may be expressed and evaluated. One method for expressing the distribution of damage is to use damage probability matrix (DPM) which was introduced by Whitman et al. (1973) and DPM were developed in ATC-13 for 42 building classes for California (ATC, 1985). Fragility curves introduced by Kennedy et al. (1980) and Kircher and McCann (1983) provide essentially the same information as DPM does. Each curve gives the probability that the indicated damage states is equaled or exceeded as a function of the intensity of ground shaking, such as those in FEMA-177 (FEMA, 1989).

FRAGILITY CURVES FOR LOW-RISE TO HIGH-RISE BUILDINGS IN HONG KONG

In this study, three types of buildings are considered: a 40-story with RC frame apartment building of the HKHA, which contains a thick RC transfer plate located between upper apartment and lower department store or parking area; a 21-story RC frame/shear building of Mei Foo Sun Chuen in Hong Kong, which was built in 1960s and consists of two lower levels of car park and 19 levels of upper residential block; and a 6-story RC government primary school. All these building are designed according to the Hong Kong wind code with no seismic provision.

The fragility curves are constructed at various levels of PGA in different site conditions (I, II, III, IV) and epicentral distance (near or far). In this paper, we only show those plots for near field earthquakes in Figs. 1-3 due to page limit. The vulnerability of these buildings depends on the site condition as well as on the building types. In general, we also observe that high-rise buildings with longer fundamental period are more vulnerable from far field earthquakes than from near field earthquakes, and on soft site than on stiff site.

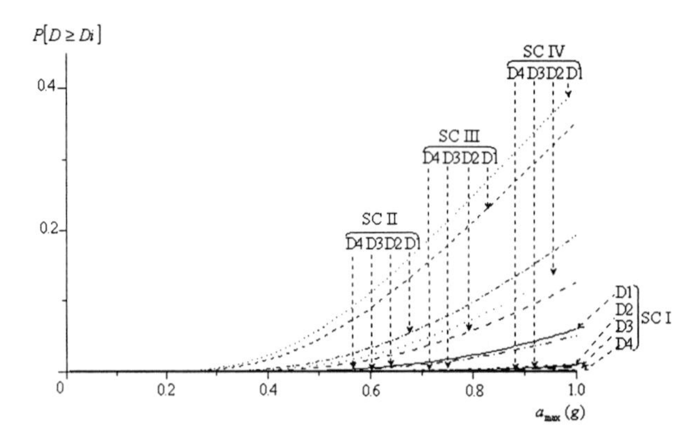

Figure 1: The fragility curves for the 6-story government school under near field earthquakes

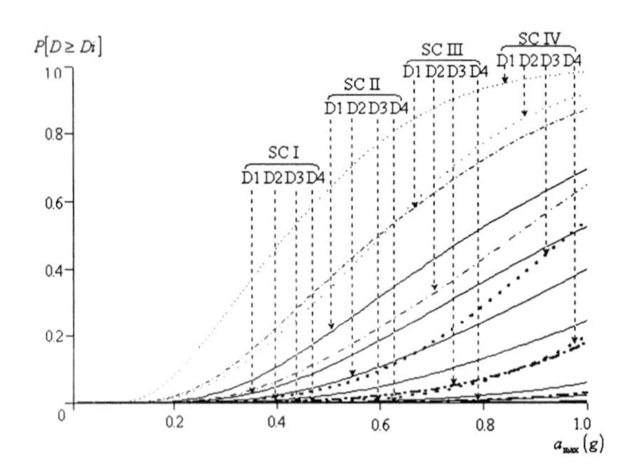

Figure 2: The fragility curves for the 42-story HKHA building under near field earthquakes

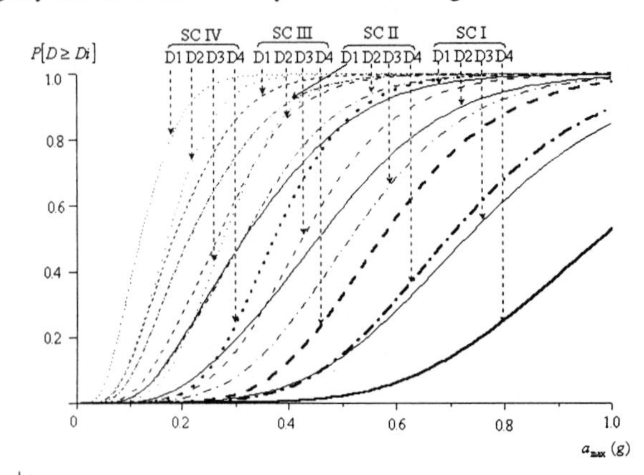

Figure 3: The fragility curves for the 21-story Mei Foo building under near field earthquakes

DISCUSSION AND CONCLUSION

We have proposed a simple model to consider the seismic vulnerability of a 6-story school, a 21-story residential building and a 42-story HKHA residential building, under various PGA input and site conditions. All these buildings are designed in compliance with the Hong Kong wind code with no seismic provisions. The fragility curves are given in terms of probability of various damage state or above versus the input PGA for various site conditions. The input acceleration spectrum is adopted from GBJ 11-89 and is site- and epicentral distance-dependent. The yield shear of each story of the building is checked against the shear force induced by various ground excitations to yield the probability of damages. The yield shear is assumed proportional to the total cross-sectional area of columns and walls at each level, while the seismic shear force is calculated based on an equivalent force method on a multi-degree-of-freedom-oscillator. In addition, the weight distribution and fundamental period of the building have also been incorporated into the fragility analysis. Uncertainty in both hazard and structural response is incorporated through an assumed log-normal distribution of the ductility of the building subject to certain ground motions.

For near field earthquakes, Figs. 1-3 demonstrate that the 6-story government school designed by the Architectural Services Department is the strongest against potential seismic shakings. The weakest building against earthquake attack is the 21-story Mei Foo Sun Chuen building, while the seismic performance of the 42-story HKHA building is between the 6-story school and the 42-story HKHA building. According to the HK wind code, if no topographic effect is considered, the design wind pressure for the 6-story school is from 1.2 to 2.2 kPa, that for 21-story building is from 1.2 to 3.0 kPa, and that 42-story building is 1.2 to 3.5 kPa. By enforcing a top floor drift of 1/500 in serviceability state under wind load, the bending stiffness EI of a typical Hong Kong building versus height L (in m) can be approximated as (Koo et al, 2002):

$$EI = 500(0.001256175L^2 + 0.2480387L - 0.45484)L^3 B \tag{10}$$

where B is the width of the building in m. Therefore, the stiffness (or the net sectional area of the columns) increases nonlinearly with the number of story of the building, but the seismic shear force also increases nonlinearly with height. Figures 1-3 illustrate that the medium-rise building is more vulnerable to seismic attack than both low-rise and high-rise buildings. In addition, we also find that site condition and epicentral distance play an important role in building damage and earthquake loss estimation, especially at high PGA. This conclusion should be true for buildings in other region of moderate seismicity in the world, at where buildings were designed according to wind loads instead of seismic loads.

ACKNOWLEDGEMENTS

The research was fully supported by ASD Project A202 and A214 of the Hong Kong Polytechnic University. Their support is gratefully acknowledged. The authors would like to thank Mr. Philip Kwok the Buildings Department of Hong Kong SAR Government in providing building information.

REFERENCES

Applied Technology Council (ATC) (1985). *ATC-13: Earthquake damage evaluation data for California.* Applied Technology Council, Redwood City, CA.

Coburn A. and Spence R. (1992). *Earthquake Protection.* Wiley, Chicheste.

Dolce M., Zuccaro G., Kappos A. and Coburn A.W. (1994). Report of the EAEE working group 3: vulnerability and risk analysis. *Proceedings of 10th European Conference on Earthquake Engineering*, 4, Vienna, Austria, 3049-3077

Federal Emergency Management Agency (FEMA) (1989). *Estimating Losses From Future Earthquakes*. FEMA –177, National Research Council (NRC), Washington, D.C.

Gou J.W. (1990). *Seismic Design of Building*. Beijing: High Education Press.

Hu Y.X. (1988). *Earthquake Engineering*. Beijing: Seismological Press (in Chinese), 1988; pP60-61.

Hu, Y.X. (1998). *Earthquake Disaster Mitigation & Engineering Seismology*. IUGG/SSB International Training Course on Continental Earthquakes and Seismic Hazard, 70pp., Beijing, China.

Hu Y.X., Liu S.C. and Dong W.M. (1996). *Earthquake Engineering*. E & FN SPON.

IAEE. (1996). *Earthquake Resistant Regulations a Word List*. International Association for Earthquake Engineering, Tokyo, Japan.

Kennedy R.P., Cornell C.A., Campbell R.D., Kaplan S. and Perla H.F. (1980). Probabilistic seismic safety study of an existing nuclear power plant. *Nuclear Engineering and Design* 59, 315-338.

Kircher C.A. and McCann M.W. (1983). *Appendix A: Development of seismic fragility curves for sixteen types of structures common to cities of the Mississippi Valley region*, J.R. Benjamin & Assoc., Mountain View, Calif.

Kircher C.A., Nassar A.A., Kustu O. and Holmes W.T. (1997). Development of building damage functions for earthquake loss estimation. *Earthquake Spectra* 13(4), 663-682.

Koo K.K., Chau K.T., Yang X., Lam, S.S. and Wong Y.L. (2001). Soil-pile-structure interactions under SH waves. *Earthquake Engineering and Structural Dynamics*, 2000, accepted, in press.

Mohraz B. (1992). Recent studies of earthquake ground motion and amplification. *Proceedings of 10th World Conference on Earthquake Engineering*, Spain, 1992, Vol.11, pp. 6695-6704.

Newmark N.M. and Hall W.J. (1982). *Earthquake Spectra and Design*. Berkeley: Earthquake Engineering Research Institute.

Newmark N.M. and Rosenblueth E.A. (1971). *Fundamentals of Earthquake Engineering*. New Jersey: Prentice-Hall.

Paz M. (1994). *International Handbook of Earthquake Engineering, Codes, Programs, and Examples*. Chapman & Hall.

Shibata A. (1980). Prediction of the probability of earthquake damage to reinforced concrete building groups in a city. *Proceedings of the Seventh World Conference on Earthquake Engineering* 1980, Istanbul, Vol. 4, 395-402.

Veletsos A.S. and Newmark N.M. (1960). Effect of inelastic behavior on the response of simple systems to earthquake motions. *Proceedings of the second World Conference of Earthquake Engineering*, Vol.2, 1960.

Wen Z.P., Hu Y.X. and Chau K.T. (2002). Site effect on vulnerability of high-rise shear wall-buildings under near and far field earthquakes. *Soil dynamics and Earthquake Engineering*, accepted, in press.

Whitman R.V., Reed J.W. and Hong S.T. (1973). Earthquake damage probability matrices. *Proceedings of 5th World Conference on Earthquake Engineering 1973*, Rome, Italy.

Yamaguchi N. and Yamazaki F. (2001). Estimation of strong motion distribution in the 1995 Kobe earthquake based on building damage data. *Earthquake Engineering and Structural Dynamics* 30, 787-801.

Yin Z. (1995). *The Method of Seismic Damage and Loss Prediction*. Beijing: Seismological Press (in Chinese).

Yin Z.Q. Li S. and Sun P. (1981). Elasto-plastic earthquake responses of multi-story frames structures. *Earthquake Engineering and Engineering Vibration* 1(2), 56-77.

Yin Z.Q., Li S. and Sun P. (1985). Relation between story displacement and yield strength in multistory framed structures and problem of controlling displacement to prevent collapse. *Earthquake Engineering and Engineering Vibration* 5(1), 33-44.

Advances in Building Technology, Volume 1
M. Anson, J.M. Ko and E.S.S. Lam (Eds.)

A PRELIMINARY STUDY OF STRUCTURAL SEMIACTIVE CONTROL USING VARIABLE DAMPERS

R. L. Yang[1], X.Y. Zhou[1] and X. H. Liu[2]

[1] Institution of Earthquake Engineering, China Academy of Building Research, Beijing 100013, China
[2] Electronic Information Center, Ministry of Information Industry, Beijing 100846, China

ABSTRACT

It's well known that the semiactive control system is very attractive due to a large control force capacity, a small power demand and its inherent stability. As a classical example the semiactive variable damper is widely used, however, the latest research results show that its performance still need to be further studied. In this paper based on the theory of a SDOF vibration-isolation system under harmonic excitation, the basic conditions and limits of the applicability of viscous variable dampers are discussed. As a numerical example a six-story base-isolated frame is considered. According to the qualitative analysis and numerical simulation, two basic conclusions are drawn. They include: (1) Only far from the resonance may reductions in both the displacement and acceleration responses be possible for structures with viscous variable dampers; (2) The effectiveness of viscous variable dampers on reducing the response of structures depends on the characteristics of the ground motion that excites structures. In contrast to that of passive dampers, the effect of variable dampers is only slightly improved.

KEYWORDS

Semiactive variable dampers, fluid viscous dampers, OTE algorithm, structural control, hybrid base-isolated system, seismic excitations

1 INTRODUCTION

Similar to the natural evolution of species, vibration control techniques have a development process from a low-grade phase to a high-grade phase, i.e., from passive control systems to active/semiactive control systems. In recent years, considerable interests are shifted to semiactive control systems because of a broad dynamic adjustable range, a small power demand and its inherent stability. As a basic type of semiacive devices, semiactive variable dampers have been widely adopted. Nevertheless, the latest research results have shown that further attention should be paid to its efficiency.

In an extensive analytical and experimental study, Symans and Constantinou[1] developed and tested a two-stage damper and a variable semiactive fluid damper. For two dampers, different algorithms were used. Their study included a single-stoery frame and a three-stoery frame under different seismic excitations. The results indicated that while variable dampers significantly reduced the response as compared to the case with no control, no reduction was observed when compared to the same device acting as a passive damper with the upper limit of the damping of corresponding variable damper. As a result their study indicates that the use of semiactive dampers in structures is inefficient in some cases when compared to passive systems. To probe into the influence of supplemental damping and structural period on the seismic response of structures, Sadek and Mohraz[2] used six-SDOF structures with periods 0.2, 1.0, 1.5, 2.0, 2.5, and 3.0s respectively, and a structural damping ratio 0.05. Two supplemental passive dampers with damping ratios 0.05 and 0.40 were considered. 20 horizontal accelerograms were chosen. The average responses for the 20 records for the six structures showed for rigid structures (structures with periods below 1.5s) increasing the supplemental damping ratio decreased both the relative displacement and the absolute acceleration, whereas for flexible structures (structures with periods beyond 1.5s) increasing the supplemental damping ratio decreased the relative displacement but increased the absolute acceleration. Sadek and Mohraz made such a conclusion: Variable dampers can be effective in reducing the acceleration response in flexible structures,such as base-isolated and tall buildings, where an increase in the damping adversely affects the acceleration response. Variable dampers, however, are not effective for rigid structures as compared to passive dampers.

Inspired by these research results, the authors' latest study demonstrates for structures with variable dampers there is a dependency of the attained reduction in response on the characteristics of the ground motion that excites the structure. Referring to the transmissibility relation of an SDOF vibration-isolation system under support excitation[3], there are some basic features: If the ratios of the disturbance frequency to the fundamental frequency of structures are below $\sqrt{2}$, increasing the damping ratio decreases both the relative displacement and the absolute acceleration, whereas if that is beyond $\sqrt{2}$, increasing the damping ratio decreases the relative displacement but increases the absolute acceleration. Therefore, for structures with variable dampers under harmonic loadings, only is the ratio of the disturbance period to the fundamental period below $\sqrt{2}$, it's possible to achieve reductions in both the relative displacement and the absolute acceleration responses. Despite seismic excitations which contain a broad rang of frequency spectrum are very complicated compared to harmonic excitations, a qualitative analysis, referring to the corresponding results of harmonic loadings, is possible.

The authors, recently, presented a new general control strategy-OTE strategy in which the control operation is considered from the viewpoint of the relations of the motion of structures, the exciting force and the control operation. Based on the OTE strategy, all the control operations in the process of leaving/approaching balancing-position should suppress the motion of structures, however, in the process of approaching balancing-position should be kept at a slightly suppressive level. OTE algorithm, including the continuous control and the discrete control, which combines OTE strategy and fuzzy control theory, is capable of acquiring an ideal result.

Based on OTE algorithm, a discrete control by means of setting the damping into a high/low binary state is used in this paper. The numerical example studies the reducing-vibration performance of a six-stoery base-isolated frame with a hybrid seismic isolation system, i.e., isolators with supplemental variable dampers, and verifies the authors' viewpoints.

2 HYBRID ISOLATION SYSTEM

The ideal of protecting a building from the damaging effects of an earthquake by introducing some type of support that isolates it from the shaking ground is an appealing one, and many mechanisms to achieve this result have been invented successively. Although the early proposals may be traced back to 100 years, it's in recent years that base isolation has become a practical technique for earthquake-resistant design. Despite wide variation in detail, there are two basic types of isolation systems: elastometric bearings (which are predominantly natural rubber) and sliders. The mostly widely adopted system is the former, which introduce a layer of low lateral stiffness between the structure and the foundation. With this isolation layer the structure has a natural period that is much longer than its fixed-base natural period. The deformation of the isolated structure primarily occurs in the isolation layer, and the superstructure basically shows itself a rigid behavior. Considering the case in further earthquakes that occur close to seismic-isolated structures, a certain level of damping is beneficial to suppress the responses of structures[4-6]. Therefore a hybrid isolation system - isolators with variable dampers is studied in order to illustrate the efficiency of using variable dampers (Figure1).

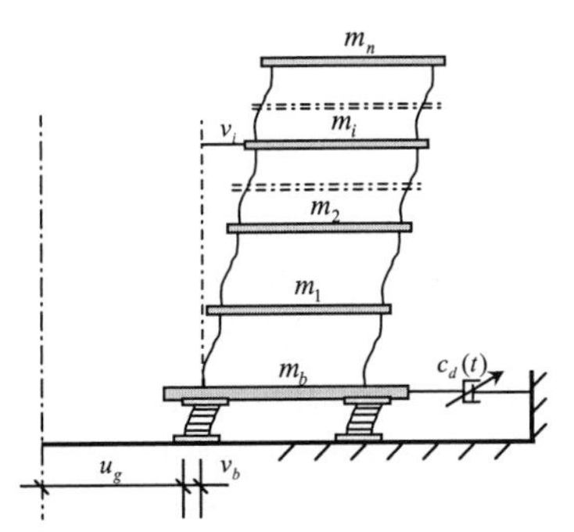

Figure 1: Multistory Hybrid Base-Isolated Structure

Assume that $\mathbf{M}$、$\mathbf{C}$、$\mathbf{K}$ are , respectively, the mass, damping and stiffness matrices of the superstructure, with n-degrees-of-freedom, m_b is the mass of the base, k_b is the total stiffness of isolators, $c_d(t)$ is the total instantaneous damping of viscous variable dampers, u_g、v_b and $\mathbf{v}$ are the ground acceleration, the relative base displacement with respect to the ground, the relative floor displacement with respect to the base respectively, $\mathbf{r}$ is a column of ones and $\mathbf{0}$ of zeros, m is the total mass of the superstructure. For this hybrid isolation system the equations of motion are as follows:

$$\mathbf{M}^*\ddot{\mathbf{v}}^* + \mathbf{C}^*\dot{\mathbf{v}}^* + \mathbf{K}^*\mathbf{v}^* = -\mathbf{M}^*\mathbf{r}^*\ddot{u}_g \qquad (1)$$

where $\mathbf{M}^* = \begin{bmatrix} m+m_b & \mathbf{r}^T\mathbf{M} \\ \mathbf{Mr} & \mathbf{M} \end{bmatrix}$，$\mathbf{C}^* = \begin{bmatrix} c_d(t) & \mathbf{0}^T \\ \mathbf{0} & \mathbf{C} \end{bmatrix}$，$\mathbf{K}^* = \begin{bmatrix} k_b & \mathbf{0}^T \\ \mathbf{0} & \mathbf{K} \end{bmatrix}$，$\mathbf{r}^* = \begin{bmatrix} 1 \\ \mathbf{0} \end{bmatrix}$,$\mathbf{v}^* = \begin{bmatrix} v_b \\ \mathbf{v} \end{bmatrix}$。

3 APPLICATION OF OTE ALGORITHM

According to OTE strategy, the damping function isn't always in favor of pushing the structure to go back to the balancing position. The characteristics of the damping function combining with the detailed motive trend of structures require being analyzed carefully. If the structure is departing from the balancing position, the instantaneous damping of variable dampers should be kept at a higher level and suppresses this trend; if the structure is approaching the balancing position, the instantaneous damping of variable dampers should be kept at a lower level and suppresses this trend slightly. Therefore in order to examine the efficiency of using variable dampers a simple on-ff control model was chose for variable dampers in this paper (Table 1).

TABLE 1
CONTROL CONDITIONS BASED ON OTE ALGORITHM

Algorithm	Control conditions	
OTE	If $x_b \dot{x}_b > 0$, then $c_d = c_{d\max}$ (ON)	If $x_b \dot{x}_b < 0$, then $c_d = c_{d\min}$ (OFF)

4 NUMERICAL EXAMPLE

A six-stoery base-isolated frame is considered as the numerical example to examine the efficiency of using variable dampers. Assume that all the floor masses are $m_i = 100$ ton, all the floor stiffnesses $k_i = 3 \times 10^5$ kN/m; the superstructure is with a damping ratio of 5 per cent in its fundamental period, and with a damping matrix proportional to its own stiffness matrix; the base mass is $m_b = 140$ ton; the stiffness of isolators is $k_b = 9 \times 10^3$ kN/m, and its damping is attached to the semiacive viscous damper; the lower and upper limits of the damping of the attached damper are 100 and 900kN.s/m respectively. All the isolators are expressed as linear models, and the fundamental period of the base-isolated frame is 1.82s.

In order to evaluate the ratio of the disturbance period to the fundamental period of the base-isolated frame, the predominant period of the acceleration response spectrum may be considered as a reference. Such a reference is approximate but meaningful by its physical meaning. It's enough to explain the authors' viewpoints. Two input records are as follows: (a) Luquan accelerogram(Luquan, Yunan, China, April 19, 1985, Station: Zhuanlong, Orientation: E-W, the predominate period of the acceleration spectra: 0.15s), (b) A S60E accelerogram synthesized by combining the two components of horizontal motion recorded at SCT during Sep.19,1985, Mexico City earthquake(the predominate period of the acceleration response spectrum: 2.0s). The numerical simulation considers three cases: off (the min. damping), on (the max. damping) and on-off control (damping adjustable). During the calculation the seismic-isolated frame is assumed to behave elastically at all times.

Figure 2 presents the peak responses of the base-isolated frame under two different seismic excitations. It's clear that

- Since the predominate period of the acceleration response spectrum of Luquan accelerogram is far below the fundamental period of the seismic-isolated frame (a non-resonant condition), an increase in the damping ratio further decreases the displacement response but increases the acceleration response.
- Since the predominate period of the acceleration response spectrum of Mexico City accelerogram is close to the fundamental period of the seismic-isolated frame (a resonant

condition), an increase in the damping ratio further decreases both the displacement response and the acceleration response.

This phenomenon is consistent with the corresponding characteristics of a SDOF vibration-isolation system subjected to harmonic excitations[7].

No matter what an algorithm is adopted, the potential effect of any semiactive control system by adjusting the damping is, undoubtfully, interposed between that of the corresponding passive control system by using the min. or max. damping. The results seen in Figure 2 also imply that

- Under non-resonant conditions, the improvement of both the displacement and acceleration response can be gained simultaneously. The semiactive variable dampers are possible to achieve reductions in both the displacement and acceleration response. However, because of small peaks of the responses the use of variable dampers is not valuable.
- Under resonant conditions, an increase in the damping is always beneficial to suppress the responses of structures. Therefore passive control systems with high damping coefficient are more suitable. The efficiency of using variable dampers is questionable.

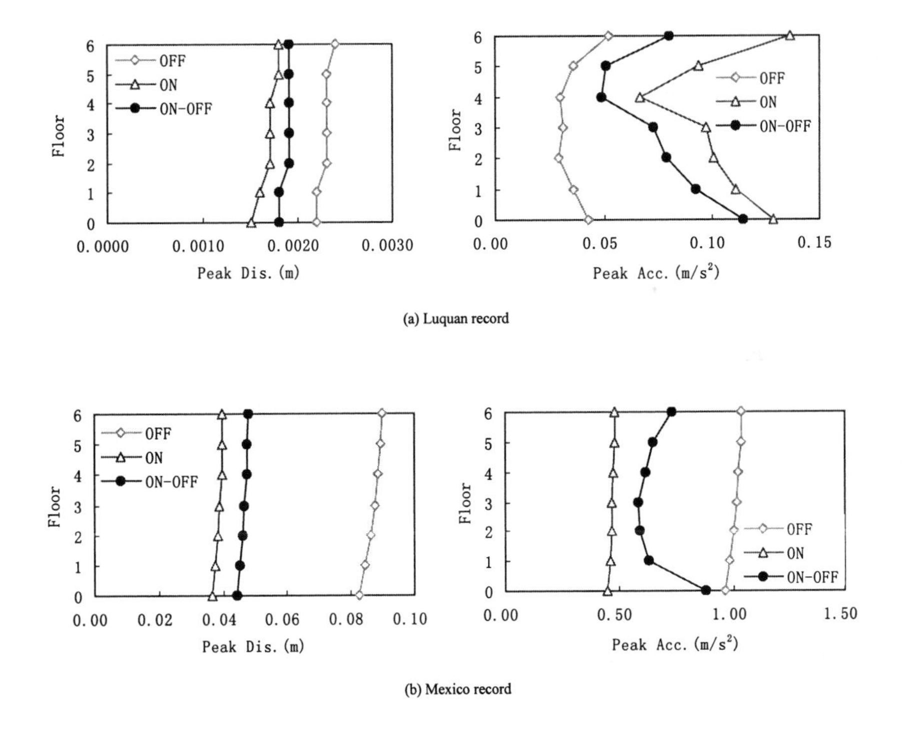

Figure 2: The peak response of the base-isolated structure

5 CONCLUSIONS

Based on OTE algorithm, the efficiency of using semiactive variable dampers is studied thoroughly in this paper. There are some basic conclusions as follows

a) Only under non-resonant conditions is using semiactive variable dampers possible to improve the control effects slightly compared to passive dampers.
b) The effectiveness of viscous variable dampers on reducing the response of structures depends on the characteristics of the ground motion that excites structures.

On the basis of known research results the authors find that considering the characteristics of the ground motion is necessary in order to study the efficiency of using variable dampers. As an index the ratios of the disturbance frequencies to the fundamental frequencies of structures is of consequence rather than the periods of structures.

6 LMITATIONS AND THE APPLICABLE RANGE

Semiactive variable dampers involved in this paper are primarily fluid viscous dampers, i.e., the damping forces are only correlative with the velocity (zero stiffness, irrespective to the displacement). Since only under the condition far from the resonance is using variable dampers possible to improve the control effects, variable dampers can be effective only in tall buildings or base-isolated structures. Meanwhile it should be pointed out that TMDs(tuned mass dampers) and TLCDs with varying-damping devices, based on the theory of dynamic absorbers, are exceptional.

REFERENCES

[1] Symans M.D. and Constantinou M.C. (1995). *Development and Experimental Study of Semi-Active Fluid Damping Devices for Seismic Protection of Structures.* Rep. No. NCEER-95-0011, State Univ. of New York at Buffalo, Buffalo, N.Y.

[2] Sadek F. and Mohraz B. (1998). Semiactive Control Algorithms for Structures with Variable Dampers. *Journal of Engineering Mechanics* **124:9,** 981-990

[3] Chopra A.K. (1995). *Dynamics of Structures: Theory and Applications to Earthquake Engineering.* Prentice Hall, New Jersey

[4] Kelly J. M. (1997). *Earthquake-Resistant Design with Rubber (Second Edition).* Springer, London, UK

[5] Nagarajaiah S. (1994). Fuzzy Controller for Structures with Hybrid Isolation System. *Proceedings of 1st World Conference on Structural Control* **2:** TA2 -67—TA2 -76

[6] Hussain S.M. and Retamal E. (1994). A Hybrid Seismic Isolation System-Isolators with Supplemental Viscous Dampers. *Proceedings of 1st World Conference on Structural Control* **3:** FA2 -53—FA2 -62

[7] Paz M. (1997). *Structural Dynamics: Theory and Computation (Fourth Edition).* Chapman&Hall, New York

TEAM PRESENTATION:

FLEXIBLE MANUFACTURING SYSTEM IN CONSTRUCTION

Advances in Building Technology, Volume 1
M. Anson, J.M. Ko and E.S.S. Lam (Eds.)

DYNAMIC MONITORING OF THE BORING TRAJECTORY IN UNDERGROUND CHANNEL FOR DRIVING COMMUNICATIONS

Prof. Dr. Eng. sc. tech. Alexej Bulgakow
Faculty of Architecture, Munich University of Technology
Arcisstr. 21, D-80333 München, Germany
Dr. Eng. Dimitry Krapivin, Dipl. Eng. Sergej Aleksyuk, Dr. Eng. Sergej Pritchin
Faculty of Mechatronik, South Russian State Technical University
Proswestchenja Str. 132, 346428 Novotcherkassk, Russia

ABSTRACT

This paper deals with high precision directional boring of underground channels, including programmable curved trajectories for enclosed digging driving communications and other purposes.

KEYWORDS

Underground channel, dynamic monitoring, driving communications.

As the working tool for boring in soft anisotropic soils a cone is taken, which is imparted a revolving motion whose angular velocity is ω_{pr}, its impact frequency f. The impact makes the revolving cone penetrate into the soil along the screw trajectory. The cone thus revolves around its symmetry axis OO_1 with angular velocity ω_K (Fig. 1) in such a way that the above axis OO_1 itself oscillates

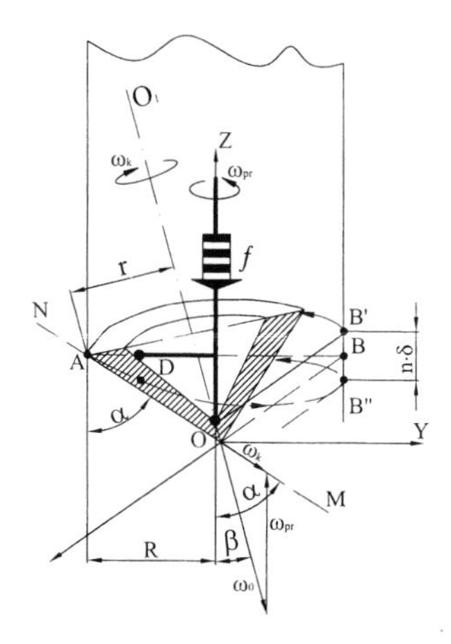

Figure 1

eccentrically around O_Z at an angular velocity ω_0 at β angle.
Acted by various factors the conditions of oscillating motion of OO_1 axis can change in such a way that the apex of the cone is displaced from the axis of the so, the canal symmetry and takes port in oscillating motion as well, its parameters being e and Δh to O_Z axis (Fig. 2). Here, the cone slips and performs friction work while its constituent OA is displaced from NM line during its turn at some angle φ to O_2 O_3 at an angular velocity ω_e.

The above parameters of this motion are found following the calculation of point O_2 location, based on the principle of the least action. With uniform distribution of the normal force

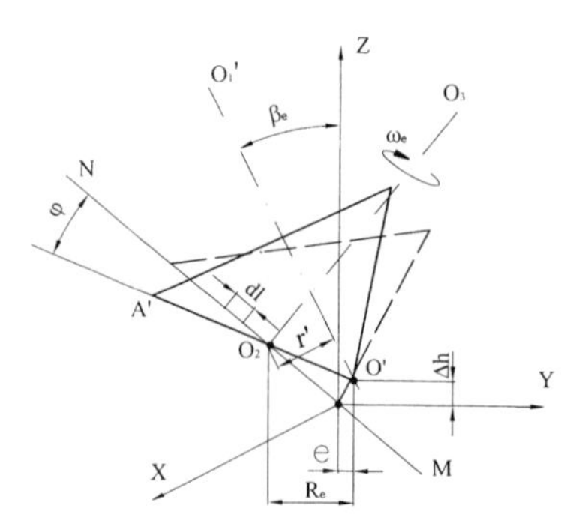

Figure 2

longitudinally the generating line of OA, the elementary work of the friction force F on $d\ell$ element will be $dA = F \cdot V \cdot d\ell$,
where $V = \omega_{\varepsilon} \cdot S$ - linear rate of $d\ell$ element, S – the distance of $d\ell$ element up to NM.

Total work of friction forces is

$$A = \frac{F}{2} \cdot \omega_e \cdot \frac{\cos\beta \cdot \sin\varphi}{\sin(\beta - \alpha)} \cdot \left[(L-\ell)^2 + \ell^2\right].$$

The minimum of the function A will be at $\frac{dA}{d\ell}$=0 or with $\ell = \frac{L}{2}$, where L=OA, i. e. when the cone constituent slips along the well face relatively to point O_2 located in the middle of cone constituent.
The slip appearance will make point A shift from the screw line B'AB". The coordinates of it point A can be derived as

$$x = (R - e) \cdot \cos\omega_{pr} t; \qquad x = (R - e) \cdot \sin\omega_{pr} t; \qquad z = n \cdot \delta,$$

where n - number of impacts,
δ - permeating in a direction of operation of impacts.

R-e value is found from $R - e = \frac{\omega_k}{\omega_{pr}} \cdot r$,
where r is the greatest cone diameter.

Thus, dynamic monitoring of the trajectory of a working cone motion is carried out by means of defining angular velocities ω_e, ω_{pr} and analysis of the amplitude readings of impacts. These indications can be used to control the directed cone motion. The shift value e at every turn can change with varying ω_{pr} and redistributing the number of impacts around the face perimeter.
When boring the channel in a uniform medium e=const. In a particular case, if e=0, the motion parameter relation is observed

$$h[n] = \frac{\omega_K \cdot r}{\omega_{pr}} \cdot \sin\beta, \tag{1}$$

Where h[n] is boring length in the preset direction within n impact - $h=n\delta$, β is the angle of screw line

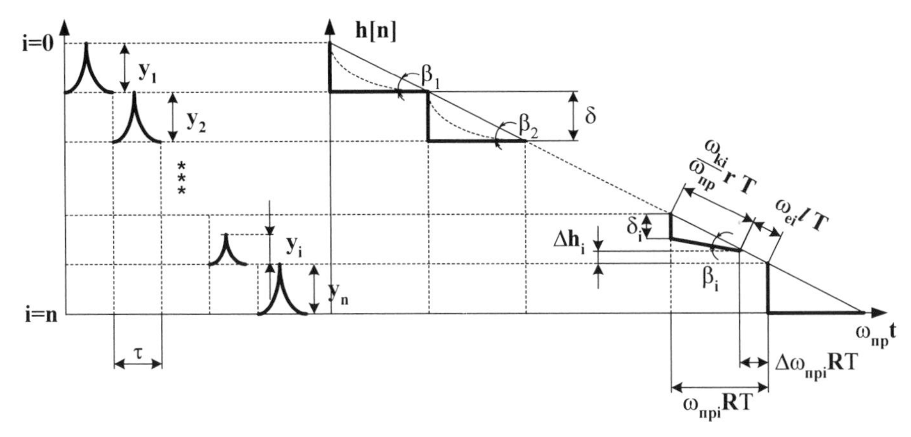

Figure 3

inclination. Figure 3 show the plane projection of a screw line for a turn.

Impact of duration τ=T and amplitude y make line B'AB" to be formed, which consists of curved portions of separate tunneling for every impact, thus resulting in screw line inclination at β angle.

If the medium is nonuniform, the parameter relationship in (1) is violated. Thus, in figure 3 at i-th stage impact amplitude $y_i << y_n$, impact duration $\tau_i < T$, by Δh_i decreases $\delta_i < \delta$, the angle $\beta_i < \beta$ changes and there appears eccentricity e_i. All the above can result in trajectory deviation due to variation of dynamic properties of tunnelling process.

To correct this deviation operatively one must have a self – adjusting system of automatic controlling the direction, which is effective if the anisotropy of medium acts casually. The system shown figure 4 has as a source of data gauge 1 of the amplitude of impact, gauge 2 of angular velocity of cone turn ω_K, gauge 3 of the angle of drive rotation velocity, a discrete correlator, a controlled filter, adaptive regulator. To find the dynamic properties of the system consisting of a drive feeding set, penetrating cone with a striker, a pulse transient function as a sequences of coordinates is being synthesized in a controlled filter as:

$$W_0 = W(0),\ W_1 = W(\tau),\ W_2 = W(2\cdot\tau),\ W_n = W(n\cdot\tau).$$

These values enter the RAM on OS_1 - OS_n.

Standard values ω^*_0–ω^*_n are recorded on $OS^*_1 - OS^*_n$ to put programmable trajectory into ROM.
On the delay line a periodic signal is put from the input unit, which is proportional to the centered correlation input – output function. Thus, the following expression is obtained on the adder output:

$$K^{*0}_{xy}(\tau) = T_m \cdot \sum_{i=0}^{n} W(i\cdot T_m)\cdot K^0_y \cdot(\tau - iT_m).$$

The difference is put to the comparator:

$$\varepsilon = K^{0*}_{xy}(\tau) - K^{0}_{xy}(\tau),$$

By adjusting the coefficients they ten fd to make it minimal. Extreme regulators are used to make the process of impact function estimate automatic.

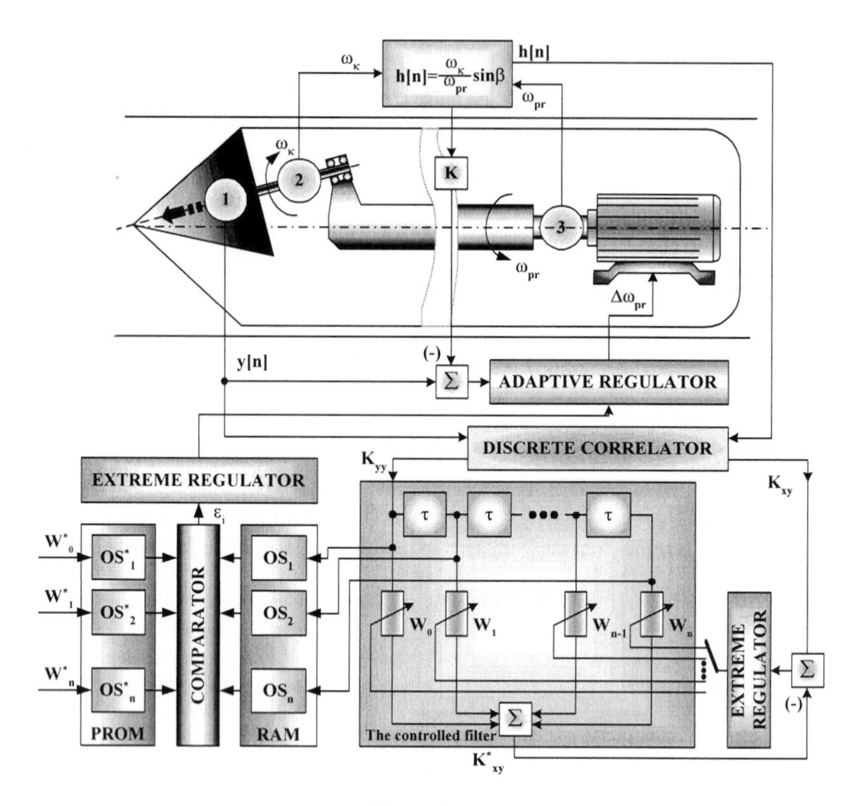

Figure 4

Thus, the procedure of trajectory monitoring envisages a complete analysis of dynamic system properties taking into account stochastic character of effect. This method of driving uses intensifying impact effects for probing anisotropic properties at the face and controlling the process of direction correction.

Advances in Building Technology, Volume I
M. Anson, J.M. Ko and E.S.S. Lam (Eds.)

Façade Cleaning Robot

Thomas Bock, Alexej Bulgakow, Shigeki Ashida
The Chair for Realization and Informatics of Construction
The Faculty of Architecture, Technical University of Munich
Arcisstr. 21 80333 Munich, Germany
Tel. +49-89-289-22100 Fax. +49-89-289-22102
Email: thomas.bock@bri.ar.tum.de : bulgakow@bri.ar.tum.de : ashida@bri.ar.tum.de

Abstract

Considering all available basic conditions, a semiautomatic façade cleaning robot can be best implemented for technical, organizational and economic reasons. All cleaning processes are executed by means of a service robot. The robot is controlled and adjusted by just one operator. The cleaning quality remains constantly excellent and the cleaning efficiency is increased significantly.

Keywords: Cleaning Robot, Façade cleaning, Semiautomatic, Vacuum suction, Nacelle traveling, Skyscraper

Introduction

Skyscrapers have to fulfill above all a representative function for the companies. The status of a façade surface and the entire appearance plays a particularly important role. The environmental leads to intensive contamination of facades, and modern building materials react quite differently to weather influences. Even high-quality natural stone facades do not remain spared from visible signs of pollution. Facades must be cleaned regularly, and the cleaning cycle can last for decades. Nevertheless the demand of the facades must be kept by the cleaning processes within maintainable boundaries. The higher the degree of pollution is, the more intensive the cleaning agents and instruments must be.
The first skyscraper with anodized aluminum facades was built in the seventies. Such facades were offered at that time as maintenance-free, and cleaning was not judged to be necessary, because the anodicl oxide coating is hard and weatherproof. Commercial criteria were a further aspect. Experience showed however that facades anodized and got dirty by weathering. A further important argument for cleaning was not considered, i.e. that the dirtier the façade, the greater is the corrosion load. In addition contamination increases the original surface area making it more absorptive so that larger quantities of aggressive air impinge on the surface. That can lead within several years to corrosion of facade surface to the extent that the original immaculate appearance cannot be repaired.
Cleaning anodized facades therefore is recommended for the following reasons:

- for good appearance
- to remove the chemical load by removing contamination
- to remove iridescence
- preservation of value

Implementation concept

Considering all available basic conditions, a semiautomatic cleaning version as described below is best implemented for technical, organizational and economic reasons. The robot is accommodated thereby on a nacelle of the traveling unit, adjusted during the whole cleaning process until completion and with the transfer of a worker (figures 1 – 2).

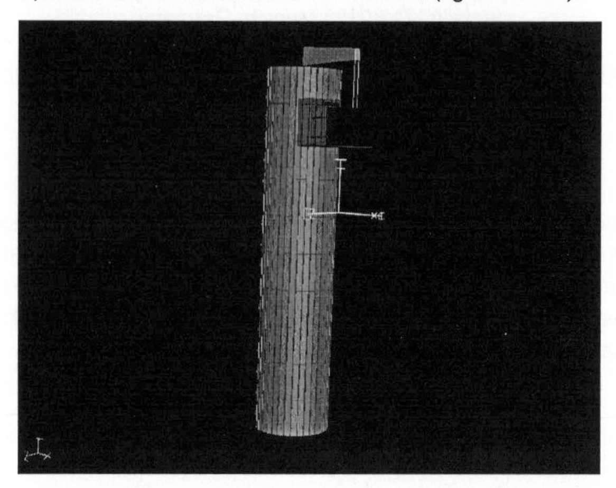

Fig. 1. Nacelle of the traveling Unit

Fig. 2. Fixing of the Robot on the facade surface with Vacuum Suction Cups

The function of the worker exists in the monitoring of the cleaning process and in the execution of the cleaning of the facade surface zones difficult to access by hand. The latter enables a simplification of the robot system and thus a cost saving.

The robot can be used also at surfaces of other types, e.g. for window surface cleaning.

Technical data of the facade cleaning robot

Type of cleaning	Brushes
Brush size	Length: 300 mm Thickness: 50 mm
Drive	Automatic guidance of the cleaning heading
Heading pressure	Manual adjusts
Dressing plant	Negative pressure system
Capacity	4,0 kW
Cleaning performance	50 m²/h
Advantages	Small personal expenditure Good and safe cleaning quality
Weight	75 kg

Application of the cleaning operation

When an electromechanical system consisting of power supply, motors, transmissions is designated for the cleaning drive, it brings the cleaning brushes into motion. The main parameters of the drive are performance, speed, precision, controllability, dimensions, weight and price. The special feature of electric drives is that they are of small dimensions, easy to control, of high precision, maximum efficiency and reliability (fig. 3).
Regardless of their relatively low performance and quite high price, electric motors are the best fit for cleaning robots of facades.

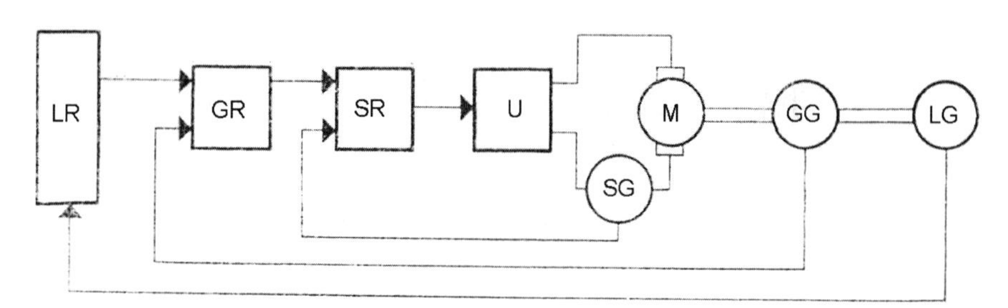

LR ; Position regulator, GR ; Speed regulator, SR ; Power regulator,
U ; Converter, M ; Motor,
LG ; Position sensor, GG ; Speed sensor, SG ; Power sensor

Fig. 3. Main Scheme of electrical operation

Application of the robot platform

The platform of the robot is equipped with a pneumatic drive. The vacuum cups serve mainly to maintain the contact pressure of the robot to the surface during the cleaning process. Additionally they permit a progressive movement up and down along the facade. The robot is held by the wire rope of the traveling unit. The vacuum force is kept normally as a function of

the entire mass of the robot, either by the degree of the vacuum or by the number of suction cups. These adjustments are made before beginning the work and do not need to be corrected during the job.
Completely differently adaptive vacuum cups are possible. At the fig. 4 an appropriate structural variant is represented. The vacuum cup consists of the pneumatic cylinder (1), the lower part of which is a suction cup in the form of an open vacuum chamber (2). The vacuum chamber is through the connecting piece (3) is connected to a vacuum source. The entire device is connected by the coupling (5) to the lifting machine. The axle (4) forms a link connection to the piston rod (6) of the cylinder. The vacuum chamber additionally has a valve (7) which connects to the atmosphere. Between the housing of the pneumatic cylinder and the coupling the torsion spring (8) is placed. When releasing the instruction for touching the object, the valve is closed, and in the vacuum chamber a vacuum is created by the compressor. If the load (e.g. because of wind conditions) exceeds the nominal value, the housing of the pneumatic cylinder shifts downward relative to the coupling, the torsion spring stretches, and the piston space of the cylinder becomes larger, so that the vacuum in the vacuum chamber corresponding to the load of the robot increases. The suction force increases, and the holding of the robot in position becomes more reliable. The greater the difference between the nominal and the actual value of the load, the larger the shifts of the housing of the cylinder, the volume of the piston space and thus the suction force of the grip arm.

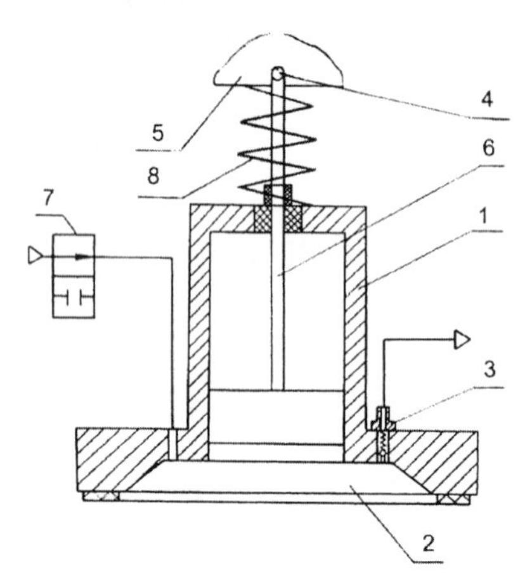

Fig. 4. Adaptable Vacuum Sucker

Conclusion

Considering all of available basic conditions of technical, organizational and economic constraints, semiautomatic cleaning version can be best implemented. The cleaning process is executed by the robot. The check of the process and adjusting the robot are implemented by an operator. The second worker can be saved. The quality of the surface remains constant and the cleaning efficiency increases significantly.

Advances in Building Technology, Volume I
M. Anson, J.M. Ko and E.S.S. Lam (Eds.)

ADVANCED CONSTRUCTION DESIGN AND TECHNOLOGY BY ROBOTICS

Prof. Dr. Dipl. Eng. Thomas Bock
Dipl. Eng. Klaus Kreupl

Faculty of Architecture, Munich University of Technology
Arcisstr. 21, D-80333 München, Germany

ABSTRACT

When a single client needs a building the normal procedure usually produces low quality. The establishment of big general contractor-companies or powerful clients leads to a loss of price transparency and variety of building types. The robot-oriented construction method will allow to satisfy all requirements of all parties who are involved in construction and achieve a robot-oriented building market. This will increase productivity and profitability because the robot-oriented construction process is characterized by minimized construction time, high variety and flexibility of design. This process will lead to more complex and coordinated building systems and will need an integrated design method to fulfil any functional requirements and regional building codes. Joining will be a key technology to reduce time of assembly. Due to the high working speed of robots, even small elements can be assembled quickly on-site . This results in a high design flexibility and minimized construction time. A robot oriented building block system requires well defined conditions, which can be achieved by the prefabrication of value added and highly finished small blocks, that have to be only finally assembled on-site. Especially if finish operations can be reduced through high degree of prefabrication the construction cost can be decreased – or profitability of buildings can be increased. Installations such as ducts and pipes can be integrated in building elements. In the early design and conception phases all conditions of construction have to be defined in their physical and geometrical quality. With a clear product hierarchy, the advantages of Industrial Robotics in the controlled environment of factories for mass-production of small and value added elements can be used thus on-site construction time can be reduced.

KEYWORDS

Robot-oriented construction, building processes, building systems, Integrated Design Method, Joining Operations, Design Flexibility, Prefabrication Systems, Robotic Assembly Systems, Product Hierarchy.

1. INTRODUCTION

Although the building industry doesn't reach the degree of automation and rationalisation which is common in other industries (e.g. the automobile industry), there are many singular implementations in use with a remarkable technological level. Industrial automated processes are only implemented in the

production of components in stationary prefabrication plants (e.g. masonry and concrete elements) in order to obtain high-quality results (Dalacker, 1997). Irregular demand, great variety and high capital costs in construction require similar strategies as already applied in the manufacturing industry. The shorter the time span between contract signing and utilization of the building the cheaper the financing of the project. Therefore robot oriented manu-facturing requires standardized planning, production and assembly systems (Bock 1988).
An overall hightech construction method for buildings would offer many benefits: Increasing quality of product and proceeding, less construction time, lower prices or higher profitability, so why are these potentials not used? The reason is, that all current solutions are isolated applications. So "Robot Oriented Construction" has to consider all relevant levels of a building project: Beginning with the market conditions and contactors structures, leading over the project organisation concerns, and finally reaching the production and assembly methods – everything has to be redesigned in view of automatisation.

2. GLOBAL STRATEGY FOR ROBOT-ORIENTED CONSTRUCTION – TRANSITION OF BUILDING MARKET TOWARDS ROBOT-ORIENTED CONSTRUCTION

Normally when a client needs a building, he will ask a company, for which architects and engineers are working, for the design, planning and disposition of the construction project. Later one or more builders will be assigned to execute the construction works (figure 1).
A major disadvantage of this model is the way of defining the quality. Since there is no sufficient quality control of works between company and builder even though specifications are usually listed. Due to the fast development of science and technology, building related specifications of up to date technology is hardly available especially to small builders. The smaller the builder is, the less successful is the definition and control of quality. Since technological know-how is related to production technology which is varying from one company to other, the quality of the building as a product is differently decided by each builder. Furthermore the architect is sometimes not aware of the state of the art of technology. This dilemma can be seen at the fact that former tasks of architects are now executed by specialized engineers. This development supported the establishment of large general contractors offering turn-key-system in construction (figure 2).
If the client is confronted to few large contractors which decide the quality and guaranty of the building, he can only choose between some companies and there is no transparency of prices. Figure 3 shows powerful public or private clients which also run their own planning office. In this case, the client can partly influence the producer, but the producer will always decide the variety of building types for the production and construction of large projects. The robot-oriented building and construction system will allow to satisfy the requirements of all parties that are involved in construction. Due to the inherent character of robotic technology, all standards of soft- and hardware have to be well defined. Suppliers have to provide the geometrical and physical qualities of building elements that are required for robot-oriented construction method (figure 4).

TRANSITIONS OF BUILDING MARKET TOWARDS ROBOT-ORIENTED CONSTRUCTION
FIGURE 1

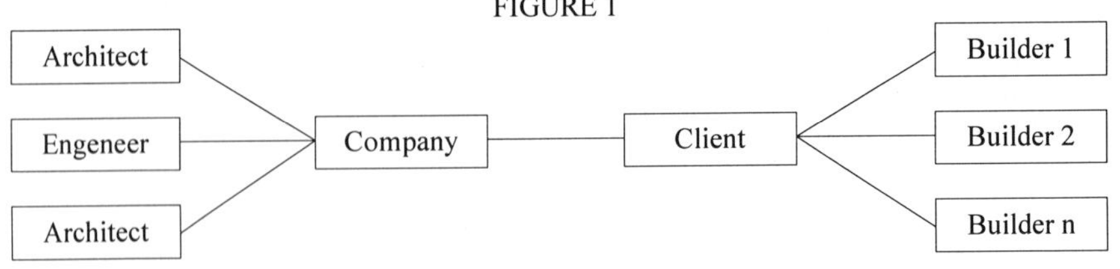

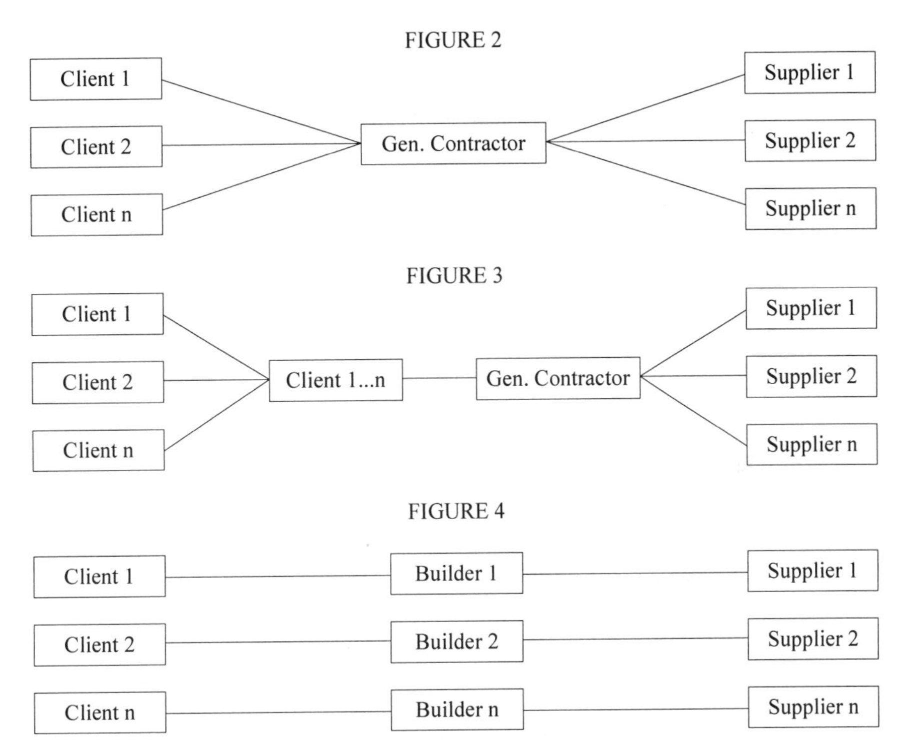

This robot-oriented building market will be achieved, if the following 15 parameters can be simultaneous fulfilled without excluding each other.

3. PARAMETERS OF ROBOT-ORIENTED CONSTRUCTION

1. Flexibility
2. Definition of cost
3. Clear marketability
4. Definition of price
5. Guaranty of price
6. Transparency of price
7. Homogeneity of production
8. Continuity of production
9. Definition of quality
10. Control of Quality
11. Transparency of quality
12. Standardization
13. Definition of construction period and time
14. Control of construction time
15. Transparency of construction time

a. Increase of Productivity and Profitability

Productivity of construction industry can be increased through production of highly differentiated building elements which are suited for many function. A major improvement in Robot-Oriented Construction is the reduction of on-site construction time through higher working speed and almost 24 hours working hour a day. For example, it takes a mason about 1 to 1½ hours to construct 1 m^2 of masonry. The same job, a robot could do 10 to 50 times faster. Thus the cost of the building will increase because the building can be utilized shortly after it has been ordered or purchased.

b. The Transition of Building Processes Towards Robot-Oriented Construction

The traditional building process is characterized by a long construction period, which is resulting in a lower profitability of the building and therefore uncertain prices but a big variety and flexibility of design (figure 5). The rationalized construction process is characterized by typification of planning and well defined supply system already before the contract is signed thus reducing the construction time but also limiting variety and flexibility of design through the use of large sized elements (figure 6). The industrialized construction process is characterized by a reduced construction time thus allowing higher profitability and security of price through homogeneous and continuous production of mass-produced building elements (figure 7). But this model is weak for fluctuations of the building market. The robot-oriented construction process will be characterized by minimized construction time, high variety and flexibility of design. The flexibility of software allows the production to adjust to fluctuations in demand. This system requires a certain product hierarchy which results in well defined and specially designed building elements for robotic construction (figure 8).

FIGURE 5

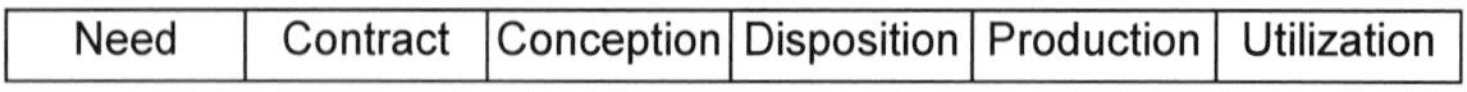

Need | Contract | Conception | Disposition | Production | Utilization

FIGURE 6

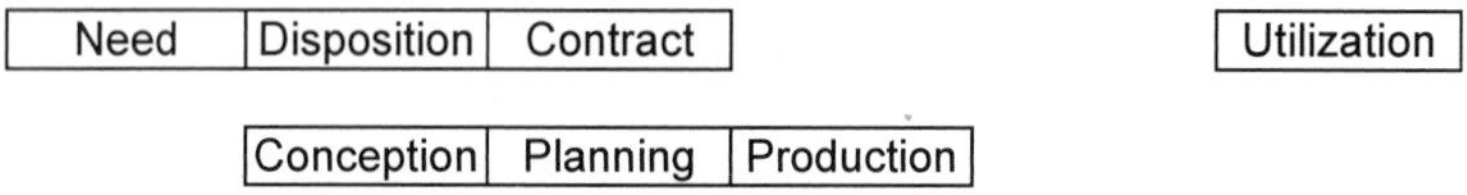

Need | Disposition | Contract   Utilization

Conception | Planning | Production

FIGURE 7

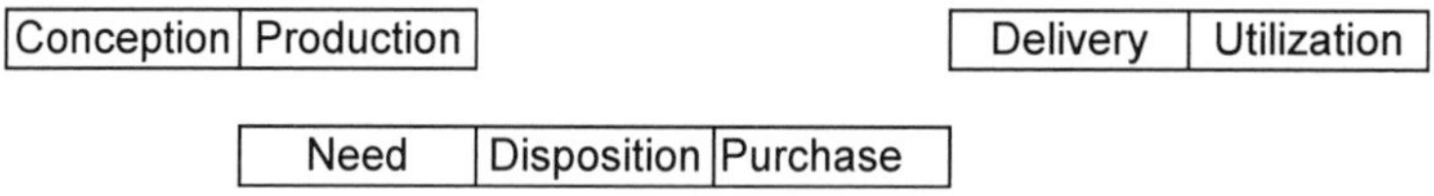

Conception | Production   Delivery | Utilization

Need | Disposition | Purchase

FIGURE 8

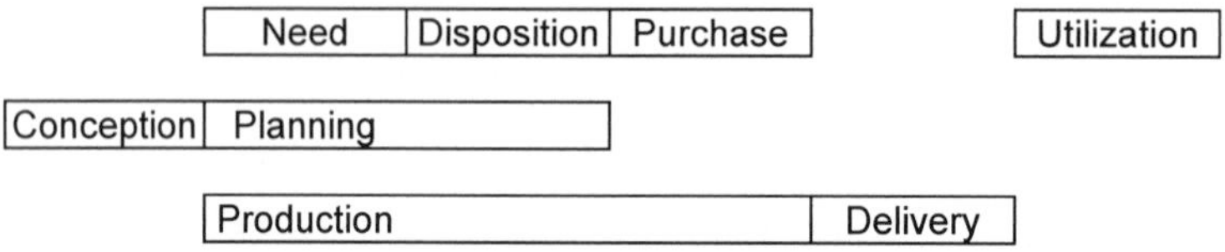

Need | Disposition | Purchase   Utilization

Conception | Planning

Production | Delivery

c. *Planning of Building Systems and Coordination of Building Systems*

The more advanced and industrialized a construction process is the more important the design stage becomes, since most building elements are designed and produced before the construction of a building is ordered. The accuracy of subsystems vary: The low accuracies of foundation works are insufficient for other construction works and the higher accuracy of floor wall or installation assemblies are not required for earth works. Different categories of accuracies should be defined for different construction works, considering with compliant interfaces of building sub-system. This is also true for the production of complex elements which are composed by e.g. wood, plastic, steel section whose tolerances also vary, affecting each other and therefore they have to be compliantly coordinated. Due to the increased complexity of modern building compared to the ancient building this kind of coordination of subsystem is required.

d. Integrated Design Method

Building systems should not only be designed for robot-oriented construction but also for usual maintainability, reliability, stability, serviceability, safety and aesthetics. This requires the coordination of production-related engineers with planners and architects. The robot oriented design of small standardized, functional and material elements are required for robotic construction in order to adjust to any project. This requires the modular coordination of smallest elements of building kit as well as modularization of robot software. Specifications and building systems have to be designed as kits which consist of highly differentiated parts which fulfill any functional requirements and regional building codes.

e. Importance of Joining Operations in Construction

In construction more than 2/3 of the production time is used for some kind of assembly operation (DIN 8593 "FUEGEN"). Therefore efficiency of construction can be mostly increased if building elements are designed for automatic assembly.

f. Increase of Construction Efficiency and Design Flexibility

Due to the high working speed of robots, even small elements can be assembled quickly on-site . This results in a high design flexibility and minimized construction time (figures 9 and 10).

FIGURE 9
COMPARISON OF ON-SITE AND FACTORY PRODUCTION TIMES OF CONSTRUCTION

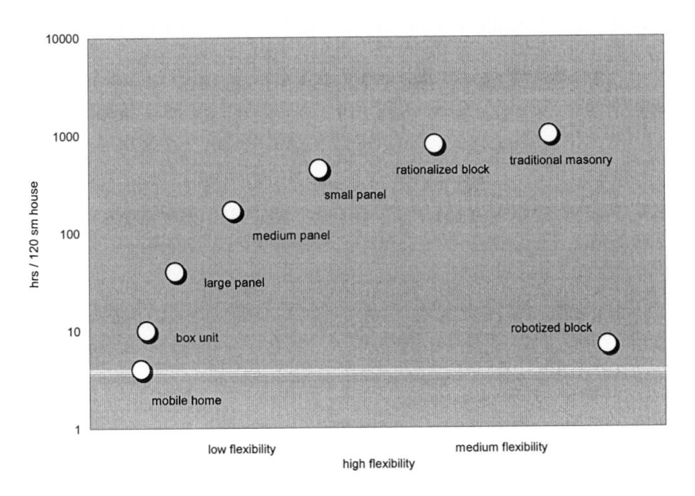

FIGURE 10
DESIGN FLEXIBILITY

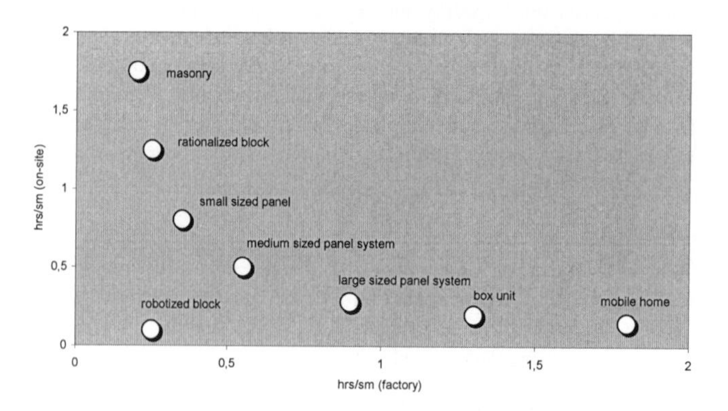

g. Increase of Construction Speed and Design Flexibility

Most commonly used prefabrication systems are based on the following principle of building production: Building elements, that are as large as possible and as prefabricated as possible, are centrally produced and finally assembled on-site. The basic idea is to minimize on-site construction operations. All existing prefabrication systems are based on this principle, which is economically oriented in order to improve quality of production and increase productivity of construction.

Among existing prefabrication systems, the box unit system allows the highest reduction of on-site construction time. The size of the box unit is decided by the conditions of transport. Other prefabrication systems consist of elements and parts that are transported to the site for further processing.

As most prefabrication systems, especially the box unit system incorporates the problem of low flexibility in time. Conventional building systems which are almost produced on-site offer an increased flexibility in time and in design. Changes in design can be realized at a later time since the size of elements and parts is smaller and allows later adjustments – during construction phases.

This disadvantage of prefabrication systems might be a reason for the stagnating market share of 10% to 20% of all construction of housing.

Through robotic technology we can combine the advantage of traditional and flexible building systems and prefabrication systems:

- Small building elements can be prefabricated centrally and shipped to the site where assembly robots construct the building at an increased assembly speed.
- The size of the building elements does not necessarily have to be large since the working speed of robots is many times faster and robots can work 3 shifts a day.

Figure 11 shows a robotic assembly system.

FIGURE 11
PROTOTYPE OF A MOBILE INTERIOR ROBOT
(FELDMANN, KOCH 2000)

The advantage of the robotized block system is increased design flexibility and easy material transportations even on narrow streets and over small bridges.
The centralized prefabrication of small block-like building elements enables the following on-site production:

- higher investments of production facilities due to long-term and continuous production
- free from bad environmental conditions (weather)
- homogeneous working conditions
- high level of organization
- due to minimized on-site construction building materials and elements are less damaged
- highly prefabricated and finished elements can be assembled on-site

A robot oriented building block system requires well defined conditions, which can be achieved by the prefabrication of value added and highly finished small blocks, that have to be only finally assembled on-site. Especially if finish operations can be reduced through high degree of prefabrication the construction cost can be decreased – or profitability of buildings can be increased. Installations such as ducts and pipes can be integrated in building elements.

h. Physical and Geometrical Definition of all Elements

During early design and concepting phases, a three dimensional space is created in which the positions of all parts are well defined. All conditions of construction have to be defined in their physical and geometrical quality.

i. Clear Product Hierarchy

The clearer the product hierarchy becomes, the easier robots can be implemented in construction. In order to rationalize construction building kits, they are split up into well defined hierarchical levels. This means a product is subdivided into sub-products according to robot-oriented requirements. If a project is realized an hierarchy of all the building parts , elements and groups necessary for the project can be shown. On construction sites, we can see all kinds of parts, elements sections and material that are transported to the site and further processed there. But in case of robotic construction, the site is just the final production stage of building production. Prefabricated, value added, small elements are shipped and installed on-site before the building is utilized. In this case, advantages of Industrial Robotics in the controlled environment of factories for mass-production of small and value added elements can be used thus on-site construction time can be reduced (figures 12 and 13).

FIGURE 12
PRODUCT STRUCTURE OF CONVENTIONAL CONSTRUCTION

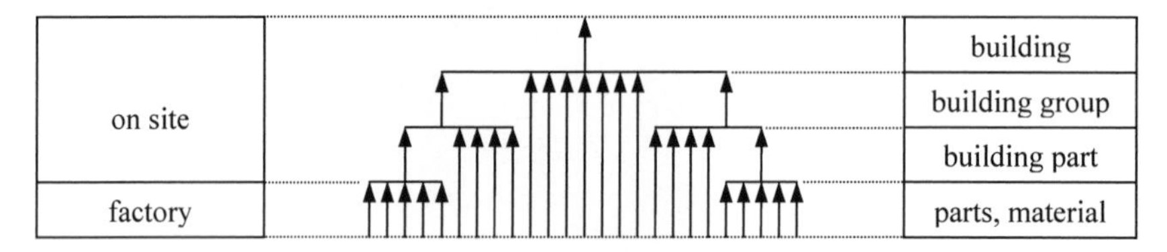

FIGURE 13
PRODUCT STRUCTURE OF ROBOT ORIENTED CONSTRUCTION

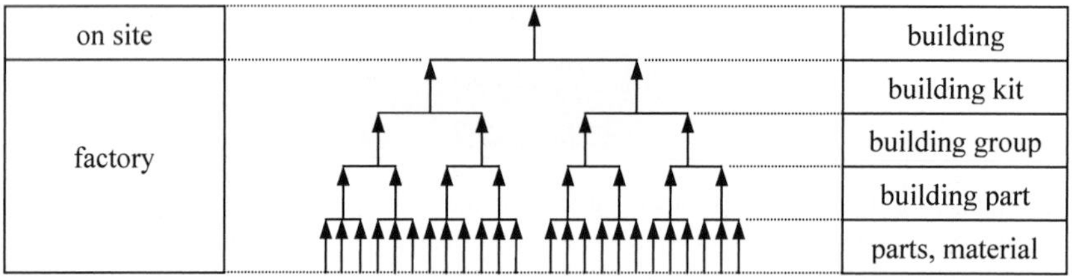

on site
factory
building
building kit
building group
building part
parts, material

4. CONCLUSION

An overall hightech construction method for buildings would offer many benefits for every party involved in building construction processes (clients, building industry, planners): Increasing quality of product and proceeding, less construction time, lower prices or higher profitability.
These potentials could only be used, if all levels of construction processes (the market structures, planning and building systems, design methods and product integration), could be aligned with the concerns of automation.

REFERENCES

Bock, Th., *Robot Oriented Design*, Shokokusha, Tokio, May 1988.

Bock, Th., Kreupl, K., Herbst, J. *"Planning aids for enhancing the implementation of a mobile robot on the construction site"*, proceedings of the 17th International Symposium on Automation and Robotics in Construction, ISARC, September 2000.

Dalacker, M. "Entwurf und Erprobung eines mobilen Roboters zur automatisierten Erstellung von Mauerwerk auf der Baustelle", *Schriftenreihe Planung, Technologie, Management und Automatisierung im Bauwesen*, Band 1, Fraunhofer IRB Verlag, Stuttgart, 1997.

Feldmann, K., Koch, M. *"Development of an Open and Modular Control System for Autonomous Mobile Building Robots with Flexible Manipulators"*, proceedings of the 17th International Symposium on Automation and Robotics in Construction, ISARC, September 2000.

Advances in Building Technology, Volume 1
M. Anson, J.M. Ko and E.S.S. Lam (Eds.)

Embedded System and Augmented Reality in Facility Management

Thomas Bock, Helga Meden, Shigeki Ashida
The Chair for Realization and Informatics of Construction
The Faculty of Architecture, Technical University of Munich
Arcisstr. 21 80333 Munich, Germany
Tel. +49-89-289-22100 Fax. +49-89-289-22102
Email: thomas.bock@bri.ar.tum.de : helga.meden@bri.ar.tum.de : ashida@bri.ar.tum.de

Abstract

The functions of facilities often have to be optimised regarding all requirements. To optimise effectively, we must analyse the present situation and refer to optimising future requirements.
For each aspect of facility management, we should use real time data, and not store data which are sometimes never used anymore.
For integrating real time data systematically, 3D scanning is simple and the best, but there are still some difficulties in such methods.
We can employ an embedded system with RFID (Radio Frequency Identification) and AR (augmented reality) system with mobile computers. With this combination, we have the following possibilities of;

- Fast and easy grasp of the basic data of facilities through RFID and virtual realizing with AR
- Visual comparison between the basic data and the present situation, and the adjustment of data
- The ability to see through obstacles
- Simulation of consequences

Both, RFID and AR systems, are already existing products, offered as mobile or wearable devices in the market.

Keywords: Facility Management, Embedded System, RFID (Radio Frequency Identification), Augmented Reality, Mobile Computers, Recognizing

1. Introduction

Facility Management is one of the disciplines, used in Architecture for the past 25 years in the US-market, 20 years in the Netherlands and Great Britain and about 10 years in Germany. Facility management aims at more effectiveness in using facilities at their maximum function at any time for any intention, instead of reconstructing a new building. Facility Management is more than a management of facilities, but a management of the function and services in facilities.
Function means convenience, satisfaction, safety, pleasure by facilities, which are for example rooms, walls, furniture, lights, machinery, papers (information) and others.
These functions should be always adapted to suit present requirements, when users change, organizations or the environment has to be rearranged. Sometimes it is better a total change of facilities in the case of new technology. Of course facilities must be maintained of waste and breakdown.
To optimising the functions of facilities effectively, the present situation and findings consolidated in the future.
3D scanning with AR is one of the best methods, but there are still some technical and practical difficulties in achieving such method. It is sometimes expensive, inexact and time for scanning is required. It is also still hard to carry or wear the equipment easily. Therefore an accepted technique with visual 3D only scanning all facilities concerned is not practical at the moment. We must change the functions of facilities when requirements change. The cost for these changes must be minimized by manager in achieving the maximum function and performance.

2. Identification technology

RFID (Radio Frequency Identification) is a popular identification method. A chip is embedded in each object containing its specific data as registered. The scanner for this system is in most of cases easy to carry or wear. AR (Augmented Reality) itself with wearable computers has also already been popular. One such application needs trials of AR for 3D scanning.
It seems becoming possible, but still many devices are required for recognizing self-position. At the moment it is better to fix the scanning position for easy and exact scanning.

3. Concept of mobile management

Our work is trying to adopt an embedded system using RFID and AR with mobile computers for recognizing of facilities.
All specific facilities data are initially registered and a RFID chip embedded. If more detail data must be stored, the identification is related to a data bank automatically (Fig. 1). The manager does not need to search, compare and select the data.

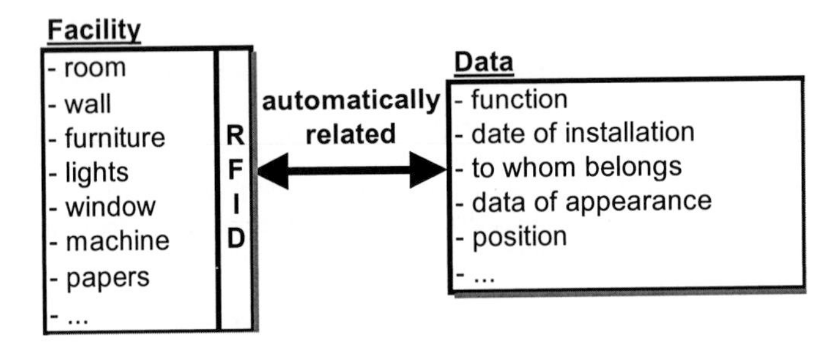

Fig. 1: Embedded data

With RFID, instead of using a bar code, all facilities within radio reach identified at one time. The manager can also take invisible objects into account. Furthermore a RFID chip can be embedded in the facility without disturbing the aesthetic appearance.
The ability of RFID for rewriting is important as well. The visual situation, which means position, wearing, shape, colour and others, is shown virtually with AR. The manager adjusts his/her position of view, and compare with the real present situation. Then the difference is rewritten on RFID system. (Fig. 2)
These devices are easy to carry, even to wear when more convenient. The manager can move to all related areas light on footwork. He/She can collect quickly only the necessary data for facilities under consideration, and the data are transferred for analysis between requirement and action. Only the necessary data, which are not expensive in quantity, are handled.
In case immediate action is required, fast and suitable decisions can be made just in time and place, because all relevant data are carried or worn by him/her. It is also possible to confer with a different person, who is in different place at the same moment, though a mobile communication tool to make better decisions whenever it is necessary.

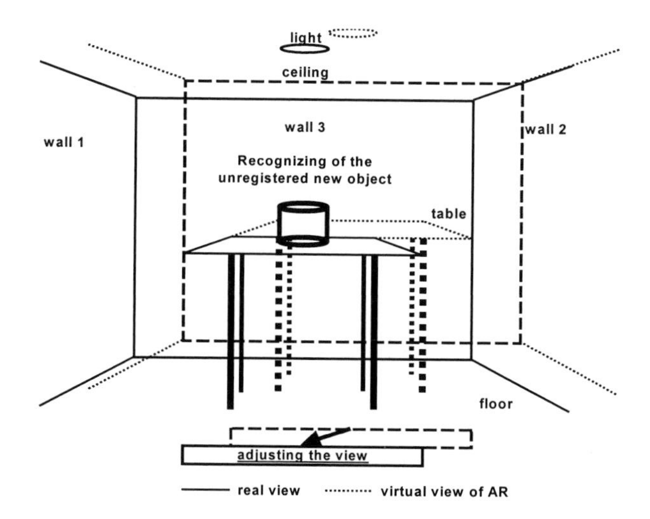

Fig. 2: Concept of recognizing and adjustment

4. Enhanced capacity by AR

With AR, it is also possible to show invisible objects. An object behind some obstacle, something inside the facility, pipes or ducts above the ceiling, electrical or communication cables under the floor, reinforcement in the walls.
A helpful view for consideration becomes visible with AR as well. The temperature, humidity, airflow, direction and level of lighting, smell, sound and others can be surveyed visually by the manager.
In the case there are some different ways to meet a requirement, the manager can compare or confirm the effect of each action through virtual simulation with AR. It is possible to select the best action, and also check the need for materials or new facilities.

5. Future work

This trial is still just a concept, and we must think furthermore about the realization. The best concrete systems should be created and combined.
Of course its efficiency must be confirmed with some case study or practical adoption, if this survey method as to be used in Facility Management. It should not only be used to analyse an existing situation, but also included as part of a total system of management if it is to be used effectively. We will create such a facility management system with RFID and AR recognizing system.

6. Conclusion

There are exciting possibilities for the Facility Management discipline through its development of AR and RFID embedded systems. Practical trials are in work; results will be published.

Advances in Building Technology, Volume 1
M. Anson, J.M. Ko and E.S.S. Lam (Eds.)

ROBOTS FOR LOCATION AND SPACING OF WOOD

Prof. Dr. Eng. Thomas Bock
Faculty of Architecture, Munich University of Technology
Arcisstr. 21, D-80333 München, Germany
Dr. Eng. Dimitry Parshin
Faculty of Mechatronik, South Russian State Technical University
Proswestschenja Str. 132, 346428 Novotcherkassk, Russia
Dipl. Oec. Aleksei Boulgakov
Faculty of building technology, Moscow State Building University
Kedrowstr. 14, H. 3, of. 405, 117036 Moscow, Russia

ABSTRACT

Construction is of great interest for usage of wood processing robotics. The structure, kinematic and dynamic models of robots for material location and spacing as well as wood processing have been considered in the paper. Taking into account specific features and dynamic characteristics of robots, recommendations on movements planning and forming laws of control have been given. In conclusion recommendations for obtaining accurate measuring data and on the composition of robot software have been provided.

KEYWORDS

Robots, wood, location, spacing, dynamic models, kinematics.

1. INTRODUCTION

When producing elements of prefabricated wooden houses a great deal of work on location and spacing of wood and shaping of holes, recesses, decorative patterns is done. The project´s individual peculiarities require the application of technological equipment with quick readjustment and opportunity to prepare control programs on the model during a short period of time. While solving this problem much attention is paid to the robotization of the operations mentioned above and the creation of robotic systems. The successful solution of robotization tasks is first connected with the development of original kinematic structures and competent structural analysis. One more significant problem of robotization of wood processing operations is setting the trajectory for cutting tool movement and provision of its purposeful movement along these trajectories with definite orientation.

2. ROBOT STRUCTURE, KINEMATIC AND DYNAMIC MODELS

When producing elements of prefabricated wooden houses a great deal of work on location and spacing of wood and shaping of holes, recesses, decorative patterns is done. The projects individual peculiarities require the application of technological equipment with quick readjustment and opportunity to prepare control programs for a model duping a short period of time.

The analysis of jobs connected with the location and spacing of wood as well as the shaping of holes and decorative patterns has shown that in the basis of the robotic system there should be a rectangular 3-coordinate gantry robot (fig.1). This cell provides the working tool movement in the plane of a working table as well as lifting, lowering and pressing of the tool. The second cell is an orienting working head providing changes in the tool position relative to the working plane or its rotation around the axis Z. The main kinematic relations defining the nature of the wood processing robot motions are presented by the system of the form

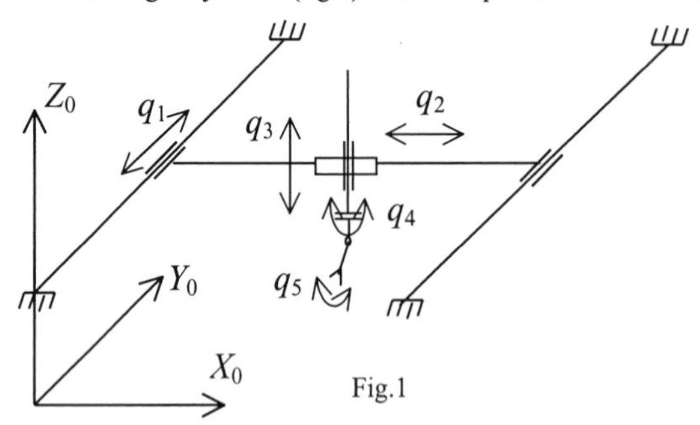

Fig.1

$$x(t) = q_1(t) + l\sin(q_4(t))\sin(q_5(t));$$
$$y(t) = q_2(t) + l\cos(q_4(t))\sin(q_5(t));$$
$$z(t) = q_3(t) + l\cos(q_5(t)),$$

where $q_i(t)$ are generalized robot's coordinates, l is the length of the end link.

The laws for changing the tool phase coordinates $x(t), y(t), z(t)$ and its orientations $\theta(t), \gamma(t)$ are determined by shape and view of the pattern being fulfilled. The angle θ specifies the tool pitch relative to the plane XY, and the angle γ - the direction of the inclination being read from the axis X. The values of these parameters their laws of change in time are formed at the stage of planning robot motions. The kinematics of orienting degrees of freedom have been chosen so that the tool pitch angles θ and the pitch directions γ are given separately by the degrees of freedom q_5 and q_4: $\theta(t) = q_5(t), \gamma(t) = q_4(t)$. In this case the laws of the generalized coordinates changes of the transportable degrees of freedom are described by the following ratios:

$$q_1(t) = x(t) - l\sin(\theta(t))\cos(\gamma(t));$$
$$q_2(t) = y(t) - l\sin(\theta(t))\sin(\gamma(t));$$
$$q_3(t) = z(t) - l\cos(\theta(t)).$$

To simplify the controlling functions and increase the accuracy of tool positioning a special-purpose orienting head has been developed, whose rotation axis of the last degree of freedom has been brought onto the plane of tool cutting. While rotating it allows the centre of the tool coordinate system on the rotation axis q_4 to be retained and obtain the main kinematic ratios of a simple kind:

$$[x(t), y(t), z(t), \theta(t), \gamma(t)] = [q_1(t), q_2(t), q_3(t), q_4(t), q_5(t)]\ .$$

Dynamic models have been used to control robots and develop control programs. For translational degrees of freedom in the basis of dynamic models the equation for the forces balance is used

$$\sum m_i \cdot \ddot{x} = F_e - \sum F_j\ ,$$

where m_i are the weights of movable parts; F_e - control force of degrees of freedom; F_j - disturbing effects incorporating frictional forces in movable parts of the mechanism and load forces on the working tool. The control force F_e is connected with a moment M. on the drive by the ratio $F_e = M \cdot i_r \cdot G / r$, where i_r is the drive gear ratio; G is the gear box efficiency; r is an effective radius of the mechanism transforming rotational motions into translatory ones. The drive dynamics for each degree of freedom in the models is presented in a linearized form by the system of equations

$$\alpha(s) = \frac{k_m}{s(T_m s + 1)} u(s),$$

where α is the angle of the motor shaft turn; u is the motor control voltage.

The degree of freedom along the coordinate Z has a double function. One is connected with tool lifting and lowering and the other with creating the necessary pressing force on the working surface. When a resolved force control is applied to the drive Z coordinate, the relationships connecting the moment and control voltage have been incorporated in the model:

$$M_m(s) = \frac{k_m s(Js + f)}{s(T_m s + 1)} \cdot u_m(s),$$

where J, T_m, k_m are inertia, a time constant and the gain factor of the drive.

3. MOVEMENT PLANNING AND CONTROL

The specific feature of controlling robots for location and spacing of wood is the necessity of forming programmed trajectories of cutting tool movement. On this basis the prediction of displacements according to the coordinates is made and control voltages for each of them are determined. At the foundation of the programmed robot control is the principle of setting movement trajectories for a cutting tool and movement program with help of the pattern being performed with the CorelDraw vector graphic editor. While composing patterns their accurate scaling in a special file is provided, this file is then used as an assigning file while carrying out control. On the pattern for material location and spacing or for performing decorative cutting the coordinates of the initial point, and the starting point of the process trajectory are assigned, transition lines between closed figures of the pattern are assigned as well. While processing each figure pattern scanning with digitization step T and read-out of information about the coordinates of the next positioning $x[kT+1], y[kT+1]$ are fulfilled. After the coordinates having been obtained control voltages are determined $u_x[kT+1]$ and $u_y[kT+1]$, and motion speeds along the coordinates x and y during the next control step are calculated as well. In order to reduce the effect of quantization as a control means in the interval $t_n \le t \le t_{n+1}$ we choose:

$$u[t_n] = 0.5[u(t_n) + u(t_{n+1})].$$

This ensures minimization of maximal deviation $\overline{u}[t_n]$ from the true value when changes $u(t)$ take place. When applying graphical means for setting movement trajectories there appears necessity to build algorithms ensuring the proper tool orientation at each point of the trajectory. To solve this problem different kinds of interpolation have been analysed and in the algorithms of robot control a parabolic interpolation in the interval $[(n-1)T, (n+1)T]$ is selected as a basic one. In this case for the time moment $t = (n+1)T$ a predictive calculation of coordinates for the point of tool position is made and according to the coordinates of three points coefficients of an interpolating equation are defined:

$$\begin{bmatrix} a_0 \\ a_1 \\ a_2 \end{bmatrix} = \begin{bmatrix} x^2[(n-1)T] & x[(n-1)T] & 1 \\ x^2[nT] & x[nT] & 1 \\ x^2[(n+1)T] & x[(n+1)T] & 1 \end{bmatrix}^{-1} \cdot \begin{bmatrix} y[(n-1)T] \\ y[nT] \\ y[(n+1)T] \end{bmatrix}$$

Next the obtained values of coefficients a_t are checked:

$$y[(n+1)T] = a_0 + a_1 x[(n+1)T] + a_2 x^2[(n+1)T]$$

and in case of identify the angle of normal to the trajectory in the point $t[(n+1)T]$ is calculated

$$\theta[(n+1)T] = 0.5\pi - \mathrm{arctg}(2a_0x[(n+1)T] + a_1 \,.$$

After determining the vector of tool position and orientation and the vector of generalized coordinates $\overline{q}[(n+1)T]$ for the next step of control the values of tool motion speeds $\dot{x}$, $\dot{y}$ and the angular speed of rotation θ are calculated. The accuracy of development of the movement trajectory depends on the accuracy of fulfilling the task mentioned above. The simplest algorithm is the linear dependence

$$v_x[(n+1)T] = (x[(n+1)T] - x[nT])/T$$
$$v_y[(n+1)T] = (y[(n+1)T] - y[nT])/T$$
$$\omega_\theta[(n+1)T] = 2\pi(\theta[(n+1)T] - \theta[nT])/360 \cdot T$$

To perform high-quality patterns we have incorporated high precision control algorithms. For this purpose the trajectory of tool movement is interpolated by a polynomial of the third degree and according to the interpolating function $S_n(T)$ the derivative $\dot{S}_n[(n+1)T]$ is determined. The obtained derivative values are applied for calculating generalized speeds:

$$v_x[(n+1)T] = \dot{q}_1[(n+1)T] = V_0 \cos(\mathrm{arctg}(\dot{S}_n[(n+1)T]))$$
$$v_y[(n+1)T] = \dot{q}_2[(n+1)T] = V_0 \sin(\mathrm{arctg}(\dot{S}_n[(n+1)T]))$$
$$\omega_\theta[(n+1)T] = \dot{q}_4[(n+1)T] = (\dot{S}_n[(n+1)T] - \dot{S}_n[nT])/T$$

Taking into account the fact that during the functioning of control algorithm we can form sampled-data functions $q[nT]$ describing the trajectories of links movement, then to determine derivatives in the points of tool movement trajectory it is advantageous to use control algorithms for the model. While simulating the process of motion differentiation of digital sequences is presented as a sum of the kind:

$$\dot{q}[n] = T^{-1}\sum_{k=1}^{m} K^{-1}\nabla^k q[n] = \sum a_i q[n-i],$$

where $\nabla q[n] = q[n] - q[n-1]$ - inverse difference; m - number of terms of a degree series; $a_i = (-1)^i \sum_{i=0}^{m} K^{-1} C_k^i$; C_k^i - binomial coefficients.

Besides the control method considered on the graphical model it is necessary to include algorithms of programmed control by the path reference point into mathematical calculations and software. In this case reference points $P[x,y]$ and angles of tool orientation in each of them are assumed. Taking into account these values we form the data base to develop the laws of generalized coordinates changes.

Motions planning is thus fulfilled on the basis of interpolation with cubic splines: $S_3(t) = \sum_{k=0}^{3} a_k t^k$.

Coefficients a_k are calculated in each section of interpolation having assumed that the trajectory is continuous and smooth. Planning of the robot's movement is carried out taking account of limitations in degrees of freedom which are preset with reference to the table of limit values for each coordinate $q_i^{\min} \le q_i \le q_i^{\max}$.

The authenticity of the measured parts is of great importance for control. In the encoding position sensors being used there may be abnormal short-term imperfect data that can lead to short-term limit accelerations and deviations from the trajectory. The casual character of some changes of these sensors makes us to introduce predictive and correcting algorithms of simple calculating structure into the algorithms of robot control. In the applied algorithms, expected values of position along the coordinates for the period of change T are predicted in each control step with help of m degree polynomial. Meanwhile we apply algorithms of single prediction, which are based on the Lagrangian interpolating polynomial

$$P(t)=\frac{1}{\tau^{m}}\sum P_{k-i}\prod_{j=0}^{m}\frac{k\tau+\tau r}{j-i}+\frac{M_{m+1}}{(m+1)!}\prod_{j=0}^{m}(k\tau+j\tau)\ ,$$

where k=1,2,3 - prediction steps.

The analysis of prediction errors has shown that algorithms of double prediction should be applied for robots and values at three points (m=3) are to be taken. For convenience in usage of predictive expressions and reductions of calculations it is better to use a recurrent form of the analysis.
A robot software solves the following problems: analog and discrete information about the parameters of robot condition is obtained, interface with drivers of analog inputs and outputs, output of discrete and analog control signals, formation of technological and emergency information, data display in real time. The programs can function in the environment of Windows and other versions and use all features of this environment. Drivers for data exchange are formed as dynamic libraries DLL. The software includes robot models, which allow to carry out check of algorithm operation.

4. CONCLUSION

The material is prepared on the basis of the authors' research, which was carried out while developing robotic systems for wood processing. The presented kinematic structure, algorithms of motions planning and control have been investigated on the models. Computer simulation of robot motions has shown the effectiveness of the described methods and algorithms. The obtained simulation results were applied while developing and designing robots for cutting and spacing of wood.

Advances in Building Technology, Volume I
M. Anson, J.M. Ko and E.S.S. Lam (Eds.)

ASSEMBLY AND DISASSEMBLY OF INTERIOR WALL

Dipl-Eng. Albrecht Hanser

Technical University Munich, chair of building realization + -informatics, Prof. T. Bock
Arcisstrasse 21, D 80333 München, albrecht.hanser@bri.ar.tum.de

Abstract

Flexible extension walls - watched on a long-term-use - have lower costs and are more economical and especially more variable compared to convenient post-and- beam-structures. Because room requirements – mainly in office and administration building – change faster and faster, an extensive potential on future markets for these technologies exists

Keywords

Flexibility, Economic Efficiency, Non-supporting removable wall-unity, Media- and electric wiring, Put-and-slide-technology, Relocate systems, Cost, Different Elements,

Figure 1: The non-supporting removable wall-unity

Present situation

Flexibility
A higher and higher flexibility of the floor plan is expected in office- and commercial building as in house building. The change of users or function often results in the wish redesigning the existing

spatial pattern. A basic floor plan flexibility is often possible and implemented by dividing the load-bearing building surface from non-structural separation walls and ceilings. A well known system of this type of walls is a wooden or metal post-and-beam-structure with mineral fibre filling and plasterboard panelling. By using this building system it is difficult to realize variations in different room structures and subdivisions by self building. Once they are cheaper to produce and to install but in the end they cause much difficulty in redesigning. They are not flexible and actually not reusable. There are reusable, demountable partition systems available, usually to expensive and to much trouble to construct.

Economic Efficiency

All partners in the construction business know exactly what construction cost means. Very often we think and act only in this kind of category.

The reason for this is the existence of two different types of budget. One for construction and a second for consequential costs during the life cycle, which are even managed by two different teams. Feedback between the two will rarely occur. So at the end of a life cycle for a building, nobody is able to find out the real cost situation which has been followed after construction time, of a product incorporated in the building. We cannot find out what the savings could have been, , even if we had constructed in prefabricated and modular systems.

Indeed thinking in a complex overall view, we realize the advantages for prefabricated and relocatable partitions are:

- Connections with its components, seperations in floor- and ceiling-areas are close in cost of these construction types.
- The finishes of gypsum board partition-walls, assembled on site, do not have the same standard of quality as industrial prefabricated partitions. Prefabricated partitions need no new coatings for 30 years. Conventional gypsum boards need it after 5 years.
- If prefabricated partitions get relocated, the difference in investment cost becomes obvious. A prefabricated partition incurs 40% of its former construction cost, while a conventional partition, which has to be demolished and reconstructed almost completely new, incurs 220%. Beyond a doubt, the payout-return depends on type of construction prefabrication and the number of relocations.
- Recovery and cleaning up belong to an ecological and economical overview; in this context reusable systems offer advantages. Savings of material and avoidance of waste have no exceeding fees for supervised dumps. These charges have been growing constantly in recent years.

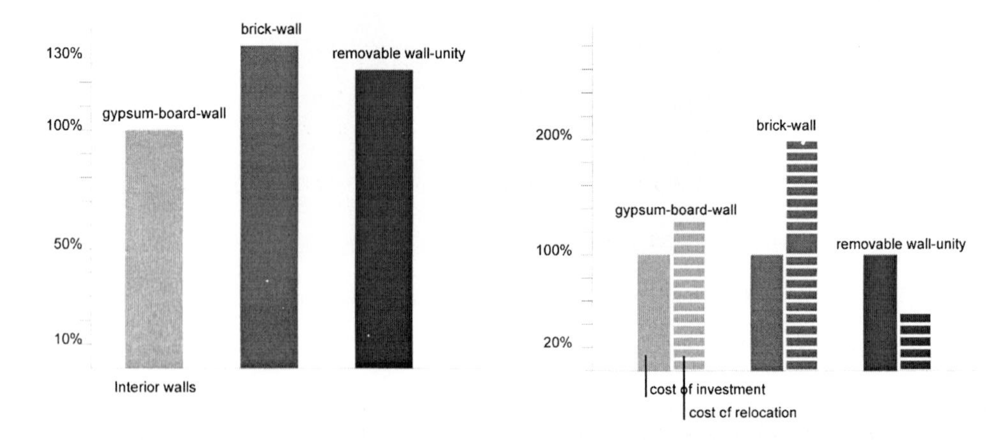

Figure 2: interior walls: cost of investment and relocation

Aim of the project

It was valid to develop a non-supporting removable wall-unity, which allows adaptable, fast and disassembling room-splitting.
Next to the standard components we have to develop elements for the connection at the existing support structure, elements for passages and doors, for windows on the inside and other glazing structures.
Media- and electric wiring are components of the removable wall-unity. The elements are delivered prefabricated and manageable on the construction site.
The connection to the available support structure is created with a wiring mechanism (Justix); the coupling of the elements and also the management of media and electric to work with a put-and-slide-technology.

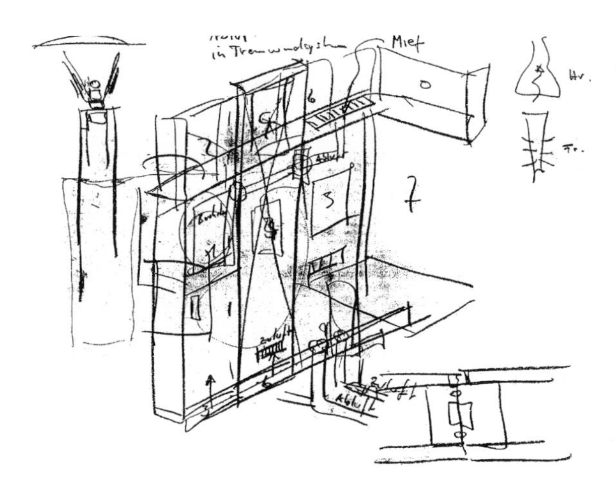

Figure 3: First sketch

Development

The system must be:

- "More than a wall", which means integrated media, such as TV, phone, flat screen, switches, bus, safe, mini-bar, shelf-systems, cupboards, completely connected up;
- Demountable and reusable without damaging the existing solid construction;
- With finish-surface on both sides, in complete lightweight construction;
- Suitable for two persons to handle and install
- Located in price between the plaster board/ post-and-beam-structures and the conventional wall systems.
- Simple construction and long-term use

Beside separating for offices, administration or housing the wall unit is supposed to serve as a room divider, half height screen wall or fair stand.
Put on wheels it can even become a mobile office wall.

One of our aims is to see the wall not only as a separating element between two areas, but to include additional functions such as screen lights, mini-bars, shelf-boards, integrated air-conditioners, even a PC or printer can be hidden in the wall;

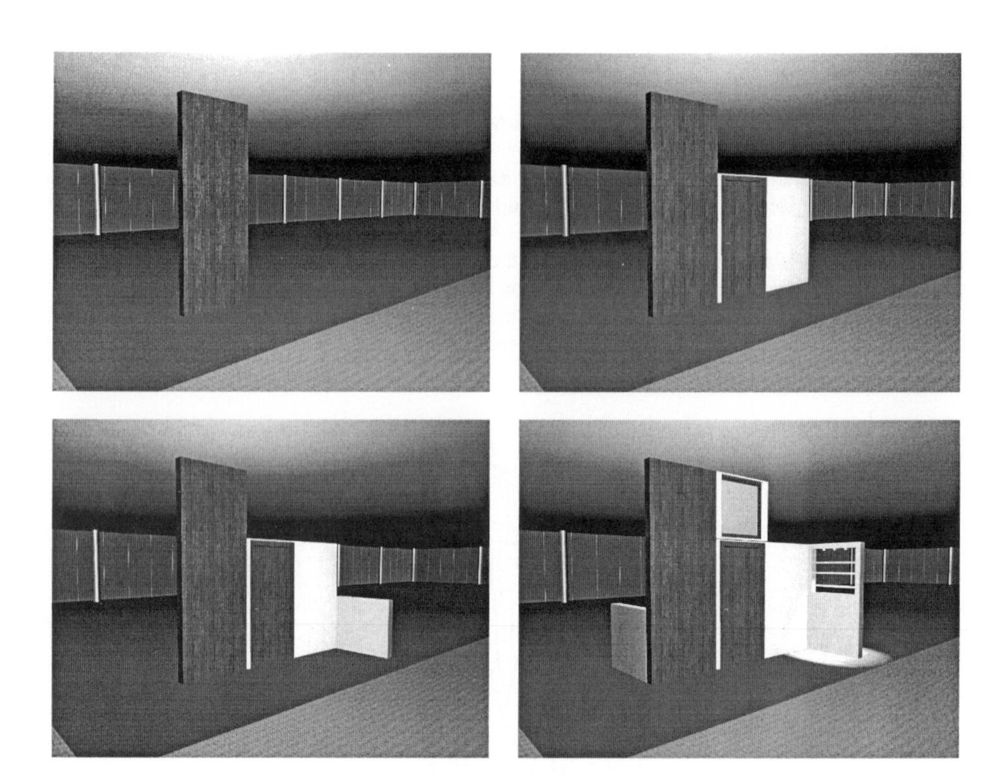

Figure 4: Simulation of the elements

Elements

Surface finished, easy-to-handle modules are basic for interior wall surfaces.
Elements are:

- Full room heightelements (size depending on height of the ceiling, standard sizes)
- Door high elements (height 210 cm)
- Elements fitting above door high elements (size depending on height of the ceiling)
- Elements for easy zoning, separating areas (height 105 cm)
- Special elements

Construction

The substructure for the single module is to be a ring-like frame. Panel materials for wall surfaces such as wooden panels, plasterboard, glass etc. serve as frame bracing, and wooden profiles but also special formed plates can be the frame material.
The connection to the load bearing wall construction and to the ceiling is designed as edging board, and the gap between the bottom of the wall and the floor is covered by a clipped baseboard. A dovetail-like joint with integrated plugs for power and data supply connects the single elements to one another and to the load bearing walls.
The single modules, therefore, are coupled by fitting and pushing the connecting elements together. The stack design is variable.

Door high elements and end elements are connected in the same way. A special processing of the frame profiles makes it possible to carry the prefabricated electronic wires through the element.
The contact from element to element becomes possible through the connecting modules just mentioned. The necessary fitting in to the surrounding construction could be succeeded by vertical adjustable foot construction, which there are two for each wall panel. After connecting the elements the foot adjustor is screwed up and pressed against a connecting profile. The gap arising at the bottom of the wall can be used for supplementary wires – especially for wire systems, that can only be added later (phone, networks). A clip on profile covers the open bottom area as a baseboard. There must be sound and fire technical measures on the surface, in the stack area and in the base and ceiling area.

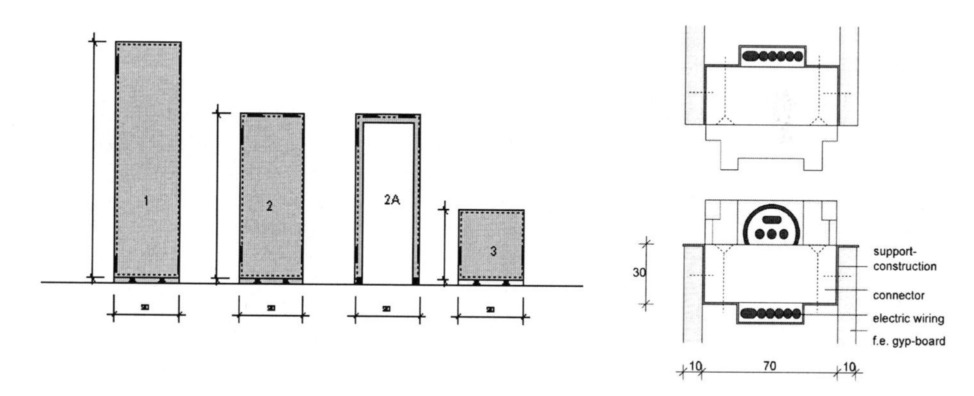

Figure 5: The elements; dimension: 90cm wide, 300cm, 210cm, 90cm high; the connecting system

Fabricating prototypes

Currently parallel extensive tests are taking place using prototype wall modules to transfer the practiced knowledge gained.

Conclusion

The system shown in this paper will find place between the low cost but not reusable gypsum board partition-walls and the industrial prefabricated partitions, which are very expensive. The combination between a prefabricated high tech system, affordable and economical, usable as flexible elements for different tasks are the main advantages of the new system.

References

Studiengemeinschaft Fertigbau (1987): Trennwandsysteme

Tichelmann / Pfau (2000): Entwicklungswandel Wohnungsbau, Vieweg, Wiesbaden

Diverse Herstellerunterlagen(1999, 2000, 2001, 2002): Lindner, Knauf, Item, Dorma, Häfele, Strähle, Hettich, u.a.

Advances in Building Technology, Volume 1
M. Anson, J.M. Ko and E.S.S. Lam (Eds.)

MUNITEC-FAST-CONNECTORS - KEY TECHNOLOGY FOR PREFAB HOUSES

Dipl. Eng. Frank Prochiner
Faculty of Architecture, Technical University of Munich, Germany

ABSTRACT

Most leading manufacturers of timber housing confirm that developments leading to a greater degree of prefabrication, and thus a shorter construction period, would significantly improve the sales of this type of building. Today, the carcass structure of so-called prefabricated houses can be erected in an extremely short period, but the subsequent fitting out can easily take a further 4 –6 weeks because of the division of work between different trades that are often dependent on each other. The site assembly of complete, large-scale elements that require no further processing is something virtually unknown at present.

KEYWORDS

Prefabrication, house, connector, process, construction, connection system.

The automobile industry provides a good example of the way an assembly process can be divided between a number of construction groups and suppliers and carried out concurrently or in an efficient sequential manner. Crucial in this respect is the coordination of the various working interfaces between different trades. This includes a strict compliance with manufacturing tolerances and the development of a simple connection system for all elements. Similarly, the advantages of any assembly system will be fully felt only if the connections can be made without the presence on site of various specialists. As part of a project being conducted at the University of Technology in Munich and supported by the Federal German Ministry for Education and Research, a quick, plug in connection unit is being developed that will facilitate the erection of housing in record time. It should be possible in future to take a building into occupation after only 24 working hours; in other words, three days after the commencement of construction. The first day would be taken up by the actual assembly work, the second day by the fitting out and finishing, and the third day by the occupants moving in.
The new quick-connection unit, for which a patent has been applied, will allow a simultaneous coupling not only of the building elements, but of all the systems integrated within them, , such as heating and electrical installations and control networks. The units are fixed to the end faces of the building elements that are to be connected.

Where a number of runs for different media have to be linked, the connection
units can be installed in a row. Where there are no installations in two adjoining elements, the connecting units will serve solely as a means of mechanical jointing. In collaboration with partners in industry, plugs and sockets have been developed for all kinds of media, including electricity, telephones, data lines, TV, and also liquid media such as heating runs. Further developments are foreseen for waste and soil pipes.
When two wall elements are pushed together, the quick-connection units provide a rough centring for the actual plug-in link. The self centring geometry of the plug then ensures a safe

connection. The connectors permit the coupling to be executed from the side or the rear, in which case the service runs can be installed in the inner skin of the wall or in the wall itself. On

MUNITEC-Fast-Connector

inserting the connectors, their wedges haped geometry causes the elements to be pressed together against the sealing strips. The dead load of the wall element thus ensures a good seal. All heating and electrical runs are installed in the walls and on the underside of the floors, so that the depth of the floor finishings is not a factor in this respect. The ends of the circuits in the walls simply have to be plugged into a central vertical services shaft.

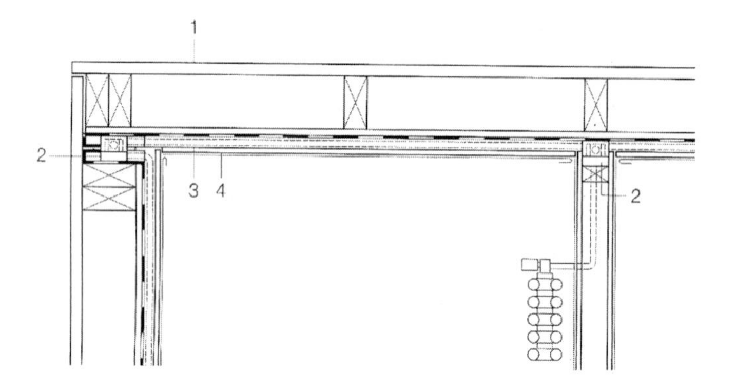

External wall connection – internal wall connection

1 – wall construction depends on manufacturer; 2 – Munitec-fast-connector; 3 – installation zone, integrated in Pavatex panel; 4 – plasterboard, wallpaper and paint finished

This system reveals a number of advantages over traditional forms of construction and assembly.

• It is faster: the construction period is reduced to three days; service installations are prefabricated and incorporated more swiftly; and building firms are less dependent on the weather and time of year.

• It is more economical: the shorter assembly period represents a big potential cost saving; and life-cycle costs are reduced as a result of the simple method of removal and disposal.

• It ensures a higher level of quality: all components and services are installed at works to a constant quality level and with comprehensive quality controls; and the risk of accidents on site is reduced.

• It is environmentally friendly: building sites are cleaner; waste products are avoided on site; and the reuse of components is simplified by the scope provided for a damage-free dismantling and removal of elements.

An adjustable foot serves to position the individual wall elements on the concrete floor slab. A threaded member integrated in the foot allows the wall unit to be raised and

Adjustable foot

Lowered until it is precisely in position. This is indicated by a sound signal from an electronic spirit level on the top of the wall. To facilitate a parallel execution of the work by the various trades, all service installations should be prefabricated to as great an extent as possible, so that these can simply be laid in the wall, floor or roof elements by the respective timber construction firm.

In the case of the electrical installation, a patented system is available – a cable harness or premoulded wiring system – that simply needs to be plugged in, in a similar technique to that used in the car industry. The planning data are coordinated in a special computer program, which also prepares a wall diagram showing the actual cable runs. The ready-wired plugs can be simply clipped into the quick-connection unit. To avoid too many transfer points between the individual building elements ,the cable runs are distributed in the wall with the aid of a newly developed subdistribution system. The conventional distribution box is

Backview of the MUNITEC fast connector with plugged in connector for data.

replaced by a power installation box (PIB).A simple coding system ensures that runs cannot be plugged into the PIB in the wrong position. In this way, the whole process of wiring up and making connections is obviated, saving time, and allowing the work to be executed by non-electricians.

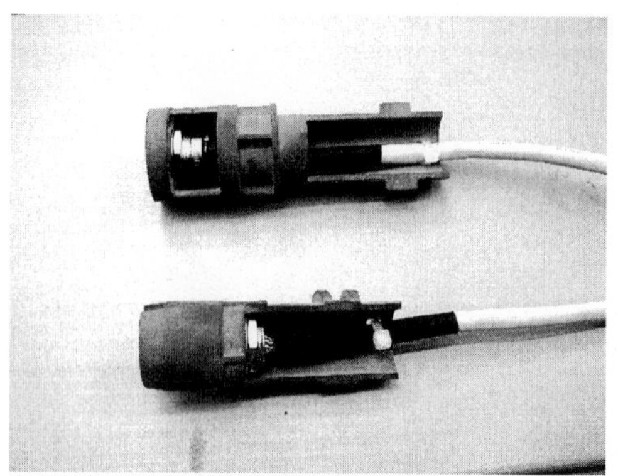

Pair of plugs for data wires.

If the pipe runs for the heating installation are installed in the inner skin of the wall as heating coils, it will be possible to obviate the need for radiators entirely. Wall heating panels of this kind make sense especially in low-energy buildings. They provide a pleasant radiant heat and are not visible. In contrast to buildings erected in traditional forms of construction ,the installation of heating coils at works avoids damage to pipes and potential leaks. A 30 mm sheet of fibreboard laid on the inner face of the structural wall elements serves as the plane for the installation. This layer allows chases to be cut for the simple fixing of electrical, water and heating runs. At the same time, the fibreboard improves the sound and thermal insulation of the wall.

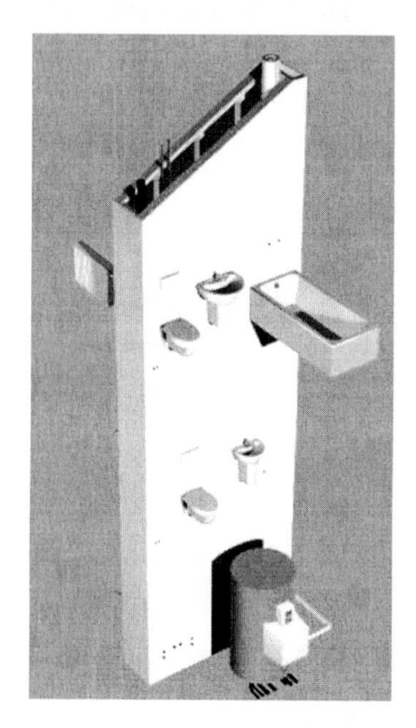

The central services core is prefabricated complete for all storeys of the building and hoisted into position on site by crane. The entire sanitary installation will possibly be incorporated in it, as well as the vertical distribution cables and runs for the electrical installation and heating.
The vertical shaft will be fitted with the appropriate connection units on all faces that abut walls. It is also conceivable that entire construction groups such as the boiler for the heating, or the sanitary fittings can be plugged in via an appropriate connector. Using this connection technology, it would be possible to achieve a new quality in building in the future. In a system analogous to that found in vehicle construction ,subcontractors and suppliers of components would work collaboratively to provide large system modules. This would also open up new perspectives for the export of prefabricated housing. At present, manufacturers and assembly firms work largely on a regional basis because of the transport costs involved and the fact that site work can require the presence of a trained assembly team for between 2 and 6 weeks. The use of a plug-in system of connection units would reduce the time required for assembly and finishing work to only 2 –3 days. Countries like Japan that are heavily dependent on the import of timber for construction purposes could order entire prefabricated housing kits. Similar markets exist in eastern Europe as well. Emergency shelters for disaster or crisis areas could also be provided in fully functioning form within a short time.
An important aspect of this system is the absolutely damage-free dismantling process once the building has served its useful life. Components or entire buildings can be reused or returned to the manufacturer for recycling – an important aspect when one considers that the building industry holds the record at present for
the creation of waste and the consumption of raw materials.
The use of quick-connection, plug-in technology need not be restricted to the building sector, however. The principle could be adapted for application in many different areas where, together with mechanical connections, service, supply or signal runs have to be linked. Easy maintenance by means of a simple exchange of
units could result, for example, in defect telephone boxes not requiring the services of qualified engineers or electricians to repair them at the place of installation. They could be simply unplugged and replaced with a new unit by service staff. Repairs and maintenance could then be carried out by qualified personnel under more pleasant conditions in the workshop.

TEAM PRESENTATION:

INDUSTRIALIZED HOUSING

Advances in Building Technology, Volume 1
M. Anson, J.M. Ko and E.S.S. Lam (Eds.)

INITIAL THOUGHTS ON INDUSTRIALISED SOLUTION FOR LOW-COST HOUSING IN CHINA'S RURAL AREA

Guan Kai and Li Shirong

Research Centre for International Construction Economics and Management,
Chongqing University
Shapingba, Chongqing, 400045, PR China

ABSTRACT

This paper deals with the industrialised solution for low-cost housing in China's rural areas. The paper firstly focuses on materials used for frames, roofs, floors, walls and ceilings. It then analyses the requirement and the construction methods for each material on fireproofing, antisepsis and waterproofing. Based on the analysis it also gives some suggestions on the detailed connections of the structural components and the surrounding components. From the structural engineer's point of view, this paper analyses the general load supporting system and the basic principles on the disposition of the structural components in both horizontal and vertical directions. It also gives an introduction an the normal steel components on the size, component, connection and methods of antisepsis and fireproofing. The last part of the paper deals with manufacture and site construction for industrialised housing in China.

KEYWORDS

Cold-formed steel frame, design, industrialised housing, China, construction, low-cost housing, poverty

THE BACKGROUND OF THE RESEARCH

The housing for poor people or low-income people is a mayor problem that every government must face. Because of the huge population and the undeveloped economy in China, it is especially a serious social problem. The traditional construction methods used for hundred of years can't fit the modern needs due to the consideration of sustainability, low-cost, quick and easy construction, high quality, easy maintenance and possibility to move. There is a need to find new approaches to solve these problems.

In addition, looking at the huge rural area in China, there are still many people who cannot afford normal housing. It's especially difficult for the people living in a mountainous area or some old, weak and disabled peasants who are less able to work.

In order to provide low cost, good quality and sustainable housing for those poor people in rural areas in China, the Chongqing University (China), in cooperation with South Bank University (UK), has set up a research team in 2000. The research team has been given financial support from both the British Council and the China of the Foreign Expert Bureau for three years (from 2001 to 2003). The research team has finished the first stage work so far which was to survey the current rural housing situation, the rural people's need and the local building resources in the rural area. This paper is mainly based on this research project.

THE TARGET GROUP IN THE RESEARCH

Not all the people living in rural area are the target group for the research. The determination of the target group is based on the policies of some local government towards poor housing in rural areas. Each year, for example, the local government provides a certain amount of money to help those people to build new houses. It focuses on those people who live in the caves and the mountain regions and have poor living conditions. There are five types:

(1) Those people who live in the mountain regions where the altitude is over 1000 metres, and also very cold with poor living conditions;
(2) Those people who live in the remote mountain regions, which are far from villages (more than 3 km). The investment for infrastructure is too big and can not to be solved in a short time;
(3) Those people who live in the regions where endemic is very difficult to cure;
(4) Those people who live in the regions where 70% of the cultivated lands slopese more than 25 degree, and lean soil less than 15 cm in depth, and the outputs of foodstuff per hectare are not more than 2kg;
(5) Those people who live in the regions where there are no water supply facilities and it is very difficult to get water without walking more than 1km and climbing is more than 100 metres.

THE PRINCIPLES OF THE SOLUTION

The aim of the research is to find a way to build low cost, easy to build and maintain, good quality and sustainable houses for the poor people in rural areas in west China. To meet the general requirement, some principles to focus the research are listed as follows:

- The layout of the house must fit the needs of the poor people's real life
- The house should keep a traditional Chinese character
- Local materials and resources should be used as much as possible
- The construction should be easy, quick and economical
- The building materials should best be manufactured locally and be usable by less skilled people
- The house should be sustainable with lower environmental impact
- The house should be cheap and affordable

THE RESULT OF THE FIRST STAGE WORK

After the first stage of the research, some useful information on housing in China's rural areas was obtained and this information is most important for the next stage of the research work.

The Cost of the House

Clay-brick and pre-stressed reinforced concrete boards are widely used to build houses in rural areas. The cost of the house is usually around 200~300 Yuan per square meter. The cost difference mainly depends on the local building products and local transport conditions.

The cost items of the house in rural area are building material costs and construction costs. The land for a farmer's house in rural area can be used after the application is accepted by the local town authority. The local rural people help each other to build houses. The construction costs then usually cover the food and accommodation for these people and the owner also invites several skilled workers to guide the work. The detailed items and the percentage of each element of work are listed in table 1.

TABLE 1
THE DETAILED COST ITEMS

Item Name	Percentage	Sub-item Name	Percentage
Building Material cost	90%	Walls	25%
		Roof & floors	30%
		Doors & windows	15%
		Foundation	20%
Construction cost	10%		

Table 1 shows the normal costs for a two-story brick-concrete house in China's rural area. If the building materials and the floors are different, the percentage of each item should be quite different. For example, to build a one story brick-concrete house, the percentage of the foundation and the roof & floors will drop while the percentage of other items will increase.

Based on the initial investigation, some initial thoughts come out for the housing design, namely:

- Looking for cheap materials to build walls, floors and roof
- Using light weight materials to reduce the load and the cost of the foundation
- Adopting new construction technology to shorten the construction period

The Local Building Resources

The investigation found that the local resources to be used could be straw, bamboo, stone and sugar cane.

In most parts of South-west China, rice is the main crop in rural area. Only 5% of the straw are used for animal food and then most of them are burned.

Bamboo is the most popular plant in South-west China in the rural areas. Normally people use bamboo for crafts such as basket or other commodities and it has not been used in housing in recent years. Bamboo grows much faster than wood and has high strength in the vertical direction. Another advantage of bamboo is its very cheap price of 0.12 Yuan per kg for retail.

Stone is easy to get for the people living in mountain areas and highlands. Even today it is still widely used for the foundation or walls of the houses. It is almost free to local people who find stone on the mountain. The only cost is to transport and shape it but people in rural areas usually do this themselves.

Sugar cane can be found in the countryside in the western part of China but not over the whole area. There is potential to use sugar cane as a building material in the future.

The number of people living in a house

The number of people living in a house is one of the key issues in the research and it is different from the number of people in a family, because some people in the family usually go out to work and don't live in the house. It has been found that most of the poor families have up to 4 people. Due to the poor conditions in the poor areas, some young people in the families usually leave the area.

THE INITIAL THROUGHT ABOUT THE SOLUTION

After the first stage of work some thoughts exist on the industrialised solution for the housing for poor people in rural area in China. This will be the main content of the next stage of work once proved to be practical and effective.

The design

The conceptual design is based on the real rural life need listed in the report of the survey on rural housing. However, the flat has a toilet and kitchen which will give much more improvement to people's life in rural areas.

The dimension of the kitchen is 3 meters in width and 4.2 meters in length, while the dimension of each room are almost uniform, about 3.6 meters in width and 4.2 meters in length. One of the reasons for doing so is that local people like a big kitchen and big rooms so that they can put some firewood in the kitchen and some working tools in rooms. Another reason is that the more uniform the dimensions are, the more convenient the manufacture and construction is.

Figure 1 shows the lay out for a house, for a couple. If there is more than a couple living in the house, the lay out can be changed as shown in Figure 2.

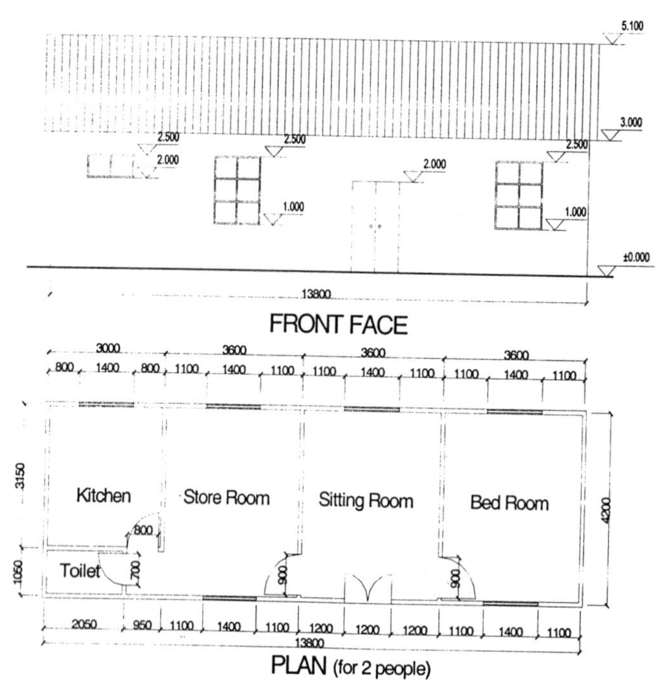

Figure 1 Lay out for two people (a couple)

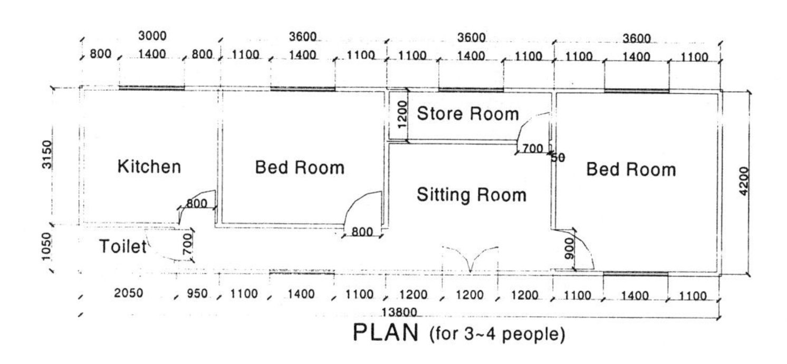

Figure 2 Lay out for 3-4 people

The lay outs shown in Figure 1 and 2 are for up to 4 people living together. If there are more than 4 people living in the house, a dwelling unit can be easily added to fit the need. On the same way, more dwelling units can be continually added for more people living together.

The wall and roof

The wall and roof are the main costs for housing. If the costs for walls and roof can be controlled successfully the total costs to build a house can be greatly reduced.

The best way to reduce the costs of walls and roof in rural area is to use straw board or bamboo board. build it. The techniques of making this board, with different thickness and different strength has already been researched.

The suggested combined board for the project is shown in Figure 3.

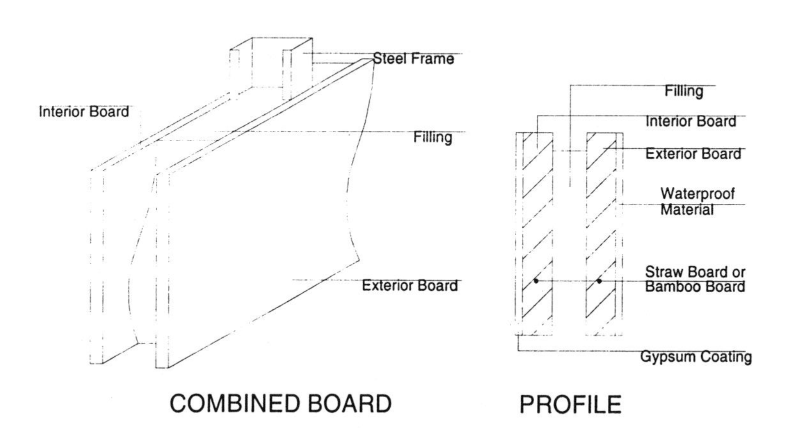

Figure 3 The housing structure

As shown in Figure 3, the interior and exterior boards can be made from straw or bamboo, shaped at high temperature and high pressure. The byproduct of the board such as the bamboo leaves and straw can be used as a filling material after it has been dried, pressed and shaped. This filling material can bring benefit both at taking full use of local resources and getting better thermal performance and sound insulation. The gypsum coating on the interior surface can help the house to be dry and comfortable. Asphaltum or paint can be used as a waterproof material on the exterior surface. The suggested thickness of these coatings is 5mm.

This board is cheap, strong, light-weight, has good performance and is local resource based.

The structure and the foundation

There are two choices of load-bearing systems identified in the research.

The first is the load-bearing wall system making some external wall and internal wall act as load-bearing to stand the vertical load. The advantage of this system is simplicity and low cost, but the disadvantage is to increase the thickness of the board to ensure that the board has enough strength to stand the load. Probably it is difficult for the local farmers to manufacture the board with simple facilities.

The second system is to use a steel frame to stand both vertical and horizontal loads and non load-bearing walls to enclose the house. The advantage is a high strength frame and board made of minimal thickness. However, the disadvantage of the system is that the cost may be higher than the load-bearing wall system.

Of these two systems, the steel frame system is more responsible. Considering overall costs, though the steel frame system will cost more than the board system, this board can be cheaper, the manufacture and construction can be easier and the equipment in the factory can be simpler and cheaper. Therefore, a steel frame system is considered as a possible solution.

Hot-formed steel is used because it cheap and has been widely used to build structures. Steel poles and beams are necessary to build a basic frame shown in Figure 4.

In this way, the house could be built to provide enough resistance at a affordable cost.

The sizes and sections of the steel elements are carefully designed to obtain the structure with best economic performance. L section hot formed steel is the primary material for the structure because of it's low cost and low skill needed to form it.

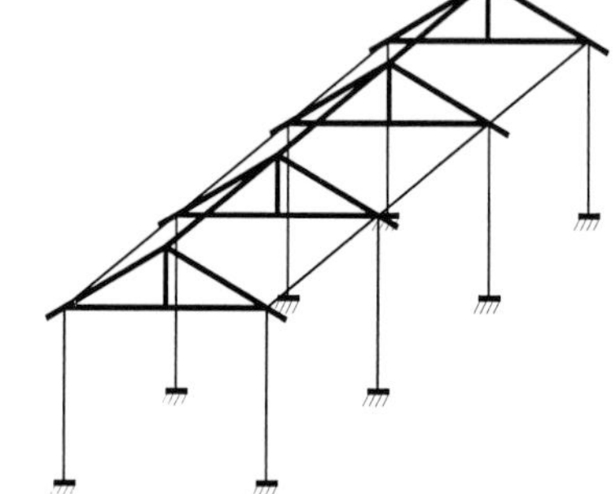

Figure 4 The structure

For resistance to rust of the steel frame, it is suggested to use paint because it is cheap, easy to get in rural area and low skill required.

Stone is a good material for the foundation because it's cheap and easy to find in rural areas. In some areas, such as the north west of China, stone may be difficult to find. Concrete should then be the foundation material. The steel poles are counter sunk into the stone foundation and cement concrete is used to secure them in place.

The financial support

If the cost of the house is kept below 200 Yuan per square meter, the total cost of the house should be no more than 11,500 Yuan. Even though this is a low cost house it is still a great deal of money for the local farmers and financial support is very important in the project. For poor people there are two ways to have their own houses.

The first is to obtain government financial support. For poor people, the Chinese government has a poverty support plan every year. The central government and local government will give special money to homeless people to get their house on the plan, at about 10,000~15,000 Yuan for each family. In Chongqing, for example, the policy is that the government pays 90% of the money and the farmer pays the rest.

The second way is to get support from other organisations. For example, the World Bank and the Asia Development Bank have fund for poor people support every year.

CONCLUSION - THE FUTURE RESEARCH

The following will be the key activities for the next stage of the research:

- Collecting detailed information from some organisations which are good at using bamboo or straw to make building materials and ensuring that there should be no technical problems on our solution.
- Setting up supply chain in both China and abroad focusing on building materials.
- Developing the detailed design, including the detailed structure, building services and connections.
- Doing further research on the local climate, people's income and related local government policy.
- Looking for possible organisations to give financial support for the poor people in rural area.

Advances in Building Technology, Volume 1
M. Anson, J.M. Ko and E.S.S. Lam (Eds.)

INDUSTRIALIZED HOUSING CONSTRUCTION – THE UK EXPERIENCE

R. Howes

Professor in Construction Management
Chairman of the Innovation and Research Committee, Construction Industry Council, UK

ABSTRACT

The historical development of industrialized housing in the United Kingdom is outlined and key milestone developments have been identified. Past mistakes have been analysed, together with the affect that this has had on the progress of advances in building technology associated with domestic house construction. The circumstances relating to the renaissance of industrialized housing have been highlighted, leading to the current state of development. The paper analyses the benefits to be derived from industrialized housing over traditional house construction, assuming favourable circumstances. Attention has been given to problems and pitfalls to be avoided in order not to repeat mistakes that have manifested themselves in the past. The paper concludes with an assessment of potential future developments and draws attention to the need for acceptance by the construction industry and future homebuyers of industrialized housing.

KEYWORDS

Industrialized, system, traditional, frame, panel, steel, concrete, timber.

INTRODUCTION

The need to provide affordable, sustainable and functionally competent housing is fundamental to the living standards and well being of individuals and families throughout the World. The United Kingdom is no exception and there has been a long history of housing shortages and substandard housing. A common perception is that the UK housing system has failed to provide adequate housing for low-income households and the generally poor maintenance of the housing stock, particularly at the low end, Gann et el (1997). A major problem is the cost of land and the limited availability of affordable urban sites especially in the South East of England. The British Government has been slow in releasing land for housing and the cost of cleaning up brown field sites has led to a situation where the demand for housing far out paces the supply. This situation has been exacerbated by skill shortages in traditional building trades and escalating wage rises. The quality of construction and the lack of design innovation have resulted in higher than necessary life cycle costs and the premature decline in the condition of property leading to higher expense to rectify defects.

In Japan, USA and Europe savings of up to 30% in cost and time have been claimed by the introduction of industrialized housing techniques where housing is constructed in factories using standard components, Fairs (1998). This has the advantage of reducing construction activity and time on site and provides the potential to assure quality and guarantee in use performance.

The introduction of prefabricated industrialized housing is not new and there have been many public and private initiatives over the past 80 years to implement industrialized techniques in support of the need to meet increasing housing demand. None of these initiatives have led to an identifiable progressive development of industrialized housing and as a consequence it only represents a small percentage of the total UK market. This paper is intended to trace the historical development of industrialized housing in the UK with the objective of identifying successes and failures for the purpose of establishing what needs to be done to increase the potential of future attempts to produce houses using factory based manufacturing.

HISTORICAL DEVELOPMENT

The review traces the development of industrialized housing and aims to identify significant systems and their degree of success or failure. It has been stated that standardization applied to building technology was borrowed from manufacturing and the intention was to increase precision and gain efficiency through repetition and continuous improvement, Finnimore (1989). These concepts were developed and it was proposed that houses should be prefabricated in factories and then transported to the site for fast assembly.

The first attempt to introduce industrialized housing occurred because of labour and material shortages caused by the First World War. Subsequent to the publication of the Tudor Report in 1918 a series of committees were established to investigate the potential for implementing new ways of building that did not rely on bricklaying, carpentry and joinery. The Committee for Standardization and New Methods of Construction was tasked with the role of looking into the potential for standardization in state aided housing schemes. In 1919 the Addison Housing Programme replaced traditional construction with pre-cast concrete and steel that employed unskilled labour. Further development work took place and principles where established for social housing, which led to the Homes Fit for Heroes Programme in 1921. Approximately 30,000 steel and concrete prefabricated homes where built and additional strains were placed on the construction industry due to a lack of proper planning or control. Unfortunately these housing programmes suffered from serious defects associated with cracking, leaking and corrosion, which were never resolved. This initiative provided the first example of hastily prepared designs that where not properly thought through and tested. The consequences manifested themselves in costly repairs and poor living environments, which gave industrialized housing a bad name.

It was not until immediately after the Second World War that industrialized housing was revisited. War damage and neglect had created a major housing crisis, which was addressed by the introduction of prefabricated homes first designed in 1944. The Portal House was based on a lightweight steel frame, however due to the shortage of steel it was never put into production. Instead manufacturers adapted the design using timber frames and aluminium clad in asbestos wall panels and roof sheets, Vale (1995). A major development occurred when the Arcan Group amalgamated the resources of materials suppliers and engineering companies to produce a range of designs and 41,000 temporary homes were provided, Gear (1967). These dwellings proved to be extremely robust and there are still examples that have survived and provide homes over 55 years after their manufacture, thus providing an important demonstration of the longevity of non-traditional construction methods.

After the Second World War the Cornish and Airy systems were designed based on small pre-cast concrete panels that could be man handled without the use of cranes, Ministry of Works (1948). It was soon established that the system incurred heavy labour costs and was found to be uneconomic resulting in a rapid decline in use. In 1948 a study by the Ministry of Works established that large pre-cast panel industrialized systems were more economical and Wates were the first UK company to produce a system using room sized load bearing external cladding panels. The Reema system followed, which combined the external and internal cladding surfaces into one load-bearing panel that was linked to internal pre-cast load bearing panels. Hence this was the start of the large panel pre-cast housing systems that dominated public housing in the UK for the next 20 years. In comparison the Reema system required less man-hours to produce than the Wates system and 20,000 houses were built using this method up to 1962.

Due to two successive World Wars the majority housing stock in the early 1950's was old, lacked amenities and was in an advanced state of decay and disrepair. The British Government initiated a housing policy of slum clearance to provide land for new homes, especially in city and urban areas. It was decided that public housing redevelopment would rely heavily on system building for both high-rise and low-rise development. In London 34% of new housing was built using an industrialized system and this figure was exceeded by County Boroughs and New Towns who achieved 44% and 43% respectively. In the 1950's the London County Council (LCC) developed a policy to employ new technologies and it instigated the use of system building in its Post War General Needs Programme, Walker (1953). Throughout the 1950's it became increasingly clear that industrialized housing would need to be used in order to meet the large demand for new homes brought about by the regeneration of the UK's industrial base, especially in the Midlands and the South. The British Government invested money into research and development aimed at deriving economy of scale through standardized components that could be used in buildings designed to a common dimensional framework. The aim was to enforce the design of public housing through a prescribed set of dimensions that in turn allowed the use of standard components, Ministry of Works (1948). Unfortunately, the commercial sponsors of the various industrialized systems ignored this initiative and to some extent the potential advantage was lost.

Towards the end of the 1950's attention turned to the severe shortage of new public housing in inner cities. The use of confined sites resulted in the need to increase the density of housing and the obvious solution was the adoption of high-rise designs. Hence, the birth of large scale inner city housing estates dominated by tower blocks. The British Government's research into industrialized building was not restricted to housing. Consortiums of local authorities were set up to produce designs for schools, hospitals and other public buildings using a national system of interchangeable components. The majority of systems developed where steel framed and used a variety of cladding components. In contrast the design of public housing was left mainly to the commercial sector that, in general, favoured the use of large pre-cast concrete panel systems.

Bison was one of the first British large pre-cast systems that were designed for tower blocks between 8 and 20 storeys. The pre-cast panels integrated external and internal finishes, services and fixings and were produced in factories and transported to the site. Because of the high demand the LCC decided to explore the possibility of adapting pre-cast panel systems from elsewhere in Europe and this resulted in the use of the Danish Nielson System being licensed to Taylor Woodrow Anglian. Other foreign systems where also used, including the Camus system. Two notable British manufactured systems where 12m Jesperson developed in 1963 by Laing and Wates Modular , which used site based factories to reduce transportation costs. Typically public housing estates comprised of 500-600 dwellings. The Larsen Nielson system alone accounted for 182 residential tower blocks constructed for the LCC.

Increasing prosperity in the UK economy of the 1960's created the view that standards of social housing were not keeping pace with the development of the economy and an investigation into desirable housing standards was implemented. The outcome was the Parker Morris Report, entitled "Homes for Tomorrow" which concluded that the quality of social housing was not keeping pace with living standards and it advised that more space and amenities be provided and new guidelines were drawn up for dwellings.
During the mid 1960's the UK construction industry was at the centre of the modernization of the nation's housing and infrastructure. In 1965 Government policy decreed that an annual programme of 500,000 new homes should be in place by 1970. Housing production continued to rise from 1966 to 1972 when house completions in system building averaged 30% of all public housing and it peaked in 1970 at 41%, Housing

Statistics GB (1972). In contrast to the public sector, private housing continued to use traditional building construction, mainly for two storey dwellings in suburban and rural and rural areas. Low to medium rise apartment developments were adopted in inner cities.

THE DECLINE

A defining event occurred in 1968 when a gas explosion on an upper floor of a Larsen Nielson system built tower block at Ronan Point in London, led to a progressive collapse of all twenty two floors of one corner of the building, killing and injuring those residents in that part of the building. The subsequent investigation into the collapse revealed a major design fault in that the walls and floors were not structurally tied together. Huge public expenditure was necessary to carry out remedial works to make other tower blocks safe. It was also revealed that in Denmark this system was not built over six storeys and in the USA this type of construction was banned over 6 storeys. The resulting scandal did much to destroy British public confidence in high-rise residential accommodation.

Critics pointed to the fact that system building solutions were born out of crisis and stated that political ideologies bypassed the need for proper research, development and testing of designs and their continual improvement over time, Russell (1981). Russell cited 'Homes for Heroes' , the post war 'prefabs' and the large panel systems of the 1960's and blamed a lack of long term planning, which was aided and abetted by politicians and civil servants. The 1967 economic crisis effectively curtailed expenditure on housing and questions were raised about the efficiency and desirability of high-rise residential blocks. In the early 1970's it was realized that the system building stock was expensive to build and there were emerging problems of technical deficiency and social behaviour and well being. As a result system public house building was abandoned in the UK in favour of low-rise high density traditional construction.

During the 1970's and 1980's the problems associated with large panel concrete systems began to reveal themselves. Common defects were the spalling and cracking of panels and water barriers incorporated in the panel joints were not adequate to stop in ingress of rainwater. Further, the panels suffered serious condensation destroying internal decoration and creating a damp living environment. Window frames rotted and services, including lifts, suffered breakdowns. Social problems created vandalism and abuse of the buildings and the resultant cost of renovation and repairs placed a heavy additional burden on public expenditure. Many local authorities decided to cut their losses by demolishing many of the worst affected tower blocks after just 10-20 years of use. The sell off of public housing by the Thatcher Government resulted in all the best housing being purchased by tenants. However, many of the worst developments are still being maintained by local authorities to satisfy social housing need. These events have been well publicized and the British public is extremely sceptical of system building. The general public perception equates to images of the worst forms of architectural design with social deprivation, poor engineering and shoddy workmanship.

DEVELOPMENTS POST 1970

An alternative to constructing traditional low rise load bearing masonry dwellings is the use of timber frame construction. Supported by the increasing supply of domestic plantation softwood, timber frame became a viable alternative to fill the gain left by the abandonment of large panel construction. Timber frame housing increased the speed of construction and improved thermal insulation. It also provided traditional brick and tile elevations, which were aesthetically pleasing to the public. Technology 'know-how' was borrowed from the USA, Sweden and Canada and the British Timber Research and Development Association conceptualized and developed design recommendations. There was growing acceptance of this technique until a disastrous BBC World in Action programme that produced a damning report of the poor workmanship adopted by Barratts, a major UK house builder specializing in timber framed housing. The report could not fault the basic design principles or the best practice adopted overseas, however it did reveal poor working practices and a complete failure in site management in quality control procedures. The outcome resulted in a rapid down turn in the sale of timber framed houses. Unlike the large panel systems that were fundamentally flawed, timber frame systems made a comeback in the early 1990's and these are now accepted as an efficient and quality means to produce housing to an acceptable standard.

Steel framed systems such as CLASP and NENK were produced in the 1960's and were applied mainly to low rise schools, libraries and other public buildings, but these also suffered defects and failures e.g. inability to limit the spread of fire. There were limited attempts to produce steel frame housing systems but these never caught on, mainly due to cost and poor aesthetic design. The decreasing price of steel and developments in design technology in the mid 1990's led to system building using lightweight cold rolled steel. McDonald's, the global fast food chain were able to demonstrate the advantages of lightweight steel systems in terms of speed of construction, quality and cost. It was quickly realized that this system could be effectively adapted to produce affordable housing.

LESSONS LEARNED

System building in the UK has suffered major failures; by far the worst was that of the large-scale panel systems in the 1960's. The British public still holds the general view that system building represents the worst of modern architecture and defects resulting in poor life cycle performance, Russell (1981). Construction professionals are still sceptical about the implementation of systems and there is a strong body of opinion that considers previous system house building to be a series of misguided attempts to introduce higher levels of construction design and manufacturing techniques. As a consequence, there has been no consistent development of system building in the UK over the past 85 years. Instead the subject has been revisited twice as the result of two World Wars and once to solve a peace time housing crisis, which resulted in almost total failure after a period of 20 years commencing in the 1950's. The main lessons to be learned from the UK experience are listed below, however underlying these is the need for a long term strategic development plan for system built housing, backed by sufficient innovation, research and development:

- The design should be subject to full value analysis, where aesthetic, performance and cost factors are given detailed consideration in the context of the planned life cycle. The selected design should be engineered and tested by means of prototypes in order to assure design quality and performance through rigorous testing procedures;
- Solutions should be sustainable in that they require low embodied and in-use energy, as well as using materials that can be recycled on completion of the life cycle;
- Systems should be open and flexible in order that standard components are based on a common approach to dimensional co-ordination. Thus interchangeability between components provides for greater design flexibility and aesthetic appearance;
- Advantages offered by standardization and economy of scale should be fully exploited; and
- New components and fabrications require independent accreditation regarding their quality and functional performance. They must also comply with the minimum standards set by building regulations and codes of practice;

CURRENT HOUSE BUILDING PRACTICE IN THE UK

New housing in the UK is produced primarily by the private sector for purchase by homeowners. Most of this construction is traditionally based on load bearing brick and block cavity walls, nevertheless timber framed has gained ground in this sector. For higher rise apartments in city centres, traditional construction in steel and concrete is in common use. Social housing is provided by housing associations and it is mainly traditionally constructed. Advances have been made in the use of timber-framed systems and there has been innovation involving the use of steel frames and concrete walling. Public housing in the UK has progressively diminished to the point where it is non-existent apart from a limited number of sheltered housing projects to accommodate old people. In the suburbs and rural areas houses are normally built to a maximum of three storeys as detached, semi-detached and terraced units. In city areas flats and apartments are normally provided up to six storeys and in a minority of cases where land is extremely expensive, high rise will be employed utilizing structural frames and cladding systems. It may therefore be concluded that the construction of housing can be classified into two prime categories, namely 'traditional' and 'system'. These categories can be further subdivided as shown in Figures 1 and 2. Although system built housing is very much in the minority there is a growing body of opinion in the UK that it offers the potential to build more efficient and better quality homes at an affordable price.

Progressive development of steel framed systems since the 1970's have lead to the advent of cold rolled lightweight steel framed panel and modular systems. This has enabled lightweight modules to be designed and constructed for feasible transportation from the factory to site by road. There are five major manufacturers of these units in the UK. Much of this production has been geared to restaurants and hotels, but recent pilot housing projects such as Murray Grove in London have proved that this technology can be successfully applied to housing.

Innovation and research into the production of timber framed housing dates back over fifty years. Houses produced in kits have been a commercial success and this has encouraged further investment. Recently development has concentrated on fabrication away from the site in factories and the feasible construction of timber framed construction above three storeys. Modular construction has proved to be economically feasible for the construction of low-rise hotels and this approach is now being developed for application in the social housing sector. Timber systems are more established than steel systems and will see further ongoing development.

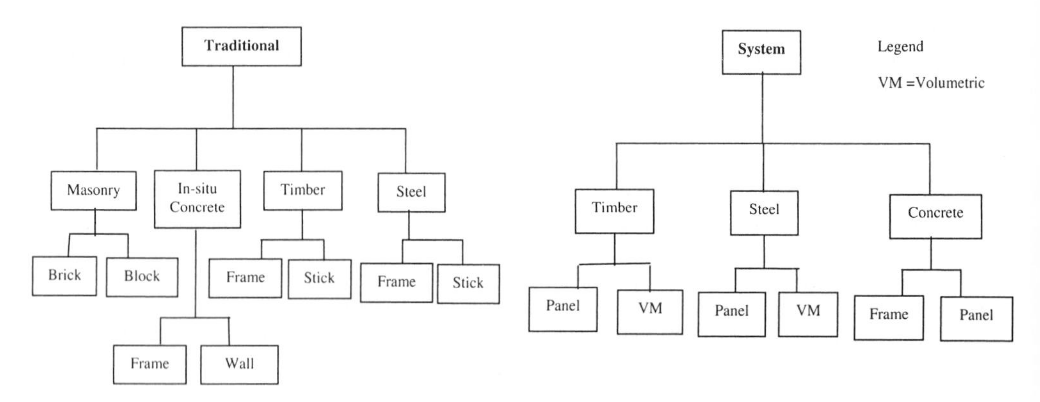

Figure 1: Classification of Traditional House Construction
Source: Kaluarachchi, Howes and Tah 2000

Figure 2: Classification of System House Construction
Source: Kaluarachchi, Howes and Tah 2000

Building on experience gained experience gained from Europe, concrete systems have been developed. These are primarily panel and frame based systems that are pre-cast in factories using casting batteries or extrusion machines. Because they are relatively heavy, volumetric construction appears not to be a feasible option. Developments in the finishing and insulation of external walls provide more design flexibility and better thermal performance. Concrete systems claim to be robust requiring only limited life cycle maintenance.

Although a wide variety of housing systems are available in the UK none of these have achieved the potential advantages of economy of scale. Traditional house building still dominates the market, but there are problems due to the cost of construction and a lack of traditional craftsmen such as bricklayers, carpenters, plumbers and plasterers. Currently there is a serious housing shortage in the UK and the number of new homes being built has dropped to only 162,900 in 2001, the lowest number for decades. Statistics released in April 2002 have shown that house prices had risen by 3.8% in March 2002, pushing the average price above £100,000. Pressures are now building for the introduction of system built homes to provide affordable quality housing for public sector workers, teachers, nurses and police officers who cannot afford to purchase housing in inner city areas.

POTENTIAL DEVELOPMENTS

The construction of housing can be considered as a largely repetitive process, involving a range of components and materials, which can be constructed in a variety of ways to produce an infinite range of designs. Traditional construction relies heavily on the effective integration of the various trades in a sequential process to achieve production on site. Research indicates that 80% of building work and repairs are of a repetitive nature, Environment Committee Report (1996). Under such circumstances there is scope to rationalize and advance the quality of house building through the use of systems based standardized components and processes where continual improvement can be facilitated. A wide range of choice and flexibility can be achieved through the use of a carefully selected supply chain to provide the materials and components necessary to support factory-based assembly. To achieve this, the construction of housing under factory conditions must be seen as a total process using integrated teams, Environment Committee Report (1996), Gann (1996). The Egan Report "Rethinking Construction" and the follow up report "Accelerating Construction" points towards the high level of waste occurring in labour, materials and the misuse of equipment which is making houses too expensive. As a consequence low income groups are being precluded from owning their own homes, Egan (1998), Strategic Forum (2002). Egan also points towards lessons that can be learned from manufacturing and the utilization of integrated supply chains.

Important progress has already been made in the production of system housing demonstration projects and prototypes. Indications are that quality has improved and construction times have been dramatically reduced, however the cost is still too high. Further development needs to take place to improve designs and to provide more flexibility and choice for house purchasers. Manufacturing efficiency requires improvement and potential purchasers need convincing that modern factory produced homes really are superior and are better value for money than the traditional alternative. Better co-operation within the industry and the acceptance of a range of standard products and components will considerably assist price reduction by means of economy of scale. A reduction in construction time can be achieved by adopting multiple assembly lines in order that modules can be constructed concurrently. At the same time site works can be scheduled so that rapid erection can commence once modules, panels, floors and roof assemblies are delivered from the factory.

CONCLUSION

The UK has experienced over 80 years of spasmodic activity in the development and use of industrialized housing. Periods of high activity have been brought about by two World Wars, which reduced available manpower, created considerable damage to the housing stock through bombing and diverted attention away from essential maintenance and repair. The most significant period of industrialized housing occurred in the 1960's, which was brought about by increased prosperity and the need to improve living standards. Political, economic and social pressure together with the misconception that living standards could be improved by untried and inadequately tested technology, directly contributed towards the major failure of concrete panel system tower blocks and urban housing estates. The public was devastated by the scandals, failures and waste associated with system building during this period and great scepticism still exists. This effectively killed off system building in the UK from the late 1970's to the mid 1990's. The exception has been the development of timber frame housing systems and kits, although this area suffered a major set back due to the BBC's World in Action programme on Barratt House Builders.

Important research and development sponsored by forward thinking clients such as McDonalds, British Airports Authority, Travel Lodge and supermarket chains eg Asda have proved that factory based construction technology can produce superior buildings for the same, and in some cases less cost, with the added bonus of much faster construction times. The demonstration and prototype-industrialized housing projects have affectively proved that the new system based technology can be transferred to housing. The UK Construction Industry now has reached the stage were new system housing has the potential to satisfy the huge demand for home ownership at an affordable price. It is now the responsibility of the wider construction industry, housing associations, building societies and banks to embrace and further develop this technology to meet the needs of future homeowners and society in general. Success or failure will be driven through private demand and not through a state housing programme. Unlike on previous occasions referred to earlier in this paper, new industrialized building technology has been carefully developed and thoroughly tested and it has been shown to be sound. In conclusion there would seem to be a promising future for system building in the UK.

REFERENCES

Egan, J.,The Construction Task Force, (1998), *Rethinking Construction*, DETR, London.
Environment Committee Report, (1996), *Housing Need*, Vol.1,Report Session 1995-96.
Fairs, M., (June 1998), *Evolve or be Damned*, Building Design, p6.
Finnimore, B., (1989), *Houses from the Factory*, Rivers Oran Press, London.
Gann, D., (1996), *Construction as a Manufacturing Process*?, Vol 14, pp 437-450, Journal of Construction Management and Economics, SPON.
Gann, D., et el, (December 1997), *Mission Statement – Flexibility and Choice in Meeting Occupants' Housing Needs,* RSA, SPRU University of Sussex.
Gear A. M. and Associates, (1967), *The Arcon Group 1943-67*, Report.
Housing Statistics Great Britain 1964-72, (1972), HMSO.
Ministry of Works, (1948), *Post War Building Studies*, No 45, HMSO.
Russell, B., (1981), *Building Systems Industrialisation and Architecture*, John Wiley & Sons, London.
Strategic Forum, (2002), *Accelerating Change*, DTI, London.
Vale, B., (1995), Prefabs; *A History of the UK Temporary Housing Programmes*, Chapman and Hall, London.
Walker, H., (1953), *Prefabrication 1*, p32.

Advances in Building Technology, Volume 1
M. Anson, J.M. Ko and E.S.S. Lam (Eds.)

A PERSPECTIVE OF INDUSTRIALISED HOUSING IN CHONGQING

Li Shirong

Research Centre for International Construction Economics and Management
Chongqing University
Shapingba, Chongqing, 400045, PR China

ABSTRACT

The aim of the paper is to provide a better understanding of Chongqing's features and the potential for promoting industrialised housing in this region. It first reviews the current development for industrialised buildings abroad. It then focuses on the features of Chongqing, the importance, the potential and the advantages for promoting industrialised housing in Chongqing. It also introduces an international cooperation project in Chongqing.

KEYWORDS

Housing, industrialised housing, China, Chongqing, construction industry, manufacture

DEVELOPMENT OF INDUSTRIALISED HOUSING

Industrialised housing development abroad

In 1950s, due to the damage made in the Second World War, there was a great demand for houses in Europe. Therefore, in order to settle the problem, many industrialised houses were assembled, which are still standing. There were complete, standard and systematic housing system in Europe. Ten years later, the wave of industrialised housing spread all over the European countries and soon to USA, Canada, Japan and other developed countries.

With the continual enhancement of science and technology in recent years, the demand for houses came through a phase of "Pay attention to: Quantity-Quantity & Quality-Quality First-Individualisation, diversification, high qualify environment". At the same time, the production of industrialised houses in many western countries reached its climax. As it is well known, Sweden is one of the most advanced countries developing industrialised housing, and 80% of the country's houses adopt generally a housing system based on universally available parts. In USA, there are 34 professional companies producing modular construction parts now. In the UK, the Egan Report in 1998 further pushes this new trend. Currently, the study of the production of industrialised housing in the UK has paid more attention to the flexibility and design variation of housing systems, in order to widen the adoptable coverage and to bring in the scale effect (Prewer, 2001).

Although industrialised housing has been developed in the past 40 years, the speed is slow. The main reasons include the techniques, user's understanding and economic conditions. For example, during the 1960s and 1970s, the UK built many industrialised residential buildings with concrete panels, which many people don't like. Therefore, even today when people hear about industrialised housing they are negative (Shirong, 2002).

To shorten the construction period is not the only objective of implementing industrialized housing. Within the report 'Rethinking Construction' provided by UK key clients who appointed by the British government in 1998, it is clearly stated the objectives to be achieved within the British construction industry through innovation, implementation, industrialized production are (Egan, 1998):

- To reduce cost 10%
- To shorted the time 10%
- To improve the predictability 20%
- To reduce the defect rate 20%
- To reduce the accident rate 20%
- To improve the production rate 10%
- Eventually, to improve the profit rate 10%

Therefore, the main purpose of industrialized housing is to absorb the principles developed in the manufacture industry into the construction industry. Thus, it is hoped to change the traditional construction methods and lead the housing industry to the development of industrialization, standardization, integration and information technology. This is a kind of revolution within the construction industry. Obviously, the principles of implementing industrialized housing are to achieve standardization through modules, and then to achieve industrialization through standardization. An example shows the importance of using an industrialised housing system. There are about 3,000 spare parts for making a car, while there are about 40,000 for building a house (Shirong, 2002). Due to the complication of building houses, it is important to use standardised parts with a manufactured approach for the building of houses. It is an important approach to improve quality and efficiency in the construction industry.

The common interesting in housing development

From a global market perspective, the following are some common interesting in housing development:

- In housing development, some key issues become more and more important, such as the quality, function and efficiency of construction, thinking of people, environment protection and sustainable development.
- There is a new revolution in construction approach focusing on manufacture-based construction.
- Partnering, project management and supply chain logistics are the key issues for success in this revolution.
- Industrialisation, Standardisation, integration and information technology are key factors for developing industrialised housing.
- Transferring the production approach of manufacturing industry into the construction industry is the breach in the traditional approach construction in order to achieve and achieving the objectives of the innovation revolution.
- The new challenges in the new century requires rethinking of the construction approach in order to meet the requirements from society in terms of high quality, zero defects, reduced waste, high speed and high efficiency, as well as sustainable development and environment protection.

- The lack of technology for industrialised housing affected the development in the past, while today the industrialised housing system has to be promoted to overcome the challenges.
- Although for many years, there were been different type of structures in industrialised housing systems, the light-steel frame structure seems to have more advantages particularly with today's requirements in sustainable development and environment protection.

THE IMPORTANCE FOR USING INDUSTRIALISED HOUSING IN CHONGQING

Compared with other major cities in China, it is more urgent and important for Chongqing to develop industrialised housing. The technology and management level in Chongqing's housing industry is comparatively lower than in other major cities in China. Along with lack of experienced human resources within the local housing industry and the low level of industrialization, it is more difficult for Chongqing to further develop the industry and to grasp the opportunities. How to adjust the industry structure, change the construction methods and fully use local resources to realize the objectives of industrialized housing and restructure the related industries is the key problem for Chongqing in the new century. Therefore, to implement the industrialized housing can not only meet the requirement of central government, but also comply with the current conditions in Chongqing.

Accelerating the development of Chongqing's construction industry

Meeting the great housing demand in Chongqing

Due to the use of standardized and manufacture-based components, procurement and production of large scale become possible which could speed up the housing production process.

Due to the huge population (30 million), the average floor space per head in Chongqing is below the national level. With the quick economic development, the demand for houses in Chongqing now is up to about 20 million square metres per year (Chongqing Construction Industry Yearbook, 2000). Although the housing sector in Chongqing has been developed in Chongqing since 1980s, it still can not meet the requirement of the society. For example, the average floor space per head for housing was 3.12 square metres in 1980, 7 square metres in 1995, and 11.59 square metres in 2001. The rapid development of urbanisation in Chongqing further increases the demand for housing in Chongqing. Based on the city planning for 2010 and 1020, the urban population will account for 37% in 2010 and 47% in 2020.

Improving construction productivity in Chongqing

Currently, traditional approaches are still being used in housing construction in Chongqing. Construction productivity in Chongqing is lower than the average level of the country and even lower than the level of western part of China, much lower than in Beijing, Tianjin and Shanghai, as shown in Figure 1.

In addition, due to the change of production when using industrialised housing approach, majority of building components could be made in factories that will improve the quality. In Chongqing, the skills of construction employees are relatively low because a large number of construction workers in Chongqing are from rural areas. Until the end of 2000, the total number of construction workers in Chongqing was 1 million but about 75% of the total number were farmers. The other three cities are much smaller than in Chognqing, such as Beijing: 14.7%, Tianjin: 32.7%, Shanghai: 24.9% (China Statistical Yearbook, 2001).

Speeding up housing construction in Chongqing

One of the main advantages of using industrialised housing is to speed up the construction process and greatly reduce the site work that also avoids the outside affects for construction. It is great desire for both developers and users to speed up housing construction in Chongqing. It happens for some customers that they have to wait for 2 years to get their new houses although they pay before. In addition, Chongqing has about 800,000 people who should be relocated due to Three-Gorge Project. In 1998, for example, about 70,319 people were relocated in Chongqing, costing almost 3 billion yuan. The total floor space was 2.25 million square metres.

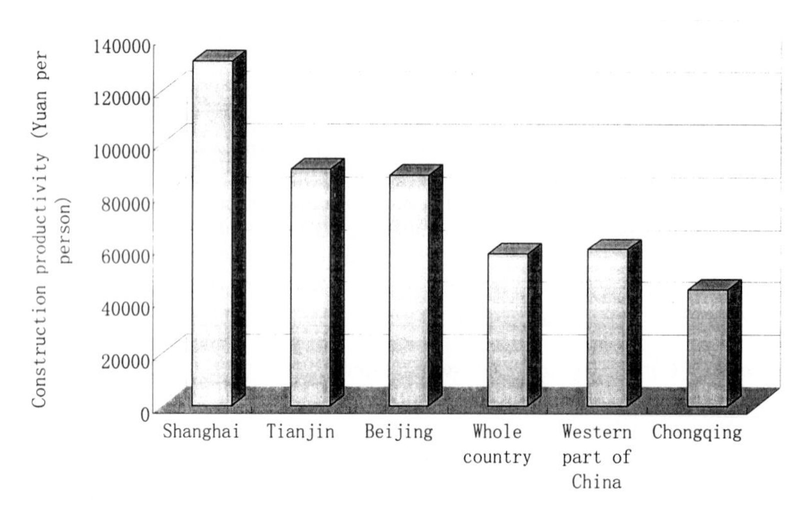

Figure 1 Construction productivity in some cities (2000)

Setting up production base for building materials in Chongqing

Though construction industry has been one of the key industries in Chongqing, the variety and the size of building materials and products are still very small. For example, about 80 percent of building materials and products for fitting out used in Chongqing are produced from other cities.

Besides, there are a lot of industrial and agricultural wastes materials in Chongqing that can be used in the production of building materials, and this'll do good to environment protection of Chongqing.

Chongqing has strong steel production, but there is small amount steel used in building structure in Chongqing.

Chongqing is also one of the key districts in agriculture with abundant agricultural byproducts which can be used for producing building materials such as light boards.

Restructuring manufactures and introducing new products

Chognqing is an old industry city with many heavy industry manufactures, such as steel and shipbuilding. Restructuring industries is one of the key tasks for Chongqing government. Many state-owned enterprises in Chongqing need to introduce new products, such as Chongqing Steel and Chuandong Shipyard. Many state-owned enterprises with traditional industry products have low productivity. For example, in 2000, the industry productivity in Chongqing was 31252 Yuan per head, while it was 63878 in Beijing, 52425 in Tianjin and 82327 in Shanghai.

Introducing foreign funds and technology to Chognqing

Economic development needs money. For many years, Chongqing has the shortage of foreign investment. For example, in 2000, FDI was 1.98 billion yuan in Beijing, 1.76 billion yuan in Tianjin, 2.84 billion yuan in Shanghai, but only 240 million yuan in Chongqing. As a lot of new technology and products will be involved in the carrying out of industrialized housing, we can introduce these and other experience from advanced countries, and may cooperate with them in setting up factories.

Re-employment of people

The promotion of industrialized housing can not only bring along many relevant industries, but also produce some new industries, which can solve the re-employment of people who are out of work. Economic reforms in Chognqing have brought about many people out of work (about 400,000 people), while the revolution in construction process with manufacture-based approach will bring new industries, particularly building materials. This will provide opportunities for unemployed people in Chongqing for new job.

Benefit for environment protection and sustainable development

With the carrying out of industrialized housing, construction site work and waste of construction materials can be reduced; some waste materials can be recycled; and new green and light materials can be put into use. All of these act a positive part in environment protection and sustainable development of Chongqing.

THE MAIN ADVANTAGES FOR USING INDUSTRIALISED HOUSING IN CHONGQING

- Chongqing's market environment provides a basis for promoting industrilaised housing in Chongqing (municipal city, Gorge project, western part of China, key economic areas along Yangze River).
- Great demand for housing in Chongqing provides huge market for industrialised housing.
- Abundant resources in Chongqing provides basic condition for establishing building material manufactures.
- Sufficient labour force in Chongqing makes it possible for establishing manufactures.
- Existing shipyard in Chongqing provides the site for housing manufacture.
- The transportation network in Chognqing (including roads, railways, and water roads) provides a good condition for establishing supply chain.

In a word, Chongqing has the potential for being one of the main manufacture centres for industrialised housing in China.

AN INTRODUCTION TO INDUSTRIALISED HOUSING IN CHONGQING

A joint research team was established at the end of 1999, including the members from UK, Hong Kong and Chongqing. The main members are from three universities (South Bank University from UK, Hong Kong Polytechnic University from Hong Kong, and Chongqing University from Chongqing). Research work has been under taken since then.

Project objectives

An industralised housing system is to be established in Chongqing, which includes housing industry and other industries such as steel, metal, building material, real estate, decoration, manufacture, environment, environment protection and information technology. This system is aimed at research on serial technologies of industralised housing and the base establishment and directed by the basic ideas of technology innovation service system. In this system the serial technologies of industralised housing serves as the main body of innovation and the goal is to establish the base for industralised housing in Chongqing.

Principles for development industralised housing system in Chongqing

The following main principles are the key drivers for the project:

- Technology transfer in the form of appropriate industrialized building systems will be adapted to conform with local requirements.
- The use of locally available material and human resources will determine the type of system to be developed.
- The proposed system must be within the potential skill capability of local professionals, technicians and operators, after they receive some training courses.
- The system will supply sufficient freedom to allow variety in design.
- The system will be suitable for local environmental condition. Energy consumption will be a major consideration. Priorities will be focused on the internal environmental factors including: air quality, damp proofing, acoustic insulation, as well as temperature and humidity control.
- The design of the dwellings will be determined by the production cost and maintenance cost.
- Quality and fitness for purpose will be considered according to cost and affordability.
- Waste will be minimized in the process and recognition will be given to issues concerning sustainability.
- Safety performance improvement will be focused on the quality control of factory production, which are the spare parts for the dwellings.
- Time will be saved and progress will be accelerated since more parts do not have to be pre-fabricated on site.

Possible systems

There would appear to be three basic approaches to industrialized steel construction, namely:

- High rise using a hot rolled steel frame to support modular or panel units consisting of lightweight cold rolled sections.
- Lightweight cold rolled steel framed panels intended for construction up to around eight stories.
- Load-bearing modular units stacked one on the other up to six stories.

Pre-fabricated bathroom and kitchen pods could be used in all three categories.

Of the three approaches, the second appeared to offer the most flexibility. It could be extended to high-rise construction through the use of hot rolled principal structural frames and could be adapted to small-scale production by using the kind of portable bending equipment.

Work stages

There are three main stages in this project, including research on technology and system, prototype, demonstration.

The first stage includes:

- Understanding local conditions.
- Analysis and valuation of industralised housing systems used in practice, finding out both advantages and disadvantages of these systems, and determining suitable systems for further analysis.
- Designing new system.

The second stage includes:

- Setting up partnering (including government, research organisations, industry people).
- building the prototype.
- Testing and showing.

The third stage includes:

- Establishing key manufactures.
- Building houses for demonstration.

The key technologies for the project

- The establishment of structure system of houses, which is suitable for industralised housing and local conditions in Chongqing which includes the index such as structure type, size and height
- The establishment of standard and specification
- New technology of building material
- The product upgrade of other relative industries. How to make the needed production for the industrialised housing by the utilization of the production ability of mental, steel and shipbuilding industries and the importation of new product and production line
- The material supply chain which is gradually formed for the housing industrialization
- The establishment of industrialised housing base which is the objectives at the first stage and the key jobs of the project

Main features of the project

Firstly due to the involvement of many different fields and departments, the project has the following characters:

- **One** significant character --- a system engineering
- **Two** important aspects --- production systems development, demonstration project
- **Three** parties' joint effort --- Government, enterprise and research organization
- **Four** approaches --- research and innovation, communication, demonstration and training
- **Five** assurances--- policies, funding, personnel, land and time

Just because of the above characters, there is specialty in the implementation of the project. However, the following must be mutually agreed for implementing the project:

- The government is the leader
- The enterprises are the main players
- The research organizations are the supporting bodies

From above the implementation of the project is big system engineering involving multi directions, multi levels, multi objectives and multi stages. As government is concerned, it involves construction, economy and agriculture commission, planning science and environment protection departments; As the enterprise is concerned, it involves the bodies such as design, consulting, developer, contractor, supplier and manufacturer; As the research organization is concerned, it involves the university and design institute; As the industry is concerned, it involves architecture, real estate, material, mental, fabrication and service.

Therefore to ensure the effective implementation of the project, a project team including government, research organization and enterprise must be established. And in the team the work of each body shall be clearly clarified, project organization shall be realistic and functional so that the advantage and resource of each participant can be fully taken and exploited.

CONCLUSION - KEYS TO SUCCESS

- It is important for Chongqing government to realise the potentials for promoting industrialized housing project in Chongqing. It is not only to improve building sector but also to improve other related industries. Compared with other cities, Chongqing has more potential for promoting this work.
- The publics should have a better understanding of industrialised housing because they are the main forces for promoting this work. Without their understanding, the work is difficult to carry out. Education and training are important approaches for reaching this objective.
- It is also important for all parties to the project to know their roles and responsibilities in implementing the project. They should also clearly see their benefits from this project. Without their understanding, it is difficult to carry out. Among them, government is a key party to make this project happen in Chongqing. The benefits that local government achieves are huge, including promoting the development of Chongqing construction industry and other related industries, helping restructuring industries.
- The most advantages for international cooperation is to attract foreign funding and introduce overseas advanced technology and management skills.
- The possibility for carrying out this project depends upon the government's active attitude, good cooperation among all parties.
- Industrialised housing systems are new in China. Many cities are eager to set up the system. Early action will bring a lot of benefits.

References

Chongqing Construction Commission (2001), *Chongqing Construction Industry Year Book*, Chongqing Publishing House, China.

Egan John. (1998), *Rethinking Construction*, DETR, UK

Genfang Wang. (2001), Perspective of Chongqing Housing Development, *Journal of Chongqing Jianzhu University*, 23:3, 6-10, China.

Ministry of Construction (2000), *China Construction Industry Yearbook*, China Construction Industry Publishing House, China.

National Bureau of Statistics (2001), *China Statistical Yearbook*, *China Statistic Press*, China.

Prewer John. (2001), Innovation and development in moudular industrialised housing, *Journal of Chongqing Jianzhu University*, 23:3, 55-66, China.

Chirong, Li. (2001), *Journal of Chongqing Jianzhu University*, 23:3, 34-42, China.

Steel Construction Institute (2000), *Value and benefits assessment of modular construction*, Steel Construction Institute, UK.

Steel Construction Institute (1999), *Modular Construction Using Light Steel Framing, An Architect's Guide*, Steel Construction Institute, UK.

Advances in Building Technology, Volume I
M. Anson, J.M. Ko and E.S.S. Lam (Eds.)

IMPACT ON STRUCTURE OF LABOUR MARKET RESULTING FROM LARGE-SCALE IMPLEMENTATION OF PREFABRICATION

C M Tam

Department of Building & Construction, City University of Hong Kong,
83 Tat Chee Avenue, Kowloon Hong Kong

ABSTRACT

The construction industry of Hong Kong is characterized by labour intensive trades such as plastering, tiling, concreting, steel bar fixing, formwork erection and bamboo scaffolding, which are considered by the public as dirty and hazardous. Traditionally, the industry has found difficulties in recruiting appropriate workers. The Construction Industry Training Authority (CITA) has set the entrance requirement of trainees at Form 3 school leavers, the lowest standard in the territory. Over the years, CITA has helped relieving the social pressure by offering these youngsters a training opportunity. In recent years, the industry has also helped to absorb the surplus labour supply, especially of the low technology sector when most of the manufacturing industries have been moving from the Hong Kong Special Administration Region (HKSAR) back to the Mainland since early 80's. The construction industry offers employment opportunities to the new immigrants from the Mainland who lack modern skills to survive in the new economy of the territory. However, this scenario is put under threat as prefabrication has become more widely accepted by the industry. Akin to the manufacturing sector, prefabrication will encourage the shifting of site casting work to factory casting in the PRC. Recent studies by the authors have indicated that the labour demand can be reduced by as much as 40%.

KEYWORDS

Construction Labour Supply, Labour Market, Prefabrication.

INTRODUCTION

The free market economy will drive capitalists to move the place of production from the costly regions to places with a lower cost of manufacturing (Lipsey and Steiner1978). The manufacturing industry has followed strictly this doctrine and hence, most factories from Hong Kong have moved to the Mainland and this has triggered the historically high unemployment rate of 6.7% in early 2002. Fortunately, the construction industry, whose products are for local consumption only, still remains very labour intensive. Owing to the public's strong objection to the importation of labour, construction workers are mainly drawn from local residents, providing a good employment opportunity to the Hong

Kong citizens. Currently, there are about 300,000 construction workers (including the interior decoration trades) recorded, contemplated least affected by the shift of factories to the Mainland.

Most of the construction trades have their strict and hard boundaries which workers will not or cannot cross over and hence induces inflexibility both for workers in seeking continuous employment and for employers in labour deployment (Tam et al 2001). This results in the lack of continuous employment for workers due to the daily pay arrangement of the workforce. The entrenched tradition of labour subcontracting is seen as an effective cost control and reduction measure in response to the cyclical nature of the construction industry (Gray and Flanagan 1989). However, this practice has been criticized by the industry and the public in general for the induced poor safety records and unacceptable quality performance.

As a result, the use of prefabrication has been strongly advocated recently in the industry, as it can help to improve site safety by providing a cleaner and tidier site environment, enhance quality by producing elements under the factory conditions and eliminate site malpractices. Further, factory production can reduce wastage and encourage re-cycling of construction waste, leading to environment protection and sustainability of the industry.

IMPEDIMENTS TO WIDE ADOPTION OF PREFABRICATION

However, prefabrication bears some limitations. Firstly, the modular units may bring about monotonous designs, which are strongly opposed by town planners and architects. Secondly, the prevailing high-rise building structures in Hong Kong lead to connection and jointing problems and concerns on structural integrity. Thirdly, the water and weather tightness at joints (rains come with typhoons in Hong Kong) also imposes barriers in the wide adoption of prefabrication.

The Hong Kong Housing Authority has adopted small scale prefabrication since 1988 (Cheung et al 2002). The prefabricated elements include precast façade units, staircases, drywall and semi-precast floor planking. Their experience in using prefabrication is positive in terms of quality, time, safety; some good responses are listed as follows:

? Site tidiness is obviously improved resulting in reduction in site accidents;
? The speed of construction can be improved by converting some critical site casting activities into precasting works;
? The external outlook of building structures can be varied by changing the combinations of modular units;
? The in-situ grouted joints can minimize the occurrence of water leakage;
? The quality of prefabrication units is much improved. The former quality breakdowns like the de-lamination of external mosaic tiles and water leakage along external window frames have been seldom recorded.

STIMULUS TO PREFABRICATION DUE TO NEW CONSTRUCTION TECHNOLOGY

As regards structural integrity, the new developments in high strength concrete and structural lightweight concrete can, to certain extent, overcome some structural difficulties. Modular flat units can be designed to sit upon each other and lumped together to form a tower block, which are supported by cast in-situ structural cores and tie beams.

This new development gives ways to the wide adoption of prefabrication. Consequently, this poses an intense threat to the local labour market. With the extensive use of prefabrication, it is anticipated that large numbers of site workers will be replaced with factory workers at the prefabrication yards. Further, some trades will diminish while some new trades may emerge. This shifting of work nature will induce significant impact on the existing labour market structure, as well as training provisions.

STUDY OF THE IMPACT ON THE LOCAL LABOUR MARKET

The authors have conducted interviews with the key staff of the major contractors to identify the impact on the labour market resulting from the widespread use of prefabrication. Eight interviewees, all at directorate levels of six major building contractors who have a market share of more than 50% of the local building industry market were met and interviewed. Prior to the interviews, information requisition forms were designed and distributed to the six contractors to collect the necessary data. Existing labour consumption data was collected from the standard information requisition forms. In the interviews, buildings with a high degree of prefabrication (70% of elements by prefabrication) were postulated and used as the basis to work out the labour consumption.

The results of the study are summarized as follows:

Comparison of Site Worker Consumption

Summarizing from the interviews, the total manpower for the prevailing residential tower block and the proposed new highly prefabricated tower block are derived. Table 1 shows the comparisons of the average site manpower requirements for the two.

TABLE 1
Comparison of HB/NCB and SPC on Site Labour Consumption

In Labour-days / Tradesmen	Average per Existing Residential Unit	Average per New Highly Prefabricated Residential Unit	Difference per unit
Foreman	1.24	0.71	-(0.53)
Assistant foreman	1.39	0.94	-(0.45)
Coolie	2.47	3.17	0.7
Steel Fixer	4.21	3.42	-(0.79)
Formworker	2.61	0.60	-(2.01)
Concretor	1.20	0.07	-(1.13)
Lightweight Partition Fixer	0.43	0.00	-(0.43)
Plasterer	5.24	1.65	-(3.59)
Joiner	0.40	0.15	-(0.25)
Metal Worker	0.27	0.35	0.08
Painter	1.73	0.60	-(1.13)
Electrician	2.00	0.94	-(1.06)
Plumber	1.36	0.54	-(0.82)
Joint Filling Worker	0.18	2.37	2.19
General Labour	5.24	1.17	-(4.07)
Mechanics	1.83	1.40	-(0.43)
Total:	31.79	18.08	-(13.71)

Truck Driver Consumption

With the widespread adoption of prefabrication, there will be a great demand on local transportation. As a result, the need for truck drivers will increase. Table 2 summarizes the truck driver consumption.

TABLE 2
Comparison of HB/NCB and SPC on Truck Drivers Consumption

	Driver-days Required per Existing Residential Unit	Driver-days Required per New Highly Prefabricated Residential Unit	Difference per unit (in Driver-days)
Truck Drivers*	2.98	3.82	0.84

*Assume each truck driver can manage two round trips per day

For a typical residential tower block, there will be a reduction of site workers of 43%, a terrifying figure to the local construction workers. The decision to promote prefabrication will become a political rather than a technical or managerial problem.

RECOMMENDATIONS ON PREVENTING FROM DRAINING OFF OF JOBS

One of the ways to retain construction jobs is to set up prefabrication yards within the Hong Kong SAR by:

Offer of Cheap Land for Prefabrication Yards

The Hong Kong SAR government can encourage contractors to set up prefabrication yards within the Hong Kong region by offering land with low rental fees.

Contractual Arrangements

To facilitate building control and to ensure that the use of materials and construction methods coupled with the performance of the finished products do comply with the Hong Kong Building Regulations, the large developers, especially the public clients, can mandate that all prefabrication yards must be located within the Hong Kong SAR or located in a place acceptable to them.

CONCLUSIONS

The shift from the current high percentage of site casting to prefabrication designs will definitely affect the local labour consumption. The study has revealed that there could be a 43% reduction in site labour consumption if the move is realized. The figure may have been exaggerated as some minor trades such as setters-out, scaffolders, marble-layers, project management staff and etc., which should remain unaffected, were not counted in the model. Although the cut in jobs seems alarming, the move towards a more advanced building technology for public housing construction is a worldwide trend, which is inevitable. If prefabrication is not widely adopted, the competitiveness of the construction industry in Hong Kong in terms of productivity, time, quality and safety performance will suffer.

References

Gray, C and Flanagan, R. (1989), The Changing Role of Specialist and Trade Contractors. The Chartered Institute of Building, Ascot.

Lipsey R.G. and Steiner P.O. (1978) Economics. Fifth Edition, Harper & Row, Publishers, Inc., N.Y.

S O Cheung, Thomas K L Tong, and Tam C.M., (2002) "Site Pre-cast Yard Layout Arrangement Through Genetic Algorithms" Journal of Automation in Construction, Vol. 11, No. 1, pp. 35-46.

Tam C.M., Tong T.K.L., Cheung S.O. and Chan A.P.C. (2001) "Genetic Algorithm Model in Optimising Labour Utilization" Journal of Construction Management and Economics, UK., Vol. 19, No. 2, pp.207-215.

Advances in Building Technology, Volume I
M. Anson, J.M. Ko and E.S.S. Lam (Eds.)

PRIORITY SETTING OF PREFERENTIAL PARAMETERS FOR HOME PURCHASE IN CHONGQING – AN ANALYTIC HIERARCHY PROCESS APPROACH

K.W. Wong [1] and M. Wu [2]

[1] Department of Building and Real Estate, Faculty of Construction and Land Use
The Hong Kong Polytechnic University, Kowloon, Hong Kong
[2] Department of Building, School of Design and Environment
The National University of Singapore, 4 Architecture Drive, Singapore, 117566

ABSTRACT

Chongqing, being the fourth metropolis ranked after Beijing, Shanghai, and Tianjin, is one of the most important cities in the western part of Mainland China. It is situated on the Yangtze River, 2500 kilometres from the coastal city of Shanghai. Chongqing, with a population about 32.5 million, has developed rapidly in recent years due mostly to major economic and infrastructure developments, including the "three gorges" project.

A prominent problem in the real estate market in Chongqing is the high vacancy rate of commodity housing. According to the Chongqing Construction Committee (2000), 1.1 million square metres of residential commodity housing have remained unsold by the end of 1999. By contrast, around a hundred thousand workers and their families were living in quarters with less than 4 square metres per capita. Thirty percent of housing unit lacked basic facilities such as kitchens and lavatories (Chongqing Municipal Bureau, 1999). A mismatch between housing demand and supply is the key to the dilemma of housing vacancies, though low affordability contributes to part of the problem, as the economic situation of the households in Chongqing has experienced continuing improvment during recent years. A Housing Preference Expert System, which contains parameters with priority vectors, are urgently needed as a tool to assist the developers to supply marketable residence, and to help homebuyers to select suitable houses.

This paper reports the preliminary findings from stage 1 of a collaborative study on affordable housing conducted by a tripartite research team from The Hong Kong Polytechnic University, the Chongqing

University, and the South Bank University. The overall research objectives are to use the concept of industrialized affordable housing to examine the feasibility and viability of its use, and to design suitable types of industrialized affordable housing for use in Chongqing, by maximising the use of indigenous resources from the region. At this stage, an investigation has been carried out to identify the basic characteristics of the construction industry in Chongqing and to set the scene for the housing research. This focuses on the priority of parameters regarding potential buyers' preference, by using an analytic hierarchy process approach.

KEYWORDS

Industrialized Housing, Analytic Hierarchy Process, Chongqing.

INTRODUCTION

Chongqing as a Key City in the Western Part of China

Chongqing is positioned strategically at the gateway to western China, bridging the western part and the coastal region (Wong, et. al., 2001). As China's most populated city, it has an area of 82.4 thousand square k.m., is 470 km long from East to West, 450 km wide from North to South, and has a total population of 32.5 million. According to the Mayor of Chongqing, the city is important in the western part of China for its rich mineral resources, its huge market, and its cheap labour (Bao, 2000). The technology, human resources, and industrial setting in Chongqing are comparatively more advanced than other cities.

There are three attributes that distinguish Chongqing from the rest of western China. Firstly, Chongqing is the sole and biggest central metropolis, being directly controlled by the central government and equipped with a highly developed infrastructure. Secondly, being situated at the upper stream of the Yangtze River, Chongqing is an excellent economic centre for transportation facilities. Thirdly, the Three Gorges Project requires the migration of one million people and the reconstruction of hundreds of towns, which will create an enormous market to encourage the development of industrialized housing in Chongqing. More than 20 provinces, in terms of capital, technology, and managerial skills, will assist the Three Gorges Project. Completion of the project will bring fundamental improvements to the supply and distribution of hydroelectric power, which in turn will benefit all industries.

Problems Encountered in the Chongqing Housing Market

One of the main problems in Chongqing's real estate market is the high vacancy ratio caused by a huge housing demand, and poor affordability. This mismatch between supply and demand is explained below.

Huge housing demand

The total amount of housing floor space sold in 1998 was 4.2 million square metres, which amounted to 4.186 billion RMB. Housing units amounting to 2.5 billion yuan were sold to individual families and the remainder was sold to work units (China Municipal Publishing House, 2000). Up to now there are still 8,000,000 families who are classified as living under difficult conditions in Mainland China (China Municipal Bureau, 2000).

The estimated demand for commodity housing in the Chongqing urban area during the year 2000 amounted to 6 million sq.m. (Chongqing Construction Committee, 2000).

High vacancy ratio

The real estate market in Chongqing has become more agile in recent years yet a considerable number of housing units in the market remain unsold, possibly due to lack of a proper feasibility study before the development stage.

	Item	Total
1998	Housing Under Construction (million sq.m)	12.24
	Housing Newly Completed (million sq.m)	5.96
	Housing Sold (million sq.m)	4.21
	Housing Unsold (million sq.m)	1.17
1999	Housing Under Construction (million sq.m)	12.68
	Housing Newly Completed (million sq.m)	4.29
	Housing Sold (million sq. m)	4.34
	Housing Unsold (million sq. m)	1.12

Source: Chongqing Construction Committee, 2000

Figure 1: Housing Supply in Chongqing 1998 - 1999

Figure 1 shows that the unsold number of housing units in Chongqing remained about the same from 1998 to 1999. In 1999, 12.68 million sq.m. of housing was under construction and only 4.29 million sq.m. was completed, of this total 1.12 million sq.m. remained unsold (Chongqing Construction Committee, 2000). Clearly a mismatch exists between supply and demand because of affordability and quality issues.

This mismatch between housing demand and supply is the key to the dilemma of housing vacancy, though low affordability contributes to part of the problem, as the economic situation of the households in Chongqing has experienced continuing improvement during recent years. To help alleviate the mismatch problem, the Housing Preference Expert System (HPES), with appropriate parameters and priority vectors, can be used to assist developers to supply marketable residence, and to assist homebuyer to select their suitable houses.

RESEARCH

Objectives

- To determine the priority vector for each parameter of the proposed Housing Preference Expert System.
- To scrutinize the factors that affect homebuyers' behavior in China.

Methodology

- Identify the housing experts in Chongqing.
- Invite the housing experts to complete an investigation form.
- Use an Analytic Hierarchy Process (AHP) approach to calculate the priority vector of each parameter.

ANALYTIC HIERARCHY PROCESS (AHP)

The Analytical Hierarchy Process is an analytical tool using a deductive approach. It is a systematic procedure to represent the elements of any problem, and is particularly useful for solving complex problems, which are in the state of 'fuzziness'. It organizes the basic rationality by breaking down a problem into its smaller constituent parts and then calls for only simple pair-wise comparison judgments, to develop priorities in each hierarchy.

One should note that to design an analytic hierarchy, like the structuring of a problem by any other method, is more of an art than a science. It necessitates substantial knowledge of the system in question. A very strong aspect of the AHP is that the knowledgeable individuals who supply judgments for the pair-wise comparisons usually also play a prominent role in specifying the hierarchy.

SETTING UP WEIGHTINGS FOR THE RESIDENCE CRITERIA

Structuring a Decision-making Hierarchy

The prioritization problem has been broken down into a hierarchy of interrelated decision levels, each of which involves a number of factors. Level 1 describes the objective (goal) in the decision-making problem whilst Level 2, 3, and 4 contain factors that are perceived as important attributes affecting the level 1 decision. Level 5 gives the decision alternatives. The factors have been distilled from a comprehensive literature review of housing preferences.

Formation of a Pair-wise Comparison Matrix

Housing experts in Chongqing were invited to make a pair-wise judgment for the selected factors. Each pair-wise comparison involved assigning the priority factor with respect to the other factors on a rational scale. The scale used in this study is based on the recommendations from Saaty (1991).

Consistency of Priorities

The results of the pair-wise comparison matrix were examined for consistency of priorities by using the following formula:

$$\text{The Index of Consistency (C.I.)} = (\lambda_{max} - n) / (n-1)$$

(Note: n= number of factors for pair-wise comparison, and λ_{max} = largest eigenvalue for size matrix)

$$\text{Consistency Ratio (C.R.)} = \text{C.I.} / \text{Random Consistency (n)}$$

Size of Matrix	1	2	3	4	5	6	7	8	9	10
Random Consistency	0.00	0.00	0.58	0.90	1.12	1.24	1.32	1.41	1.45	1.49

Figure 2: Random Consistency for Different Matrix Sizes (Saaty, 1982)

THE PRIORITY VECTOR OF RESIDENCE PREFERENCE PARAMETERS

Totally six experts on housing issues in Chongqing completed the questionnaire. They comprised one general manager and one director from the China Construction Science & Technology (Chongqing) Corporation, two senior architects from the Chongqing Municipal Architecture Design Institute, and two professors from the Chongqing University.

The overall housing preference parameters' priority vectors

Level 1 Matrix							
	A & S	Facility	Trans.	Price	Environ.	Decor.	P.M.
Architecture & Structure	1	1	1/3	1/5	1/4	1	1/2
Facility	1	1	1/3	1/5	1/4	1	1/2
Transportation	3	3	1	3/5	3/4	3	3/2
Price	5	5	5/3	1	5/4	5	5/2
Environment	4	4	4/3	4/5	1	4	2
Decoration	1	1	1/3	1/5	1/4	1	1/2
Property Management	2	2	2/3	2/5	1/2	2	1
Priority Vector	0.059	0.059	0.176	0.294	0.235	0.059	0.118

Figure 3: Results of Priority Vectors

Figure 3 provides the consolidated pair-wise comparison results from the housing experts, and Table 1 and Table 2 provide the breakdown details.

Table 1
Pair-wise Comparison Matrix

	Housing preference – pair-wise comparison matrix	**Priority vector**
Architecture & Structure	7 √ 1 x 1 x 1/3 x 1/5 x 1/4 x 1 x 1/2 = 7 √ 0.0083 = 0.5043	0.5043 / 8.5778 = 0.059
Facility	7 √ 1 x 1 x 1/3 x 1/5 x 1/4 x 1 x 1/2 = 7 √ 0.0083 = 0.5043	0.5043 / 8.5778 = 0.059
Transportation	7 √ 3 x 3 x 1 x 3/5 x 3/4 x 3 x 3/2 = 7 √ 18.225 = 1.5139	1.5139 / 8.5778 = 0.176
Price	7 √ 5 x 5 x 5/3 x 1 x 5/4 x 5 x 5/2 = 7 √ 651.0416 = 2.5232	2.5232 / 8.5778 = 0.294
Environment	7 √ 4 x 4 x 4/3 x 4/5 x 1 x 4 x 2 = 7 √ 136.53 = 2.0185	2.0185 / 8.5778 = 0.235
Decoration	7 √ 1 x 1 x 1/3 x 1/5 x 1/4 x 1 x 1/2 = 7 √ 0.0083 = 0.5043	0.5043 / 8.5778 = 0.059
Property Management	7 √ 2 x 2 x 2/3 x 2/5 x 1/2 x 2 x 1 = 7 √ 1.0667 = 1.0093	1.0093 / 8.5778 = 0.118
	Total = 8.5778	1.0000

Table 2
Multiplication of the Matrix by the Priority Vector

Architecture & Structure	(1x0.0059)+(1x0.0059)+(1/3x0.176)+(1/5x0.294) +(1/4x0.235)+(1x0.059)+(1/2x0.118)=	0.413
Facility	(1x0.0059)+(1x0.0059)+(1/3x0.176)+(1/5x0.294) +(1/4x0.235)+(1x0.059)+(1/2x0.118)=	0.413
Transportation	(3x0.059)+(3x0.059)+(1x0.176)+(3/5x0.294) +(3/4x0.235)+(3x0.059)+(3/2x0.118)=	1.236
Price	(5x0.059)+(5x0.059)+(5/3x0.176)+(1x0.294) +(5/4x0.235)+(5x0.059)+(5/2x0.118)=	2.061
Environment	(4x0.059)+(4x0.059)+(4/3x0.176)+(4/5x0.294) +(1x0.235)+(4x0.059)+(2x0.118)=	1.649
Decoration	(1x0.0059)+(1x0.0059)+(1/3x0.176)+(1/5x0.294) +(1/4x0.235)+(1x0.059)+(1/2x0.118)=	0.413
Property Management	(2x0.059)+(2x0.059)+(2/3x0.176)+(2/5x0.294) +(1/2x0.235)+(2x0.059)+(1x0.118)=	0.825
	Largest eigenvalue for size matrix λ_{max} =	7.01

The results of the pair-wise comparison matrix from Table 1 and Table 2 were examined for consistency of priorities:

$$\text{The Index of consistency (C.I.)} = (\lambda_{max} - n) / (n-1)$$

$$= (7.01-7) / (7-1) = 0.01 / 6 = 0.0017$$

$$\text{Consistency Ratio (C.R.)} = \text{C.I.} / \text{Random Consistency (n)}$$

$$= 0.0017 / 1.32 = 0.0013$$

The value should be in the region of 0.1 or less. Since the consistency ratio in this case is at 0.0013, which is less than 0.1, the judgment is proved to be numerically consistent. A similar approach could be used to determine the priority vectors of sub-parameters at the other levels.

CONCLUSIONS

The priority setting of preferential parameters for home purchase in Chongqing was determined as follows (priority vector): price (2.061), environment (1.649), transportation (1.236), property management (0.825), architecture & structure (0.413), facility (0.413) and decoration (0.413).

To conclude, the Analytic Hierarchy Process (AHP) approach can be used as an effective analytical tool to determine complex issues, such as preferential parameters for home purchase. With appropriate parameters and priority vectors, a Housing Preference Expert System (HPES) could be developed, to assist potential homebuyers to select suitable properties. However, caveat is that such an analysis will very much depend on the comprehensiveness of the literature review to determine the appropriate parameters to be adopted in the study, and the professional judgment of the experts asked to make the pair-wise comparisons. An obvious limitation of this approach is that there is a certain degree of subjectivity in prioritising the parameters. Finally, it is recommended that similar studies be conducted in other major cities in China Mainland, so that comparisons between major cities and regions could be made in the future.

ACKNOWLEDGEMENTS

The Hong Kong Polytechnic University's Institutional Research Fund funds this study. The research title is 'A Study of Manufactured Affordable Housing in Chongqing', and the Project Account Code is G – YC 32.

REFERENCES

Bao Xu Ding (2000) Western Exploration Hundreds of Questions, *Chongqing Municipal Party Committee Propaganda Department* (in Chinese).
China Municipal Bureau (2000) 2000 *China Statistical Yearbook* (in Chinese).

China Municipal Publishing House (2000) *PRC real Estate Statistical Year Book.*
Chongqing Construction Committee (2000) *Chongqing Construction Industry Report in Year 2000.*
Chongqing Municipal Bureau (1999) *Chongqing Statistical Yearbook* (in Chinese).
Li Ye Heng Zheng Estate (2000) *Report on Housing Demands at 1999 Chongqing Trade Fair of Real Estate.*
Saaty, T.L. (1980) *The Analytic Hierarchy Process*, McGraw-Hill.
Saaty, T.L. (1982) The Analytic Hierarchy Process: A New Approach to Deal with Fuzziness in Architecture, *Architectural Science Review*, Vol.25, No.3, September, pp. 64-69.
Saaty, T.L. (1997) A Scaling Method for Priorities in Hierarchical Structures, *Journal of Mathematical Psychology*, Vol.15, pp.234-281.
Saaty, T.L., & Beltran, M. (1980) Architectural Design by the Analytic Hierarchy Process, *Design Methods and Theories*, Vol. 14, Nos. 3/4, pp. 124-134.
Saaty, T.L. & Erdener, E. (1979) A New Approach to Performance Measurement - The Analytic Hierarchy Process, *Design Methods and Theories*, Vol. 13, No. 2, pp. 64-72.
Saaty, T.L., & Vargas, L. (1991) The Logical of Priorities, *Kluwer Nijhoff Publishing.*
Shen, Q.P. & Lo, K.K. (1999) Priority Setting in Maintenance Management – An Analytic Approach, *The Hong Kong Polytechnic University*, 62 pp.
Wong, K.W., Hui, C.M., Li, S.R., Howes, R., & Wu, M. (2001) A Study of Manufactured Affordable Housing in Chongqing, Research Monograph, *The Hong Kong Polytechnic University*, 77pp.

TEAM PRESENTATION:

RESEARCH AT THE ATLSS CENTER ON NEW MATERIALS AND SYSTEMS TO RESIST EARTHQUAKES AND WIND

Advances in Building Technology, Volume 1
M. Anson, J.M. Ko and E.S.S. Lam (Eds.)

STEEL MOMENT FRAME CONNECTIONS THAT ACHIEVE DUCTILE PERFORMANCE

John W. Fisher[1], J.M. Ricles[1], Le-Wu Lu[1], C. Mao[1], Eric J. Kaufmann[1]

[1]ATLSS Engineering Research Center, Lehigh University, 117 ATLSS Drive,
Bethlehem, PA 18015-4729, USA

ABSTRACT

An extensive research program was conducted at the ATLSS Center, Lehigh University, to investigate the causes of brittle failures observed in welded connections during the 1994 Northridge earthquake and ways to improve their ductility. Tests on pre-Northridge connection details verified several weak links in the connection design and fabrication. Full-scale connection tests were carried out to develop improved details for welded unreinforced connections. These test results are presented, along with recommendations to insure ductile connection performance.

KEYWORDS

Continuity plates, ductility, fracture, inelastic rotation, notch toughness, panel zone, weld access hole, welded moment connection.

INTRODUCTION

The 1994 Northridge earthquake in California resulted in widespread damage to welded beam-to-column connections in moment resisting framed structures in the Los Angeles area [1]. The moment connection used in these structures was the popular welded flange and bolted web type with beam flanges welded to the column flange with full penetration groove welds with backing bars and the beam web is bolted to a shear tab. This connection was "prequalified" as a ductile connection based on the results of tests on moderate size laboratory specimens [2], and had been widely adopted since the late 1960's. Shortly after the earthquake damage was found in more than 100 low- and high-rise buildings in the Los Angeles area.

This paper summarizes the results of a comprehensive research program that was conducted at Lehigh University following the 1994 Northridge earthquake. This research program included investigations to study causes of the connection fractures that occurred during the earthquake as well as experimental and analytical studies to develop improved performance of welded connections under inelastic cyclic loading.

FAILURE ANALYSIS OF FRACTURED PRE-NORTHRIDGE CONNECTIONS

Following the 1994 Northridge earthquake a survey of 1290 floor-frames in 51 steel buildings was conducted by Youssef at al. [1]. The buildings were from a variety of locations in Los Angeles, of different sizes, frame configuration and construction type.

Youssef et al. [1] reported that the most common form of damage found was partial or complete fracture of the beam flange groove welds. Damage to the base metal was found to occur most frequently as a fracture of the column flange adjacent to the beam bottom flange weld. The observed column flange and weld fracture patterns strongly suggested that the damage was related to the condition at the weld root. The welded joints appeared to be brittle, with little evidence of inelastic behavior in the weldment or base metal. The fractures were attributed to a variety of other factors, including the weld process, quality of workmanship, base metal properties, and connection design [3].

A study was undertaken at Lehigh University [4] in order to characterize the origin of the fracture and material properties of failed connections. A detailed analysis was performed of 19 fractures removed from five different buildings damaged during the Northridge earthquake. The fracture surfaces of all of the samples were analyzed in order to characterize the fractures. The size and source of the originating defect was recorded.. The details are presented by Kaufmann et al. [4]. Cleavage fracture was found to be the mechanism of crack propagation in all of the samples regardless of crack propagation path. Brittle fracture initiated close to the mid-width of the weld at the column centerline usually from a weld root incomplete fusion flaw arising from entrapped slag or porosity at the back-up bar.

Another common factor among all of the fractured connections that were studied is that the welds were deposited with E70T-4 electrodes which were in common use. Fracture toughness tests of weld metal removed from sixteen connection samples, other damaged building connections and laboratory tests [4,5] have consistently shown that the E70T-4 weld metal provides a low level of fracture toughness. Figure 1 shows CVN values of various self-shielded, flux core electrode E70T-4, E70T-7, E71T-8 and E70TG-K2 and E7018 a shielded-metal-arc electrode.

The dynamic fracture toughness, K_{ID}, of the weld metal obtained by applying the correlation $K_{ID} = \sqrt{0.64E(CVN)}$, to the Charpy V-notch data, where E and CVN are the modulus of elasticity (in MPa), and the measured Charpy V-notch energy (in Joules). Weld metal compact tension tests of E70T-4 weld metal tested statically and at intermediate loading rates similar earthquakes [4] provided fracture toughness of 52 to 54 MPa $\sqrt{m}$. A strain shift rate of $0.75T_S$ shows good agreement between the lower bound dynamic toughness and the intermediate rate K_{IC} tests. This indicated that E70T-4 weld metal toughness at intermediate loading rates was 50 MPa$\sqrt{m}$ to 75 MPa$\sqrt{m}$ at room temperature [4].

PRE-NORTHRIDGE CONNECTION TESTS AND OTHER PRIOR STUDIES

Following the Northridge earthquake, numerous experimental studies were conducted to investigate the various aspects that were believed to be associated with the failures observed in the pre-Northridge connection and to improve connection performance. Two full-scale beam-and-column assemblies with a "pre-Northridge" connection detail were dynamically tested. The column was a W14x311 made of A572 Gr. 50 steel. It was supported by a pin at its bottom and a roller at its top. The beam was an A36 steel W36x150 section, with a flange yield strength equal to 262 MPa. The beam was connected to the column at mid-height. Dynamic load cycles were applied at the free end of the beam by two actuators.

The first specimen tested was Specimen A-1, which was typically pre-Northridge practice. The second specimen, Specimen A-2, was the same as Specimen A-1, but with the back-up bar and weld tabs removed and a 9.5-mm reinforcing fillet weld (E71T-8) added to its weld root. E70T-4 was used as the filler metal in both specimens. The groove welds were ultrasonically tested. In both specimens ten A325 bolts, 25.4-mm in diameter, were used to connect the beam web to a shear tab.

The applied beam load-beam tip displacement responses of Specimens A-1 and A-2 are shown in Figure 2. Specimen A-1 fractured at the bottom flange connection in a brittle manner. The cracks started at the root of the weld, extended into the column flange and caused a divot type fracture (Figure 3). The maximum bending moment achieved in the beam during the test was only $0.87M_p$. The fracture occurred when both the beam and column were still in the elastic range. The applied stress at the groove welds was approximately 262 MPa. The critical value for the stress intensity K is about 58 MPa$\sqrt{m}$. This value is compared with the estimated weld fracture toughness (point A-1) in Figure 4. Also shown are the results of static tests on other large-size connection specimens fabricated with E70T-4 welds [10] that fractured without significant yielding of the beam.

The removal of the back-up bars in Specimen A-2 slightly improved the connection performance, but brittle fracture of the flange welds again led to failure. Other than the limited yielding at the weld access holes, the connection behaved elastically when the welds of the top and bottom flanges fractured almost simultaneously during reversed loading. The fracture surface was confined entirely in the weld. The maximum bending moment resisted by the connection was $0.92M_p$. The inferior performance of Specimens A-1 and A-2, with E70T-4 filler metal, was mainly due to the low fracture resistance of the E70T-4 weld metal. A Charpy V-notch fracture toughness of 27 Joules at –29°C was imposed by the AISC in 1997 [6].

Since the Northridge Earthquake several experimental studies were conducted in an attempt to develop improved details for ductile welded unreinforced moment connections. These studies included 13 tests conducted by Lu et al. [7], Stojadinovic et al. [8], and Engelhardt [9]. The test specimens consisted of a beam-to-column exterior connection, where the beam and column sections ranged in size from W24x68 to W36x150 and W14x120 to W14x455, respectively. The complete penetration groove beam flange welds were made for all specimens using the E70TG-K2 electrode, except for Specimen A3 tested by Lu et al. which used an E7018 electrode. The bottom flange the backing bar was removed; the root of the weld was gouged rewelded and reinforced with a fillet weld using the E71T-8 electrode. On the top flange the backing bar was left in place and the bottom side facing the column was reinforced with an E71T-8 fillet weld. In the two tests by Lu et al. and the one test by Engelhardt, both the top and bottom flange backing bars were removed. The beam webs of all of the specimens were bolted, except for the tests by Lu et al. which had the beam webs attached to the column flange with a groove weld. All specimens had continuity plates, except Specimen NSF3. The access holes for the specimens are shown in Figure 5 and represented typical fabrication practice.

All of the specimens developed inelastic deformation in both their beams and panel zones, except for two specimens tested by Stojadinovic et al. [8] and Specimen NSF3 tested by Engelhardt which had strong panel zones. A summary of the total plastic story drift, θ_p, developed in the specimens and the contribution from the beam and panel zone is given in Figure 5. Of the 13 tests, only one successfully achieved a plastic rotation of 0.03 radians or more. The typical failure mode in all of the specimens was a fracture in the base metal of the beam flange that originated at the access hole. Some of the specimens developed a through thickness crack that was a low-cycle fatigue failure [10].

NEW STUDIES TO IMPROVE CONNECTION DUCTILITY

The numerous causes that contributed to the poor performance of moment connections motivated additional analytical and experimental studies by the authors which were conducted under the second phase of the SAC Steel Project [11]. In these studies the effects that various connection details have on inelastic cyclic performance of welded unreinforced moment connections were investigated. The details investigated included: (1) geometry and size of the weld access hole, (2) beam web attachment, and (3) panel zone capacity. The analytical study involved a parametric study using nonlinear finite element models of a welded moment flange connection [12]. These design details were verified by testing 11 full-scale specimens, and shown to produce ductile connections.

Test Program

The experimental study consisted of six exterior moment connections and five interior moment connections. All of the specimens had complete penetration beam flange groove welds using the E70TG-

K2 filler metal. The top flange backing bar was left in place, and a fillet weld reinforcement was provided between the backing bar and the column flange. The beam bottom flange backing bar was removed, back gouged and reinforced with a fillet weld. The beam web attachment detail, panel zone strength, and use of continuity plates were varied among the specimens.

Specimens T1 to T6 were exterior connections. The beam and column were A572 Gr. 50 W36x150 and W14x311 sections. The beam web attachment and panel zone strength were the main variables. Five different beam web attachment details were tested. Specimen T1 beam web was welded directly to the column flange with a groove weld and supplementary fillet weld around the shear tab. In Specimen T2, the weld details are the same as T1 except no supplementary fillet weld. In Specimen T3, a fillet weld connected the shear tab. In Specimen T4, the beam web was bolted to the shear tab. Specimens T1 through T4 had continuity plates, but no doubler plates because their panel zone strength satisfied the 1997 AISC Seismic Provisions. Specimens T5 and T6 had stronger panel zones. Specimen T5 was similar to Specimen T1 except that it had a 12mm thick doubler plate in the column panel zone and no continuity plates. Specimen T6 was also similar, except that it had a 12 mm doubler plate and a heavy shear tab welded to the column face and no continuity plates. The shear tab and welds for Specimen T6 were designed to resist the plastic moment capacity of the beam web. The design of the doubler plates for Specimens T5 and T6 was based on the AISC LRFD provisions, but the panel zone was designed to resist the expected beam capacity M_{pe}=1.1M_p, as opposed to 0.8M_{pe} per AISC [6].

The interior connection specimen consisted of two W36x150 A572 Grade 50 beams with either a W14x398 A572 Grade 50 column (Specimens C1, C2, and C5) or a W27x258 A572 Grade 50 column (Specimens C3 and C4). Specimens C1, C3, and C5 did not have continuity plates. The W14x398 column section (Specimen C1, C2, and C5) has a 72- mm thick flange and a flange width-to-thickness ($b_{f,col}/t_{f,col}$) ratio of 5.83. The W27x258 section, having a 45-mm thick flange, was selected as the column for Specimens C3 and C4 because the ($b_{f,col}/t_{f,col}$) ratio for the section is 8.06.

The test setup for both exterior and interior connections was designed to simulate the boundary conditions of a beam-to-column connection in a MRF subjected to lateral loading. The ends of the beams were pin connected to rigid links using cylindrical bearing to simulate beam points of inflection in the prototype frame. The base of the column was pinned to the laboratory floor. At the top of the column a pair of parallel horizontal actuators were pin connected to the column to impose story drift to the specimen. Lateral bracing was provided and conformed to the requirements of the 1997 AISC Seismic Provisions. Each specimen was instrumented with strain gages and displacement transducers.

Test Results

The normalized beam moment at the column face-total plastic connection rotation (M/M_p - θ_p) relationship for two exterior connections is shown in Figure 6, where M_p is the plastic beam moment capacity based on measured material properties and section dimensions. The behavior of the exterior connection specimens varied, depending on the web attachment detail and strength of the panel zone.

Specimen T1 developed 2% total story drift when cracks formed at the fusion line of the beam web groove weld at the top and bottom of the web. While yielding occurred in the beam, along with minor local beam flange buckling, extensive yielding occurred in the panel zone. Consequently, the M/M_p-θ_p response showed a progressive increase in capacity with each cycle, until at 5% story drift fracture of the beam bottom flange occurred. In Specimen T2 the beam top flange fractured at the fusion line of the groove weld during the second cycle of 4% total story drift. The omission of the supplemental fillet weld in Specimen T2 reduced the ductility of the connection. Specimen T3 had the fillet welded shear tab separate from the column face, which led to a fracture of the beam bottom flange base metal at 3% total story drift. The fillet weld was not strong enough to resist the moment in the beam web. In Specimen T4, slip occurred in the bolted shear tab. This lead to a fracture of the beam top flange in the HAZ at 4% total story drift. During the testing of Specimen T5, which had a stronger panel zone, yielding developed in the panel zone followed by yielding of the beam flanges and web, before significant local buckling of the beam flanges and web occurred at 3% total story drift. The beam bottom flange fractured at a total story drift of 6% in the region of local flange buckling. Specimen T6 did not fracture, and the test was stopped at a total story drift of 6%. The performance of the access hole was exceptional since in all of the specimens no cracking or fractures were observed to have initiated in the access hole region.

The behavior of the interior connection specimens were all similar, where pronounced yielding occurred in the beam flanges and web, leading to local flange and web buckling (see Figure 7). The location was about one-half the beam depth from the face of the column. After local beam buckling the capacity of the specimens then began to deteriorate with further cyclic displacements. A fracture of the beam flange eventually occurred in all specimens except Specimen C4. The maximum panel zone inelastic deformation occurred when specimen peak capacity developed. The total story drift at which flange fracture occurred ranged from 4% to 6%. The fracture in Specimen C5, which had the composite slab and shear studs welded to the beam top flange, occurred suddenly in the east beam at a shear stud located 533 mm from the column face where cyclic local buckling developed.

A summary of the contribution of the beam, column and panel zone to the total maximum plastic story drift θ_p for each specimen is given in Figure 8. These results are based on the last successful cycle that was achieved prior to fracture, or termination of the test. With the exception of Specimen C5, specimens that had dominant beam plastic rotation exceed the total plastic rotation of the specimens that had mainly panel zone deformation. Only the specimens with either the complete penetration groove welded beam web and supplement fillet welds or the heavy shear tab fillet welded to the beam web achieved a maximum plastic rotation of at least 0.03 radians.

CONCLUSIONS AND DESIGN RECOMMENDATIONS

Based on the research presented, the following conclusions and design recommendations are given.

1. Full-scale dynamic cyclic tests of pre-Northridge connections show failure modes similar to those observed in the field. Brittle fracture developed in the elastic range of response from flaws in the low toughness E70T-4 weld metal and geometric conditions.

2. In the connection tests with a notch tough filler metal and access holes with conventional geometry, ductile fracture initiating at weld access hole was observed, degrading the performance of the connection. The modified access hole geometry shown in Figure 8 was found to result in good performance and has been adopted by FEMA 350.

3. The web attachment detail can significantly affect the ductility of a connection which has been fabricated using a notch tough filler metal. A beam web attachment detail consisting of a complete penetration groove welded web and supplemental fillet welds around the beam shear tab, or a heavy shear tab welded to the column with a complete penetration groove weld and fillet welded to the beam web produced the best results.

4. Test specimens with a weak panel zone and notch tough filler metal developed significant inelastic shear deformation in the panel zone and increased the plastic strain demand on the beam web welds. Better performance was achieved when the panel zone strength was based on the web alone.

5. Shear studs welded to the beam flange in the region of inelastic cyclic local flange bucking caused cracks to initiate and deteriorate plastic rotation. Test results also indicate that continuity plates are not always required. All specimens without continuity plates and shear studs had a plastic story drift greater than 0.041 radians.

ACKNOWLEDGMENTS

This research described in this paper was supported by grants from the U.S. National Science Foundation, the National Institute of Standards, the SAC Joint Venture, and the Pennsylvania Infrastructure Technology Alliance.

REFERENCES

[1] Youssef, N. F. G, Bonowitz, D., and Gross, J. L., "A Survey of Steel Moment Resisting Frame Buildings Affected by the 1994 Northridge Earthquake," NIST, Report No. NISTIR 5625, 1995, Gaithersburg, MD.

[2] Popov, E. P., and Pinkney, R. B., "Cyclic Yield Reversal in Steel Building Connections," *Journal of the Structural Division*, ASCE, Vol. 95, No. ST3, 1969, pp. 327-353.

[3] Campbell, H.H., "The Northridge Fractures: Are We Learning the Right Lessons?" *Civil Engineering*, March, 1995.

[4] Kaufmann, E.J., Fisher, J.W., Di Julio Jr. R.M., and Gross, J.L., "Failure Analysis of Welded Steel Moment Frames Damaged in the Northridge Earthquake," National Institute of Standards and Technology, Report No. NISTIR 5944, January, 1997.

[5] Xue, M., Kaufmann, E. J., Lu, L-W, and Fisher, J. W., "Achieving Ductile Behavior of Moment Connections-Part II," Modern Steel Construction, June, 1996.

[6] AISC, "Seismic Provisions for Structural Steel Buildings," AISC, Chicago, Illinois, 1997.

[7] Lu, L.W, Xue, M., Kaufmann, E. J. and Fisher, J. W., "Cracking, Repair And Ductility Enhancement of Welded Moment Connections." *Proceedings, NEHRP Conference and Workshop on Research on the Northridge, California Earthquake of January 17, 1994*, Vol. III, 1997, pp. 637-646, Los Angeles.

[8] Stojadinovic, B., Goel, S., Lee, K., Margarian, A.G., Ghoi, J.H., "Parametric Tests on Unreinforced Steel Moment Connections," *Journal of Structural Engineering*, ASCE, Vol. 126, No. 1, 2000, pp. 40-49.

[9] Engelhardt, M., "Cyclic Testing of Welded Unreinforced Moment Connections," Unpublished report, University of Texas, Austin, 2000.

[10] Barsom, J., "Failure Analysis of Welded Beam to Column Connections," Report No. SAC/BD-99/23, SAC Joint Venture, Sacramento, California, 1999.

[11] Ricles, J. M., Mao, C., Lu, L.W. and Fisher, J.W., "Development and Evaluation of Improved Ductile Welded Unreinforced Flange Connections," Report No. SAC/BD-00/24, SAC Joint Venture, Sacramento, California, 2000.

[12] Mao, C., Ricles, J., Lu, L-W, Fisher, J. W., "Effect of Local Details on Ductility of Welded Moment Connections," *Journal of Structural Engineering,* Vol. 127, No. 9, Sept. 2000.

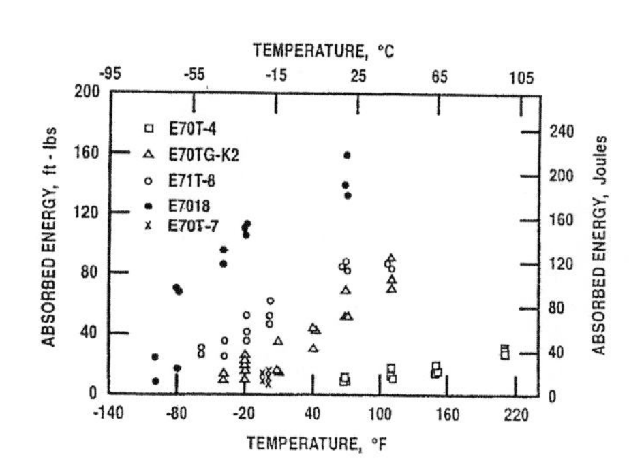

Figure 1 CVN test data for various filler metals [6]

Figure 3 Beam flange weld fracture, Specimen A-1

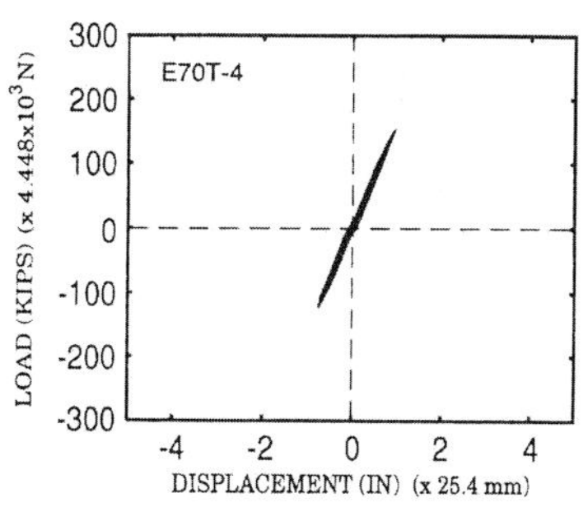

(a) Specimen A-1

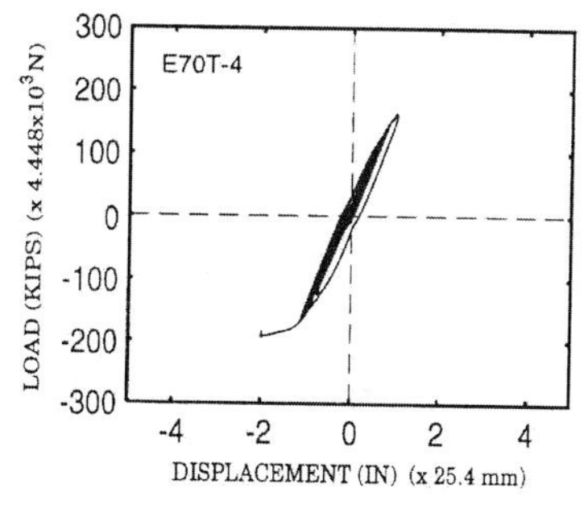

(b) Specimen A-2

Figure 2 Applied load-beam tip displacement response

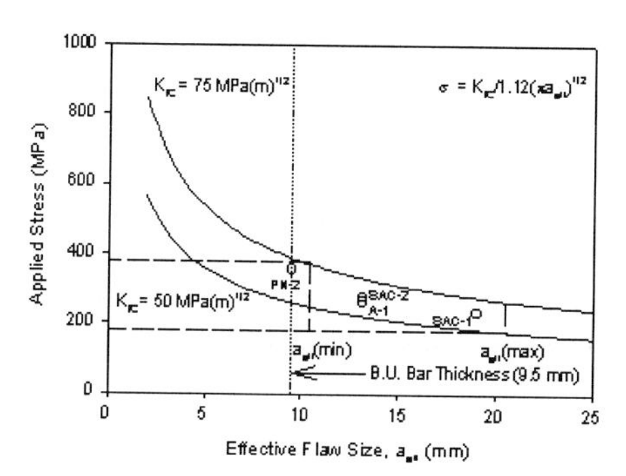

Figure 4 Critical applied stress vs. effective flaw depth
for E70T-4 WSMF Connections

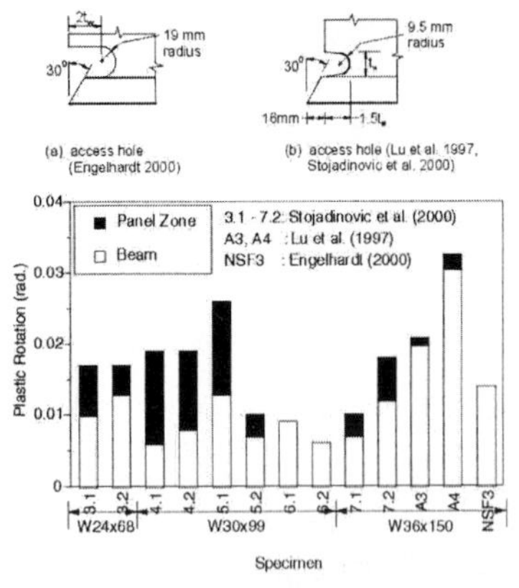

Figure 7 Interior Specimen C4 at completion of testing connection test

Figure 5 Summary of plastic story drift of prior Post-Northridge

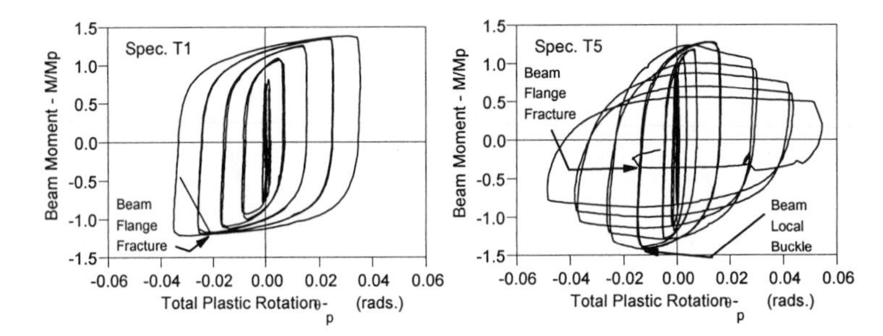

Figure 6 Exterior connection beam moment-plastic rotation response for a weak and strong panel zone

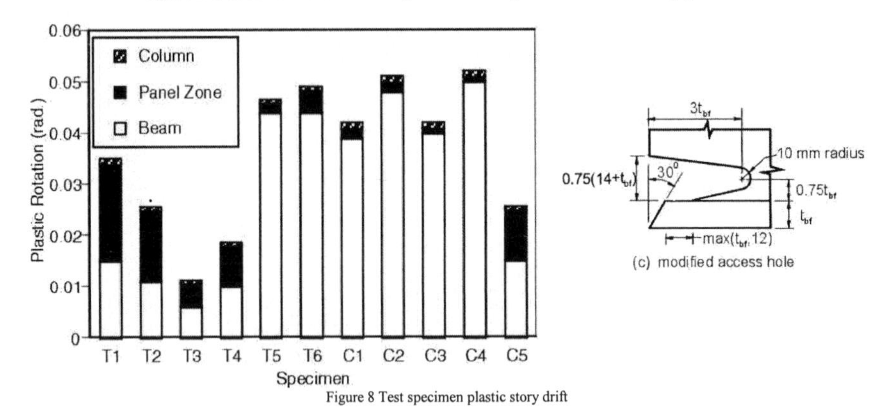

Figure 8 Test specimen plastic story drift

Advances in Building Technology, Volume 1
M. Anson, J.M. Ko and E.S.S. Lam (Eds.)

LATERAL LOAD BEHAVIOR OF UNBONDED POST-TENSIONED PRECAST CONCRETE WALLS

F.J. Perez, R. Sause, S. Pessiki, and L.-W. Lu

ATLSS Research Center, Department of Civil and Environmental Engineering, Lehigh University, Bethlehem, PA 18015, USA

ABSTRACT

Unbonded post-tensioned precast walls are constructed by post-tensioning precast panels across horizontal joints with the post-tensioning steel unbonded to the panels over a significant portion of its length. The lateral load behavior of these walls differs significantly from that of conventional cast-in-place reinforced concrete walls and precast concrete walls designed to emulate cast-in-place walls. An unbonded post-tensioned wall can be designed with strength and initial stiffness similar to those of a conventional cast-in-place reinforced concrete wall. In addition, however, the unbonded post-tensioned wall can be designed to undergo significant nonlinear lateral drift without significant damage. The nonlinear behavior results from the opening of gaps along the horizontal joints. Subsequent closing of these gaps, as a result of the post-tensioning, results in a self-centering behavior that eliminates residual lateral drift. The paper reports on the experimentally and analytically observed lateral load behavior of unbonded post-tensioned precast concrete walls. The limit states in the behavior are outlined. Results of monotonic and cyclic lateral load tests of large-scale wall specimens are presented and compared with analytical results. Conclusions regarding the suitability of the observed behavior are given.

KEYWORDS

Walls, Precast, Unbonded, Prestressed, Concrete, Seismic, Earthquake

INTRODUCTION

Past earthquakes have demonstrated the superior seismic performance of buildings with reinforced concrete walls as the primary lateral load resisting system (Fintel, 1995). However, the provisions of current U.S. building codes limit the use of precast concrete walls as a primary lateral load resisting system in seismic regions. Precast concrete systems that can be designed in the U.S. under current building codes use connections made with cast-in-place concrete, and the resulting wall emulates a cast-in-place wall. However, these emulative precast walls do not have all the economic advantages of precast construction as a result of the cast-in-place concrete and expensive details in the connections.

Alternative non-emulative precast seismic structural systems have been developed that take advantage of natural discontinuities at the connections of precast systems. In the U.S., these alternative systems have been the focus of the PRESSS (PREcast Seismic Structural Systems) program, initiated in 1990. PRESSS research has shown that the use of unbonded post-tensioned connections of precast elements results in precast concrete systems with excellent load-deformation hysteretic behavior (Cheok et al., 1993, MacRae et al., 1994, Priestley et al., 1993). An analytical study of the seismic response of unbonded post-tensioned precast concrete walls with horizontal joints was carried out at Lehigh University by Kurama et al. (1999). These unbonded post-tensioned precast walls are constructed by post-tensioning precast wall panels across horizontal joints at the floor levels (Figure 1). Grout is used between the joints for construction tolerances. The analytical study shows that, as a result of unbonding, large nonlinear lateral drifts can be achieved in a wall without yielding or fracturing the post-tensioning steel. Dynamic analysis results indicate that unbonded post-tensioned precast walls have large flexural ductility and self-centering capacity without sustaining significant damage and excessive drift under moderate-to-severe earthquakes.

An experimental program to verify the results of this analytical study is currently underway at Lehigh University. This paper discusses the behavior of unbonded post-tensioned precast concrete walls subjected to static monotonic and cyclic lateral loads. The experimental results obtained to date are compared with results obtained using the analytical model developed by Kurama et al. (1996).

BEHAVIOR

This paper reports on the experimentally and analytically observed lateral load behavior of unbonded post-tensioned precast concrete walls. The walls considered in this study are shown schematically in Figure 1. Each wall is comprised of six one-story precast panels that are connected along horizontal joints using unbonded post-tensioned steel, which is anchored at the roof and within the foundation (Figure 1(a)). The bottom panel has regions that are confined, as shown in Figure 1(b). This confinement enables the base panel to sustain the large compressive strains that develop as a result of gap opening displacements that develop along the base of the wall due to lateral loads.

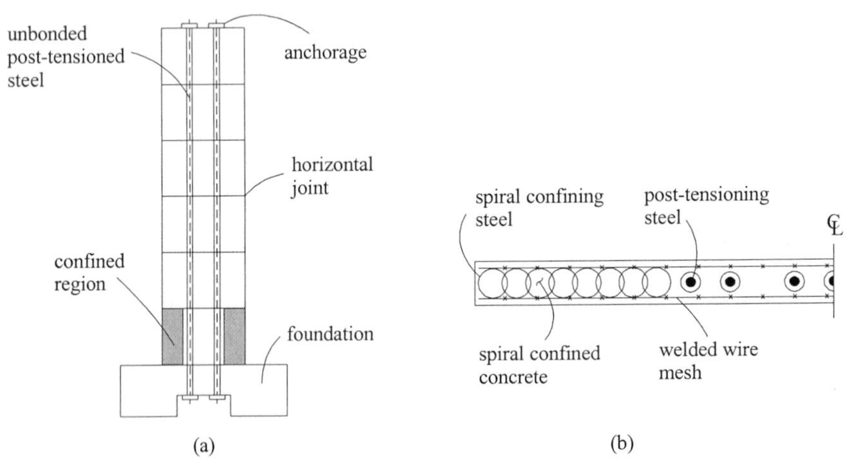

Figure 1: Unbonded post-tensioned wall.

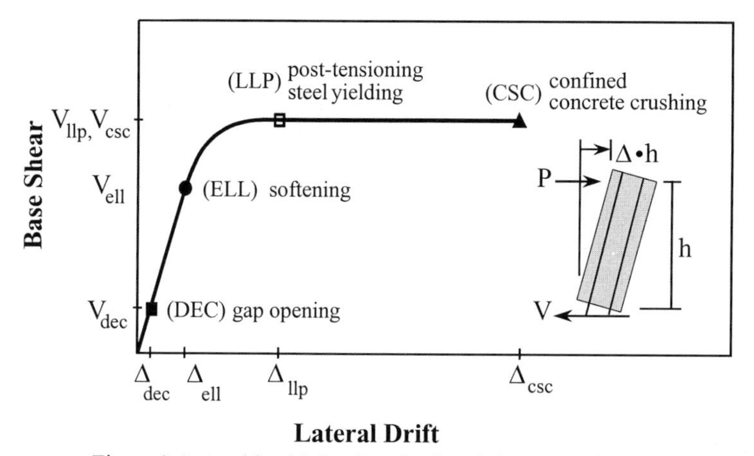

Figure 2: Lateral load behavior of unbonded post-tensioned walls.

As shown in Figure 2, the lateral load behavior of unbonded post-tensioned walls is characterized by several limit states: (1) decompression at the base of the wall; (2) significant reduction in lateral stiffness (i.e., softening) due to gap opening in flexure along the wall-to-foundation connection; (3) yielding of the post-tensioning steel; (4) base shear capacity; and (5) crushing of confined concrete. The limit states are described below.

Decompression at the base of the wall occurs when the overturning moment at the base of the wall causes one end of the wall to begin to decompress. Under a specified lateral load distribution, decompression of the wall can be related to a specific level of base shear and lateral drift, V_{dec} and Δ_{dec} respectively. Decompression is accompanied by the *initiation* of gap opening along the horizontal joint between the base panel and the foundation. The point on the base-shear-lateral-drift response of a wall when the lateral stiffness of the wall begins to reduce significantly is called the effective linear limit. This reduction in lateral stiffness (called softening) is caused by significant gap opening along the horizontal joint between the base panel and the foundation. The base shear and lateral drift corresponding to the effective linear limit are denoted as V_{ell} and Δ_{ell} respectively. Softening usually occurs in a smooth and continuous manner in a precast wall with unbonded post-tensioned horizontal joints (Kurama et al., 1999). Hence, the term *effective* linear limit is used to describe this point on the base-shear-lateral-drift relationship.

Yielding of the post-tensioning steel occurs when the strain in the steel reaches the linear limit of the steel. The base shear and lateral drift corresponding to the linear limit strain of the post-tensioning steel are denoted as V_{llp} and Δ_{llp}. Since the post-tensioning steel is unbonded, this limit state is reached after large nonlinear lateral drift has occurred, providing the wall with significant drift capacity before significant inelastic deformation develops (Kurama et al., 1999). Failure of the wall occurs when the confined concrete at the base of the wall fails in compression. Thus the drift capacity of the wall is controlled by the compression strain capacity of the confined concrete. Figure 1 refers to spiral confined concrete, although other confining reinforcement details may be used. The base shear and lateral drift corresponding to failure of the confined concrete are denoted as V_{csc} and Δ_{csc}. In a well-designed wall Δ_{csc} is significantly larger than Δ_{llp}. Also, the base shear capacity of the wall is V_{llp}, which is approximately equal to V_{csc}. Thus, in a well-designed wall, the base shear capacity is governed by axial-flexural (overturning) capacity of the wall, rather than the shear sliding capacity.

EXPERIMENTAL PROGRAM

An experimental program is currently underway at Lehigh University to study the behavior of unbonded post-tensioned precast concrete walls with horizontal joints under static monotonic and cyclic lateral loads. The walls considered in this study are 5/12-scale models of a 6-story full-scale wall similar to that shown in Figure 1. An elevation view of the test set-up is shown schematically in Figure 3. At the base is a large precast concrete foundation with a manhole (not shown) to provide access to the inside of the foundation, where the post-tensioning bars (shown dashed) are anchored. The test specimen is comprised of four wall panels, a loading block, and two extension panels. The bottom two panels have confinement on both ends of the wall, as shown by the shaded regions in Figure 4. The confinement is the same as that shown in Figure 1(b). It is noted that the confinement in the second panel is provided to preserve its integrity and thus make it reusable throughout the test program.

The loading block, which rests on the fourth panel, is used to apply gravity and lateral loads. The gravity load is applied by stressing an external bar on either side of the wall using a hydraulic cylinder. The gravity load remains constant throughout the experiment. The lateral load actuator is connected to the loading block at a height of 7.14 m from the base of the wall. This height represents the resultant of a triangular inertia force profile with zero load at the base and the maximum load at the roof. The two extension panels above the loading block provide a total unbonded height of 9.91 m, which is the unbonded height of the full-scale 6-story wall, scaled by 5/12.

The effect of five parameters on the flexural behavior of the walls is investigated. As summarized in Table 1, these parameters are: (1) area of post-tensioning steel, A_p; (2) initial stress in post-tensioning steel, f_{pi} (normalized with respect to the maximum stress of the post-tensioning steel, f_{pu}); (3) initial stress in concrete due to post-tensioning, $f_{ci,p}$; (4) eccentricity of post-tensioning steel, e_p; and (5) confining reinforcement details in the base panel. To date, Tests 1 and 2 have been performed (See Table 1) to compare the effects of monotonic versus cyclic loading on the flexural behavior of the walls. Results from these tests are discussed next.

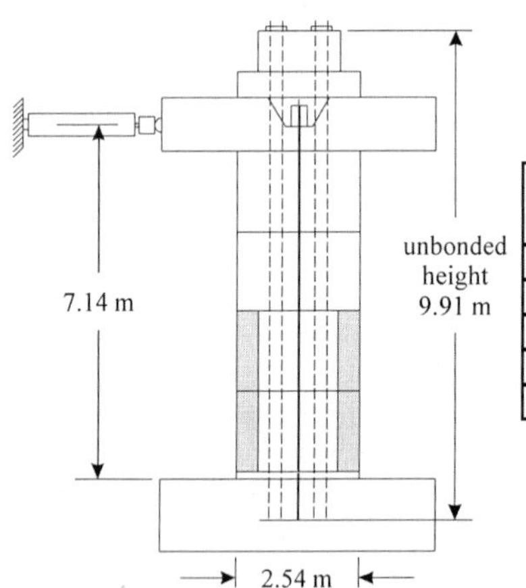

Figure 3: Test set-up.

TABLE 1
EXPERIMENTAL PARAMETER STUDY

Test	Loading	A_p (mm^2)	f_{pi}/f_{pu}	$f_{ci,p}$ (N/mm^2)	e_p (mm)	Confinement Type	Confinement Ratio (%)	
							Volumetric	Area
1	monotonic	4839	0.59	6.07	508	spirals	7.3	-
2	cyclic	4839	0.59	6.07	508	spirals	7.3	-
3	cyclic	2419	0.59	3.03	508	hoops	-	1.75
4	cyclic	4839	0.295	3.03	508	hoops	-	1.75
5	cyclic	2419	0.59	3.03	127	hoops	-	1.75

EXPERIMENTAL RESULTS

Figure 4 shows three plots of base shear versus lateral drift for Test 1. The base shear corresponds to the lateral load applied by the horizontal actuator shown in Figure 3. The lateral drift is taken as the horizontal displacement at the level of the actuator, divided by the height of the actuator relative to the base of the wall (i.e., 7.14 m). The two smooth plots in Figure 4 are obtained analytically using the wall model proposed by Kurama et al. (1996). The third plot, with the repeated load drops, is the experimental results. Each load drop is a result of a temporary halt in the loading of the specimen. The analytical result plotted with the darker line is from a pre-test prediction. It can be seen that the analytical model predicts the results quite accurately in the elastic range and through the effective linear limit point (ELL). However, the peak base shear and the lateral drift at failure do not agree. After Test 1 was completed, post-tensioning steel samples were tested in tension to obtain the stress-strain behavior of the material. These samples were taken from bars that did not yield during the test. It was observed that the yield strength of the post-tensioning steel was 15 percent larger than the pre-test model. A post-test analytical model was created from the pre-test model by increasing the post-tensioning steel yield strength, and the analysis was repeated. The post-test analytical result is shown in the lighter curve in Figure 4. It can be seen that (except for the lateral drift at failure) there is an excellent correlation between the wall analytical model and the experimental result. The values obtained for various key limit states from the second analytical model and the experiment are shown in tabular form in Figure 4.

At this time, two explanations are proposed for the difference in the lateral drift at failure (i.e., lateral drift corresponding to crushing of spiral confined concrete, Δ_{csc}). First, the strain capacity of the confined concrete in the experiment may be larger than the strain capacity in the analytical model, which was obtained using the stress-strain relationship for confined concrete proposed by Mander et al. (1988). Second, the compressive strain demand in the concrete, as a function of the lateral drift of the wall, may be smaller in the experiment than in the analytical model. That is, the model, which assumes plane sections in the concrete, overestimates the compressive strains at the base of the wall, and thus predicts an early failure when compared to the experiment.

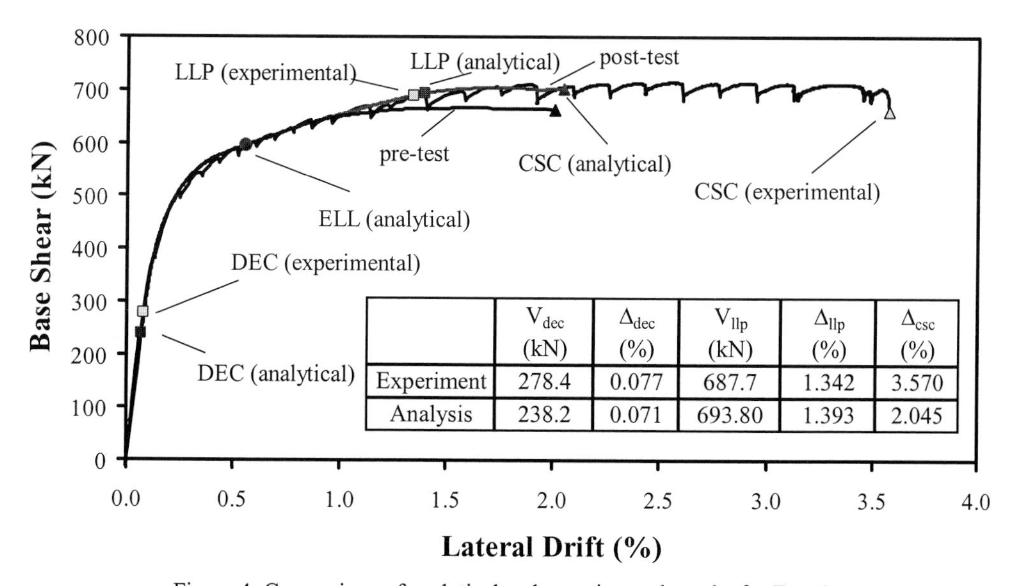

	V_{dec} (kN)	Δ_{dec} (%)	V_{llp} (kN)	Δ_{llp} (%)	Δ_{csc} (%)
Experiment	278.4	0.077	687.7	1.342	3.570
Analysis	238.2	0.071	693.80	1.393	2.045

Figure 4: Comparison of analytical and experimental results for Test 1.

The base shear-lateral drift results for Tests 1 and 2 (see Table 1) are plotted in Figure 5. Test 1 was conducted under a static monotonic lateral load, while Test 2 was performed under a static cyclic lateral load. As shown in Figure 5 and in Table 2, the lateral load results for the two tests are in good agreement. Test 2 shows that an unbonded post-tensioned precast concrete wall exhibits a nearly nonlinear elastic load-deformation response with a small amount of energy dissipation per cycle of loading. Before yielding of the post-tensioning steel, the nonlinearity results from gap opening that occurs between the base panel-to-foundation connection as the precompression due to prestressing is overcome by the moment due to lateral load. In addition, it can be observed that the wall exhibits excellent self-centering behavior. The self-centering capacity of the wall is only slightly compromised when it is loaded past yielding of the post-tensioning steel. Therefore, preliminary experimental results indicate that an unbonded post-tensioned precast concrete walls exhibit cyclic lateral load behavior that make them good candidates for use in earthquake-resistant building structures.

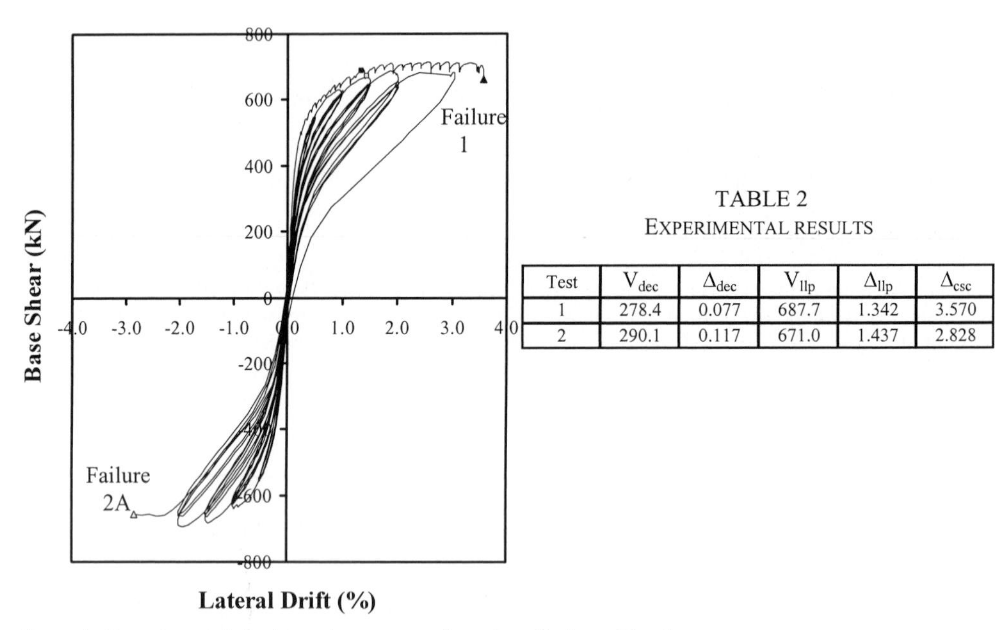

TABLE 2
EXPERIMENTAL RESULTS

Test	V_{dec}	Δ_{dec}	V_{llp}	Δ_{llp}	Δ_{csc}
1	278.4	0.077	687.7	1.342	3.570
2	290.1	0.117	671.0	1.437	2.828

Figure 5: Experimental results under monotonic and cyclic lateral loads.

FAILURE MODES

An unbonded post-tensioned precast concrete wall that is well designed and adequately detailed is expected to fail when the confined concrete at the base of the wall fails in axial-flexural compression. This failure occurs when the confining steel fractures and the lateral confining pressure in the confined concrete is lost. As noted earlier, the lateral drift when crushing of the confined concrete occurs is denoted as Δ_{csc}. This limit state is shown in Figure 5 for Tests 1 and 2 and is labeled as "Failure 1" and "Failure 2A," respectively. The observed failure at the base of the wall for Test 1 is shown in Figure 6. Figure 6(a) shows the base panel after failure. After excavating the concrete at the base of the compression edge of the wall, it was observed that the spirals fractured in the through-thickness direction of the wall, as shown in Figure 6(b). The same mode of failure was observed for Test 2 (Failure 2A). After Failure 2A occurred at one end of the wall for Test 2, the specimen was loaded in the opposite direction for additional cycles, as shown in Figure 7(a) until that side of the wall failed (labeled "Failure 2B"). As shown in Figure 7(b), Failure2B was a sudden local buckling failure of the confined concrete region of the base panel. It is important to emphasize that only the confined concrete region buckled, while the rest of the base panel did not, as shown in Figure 7(b) by the shiny vertical conduit which remained attached to the unbuckled portion of the panel. This failure mode can be explained as follows. During Test 2, as the wall was subjected to cyclic lateral loads, numerous short diagonal cracks formed in a narrow vertical band along the height of the base panel. These cracks formed adjacent to the vertical conduit shown in Figure 8(b), which was the outermost conduit, next to the highly confined region (see also Figure 1(b)). The development of these cracks ultimately permitted the confined concrete region to act independently from the rest of the wall. As a result, the confined region, which was carrying the compressive load, buckled after it was no longer restrained from buckling by the rest of the wall. This mode of failure is undesirable as the gravity load capacity of the wall is jeopardized. Therefore, the reinforcing steel in the base panel is being redesigned to better control cracking and to prevent this mode of failure.

(a) (b)

Figure 6: Observed failure mode for Test 1.

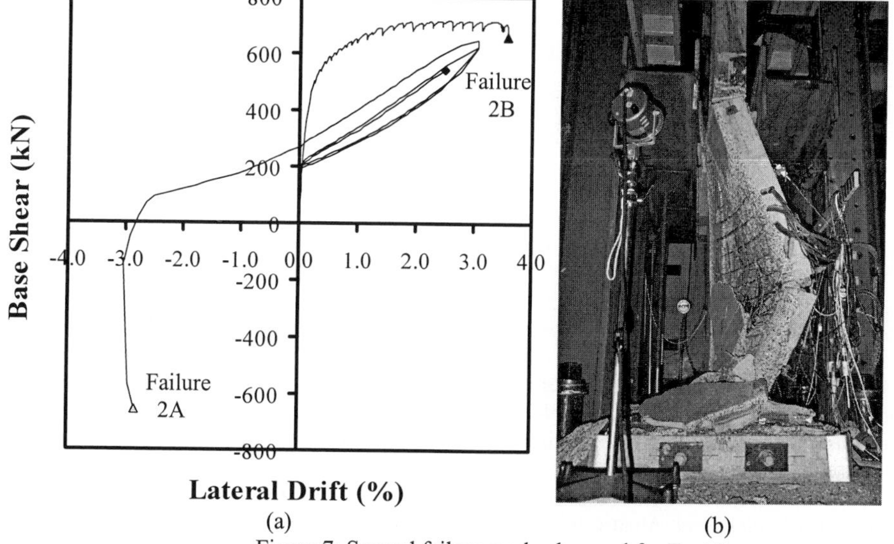

(a) (b)

Figure 7: Second failure mode observed for Test 2.

CONCLUSIONS

This paper reports on the experimentally and analytically observed lateral load behavior of unbonded post-tensioned precast concrete walls. The limit states that characterize the lateral load behavior are outlined. Results of monotonic and cyclic lateral load tests of large-scale wall specimens are presented and compared with analytical results. The results show that the limit states that characterize the lateral load behavior occur as anticipated in the design of the walls and at force and drift levels predicted by the analytical model, except that the experimentally observed drift capacity significantly exceeds the drift capacity predicted by the analytical model. The results demonstrate that unbonded post-tensioned precast walls can be designed to undergo significant nonlinear lateral drift without significant damage, and to retain their ability to self-center, thereby eliminating residual lateral drift. The cyclic lateral load behavior of the walls is nearly nonlinear elastic, with only a small amount of energy dissipation per cycle of loading. As a result, larger lateral drifts can be expected under earthquake loading. However dynamic analysis results (Kurama et al. 1999) show that these walls can be designed to avoid sustaining excessive drift under moderate-to-severe earthquakes.

ACKNOWLEDGMENTS

This research is funded by the National Science Foundation (NSF) under Grant No. CMS-9612165 as part of the PREcast Seismic Structural Systems (PRESSS) program. The support of NSF program director S.C. Liu is gratefully acknowledged. The research is also supported by the Pennsylvania Infrastructure Technology Alliance through a grant from the Pennsylvania Department of Community and Economic Development. Numerous individuals from the U.S. Precast Concrete Industry have also provided assistance. The opinions, findings, and conclusions in this paper are those of the authors and do not necessarily reflect the views of those acknowledged herein.

REFERENCES

Cheok, G., Stone, W., and Lew, H. (1993). Model Precast Concrete Beam-to-Column Connections Subject to Cyclic Loading. *PCI Journal*, Precast/Prestressed Concrete Institute **38:4,** 80-92.

Fintel, M. (1995). Performance of Buildings with Shear Walls in Earthquakes of the Last Thirty Years. *PCI Journal*, Precast/Prestressed Concrete Institute **40:3,** 62-80.

Kurama, Y., Pessiki, S., Sause, R., Lu, L. W., and El-Sheikh, M. (1996). Analytical Modeling and Lateral Load Behavior of Unbonded Post-Tensioned Precast Concrete Walls. Research Report No. EQ-96-02, Department of Civil and Environmental Engineering, Lehigh University, Bethlehem, PA.

Kurama, Y., Pessiki, S., Sause, R., and Lu, L. W. (1999). Seismic Behavior and Design of Unbonded Post-Tensioned Precast Concrete Walls. *PCI Journal*, Precast/Prestressed Concrete Institute **44:3,** 72-89.

MacRae, G., and Priestley, M. (1994). Precast Post-Tensioned Ungrouted Concrete Beam-Column Subassemblage Tests. Report No. PRESSS – 94/01, Department of Applied Mechanics and Engineering Sciences, Structural Systems, University of California, San Diego, CA.

Mander, J., Priestley, M., and Park, R. (1988). Theoretical Stress-Strain Model for Confined Concrete. *Journal of Structural Engineering*, American Society of Civil Engineers **114:8**, 804-1826.

Priestley, M., and Tao, J. (1993). Seismic Response of Precast Prestressed Concrete Frames with Partially Debonded Tendons. *PCI Journal*, Precast/Prestressed Concrete Institute **38:1**, 58-69.

Advances in Building Technology, Volume I
M. Anson, J.M. Ko and E.S.S. Lam (Eds.)

DESIGN OF HIGH STRENGTH SPIRAL REINFORCEMENT IN CONCRETE COMPRESSION MEMBERS

Stephen Pessiki[1], Benjamin Graybeal[2] and Michael Mudlock[3]

[1] Department of Civil and Environmental Engineering, Lehigh University
Bethlehem, PA 18015, USA
[2] PSI Inc., FHWA Turner-Fairbank Highway Research Center
Mclean, VA 22101, USA
[3] Simpson, Gumpertz & Heger Inc., Arlington, MA 02474, USA

ABSTRACT

Current provisions of the ACI 318 Code and AASHTO Design Specification limit the nominal yield stress of spiral reinforcement for compression members to 414 MPa (60 ksi). A procedure for the design of high strength spiral reinforcement for compression members is described. The procedure utilizes useable stress values, rather than nominal yield stress values, for spiral design. To evaluate the design procedure, concentric axial load tests were performed on six spirally reinforced concrete compression members with spiral reinforcement yield stress values ranging from 538 MPa (78 ksi) to 1345 MPa (195 ksi), and useable stress values ranging from 545 MPa (79 ksi) to 1131 MPa (164 ksi). It was found that, for the compression member geometries and material properties treated in this study, the proposed design procedure satisfactorily predicted the behavior of the compression members made from spirals with useable stress values up to 758 MPa (110 ksi).

KEYWORDS

compression, concrete, confinement, design, high strength steel, spiral reinforcement

INTRODUCTION

In current U.S. practice, spiral reinforcement in compression members is designed so that, should the compression member be overloaded, confinement by the spiral reinforcement provides enough strength enhancement to the concrete core to replace the strength lost by the spalling of the concrete cover. For a given compression member cross-section, a higher strength concrete requires a greater amount of confinement. Current provisions of the ACI 318 Building Code Requirements for Structural Concrete (1995) and the American Association of State Highway and Transportation Officials LRFD Bridge Design Specifications (1994) limit the spiral reinforcement yield stress to 414 MPa (60 ksi). As concrete strengths have increased over the years, the required amount of spiral reinforcement has also increased. To satisfy the need for a greater amount of spiral reinforcement, either the use of larger diameter spiral wires or decreased pitch can be employed. However, fabrication problems and concrete placement difficulty based on practical spacing limits and code requirements may result. One alternative to the above practices is to use higher strength spiral reinforcement (in excess of current code provisions), wherein a smaller amount of higher strength spiral steel is used to provide a given required amount of confining pressure to the concrete. This paper presents a procedure for the design of high strength spiral reinforcement and the results of tests to evaluate this procedure. Complete details of the research presented in this paper are given in Graybeal and Pessiki (1998) and Mudlock and Pessiki (1999).

BACKGROUND

In the early 1900's Considère (1903) found that the compressive strength of concrete was increased by transverse confining pressure. This confining pressure, whether active (i.e. applied externally by a pressurized fluid) or passive (i.e. created by lateral expansion of the axially compressed concrete against a confining material), worked to resist the lateral expansion of concrete as it was loaded axially. Richart, Brandzaeg, and Brown (1934) presented Eqn.1 to describe the relationship between the unconfined concrete strength f_{co}, lateral pressure $f_{2\text{-}2}$, and the confined concrete strength f_{c2}:

$$f_{c2} = f_{co} + 4.1 f_{2-2} \tag{1}$$

Richart et al. also proposed Eqn. 2 to relate the longitudinal strain at the confined concrete strength, ε_{c2}, to the confined strength, f_{c2}, the unconfined strength, f_{co}, and the corresponding unconfined longitudinal strain, ε_{co}:

$$\varepsilon_{c2} = \varepsilon_{co}\left(5\frac{f_{c2}}{f_{co}} - 4\right) \tag{2}$$

Eqns. (1) and (2) are used as part of the proposed design procedure.

Current Design Requirements

The ACI 318 Code and the AASHTO Design Specification both state that the nominal concentric axial load capacity, P_o, of a reinforced concrete compression member is given by:

$$P_o = 0.85 f_{co}\left(A_g - A_{st}\right) + A_{st} f_y \tag{3}$$

where A_g is the gross cross-sectional area of the compression member, A_{st} is the area of the longitudinal reinforcement, and f_y is the yield stress of the longitudinal reinforcement. The volumetric ratio of spiral reinforcement in a compression member, ρ_{sp}, defined as the ratio of the volume of the spiral to the volume of the concrete core, is computed as:

$$\rho_{sp} = \frac{4A_{sp}}{d_{sp}s} \tag{4}$$

where the spiral wire cross-sectional area, pitch, and out-to-out diameter are A_{sp}, s, and d_{sp}, respectively. The confining pressure $f_{2\text{-}2}$ which is applied to the concrete core by spiral reinforcement is calculated as:

$$f_{2-2} = \frac{1}{2}\rho_{sp} f_{sp} \tag{5}$$

where f_{sp} is the stress in the spiral reinforcement. Spiral reinforcement is designed according to the philosophy that the strength reduction caused by spalling of the concrete cover should equal the strength gain of the concrete core due to confinement. In design codes, this leads to the following equation for the required amount of spiral reinforcement:

$$\rho_{sp} \geq 0.45\frac{f_{co}}{f_{sy}}\left(\frac{A_g}{A_{core}} - 1\right) \tag{6}$$

In Eqn. (6), the stress in the spiral reinforcement, f_{sp}, is assumed to equal f_{sy}, the yield stress of the spiral reinforcement.

PROPOSED DESIGN PROCEDURE

In the proposed design procedure, spiral reinforcement design is based upon a useable stress $f_{sp2,dsgn}$ instead of the nominal yield stress of the spiral reinforcement. The proposed design procedure provides a means to calculate the useable strain $\varepsilon_{sp2,dsgn}$ in the spiral reinforcement, which in turn is used to determine the useable stress from the stress-strain curve of the particular spiral wire.

Step 1 in the procedure determines the level of confined compressive strength $f_{c2,dsgn}$ required for the core concrete in the compression member. Using an approach consistent with the ACI 318 Code and AASHTO Design Specification, this core strength enables the member to carry the same axial load after the loss of the concrete cover that it carried prior to concrete cover failure:

$$f_{c2,dsgn} = \frac{f_{co}(A_g - A_{st})}{A_{core}} \tag{7}$$

where A_{core} is the area of the core concrete. Step 2 calculates the confining pressure, $f_{2\text{-}2,dsgn}$, which is required for the core concrete to reach its desired strength. From Eqn. (1), $f_{2\text{-}2,dsgn}$ is calculated as:

$$f_{2-2,dsgn} = \frac{1}{4.1}(f_{c2,dsgn} - f_{co}) \tag{8}$$

Step 3 determines the longitudinal strain $\varepsilon_{c2,dsgn}$ using Eqn. (9), which is taken from Eqn. (2):

$$\varepsilon_{c2,dsgn} = \varepsilon_{co}\left(5\frac{f_{c2,dsgn}}{f_{co}} - 4\right) \tag{9}$$

Step 4 determines the transverse strain in the concrete core, $\varepsilon_{ct2,dsgn}$, which occurs at fracture of the spiral reinforcement. The value of $\varepsilon_{ct2,dsgn}$ is calculated as:

$$\varepsilon_{ct2,dsgn} = 0.41\varepsilon_{c2,dsgn} - 0.105\varepsilon_{co} \tag{10}$$

The development of Eqn. (10) is discussed in Graybeal and Pessiki (1998). In Step 5, from strain compatibility, the useable strain in the spiral reinforcement $\varepsilon_{sp2,dsgn}$ is the same as the transverse strain in the concrete core $\varepsilon_{ct2,dsgn}$. This relationship applies at all strains, including those at spiral fracture. This relationship is expressed as:

$$\varepsilon_{sp2,dsgn} = \varepsilon_{ct2,dsgn} \tag{11}$$

Thus the value of useable strain is determined by the compression member geometry and concrete material properties. The useable strain is not dependent upon the stress-strain curve of the particular spiral wire that is used. The chosen spiral wire must possess sufficient ductility to achieve this strain value.

Step 6 determines the useable stress $f_{sp2,dsgn}$ from the useable strain $\varepsilon_{sp2,dsgn}$ and the stress strain curve for the particular spiral wire under consideration. Thus, while the useable strain is independent of the stress-strain curve of the spiral, the useable stress does depend upon the stress-strain curve. The ideal situation is to utilize the stress-strain curve of the spiral in its in-situ, or spiraled, state. However, it is more likely that only a stress-strain curve from a tension test of an unspiraled wire will be available, which may not accurately represent the behavior of the spiraled wire. In this case, the value obtained may need to be modified in order to ensure a proper design. This is discussed more fully in Mudlock and Pessiki (1999).

Step 7 determines the required volumetric ratio of spiral reinforcement $\rho_{sp,dsgn}$ from Eqn. (12), which is a rearranged form of Eqn. (5).

$$\rho_{sp,dsgn} = \frac{2f_{2-2,dsgn}}{f_{sp2,dsgn}} \tag{12}$$

Finally, Step 8 computes the required pitch, s_{dsgn}, from Eqn. (13), which is a rearranged form of Eqn. (4).

$$s_{dsgn} = \frac{4A_{sp}}{d_{sp}\rho_{sp,dsgn}} \tag{13}$$

EXPERIMENTAL PROGRAM

Concentric axial load compression tests were performed to evaluate the design procedure presented above. The test program, described in Table 1, included six 356 mm (14 in.) diameter compression member specimens made with six different spiral wires. Each specimen was designed using an unconfined concrete compressive strength of 55.2 MPa (8.0 ksi). The nominal yield stress of the spiral reinforcement varied from 538 MPa (78 ksi) to 1345 MPa (195 ksi), and the useable stress varied from 545 MPa (79 ksi) to 1131 MPa (164 ksi). The specimens all measured 1.42 m (56 in.) in height and had a 51 mm (2 in.) concrete cover to the outside of the spiral reinforcement.

The design procedure described above was used to design the spiral for each specimen. The calculation of useable strain proceeded as follows. First, $f_{c2,dsgn}$ was calculated from Eqn. (7) as 109 MPa (15.8 ksi). From Eqn. (8), the confining pressure $f_{2-2,dsgn}$ was calculated as 13.0 MPa (1.88 ksi). Next, $\varepsilon_{c2,dsgn}$ was calculated by Eqn. (9) as 0.0158, and $\varepsilon_{ct2,dsgn}$ was calculated from Eqn. (10) to be 0.0062. Finally, Eqn. (11) was used to obtain a value for $\varepsilon_{sp2,dsgn}$ equal to 0.0062. This was the useable strain value used in the design of the specimens. The useable stresses $f_{sp2,dsgn}$ were determined for each wire at that strain, and Eqns. (12) and (13) were used to calculate $\rho_{sp,dsgn}$ and finally the spiral pitch s_{dsgn}.

Extra spiral reinforcement was provided in each specimen over one diameter of height from each end. The test region was the portion of the specimen between the more heavily confined end regions. Two wires were bundled to create the spiral in specimens14-A' and 14-B'. This is shown in Table 1, where n_{sp} is the number of wires in the bundle. Four 12.7 mm (0.5 in.) diameter mild steel longitudinal reinforcement held the spiral in position during fabrication. Each specimen was tested under concentric axial compression in a 22.2 MN (5000 kip) capacity universal testing machine. Twelve electrical resistance strain gages were used to monitor strains in both the spiral and longitudinal reinforcement within the test region, and linear

TABLE 1
DESCRIPTION OF TEST SPECIMENS

Specimen		Spiral Reinforcement						
I.D.	d_c (mm)	$f_{sy,nom}$ (MPa)	$f_{sp2,dsgn}$ (MPa)	d_{sw} (mm)	n_{sp}	A_{sp} (mm^2)	s (mm)	ρ_{sp} %
14-A'	356	538	545	8.9	2	124	41	4.76
14-B'	356	738	717	8.9	2	124	54	3.62
14-C'	356	834	745	8.9	1	62	28	3.48
14-D'	356	965	758	8.9	1	62	29	3.42
14-E'	356	1345	1131	9.1	1	65	44	2.29
14-F'	356	1276	1110	10.9	1	93	62	2.34

variable differential transformer displacement transducers were used to measure overall axial shortening. Based on material testing, the unconfined concrete compressive strength f_{co} for the Series 2 tests was taken as 52.1 MPa (7.55 ksi), and the axial concrete strain corresponding to the unconfined concrete strength, ε_{co}, was taken as 0.0027. Due to a fabrication error, Specimen 14-B' failed prematurely during testing in the heavily confined region at the top of the specimen. The results from this specimen are excluded from the remaining discussions.

RESULTS AND DISCUSSION

The axial load-axial shortening response of each specimen is shown in Figure 1. Failure of all eight specimens was defined by rupture of one or more turns of wire of the spiral reinforcement and a subsequent loss of confining pressure on the concrete core. This resulted in a significant decrease in the load carrying capacity of the specimen. However, two observations indicate that the failures were not due to the spiral reinforcement reaching its ultimate strain as caused by lateral expansion of the core. First, for all specimens, the measured strain in the spiral reinforcement just before failure was less than one-third of the spiral strain at ultimate stress, ε_{su}. Typically, the maximum strains measured during the tests were less than 0.01. Second, for most of the specimens, multiple spiral fractures occurred simultaneously along a well defined inclined plane throughout the test region. This suggests that the plane formed first, and that the spirals fractured as a result of relative movement of the concrete along this plane.

Longitudinal Strains

Tables 2 and 3 show the experimentally obtained values and design values of ε_{c2}, and Table 4 shows the ratio of these two values, $R(\varepsilon_{c2})_{exp/dsgn}$. The experimental value $\varepsilon_{c2,exp}$ was obtained as the average measured strain in the longitudinal reinforcement at failure. Table 4 shows that the experimental values exceeded the design values in every case except Specimen 14-A', where the value of $R(\varepsilon_{c2})_{exp/dsgn}$ was 0.97. This indicates that the design procedure gives a reasonable to conservative prediction of longitudinal strain at failure.

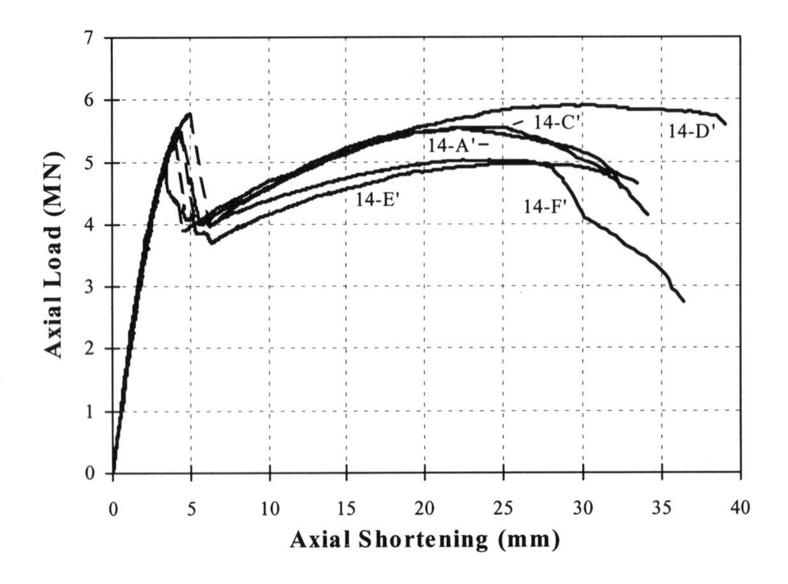

Figure 1: Axial load-axial shortening responses

Spiral Strains and Stresses

Tables 2 and 3 show the experimentally obtained and design values of ε_{sp2}, and Table 4 shows the ratio of these two values $R(\varepsilon_{sp2})_{exp/dsgn}$. Table 4 shows that, in all cases, the measured strains were equal to or greater than the design values. Tables 2, 3 and 4 show similar results for the experimental and design values of f_{sp2}. From the foregoing results, it is clear that the design procedure provides a reasonable to conservative estimate of the values of the spiral strains and stresses, and thus a reasonable to conservative estimate of the confining pressure provided by the spiral reinforcement.

Confined Concrete Strength

An experimental value for the increase in compressive strength of the confined concrete core, $\Delta f_{c12,exp}$ was computed from the experimental results. This experimental value is compared to the design value of the increase in strength of the concrete core, $\Delta f_{c12,dsgn}$. Tables 2 and 3 show the values of $\Delta f_{c12,exp}$ and $\Delta f_{c12,dsgn}$, and Table 4 shows the ratio of these values, $R(\Delta f_{c12})_{exp/dsgn}$. Table 4 shows that Specimens 14-A', 14-C' and 14-D' either achieved or exceeded the design values of core strength enhancement based on the spiral design useable stress. For these specimens, $R(\Delta f_{c12})_{exp/dsgn}$ ranged from 1.00 to 1.14. In contrast, the core strength enhancement in Specimens 14-E' and 14-F' fell approximately 20 percent below the design values.

Table 3 also shows the Richart et al. predicted values of $\Delta f_{c12,Rich}$, and Table 4 shows $R(\Delta f_{c12})_{exp/Rich}$, which is the ratio of $\Delta f_{c12,exp}$ to $\Delta f_{c12,Rich}$. Table 4 shows that Specimens 14-A' and 14-C' satisfied the Richart et al. prediction of increased compressive strength of the concrete core. Specimen 14-D' nearly satisfied the prediction, falling short by only 5 percent. However, Specimens 14-E' and 14-F' achieved only 67 percent and 69 percent of the Richart et al. predictions. Thus Specimens 14-A',14-C' and 14-D' exhibited experimental core strength increases consistent with the stresses observed in the spiral steel, but Specimens 14-E' and 14-F' did not. This suggests that the Richart et al. equation may have a limit on its range of applicability for higher strength steels.

TABLE 2
SUMMARY OF DESIGN VALUES

I.D.	$\varepsilon_{c2,dsgn}$	$\varepsilon_{sp2,dsgn}$	$f_{sp2,dsgn}$ (MPa)	$\Delta f_{c12,dsgn}$ (MPa)
14-A'	0.0158	0.0062	545	53.2
14-C'	0.0158	0.0062	745	53.1
14-D'	0.0158	0.0062	758	53.2
14-E'	0.0158	0.0062	1131	53.1
14-F'	0.0158	0.0062	1110	53.2

TABLE 3
SUMMARY OF EXPERIMENTAL VALUES

I.D.	$\varepsilon_{c2,exp}$	$\epsilon_{sp2,exp}$	$f_{sp2,exp}$ (MPa)	$f_{c2,exp}$ (MPa)	$\Delta f_{c12,exp}$ (MPa)	$\Delta f_{c12,Rich}$ (MPa)
14-A'	0.0153	0.0063	545	105.4	53.4	53.2
14-C'	0.0219	0.0063	752	105.9	53.8	53.6
14-D'	0.0344	0.0091	917	112.8	60.7	64.3
14-E'	0.0231	0.0122	1338	94.1	42.1	62.8
14-F'	0.0206	0.0086	1310	95.4	43.3	62.8

TABLE 4
RATIOS OF KEY VALUES

I.D.	$R(\epsilon_{c2})_{exp/dsgn}$	$R(\epsilon_{sp2})_{exp/dsgn}$	$R(f_{sp2})_{exp/dsgn}$	$R(\Delta f_{c12})_{exp/dsgn}$	$R(\Delta f_{c12})_{exp/Rich}$
14-A'	0.97	1.02	1.00	1.00	1.00
14-C'	1.39	1.02	1.01	1.01	1.00
14-D'	2.18	1.47	1.21	1.14	0.95
14-E'	1.46	1.97	1.18	0.79	0.67
14-F'	1.30	1.39	1.18	0.81	0.69

CONCLUSIONS

The conclusions of this research are as follows:

1. The proposed design procedure provides satisfactory spiral designs for spiral steels with useable stress values up to 758 MPa (110 ksi). This conclusion is applicable to the specimen geometries and material strengths treated in this research.

2. From the conclusion above, it follows that current design requirements that limit the design yield stress of spiral reinforcement in compression members to 414 MPa (60 ksi) may be modified to permit higher stresses. Spiral steel stresses in excess of 414 MPa (60 ksi) can be used to design spiral reinforcement in compression members similar to the specimens treated in this research.

3. Three key relationships used in the design procedure provide acceptable results for the design purposes. First, Eqn. (9) (Eqn. (2) from Richart et al.) provides reasonable to conservative estimates of the longitudinal strain at failure. Second, Eqn. (10), the tangent dilation ratio relationship proposed by Graybeal et al., provides reasonable to conservative estimates of transverse strains in the concrete, and thus strains in the spiral reinforcement at failure. Third, Eqn. (8) (Eqn. (1) from Richart et al.) which relates confining pressure to axial strength enhancement, is valid for compression members made with useable stress values up to 758 MPa (110 ksi).

ACKNOWLEDGMENTS

This research was funded by the Precast/Prestressed Concrete Institute, through a Daniel P. Jenny Fellowship, and by the Center for Advanced Technology for Large Structural Systems at Lehigh University. Additional support was provided by Concrete Technology Corporation, Sumiden Wire Products Corporation, Florida Wire and Cable, Inc., and Neturen Co., Ltd. The findings and conclusions presented in this report are those of the authors.

REFERENCES

ACI Committee 318. (1995). *Building Code Requirements for Structural Concrete (ACI 318-95) and Commentary (ACI 318R-95),* American Concrete Institute, Farmington Hills, MI, USA.

AASHTO. (1994). *AASHTO LRFD Bridge Design Specifications,* American Association of State Highway and Transportation Officials, Washington, D.C., USA, First Edition.

Graybeal, B.A., and Pessiki, S. (1998). *Confinement Effectiveness of High Strength Spiral Reinforcement in Prestressed Concrete Piles.* Report No. 98-01, Center for Advanced Technology for Large Structural Systems, Lehigh University, 164 pp.

Mudlock, M., Pessiki, S. (1999). *Design of High Strength Spiral Reinforcement for Prestressed Concrete Piles.* Report No. 99-12, Center for Advanced Technology for Large Structural Systems, Lehigh University, 134 pp.

Considère, A. (1903). Rèsistance à la compression du bèton armè et du bèton frettè. *Gènie Civil,* 1903.

Richart, F.E., and Brown, R.L. (1934). *An Investigation of Reinforced Concrete Columns.* University of Illinois Bulletin, Vol. XXVI, No. 40, 94 pp.

Advances in Building Technology, Volume 1
M. Anson, J.M. Ko and E.S.S. Lam (Eds.)

MODELING AND BEHAVIOR OF COMPOSITE MRFs WITH CONCRETE FILLED STEEL TUBULAR COLUMNS SUBJECT TO EARTHQUAKE LOADING

J.M. Ricles[1], R. Sause[1], T. Muhummud[1], R. Herrera[1]

[1]Department of Civil and Environmental Engineering, Lehigh University,
117 ATLSS Drive, Bethlehem, PA 18015-4729, USA

ABSTRACT

In conjunction with the U.S.-Japan Cooperative Research Program in Earthquake Engineering on Composite and Hybrid Structures, analytical studies on the seismic behavior of steel moment resisting (MRFs) with concrete filled steel tubular (CFT) columns were performed. The CFT columns consisted of normal strength concrete with a high strength steel tube. In these studies the seismic performance of this type of composite structure was investigated using a fiber element in order to account for non-linear behavior due to material yielding, local buckling, and concrete cracking and crushing. Using the IBC 2000 provisions several frames were designed and analyzed. It was found that the frames are likely to perform adequately under both the design level and maximum considered earthquakes.

KEYWORDS

composite construction, concrete filled steel tubular column, earthquake, high strength steel, nonlinear analysis, structural steel.

INTRODUCTION

Composite construction is becoming more common for designing the structural systems of buildings. This is due to the numerous advantages that composite construction offers, including: the inherent mass, stiffness, damping, and economy of reinforced concrete; and the speed of construction, strength, long-span capability, and light weight of structural steel. Composite structures for buildings often include steel moment resisting or braced frame systems with steel-concrete composite columns to help control lateral drift. The composite columns may be structural shapes encased in reinforced concrete (SRC), or circular or rectangular concrete filled steel tubes (CFT). CFT columns have advantages over SRC columns because the steel tube serves as formwork and offers confinement to the concrete, thus improving its ductility. Additionally, the concrete infill delays the local buckling of the rectangular steel tubes and increases the flexural stiffness of the column. This enables the inter-story drift requirements to be met.

This paper presents the results of research that was conducted in conjunction with the U.S.-Japan Cooperative Research Program in Earthquake Engineering on Composite and Hybrid Structures. A composite moment resisting frame (MRF) was designed with high strength CFT columns. The members and connections of the structure were modeled to account for the effects of local buckling, concrete crushing and cracking, connection deformations, as well as P-delta effects. The results of static pushover and non-linear time history analysis involving the design basis earthquake (DBE) and maximum considered earthquake (MCE) records are presented.

PROTOTYPE FRAME

Description of Frame

The six-story five-bay frame shown in Figure 1 was designed as a special MRF in accordance with the International Building Code (IBC) 2000 provisions ("*International*" 2000). Drift was found to control the design of the MRF. The dimensions of the frame were based on the U.S.-Japan theme structure (Goel and Yamanouchi 1992). The beams were A992 steel wide flange sections with a nominal 345 MPa yield strength. The CFT columns were high strength steel tubes (with a nominal 552 MPa yield strength) filled with a 55 MPa nominal strength concrete. The beam-to-column moment connections are shown in Figure 2, and consisted of split tees that were fillet welded (in the field) to the beam flanges and bolted to the column. No shear tab was used. The design of these connections was based on the research conducted by Peng et al. (2001), which demonstrated that good strength and ductility can be developed with this detail under inelastic cyclic loading. The members were proportioned in order to satisfy a weak beam-strong column design in accordance with FEMA Provisions (FEMA 2000). The required column moment capacity was based on the expected plastic beam moment developing at a distance of one-third the beam depth from the end of the tee stems in the connection. The inflection points in the column were assumed to be located at the far end of the column above and below the joint when determining the required column moment capacity.

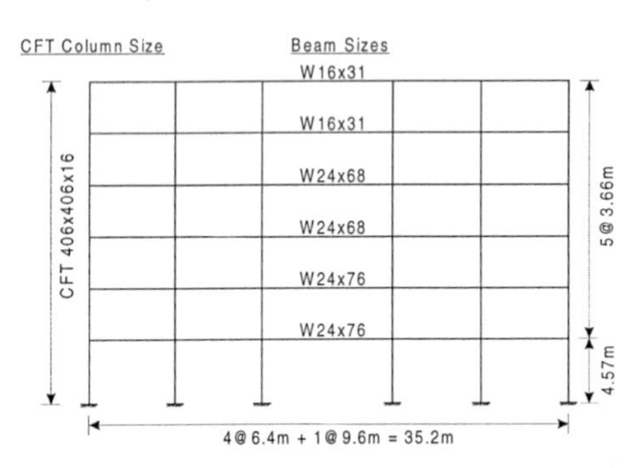

Figure 1: MRF with CFT columns.

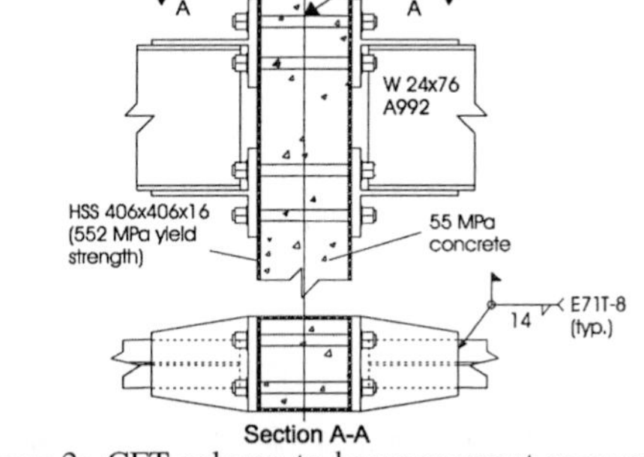

Figure 2: CFT column-to-beam moment connection.

Modeling of Frame

The frame was analyzed using the DRAIN-2DX computer program (Prakash et al. 1993, Herrera et al. 2001). To account for the possibility of concrete crushing, concrete cracking, steel yielding, and steel local buckling (which can lead to a deterioration of stiffness and strength under cyclic loading) the CFT columns in the frame were modeled using a fiber element, as shown in Figure 3. The cross section of the column was discretized using 22 fibers (12 for the concrete and 10 for the steel). Fibers with local buckling incorporated into their stress-strain relationship were located at the ends of the column, as shown in Figure 3(b), to account for the effects of any local buckling that may occur in the steel tube, particularly at the base of the ground floor column where a plastic hinge is expected to form (Varma et al. 2001). Shear springs were prescribed to transfer the transverse shear through the plastic hinge zone. The stress-strain curves for the steel and concrete fibers were developed using the results of finite element analysis of the CFT section under monotonic axial load to account not only for the effects of local buckling in the steel tube wall but also confinement on strength, stiffness and ductility of the concrete. Representative steel and concrete fiber stress-strain curves are shown in Figure 3(c), where the cyclic unloading and reloading rules adopted for the analysis are illustrated. Existing test data was used to validate the CFT column model, with representative results shown in Figure 4.

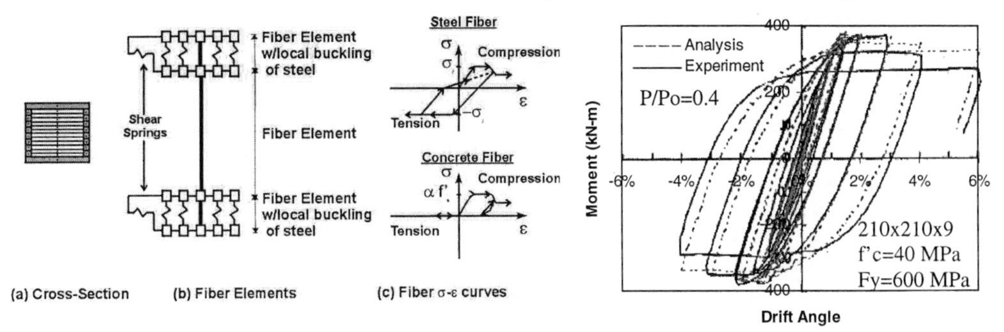

Figure 3: Modeling of CFT columns.

Figure 4: Comparison of CFT column analysis and experimental results.

The wide flange steel beams were modeled using steel fibers located in the plastic hinge region of the beam to capture the effects of cyclic yielding, strain hardening, and local buckling. Outside the plastic hinge region the beam was modeled using an elastic beam element. The cyclic stress-strain relationship for the fibers was based on experimental data from Ricles et al. (2002). A comparison of the moment-plastic rotation cyclic response predicted by the beam model and test data is shown in Figure 5. The connection was carefully modeled to account for flexibility and any inelastic behavior that may develop. A set of master and slaved nodes was used to define the panel zone kinematics that considered the depth of the column and beams at the joint, as shown in Figure 6. The shear-deformation relationship for the panel zone was defined by specifying the properties of the two parallel rotational springs ($k_{\gamma 1}$ and $k_{\gamma 2}$) shown in Figure 6. Tests on CFT panel zones conducted by Koester (2000) were used to define the spring properties. A comparison of measured panel zone cyclic behavior with predicted response is shown in Figure 7.

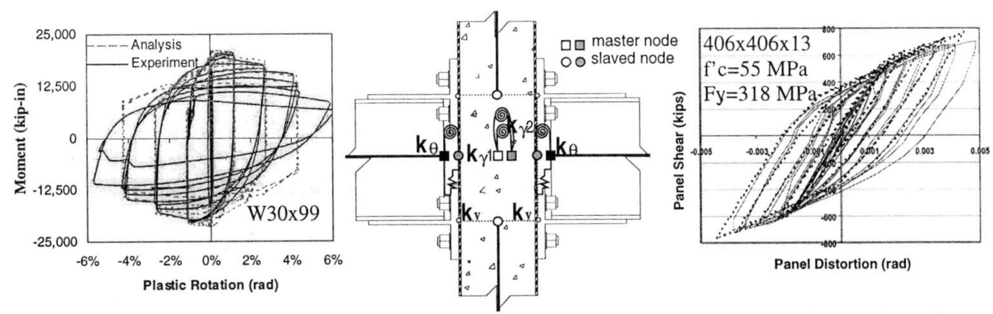

Figure 5: Analysis of beam with local buckling.

Figure 6: Connection modeling.

Figure 7: Comparison of analysis and experimental results.

The rotational and transverse flexibility of the tee stems was considered through the use of a rotational (k_θ) and transverse spring (k_v) at each beam-column interface as shown in Figure 6. The properties of the tee-stem springs were based on research conducted by Peng et al. (2001) on split-tee moment connections. P-delta effects from the interior gravity frames were incorporated into the frame model using a leaning column (not shown in Figure 1).

ANALYSIS RESULTS

Static Pushover Analysis

The frame model was used to perform a nonlinear static pushover analysis of the composite MRF. The lateral loads were distributed over the height of the frame in accordance with the IBC 2000 provisions. Figure 8 shows the relationship between the normalized base shear and the roof drift index (RDI). The normalized base shear is defined as the base shear divided by the seismic dead weight of the building, W, whereas the RDI is the roof displacement divided by the building height. The analysis results indicated that the frame first yields at a base shear of 0.15W at the fifth floor beams at a corresponding RDI of 1.1%. The maximum overstrength of the MRF is 3.29 (0.23W) at a corresponding RDI of 2.3%, where the overstrength is defined as the base shear divided by the IBC 2000 design base shear. In the static pushover that maximum story drift was 5.26% and occurred at the first story. The inelastic deformations in the structure are concentrated in all the floor beams and at the base of the ground floor columns. The panel zones in the interior columns at the first and second floors developed minor inelastic response, illustrating that they are relatively strong in shear compared to the demand imposed on them.

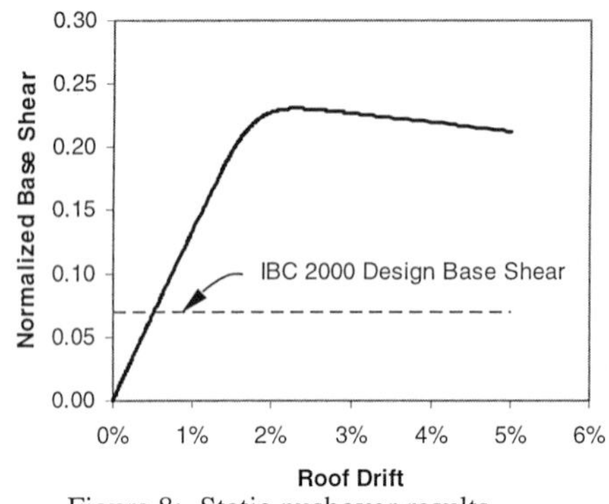

Figure 8: Static pushover results.

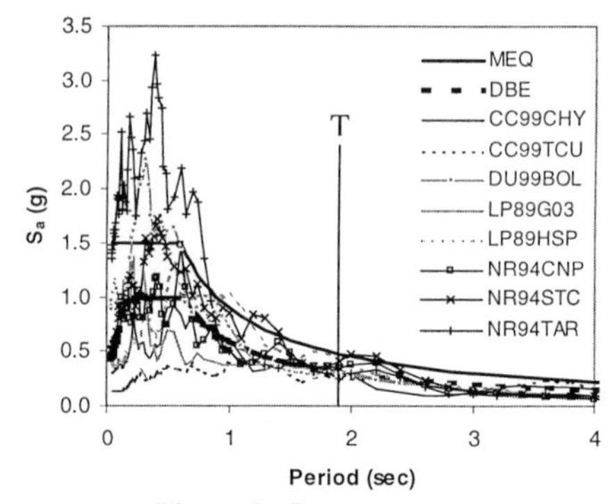

Figure 9: Response spectra.

Time History Analysis

Nonlinear time history analyses were conducted using an ensemble of eight earthquake records. The eight earthquake records are summarized in Table 1. Two sets of analyses were conducted. The first set involved scaling the eight earthquake records to the Design Basis Earthquake (DBE) and evaluating the response of the frame to these scaled records. The scale factor for each record was determined using the IBC 2000 design response spectrum as the target spectra and the scaling procedure developed by Somerville (1997). The second set of analysis involved scaling the same eight earthquake records to the Maximum Considered Earthquake (MCE) and evaluating the response of the frame to these scaled records. The scale factors corresponding to the MCE were determined using the IBC 2000 maximum considered earthquake response spectrum in conjunction with the procedure of Somerville (1997). The IBC 2000 design and maximum considered earthquake response spectra are compared to the response spectra for the eight un-scaled records in Figure 9. The scale factors for each record corresponding to the DBE and MCE are summarized in Table 1. The IBC 2000 DBE has a return period range from 300 to 600 years in areas of high seismicity (approximately a 8% to 16% probability of exceedance in 50 years) while the MCE has a 2% probability of exceedance in 50 years (2/50). The fundamental elastic period, T_1, of the frame was 1.93 seconds.

TABLE 1
TIME HISTORY ANALYSIS RESPONSE RESULTS

Ground Motion (name, year, component)	Design Basis Earthquake				Max. Considered Earthquake			
	Scale Factor	Max. Roof Disp. (mm)	Max. IDI (%)	Max. Base Shear (%W)	Scale Factor	Max. Roof Disp. (mm)	Max. IDI (%)	Max. Base Shear (%W)
Chi-Chi 1999, CHY036-W	1.04	518	3.7	24.3	1.55	582	4.4	25.4
Chi-Chi 1999, TCU123-N	1.34	480	2.3	25.2	2.01	662	3.4	26.9
Duzce 1999, BOL000	1.08	392	2.4	25.8	1.61	523	4.1	26.9
Loma Prieta 1989, G03090	1.49	578	2.9	25.5	2.24	833	4.1	25.6
Loma Prieta 1989, HSP000	0.95	430	2.7	26.4	1.42	709	4.3	26.9
Northridge 1994, CNP196	1.27	581	3.1	25.1	1.91	906	5.0	25.9
Northridge 1994, STC180	0.89	501	3.1	26.2	1.33	636	4.0	27.7
Northridge 1994, TAR360	1.09	450	4.1	28.1	1.64	653	4.5	28.7

Table 1 summarizes the maximum roof displacement, inter-story drift (IDI), and base shear for both sets of analysis. It can be seen that the response increases for the MCE compared to the DBE. Furthermore, the maximum base shear developed during the DBE and MCE both exceed that developed in the static pushover (0.23W). This effect is due to the participation of the higher vibration modes in the response to the earthquake excitation.

The maximum story drifts at each floor level are shown in Figure 10, where the mean and mean plus one standard deviation are noted. The mean plus one standard deviation results are approximately the 84th percentile, assuming that the results are lognormal distributed. The results in Figure 10 indicate there is a greater drift developed in the upper floors for this design, particular under the MCE. The maximum mean and mean+standard deviation for the maximum drift are 2.9% and 3.6%, respectively, for the DBE, and 4.0% and 4.5%, respectively, for the MCE. FEMA 273 (1997) recommends that the drift not exceed 2.5% and 5.0% under the DBE and MCE, respectively. Although under the DBE the drift exceeded 2.5% in some analysis, a stability analysis indicated that the structure was not on the verge of collapse, and the plastic rotation demand on the members was not excessive.

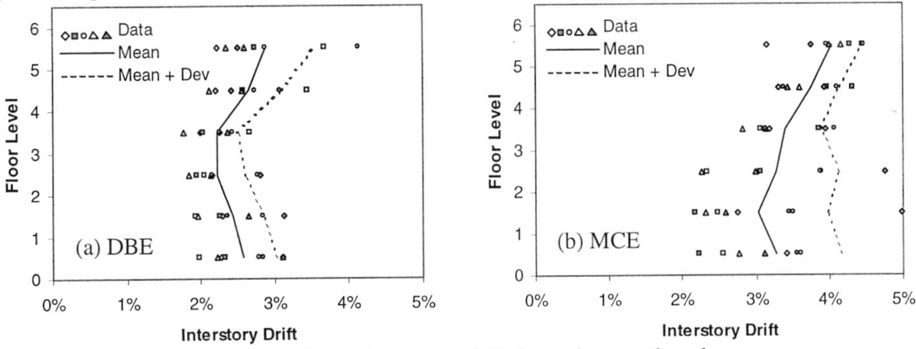

Figure 10: Maximum inter-story drift for various earthquakes.

The maximum plastic rotation in the beams at each floor level is shown in Figure 11, where also the mean and mean plus one standard deviation are shown. Figure 11 indicates statistically that the maximum plastic beam rotation develops in the upper floors for this design, although under the MCE a first-floor beam developed 4.79% radians of plastic rotation. The maximum mean and mean+standard deviation of maximum plastic rotation are 1.6% and 2.1% radians, respectively, for the DBE, and 2.7% and 3.0% radians, respectively, for the MCE. Tests by Peng et al. (2001) indicated that split-tee connections could enable a plastic rotation in excess of 5.0% radians to develop in the beam under cyclic loading before failure. An examination of the hysteretic stress-strain response of the fibers in the plastic hinge zone indicated that the beam flanges developed local buckling, however the web did not develop enough appreciable local buckling to cause any appreciable degradation in beam capacity.

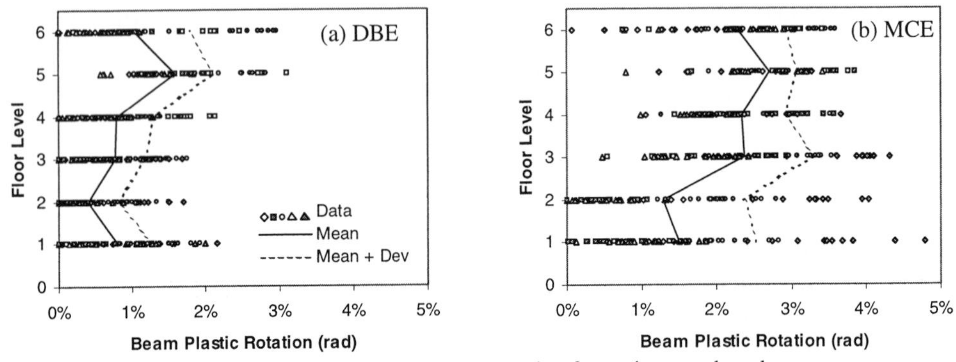

Figure 11: Maximum beam plastic rotation for various earthquakes.

The maximum plastic rotation in the columns at each floor level is shown in Figure 12, where also the mean and mean plus one standard deviation are shown. Yielding in the columns occurred only at the base of the ground floor columns. The maximum mean and mean+standard deviation of maximum plastic rotation are 1.1% and 1.7% radians, respectively, for the DBE, and 1.8% and 2.7% radians, respectively, for the MCE. An examination of the hysteretic stress-strain response in the plastic hinge zone indicated that both the flanges and webs developed local buckling, however the latter did not develop an appreciable amount to cause any significant amount of degradation in column capacity. Because of concrete cracking, however, there was degradation in the column elastic flexural stiffness. The fact that the columns yielded only at the base at the ground floor indicates that the approach used to proportion the column and beam sizes ensured a weak beam-strong column design. The accumulative plastic strain in the ground floor columns was less than that to cause low cycle fatigue failure for high strength CFT columns (Varma et al. 2001).

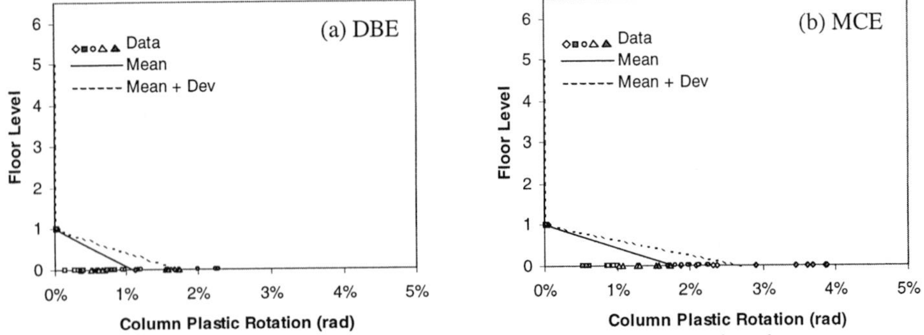

Figure 12: Maximum column plastic rotation for various earthquakes.

The maximum plastic shear deformation in the panel zones at each floor level is shown in Figure 13, where also the mean and mean plus one standard deviation are shown. Yielding in the panel zones occurred at only the first and second floors. The maximum mean and mean+standard deviation of panel zone maximum plastic deformation are 0.13% and 0.14% radians, respectively, for the DBE, and 0.15% and 0.17% radians, respectively, for the MCE. An examination of the hysteretic stress-strain response of the panel zones with plastic deformation indicated that they developed the onset of inelastic response, and had significant reserve strength (i.e., they did not achieve their ultimate capacity).

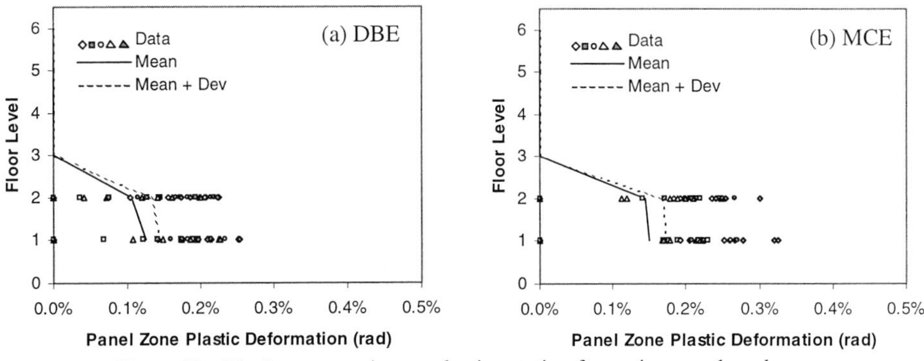

Figure 13: Maximum panel zone plastic rotation for various earthquakes.

The tee stems were found to remain elastic throughout all of the time history analyses.

SUMMARY AND CONCLUSIONS

A steel MRF with high strength CFT composite columns was designed and analyzed for response to the Design Basis and Maximum Considered Earthquakes. The modeling of the CFT columns and steel beams involved the use of fiber elements to account for nonlinearities due to steel yielding, local buckling, concrete cracking and crushing. Joint flexibility and nonlinearities were considered by modeling the panel zones and connection elements. Based on the results of the analysis, and a comparison of the demand with resistance of test specimens, it appears that the frame performed satisfactory. An experimental program is currently in progress at Lehigh University involving large-scale testing of steel MRFs with CFT columns subject to earthquake loading. The test results will provide data to verify the analysis results.

ACKNOWLEDGEMENTS

The research reported herein was supported by the National Science Foundation (Dr. Shih-Chi Liu cognizant program official), by a grant from the Pennsylvania D.C.E.D. through the PITA program, and a scholarship to the third author through the Royal Thai Government. The opinions expressed in this paper are those of the authors and do not necessarily reflect the views of the sponsors.

REFERENCES

American Institute of Steel Construction (1997) "Seismic Provisions For Structural Steel Buildings," Chicago, Illinois.

FEMA (1997) "NEHRP Guidelines for the Seismic Rehabilitation of Buildings," *Report No. FEMA 273*, FEMA, Washington, D.C.

FEMA (2000) "Recommended Seismic Design Criteria for New Steel Moment-Frame Buildings," *Report No. FEMA 350*, FEMA, Washington, D.C.

Goel, S. and K. Yamanouchi (1992) "Recommendations for U.S.-Japan Coop. Research Program – Phase 5: Composite and Hybrid Structures," Report No. UMCEE 92-29, Department of Civil and Environmental Engineering, University of Michigan, Ann Arbor, Michigan.

Herrera, R., Sause, R., and J. Ricles, (2001) "Refined Connection Element (Type 05) for DRAIN-2DX, Element Description and User Guide," ATLSS Report No. 01-07.

"International Building Code" International Code Council (ICC) (2000) Falls Church, Virginia.

Koester, B. (2000) " Panel Zone Behavior of Moment Connections Between Rectangular Concrete-Filled Steel Tubes and Wide Flange beams" Ph. D. Dissertation, Department of Civil and Environmental Engineering, University of Texas, Austin, Texas.

Peng, S.W., Ricles, J.M., and L.W. Lu, (2001) "Seismic Resistant Connections for Concrete Filled Column-to-WF Beam Moment Resisting Frames," ATLSS Engineering Research Center, Report No. 01-08, Lehigh University, Bethlehem, Pennsylvania.

Prakash, V., Powell, G., and S. Campbell, (1993). "DRAIN-2DX Base Program Description and User Guide, Version 1.10," Report No. UCB/SEMM-93/17 & 18, Structural Engineering Mechanics and Materials, Department of Civil Engineering, University of California, Berkeley, California.

Ricles, J.M., Mao, C., Lu, L.W., and J.W. Fisher, (2002) "Inelastic Cyclic Testing of Welded Unreinforced Moment Connections," Journal of Structural Engineering, ASCE, Vol. 128, No. 4.

Somerville, P. (1997). "Development of Ground Motion Time Histories for Phase 2 of the FEMA/SAC Steel Project," Report No. SAC/BD-97/04.

Varma, A.H., Ricles, J.M., Sause, R., and L.W. Lu, (2001) "Seismic Behavior, Analysis, and Design of High Strength Square Concrete Filled Steel Tube (CFT) Columns," ATLSS Engineering Research Center, Report No. 01-02, Lehigh University, Bethlehem, Pennsylvania.

Advances in Building Technology, Volume 1
M. Anson, J.M. Ko and E.S.S. Lam (Eds.)

THE USE OF POST-TENSIONING TO REDUCE SEISMIC DAMAGE IN STEEL FRAMES

J. M. Ricles[1], R. Sause[1], Pedro Rojas[1], and M. M. Garlock[1]

[1]Department of Civil and Environmental Engineering, Lehigh University,
117 ATLSS Drive, Bethlehem, PA 18015-4729, USA

ABSTRACT

A post-tensioned friction damped connection for earthquake resistant steel moment resisting frames (MRFs) is introduced. The connection includes friction devices with post-tensioned high strength strands running parallel to the beam. The connection has excellent ductility, minimizes inelastic deformation to the components of the connection, requires no field welding, and returns the structure to its pre-earthquake position. Inelastic analyses were performed on a 6-story, 4-bay post-tensioned steel MRF to study its response to strong ground motions. Results show good energy dissipation, strength, and ductility in the post-tensioned system. The analyses indicate that the seismic performance of a post-tensioned steel frame can exceed that of a frame with conventional moment resisting connections.

KEYWORDS

earthquake, moment resisting connection, moment resisting frame, nonlinear analysis, post-tensioning, seismic damage mitigation, structural steel.

INTRODUCTION

The inadequate performance of steel moment resisting frames (MRFs) in recent earthquakes has caused much concern. During the 1994 Northridge Earthquake, many steel-framed buildings suffered unexpected premature fractures in their welded beam-to-column connections. The conventional earthquake resistant steel moment connection is composed of full penetration welds between the beam flanges and the column face, with the beam web bolted to a shear tab extending from the column. The occurrence of premature fractures in the beam flange welds between the beams and columns indicates an immediate need to revise current design and construction standards.

As an alternative to welded construction the authors developed a post-tensioned friction damped moment connection (PFDC) for use in seismic resistant steel MRFs. The connection utilizes high strength steel strands that are post-tensioned after friction devices are installed (Figure 1(a)). The post-tensioning strands run through the column, and are anchored outside the connection region (Figure

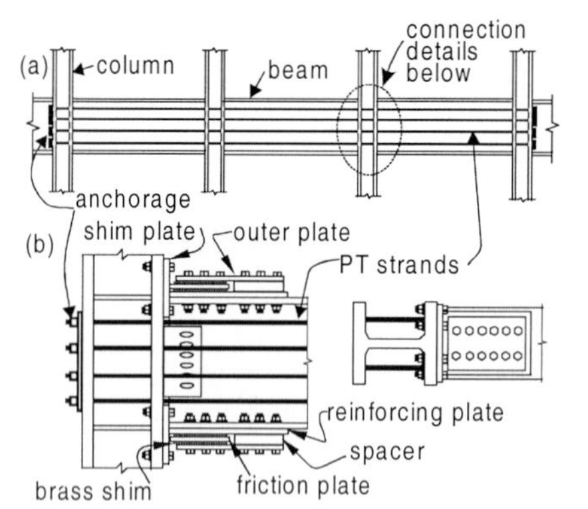

Figure 1(a) Schematic elevation of one floor of a frame with PFDCs and (b) connection details.

1(b)). A properly designed PFDC has several advantages: (1) field welding is not required; (2) the connection is made with conventional materials and skills; (3) the connection has an initial stiffness similar to that of a typical welded connection; (4) the connection is self-centering without residual deformation, thus the MRF will not have residual drift after an earthquake if significant residual deformation does not occur at the base of the ground floor columns; and (5) the beams and columns remain essentially elastic while the friction plates provide energy dissipation.

This paper describes the details, behavior, and analytical studies of a steel MRF with PFDCs. Experimental test results were used to calibrate a fiber element connection model, which was in turn used to analyze a 6-story, 4 bay steel MRF with PFDCs. Nonlinear static pushover and seismic time history analysis were performed on the frame model and compared to the response of the same MRF with standard fully restrained welded connections (FR MRF).

POST-TENSIONED FRICTION DAMPED CONNECTION

Connection Details

The post-tensioned high strength strands run parallel to the beam and are anchored outside of the connection. Due to the initial post-tensioning force applied to the strands, the beam flanges are compressed against the column flanges (Figure 1(a)). As shown in Figure 1(b), reinforcing plates are placed on the beam flanges in order to limit beam compression yielding and thus, to minimize structural damage. Shim plates are placed between the column flange and the beam flanges so that only the beam flanges and reinforcing plates are in contact with the column. This enables good contact to be maintained between the beam flanges and column face, while protecting the beam web from yielding under bearing.

Friction devices are located at the beam flanges, consisting of a friction plate sandwiched by two brass shim plates that are inserted between the beam flange reinforcing plate and an outer plate. All plates are bolted to the beam flanges. Long slotted holes are drilled through the friction plate. The shim plate is bolted to the column flange and serves as a tee flange that the friction plate is attached to. Friction is generated when the beam flanges and outer plate slide against the stationary friction plate when the beam rotates about the center of rotation situated at the mid-depth of the reinforcing plates (see Figure 2 insert). The brass shim plates are used to produce a stable friction force and to control the energy capability of the PFDC (Grigorian et al. 1993, Petty 1999). A shear tab with slotted holes is bolted to the beam web and welded to the column flange to transmit the shear forces.

Flexural Behavior

The moment-relative rotation (M-θ_r) curve for a PFDC when subjected to cyclic loading is shown in the schematic given in Figure 2. The behavior is characterized by a gap opening and closing at the beam-column interface. The total moment resistance of the connection is provided by the moment due to the initial post-tensioning force in the strands, friction force, and an additional force developed due to elongation of the strands. For simplicity, the post-tensioned forces are assumed to be acting at the centroidal axis of the beam while the friction forces are assumed to be acting at the mid-depth of top and bottom friction plates.

Under applied moment, the connection initially behaves as a fully restrained connection, where the initial stiffness is similar to that of a fully restrained welded moment connection when θ_r is equal to zero (events 0 to 2 in Figure 2). Once the magnitude of the applied moment reaches the moment resistance due to the initial post-tensioning force in the strands, decompression of the beam from the column face occurs. The moment at which this occurs (event 1) is called the decompression moment. The applied moment continues to increase between events 1 and 2 as the rotation of the beam is restrained by the resistance of the friction component. At event 1 the friction force is minimal and increases gradually up to its maximum value at point 2, which is the point of incipient rotation. The maximum value of the friction force is computed using classical friction Coulomb's theory.

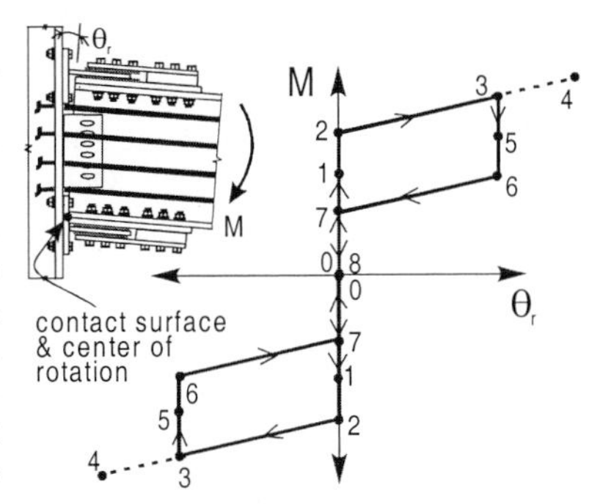

Figure 2. Moment-relative rotation behavior.

The stiffness of the connection after gap opening is associated with the elastic axial stiffness of the post-tensioned strands. With continued loading, the strands elongate producing an additional force, which contributes to resist the total applied moment. Yielding of the strands eventually may occur at event 4. Upon unloading (event 3), θ_r remains constant. At event 5, the friction force is zero. Between events 5 and 6 the friction force changes direction and starts increasing until reaching its maximum value at event 6. Between events 6 and 7, the beam rotates until the beam top flange is back in contact with the shim plate, but not compressed. Between events 7 and 8 the value of the friction force decreases with the beam being compressed against the shim plates and M equal to zero at event 8. A complete reversal in the applied moment will result in a similar connection behavior occurring in the opposite direction of loading, as shown in Figure 2. Because a residual friction force exists at event 8, the forces in the system are indeterminate until event 2 is again reached. Thus, there is no clear point of decompression on the curve following the first half-cycle.

As long as the strands remain elastic, and there is no significant beam yielding, the post-tensioning force is preserved and the connection will self-center upon unloading (i.e., θ_r returns to zero rotation upon removal of the connection moment and the structure returns to its pre-earthquake position). The energy dissipation capacity of the connection is related to the force developed between the friction surfaces.

Main Design Considerations

The connection behavior described above is the basis for the design of a PFDC. Preliminary design considerations were developed using experimental results and nonlinear static pushover analyses of beam-column subassemblies (Garlock et al. 2002), in addition to nonlinear time history analysis of MRFs with PT connections having top and seat angles (Ricles et al. 2001). For design it was assumed that the connection developed a resistance of at least $0.93M_p$ (where M_p is the plastic nominal moment of the beam) corresponding to a relative rotation, θ_r, of 0.0225 radians. The connections were designed with a moment of at least $0.60M_p$ when gap opening becomes imminent.

ANALYTICAL MODELING OF A PFDC

Description of Model

The computer program DRAIN-2DX (Prakash et al. 1993) was used to develop a model of the PFDC and the associated beams and columns of the MRF. This computer program is well suited for this purpose since it accounts for geometric and material non-linearities in static and dynamic two-dimensional structural analysis. A brief description of the model is presented. More detailed information on the modeling of PFDCs is given in Rojas (2002).

The beams and columns are modeled using fiber elements. Each fiber element is divided into a number of segments along the element length. Only one segment is used in those elements were elastic behavior is expected. The cross-section of each segment is comprised of several fibers. A material stress-strain relationship, a cross sectional area, and a distance from the longitudinal reference axis of the member characterize each fiber. The beam fiber elements adjacent to the columns are used to model connection gap opening. The fibers of the beam cross section initially in contact with the shim plates are assigned a stress-strain relationship that has stiffness in compression, but none in tension. Fibers are omitted from the cross section not in contact with the shim plates.

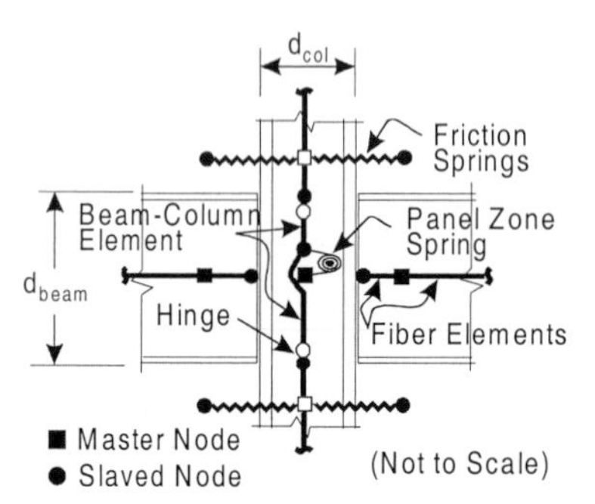

Figure 3. Post-tensioned friction damped connection analysis model.

To properly account for the depth of the beam and the size of the panel zone in the connection region, it was necessary to utilize a set of master-slave nodes as shown in Figure 3. The flexibility of the panel zone was modeled by placing a rotational spring with moment-rotational characteristics determined from the shear stiffness and strength of the panel zone. The strands were modeled using truss elements. The friction components were modeled with spring elements having a bilinear force deformation relationship.

A model was also developed for the FR MRF. Beams, columns, and panel zones were modeled as described in the model with PFDCs. However, one of the limitations of DRAIN-2DX is that it cannot model local buckling of members. To overcome this difficulty Herrera (2001) developed a fiber element to include the effects of local buckling. Ojeda and Muhummud (2002) implemented a beam plastic hinge model that includes strength degradation and strain hardening in order to model beam local buckling. P-delta effects from the interior gravity frames were incorporated into the FR MRF and MRF with PFDC models using a leaning column.

SEISMIC BEHAVIOR OF A MRF WITH PFDCS

Description of Frames

To study the seismic behavior of steel frames with PFDCs, a 6-story perimeter MRF with 4 bays was designed as a special MRF (SMRF) in accordance with the 2000 International Building Code (IBC 2000) provisions (ICC 2000). The beam and column sections were designed based on the following assumptions: 1) connections were assumed to be rigid; 2) the structure is an office building located on stiff soil in the Los Angeles area; 3) the design accelerations were determined using the deterministic limit of the site-specific procedure; and 4) A992 steel was used. The resulting member sizes are shown in Figure 4. Two frame models were created with these members where one model had rigid connections and the other had PDFCs. The PFDCs were designed for $0.93M_p$ in the first three stories and increased to M_p in the last three stories corresponding to a θ_r=0.0225 radians. The reinforcing plates were 1778 mm long. Figure 4 shows the initial post-tensioning (T_o) and friction forces (F_f) provided in the MRF with PFDCs. Additional details can be found in Rojas (2002).

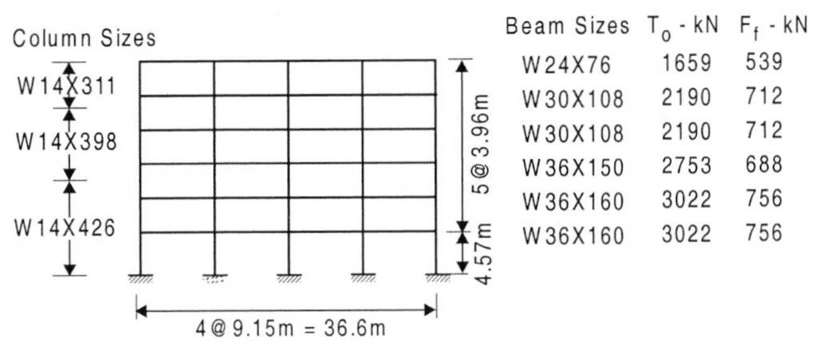

Figure 4. Frame used in analytical study.

Static Pushover Analysis

A nonlinear static pushover analysis of both frames was conducted. The lateral loads were distributed over the height of the frame in accordance with the IBC 2000 provisions. Figure 5 shows the relationship between the normalized base shear and the roof drift index (RDI) for both frames. The normalized base shear is the base shear divided by the seismic dead weight of the building, W. The analysis results for the FR MRF show that first yielding occurs when the applied base shear is 0.14W at a corresponding RDI of 0.93%. The maximum overstrength of the FR MRF is 3.16 (0.21W) at a corresponding RDI of 2.45%. The overstrength is defined as the base shear divided by the IBC 2000 design base shear. This value is close to the one anticipated in the IBC 2000 code (3.00) for SMRF systems. The maximum interstory drift index (IDI) was 5.91% and occurred at the first story. The inelastic behavior in the structure is concentrated at the girders of all floors and at the base of the ground floor columns. Panel zone yielding also occurred in most of the interior columns of the frame.

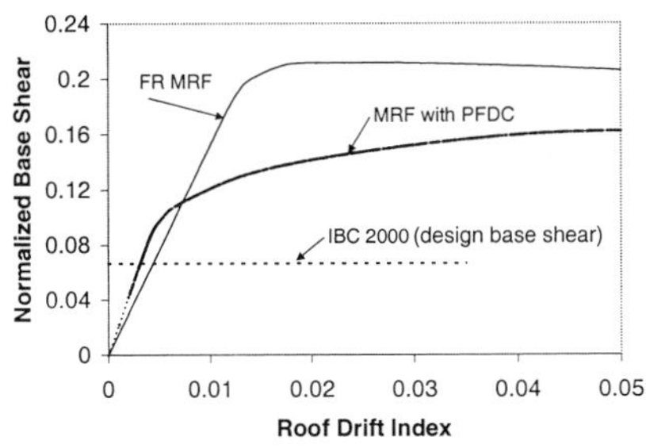

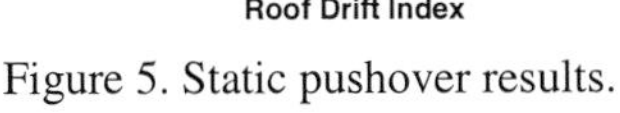

Figure 5. Static pushover results.

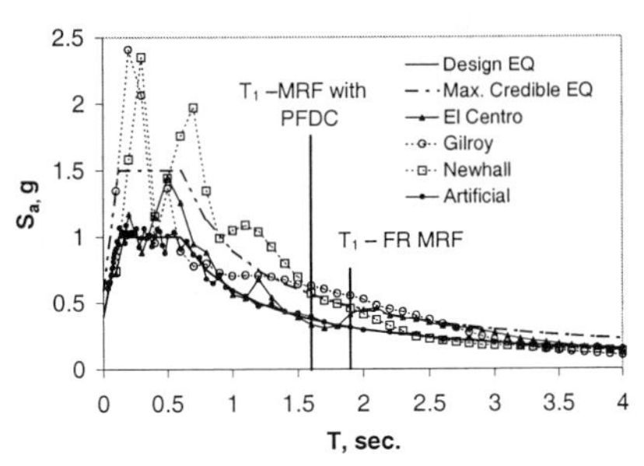

Figure 6. Response spectra.

The frame with PFDCs has a larger initial stiffness than the FR MRF due to the reinforcing plates. Gap opening becomes imminent at a base shear of 0.098W, corresponding to a RDI of 0.5%. After gap opening, the lateral stiffness of the MRF with PFDCs decreases. In the MRF with PFDCs, first yielding occurs at the base of the ground floor columns when the applied base shear is 0.12W and the RDI is 0.97%. The maximum overstrength of the MRF with PFDCs is 2.42 (0.162W) when the RDI is 5%. The maximum interstory drift was 5.45% and occurred at the second story. The maximum connection relative rotation, θ_r, was 0.055 radians and occurred at the first floor. Beam compression yielding at the end of the reinforcing plates begins when the RDI is 1.67%. The panel zones remain elastic.

Time History Analysis

Nonlinear dynamic time history analyses were conducted using an ensemble of four accelerograms to investigate the seismic behavior of the MRF with PFDCs and the FR MRF. These accelerograms include three records classified as ground motions with 10% probability of exceedance in 50 years (10/50) according to the SAC project (Somerville 1997) and an artificial ground motion generated to be compatible with the IBC 2000 design response spectrum for the seismicity conditions noted previously. The SAC records are: (1) a scaled component rotated 45° from fault normal of the 1940 Imperial Valley Earthquake recorded at El Centro, CA; (2) a scaled component rotated 45° from fault parallel of the 1994 Northridge Earthquake recorded at Newhall, CA; and (3) a scaled component rotated 45° from fault normal of the 1989 Loma Prieta recorded at Gilroy, CA. The peak ground accelerations for the three SAC ground motions and the artificial ground motion were 0.46g, 0.66g, 0.67g, and 0.40g, respectively.

Figure 6 shows the IBC 2000 design and maximum credible earthquake response spectra, as well as the response spectra of the four ground motions. The IBC 2000 design response spectrum has approximately a 10% probability of exceedance in 50 years while the IBC 2000 maximum credible earthquake response spectrum has a 2% probability of exceedance in 50 years. Scale factors for the three SAC ground motions were computed (according to the procedure described by Somerville (1997)) using the IBC 2000 response spectra as the target spectra and compared with the scale factors provided by Somerville (1997). From these new scale factors it was concluded that the as-scaled El Centro ground motion may be classified as approximately a 10/50 ground motion, even though at some period ranges it exceeds the IBC 2000 design spectrum. The as-scaled Newhall and Gilroy ground motions lie between the 2/50 and 10/50 ground motions, with the Newhall ground motion closer to a ground motion with a 2% probability of being exceeded in 50 years (2/50). However, it is interesting to note that between 1.7 and 3.0 seconds, the as-scaled Gilroy ground motion exceeds the as-scaled Newhall ground motion. The fundamental elastic period, T_1, of the frames are 1.87 and 1.60 seconds for the FR MRF and MRF with PFDCs, respectively.

Table 1 presents a summary of the seismic response for both frames. It can be seen that the MRF with PFDCs develops a smaller magnitude of displacement response than the FR MRF for the artificial and El Centro ground motions. The maximum residual IDI is negligible in both frames. Both frames perform satisfactorily for this ground motion, having minimal yielding.

TABLE 1
RESPONSE OF MRF WITH PFDCS AND FR MRF

		Max. Roof Disp. (mm)		Max. IDI (%)		Max. Base Shear (%W)		Max. Res. IDI (%)	
Ground Motion	$\theta_{r,max}$ (rads)	MRF-PFDC	FR-MRF	MRF-PFDC	FR-MRF	MRF-PFDC	FR-MRF	MRF-PFDC	FR-MRF
Artificial	0.0168	303.0	365.4	1.50	2.01	17.7	23.1	0.06	0.09
El Centro	0.0222	432.7	456.7	2.31	2.63	17.1	25.3	0.20	0.50
Newhall	0.0232	521.1	478.0	2.67	2.55	19.0	25.8	0.05	0.24
Gilroy	0.0347	673.3	640.5	3.51	3.06	17.3	24.6	0.07	0.70

The response envelope of lateral displacement and story shear for the MRF with PFDCs and the FR MRF to the El Centro ground motion is shown in Figure 7(a). It can be seen that the maximum floor displacements in the MRF with PFDCs is less than the FR MRF. Also, the maximum story shears are shown in Figure 7(b) to be less for the frame with PFDCs. The roof displacement time-history is given in Figure 7(c). The girders and panel zone remain elastic in the MRF with PFDCs. The FR MRF has significant permanent deformation in the lower stories. Inelastic deformation developed at the base of the ground floor columns, and the girders and panel zones in almost all stories of the FR MRF. The maximum residual displacement for the MRF with PFDCs is 19.5 mm while for the FR MRF it is 47.3 mm (see Figure 7(d)). The residual deformation in the MRF with PFDCs is due to some inelastic hinging at the base of the ground floor columns. This can be avoided by using a PFDC at the base of the ground floor columns (beyond the scope of this paper).

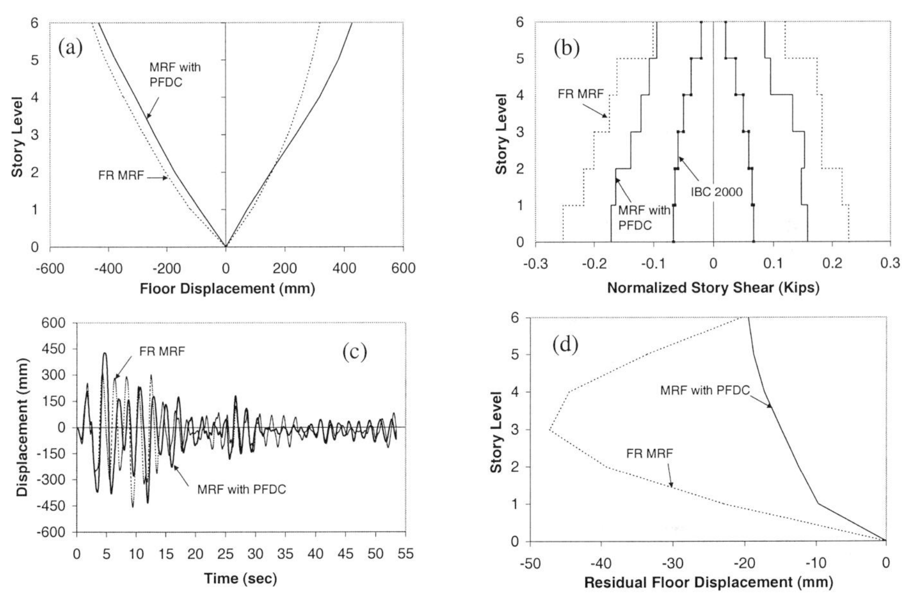

Figure 7. Response of frames to El Centro earthquake.

In the cases studied the MRF with PFDCs did not always develop a smaller magnitude of maximum drift response to the selected ground motion, as seen in Table 1. However, even in the cases where the FR MRF exhibited a smaller magnitude of response, the MRF with PFDCs exhibited superior behavior that included self-centering capability with minimal inelastic deformation to the main structural components of the system. Therefore, after a major earthquake the MRF with PFDCs will have little damage and its post-earthquake position will be very close to the undeformed configuration. The maximum values for θ_r in Table 1 are close to the design value of 0.0225 radians except the Gilroy record (which is however larger than the design earthquake).

SUMMARY AND CONCLUSIONS

An innovative connection for seismic resistant design of steel structures that requires no field welding has been presented. Combining friction devices with high strength post-tensioned steel strands results in a connection with excellent ductility and an initial stiffness that is similar to fully welded moment resisting connections. In addition, the connection has a self-centering capability, resulting in minimal permanent interstory drift in a building following a severe earthquake. Adequate design of reinforcing plates in beam flanges and doubler plates in panel zones will avoid damage in these elements. The frame however must be properly designed to avoid permanent rotational deformation at the base of the ground floor columns. An analytical model was presented that was based on fiber elements. Results from the analytical studies indicate that moment resisting frames with PFDCs may perform better than moment resisting frames with conventional welded moment connections.

ACKNOWLEDGEMENTS

The research reported herein was supported by the National Science Foundation (Dr. Ken Chong and Dr. Ashland Brown cognizant program officials) and by a grant from the Pennsylvania DCED through the PITA program. The opinions expressed in this paper are those of the authors and do not necessarily reflect the views of the sponsors.

REFERENCES

Garlock, M., Ricles, J., and Sause, R. (2002). "Experimental Studies on Full-Scale Post-Tensioned Seismic-Resistant Steel Moment Connections," *Proceedings of the 7th U.S. National Conference on Earthquake Engineering (NCEE), Boston, MA.*

Grigorian, C.E., Yang, T.S., and Popov, E.P. (1993). "Slotted Bolted Connection Energy Dissipators," Earthquake Spectra, 9(3), pp. 491-504.

Herrera, R., Sause, R., and Ricles, J. (2001). "Refined Connection Element (Type 05) for DRAIN-2DX, Element Description and User Guide," ATLSS Report No. 01-07.

ICC (2000) "International Building Code" International Code Council (ICC), Falls Church, VA.

Ojeda, J. and Muhummud, T. (2002). "Beam Strength Degradation Model for DRAIN-2DX," ATLSS report in progress.

Petty, G.D. (1999). "Evaluation of a Friction Component for a Post-Tensioned Steel Connection," M.S. Thesis, Lehigh University, 328pp.

Prakash, V., Powell, G., and Campbell, S., (1993). "DRAIN-2DX Base Program Description and User Guide, Version 1.0," Report No. UCB/SEMM-93/17 & 18, Structural Engineering Mechanics and Materials, Department of Civil Engineering, University of California, Berkeley, CA.

Ricles, J., Sause, R., Garlock, M, and Zhao, C. (2001). "Post-Tensioned Seismic Resistant Connections for Steel Frames," *Journal of Structural Engineering,* ASCE, 127(2), 113-121.

Rojas, P. (2002). "Seismic Analysis and Design of Post-Tensioned Friction Damped Connections for Steel Moment Resisting Frames," *Ph.D. Dissertation*, Civil and Environmental Engineering Dept., Lehigh University, Bethlehem, PA.

Somerville, P. (1997). "Development of Ground Motion Time Histories for Phase 2 of the FEMA/SAC Steel Project," Report No. SAC/BD-97/04.

TEAM PRESENTATION:

SMART STRUCTURES FOR DYNAMIC HAZARD MITIGATION

Advances in Building Technology, Volume 1
M. Anson, J.M. Ko and E.S.S. Lam (Eds.)

Stochastic Approach to Control and Identification of Smart Structures

James L. Beck and Ka-Veng Yuen

Division of Engineering and Applied Science, California Institute of Technology

ABSTRACT. To fully exploit new technologies for response mitigation and structural health monitoring, improved design methodologies are desirable. In this paper, a stochastic framework is presented for robust control and identification of structural systems under dynamical loads, such as those induced by wind or earthquakes. A reliability-based stochastic robust control approach is used to design the controller for an active or semi-active control system. Feedback of the incomplete response at earlier time steps is used, without the need for any state estimation. The optimal controller is chosen by minimizing the robust failure probability over a set of possible models for the system. Here, failure means excessive levels of one or more response quantities representative of the performance of the structure and the control devices. When calculating the robust failure probability, the plausibility of each model as a representation of the system's dynamic behavior is quantified by a probability distribution over the set of possible models; this distribution is initially based on engineering judgment but it can be updated using Bayes' Theorem if dynamic data become available from the structure. For this purpose, a probabilistic system identification technique is presented for model updating using incomplete noisy measurements only. This method allows for updating of the uncertainties associated with the values of the parameters controlling the dynamic behavior of the structure by using only one set of stationary or nonstationary response data. This updated probabilistic description of the system can be used to modify the controller for improved performance of the system. It can also be used for structural health monitoring. An example is presented to illustrate the proposed stochastic framework for identification and control of smart structures.

KEYWORDS: Bayesian inference, model updating, robust control, robust reliability, smart structures, system identification.

1 STOCHASTIC RESPONSE ANALYSIS

Consider a linear model of a structural system with N_d degrees-of-freedom (DOFs) and equation of motion:

$$\mathbf{M}(\boldsymbol{\theta}_s)\ddot{\mathbf{x}}(t) + \mathbf{C}(\boldsymbol{\theta}_s)\dot{\mathbf{x}}(t) + \mathbf{K}(\boldsymbol{\theta}_s)\mathbf{x}(t) = \mathbf{T}\cdot\mathbf{f}(t) + \mathbf{T}_c\cdot\mathbf{f}_c(t) \quad (1)$$

where $\mathbf{M}(\boldsymbol{\theta}_s)$, $\mathbf{C}(\boldsymbol{\theta}_s)$ and $\mathbf{K}(\boldsymbol{\theta}_s)$ are the $N_d \times N_d$ mass, damping and stiffness matrix, respectively, parameterized by the structural model parameters $\boldsymbol{\theta}_s$ for the system; $\mathbf{f}(t) \in \mathbb{R}^{N_f}$ and $\mathbf{f}_c(t) \in \mathbb{R}^{N_{fc}}$ are the external excitation and control force vector, respectively, and $\mathbf{T} \in \mathbb{R}^{N_d \times N_f}$ and $\mathbf{T}_c \in \mathbb{R}^{N_d \times N_{fc}}$ are their distribution matrices. A control law is given later that specifies $\mathbf{f}_c$ by feedback of the measured output.

The uncertain excitation $\mathbf{f}(t)$ could be earthquake ground motions or wind forces, for example, and it is modeled by a zero-mean stationary filtered white-noise process described by

$$\begin{aligned}\dot{\mathbf{w}}_f(t) &= \mathbf{A}_{wf}(\boldsymbol{\theta}_f)\mathbf{w}_f(t) + \mathbf{B}_{wf}(\boldsymbol{\theta}_f)\mathbf{w}(t) \\ \mathbf{f}(t) &= \mathbf{C}_{wf}(\boldsymbol{\theta}_f)\mathbf{w}_f(t)\end{aligned} \tag{2}$$

where $\mathbf{w}(t) \in \mathbb{R}^{N_w}$ is a Gaussian white-noise process with zero mean and unit spectral intensity matrix; $\mathbf{w}_f(t) \in \mathbb{R}^{N_{wf}}$ is an internal filter state and $\mathbf{A}_{wf}(\boldsymbol{\theta}_f) \in \mathbb{R}^{N_{wf}\times N_{wf}}$, $\mathbf{B}_{wf}(\boldsymbol{\theta}_f) \in \mathbb{R}^{N_{wf}\times N_w}$ and $\mathbf{C}_{wf}(\boldsymbol{\theta}_f) \in \mathbb{R}^{N_f\times N_{wf}}$ are the parameterized filter matrices governing the properties of the filtered white noise. A vector $\boldsymbol{\theta}$ is introduced, which combines the structural model and excitation model parameter vectors, i.e. $\boldsymbol{\theta} = [\boldsymbol{\theta}_s^T, \boldsymbol{\theta}_f^T]^T \in \mathbb{R}^{N_\theta}$. The dependence on $\boldsymbol{\theta}$ will be left implicit hereafter in this section.

By introducing the state vector $\mathbf{y}(t) = [\mathbf{x}(t)^T, \dot{\mathbf{x}}(t)^T]^T$, (1) can be rewritten in state-space form as:

$$\dot{\mathbf{y}}(t) = \mathbf{A}_y\mathbf{y}(t) + \mathbf{B}_y\mathbf{f}(t) + \mathbf{B}_{yc}\mathbf{f}_c(t) \tag{3}$$

where $\mathbf{A}_y = \begin{bmatrix} \mathbf{0}_{N_d\times N_d} & \mathbf{I}_{N_d} \\ -\mathbf{M}^{-1}\mathbf{K} & -\mathbf{M}^{-1}\mathbf{C}\end{bmatrix}$, $\mathbf{B}_y = \begin{bmatrix}\mathbf{0}_{N_d\times N_f} \\ \mathbf{M}^{-1}\mathbf{T}\end{bmatrix}$ and $\mathbf{B}_{yc} = \begin{bmatrix}\mathbf{0}_{N_d\times N_{fc}} \\ \mathbf{M}^{-1}\mathbf{T}_c\end{bmatrix}$. Here, $\mathbf{0}_{a\times b}$ and $\mathbf{I}_a$ denote the $a \times b$ zero matrix and $a \times a$ identity matrix, respectively.

In order to allow more choices of the output to be fed back or to be controlled, an output vector $\mathbf{y}_f \in \mathbb{R}^{N_{yf}}$ is introduced that is modeled by the following state equation:

$$\dot{\mathbf{y}}_f(t) = \mathbf{A}_{yf}\mathbf{y}_f(t) + \mathbf{B}_{yf}\mathbf{y}(t) \tag{4}$$

where $\mathbf{A}_{yf} \in \mathbb{R}^{N_{yf}\times N_{yf}}$, $\mathbf{B}_{yf} \in \mathbb{R}^{N_{yf}\times 2N_d}$ are the matrices that characterizes the output filter. Note that the output vector can represent many choices of feedback. For example, it can handle displacement, velocity or acceleration measurements if the matrices $\mathbf{A}_{yf}$ and $\mathbf{B}_{yf}$ are chosen appropriately. Accelerations can be obtained approximately by passing the velocities in the state vector through a filter with the transfer function $H_d(s) = \omega_0^2 s(s^2 + \sqrt{2}\omega_0 s + \omega_0^2)^{-1}$: This filter can approximate differentiation very accurately if ω_0 is chosen larger than the upper limit of the frequency band of interest. On the other hand, one can model the sensor dynamics for displacements or velocities measurements by using a low-pass filter with the transfer function $H_l(s) = \omega_0^2(s^2 + \sqrt{2}\omega_0 s + \omega_0^2)^{-1}$.

If the full state vector $\mathbf{v}(t) = [\mathbf{w}_f(t)^T, \mathbf{y}(t)^T, \mathbf{y}_f(t)^T]^T$ is introduced, then (2) - (4) can be combined as follows:

$$\dot{\mathbf{v}}(t) = \mathbf{A}\mathbf{v}(t) + \mathbf{B}\mathbf{w}(t) + \mathbf{B}_c\mathbf{f}_c(t) \tag{5}$$

where the matrices $\mathbf{A}$, $\mathbf{B}$ and $\mathbf{B}_c$ are given by

$$\mathbf{A} \equiv \begin{bmatrix} \mathbf{A}_{wf} & \mathbf{0}_{N_{wf}\times 2N_d} & \mathbf{0}_{N_{wf}\times N_{yf}} \\ \mathbf{B}_y\mathbf{C}_{wf} & \mathbf{A}_y & \mathbf{0}_{2N_d\times N_{yf}} \\ \mathbf{0}_{N_{yf}\times N_{wf}} & \mathbf{B}_{yf} & \mathbf{A}_{yf}\end{bmatrix}, \quad \mathbf{B} \equiv \begin{bmatrix}\mathbf{B}_{wf} \\ \mathbf{0}_{2N_d\times N_w} \\ \mathbf{0}_{N_{yf}\times N_w}\end{bmatrix} \text{ and } \mathbf{B}_c \equiv \begin{bmatrix}\mathbf{0}_{N_{wf}\times N_{fc}} \\ \mathbf{B}_{yc} \\ \mathbf{0}_{N_{yf}\times N_{fc}}\end{bmatrix} \tag{6}$$

By treating $\mathbf{w}$ as constant over each subinterval $[k\Delta t, k\Delta t + \Delta t)$, where Δt is the sampling time interval that is small enough to capture the dynamics of the structure, (5) yields the following discrete-time equation:

$$\mathbf{v}[k+1] = \bar{\mathbf{A}}\mathbf{v}[k] + \bar{\mathbf{B}}\mathbf{w}[k] + \bar{\mathbf{B}}_c\mathbf{f}_c[k] \tag{7}$$

where $\mathbf{v}[k] \equiv \mathbf{v}(k\Delta t)$, $\bar{\mathbf{A}} \equiv e^{\mathbf{A}\Delta t}$, $\bar{\mathbf{B}} \equiv \mathbf{A}^{-1}(\bar{\mathbf{A}} - \mathbf{I}_{N_{wf}+2N_d+N_{yf}})\mathbf{B}$ and $\bar{\mathbf{B}}_c \equiv \mathbf{A}^{-1}(\bar{\mathbf{A}} - \mathbf{I}_{N_{wf}+2N_d+N_{yf}})\mathbf{B}_c$, and $\mathbf{w}[k]$ is Gaussian discrete white noise with zero mean and covariance matrix $\boldsymbol{\Sigma}_w = \frac{2\pi}{\Delta t}\mathbf{I}_{N_w}$.

Assume that discrete-time response data, with sampling time interval Δt, is available for N_o components of the output state, that is, the measured output is given by

$$\mathbf{z}[k] = \mathbf{L}_o\mathbf{v}[k] + \mathbf{n}[k] \tag{8}$$

where $\mathbf{L}_o \in \mathbb{R}^{N_o\times(N_{wf}+2N_d+N_{yf})}$ is the observation matrix and $\mathbf{n}[k] \in \mathbb{R}^{N_o}$ is the uncertain prediction error which accounts for the difference between the actual measured output from the structural system and the predicted output given by the model defined by (7); it includes both modeling error and measurement noise. The prediction error is modeled as a stationary Gaussian discrete white noise process with zero mean and covariance matrix $\mathbf{\Sigma}_n$; this choice gives the maximum information entropy (greatest uncertainty) in the absence of any additional information about the unmodeled dynamics or output noise.

Now, choose a linear control feedback law using the current and the previous N_p output measurements,

$$\mathbf{f}_c[k] = \sum_{p=0}^{N_p} \mathbf{G}_p \mathbf{z}[k-p] \tag{9}$$

where $\mathbf{G}_p, p = 0, 1, \ldots, N_p$ are the gain matrices, which will be determined in the next section. It is worth noting that if the matrices $\mathbf{G}_p, p = 0, \ldots, N_p^\star$ ($N_p^\star < N_p$) are fixed to be zero, the controller at any time step only utilizes output measurements from time steps that are more than $N_p^\star \Delta t$ back in the past. Furthermore, by choosing a value of $N_p^\star$ such that $N_p^\star \Delta t$ is larger than the reaction time of the control system (data acquisition, online calculation of the control forces and actuator reaction time), it is possible to avoid any instability problem caused by time-delay effects.

Substituting (9) into (7):

$$\mathbf{v}[k+1] = (\bar{\mathbf{A}} + \bar{\mathbf{B}}_c \mathbf{G}_0 \mathbf{L}_o)\mathbf{v}[k] + \bar{\mathbf{B}}\mathbf{w}[k] + \bar{\mathbf{B}}_c \sum_{p=1}^{N_p} \mathbf{G}_p \mathbf{z}[k-p] + \bar{\mathbf{B}}_c \mathbf{G}_0 \mathbf{n}[k] \tag{10}$$

Now, define an augmented vector $\mathbf{U}_{N_p}[k]$ as follows:

$$\mathbf{U}_{N_p}[k] \equiv [\mathbf{v}[k]^T, \mathbf{z}[k-1]^T, \ldots, \mathbf{z}[k-N_p]^T]^T \tag{11}$$

Then, (10) can be rewritten as follows:

$$\mathbf{U}_{N_p}[k+1] = (\bar{\mathbf{A}}_u + \bar{\mathbf{B}}_{uc})\mathbf{U}_{N_p}[k] + \bar{\mathbf{B}}_u \bar{\mathbf{f}}[k] \tag{12}$$

where

$$\bar{\mathbf{f}}[k] \equiv [\mathbf{w}[k]^T \quad \mathbf{n}[k]^T]^T \tag{13}$$

and $\bar{\mathbf{A}}_u$, $\bar{\mathbf{B}}_u$ and $\bar{\mathbf{B}}_{uc}$ are given by

$$\bar{\mathbf{A}}_u \equiv \begin{bmatrix} \bar{\mathbf{A}} & & \mathbf{0}_{(N_{wf}+2N_d+N_{yf})\times N_p N_o} \\ \mathbf{L}_o & & \mathbf{0}_{N_o \times N_p N_o} \\ \mathbf{0}_{(N_p-1)N_o \times (N_{wf}+2N_d+N_{yf})} & \mathbf{I}_{(N_p-1)N_o} & \mathbf{0}_{(N_p-1)N_o \times N_o} \end{bmatrix}$$

$$\bar{\mathbf{B}}_{uc} \equiv \begin{bmatrix} \bar{\mathbf{B}}_c \mathbf{G}_0 \mathbf{L}_o & \bar{\mathbf{B}}_c \mathbf{G}_1 & \cdots & \bar{\mathbf{B}}_c \mathbf{G}_{N_p} \\ \multicolumn{4}{c}{\mathbf{0}_{N_p N_o \times (N_{wf}+2N_d+N_{yf}+N_p N_o)}} \end{bmatrix}, \quad \bar{\mathbf{B}}_u \equiv \begin{bmatrix} \bar{\mathbf{B}} & \bar{\mathbf{B}}_c \mathbf{G}_0 \\ \mathbf{0}_{N_o \times N_w} & \mathbf{I}_{N_o} \\ \mathbf{0}_{(N_p-1)N_o \times N_w} & \mathbf{0}_{(N_p-1)N_o \times N_o} \end{bmatrix} \tag{14}$$

Therefore, the covariance matrix $\mathbf{\Sigma}_u \equiv E[\mathbf{U}_{N_p}[k]\mathbf{U}_{N_p}[k]^T]$ of the augmented vector $\mathbf{U}_{N_p}$ is readily obtained:

$$\begin{aligned} \mathbf{\Sigma}_u &= (\bar{\mathbf{A}}_u + \bar{\mathbf{B}}_{uc})\mathbf{\Sigma}_u(\bar{\mathbf{A}}_u + \bar{\mathbf{B}}_{uc})^T + \bar{\mathbf{B}}_u \mathbf{\Sigma}_{\bar{f}} \bar{\mathbf{B}}_u^T \\ \mathbf{\Sigma}_{\bar{f}} &= \begin{bmatrix} \mathbf{\Sigma}_w & \mathbf{\Sigma}_{wn} \\ \mathbf{\Sigma}_{wn}^T & \mathbf{\Sigma}_n \end{bmatrix} \end{aligned} \tag{15}$$

where $\mathbf{\Sigma}_{\bar{f}}$ denotes the covariance matrix of the vector $\bar{\mathbf{f}}$ in (13). Note that (15) is a standard stationary Lyapunov covariance equation in discrete form. The system response is a stationary Gaussian process with zero mean and covariance matrix that can be readily calculated using (15). These properties are used to design the optimal robust controller for the structure.

2 OPTIMAL CONTROLLER DESIGN

The optimal robust controller is defined here as the one which maximizes the robust reliability (Papadimitriou et al. 2001) with respect to the feedback gain matrices in (9), that is, the one which minimizes the robust failure probability for a structural model with uncertain parameters representing the real structural system. Failure is defined as the situation in which at least one of the performance quantities (structural response or control force) exceeds a given threshold level. This is the classic 'first passage problem', which has no closed form solution. Therefore, the proposed method utilizes an approximate solution based on Rice's 'out-crossing' theory.

2.1 *Conditional Failure Probability*

Use $\mathbf{q}[k] \in \mathbb{R}^{N_q}$ to denote the control performance vector of the system at time $k\Delta t$. Its components may be structural interstory drifts, floor accelerations, control forces, etc. The system performance is given by $\mathbf{q}[k] = \mathbf{P}_0\mathbf{U}_{N_p}[k] + \mathbf{m}[k]$, where $\mathbf{P}_0 \in \mathbb{R}^{N_q \times (N_{wf}+2N_d+N_{yf}+N_pN_o)}$ is a performance matrix which multiplies the augmented vector $\mathbf{U}_{N_p}$ from (11) to give the corresponding performance vector of the model. In order to account for the unmodeled dynamics, the uncertain prediction error $\mathbf{m} \in \mathbb{R}^{N_q}$ is introduced because the goal is to control the system performance, not the model performance; it is modeled as discrete white noise with zero mean and covariance matrix $\mathbf{\Sigma}_m$.

For a given failure event $F_i = \{|q_i(t)| > \beta_i \text{ for some } t \in [0,T]\}$, the conditional failure probability $P(F_i|\boldsymbol{\theta})$ for the performance quantity q_i based on the structural model and excitation model specified by $\boldsymbol{\theta}$ can be estimated using Rice's formula:

$$P(F_i|\boldsymbol{\theta}) \approx 1 - \exp[-\nu_{\beta_i}(\boldsymbol{\theta})T] \tag{16}$$

where $\nu_{\beta_i}(\boldsymbol{\theta}) = \frac{\sigma_{\dot{q}_i}}{\pi\sigma_{q_i}} \exp(-\frac{{\beta_i}^2}{2\sigma_{q_i}^2})$ is the mean out-crossing rate for the threshold level β_i, and where σ_{q_i} and $\sigma_{\dot{q}_i}$ are the standard deviations for the performance quantity q_i and its derivative $\dot{q}_i$, respectively. In implementation, $\dot{q}_i$ must be included in $\mathbf{y}_f$ in (4) if it is not already part of $\mathbf{y}$.

Now consider the failure event $F = \cup_{i=1}^{N_q} F_i$, that is, the system fails if any $|q_i|$ exceeds its threshold β_i. Since the mean out-crossing rate of the system can be approximated by $\nu = \sum_{i=1}^{N_q} \nu_{\beta_i}$ (Veneziano et al. 1977), the probability of failure $P(F|\boldsymbol{\theta})$ of the controlled structural system is given approximately by

$$P(F|\boldsymbol{\theta}) \approx 1 - \exp[-\sum_{i=1}^{N_q} \nu_{\beta_i}(\boldsymbol{\theta})T] \tag{17}$$

where N_q denotes the number of performance quantities considered.

2.2 *Robust Failure Probability*

No matter what technique is used to develop a model for a structural system, the values of the structural model parameters that best represent the system are always uncertain to some extent. Furthermore, the excitation model is uncertain as well. Therefore, a probabilistic description is used to describe the uncertainty in the model parameters $\boldsymbol{\theta}$ defined earlier. Such probability distributions can be specified using engineering judgement or they can be obtained using system identification techniques. This leads to the concept of the robust failure probability given by the theorem of total probability (Papadimitriou et al. 2001):

$$P(F|\Theta) = \int_{\Theta} P(F|\boldsymbol{\theta})p(\boldsymbol{\theta}|\Theta)d\boldsymbol{\theta} \tag{18}$$

which accounts for modeling uncertainties in deriving the failure probability. This robust failure probability is conditional on the probabilistic description $p(\boldsymbol{\theta}|\Theta)$ of the parameters that is specified over the set of possible models Θ. Since this high-dimensional integral is difficult to evaluate numerically, an asymptotic expansion is utilized (Papadimitriou et al. 1997). Denote the integral of interest by $I = \int_{\Theta} e^{l(\boldsymbol{\theta})}d\boldsymbol{\theta}$, where $l(\boldsymbol{\theta}) = \ln[P(F|\boldsymbol{\theta})] + \ln[p(\boldsymbol{\theta}|\Theta)]$. The basic idea is to fit a Gaussian density centered at the 'design point' at which $e^{l(\boldsymbol{\theta})}$, or $l(\boldsymbol{\theta})$, is maximized. It is assumed here that there is a unique design point; see Au et al. (1999) for a more general case. Then, this integral is approximated by

$$I \equiv P(F|\Theta) \approx (2\pi)^{\frac{N_\theta}{2}} \frac{P(F|\boldsymbol{\theta}^\star)p(\boldsymbol{\theta}^\star|\Theta)}{\sqrt{\det \mathbf{L}(\boldsymbol{\theta}^\star)}} \tag{19}$$

where $\boldsymbol{\theta}^\star$ is the design point at which $l(\boldsymbol{\theta})$ has a maximum value and $\mathbf{L}(\boldsymbol{\theta}^\star)$ is the Hessian of $-l(\boldsymbol{\theta})$ evaluated at $\boldsymbol{\theta}^\star$. The optimization of $l(\boldsymbol{\theta})$ to find $\boldsymbol{\theta}^\star$ can be performed, for example, by using MATLAB subroutine 'fmins'.

The proposed control design can be summarized as follows: By solving (15), the covariance matrix of the structural response can be obtained. Then, the robust failure probability can be calculated using the asymptotic expansion formula in (19) along with (16) - (17). The optimal robust controller is obtained by minimizing the robust failure probability over all possible controllers parameterized by their gain matrices, which again can be performed, for example, using MATLAB subroutine 'fmins'.

3 SYSTEM IDENTIFICATION

The optimal controller can be readily updated when dynamic data $\mathbf{Z}_{1,N}$ is available from the system (Beck and Katafygiotis 1998; Yuen et al. 2002). Here, $\mathbf{Z}_{m,p}$ denotes the set of the measurements from time $m\Delta t$ to $p\Delta t$. In this case, Bayes' Theorem is used to get an updated PDF (probability density function) $p(\boldsymbol{\theta}|\mathbf{Z}_{1,N},\Theta)$ that replaces $p(\boldsymbol{\theta}|\Theta)$ in (18) and hence the updated robust failure probability $p(F|\mathbf{Z}_{1,N},\Theta)$ (Papadimitriou et al. 2001) is minimized to obtain the optimal control gains.

Using Bayes' Theorem, the expression for the updated PDF of the parameters $\boldsymbol{\theta}$ given some measured response $\mathbf{Z}_{1,N}$ is:

$$p(\boldsymbol{\theta}|\mathbf{Z}_{1,N},\Theta) = c_1 p(\boldsymbol{\theta}|\Theta)p(\mathbf{Z}_{1,N}|\boldsymbol{\theta},\Theta) \tag{20}$$

where c_1 is a normalizing constant such that the integral of the right hand side of (20) over the domain of $\boldsymbol{\theta}$ is equal to unity. The factor $p(\boldsymbol{\theta}|\Theta)$ in (20) denotes the prior PDF of the parameters and is based on previous knowledge or engineering judgement. As long as it varies smoothly over the parameter space Θ, $p(\mathbf{Z}_{1,N}|\boldsymbol{\theta},\Theta)$ is the dominant factor on the right-hand side of (20) and the prior PDF has little influence. To evaluate this factor, a recently developed approach (Yuen et al. 2002) is used that can handle stationary or nonstationary response. This approach is based on the approximation:

$$p(\mathbf{Z}_{1,N}|\boldsymbol{\theta},\Theta) \simeq p(\mathbf{Z}_{1,N_p}|\boldsymbol{\theta},\Theta) \prod_{k=N_p+1}^{N} p(\mathbf{z}(k)|\boldsymbol{\theta},\Theta,\mathbf{Z}_{k-N_p,k-1}) \tag{21}$$

where the conditional probabilities depending on more than N_p previous data points are approximated by conditional probabilities depending on only the last N_p data points. The validity of this approximation was verified by numerous simulations. The factors $p(\mathbf{Z}_{1,N_p}|\boldsymbol{\theta},\Theta)$ and $p(\mathbf{z}(k)|\boldsymbol{\theta},\Theta,\mathbf{Z}_{k-N_p,k-1})$, $k = N_p+1,...,N$ can be calculated using fundamental time series theory (Yuen et al. 2002).

4 ILLUSTRATIVE EXAMPLE

A three-bay four-story structural frame (Figure 1) with an active mass damper on the roof is used to demonstrate the proposed framework. Mass is assumed uniformly distributed over the floor slabs and along the columns, but the mass per unit length of the slabs is twice as much as that of the columns. The stiffness-to-mass ratio is taken as $EI_{c1}/m = 1.25 \times 10^5 \mathrm{m}^4\,\mathrm{sec}^{-2}$, where m is the mass per unit length of the columns. Furthermore, $[EI_{c2}, EI_{c3}, EI_{c4}] = [0.9, 0.8, 0.7]EI_{c1}$. The rigidity of the slab is taken to be $EI_s/m = 2.00 \times 10^6 \mathrm{m}^4\,\mathrm{sec}^{-2}$. The first four natural frequencies of the structure are 4.108Hz, 11.338Hz, 17.26Hz and 21.50Hz. Rayleigh damping is assumed, so the damping matrix $\mathbf{C}$ is given by $\mathbf{C} = \alpha_m\mathbf{M} + \alpha_k\mathbf{K}$, where $\mathbf{M}$ and $\mathbf{K}$ are the mass and stiffness matrices of the system; and $\alpha_m = 0.376\,\mathrm{sec}^{-1}$, $\alpha_k = 2.07 \times 10^{-4}\,\mathrm{sec}$, which gives 1.0% damping for the first two modes.

Assume that absolute accelerations measurements are available at the 2nd and 4th DOFs for 30 sec with a sampling frequency 200 Hz. These data are simulated using the actual model with 10% rms noise added. Furthermore, assume that the system is subjected to a white noise ground motion with spectral intensity $S_{f0} = 1.0 \times 10^{-3}\mathrm{m}^2\,\mathrm{sec}^{-3}$.

In order to have better scaling, the rigidities are parameterized as follows: $EI_j = \theta_j \widetilde{EI}_j, j = 1,2,3,4$, where EI_j denotes the jth story rigidity and $\widetilde{EI}_j$ is its nominal value. The rigidity parameters θ_j are

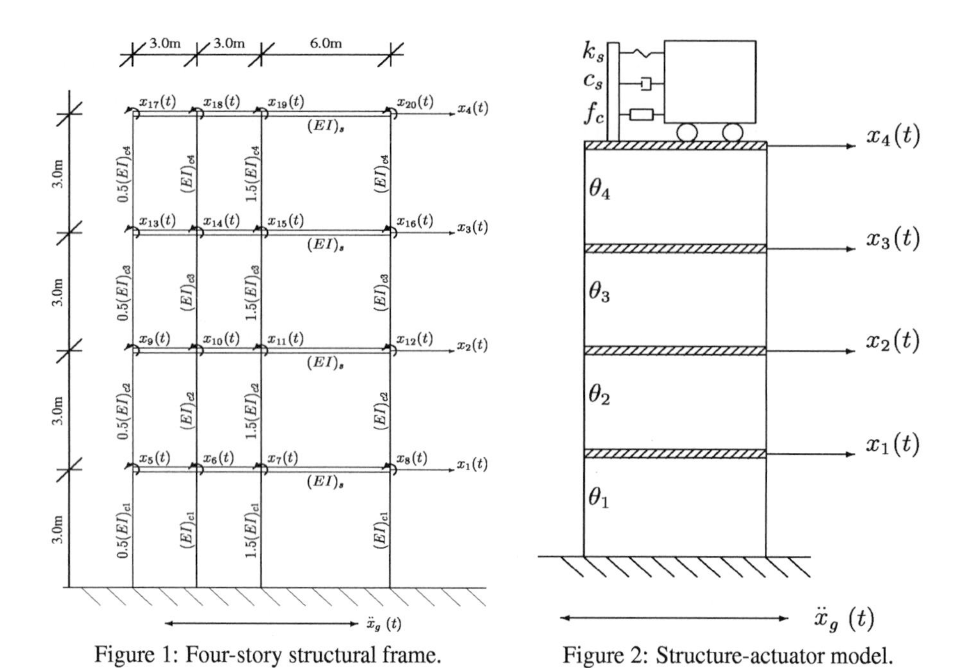

Figure 1: Four-story structural frame.

Figure 2: Structure-actuator model.

considered unknown and are determined by identification. Table 1 shows the exact values and the identified rigidity parameters and damping ratios for the two classes of models. Note that the damping ratios are presented in percentages. Table 2 shows the corresponding frequencies for the actual and the optimal models.

Table 1: Optimal structural parameters for the model class representing the structural frame.

Parameter	θ_1	θ_2	θ_3	θ_4	ζ_1	ζ_2	ζ_3	ζ_4
Exact	—	—	—	—	1.000	1.000	1.296	1.537
Identified	0.908	0.987	0.881	1.040	1.128	0.563	2.342	0.623

Table 2: Natural frequencies (in Hz) of the best model in each class.

Mode	f_1	f_2	f_3	f_4
Actual	4.108	11.34	17.26	21.50
Identified	4.158	11.35	17.40	21.02

The updated PDF of the rigidity parameters is used for calculating the robust failure probability in the controller design. First, the identified model is used to design the stiffness and damping of the AMD (active mass damper). The AMD mass M_s is chosen to be 10% of the mass of the building and the AMD stiffness and damping parameters are taken as: $k_s/M_s = 500\,\text{sec}^{-2}$ and $c_s/M_s = 15.0\,\text{sec}^{-1}$ to give a natural frequency and damping ratio approximately equal to that of the identified fundamental mode of the structure.

In Dyke et al. (1995), hydraulic actuators are modeled as follows: $\dot{f}_c = A_f f_c + B_f \dot{x}_a + B_{fu} u$, where f_c is the control force applied by the actuator; $\dot{x}_a$ is the actuator velocities; u is the signal given to the actuator; and A_f, B_f and B_{fu} are given by $A_f = -\frac{2\beta k_a}{V}$, $B_f = -\frac{2\beta A^2}{V}$ and $B_{fu} = \frac{2\beta A k_q}{V}$, where β

is the bulk modulus of the fluid; k_a and k_q are device constants; V is the characteristic hydraulic fluid volume of the actuator; and A is the cross-sectional area of the actuator. Here, these parameters are taken from Dyke et al. (1995): $2\beta k_a/V = 66.7\,\mathrm{sec}^{-1}$, $2\beta A^2/V = 250$ N/m, $2\beta A k_q/V = 1670\mathrm{Nsec}^{-1}$. The structure-actuator system is shown schematically in Figure 2.

The output vector $\mathbf{y}_f$ in (4) is comprised of: $\mathbf{y}_f = [f_c, \tilde{f}_c, \tilde{\dot{f}}_c]^T$, where $\tilde{f}_c$ and $\tilde{\dot{f}}_c$ are the state vectors for a low-pass filter with input f_c that approximates differentiation of f_c. Note that $\tilde{\dot{f}}_c$ is used to estimate the out-crossing rate of the control force and it is not used in the control system feedback. The full state vector equation is:

$$\dot{\mathbf{v}}(t) = \mathbf{A}\mathbf{v}(t) + \mathbf{B}w + \mathbf{B}_c u \tag{22}$$

where $\mathbf{v} = [\mathbf{x}^T, \dot{\mathbf{x}}^T, \mathbf{y}_f^T]^T$; w is the ground motion; and u is the signal given to the actuator which is discussed in more detail later. The matrices $\mathbf{A}$, $\mathbf{B}$ and $\mathbf{B}_c$ are given by

$$\mathbf{A} = \begin{bmatrix} \mathbf{0}_{5\times5} & \mathbf{I}_5 & \mathbf{0}_{5\times3} \\ -\mathbf{M}^{-1}\mathbf{K} & -\mathbf{M}^{-1}\mathbf{C} & \mathbf{M}^{-1}\mathbf{A}_{12} \\ \multicolumn{2}{c}{\mathbf{A}_{21}} & \mathbf{A}_{22} \end{bmatrix} \tag{23}$$

$$\mathbf{B} = -[0,0,0,0,0,1,1,1,1,1,0,0,0]^T, \quad \mathbf{B}_c = [\mathbf{0}_{1\times10}, B_{fu}, 0, 0]^T$$

where $\mathbf{A}_{12}$ and $\mathbf{A}_{21}$ and $\mathbf{A}_{22}$ are given by

$$\mathbf{A}_{12} = \begin{bmatrix} & \mathbf{0}_{3\times3} & \\ -1 & 0 & 0 \\ 1 & 0 & 0 \end{bmatrix}, \mathbf{A}_{21} = \begin{bmatrix} \mathbf{0}_{1\times8} & -B_f & B_f \\ & \mathbf{0}_{2\times10} & \end{bmatrix}, \mathbf{A}_{22} = \begin{bmatrix} A_f & 0 & 0 \\ 0 & 0 & 1 \\ \omega_c^2 & -\omega_c^2 & -\sqrt{2}\omega_c \end{bmatrix}, \omega_c = 10.0\mathrm{Hz} \tag{24}$$

In analogy with (5), (22) can be transformed to the discrete-time augmented state equation (7) where $\bar{\mathbf{A}} \equiv e^{\mathbf{A}\Delta t}$, $\bar{\mathbf{B}} \equiv \mathbf{A}^{-1}(\bar{\mathbf{A}} - \mathbf{I}_{13})\mathbf{B}$ and $\bar{\mathbf{B}}_c \equiv \mathbf{A}^{-1}(\bar{\mathbf{A}} - \mathbf{I}_{13})\mathbf{B}_c$.

The absolute acceleration measurements at the 2nd and 4th DOF are modeled by

$$\mathbf{z}[k] = -\mathbf{L}_o\mathbf{M}^{-1}\mathbf{K}\mathbf{x}[k] - \mathbf{L}_o\mathbf{M}^{-1}\mathbf{C}\dot{\mathbf{x}}[k] + \mathbf{n}[k] \tag{25}$$

where $\mathbf{L}_o$ is an observation matrix which picks up the second and fourth row and $\mathbf{n}[k]$ is discrete white noise with zero mean and standard deviations 0.01g, which models the prediction error, i.e. measurement noise, the differentiator errors and modeling error.

The signal given to the AMD actuator is given by

$$u[k] = \mathbf{G}_1\mathbf{z}[k-1] + \mathbf{G}_2\mathbf{z}[k-2] \tag{26}$$

so the controller feeds back only the previous two time steps. Here, the gains $\mathbf{G}_p \in \mathbb{R}^{1\times2}, p = 1,2$ are design parameters. Substituting (25) and (26) into the discrete-time augmented state equation, one can obtain the augmented vector equation:

$$\mathbf{U}[k+1] = \bar{\mathbf{A}}_u\mathbf{U}[k] + \bar{\mathbf{B}}_u\bar{\mathbf{f}}[k] \tag{27}$$

where $\mathbf{U}[k] = [\mathbf{v}[k]^T, \mathbf{z}[k-1]^T, \mathbf{z}[k-2]^T]^T$; $\bar{\mathbf{f}} = [w[k], \mathbf{n}[k]^T]^T$; and $\bar{\mathbf{A}}_u$ and $\bar{\mathbf{B}}_u$ are given by

$$\bar{\mathbf{A}}_u = \begin{bmatrix} \multicolumn{3}{c}{\bar{\mathbf{A}}} & \bar{\mathbf{B}}_c\mathbf{G}_1 & \bar{\mathbf{B}}_c\mathbf{G}_2 \\ -\mathbf{L}_o\mathbf{M}^{-1}\mathbf{K} & -\mathbf{L}_o\mathbf{M}^{-1}\mathbf{C} & \mathbf{0}_{2\times3} & \mathbf{0}_{2\times2} & \mathbf{0}_{2\times2} \\ \multicolumn{3}{c}{\mathbf{0}_{2\times13}} & \mathbf{I}_2 & \mathbf{0}_{2\times2} \end{bmatrix}, \quad \bar{\mathbf{B}}_u = \begin{bmatrix} \bar{\mathbf{B}} & \mathbf{0}_{13\times2} \\ \mathbf{0}_{2\times1} & \mathbf{I}_2 \\ \mathbf{0}_{2\times1} & \mathbf{0}_{2\times2} \end{bmatrix} \tag{28}$$

Then, the covariance matrix $\boldsymbol{\Sigma}_u$ of the augmented vector is the solution of the following Lyapunov equation in discrete form: $\boldsymbol{\Sigma}_u = \bar{\mathbf{A}}_u\boldsymbol{\Sigma}_u\bar{\mathbf{A}}_u^T + \bar{\mathbf{B}}_u\boldsymbol{\Sigma}_{\bar{f}}\bar{\mathbf{B}}_u^T$, where $\boldsymbol{\Sigma}_{\bar{f}}$ is the covariance matrix of $\bar{\mathbf{f}}$. The threshold levels for the interstory drifts, actuator stroke and the actuator acceleration are 1.0cm, 20cm and 1.0g, respectively. The optimal gain coefficients are found to be $\mathbf{G}_1 = [0.271, 0.305]$ and $\mathbf{G}_2 = [-0.395, -0.350]$. Tables 3 and 4 show the performance quantities, including the interstory drifts, AMD actuator stroke (x_s) and control force (f_{cn}), for the cases of a random excitation sample and the 1940 El Centro earthquake record, respectively. In these tables, σ denotes the rms value of a quantity. It can be seen that the interstory drifts are significantly reduced in both cases when the AMD is installed.

Table 3: Performance quantities under random excitation sample.

Performance quantity	Threshold	Uncontrolled	Controlled
σ_{x_1} (m)	—	0.0072	0.0022
$\sigma_{x_1-x_2}$ (m)	—	0.0073	0.0022
$\sigma_{x_2-x_3}$ (m)	—	0.0061	0.0020
$\sigma_{x_3-x_4}$ (m)	—	0.0037	0.0015
$max\|x_1\|$ (m)	0.01	0.0222	0.0091
$max\|x_1 - x_2\|$ (m)	0.01	0.0217	0.0086
$max\|x_2 - x_3\|$ (m)	0.01	0.0212	0.0084
$max\|x_3 - x_4\|$ (m)	0.01	0.0133	0.0056
σ_{x_s} (m)	—	—	0.0075
$\sigma_{f_{cn}}$ (g)	—	—	0.1256
$max\|x_s\|$ (m)	0.20	—	0.0288
$max\|f_{cn}\|$ (g)	1.00	—	0.4990

Table 4: Performance quantities under 1940 El Centro earthquake record.

Performance quantity	Threshold	Uncontrolled	Controlled
σ_{x_1} (m)	—	0.0012	0.0005
$\sigma_{x_1-x_2}$ (m)	—	0.0013	0.0005
$\sigma_{x_2-x_3}$ (m)	—	0.0010	0.0005
$\sigma_{x_3-x_4}$ (m)	—	0.0006	0.0003
$max\|x_1\|$ (m)	0.01	0.0053	0.0037
$max\|x_1 - x_2\|$ (m)	0.01	0.0053	0.0037
$max\|x_2 - x_3\|$ (m)	0.01	0.0043	0.0032
$max\|x_3 - x_4\|$ (m)	0.01	0.0024	0.0020
σ_{x_s} (m)	—	—	0.0018
$\sigma_{f_{cn}}$ (g)	—	—	0.0236
$max\|x_s\|$ (m)	0.20	—	0.0125
$max\|f_{cn}\|$ (g)	1.00	—	0.1693

References

Au, S. K., C. Papadimitriou, and J. L. Beck (1999). Reliability of uncertain dynamical systems with multiple design points. *Structural Safety 21*, 113–133.

Beck, J. L. and L. S. Katafygiotis (1998). Updating models and their uncertainties. I: Bayesian statistical framework. *Journal of Engineering Mechanics 124(4)*, 455–461.

Dyke, S. J., B. Spencer, Jr, P. Quast, and M. K. Sain (1995). Role of control-structure interaction in protective system design. *Journal of Engineering Mechanics 121(2)*, 322–338.

Papadimitriou, C., J. L. Beck, and L. S. Katafygiotis (1997). Asymptotic expansions for reliability and moments of uncertain systems. *Journal of Engineering Mechanics 123(12)*, 1219–1229.

Papadimitriou, C., J. L. Beck, and L. S. Katafygiotis (2001). Updating robust reliability using structural test data. *Probabilistic Engineering Mechanics 16(2)*, 103–113.

Veneziano, D., M. Grigoriu, and C. A. Cornell (1977). Vector-process models for system reliability. *Journal of Engineering Mechanics 103(EM3)*, 441–460.

Yuen, K.-V., J. L. Beck, and L. S. Katafygiotis (2002). Probabilistic approach for modal identification using non-stationary noisy response measurements only. *Earthquake Engineering and Structural Dynamics 31(4)*, 1007–1023.

Advances in Building Technology, Volume I
M. Anson, J.M. Ko and E.S.S. Lam (Eds.)

BUILDING HAZARD MITIGATION WITH PIEZOELECTRIC FRICTION DAMPERS

G. D. Chen[1] and C. Q. Chen[1]

1. Department of Civil Engineering, University of Missouri-Rolla, Rolla, MO 65409-0030, USA

ABSTRACT

Semi-active piezoelectric friction dampers have been recently introduced to reduce the peak responses of buildings. They are regulated in real time with applied voltages according to a simple yet effective control algorithm that combines the viscous and nonlinear Reid damping mechanisms. This paper is aimed at further addressing some performance-related issues such as the stick and sliding features of a friction damper, optimum ratio of Reid and viscous damping in the control algorithm, performance comparison with Coulomb dampers, and optimal placement of friction dampers in a multi-story building. Numerical simulations of a single- and a 20- story building (statistical analysis under 10 earthquake ground motions) indicate that Coulomb dampers are effective when the external excitation on the building is known a *prioror*. Semi-active friction dampers, however, are effective in response reduction of a building subjected to excitations of various intensities. They have been applied to effectively mitigate the responses of buildings under both near-fault and far-field ground motions. The optimum ratio of control gain factors for the Reid and viscous damping is equal to $2/\pi$ times the excitation frequency. The application of piezoelectric actuators in civil engineering relies upon the distribution effect of multiple dampers. A sequential sub-optimal procedure for damper placement is developed for practical applications. Due to their adaptability to external disturbances, semi-active dampers can greatly enhance the multi-objective performance of buildings under multi-level excitations and will play an important role in performance-based engineering.

KEYWORDS:

Piezoelectric actuators, active friction dampers, semi-active friction dampers, Coulomb dampers, seismic effectiveness, performance evaluation, near-fault effect, performance-based engineering

INTRODUCTION

Friction dampers guarantee the dissipation of energy by friction and therefore do not cause instability of a structure being controlled. In this paper, piezoelectric actuators are used to modulate the clamping force of a fiction damper to make the damper adaptive to external disturbances. Such a device is referred to as a piezoelectric friction damper. Piezoelectric materials offer such unique features as effectiveness over wide frequency bands, high-speed actuation, low power consumption, simplicity, reliability, and compactness as demanded in civil engineering applications (Housner *et al.*, 1994). However, these materials work under a small strain level and therefore are limited in their loading capacity. To generate a sufficient control force with piezoelectric materials in civil engineering applications must rely on the distribution effect of many actuators.

Piezoelectric friction dampers have recently been used to mitigate the peak responses of elastic and inelastic buildings (Chen and Chen, 2000; 2002) under dynamic loading. A new control algorithm has been developed to command friction dampers. It has been demonstrated very effective in suppressing the harmonic responses of single-story structures and seismic responses of multi-story buildings. In this paper, some issues related to the control algorithm and optimal placement of multiple dampers in a building are addressed.

SEMI-ACTIVE PIEZOELECTRIC FRICTION DAMPERS

A prototype damper and its schematic representation are shown in Figure 1. The schematic consists of two U-shaped bodies, one sliding against the other. The outer body is assumed to be rigid. The clamping force, $N(t)$, acting on the sliding surfaces is controllable with four PZWT100 stack actuators. Each actuator consists of 24 discs with 0.02 in. thick each; they are connected mechanically in series and electrically in parallel. The clamping force on two friction surfaces is determined by

$$N(t) = N_{pre} + \frac{4EAd_{33}V(t)}{h} \tag{1}$$

in which N_{pre} is the pre-load on the stack actuators required for the generation of the passive friction force, E is the Young's Modulus of the PZWT material, A is the area of cross section of the stacks, h is the thickness of each layer, d_{33} is the piezoelectric strain coefficient and $V(t)$ is the applied voltage on the stack actuators. When $V(t)$ in Eqn.1 is equal to zero, the clamping force is constant $(= N_{pre})$ and it corresponds to a passive Coulomb friction damper. When N_{pre} is negligible, the clamping force is proportional to the applied voltage $V(t)$, which is fully adaptive to the feedback of the damper slippage and thus corresponds to an active friction damper. When both pre-load and voltage are applied, the damper is semi-active and requires less energy than an active damper to operate.

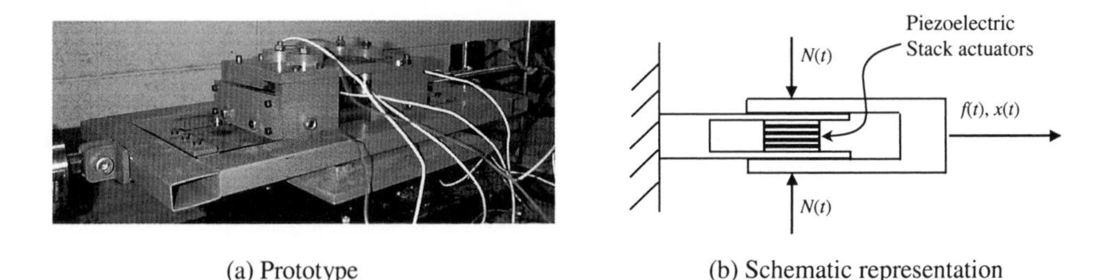

(a) Prototype (b) Schematic representation

Figure 1: A piezoelectric friction damper

Control Algorithm

For a damper installed in the ith story of a building, a semi-active control strategy is proposed to regulate the damper with an applied voltage such that the following friction force is generated:

$$f_i(t) = \begin{cases} 2\mu N_{pre} sgn[\dot{x}_i(t)], & when\ e|x_i(t)| + g|\dot{x}_i(t)| \le N_{pre} \\ 2\mu e|x_i(t)| sgn[\dot{x}_i(t)] + 2\mu g \dot{x}_i(t), & when\ e|x_i(t)| + g|\dot{x}_i(t)| > N_{pre} \end{cases} \quad (2)$$

in which e and g are positive gain factors, $|x_i(t)|$ and $|\dot{x}_i(t)|$ are respectively the absolute values of the drift and drift rate of the ith building story, μ is the coefficient of friction and $sgn[\ ]$ represents the sign of the argument in the bracket. Note that the factor of 2 is used to account for two friction surfaces. The first expression represents the passive damper mechanism when structural responses are relatively small. When structural responses become significant, the active damper mechanism represented by the second expression is activated. Eqn. 2 indicates that, for relatively large structural responses, the active component of the proposed semi-active control strategy essentially combines both the viscous and the non-linear Reid damping mechanisms, which has been proved to be more effective than individual damping mechanisms (Chen and Chen, 2000).

Optimum Gain Factor Ratio

The active component of the proposed semi-active control strategy involves two gain factors, e and g, representing the weighting effect on the Reid and viscous damping mechanisms. Since the Reid damping force reaches its maximum when the viscous damping force is zero, there must exist an optimum gain ratio, e/g, which corresponds to the minimal responses of a structure under the same maximum control force. Consider the single-story frame structure with mass, natural frequency, and damping ratio respectively equal to $m=2.92\ kN\text{-}sec^2/m$, $\omega_0=3.47\ Hz$, and $\xi=1.24\%$ (Soong, 1990). The structure is controlled with an active friction damper ($N_{pre}=0$). The proposed control algorithm is homogeneous to the first degree and is linearized, leading to an equivalent structure-damper system:

$$\begin{aligned} &m\ddot{x}(t) + m2\xi\omega_0\dot{x}(t) + m\omega_0^2 x(t) + C_e\dot{x}(t) = F_0 \sin(\omega t) \\ &C_e = \mu g(1 + 2e/\pi g\omega) \end{aligned} \quad (3)$$

in which C_e is the damping coefficient of the equivalent viscous damper for the active control force (Chen and Chen, 2000a). The active control force in the second expression of Eqn. 2 (i=1) can then be approximately represented by the stead state response from Eqn. 3 and its maximum value is

$$f_{1\max} = \mu g A\sqrt{\omega^2 + (e/g)^2} \quad (4)$$

where A is the displacement amplitude of the structure described by Eqn. 3. After solving for μg from Eqn. 4 and plugging it into C_e of Eqn. 3, the equivalent damping coefficient C_e can be expressed as a function of f_{1max}, A, and e/g, and A satisfies the following equation:

$$[(1-\beta^2)^2 + (2\xi\beta)^2](\frac{Am\omega_0^2}{F_0})^2 + 4\xi\beta\frac{f_{1\max}}{F_0}\frac{1+2e/\pi g\omega}{\sqrt{1+(e/g\omega)^2}}\frac{Am\omega_0^2}{F_0} + \frac{f_{1\max}^2}{F_0^2}\frac{(1+2e/\pi g\omega)^2}{1+(e/g\omega)^2} - 1 = 0 \quad (5)$$

in which β is the ratio of excitation and natural frequencies. When f_{1max} is constant, A can be solved from the above equation as a function of e/g as well as structural and excitation parameters. The displacement amplitude reaches its minimum value when e/g is approximately equal to $2\omega/\pi$. To obtain a real solution for A in Eqn. 5, the maximum control force must satisfy

$$f_{1\max}/F_0 \le \sqrt{(1-\beta^2)^2 + (2\xi\beta)^2}/|1-\beta^2|\sqrt{1+4/\pi^2} \quad (6)$$

Roles of Passive and Active Mechanisms in the Proposed Control Algorithm

The same single-story frame structure is studied. When Coulomb damping is considered, the displacement and acceleration ratio of the controlled and uncontrolled structure are respectively shown in Figures 2(a, b). Their corresponding ratios when active damping is considered are presented in Figures 3(a, b), respectively. It can be observed from Figure 2 that, as the friction force increases or the excitation decreases, the displacement of the structure controlled with a Coulomb damper always decreases. However, its associated acceleration increases when β is smaller than *1.0* since the stick nature of the friction damper results in a sudden movement at the beginning of sliding stages. On the other hand, Figure 3 indicates that the reduction in both displacement and acceleration is nearly independent of the intensity of excitation due to the adaptability of an active damper.

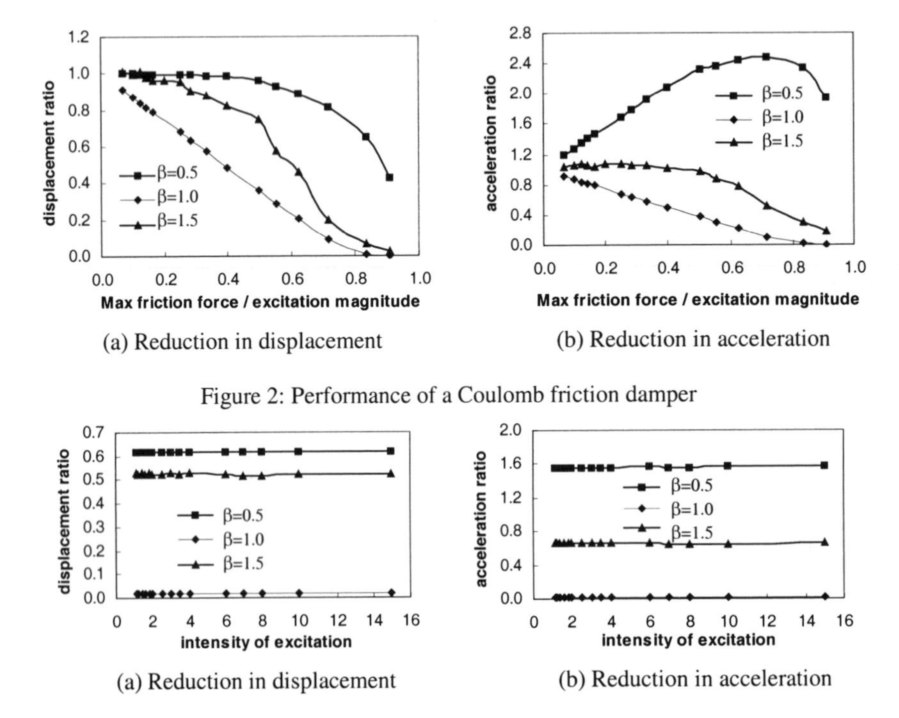

(a) Reduction in displacement (b) Reduction in acceleration

Figure 2: Performance of a Coulomb friction damper

(a) Reduction in displacement (b) Reduction in acceleration

Figure 3: Performance of an active damper with the proposed algorithm

In building designs, both story drift and acceleration are important measures of building performance. Therefore, it would be beneficial to implement a semi-active control strategy for performance-based designs of buildings that are subjected to dynamic loads with uncertainties in magnitude and phase.

OPTIMAL PLACEMENT OF DAMPERS IN MULTI-STORY BUILDINGS

The 20-story steel frame structure and 10 earthquake ground motions studied by Ohtori *et al.* (2000) are used to demonstrate the distribution and seismic effectiveness of piezoelectric friction dampers. For practical applications, a heuristic approach is taken to develop a simple and effective procedure for damper placement. The optimum gain factor ratio, *e/g*, is determined with the average dominant frequency ($\omega_u \cong 4.73$ *rad/sec*) of the uncontrolled displacement responses under the 10 earthquake inputs. The gain factor g is taken to be 1.9×10^5 *N-sec/m* to ensure that the dampers can generate the maximum friction force nearly equal to their capacity (*93.1 kN*) without saturation under the earthquake excitations. The pre-load N_{pre} is selected to provide a friction force equal to *10%* of the damper capacity. This algorithm is implemented in MATLAB (Chen and Chen, 2002).

The following sequential procedure is used for the sub-optimal placement of dampers in a building:

Step 1. Select an optimization index such as peak story drift or peak floor acceleration of the building.

Step 2. Place one or several dampers on the structure toward the reduction of the optimization index. Specifically, place one damper on each permissible location of the building and evaluate the performance index. If the index is reduced with the installation of the damper at various locations, the damper is placed at the location corresponding to the maximum reduction. If adding one damper at a time cannot further reduce the performance index, two or more dampers should be placed on the structure simultaneously.

Step 3. Repeat Step 2 until all the dampers have been placed on the structure.

The peak story drift and the peak floor acceleration are used as potential optimization indices to facilitate the placement of dampers. To determine which index is more effective in reducing the overall building responses, both peak story drift and peak acceleration of the building under six earthquake ground motions are presented in Figures 4(a, b) and 5(a, b) as the number of dampers on the structure increases. It can be observed from Figure 4 that the use of peak floor acceleration as an optimization index results in comparable reduction in both acceleration and drift of the structure except for the Kobe earthquake. On the other hand, if the peak drift ratio is used as an index as indicated in Figure 5, it is very difficult to reduce the peak floor acceleration simultaneously. Sometimes it even increases the acceleration as illustrated in Figure 5(b). Therefore, it is recommended that the peak floor acceleration be used as the optimization index for optimal damper placement.

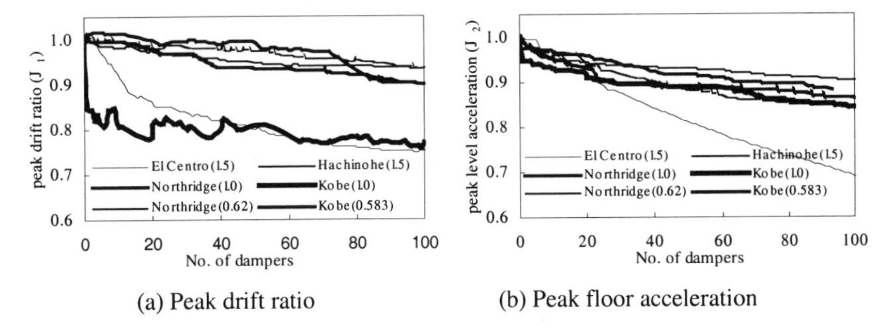

(a) Peak drift ratio (b) Peak floor acceleration

Figure 4: Effectiveness of various numbers of dampers with acceleration-based damper profiles

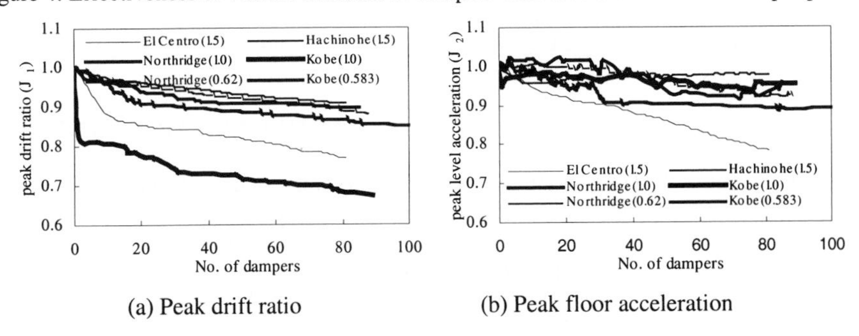

(a) Peak drift ratio (b) Peak floor acceleration

Figure 5: Effectiveness of various numbers of dampers with drift-based damper profiles

Figures 4 and 5 indicate the comparable reductions in both story drift and floor acceleration under the Northridge earthquake of different modification factors of 62% and 100%. However, under the Kobe earthquake, dampers can suppress significantly more story drift under the high excitation (100%) than that under the low excitation (58.3%). This result implies the occurrence of substantial plastic deformation at several locations under the 100%Kobe earthquake. Considering various earthquakes of the same intensity, e.g. 150%El Centro vs. 62%Northridge earthquake, it is observed from Figures 4 and 5 that both story drift and floor acceleration can be reduced more effectively under the El Centro earthquake than those under the Northridge earthquake. Those under the 150%Hachinohe are mitigated slightly less effectively than under the 58.3%Kobe earthquake. In general, a structure subjected to far-field ground motions, such as the El Centro earthquake, can be more easily controlled with dampers than that under near-fault ground motions such as the Northridge and Kobe earthquakes. However, it is not difficult to control the excessive inelastic story drift of a structure under strong near-fault earthquakes such as the 100%Kobe.

To select the optimal distribution of the 80 dampers over the building height for overall performance under various earthquakes, six damper profiles were determined using the proposed sequential procedure with the peak floor acceleration as the optimization index. Each profile corresponds to one of the six earthquake inputs used in Figures 4 and 5. These profiles were then employed to calculate the peak accelerations at all floors of the building subjected to the 10 earthquake excitations (Ohtori *et al.*, 2000). The maximum value of the peak accelerations at all floors was normalized with that of the uncontrolled structure under each of the 10 earthquake excitations. The statistical characteristics of the 10 samples for peak acceleration ratio (r_a) such as the maximum value [$max(r_a)$], average ($\bar{r}_a$), and standard deviation [$s(r_a)$] are shown in Table 1. It can be observed from the table that Profile8 has the lowest maximum and average accelerations. The standard deviation associated with this profile is also very small, indicating the consistent damper performance under the 10 earthquake ground motions. Thus, Profile8, determined under the Northridge earthquake, achieves the best overall performance in reducing the peak floor acceleration of the structure. The 80 dampers in this case are distributed along the building height from top to bottom as: [26, 1, 3, 0, 7, 0, 0, 1, 7, 0, 7, 5, 4, 2, 1, 0, 6, 1, 0, 9].

TABLE 1
PEAK ACCELERATION RATIOS WITH 10% CAPACITY FROM THE PASSIVE COMPONENT

Damper Profile	$max(r_a)$	$\bar{r}_a$	$s(r_a)$
Profile3 (150%El Centro EQ)	1.111	0.882	0.122
Profile6 (150%Hachinohe EQ)	0.985	0.891	0.059
Profile8a (62%Northridge EQ)	1.085	0.882	0.110
Profile8 (100%Northridge EQ)	**0.989**	**0.878**	**0.081**
Profile10a (58.3%Kobe EQ)	1.004	0.885	0.085
Profile10 (100%Kobe EQ)	1.067	0.882	0.104

PERFORMANCE OF PASSIVE, SEMI-ACTIVE, AND ACTIVE FRICTION DAMPERS

To understand the pre-load effect on the damper performance, the statistical characteristics similar to those in Table 1 under the Northridge earthquake are given in Table 2 for various N_{pre}. It can be clearly seen from the table that all statistical characteristics of the peak floor acceleration increases as the pre-load increases. This means less effectiveness of dampers in reducing the structural acceleration when the percentage of the passive friction force increases in the proposed semi-active strategy. Similarly, more pre-load on the passive component of the semi-active damper results in the increasing maximum and average peak drift ratio. On the other hand, Table 2 indicates that the required maximum active control force (normalized by the total building weight, 1.09×10^5 *kN*) significantly drops as the pre-load increases. In conclusion, the introduction of more passive friction force in the semi-active friction dampers significantly degrades the performance of the dampers but reduces the active control force. To maintain the effectiveness of dampers with reasonable power demand, a pre-load corresponding to 10% of the damper capacity was used in the previous section.

TABLE 2
PERFORMANCE COMPARISON BETWEEN ACTIVE AND SEMI-ACTIVE DAMPERS

Damper Profile	Response Quantity	$max(r)$	$\bar{r}$	$s(r)$
Profile8b ($N_{pre}\Rightarrow$0% capacity) (active)	Story drift ratio	0.957	0.877	0.718
	Acceleration ratio	0.944	0.872	0.044
	Control force ($\times10^{-4}$)	8.186	4.228	2.099
Profile8 ($N_{pre}\Rightarrow$10% capacity) (semi-active)	**Story drift ratio**	**0.962**	**0.875**	**0.078**
	Acceleration ratio	**0.980**	**0.878**	**0.081**
	Control force ($\times10^{-4}$)	**7.470**	**3.362**	**2.174**
Profile8d ($N_{pre}\Rightarrow$30% capacity) (semi-active)	Story drift ratio	0.976	0.860	0.095
	Acceleration ratio	1.428	0.976	0.191
	Control force ($\times10^{-4}$)	5.339	1.799	1.827
Profile8f ($N_{pre}\Rightarrow$50% capacity) (semi-active)	Story drift ratio	0.982	0.854	0.100
	Acceleration ratio	2.255	1.211	0.481
	Control force ($\times10^{-4}$)	4.164	0.847	1.422

To compare the performance between Coulomb dampers and semi-active friction dampers, Table 3 lists the statistical characteristics of the peak story drift and peak floor acceleration ratios associated with the optimal damper placement, Profile8, determined under the Northridge earthquake. For passive dampers, different levels of friction forces have been used in numerical simulations. It is seen from Table 3 that the maximum values of both story-drift and floor acceleration ratios can be reduced more by semi-active dampers. As the friction force of Coulomb dampers increases, both the maximum and average drifts decreases and their standard deviation increases due to stick effects. All the statistical quantities of the floor acceleration substantially increase with the friction force of Coulomb dampers, indicating degrading performance. These results indicate that semi-active dampers outperform Coulomb dampers for the multi-level earthquake design of buildings.

TABLE 3
PERFORMANCE COMPARISON BETWEEN SEMI-ACTIVE AND PASSIVE DAMPERS

Control Strategy	Response Quantity	$max(r)$	$\bar{r}$	$s(r)$
Semi-active ($N_{pre}\Rightarrow$10% capacity)	**Story drift ratio**	**0.962**	**0.875**	**0.078**
	Acceleration ratio	**0.980**	**0.878**	**0.081**
Passive ($N_{pre}\Rightarrow$10% capacity)	Story drift ratio	1.012	0.932	0.078
	Acceleration ratio	1.036	0.938	0.074
Passive ($N_{pre}\Rightarrow$50% capacity)	Story drift ratio	0.986	0.833	0.101
	Acceleration ratio	2.249	1.173	0.431
Passive ($N_{pre}\Rightarrow$100% capacity)	Story drift ratio	0.964	0.778	0.109
	Acceleration ratio	4.224	1.850	0.994

The optimum ratio of control gain factors, *e/g*, was derived from a single-story building subjected to harmonic loading. To see how sensitive the dampers' performance is to the change of the gain factor ratio, Table 4 shows the statistical results of peak story-drift and peak acceleration ratios when ±10%

uncertainties in the determination of the gain factor ratio is introduced. Clearly, it is observed that the changes in maximum and average story-drift ratios or acceleration ratios are all within 1%. The optimum ratio of two gain factors derived in this study is thus valid for the practical design of dampers.

TABLE 4
PERFORMANCE SENSITIVITY TO OPTIMUM GAIN FACTOR RATIO

Excitation Frequency	Response Quantity	*max*(r)	$\bar{r}$	$s(r)$
$0.9\omega_u$	Story drift ratio	0.965	0.880	0.076
	Acceleration ratio	0.978	0.880	0.081
$\mathbf{1.0\omega_u}$	**Story drift ratio**	**0.962**	**0.875**	**0.078**
	Acceleration ratio	**0.980**	**0.878**	**0.081**
$1.1\omega_u$	Story drift ratio	0.959	0.870	0.082
	Acceleration ratio	0.980	0.874	0.079

CONCLUDING REMARKS

Based on the extensive numerical simulations in this study, it is concluded that Coulomb dampers can be designed to effectively reduce the dynamic responses of a building when the intensity of an external disturbance exerted on the building is known during the design period. Semi-active friction dampers, however, are effective in response reduction of a building subjected to excitations of various intensities. The proposed algorithm is effective in mitigating the inelastic responses of buildings under near-fault ground motions. The optimum ratio of control gain factors for the Reid and viscous damping is equal to $2/\pi$ times the excitation frequency. The application of piezoelectric actuators in civil engineering relies upon the distribution effect of multiple dampers. The proposed sequential procedure for damper placement is very efficient.

ACKNOWLEDGEMENTS

This study was sponsored by the U.S. National Science Foundation under Award No. CMS-9733123 with Drs. S. C. Liu and P. Chang as Program Directors. The results, opinions and conclusions expressed in this paper are solely those of the authors and do not necessarily represent those of the sponsor.

REFERENCES

Chen, G. D. and Chen, C. C. (2000). Behavior of Piezoelectric Friction Dampers under Dynamic Loading. *Proc. SPIE Symposium on Smart Structures and Materials: Smart Systems for Bridges, Structures, and Highways*, Newport Beach, CA, Vol. **3988**, 54-63.

Chen, G. D. and Chen, C. C. (2002). Semi-Active Control of the 20-Story Benchmark Building with Piezoelectric Friction Dampers. *Journal of Engineering Mechanics*, ASCE (accepted for publication).

Housner, G. W., Soong T. T. and Masri, S. F. (1994). Second Generation of Active Structural Control in Civil Engineering. *Proc.1st World Conference on Structural Control*, Los Angeles, CA, Panel: 3-18.

Ohtori, Y., Christenson, R. E. and Spencer, Jr., B. F. (2000), "Benchmark control problems for seismically excited nonlinear buildings", *http://www.nd.edu/~quake/nlbench.html.*

Soong T.T. (1990). *Active Structural Control: Theory & Practice*, Longman & Scientific Technical, New York.

Advances in Building Technology, Volume I
M. Anson, J.M. Ko and E.S.S. Lam (Eds.)

A RESEARCH AND DEVELOPMENT OF SMART BUILDING STRUCTURES BY MAGNETO-RHEOLOGICAL DAMPER

H. Fujitani [1], Y. Shiozaki [2], T. Hiwatashi [3], K. Hata [4], T. Tomura [4], H. Sodeyama [5] and S. Soda [6]

[1] Department of Structural Engineering, Building Research Institute
Tsukuba, Ibaraki 305-0802, Japan
[2] Nishimatsu Construction Co. Ltd. (Visiting Researcher of Building Research Institute)
Tsukuba, Ibaraki 305-0802, Japan
[3] TOA Technical Research Institute (Visiting Researcher of Building Research Institute)
Tsukuba, Ibaraki 305-0802, Japan
[4] Central Research & Development Division, Bando Chemical Industries, Ltd.
Ashihara-dori, Hyogo-ku, Kobe 652-0882, Japan
[5] Manager, Test & Research Department, Sanwa Tekki Corporation
Naka-okamoto, Kawachi-machi, Tochigi 329-1192, Japan
[6] Department of Architecture, School of Science & Engineering, Waseda University
Okubo, Shinjuku-ku, Tokyo 169-8555, Japan

ABSTRACT

Magneto-rheological damper (MR damper) has been expected to control the response of building structures in recent years, because of its large force capacity and variable force characteristics. Authors developed some MR dampers and conducted shaking table tests for the improvement the performance of building structures subjected by earthquake. A three-story large-scaled test frame supposing both base-isolated and normal structure was developed. As for base-isolated building, 40kN Magneto-Rheological damper was installed at base-isolation. The effectiveness of MR damper and proposed control algorithm was verified. In the other hand, a new Magneto-Rheological fluid is also developed in order to apply to the MR damper. The new MR fluid had a good property concerning the settlement of particles in dampers. The settlement was one of serious problems to be solved in handling of MR fluid in case of controlling the earthquake response of building structures. Fourth author developed a new MR fluid, which keeps the particles in the fluid adequately enough for usual use of MR damper

KEYWORDS

MR Fluid, MR damper, Test frame, Shaking table test, Semi active control, Base-isolation, Earthquake response control

INTRODUCTION

The Building Research Institute (BRI) of Japan and the U.S. National Science Foundation (NSF) initiated the U.S.-Japan Cooperative Research Program on Auto-adaptive Media (Smart Structural Systems) in 1998 (Otani et al (2000)), under the aegis of the U.S.-Japan Panel on Wind and Seismic Effects of the U.S.-Japan Cooperative Program in Natural Resources. At the Joint Technical Coordinating Committee (JTCC) meeting, research items and plans were discussed in detail for three research thrusts: (1) structural systems, (2) sensing and monitoring technology, and (3) effecter technology.

BRI planned a series of large-scale tests to verify some smart systems developed in this project. The effectiveness of "Semi-active control by MR dampers", "Damage detection system" and "Rocking energy dissipation system" was confirmed. This paper outlines the development of MR fluid and MR damper, the characteristics of the large-scale test frame and test results of response control of base-isolated building by MR damper.

LARGE-SCALE TEST OF RESPONSE CONTROL BY MR DAMPER

The passive control has a limitation of damping effect to a certain range with frequency and input level in general. Semi-active control reduces both response displacements and accelerations. MR damper generates damping force, which does not depend on the piston speed so much (Fujitani et al (2000)). The target of this subject is to improve the safety and functionality and habitability by controlling the response displacements and accelerations by using MR dampers. For this purpose, a MR damper and a control algorithm have been developed and their validity is discussed by an analytical study and shaking table tests.

The basic test frame used for the large-scale test is a 3-story steel frame (Figure 1). It is 3m (1 span) x 4m (2m x 2 span) in plan and 6m high, weighs a total of 20.23 ton (including the weight of the 1st floor), and has a natural period of about 0.546 sec. (TABLE 1). The shaking table of the National Research Institute for Earth Science and Disaster Prevention (NIED) was used. The shaking table has a maximum displacement of 22 cm, a maximum velocity of 75 cm/s, and a maximum weight of 500 ton.

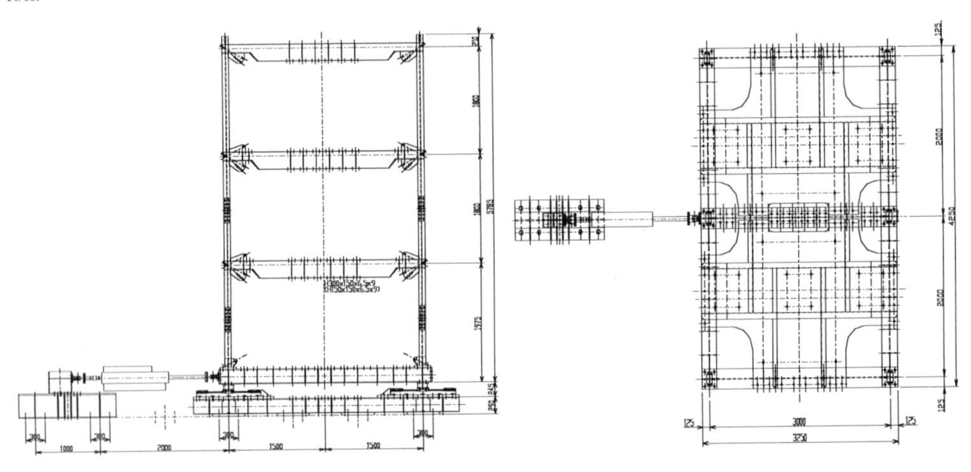

(a) Central frame of span direction (b) First floor of base isolation system

Figure 1: Test frame for response control by MR dampers

TABLE 1
DATA OF TEST FRAME

(a) Masses (ton),(b) Stiffness and Elastic displacement of the frame,(c) Characteristics as base-isolation

Roof	4.67
3rd Floor	4.78
2nd Floor	4.78
1st Floor	6.00
Total	20.23

	Stiffness(kN/cm)	Elastic Displacement(cm)
3F	27.6	1.74
2F	28.4	3.00
1F	35.4	2.27

Stiffness(kN/cm)	1.01
Friction Force(kN)	0.692
Damping Coefficient(kNs/cm)	0.0333
Damping Ratio	0.037
Natural Period(sec)	2.84

DEVELOPMENT OF MR FLUID

Trial product of MR fluid "#230", which was made on an experimental basis by the fourth and fifth authors of this paper, was applied for MR damper. The "#230" is oil–based fluid, and has a good stability. Properties of "#230" are shown in TABLE 2.

TABLE 2
PROPERTIES OF MR FLUID "#230"

Properties	#230
Base fluid	Oil
Density	$3.3*10^3$ kg/m^3
Stability *)	Approximately 98%(by volume) after $2*10^4$ min.

*):Stability(%)=V(t) / V(0) *100
V(t):sedimental volume after t min
V(0):original sedimental volume

DEVELOPMENT OF MR DAMPER

Design of MR damper

The sixth author of this paper developed a MR damper. The design of the bypass-flow-type MR damper for base isolation used in the experiment is shown in Figure 2 and TABLE 3. It is composed of three parts: bypass-flow and pressure chambers and a reservoir. In bypass-type hydraulic dampers

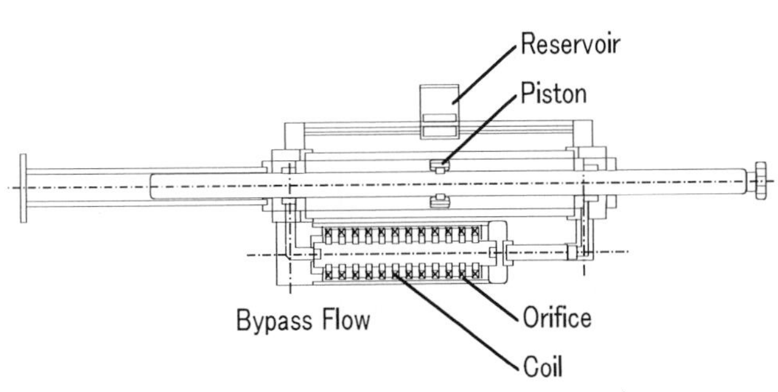

Figure 2: Structure of MR damper

TABLE 3
DESIGN SPECIFICATION

Max.Force		40kN
Stroke		±295mm
Cylinder Bore		φ90mm
Orifice	Section	Outer diameter φ48mm
		Gap 3.0mm
	Length	420mm
Electromagnet	Inductance	37.4mH × 3
	Resistance	12Ω × 3
	Max.Current	3A
MR fluid		Bando:#230

such as this, the cylinder is divided into two pressure chambers by a piston with rubber O-rings. The bypass flow portion is a passage for MR fluids connecting two pressure chambers. The bypass flow portion has an orifice for effectively magnetizing the fluid. The uniform magnetic field is applied perpendicularly to the MR fluid flow at the annular orifice. The thermal expansion due to the temperature rise of the MR fluid is absorbed by the reservoir.

Mechanical characteristic of MR damper by cyclic loading

Cyclic loading tests were conducted to clarify the performance of the MR damper. Figure 3 shows the force-displacement relationship. The electric current to the electromagnet was set at the constant value at 0A～3A in the 0.3A interval. The hysteretic loop shows like rigid-plastic characteristics caused by friction force of MR fluid with magnetic field. The force increases almost proportionally for the increase in the electric current. Figure 4 shows the force-velocity relationship. The velocity is the maximum velocity of piston in case of sinusoidal loading in this figure. This figure also shows that the force increases almost proportionally for the increase in the electric current and the force does not depend on the piston velocity so much.

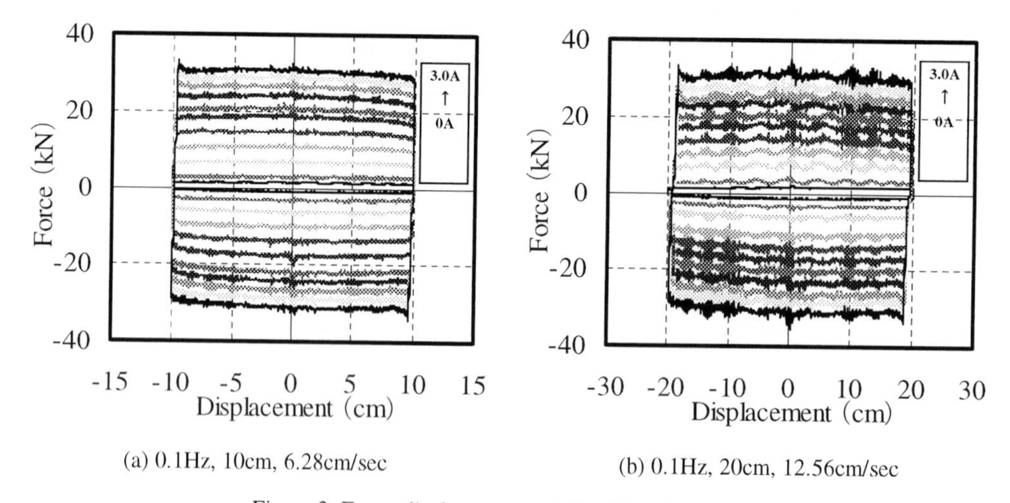

(a) 0.1Hz, 10cm, 6.28cm/sec

(b) 0.1Hz, 20cm, 12.56cm/sec

Figure 3: Force-displacement relationship of MR damper

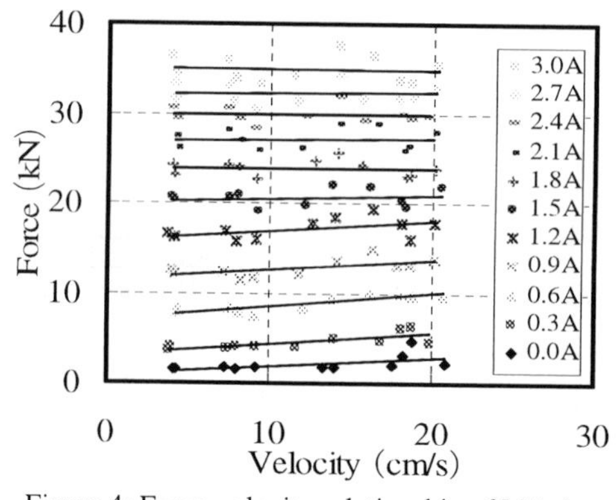

Figure 4: Force-velocity relationship of MR damper

SHAKING TABLE TEST

Testing Facilities

The base-isolation system is constituted of six roller bearings with four laminated rubber bearings (Photo 1). The laminated rubber bearing gives the restoring force, and the roller bearing supports the vertical load. The input wave for the shaking table test were sweep sinusoidal wave, white noise wave, and five earthquake waves of El Centro 1940 NS, Hachinohe 1968 NS, JMA Kobe 1995 NS and Taft 1952 EW standardized at a maximum velocity of 50cm/s. Figure 5 shows the performance of the shaking table.

Photo 1: Test frame

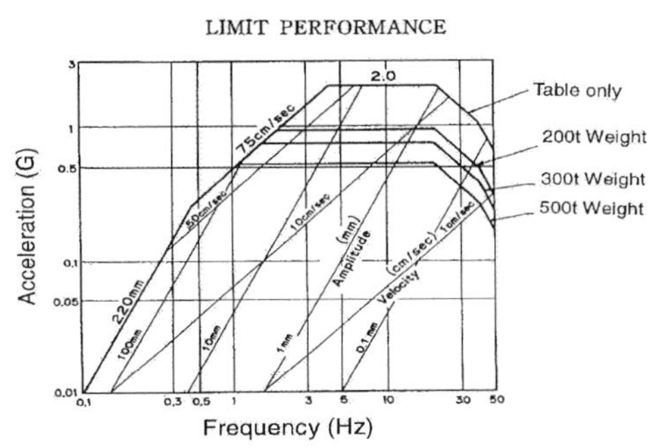

Figure 5: Performance of the shaking table

Control System

The measurement and control system in the shaking table test is shown in Figure 6. Accelerometers, transducers for measuring displacement and strain gauges were installed at each part of the test frame, and the data were recorded. There were 256 recording channels, and the data sampling was 2kHz. The data were recorded by an EWS for the measurement, which served as D/A converter of input waveform to the shaking table. The control signal incorporated the signal from the sensor in the amplifier, and it took in a voltage signal output from the amplifier to the A/D converter of the control PC. The control signal of electric current supplied to the MR damper was output from the D/A converter of the control PC to the high-speed DC power supplies.

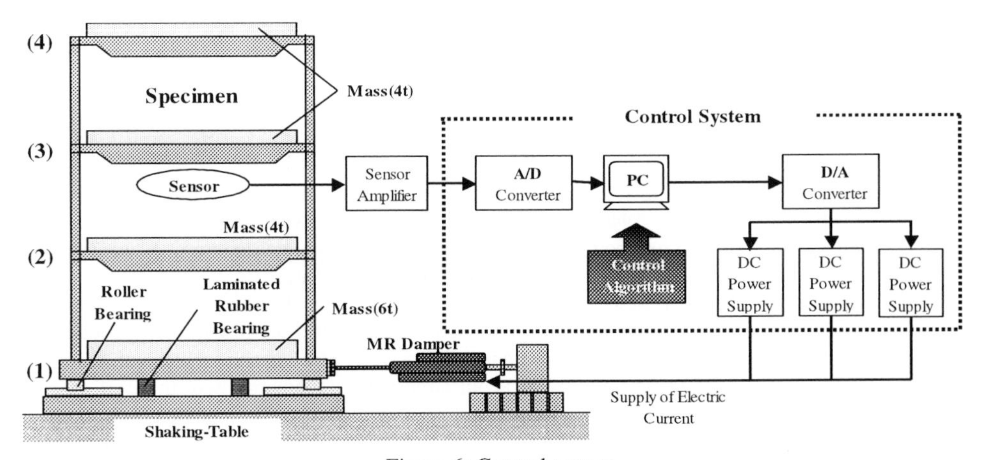

Figure 6: Control system

Control Algorithm

Although response displacements are reduced with passive dampers, response accelerations may become larger. Thus, the following two control objectives were considered.

(1)To reduce response displacement by semi-active control as much as by a passive damper.

(2)To reduce absolute acceleration response by semi-active control to less than that achieved by a passive damper.

The second author proposed a very simple semi-active control algorithm formulated in Eqn. 1 and shown in Figure 7, which aims the above two control objectives (Shiozaki et al. (2002)), by considering the base-isolated structure to be a one-degree-of-freedom system. In Eqn. 1, F is force of MR damper, k is the stiffness of base-isolation, u is relative displacement of base-isolation, ω is natural circular frequency, λ_{opt} is optimized constant.

$$F = \begin{cases} \mathrm{sgn}(\dot{u})k\left(\sqrt{\left(\dfrac{\lambda_{opt}\dot{u}}{\omega}\right)^2 + u^2} - |u|\right) & (u\dot{u} \geq 0) \\ k\dfrac{\lambda_{opt}\dot{u}}{\omega} & (u\dot{u} < 0) \end{cases} \qquad (1)$$

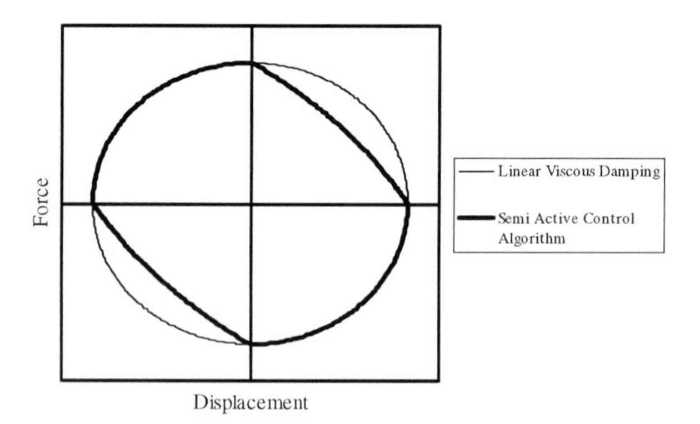

Figure 7: Concept of semi-active control

TEST RESULTS

Figures 8.1-8.3 show the maximum response accelerations of each story of test frame subjected to earthquake excitation of El Centro 1940 NS 50cm/sec, with constant electric current applied to MR damper. Applying high ampere of electric current reduced the maximum response displacements; in other hand, the maximum response accelerations became larger.

Figures 9.1-9.3 show the maximum response accelerations of each story of test frame controlled by proposed semi-active algorithm, comparing with those of passive control as shown in Figure 7. The maximum values of accelerations are reduced at about largest 20% compared with those of passive control. The RMS values of accelerations are reduced at about largest 30% compared with those of passive control. The efficacy of the proposed semi-active control is sufficiently demonstrated.

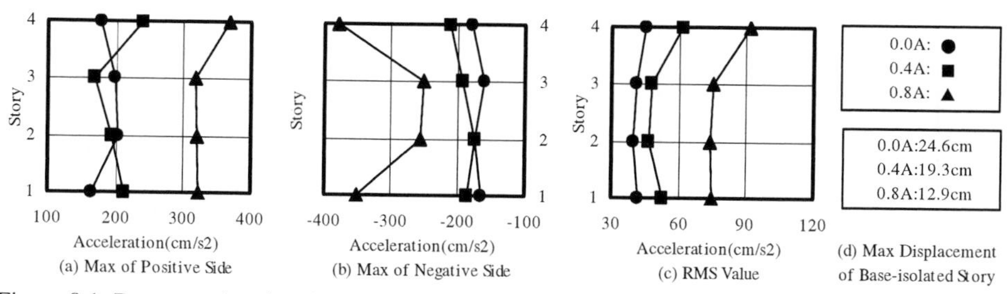

Figure 8.1: Response Acceleration of each story in comparison among three constant electric current values of 0.0A, 0.4A, 0.8A (El Centro 1940 NS 50cm/sec)

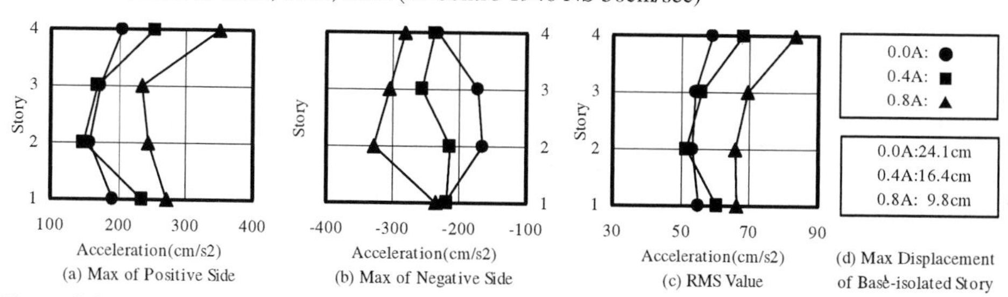

Figure 8.2: Response Acceleration of each story in comparison among three constant electric current values of 0.0A, 0.4A, 0.8A (Hachinohe 1968 NS 50cm/sec)

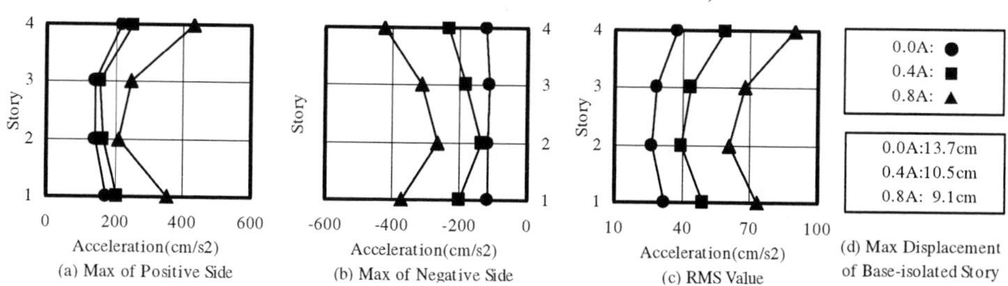

Figure 8.3: Response Acceleration of each story in comparison among three constant electric current values of 0.0A, 0.4A, 0.8A (JMA Kobe 1995 NS 50cm/sec)

CONCLUSIONS

The MR fluid showed the good results in the point of view that the force-displacement relationship of the MR damper is like rigid-plastic characteristics caused by friction force. The basic viscosity is low enough so that the MR damper generates small force without electric current. Authors have to evaluate the performance of the MR fluid to resist against the settlement of articles.

MR damper developed for this large-scale test showed the good results, too. The force-displacement relationship shows that the MR damper responds quickly to the inversion of the movement of the piston.

The maximum values of response accelerations are reduced by proposed semi-active control at about largest 20% compared with those of passive control. The RMS values are reduced at about largest 30% compared with those of passive control. The effectiveness of the semi-active control for the passive control was clarified.

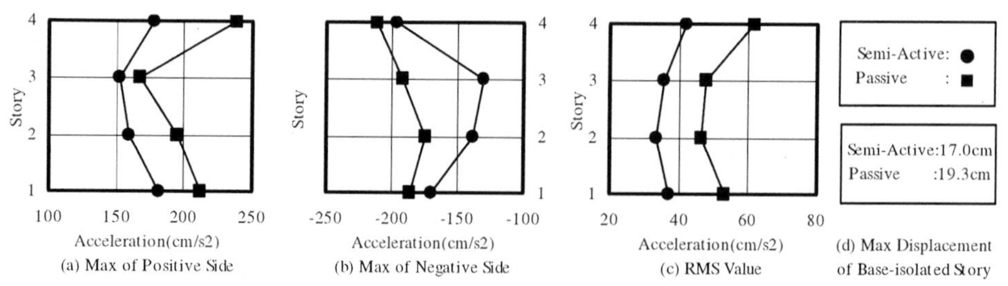

Figure 9.1: Response Acceleration of each story in comparison between proposed semi-active control and passive control (El Centro 1940 NS 50cm/sec)

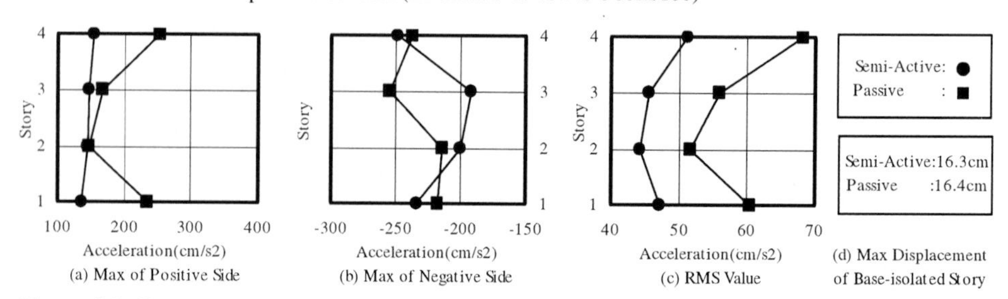

Figure 9.2: Response Acceleration of each story in comparison between proposed semi-active control and passive control (Hachinohe 1968 NS 50cm/sec)

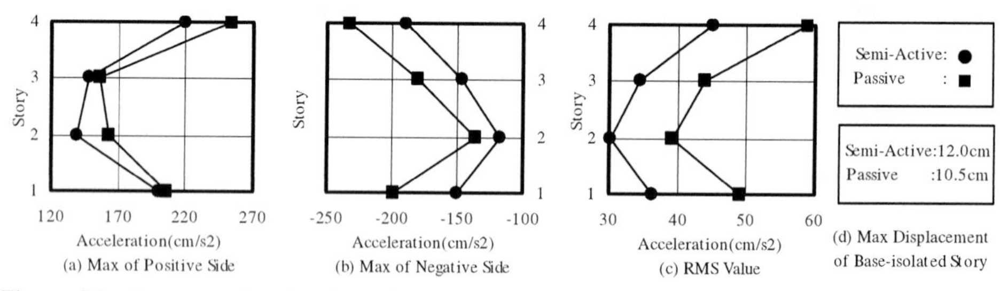

Figure 9.3: Response Acceleration of each story in comparison between proposed semi-active control and passive control (JMA Kobe 1995 NS 50cm/sec)

Acknowledgement

This work has been carried out under the US-Japan cooperative structural research project on Smart Structural Systems. The authors wish to thank the members of ER/MR working group of the project for their fruitful discussions and thank Mr. Chikahiro Minowa of NIED. This research was partly supported by 2001 Grant-in-Aid for Scientific Research (B).

References

Fujitani H., et al. (2000). Dynamic Performance Evaluation of Magneto-Rheological Damper, Proceedings of Advances in Structural Dynamics 2000, Vol.1, pp.319-326.

Otani S., et al. (2000). Development of Smart Systems for Building Structures (Invited paper), Proc.of SPIE's 7th. International Symposium on Smart Structures and Materials, Paper No. 3988-01

Shiozaki Y. et al. (2002). Study of response control by MR Damper for base-isolated building, Proc. of 3WCSC.

Advances in Building Technology, Volume I
M. Anson, J.M. Ko and E.S.S. Lam (Eds.)

STRUCTURAL HEALTH MONITORING SYSTEM USING SUPPORT VECTOR MACHINE

Hiromi Hagiwara and Akira Mita

Department of System Design Engineering, Keio University
3-14-1 Hiyoshi, Kohoku-ku, Yokohama 223-8522, Japan

ABSTRACT

A structural health monitoring system for a building structure utilizing Support Vector Machine (SVM) is proposed. The SVM is a new machine learning technique for pattern recognition. In our proposed system, modal frequencies of a structure are used for pattern recognition. As our system does not require modal shapes, typically only two vibration sensors detecting an input signal and an output signal for a structural system are enough to define damage. Changes in moral frequencies normalized by original modal frequencies before suffering any damage are used as feature vectors for feeding into the SVMs. The proposed system is capable of identifying the damaged locations even when multiple stories suffered damages.

KEYWORDS

Damage assessment, support vector machine, health monitoring, machine learning, modal frequency

1. INTRODUCTION

Many methods and approaches for health monitoring of a structure have been proposed targeting at reducing the maintenance cost or improving the structural performance. In those methods or approaches, one of the critical issues is the number of sensors available for monitoring. Though it is theoretically possible to obtain the detailed health information of a structure using unlimited number of sensors, it is not the case for most practical situations. The possible financial allocation for a structural health monitoring system is not so large and should be lower than the expected merit by introducing the system. For example, there is little rationale at this point to install a health monitoring system that measures acceleration of all floors.

The purpose of this study is to propose a new structural health monitoring system that can obtain the detailed damage information from the minimum number of sensors by utilizing the SVM. The feature vectors fed to the SVMs are formed based on the physical relationship between damages and modal frequency changes. The relationship is established by a sensitivity analysis.

2. SUPPORT VECTOR MACHINE

The Support Vector Machine (SVM) is a mechanical learning system introduced by Vapnik and his co-workers. The SVM uses a hypothesis space of linear functions in a high dimensional feature space[1), 2)]. The simplest SVM model is the so-called Linear SVM (LSVM). It works only for the case where data are linearly separable in the original feature space. Hence, applicable problems in the real-world are limited. In the early 1990s, nonlinear classification in the same procedure as LSVM became possible by introducing a nonlinear function called a kernel function, without being conscious of the mapped high-dimensional space. The machine extended to nonlinear feature spaces is called Nonlinear SVM (NSVM).

In what follows, we assume a training sample S consisting of vectors $\boldsymbol{x}_i \in R^n$ with $i = 1,\ldots,N$.

$$S = ((\mathbf{x}_1, y_1),\ldots,(\mathbf{x}_N, y_N)) \tag{1}$$

Each vector $\boldsymbol{x}_i$ belongs to either of two classes and thus is given a label $y_i \in \{-1,1\}$. The pair of (w,b) defines a hyper-plane of equation

$$(\mathbf{w} \cdot \mathbf{x}) + b = 0 \tag{2}$$

This hyperplane is named separating hyperplane (see Fig.1 (a)).

We now need to establish the optimal separating hyperplane (OSH) that divides S leaving all the points of the same class on the same side while maximizing the margin which is the distance of the closest point of S (see Fig.1 (b)). This closest vector $\mathbf{x}_i$ is called the support vector.

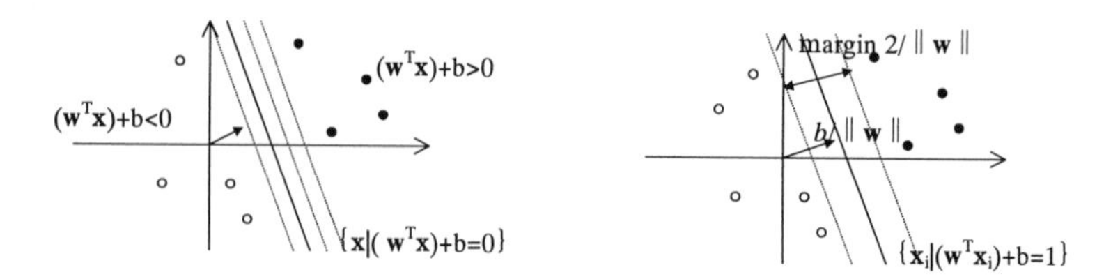

Figure 1: (a) separating hyperplane (b) OSH (the dashed lines in (b) define the margin)

For the reasons mentioned above, the OSH (w,b) can be determined by solving an optimization problem defined by

$$\begin{aligned} &\text{minimize} \quad \text{margin } d(\mathbf{w}) = \frac{1}{2}(\mathbf{w} \cdot \mathbf{w}), \\ &\text{subject to} \quad y_i((\mathbf{w} \cdot \mathbf{x}_i) + b) \geq 1, \quad i = 1,2,\ldots,N \end{aligned} \tag{3}$$

The resulting SVM is called Hard Margin SVM. However, applicable problems are still limited. In order to relax the situation, we allow a small number of misclassified feature vectors. The previous analysis (Eqn.3) is generalized by introducing N nonnegative variables $\xi = (\xi_1, \xi_2, \ldots, \xi_N)$ such that

$$\begin{aligned} &\text{minimize} \quad \text{margin } d(\mathbf{w}) = -\frac{1}{2}(\mathbf{w} \cdot \mathbf{w}) + C\sum \xi_i, \\ &\text{subject to} \quad y_i((\mathbf{w} \cdot \mathbf{x}_i) + b) \geq 1 - \xi_i, \quad i = 1,2,\ldots,N, \ \xi \geq 0. \end{aligned} \tag{4}$$

This SVM defined by Eqn. 4 is called Soft Margin SVM. The purpose of the term $C\Sigma\,\xi_i$, where the sum is for i=1,2, ...,N is to keep under control the number of misclassified vectors. The parameter C can be regarded as a regularization parameter. The OSH tends to maximize the minimum distance $1/w$ for small C, and minimize the number of misclassified vectors for large C.

We described the case of linear decision surfaces in the above. To allow much more general decision surfaces, we have to introduce nonlinear transformation to a set of original feature vectors x_i into a high-dimensional space by a map $\mathbf{F}: \mathbf{x}_i \mapsto \mathbf{z}_i$ as in Fig. 2. Then a linear separation becomes possible in the high-dimensional space there. However the computation of inner products $(\mathbf{F}(\mathbf{x}) \cdot \mathbf{F}(\mathbf{x}_i))$ in a high-dimensional space is prohibitively time-consuming. Therefore, we are interested in cases where these expensive calculations can be significantly reduced by using a kernel function which satisfies the Mercer's theorem such that

$$(\mathbf{F}(\mathbf{x}) \cdot \mathbf{F}(\mathbf{x}_i)) = K(\mathbf{x}, \mathbf{x}_i) \tag{5}$$

The polynomial kernel and the Gaussian kernel are typical kernel functions. Each kernel function is shown in the following formula.

$$K(\mathbf{x}, \mathbf{x}_i) = (\mathbf{x}, \mathbf{x}_i)^d \quad \text{(polynomial kernel)} \tag{6}$$

$$K(\mathbf{x}, \mathbf{x}_i) = \frac{\exp(-\|\mathbf{x} - \mathbf{x}_i\|)^2}{\sigma} \quad \text{(Gaussian kernel)} \tag{7}$$

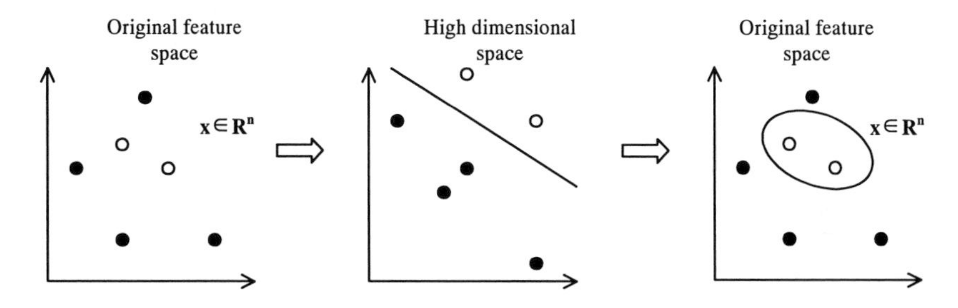

Figure 2: Nonlinear SVM

3. DAMAGE DETECTION IN A BUILDING USING SVM

3.1 Relation between stiffness change and natural frequency change

A natural frequency change associated with a certain mode does not provide any spatial information of any structural damage. However, multiple natural frequency changes may provide the information on the location of damaged stories.[3), 4)]

In a multi-mass shear system shown in Fig.3, the equilibrium equation for an undamped structure is given by

$$\left(-\omega_r^2[M]+[K]\right)\{\phi\}_r = \{0\} \tag{8}$$

where r = 1,2, ...,N, $[M]$ and $[K]$ = mass and stiffness matrices, respectively. $\{\phi\}_r$ = r-th mode shape corresponds to the natural frequency ω_r and is normalized by $\{\phi\}_r^T[M]\{\phi\}_r = 1$.

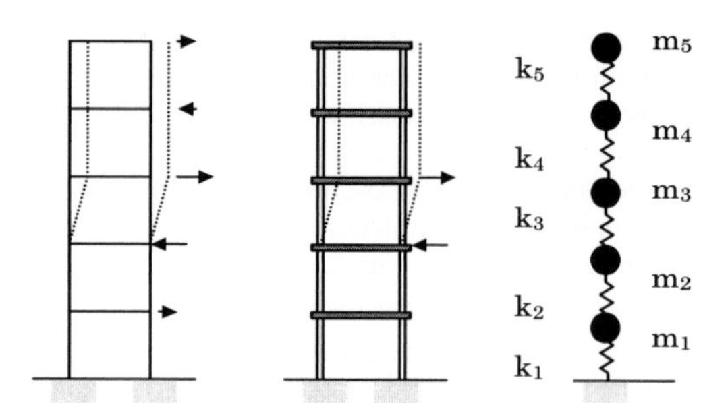

Figure 3: Multi-mass shear system

The sensitivity coefficient of the r-th natural frequency in terms of k_{ij} is defined by the derivative of Eqn.8 with respect to k_{ij} [3].

$$\frac{\partial \omega_r}{\partial k_{ij}} = \frac{1}{2\omega_r}\{\phi\}_r^T \frac{\partial [K]}{\partial k_{ij}}\{\phi\}_r \tag{9}$$

If we take into consideration the symmetry of stiffness matrix, the sensitivity coefficients of the natural frequencies can be rewritten as

$$\frac{\partial \omega_r}{\partial k_{ij}} = \begin{cases} \dfrac{1}{\omega_r}\phi_{ir}\phi_{jr}, & i \neq j \\ \dfrac{1}{2\omega_r}\phi_{ir}^2, & i = j \end{cases} \tag{10}$$

where $\{\phi\}_{ir}$ is i -th component of the r-th mode.

This equation for the natural frequencies may be expanded into first-order Taylor's series. The resulting series represents the change in natural frequency as

$$\Delta\omega_r = \sum_{i=1}^{N}\sum_{i=1}^{N}\frac{\partial \omega_r}{\partial k_{ij}}\Delta k_{ij} \tag{11}$$

For the multi-mass shear system, when the i-th story stiffness is reduced, only k_{ii}, $k_{(i-1)(i-1)}$, $k_{i(i-1)}$ and $k_{(i-1)i}$ are altered in the stiffness matrix. Hence, Eqn.11 is simplified into

$$\frac{\Delta\omega_r}{\omega_r} = \frac{\Delta k_i}{2\omega_r^2}\left(\phi_{ir} - \phi_{(i-1)r}\right)^2 \tag{12}$$

We define the i-th vector of natural frequency change associated with the i-th story stiffness change in Eqn.12 as.

$$\{\gamma_i\} = \left[\frac{\Delta\omega_{1i}}{\omega_1}, \frac{\Delta\omega_{2i}}{\omega_2}, \ldots, \frac{\Delta\omega_{Ni}}{\omega_N}\right] \quad (i = 1, 2, \cdots, N) \tag{13}$$

where $\Delta\omega_{ri}$ is i–th component of the r-th natural frequency. The vectors have typical patterns depending on the location of damaged story. In this study, the above vectors of natural frequency change are used as the basis of feature vectors.

Under the assumption that the 5-mass shear system model consists of the same stiffness and the same mass at each story, vectors defined by Eqn. 13 are typically plotted in Fig.4. The stiffness of each story is varied from 95% to 15% of the original stiffness for simulating damages. From Fig.4, we could observe the vectors are showing characteristic shapes depending of the location of damage. This fact indicates us recognition of damage location may be possible from these vectors.

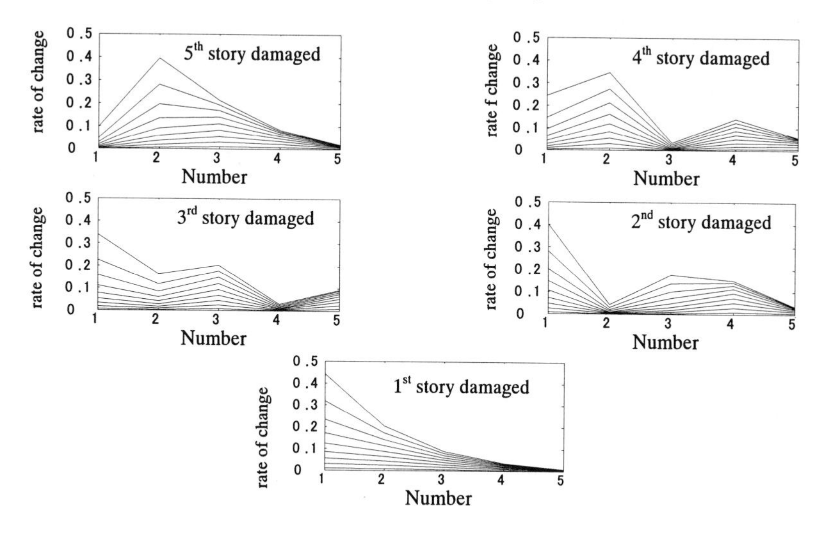

Figure 4: Basis of feature vectors

3.2 Support vector machines for 5-story building

The number of classes that should be distinguished by the SVMs is 6. Therefore, we need six SVMs defined by undamaged (SVM0), 1st story damaged (SVM1), 2nd story damaged (SVM2), 3rd story damaged (SVM3), 4th story damaged (SVM4) and 5th story damaged (SVM5). The feature vector is slightly modified from the vector defined by Eqn. 13. The vector is normalized so that the minimum component is set to zero. As the feature vectors can not be classified using LSVM, a Gaussian kernel function is used to introduce the NSVM. In addition, slack variables are used for softening the condition.

The parameters to be determined for each NSVM are σ as in Eqn.7, and C as in Eqn.4. These parameters are determined to minimize misclassified data. The list of the parameters used for SVM is shown in Table.1. The *l-o-o* (leave-one-out bounds) shown in Table.1 represents the probability that the data does not exist in a margin ($0<f<1$). Thus, after tuning the SVM to classify the data into two classes by the boundary of $f=0$ (correctness=100%), discernment accuracy becomes good as *l-o-o* is close to 100%. The column SV represents the number of support vectors. It is observed that about half of the training feature vectors should be used as support vectors for each classification.

The SVM0 that classifies the data of undamaged structure is shown in the top of Table 1. This SVM can be used to know whether any damage exists in the structure. The SVM5 shown in the next line of Table 1 classifies the data with some damage in the 5th story. Therefore, this SVM finds the case when the damage is expected in the 5th story. Other SVMs have similar meaning.

TABLE 1
SVM FOR DAMAGE DETECTION OF THE 5-STORY BUILDING

SVM	σ	C	SV	l-o-o[%]
SVM0	60	20	21	88.5
SVM5	90	90	45	81.3
SVM4	90.	80	48	83.3
SVM3	90	80	47	82.3
SVM2	90	80	48	86.5
SVM1	90	80	48	78.1

3.3 Performance evaluation

In the following, ■ indicates it is damaged. □ it may be damaged, and ■ it is undamaged. By observing outputs from each SVM (as shown Fig.5 (b)), we can judge whether the structure is damaged or not. And if it is damaged, the SVM will tell the location. Moreover, the magnitude of an output represents the distance from the margin.

The performance of SVMs was tested using 46 feature vectors generated by a simulation. In Fig.5, The data No. 1 is for an undamaged structure. The next 9 data (No.2~10) are those with damage in the 1st story. The story stiffness is reduced to 95% to 15% of the original stiffness with a pitch of 10%. The next 9 data are similarly for the 2nd story damage. The last 9 data are for the 5th story damage.

The SVM0 is observed to give correct outputs for all data sets. However, SVM2 to SVM5 failed to give correct outputs for the damage with 95% stiffness. In addition, the outputs are in the margin for 35%, 25%, 15% stiffness data sets.

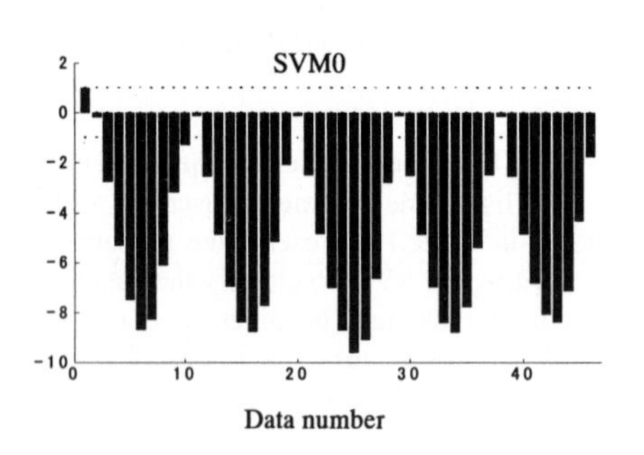

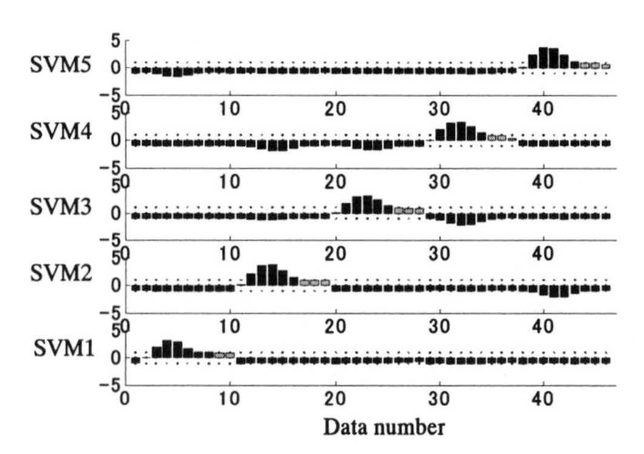

Figure 5: Damage detection using simulated data

Using the SVMs taught for single story damage, the possibility of detecting the damage in multiple stories was evaluated. Assuming both the 3rd story and the 5th story are damaged as listed in Table 2, outputs from the SVMs were plotted in Fig. 6. The outputs shown in Fig.6 indicate that the proposed

SVMs can detect the multiple damages. However, accuracy is less for small stiffness reduction and for very large stiffness reduction as was the case for single damage story.

TABLE 2

DATA NUMBER FOR CASES WITH DAMAGE IN 3RD AND 5TH STORIES

The 5th story / The 3rd story	$0.9k_0$	$0.8k_0$	$0.7k_0$	$0.6k_0$	$0.5k_0$
$0.9k_0$	No.1	No.2	No.3	No.4	No.5
$0.8k_0$	No.6	No.7	No.8	No.9	No.10
$0.7k_0$	No.11	No.12	No.13	No.14	No.15
$0.6k_0$	No.16	No.17	No.18	No.19	No.20
$0.5k_0$	No.21	No.22	No.23	No.24	No.25

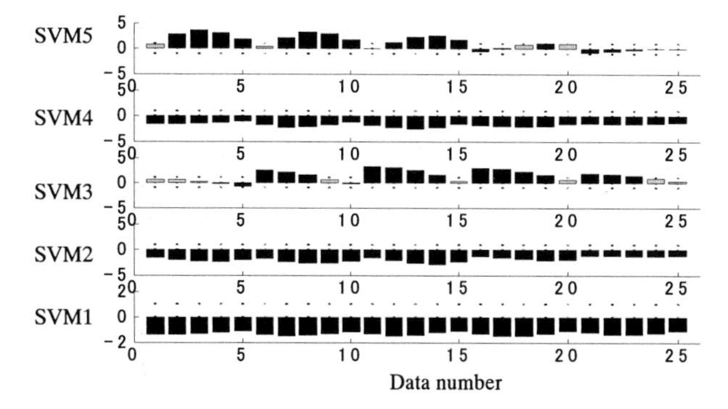

Figure 6: SVM outputs for damaged structure in 3rd and 5th stories

4. EXPERIMENTAL VERIFICATION

A series of experiments were performed to verify the performance of our proposed approach. The model structure is depicted in Fig. 7(a). The damage was introduced by replacing columns by weak columns. By replacing two columns (Fig. 7(b)), the story stiffness is reduced to 60% of the original stiffness. Although acceleration sensors are installed at all stories, only the top sensor and the bottom sensor were used for obtaining modal frequencies. The modal frequencies were calculated from time histories of the vibration experiment by applying the subspace identification method.

Fig.8 shows the outputs from the SVMs. Fig. 8 (a) shows the results for cases with single story damage. Fig. 8 (b) gives the results for cases in which two stories are damaged. The combinations are (0, 0)(no damage), (1st,2nd), (1st,3rd), (1st,4th), (1st,5th), (2nd,3rd), (2nd,4th), (2nd,5th), (3rd,4th), (3rd,5th), (4th,5th). From these figures, we may conclude that the method is indeed applicable to realistic problems. However, the existence of uncertainty due to the sensor noises degraded the accuracy compared to the simulated ones.

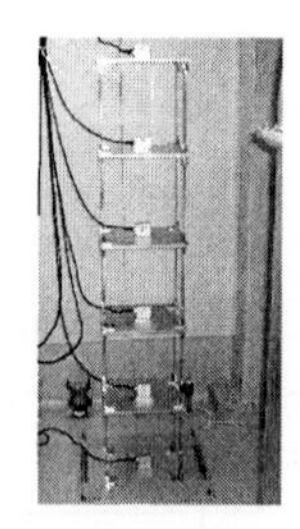

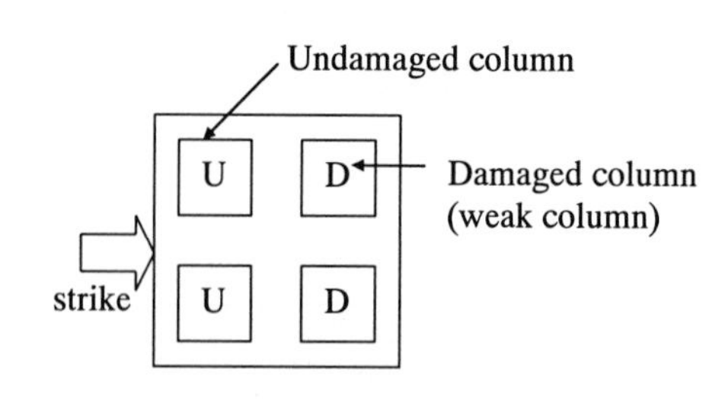

Figure 7: (a) Model structure (b) Plan of damaged story
(Story stiffness is 60% of the original)

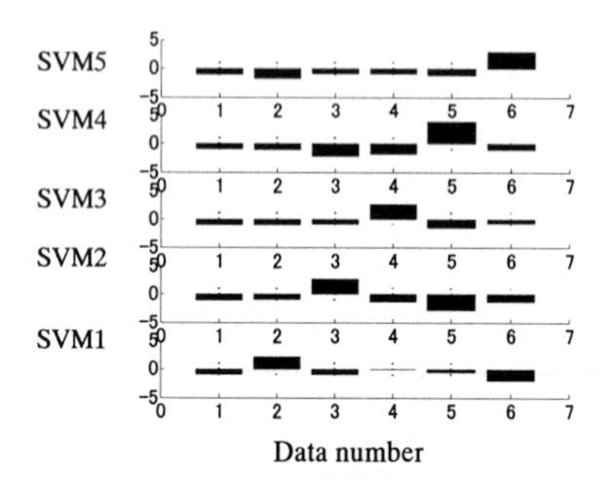

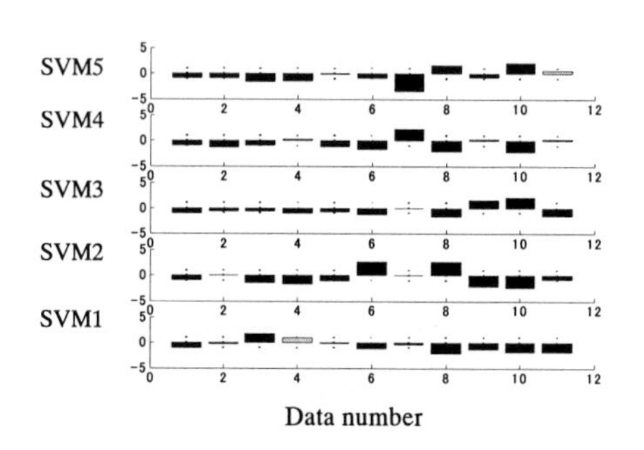

Figure 8: (a) Single story damage (b) 2 stories damage

5. CONCLUDING REMARKS

A structural health monitoring system for a building structure utilizing Support Vector Machine (SVM) was proposed. The proposed system uses modal frequencies of a structure as feature vectors fed to the SVMs for pattern recognition. As our approach does not require modal shapes, minimum number of sensors is necessary for detecting the damage. Typically only two vibration sensors detecting an input signal and an output signal for a structural system are enough to define modal frequencies. The SVMs taught by the data with single damage in a story was found to be also capable of detecting the damage in multiple stories.

6. REFERENCES

1) Vapnik, V.N. (1995). *The Nature of Statistical Learning Theory*, Springer.
2) Christianini, Nello and John Shawe-Taylor (200). *An Introduction to Support Vector Machines*.
3) Zhao, J. and T. DeWolf (1999). Sensitivity Study for Vibrational Parameters Used in Damage Detection. *J. of Structural Engineering, ASCE*, vol.125, NO.4,410-416.
4) Morita, K., M. Teshigawara, H. Isoda, T. Hamamoto and A. Mita (2001). Damage Detection Tests of Five-Story Frame with Simulated Damages. *Proc. of the SPIE* vol.4335, Advanced NDE Methods and Applications.

Advances in Building Technology, Volume 1
M. Anson, J.M. Ko and E.S.S. Lam (Eds.)

PERFORMANCE OF STRUCTURES WITH PASSIVE DAMPERS SUBJECT TO NEAR-FAULT EARTHQUAKES

W. L. He[1] and A.K. Agrawal[2]

Graduate Student Researcher[1] and Assistant Professor[2], Department of Civil Engineering, The City College of New York, New York, NY 10031.

ABSTRACT

In this paper, performance of passive viscous and friction dampers subject to near-fault ground motion velocity pulses for elastic and inelastic structures has been investigated through numerical simulations. The objective of the investigation is to determine the guidelines for types of structures where a passive damper may be effective in reducing damages during a near-fault ground motion. Simulation results show that viscous dampers are most effective around $T/T_g \approx 1$ for elastic structures, whereas they are most effective for inelastic structures in range of natural periods $T/T_g < 1.0$ when both the input energy and hysteretic energy (damage) decrease and dissipated viscous energy increases. Passive friction dampers have good performance for elastic structures around $T/T_g \approx 1$ where input energy decreases and energy dissipated by the damper increases. For inelastic structures with μ=4, friction dampers have good performance in the range of for $0.3 < T/T_g < 2$.

KEYWORD: Near-fault ground motions, long period velocity pulses, passive viscous dampers, passive friction dampers, supplemental energy dissipation systems, structural response control

INTRODUCTION

The traditional approach for seismic hazard mitigation is to design structures with sufficient strength capacity and the ability to deform in a ductile manner. Such design philosophy is based on the assumption that the structure dissipates input seismic energy as hysteretic energy through damage to structural members. Observations during recent seismic destructive events show that displacement ductility ratios higher than 4 imply unacceptable levels of damage to the structure. An alternative to the traditional seismic design is the use of supplemental energy dissipation devices. The objective of using these devices is to dissipate or reduce the input seismic energy of the structures. If a structure is equipped with energy dissipation devices, the primary members of structures are expected to be protected during strong earthquakes, since a part of the energy input from ground motion will be dissipated by the supplementary energy dissipation elements. Iwan (1980) investigated the input energy dissipated due to the inelastic deformation during earthquakes by approximating the inelastic energy spectrum with an elastic structure using an equivalent viscous damping and an equivalent natural period. The concept of input energy and energy balance in structures was systematically investigated by Uang and Bertero (1990). Fajar and Vidic (1994) constructed the E_H/E_I spectra, where E_H is the hysteretic energy (damage) and E_I is the input energy, for inelastic structures as a function of the damping, the ductility factor and the hysteretic behavior of the structures. Recently, Goel (1997) and Fu and Kasai (1998) discussed the idea of energy balance for designing supplementary passive energy devices. In this paper, the energy transfer between supplemental energy dissipation devices and hysteresis actions in structural members, and the performance of supplemental energy dissipation devices in preventing damage to the structural members is investigated.

MATHEMATICAL FORMULATION

Consider a single-degree-of-freedom (SDOF) system with the governing equation

$$\ddot{x}(t) + 2\zeta\omega\dot{x}(t) + f_r(x) + u = -\ddot{x}_g(t) \tag{1}$$

where $x(t)$, $\dot{x}(t)$ and $\ddot{x}(t)$ are the relative displacement, velocity and acceleration of the structure, respectively, ζ is inherent viscous damping ratio, ω is the natural frequency of the structure, $f_r(x)$ is the restoring force per unit mass (e.g., $f_r(x) = \omega^2 x$, if the structures is in elastic range), u is unit mass control force, and $\ddot{x}_g(t)$ is the earthquake ground motion. For passive energy dissipation systems that are functions of the velocity across the damper, the control force u is given as

$$u(t) = c\dot{x} \tag{2}$$

Likewise, the control force for a pure Coulomb friction damper is given is given as

$$u = \mu N(t)\,\mathrm{sgn}(\dot{x}) \tag{3}$$

where μ is the coefficient of friction and N(t) is the normal force across the damper. For passive friction dampers, N is constant and it is variable for semi-active friction dampers [e.g., He et al (2002a)]. The control force in Equations (2) or (3) is a function of structural response, which depends on the characteristics of earthquake ground motions. Therefore, the performance of passive energy dissipation devices depends on the nature of earthquake ground motions. By assuming the initial displacement and velocity of the structure to be zero, multiplying Eq.(1) by dx and integrating it from time t_0 to t_1, the energy balance from time $t_0 = 0$ to t_1 is obtained as

$$\left[\frac{1}{2}\dot{x}^2\right]_{t=t_1} + \left[\int_0^{t_1} f_r\dot{x}dt\right] + \left[\int_0^{t_1} 2\zeta\omega\dot{x}^2 dt\right] + \left[\int_0^{t_1} u\dot{x}dt\right] = \left[-\int_0^{t_1} \ddot{x}_g\dot{x}dt\right] \tag{4}$$

The terms in brackets in the left side of equation are kinetic energy, strain energy, damping energy, and control force energy, respectively. The term in the right hand side is the input seismic energy. In Eq.(4), various energy terms are expressed in per unit mass of the structure. It is noted that the energy terms in equation (4) are for relative energy, which is inappropriate for long period structures. However, it has been pointed out by Uang and Bertero (1990) that the use of relative energy form is valid for structures in realistic period ranges. For any supplemental velocity dependent damper in Eq.(2), the fourth term in Eq.(4) is always positive definite for any positive value of c. A similar argument applies for the friction damper in Eq.(3) since N(t) > 0. Hence, such systems are always dissipative in nature and they continuously dissipate energy.

Consider a Single-Degrees-of-Freedom (SDOF) elastic-perfectly-plastic structure in Eq.(1) equipped linear viscous damper in Eq (2) or a friction damper in Eq. (3). For a general hysteretic energy dissipation system, the nonlinear damper force can be represented by the Bouc-Wen model,

$$u(t) = \alpha K_e x(t) + (1-\alpha)K_e u_y z(t) \tag{5}$$

where x(t) is the displacement across the damper, K_e is a reference stiffness, α is a parameter controlling post-yielding stiffness, u_y is the value of yield displacement and z(t) is a dimensionless hysteretic quantity given by

$$u_y \dot{z} + \gamma |\dot{x}(t)| z |z|^{n-1} + \beta \dot{x}(t) |z|^n - A\dot{x}(t) = 0 \tag{6}$$

where β γ n, A are dimensionless quantities that control the shape of the hysteretic loop. Equations (5) and (6) can be used to model various types of nonlinear energy dissipation systems. For example, appropriate values of parameters β, γ, n and A can be selected to model the elasto-plastic type of hysteretic loops of viscoelastic elements. For a pure Coulomb type of friction damper, the rigid plastic behavior can be simulated by assuming A = 1, α = 0, u_y a sufficiently small quantity and $K_e u_y = \mu N$. Hence, a general form of hysteretic controller that represents a wide variety of elastic-plastic behavior can be written as

$$u(t) = \varepsilon \cdot (\ddot{x}_g)_{max} z(t) \tag{7}$$

where $\varepsilon = \left| u/\ddot{x}_{g\,max} \right|$ is the ratio of magnitude of control force to the peak shaking force per unit mass generated by ground motion, and u is the magnitude of the control force.

The efficiency of the supplement energy dissipation devices depends on the structures, frequency content and energy distribution of ground motions and energy dissipation mechanism of passive dampers. Since parameters of ground motions are highly stochastic in nature, no attempt has been made to perform rigorous statistical analysis using a large samples of ground motions. Instead of using a large number of ground motions, a closed-form approximation which is demonstrated to be capable of capturing kinematic characteristics of near-fault ground motion pulses has been recently proposed by the authors [He and Agrawal (2002b)]. In this formulation, the predominant velocity pulse in near-fault ground motions has been modeled by decaying sinusoidal functions,

$$\dot{x}_g = s e^{-\zeta_g \omega_g t} \sin \omega_g \sqrt{1-\zeta_g^2}\, t \tag{8}$$

where ζ_g is the ground motion decaying factor, ω_g is the predominant frequency of the pulse of ground motions, and s is the initial amplitude of the pulse. Expressions for acceleration for the velocity pulses in Eq.(8) can be obtained by differentiating Equation (8), i.e.,

$$\ddot{x}_g = s e^{at} [a \sin bt + b \cos bt] \tag{9}$$

where the coefficient a and b are obtained as

$$a = -\zeta\ \omega_p\ ;\ b = \omega_g \sqrt{1-\zeta_g^2} \tag{10}$$

The displacement u_g of the pulse can be obtained by integrating Equation (8) as follows

$$x_g = \frac{s e^{at}}{\omega_g^2} [a \sin bt - b \cos bt] + \frac{sb}{\omega_g^2} \tag{11}$$

Makris and Chang (1998) have modeled near-fault ground motions pulses by three distinct types of pulses, e.g., Type A, Type B and Type C_n. The authors have investigated the pulses in several ground motions recorded during Northridge, Kobe, Turkey and Chi-Chi earthquakes, and have demonstrated that actual ground motion pulses behave like decaying sinusoidal waveforms rather than distinct types of pulses modeled by Type A, Type B and Type C_n. The correlation of the decaying sinusoidal formulation with recorded ground motions pulses has been demonstrated in He and Agrawal (2002).

In this paper, a ground motion pulse with peak acceleration of 0.5g, a decay ratio of $\zeta_g = 0.2$ and $\omega_g = 2\pi$ (i.e., $T_g = 1.0$ sec) is used to study the effectiveness of passive dampers.

STRUCTURE WITHOUT DAMPERS

A SDOF elastic-perfectly-plastic structure with inherent viscous damping ratio of 5% and the natural period of the elastic structure varying from 0.05 to 5 s subject to the velocity pulse in Eq.(8) is investigated. The plots of seismic input energy of the structure with displacement ductility μ=1, 2, 3, 4, 6 and 8 are shown in Figure 1(a). In this figure, μ=1 represents the elastic structure. The input energy of SDOF inelastic systems is approximately independent of the strength of the structure in short (i.e., T < 0.25 sec.) and long (T > 2.5 sec.) period ranges, since the plots for structures with different ductilities almost coincide. The peak of the input energy spectrum corresponds to the predominant period T_g of the ground motions and it shifts slightly towards smaller periods as the ductility is increased. Input energy increases slightly with ductility for structures with $T/T_g < 0.7$ and it decreases with ductility ratio for longer period structures.

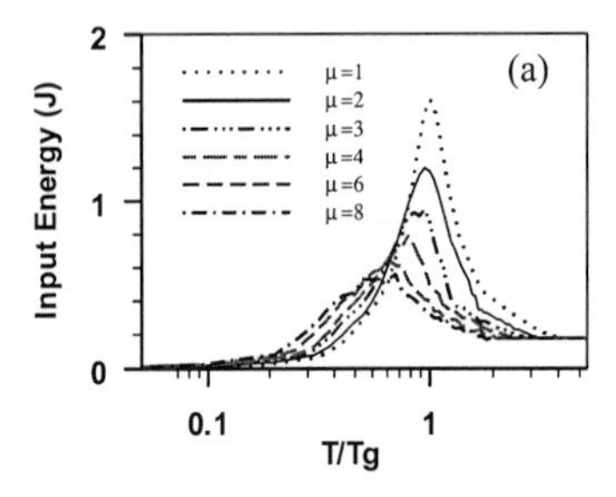

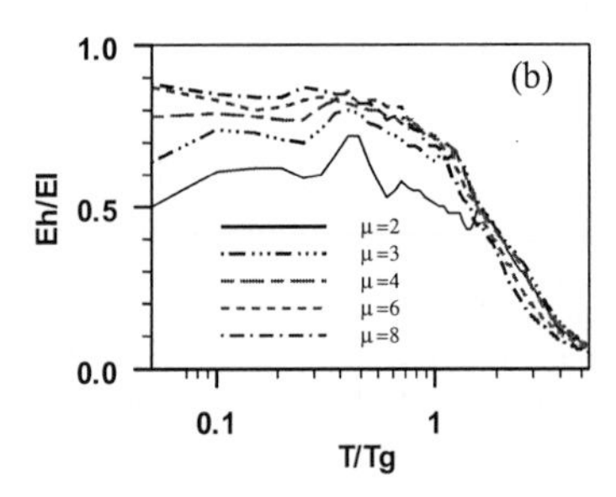

Figure 1 (a) Input Energy; (b) Eh/EI ratio of Structures Without Dampers.

The input energy is completely dissipated by viscous damping at the end of the ground motion in an elastic structure. The total input energy is partly dissipated by viscous damping mechanism and partly through hysteretic damage to the structure during inelastic deformation. Figure 1(b) shows the plots of E_H/E_I ratio for the inelastic structure with different ductilities, where E_H is the hysteretic energy of the structure and E_I is the input energy. It is observed that the E_H increases with μ from 2 to 4 and the increase in E_H (i.e., damage to the structure) is smaller for $\mu > 4$ for structure in the period range $T/T_g < 1$. The E_H/E_I ratio is approximately stable for $T/T_g < 1$ for a given ductility demand μ. This characteristics of the E_H/E_I spectra is similar to that of presented by Fajar and Vidic (1994). For structures with $T/T_g > 1$, the E_H/E_I ratio decreases gradually to zero for all displacement ductilities since the present study uses relative input energy.

EFFECTS OF PASSIVE VISCOUS DAMPER

The SDOF structures with displacement ductilities of μ=1 and 4 are assumed to be equipped with supplemental passive viscous damper in Eq. (2). Two cases of supplemental dampers are designed such that the total damping ratios (i.e., supplemental +inherent) of the structure are 15% and 30%. Such design of dampers is practical using commercially available passive dampers. Figure 2 shows input energy spectra of the structures with μ=1 and 4. It is observed from Figure 2(a) that the input energy decreases significantly as the viscous damping is increased for elastic structures of natural

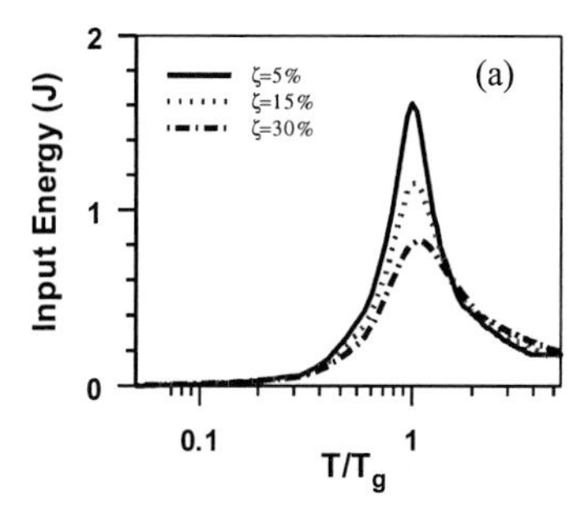

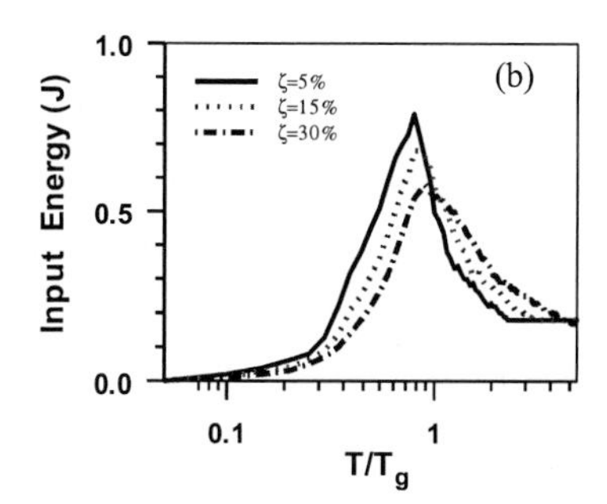

Figure 2: Input energy spectra of the structure with supplemental viscous dampers: (a) μ=1; (b) μ=4.

period around T_g, i.e., $T/T_g \approx 1$. For the inelastic structure with μ=4, it is observed from Figure 2(b) that the peak input energy without any supplemental damper is 0.8 J as compared to 1.6 J for the elastic structure. The reduction in input seismic energy for the inelastic structure without dampers is achieved at the expense of damage to the structure. An increase in damping ratio of the inelastic structure through supplemental viscous dampers has a lesser effect in reducing the peak input energy. In fact, increasing the damping ratio of the inelastic structure reduces the input energy for structures with $T/T_g < 1$ and it increases the input energy for long period structures with $T/Tg > 1.0$.

Figure 3 shows plots for hysteretic and viscous energy dissipated for the inelastic structure with μ=4 equipped with supplemental viscous damper. It is observed from Figure 3(a) that the viscous dampers are most effective in reducing hysteretic damage to the structures of natural periods $T/T_g < 1.5$. For longer period structures with periods $T/T_g > 1.5$, supplemental viscous dampers dissipate more viscous energy [as shown in Figure 3(b)] since the input energy is also increased because of addition of viscous dampers [as shown in Figure 2(b)]. They are, however, ineffective in reducing hysteretic

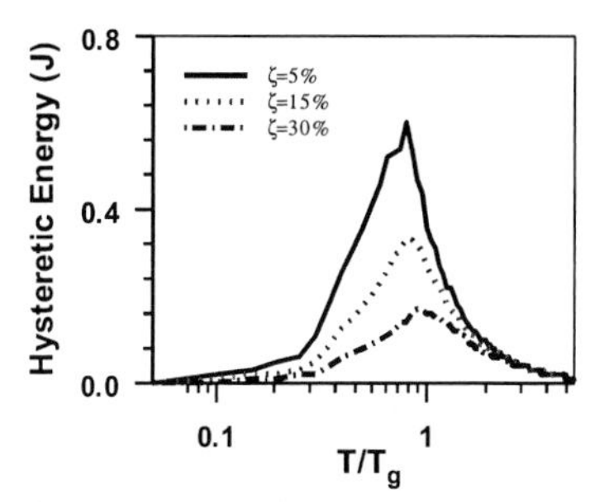

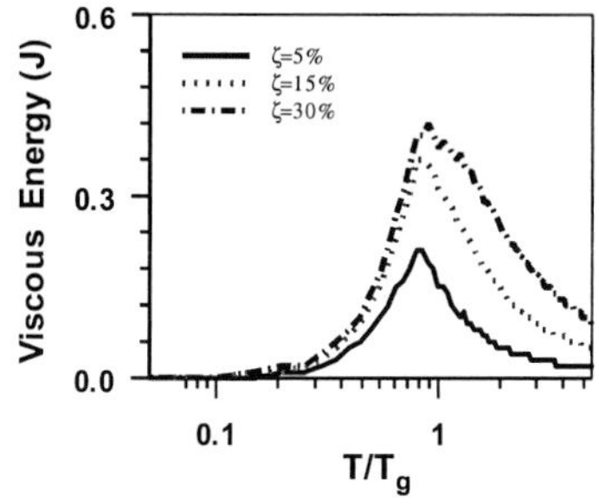

Figure 3: Hysteretic and viscous energy spectra for the inelastic structure with original displacement ductility μ=4.

damage to the longer period ($T/T_g < 1.5$) inelastic structure, as shown in Figure 3(a). Viscous dampers are most effective for structures in the range of natural periods $T/T_g < 1.0$ when both the input energy and hysteretic energy (damage) decrease and dissipated viscous energy increases.

Figure 4 shows the displacement ductility spectra of the elastic (μ=1) and the inelastic (μ=4) structures with supplemental viscous dampers. It is observed from Figure 4(a) that the supplemental viscous dampers can more effectively reduce the displacement of elastic structures in the period range 0.3< T/T_g < 2, although displacements decrease in the entire period range of $0 \le T/T_g \le 5$. For inelastic structures with μ = 4, supplemental viscous dampers are more effective for structures in the range of T/T_g < 1, as shown in Figure 4(b). It is further observed from Figure 4(b) that supplemental dampers with total structural damping ratio (inherent + supplemental) of 30% are capable of reducing the displacement ductility of structures with T/Tg = 0.6 from μ=4 to approximately μ=1.25, implying that the structure will undergo significantly smaller damage than the original structure. For long period structures with T/Tg > 1.0, the displacement ductility decreases to 3 and 2.5 for 15% and 30% damping ratios, respectively. The effects of supplemental viscous dampers on the acceleration of elastic and inelastic structures has also been investigated. It has been observed that the acceleration decreases for all period ranges except for a slight increase in long period ranges (T/T_g > 1) for elastic structures. For inelastic structures, an increase in damping ratio increases the acceleration of the structure in all period ranges.

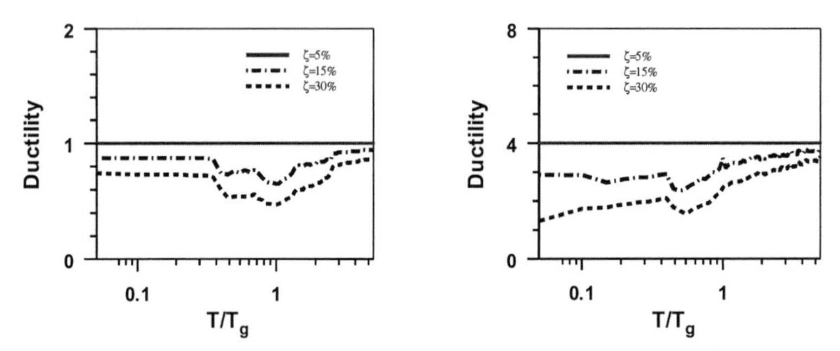

Figure 4: Displacement ductility spectra of the structure with viscous dampers: (a) elastic structure (μ=1); (b) inelastic structure (μ=4).

EFFECTS OF PASSIVE FRICTION DAMPERS

Next, a coulomb friction damper is used as a supplemental damping device in a SDOF structure with displacement ductility μ=1 and μ=4. The friction damper is installed in a building with 5% inherent viscous damping and sliding force ε =10% and 20% of the maximum seismic inertial force. Figure 5 shows input energy spectra for elastic and inelastic structures with supplemental friction dampers. It is observed from Figure 5(a) that the friction damper is effective in reducing input seismic energy significantly for T/T_g < 2 for the elastic structures. However, the input energy increases slightly for T/T_g > 2. For inelastic structures in Figure 5(b), the friction damper reduces the input seismic energy for T/T_g < 1 (more predominantly around $T/T_g \approx 1$) and it increases the input energy for T/T_g > 1. Since the energy is dissipated both by friction damper and inelastic behavior (hysteretic) of structure, a comparative study is done for the two types of hysteretic energy dissipation mechanisms in Figure 6. Fig 6(a) shows hysteretic energy dissipated by the friction damper for both elastic and inelastic structures. For the elastic structure, significant amount of hysteretic energy is dissipated by the friction damper. However, the hysteretic energy doesn't increase significantly by increasing the friction force from ε = 10% to 20%. This may happen because of sticking in the friction damper at the higher friction force. For the inelastic structure, a higher amount of energy is dissipated by the larger frictional force. Figure 6(b) shows the energy dissipated because of inelastic behavior of the structure. It is observed that the hysteretic energy (damage) decreases drastically as friction force is increased, clearly indicating a lesser degree of damage in the structure using a larger friction force. The friction damper is more effective in reducing hysteretic damage to the structure for 0.3 <T/T_g <2.

Figure 7(a) and (b) show displacement ductility spectra for elastic (μ=1) and inelastic (μ=4) structures. For a comparative study, ductility spectra for viscous dampers with total damping of 15% and 30% are also shown in these figures. It is observed from Figure 7(a) that friction dampers are more effective than viscous dampers for long period elastic structures. For inelastic structures (μ=4), while friction

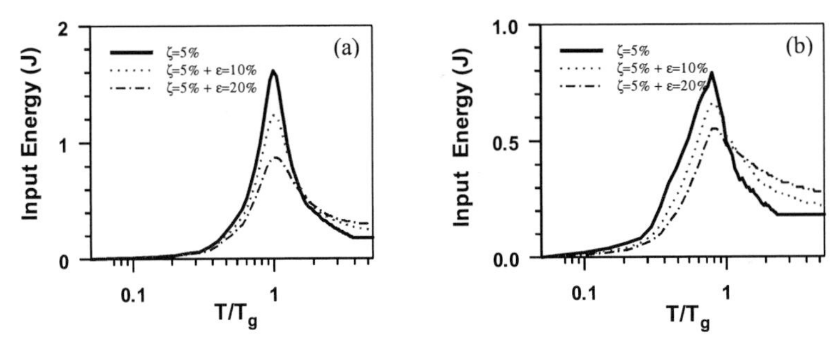

Figure 5: Input Energy spectra of structure with friction damper: (a) elastic structures (μ=1); (b) Inelastic structures (μ=4).

dampers are effective over the period range of 0.2 <T/T_g <5, viscous dampers are more effective for T/T_g < 1. Through acceleration spectra for elastic and inelastic structures for both friction damper and viscous dampers, it has been observed that acceleration of structures with friction damper is generally higher than those of structure with viscous dampers.

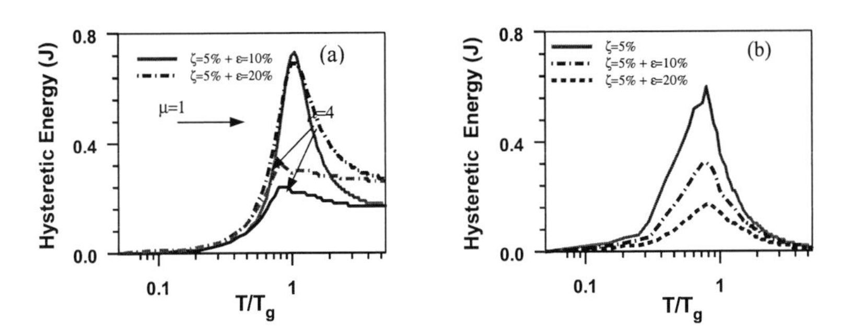

Figure 6: Hysteretic energy dissipated by: (a) Friction damper; (b) Structural damage (μ=4).

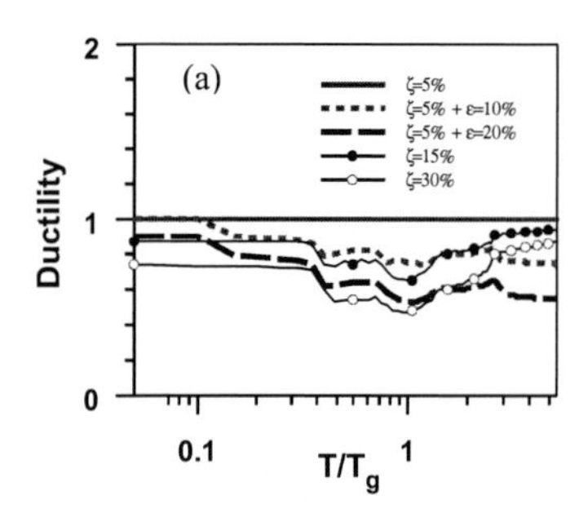

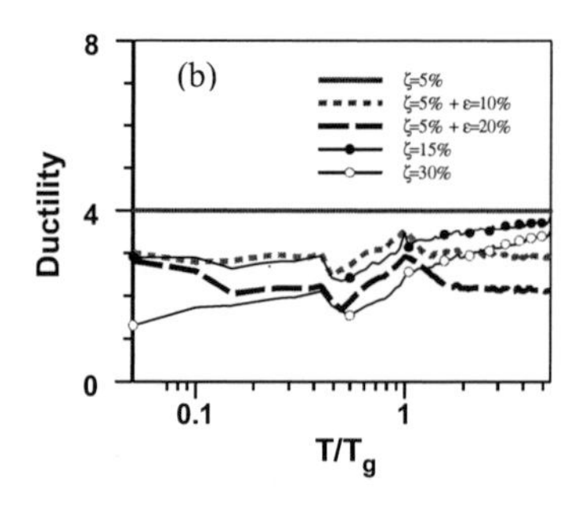

Figure 7: Displacement ductility spectra of the structure with friction dampers: (a) Elastic structure (μ=1); (b) Inelastic structure (μ=4).

CONCLUSIONS

In this study, the performance of the supplemental energy dissipation devices for a SDOF elastic or elastic-perfectly-plastic structure subject to near-fault velocity pulses is investigated. By adding the supplement energy dissipation devices, the input energy and the hysteretic energy dissipated by the structure, which is directly related to the damage of the structure, as well as the displacement ductility are investigated in detail. For elastic structure, viscous dampers are most effective around $T/T_g \approx 1$, whereas they are most effective for inelastic structures in range of natural periods T/Tg < 1.0 when both the input energy and hysteretic energy (damage) decrease and dissipated viscous energy increases. Passive friction dampers have good performance for elastic structures around $T/T_g \approx 1$ where input energy decreases and energy dissipated by the damper increases. For inelastic structures with μ=4, friction dampers have good performance in the range of for $0.3 < T/T_g < 2$. The present investigation is a preliminary effort to address the design of passive supplement dampers from input seismic energy point of view. More detailed investigation, including the verification using actually recorded ground motion is underway and it will be reported in future publications.

ACKNOWLEDGMENT

The research presented in this paper is sponsored by the Grant Number CMS 0099895 by the National Science Foundation.

REFERENCES

Fajfar P and Vidic T. (1994), Consistent inelastic design spectra: hysteretic and input energy. *Earthquake Engineering and Structural Dynamics*, **23**, 523-532.

Fu, Y. and Kasai, K. (1998). Comparative study of frames using viscoelastic and viscous dampers. *J. Struct. Eng.,* **124:5**, 513–552.

Goel, R. K. (1997). Seismic response of asymmetric systems: Energy based approach. *J. Struct. Eng.,* **123:11**, 1444–1453.

He, W.-L. and Agrawal, A.K. (2002a). A Novel Semi-Active Friction Controller for Linear Structures Against Earthquakes. Accepted for publication in *Journal of Structural Engineering*, ASCE.

He, W.-L. and Agrawal, A.K. (2002b). A closed-form approximation of near-fault ground motion pulse for flexible structures. Submitted to ASCE *Journal of Engineering Mechanics*.

Iwan, W. D. (1980). Estimating inelastic response spectra from elastic spectra. *Earthquake Eng. Struct. Dyn.* **8**: 375–388.

Uang C.M., Bertero V.V. (1990), Evaluation of seismic energy in structures. *Earthquake Engineering and Structural Dynamics*. **19**: 77-90.

Advances in Building Technology, Volume 1
M. Anson, J.M. Ko and E.S.S. Lam (Eds.)

RELIABILITY AND CONTROL PERFORMANCE VERIFICATION OF AN ACTUAL SEMI-ACTIVE STRUCTURAL CONTROL SYSTEM

Narito KURATA

Kobori Research Complex, Kajima Corporation,
KI Building, 6-5-30, Akasaka, Minato-ku, Tokyo 107-8502, Japan

ABSTRACT

In the past 15 years, active and semi-active structural control systems have been applied to actual buildings and bridges. However, a reliability of these systems was not discussed enough in previous stages. This can be secured by a system's health monitoring function and a fail-safe function in unpredicted emergency cases. In this paper, the reliability and control performance about the following practice are presented. After the theoretical and experimental study, a semi-active structural control system using a continuous-variable semi-active hydraulic damper with a maximum damping force of 1,000 kN, was applied to an actual 5-story building in 1998. A consideration of health monitoring for the system, fail-safe function, treatment at the time of power failure and trouble of the sensor, control device, computer and communication system, etc., are required. A forced vibration test was carried out to recognize the required functions. These results show the effectiveness of system's functions. Finally, it is presented that the reliability of the system will be secured as well as the control performance.

KEYWORDS

semi-active structural control, semi-active hydraulic damper, reliability, fail-safe, health monitoring, power failure

INTRODUCTION

In the semi-active structural control systems, a large control force can be obtained only by adjusting the characteristics of the control device itself with a small external energy supply. This enables a structural control system to be realized that is highly effective over the range from small to large earthquakes, and that is highly reliable and economical. With this in mind, the research of a semi-active control has been carried out actively from the middle of the 1990's (Kobori, 2000; Spencer, 2000; Soong and Spencer, 2000). For example, Kurata et al. (1998) and Matsunaga et al. (1998) have developed a semi-active hydraulic damper with a maximum damping force of 2000 kN and demonstrated the effectiveness of this semi-active control system through simulation analyses of a semi-actively controlled building in severe earthquakes. Spencer, Jr. et al. (1996, 1998) proposed a Magneto-rheological (MR) fluid damper for the semi-active structural control and developed a full-scale damper. Dyke et al. (1996) and Jansen and Dyke (2000) developed a control algorithm for investigating the use of the MR dampers. Furthermore, there have been three applications of an on-off variable hydraulic device to two actual buildings (Kobori et al. 1991; Kobori, 2000) and a bridge (Patten et al. 1999), and an application of a multi-stage variable hydraulic damper to a base isolated building (Yoshida and Fujio, 2000).

We developed a continuously variable semi-active hydraulic damper (SHD) that can produce a maximum damping force of 1000 kN with an electric power of 70 watts, and installed it with a computer system in an actual five-story building in 1998 (Kurata et al. 1999, 2000; Kurata, 2001) after basic studies (Kurata et al. 1994, 1998). This was the world's first application of the semi-active structural control system to civil structure in terms of the following two points.

(i) Application of a continuously variable semi-active control device;

(ii) Full support for large earthquakes. It satisfied the following criteria for an input motion equivalent to the "large earthquake" referred in design of high-rise buildings in Japan. Here, the criteria are that the structures should be elastic and that the story drift angle should be within 1/200.

For this application, consideration of health monitoring for the system, fail-safe function, treatment at the time of power failure and trouble of the sensor, control device, computer and communication system, etc., are required. In this paper, the reliability of this semi-active structural control system is discussed. Detail of a system's health monitoring system and a fail-safe semi-active control are described. After the system installation, a forced vibration fail-safe test for the building with semi-active control was carried out. In the test case, the electrical power supply for the 1st floor SHDs was cut off during the semi-active control in the forced vibration. These results show the effectiveness of system's fail-safe functions. Finally, it is presented that the reliability of the system will be secured as well as the control performance.

SEMI-ACTIVE STRUCTURAL CONTROL SYSTEM

The office building in which the SHD system is installed is a five-story steel structure with PC curtain walls and a basement, and it is located in Shizuoka City, Japan. A configuration of the SHD system is shown in Figure 1. The system consists of velocity sensors, SHDs, computers and an uninterruptible power supply unit. The sensors, which are placed on each floor, measure the absolute acceleration and are able to output the absolute velocity by the analog integration. The eight SHDs are installed transversely between a steel brace and a beam on each story from the first to the fourth on both gable sides. The computers are placed in the control room on the first floor. The elasto-plastic steel dampers are installed in the longitudinal direction.

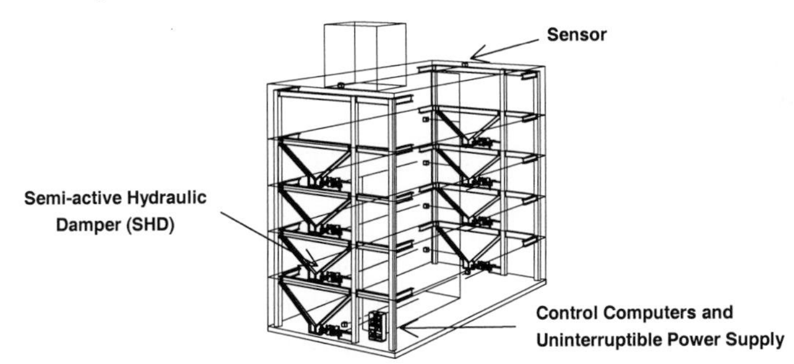

Figure 1: Semi-active damper system configuration

A full-size SHD that can produce a maximum damping force of 1000 kN with an electric power of only about 70W has been developed (see Figures 2 and 3). It is compact and requires no special installation space, so the required number of them can easily be installed in a building. It comprises a flow control valve, a check valve and an accumulator. A relief valve that opens at a set pressure is installed in parallel with the flow control valve, so that the load cannot cause the design stress of the SHD to be exceeded. It is activated, and this restricts the damping force to 900 kN when an excessive command is given. Furthermore, a solenoid valve that opens in case of an interruption of the electrical service is provided as a fail-safe to an unexpected system fault or power failure. When it opens, the oil flows through the orifice and the SHD works as a passive damper.

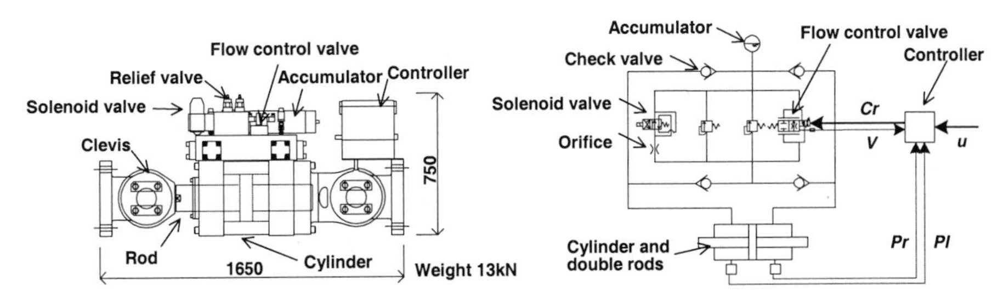

Figure 2: Outline of SHD

Figure 3: Hydraulic circuit of SHD

The functions of system management, control computation and command transmission are assigned to four computers in the control room on the 1st floor. It is possible to achieve real-time control with a 5-millisecond sampling time with high reliability. The control procedure is as follows: (i) the sensors measure the building's responses; (ii) the computer calculates the damping force command to minimize the response based on the measured data; and (iii) the SHDs generate the damping forces according to the computer's command. By this control procedure, the damping force is optimized and the semi-active control system can achieve a high response reduction performance that is not obtained with a passive device.

The damping force f_{vi} of the i-th SHD can be expressed by the following equation.

$$f_{vi} = \begin{cases} f_{\max} \cdot sign(v_i) & u_i \cdot v_i > 0, \ \left|u_i\right| > f_{\max} \\ c_{\max} \cdot v_i & u_i \cdot v_i > 0, \ \left|\frac{u_i}{v_i}\right| > c_{\max}, \ \left|u_i\right| \le f_{\max} \\ c_i(t) \cdot v_i = u_i & u_i \cdot v_i > 0, \ \left|\frac{u_i}{v_i}\right| \le c_{\max}, \ \left|u_i\right| \le f_{\max} \\ 0 & u_i \cdot v_i \le 0 \end{cases} \quad (1)$$

where u_i is the damping force command from the computer to the i-th SHD and v_i is the i-th SHD's velocity. $f_{\max}$ and $c_{\max}$ are the upper limit values of the damping force and the damping coefficient of the SHD, respectively. They were determined through a dynamic loading test (Kurata et al. 1999). The damping force command u_i is the optimal control force, which is designed to minimize the building's response by using the relative velocity feedback law based on a Linear Quadratic Regulator (LQR). The optimal control force is:

$$U(t) = \{u_i(t)\}^T = -G_{j,k} X(t) \quad , j=1,10; \ k=1,15 \quad (2)$$

where $G_{j,k}$ is the velocity feedback gain. The index j and k, which are the variable gain number, relates response level and fail-safe mode, respectively.

HEALTH MONITORING AND FAIL-SAFE SEMI-ACTIVE CONTROL

The computer system configuration is shown in Figure 4. The required function of the control system is real-time control and system management. The former contains control trigger determination, real time computation with several milliseconds and communication between the SHD and computers. The latter contains measurement of a lot of information, health monitoring of the system, fail-safe, and emergency alarm. Each function was realized by a separate computer. A control computer (Con-C) determines the damping force command to the SHD from the building's response measured by the sensors and outputs it to the communication computer (Com-C) with a sampling time of five milliseconds. One Com-C outputs the damping force command to four SHDs. A management computer (Man-C) monitors the condition of each SHD and communication situation of each computer, and determines whether or not it is normal. If there is a fault with the system, the fail-safe function corresponding to each abnormal mode becomes effective and an alarm is sent through the telephone line to the remote computer (Rem-C) located in the Kobori Research Complex, Akasaka, Tokyo. The sampling time of the Man-C is 0.02 seconds. The remote operation of the Man-C and the starting and shutdown of the whole system can be done from the Rem-C. Furthermore, the control can be continued, even if the function of the Man-C is lost. The computers are in real time monitoring the health of the sensors, the SHDs and all communication lines. At the same time, the SHDs and the computers have a self-diagnosis function and a mutual surveillance function. It is possible to make the fail-safe function corresponding to various abnormalities work effectively by these health monitoring.

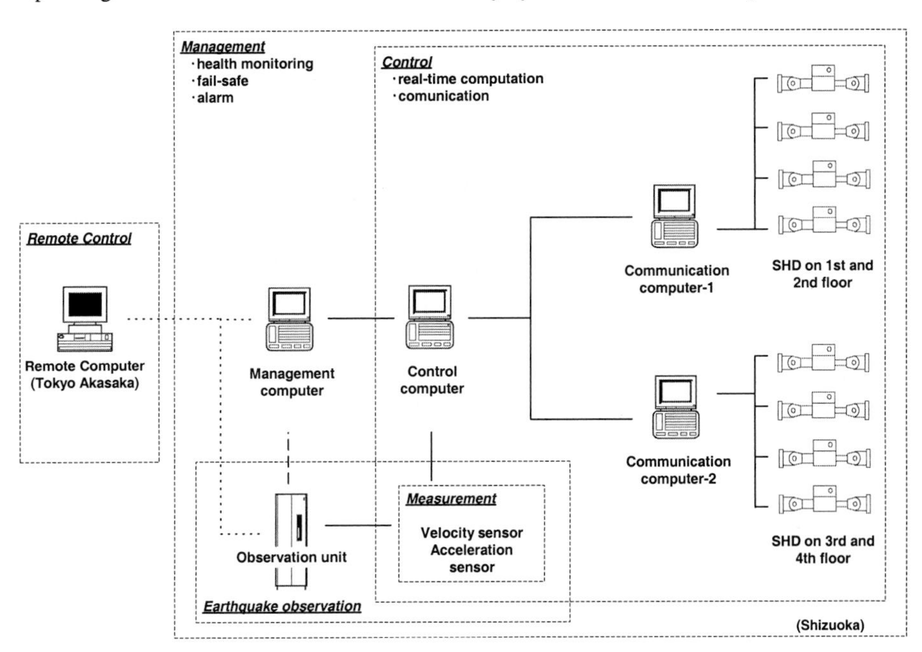

Figure 4: Computer system configuration

Waiting, control and fail-safe system modes are provided. Usually, in the waiting mode, all the SHDs function as passive dampers. Then, the semi-active control mode starts when the trigger by the control computer turns ON. Next, it returns in the waiting mode when an earthquake finishes and the trigger turns OFF. The trigger turns ON when the first floor acceleration exceeds five cm/sec^2. The fail-safe mode is set for emergency situations. Table 1 shows the semi-active control under fail-safe modes. The SHDs corresponding to an abnormal part changes to a passive damper for every floor unit and optimal semi-active çontrol using variable gains is carried out under this situation. Accordingly, this system secures high control performance as a hybrid structural control system, even if the power supply is cut off. The conditions of the SHD's hydraulic circuits are the same in the waiting mode and the fail-safe mode. However, all the SHDs are in the passive condition in the waiting mode.

TABLE 1

FAIL-SAFE SEMI-ACTIVE CONTROL

State	Number of semi-active controlled stories	Gain	Story			
			1	2	3	4
Semi-active control	4	$G_{j,1}$	S	S	S	S
Fail-safe semi-active control	3	$G_{j,2}$	S	S	S	P
		$G_{j,3}$	S	P	S	S
		$G_{j,4}$	S	S	P	S
		$G_{j,5}$	P	S	S	S
	2	$G_{j,6}$	S	S	P	P
		$G_{j,7}$	S	P	S	P
		$G_{j,8}$	S	P	P	S
		$G_{j,9}$	P	S	S	P
		$G_{j,10}$	P	S	P	S
		$G_{j,11}$	P	P	S	S
	1	$G_{j,12}$	S	P	P	P
		$G_{j,13}$	P	S	P	P
		$G_{j,14}$	P	P	S	P
		$G_{j,15}$	P	P	P	S
Passive control	–	–	P	P	P	P

FORCED VIBRATION FAIL-SAFE TEST

A forced vibration fail-safe test for the building using an exciter was carried out to recognize an actual fail-save functions of this system. A large-scale synchronous exciter with a maximum force of 100 kN was used. This exciter generates a sinusoidal force by rotating two pairs of eccentric masses. It was placed at the center of the building's roof floor, and a transverse forced excitation was applied. The tests were carried out under a constant excitation force amplitude. Displacement sensors were placed transversely on each floor from the basement to the roof at both ends. In addition, sensors for semi-active control were placed in each floor as shown in Figure 1. For each SHD, the damping force command u, the hydraulic pressures Pl and Pr at the inside left and right parts of the cylinder, and the actual valve opening V of the flow control valve and the control current Cr, shown in Figure 3, were measured. A constant harmonic excitation force of 16 kN with a frequency of 1.46 Hz was applied to the semi-active controlled building. In the test case, the electrical power supply for the 1st floor SHDs was cut off during the semi-active control in the forced vibration. The damping force and the valve opening of the SHD located on the 1st and 3rd floor, and the velocity at the roof floor, are shown in Figure 5. At the same time, two SHDs located on the 1st floor switched to the fail-safe passive state. The opening of the flow control valve went to zero, the orifice was activated and the passive damping force generated. However, other SHDs continued to be semi-actively controlled. The velocity at the roof floor became large a little after the switching.

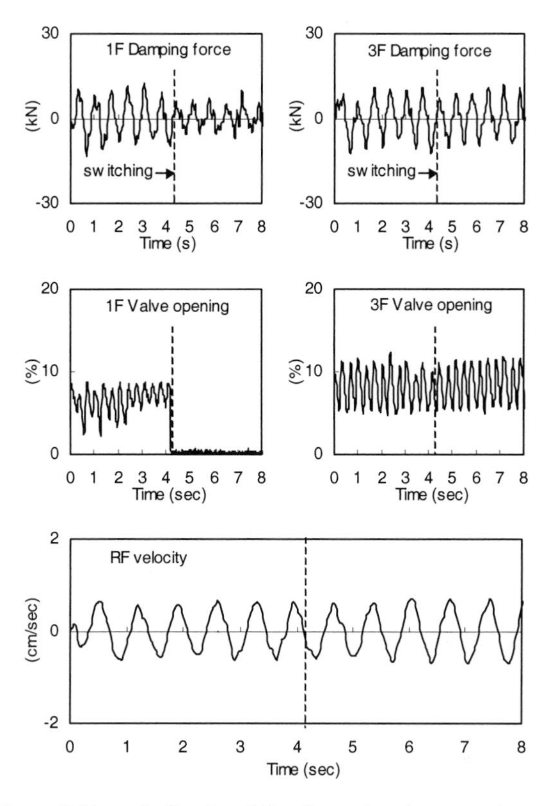

Figure 5: Forced vibration fail-safe semi-active control test

CONCLUSIONS

In this paper, the reliability of a semi-active structural control system is discussed. First, the outline of a building with Semi-active Hydraulic Damper (SHD) system was presented. Next, the health monitoring and fail-safe semi-active control was described. Then, the actual fail-safe performance of this system was confirmed through the forced vibration test. The elements that composed the SHD system demonstrated the specified performance and the whole system operated well. Finally, it is presented that the reliability of the system will be secured as well as the control performance.

ACKNOWLEDGEMENTS

The author would like to express his gratitude to Messrs. Y. Matsunaga, N. Niwa and H. Midorikawa of Kajima Corporation, and to Kawasaki Heavy Industry Ltd. for developing the SHD system.

REFERENCES

Dyke S.J. and Spencer Jr. B.F. (1996). Seismic response control using multiple MR dampers. *Proc., 2nd Int. Workshop on Struct. Control*, 163-173.

Jansen L.M. and Dyke S.J. (2000). Semiactive control strategies for MR dampers: comparative study. *J. Engrg. Mech.*, ASCE, **126:8**, 795-803.

Kobori T. et al. (1991). Shaking Table Experiment and Practical Application of Active Variable Stiffness (AVS) System. *Proc., 2nd Conf. on Tall Buildings in Seismic Regions*, 213-222.

Kobori T. (2000). Stream of structural control for large earthquakes - Semiactive control system -. *Proc., 2nd European Conf. on Struct. Control*, in press.

Kurata N. et al. (1994). Shaking table experiment of active variable damping system. *Proc., 1st World Conf. on Struct. Control* **2**, USC Publications, TP2 108-117.

Kurata N. et al. (1998). Semi-active damper system in large earthquakes. *Proc. 2nd World Conf. on Struct. Control* **1**, Wiley, West Sussex, U.K., 359-366.

Kurata N. et al. (1999). Actual seismic response controlled building with semi-active damper system. *Earthquake Engrg. Struct. Dyn.* **28**, 1427-1447.

Kurata N. et al. (2000). Forced vibration test of a building with semi-active damper system. *Earthquake Engrg. Struct. Dyn.* **29**, 629-645.

Kurata N. (2001). Actual seismic response control building with semi-active damper system. *Proc. Struct. Congr. 2001*, ASCE, CD-ROM.

Matsunaga Y. et al. (1998). Development of actual size semi-active hydraulic damper for large earthquake. *Proc., 2nd World Conf. on Struct. Control* **2**, Wiley, West Sussex, U.K., 1615-1622.

Patten W. N. et al. (1999). Field test of an intelligent stiffener for bridges at the I-35 Walnut Creek Bridge. *Earthquake Engrg. Struct. Dyn.* **28**, 109-126.

Soong T.T. and Spencer Jr. B.F. (2000). Active, Semi-active and hybrid control of structures. *Proc., 12th World Conf. on Earthquake Engrg.*, No. 2834.

Spencer Jr. B.F. et al. (1996). Magnetorheological dampers: a new approach to seismic protection of structures. *Proc. Conf. on Decision and Control,* 676-681.

Spencer Jr. B.F. et al. (1998). "Smart" dampers for seismic protection of structures: a full-scale study. *Proc., 2nd World Conf. on Struct. Control* **1**, Wiley, West Sussex, U.K., 417-426.

Spencer Jr. B.F. (2000). Advanced in semi-active control of civil engineering structures. *Proc., 2nd European Conf. on Struct. Control*, in press.

Yoshida K. and Fujio T. (2000). Semi-active base isolation control of a building using variable oil damper. *Proc., 3rd Int. Workshop on Struct. Control*, World Scientific, Singapore, 567-575.

Advances in Building Technology, Volume 1
M. Anson, J.M. Ko and E.S.S. Lam (Eds.)

HYBRID RESPONSE CONTROL OF CLOSELY SPACED BUILDINGS

George C. Lee, Mai Tong[1] and Yihui Wu[2]

(1) Multidisciplinary Center for Earthquake Engineering Research,
State University of New York at Buffalo, Buffalo, NY 14261, USA
(2) Maguire Group Inc., 225 Chapman St. Providence, RI 02852, USA

ABSTRACT

Seismic retrofit of buildings in urban centers often encounters the challenge of limiting the displacement between closely spaced buildings in addition to acceleration reduction. Recent advances in energy dissipation technologies have enabled structural engineers to control deformational responses to desired levels. However, in order to handle responses of various magnitude, particularly those associated with large inelastic deformations, energy dissipation devices are not effective. This paper describes a hybrid control strategy for reducing building seismic responses. By switching between damping and stiffness devices, the control system is adaptive to different levels of building displacements. An example of controlling the roof displacement of two closely spaced buildings is given to compare the tradeoff of the control effect using a combination of passive damping with/without variable stiffness.

KEYWORDS

Semi-active control, effective damping, structural responses, response reduction.

PASSIVE DAMPING DEVICES AND THEIR APPLICATIONS IN BUILDING SEISMIC PROTECTION

The use of energy dissipation devices (EDD), in particular the use of viscous dampers, as a retrofit strategy for building structures has increased in recent years. Some design procedures for EDD are available in FEM 273/274 *"Guidelines for the Seismic Rehabilitation of Buildings"* and FEMA356, "*Prestandard and commentary for the seismic rehabilitation of buildings*". Design of dampers under performance based design considerations have also been studied (Elhassan et al, 1998).

The current practice typically follows a linear elastic response analysis approach, as described in FEMA guidelines. In FEMA 273, the corresponding effective damping due to the added supplemental energy dissipation devices is to be calculated according to the formula:

$$\beta_{eff} = \beta + \frac{\sum_j C_j \cos^2 \theta_j \phi_{rj}^2}{2\omega \sum_i \left(\frac{w_i}{g}\right) \phi_i^2} \tag{1}$$

where C_j is the damping constant of device j, θ_j is the angle of inclination of device j to the horizontal direction, ϕ_{rj} is the first mode relative displacement between the ends of device j in the horizontal direction, ω is the fundamental frequency of the rehabilitated building including the stiffness of the velocity dependent devices, w_i is the reactive weight of floor level i, and ϕ_i is the first mode displacement at floor level i.

The effective damping β_{eff} is then used to modify the design spectrum coefficient B_s (short period design spectral coefficient) and B_l (long period design spectral coefficient). Figure 1 shows the changes of the level of design spectrum as the effective damping is increased.

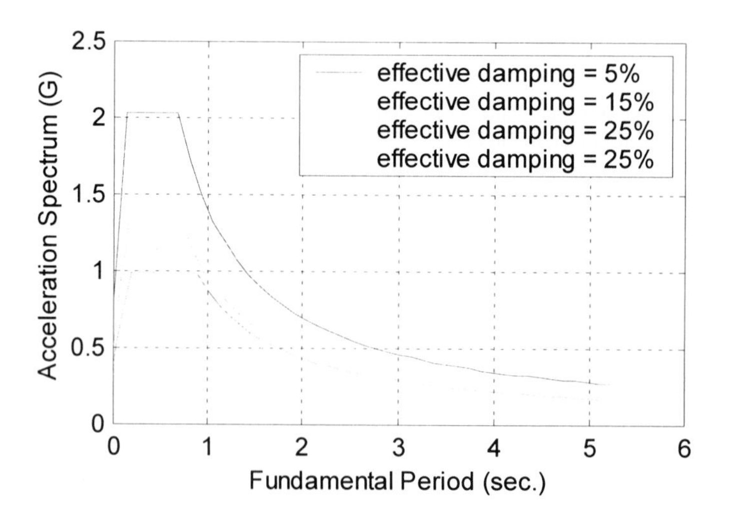

Figure 1: Acceleration Spectrum for LA area per FEMA 273/356

Figure 1 is based on an assumption of 5% design damping. For destructive earthquakes, most structures will experience inelastic deformation, in which structures will have undergone of their ductility capacity to various degrees. Response reduction factors in design codes typically reflect such effect (Bertero, 1986, 1991). The equivalent damping from the ductility will largely overpass that of the energy dissipation devices. As it is shown in Figure 1, when the effective damping level increases, the same percentage increase of damping will produce less response reduction. On the other hand, as structural responses enter the inelastic range, the fundamental period will increase, which will result in reduced acceleration response, but increased displacement response. Therefore, displacement response level will be the major concern for large earthquakes.

COMBINED RSPM AND PASSIVE DAMPING HYBRID CONTROL SYSTEM

The Real-time Structural Parameter Modification (RSPM) technology is a semi-active nonlinear control system for reducing structural seismic responses (Liang, Tong and Lee, 1999a,b; Lee et al, 1997). This semi-active nonlinear system can effectively reduce seismic responses, especially when the objective is to reduce displacement or story drift. A combined RSPM and passive damping control approach can achieve better seismic reduction than the passive damping alone. The combined system can reduce the acceleration response very effectively because of the added damping. Figure 2 shows this hybrid system which consists a passive damper and a controlled stiffness unit. The passive damper will always contribute to the dissipation of energy of the system which the stiffness unit the deformation reaches a threshold value—termed as the open distance. If the relative displacement (the absolute value) is larger than the open distance, the stiffness unit will contribute to reduce the response. If, at any instant, the displacement (absolute value) becomes smaller than the threshold, the RSPM nonlinear stiffness will be automatically switched off. It will be switched on only when the displacement exceeds the threshold, and the RSPM control mechanism is activated only when the stiffness is engaged. The devices are combined in pairs of a tension and a compression units working as a push-and-pull set.

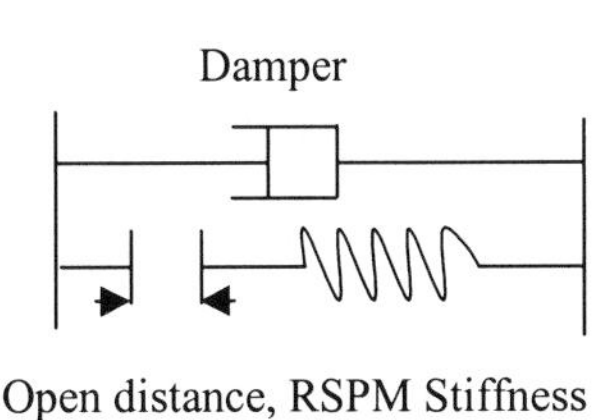

Figure 2: Combined RSPM and passive damping hybrid control system

This hybrid control system has been studied and designed fro two adjacent existing buildings in LA. The weaker building is a seven story steel structure (Building I) with full moment resisting connections. The seismic retrofit problem is to avoid collision of this building with the much stiffer adjacent building II. The gap between these two buildings at the roof level is only six inches (see Figure 3). The combined RSPM and passive damping devices are considered for implementation between the two buildings at the roof level A secondary requirement of this retrofit project is to reduce the acceleration response level, for which pure damping device can be effectively used. The combined RSPM and damping approach therefore is developed to take advantage of their respective strengths.

The hybrid system consists of a linear viscous damper with C=6kips/inch, which is able to provide 30% critical damping for the structure. RSPM is controlled in seven different open distances (1,1.5, 2.0, 2.5, 3.0, 4.0, 5.0 inch) and eight stiffness levels (10, 20, 50, 100, 150, 200, 300, 500 kips/inch). Ground motions used in the analysis include the San Fernando earthquake and Kobe earthquake. The San Fernando earthquake ground acceleration is considered as the design earthquake. Under this excitation, the roof displacement response reaches up to 7.41 inches if without any response control devices.

Figure 4 shows the different roof response levels under different control schemes. Figure 5 shows the roof displacement response and acceleration response for the different level of controlled open distance under Kobe earthquake. It is seen that the displacement and acceleration responses have a tradeoff effect. As the open distance increases, the acceleration response will be reduced more effectively in the tradeoff of the displacement response reduction.

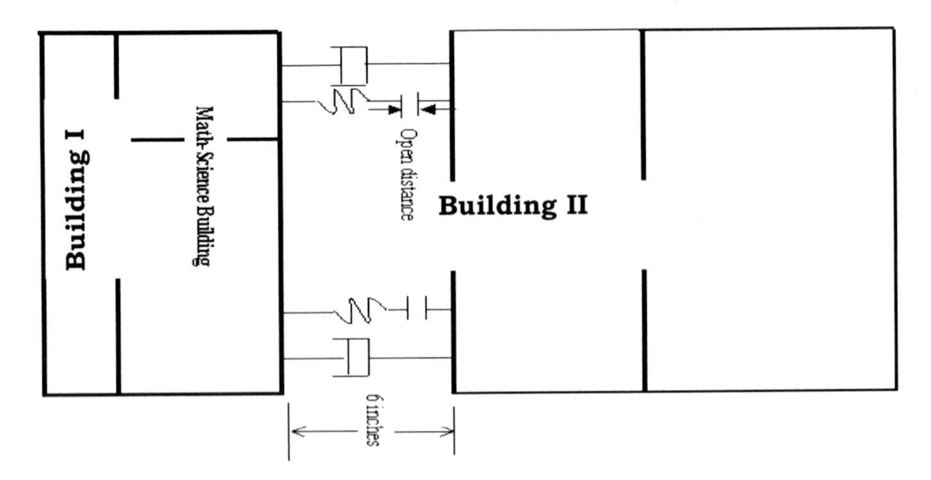

Figure 3: Application to two adjacent buildings

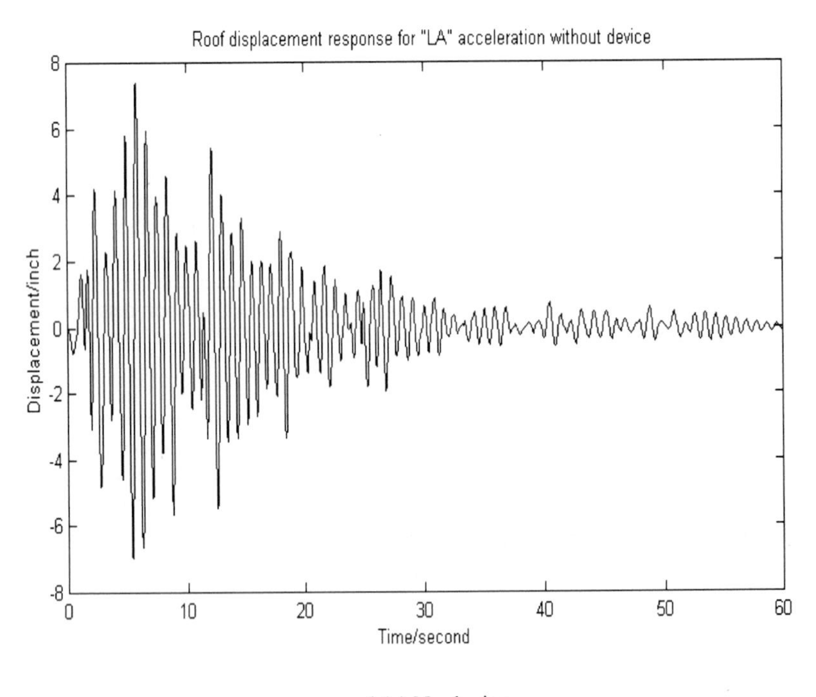

4 (a) No device

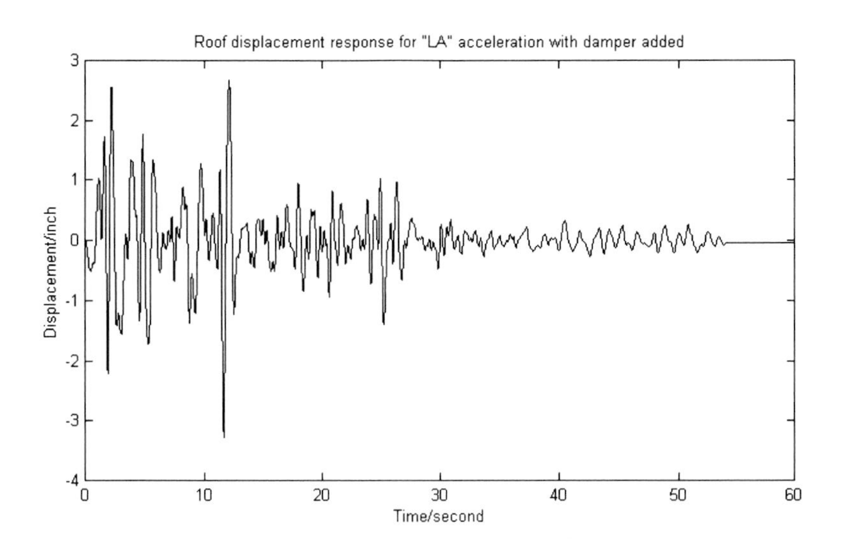

4 (b) Pure damping

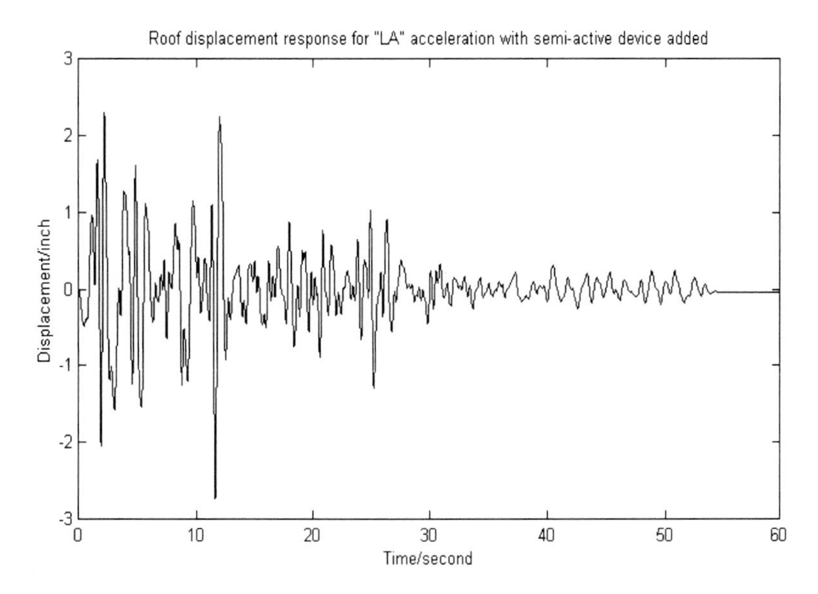

4 (c) Combined RSPM

Figure 4: Roof displacement response

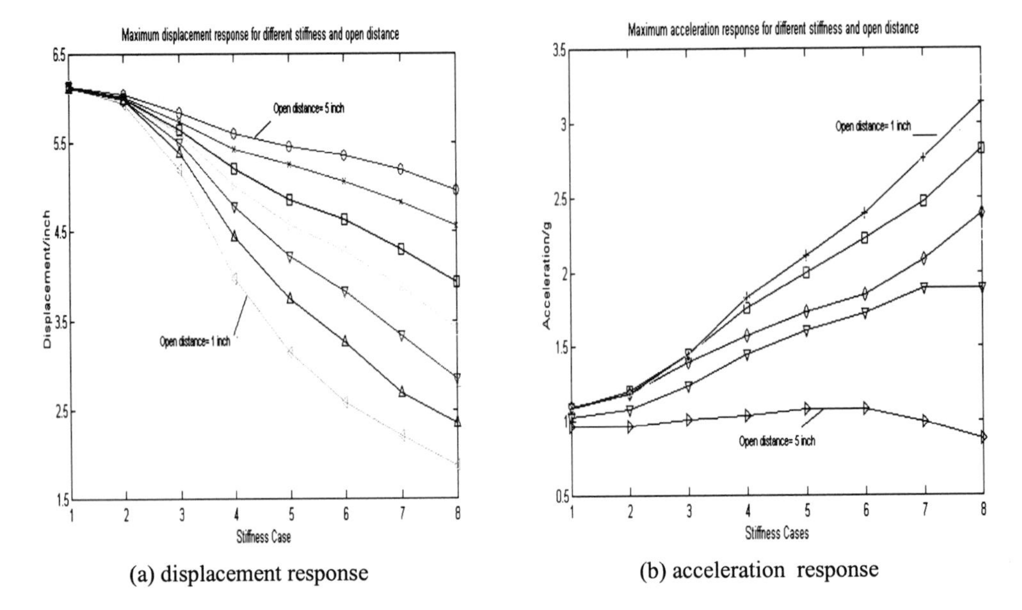

(a) displacement response

(b) acceleration response

Figure 5: Roof response reduction

COMPARISON OF DAMPING AND HYBRID CONTROL EFFECTS FOR INELASTIC STRUCTURAL RESPONSE

To compare the effectiveness of the hybrid system under various ductility demands, a single DOF nonlinear model with perfect plastic stiffness and 5% hardening spring is considered. The mass and linear stiffness of the spring are equivalent to the UCLA math science building. A 4% of critical damping is assumed for the structure. The bare structure is subject to ductility demands of 1-2 for post yield, 3-10 for life safety level and 10-16 for near collapse stage. An equivalent effective damping of 5% to 40% is added. The nonlinear time history analysis is carried out under the same design earthquake ground acceleration used in the linear analysis, but the amplitude is scaled to match the ductility demand levels of the bare structure (that is at each ductility demand level, the acceleration record is scaled up to the level to produce the corresponding ductility demand for the bare structure). The response reduction with pure damping is illustrated in Figure 6. As the ductility demands increases, the damping effectiveness decreases. It can be seen from Fig.6, for the same 15% pure damping, the structural response is reduced by about 15% for ductility demand of 15, 20% for ductility demand of 10, and 30% for ductility demand of 2.

In a separate comparison of the response reduction effect with combined RSPM and passive damping device, three RSPM stiffness levels of 10%, 20% and 30% of the structural stiffness are considered respectively. A 15% effective passive damping ratio is assumed for the hybrid system. The open distance is selected as 50% of the yield displacement. The results are illustrated in Figure 7. The response reduction (dashed line) of the 15% pure damping is also shown in the figure. It can be seen that the hybrid system has more response reduction than pure damping only. The response reduction effect increases as the ductility demand increases until 10. Higher RSPM will result in more response reduction. In the near collapse stage of ductility demand (11-16), because the RSPM stiffness is limited to less than 30% of the total stiffness, the response reduction effect become smaller. Overall, the most effective response reduction is in the life safety range (3-10).

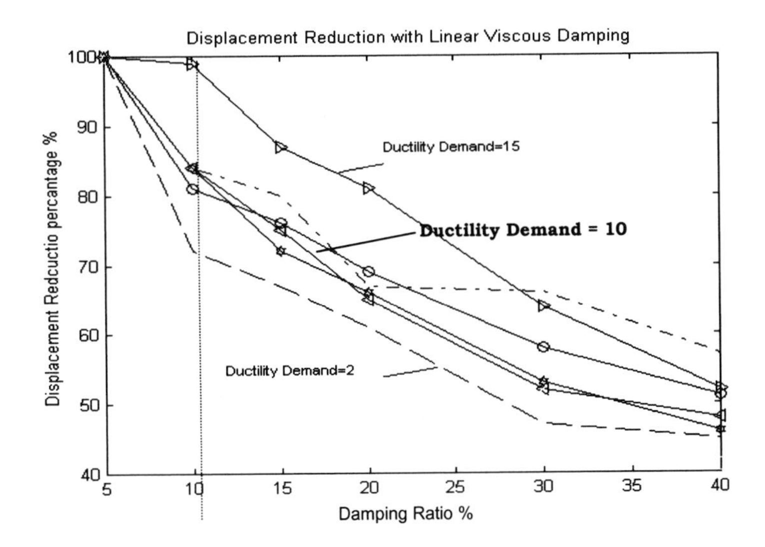

Figure 6. Nonlinear displacement reduction under combined damper

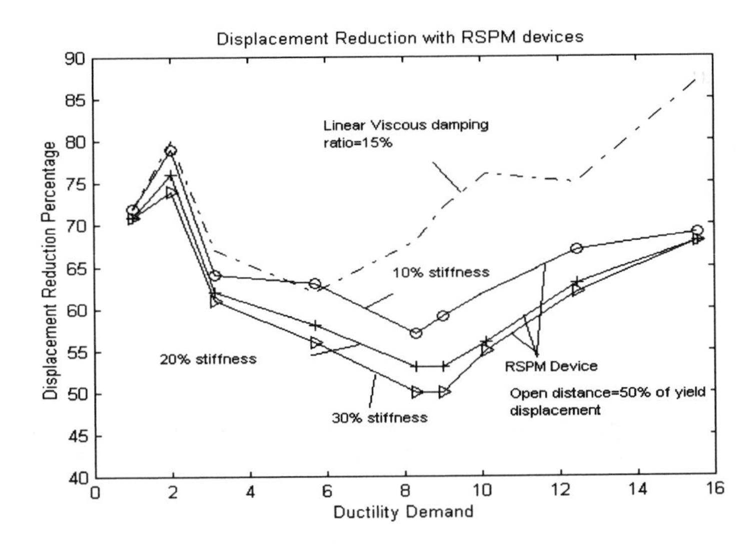

Figure 7: Nonlinear displacement reduction under combined RSPM

CONCLUSION

A combined RSPM and passive damping control system approach takes advantage of the RSPM effectiveness in reducing displacement and acceleration by passive damping. It can be used to effectively reduce the displacement and acceleration. By optimizing the control parameter, the roof displacement of s building (I) can be reduced effectively and the acceleration response can also be controlled at the same time. Inelastic response analysis indicates that the combined RSPM and passive damping hybrid control system can have 20% more response reduction for 15% damping and 30% stiffness level. The system has the greatest reduction in the ductility demand range of 5 to 10, which is the life safety performance ductility range for most building structures. In general, the nonlinear stiffness part of the RSPM technology alone can provide damping up to 30%; however, the size and capacity are normally balanced with other considerations such as the level of damping and the capacities of the supporting load bearing structural members.

ACKNOWLEDGEMENT

This study is performed by a grant from the National Science Foundation through the Multidisciplinary Center for Earthquake Engineering Research (Grant No. ECC-9701471).

REFERENCES

Bertero, V. V. (1986). Evaluation of Response Reduction Factors Recommended by ATC and SEAOC. *Proceedings of the 3rd US National Conference on Earthquake Engineering*, 1663-1673.

Bertero V. V. (1991). Design Guidelines for Ductility and Drift Limits: Review of State-of-the-Practice and State-of-the-Art in Ductility and Drift-Based Earthquake-Resistant Design of Buildings. UCB/EERC No.91/15.

Bunce, B.T. (1970). *Two and Three Dimensional Modeling and Dynamic Analysis of the Five Story Math-science Addition.* Master Thesis, UCLA, 1970.

Elhassan, R.M., Hart, G.C. and Liu, X. (1998). Viscous dampers in performance-based seismic rehabilitation of buildings. *Structural Engineering World Wide*, Sponsored by SEI-ASCE (CD-Rom), 1998.

Lee, G.C., Tong, M., Liang, Z., Houston, B., Taylor, S. Clinard, R.L., Tomita, K. and Gottwald, W. (1997). Structural Vibration Reduction with ISMIS Control Technology. *Proceedings of the 68th Shock and Vibration Symposium.*

Liang, Z., Tong, M. and Lee, G.C. (1999a). A Real-time Structural Parameter Modification (RSPM) Approach for Random Vibration Reduction: Part I - Principle. *J. Probabilistic Engineering*, June 1999, Vol. 14, 349-362.

Liang, Z., Tong, M. and Lee, G.C. (1999b). A Real-time Structural Parameter Modification (RSPM) Approach for Random Vibration Reduction: Part II - Experimental Verification. *J. Probabilistic Engineering*, Vol. 14, 362-385.

Advances in Building Technology, Volume 1
M. Anson, J.M. Ko and E.S.S. Lam (Eds.)

MODEL TESTING OF A NONLINEAR BASE ISOLATION CONCEPT

D. Michael McFarland[1], Yumei Wang[2], Alexander F. Vakakis[3] and Lawrence A. Bergman[1]

[1]Department of Aeronautical and Astronautical Engineering,
University of Illinois, Urbana, IL, 61801, USA
[2]Department of Civil and Environmental Engineering,
University of Illinois, Urbana, IL, 61801, USA
[3]Division of Mechanics, National Technical University of Athens,
GR–157 10 Zografos, Athens, Greece

ABSTRACT

We present a unique response reduction strategy for complex structures subjected to base or ground shock-induced vibrations. The primary focus of our efforts has been the development of a new type of passive nonlinear shock isolation system for protection of flexible structures that effectively and robustly isolates the structure from transient ground motions. The proposed design is based on the concept of nonlinear localization whereby induced vibrational energy is passively confined to a preassigned secondary system and away from the primary structure to be isolated. In this paper, we demonstrate the performance of the system using a simple, spatially one-dimensional model. The structure to be isolated is a two-degree-of-freedom system, which is weakly and linearly coupled to a single-degree-of-freedom intermediate or secondary system. This, in turn, is connected to ground through a hardening spring which, here, is produced by the parallel combination of two linear springs, one having a clearance nonlinearity. The structural system to be protected is weakly coupled to an intermediate system consisting of a subfoundation and linear spring which is tuned to a mode of the structural system. A clearance spring acts to excite the appropriate nonlinear normal mode, the clearance being a design parameter which determines the onset of nonlinear behavior as a function of the input to the system. The performance of the system is examined for pulse-type inputs using simulation.

KEYWORDS

Base isolation, nonlinear dynamics, nonlinear normal mode, mode localization

INTRODUCTION

The vulnerability of flexible structures to large-magnitude shock inputs is well documented. Various methods have been proposed, and in some cases implemented, through which the reliability of these structures can be enhanced. Implementations include passive isolation systems and auxiliary damping devices, while proposed solutions include various manifestations of active and semi-active protective systems. Generally speaking, active systems have not been readily embraced due to considerations such as reliability, power requirements and cost, and semi-active systems are still in the research stage, while passive devices and systems have been employed for many years with great success despite some performance limitations that we seek to address with this technology.

The primary focus of our efforts has been the development of a new type of passive nonlinear shock isolation system for protection of flexible structures that, when implemented, effectively and robustly isolates the structure from transient ground motions, thus reducing the potential for damage and extending useful life. The proposed design is based on the concept of nonlinear localization whereby induced vibrational energy is passively confined to a preassigned secondary system and away from the primary structure to be isolated. We show that such a robust nonlinear localization phenomenon can be effected by: (a) tuning the shock isolation system so that a 1:1 internal resonance exists between the secondary substructure and a mode of the primary structure; and (b) inducing a strongly localized nonlinear normal mode in which vibration is predominantly confined to the secondary substructure. In this paper, we demonstrate the performance of the system using a simple, spatially one-dimensional model.

PROBLEM STATEMENT

A conventional base isolation system consists of linear stiffness and damping elements arranged to permit relative motion between the ground and a building's foundation. Its tuning is limited to the adjustment of the values of these components with the goal of decoupling the structure to be isolated from the ground supporting it. This decoupling can be achieved by making the stiffness of the foundation very small, with the result that small disturbances may produce large displacements while larger ground motions may use up all the travel available in the isolation system without adequately protecting the primary structure. On the other hand, a linear isolation system sized to function properly in response to a strong ground input may be too stiff and highly damped to offer the desired protection (decoupling) during a lesser disturbance. The design process typically includes trading off the reduction of "rattle space" to simplify utility connections, the minimization of the absolute accelerations of the superstructure to preserve occupant comfort and confidence, and the minimization of inter-story displacements to limit potentially damaging shear stresses.

The design goals for the system we propose are the same, but the introduction of an additional, nonlinear degree of freedom offers new options in meeting them. In particular, it becomes possible to preferentially direct energy to an auxiliary mass and confine it there by designing the isolation system so that the entire structure has a nonlinear normal mode of vibration in which motion is effectively limited to this degree of freedom. It will be seen that this tuning can be achieved through the linear components of the isolation system, leaving the nonlinear elements (a clearance and an additional linear stiffness) as additional parameters which may be adjusted to improve the overall response. One use of this additional capability is to create a structural system that responds appropriately to both large and small ground motions.

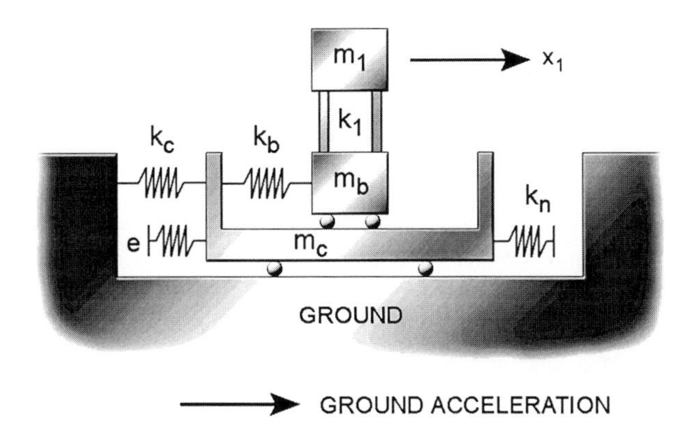

Figure 1: Nonlinear base isolation system with SDOF superstructure.

MECHANICAL MODEL AND ANALYSIS

The nonlinear base isolation system to be analyzed is shown schematically in Figure 1. The structure to be isolated is a two-degree-of-freedom system, which is weakly and linearly coupled to a single-degree-of-freedom intermediate or secondary system. This, in turn, is connected to ground through a hardening spring which, here, is produced by the parallel combination of two linear springs, one having a clearance nonlinearity. The structural system consisting of the foundation m_b, superstructure m_1 and column stiffness k_1 is weakly coupled to the subfoundation m_c through the spring k_b. The intermediate system consisting of the subfoundation m_c and linear spring k_c is tuned to a mode of the structural system. The clearance spring k_n acts to excite the appropriate nonlinear normal mode, while the clearance e is a design parameter which determines the onset of nonlinear behavior as a function of the input to the system.

Equations of Motion

We denote the restoring force provided by the nonlinear spring-gap element by

$$f_n(y_c) = \begin{cases} k_n(y_c - e), & y_c > e, \\ k_n(y_c + e), & y_r < e, \\ 0, & \text{otherwise} \end{cases} \tag{1}$$

and adopt the convention that x represents displacement with respect to a fixed reference frame, y, displacement with respect to the ground, and z, displacement with respect to the base mass m_b. Hence

$$x = y + x_g = z + y_b + x_g \tag{2}$$

and the equations of motion of the structure may conveniently be written as

$$\ddot{y}_c = \frac{1}{m_c}[-(c_c + c_b)\dot{y}_c + c_b\dot{y}_b - (k_c + k_b)y_c + k_b y_b - f_n] - \ddot{x}_g, \tag{3}$$

$$\ddot{y}_b = \frac{1}{m_b}[-c_b\dot{y}_b + c_b\dot{y}_c + c_1\dot{z}_1 - k_b y_b + k_b y_c + k_1 z_1] - \ddot{x}_g \tag{4}$$

plus an equation representing the dynamics of the superstructure,

$$m_1\ddot{z}_1 + c_1\dot{z}_1 + kz_1 = -m(\ddot{x}_g + \ddot{y}_b). \tag{5}$$

The input to the system in all cases examined here was a ground acceleration shock pulse of amplitude A_g and duration T_g. This ground motion was further assumed to be either a rectangular pulse,

$$\ddot{x}_g(t) = \begin{cases} A_g, & 0 < t < T_g, \\ 0, & \text{otherwise}, \end{cases} \tag{6}$$

or a half-sine pulse,

$$\ddot{x}_g(t) = \begin{cases} A_g \sin \dfrac{\pi t}{T_g}, & 0 < t < T_g, \\ 0, & \text{otherwise}, \end{cases} \tag{7}$$

both of which provide a large amount of energy to the system in a short time.

Numerical Simulation

The response of the system was simulated using Matlab and Simulink, where the isolation system was represented by a detailed block diagram while the building was described by the state-space model

$$\dot{\xi} = \mathbf{A}\xi + \mathbf{Bu}, \tag{8}$$

$$\eta = \mathbf{C}\xi + \mathbf{Du} \tag{9}$$

with

$$\xi = \begin{Bmatrix} z_1 \\ \dot{z}_1 \end{Bmatrix}, \quad \mathbf{A} = -\mathbf{A}_1^{-1}\mathbf{A}_2, \quad \mathbf{B} = \mathbf{A}_1^{-1}\mathbf{B}_1, \quad \mathbf{C} = \mathbf{I}, \quad \mathbf{D} = \mathbf{0} \tag{10}$$

where

$$\mathbf{A}_1 = \begin{bmatrix} c_1 & m_1 \\ m_1 & 0 \end{bmatrix}, \quad \mathbf{A}_2 = \begin{bmatrix} k_1 & 0 \\ 0 & -m_1 \end{bmatrix}, \quad \mathbf{B}_1 = \begin{Bmatrix} -m \\ 0 \end{Bmatrix}, \quad u = \ddot{x}_g. \tag{11}$$

No difficulty was encountered in solving these equations numerically using either fixed-step or adaptive integrators (e.g., Matlab's functions `ode4` and `ode45`, respectively). Numerical results obtained using Matlab were compared to the analytical solution for a simpler, SDOF model exhibiting the same sort of dead-band nonlinearity to verify that the potentially abrupt change in stiffness associated with the subfoundation mass m_c contacting the spring k_n posed no unusual numerical problems. In general, the adaptive integrator performed well, and it was used for all the simulations reported herein. The fixed-time step solver was found to be considerably slower for these equations, but has the advantage that its output is more amenable to some forms of post-processing.

TUNING ALGORITHM

The design of the isolation system begins with knowledge of the superstructure mass and stiffness, m_1 and k_1. The masses of the base, m_b, and the intermediate subsystem, m_c, are then chosen. These are in principle arbitrary but in any given application the total foundation and auxiliary mass will be expected to lie in a range familiar to designers; here we have chosen m_b for compatibility with existing experimental apparatus, then selected m_c as a practical fraction of the total mass of the isolated structure. Tuning of the isolation system is then accomplished through the selection of appropriate values of the springs k_c, k_b and k_n. This process is simplified by fixing the ratios of the other two stiffnesses to k_c, that is, by setting

$$k_b = \frac{1}{10} k_c, \tag{12}$$

$$k_n = 10 k_c. \tag{13}$$

TABLE 1
PHYSICAL PARAMETERS OF MODEL SYSTEM

Superstructure	
m_1	0.2 kg
k_1	3500 N/m
c_1	0 N/m/s

Base	
m_b	0.3 kg
k_b	583.1 N/m
c_b	0 N/m/s

Intermediate System	
m_c	0.2 kg
k_c	5831 N/m
c_c	0 N/m/s
k_n	58310 kg
e	Variable

The first of these equations reflects the requirement that the coupling between the intermediate SDOF system and the base mass be weak, and allows us to use a very simple formula to determine the single remaining parameter, k_c.

The desired response of the complete structure to ground motion is a mode in which vibration is almost entirely confined to the intermediate subsystem—a localized mode in which the base and superstructure participate very little. With the above assumption of weak coupling, and assuming for the moment that the displacement of the intermediate mass with respect to ground is smaller than the gap e so that the nonlinear element may be ignored, an approximate expression for the resonant frequency of the isolator's degree of freedom is seen to be

$$\omega_c = \sqrt{k_c/m_c}. \tag{14}$$

The determination of k_c, and hence k_b and k_n, is thus reduced to the specification of the resonant frequency ω_c. On physical grounds, this is chosen to be equal to the frequency of the first mode in which the base-superstructure combination would experience significant inter-story displacements. Considering the linear system consisting of masses m_b and m_1 and the spring k_1, but neglecting k_b, we can readily compute its natural frequencies. The lowest of these will be zero, corresponding to a rigid-body mode, but the second will be non-zero and will correspond to the mode to which we seek to tune the isolator. Denoting these frequencies

$$\tilde{\omega}_0 = 0, \quad \tilde{\omega}_1 > 0, \tag{15}$$

we can state the tuning condition as

$$\omega_c = \tilde{\omega}_1, \tag{16}$$

from which we immediately obtain

$$k_c = m_c \tilde{\omega}_1^2. \tag{17}$$

The values in Table 1 were calculated using this algorithm. The only parameter of the isolation system not now uniquely determined is the clearance e, the effect of which will be investigated below.

An experiment is planned which will utilize existing hardware to the extent practical while demonstrating the efficacy of the proposed isolation system and tuning algorithm. The values shown in Table 1 have been selected for the experimental structure to be used. These reflect previously identified or readily available components and are consistent with the tuning criteria for the isolator as outlined above. The most significant departure from similitude to a realistic civil structure is in the relatively large base mass m_b. This will be mollified in later experiments by increasing the number of floors in the superstructure and, thus, its total mass. However, the present values are adequate for proof-of-concept work, and the SDOF superstructure is retained here for simplicity.

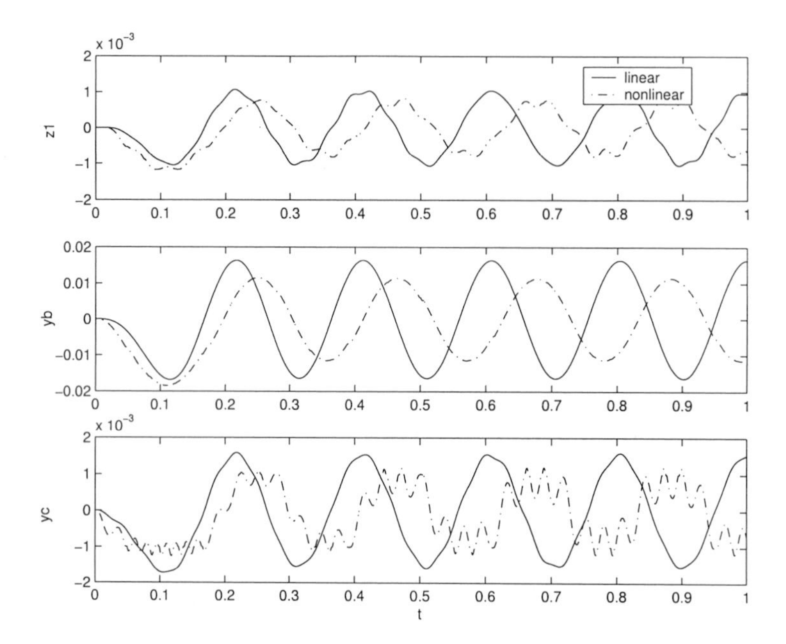

Figure 2: Time responses of the linear ($k_n = 0$) and nonlinear ($k_n = 58\,310$ N/m, $e = 0.001$ m) systems.

SIMULATION RESULTS

With the isolation system tuned as just described, a number of numerical simulations were made to assess its potential effectiveness. This section presents some typical results of these runs, then summarizes the influence of the design parameter e, the clearance between the spring k_n (tied to ground) and the intermediate mass m_c.

Time Histories of Response

Time history results from two cases are shown in Fig. 2, and clearly demonstrate the existence of the desired localized mode in the nonlinear system. The larger response plotted in this figure is that of the linear system obtained when $k_n = 0$. It depicts the interaction of three nonlocalized modes; the intermediate degree of freedom introduced by the isolator is ineffective in limiting the response of the superstructure. In the other simulation results shown on the same axes, the nonlinear spring coefficient has been returned to its nominal value of $k_n = 58\,310$ N/m and the clearance set to $e = 0.001$ m (a large gap in that it is approximately two-thirds of the peak value of y_c from the linear analysis). The response plots indicate the isolator is beginning to function as intended, with the building and base mass moving less than in the linear case while the intermediate mass m_c responds more strongly than before and with additional, higher-frequency motion.

Shock Spectra

In order to compare the simulation results obtained with different nonlinear isolation system designs, we have found it useful to adopt the familiar concept of the *shock spectrum* of a mechanical system (Meirovitch, 1986), in which the maximum value of a response quantity is plotted against the shock pulse duration T_g. Conventionally, both pulse duration and response amplitude are normalized with respect to

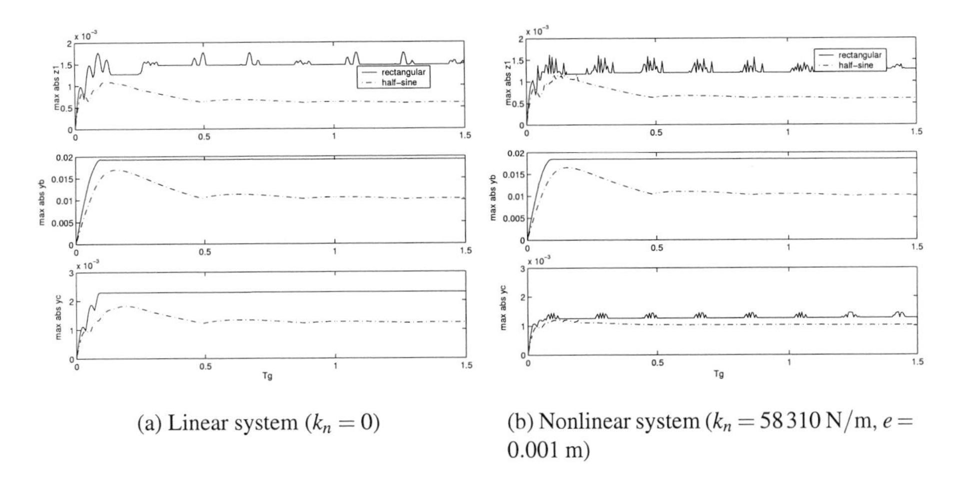

(a) Linear system ($k_n = 0$)

(b) Nonlinear system ($k_n = 58\,310\ \mathrm{N/m}$, $e = 0.001$ m)

Figure 3: Shock spectra of responses to rectangular and half-sine ground accelerations.

combinations of the system parameters, but in dealing with nonlinear systems it is more convenient to work directly with dimensional quantities.

The shock spectra for the two systems of the previous section are shown in Fig. 3. In each case, two curves are plotted for each of the three degrees of freedom, corresponding to rectangular and half-sine shocks. Those familiar with the classical shock spectra of an SDOF system will notice these are less smooth due to the modal interactions of the multiple degrees of freedom present. The isolator's effectiveness can be seen most clearly by comparing the peak superstructure responses at several values of the pulse duration. With the clearance used here ($e = 0.001$ m), the response spectrum resulting from the half-since shock is almost the same for the nonlinear system as for the linear. However, the peak response z_1 due to the rectangular pulse is significantly reduced for all pulse durations. Although not shown by these plots, a similar response improvement occurs for the half-sine input when the clearance is reduced or the shock amplitude increased.

Effect of the Isolator Clearance

With the other design parameters fixed, the clearance (gap) e is still available with which to optimize the response of the isolated structure. Its influence is indicated in Fig. 4, where the maximum value of the shock spectra are plotted against e. In constructing these plots, the amplitude of the ground acceleration was held constant at $A_g = 1$ g. In an actual application, there is a minimum level of ground motion that must occur before the nonlinearity comes into play, and the larger the gap is made the greater this ground acceleration threshold becomes. In general, however, it can be said that there is an optimum gap from the standpoint of reducing the response of the isolated structure and its base to strong ground motions, although for the present problem it appears that optimum lies outside the range of Fig. 4.

CONCLUSION

A novel base isolation system depending for its operation on the tuning and excitation of a highly localized nonlinear mode has been described and demonstrated through numerical simulations. Isolation is

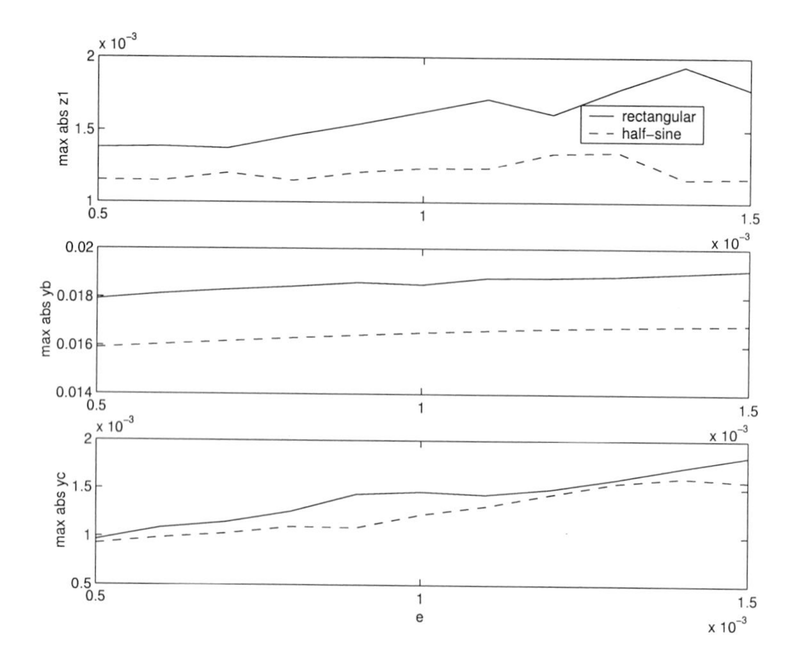

Figure 4: Effect on the maximum value of the shock spectrum of varying the clearance e.

achieved by confining response energy to an auxiliary mass, vibration of which dominates this localized mode. An effective tuning algorithm has been reviewed and the influence of the remaining free design parameter, a clearance, has been investigated through the computation of shock spectra.

REFERENCES

Gendelman O. (2001). Transition of energy to a nonlinear localized mode in a highly asymmetric system of two oscillators. *Nonlinear Dynamics* **25:1–3**, 237–253.

Ma X., Nayfeh T.A., Vakakis A.F. and Bergman L.A. (2000). Experimental verification of shock reduction achieved through non-linear localization. *Journal of Sound and Vibration* **230:5**, 1177–1184.

Meirovitch L. (1986). *Elements of Vibration Analysis*, 2nd ed., McGraw-Hill, New York.

Vakakis A.F. (2001). Shock isolation through the use of nonlinear energy sinks. Submitted to the *Journal of Vibration and Control*.

Vakakis A.F., Kounadis A.N. and Raftoyiannis I.G. (1999). Use of non-linear localization for isolating structures from earthquake-induced motions. *Earthquake Engineering and Structural Dynamics* **28**, 21–36.

Vakakis A.F., Manevitch L.I., Mikhlin Y.V., Pilipchuk V.N., Zevin A.A. (1996). *Normal Modes and Localization in Nonlinear Systems*, John Wiley & Sons, New York.

Advances in Building Technology, Volume I
M. Anson, J.M. Ko and E.S.S. Lam (Eds.)

RISK CONTROL OF SMART STRUCTURES USING DAMAGE INDEX SENSORS

Akira Mita and Shinpei Takahira

Department of System Design Engineering, Keio University
3-14-1 Hiyoshi, Kohoku-ku, Yokohama 223-8522, Japan

ABSTRACT

Simple and inexpensive passive sensors that can monitor the peak strain or displacement of a critical structural member were developed. The developed sensors have an ability to quickly assess the degree of damage in a structure when a checkup is needed. The sensors need no power supply for monitoring. The peak values can be retrieved wirelessly if desired. In addition, they can be easily modified to measure other damage indexes such as maximum acceleration and force. The mechanism to memorize the peak strain or displacement values relies on the pure plastic extension of sensing section. The pure plastic extension of the sensing section is made possible by introducing elastic buckling. The peak value is detected by measuring a change in electric resistance, inductance or capacitance. In addition, introduction of an LC circuit into the sensor enabled wireless retrieval of the data. Theoretical and experimental studies exhibit the feasibility of the developed sensors for structural health monitoring of smart structures.

KEYWORDS

Damage index, smart structure, health monitoring, peak sensor, wireless, damper

INTRODUCTION

A health monitoring system is getting strong attention for controlling or reducing risks associated with natural hazards such as large earthquakes. This trend has been accelerated after the 1994 Northridge Earthquake and the 1995 Hyogo-Ken Nanbu (Kobe) Earthquake[1)]. Both occurred in the populated areas so that many infrastructures and engineered buildings experienced unexpected damages. For example, many steel buildings suffered sever damages mainly at their beam-column joints. It was a

surprising fact that the damages were not identified until removing fire-protection coatings on beam-column joints. In most cases, it was not possible to find the correct degree of the damages by a simple eye-inspection of the structure surface because there were no major visible damages.
One of the key issues raised was the difficulty to find and quantify damages in piles and foundations of a building. If the building were inclined or tilted, it would be rather easy to notice the structural damage. However, as was the case for beam-column joints in steel structures, a simple eye-inspection can not identify the damages in piles and foundations in most cases. Even when the damage was identified to be in a pile system, it is necessary to excavate soils around the foundation for accurate assessment. Thus, an efficient health monitoring system for piles and foundations is indeed useful.

In the course of developing a health monitoring system for steel structures, piles and others, it was found that the current sensor and network technologies had certain limitations for proper assessment. For example, nondestructive damage assessment of beam-column joints in a tall steel building is very difficult without removing coating materials for fire-protection. Evaluation of embedded piles is further difficult since excavation of foundation soil is necessary for installing sensors. Though several nondestructive methods were proposed, no single method could satisfy all the requirements for correct assessment of damages.

In addition to the beam-column steel joints and piles, damper devices for smart structures are becoming sources of needs for health monitoring systems. Typical building structures equipped with damper devices are depicted in Fig. 1. The number of devices installed into a tall steel building may easily exceed one hundred. The current inspection practice depends solely on eye-inspections so that a certain number of wall panels covering damper devices have to be removed when the building is hit by a large earthquake. This inspection work will be not only time-consuming but also costly.

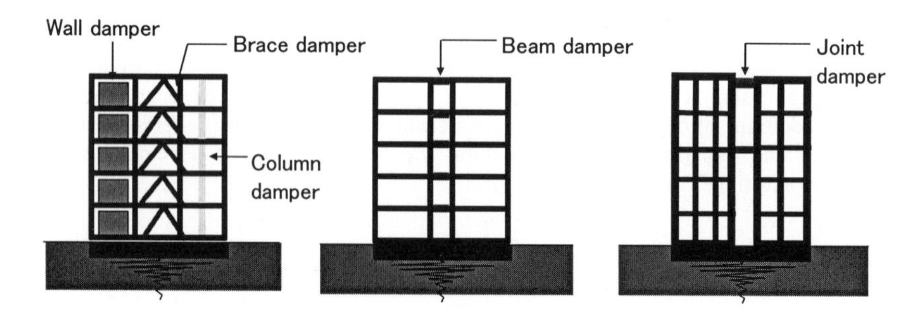

Figure 1: Smart structures with dampers

For a conventional building structure, introducing sensors to detect damages may not be practical as the damage scenario is complicated due to high degree of nonlinearity and statical indeterminacy. However, a smart structure that is equipped with passive or active dampers as shown in Fig. 1 has a clearer scenario for the damage. The input power due to earthquakes or strong winds is guided to be absorbed by dampers. Therefore, installing sensors to identify the condition of dampers may become feasible. For a ten-storey building, we may need to know the health of one hundred dampers for assuring the structural health. As a realistic and economical means, employing damage index sensors to the dampers will be attractive if the cost for sensors is reasonable. A damage index is defined as a physical value that is well correlated to a critical damage in a structure. Typical damage indexes include peak strain, peak displacement, peak acceleration, story drift, absorbed energy, accumulated plastic deformation, and so on. They will only memorize the indexes that are well correlated to the damage without any electrical power supply. Therefore, the maintenance cost for such sensor systems is minimal.

In the past few years, several peak strain sensors have been proposed and developed. The peak strain sensor is one of the most promising damage index sensors. The TRIP (TRansformation Induced Plasticity) steel was used as a sensor head as it is magnetized when large strain is applied[2]. However, the TRIP sensor may not be reused once the sensor experiences a large strain. In addition, detection of the magnetization level is rather difficult. In another research, the relationship between the electric resistance change and the peak strain was studied for CFGFRP (Carbon Fiber Glass Fiber Reinforced Plastics)[3]. It was concluded that the change of the electric resistance in the CFGFRP material would be well correlated to the peak strain. This feature is unique as the material itself can function as a sensor. Unfortunately, however, the electric resistance is correlated not only to the peak strain but also to the residual strain so that isolation of the peak strain is difficult. In the other studies, the electric resistance of SMA (Shape Memory Alloys) was used in the hope of using it as a peak strain sensor[4]. However, the characteristics of the sensor output was not so simple.

In this paper, a new concept is presented to memorize the peak strain or the peak displacement. The mechanism utilizes elastic buckling of a thin wire. In addition, the memorized value can be retrieved wirelessly.

MECHANISM OF DAMAGE INDEX SENSOR

One of the most useful damage indexes is the peak strain. In the following, the mechanism of a sensor that can memorize the peak strain is explained. Although the mechanism is to memorize the maximum strain only, the same mechanism is applicable to obtain other damage indexes with minor modifications. The ideal response of a peak strain sensor is plotted in Fig. 2 compared with a conventional strain sensor. The peak strain sensor keeps the peak strain value even when the strain in the object material is released. This feature can be realized by using a material that has a pure plastic response against applied load for a sensor element. We propose to use the elastic buckling of a thin wire to introduce such plastic response as shown in Fig. 3.

In Fig. 3, the right end of a thin wire is attached to a conductive block. The left side of the wire is sandwiched by a conductive block resulting in a certain level of friction force. At the initial phase, no tension is applied to the wire. When the left block is pulled to the left direction, the wire will be stretched. Under the condition that the tension force in the wire reaches beyond the static friction force, the wire is pulled out from the left conductive block. When the tension force is removed, the wire may keep the extended length if the static friction force is larger than the elastic buckling force for the extended wire. Thus the peak strain is obtained by measuring the length of the wire. If the wire is electrically resistive, the length is linearly related to the change of the electric resistance.

A slightly different mechanism is shown in Fig. 4. In this case, the plastic behavior is achieved by a variable element such as a variable capacitor, a variable inductor or a variable resistance. Therefore, the length of the thin wire in this case is unchanged except for elastic elongation. The plastic deformation is kept in the variable element by utilizing the inherent friction. The sensor is sensitive in only one direction. In the opposite direction, the sensor is insensitive by the buckling of the thin wire.

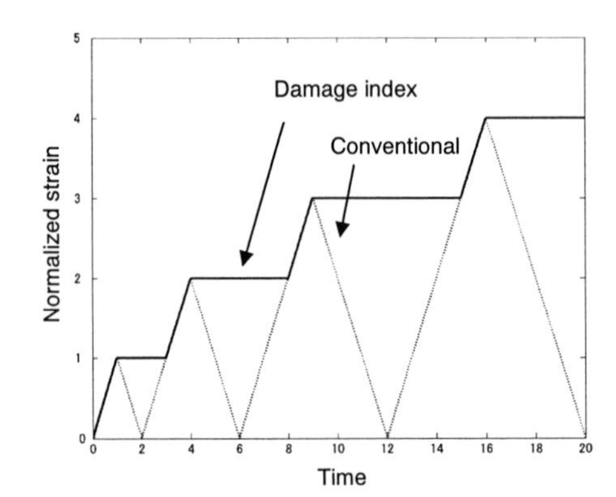

Figure 2: Ideal response of damage index sensor

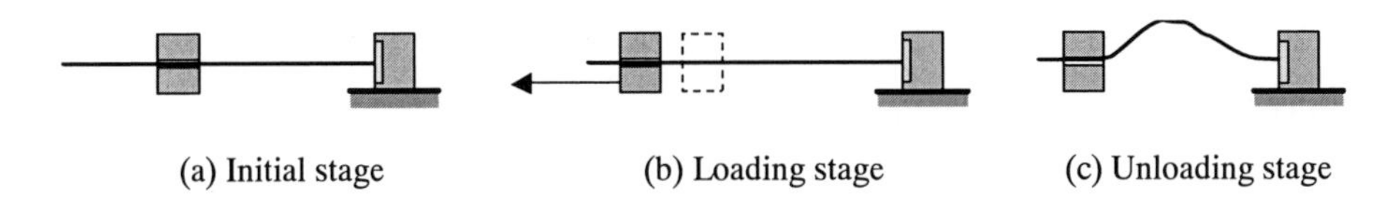

(a) Initial stage (b) Loading stage (c) Unloading stage

Figure 3: Mechanism of memory using elastic buckling of a thin wire

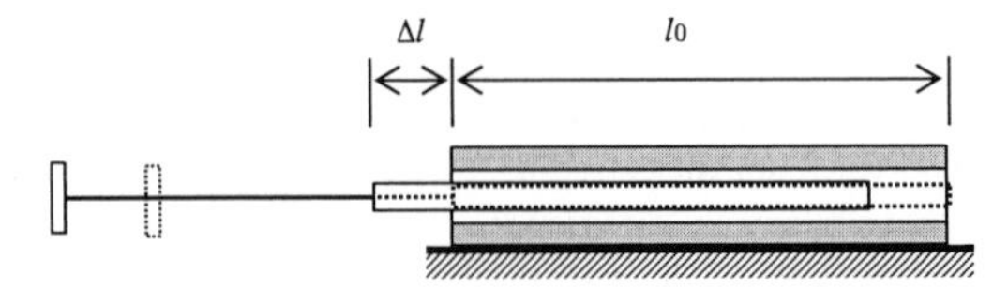

Figure 4: Mechanism of a damage index sensor consisting of a variable element and a thin wire

The buckling force P_{cr} to induce the first buckling mode in a thin wire with fixed-end boundary conditions at both ends and of its length of l is given by

$$P_{cr} = \frac{4\pi^2 EI}{l^2} \tag{1}$$

where E is Young's modulus and I is the moment of inertia of an area. The corresponding buckling stress and strain in the wire are given by

$$\sigma_{cr} = \frac{4\pi^2 EI}{l^2 A}, \quad \varepsilon_{cr} = \frac{4\pi^2 I}{l^2 A} \tag{2}$$

where A is the cross sectional area of the wire. The condition required for the peak strain sensor is to keep this buckling force smaller than the static friction force so that the extended length of the wire should be kept. In addition, the buckling strain should be as small as possible since it is associated with the maximum elastic strain induced in the sensor. The elastic component in the response of the sensor should be minimized.

For a peak strain sensor depicted in Fig. 2 of its initial length l_0 for the resistant wire, the output from the sensor as a change of resistance ΔR is given by.

$$\frac{\Delta R}{R_0} = \frac{\Delta l}{l_0} + 2\nu \frac{\Delta l^e}{l_0} \tag{3}$$

where ν is Poisson's ratio. The increase of the length is divided into two components.

$$\Delta l = \Delta l^e + \Delta l^p \tag{4}$$

The superscript e represents elastic elongation. The superscript p indicates the component associated with plastic or permanent elongation beyond the friction force. The typical but exaggerated response of the sensor is shown in Fig. 5. In this figure, the dynamic friction coefficient was assumed to be half of the statical friction coefficient. The buckling stress was assumed to be negligibly small. In the segment 1 in Fig. 5, only elastic elongation occurs until the stress in the thin wire exceeds the static friction force. In the segment 2, the friction is reduced to the dynamic friction. The increase of the resistance is due to the plastic elongation. In the segment 3, the elastic elongation is reduced to zero. In the segment 4, the thin wire is buckled so that the same output is kept.

For a peak strain sensor using a variable capacitor and a thin wire depicted in Fig. 4, the expected response is slightly different from the above. The response is typically given by

$$\frac{\Delta C}{C} = \frac{\Delta l}{l_0} \tag{5}$$

where C is capacitance. From Eqn. 5, it is clearly understood that the response of this sensor is only due to the change of variable element and not due to the elastic deformation of the thin wire. A simulated response of the sensor consisting of a thin wire and a capacitor is depicted in Fig. 6. In the segment 1, due to the static friction force, the sensor will not give any output before exceeding a certain strain where the resulting stress becomes larger than the static friction. In the segment 2, the tension force is larger than the dynamic friction force so that the variable capacitor changes its capacitance. In the segment 3, the tension force is removed from the thin wire so that the capacitance is kept at the maximum value.

The damage indexes are obtained by measuring changes in resistance, capacitance or inductance. Although the simplest configuration would be using a variable resistor, the use of a variable capacitor or a variable inductor has an additional benefit, that is, wireless retrieval capability. When a variable capacitor or a variable inductor is used, the sensor can be easily modified to form a closed LC circuit by adding an inductor or a capacitor. When an LC circuit consisting of a capacitor C and an inductor

L exists, the natural frequency of the circuit is given by

$$f = \frac{1}{2\pi\sqrt{LC}} \tag{6}$$

This natural frequency can be detected without touching the wire. The simplest way to measure the frequency would be using a dip meter. A dip meter generates radio waves of various frequencies and detects the frequency at which the energy is dipped by the nearby LC circuit. The frequency at which the energy is absorbed can be considered to be the natural frequency of the nearby sensor. The desired peak strain or displacement value is therefore retrieved wirelessly from the measured natural frequency. This feature is extremely useful, when the sensor should be covered by a fire-protection material or a cosmetic wall.

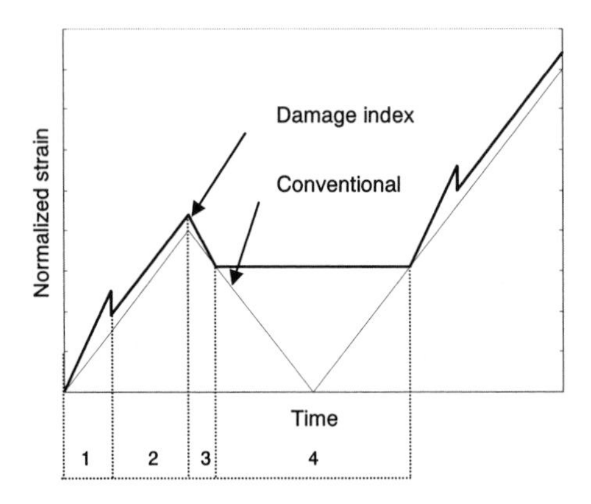

Figure 5: Simulated response of damage index sensor using a resistant thin wire

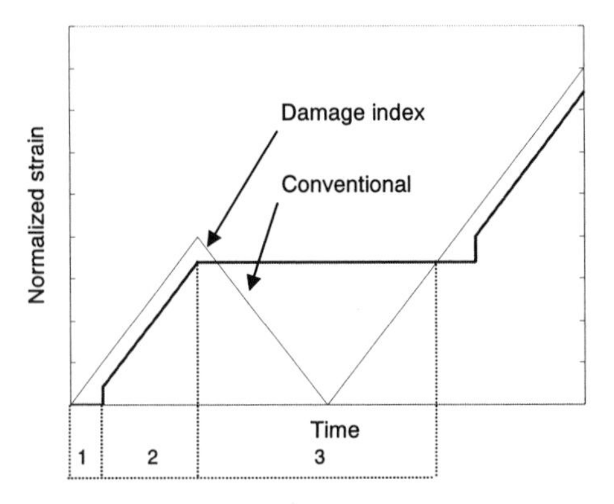

Figure 6: Simulated response of a damage index sensor using a variable capacitor and a thin wire

EXPERIMENTAL VERIFICATION

In the following experiment, a peak strain sensor consisting of a thin wire and a variable capacitor depicted in Fig. 4 was chosen for verification as shown in Fig. 7.

A variable capacitor used here is depicted in Fig. 8. It is made of two aluminum cylinders that are separated by a non-conductive material as a separator. Therefore, the capacitance of this system can be varied by changing the overlapping length of two aluminum cylinders. The measured initial capacitance was 217 pF. An inductor added to the variable capacitor to form a closed LC circuit was measured to be about 25 μH. Therefore, the natural frequency of the prototype sensor at its initial condition is 2.16 MHz at its initial position. The buckling wire is made of a fluorocarbon line to assure high Young's modulus. The diameter of the wire is 219 μm. The output was compared with the data taken by the laser sensor. The data were retrieved in the form of natural frequency change.

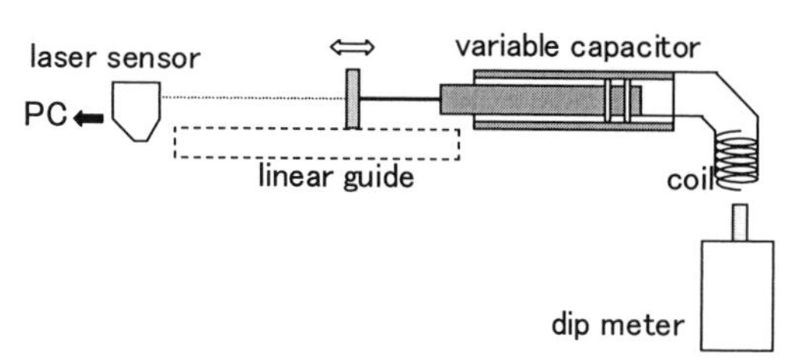

Figure 7: Experimental setup

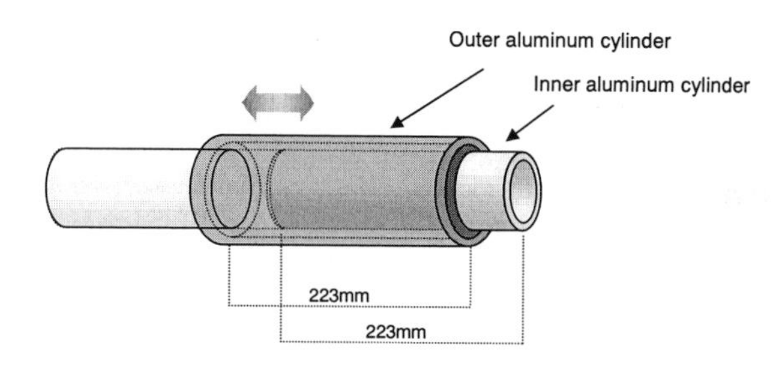

Figure 8: Variable capacitor consisting of aluminum cylinders

The measured results are shown in Fig. 9. From this figure, it is clearly shown that the peak response is indeed memorized. The response was measured using a dip meter. The initial position corresponds to 2.1637 MHz. The largest value of the peak strain sensor in Fig. 9 corresponds to 2.4177 MHz. If this sensor is bonded to a material with the initial sensor length of 400 mm, the range tested here corresponds to 0 μstrain to 52,750 μstrain. As the mechanism employed here is very flexible, the measurement range can be easily narrowed or widened by changing the length of variable capacitor and the length of wire. The size of the sensor can be as small as a conventional strain gauge or as large as the largest displacement sensor. The mechanism employed here to memorize the peak strain or displacement value can be extended to memorize other physical values such as force, stress, acceleration, velocity, accumulation of plastic deformation and so on.

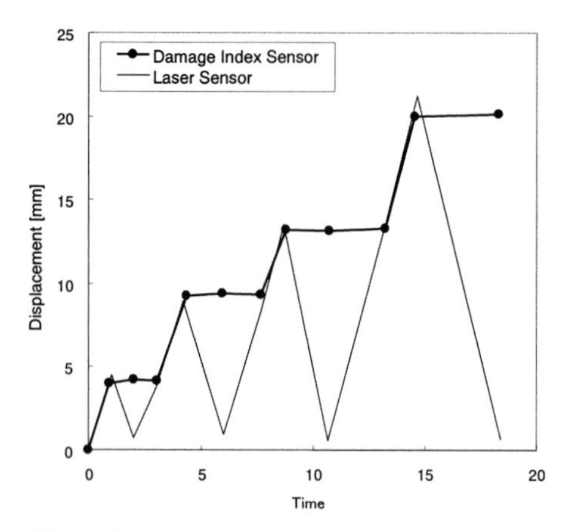

Figure 9: Response of damage index sensor

CONCLUDING REMARKS

A damage index is defined as a physical value that can be well correlated to a critical damage in a structure. Typical damage indexes include peak strain, peak displacement, peak acceleration, absorbed energy, accumulated plastic deformation, and so on. A mechanism of simple and inexpensive passive sensors that can monitor such damage indexes was developed. The mechanism to memorize the damage indexes relies on the pure plastic deformation of the sensing section. The pure plastic deformation of the sensing section is made possible by introducing a thin wire that is easily buckled by a small compressive force. The change in the length of the sensing section is detected by a change in resistance, inductance or capacitance. The sensors need no power supply for monitoring. In addition, introduction of an LC circuit into the sensor enabled wireless reading of the damage index. The damage index sensors are the most promising tools for controlling risks associated with natural hazards for smart structures with passive or active dampers by assessing the quantitative health conditions within a reasonable cost.

REFERENCES

1. Mita, A. (1999). Emerging Needs in Japan for Health Monitoring Technologies in Civil and Building Structures. *Proc. Second International Workshop on Structural Health Monitoring*, Stanford University, Sept. 8-10, 56-67.
2. Westermo, B.D. and L. Thompson (1994). Smart Structural Monitoring: A New Technology. *International Journal of Sensors*, 15-18.
3. Muto, N., H. Yanagida, T. Nakatsuji, M. Sugita, Y. Ohtsuka and Y. Arai (1992). Design of Intelligent Materials with Self-Diagnosing Function for Preventing Fatal Fracture. *Smart Materials and Structures*, **1**, 324-329.
4. Kakizawa, T. and S. Ohno (1996). Utilization of Shape Memory Alloy as a Sensing Material for Smart Structures,. *Proc. Advanced Composite Materials in Bridges and Structures*, 67-74.

Advances in Building Technology, Volume 1
M. Anson, J.M. Ko and E.S.S. Lam (Eds.)

Seismic Protection of Base-isolated Structures Using Semi-active MR Dampers

T. Tse[1], and C.C. Chang[2]

[1] Research Assistant, Department of Civil Engineering, Hong Kong University of Science and Technology, Clear Water Bay, Kowloon, Hong Kong.
[2] Associate Professor, Department of Civil Engineering, Hong Kong University of Science and Technology, Clear Water Bay, Kowloon, Hong Kong.

ABSTRACT

Magnetorheological (MR) damper is one semi-active device that shows a great potential for vibration suppression and hazard mitigation of civil structures. One special advantage of the damper is that its damping characteristics can be adaptively and quickly altered by a varying magnetic field. This semi-active damper has been shown to be more effective than passive dampers in reducing the seismic response of flexible structures with a dominant natural period of 1.5 sec or more. Examples of these structures include base-isolated structures, cables of long-span bridges and tall buildings. In this study, the effectiveness of MR dampers for seismic protection of base-isolated structures is investigated. The damper is commanded by an inverse Bouc-Wen model together with a linear quadratic regulator (LQR) algorithm. An elevated highway bridge, modeled as a 2-degree-of-freedom system, is used to illustrate the effectiveness of the MR damper in protecting a base-isolated structure.

KEYWORDS

MR damper, Based-isolated Structure, Inverse Dynamic Model, LQR, Semi-active Control

INTRODUCTION

Magnetorheological (MR) damper is a newly developed semi-active device that shows great promises for civil engineering applications. The damper utilizes special characteristics of the MR fluid that possesses magnetically controllable yield strength and can reversibly change from free-flow to semi-solid in milli-seconds. This semi-active damper has been shown to be more effective than passive dampers in reducing the seismic response of flexible structures with a dominant natural period of 1.5 sec or more. A clipped-optimal control algorithm proposed by Dyke et al. (1996) has been adopted to adjust the command voltage of the MR damper (Dyke et al. 1996, Dyke et al. 1998, Johnson et al. 2000). Analytical and experimental studies demonstrated that the MR damper, used in conjunction with the clipped optimal control algorithm, was effective for controlling a multi-story building model.

Alternatively, Chang and Tse (2001) proposed an inverse dynamic model to command the MR damper. The inverse dynamic model could directly estimate voltage that needed to be input into the damper in order to produce a desired damper force. The inverse dynamic model was derived from the Bouc-Wen model based on the following two assumptions: (i) the external displacement is much larger than the internal displacement, and (ii) the MR fluid always behaved in the post-yielding region. Under these assumptions, it was possible to directly link the damper force to the command voltage. Experimental and numerical results showed that the inverse model can provide a smoother tracing of target force as compared with the clipped-optimal model.

In this study, the effectiveness of MR damper for seismic protection of base-isolated structure is investigated. The damper is commanded by an inverse Bouc-Wen model together with a linear quadratic regulator (LQR) algorithm. An elevated highway bridge modeled as a 2-degree-of-freedom system is used to illustrate the effectiveness of MR damper in protecting this bridge under three different ground excitations: the El Centro, the Northridge, and the Kobe earthquake.

BOUC-WEN HYSTERESIS MODEL

A phenomenological model proposed by Spencer et al. (1997) to portray the hysteretic behavior of a prototype MR damper was shown to be accurate over a wide range of operation. The governing equations of the model are:

$$f = c_1\dot{y} + k_1(x - x_0) \tag{1}$$

$$\dot{y} = \frac{1}{c_0 + c_1}\left[\alpha z + c_0\dot{x} + k_0(x - y)\right] \tag{2}$$

$$\dot{z} = -\gamma|\dot{x} - \dot{y}|z|z|^{n-1} - \beta(\dot{x} - \dot{y})|z|^n + A(\dot{x} - \dot{y}) \tag{3}$$

$$\alpha = \alpha_a + \alpha_b u \tag{4}$$

$$c_0 = c_{0_a} + c_{0_b} u \tag{5}$$

$$c_1 = c_{1_a} + c_{1_b} u \tag{6}$$

$$\dot{u} = -\eta(u - v) \tag{7}$$

where x is the external displacement of the MR damper; f is the force generated by the MR damper; y is the internal displacement of the damper; c_0 and c_1 control the viscous damping at large and low velocities, respectively; k_0 and k_1 represent the stiffness at large velocity and the accumulator stiffness, respectively; x_0 is the initial displacement of spring k_1 associated with the nominal damper force due to the accumulator; α is a parameter associated with the evolutionary variable z; γ, β and A are the hysteresis parameters for the yield element; u and v are the output of a first-order voltage filter and the command voltage, respectively. The optimal values of the 14 parameters (k_0, k_1, γ, β, A, n, x_0, α_a, α_b, α_c, c_{0_a}, c_{0_b}, c_{1_a}, c_{1_b}, and η) were determined by solving a constrained nonlinear optimization with experimental data of a prototype MR damper. While large-scale MR dampers that are capable of producing forces in the order of 20 tons are currently under study, the force ranges for commercially available MR dampers are too small for practical application. For numerical demonstration, Erkus et al. (2002) suggested multiplying some of the parameters by a modification factor MF to scale up the force of the prototype MR damper. The parametric values and the MF factor used in this study are summarized in Table 1. By setting the MF factor at 1264, the force range of the MR damper will be in the order of a few hundred tons.

INVERSE DYNAMIC MODEL

It has been suggested that the force generated by MR damper cannot be directly commanded. Only the voltage v supplied to the current driver for the MR damper can be directly changed. A clipped-optimal control algorithm proposed by Dyke et al. (1996), based on acceleration feedback, is thus used along with MR damper to approximately generate the optimal control force f_c. The command voltage is determined based on how the damper force f compared with the desired control force f_c,

$$v = v_{max} H[(f_c - f)f] \tag{8}$$

where v_{max} is the maximum voltage to the current driver associated with saturation of the magnetic field in the MR damper and H(•) is the Heaviside step function. However it was found that the force generated by the MR damper using the clipped algorithm oscillated about the target control force. This was primarily due to the switching of voltage at either its minimum or maximum value.

Chang and Tse (2001) proposed an approximate formula based on the Bouc-Wen hysteretic model with the following two assumptions: (i) the external displacement x was much larger than the internal displacement y; and (ii) the evolutionary variable z was assumed to be equal to its ultimate hysteretic strength z_u (Spencer 1986), i.e.,

$$z \cong z_u = \mathrm{sgn}(\dot{x} - \dot{y})\left(\frac{A}{\gamma + \beta}\right)^{1/n} \cong \mathrm{sgn}(\dot{x})\left(\frac{A}{\gamma + \beta}\right)^{1/n} \tag{9}$$

Based on these two assumptions, the damper force f can then be approximated as,

$$f \cong \frac{(c_{1_a} + c_{1_b} u)}{(c_{0_a} + c_{1_a}) + (c_{0_b} + c_{1_b})u}\left[(\alpha_a + \alpha_b u)z_u + (c_{0_a} + c_{0_b} u)\dot{x} + k_0 x\right] + k_1(x - x_0) \tag{10}$$

The evolutionary variable z is found to be equal to the ultimate hysteretic strength for most regions. This suggests that the MR fluid operates in its post-yielding region for most of the time. There is some discrepancy for forces at low velocity, which is attributed to the second assumption where no intermediate values are allowed for z. The discrepancy however does not affect the control performance since the force magnitudes in these regions are small and the time duration is short.

It can be seen from Eqn. 10 that the damper force is a function of the voltage filter output u, the external displacement x and velocity $\dot{x}$. If the magnitude of the damper force f is specified (i.e. optimal control force f_c determined from control algorithm), Eqn. 10 can be rearranged to a quadratic function in u:

$$p_2 u^2 + p_1 u + p_0 = 0 \tag{11}$$

$$p_2 = c_{1_b} \alpha_b z_u + c_{0_b} c_{1_b} \dot{x} \tag{12}$$

$$p_1 = (c_{1_a} \alpha_b + c_{1_b} \alpha_a) z_u + (c_{0_a} c_{1_b} + c_{1_a} c_{0_b})\dot{x} + c_{1_b} k_0 x + (c_{0_b} + c_{1_b})[k_1(x - x_0) - f] \tag{13}$$

$$p_0 = c_{1_a} \alpha_a z_u + c_{0_a} c_{1_a} \dot{x} + c_{1_a} k_0 x + (c_{0_a} + c_{1_a})[k_1(x - x_0) - f] \tag{14}$$

It is seen that u at any time instance can be solved rather easily and the corresponding command voltage v can then be numerically obtained from Eqn. 7. Two experiments were conducted to illustrate how the damper forces could be commanded to trace desired target forces. Results showed that this inverse model could provide a smoother tracing of target forces as compared to the clipped algorithm.

NUMERICAL EXAMPLE

Modeling of elevated highway bridge

An elevated highway bridge modeled as a 2-degree-of-freedom system (see Fig. 1) proposed by Erkus et al. (2002) is used here to illustrate the effectiveness of the MR damper. The MR damper is rigidly connected between the pier and the bridge deck. The equations of motion for this system and the MR damper are

$$\boldsymbol{M_s}\ddot{\boldsymbol{s}} + \boldsymbol{C_s}\dot{\boldsymbol{s}} + \boldsymbol{K_s}\boldsymbol{s} = \boldsymbol{\Gamma}\mathrm{f_c} - \boldsymbol{M_s}\Lambda\ddot{\mathrm{s}}_\mathrm{g} \tag{15}$$

where $\boldsymbol{s}$ and $\ddot{\mathrm{s}}_\mathrm{g}$ are the displacement vector of the building model and the earthquake ground acceleration, respectively; $\mathrm{f_c}$ is the control force; $\boldsymbol{\Gamma}$ and Λ represent the location of the MR damper and the excitation, respectively; and $\boldsymbol{M_s}$, $\boldsymbol{C_s}$ and $\boldsymbol{K_s}$ are the mass matrix, the damping matrix and the stiffness matrix with the following values:

$$\boldsymbol{M_s} = 10^3\begin{bmatrix}100 & 0\\ 0 & 500\end{bmatrix}\ \mathrm{kg};\quad \boldsymbol{C_s} = 10^3\begin{bmatrix}321.6 & -196\\ -196 & 196\end{bmatrix}\ \mathrm{N\cdot sec/m};\quad \boldsymbol{K_s} = 10^6\begin{bmatrix}23.5 & -7.7\\ -7.7 & 7.7\end{bmatrix}\ \mathrm{N/m}$$

For semi-active control, the mass and the stiffness matrices remain the same, while the damping coefficient of bearing c_2 is set to 0. The state-space form of Eq. 16 is

$$\dot{\boldsymbol{x}} = \boldsymbol{A}\boldsymbol{x} + \boldsymbol{B}\mathrm{f_c} + \boldsymbol{E}\ddot{\mathrm{s}}_\mathrm{g} \tag{16}$$

where $\boldsymbol{x}$ is the state vector. The matrices $\boldsymbol{A}$, $\boldsymbol{B}$, and $\boldsymbol{E}$ can be written as,

$$\boldsymbol{A} = \begin{bmatrix}\boldsymbol{0} & \boldsymbol{I}\\ -\boldsymbol{M_s}^{-1}\boldsymbol{K_s} & -\boldsymbol{M_s}^{-1}\boldsymbol{C_s}\end{bmatrix};\quad \boldsymbol{B} = \begin{bmatrix}\boldsymbol{0}\\ \boldsymbol{M_s}^{-1}\boldsymbol{\Gamma}\end{bmatrix};\quad \boldsymbol{E} = -\begin{bmatrix}\boldsymbol{0}\\ \Lambda\end{bmatrix}; \tag{17a-c}$$

The control force $\mathrm{f_c}$ depends on the gain matrix $\boldsymbol{G}$ as follows,

$$\mathrm{f_c} = -\boldsymbol{G}\boldsymbol{x} \tag{18}$$

The gain matrix $\boldsymbol{G}$ is a constant matrix depending on the control algorithm used.

Control Algorithm

In this study, the linear-quadratic regulator (LQR) algorithm is used for the semi-active control. The control force $\mathrm{f_c}$ is so determined that it minimizes the following quadratic objective function,

$$\mathrm{J} = \int_0^\infty (\boldsymbol{x}^\mathrm{T}\boldsymbol{W}\boldsymbol{x} + \mathrm{f_c}^\mathrm{T}\boldsymbol{R}\mathrm{f_c})\mathrm{dt} \tag{19}$$

where $\boldsymbol{W}$ and $\boldsymbol{R}$ are the weighting matrices. By adjusting these weighting matrices, one can reduce the structural responses with different objectives, such as reducing the pier displacement, the deck displacement, or the inter-story drift. The weighting matrices used in this study are assumed to be,

$$\boldsymbol{W} = \frac{1}{2}\begin{bmatrix} rk_1 + k_2 & -k_2 & 0 & 0 \\ -k_2 & k_2 & 0 & 0 \\ 0 & 0 & rm_1 + m_2 & -m_2 \\ 0 & 0 & -m_2 & m_2 \end{bmatrix}; \quad R = 10^{-9} \qquad (20a,b)$$

where k_1 (=15.8 MN/m) and k_2 (=7.7 MN/m) are the stiffness of pier and deck respectively; m_1 (=100 tons) and m_2 (=500 tons) are the mass of pier and deck respectively; the weighting parameter r defines the relative importance of the pier and the deck responses. A larger value of r would decrease the pier response and a smaller value would decrease the deck response.

Numerical Results

To quantify the control effect, three performance indices are introduced.

$$J_1 = \max(s_1); \; J_2 = \max(s_2); \; J_3 = \max(s_2 - s_1) \qquad (21a\text{-}c)$$

where s_1 and s_2 are the pier and the deck displacement respectively. The weighting parameter r is set to be 0.1, which means the deck response is to be minimized. The time increment is set to be 0.005 sec. It is also assumed that an ideal actuator that can produce exactly the desired control forces specified by the control algorithm is available for comparison. Assume that the elevated bridge is excited by ground motion of three earthquakes: the El Centro, the Northridge and the Kobe.

Figures 2-4 show the simulation results for the three different ground excitations. It can be seen from the figures that the inter-story drift $s_1 - s_2$ is substantially reduced. Also the responses of the bridge controlled by the ideal actuator nearly coincide with those controlled by the MR damper. The MR damper forces are seen to coincide with the forces of the ideal actuator for all three earthquakes. These results demonstrate that the inverse dynamic model can be used effectively for commanding the MR damper in the control application. The three performance indices for the bridge under the three earthquake excitations are summarized in Table 2. Although the pier response J_1 appears to increase slightly under the El Centro and the Kobe earthquake, it can be seen clearly that the deck response J_2 and the drift between the deck and the pier J_3 are both reduced significantly for all three earthquakes. The control effect of installing an MR damper between the deck and the pier is quite vividly demonstrated in this example.

CONCLUDING REMARKS

The effectiveness of MR dampers for seismic protection of base-isolated structures is investigated in this study. The damper is commanded by an inverse Bouc-Wen model together with a linear quadratic regulator (LQR) algorithm. This inverse Bouc-Wen model provides a direct relationship between the target control force computed from the LQR algorithm and the required voltage input to the damper. An elevated highway bridge, modeled as a 2-degree-of-freedom system, is used to illustrate the applicability of the MR damper. Results show that the MR damper can effectively reduce the deck response and the drift between the deck and the pier without significantly increase the pier response.

ACKNOWLEDGEMENT

This work is supported by the Hong Kong Research Grant Council Competitive Earmarked Research Grant HKUST 6218/99E.

REFERENCE

Chang, C.C. and Tse, T. (2001). A Bouc-Wen Based Inverse Dynamic Model for Ccommanding MR Dampers. Submitted to *Earthquake Engineering and Structural Dynamics*.
Dyke, S.J., Spencer, Jr. B.F., Sain, M.K., and Carlson, J.D. (1996). Modeling and Control of Magnetorheological Dampers for Seismic Response Reduction. *Smart Materials and Structures*, **5**, 565-575.
Dyke, S.J., Spencer, Jr. B.F., Sain, M.K., and Carlson, J.D. (1998). Experimental Study of MR Dampers for Seismic Protection. *Smart Materials and Structures: Special Issue on Large Civil Structures*. **7:5**, 693-703.
Erkus, B., Abe, M., and Fujino Y. (2002). Investigation of Semi-active Control for Seismic Protection of Elevated Highway Bridges. *Engineering Structures*, **24:3**, 281-293.
Johnson, E.A., Christenson, R.E. and Spencer, Jr. B.F. (2000). Semiactive Damping of Cables with Sag." Proceedings of the International Conference on *Advances in Structural Dynamics*, Eds. J.M. Ko and Y.L Xu, 13-15 Dec. 2000, Hong Kong, Elsevier Service Ltd. Vol. I, 327-334.
Spencer Jr., B.F., Dyke, S.J., Sain, M.K. and Carlson, J.D. (1997). Phenomenological Model for Magnetorheological Dampers. *Journal of Structural Engineering*, ASCE, **123:3**, 230-238.
Spencer, Jr., B.F. (1986). *Reliability of Randomly Excited Hysteretic Structures*. Springer-Verlag, New York.

TABLE 1
PARAMETERS FOR THE MR DAMPER

Parameter	Value	Parameter	Value
MF	1264	α_a	140 MF N/cm
c_{0_a}	21 MF N·sec/cm	α_b	695 MF N/cm·V
c_{0_b}	3.5 MF N·sec/cm·V	γ	363 cm^{-2}
k_0	46.9 MF N/cm	β	363 cm^{-2}
c_{1_a}	283 MF N·sec/cm	A	301
c_{1_b}	2.95 MF N·sec/cm·V	n	2
k_1	5.0 MF N/cm	η	190 sec^{-1}
x_0	14.3 cm		

TABLE 2
PERFORMANCE INDICES J_1, J_2, AND J_3 OF THE ELEVATED HIGHWAY BRIDGE

	J_1 (cm)			J_2 (cm)			J_3 (cm)		
	No Control	Ideal	MR	No Control	Ideal	MR	No Control	Ideal	MR
El Centro	6.91	7.99	8.08	20.05	8.26	8.33	13.73	2.03	1.95
Northridge	18.03	17.62	16.68	52.80	18.86	18.85	34.89	4.53	4.81
Kobe	16.77	25.93	22.73	42.75	25.46	26.21	29.92	6.00	7.20

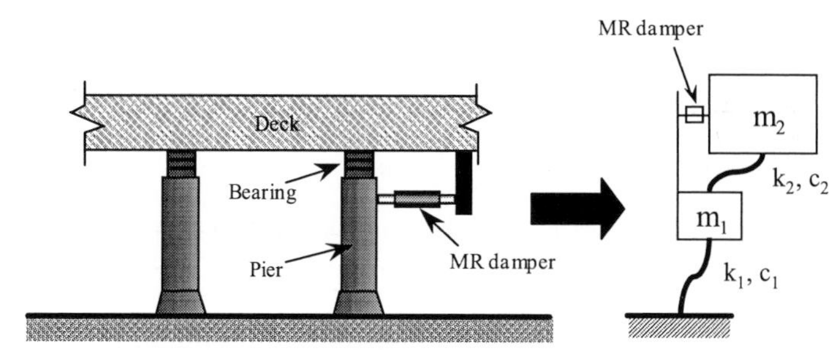

Figure 1: The elevated bridge and its mathematical model

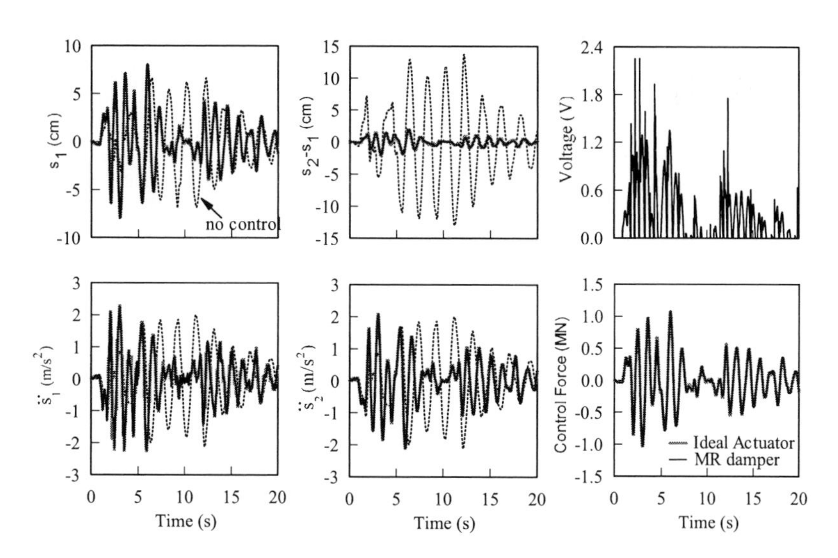

Figure 2: Results under the El Centro ground excitation

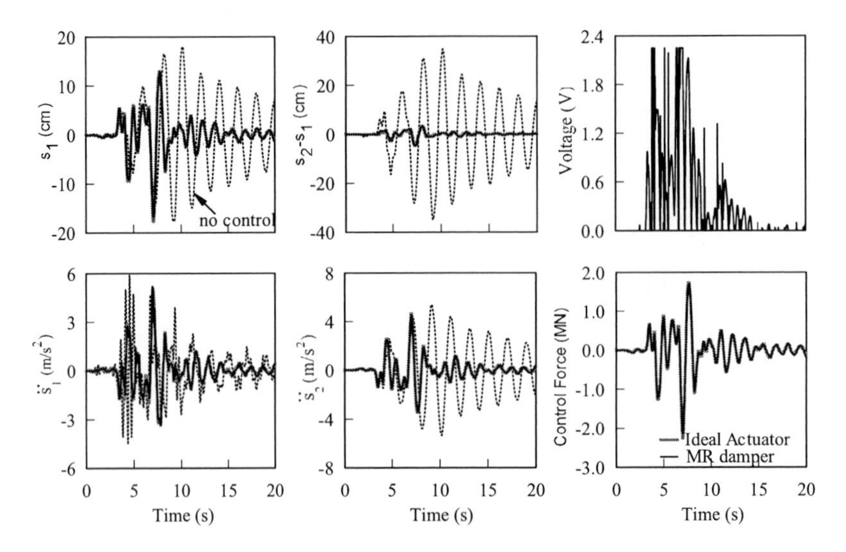

Figure 3: Results under the Northridge ground excitation

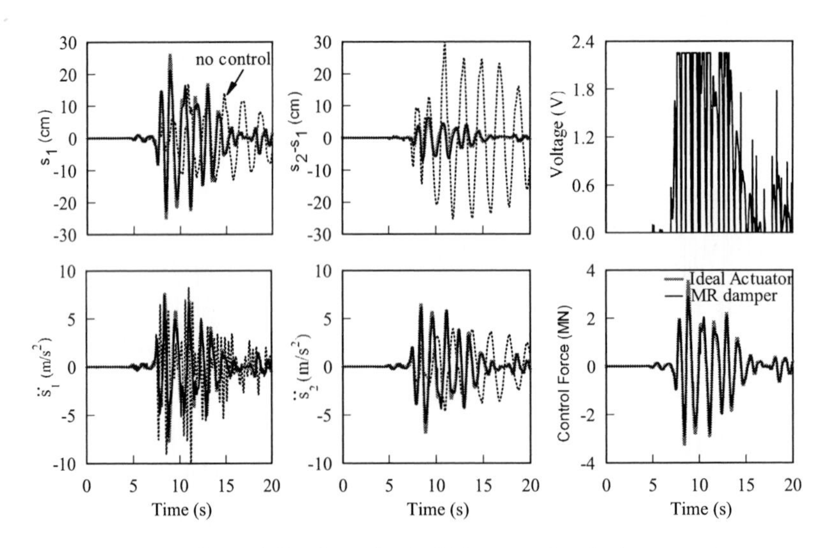

Figure 4: Results under the Kobe ground excitation

Advances in Building Technology, Volume 1
M. Anson, J.M. Ko and E.S.S. Lam (Eds.)

SEMI-ACTIVE TLCDS USING MAGNETO-RHEOLOGICAL FLUIDS FOR VIBRATION MITIGATION OF TALL BUILDINGS

J. Y. Wang[1], Y. Q. Ni[1], J. M. Ko[1], B. F. Spencer, Jr.[2]

[1]Department of Civil and Structural Engineering, The Hong Kong Polytechnic University, Hung Hom, Kowloon, Hong Kong
[2]Department of Civil and Environmental Engineering, University of Illinois at Urbana-Champaign, Urbana, Illinois 61801, USA

ABSTRACT

A semi-active tuned liquid column damper using magneto-rheological fluid (MR-TLCD) is proposed for wind-induced vibration mitigation of high-rise buildings. The magneto-rheological (MR) fluids can reversibly change from a free-flowing, linear viscous fluid to a semi-solid with controllable yield strength in milliseconds when exposed to a magnetic field. They are used as damping fluids to devise semi-active MR-TLCDs with alterable fluid viscosity. The sharply alterable fluid viscosity results in adjustable and controllable damping forces that achieve real-time structural vibration control. By applying a magnetic field to the MR fluid at the bottom part of a U-shaped tube, the damping effects of an MR-TLCD can be changed continuously through altering the magnetic field strength. Making use of the equivalent linearization technique, a stochastic response analysis for tall buildings incorporating semi-active MR-TLCDs subjected to random wind excitation is formulated. A 50-storey tall building using MR-TLCDs for wind-induced vibration mitigation is considered as a case study. The control effectiveness using MR-TLCDs is evaluated by comparing the RMS acceleration and deflection of the structure with an MR-TLCD damping system and with a conventional TLCD system.

KEYWORDS

Tuned liquid column damper (TLCD), magneto-rheological (MR) fluid, tall building, wind-induced excitation, random response analysis, vibration mitigation.

INTRODUCTION

The tuned liquid column damper (TLCD) is a supplemental energy dissipation system that relies on the motion of the liquid column in a U-shaped tube to counteract the action of external forces acting on the structure. Due to its cost-effectiveness, simplicity in installation, and low maintenance costs, the TLCD has attracted significant interest for researchers and engineers (Xu et al. 1992; Belendra et al. 1995, 1999).

The dynamic property and damping effect for conventional TLCDs are determined from the geometric construction of the liquid column. Therefore, the damping coefficient is unchangeable. However, the wind and earthquake forces are random in nature, and the extent and frequency content of the external forces acting on a structure are different at different times. In this sense, a damping-variable TLCD is desirable to achieve optimal damping under various load conditions, for example, using a controllable valve to adjust the TLCD orifice opening (Yalla et al. 2001).

In this study, a semi-active tuned liquid column damper using magneto-rheological fluid (MR-TLCD) is proposed, as shown schematically in Figure 1. The essential characteristic of MR fluids is their ability to reversibly change from a free flowing, linear viscous liquid to a semi-solid having a controllable yield strength in milliseconds when exposed to a magnetic field. By adjusting the strength of the applied magnetic field, the yield stress of the MR fluid, and thus the controllable damping of the MR-TLCD, can be developed.

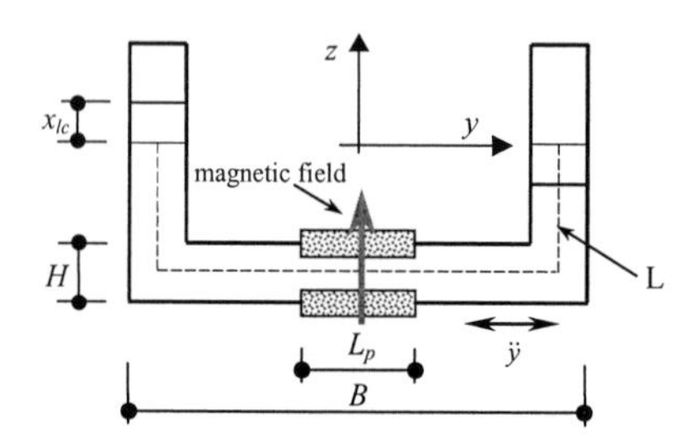

Figure 1: Schematic of MR-TLCD

STRUCTURAL SYSTEM WITH MR-TLCDS

A high-rise building is generally considered as a lumped-mass system with n degrees-of-freedom (DOF) corresponding to the n floor stories. Wind loading in the along-wind direction is applied to discrete lumped masses. To derive the analytical formulation, the case of installing p MR-TLCDs on selected floors is considered. It is assumed that the wind load does not act on the dampers directly.

Mechanics of MR Fluid in a Rectangular Duct

As shown in Figure 2, the motion of the MR fluid between fixed poles can be modeled by the parallel-plate model (Yang et al. 2002), The pressure drop developed along a pressure-driven flow can be assumed as the sum of a viscous component Δp_η and a component Δp_τ due to the field-dependent yield stress (Phillips 1969; Jolly et al. 1999).

$$\Delta p = \Delta p_\eta + \Delta p_\tau = \frac{12\eta Q L_p}{h^3 w} + \frac{c\tau_y L_p}{h} \tag{1}$$

where L_p, h and w are the length, depth and width of the flow between the fixed poles; Q is the volumetric flow rate; η is the field-independent viscosity and τ_y is the yield stress developed by the applied field; c is a function of the flow velocity and has a value ranging from 2.07 to 3.07 (Spencer et al. 1998; Yang et al. 2002).

The head loss and the pressure drop are related through

$$h_p = \frac{\Delta p}{g\rho} = \frac{12\eta Q L_p}{g\rho h^3 w} + \frac{c\tau_y L_p}{g\rho h} \tag{2}$$

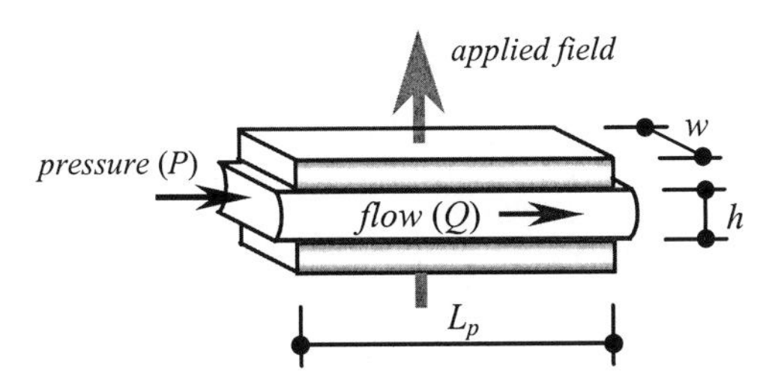

Figure 2: Operational model for pressure-driven fluid

The first term in the previous equation is the head loss due to the viscosity, which can be expressed as

$$h_{pv} = \frac{\delta_v V^2}{2g} \tag{3}$$

in which the head loss coefficient δ_v is obtained by

$$\delta_v = \frac{48}{\mathrm{Re}_H (1+\frac{h}{w})^2} \frac{L_p}{h}, \qquad \mathrm{Re}_H = \frac{2\rho V(w+h)}{\eta w h} \tag{4}$$

where Re_H is the Reynold number for rectangular duct. Then the total head loss along the complete fluid column can be computed from

$$h_w = \frac{\delta V^2}{2g} + \frac{c\tau_y L_p}{g\rho h}, \qquad \delta = \frac{48}{\mathrm{Re}_H (1+\frac{h}{w})^2} \frac{L}{h} + \sum_j \zeta_j \tag{5}$$

where δ is referred as the overall head loss coefficient, V is the fluid velocity, ζ_j is the coefficient of minor loss, including elbows and orifices.

Governing Equation of MR-TLCD

The governing equation of an MR-TLCD is given by

$$\rho A L \ddot{x}_{lc} + \frac{1}{2}\rho A \delta |\dot{x}_{lc}| \dot{x}_{lc} + \frac{c\tau_y A L_p}{h} \frac{\dot{x}_{lc}}{|\dot{x}_{lc}|} + 2\rho A g x_{lc} = -\rho A B \ddot{y} \tag{6}$$

where A, L and B denote the cross-sectional area, the total length, and the horizontal length of the liquid column; ρ is the liquid density; and $\ddot{y}$ is the acceleration of the structure. The MR-TLCD has a natural frequency of $\omega_{lc} = \sqrt{2g/L}$, which is not affected by the applied field.

It is noted that the above governing equation is nonlinear. To simplify the computation, the equivalent linearization technique is adopted in dealing with the nonlinear damping terms. Let the equivalent linearized equation be

$$\rho A L \ddot{x}_{lc} + c_{lc}\dot{x}_{lc} + 2\rho A g x_{lc} = -\rho A B \ddot{y} \tag{7}$$

Assuming that the external excitation can be modeled as a zero-mean stationary Gaussian process, the equivalent damping coefficient, c_{lc}, can then be derived by minimizing the mean square value of the equation error (Iwan & Yang 1972; Wen 1980):

$$c_{lc} = \sqrt{\frac{2}{\pi}}\rho A\delta\sigma_{\dot{x}_{lc}} + \sqrt{\frac{2}{\pi}}\frac{c\tau_y AL_p}{h\sigma_{\dot{x}}} \tag{8}$$

The corresponding equivalent damping ratio is

$$\xi_{lc} = \frac{c_{lc}}{2\rho AL\omega_{lc}} = \frac{1}{2\sqrt{\pi gL}}(\delta\sigma_{\dot{x}} + \frac{c\tau_y L_p}{\rho h\sigma_{\dot{x}}}) \tag{9}$$

where $\sigma_{\dot{x}}$ is the standard deviation of the fluid velocity. Since the value of $\sigma_{\dot{x}}$ is unknown *a priori*, an iterative procedure is necessary in the analysis.

Analytical Model

By taking the motion of dampers as the control force, the governing equations for vibration of the coupled MR-TLCD-structure system can be expressed as

$$M_j\ddot{y}_j + C_j\dot{y}_j + K_j y_j = F_j(t) + F_{dj}(t) \qquad (j = 1,2,\cdots,n) \tag{10}$$

$$m_{lc,q}\ddot{x}_{lc,q} + c_{lc,q}\dot{x}_{lc,q} + k_{lc,q}x_{lc,q} = -\lambda_q m_{lc,q}\ddot{\bar{y}}_q \qquad (q = 1,2,\cdots,p) \tag{11}$$

where M_j, C_j, K_j, $F_j(t)$ are the mass, damping, stiffness, and external force of the structure at the jth floor; $m_{lc,q}$, $c_{lc,q}$, $k_{lc,q}$ are the mass, equivalent damping, and stiffness of the qth MR-TLCD, respectively. The control force of the qth MR-TLCD, which is installed at the jth floor, is obtained as

$$F_{dj}(t) = m_{lc,q}(\frac{1}{\lambda_q}((1.0 - \lambda_q^{\ 2})\ddot{x}_{dq} + 2\xi_{dq}\omega_{dq}\dot{x}_{dq} + \omega_{dq}^2 x_{dq}))^T, \ \lambda_q = \frac{B_q}{L_q} \tag{12}$$

In the frequency domain, the responses can be written as

$$[S_r(\omega)] = [H(i\omega)][S_F(\omega)][H^*(i\omega)]^T \tag{13}$$

in which, the transfer function matrices corresponding to cases without and with the dampers being installed are

$$[\widetilde{H}(i\omega)] = diag[\omega_1^2 - \omega^2 + 2i\xi_1\omega_1\omega \quad \omega_2^2 - \omega^2 + 2i\xi_2\omega_2\omega \quad \cdots \quad \omega_n^2 - \omega^2 + 2i\xi_n\omega_n\omega] \tag{14}$$

$$[H(i\omega)] = ([\widetilde{H}(i\omega)]^{-1} - [\Phi]^T[m_{lc}][D][\widetilde{H}_d(i\omega)][\Phi])^{-1} \tag{15}$$

where $[\widetilde{H}_d(i\omega)]$ is the transfer function matrix of MR-TLCDs,

$$[\widetilde{H}_d(i\omega)] = diag\left[\frac{\lambda_1\omega^2}{\omega_{d1}^2 - \omega^2 + 2i\xi_{d1}\omega_{d1}\omega} \quad \frac{\lambda_2\omega^2}{\omega_{d2}^2 - \omega^2 + 2i\xi_{d2}\omega_{d2}\omega} \quad \cdots \quad \frac{\lambda_p\omega^2}{\omega_{dp}^2 - \omega^2 + 2i\xi_{dp}\omega_{dp}\omega}\right] \tag{16}$$

and $[D]$ denotes the factor of the control force having a form of

$$D_q = \frac{1}{\lambda_q}((\lambda_q^{\ 2} - 1)\omega^2 + \omega_{dq}^2 + 2i\xi_{dq}\omega_{dq}\omega) \qquad (q = 1,2,\cdots,p) \tag{17}$$

The response quantities of displacement spectrum at jth DOF can be calculated by

$$\widetilde{S}_{y,j}(\omega) = \{\Phi\}_j[\widetilde{H}(i\omega)][\overline{M}]^{-1}[\Phi]^T[S_F(\omega)][\Phi]([\overline{M}]^{-1})^T[\widetilde{H}^*(i\omega)]\{\Phi\}_j^T \quad \text{(without dampers)} \tag{18}$$

$$S_{y,j}(\omega)=\{\Phi\}_j[H(i\omega)][\overline{M}]^{-1}[\Phi]^T[S_F(\omega)][\Phi]([\overline{M}]^{-1})^T[H^*(i\omega)]\{\Phi\}_j^T \quad \text{(with MR-TLCDs)} \tag{19}$$

The root-mean-square (RMS) displacement responses are

$$\tilde{\sigma}_{y,j}^2=\int_0^{+\infty}\tilde{S}_{y,j}d\omega \quad \text{(without dampers)} \tag{20}$$

$$\sigma_{y,j}^2=\int_0^{+\infty}S_{y,j}d\omega \quad \text{(with MR-TLCDs)} \tag{21}$$

Similarly, the root-mean-square (RMS) acceleration responses can be obtained from

$$\tilde{\sigma}_{\ddot{y},j}^2=\int_0^{+\infty}\omega^4\tilde{S}_{y,j}d\omega \quad \text{(without dampers)} \tag{22}$$

$$\sigma_{\ddot{y},j}^2=\int_0^{+\infty}\omega^4 S_{y,j}d\omega \quad \text{(with MR-TLCDs)} \tag{23}$$

The effectiveness of the MR-TLCD in mitigating the wind-induced vibration is quantified using the following indicators:

$$\alpha_{y,j}=(1-\frac{\sigma_{y,j}}{\tilde{\sigma}_{y,j}})\times 100, \quad \alpha_{\ddot{y},j}=(1-\frac{\sigma_{\ddot{y},j}}{\tilde{\sigma}_{\ddot{y},j}})\times 100 \tag{24}$$

Wind Load Spectrum

The spectrum of longitudinal turbulence proposed by Davenport (1961) has been used to simulate the wind loading in this study. The cross-spectral density representation of along-wind force in the frequency domain is given as

$$S_{F,ij}(\omega)=\rho_a^2C_D^2V_{10}^2A_iA_j(\frac{z_iz_j}{100})^{\alpha}coh(z_i,z_j,\omega)\frac{S(\omega)}{2\pi} \tag{25}$$

where the coherence function, $coh(z_i,z_j,\omega)$, reflecting the spanwise correlation of the fluctuating force, is calculated by

$$coh(z_i,z_j,\omega)=\exp(-\frac{C_z\cdot\omega\cdot(z_i-z_j)}{2\pi V_{10}}) \tag{26}$$

The wind velocity spectral density is given by

$$S(\omega)=8\pi K_DV_{10}^2\frac{n^2}{\omega(1+n^2)^{4/3}}, \quad n=\frac{600\omega}{V_{10}\pi} \tag{27}$$

in which K_D is the ground coarse coefficient; α is a power-law exponent; C_D is the drag coefficient; C_z is the exponential decay constant; ρ_a is the air density; ω is circular frequency in rad/s; V_{10} is the mean wind velocity at 10 m above the ground; z_i and z_j are heights of ith and jth nodes; A_i; and A_j are the equivalent projection areas at z_i and z_j, respectively.

NUMERICAL EXAMPLE

A 50-storey building, 162m in height, is used as an example to demonstrate the effectiveness of MR-TLCD system in mitigating wind-induced vibration. The total mass of the building, $\tilde{M}$, is 27,774 ton.

Modal properties obtained from a three-dimensional finite element model in the along-wind direction are listed in Table 1. Damping ratios for all modes are assumed to be 4.0%. The first three modes, which contribute over 85% in the along-wind vibration are taken into account in the analysis. Two cases, i.e., installation of one MR-TLCD on the top-floor and one at the top and another at the 25th floor, are considered. The following values of the MR-TLCD and wind loading parameters have been adopted in this example: ρ=2500kg/m^3, c=2.1, δ=30, λ=0.9, L_p=0.5m, h=0.3m, ρ_a=1.28kg/m^3, α=0.19, V_{10}=45.3 m/s, K_D=0.02, C_D=1.2, C_z=10.

Tables 2 and 3 give the mitigation indicators at the top floor when the mass ratio ($\mu = m_{lc} / \widetilde{M}$) is equal to 1.0% and 2.5%, respectively. The corresponding frequency ratio ($\mu_f = \omega_{lc} / \omega$) is set to be 1.0 and 0.95. It is noted when no magnetic field is applied, i.e. $\tau_y = 0$, the damper(s) has little effect. As the magnetic field, consequently the fluid yield stress, increases, a larger response reduction is achieved. However, it should be pointed out that there exists an optical value to make the MR-TLCD have an ideal performance. The installation of the second MR-TLCD at the 25th floor contributes a little in vibration mitigation.

Figure 3 shows the displacement and acceleration spectra of the building with an MR-TLCD installed on the top. The resonant peak responses are significantly reduced. Figures 4 and 5 illustrate the

TABLE 1

MODAL PROPERTIES OF BUILDING

Mode order	1	2	3	4	5
f (Hz)	0.216	0.940	2.278	3.941	5.932
ω (rad/s)	1.355	5.904	14.314	24.762	37.269
Mass factor (%)	66.82	13.08	5.51	3.11	1.92

TABLE 2

DISPLACEMENT AND ACCELERATION MITIGATION INDICATORS (μ=1.0%, μ_f=1.0)

	$p=1$			$p=2$		
τ_y	$\alpha_{y,51}$	$\alpha_{\ddot{y},51}$	ξ_{lc}	$\alpha_{y,51}$	$\alpha_{\ddot{y},51}$	ξ_{lc}
0.0	1.11	4.58	0.003	0.73	4.35	0.003
10.0	9.57	17.17	0.020	10.46	18.94	0.020
20.0	14.06	23.94	0.039	14.88	25.64	0.039
40.0	16.44	27.48	0.078	16.98	28.73	0.079
60.0	16.03	26.75	0.114	16.47	27.83	0.116
100.0	14.02	23.59	0.181	14.38	24.53	0.183

TABLE 3

DISPLACEMENT AND ACCELERATION MITIGATION INDICATORS (μ=2.5%, μ_f=0.95)

	$p=1$			$p=2$		
τ_y	$\alpha_{y,51}$	$\alpha_{\ddot{y},51}$	ξ_{lc}	$\alpha_{y,51}$	$\alpha_{\ddot{y},51}$	ξ_{lc}
20.0	18.20	28.32	0.041	19.73	31.64	0.043
40.0	27.34	43.91	0.098	28.50	47.06	0.103
60.0	30.63	50.55	0.161	31.22	52.80	0.166
80.0	31.55	52.91	0.220	31.94	54.68	0.226
100.0	31.62	54.49	0.275	31.92	55.01	0.281
150.0	30.65	52.38	0.397	30.87	53.63	0.403

displacement and acceleration root-mean-square (RMS) response profiles. The MR-TLCD is found to be capable of reducing responses along the entire height of the building. However, the reduction is not uniformly distributed.

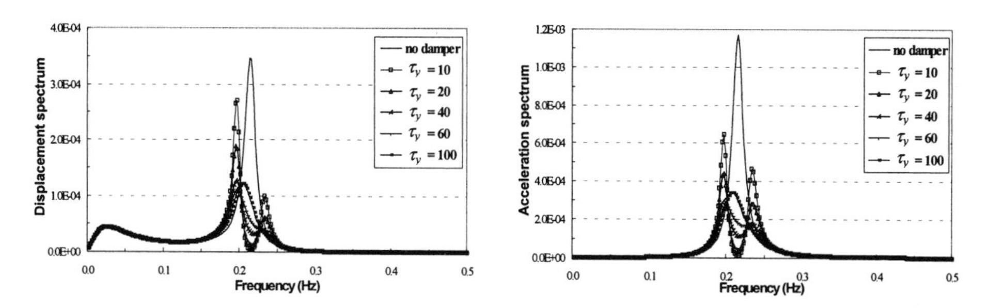

Figure 3: Displacement and acceleration spectra (p=1, μ=1.0%, μ_f=1.0)

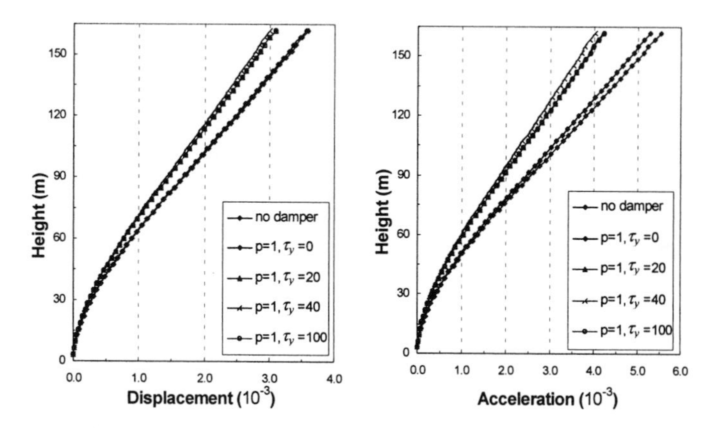

Figure 4: RMS displacement and acceleration response profiles (μ=1.0%, μ_f=1.0)

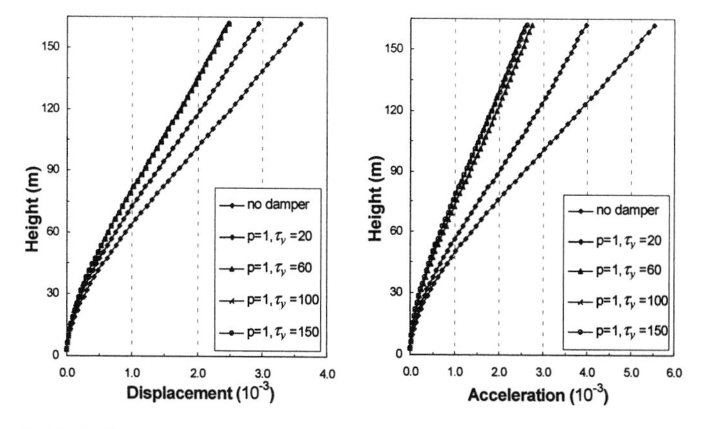

Figure 5: RMS displacement and acceleration response profiles (μ=2.5%, μ_f=0.95)

CONCLUSION

A semi-active tuned liquid column damper using magneto-rheological fluid (MR-TLCD) is proposed in the present study. Making use of the equivalent linearization technique, a random response analysis for an MR-TLCD-structure system under stochastic wind excitation is conducted. Numerical analysis of a 50-storey building demonstrates that using a larger mass ratio provides a larger reduction in vibration response and MR-TLCDs are able to supply the optimal damping requirement facilely. Further study aims to evaluate the effectiveness of MR-TLCDs for structures with varying dynamic characteristics and to use MR-TLCDs in hybrid damping systems, as well as to develop a real-time control strategy for semi-active seismic response mitigation.

ACKNOWLEDGEMENTS

The work presented in this paper was supported by a grant from The Hong Kong Polytechnic University through the Area of Strategic Development Programme (Research Centre for Urban Hazard Mitigation). This support is gratefully acknowledged.

References

Balendra T., Wang C.M. and Cheong H.F. (1995). Effectiveness of tuned liquid column damper for vibration control of towers. *Engineering Structures* **17:9,** p668-675.

Balendra T., Wang C.M. and Rakesh G. (1999). Effective of TLCD on various structural systems. *Engineering Structures* **21,** 291-305.

Davenport A.G. (1961). The application of statistical concepts to the wind loading of structures. *Proc., Inst. of Civil Engineer* **19,** 449-472.

Iwan W.D. and Yang I.M. (1972). Application of statistical linearization techniques to nonlinear multidegree-of-freedom systems. *Journal of Applied Mechanics* **39,** 545-550.

Jolly M.R., Bender J.W. and Carlson J.D. (1999). Properties and applications of commercial magnetorheological fluids. *Journal of Intelligent Material Systems and Structures* **10,** 5-13.

Phillips R.W. (1969). Engineering applications of fluids with variable yield stress, PhD Thesis, University of California, Berkeley, USA.

Spencer Jr. B.F., Yang G., Carlson J.D. and Sain M.K. (1998). Smart dampers for seismic protection of structures: a full-scale study. *Proceedings of 2nd World Conference on Structural Control* **1,** Kyoto, Japan, 417-426.

Wen Y.K. (1980). Equivalent linearization for hysteretic systems under random excitation. *Journal of Applied Mechanics* **47,** 150-154.

Yalla S.K., Kareem A. and Kantor J.C. (2001). Semi-active tuned liquid column dampers for vibration control of structures. *Engineering Structures* **23,** 1469-1479.

Yang G., Spencer Jr. B.F., Carlson J.D. and Sain M.K. (2002). Large-scale MR fluid dampers: modeling and dynamic performance considerations. *Engineering Structures* **24,** 309-323.

Xu Y.L., Samali B. and Kwok K.C.S. (1992). Control of along-wind response of structures by mass and liquid dampers. *Journal of Engineering Mechanics* **118:1,** 20-39.

Advances in Building Technology, Volume I
M. Anson, J.M. Ko and E.S.S. Lam (Eds.)

PHENOMENOLOGICAL MODEL OF LARGE-SCALE MR DAMPER SYSTEMS

G. Yang,[1] B.F. Spencer, Jr.,[1] H.J. Jung,[2] and J.D. Carlson[3]

[1] Department of Civil and Environmental Engineering, University of Illinois at Urbana-Champaign
205 North Matthews Ave., Urbana, IL 61801, USA

[2] Dept. of Civil Engineering, Korea Advanced Institute of Science and Technology,
Taejon 305-701, Korea

[3] Material Division, Lord Corporation, 111 Lord Drive, Cary, NC 27512, USA

ABSTRACT

Magnetorheological (MR) dampers are one of the most promising new devices for structural vibration mitigation. Because of their mechanical simplicity, high dynamic range, low power requirements, large force capacity, and robustness, these devices have been shown to mesh well with demands and constraints of earthquake and wind engineering applications. Quasi-static models of MR dampers have been investigated by researchers. Although useful for damper design, these models are not sufficient to describe the MR damper behavior under dynamic loading. This paper presents a new dynamic model of the overall MR damper system which is comprised of two parts: (i) a dynamic model of the power supply, and (ii) a dynamic model of the MR damper. Because previous studies have demonstrated that a current-driven power supply can substantially reduce the MR damper response time, this study employs a current driver to power the MR damper. The operating principles of the pulse-width modulated current driver, and an appropriate dynamic model are provided. Subsequently, MR damper force response analysis is performed, and a phenomenological model based on the Bouc-Wen model is proposed to predict the MR damper behavior under dynamic loading. This model accommodates the MR fluid stiction phenomenon, as well as fluid inertial and shear thinning effects. Compared with other types of models based on the Bouc-Wen model, the proposed model has been shown to be more effective, especially in describing the force roll-off in the low velocity region, force overshoots when velocity changes in sign, and two clockwise hysteresis loops at the velocity extremes.

KEYWORDS: MR dampers, Smart damping devices, Hysteresis model, Parameter estimation, System identification, Rheological technology

INTRODUCTION

MR fluid dampers are new semi-active devices that utilize MR fluids to provide controllable damping forces. These devices overcome many of the expenses and technical difficulties associated with semi-active devices previously considered (Carlson and Spencer 1996; Yang 2001). Recent studies have shown that the semi-active dampers can achieve the majority of the performance of fully active systems, thus allowing for the possibility of effective response reduction during both moderate and strong seismic activity (Dyke et al. 1996; Spencer et al. 2000). For these reasons, significant efforts have been devoted to the development and implementation of MR devices. Indeed, the first full-scale implementation of MR dampers in building was realized in the National Museum of Emerging Science and Innovation of Tokyo, Japan in 2001. In this building, two 30-ton MR fluid dampers built by the Sanwa Tekki Corporation were installed between the 3rd and 5th floors; the MR fluid inside the damper is manufactured by the Lord Corporation.

To investigate the salient features of large-scale MR devices, a 20-ton MR fluid damper has been designed and built (Carlson and Spencer 1996; Yang et al. 2002). At that time, this device constituted the largest MR damper ever produced. Fig. 1a shows the schematic of the MR damper tested in this paper. Detail information about this damper can be found at Yang (2001). Fig. 1b shows the experimental setup at the University of Notre Dame for this large-scale MR fluid damper.

Because of quasi-static models of MR dampers are not sufficient to describe the damper nonlinear behavior under dynamic loading (Yang 2001; Yang et al. 2001, 2002), this paper presents a new dynamic model of the overall MR damper system which is comprised of two parts: (i) a dynamic model of the power supply, and (ii) a dynamic model of the MR damper. Because previous studies have demonstrated that a current-driven power supply can substantially reduce the MR damper response time (Yang et al. 2001), a dynamic model of the pulse-width modulated current driver is provided. Subsequently, MR damper force response analysis is performed, and a phenomenological model based on the Bouc-Wen model is proposed to predict the MR damper behavior under dynamic loading. This model accommodates the MR fluid stiction phenomenon, as well as inertial and shear thinning effects. Experimental verification has shown that the proposed dynamic model of the MR damper system predicts the experimental results very well.

DYNAMIC MODEL OF CURRENT DRIVERS

Previously study has demonstrated that the exponential current response in damper coil when driven by a voltage source is insufficient for many practical applications (Yang et al. 2000, 2002). Several approaches can be considered to decrease the response time of the MR damper's electromagnetic circuit. One of them is to use a current driver instead of a voltage-driven power supply. The schematic of the current driver based on a PWM amplifier are given in Fig. 2. Usually, a PI controller is employed in the feedback loop to regulate the duty cycle based on the error between the measured and desired current. Assuming that the duty cycle is proportional to the controller output u_c and that u_c is not satu-

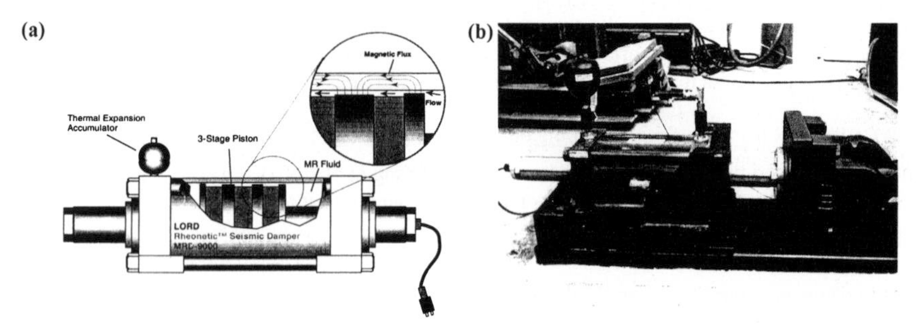

Figure: 1 (a) Schematic of the large-scale 20-ton MR fluid damper; (b) experimental setup

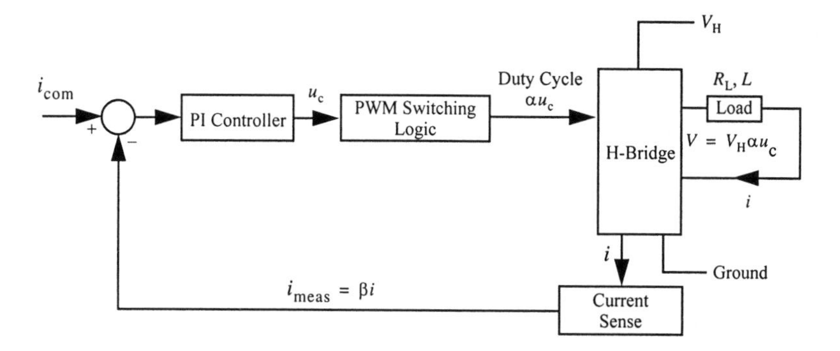

Figure: 2 (a) Schematic of the current driver based on the PWM servo amplifier

rated, the governing equation for current driver is given as (Yang 2001)

$$\frac{L}{\beta\gamma}\frac{d^2 i}{dt^2} + \frac{(R_L + \beta\eta)}{\beta\gamma}\frac{di}{dt} + i = \frac{i_{com}}{\beta} + \frac{\eta}{\beta\gamma}\frac{di_{com}}{dt} \tag{1}$$

where $\gamma = \alpha K_i V_H$; $\eta = \alpha K_p V_H$; i_{com} = reference input signal; β = sensitivity of the current sensing; and K_p, K_i = controller proportional and integral gains. The steady state current is given by

$$I_s = \frac{i_{com}}{\beta} \tag{2}$$

which depends only on the input reference signal i_{com} and the sensitivity of the current sensing β. The load resistance R_L and the bus voltage V_H have no effect on the steady state current.

Fig. 3 shows a typical current driver response to a step reference signal. At the beginning of the response, the error signal is large. The power supply applies maximum voltage to facilitate the current increase. The current increase follows the same path as that of the 100% duty cycle. As the current increases, the error signal decreases. The controller output u_c is no longer saturated, and the current is governed by Eq. (1). The controller regulates the current to a steady state current of i_{com}/β. As compared with the exponential current response using a voltage source, the current driver can substantially reduce the current response time which is readily seen in Fig. 3. To experimentally verify the effectiveness of the current driver, currents in the 20-ton MR fluid damper coil (connected in series) due to a step input command generated by both a constant voltage power supply and a current driver are compared. The experimental results are shown in Fig. 4 (Yang et al. 2002). The constant voltage case corresponds to the scenario where a voltage-driven power supply is attached to the damper coils. The time constant L_0/R_0 for the coils of the 20-ton MR fluid damper arranged in series is 0.3 sec. Therefore, as shown in Fig. 4, it takes about 1 sec for the current to achieve 95% of the final value, which is unacceptable for most of civil engineering applications. Alternatively, using a current driver, the 5% error range is achieved within 0.06 sec.

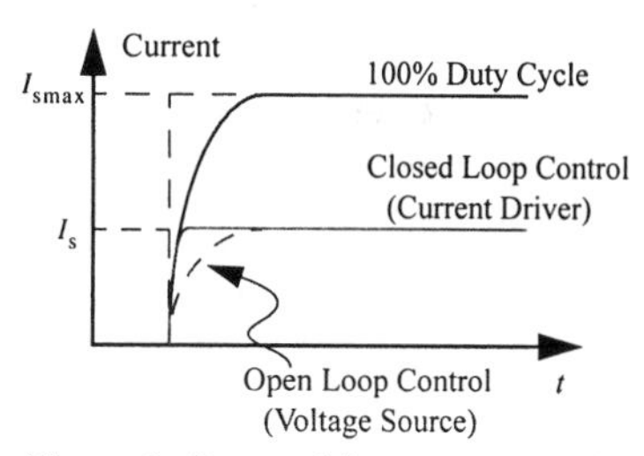

Figure: 3 Current driver response under a step reference signal

Based on Eq. (1), the dynamic model of the current driver used in the experiment is identified. The transfer function between the input reference signal i_{com} and current i is given by (Yang et al. 2001; Yang 2001)

$$i(s) = \frac{1001.45s + 1016.1}{s^2 + 503.7s + 508.05} i_{com}(s) \tag{3}$$

A comparison between the measured and predicted current is provided in Fig. 5; close agreement is observed.

Figure: 4 Comparison between the coil current driven by constant voltage and current driver (Yang et al. 2002)

DYNAMIC MODEL OF MR DAMPERS

In this section, following the MR damper force response analysis, a phenomenological model of MR dampers based on the Bouc-Wen hysteresis model is proposed, which accommodates the MR fluid stiction phenomenon, as well as inertial and shear thinning effect. The experimental verification of the proposed model under various types of input current and loading conditions is also provided.

MR damper force response analysis

For the purposes of this discussion, the damper response in Fig. 6a can be divided into three regions. At the beginning of region I, the velocity changes in sign from negative to positive, the velocity is quite small and flow direction reverses. At this stage, the MR damper force is below the yield level, and the MR fluid operates mainly in the pre-yield region, i.e., not flowing and having very small elastic deformation. After the damper force exceeds the yield level, a damper force loss occurs during the transition from the pre-yield to post-yield region due to the stiction phenomenon of MR fluids (Pignon et al. 1996; Powell 1995; Weiss et al. 1994), resulting in an overshoot type of behavior in force. By definition, stiction is a particle jamming or a mechanical restriction to flow that is highly dependent upon both particle size and shape, as well as the prior electric field and flow history of the material. The illustration of stiction phenomenon is shown in Fig. 6b, which is similar to the Coulomb friction. As shown in this figure, the MR fluid stress increases in the pre-yield region when strain increases. As the strain exceeds the critical strain, the MR fluid changes from pre-yield to post-yield region and begins to flow; consequently, the elastic stress is released, and stress loss is observed (Weiss et al. 1994; Pignon et al. 1996). Note that due to the stiction phenomenon, the displacement measurement lags the command signal during the force transition from the pre-yield to the post yield region, as shown in Fig. 6a. Because the

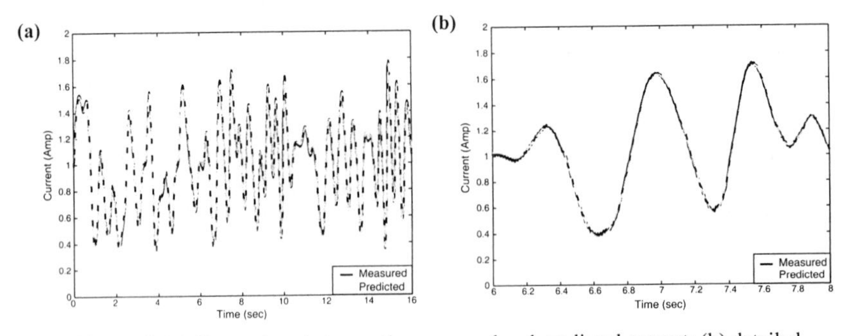

Figure: 5 (a) Comparison between the measured and predicted current; (b) detailed comparison in current fast-changing region

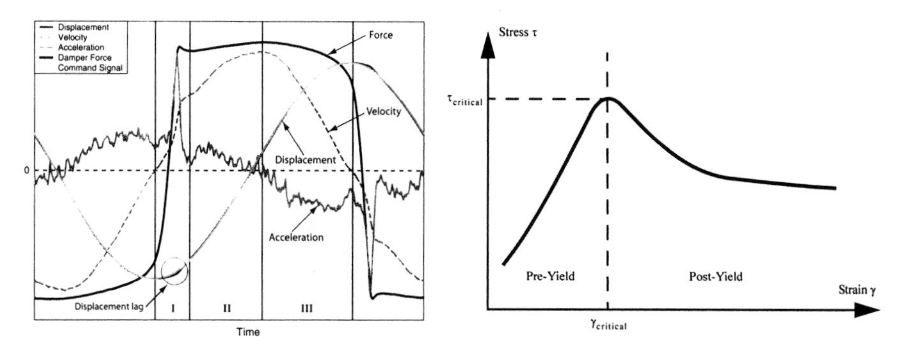

Figure: 6 (a) MR damper force response analysis; (b) Illustration of stiction phenomenon of MR fluids (Weiss et al. 1994)

servo controller uses displacement feedback, the controller tends to command a large valve opening to facilitate the damper movement. Therefore, a substantial increase in acceleration is observed. After the damper force exceeds the yield level, the acceleration quickly drops to its normal sinusoidal trajectory, as shown at the end of region I. Because the fluid inertial force is related to the acceleration, an additional force overshoot due to the dynamics of the experimental setup may be introduced, and may increase the degree of force overshoot caused by the stiction phenomenon of MR fluids. However, the experimental results have been shown that the force overshoot resulting from the inertial force is much smaller than the force due to the stiction phenomenon. This indicates that the stiction phenomenon dominates the force overshoot (Yang 2001).

In region II, the velocity continues to increase while still remaining positive. Therefore, the plastic-viscous force increases, and a damper force increase is observed.

In region III, the velocity decreases. The damper velocity approaches zero at the end of this region, and the plastic viscous force drops more rapidly due to the fluid shear thinning effect. Therefore, a force roll-off is observed. Note that the stiction phenomenon or the damper force loss after yielding is irreversible (Powell 1995), the force overshoot does not occur in this region; therefore, two clockwise loops are observed in the force-velocity plot.

Dynamic model of MR dampers

Two types of dynamic models of controllable fluid damper have been investigated by researchers: non-parametric and parametric models. Ehrgott & Masri (1992) and Gavin (1996) presented a non-parametric approach employing orthogonal Chebychev polynomials to predict the damper resisting force using the damper displacement and velocity information. Chang & Roschke (1998) developed a neural network model to emulate the dynamic behavior of MR dampers. However, the non-parametric damper models are often quite complicated. Gamato & Filisko (1991) proposed a parametric viscoelastic-plastic model based on the Bingham model. Wereley et al. (1998) developed a nonlinear hysteretic biviscous model, which is an extension of the nonlinear biviscous model having an improved representation of the pre-yield hysteresis. Spencer et al. (1997) proposed a mechanical model based on the Bouc-Wen model.

Based on the damper response analysis presented in the previous section, a phenomenological model considering the MR fluid stiction phenomenon, as well as inertial and shear thinning effects. The schematic of the model is shown in Fig. 7. The damper force is given by

$$f - f_0 = m\ddot{x} + c(\dot{x})\dot{x} + kx + \alpha z \qquad (4)$$

where the evolutionary variable z is governed by

$$\dot{z} = -\gamma|\dot{x}|z|z|^{n-1} - \beta\dot{x}|z|^n + A\dot{x} \qquad (5)$$

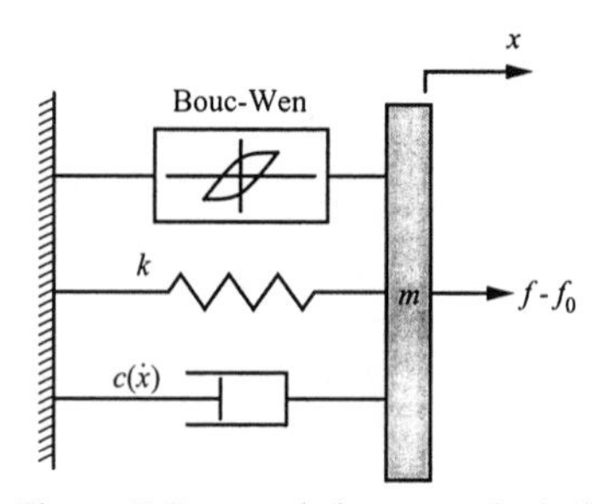

Figure: 7 Proposed phenomenological model of MR dampers

In this model, m = equivalent mass which represents the MR fluid stiction phenomenon and inertial effect; k = accumulator stiffness and MR fluid compressibility; f_0 = damper friction force due to seals and measurement bias; and $c(\dot{x})$ = post-yield plastic damping coefficient. To describe the MR fluid shear thinning effect which results in the force roll-off of the damper resisting force in the low velocity region, the damping coefficient $c(\dot{x})$ is defined as a mono-decreasing function with respect to absolute velocity $|\dot{x}|$. In this paper, the post-yield damping coefficient is assumed to have a form of (Yang et al. 2001)

$$c(\dot{x}) = a_1 e^{-(a_2|\dot{x}|)^p} \qquad (6)$$

where a_1, a_2 and p are positive constants.

To determine a model which is valid under the fluctuating input current, the functional dependence of the parameters on the input current must be determined. Since the fluid yield stress is dependent on input current, α can then be assumed as a function of the input current i. Moreover, from the experiment results, a_1, a_2, m, n, and f_0 are also functions of the input current. In order to obtain the relationship between the input current i and damper parameters α, a_1, a_2, n, m and f_0, the damper was driven by band-limited random displacement excitations with a cutoff frequency of 2 Hz at various constant current levels. A constrained nonlinear least-squares optimization scheme based on the trust-region and preconditioned conjugated gradients (PCG) methods is then used. The results are shown in Table 1. A linear piecewise interpolation approach is utilized to estimate these damper parameters for current levels which are not listed in the above table. The rest damper parameters which are not varied with input current are chosen to be $\gamma = 25179.04\ \text{m}^{-1}$, $\beta = 27.1603\ \text{m}^{-1}$, $A = 1377.9788\ \text{m}^{-1}$, $k = 20.1595\ \text{N/m}$, and $p = 0.2442$. Note that a first order filter needs to be used to accommodate the dynamics involved in the MR fluid reaching rheological equilibrium

$$H(s) = \frac{31.4}{s + 31.4} \qquad (7)$$

Table 1. Damper parameters at various current levels under random displacement excitation

Current (A)	α (10^5 N)	a_1 (N · sec/m)	a_2 (sec/m)	m (kg)	n	f_0 (N)
0.0237	1.3612	4349000	862.03	3000	1.000	1465.82
0.2588	2.2245	24698000	3677.01	11000	2.0679	2708.36
0.5124	2.3270	28500000	3713.88	16000	3.5387	4533.98
0.7625	2.1633	32488000	3849.91	18000	5.2533	4433.08
1.0132	2.2347	24172000	2327.49	19500	5.6683	2594.41
1.5198	2.2200	38095000	4713.21	21000	6.7673	5804.24
2.0247	2.3002	35030000	4335.08	22000	6.7374	5126.79

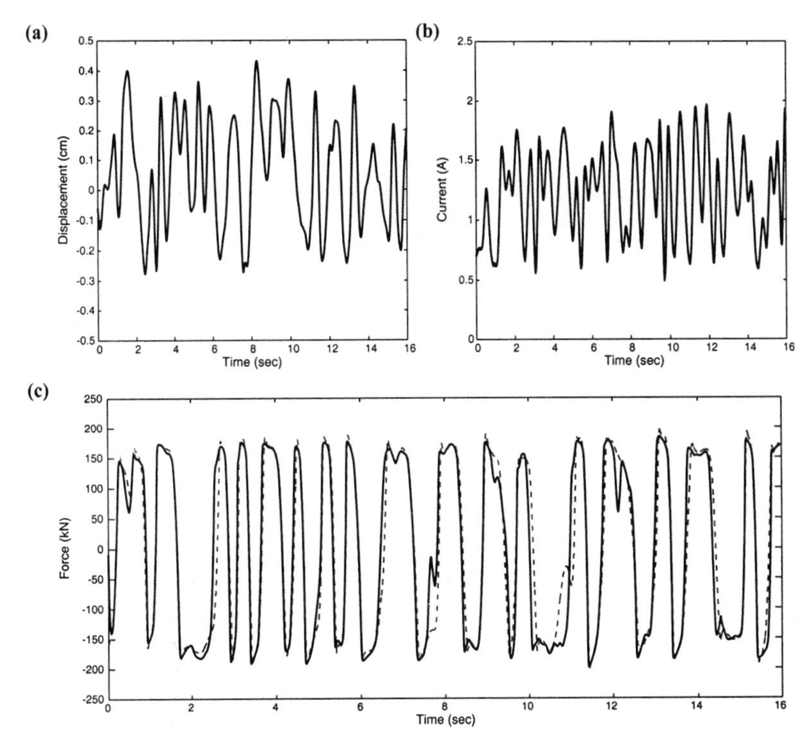

Figure: 8 Comparison between predicted and experimentally-obtained damper responses under random displacement excitation with random input current, (a) displacement vs. time; (b) current vs. time; (c) force vs. time

Fig. 8. provides the comparison between the predicted force and experimental data when the damper is subjected to a fluctuating current input. In this test, a band-limited random displacement excitation is applied. The displacement excitation and input current is also shown in Fig. 8. Excellent agreement is observed between the experimental and model responses.

CONCLUSIONS

Magnetorheological (MR) fluid dampers provide a level of technology that has enabled effective semi-active control in a number of real world applications. This paper provides fundamental insight into the dynamic behavior of a 20-ton MR damper capable of providing semi-active damping for full-scale civil engineering applications. A phenomenological model based on the Bouc-Wen hysteresis model is proposed to describe the dynamic behavior of this large-scale MR damper. The model considers the MR fluid stiction phenomenon, as well as shear thinning and inertial effects, which are observed in the experimental data. Moreover, experimental result shows that use of a pulse-width modulated current driver can dramatically reduce the MR damper response time. The operational principle of the current driver is also discussed, and its dynamic model is given. This dynamic model is then combined with the phenomenological model proposed for the MR damper to obtain the dynamic model of the overall MR damper system. The predicted results compare well with the experimental data, and effective predictions are obtained. This model provides a new tool for analysis and synthesis of structures employing such large-scale MR dampers.

ACKNOWLEDGEMENT

The authors gratefully acknowledge the support of this research by the National Science Foundation under grant CMS 99-00234 (Dr. S.C. Liu, Program Director).

REFERENCES

Carlson J.D. and Spencer Jr. B.F. (1996). Magneto-Rheological Fluid Dampers for Semi-active Seismic Control. *Proc. 3rd International Conference on Motion and Vibration Control*, vol. 3, Chiba, Japan, pp. 35–40.

Chang C.C. and Roschke P. (1998). Neural Network Modeling of a Magnetorheological Damper. *J. of Intelligent Material Systems and Structures* **9**, 755–764.

Dyke S.J., Spencer Jr. B.F., Sain M.K. and Carlson, J.D. (1996). Modeling and Control of Magnetorheological Dampers for Seismic Response Reduction. *Smart Materials and Structures* **5**, 565–575.

Ehrgott R.C. and Masri S.F. (1992). Modeling the Oscillatory Dynamic Behavior of Electrorheological Materials in Shear. *Smart Materials and Structures* **1**, 275–285.

Gamota D.R. and Filisko F.E. (1991). Dynamic Mechanical Studies of Electrorheological Materials: Moderate Frequencies. *J. of Rheology* **35**, 399–425.

Gavin H.P., Hanson R.D. and Filisko, F.E. (1996). Electrorheological Dampers, Part 2: Testing and Modeling. *J. of Applied Mechanics* ASME **63:9**, 676–682.

Pignon F., Magnin A. and Piau J-M. (1996). Thixotropic cOlloidal Suspensions and Flow Curves with Minimum: Identification of Flow Regimes and Rheometric Consequences. *J. Rheology* **40:4**, 573–587.

Powell J.A. (1995). Application of a Nonlinear Phenomenological Model to the Oscillatory Behavior of ER Materials. *J. Rheology* **39:5**, 1075–1094.

Spencer Jr. B.F., Dyke S.J., Sain M.K. and Carlson. J.D. (1997). Phenomenological Model of a Magnetorheological Damper. *J. of Engineering Mechanics* ASCE **123:3**, 230–238.

Spencer Jr. B.F., Johnson E.A. and Ramallo J.C. (2000). “Smart” Isolation for Seismic Control. *JSME International Journal* JSME **43:3**, 704–711.

Weiss K.D., Carlson J.D., and Nixon D.A. (1994). Viscoelastic Properties of Magneto- and Electrorheological Fluids. *J. Intelligent Material Systems and Structures* **5:11**, 772–775.

Wereley N.M., Pang L. and Kamath G.M. (1998). Idealized Hysteresis Modeling of Electrorheological and Magnetorheological Dampers. *J. of Intelligent Material, Systems and Structures* **9:8**, 642–649.

Yang G., Ramallo J.C., Spencer Jr. B.F., Carlson J.D. and Sain M.K. (2000). Large-Scale MR Fluid Dampers: Dynamic Performance Considerations. *Proceedings of International Conference on Advances in Structure Dynamics*, Hong Kong, China. pp. 341–348.

Yang G., Jung H.J., and Spencer Jr. B.F. (2001). Dynamic Modeling of Full-Scale MR Dampers for Civil Engineering Applications. *US-Japan Workshop on Smart Structures for Improved Seismic Performance in Urban Region,* Seattle, WA., Aug. 14. pp. 213–224.

Yang G. (2001). *Large-scale magnetorheological fluid damper for vibration mitigation: modeling, testing and control.* Ph.D dissertation, University of Notre Dame, IN.

Yang G., Spencer Jr. B.F., Carlson J.D. and Sain M.K. (2002). Large-Scale MR Fluid Dampers: Modeling, and Dynamic Performance Considerations. *Engineering Structures* **24**:309–323.

TEAM PRESENTATION:

STRUCTURAL ANALYSIS RESEARCH AT THE BEIJING POLYTECHNIC UNIVERSITY

Advances in Building Technology, Volume I
M. Anson, J.M. Ko and E.S.S. Lam (Eds.)

THE PRACTICAL ANTI-EARTHQUAKE OPTIMUM DESIGN OF CIVIL INFRASTRUCTURE NETWORK SYSTEM

Chen Yanyan[1] and Wang Guangyuan[2]

[1] College of Civil Engineering, Beijing Polytechnic University
Beijing, 100022, P.R.China
[2] College of Civil Engineering, Harbin Polytechnic University
Harbin, 150090, P.R.China

ABSTRACT

The optimum anti-earthquake design of civil infrastructure system (CIS) is a problem with numerous variables, complex constraints and high dimensions. In this paper, we try to alleviate the difficulty by dividing the design into two stages, i.e. system optimization and structure optimization. Taking the structure design load or reliability as bridge, the two stages are joined together. Considering the future loss expectation, the global optimum objective is to minimize the total investment cost (the sum of system original cost and future loss expectation). Through changing the system optimum variables into fewer discrete ones, an effective optimum method for CIS anti-earthquake design is suggested.

KEYWORDS

Anti-earthquake optimum design, civil infrastructure system, Reliability, Life-cycle cost

INTRODUCTION

Many civil infrastructure system (CIS) are large building-lifeline interrelated systems. It includes various kinds of buildings and lifeline segments and can be modeled as a network of interconnected nodes and links. It's important to undertake anti-earthquake design for the system because seismic damages may cause extraordinary loss. In the meantime, there is clearly a need for an optimum design procedure that is able to strike an acceptable compromise between safety and economy.

In the past, structure optimum design is often independent and only aims at minimum cost. And the optimized variable is the structure dimension, with which it is difficult to express system loss expectation. In this paper, the optimum design of CIS is divided into two stages, system optimization and structure optimization, that is:

(1) system optimization design: i.e. optimization of structure design intensity. The optimization objective is:

$$W_S[\bar{x}_n(I_{dn})] = C_S[\bar{x}_n(I_{dn})] + M_S(\bar{x}_n) + L_S[\bar{x}_n(I_{dn})] \rightarrow \min \quad (1)$$

Here, structure design intensity I_{dn} is taken as the parameter to represent the nth structure anti-earthquake level. $\bar{x}_n$ is the design vector of the nth structure. W_S is CIS lifecycle cost, which includes system initial cost C_S, maintenance cost M_S and loss expectation L_S. Because routine maintenance cost is irrelevant to the structure anti-earthquake level, so it may be omitted in the anti-earthquake design.

(2) structural optimum design: i.e. optimization of structure design vector. The optimization objective is:

$$C_n[\bar{x}_n(I^*_{dn})] \rightarrow \min \quad (2)$$

Here, I^*_{dn} is the optimum design intensity got in the first stage. C_n is the cost of the nth structure.

THE COMPONENT VULNERABILITY ANALYSIS

In the building design code of P.R.China (1994), five damage states are considered for building structure. If the ith damage states is denoted by B_i, then there is

$[B_1, B_2, B_3, B_4, B_5]$=[intactness, minor damage, medium damage, major damage, collapse]

According to the three-level anti-earthquake design norm in the design code of P.R.China (1994), which requires the structure to be intact ($=B_1$) under minor earthquake (written as I^L, lower by 1.55 degree than design intensity I_d), repairable($=B_3$) under medium earthquake (equal to I_d), not collapse($=B_4$) under strong earthquake(written as I^U, higher by 1 degree or so than I_d), building structure is designed under design intensity (written as I_d). So under the single failure mode, for given earthquake intensity s, the probability of the structure with design intensity I_d being damaged over (more seriously than) B_i state (written as $P_f[B_i^*, \bar{x}(I_d)/s]$) can be shown in fig.1 (Wang guangyuan, 1999).

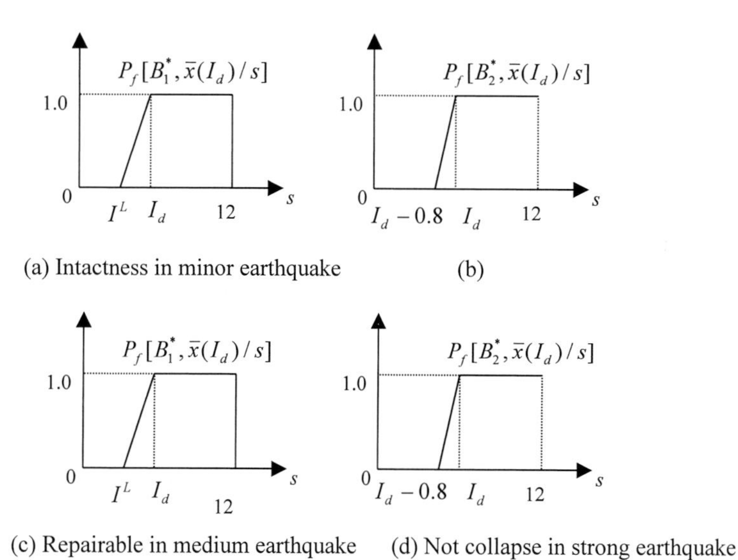

(a) Intactness in minor earthquake (b)

(c) Repairable in medium earthquake (d) Not collapse in strong earthquake

Fig.1: The damage probability curve of structure for different intensity s

Fig.1 (a), fig.1 (c), fig.1 (d) correspond to the three-level anti-earthquake design norm separately, while fig1. (b) is decided approximately by fig1. (a) and fig.1(c).

Depending on the probability curve of the biggest earthquake intensity in design terms, the structure damage probability $P_f[B_i^*, \bar{x}(I_d)]$ can be obtained from

$$P_f[B_i^*, \bar{x}(I_d)] = \int_0^{12} f_s(s) P_f[B_i^*, \bar{x}(I_d)/s] ds \tag{3}$$

Here, $f_s(s)$ is the probability density function of the biggest earthquake intensity in design terms.

It can easily be seen that, the structure damage probability $P_f[B_i^*, \bar{x}(I_d)]$ is only a function of design intensity and damage degree, and has no relationship with the design dimension vector.

Suppose the component in CIS has only two states, connective and failure. Failure often happens only when building structure being damaged over B_3 level. Then the connective reliability R of building structure component can be described by

$$R = R(I_d) = 1 - P_f[(B_3^{\bullet}, \bar{x}(I_d)] \tag{4}$$

For the link component (lifeline segment), we can suppose the damage-place number conforming to Poisson distribution, then the component connective reliability can be described by

$$R = R(\bar{x}) = 1 - e^{-\alpha l} \tag{5}$$

Here, α is the damage rate (the number of damage places per unit length), which is related with design vector $\bar{x}$, l is the link component length.

THE SYSTEM LIFECYCLE COST

The key problem of system anti-earthquake optimum design is the decision on the appropriate objective function. According to discussion above, the objective should be minimizing system life-cycle cost, which includes system original cost C_S and future loss expectation L_S. The system original cost could be expressed by the following formula:

$$C_S = \sum_{n=1}^{N_1} C_n[\bar{x}_n(I_{dn})] + \sum_{n=1}^{N_2} C_n(\bar{x}_n) \tag{6}$$

Here, $C_n[\bar{x}(I_{dn}]$ is the cost of the nth building component with design intensity I_{dn}, N_1 is the number of the buildings in the system . $C_n(\bar{x}_n)$ is the cost of the nth lifeline segment component with design vector $\bar{x}_n$, N_2 is the number of the lifeline segment components.

In fact, there will be different design vectors under the same design intensity, which means, there will be different cost under the same design intensity. But the optimum design should be the most economic design, i.e. $C_n[\bar{x}(I_{dn}]$ should be the lowest cost $C_{\min,n}[\bar{x}(I_{dn}]$. Approximately, in the practical application, we can take the cost of a reasonable plan with design intensity I_{dn} as $C_{\min,n}[\bar{x}(I_{dn})]$.

Under the earthquake action, the loss expectation L_S of CIS could be divided into two parts, one results from the component structural damage, the other results from system service interruption, which can be described by

$$L_S = L^{(e)} + L^{(s)} \tag{7}$$

Here, $L^{(e)}$ is the structural loss expectation resulted from the component structural damage, $L^{(s)}$ is the system loss expectation resulted from system service interruption.

$L^{(e)}$ is the sum of all component structural loss expectation, through the component vulnerability analysis, $L^{(e)}$ can be described by

$$L^{(e)} = \sum_{n=1}^{N_1} L_n^{(e)}(\mathrm{I}_{\mathrm{dn}}) + \sum_{n=1}^{N_2} L_n^{(e)}(\bar{x}_n) \tag{8}$$

For the building structure

$$L_n^{(e)}(\mathrm{I}_{\mathrm{dn}}) = \sum_{i=1}^{5} P_{fn}^{(e)}(B_i, I_{dn}) D_{ni}^{(e)} \tag{9}$$

Here, $L_n^{(e)}(I_{dn})$ is the structural loss expectation of the nth component, $P_{fn}^{(e)}(B_i, I_{dn})$ is the probability of the nth building component designed under intensity I_{dn} in damage state B_i, $D_{ni}^{(e)}$ is the structural loss of the nth structure in damage state B_i.

$L_n^{(e)}(\bar{x}_n)$ is the structural loss expectation of the nth lifeline component with design vector $\bar{x}_n$. For the lifeline segment, the damage degree is often expressed by the damage rate (the number of damage places per unit length). Then the component loss expectation can be determined by

$$L_n^{(e)}(\bar{x}_n) = \int_0^{\lambda_n(\bar{x}_n)l_n} D_n^{(e)}(m)dm \tag{10}$$

Here, m is the variable of expectation number of damage places. $\lambda_n(\bar{x}_n)$ is the damage rate expectation of the nth component with design vector $\bar{x}_n$. $D_n^{(e)}$ is the component loss of the nth structure, which is the function of damage-place number.

The system loss is resulted from the failure in connection from source nodes to terminal nodes, which can be expressed as:

$$L^{(s)} = \sum_{k=1}^{K} L_k^{(s)} = \sum_{k=1}^{K} (1 - R_S^{(k)}) D_k^{(s)} \tag{11}$$

Here, $L_k^{(s)}$ is the system loss expectation of the kth source-terminal pair. K is the number of source-terminal pairs. $D_k^{(s)}$ is the system loss of the kth source-terminal pair when system service being interrupted. $R_S^{(k)}$ is the connective reliability of the kth source-terminal pair, which can be obtained by Monte Carlo simulation or Analysis algorithm(Li Guiqing,1994).

Then the anti-earthquake optimum objective of CIS is:

$$W_S = C_S + L_S^{(e)} + L_S^{(s)} \rightarrow \min \tag{12}$$

THE PRACTICAL OPTIMUM ANTI-EARTHQUAKE DESIGN FOR CIS

The global optimum anti-earthquake design of CIS is to find the optimum design intensity I_d for each building component and design vector $\bar{x}$ for each lifeline segment component to minimize the system lifecycle cost subject to the system original cost constraints and system function constraints.

In design practice, the design intensity can be taken as discrete values. In any case, the reasonable designs of the lifeline segment are relatively few because of the limits to structure type, function, flux and coded spare part size. So in system optimum design, depending on the earthquake risk analysis, component importance, function requirement and so on, the designer can choose several possible discrete values of design intensity for each building and several possible design vectors for each lifeline segment. Combining every possible selection through enumeration, and computing system objective and constraints in each possible combination, then the optimum result meeting system global optimization requirements can be chosen. Of course, other discrete variable optimization algorithms, such as GA can be taken. Finally, under optimum design intensity, minimum cost design or traditional design of the structure can be undertake.

EXAMPLE

To check the application of the proposed method, an underground pipeline network to be built is considered here. The simplified network model is shown in fig.2.

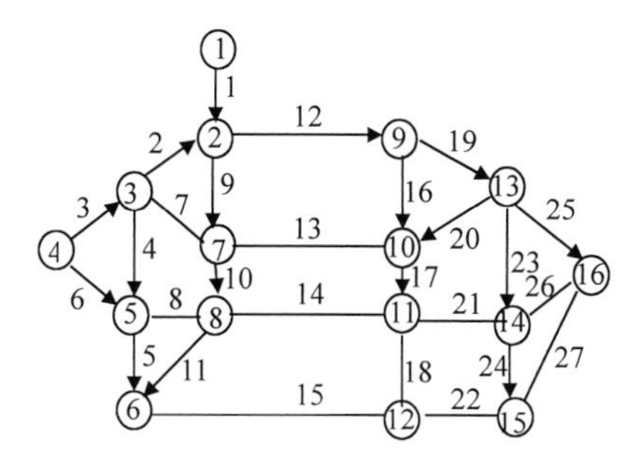

Fig.2: a pipeline network

The optimum design of the network can be made through the following procedure:

a. Decide the importance of components.

b. Decide some relatively reasonable design intensity values for the plant building and pumping plant building or design vectors (here we only consider the joint type variables) for each pipeline segment components according to their importance, requirement of function, seismic risk analysis of the region and so on.

c. Calculate the corresponding original cost, damage probability and future structural loss expectation of each component under various design intensity values or design vectors.

d. Combine every possible selection through enumeration, and computing system objective and constraints in each possible combination.

e. Under the original cost constraint, joint type (flexible joint or rigid joint) for each lifeline segment and optimum design intensity for pumping plant building can be decided to minimize the system lifecycle cost.

f. Under optimum design intensity, minimum cost design of the structure can be made for the pumping plant building. The optimum result is that pipeline [1][2][3][4][5][6][14][15][22][23][27] use flexible joints, the design intensity of water plant building and pump building are 9 and 8 separately.

CONCLUSION

A two-stage optimization method is presented for CIS anti-earthquake design. A simplified method for component and system seismic risk analysis is suggested. The optimum objective allowing for future loss expectation can meet the requirement of system global optimization. Through changing the system optimum variables into fewer discrete ones, the anti-earthquake optimum design of CIS becomes simple and practical.

References:

1. Ministry of construction department of P.R.China. (1994). *Design code of civil engineering*, Press of the Earthquake, Beijing, P.R.China
2. Wangguangyuan.(1999), Practical Methods of Optimum Aseismic Design for Engineering Structure and System, Building Factory Press, Beijing, P.R.China
3. Li Guiqing, Huo Da &Wang Dongwei. (1994), *The Anti-earthquake Reliability Analysis of Urban Civil engineering network system*, Press of the Earthquake, Beijing, P.R.China

Announcement

Thanks for the sponsor of China Natural Science Fund Committee.

Advances in Building Technology, Volume 1
M. Anson, J.M. Ko and E.S.S. Lam (Eds.)

Earthquake Analysis of Arch Dam with Joints Including Dam-Foundation Rock Nonlinear Dynamic Interaction

Xiuli Du[1] Jin Tu[2]

[1]Department of Science and Technology, Beijing Polytechnic University,
No. 100, Pingleyuan Road, Beijing 100022, China
[2] China Institute of Water Resources and Hydropower Research,
No.20, Chegongzhuan Road, Beijing 100044, China

ABSTRACT

The influence of contraction joints in arch dams on the seismic response analysis under strong earthquake is all very obvious. Existing models simulating the contact process of joints used in the seismic response analysis of arch dams, such as Fenves' model and Dowlings' model, are based on the penalty parameter method. It is difficult to determine the penalty parameters and in addition, the effect of dam-foundation dynamic interaction has not yet been considered in the analyses methods by the models stated above, and it is difficult to consider the two effects together by the frequency domain methods. Therefore, a combination method in the time domain of the explicit FEM and a transmitting boundary have been proposed for the earthquake analysis of arch dams by X.L., Du et al (2002). In this paper, the first effect stated above is studied by introducing a contact force model into the combination method. Finally, some results have been obtained by using this approach for the designed Xiaowan arch dam with maximum height of 292m in China.

KEYWORDS

Arch Dam, Foundation, Contraction Joint, Dynamic Interaction, Nonlinear Seismic Response

INTRODUCTION

The influences of energy radiation in an infinite foundation on the seismic response of arch dams have been studied by O. Maeso, J. Dominguez (1993), J. Dominguez O. Maeso (1993), H.Tan and A.K. Chopra (1995a, b) through the combination method of BEM and FEM in frequency domain, which is not suitable for nonlinear problems and is complicated in calculation. Taking consideration of the effect of faults and cracks existing in near-field foundation on the seismic response of arch dams, X.L. Du et al (2002) studied the nonlinear seismic response of arch dam-foundation systems by use of the explicit FEM and a transmitting boundary in the time domain. In this way some advantages are very obvious for solving nonlinear problems with large degrees of freedom due to the uncoupling feature. As far as the influence of simulating the opening and closing of contactable joints in arch dams on the seismic response of the arch dams during strong earthquake is concerned, some developments have been achieved. D. Row and V. Schricker (1984) proposed a contact element model with two nodes. G.L. Fenves, S. Mojthahedi and R.B. Reimer (1989) used a block contact element model without thickness. M.J. Dowling (1989) presented a three parameter contact model with rotational angle spring, transational spring and its position. The models mentioned above are based on the penalty parameter method. J.F. Hall (1996) used a smear crack method for simulating the influences of contactable joints on the seismic response of arch dams, which is suitable for simulating the stress response of arch dams. In this paper, based on the method suggested by X.L. Du et al (2002), a dynamic contact force model proposed by J.B. Liu, S. Liu and X.L. Du (1999) is used to analysis the seismic responses of arch dams with contactable joints, meanwhile, the energy radiation in infinite foundation and the local topographical and geological features in the near-field foundation are also considered.

WAVE MOTION SIMULATION METHOD

From the explicit FEM, the equation of motion in the j direction at any node i in the discrete model of arch dam-foundation systems can be written as

$$m_i \ddot{u}_{ij} + \sum_{l=1}^{n_e} \sum_{k=1}^{n} c_{ijlk} \dot{u}_{lk} + R_{ij} = F_{ij} \tag{1}$$

Where, m_i is the mass of the node i, $\ddot{u}_{ij}$ is the acceleration of j direction at the node i, $\dot{u}_{lk}$ is the velocity in k direction at some node l which is around the node i and belongs to the same element with the node i, R_{ij} and F_{ij} are the restoring force and external force in j direction at the node i respectively, c_{ijlk} is the damping influence coefficient of k direction at the node l on j direction at the node i, n_e is the nodal number related to the node i, n is the number of degrees of freedom, R_{ij} may be determined by an elastic-plastic constitutive model. If R_{ij} is an elastic restoring force, it is expressed by

$$R_{ij} = \sum_{l=1}^{n_e} \sum_{k=1}^{n} k_{ijlk} u_{lk} \tag{2}$$

In which, u_{lk} is the displacement in k direction at the node l, k_{ijlk} is the stiffness coefficient of k direction at the node l on j direction at the node i. By introducing an explicit integration formula with the accuracy of 2-order ($O(\triangle t^2)$) given by X.J. Li, et al (1992), we have

$$u_{ij}^{t+\Delta t} = \left(F_{ij}^t - R_{ij}^t - \sum_{l=1}^{n_e} \sum_{k=1}^{n} c_{ijlk} \dot{u}_{lk}^t \right) \Delta t^2 / 2m_i + u_{ij}^t + \Delta t \cdot \dot{u}_{ij}^t \tag{3}$$

$$\dot{u}_{ij}^{t+\Delta t} = \dot{u}_{ij}^t + \Delta t \left(F_{ij}^{t+\Delta t} - R_{ij}^{t+\Delta t} + F_{ij}^t - R_{ij}^t \right) / 2m_i - \left(\sum \sum c_{ijlk} \left(u_{lk}^{t+\Delta t} - u_{lk}^t \right) \right) / 2m_i \tag{4}$$

DYNAMIC CONTACT FORCE MODEL

A contactable interface S is shown in Fig.1, the two side surfaces of S are called S^+ and S^-, they are coincident with each other, and assuming that S is a smooth curve or curve surface. The finite element nodes between S^+ and S^- are a pair of nodes. The sliding deformation along the direction of interface of the contactable joints is too little to be considered. Hence, the contact process between the surfaces of the contactable joints may be simulated by the contact process between the two nodes of the pairs of nodes.

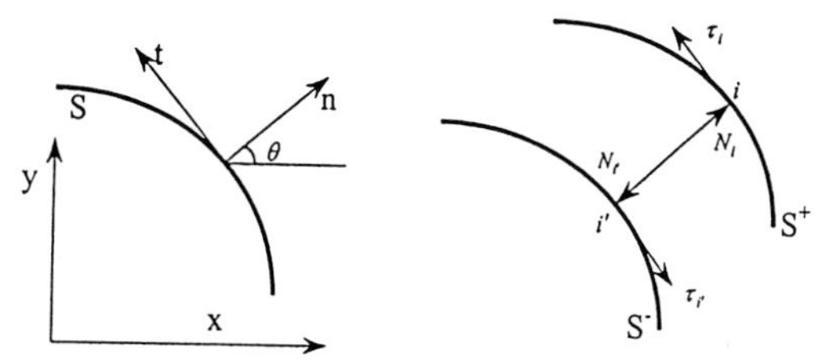

Fig1:Model with Joints and Two Side of Contactable Interface

For any pair of nodes, i and i', the node i is located on the side of S^+ and the node i' is located on the side of S^-, $\vec{N}_i$ and $\vec{\tau}_i$ are a normal contact force and a tangential contact force, respectively. Here, it is assumed that $\vec{n}$ in the normal direction to S^+ is positive. From the formula (3), by neglecting the effects of damping force, the displacement in j direction at the node i can been written as

$$u_{ij}^{t+\Delta t} = ((F_{ij}^t + N_{ij}^t + \tau_{ij}^t) - R_{ij}^t)\Delta t^2 / 2m_i + u_{ij}^t + \Delta t \dot{u}_{ij}^t \tag{5}$$

In which, N_{ij} and τ_{ij} are a projection of $\vec{N}_i$ and $\vec{\tau}_i$ on j direction, respectively. From the motion of equilibrium equations of j direction at the node i at time t, it can be known that $u_{ij}^{t+\Delta t}$ can not be got directly from the formula (5) in which N_{ij}^t and τ_{ij}^t are the functions of motion state at the time t and the time t+△t and unknown because $u_{ij}^{t+\Delta t}$ is unknown. Usually, the formula (5) may be solved by the iteration method. Fortunately, because of the characteristic of the explicit method, the tedious iterative process may be avoided through determining the contact state between the interfaces of the contactable joints at the time t. For convenience to derive, the formula (5) is written as:

$$u_{ij}^{t+\Delta t} = \overline{u}_{ij}^{t+\Delta t} + \Delta u_{ij}^{t+\Delta t} + \Delta v_{ij}^{t+\Delta t} \tag{6}$$

$$\overline{u}_{ij}^{t+\Delta t} = (F_{ij}^t - R_{ij}^t)\Delta t^2 / 2m_i + u_{ij}^t + \Delta t \dot{u}_{ij}^t \tag{7}$$

$$\Delta u_{ij}^{t+\Delta t} = N_{ij}^t / M_i \tag{8}$$

$$\Delta v_{ij}^{t+\Delta t} = \tau_{ij}^t / M_i \tag{9}$$

$$M_i = 2m_i / \Delta t^2 \tag{10}$$

Similarly, for the node i' there are the formulas as follow:

$$u_{i'j}^{t+\Delta t} = \overline{u}_{i'j}^{t+\Delta t} + \Delta u_{i'j}^{t+\Delta t} + \Delta v_{i'j}^{t+\Delta t} \tag{11}$$

$$\overline{u}_{i'j}^{t+\Delta t} = (F_{i'j}^t - R_{ij}^t)\Delta t^2 / 2m_{i'} + u_{i'j}^t + \Delta t \dot{u}_{i'j}^t \tag{12}$$

$$\Delta u_{i'j}^{t+\Delta t} = N_{i'j}^t / M_{i'} \tag{13}$$

$$\Delta v_{i'j}^{t+\Delta t} = \tau_{i'j}^t / M_{i'} \tag{14}$$

$$M_{i'} = 2m_{i'} / \Delta t^2 \tag{15}$$

$$N_{ij}^t = -N_{i'j}^t \tag{16}$$

$$\tau_{ij}^{t} = -\tau_{i'j}^{t} \tag{17}$$

Next, the computation methods for N_{ij}^{t} and τ_{ij}^{t} are introduced, respectively.

Computation of N_{ij}^{t}

N_{ij}^{t} is dependent on the contact state between the interfaces of the contactable joints. By the geometric conditions of not intruding into each other, there is

$$\vec{n}_i^T \cdot (\vec{u}_{i'}^{t+\Delta t} - \vec{u}_i^{t+\Delta t}) = 0 \tag{18}$$

The formula (18) represents the normal contact condition between the pair of the nodes i and the node i'. While to satisfy

$$\vec{n}_i^T \cdot (\vec{u}_{i'}^{t+\Delta t} - \vec{u}_i^{t+\Delta t}) \geq 0 \tag{19}$$

The contacting occurs between the nodes i and the node i', otherwise, the node i and the node i' keep separating. When separation between the node i and the node i' occurs, we have

$$N_{ij}^{t} = N_{i'j}^{t} = 0 \quad , \quad \tau_{ij}^{t} = \tau_{i'j}^{t} = 0 \tag{20}$$

$$u_{ij}^{t+\Delta t} = \overline{u}_{ij}^{t+\Delta t} \quad , \quad u_{i'j}^{t+\Delta t} = \overline{u}_{ij}^{t+\Delta t} \tag{21}$$

When the contact between the node i and the node i' occurs, there is the formula (18). Substituting formula (6) and (11) into formula (18), we have

$$\vec{n}_i^T \cdot (\vec{\overline{u}}_{i'}^{t+\Delta t} + \Delta\vec{u}_{i'}^{t+\Delta t} + \Delta\vec{v}_{i'}^{t+\Delta t} - \vec{\overline{u}}_i^{t+\Delta t} - \Delta\vec{u}_i^{t+\Delta t} - \Delta\vec{v}_i^{t+\Delta t}) = 0 \tag{22}$$

Because the displacement in the normal direction induced by the tangential force is zero, we have

$$\vec{n}_i^T \cdot \Delta\vec{v}_i^{t+\Delta t} = \vec{n}_i^T \cdot \Delta\vec{v}_{i'}^{t+\Delta t} = 0 \tag{23}$$

Substituting formula (23) into formula (22), we obtain

$$\vec{n}_i^T \cdot \left(\Delta\vec{u}_i^{t+\Delta t} - \Delta\vec{u}_{i'}^{t+\Delta t}\right) = \vec{n}_i^T \cdot \left(\Delta\vec{\overline{u}}_{i'}^{t+\Delta t} - \Delta\vec{\overline{u}}_i^{t+\Delta t}\right) \tag{24}$$

and we can define

$$\Delta_{1i} = \vec{n}_i^T \cdot \left(\Delta\vec{u}_i^{t+\Delta t} - \Delta\vec{u}_{i'}^{t+\Delta t}\right) = \vec{n}_i^T \cdot \left(\Delta\vec{\overline{u}}_{i'}^{t+\Delta t} - \Delta\vec{\overline{u}}_i^{t+\Delta t}\right) \tag{25}$$

Substituting formule (8) and (13) into formula (24) again, we obtain

$$\vec{N}_i^t = \left(M_i M_{i'} / (M_i + M_{i'})\right)\Delta_{1i} \cdot \vec{n}_i \tag{26}$$

Computation of τ_{ij}^{t}

When the contact between node i and node i' occurs, the sliding between node i and node i'

along the tangential direction may or may not be generated or not be generated. When there is no relative sliding, a static friction state is generated. When there is relative sliding, the dynamic friction state is generated. μ_s and μ_d are the static friction coefficient and the dynamic friction coefficient, respectively.

For the static friction state, we have

$$\vec{t}_i^{\,T} \cdot \left(\vec{u}_i^{\,t+\Delta t} - \vec{u}_{i'}^{\,t+\Delta t}\right) = \vec{t}_i^{\,T} \cdot \left(\vec{u}_i^{\,t} - \vec{u}_{i'}^{\,t}\right) \tag{27}$$

Where, $\vec{t}_i$ is the tangential vector at the node i on S^+. Substituting formula (6), (11) into formula (27), we have

$$\vec{t}_i^{\,T} \cdot \left(\vec{\bar{u}}_i^{\,t+\Delta t} + \Delta\vec{u}_{i'}^{\,t+\Delta t} + \Delta\vec{v}_{i'}^{\,t+\Delta t} - \vec{\bar{u}}_i^{\,t+\Delta t} - \Delta\vec{u}_i^{\,t+\Delta t} - \Delta\vec{v}_i^{\,t+\Delta t}\right) = \vec{t}_i \cdot \left(\vec{u}_{i'}^{\,t} - \vec{u}_i^{\,t}\right) \tag{28}$$

Noting the condition, $\vec{t}_i^{\,T} \cdot \Delta\vec{u}_{i'}^{\,t+\Delta t} = \vec{t}_i \cdot \Delta\vec{u}_i^{\,t+\Delta t} = 0$, and assuming that $\Delta_{2i} = \vec{t}_i^{\,T} \cdot \left[\left(\vec{\bar{u}}_{i'}^{\,t+\Delta t} - \vec{\bar{u}}_i^{\,t+\Delta t}\right) - \left(\vec{u}_{i'}^{\,t} - \vec{u}_i^{\,t}\right)\right]$, from formula (28), we can obtain

$$\vec{t}_i^{\,T} \cdot \left(\Delta\vec{v}_i^{\,t+\Delta t} - \Delta\vec{v}_{i'}^{\,t+\Delta t}\right) = \Delta_{2i} \tag{29}$$

Substituting formula (9) and (14) into formula (29), we have

$$\vec{\tau}_i^{\,t} = -\vec{\tau}_{i'}^{\,t} = \left(M_i M_{i'} / \left(M_i + M_{i'}\right)\right)\Delta_{2i} \cdot \vec{t}_i \tag{30}$$

For the dynamic contact state, we have

$$\left|\vec{\tau}_i^{\,t}\right| = \mu_d \left|\vec{N}_i^{\,t}\right| \tag{31}$$

During the calculation process, it is first assumed that the friction state between node i and node i' is the same as the former time step. If the friction state at the former time step is static, $\vec{\tau}_i^{\,t}$ can be solved by formula (30), then, the assumption is examined

$$\left|\vec{\tau}_i^{\,t}\right| \le \mu_s \left|\vec{N}_i^{\,t}\right| \tag{32}$$

If the formula (32) is not satisfied, $\left|\vec{\tau}_i^{\,t}\right|$ is calculated by formula (31), the sign of $\vec{\tau}_i^{\,t}$ is determined by Δ_{2i}. If the contact state between the node i and the node i' is dynamic, the next sufficient and necessary conditions are observed

$$\text{If } \operatorname{sgn}(\Delta 2i) = 1, \; \vec{t}_i^{\,T} \cdot (\vec{u}_{i'}^{\,t+\Delta t} - \vec{u}_i^{\,t+\Delta t}) > \vec{t}_i \cdot (\vec{u}_{i'}^{\,t} - \vec{u}_i^{\,t}) \tag{33}$$

$$\text{If } \operatorname{sgn}(\Delta 2i) = -1, \; \vec{t}_i^{\,T} \cdot (\vec{u}_{i'}^{\,t+\Delta t} - \vec{u}_i^{\,t+\Delta t}) < \vec{t}_i \cdot (\vec{u}_{i'}^{\,t} - \vec{u}_i^{\,t}) \tag{34}$$

The calculation accuracy of the dynamic contact force model has been verified by some typical examples with an analytic solution case in the research work of J.B. Liu, S. Liu, and X.L. Du (1999). In the dynamic contact force model, only the equilibrium conditions of mechanics and the contact conditions of geometry are considered. It is also suitable for the contact problems with nonlinear material media.

For a three-dimensional problem, S is a curved surface. The tangential sliding direction in the tangential plane at any contact point at t time is unknown. Therefore, the tangential sliding direction at any contact point should firstly be determined, then, the tangential friction force is determined by the same method as that for the two-dimensional problem. The tangential sliding direction at the time t can be determined through neglecting the friction force at the time t.

ANALYSIS EXAMPLE OF THE SEISMIC RESPONSE OF AN ARCH DAM WITH JOINTS

Xiaowan arch dam with a height of 292m will be located at the upper reach of the Lanchuang River in Yunnan province in China, the descriptions of the finite element model of the arch dam-foundation system and the earthquake input are seen in Ref. Given by X.L. Du et al (2002)

The four different kind of analytical case considered here are illustrated in Table.1. For simplicity, the influence of arch dam-reservoir water dynamic interaction was modeled through added mass with neglect of the compressibility of reservoir water. The influences of gravity load, temperature load, hydrostatic load, and sediment load have been considered. Five strips of joints are set up in the arch dam for Case A, C, D.

TABLE 1
THE FOUR DIFFERENT KINDS OF ANALYTICAL CASES

Case	Water Lever	Strips of Joints	Bottom Crack
A	Low water lever	5	Without
B	Low water lever	Without	Without
C	Normal water lever	5	With
D	Normal water Lever	5	Without

The Influence of the Contraction Joints on the Seismic Response of the Arch Dam

The comparison of X direction stresses and Z direction stresses at the different heights on the surface of the arch dam with joints and without joints are shown in Table.2 and Table.3 respectively. It can be seen that the effect of the joints make X direction stresses (approximate arch stresses) lower obviously, Z direction stress (Cantilever stresses) increase roughly.Table.4 shows the maximum degree of opening for the joints for the case A. Table 5 shows the maximum dynamic stresses at the different heights on the surface of the crown cantilever given by the paper method. The maximum opening on the upstream and downstream surfaces of the joints are shown in Table 6.

TABLE 2
THE INFLUENCES OF TRANSVERSE JOINTS TO STRESSES OF X DIRECTION IN MAIN LOCATION (MPa)

Location			Dam Crest			Mid Dam	Dam Bottom
			Left Abutment	Crown	Right Abutment	Crown	Crown
Maximum	Case A	US	1.21	0.76	-0.17	-1.69	0.44
		DS	0.42	0.75	0.63	-0.18	--
	Case B	US	1.20	2.20	-0.28	-1.45	0.30
		DS	0.47	3.10	0.68	-0.50	--
Minimum	Case A	US	-0.33	-6.10	-1.88	-6.23	-1.96
		DS	-0.43	-5.69	-0.18	-2.86	--
	Case B	US	-0.20	-6.02	-1.76	-5.96	-2.05
		DS	-0.42	-5.70	-0.14	-2.80	--

Note: US =Upstream Surface, DS =Downstream Surface

TABLE 3
THE INFLUENCES OF TRANSVERSE JOINTS TO STRESS OF Z DIRECTION IN MAIN LOCATION (MPa)

Location			Dam Crest			Mid Dam	Dam Bottom
			Left Abutment	Crown	Right Abutment	Crown	Crown
Maximum	Case A	US	0.16	1.07	-0.67	-1.40	2.75
		DS	-0.58	0.31	0.07	0.79	--
	Case B	US	0.15	1.10	-0.64	-1.30	2.50
		DS	-0.58	0.05	0.10	0.75	--
Minimum	Case A	US	-1.08	-2.01	-2.17	-5.25	-6.99
		DS	-1.36	-0.93	-0.47	-2.68	--
	Case B	US	-1.00	-1.60	-2.12	-5.25	-7.20
		DS	-1.35	-0.84	-0.47	-2.75	--

Note: US =Upstream Surface, DS =Downstream Surface

TABLE 4
THE MAXIMAL OPENING DEGREE OF JOINTS (mm)

Location	Joint 1	Joint 2	Joint 3	Joint 4	Joint 5
US	4.06	4.67	11.05	3.56	8.65
DS	5.05	3.65	10.25	4.22	6.21

Note: US =Upstream Surface, DS =Downstream Surface

TABLE 6
EFFECTS OF RADIANT DAMPING TO THE OPENING DEGREE OF JOINTS (mm)

Case	Location	Joint 1	Joint 2	Joint 3	Joint 4	Joint 5
A	US	5.05	3.65	10.25	4.22	6.21
	DS	4.06	4.67	11.05	3.56	8.65
B	US	0.70	0.00	0.00	0.09	0.75
	DS	0.47	0.67	0.98	0.04	0.00

Note: US =Upstream Surface, DS =Downstream Surface

TABLE 5
EFFECTS OF RADIANT DAMPING TO MAXIMAL DYNAMIC STRESS (MPa)

Altitude	Location	Arch stress	Cantilever Stress
▽1210	US	3.186	1.176
	DS	1.900	0.827
▽1170	US	3.034	1.559
	DS	2.008	1.654
▽1130	US	2.579	1.750
	DS	1.761	1.958
▽1090	US	2.024	1.664
	DS	1.450	1.774
▽1050	US	1.441	1.646
	DS	1.215	1.361
▽1010	US	0.883	2.055
	DS	0.960	1.347
▽975	US	0.818	3.460
	DS	0.639	1.382
▽953	US	1.409	5.196
	DS	0.359	1.378

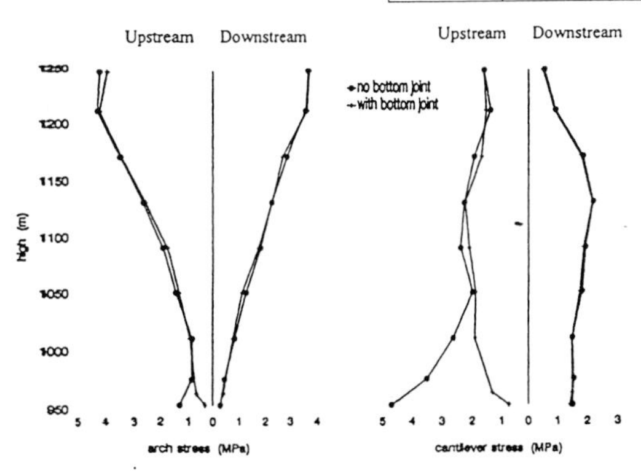

Fig.2 Arch and Cantilever Stress at Crown of Arch Dam with and without Bottom Crack

The Influence of Bottom Joint on the Seismic Response of Arch Dam

A strip of joint at the bottom of the dam was set up for comparison of the influence of the bottom joint on the seismic responses of arch dams. The comparison between dynamic arch and cantilever stresses at the crown of the arch dam with bottom joint and without bottom joint for the normal level of water is shown in Fig.2. It can be seen that the bottom joint makes the cantilever dynamic stresses lower as would be expecting.

CONCLUSION

In this paper, a calculation method has been established for solving the nonlinear seismic responses of the arch dam-foundation systems by combining the dynamic contact force model with the explicit FEM. As an example, the influences of contraction joints and bottom joints on the seismic response of Xiaowan arch dam are calculated by the use of the proposed method in this paper, and some results are obtained for reference to engineering design.

ACKNOWLEDGEMENTS

The study was supported by CNSF grant (No. 59739180)

REFERENCES:

1. Dominguez, J., and Maeso, O. (1993). "Earthquake Analysis of Arch Dams, II . Dam-Water-Foundation Interaction." *J. Eng. Mech.*, ASCE, 119(3), 513-530.
2. Dowling, M. J. (1987). "Nonlinear Seismic Analysis of Arch Dams." *Report No. EERL-87/03*, Earthquake Engineering Research Laboratory, California Institute of Technology, Pasadena, California..
3. Du, X. L., Zhang, B. Y., Tu, J. and Zhang, Y. H. (2002). "Earthquake Analysis of Arch Dam Including Dam-Foundation Nonlinear Dynamic Interaction." International Conf. On Advances in Building Technology.
4. Fenves, G. L., Mojthahedi, S., and Reimer, P. B. (1989). "ADAD-88: A Computer Program for Nonlinear Earthquake Analysis of Concrete Arch Dams." *Report No. UCB/EERC-89/12*, Earthquake Engineering Research Center University of California, Berkeley, California.
5. Hall, J. F. (1996). "Efficient Nonlinear Seismic Analysis of Arch Dams-Smeared Crack Arch Dam Analysis." *California Institute of Technology, Pasadena, California*, April.
6. Li, X. J., Liao, Z. P., and Du, X. L. (1992). "An Explicit Finite Difference Method for Viscoelastic Dynamic Problem." *Earthquake Engineering and Engineering Vibration*, 12(4), 74-80.
7. Liu, J. B., Liu, S., and Du, X. L. (1999). "A Method for the Analysis of Dynamic Response of Structure Containing Non-Smooth contactable Interfaces." *ACTA Mechanic Sinica* (English Series), 15(1), 63-72.
8. Maeso, O., and Dominguez, J (1993). "Earthquake Analysis of Arch Dams, I , Dam-Foundation Interaction." *J. Eng. Mech.*, ASCE, 119(3), 496-512.
9. Row, D., and Schricker, V. (1984). "Seismic Analysis of Structures With Localized Nonlinearities." *Proc. 8th WCEE.*, San Francisco, California.
10. Tan, H., and Chopra, A. K. (1995a). "Earthquake Analysis of Arch Dams Including Dam-Water-Foundation Rock Interaction." *Earthquake Eng. Struct. Dyn.* , 24(11), 1453-1474.
11. Tan, H., and Chopra, A. K. (1995b). "Dam-Foundation Rock Interaction Effects in Frequency-Response Function of Arch Dams." *Earthquake Eng. Struct. Dyn.* , 24(11), 1475-1489.

Advances in Building Technology, Volume 1
M. Anson, J.M. Ko and E.S.S. Lam (Eds.)

Earthquake Analysis of Arch Dam Including Dam-Foundation Rock Nonlinear Dynamic Interaction

Xiuli Du[1] Boyan Zhang[2] Jin Tu[2] Yanhong Zhang[2]

[1]Department of Science and Technology, Beijing Polytechnic University,
No. 100, Pingleyuan Road, Beijing 100022, China
[2] China Institute of Water Resources and Hydropower Research,
No.20, Chegongzhuan Road, Beijing 100044, China

ABSTRACT

An analysis method in the time domain is proposed by combining the explicit finite element method and the transmitting boundary, and by using a dynamic analysis method the static responses of arch dams is solved and a combination method is also proposed to calculate the nonlinear responses of arch dams under the actions of static and dynamic load. The influences of the arch dam-foundation dynamic interaction with energy dispersion and its local material nonlinear and non-homogeneous behavior on the seismic response of arch dams have also been studied for the designed Xiaowan arch dam with maximum height of 292m in China.

KEYWORDS

Arch dam, Foundation Rock, Nonlinear Dynamic Interaction, Earthquake

INTRODUCTION

Prof. R.W. Clough (1973) developed one of the earliest computer programs, ADAP, based on the finite element method for calculating the seismic response of arch dams, in which a massless foundation rock model with fixed boundaries was adopted and the influences of impounded water on the seismic response of arch dams were not considered. Based on ADAP, J. S. –H. Kuo (1982) and Y. Ghanaat and R.W. Clough (1989) studied the influences of the hydrodynamic effects with added mass approximation on the seismic response of arch dams and developed a computer program EADAP. In order to consider the hydrodynamic effects of impounded water with compressibility and the reservoir boundary absorption effects, a substructure technique in the frequency domain was developed by J.F. Hall and A.K. Chopra (1983) and K. –L. Fok (1986a, 1986b, 1986c,, 1987). O. Maeso and J. Dominguez (1993) and J. Dominguez and O. Maeso (1993) used the boundary element method in the frequency domain for considering the arch dam-foundation rock dynamic interaction and the travelling-waves effects of arch dam systems under the action of the harmonic seismic motion input. H. Tan and A.K. Chopra (1995a, b) also used the boundary element method to calculate the foundation impedance matrix and developed a substructure technique considering the foundation rock-arch dam dynamic interaction, impounded water with compressibility-arch dam dynamic interaction and the reservoir boundary absorption effects. There are still the some difficulties to consider the foundation rock-arch dam dynamic interaction and the effects of the local nonlinear and non-homogenous behavious in foundation rock in the meanwhile. It is the reason that the methods used for analysis of the arch dam-foundation rock dynamic interaction in the frequency domain are not suitable to the problem with nonlinear behavious. In this paper, an analytical technique in the time domain has been proposed for considering the foundation rock-arch dam dynamic interaction and the nonlinear effects of materials mentioned above.

At present, there are mainly the two classes of method to simulate scattering wave propagation in infinite media, the one class satisfies all the differential equations of motion and the radiation conditions at infinite region in the exterior of discrete models exactly, such as the boundary element method. They couple in space and time, and are usually formulated in the frequency domain, in which the shortcomings are that computing costs are high and it is difficult to simulate the local nonlinear and non-homogeneous features in the foundation near structures. The other class satisfies approximately the non-reflection condition at the artificial boundary of the finite discrete models to be used to simulate the effects of infinite foundation, in which the assumption of outgoing plane waves is mostly made, such as a transmitting boundary suggested by Z.P. Liao et al (1984), an absorbing boundary suggested by R. Clayton and B. Engquist (1977), a viscous boundary suggested by J. Lysmer and R.L. Kuhlemeyer (1969), they are all local non-coupling in space and time, and have the advantages for solving of the near field wave motion problems with local nonlinear behaviors. J.P. Wolf (1986) compared the accuracy of these artificial boundary methods for the simple one-dimensional case and pointed out that the well-known viscous boundary and the transmitting boundary lead to good accuracy. Due to the feature of multi-times transmitting, the high-order transmitting boundary methods lead to good accuracy for two-and-three dimensional cases also.

In this paper, an analysis method in the time domain used for computing the seismic responses of arch dams with consideration of the dam-foundation rock dynamic interaction and the local nonlinear behavior in the near-field foundation rock has been developed by combining the explicit finite element method with the transmitting boundary proposed by Z.P. Liao, et al (1984). It is noticed that there is a high frequency unstable phenomenon at artificial boundaries by use of the transmitting boundary. A method to eliminate the phenomenon has been proposed by Z.P. Liao (1996), in which the damping forces of media are related to the stiffness of media. Therefore, Rayleigh damping assumption was used here.

FUNDAMENTAL APPROACH

In this paper, an arch dam-foundation system is divided into an interior region with dam body and its adjacent near-field foundation with important local topographical and geological features and an infinite far-field foundation region with homogenous and linear features. The interior region is meshed by using the finite element method. The infinite exterior region is simulated by setting a set of artificial boundaries to simulate the propagation of outgoing scatter waves from the interior region to the infinite exterior region. The motions of the nodes of the discrete finite element model in the interior region are calculated by the explicit finite element method and the motions of the nodes at the artificial boundaries are solved by the transmitting boundary. The detailed introduction on the accuracy and the efficiency of the transmitting boundary can be seen in the references given by Z.P. Liao (1996). Next, the brief descriptions of the explicit finite element method are given.

From the explicit FEM, the motion of equation of j direction at any node i in the discrete model of arch dam-foundation systems can be written as

$$m_i \ddot{u}_{ij} + \sum_{l=1}^{n_e} \sum_{k=1}^{n} c_{ijlk} \dot{u}_{lk} + R_{ij} = F_{ij} \tag{1}$$

Where, m_i is the mass of the node i, $\ddot{u}_{ij}$ is the acceleration of j direction at the node i, $\dot{u}_{lk}$ is the velocity of k direction at the node l which is around the node i and belongs to the same element with the node i, R_{ij} and F_{ij} are the restoring force and the external force of j direction at the node i respectively, c_{ijlk} is the damping influence coefficient of k direction at the node l on j direction at the node i, n_e is the nodal number related to the node i, n is the number of freedom degree, R_{ij} may be determined by an elastic-plastic constitution model.

If R_{ij} is an elastic restoring force, it is expressed by

$$R_{ij} = \sum_{l=1}^{n_e} \sum_{k=1}^{n} k_{ijlk} u_{lk} \tag{2}$$

In which, u_{lk} is the displacement of k direction at the node l, k_{ijlk} is the stiffness coefficient of k direction at the node l on j direction at the node i.

Assuming that the damping force of any direction at the node i is only relative to itself velocities at the node i, thus

$$\begin{cases} c_{ijlk} = 0 & l \neq i \\ c_{ijlk} = c_{ij} & l = i \end{cases} \tag{3}$$

When $c_{ij} = c_{ik} = c_i$, the formula (1) may by written as

$$m_i \ddot{u}_{ij} + c_i \dot{u}_{ij} + R_{ij} = F_{ij} \tag{4}$$

It is known that formula (4) is a special example of formula (1), this is, when the damping force is only relative to the velocity of the node i and every direction damping coefficient is the same, there is the formula (4). Substituting the central difference formulations for acceleration and displacement into the formula (4), the explicit integration formula of Eq. (4) can by obtained. Generally, the assumption from the formula (1) to the formula (4) is not reasonable. Thus, an explicit integration formula can not be obtained by introducing the central difference method. X.J. Li, et al (1992) derived a explicit integration formula with the accuracy of 2-order ($O(\triangle t^2)$), which can be expressed as

$$u_{ij}^{t+\Delta t} = \left(F_{ij}^t - R_{ij}^t - \sum_{l=1}^{n_e} \sum_{k=1}^{n} c_{ijlk} \dot{u}_{lk}^t \right) \Delta t^2 / 2m_i + u_{ij}^t + \Delta t \cdot \dot{u}_{ij}^t \tag{5}$$

$$\dot{u}_{ij}^{t+\Delta t} = \dot{u}_{ij}^t + \left(F_{ij}^{t+\Delta t} - R_{ij}^{t+\Delta t} + F_{ij}^t - R_{ij}^t \right) \Delta t / 2m_i - \left(\sum_{l=1}^{n_e} \sum_{k=1}^{n} c_{ijlk} \left(u_{lk}^{t+\Delta t} - u_{lk}^t \right) \right) / 2m_i \tag{6}$$

The formulations introduced above is conditional stable, the determination of the time step Δt is seen in Ref. Given by X.J.Li, et al (1992).

COMPUTING OF THE COMBINATIVE RESPONSES OF ARCH DAM-FOUNDATION SYSTEMS DURING STATIC LOADS AND EARTHQUAKE

In this paper, the faults, cracks etc are treated as elastic-plastic media with a nonlinear constitutive relation with Druker-Prager Cap model adopted here. The nonlinear seismic response of arch dams is related to the static response of ones subjected by static loads. Such as self-weight, impound water and sediment etc. A techniquefor calculating the static responses by the dynamic method has been adopted here, in which the static loads are regarded as the step loads exerted at the initial time and the artificial damping are exerted during the step loads. Because the attenuation phenomenon of outgoing scattering waves in the infinite domain can not be considered in the transmitting boundary, the constrained effect of the stiffness supplied by the infinite domain for the finite model of the near-field medium is neglected. While the elastic restore behavior of the artificial boundaries of the finite model is needed in order to keep the equilibrium of the calculation model, the transmitting boundary may lead to drift instability. In order to overcome the instability, a constrained artificial boundary on which the elastic coefficients are given by experiences is introduced for computing the static response of the problem.

APPLICATION TO ARCH DAM-FOUNDATION SYSTEMS

The proposed method is used to calculate the seismic dynamic responses of an arch with height of 292 m called as Xiaowan arch dam located in the upper reach of the Lanchuang River in Yunnan province in China. The design earthquake intensity for the arch dam is IX degree with a horizontal peak acceleration of 0.290g, while the vertical peak acceleration is 2/3 of the horizontal one. The finite element model containing the arch dam and the foundation region to be calculated has the total element number of 20107 and the total nodal number of 22878, and there are the material kinds of 22 in the model. The sketch of the finite element discrete model except the artificial boundary region is shown in Fig.1. The altitude at the top of the dam is 1245 m. The arc length at the top of the dam is 935 m, the thickness at the top of the dam is 12 m, and the thickness at the bottom of the dam is 73 m. The dynamic elastic modulus is 27.3GPa, the density is 2400 N/m^3, and the Poisson ratio is 0.189 in the concrete of dam body. The material constants in the foundation region are shown in Table 1. Here, the dynamic elastic modulus is 1.3 times the static elastic modulus. For simplicity, the influence of arch dam-reservoir water dynamic interaction was modeled through added mass with neglect of the compressibility of reservoir water. The acceleration histories of incident seismic waves are shown in Fig.2 (incident S waves for stream-wise component and cross-stream component, incident P wave for vertical component). In the model the most liked fault location apart from the dam heel is also considered with a nonlinear constitutive relation with Druker-Prager Cap model adopted here. In the fault, the two groups material parameters are given out according to the distance from the valley site. In group 1, the static elastic modulus is 3500 MPa, the Poisson ratio is 0.3, the density is 1900 N/m^3, the friction coefficient f=1.00, the cohesive force C=0.5 MPa. In group 2, the static elastic modulus is

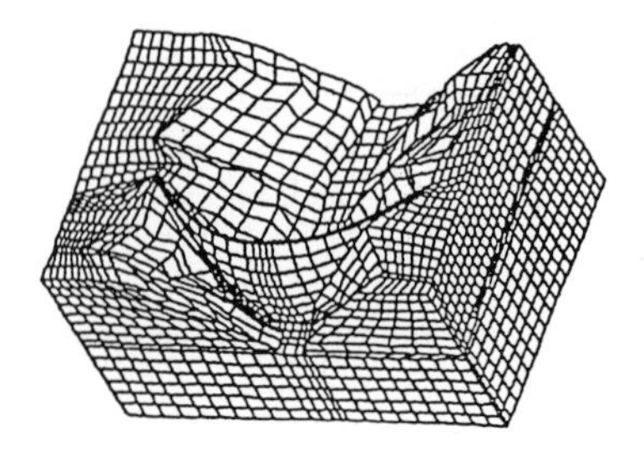

Fig.1 Finite Element Discrete Model of Xiaowan Arch Dam-foundation

1400 MPa, the Poisson ratio is 0.31, the density is 1900 N/m^3, the friction coefficient f=0.85, the cohesive force C=0.35 MPa. In the finite element discrete model, the minimum element size is 3 m, Δt is determined as 0.0004 second according to the computational requirements given by X.J.Li, et al (1992). The stress and displacement fields induced by gravity load, temperature load and hydrostatic load of normal water lever are considered during the calculation process of seismic action.

TABLE 1
MATERIAL CONSTANTS OF THE FOUNDATION REGION MEDIA (SI units)

Group No.	Density (N/m^3)	Dynamic elastic modulus ($\times 10^{11}$Pa)	Poisson ratio	Group No.	Density (N/m^3)	Dynamic elastic modulus ($\times 10^{11}$Pa)	Poisson's ratio
1	2630.00	0.3250	0.2200	12	2630.00	0.2210	0.2600
2	2630.00	0.2600	0.2500	13	2630.00	0.7150	0.3000
3	1900.00	0.4550	0.3000	14	2630.00	0.2990	0.2300
4	1900.00	0.1820	0.3100	15	2630.00	0.3250	0.2200
5	2000.00	0.5070	0.3000	16	2630.00	0.2600	0.2500
6	2630.00	0.1170	0.2650	17	2630.00	0.2860	0.2400
7	1900.00	0.2600	0.3000	18	2630.00	0.3380	0.2100
8	1900.00	0.1326	0.3100	19	2630.00	0.2730	0.2500
9	1900.00	0.3900	0.3500	20	2630.00	0.2470	0.2500
10	1900.00	0.7150	0.3000	21	2630.00	0.2340	0.2500
11	1900.00	0.1040	0.2900				

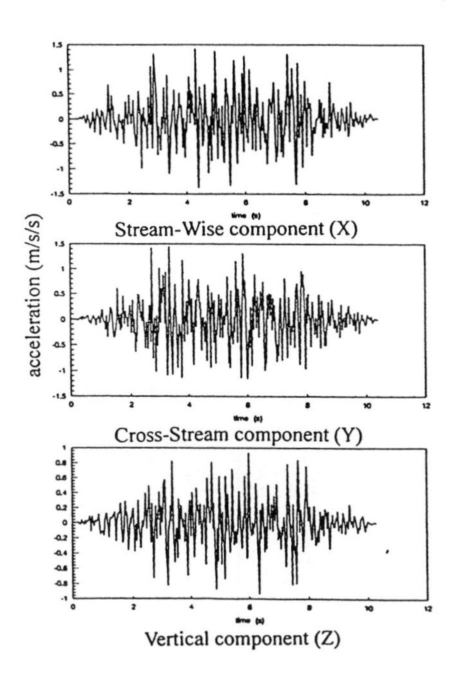

Fig.2 Incident Seismic Waves of Xiaowan Arch Dam

The comparison between the proposed method and the conventional method

The arch and cantilever peak dynamic stress absolute values at the different heights of the crown cantilever calculated by the proposed method and the conventional method are shown in Table2 and Table3. We can see that the peak dynamic stresses obtained by the proposed method with consideration of energy dispersion in infinite foundation through the transmitting boundaries are more less than the ones obtained by the conventional method with the assumptions of a massless foundation with fixed boundaries and uniform input seismic motion along the interface of dam-foundation on an average. The principal stresses of the upstream and downstream surfaces at the top altitude near the abutments are illustrated in Table 4. It can be seen that the maximum principal stresses by the two methods are roughly identical. The minimum principal stress absolute values by the proposed method are obviously larger than the ones by the conventional method.A and B represent the results of the conventional method and the proposed method here, respectively.

TABLE2
DYNAMIC ARCH STRESS OF CROWN CANTILEVER OF XIAOWAN ARCH DAM (MPa)

Altitude (m)		▽1210	▽1170	▽1130	▽1090	▽1050	▽1010	▽975	▽953
Upstream	B	4.18	3.47	3.05	2.39	1.83	1.11	0.79	1.25
	A	4.70	4.02	3.28	2.53	1.85	0.95	0.51	0.72
Downstream	B	3.38	2.39	1.62	1.10	0.80	0.77	0.41	0.26
	A	3.80	2.88	2.25	1.83	1.27	0.85	0.46	0.29

TABLE3
DYNAMIC CANTILEVER STRESS OF CROWN CANTILEVER OF XIAOWAN ARCH DAM (MPa)

Altitude (m)		▽1210	▽1170	▽1130	▽1090	▽1050	▽1010	▽975	▽953
Upstream	B	1.29	1.86	2.19	2.32	1.92	2.57	3.43	4.82
	A	1.32	2.52	2.92	2.77	2.45	2.58	3.52	4.36
Downstream	B	0.94	2.03	2.26	2.04	1.83	1.51	1.55	1.65
	A	1.15	2.21	2.44	2.16	1.87	1.72	2.52	3.09

TABLE4
DYNAMIC PRINCIPAL STRESSES NEAR THE ABUTMENTS (MPa)

			Right Abutments	Left Abutments
Upstream surface	Maximum	A	1.82	1.40
		B	1.93	1.13
	Minimum	A	-0.13	-0.12
		B	-1.85	-1.14
Downstream surface	Maximum	A	2.10	1.80
		B	1.49	1.56
	Minimum	A	-0.10	-0.08
		B	-1.53	1.17

The effects of the fault

In order to study the effects of the fault with the different type of mechanical behavious, comparison calculations have been carried out for the two foundation models with the fault with linear and nonlinear behavious. The results are illustrated in Table 5 and Table 6, respectively. It can be seen that with consideration of the nonlinear behavious of the fault, the influences on the arch and cantilever stresses at the different parts upstream and downstream of the crown cantilever are less. The reason may be reason that the fault located not at the dam heel is not across the interface between the arch dam and the rock foundation.

TABLE5
NONLINEAR AND LINEAR EFFECTS OF THE FAULT ON PEAK SYNTHETICAL ARCH STRESS OF CROWN CANTILEVER OF XIAOWAN ARCH DAM (MPa)

Altitude (M)		▽1210	▽1170	▽1130	▽1090	▽1050	▽1010	▽975	▽953
Upstream	Linear Fault	11.45	11.25	10.38	9.04	7.64	5.75	3.64	1.58
	Nonlinear Fault	11.61	11.09	10.32	9.17	7.74	5.88	3.70	1.57
Downstream	Linear Fault	8.42	6.04	4.86	4.18	3.21	2.15	1.32	1.63
	Nonlinear Fault	8.77	6.28	4.99	4.20	3.11	1.99	1.24	1.53

TABLE6
NONLINEAR AND LINEAR EFFECTS OF THE FAULT ON PEAK SYNTHETICAL CANTILEVER STRESS OF CROWN CANTILEVER OF XIAOWAN ARCH DAM (MPa)

Altitude (M)		▽1210	▽1170	▽1130	▽1090	▽1050	▽1010	▽975	▽953
Upstream	Linear Fault	2.36	3.64	4.42	4.62	4.72	4.12	3.30	8.59
	Nonlinear Fault	2.58	3.87	4.55	4.66	4.77	4.26	3.29	8.52
Downstream	Linear Fault	1.35	2.79	3.33	3.42	3.70	5.35	9.10	10.26
	Nonlinear Fault	1.40	2.79	3.41	3.56	3.79	5.25	8.95	10.11

CONCLUSION

Some conclusions for the analysis of the seismic responses of the Xiaowan arch dam can be drawn as follows:

1. Allowing for energy dispersion in the infinite foundation and the effects of non-uniform motion at the dam base along the canyon, the arch and cantilever stresses in the central and upper part of the arch dam are obviously reduced by (20%~40%).
2. The influence of the fault with nonlinear behavious on the seismic stresses in the arch dam is very little.

ACKNOWLEDGEMENTS

This research has been supported by CNSF grant (No.59739180).

REFERENCES

1. Clayton, R., and Engquist, B. (1977). "Absorbing Boundary Conditions for Acoustic and Elastic Wave Equations." *BSSA*, 67(6), 1529-1540.
2. Clough, R. W., Raphael, J. M., and Mojtahedi, S. (1973). "ADAP-A Computer Program for static and Dynamic Analysis of Arch Dams." *Report No. EERC 73-14*, Earthquake Engineering Research Center, University of California, Berkeley, California
3. Dominguez, J., and Maeso, O. (1993). "Earthquake Analysis of Arch Dams Ⅱ. Dam-Water-Foundation Interaction." *J. Eng. Mech.*, ASCE, 119(3), 513-530.
4. Fok, K. -L., and Chopra, A. K. (1986a). "Earthquake Analysis of Arch Dams Including Dam-Water Interaction, Reservoir Boundary Absorption and Foundation Flexibility." *Earthquake Eng. Struct. Dyn.*, 14(2), 155-184.
5. Fok, K. -L., and Chopra, A. K. (1986b). "Frequency Response Functions for Arch dams: Hydrodynamic and Foundation Flexibility Effects." *Earthquake Eng. Struct. Dyn.,* 14(5), 769-795.
6. Fok, K. -L., and Chopra, A. K. (1986c). "Hydrodynamic and Foundation Flexibility Effects in Earthquake Response of Arch Dams." *J. Struct. Eng.*, ASCE, 112, 1810-1828.
7. Fok, K. -L., and Chopra, A. K. (1987). "Water Compressibility in Earthquake Response of Arch Dams." *J. Struct. Eng.*, ASCE, 113(5), 958-975.
8. Ghanaat, Y., and Clough, R. W. (1989). "EADAP Enhanced Arch Dam Analysis Program, User's Manual." *Report No. UCB/EERC-89/07*, Earthquake Engineering Research Center, University of California, Berkeley, California.
9. Hall, J. F., and Chopra, A. K. (1983). "Dynamic Analysis of Arch Dams Including Hydrodynamic Effects." *J. Eng. Mech.*, ASCE, 109(1), 149-167.
10. Kuo, J. S. -H. (1982). "Fluid-Structure Interaction: Added Mass Computations for Incompressible Fluid." *Report No. UCB/EERC-82/09*, Earthquake Engineering Research Center, University of California, Berkeley, California.
11. Li, X. J., Liao, Z. P., and Du, X. L. (1992). "An Explicit Finite Difference Method for Viscoelastic Dynamic Problem, Earthquake Engineering and Engineering Vibration." 12(4), 74-80 (In Chinese).
12. Liao, Z. P. (1996). "An Introduction to Wave Motion Theories in Engineering." Beijing, China, Academic Press, 165-322 (In Chinese).
13. Liao, Z. P., Wong, H. L., Yang, B. P. and Yuan, Y. F. (1984). "A Transmitting Boundary for the Numerical Simulation of Elastic Wave Propagation." *Soil Dyn, Earthq. Eng.,* 3(4), 174-183.
14. Lysmer, J., And Kuhlemeyer, R. L. (1969). "Finite Dynamic Model for Infinite Media." *J. Eng. Mech. Div.*, ASCE, 95(4), 859-877.
15. Maeso, O., and Dominguez, J. (1993). "Earthquake Analysis of Arch Dams, Ⅰ. Dam-Foundation Interaction." *J. Eng. Mech.*, ASCE, 119(3), 496-512.
16. Tan, H., and Chopra, A. K. (1995). "Earthquake Analysis of Arch Dams Including Dam-Water-Foundation Rock Interaction." *Earthquake Eng. Struct. Dyn.*, 24(11), 1453-1474 .
17. Tan, H., and Chopra, A. K. (1995). "Dam-Foundation Rock Interaction Effects in Frequency-Response Functions of Arch Dams." *Earthquake Eng. Struct. Dyn.*, 24(11), 1475-1489 .
18. Wolf, J. P. (1986). "A Comparison of Time-Domain Transimitting Boudaries." *Earthq. Eng. Struct. Dyn.*, 14(4), 655-673.

Advances in Building Technology, Volume I
M. Anson, J.M. Ko and E.S.S. Lam (Eds.)

Stability Analysis of Large Complicated Underground Excavations

Tao Lianjin[1] Li Panfeng Zhang Zhuoyuan[2], Zhang Lihua[3] and Li Xiaolin[1]

College of Architecture and Civil Engineering, Beijing Polytechnic University, Beijing, 100022, China
Institute of Engineering Geology, Chengdu University of Technology, Chengdu, 610059, China
Civil Engineering Department, North China Institute of Technology, Sanhe, China

ABSTRACT

Based on field investigation and data analysis, the geostress field is fitted to a three-dimension model and then a model to evaluate the stability of powerhouses is built and computed. Some suggestions are proposed based on the analysis.

KEY WORDS
underground powerhouses, surrounding rockmass, discontinuities, stability analysis, numerical modeling

INTRODUCTION

The planed Xiluodu Hydropower station is at the third stage of the principal stream of Jinshajiang River between Panzhihua City and Yibin City. The station is located in the deep downcutting valley at the border of Leibo County, Sichuan Province and Yongshan County, Yunnan Province as shown in Fig. 1.

According to the feasibility design scheme, underground powerhouses are adopted and all of the powerhouses are located along the two banks in the deep downcutting valley. The underground chambers mainly include three large chambers, i.e. main chamber with size 436×33.8×57.4m, alternative chamber 43×17×31.6m and the tailrace 300×26 ×93.5m. The three main chambers are all located in the Permian Emei Mountain Basalt strata. As a whole, the rock mass quality is quite sound and the geostress is medium. However many discontinuities can be found such as interlayer shearing belts, intrastratal shearing belts, joints and protogenesis joints and discontinuities caused by epigenetic recreation. Especially the intrastratal shearing belts randomly distributing in the strata, make the rock mass strength and quality worse. The powerhouses are characterized by their large cross-section, complicated space intersection. Therefore it is very important to properly assess the effects of discontinuities and interaction of chambers.

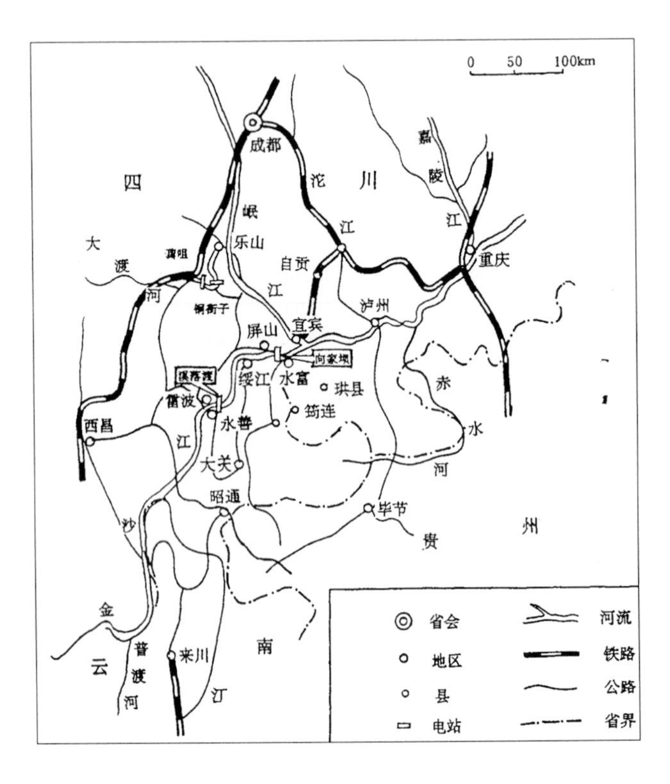

Figure 1 The sketch illustrating the location of Xiluodu Hydropower Station

GEOSTRESS FIELD FITTING

Geological Conditions

The three underground chambers(turbine, transformer and tail water chamber) on both bank located in Permian Emei Mountain Basalt strata $P2\beta 4$ to $P2\beta 6$. The rock mass quality is quite sound. However because of several times of active tectonism, many discontinuities can be found in the rockmass at the damsite. Especially the intrastratal shearing belts randomly distributing in the strata, worsen the rockmass strength and quality.

Three types of structural plane mainly distributed surrounding the underground powerhouse zone are described as follows:

Interlayer shearing belts(C for short): Interlayer shearing belts, as consequence of the interface in the tectonic rebuilding process of terrane, are the I class structural plane at dam site and extend continuously and steadily. They are gently inclined and the tilt angle is between 4 and 6 degrees. Intrastratal shearing belts (Lc for short): Intrastratal shearing belts, a suite of tectonic shearing plane distributing in terranes, are the II class structural plane. There exist a lot of intrastratal shearing belts in rock mass and some of them are bad in the view of engineering geology. Steep dip joints (J for short): The joints with steep dip angle and distributing widely and randomly in dam site, are the III class structural planes. The trace length of joints in the rock mass surrounding the powerhouses is about 2~4 meters; some of the joints are quite long and so they are called long joints.

Geostress measurement

So far a lot of geostress measurements have been carried out in the dam area, which includes six groups Kaiser effects in lab, 18 groups of stress relief borehole in-situ measurements and 6 groups of hydraulic fracturing measurements. The measurement points distribute at the bottom of the valley and in the shallow rockmass along the valley bank.

All the results are analyzed and compared with each other, which shows considerable derivation both in amount and orientation. Moreover due to limited measuring samples available, it is far insufficient to satisfy the demand of stress and deformation analysis of the excavations. It is therefore necessary to make further analysis to the original stress field and then fit it by fitting and simulating the generation process of the stress field with the approach of tectonic theory.

For this purpose, a 3-D finite element model is built to model the generation of the stress field as shown in Figure 2.

According to structural geology analysis, the present stress field is the result of deep down cutting of the valley. Therefore only the three-phase down cutting process and the main rockmass features are considered. The finite element model consists of 11842 elements and 51807 nodes.

The results of the general stress distribution are: the direction: middle stress – almost horizontal; the maximum and minimum are nearly flat; the amount: the maximum stress is 14~18 MPa in the left bank and 16~20 MPa in the right bank, the middle stress is 10~15 MPa, the minimum stress is little different in both banks, 4~6 MPa.

The modeling results have ideal consistency with the in-situ measurements in value, and some derivation in stress orientation, which has been corrected, based on the stress measurement and tectonic analysis.

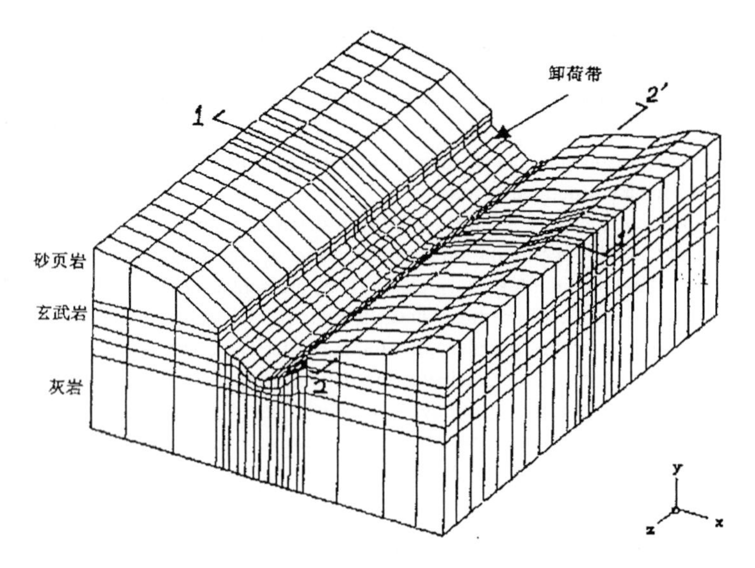

Figure 2 Finite element model to analyzing the initial stress field

TABLE 1

COMPARISON OF MEASURED AND MODELED GEOSTRESS RESULTS IN THE RIGHT BANK

Measuring point number	σ_1 (MPa)		σ_2 (MPa)		σ_3 (MPa)	
	measured	modeled	measured	modeled	measured	modeled
σ_{45-1}	15.9	16.3	10.0	12.5	5.37	5.4
σ_{45-2}	18.4	17.5	15.8	13.8	4.23	5.2
σ_{45-3}	18.2	17.0	12.3	12.8	6.51	5.7
σ_{45-4}	20.5	18.2	15.5	13.9	6.73	5.5
σ_{45-5}	18.2	17.6	12.7	12.5	6.05	5.4
σ_{45-6}	17.3	17.4	15.6	13.7	5.01	5.3

FINITE ELEMENT MODEL FOR THE UNDERGROUND EXCAVATIONS

In order to eliminate the boundary effects, the size of the 3-D model is: 840m long in Z direction, orientation N70° W; 425m wide in X direction, orientation S20° W; 300m high in Y direction. The model is divided into 16275 3-D elements, 67923 nodes. The upper load on the model is the gravitational load and the boundary load in the horizontal direction is determined by the geostress fitting results. The uniform distributed loads σ_1、

σ_3 are applied at 17MPa and 5MPa respectively. Due to the considerably large elastic modulus and brittle failure mode of the basalt, it is assumed that the rockmass is in elastic condition. However approximate yielding and failure analysis is made with elastic stress. The effects of the discontinuities are modeled by the strength equivalent approach. The computational parameters are listed in table 2.

TABLE 2 MECHANICAL PROPERTIES OF THE ROCKMASS

Rock type	Deformation modulus (MPa)	Poisson ration	Density (MN/m^3)	Cohesion (MPa)	Friction angle (^{0})	Tensile strength (MPa)
Compact basalt	13500	0.2	0.0285	2.5	55	0.2
Dotted basalt	13200	0.205	0.028	2.4	53	0.2
Porphyritic basalt	13000	0.208	0.027	2.35	52	0.2
Breccia latite	11000	0.21	0.0268	1.7	50	0.15
Discontinuities	2000	0.27	0.026	0.25	43	0
Limestone	12000	0.212	0.0265	1.8	50	0.18

RESULTS ANALYSIS

The above-mentioned model was analyzed with an FEM package 3D-σ and the excavation steps follow the design scheme.

The induced secondary stress after excavation

The distribution of secondary stress in the cross section located at No. 1 turbine is shown in Figure 3

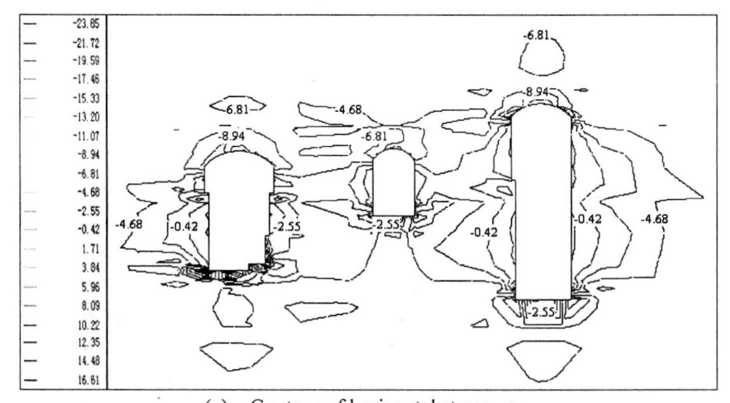

(a) Contour of horizontal stress σ_{xx}

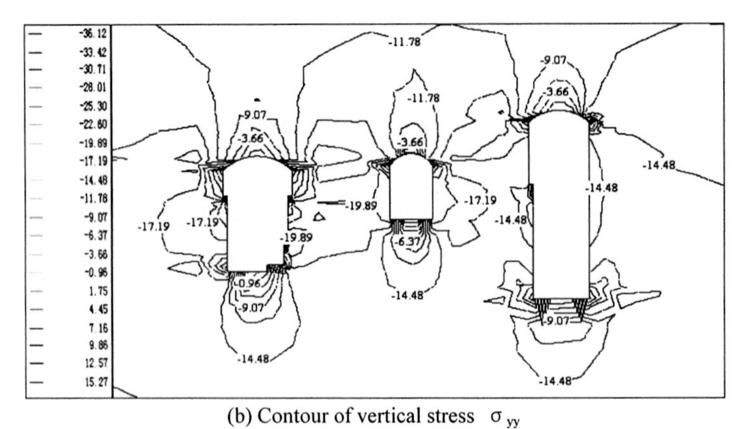

(b) Contour of vertical stress σ_{yy}

Figure 3 Stress distributions after excavation

The displacement

The displacement contour located at the No.1 turbine is shown in Figure 4.

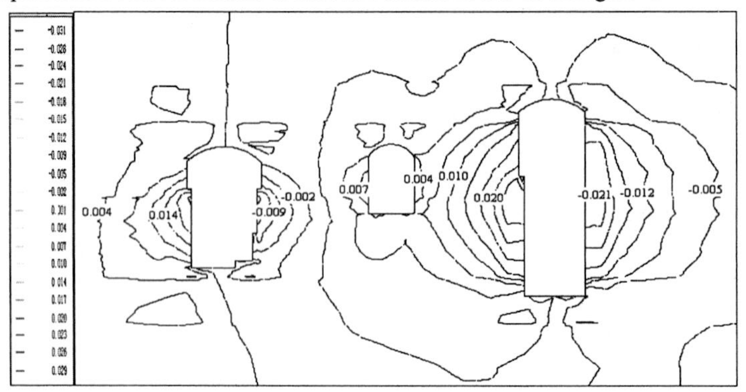

(a) Contour of displacement in X direction

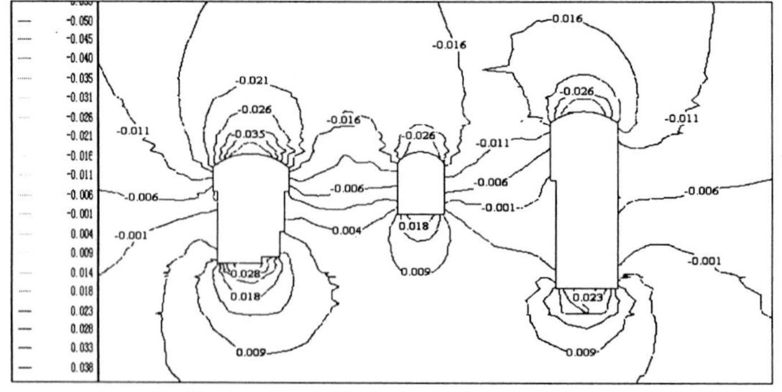

(b) Contour of displacement in Y direction

Figure 4 Displacement distributions in a typical cross section

The displacement value at some typical points is listed in Table 3

TABLE 3
DISPLACEMENT AT SOME TYPICAL POINTS IN THE CENTERLINE CROSS SECTION LOCATED AT TURBINE NO.1 (unit: cm)

Point No	Disp X	Disp Y	Point No	Disp X	Disp Y	Point No	Disp X	Disp Y
1	0.15	-4.38	7	-0.03	3.403	13	-1.37	0.183
2	0.45	-2.07	8	-0.052	3.414	14	-0.9	0.22
3	1.35	-0.25	9	-0.006	3.056	15	-0.8	-0.02
4	1.45	0.21	10	-0.08	1.666	16	-0.15	-2.07
5	1.783	0.004	11	-0.60	1.416			
6	0.2735	1.2371	12	0.8287	-0.973			

The failure zone

The failure zone is enclosed by the Coulomb-Mohr failure criteria. Figure 5 shows the failure zone in cross section located at No. 1 turbine room. The following features can be seen: (1) The shape of the failure zone at different cross section is similar; (ii) There exists considerable failure zone around the excavations, serious damage at the floor and sidewalls, the failure zone is deep in the sidewall. It should be noticed that the failure zone may connect between the alternator room and the tailwater chamber along the interlayer shearing belt C3; (iii) The damage degree and the shape of the failure zone of the surrounding rock are controlled by the discontinuities (interlayer shearing belts and intrastratal shearing belts).

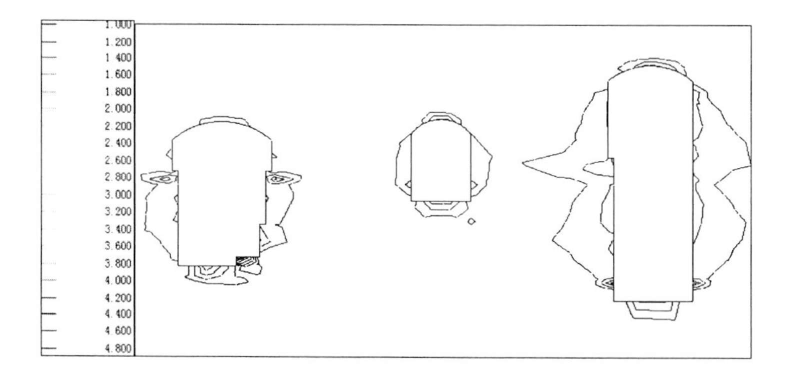

Figure 5 Failure zone around the excavations at cross section in turbine No.1

CONCLUSIONS AND SUGGESTIONS

Based on geological investigation, geostress measurement and fitting, 3-D finite element analysis, the following conclusions and suggestions may be drawn:

(1) There exists stress concentration after excavation with maximum 36.04 MPa at tailwater chamber; there is also some tensile stress zone in the surrounding rock unfavorable to the stability of powerhouses.

(2) Displacement: The maximum horizontal displacements are 1.78 cm and 1.59 cm at the sidewall of main turbine chamber; 0.90 cm and 0.25 cm at the sidewall of main alternator chamber; 2.29 cm and 2.07 cm at the sidewall of tailwater chamber. The maximum vertical displacements are 4.38 cm at the crown of the main turbine chamber; 2.96 cm at the alternator chamber; 3.69 cm at the tailwater chamber.

(3) Failure zone: There exists different failure zones around the excavations. The main turbine chamber: serious damage in the floor but the failure range is small; the damage in the sidewall is slight but deep. The failure zone between the main alternator chamber and the tailwater chamber almost connect, which is the result of the effects of the high sidewall of the tailwater chamber to the main alternator chamber, and the results of the discontinuities.

(4) The effects of the discontinuities: all the discontinuities, especially those exposed during excavation, have significant effects on the stress distribution, displacement and the failure zone of the surrounding rock.

REFERENCES

1. Chengdu Institute of Hydropower Investigation and Design, Feasibility research report on Xiluodu Hydropower, Jinshajiang River, 1998
2. Chengdu Institute of Hydropower Investigation and Design, Outline and guideline of feasibility study on Xiluodu Hydropower, Jinshajiang River, 1999
3. Hu Bin, Zhang Zhuoyuan and Tao Lianjin, Statistical analysis of geostress measurements in deeply downcutting valley, J of Geohazards Prevention and Geoenvironment Protection, 2001(3)
4. Hu Bin, Zhang Zhuoyuan and Tao Lianjin, Variation feature of geostress in a deep downcutting valley, J of World Earthquake Information, 2001(2)

ACKNOWLEDGEMENTS

This research is supported by the Young Talented Teacher Program of the Ministry of Education of China, the Natural Science Foundation of Beijing, the project of Beijing New SciTech Star and the Geohazard Prevention Openlab Foundation.

Advances in Building Technology, Volume 1
M. Anson, J.M. Ko and E.S.S. Lam (Eds.)

Evaluation of Ground Vibration Induced by Subway

Tao Lianjin [1], Li Xiaolin [1] and Zhang Dingsheng [2]

[1] College of Architecture and Civil Engineering, Beijing Polytechnic University
Beijing, 100022, China
[2] Beijing Design Institute of City Construction

ABSTRACT

Based on field monitoring and wheel-rail model analysis, a finite element model is built to evaluate the vibration in the subway tunnels between subway stations. A spring and viscous boundary is adopted in the model and transient approach is used to model the dynamic effects of the train with ANSYS. The simulation results show good consistency with the monitored data in situ, and the following conclusions are drawn according to the analysis and monitoring:

(1) The horizontal vibration just above the tunnel is minimum and the vertical vibration is maximum;
(2) The vertical ground vibration caused by subway train is more noticeable than the horizontal one;
(3) On the ground, the vibration decreases with the increase of the distance away from the subway tunnel, and that the attenuation in the vertical direction is faster than in horizontal direction.

KEYWORDS

ground vibration, wave transmission, finite element model

1. INTRODUCTION

The subway circuit in a city generally passes through the downtown area where the building is multitudinous. The embedded depth of subway tunnel is generally shallow, and the vibration that the running subway train gives rise to will be delivered to the lining through the track structure, and be delivered to the ground through the rock soil medium outside the lining structure, causing the ground vibration. This kind of vibration will influence building on the ground. At present, the environment problem initiated by engineering construction is arousing more and more attention.

In recent years, the environment questions such as vibration caused by subway are more and more noticeable owing to the great development of city tube traffic systems. It is meaningful to evaluate the effects of subway vibration to the environment.

2. CURRENT RESEARCH

The research on vibration caused by the subway chiefly includes two aspects at present: (1) vehicle-track system dynamics model, and finding the exciting force the running subway train erects on the subway tunnel by computing [1][2]. (2) from dynamic model of tunnel structure-strata, obtaining the kinetic equation of the model, and adopting a numerical method to solve the equation and so obtaining the response of the whole model. Wang Fengchao[5], Zhang Yue and Pan Changshi [6] and Sun Juns [7] made some progress in this aspect.

3. COMPUTING MODEL AND RESULTS

3.1. General Engineering Situation

The soil in Beijing generally can be divided into miscellaneous filling soil (1~3m), clay silt, silty clay, sand soil (total thickness > 20 meters) and cobble. Because the embedded depth of subway tunnel does not generally surpass 10 meters, it is considered that the subway tunnel passes through clay silt, silty clay, and sand soil. The tunnel cross section in the region between subway stations is a rectangle, and reinforced concrete lining is used. A sketch of the tunnel cross section is shown in figure 1.

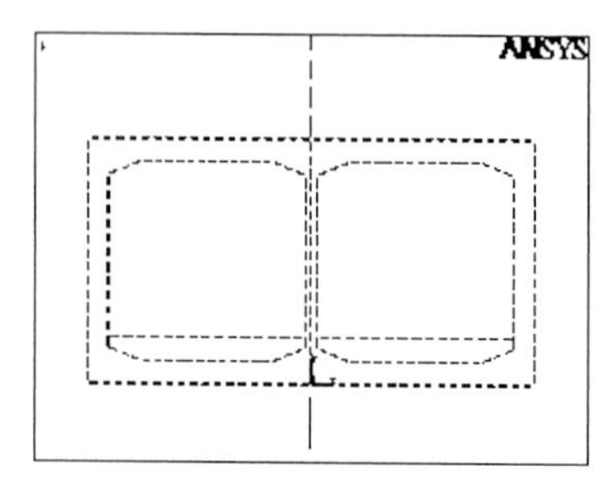

Fig. 1. Tunnel cross section

3.2. Effect of Impact of Train wheel to The Track

According to the testing data of Pan Changshi [7] to the vibration in a Beijing subway tunnel , we used spectral analysis to obtain the expression for the acceleration of the rail bottom:

$$x(t) = \sum_{k=-N/2}^{N/2} (A_k \cos k\omega t + B_k \sin k\omega t) \tag{1}$$

Where

$$A_k = \frac{2}{N}\sum_{j=0}^{N-1} x_j \cos\frac{2\pi kj}{N} \tag{2}$$

ω---- circle frequency, $\omega = \dfrac{2\pi}{N\Delta t}$

N ---- number of sampling, $N = 2048$

$$B_k = \frac{2}{N}\sum_{j=0}^{N-1} x_j \sin\frac{2\pi kj}{N} \tag{3}$$

Δt ----time interval for sampling, $\Delta t = \dfrac{s}{300}$

The model of interaction between a simplified wheel system of the subway train and the rail is shown in figure 2.

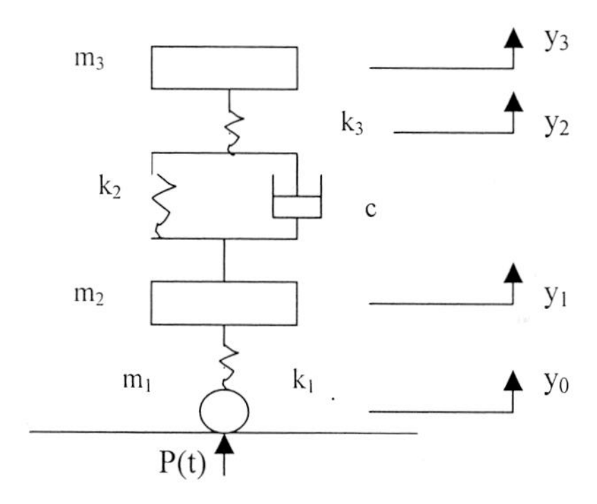

Fig 2. The interaction model of wheel and rail

In Fig. 2, m_1 = 1700.0 Ns²/m, m_3 = 9050.3 Ns²/m, $k_1 = 2.45\times10^6$ N/m, $k_3 = 1.90757\times10^6$ N/m, $c = 0.240884\times10^6$ Ns/m. The interaction force is marked as $P(t)$, the balance seat coordinates of m_1, m_2, k_3, m_3 are marked as y_0, y_1, y_2, y_3. According to the D'Alembert principle, the differential equation of motion of the train wheel system can be attained as follows:

$$P(t) = 125500 + \sum_{k=-N/2}^{N/2} \begin{Bmatrix} [128.0669A_k - 165.7383(k\omega)^2 \bullet F_k - 24.7128(k\omega)^3 \bullet E_k \\ +0.1119(k\omega)^4 \bullet F_k + 0.0103(k\omega)^5 \bullet E_k]\cos k\omega t \\ +[128.0669B_k - 165.7383(k\omega)^2 \bullet E_k - 24.7128(k\omega)^3 \bullet F_k \\ +0.1119(k\omega)^4 \bullet E_k + 0.0103(k\omega)^5 \bullet F_k]\sin k\omega t \end{Bmatrix} (N) \tag{4}$$

where, A_k, B_k are coefficients. The vertical train load can be calculated from the expression:

$$F(t) = \frac{K \bullet n \bullet N \bullet P(t)}{L} \qquad \text{(N/m)} \tag{5}$$

In this expression,

K ---- coefficient of correction
N --- -number of double wheel of each railway carriage
n ---- number of railway carriages
L ---- length of the subway train

The dynamic load induced by the subway train can be obtained when N=5, n=2, L=5*19.2=96m, K=1. We simplified the result, and produced the load-time curve as shown in figure 3.

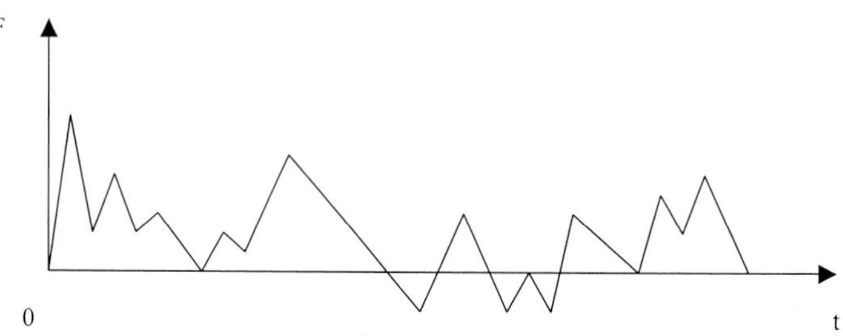

Fig 3. Load-time curve of wheel to rail

3.3. Computational Model and Method

3.3.1. Geometry model

The computational model is shown in figure 4.

Analyzing the vibrating influence question that the subway train causes, is strictly a 3-D question, which would require a long time for the computer to perform. When it is simplified to a 2D plane strain question, the time spent on computing can be shortened greatly, and the precision can still satisfy the engineering needs[8][9].

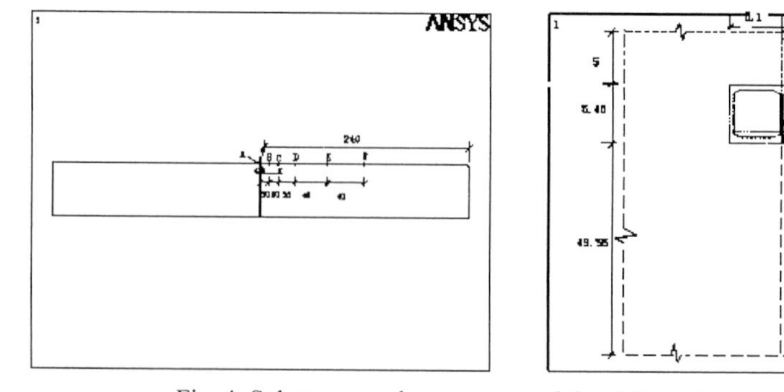

Fig. 4. Subway tunnel geometry model and its partly enlarged view

3.3.2. The computing parameters

The subway railway roadbed is of the crushed stone railway roadbed, and is considered as linear elastic material. The computing parameters are:
Young's modulus:E=1.3 × 10^3 MPa Poisson ratio:$\mu = 0.2$ Density:$\rho = 2000\ Kg / m^3$

The reinforced concrete lining is considered as linear elastic. The computing parameters are:
Young's modulus:E=2.1 × 10^4 MPa Poisson ratio:$\mu = 0.2$ Density:$\rho = 2400\ Kg / m^3$

The surrounding soil is considered as linear elastic. The computing parameters are:
Young's modulus:E=30 MPa Poisson ratio:$\mu = 0.3$ density:$\rho = 2000\ Kg / m^3$

3.3.3. Computational method

The finite element method with ANSYS is adopted. The Newmark time integral method at dispersed time points was used to solve the above-mentioned equation. The dynamic load shown in figure 2 is adopted as the computing load, and initial stress and the gravity load of all the structure are considered. The analysis time is 7 seconds, the length of step time is 0.05 second, and 140 steps were computed. The 8 nodes plane quadrilateral isoparametric element was adopted. The number of the elements was 2420, having 7413 nodes. An elasticity boundary condition was adopted at the two sides and at the bottom of the model.

The finite element mesh is shown in figure 5.

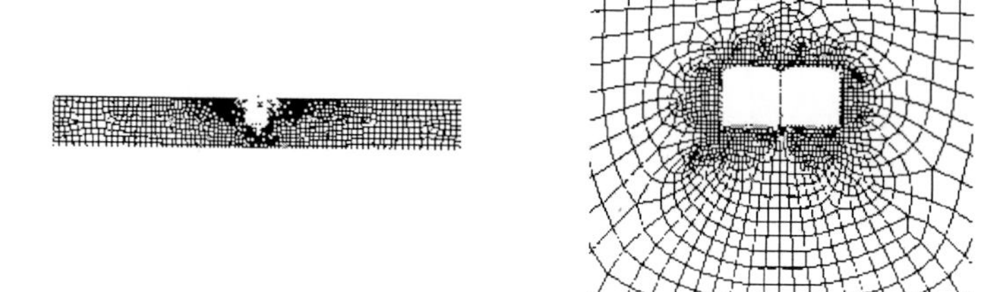

Fig. 5. The finite element mesh (whole and the close-up central part)

3.4. Part of Computational Results and the Analysis

3.4.1. Part of computational results

Six reference points were chosen in computing: A, B, C, D, E and F. The positions of each reference point are shown in figure 4. The vibrations in horizontal direction and vertical direction are computed respectively.

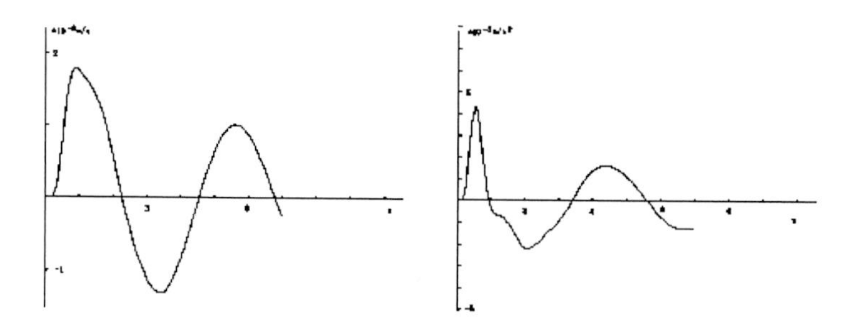

Fig. 6. Velocity-time curve (left) and acceleration-time curve (right) at point A

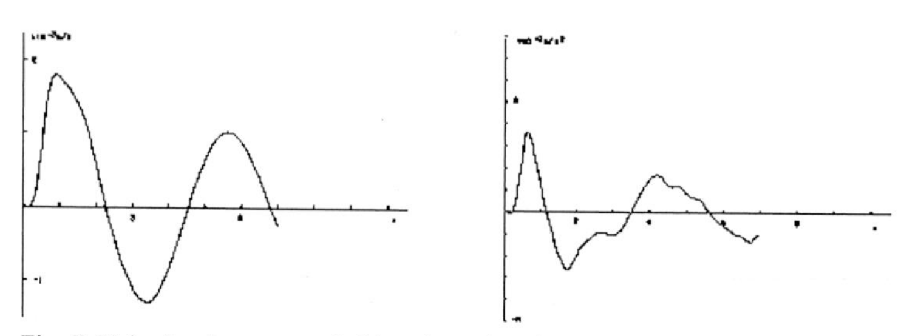

Fig. 7. Velocity-time curve (left) and acceleration-time curve (right) at point E

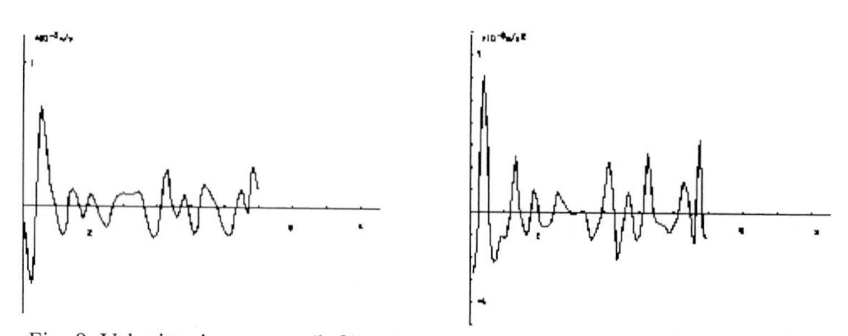

Fig. 8. Velocity-time curve (left) and acceleration-time curve (right) at point A

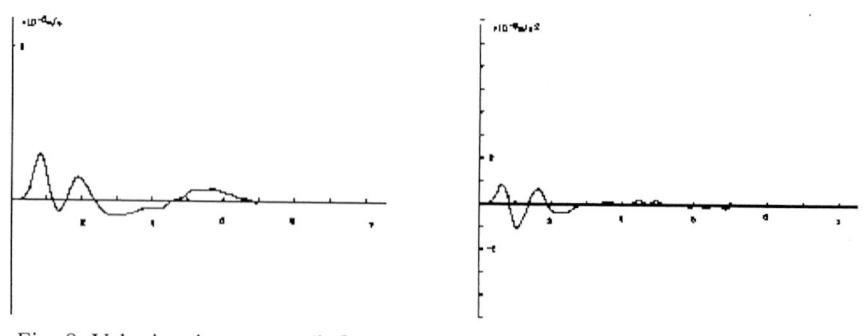

Fig. 9. Velocity-time curve (left) and acceleration-time curve (right) at point E

The maximal values of the results at each reference point are shown in in figure 10 and 11.

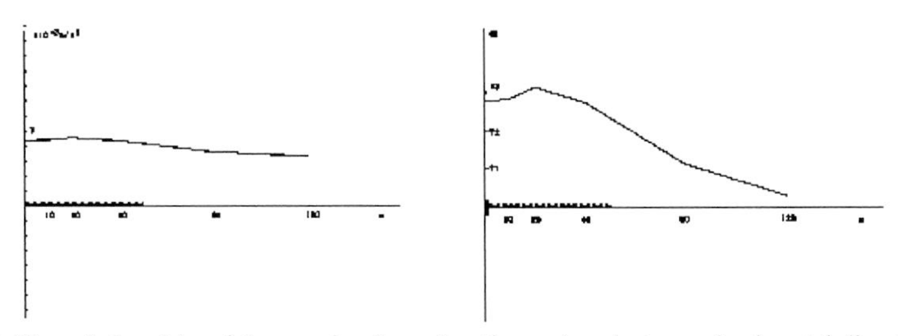

Fig.10. The relationship of the maximal acceleration values between horizontal direction and distance

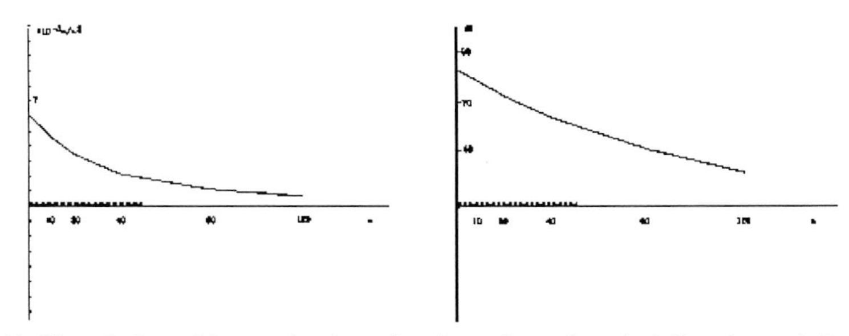

Fig.11. The relation of the maximal acceleration values of vertical direction and distance

4. CONCLUSIONS

According to the results given above, some conclusions can be drawn:

1. The horizontal vibration of the ground just above the subway tunnel is the minimum, while the vertical direction vibration is maximum;

2. The vertical vibration caused by the subway train is more significant than the vibration in the horizontal direction;

3. The ground vibration attenuates to some extent with increase of the distance from the subway tunnel, and attenuation of vibration in the vertical direction is faster than in horizontal direction.

Acknowledgements

This research is supported by the Young Talented Teacher Program of the Ministry of Education of China, the Natural Science Foundation of Beijing and the Project of Beijing New

SciTech Star.

References

[1] Zhang Yue, Bai Baohong. (2000). The method of identifying train vibration load acting on subway tunnel structure. *Journal of Vibration and Shock* **19:3**, 68-76

[2] Li Dewu, Gao Feng. (1999). In-situ measurement and frequency-spectrum analysis for the traffic vibration of foundationstructure of the tunnel. *Journal of Gansu Sciences* **11:1**, 52-54

[3] Liu Weining, Xia He, Guo Wenjun. (1996). Study of vibration effects of underground trains on surrounding environments. *Chinese Journal of Rock Mechanics and Engineering* **15**, 586-593

[4] Chen Shi, Xu Guobin,Gao Ri. (1998). Dynamic analysis of the impact of train vibration acting on the structure along the railway. *Journal of Northern Jiaotong University* **22:4**, 57-60

[5] Wang Fengchao, Xia He, Zhang Hongru. (1999). Vibration effects of surrounding buildings. *Journal of Northern Jiaotong University* **23:5**, 45-48

[6] Zhang Yue, Bai Baohong, Pan Changshi. (1997). Evaluation of the impact of subway train vibration acting on surroundings. *Control for yawp and vibration* **2**, 37-41

[7] Sun Jun. (1996). *Theory and practice of underground engineering design*, Science and Technology Publishing House of Shanghai

[8] DRJ·Irving. Zheng Guoping, Liu Zhong, Xu Jiali translated. (1989). *Theory and application of finite element method of plastic mechanics*, Weapon Industry Publishing House

[9] Pan Changshi, Zhang Mi, Wu Hongqing. (1995). *Numerical method of tunnel mechanics*, Railway Publishing House of China

Advances in Building Technology, Volume I
M. Anson, J.M. Ko and E.S.S. Lam (Eds.)

NONSTATIONARY RESPONSE OF SPATIAL LATTICE SHELLS UNDER MULTIPLE SEISMIC INPUTS

X.S. Wang, S.D. Xue and Z. Cao

The College of Architecture and Civil Engineering
Beijing Polytechnic University, Beijing 100022, China

ABSTRACT

This paper presents a nonstationary random vibration study of spatial lattice shells under multi-support and multi-dimensional seismic excitations. The multi-dimensional pseudo excitation method for random vibration analysis of structures is first developed. It is found that the derived method is particularly suitable for random vibration analysis of long span spatial structures, which are usually characterized with closely spaced natural frequencies and coupled vibration modes. Based on the method, a computer program is developed. The seismic performance of a spatial lattice shell under different earthquake inputs is investigated. Some observations are made. It is shown that the developed method is a highly efficient algorithm, which can easily solve the nonstationary problem of large structures under multi-support and multi-dimensional seismic excitations.

KEYWORDS

Random vibration, nonstationary, seismic response, lattice shell, spatial structure, pseudo excitation method, multiple seismic inputs, multi-support earthquake excitation, 3-D random analysis.

INTRODUCTION

Lattice shell is a type of popular spatial structure. At present, lattice shells have been widely used in the construction of long span sports buildings, exhibition halls, and hangars, etc. The potential social and economical influence and impacts resulting from the failure of this type of structure make it mandatory that special care be taken in the design of lattice shells, in which the seismic performance of the structure is an important concern. In recent years, dynamic analyses of lattice shells to earthquake excitations have been performed by many researchers (Cao and Zhang 2000; Shen *et al* 2000). However, most of the studies in the past focused only on single-dimensional and deterministic analysis. Since lattice shell is a type of spatial structure whose vibrations are dominated by both horizontal and vertical motion, its seismic performance to multiple earthquake inputs needs to be investigated. To this end, this paper presents a nonstationary random vibration analysis of lattice shells subjected to multi-support and multi-dimensional earthquake excitations.

The earthquake ground motion can be regarded as a random process (Houser 1947). At present, random vibration analysis method has been widely accepted for analyzing the stochastic responses of structures to earthquake excitations. Many scholars engaged in random vibration study of structures in past few years (Gasparini 1979; Harichandran and Vanmarcke 1986). Recently, structural responses to multiple seismic excitations have attracted much attention by researchers (Li 1998; Lopez *et al* 2000). Two major methods, response spectrum method and random analysis method, have been investigated. However, it is found that these methods are difficult to apply to practical large structures due to their considerable computation efforts. In order to reduce the computational effort, the cross-correlation items between the participant modes are generally neglected, i.e., the accurate CQC method is approximately replaced by the SRSS method (Clough and Penzien 1993). However, this is acceptable only for structures with small damping and sparsely spaced eigen-frequencies. For most spatial structures, it is well known that the participant frequencies are often closely spaced and the vibration modes may couple together. Therefore, the SRSS method may produce large errors. Recently, Lin *et al* (1993; 1994) proposed a highly efficient algorithm called the Pseudo Excitation Method (PEM). This method is a fast and accurate CQC algorithm, in which all the cross-correlation terms between the participant modes are involved, therefore, it is more suitable for dynamic analysis of structures which are characterized with closely spaced natural frequencies and coupled vibration modes. Using the concept of PEM, a multi-dimensional pseudo excitation method was further developed by Xue *et al* (2000) in order to investigate the random responses of lattice shells under multi-component seismic excitations. However, in the researches of Xue *et al* (2000), the seismic excitations were assumed to be stationary and no phase lags between the excitations were taken into account. In this paper, the multi-dimensional pseudo excitation method is further developed for nonstationary random vibration analysis of structures under multi-support and multi-dimensional earthquake excitations, in which the phase lags between the excitations have been involved. The formulation is first derived. Then, a random vibration analysis is performed for a spatial lattice shell to investigate its seismic performance under multiple seismic input, from which some conclusions are drawn.

FORMULATION

Selecting a global coordinate system O-XYZ with Z-axis vertically upwards. The system is assumed to be static relative to the center of the earth. For a lumped mass structure having P ground joints and N free joints, the equation of motion can be expressed by

$$\begin{bmatrix} M_{ss} & 0 \\ 0 & M_{mm} \end{bmatrix} \begin{Bmatrix} \ddot{U}_s + \ddot{U}_r \\ \ddot{U}_m \end{Bmatrix} + \begin{bmatrix} C_{ss} & C_{sm} \\ C_{ms} & C_{mm} \end{bmatrix} \begin{Bmatrix} \dot{U}_r \\ 0 \end{Bmatrix} + \begin{bmatrix} K_{ss} & K_{sm} \\ K_{ms} & K_{mm} \end{bmatrix} \begin{Bmatrix} U_s + U_r \\ U_m \end{Bmatrix} = \begin{Bmatrix} 0 \\ P_m \end{Bmatrix} \tag{1}$$

where $[M_{ss}]$, $[C_{ss}]$, $[K_{ss}]$ are, respectively, mass, damping, and stiffness matrices of the structural free joints; $[M_{mm}]$, $[C_{mm}]$, $[K_{mm}]$ are, respectively, mass, damping, and stiffness matrices of the ground joints; $[C_{sm}]$, $[C_{ms}]$ and $[K_{sm}]$, $[K_{ms}]$ are respectively coupled damping and stiffness matrices between the structural free joints and ground joints; $\{U_s\}$ and $\{U_r\}$ are respectively the quasi-static and quasi-dynamic displacement vectors caused by the ground motion; $\{U_m\}$ is the enforced displacement vector of the ground joints; $\{P_m\}$ represents the forces exerted on the structure by ground motion. After simplification (Lin *et al* 1994; 2000), Eqn. 1 becomes

$$[M_{ss}]\{\ddot{U}_r\} + [C_{ss}]\{\dot{U}_r\} + [K_{ss}]\{U_r\} = -[M_{ss}]\{\ddot{U}_s\} \tag{2}$$

$$\{U_s\} = -[K_{ss}]^{-1}[K_{sm}]\{U_m\} \tag{3}$$

Substituting Eqn. 3 into Eqn. 2 leads to

$$[M_{ss}]\{\ddot{U}_r\}+[C_{ss}]\{\dot{U}_r\}+[K_{ss}]\{U_r\}=[M_{ss}][K_{ss}]^{-1}[K_{sm}]\{\ddot{U}_m \quad (4)$$

Assuming the earthquake wave reaches the *j*th ground joint (X_j, Y_j) at $t = T_j$

$$T_j = (X_j \cos\theta + Y_j \sin\theta)/v \quad (5)$$

where v is the equivalent phase velocity of the earthquake wave, θ is the angle between v direction and X-axis direction. Thus, the acceleration of ground joints can be expressed by

$$\{\ddot{U}_m(t)\}=[G(t)]\{\ddot{U}_g(t) \quad (6)$$

where $[G(t)]$ is a specified time-lags envelope function matrix; $\{\ddot{U}_g(t)$ is a time-lags zero-mean-valued stationary random process vector

$$[G(t)]=\begin{bmatrix} G_1(t) & & & & \\ & \ddots & & 0 & \\ & & G_j(t) & & \\ & 0 & & \ddots & \\ & & & & G_p(t) \end{bmatrix} \quad (7)$$

$$[G_j(t)]=\begin{bmatrix} g_x(t-T_j) & & 0 \\ & g_y(t-T_j) & \\ 0 & & g_z(t-T_j) \end{bmatrix} \quad (8)$$

$$\{\ddot{U}_g(t)\}=[\ddot{U}_{g1}(t) \quad \cdots \quad \ddot{U}_{gj}(t) \quad \cdots \quad \ddot{U}_{gp}(t)]^T \quad (9)$$

$$\{\ddot{U}_{gj}(t)\}=[X(t-T_j) \quad Y(t-T_j) \quad Z(t-T_j)]^T \quad (10)$$

In practical calculation, we usually assume that $g_x(t)=g_y(t)=g_z(t)=g(t)$, $g(t)=0$ when $t<0$. For linear structure with orthogonal damping matrix, the structural responses are usually determined by mode-superposition scheme, the displacement of structural free joints $\{U_r\}$ can be expressed as the linear composition of the first q mode shapes ($q << 3n$)

$$\{U_r\}=\sum_{j=1}^{q}\{\phi_j\}\mu_j=[\Phi]\{\mu\} \quad (11)$$

Eqn. 4 can be decomposed into

$$\{\ddot{\mu}\}+diag[2\xi_j\omega_j]\{\dot{\mu}\}+diag[\omega_j^2]\{\mu\}=[\beta]\{\ddot{U}_m\} \quad (12)$$

where ξ_j and ω_j are, respectively, the *j*th damping ratio and frequency of the *j*th mode; $[\beta]=[\Phi]^T[M_{ss}][K_{ss}]^{-1}[K_{sm}]$, in which the following relation is utilized

$$[\Phi]^T[M][\Phi]=[I] \tag{13}$$

The input power spectral density matrix can be written as

$$[S_{\ddot{U}_m\ddot{U}_m}]=[G(t)]^*[S_{\ddot{U}_g\ddot{U}_g}][G(t)]^T \tag{14}$$

$$\left[S_{\ddot{U}_g\ddot{U}_g}(\omega)\right]=\begin{bmatrix}[I]_3e^{i\omega T_1}\\ [I]_3e^{i\omega T_2}\\ \vdots\\ [I]_3e^{i\omega T_p}\end{bmatrix}\begin{bmatrix}S_{xx}(\omega) & S_{xy}(\omega) & S_{xz}(\omega)\\ S_{yx}(\omega) & S_{yy}(\omega) & S_{yz}(\omega)\\ S_{zx}(\omega) & S_{zy}(\omega) & S_{zz}(\omega)\end{bmatrix}\left[[I]_3e^{-i\omega T_1}\quad [I]_3e^{-i\omega T_2}\quad \cdots\quad [I]_3e^{-i\omega T_p}\right] \tag{15}$$
$$=[V]^*[S(\omega)][V]^T$$

where $[I]_3=\begin{bmatrix}1 & & 0\\ & 1 & \\ 0 & & 1\end{bmatrix}$; $[S(\omega)]$ is the power spectral density matrix of three-dimensional earthquake inputs; Considering the coherence between ground motions, $[S(\omega)]$ is a Hermitian's matrix. According to the spectral decomposition scheme of the Hermitian matrix, $[S(\omega)]$ can be decomposed into

$$[S(\omega)]=\sum_{j=1}^{r}\alpha_j\{\psi_j\}^*\{\psi_j\}^T \tag{16}$$

where r is the rank of the matrix $[S(\omega)]$; α_j and $\{\psi_j\}$ are respectively the jth eigenvalue and normalized eigenvector of $[S(\omega)]$. Substituting Eqns. 15 and 16 into Eqn. 14 gives

$$\left[S_{\ddot{U}_m\ddot{U}_m}\right]=\sum_{j=1}^{r}[G(t)]^*[V]^*\alpha_j\{\psi_j\}^*\{\psi_j\}^T[V]^T[G(t)]^T \tag{17}$$

According to the principle of pseudo excitation method, taking a pseudo ground acceleration vector

$$\{\ddot{U}_{m_j}(t,\omega)\}=\sqrt{\alpha_j}\,[G(t)][V]\{\psi_j\}e^{i\omega t} \tag{18}$$

The solution for Eqn. 11 under the pseudo excitation $\{\ddot{U}_{m_j}(t,\omega)$ is given by

$$\{\ddot{U}_{r_j}(t,\omega)\}=\int_0^t[\Phi][h(t-\tau)][\beta]\sqrt{\alpha_j}\,[G(\tau)][V]\{\psi_j\}e^{i\omega\tau}\,d\tau \tag{19}$$

where

$$[h(t-\tau)]=diag[h_i(t-\tau)] \tag{20}$$

$$h_i(t-\tau)=\frac{1}{\omega_{di}}e^{-\xi_i\omega_i(t-\tau)}\sin\omega_{di}(t-\tau) \tag{21}$$

in which $\omega_{di}=\omega_i\sqrt{1-\xi_i^2}$. According to the concept of pseudo excitation method, the power spectral density matrix for structural quasi-dynamic displacement is developed by

$$\left[S_{U_rU_r}(t,\omega)\right]=\sum_{j=1}^{r}\left\{U_{r_j}(t,\omega)\right\}^{*}\left\{U_{r_j}(t,\omega)\right\}^{T} \tag{22}$$

The quasi-static displacement vector can be obtained by substituting Eqn. 18 into Eqn. 3,

$$\left\{U_{s_j}(t,\omega)\right\}=-\left[K_{ss}\right]^{-1}\left[K_{sm}\right]\int_0^t\left(\int_0^{t'}\left[G(\tau)\right]\left[V\right]\sqrt{\alpha_j}\left[F_m\right]\left\{\psi_j\right\}e^{i\omega\tau}\,d\tau\right)dt' \tag{23}$$

Then, the power spectral density matrix for structural quasi-static displacement is given by

$$\left[S_{U_sU_s}(t,\omega)\right]=\sum_{j=1}^{r}\left\{U_{s_j}(t,\omega)\right\}^{*}\left\{U_{s_j}(t,\omega)\right\}^{T} \tag{24}$$

The power spectral density of total displacement is written as

$$\left[S_{U_{ss}U_{ss}}(t,\omega)\right]=\sum_{j=1}^{r}\left\{U_{ss_j}(t,\omega)\right\}^{*}\left\{U_{ss_j}(t,\omega)\right\}^{T} \tag{25}$$

$$\left\{U_{ss_j}(t,\omega)\right\}=\left\{U_{s_j}(t,\omega)\right\}+\left\{U_{r_j}(t,\omega)\right\} \tag{26}$$

The power spectral density matrix for internal forces can be obtained through translation matrix $\left[Z_N\right]$

$$\left[S_{NN}(t,\omega)\right]=\sum_{j=1}^{r}\left\{N_j(t,\omega)\right\}^{*}\left\{N_j(t,\omega)\right\}^{T} \tag{27}$$

$$\left\{N_j(t,\omega)\right\}=\left[Z_N\right]\left\{U_{ss_j}(t,\omega)\right. \tag{28}$$

It can be easily verified that Eqns. 22, 24, 25 and 27 are mathematically identical to expressions given by conventional CQC method.

The time-dependent variance for internal force of the *i*th member can be assessed by corresponding component of $\left[S_{NN}(t,\omega)\right]$

$$\sigma_{N_i}^2(t)=\int_{-\infty}^{+\infty}S_{N_iN_i}(t,\omega)d\omega \tag{29}$$

The mean value of peak response can be estimated by the variance (Jiang and Hong 1984).

ANALYSIS

To study the seismic performance of spatial lattice shells under multiple nonstationary seismic excitations, a double-layer cylindrical lattice shell is selected for the random vibration study. Figure 1 shows a finite element model of the structure, where B_S and B_X represent respectively the upper and the lower chord members in B direction; Z_S and Z_X represent respectively the upper and the lower chord members in L direction. The boundary supports of the lattice shell are assumed to be fixed hinge joints. The structural parameters are taken as follows: span length L=30 m; width B=25 m; shell thickness d=1.0 m; the ratio of depth to span f/B=0.3; the basic grid size is taken as 2×2 m. The members are modeled as bar elements in the analysis. The roof load is taken as 1kN/m^2. All the loads and self-weights of the members are treated as lumped masses concentrated at the joints.

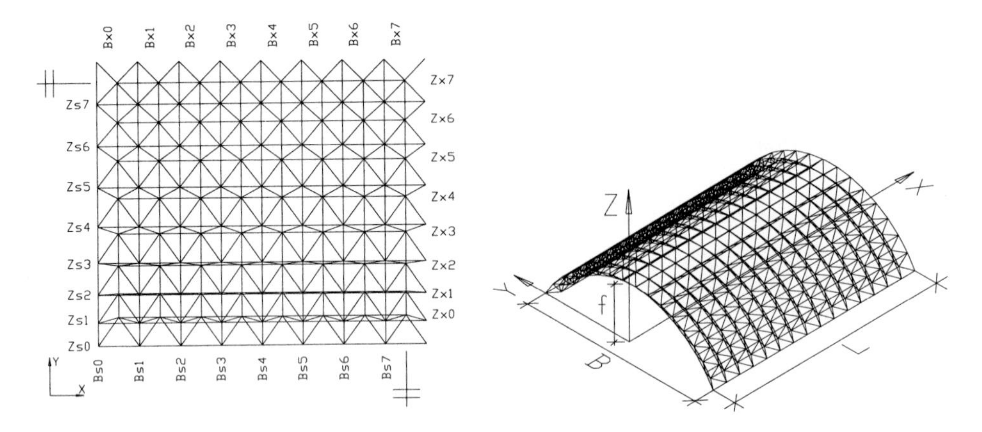

Figure 1 Finite element model of double-layer cylindrical lattice shell

The earthquake power spectrum for horizontal ground motion is represented by revised Kanai-Taijimi expression (Clough and Penzien 1993)

$$S_{xx}(\omega) = S_{yy}(\omega) = \frac{\omega_g^4 + 4\xi_g^2\omega_g^2\omega^2}{(\omega_g^2 - \omega^2)^2 + 4\xi_g^2\omega_g^2\omega^2} \cdot \frac{\omega^4}{(\omega_f^2 - \omega^2)^2 + 4\xi_f^2\omega_f^2\omega^2} S_0 \tag{30}$$

where ω_g is the natural ground frequency; ξ_g is the damping ratio of the ground; S_0 is the intensity parameter; ξ_f, ω_f are revised parameters; generally, it is taken that $\xi_f = \xi_g$, $\omega_f = 0.1 - 0.2\omega_g$ (Chen *et al* 1999). The earthquake power spectrum for vertical ground motion is considered to have similar spectral distribution of Eqn. 30, but the parameters is modified as suggested by the authors (Xue *et al* 2000; Wang *et al* 2001), $\omega_{gV} = 1.58\omega_{gH}$; $\xi_{gV} = \xi_{gH}$; $S_{0V} = 0.281S_{0H}$.

The cross-spectrum between the excitations are assumed to have the form: $S_{xy}(\omega) = \sqrt{S_{xx}(\omega)S_{yy}(\omega)}$ (Huang and Liu 1987); $S_{xz}(\omega) = 0.6\sqrt{S_{xx}(\omega)S_{zz}(\omega)}$; $S_{yz}(\omega) = 0.6\sqrt{S_{yy}(\omega)S_{zz}(\omega)}$ (Xue *et al* 2000). The following envelope function is adopted (Amin and Ang 1968)

$$f(t) = \begin{cases} (t/t_1)^2 & 0 \le t < t_1 \\ 1 & t_1 \le t \le t_2 \\ e^{-c(t-t_2)} & t > t_2 \end{cases} \tag{31}$$

The parameters used for the input spectra are respectively: ξ_g=0.8; ω_{gH}=15.71 rad/s; S_{0H}=8.6409 cm^2/s^3; $t_1 = 1.2$; $t_2 = 9.0$; $c = 0.5$; assuming $\theta = 0$, $v = 500$, 1000, 2000, 3000 m/s.

Based on the formulation, a computer program has been developed for nonstationary random vibration analysis of lattice shells under multiple seismic inputs. The random responses of the double-layer cylindrical lattice shell are obtained using the developed program. Some of the results are shown in Figures 2 to 5, respectively, where the response of dynamic internal force represents the mean value of peak response. Following observations can be drawn from the results:

(1) From Figures 2 and 3 we can observe that the seismic responses under three-dimensional

earthquake inputs (XYZ input) are always greater than those under single-dimensional earthquake input (Y input), whether or not the phase lags are considered.

(2) The results shown in Figure 4 indicate that stationary input produces larger seismic responses than nonstationary input.

(3) It is observed from Figure 5 that the change of wave speed may result in significant different responses in different members, indicating the effects of phase lags between excitations. It is also observed that the earthquake wave with low speed may produce quite larger responses in some members (see member 18 in Figure 5).

(4) The analyses of the example show that the developed method is a highly efficient algorithm, which can easily solve the nonstationary problem of large structures under multi-support and multi-dimensional seismic excitations.

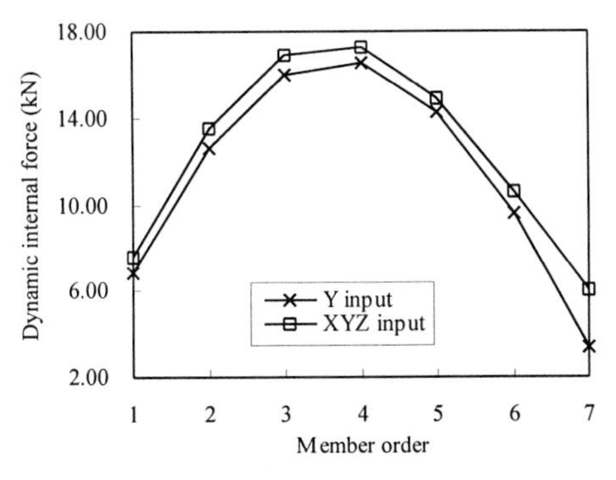

Figure 2 Dynamic internal force of Bx4 (nonstationary without phase lag, $v=\infty$)

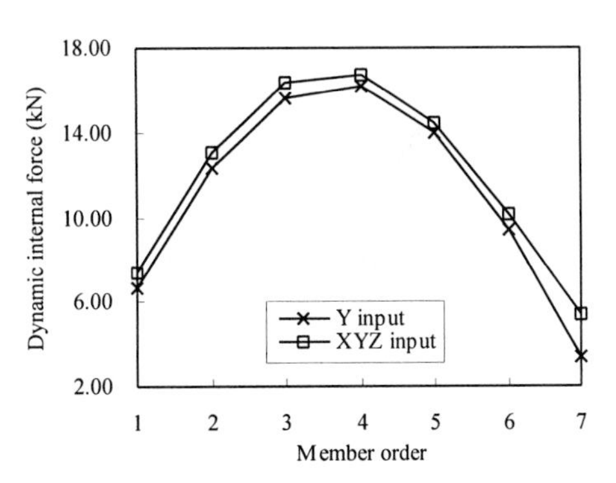

Figure 3 Dynamic internal force of Bx4 (nonstationary with phase lag, v=500m/s)

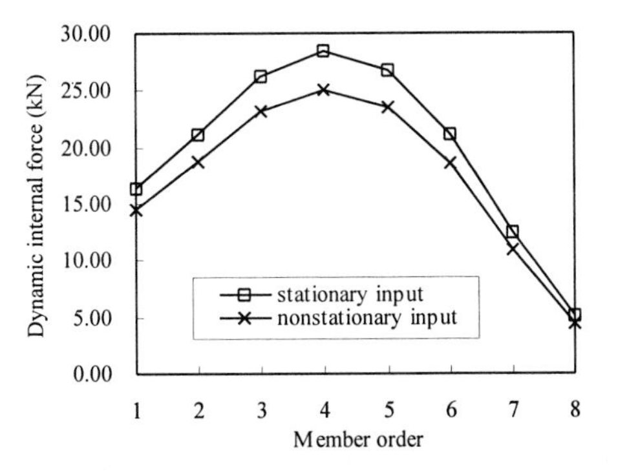

Figure 4 Dynamic internal force of Bs7 (XYZ input, without phase lag)

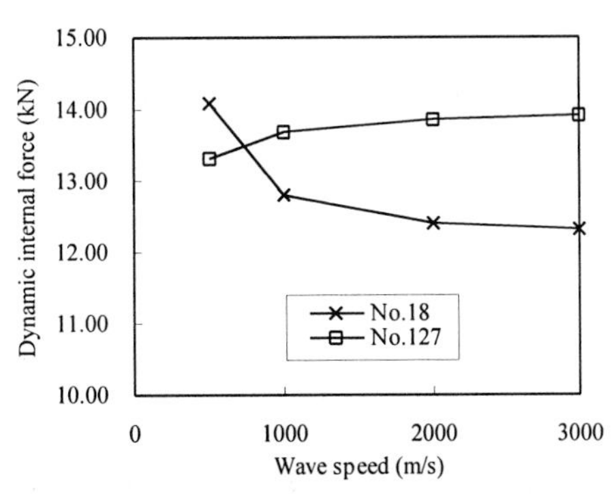

Figure 5 Dynamic internal force of member 18 and 127 at nonstationary XYZ input

CONCLUSIONS

A nonstationary random vibration study of spatial lattice shells under multi-support and multi-dimensional seismic excitations has been performed in this paper. The multi-dimensional pseudo excitation method for random vibration analysis of structures was developed. It is found that the derived method is particularly suitable for random vibration analysis of long span spatial structures, which are usually characterized with closely spaced natural frequencies and coupled vibration modes. Based on the method, a computer program has been developed. The seismic performance of a cylindrical lattice shell under different earthquake inputs was investigated. Some observations were made. It is shown that the developed method is a highly efficient algorithm, which can easily solve the nonstationary problem of large structures under multi-support and multi-dimensional seismic excitations.

ACKNOWLEDGEMENTS

The authors are grateful for the support from the National Natural Science Foundation of China.

REFERENCES

Amin M. and Ang A. (1968). Nonstationary Stochastic Models of Earthquake Motions. *J. Engineering Mechanics Division, ASCE.* **94:EM2,** 559-584.

Cao Z. and Zhang Y.G. (2000). A Study on the Seismic Response of Lattice Shells. *International Journal of Space Structures*, **15:3&4**, 243-247.

Chen G.X, Jin Y.B. and Zai J.M. (1999). A Simple Method of Random Earthquake Response Analysis for High-Rise Buildings. *Journal of Nanjing Architectural and Civil Engineering Institute*. No.1 (Sum No.48), 29-37.

Clough R.W. and Penzien J. (1993). *Dynamics of Structures*. McGraw-Hill, Inc., 2nd ed., New York.

Gasparini D.A. (1979). Response of MDOF Systems to Nonstationary Random Excitation. *J. Engineering Mechanics, ASCE.* **105:EM1,** 13-28.

Harichandran R.S. and Vanmarcke E.H. (1986). Stochastic Variation of Earthquake Ground Motion in Space and Time. *J. Engineering Mechanics, ASCE*, **112:2,** 154-175.

Houser G.W. (1947). Characteristics of strong motion earthquakes. *Bull. Seism. Soc. Am.*, **37**, 17-31.

Huang Y.P. and Liu J. (1987). The Spatial Correlation of Two Horizontal Direction Earthquake. *Journal of Harbin Architectural and Civil Engineering Institute*. **3**, 10-14.

Jiang J.R. and Hong F. (1984). Conversion Between Power Spectrum and Response Spectrum and Artificial Earthquakes. *Earthquake Engineering and Engineering Vibration*. **4:3,** 89-94.

Li H.N. (1998). *Theoretical Analysis and Design of Structures to Multiple Earthquake Excitations*. Science Press, Beijing.

Lin J.H., Williams F.W. and Zhang W.S. (1993). A New Approach to Multiphase-Excitation Stochastic Seismic Response. *Microcomputers in Civil Engineering*. **8,** 283-290.

Lin J.H., Zhang W.S. and Williams F.W. (1994) Pseudo-excitation Algorithm for Nonstationary Random Seismic Responses. *Engineering Structures*. **16:4,** 270-276.

Lin J.H., Zhong W.X. and Zhang Y.H. (2000). Seismic Analysis of Long Span Structures by Means of Random Vibration Approach. *Journal of Building Structures*. **21:1,** 29-36.

Lopez O.A., Chopra A.K. and Hernandez J.J. (2000). Critical Response of Structures to Multicomponent Earthquake Excitation. *Earthquake Engineering and Structural Dynamics*, **29**, 1759-1778.

Shen Z.Y., Zhou D. and Gong M. (2000). Analysis of the Tower-column Element and Cable Element as Well as the Dynamic Properties for Cable-stayed Reticulated Shell Structures. *Journal of Tongji University* **28:2,** 128-133.

Wang X.S., Xue S.D. and Cao Z. (2001). Review and Prospects of the Research on Multi-components Earthquake Excitation for Structures (I): Seismic Inputs. *World Information on Earthquake Engineering*. **17:4,** 27-33.

Xue S.D., Li M.H., Cao Z. and Zhang Y.G. (2000). Random Vibration Analysis of Lattice Shells Subjected to Multi-dimensional Earthquake Inputs. *Advances in Structural Dynamics*, Vol. I, JM Ko and YL Xu (Eds.), Hong Kong, 777-784.

Advances in Building Technology, Volume I
M. Anson, J.M. Ko and E.S.S. Lam (Eds.)

INTERACTION BETWEEN PILE AND RAFT IN PILED RAFT FOUNDATION

W. J. Wu[1], J. C. Chai[2] and J. Z. Huang[3]

[1]College of Architecture and Civil Engineering, Beijing Polytechnic University, Beijing, China
[2]Department of Civil Engineering, Saga University, Saga, Japan
[3]Wuhuan Chemical Engineering Corporation, Wuhan, China

ABSTRACT

The interaction between pile and raft in a piled raft foundation has been investigated by model test and finite element analysis. A method is presented to predict the complete load-settlement response of a piled raft foundation in soft clay from those of its component parts, the pile and the raft, based on interaction between pile and raft. The efficiency of a piled raft foundation and independent partial-safety factors for pile and raft in a piled raft foundation are also discussed.

KEYWORDS

piled raft, foundation, soft clay, model test, bearing capacity, finite element method

INTRODUCTION

Being one of the most sensitive clays, Ariake clay has a natural water content of 70% to 150% and has a thickness of 10 to 30m in the Saga Plain, Northern Kyushu, Japan. In this plain, most of the lightweight structures have been constructed on end bearing piles supported by a gravel layer below the soft clay layer. In this case, a large differential settlement between structure and surrounding ground can occur because of subsidence due to ground water pumping. In order to mitigate the differential settlement, a floating type of piled raft foundation is found to be a suitable alternative, since the floating pile can settle with the surrounding ground (Wu and Miura 1995). However, there is no design standard in Japan for a piled raft foundation in soft clay.

The most important thing in designing the piled raft foundation is to consider the interaction between pile and raft. Some design procedures based on elastic or non-linear elastic analysis are proposed (Randolph, 1983; Clancy and Randolph, 1993; 1996; etc). However, the mechanism of the interaction between pile and raft are not fully understood. In order to consider these factors in design, more researches are needed. In this study, laboratory tests were carried out for investigating the behavior of a piled raft foundation in soft clay. Numerical analysis (FEM) was conducted to study the interaction between the pile and raft. This indicates that due to interaction between pile and raft, the bearing capacity of a piled raft foundation will be less than the sum of the bearing capacities of piles and raft taken separately. A simple approach is presented to predict the behavior of a piled raft foundation in soft clay, based on the interaction between the pile and raft. Finally, the bearing capacity of piled raft foundation and independent partial-safety factors for pile and raft in a piled raft foundation are also discussed.

INTERACTION BETWEEN PILE AND RAFT

Laboratory Test

The model ground for the laboratory test was made of Ariake clay in a concrete soil tank with a bottom area 1.5mx1.5m and a depth, 1.0m. Sampling of Ariake clay took place at a depth of 2m in the suburb of Saga City, Japan. The physical properties of the clay are as follows: w_L=89%, I_P=50, w_n=147%. Ariake clay in a slurry state was placed in the tank. A uniform pressure of 5kPa was applied over the entire area and maintained for about 5 months for consolidation (Miura et al, 1995).

The model pile was made of timber with a diameter ($2r_p$) of 25mm. The length (L) of the pile varied from 125mm to 375mm with a length to diameter ratio ($L/2r_p$) of 5 to 15. The model raft was made of transparent plastic plate. In this case, the conduct condition between raft and soil can be checked easily. The diameter ($2r_r$) of the raft ranged from 75 to 225mm, which resulted in a raft radius to pile radius ratio (r_r/r_p) of 3 to 9. For the piled raft model, a load cell was mounted on top of the pile to measure the load transferred to the pile.

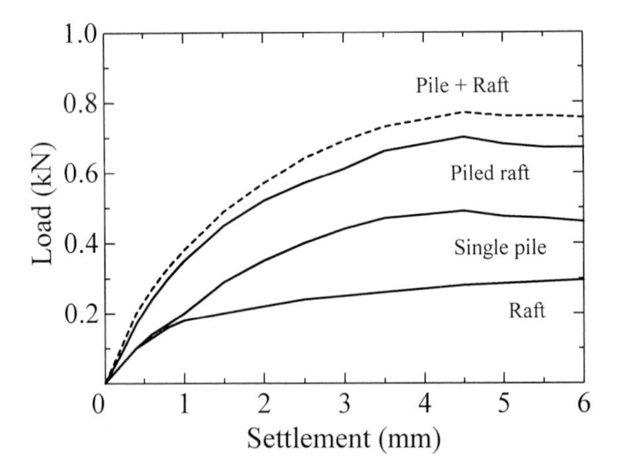

Figure 1: Load settlement diagram for model pile in soft clay

The model pile was installed in the model ground and was left for 28 days to allow the recovery of the disturbed zone around the pile. Three types of model test were carried out: (a) single pile, (b) raft only, and (c) piled raft system. The load was applied by dead load. The time interval between load increments was 30min. for each test, 10 to 20 load increments were applied. For the piled raft system, the load transferred to the pile was monitored by a load cell mounted on top of the pile. The settlement was measured by dial gages.

The load-settlement curves of model pile, raft and piled raft are presented in Figure 1. To compare the behavior of pile, raft, and piled raft, a load-settlement curve of (P+R) is also plotted in this figure, which is drawn by adding up the load-settlement curves for the pile and raft. It can be seen in Figure 1 that the curve for the piled raft (PR) and curve for (P+R) show a similar trend, but (PR) curve indicates less capacity than the (P+R) curve. Figures 2, 3 show the normalized load-normalized settlement responses of a single pile and a pile in piled raft system, raft only and raft in piled raft system, respectively.

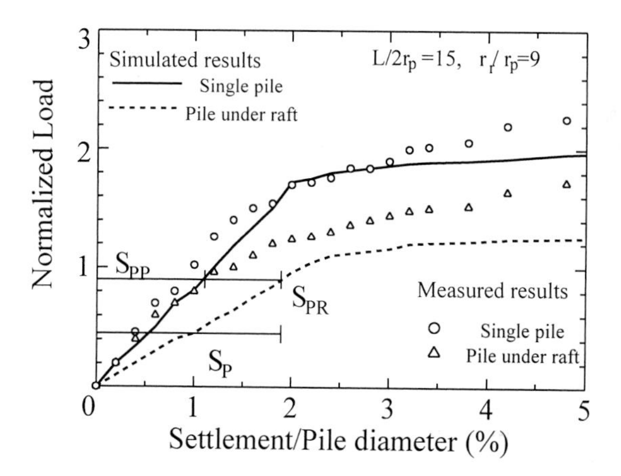

Figure 2: Load-settlement curves for pile

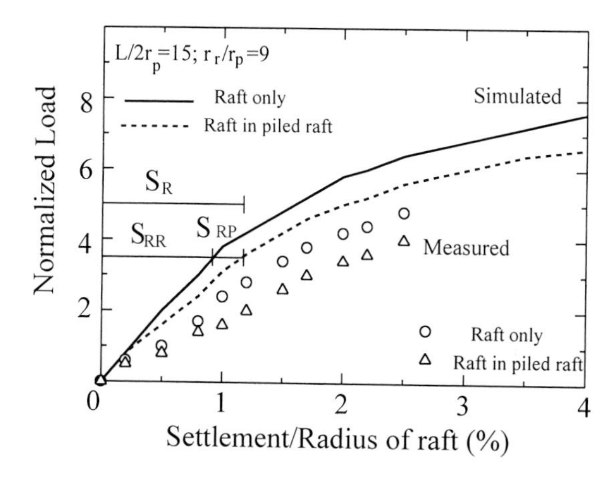

Figure 3: Bearing stress-settlement curves for raft

Numerical Analysis

For FE analysis, an eight node quadrilateral element with reduced integration scheme is used. The sub-stepping technique with error control is adopted to integrate the elasto-plastic stress-strain relationship. To illustrate the effectiveness of the numerical method used, a circular plate on a uniform foundation is analyzed and compared with the theoretical value. The foundation is assumed weightless and with an isotropic initial stress of 10 kPa and represented by a modified Cam clay model. The model parameters and the initial stress condition are estimated to represent the behavior of the model ground. The behavior of soil is represented by a modified Cam clay model. The soil parameters are λ =0.29, κ =0.03, and the coefficient of permeability, k, is determined from 1-D consolidation test results: k=1.1x10^{-4}m/day and initial void ratio, e_0=2.3. The friction parameter M, and Poisson's ratio, υ , are determined empirically: M=1.2, υ =0.3. Model pile and raft are assumed as elastic material.

The numerical results are shown in Figures 2, 3 with the results of model tests. Both measured and numerical results indicate that the curve of pile (raft) in the piled raft system shows larger settlement under any given load than for the single pile (raft only). In piled raft system, due to interaction, the load transmitted by the pile causes an additional settlement on the raft (S_{RP}) and in a similar manner; the stress from the raft to the soil generates an additional pile displacement (S_{PR}). At given load, the settlement of pile, S_P (or raft, S_R) in piled raft is the sum of the settlement (S_{PP}, S_{RR}) due to their own loads and those (S_{PR}, S_{RP}) due to stresses from the raft (or pile). Fig. 4 shows the shear strain and shear stress level contours of piled raft system ($L/2r_p$=15; r_r / r_p =9), at a settlement of 2% of raft radius. This settlement is about 18% of pile diameter, which implies that the soil/pile interface strength is mobilized at a much smaller value of settlement.

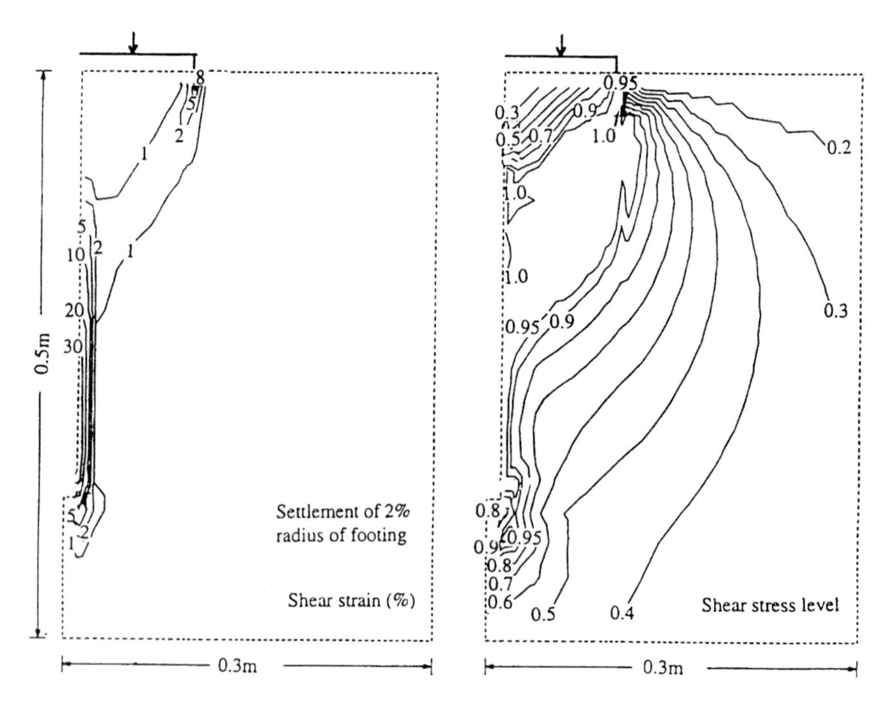

Figure 4: Shear strain and shear stress level contours for piled raft

Pile Raft Interaction

Due to the interaction between pile and raft, the total resistance of piled raft system will be less than the sum of the resistance of single pile and raft. Figures. 2 and 3 compare the resistance of single pile and raft with those in piled raft system, respectively. It can be seen that due to interaction, the bearing stress is reduced. Firstly, the effect of raft on pile behavior is to reduce the relative movement at soil/pile interface, and therefore the mobilized resistance is also reduced. This effect is most significant for that part of the pile close to the raft. As shown in Figure 4, at a depth of about the raft radius, the shear strain is less than 1%, which indicates the small relative movement at the soil/pile interface. For a given raft and pile radius ratio, r_r/r_p, the pile efficiency increases with the increase of the pile length. Then, the effect of pile on raft behavior is that the pile will drag the soil adjacent to it to move downward, which will reduce the mobilized bearing resistance under the raft. It has been found that this effect is mainly a function of r_r/r_p ratio. The larger the r_r/r_p, the less the effect of the pile on raft bearing resistance is as shown in Figure 5. The efficiency of the piled raft system is defined as the bearing resistance of piled raft system over the sum of single pile and raft resistances, which is also shown in Figure 5. The efficiency of a piled raft system is an index to show the degree of pile raft interaction. The larger the r_r/r_p ratio, the more effective is the system because the bearing resistance will be mainly controlled by the raft. For a fixed r_r/r_p ratio, the efficiency is slightly increasing with an increase of $L/2r_p$ ratio.

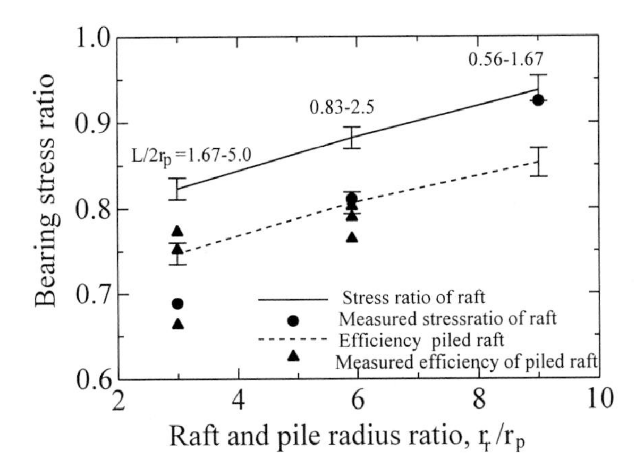

Figure 5: Efficiency of raft and piled raft

LOAD-SETTLEMENT RESPONSES OF PILED RAFT SYSTEM

The load-settlement responses of foundations can be approximated by a hyperbola. Based on the hyperbolic relations, the prediction of the load-settlement response of a piled raft foundation is discussed as follows. Eqns. 1 and 2 can express the load versus settlement responses for the single pile and the raft foundation, respectively,

$$S_{PP}=P_P/\{k_P(1-(P_P/P_{Pu}))\} \tag{1}$$

$$S_{RR}=P_R/\{k_R(1-(P_R/P_{Ru}))\} \tag{2}$$

where k_P (Randolph, 1983) and k_R (Madhav, 1980) are the initial slopes of the load-settlement curves of pile and raft, respectively. P_{Pu}, and, P_{Ru} are the ultimate bearing capacities of the pile and raft, respectively. Eqns. 1 and 2 were rewritten in incremental form as follows (Madhav et al, 1993):

$$\Delta S_{PP}=\Delta P_P/\{k_P(1-(P_P/P_{Pu}))^2\} \quad (3)$$

$$\Delta S_{RR}=\Delta P_R/\{k_R(1-(P_R/P_{Ru}))^2\} \quad (4)$$

As mentioned previously, the settlements of pile and raft in the piled raft foundation are the sum of the settlements due to their own loads and those due to stresses from the raft and the pile, respectively (Figures 2, 3). It is assumed that the pile or raft settlements due to their own loads or stresses and the interaction effects are proportional to these settlements. The average incremental settlement of the raft, ΔS_{RP}, due to an increment of load in the pile, ΔP_P, is $\Delta S_{RP}=I_{RP}\times S_{PP}$. The average incremental settlement of the pile, ΔS_{PR}, due to increment of stress, transmitted by the raft to the soil, is $\Delta S_{PR}=I_{PR}\times S_{RR}$, where, I_{RP} and I_{PR} are the interaction coefficients, respectively (Clancy and Randolph, 1993).

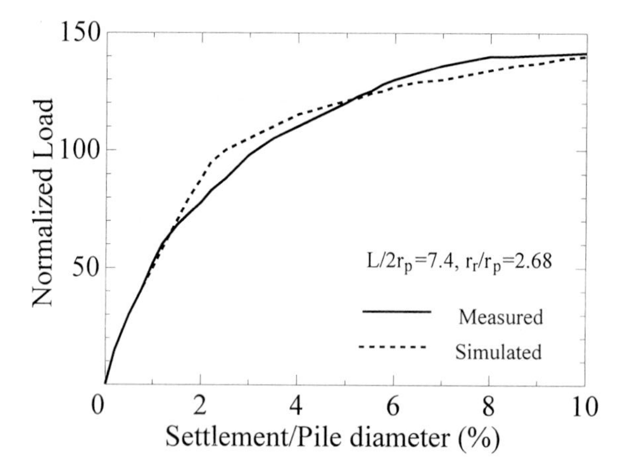

Figure 6: Measured and predicted results

In the piled raft foundation, the load on the piled raft, P_{PR}, is shared by the pile and the raft, making each share P_P and P_R ($P_{PR}=P_P+P_R$). The incremental load versus incremental settlement responses for the pile and the raft in piled raft foundation can be expressed as Eqns. 5 and 6, respectively.

$$\Delta S_P=\Delta P_P/\{k_P(1-(P_P/P_{Pu}))^2\}+I_{PR}\times\Delta P_R/k_R \quad (5)$$

$$\Delta S_R=\Delta P_R/\{k_R(1-(P_R/P_{Ru}))^2\}+I_{RP}\times\Delta P_P/k_P \quad (6)$$

Because the settlement increment for the pile and the raft in the piled raft foundation are the same, it is possible to show the relation by this form $\Delta S_P=\Delta S_R$. Solving Eqns. 5, 6, the incremental settlement of the piled raft and the loads shared by the pile and raft are obtained, and by integrating the incremental settlement the complete load settlement response of the piled raft can be determined. The test results are compared with the predicted ones obtained by the above approach as shown in

Figure 6. It shows that there exists a good agreement between the measurement and the prediction.

BEARING CAPACITY AND SAFETY FACTOR

The capacity of a piled raft foundation can be estimated from the summation of the calculated capacities of pile and raft by introducing an efficiency factor, η. It has been found that the efficiency factor, η ranges from 0.7 to 0.9.

Figure 7 shows the load-settlement curves of an idealized piled raft foundation. The ultimate bearing capacity of piled raft foundation R_{PRu}, ultimate bearing capacity of raft R_{Ru}, and ultimate bearing capacity of pile R_{Pu} are shown by AD, AB, AC in this figure, respectively. If we take a safety factor F=3, the allowable bearing capacity will be EF. The bearing capacity of pile that corresponds to this and the bearing capacity of the raft are not LK and IJ, respectively, but in fact they will be EG and EH respectively. It can be concluded that $F_R \neq F_P$ then the differential safety factor should be adopted. When the total safety factor 3 is considered, the pile safety factor and raft safety factor should be proposed separately on the basis of total load on the piled raft foundation. The allowable bearing capacity of the piled raft foundation, Ra, can be calculated by the following equations.

$$Ra=\eta\ (R_{Pu}+R_{Ru})/F \qquad (7)$$

and

$$Ra=\eta\ (R_{Pu}/F_P+ R_{Ru}/F_R) \qquad (8)$$

where, F is the safety factor of the piled raft; F_P and F_R are partial safety factors for pile and raft, respectively. Miura et al (1993) pointed out that in the usual case F is equal to 3, F_P=1.5 to 2 and F_R>3.

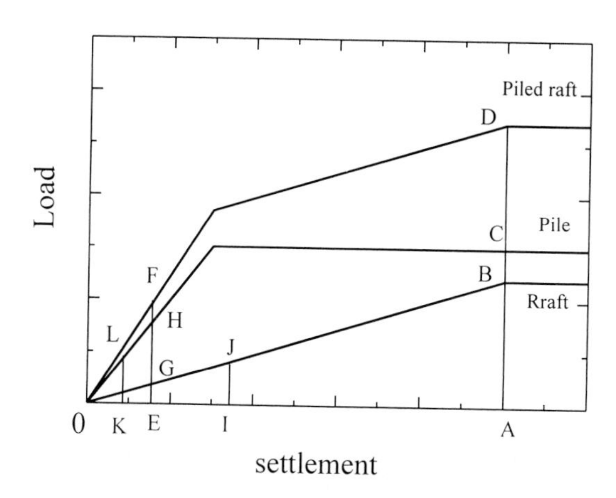

Figure 7: Idealized load vs. settlement response for pile, raft and piled raft

CONCLUSIONS

Based on the above test results and analysis, the following conclusions are summarized:

The model test and numerical analysis results indicate that due to pile raft interaction, the resistance of piled raft system is less than the sum of the resistances of pile and raft considered separately. This interaction is expressed as a piled raft system efficiency factor which is a function of raft/pile radius ratio, r_r/r_p.

The test and analysis results show that the ultimate bearing capacity of piled raft foundation can be estimated from the summation of the calculated capacities of pile and raft modified by the efficiency factor.

A method to predict the load-settlement responses of a piled raft foundation is presented based on the interaction between pile and raft. The predicted result compares well with the measured one.

Independent partial-safety factors for pile and raft foundation were discussed. If the total safety factor 3 is considered for the piled raft foundation, the safety factor for the pile in the piled raft foundation is proposed to be 1.5 to 2.

ACKNOWLEDGEMENT

The authors would like to express their sincere gratitude to professor N. Miura (Saga University, Japan) and Professor M. R. Madhav (Indian Institute of Technology, Indian), for their continuous guidance.

References

Clancy P. and Randolph M. F. (1993). An Approximate Analysis Procedure for Piled Raft Foundations, *International Journal for Numerical and Analytical Methods in Geomechanics,* **17:12,** 849-869.

Clancy P. and Randolph M. F. (1996). Simple Design Tools for Piled Raft Foundations, *Geomechanics,* **46:2,** 313-328

Madhav M. R. (1980). Settlement and Allowable Pressures for Rigid or Annular Footings, *Indian Geotechnical Journal,* **23:4,** 113-118.

Madhav M. R., Miura N. and Wu W. J. (1993). Settlement and Efficiency of Piled Raft Foundation, *The 28th Japan National Conference on SMFE,* 1647-1648.

Miura N., Wu W. J., Madhav M. R. and Nagaike S. (1993). Settlement Behavior and Bearing Capacity of Piled Raft Foundations, *Reports of Faculty of Science and Engineering, Saga University,* **22:1,** 143-149.

Miura N., Wu W. J., Nakamura R. and Ichinose T. (1995). Experimental Study on Skin Resistance of Pile in Soft Clay, (in Japanese), *JSCE, Journal of Geotechnical Engineering,* **517:3-31**, 63-72.

Randolph M. F. (1983). Design of Piled Raft Foundation, *Proc. Intl. Symposium on Recent Developments in Laboratory and Field Tests and Analysis of Geotechnical Problems*, 525-537.

Wu W. J. and Miura N. (1995), Vertical Loading Test for Evaluating Skin Resistance of Timber Pile in Clay, (in Japanese), *Tsuchi-to-Kiso,* **43:5,** 40-42.

Advances in Building Technology, Volume I
M. Anson, J.M. Ko and E.S.S. Lam (Eds.)

THE METHOD OF CALCULATING THE INTERIOR FORCE OF SOIL NAILS AND CASE ANALYSIS

Zhang Qinxi[1] Yang Jingrong[2] Huo Da[1]

[1] The College of Architectural and Civil Engineering, Beijing Polytechnic University, Beijing, 100022, China
[2] Institute of Geography Science and Natural Resource, The Chinese Academy of Science, Beijing, 100101, China

ABSTRACT

On the basis of analyzing the current methods in the technical codes of the PRC, this paper discusses a reasonable design approach for soil-nailing and put forwards a new design method, especially a more reasonable method of calculating the interior forces in soil nails.
To check the reliability of the proposed method, an in-situ measuring test has been performed in an excavation project 10 meters deep, and the measurement results show that the method proposed in this paper can achieve most satisfactory results.

KEYWORDS

excavation, soil-nailing, design method, interior forces in nails, in-situ test, case analysis

1 INTRODUCTION

Because of the advantages such as lower price, faster construction speed, and the absence of heavy construction machines, soil-nailing, as a kind of retaining structure in deep foundation pits, has been widely used in Beijing. Due to the complexity of the interaction mechanism between nails and soils, the estimation of the interior force in nails during the working process has not been ideally solved so far. At the same time, the estimation of interior force in nails is very important for soil-nailing design. So, it is meaningful to study and discuss how to better calculate the interior force in nails.

2 CURRENT METHODS FOR ESTIMATING THE INTERIOR FORCE IN NAILS

In China, the methods for estimating the interior forces in nails are all based on the active earth pressure theory (Rankine or Coulomb's). That is to say, the maximum force in a nail is equal to earth pressure times the area on which a nail acts, which can be seen in Fig.1and Eqn.1.

$$T = P_a S_v S_h / \cos\theta \tag{1}$$

where:
T=the maximum interior force in soil nails, kN
Sv and Sh=the vertical and horizontal space between nails, m
θ =the angle between nails and horizontal level, degree
Pa=working earth pressure, kPa.

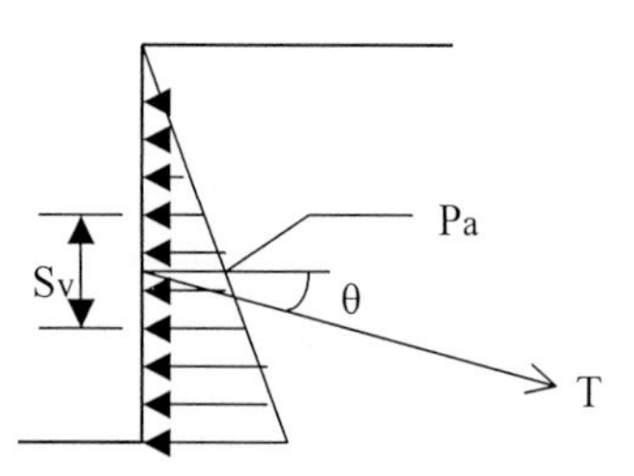

Fig.1 Model of calculating T

There are two methods to calculate the working pressure Pa at present in China, which are as follows.

2.1 Triangular Model [1]

It is supposed that the distribution of the working earth pressure is triangular as Fig.2 (a) shows:

$$P_a = \xi P_A \tag{2}$$

where:
P_A= active earth pressure of Rankine, kPa
ξ =coefficient of discount, see Eqn.3

$$\xi = tg[(\beta+\phi)/2]\{1/tg[(\beta+\phi)/2]-1/tg\beta\}/tg^2(45^\circ-\phi/2) \tag{3}$$

where:
β =the angle of slope(see Fig.3), degree
ϕ =the friction angle of soils, degree

2.2 Trapezium Model[3]

The supposed distribution of Pa is shown in Fig.2 (b). Where Pm is the maximum value of Pa and Pm can be determined according to the following stipulations:

(1) For sands and silts , $C/(\gamma H) \leqslant 0.05$:

$$Pm=0.55Ka\,\gamma H \tag{4}$$

(2) For cohesive soils, $C/(\gamma H) > 0.05$:

$$Pm=Ka\,[1-2C/(\gamma H\sqrt{Ka})]\,\gamma H$$

$$\text{and}\quad 0.2\gamma H \leqslant Pm \leqslant 0.55Ka\,\gamma H \tag{5}$$

where:
γ =unit weight, kN/m^3
H =depth of foundation pit, m
Ka=coefficient of active earth pressure of Rankine
C=cohesion of soils, kPa

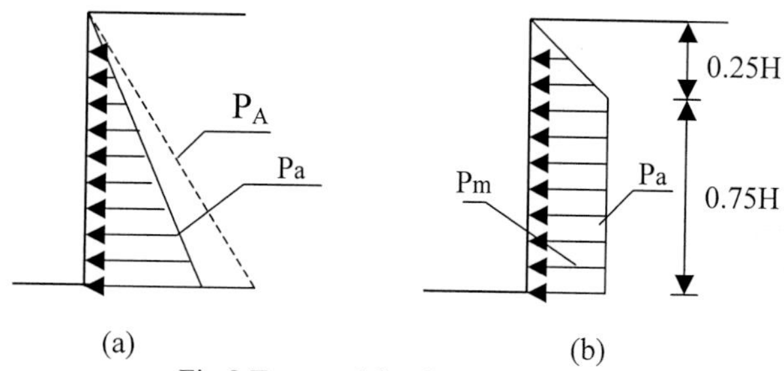

Fig.2 Two models of Pa in China

2.3 Problems Existing in the Two Models Above

(1) For the triangular model, Pa increases linearly with depth, so the nail which bears maximum force is at the bottom, but in-situ measurement tests show that this is not true. Further more, the discount coefficient ξ lacks any basis in theory;

(2) The trapezium model can not reflect the fact that Pa decreases with the decrease of β. So the interior force in nails calculated based on this model are not reasonable, especially when β is small.

3 A NEW METHOD OF CALCULATING THE INTERIOR FORCE IN SOIL NAILS

According to the analysis above and to avoid the vagueness of earth pressure, a new method called sliding wedge-mass limit analysis is proposed for estimating the interior force in nails[2].
Suppose that the failure zone is a plane and the angle between the failure plane and horizontal plane is $(\beta+\phi)/2$ (see Fig.3), we can get safety factor K from the equilibrium equation of wedge ABC:

$$K=[T\cos(\alpha+\theta)+T\sin(\alpha+\theta)\,tg\,\phi+W\cos\alpha\,tg\,\phi+CLS_h]/W\sin\alpha \tag{6}$$

then:

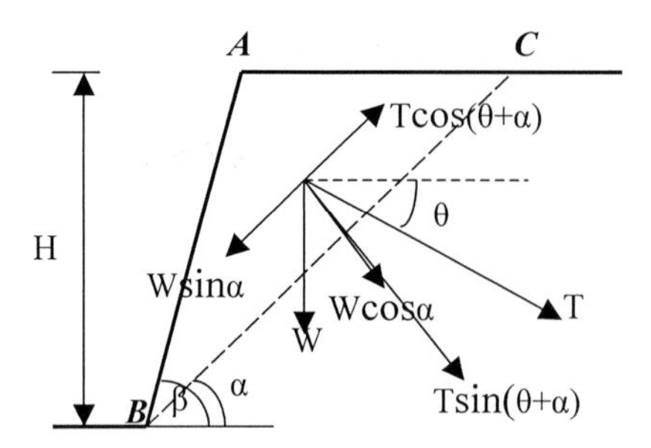

Fig.3 Model of wedge-mass analysis

$$T=[KW\sin(\alpha+\theta)-CLS_h-W\cos\alpha\, tg\phi]/[\cos(\alpha+\theta)+\sin(\alpha+\theta)\, tg\phi] \tag{7}$$

where:
K=safety factor
T=the total force provided by nails, kN
W=the weight of wedge ABC, kN
L=the length of failure plane, m
S_h=the horizontal space of nails, m
α =the angle between the failure plane and horizontal plane, $\alpha=(\beta+\phi)/2$ (see Fig.3) , degree
θ =the angle between nails and horizontal plane, degree
To get the interior force of each nail, we can work out the horizontal and vertical space between nails based on experience and soil properties, so we get "n", the number of rows along vertical, and refer to Fig.2 (b), we may suppose that the force in each row of nails are equal. So the interior force in each row of nails is:

$$T_i=T/n \tag{8}$$

where:
T_i= the interior force in each row of nails, kN
n= the number of rows along vertical direction

4 CASE ANALYSIS

To check the reliability of the method proposed above, an in-situ test has been made in a deep foundation pit from Oct. to Nov. in 2001. The measurement results are introduced as follows.

4.1 General Situation

Beijing Hai-Run International Apartment is located in Jiangtai Road, Chaoyang District, Beijing. The building complex includes 5 tower blocks with 17~28 stories and 1 public building with 2~3 stories. There is a large underground parking lot with 2 stories under the whole building complex. The depth of the foundation pit is 9.75 m, the area of excavation is about 10000m^2.

According to the geological survey report, properties of the soil layers in the field are summarized in Table1.

TABLE 1
GEOTECHNICAL PROPERTIES OF DIFFERENT SOILS IN THE CONSTRUCTION FIELD

Name of soil	Depth (m)	Thickness (m)	γ (kN/m^3)	C (kPa)	φ (degree)
Fill (silty clay)	0~2	2	20	15	15
Silty clay or clayey silt	2~10	8	20	20	25
Medium sand	10~13	3	20	0	32
Silty clay	13~17.5	4.5	20	20	20

According to the geological survey report, the underground water level is at 2~4m under the ground level. So dewatering wells surrounding the foundation pit are designed to lower the water level.
According to the technical codes concerned in China, the soil nails are designed as shown in Fig.4 and Table 2 .

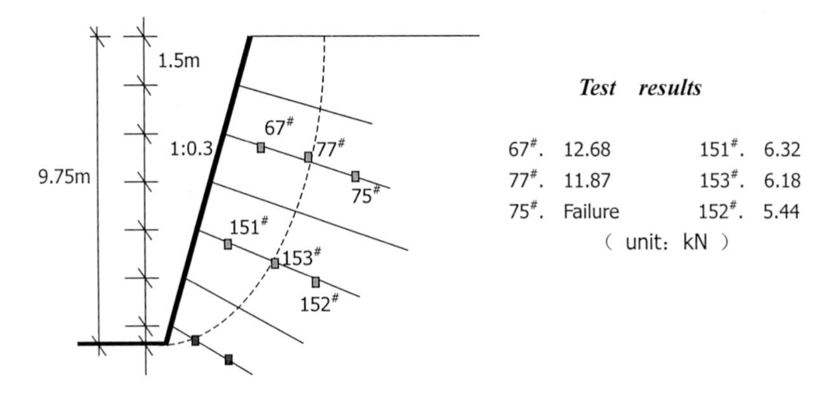

Fig.4 Location of measurement points

TABLE 2
PARAMETERS OF SOIL NAILS

No. of row	Length (m)	Steel bar	Diameter of nails(mm)
1	7	1Φ16	110
2	8	1Φ16	110
3	9	1Φ18	110
4	7	1Φ18	110
5	6	1Φ22	110
6	5	1Φ22	110

4.2 In-situ Measure Test Plan

It is planed to embed 8 pieces of steel bar stress-meter in nails, and the location of these measurement points are shown in Fig.4. Due to the changes of construction plan, 6 pieces of stress-meter are embedded actually, and their locations are shown in Fig.4. There is 1 stress-meter ($75^{\#}$) failed during test due to its quality.

4.3 Analysis of Test Results

Throughout the whole excavation process, the interior force in nails is measured every one day. The test results are shown in Fig. 5.

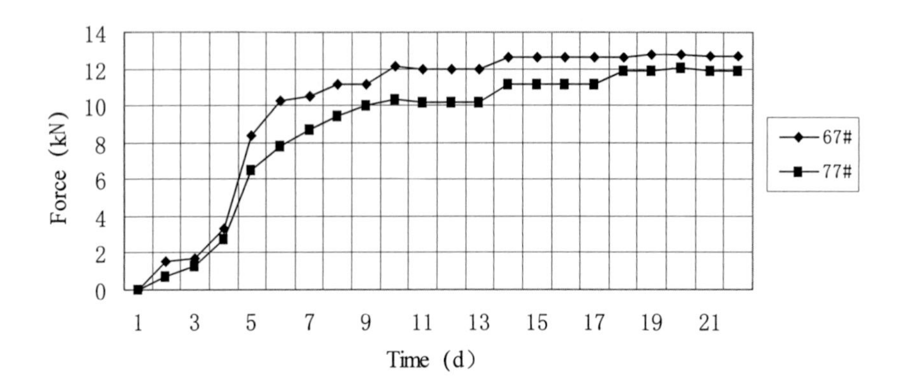

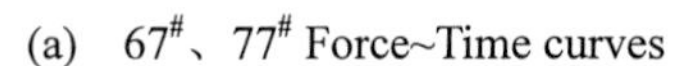

(a) $67^{\#}$、$77^{\#}$ Force~Time curves

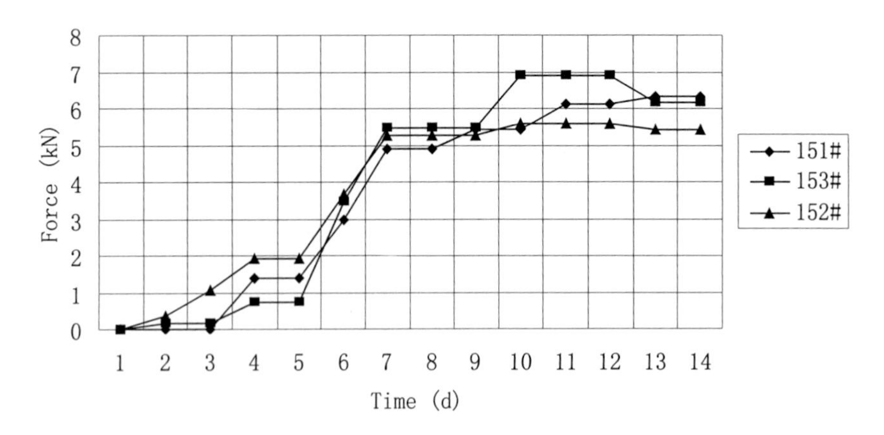

(b) $151^{\#}$、$153^{\#}$ 、$152^{\#}$ Force~Time curves

Fig.5 Force~Time curves of the measurement points

Fig.5 shows that:

(1) The interior force in the second row nails is about 12kN, smaller than the designed interior force 21.72 kN;

(2) The interior force in 4th row nails is about 6 kN, much smaller than the designed interior force 68.82 kN.

Table3 shows the results calculated with different methods, it shows that the method proposed in this paper is closest to the measurement results.

TABLE 3

RESULTS GETTING FROM DEFERENT METHODS (Unit: kN)

Location of nails	Model [1]	Model [2]	Method proposed by authors	Results measured
The second row	23.7	91.4	10	11.87~12.68
The fourth row	47.5	91.4	10	5.44 ~ 6.32

5 CONCLUSIONS

From above analysis and measurement results, we derive some conclusions as follows:

(1) The forces in soil nails are very small generally; the methods of earth pressure are not able to estimate the interior forces in nails correctly;

(2) In-situ measurement tests and calculations show that the method proposed in this paper can achieve very satisfactory results;

(3) The concept of earth pressure may not be suitable for soil nails, and the development of new methods is very important.

References

Ministry of Construction, PRC. (1999), *Technical Specification for Retaining and Protection of Building foundation Excavations,* Press of Construction Industry, Beijing, China

Q.X.Zhang. (1999). Designing Soil Nailing by Sliding Sphenoid-Mass Equilibrium Method. *Foundation Engineering* **9:1**,32-39.

Z.Y.Chen. (1997). *Specification for Soil Nailing in Foundation Excavations,* Press of China Association for Engineering Construction Standardization, Beijing, China

TEAM PRESENTATION:

STRUCTURAL BAMBOO AND BAMBOO SCAFFOLDING IN BUILDING CONSTRUCTION

Advances in Building Technology, Volume I
M. Anson, J.M. Ko and E.S.S. Lam (Eds.)

STABILITY OF MODULAR STEEL SCAFFOLDING SYSTEMS — THEORY AND VERIFICATION

Chu A.Y.T.[1], Chan S.L.[2] and Chung K.F.[2]

[1]*Buildings Department, Hong Kong, China*
[2]*Department of Civil and Structural Engineering, the Hong Kong Polytechnic University, Hong Kong, China*

ABSTRACT

In the design of modular steel scaffolding systems, it is always difficult to assess the instability behaviour of these systems as the effective length coefficient of each member depends on many factors: number of bays, storey height, joint stiffness between steel scaffolding frames, and bracing details. This paper presents the investigation on both the structural modelling and the stability analyses of modular steel scaffolding systems. Based on the second order non-linear analysis approach, a finite element model was developed, and physical tests were carried out to provide test data to verify the findings of the theoretical work obtained from the finite element modelling.

KEYWORDS

Stability analysis, effective length coefficient, column buckling, modular steel scaffolds

INTRODUCTION

Modular steel scaffolding systems have the advantages of easy fabrication, installation and dismantling. However, major problems are encountered in designing such modular steel scaffolding systems (Yen 1993) as most of the steel design codes consider building frames with beams and columns in conventional framing systems and configurations, and thus, their provisions are not strictly applicable to steel scaffolding frames. In general, collapses of steel scaffolding frames are usually accompanied by large deformations with material yielding. The deformations of these steel frames increase the lateral deflections (P-δ effect) of the posts, and, at the same time, increase the drift due to the applied load (P-Δ effect) of the entire structures. These preceding P-δ and P-Δ effects do not frequently occur in most building frames, but this can happen in modular steel scaffolding systems under large vertical loads as such destabilizing effects reduce the buckling resistances of these steel scaffolding systems significantly. According to Lightfoot and Oliveto (1977), experimental results showed that for articulated structures of a minimal and lightweight nature such as falsework, the stability of the whole and the elements of the structure was suspected when using effective lengths normally associated with permanent works. The design of steel scaffolding systems should be based on experiments with judgement, rather than standard design rules.

The objective of this paper is to investigate the instability behaviour of modular steel scaffolding systems. Moreover, experiments on similar steel scaffolding systems are carried out to provide test results to verify the findings of the theoretical work obtained from the finite element model. Through the application of an advanced integrated design and analysis software, NIDA, it is not necessary to pre-determine the effective lengths of the members in these systems as they are analysed nonlinearly. Unlike the eigenvalue analysis which only provides an upper bound solution, the load-carrying capacities of these systems are evaluated using the load-deflection path tracing technique. Consequently, the use of the empirical Merchant-Rankine formula is not needed.

ASSUMPTIONS

In order to study the structural behaviour of modular steel scaffolding systems, the following assumptions are made for the theoretical study (Chu, 1994 and Peng, 1994):

- The jack bases and the sleeve connections joining between modular steel scaffolding frames are assumed to be pinned supports and rigid joints respectively.
- The cross sections of the steel members are symmetrical about both the principal x and y axes.
- The structures are primarily under vertical loads while a horizontal notional force of 1% of the vertical loads is also applied according to codified design rules (BS5950, 2000).

Figure 1 presents the typical form and dimensions of the steel frames adopted in modular steel scaffolding systems.

STRUCTURAL FORMULATION

The structural analysis software NIDA is adopted to analyse the structural behaviour of modular steel scaffolding systems. Second-order elastic analyses with geometrical non-linearity are performed; all materials are assumed to be linear elastic. The analysis on of the elements of the modular systems is based on the self-equilibrium concept (Chan, 1992 and Chan & Zhou, 1998) in which the internal moment balances the external moment at the mid-span of the elements, as shown in Figure 2. The condition can be achieved by the following equations:-

For compatibility

At $x = -L/2$; $\dot{v} = \theta_1$

At $x = L/2$; $\dot{v} = \theta_2$

At $x = \pm L/2$; $v = 0$

For equilibrium

At $x = 0$; $$E\,I\,\ddot{v} = P\,v + \frac{M_2 + M_1}{L}\left(\frac{L}{2} + x\right) - M_1 \quad [1]$$

At $x = 0$; $$E\,I\,\dddot{v} = P\,\dot{v} + \frac{M_2 + M_1}{L} \quad [2]$$

where

EI	EI is the flexural rigidity;
M_1 and M_2	are the applied nodal moments at nodes 1 and 2;
L	is the undeformed member length;
v	is the lateral displacement with the dots on its top representing a differentiation with respect to the distance along element length, x; and
θ_1 and θ_2	are the nodal rotations at two ends.

A fifth order polynomial for the lateral displacement is assumed as follows:

$$v = a_0 + a_1 x + a_2 x^2 + a_3 x^3 + a_4 x^4 + a_5 x^5 \qquad [3]$$

By eliminating the coefficients in Equation [3], the lateral displacement can be expressed as follows:

$$v = [\, N_1 \quad N_2 \,]\, [\, L\,\theta_1 \quad L\,\theta_2 \,]^{T} \qquad [4]$$

in which the shape functions, N_1 and N_2 are given by

$$N_1 = A / H_1 + B / H_2$$

$$N_2 = A / H_1 - B / H_2$$

Refer to Chan and Zhou (1998) for detailed expressions of parameters A, B, H_1, H_2 and θ.

Using the same procedure for y axis and similar equations for three dimensional analysis, the displacement function for z axis can be evaluated and its displacement function, v, in Equation 3 is replaced by $(v + v_0)$ in which v_0 is given by

$$v_0 = v_{m0}\,(1 - t^2) \qquad [5]$$

$$t = 2\,x / L \qquad [6]$$

where v_{m0} is the deflection at mid-span of the element and t is a non-dimensional parameter.

With the displacement functions in Equation 5 and the revised form allowing for the initial imperfection in Equation 6, the tangent and secant stiffness relations can be obtained via a standard energy equation:-

$$M_1 = \frac{EI}{L}\left[c_1(\theta_1 + \theta_2) + c_2(\theta_1 - \theta_2)\right]$$

$$M_2 = \frac{EI}{L}\left[c_1(\theta_1 + \theta_2) - c_2(\theta_1 - \theta_2)\right]$$

TANGENT STIFFNESS MATRIX AND NUMERICAL PROCEDURE

The tangent stiffness matrix relates the incremental change in forces to a corresponding change in displacements and the tangent stiffness matrix can be formulated via a standard energy equation (Lightfoot and Oliveto,1977). The formulation of the analysis method is simple and the method follows a standard finite element procedure with only minimal amount of mathematical manipulation.

Both the load-deflection curve and the load-stress path for the structure analysed can be obtained by the derived elements using a numerical scheme suggested by Chan. With these plots, the load levels at which the deflection limit and the induced stress being equal to the yield strength can be determined. This load is of great interest to practising engineers, and it depends on the yield strength of materials used. It should be noted that the load calculated by the proposed non-linear theory is significantly different from the loads predicted by conventional linear analysis.

EXPERIMENTAL RESULTS

Three single storey double bay modular steel scaffolding systems were tested to provide data for comparison with the theoretical results obtained from the finite element method. The test specimens comprise of two 1930 mm × 1219 mm steel scaffolding frames which are positioned at 1819 mm apart with two cross-bracings at two planes. The jack extensions at the bases and the U-heads are adjusted so that the overall height of the specimens is 2025 mm. The specimens are mounted onto a loading frame and set in a vertical position. An axial compression load is applied progressively with a 50 ton hydraulic jack until unloading occurs. Figure 3 illustrates the general test set-up, and also the overall view of the test specimens. After the test, a proportion of the failed frame is taken out to perform a tensile test in order to determine the mechanical properties of the frame. Typical deflected shapes of the test specimens are shown in Figure 4. Large horizontal displacements were observed out of the plane of the modular units while little displacement was measured in the direction of the plane. Table 1 summarises the test results of the three test specimens.

TABLE 1
LOADING TESTS ON SPECIMENS

Test	Dimensions			Maximum Applied load per leg, P_{max}	Maximum Applied stress, σ	Yield strength, σ_y	σ / σ_y
	D	t	A				
	(mm)	(mm)	(mm^2)	(kN)	(N/mm^2)	(N/mm^2)	
1	43.2	2.63	335.2	60.6	180.8	412	0.44
2	43.2	2.63	336.0	63.5	188.8	399	0.47
3	43.35	2.76	351.9	66.1	187.8	409	0.46

It was shown that due to high slenderness in the steel members, the members are working at less than 50% of the yield strength. Moreover, the test results also suggested that substantial restraints were provided to the test specimens by the knee braces through fixings, and also by the loading beam through attachments.

RESULTS FROM FINITE ELEMENT METHOD

A three dimensional finite element model was established to study the structural behaviour of the modular steel scaffolding systems using NIDA, as shown in Figure 5. In order to provide realistic evaluation on the load carrying capacities of the systems, initial imperfection in the form of the first eigenmode of the system was provided in the model, also shown in Figure 5. The maximum out-of-straightness in the steel members was assigned to be *1/1000* of the overall system height. Due to the presence of restraints in both the jack bases and the loading beam, finite element models with four different boundary conditions with different degrees of both positional and rotational restraints onto the systems were analysed. The predicted deformations of the systems are presented in Figure 6 together with respective predicted load carrying capacities.

It is shown that while both the '*Pinned-Fixed*' condition and the '*Pinned-Pinned*' condition over-estimate the load carrying capacities significantly, the '*Free-Pinned*' condition is considered not to be sufficient. The '*Free-Fixed*' condition is found to give a load carrying capacity of 56.9 kN which compares favourably with the test results of 60.6 kN. Moreover, the predicted deformation at failure also follows broadly with the observed deformation of the test specimens.

In reality, the loading beam is considered to provide certain amount of positional restraints to the top of the steel scaffolding frames while the jacket base cannot provide fully fixed rotational restraints to the bottom of the frames. As shown in the finite element results, these two partial restraints somehow provide certain restraining effects to the test specimens which may be best approximated as the '*Free-Fixed*' condition.

CONCLUSIONS

This paper presents the investigation on both the structural modelling and the stability analyses of modular steel scaffolding systems. Based on the second order non-linear analysis approach, a finite element model was developed, and physical tests were carried out to provide test data to verify the findings of the theoretical work obtained from the finite element modelling. The instability behaviour of modular steel scaffolding systems was successfully modelled through the non-linear analysis using *NIDA*. Both the deformation and the load carrying capacities of the systems from the non-linear analysis are found to agree well with the experimental results. However, any external positional and rotational restraints will affect the structural behaviour of these systems significantly, and these restraining effects should be incorporated into the finite element model carefully to yield safe and reliable results.

REFERENCES

British Standard Institution (2000). *BS5950, Part 1 Structural Use of Steelwork in Buildings*, BSI, London, England.

Chan S.L. (1992). Large Deflection Kinematics Formulation for Three Dimensional Framed Structures, *Comp. Methods in Appl. Mech. and Engineering*, 95, 17-36.

Chan S.L. and Zhou Z.H. (1998). On the development of a robust element for second-order 'non-linear integrated design and analysis (NIDA)'. *Journal of Constructional Steel Research*, 47, 169-190.

Chu Y.T. (1994). *A Study on Elastic Stability of Steel Scaffolding*, thesis submitted to the Department of Civil and Structural Engineering, the Hong Kong Polytechnic University for partial fulfilment for the Degree of Master of Science in Civil Engineering.

Lightfoot E. and Oliveto G. (1977). The Collapse Strength of Tubular Steel Scaffold Assemblies. *Proceeding of Institution of Civil Engineers*, Part 2, June, 311-329.

Peng J.L. (1994). *Analysis Models and Design Guidelines for High Clearance Falsework Systems*. Ph.D. dissertation, School of Civil Engineering, Purdue University, West Lafayette, Ind.

Yen T. (1993). *Research in Preventing Collapses of Modular Steel Falsework in Construction*. Res. Rep., Council of Labor Affairs, Taipei, Taiwan.

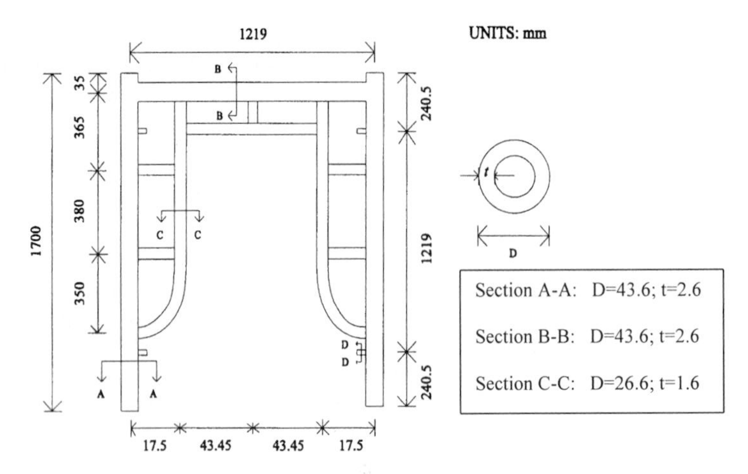

Figure 1: Typical form and dimensions of modular steel scaffolding frame

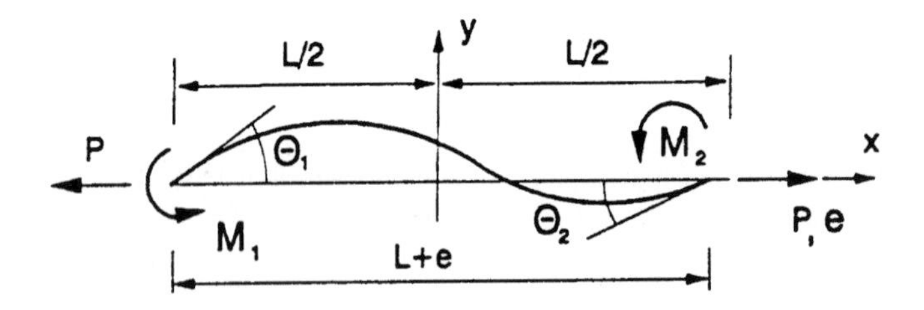

Figure 2: Basic force versus displacement in high order beam element

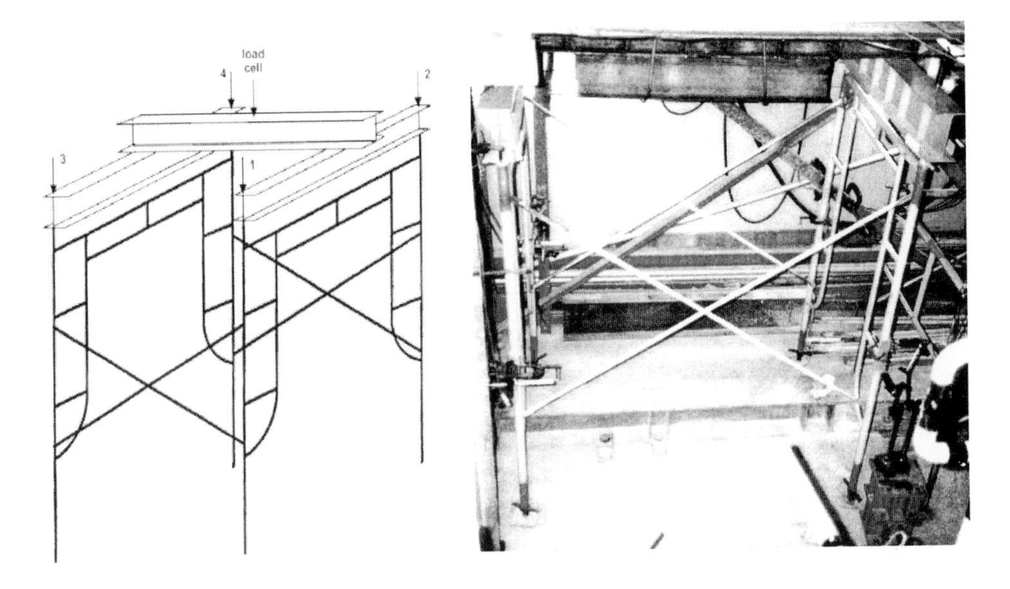

Figure 3: General view of test set-up for modular steel scaffolding systems

Figure 4: Typical failure mode in modular steel scaffolding systems – column buckling

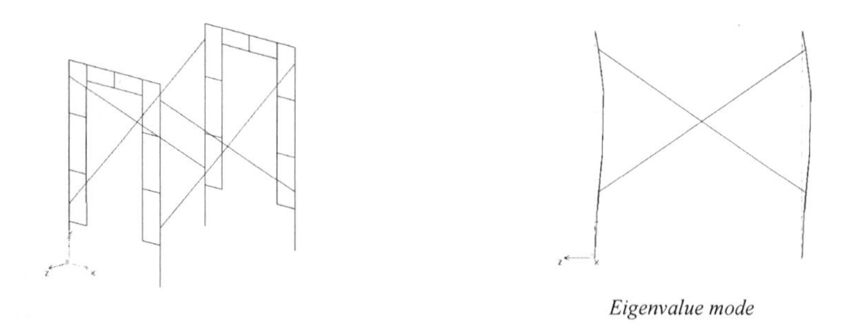

Figure 5: Geometry and initial imperfection of the finite element model

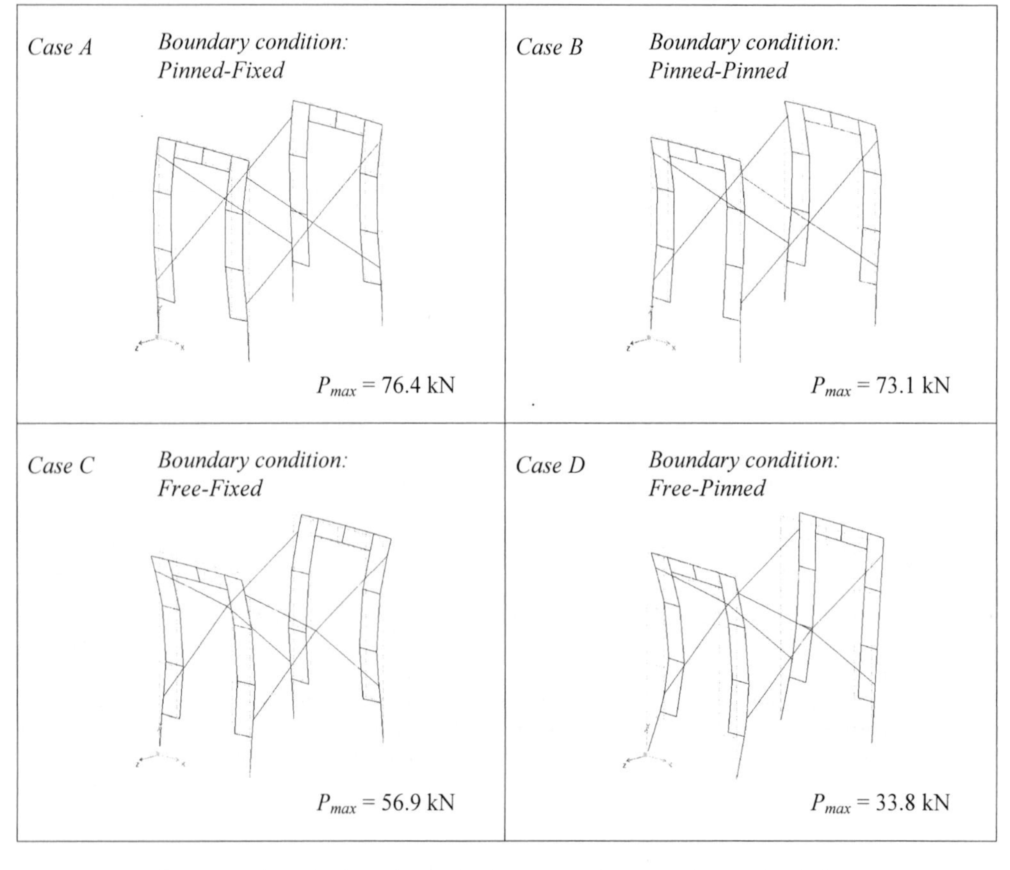

Figure 6: Numerical results obtained from non-linear analyses using *NIDA*

Advances in Building Technology, Volume 1
M. Anson, J.M. Ko and E.S.S. Lam (Eds.)

RECENT DEVELOPMENTS ON BAMBOO SCAFFOLDING IN BUILDING CONSTRUCTION

K.F. Chung, S.L. Chan and W.K. Yu

Department of Civil and Structural Engineering, the Hong Kong Polytechnic University, Hung Hom, Kowloon, Hong Kong

ABSTRACT

Bamboo scaffolds have been used in China for over a few thousand years, and both the basic framing systems and the erection methods for bamboo scaffolds in building construction were established through practice about two thousand years ago. Bamboo scaffolds provide temporary access, working platforms for workers and supervisory staff, and also prevents construction debris from falling onto passers-by. In order to promote the effective use of bamboo scaffolds in building construction, a research and development project titled '*Bamboo Scaffolds in Building Construction*' was undertaken at the Research Center for Advanced Technology in Structural Engineering of the Hong Kong Polytechnic University in 1999. This paper aims to present the major deliverables and findings of the project.

KEYWORDS

Bamboo scaffolds, green construction, structural bamboo, mechanical properties, design development.

INTRODUCTION

Bamboo scaffolds have been widely used in construction applications in South East Asia, in particular, Hong Kong for many years. Because of their high adaptability and low construction cost, bamboo scaffolds can be constructed in different shapes to follow any irregular architectural features of a building within a comparatively short period of time. In general, bamboo scaffolds are mainly used to provide access of workers to different exposed locations to facilitate various construction and maintenance process. Besides widely erected on construction sites, bamboo scaffolds are also used in signage erection, decoration work, demolition work and civil work.

In 1999, a research and development project titled '*Bamboo Scaffolds in Building Construction*' was undertaken at the Research Center for Advanced Technology in Structural Engineering (*RCATISE*) of the Hong Kong Polytechnic University with the support of the International Network of Bamboo and Rattan (*INBAR*). This was a high-level contracted research project on the technology transfer of bamboo scaffolding, and the primary objective was to promote the effective use of bamboo scaffolds in building construction.

SCOPE OF WORK

This paper aims to present major deliverables and findings derived from the project, and three major tasks are undertaken as follows:

Task 1 ***Dissemination of Established Bamboo Scaffolding Technology***
To promote the effective use of bamboo scaffolding in building construction through dissemination of technical information. Both the established knowledge and the proven practice of bamboo scaffolding in Hong Kong have been formalized and documented:

Task 1A Design Guide entitled '*Design of Bamboo Scaffolds*' covering material requirements, typical usage, structural principles and safety requirements for structural engineers.

Task 1B Erection Manual entitled '*Erection of Bamboo Scaffolds*' covering material selection, typical configurations with details, and erection procedures for scaffolding practitioners.

Task 2 ***Development of Engineered Bamboo Scaffolding Systems***

Task 2A To develop engineered bamboo scaffolding for typical applications with improved structural forms through advanced analysis and design.

Task 2B To establish connection resistances of conventional and engineered connections for practical design of bamboo scaffolds.

Task 3 ***General Technical Information of Structural Bamboo***
In order to facilitate wide engineering applications of bamboo, a document titled '*Engineering and Mechanical Properties of Structural Bamboo*' is compiled to provide general technical information on bamboo as a constructional material.

DISSEMINATION OF ESTABLISHED BAMBOO SCAFFOLDING TECHNOLOGY

Task 1A Design Guide

A '*Design Guide*' is complied [1] which comprises of the following chapters:

- Chapter 1 introduces the background on structural bamboo and also on bamboo scaffolds commonly used in South East Asia, in particular, in Hong Kong and the Southern China, namely, the Single Layered Bamboo Scaffold (SLBS) and the Double Layered Bamboo Scaffold (DLBS) as shown in Figures 1 and 2 respectively. The modern design philosophy based on limit states is also presented.
- Chapter 2 presents an experimental investigation [2] on the mechanical properties of two bamboo species, namely *Bambusa Pervariabilis* (or Kao Jue) and *Phyllostachys Pubescens* (or Mao Jue), which are commonly used in access scaffoldings in the South East Asia, in particular, in Hong Kong and the Southern China. Based on systematic compression and bending testing on a large number of test specimens, characteristic values of both the strengths and the Young's moduli of each bamboo species for limit state structural design are provided for practical design [3].
- In general, column buckling is considered to be one of the critical modes of failure in bamboo scaffolds, leading to overall structural collapse. Chapter 3 presents the design development of a limit state design method against column buckling of structural bamboo. A total of 72 column buckling tests for both Kao Jue and Mao Jue are executed [4] to provide test data, and a design method against column buckling for both Kao Jue and Mao Jue is proposed for general design.

- Chapter 4 presents an experimental investigation on the connection resistances in bamboo scaffolds where the connections are formed using either bamboo strips or plastic strips [5]. After statistical analysis, the characteristic connection resistances of the beam-column connections and the column splices with Kao Jue and Mao Jue are presented for practical design.
- Chapter 5 presents the basic configurations of typical bamboo scaffolds using both Kao Jue and Mao Jue, namely, the Single Layered Bamboo Scaffold (SLBS) and the Double Layered Bamboo Scaffold (DLBS). The results of a parametric study on the structural behaviour of bamboo scaffolds with different practical arrangements of lateral restraints are presented, and recommendations on the effective length coefficients for posts in both SLBS and DLBS with different practical arrangements of lateral restraints are also provided.
- Chapter 6 presents the general design principles of bamboo scaffolds. The design procedures of a number of typical structural bamboo members and scaffolding systems are fully presented through worked examples [6]. Moreover, the design of both SLBS and DLBS are also presented with careful consideration of various arrangements of lateral restraints for practical applications.

With the availability of rational and scientifically based design methods, structural engineers are encouraged to take the advantage offered by bamboo to build light and strong bamboo scaffolds to achieve enhanced economy and buildability.

Task 1B Erection Manual

An '*Erection Manual*' is complied [7] which comprises of the following chapters:

- Chapter 1 presents the advantages of bamboo scaffolds in building construction, and a number of practical issues such as field tests, loading requirements, safety and inspection are also presented.
- Chapter 2 introduces basic understandings on the structural forms of bamboo scaffolds, as shown in Figures 3 and 4, together with relevant requirements on supports against overall stability.
- Chapter 3 describes a total of seven types of common bamboo scaffolds, and they are fully illustrated with photographs in Chapter 4.
- Chapter 5 illustrates the structural forms and the erection sequences of various bamboo scaffolds.
- Chapter 6 introduces the basic forms of fastenings with plastic strips between bamboo members, and tightening sequences of knots for members in practical orientations are also illustrated.

With the establishment of the structural forms of the bamboo scaffolds through structural design, the conventional erection techniques are thus rationalized and simplified.

DEVELOPMENT OF ENGINEERED BAMBOO SCAFFOLDING SYSTEMS

Task 2A To develop engineered bamboo scaffolding

In order to achieve overall structural stability of the bamboo scaffolds with high structural efficiency, lateral restraints should be provided at close intervals whenever feasible. However, due to site restrictions such as availability of strong supports in proximity, it may not be practical or even possible to provide lateral restraints in every main post – main ledger connection of the bamboo scaffolds. In the absence of sufficient lateral restraints, the effective length of bamboo columns will be larger than their member lengths between ledgers, reducing the axial buckling resistance of the bamboo columns significantly. Thus, it is necessary to provide design guidance on practical arrangements of lateral restraints.

An advanced non-linear finite element analysis using *NIDA* was carried out to investigate the column buckling behaviour of bamboo scaffolds. Both SLBS and DLBS with practical arrangements of lateral restraints are examined [8], and the configurations of the systems are presented in Figure 5. Both the local buckling of bamboo posts between ledgers and the global instability of the entire structures with practical lateral restraints are accurately incorporated as shown in Figure 6. Based on the results of the parametric studies, a set of design rules for both SLBS and DLBS was proposed for practical support conditions. Moreover, a total of four full-scale tests on double layered bamboo scaffolds were carried out [9], and they provide test data for the calibration of the proposed design methods.

Task 2B *Connection resistances of conventional and engineered connections*

In order to enable bamboo scaffolds to compete directly with metal scaffolds, it is necessary to rationalize the existing bamboo scaffolding systems with simple, standardized and reliable connections for quick installation. In the past, bamboo strips were commonly used to form knots in joining members together while plastic strips are widely used instead after 1970s. The major advantage of plastic strips is quick erection of bamboo scaffolds by hands with only simple tools. However, the strength and stiffness of these connections depend primarily on the skills of the workers, and these connections tend to be weakened over time due to weathering on both the plastic strips and the bamboo members. It is highly desirable to develop reliable engineered connections for bamboo scaffolds with prolonged usage. An experimental investigation was carried out to establish connection resistances of conventional and engineered connections for practical design of bamboo scaffolds, and full details of the experimental investigation are presented in the Technical Report '*Connections in Bamboo Scaffolds*' [5]. The investigation may be divided into the following two parts:

Part I *Conventional connections using bamboo and plastic strips*

- Destructive tests on practical beam-column connections with typical member configurations, as shown in Figure 7. Material tensile tests on both bamboo and plastic strips were also carried out.

Part II *Engineered connections using steel bolts and screws*

- Destructive tests on beam-column connections with steel strips, wooden wedges and screws, as shown in Figure 8.
- Destructive tests on column-column connections with steel bolts and screws, as shown in Figure 8.

A number of different connection configurations were proposed, and connections using plastic strips, steel wires, steel bars, thin gauge steel strips, steel couplers, nails, and self drilling self tapping screws were made as samples. For those connections with high buildability and sufficient strength and stiffness, destructive tests were carried out to assess their structural adequacy, and material tests were also performed as appropriate. The results of all the test series are fully presented in the Technical Report [5] together with statistical analysis.

BAMBOO AS A CONSTRUCTIONAL MATERIAL

Task 3 *General technical information of structural bamboo*

In order to facilitate wide engineering applications of bamboo, a document titled '*Engineering and Mechanical Properties of Structural Bamboo*' is compiled [10]. The document aims to provide general technical information to readers who are considering to use bamboo as a constructional materials, to help their understanding on bamboo and to select the most suitable species for their purposes. It also covers relevant and up-to-date technical guidance to structural engineers to exploit bamboo as load bearing structural members.

CONCLUSIONS

This paper presents the major deliverables and findings of a research and development project which aims to provide technical solutions against various hinderances of wide applications of bamboo scaffolds. With the availability of modern design methods and engineering data of the mechanical properties of structural bamboo, engineers are encouraged to take the advantage offered by bamboo to build light and strong scaffolds to achieve enhanced economy and buildability.

ACKNOWLEDGEMENTS

The research project is supported by both the Research Committee of the Hong Kong Polytechnic University Research (Project No. G-V849), and the International Network for Bamboo and Rattan (Project No. ZZ04). The authors would like to thank Professor J.M. Ko of the Hong Kong Polytechnic University and Professor J.J.A. Janssen of the Eindhoven University of Technology, co-chairmen of the Steering Committee of the INBAR project, for their general guidance and technical advice. The technical support from the Construction Industry Training Authority of the Government of Hong Kong Special Administration Region on erection of bamboo scaffolds is also gratefully acknowledged.

REFERENCES

1. Chung K.F. and Chan S.L.: *Bamboo Scaffolds in Building Construction. Design of Bamboo Scaffolds*. Joint Publication, the Hong Kong Polytechnic University and International Network for Bamboo and Rattan (in press).
2. Chung K.F. and Yu W.K.: *Mechanical properties of structural bamboo for bamboo scaffolding*. Engineering Structures 24 (2002): 429-442.
3. Chung K.F., Yu W.K. and Chan S.L.: *Mechanical properties and engineering data of structural bamboo*. Proceedings of International Seminar 'Bamboo Scaffolds in Building Construction' - An Alternative and Supplement to Metal Scaffolds, May 2002, pp1-23.
4. Yu W.K., Chung K.F. and Chan S.L.: *Column buckling of structural bamboo*. Engineering Structures (submitted for publication).
5. Chung K.F., Yu W.K. and Ying K.: *Connections in Bamboo Scaffolds*. Research Centre for Advanced Technology in Structural Engineering, the Hong Kong Polytechnic University, 2002.
6. Chung K.F., Chan S.L. and Yu W.K.: *Practical design of bamboo scaffolds*. Proceedings of International Seminar 'Bamboo Scaffolds in Building Construction' - An Alternative and Supplement to Metal Scaffolds, May 2002, pp65-88.
7. Chung K.F. and Siu Y.C.: *Bamboo Scaffolds in Building Construction. Erection of Bamboo Scaffolds*. Joint Publication, the Hong Kong Polytechnic University and International Network for Bamboo and Rattan (in press).
8. Chan S.L. and Chung, K.F.: *Stability design of hollow timber section - bamboo*. Proceedings of International Seminar 'Bamboo Scaffolds in Building Construction' - An Alternative and Supplement to Metal Scaffolds, May 2002, pp55-64.
9. Yu W.K., Chung K.F. and Tong Y.C.: *Full scale tests of bamboo scaffolds for design development against instability*. Proceedings of International Conference on Advances in Building Technology, December 2002, Hong Kong (in press).
10. Chan S.L. and Xian X.J.: *Engineering and Mechanical Properties of Structural Bamboo*. Research Centre for Advanced Technology in Structural Engineering, the Hong Kong Polytechnic University, 2002 (in press).

Figure 1: Typical bamboo scaffold – *Single Layered Bamboo Scaffold (SLBS)*

Figure 2: Typical bamboo scaffold – *Double Layered Bamboo Scaffold (DLBS)*

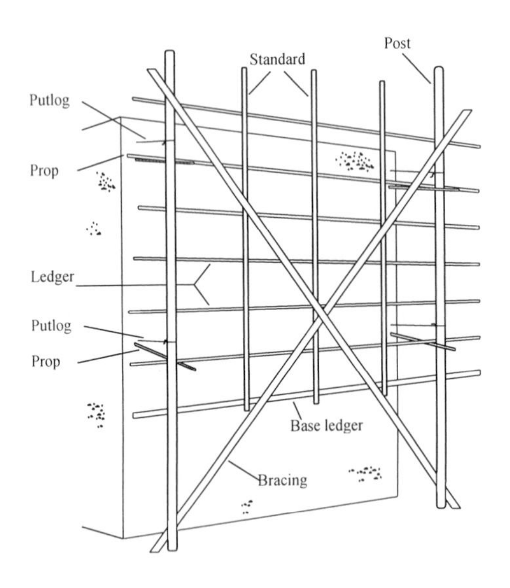

Figure 3: Typical configuration of *SLBS*

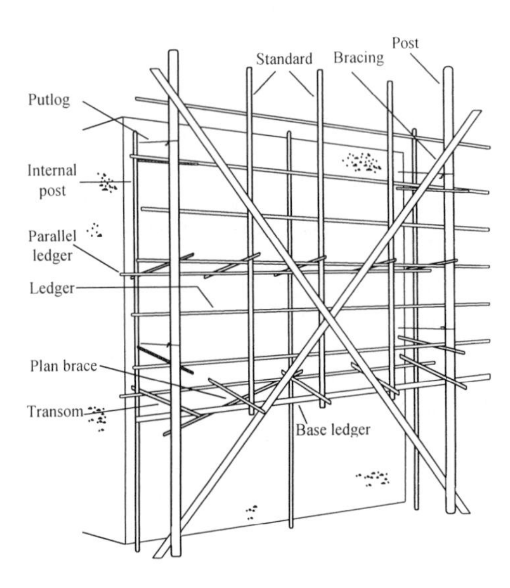

Figure 4: Typical configuration of *DLBS*

a) *SLBS* using Mao Jue

L = 1.5m ~ 2.4m
h = 1.8m ~ 2.25m

Member	Member type
Main post	*Mao Jue*
Base ledger	*Mao Jue*
Standard	*Kao Jue*
Ledger	*Kao Jue*
Diagonal	*Mao Jue*

b) *SLBS* using Kao Jue

L = 1.2m ~ 1.8m
h = 1.8m ~ 2.25m

All members using Kao Jue.

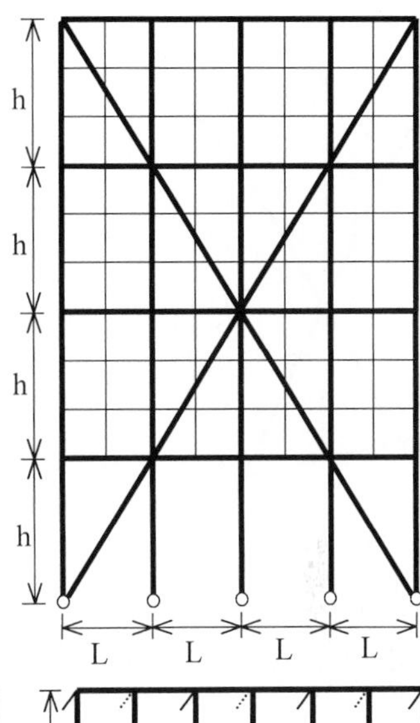

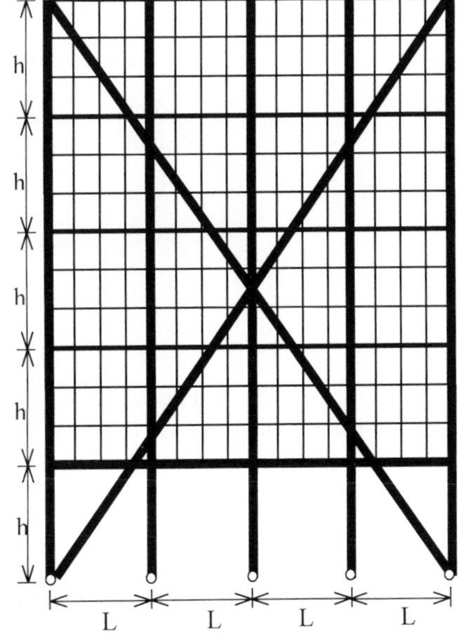

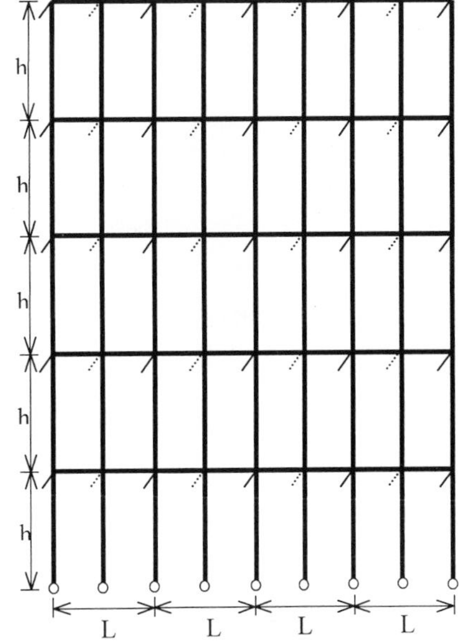

c) *DLBS* using Mao Jue and Kao Jue

L = 1.5m ~ 2.4m
h = 1.8m ~ 2.25m

Outer layer	Member type
Main post	*Mao Jue*
Base ledger	*Mao Jue*
Standard	*Kao Jue*
Ledger	*Kao Jue*
Diagonal	*Mao Jue*

Lateral restraints connected to main posts of outer layer.

Lateral restraints connected to ledgers.

All members in the Inner layer and transoms using Kao Jue.

Figure 5: Typical configurations of *SLBS* and *DLBS*

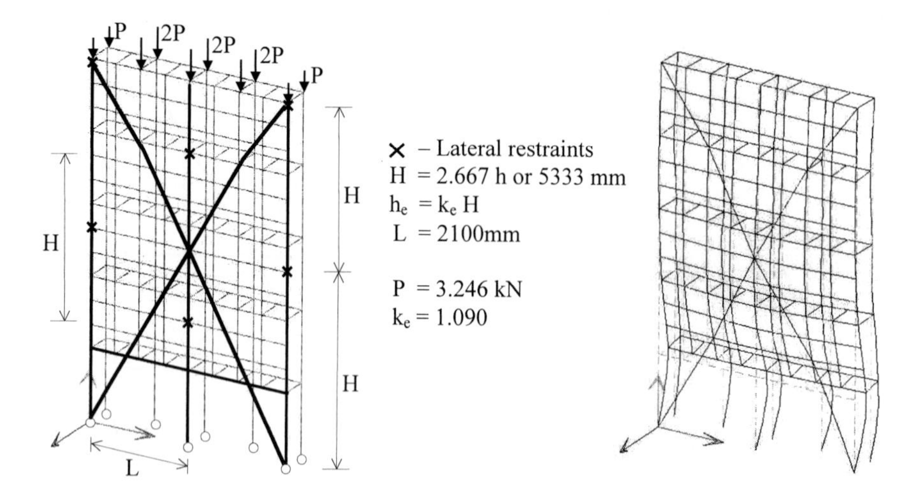

Figure 6: Advanced non-linear analysis with NIDA on *DLBS* under practical supports

a) Bamboo strips b) Plastic strips

Figure 7: Conventional beam-column connections

a) Beam-column connection using wooden wedge and steel screws

b) Column-column connection using steel bolts

c) Column-column connection using steel screws

Figure 8: Engineered connections

Advances in Building Technology, Volume 1
M. Anson, J.M. Ko and E.S.S. Lam (Eds.)

RELIABILITY ANALYSIS FOR MECHANICAL PROPERTIES OF STRUCTURAL BAMBOO

Ming Lu, W.K. Yu and K.F. Chung

Department of Civil and Structural Engineering, the Hong Kong Polytechnic University,
Hung Hom, Kowloon, Hong Kong

ABSTRACT

Some of the bamboo species are very stiff and strong, hence are considered as ideal natural structural materials suitable for many construction applications, such as low-rise houses, short span footbridges, long span roofs and assess scaffolds. On the other hand, bamboo - a natural non-homogenous organic substance, often exhibits large variations of physical properties along the length of its members, such as external and internal diameters, dry density and moisture content. While structural engineers expect variations in the mechanical properties of bamboo, they tend to accept that the mechanical properties of bamboo are likely to be more consistent than those of concrete, probably similar to structural timber.

This paper presents a series of experimental investigations on structural bamboo species, together with a neural network (NN) based approach for sensitivity analysis and reliability analysis of the experimental data. Through an optimized NN model for characterizing the mechanical properties from the physical ones for Mao Jue using the pilot test results, we have gained insight on how physical properties of bamboo influence the compressive strength and Young's modulus of structural bamboo. Further, the NN model also serves as a "virtual lab" for carrying out controlled experiments and conducting reliability analysis on the mechanical properties of bamboo. The NN-produced 5th percentile compressive strength is compared against present knowledge on structural Mao Jue, coming close to the characteristic value of Mao Jue obtained in previous findings. This new NN-based approach, as a reliable and cost-effective means to study structural bamboo, can facilitate structural engineers to understand the properties of structural bamboo and design bamboo structures at a known level of confidence against failure.

KEYWORDS

Structural bamboo, basic mechanical properties, neural networks, computer modeling, design development.

INTRODUCTION

Bamboo is a good natural structural material, and there are over 1500 different botanical species of bamboo in the world. Some of the bamboo species are considered to be effective structural materials for many construction applications such as structural members in low-rise houses, short span foot bridges, long span roofs and construction platforms in countries with plentiful bamboo resources. In general, it is considered that the mechanical properties of bamboo are likely to be at least similar, if not superior, to those of structural timber. Furthermore, as bamboo grows very fast and usually takes three to six years to harvest, depending on the species and the plantation, there is a growing global interest in developing bamboo as a substitute of timber in construction. It is generally expected that the effective use of structural bamboo will mitigate the pressures on the ever-shrinking natural forests in developing countries, and thus, facilitate the conservation of the global environment.

On the other hand, bamboo - a natural non-homogenous organic substance, often exhibits large variations of physical properties along the length of its members, such as external and internal diameters, dry density and moisture content. While structural engineers also expect variations in the mechanical properties of bamboo, they tend to accept that the mechanical properties of bamboo are likely to be more consistent than those of concrete, and probably similar to structural timber. So far, a major constraint to the development of bamboo as a modern construction material is the lack of mechanical properties and engineering data for structural bamboo. Moreover, there were only limited design rules available for the general design of structural bamboo, and further design development is highly desirable to provide guidance on the structural behaviour of bamboo members under practical loading and support conditions.

Bamboo scaffolds

Bamboo scaffolds have been used in building construction in China for over a few thousand years. Nowadays, in spite of open competition with many metal scaffolding systems imported from all over the world, bamboo scaffolds remain to be the preferred systems for access in building construction in Hong Kong and the neighbouring areas, according to Fu [1].

As reported by So and Wong [2], bamboo scaffolds are commonly employed in building construction to provide temporary access, working platforms for construction workers and supervisory staff. Owing to their high adaptability and low construction cost, bamboo scaffolds can be constructed in any layout to follow various irregular architectural features of a building within a comparatively short period of time. Moreover, they are also erected to prevent construction debris from falling onto passers-by. In 1995, an industrial guide on safety of bamboo scaffolds was issued by the Hong Kong Construction Association [3]. Typical usage of bamboo scaffolds in building construction in Hong Kong [4] include:

- Single Layered Bamboo Scaffolds (SLBS) for light duty work. It is highly adaptable to site conditions with both easy erection and dismantling.
- Double Layered Bamboo Scaffolds (DLBS) with working platform for heavy duty work. It provides safe working platforms for construction activities to be carried out at heights.
- Signage erection, decoration work, demolition work and civil work [5].

Recent research in structural bamboo

Structural bamboo has been traditionally used in China, Philippines, India, and Latin America for centuries, but little research was reported in the past. Recent scientific investigations on bamboo as a construction material were reported by Au *et. al.* [6] in Hong Kong, and also by Janssen in 1981 [7] in Holland. A large amount of data of the mechanical properties for various bamboo species all over the world were also reported by Janssen in 1991 [8]. While these sets of data provide typical values of compressive, bending and shear strengths of various bamboo species, no characteristic strengths for structural design were provided. A series of experimental studies on structural bamboo were reported by Arce-Villalobos in 1993 [9] and practical connection details for bamboo trusses and frames were also proposed and tested. Moreover, Gutierrez in 1999 [10] conducted a recent study on the traditional design and construction of bamboo in low-rise housing in Latin America, and innovative applications of bamboo in building construction in India was also reported by INBAR [11]. More recently, Amada *et. al.* [12] classified bamboo as a smart natural composite material with optimized distribution of fibers and matrices, both across cross sections and along member lengths, in resisting environmental loads in nature.

EXPERIMENTAL INVESTIGATIONS

In order to promote the safe use of structural bamboo, we carried out a series of experimental investigations on the mechanical properties of two bamboo species, namely, *Bambusa Pervariabilis* (or Kao Jue) and *Phyllostachys Pubescens* (or Mao Jue). Selection of the two bamboo species for study is attributed to their common use in Hong Kong and the Southern China as bamboo scaffolds. The mechanical properties for both Kao Jue and Mao Jue are investigated [13,14,15], aimed to establish basic design data (like characteristic strength) for evaluating their basic compression and bending capacities of structural bamboo. Due to the variations of physical properties in structural bamboo, it is necessary to carry out a large number of tests to assess the variations of the test results, and to provide engineering data based on statistical analysis of the test results. For details of the testing procedures, please refer to the INBAR documents compiled by Janssen [16,17].

PILOT STUDY ON SHORT BAMBOO MEMBERS UNDER COMPRESSION

A pilot study was first carried out to examine the variation of compressive strength against the following physical properties of bamboo members:

- External diameter, D ,
- Wall thickness, t , (both D and t are used to evaluate cross-sectional area, A)
- Dry density, ρ , and
- Moisture content, *m.c.*

Some of these physical parameters vary significantly along the length of bamboo members, and it is highly desirable to establish any correlation between the physical and the mechanical properties of bamboo members. Three dry bamboo members were tested, and all of them were mature with an age of at least three years and no visual defect. The height to diameter ratio of the short bamboo members was kept approximately at two, and the maximum height of the short members was 150 mm.

Figure 1 illustrates the general set-up of the compression tests, and both the applied loads and the axial shortening of the test specimens were measured during the tests. Two failure modes, namely *End bearing* and *Splitting*, were identified, as also shown in Figure 1. It was found that most specimens failed in *End Bearing*, especially for those specimens with high moisture contents. As the moisture content decreased, cracks along fibers were often induced, leading to *Splitting*.

Figure 2 presents the variations of the physical properties along the length of the bamboo members for Mao Jue. The variations of the ultimate compressive strengths and the Young's modulus against compression can also be found in Figure 2 for comparison.

For Mao Jue, two observations are made:

1. The physical properties of the three members are found to be significantly different from one anther. From the bottom to the top of the members, the external diameter typically decreases from 80 mm to 50 mm. The wall thickness varies from 10 mm at the bottom to 6 mm at the top of the members.
2. Contrary to the physical properties, the mechanical properties of the three members are found to be broadly similar. The compressive strength is found to be 50 N/mm^2 at the bottom of the members and increases steadily to 70 N/mm^2 at the top. The Young's modulus against compression is found to vary steadily from 5 kN/mm^2 to 10 kN/mm^2 from the bottom to the top of the members.

In brief, representative values of mechanical properties may be obtained through systemic testing on Mao Jue regardless of large variations in external diameter, wall thickness and dry density. Chung and Yu [18] conducted systemic testing on Mao Jue and obtained the characteristic values of both the compressive strengths and the Young's modulus of Mao Jue as given in Table 1. It should be noted that the characteristic values of the mechanical properties of Mao Jue are shown to be superior to common structural timber, and probably also to concrete. The material partial safety factor for structural bamboo, γ_m , is proposed to be 1.5.

SENSITIVITY AND RELIABILITY ANALYSIS THROUGH NN MODELING

The back-propagation neural network (BPNN) has been researched and applied as a convenient decision-support tool in a variety of application areas in civil engineering [19]. The BPNN is also successfully applied to analyze the relationship between the physical and the mechanical properties for Mao Jue. A total of 40 pilot test data are complied and used to train a neural network model under control of 10 randomly selected test data. The NN has 4 input parameters (i.e. external diameter, wall thickness, moisture content and dry density) and 2 output values (i.e. compressive strengths and corresponding Young's Modulus). The internal architectural and the learning parameters of NN are determined using an optimized NN modeling approach suggested in Lu and Shi (2002) [20] as: four hidden nodes are on the one hidden layer, the learning rate is equal 0.5 and the momentum term is equal 0.25, the initial random seed is equal to 60, and the number of learning iterations is equal to 19387. The trained NN model is further validated with an independent production set of 8 records, representing eight additional experimental tests.

The Monte Carlo simulation based sensitivity analysis [19] was conducted on the validated NN model and the results are shown in Figure 3. The horizontal axis represents the relative input sensitivity, that is, the output response (negative or positive in it own unit of measure) with a change of 10% relevant range in an input parameter. The vertical axis is the baseline corresponding to no output response or zero change in output. Five short vertical bars correspond to each input parameter, representing respectively the five percentiles (10th, 25th, 50th, 75th, and 90th) from left to right and reflecting the central trend, the spread and the shape of the observed slope data distribution obtained from simulation. It is observed both the diameter and the dry density of Mao Jue exhibit a positive correlation with compressive strength and Young's modulus while the thickness exhibits a negative correlation. The sensitivity of moisture content is found to be not as significant as the other parameters, which is due to the relatively small range of moisture content in the pilot test data.

Next, the NN model serves as a "virtual lab" for carrying out controlled experiments and conducting reliability analysis on the mechanical properties of Mao Jue and evaluating its design data. Over 1000 specimens for each typical combination of diameter-thickness dimensions at a constant moisture content of 20% were simulated by querying the NN model, with the dry density as a random variable uniformly distributed over the range from 600 to 850 kg/m^3. Figure 4 presents the results of the reliability analysis of compressive strength, showing that the compressive strength at fifth percentile is about 49 N/mm^2 regardless of the moisture content and the dimension combination. Note that the simulated compressive strength is found to be close to the corresponding characteristic compressive strength obtained from the systematic laboratory tests (44 N/mm^2 as read from Table 1).

CONCLUSIONS

A pilot study was carried out to examine the variation of compressive strength against the following physical properties of bamboo members, i.e. external diameter (D), wall thickness (t) dry density (ρ), and moisture content (*m.c.*). Some of these physical parameters vary significantly along the length of bamboo members, and it is highly desirable to establish any correlation between the physical and the mechanical properties of bamboo members.

The back-propagation neural network (BPNN) has been successfully applied to analyze the relationship between the physical and the mechanical properties (i.e. compressive strengths and corresponding Young's Modulus) for Mao Jue. A total of 40 pilot test data were complied and used to train a neural network model under the control of 10 randomly selected test data. The internal architectural and learning parameters of NN were decided using an optimized NN modeling approach. Following further validation of the trained NN model with an independent production set, the sensitivity and reliability analyses were performed on the NN model to (1) gain insight on how physical properties of bamboo influence its compressive strength and Young's modulus, and to (2) facilitate the design of bamboo structures at a known level of confidence against failure. This new NN-based approach has been proven to be a reliable and cost-effective means for studying mechanical properties of structural bamboo and facilitating the development of associated design methods.

ACKNOWLEDGEMENTS

The research project leading to the publication of this paper is supported by the International Network for Bamboo and Rattan (Project No. ZZ04), and also by the Research Committee of the Hong Kong Polytechnic University Research (Project No. G-V849). The authors would like to thank Professor J.M. Ko of the Hong Kong Polytechnic University and Professor J.J.A. Janssen of the Eindhoven University of Technology, Co-chairmen of the Steering Committee of the INBAR project, for their general guidance and technical advice.

REFERENCES

1. Fu, W.Y. (1993). *Bamboo scaffolding in Hong Kong.* The Structural Engineer, Vol. 71, No. 11, pp202-204.
2. So, Y.S. and Wong, K.W. (1998). *Bamboo scaffolding development in Hong Kong – A critical review.* Proceedings of the Symposium on 'Bamboo and Metal Scaffolding', the Hong Kong Institution of Engineers, October 1998, pp63-75.
3. The Hong Kong Construction Association (1995). *Practice Guide for Bamboo Scaffolding Safety Management.* Labour and Safety Committee.
4. Chung, K.F. and Siu, Y. C. (2002). *Bamboo Scaffolds in Building Construction: - Erection of Bamboo Scaffolds.* Joint Publication, the Hong Kong Polytechnic University and International Network for Bamboo and Rattan (in press).
5. Tong, A.Y.C. (1998). *Bamboo Scaffolding - Practical Application.* Proceedings of the Symposium on 'Bamboo and Metal Scaffolding', the Hong Kong Institution of Engineers, pp43-62.

6. Au, F., Ginsburg, K.M., Poon, Y.M., and Shin, F.G. (1978). *Report on study of bamboo as a construction material*. The Hong Kong Polytechnic.
7. Janssen, J.J.A. (1981). *Bamboo in building structures*. Ph.D. thesis, Eindhoven University of Technology, Holland, 1981.
8. Janssen, J.J.A. (1991). *Mechanical properties of bamboo*, Kluwer Academic Publisher.
9. Arce-Villalobos, O. A. (1993). *Fundamentals of the design of bamboo structures*. Ph.D. thesis, Eindhoven University of Technology, Holland.
10. Gutierrez, J.A. (1999). *Structural adequacy of traditional bamboo housing in Latin-America*. Technical Report No. 321-98-519. Laboratorio Nacional de Materiales y Modelos Estructurales, Universidad de Costa Rica, Costa Rica.
11. INBAR (1999). Society for Advancement of Renewable Materials and Energy Technologies, *Technical Report on 'Buildings with bamboo arch roof at Harita Ecological Institute - A review report and recommendations for design upgradation'*, International Network for Bamboo and Rattan, Beijing, China.
12. Amada, S., Munekata, T., Nagase, Y., Ichikawa, Y., Kirigai, A. and Yang, Z. (1997). *The mechanical structures of bamboos in viewpoint of functionally gradient and composite materials*. Journal of Composite Materials, Vo. 30, No.7, 1997, pp800-819.
13. Yu, W.K. and Chung, K.F. (2000a). *Qualification tests on Kao Jue under compression and bending*. Technical Report, Research Centre for Advanced Technology in Structural Engineering, the Hong Kong Polytechnic University.
14. Yu, W.K. and Chung, K.F. (2000b). *Qualification tests on Mao Jue under compression and bending*. Technical Report, Research Centre for Advanced Technology in Structural Engineering, the Hong Kong Polytechnic University.
15. Yu, W.K. and Chung, K.F. (2001). *Mechanical properties of bamboo for scaffolding in building construction*. Proceedings of International Conference on Construction, Hong Kong, June 2001, pp262-272.
16. Janssen, J.J.A. (1999a). *An international model building code for bamboo*. The International Network for Bamboo and Rattan (draft document).
17. Janssen, J.J.A. (1999b). *INBAR standard for determination of physical and mechanical properties of bamboo*. The International Network for Bamboo and Rattan (draft document).
18. Chung, K.F. and Yu, W.K. (2002). *Mechanical properties of structural bamboo for bamboo scaffoldings*. Engineering Structures, 24, pp429-442.
19. Lu, M., AbouRizk, S.M. and Hermann, U.H. (2001). *Sensitivity analysis of neural networks in spool fabrication productivity studies*. Journal of Computing in Civil Engineering, ASCE, Vol. 15, No. 4, 2001, pp299-308.
20. Lu, M. and Shi, W.Z. (2002). *Optimized Neural Network Model for Characterizing Geographical Terrain Slope*. Journal of Computing in Civil Engineering, ASCE, accepted

TABLE 1 Proposed mechanical properties for structural bamboo

Phyllostachys Pubescens (Mao Jue)	Dry (under 5% *m.c.*)	Wet (over 30% *m.c.*)
Characteristic compressive strength, $f_{c,k}$ (N/mm^2) (at fifth percentile)	117	**44**
Design compressive strength, $f_{c,d}$ (N/mm^2) ($\gamma_m = 1.5$)	78	29
Design Young's Modulus against compression, $E_{c,d}$ (kN/mm^2) (average value)	9.4	6.4

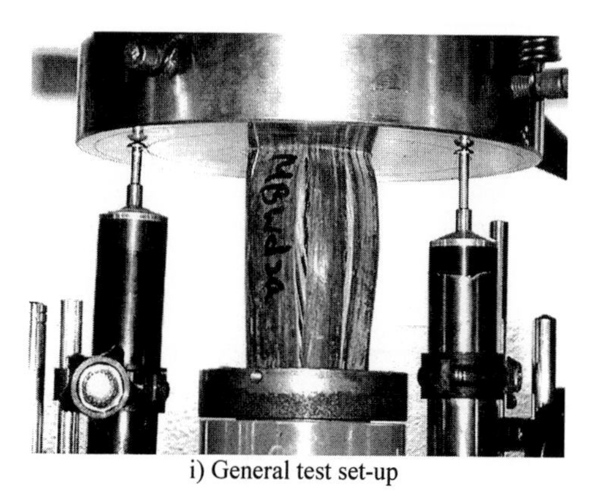

i) General test set-up

ii) Typical failure mode - End bearing

iii) Typical failure mode – Splitting

Figure 1: Compression test

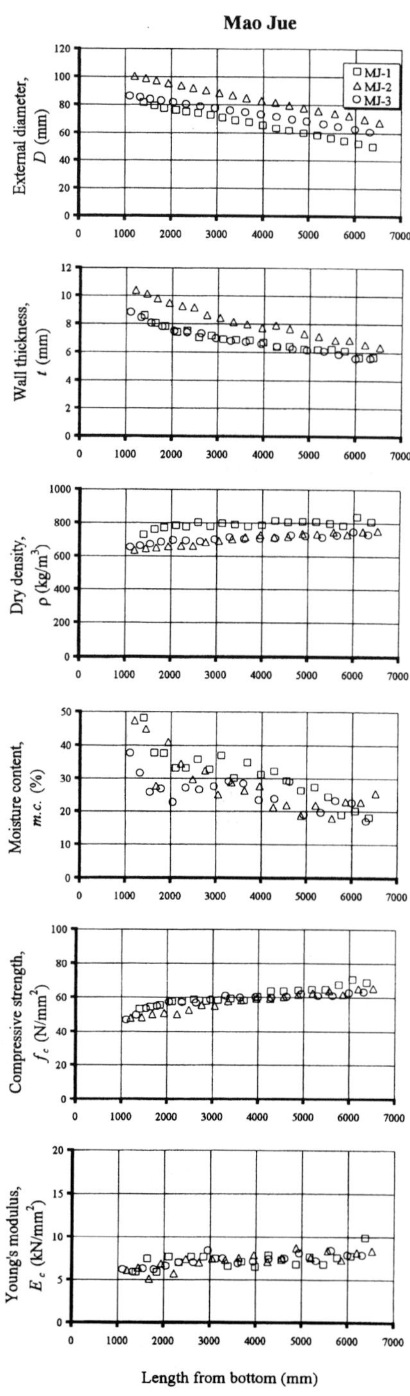

Figure 2: Variations along bamboo length

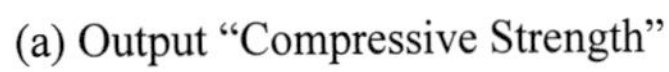
(a) Output "Compressive Strength"

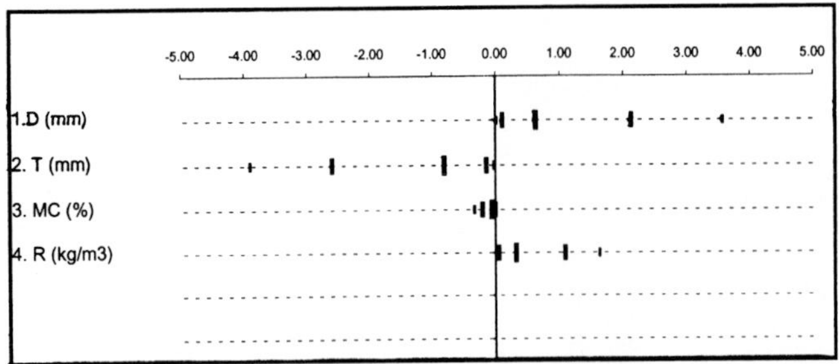

(b) Output "Young's Modulus"

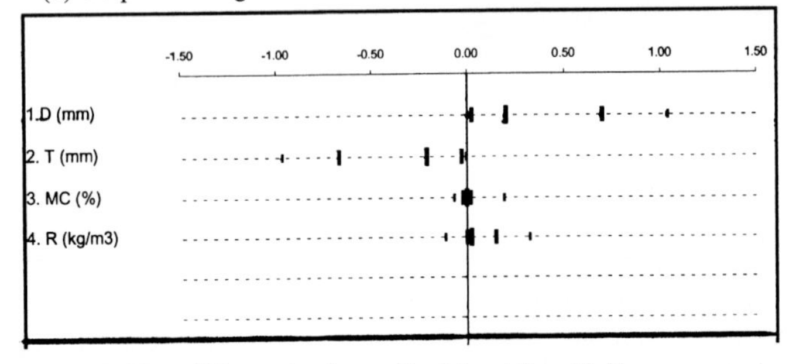

Figure 3: Neural Networks Generalized Input Sensitivity

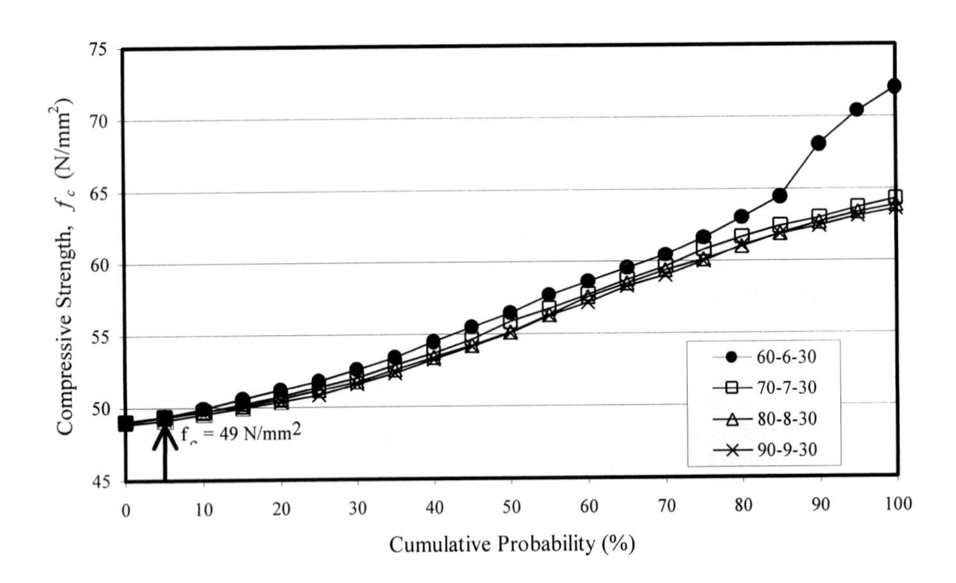

Figure 4: Reliability Analysis of Compressive Strength through Monte-Carlo simulation on the Trained NN model.

Advances in Building Technology, Volume 1
M. Anson, J.M. Ko and E.S.S. Lam (Eds.)

ERECTION OF BAMBOO SCAFFOLDS AND THEIR RECENT DEVELOPMENTS

Mr. Francis So

Wui Loong Scaffolding Works Co., Ltd.

SYNOPSIS

Bamboo scaffold has a long history of application in the construction industry in Hong Kong and Mainland China. Bamboo material is environmental-friendly and low-cost. It has been widely used in the construction of new buildings, renovation, repair works, slope maintenance and neon signage works. Basically, it is a kind of temporary working platform and a means of access and egress for workers at height. However, the trend of substantial increased in height of building, Bamboo scaffold has been partly substituted at the bottom level by Metal scaffold or Metal Bamboo Matrix System Scaffolds in some projects to provide better foundation of overall scaffolding structure.

INTRODUCTION

Most of the construction operations require the use of scaffolding and material is an important factor which affects the competitiveness as well as its overall investment.

Most experienced engineers and contractors in Hong Kong prefer using bamboo instead of steel in scaffolding works because it is easier to fabricate, light so that load on supports like steel brackets can be smaller, environmental-friendly and more economical. In addition, its characteristics of significant deflection before failure offer sufficient warning to the users that cannot be found in steel scaffolding.

COST EVALUATION

Bamboo is much cheaper than steel sections. The cost of a bamboo member of 6m length is about HK$10, compared with a steel circular hollow section of similar size of HK$80.

STRENGTH AND STIFFNESS

The design strength for typical bamboo species *Phyllostachys pubescens* is 40 N/mm2, which is the lower end of the average value of the design stress of bamboo because of its high variation. Compared to steel of grade 43, bamboo has a smaller design strength and Young's modulus of elasticity, which are two major basic parameters for engineering materials.

TABLE 1
COMPARISON OF ENGINEERING PROPERTIES BETWEEN STEEL AND BAMBOO

Material	Specific Density	Design stress (MPa)	E_t (GPa)	Poisson's ratio
Bamboo	0.8	40 - 60	12	0.35
Steel	7.8	275	205	0.3

Note : Bamboo specimen is Phyllostachys pubescens and steel grade is S275 or grade 43

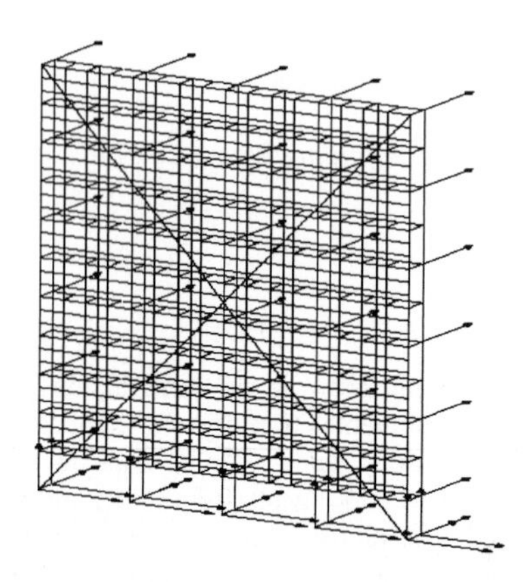

With the above data, structural analysis is carried out using computer model for bamboo scaffolding system for checking the stability of the complete system and buckling strength of individual members.

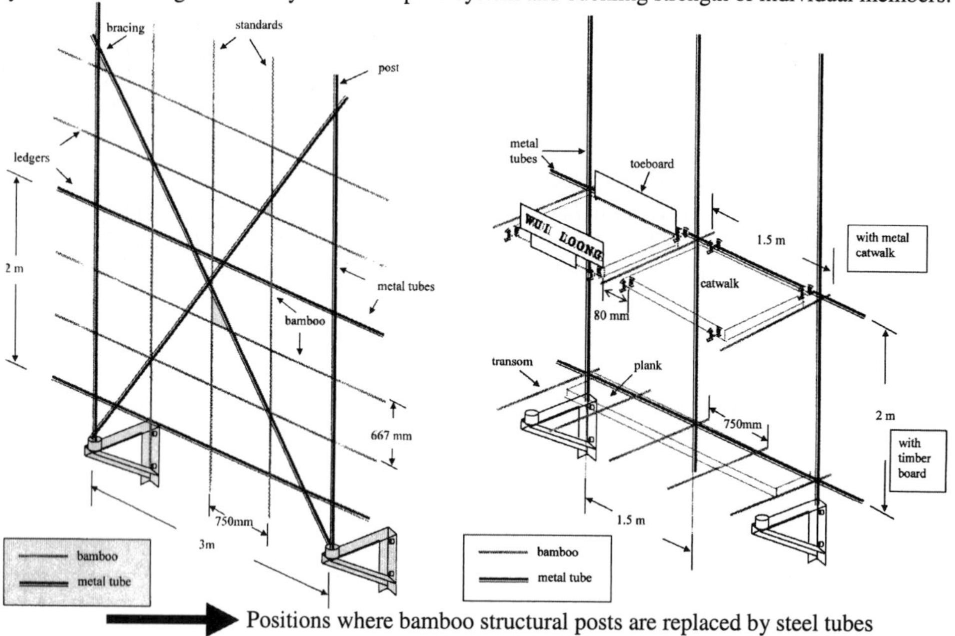

Figure 1: External layer for MBMSS

Figure 2: Internal layer for MBMSS

Owing to the difficulty in obtaining larger size and stronger bamboo posts and to the failure of bamboo itself due to various reasons such as degradation, natural disasters, destruction by other trade works, etc., the author developed a new system called the Metal Bamboo Matrix System Scaffolds (MBMSS). To increase the load carrying capacity of the system, the heavily loaded members can be replaced by steel circular sections. The system was further analyzed and checked by Professor S.L. Chan and Dr. K.F. Chung of the Hong Kong Polytechnic University for structural adequacy.

The replacement of partly bamboo structural posts by steel tubes can strengthen the stability of overall scaffolding structure. However, the financial burden to such huge investment for the whole building has limited the application of this mixed scaffolding structure partly. (most popular at one half or a third of the overall height of building instead)

In fact, light duty scaffolding structures such as traditional drama shed, neon signage or suspended working platform are still the first choice for bamboo scaffold as compare to structures like residential building etc. It is not easy to find the application of metal or mixed scaffolding on those light duty scaffolding structures.

Photo 1: Metal Bamboo Matrix System Scaffolds (MBMSS) can provide better supporting load for the scaffolding structure at the top and an ideal place for anchorage of safety harness at the connection of metal tubes.

Photo 2: The most popular usage of bamboo scaffolding in new building under construction for formwork fixing and plastering

Photo 3: The usage of bamboo scaffolding in renovation works of old building.

Photo 4: Bamboo scaffolding application in slope maintenance

Photo 5: Bamboo scaffolding application in neon signage installation

Photo 6: Suspended bamboo scaffolding working platform

Photo 7: Bamboo scaffolding application in traditional Chinese drama shed

SAFETY ISSUE

Bamboo scaffolding has a public image of being prone to collapse, causing higher casualties than steel scaffolding, which is, in fact, subject to flaw. There have been many instances of collapse in steel scaffolding around the world. For instance, there are more than 100 reported cases per annum in UK as revealed in the engineering journals. In Hong Kong, there were 2 known cases of metal scaffolding collapse during the construction stage of the KCRC West Rail Project.

The layman often has the impression that bamboo scaffolding is less safe than steel scaffolding and think that "Even when steel scaffold collapses, we cannot do much about it since we have used the strongest building material-steel". When using bamboo scaffolding, the layman may wonder why steel, a stronger material, is not used instead. I am not a defender of bamboo, but to criticize the application of traditional bamboo scaffold as unsafe and contributing to the high accident rate is really a biased, irresponsible and unprofessional allegation. It should not be made by any sensible man in construction industry. To distort the usage of bamboo in scaffolding has only one excuse – "This method is not westernized".

From the engineering point of view, safety depends on the calculated factors of safety, quality control and site practice since stronger material normally has lesser dimensions and construed as more slender in this competitive world.

CONCLUSIONS

The existence of bamboo scaffold remains as the first option of choosing scaffolding structure by contractors due to its popularity, highly competitive in economical consideration and greater flexibility. With the emphasis of safety in construction industry in Hong Kong, the quality of bamboo material to sustain high durability becomes extremely important for high-rise building construction. Therefore, the mixed system scaffolding that metal tubes serve as structural support can relieve any worries arise from the safety of scaffold caused by intentional damages by other workers and self-degradation itself. On the other hand, the usage of bamboo scaffolding structures on light duty works such as signage installation and drama shed can demonstrate this craft and art, must be inherited continuously to attract public interests.

ACKNOWLEDGEMENT

The author acknowledges the valuable comments of Professor S.L. Chan on the draft of this paper.

Advances in Building Technology, Volume I
M. Anson, J.M. Ko and E.S.S. Lam (Eds.)

STRUCTURAL DESIGN OF THE BAMBOO PAVILLION, BERLIN

David G Vesey

Ove Arup & Partners Hong Kong Limited

ABSTRACT

A small theatre pavillion was constructed in Berlin in August 2000 and then in Hong Kong in November as part of an Arts Festival jointly organised by the Haus der Kulturen der Welt and the Hong Kong Institute of Contemporary Culture. The Architect was Rocco Yim with Arup Hong Kong providing the Structural Design and Gammon as Builders. The design required the use of this traditional material in new ways to provide a light and elegant structure. Large sizes of bamboo were sourced in China, traditional joints were tested and relatively new types of joint devised.

KEYWORDS

Bamboo, Engineered Structure, Traditional material used in a modern way, Pavillion, Joints, Testing

INTRODUCTION

As part of the Millennium celebrations the Hong Kong Institute of Contemporary Culture and the Haus der Kulturen der Welt of Berlin jointly hosted a festival called "One Vision Two Cities, Festival 2000 Hong Kong In Berlin and Berlin in Hong Kong" The first half of the festival was held in Berlin in August 2000 and the second part held in Hong Kong in November.

Part of the Festival included the construction of a bamboo theatre artwork showing contemporary usage and artistic interpretations of this structural material so traditional in Chinese culture.

The vision of Rocco Yim, the Architect was to have a more open structure than the traditional bamboo lattice work of closely spaced horizontal and vertical members. In order to achieve this the structural design is interesting in that it combines the traditional material and method of jointing bamboos with modern structural technology using fabricated structural steel connections and high strength bracing cables.

The Pavillion uses traditional Chinese bamboo construction methods. It is in the form of an open theatre building, see figures 1, 2. It was erected in an ornamental pool at the Haus der Kulturen der Welt in Berlin where it stood from July 2000 to Spring 2001. The design was re-enacted at the Tamar Site in Hong Kong in November.

The Pavillion won the 2002 Award for Members work Outside of Hong Kong, given by the Hong Kong Institute of Architects.

BAMBOO AS A STRUCTURAL MATERIAL

Bamboo has been widely used in Hong Kong and China for scaffolding because of its light weight, structural efficiency and cost. It is still the material of choice for many Hong Kong Projects, see figure 3. It is also used for quite large temporary buildings at various festivals in Hong Kong, see figures 4, 5. It is used for housing in various parts of the world. The Architect Simon Velez has used it to produce some large and beautiful buildings in South America as described in the book "Grow your own House" published by Vitra Design Museum.

The bamboo for the Pavillion was sourced from Guan Xi in Southern China. Large poles over 120mm diameter and 8m long were required. They were treated in Hong Kong and shipped to Hamburg.

A team from Australia had designed and built a bridge over the river Spree in Berlin as part of a previous festival organized by Haus der Kulturen der Welt, see figure 6.

STRUCTURE

The bamboo theatre artwork is an open ended pitched roof structure occupying a footprint plan area of 22m by 22m. It consists of transverse frames of bamboo and is about 9m high at its highest point, the roof ridge. Poles at a pitch of about 25° span from the ridge onto the inner upper ends of side rakers and thence onto outward sloping palisade walls. The palisade walls form convex curves in plan and the roof poles fan outward from the ridge to match, see figures 1, 2, 7. The whole bamboo artwork springs from structural steel "shoes", which in turn sit on concrete plinths founded on the strong base of an ornamental pool, see figure 8. A timber floor is supported separately from the pool base.

Loads arise from the self weight of the bamboo and from wind pressure on the open lattice bamboo framework. The forces which can arise during storms or typhoons govern the design.

Although the bamboo theatre artwork was a temporary sculpture it was quite large and the public could walk and sit under it. It was to be constructed generally using traditional materials and jointing methods. This led to requirements for the design as follows:

- The artwork is a temporary structure and public safety must be ensured.
- As an artwork it is important to realize the Artist's vision
- Bamboo constructions have been successfully used for many years but are difficult to analyse and design precisely. This is recognized in Hong Kong and such structures are permitted without normal buildings approval so long as they are constructed by experienced bamboo masters.

The structural design method, or philosophy, needs to respect these requirements. The approach taken was therefore:

- Define the structural system, loadpaths and structural diagrams and illustrate these clearly with sketches, see figure 9.
- Adopt reasonable minimum loads for the artwork, for example do not design for roof access, other than the point load of an adventurous youth!
- Carry out relatively simple and clear calculations to establish reasonable member design forces and structural behaviour.
- Compare design forces with calculated bamboo pole capacities.
- Substantiate required connection joint capacities with simple load tests.
- Carry out and record a trial erection.

JOINTS AND CONNECTIONS

The bamboo jointing method was of particular interest because the German Building Authorities required a quite rigorous justification.

In the past traditional bamboo joints were made by soaking strips of bamboo and lashing around the poles. The arrangement and number of lashes was decided by the bamboo master based on experience and rules of thumb. More recently galvanized mild steel wire was used and today the lashing material is a 6mm wide flat strip of black polypropylene – polyethylene (PP-PE).

Initially a series of load tests were carried out on joins made with the PP-PE ties, and these satisfied strength requirements. However concerns on fire resistance were raised and led to a change to use mild steel wire ties. Further tests were also carried out on these. See figure 10, 11.

The base connection joint was made using a short steel tube welded to a base plate, see figure 8. The large ends of the bamboo poles were grouted into these sockets. Where required fin plates were also attached to these steel shoe to allow the tension cables to be anchored. The tops of the poles forming the walls were connected in pairs to increase lateral stability under the loads from bracing cables, see figure 12. Figure 13 shows connection between the roof and palisade walls.

CONSTRUCTION

A trial erection was carried out in June 2000 in Hong Kong and provided a useful learning experience for the design and construction team. Photographs and video recordings were sent to Berlin to assist the Proof Engineer in demonstrating the feasibility to the Building Authorities. The trial erection helped to confirm that additional pairs of guy cables from the ridge to ground at each end were not required. See figure 14. Figure 15 shows the Pavillion during the opening ceremony in Berlin.

ACKNOWLEDGEMENTS

Rocco Design Ltd: The Architects

Gammon Construction Ltd: Who supplied the bamboo, tested joints and built the pavilion.

World Pacific Scaffolding Works Ltd: The specialist bamboo erectors.

Ove Arup & Partners Hong Kong Ltd: Who carried out structural analysis and design, devised and supervised joint tests and liased with HKW in Berlin.

Haus Der Kulturen Der Welt: Who organized the Berlin end.

Bugler und Jaeck: Who provide the structural proof engineering service in Berlin.

Festival of Vision Hong Kong Berlin: Who organized the Festival at the Hong Kong end.

FIGURE 1

FIGURE 2

FIGURE 3

FIGURE 4

FIGURE 5

FIGURE 6

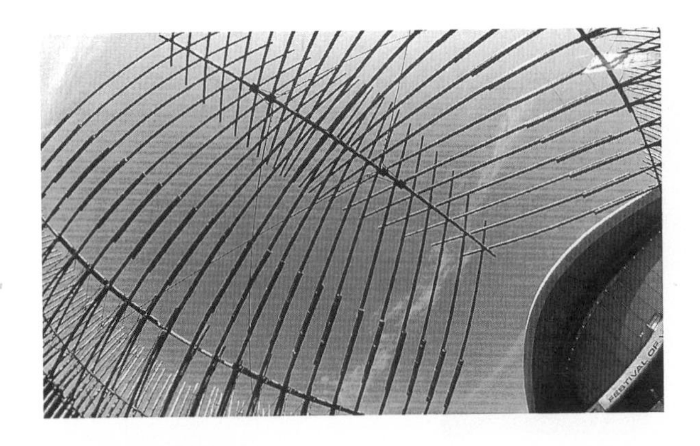

FIGURE 7

FIGURE 8

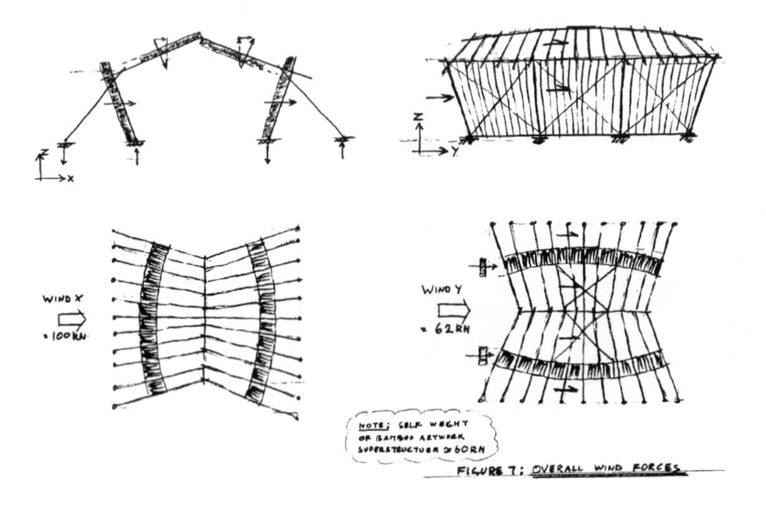

FIGURE 9

FIGURE 10

FIGURE 11

FIGURE 12

FIGURE 13

FIGURE 14

FIGURE 15

Advances in Building Technology, Volume I
M. Anson, J.M. Ko and E.S.S. Lam (Eds.)

FULL SCALE TESTS OF BAMBOO SCAFFOLDS FOR DESIGN DEVELOPMENT AGAINST INSTABILITY

Yu WK.[1], Chung KF.[1] and Tong YC.[2]

[1] *Department of Civil and Structural Engineering, the Hong Kong Polytechnic University, Hung Hom, Kowloon, Hong Kong*
[2] *Construction Industry Training Authority, Aberdeen, Hong Kong.*

ABSTRACT

Bamboo scaffolding is widely used in construction in the South East Asia, in particular, the Southern China and Hong Kong for many decades. However, bamboo scaffolds are generally erected by scaffolding practitioners based on their intuition and experiences without any structural design. In order to standardize the design and erection of bamboo scaffolds and to improve the safety of bamboo scaffolds, a research and development program to promote the effective use of structural bamboo was carried out. As part of the program activities, four full-scale bamboo scaffolds were built and tested to failure in order to examine the structural behaviour of bamboo scaffolds in site condition. The load carrying capacities of these bamboo scaffolds with different member configurations and lateral restraint arrangements were assessed and compared against practical load requirements and support conditions. Details of the full-scale tests and their test results are presented in this paper.

KEYWORDS

Bamboo scaffolds, structural bamboo, bamboo columns, full-scale tests, load proportion

INTRODUCTION

Bamboo scaffolds are widely used in construction in the South East Asia, in particular, the Southern China and Hong Kong for many decades. It is believed among Chinese that the first bamboo scaffold was built some 5000 years ago while the basic framing systems and the erection methods were established through practice about two thousand years ago. Bamboo scaffolds are traditionally erected by specialized scaffolding practitioners without any structural design, the structural adequacy and thus the safety of the bamboo scaffolds depend primarily on the skills of individual practitioners. The knowledge is passed onto young workers through an apprentice system, mostly through on-the-job training.

Bamboo scaffolds
Bamboo scaffolds provide temporary access, working platforms for construction workers and supervisory staff, and also prevent construction debris from falling onto passers-by. Figures 1 and 2

illustrate typical applications of bamboo scaffolds in Hong Kong [1], namely, Single Layered Bamboo Scaffolds (SLBS) and Double Layered Bamboo Scaffolds (DLBS) respectively. Both Kao Jue and Mao Jue are commonly used as the vertical members, or the posts, and the horizontal members, or the main ledgers, in bamboo scaffolds. They are light enough for one person to handle a single member at a time. Due to the ease of handling, bamboo scaffolds are erected and dismantled quickly; compared to steel scaffolds, where installation and dismantling take the same amount of time, bamboo scaffolds can be dismantled in one tenth of the time it takes to install. No machinery, power-driven tools and tightening equipment are required as only simple hand tools and nylon or wire ties are used.

Consequently, in spite of open competition with many metal scaffolding systems imported from all over the world, bamboo scaffold remains to be one of the most preferred access scaffolding systems in building construction in Hong Kong and the neighbouring areas. At present, the typical height of bamboo scaffolds is 15 m, and installation of steel bracket supports at regular intervals allows full coverage of the height of a building.

In Hong Kong, the usage of DLBS [2] is to provide access and safe working platforms for workers to carry out simple or *'light duty'* construction activities, such as plastering, and the construction load at each working platform should not be less than 1.5 kPa for design purpose [3].

Structural bamboo

As bamboo columns are natural non-homogenous organic materials, large variations of physical properties along the length of bamboo members are apparent: external and internal diameters, dry density and moisture content. Two bamboo species, namely Kao Jue and Mao Jue, are commonly used in access scaffolds in Hong Kong. A pilot study was carried out by Chung and Yu [4] to examine the variation of compressive strength against various physical properties along the length of bamboo culms for both Kao Jue and Mao Jue. Moreover, systematic test series with a large number of compression and bending tests were also executed [5,6] to establish characteristic values of both the strengths and the Young's moduli of each bamboo species for limit state structural design. As shown in Table 1, both Kao Jue and Mao Jue are good constructional materials with excellent mechanical properties against compression and bending.

OBJECTIVES AND SCOPE OF STUDY

In order to achieve overall structural stability of bamboo scaffolds with high structural efficiency, lateral restraints should be provided at close intervals. However, in practice, it may not be practical or even possible to provide lateral restraints to all connections between the posts and the main ledgers of the bamboo scaffolds. In the absence of sufficient lateral restraints, the effective length of bamboo columns will be larger than their member lengths between ledgers, and thus, the axial buckling resistances of the bamboo columns are reduced significantly.

Consequently, four full-scale bamboo scaffolds were built and tested to failure in order to examine the structural behaviour of bamboo scaffolds with practical lateral restraint configurations. The load carrying capacities of these bamboo scaffolds with different lateral restraint configurations will be assessed and compared against practical load requirements. Details of the full-scale tests and their test results are presented in this paper.

TEST PROGRAM

A total of four 6 m wide by 9 m high DLBS were erected in the Aberdeen Centre of the Construction Industry Training Authority, a statutory public body of the of Hong Kong SAR. The first two tests were performed on DLBS with typical configurations commonly erected in building sites in Hong

Kong, where Kao Jue and Mao Jue were used as the main posts of the inner and the outer layers respectively. The only difference between the first two tests was the number of putlogs provided to the posts of the outer layer. The first two tests were designated as *MJ3R* and *MJ2R* as both of them used Mao Jue as the posts of the outer layers, but with three and two putlogs provided to Mao Jue respectively. Refer to Figure 3 for details of the lateral restraint arrangements. For easy reference, the lateral restraint arrangement with three putlogs per post (over a height of 9 m) is referred as the full restraint condition while that with two putlogs per post is referred as the practical restraint condition.

In the third and the fourth full scale tests, it was decided to use Kao Jue as the posts of the outer layers with three and two lateral restraints provided to Kao Jue, and thus, they were designated as *KJ3R* and *KJ2R* respectively. Such bamboo scaffolds were considered to be useful in situations where there was only one kind of structural bamboo available. All bamboo scaffolds were loaded with sand-bags at the top level working platform in close proximity of the central posts, and the reaction forces at the bottom of both the inner and the outer central posts of the DLBS were measured using load cells. The general view of the test specimens is presented in Figure 4.

In order to provide practical results for general application, all bamboo culms were air-dried for at least 3 months before testing in order to allow the bamboo members to achieve typical moisture contents commonly found in practice. Moreover, after each test, at least three short bamboo culms were cut out from the failed members, and compression tests on these short culms were carried out to evaluate their compressive strengths. The length of each compression test specimen was about twice the external diameter of the bamboo culm, but not larger than 75 and 150 mm for Kao Jue and Mao Jue respectively.

TYPICAL LOAD CARRYING CAPACITIES OF BAMBOO SCAFFOLDS

From the four tests, the structural performance of all the DLBS was found to be satisfactory in terms of strength and deformation. Despite of large deformation in the posts (Kao Jue) due to axial buckling and flexural failure in the main ledgers (Kao Jue) over supports under large hogging moment, the scaffolds were found to be able to sustain additional loading without global collapse, as shown in Figure 5a. Moreover, two typical modes of failure were identified as shown in Figure 5b and 5c, namely, the overall buckling of Kao Jue between two working platforms and the slippage of the post – main ledger connection.

The failure load in test *MJ3R* was found to be 4.07 kN with apparent buckling of Kao Jue in the inner layer. As the effective loaded area of each post is equal to 1.5 m × 0.3 m or 0.45 m^2, the load carrying capacity of the DLBS with Mao Jue as posts and three lateral restraints per post is equal to 9.04 kPa.

Similarly, the failure load in test *MJ2R* was found to be 2.86 kN with apparent buckling of Kao Jue in the inner layer, and thus, the load carrying capacity of the DLBS with Mao Jue as posts and two lateral restraints per post is equal to 6.35 kPa.

For tests *KJ3R* and *KJ2R*, the failure loads were found to be 2.98 and 2.53 kN respectively, and thus, the load carrying capacities of the DLBS with Kao Jue as posts and with three and two lateral restraints per post are equal to 6.62 and 5.62 kPa respectively.

TESTS RESULTS AND DISCUSSIONS

Table 2 summarizes the load carrying capacities of the bamboo scaffolds. As the safe load carrying capacities in tests *MJ3R* and *MJ2R* are found to be 5.65 and 3.98 kPa respectively, a 30% reduction in the safe load carrying capacities of the typical DLBS should be allowed for when the full lateral restraint condition is modified to the practical restraint condition. Moreover, as the safe load carrying capacities in tests *KJ3R* and *KJ2R* are found to be 4.14 and 3.51 kPa respectively, there is only 15% reduction in the safe load carrying capacity when the full restraint condition is modified to the practical restraint condition. Consequently, it is shown that the reduction in the safe load carrying capacity of DLBS due to the change of lateral restraint arrangement is related to the member configurations of the bamboo scaffolds.

It should be noted that as the minimum load requirement for a *'light duty'* assess scaffold is 1.5 kPa (unfactored), all of the four bamboo scaffolds are shown to be structurally adequate to support two fully loaded working platforms. Only the bamboo scaffold in test *MJ3R* is shown to be structurally adequate to support three fully loaded working platforms.

In order to examine the load distribution between the end and the central posts, the support reactions of the central posts in both the inner and the outer layers were measured, and the support reactions are summarized in Table 3. For test *MJ3R*, it is found that only 40% of the applied load (from the top working platform) is resisted by the central inner and the central outer posts. There is a significant spreading of loads from the central posts to the end posts through connected ledgers and diagonals. Similar conclusions may be drawn in test *KJ3R* as about 50% of the applied load is distributed away from the inner and the outer central posts.

Furthermore, it is interesting to compare the load distribution between the inner and the outer posts. Table 3 shows that in test *MJ3R*, due to the high in-plane stiffness of the outer layer provided by Mao Jue as posts and diagonals, the support reaction of the outer post is 2 to 2.75 times to that of the inner post. However, in test *KJ3R*, the support reactions of the inner and the outer posts are practically the same. Consequently, the applied loads may be conservatively assumed to be equally distributed between the inner and the outer layers of bamboo scaffolds.

Last but not the least, although the bamboo scaffolds were declared to be failed when buckling of bamboo members was apparent, it should be noted that there was no overall collapse of the bamboo scaffolds even after all the sand-bags were placed (i.e. a total of 60 sandbags with each of them weighted 20 kg); though significant local deformation in one or two bamboo members was observed. Consequently, it is shown that as the configuration of the bamboo scaffolds allows many members to be connected through standards and ledgers, loading will always be effectively re-distributed away from any failed member, and thus, the bamboo scaffolds can take up additional loads after local failure, i.e. *multiple load paths with high structural redundancy*. This is one of the most important features of bamboo scaffolds in providing sufficient safety margin against overall structural failure. Meanwhile, lateral restraints to the bamboo scaffolds should always be sufficient and secure. Otherwise, the load carrying capacities of bamboo scaffolds will be reduced considerably.

It should be noted that the load carrying capacities of bamboo scaffolds depend on the following:

- mechanical properties of bamboo culms,
- member configurations of bamboo scaffolds including bracing members,
- support conditions of bamboo scaffolds including lateral restraint arrangements,
- buckling of bamboo columns, and
- resistances of post – main ledger connections.

For details of rational design and erection of bamboo scaffolds, refer to the design guide [7] and the erection manual [8] jointly published by the Research Centre for Advanced Technology in Structural Engineering of the Hong Kong Polytechnic University and the International Network for Bamboo and Rattan, INBAR. The results obtained from the four full-scale tests have been used to calibrate the design rules proposed in the design guide.

CONCLUSIONS

Based on the full-scale tests of four Double Layered Bamboo Scaffolds (DLBS), the typical bamboo scaffolds are shown to be structurally adequate to support two to three *'light duty'* working platforms. Due to the high structural redundancy in bamboo scaffolds, loading will be effectively re-distributed away from any failed member, and thus, the bamboo scaffolds can take up additional loads after local failure in bamboo members. This is one of the most important features of bamboo scaffolds in providing sufficient safety margin against overall structural failure.

Through rational design and erection, bamboo scaffolds will be transformed into modern building systems with high structural safety and buildability. They will continue to serve the construction industry in Hong Kong and the neighbouring areas with reliable and innovative applications.

ACKNOWLEDGEMENTS

The research and development project leading to the publication of this paper is partially supported by the International Network for Bamboo and Rattan (Project No. ZZ04), and also by the Research Committee of the Hong Kong Polytechnic University Research (Project No. G-V849). The authors would like to thank Professor J.M. Ko of the Hong Kong Polytechnic University and Professor J.J.A. Janssen of the Eindhoven University of Technology, co-chairmen of the Steering Committee of the INBAR program, for their general guidance and technical advice. Moreover, the authors would like to express their gratitude to the Construction Industry Training Authority for the execution of the tests. Technical advice on practical erection of bamboo scaffolds from Mr Y.C. Siu is also gratefully acknowledged.

REFERENCES

1. Tong, Albert Y.C. (1998). *Bamboo scaffolding - Practical Application.* Proceedings of the Symposium on '*Bamboo and Metal Scaffolding*', the Hong Kong Institution of Engineers, October 1998, pp43-62.
2. Labour Department of the Government of Hong Kong, 1995. *Code of practice for scaffolding safety*: Factory Inspectorate Division.
3. *BS5973: 1993: Code of practice for assess and working scaffolds and special scaffold structures in steel*: the British Standards Institution.
4. Chung KF and Yu WK. (2002). Mechanical properties of structural bamboo for bamboo scaffoldings. *Engineering Structures* 24, 429-442.
5. Yu WK. and Chung KF. (2000). *Qualification tests on Kao Jue under compression and bending.* Technical Report, Research Centre for Advanced Technology in Structural Engineering, the Hong Kong Polytechnic University.
6. Yu WK. and Chung KF. (2000). *Qualification tests on Mao Jue under compression and bending.* Technical Report, Research Centre for Advanced Technology in Structural Engineering, the Hong Kong Polytechnic University.
7. Chung KF. and Chan SL. (2002). *Bamboo Scaffoldings in Building Construction: Design of Bamboo Scaffolds.* Joint Publication, the Hong Kong Polytechnic University and International Network for Bamboo and Rattan (in press).
8. Chung KF. and Siu YC. (2002). *Bamboo Scaffoldings in Building Construction: Erection of Bamboo Scaffolds.* Joint Publication, the Hong Kong Polytechnic University and International Network for Bamboo and Rattan (in press).

TABLE 1 Proposed mechanical properties for structural bamboo

	Kao Jue		Mao Jue	
	Dry	Wet	Dry	Wet
Characteristic compressive strength, $f_{c,k}$ (N/mm^2) (at fifth percentile)	79	35	117	44
Design compressive strength, $f_{c,d}$ (N/mm^2) (γ_m = 1.5)	53	23	78	29
Design Young's Modulus against bending, $E_{b,d}$ (kN/mm^2) (average value)	22.0	16.4	13.2	9.6

TABLE 2 Summary of test results

Test	Measured P_i with apparent column buckling (kN)	Load carrying capacity (factored) (kPa)	Safe load carrying capacity (unfactored) (kPa)
MJ3R	4.07	9.04	5.65
MJ2R	2.86	6.36	3.98
KJ3R	2.98	6.62	4.14
KJ2R	2.53	5.62	3.51

TABLE 3a Summary of measured support reactions in test *MJ3R* – Initial ioading

Total applied load P_T (kg)	Total applied load P_T (kN)	P_o (kN)	P_i (kN)	P_o (corrected) (kN)	P_i (corrected) (kN)	$P_o + P_i$ (kN)	$\frac{P_o + P_i}{P_T}$	$\frac{P_o}{P_i}$
0	0	1.233	0.147	0.000	0.000	0.000	-	-
250	2.45	1.818	0.499	0.585	0.352	0.937	0.38	1.66
330	3.24	2.096	0.656	0.863	0.509	1.372	0.42	1.70
410	4.02	2.300	0.730	1.067	0.583	1.650	0.41	1.83
490	4.81	2.754	0.776	1.521	0.629	2.150	0.45	2.42
570	5.59	3.023	0.804	1.790	0.657	2.447	0.44	2.72
650	6.38	3.265	0.887	2.032	0.740	2.772	0.43	2.75

TABLE 3b Summary of measured support reactions in test *KJ3R*– Initial ioading

Total applied load P_T (kg)	Total applied load P_T (kN)	P_o (kN)	P_i (kN)	P_o (corrected) (kN)	P_i (corrected) (kN)	$P_o + P_i$ (kN)	$\frac{P_o + P_i}{P_T}$	$\frac{P_o}{P_i}$
0	0	0.545	0.102	0.000	0.000	0.000	-	-
240	2.35	0.822	0.510	0.277	0.408	0.685	0.29	0.68
320	3.14	1.229	0.769	0.684	0.667	1.351	0.43	1.03
400	3.92	1.562	0.992	1.017	0.890	1.907	0.49	1.14
480	4.71	1.729	1.437	1.184	1.335	2.519	0.53	0.89
560	5.49	2.042	1.558	1.497	1.456	2.953	0.54	1.03
640	6.28	2.311	1.743	1.766	1.641	3.407	0.54	1.08

Notes: P_T = Total applied load due to the self-weight of sand-bags.
P_o = Measured reaction of the post (Mao Jue or Kao Jue) at the outer layer.
P_i = Measured reaction of the post (Kao Jue) at the inner layer.

Figure 1: Single Layered Bamboo Scaffold (SLBS)

Figure 2: Double Layered Bamboo Scaffold (DLBS)

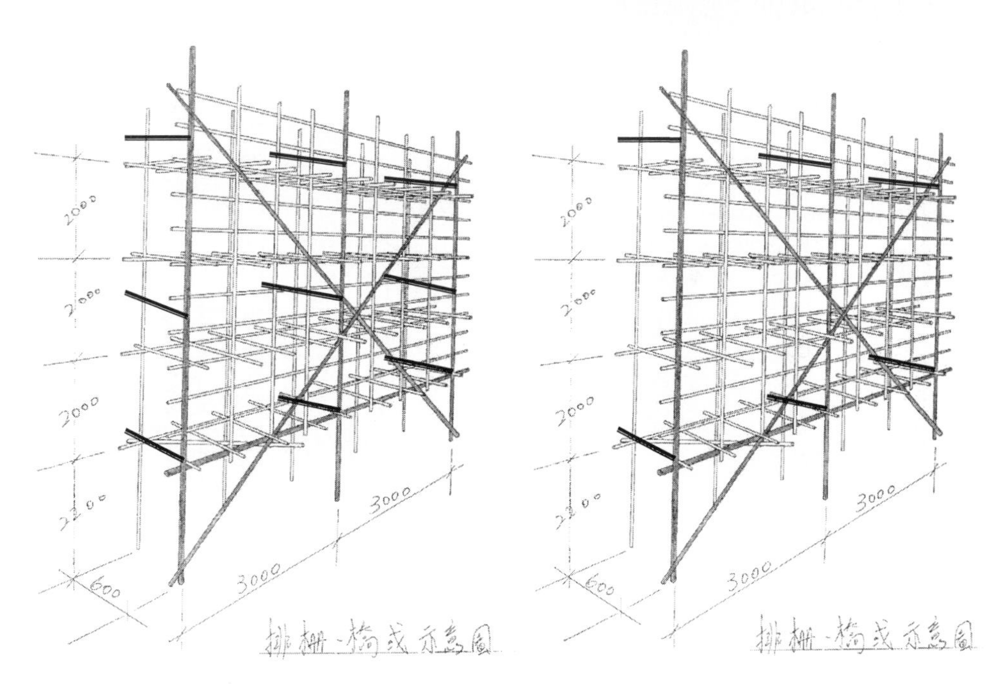

a) Three putlogs per post are provided

b) Two putlogs per post are provided

Figure 3: General layout of DLBS with typical lateral restraint arrangements

Remarks: ═══ denotes the position of a putlog

Outer layer: main posts at a horizontal spacing of 3.0 m and main ledgers at a vertical spacing of 2.0 m.

Inner layer: main posts at a horizontal spacing of 1.5 m and main ledgers at a vertical spacing of 2.0 m.

Diagonals: Mao Jue was used.

a) Typical arrangement of DLBS

a) Flexural failure of main ledgers initiating load re-distribution at large deformation – Test *MJ2R*

b) Supports to posts

b) Buckling of a Kao Jue (inner layer) – Test *KJ2R*

c) Typical arrangement of a putlog

c) Slippage in a post-main ledger connection – Test *MJ2R*

Figure 4: General view of DLBS

Figure 5: Typical failure modes in DLBS

CONSTRUCTION TECHNOLOGIES:

CONSTRUCTION TECHNOLOGIES/PLANTS

Advances in Building Technology, Volume 1
M. Anson, J.M. Ko and E.S.S. Lam (Eds.)

CAUSES OF LUMPS AND BALLS IN HIGH SLUMP TRUCK-MIXED CONCRETE: AN EXPERIMENTAL INVESTIGATION

Irtishad Ahmad, Salman Azhar and Ivan R. Canino

Department of Civil and Environmental Engineering,
Florida International University, Miami, Florida, USA

ABSTRACT

In high slump truck-mixed concrete, mostly used for drilled shafts (deep foundation system for heavy loads), lumps and balls typically the size of a lemon to a baseball, are frequently found. Such lumps and balls jeopardize the structural integrity of concrete by forming weakened zones and by increasing the potential for soil intrusion. Florida Department of Transportation (FDOT) requires that the concrete should be free from lumps. Hence when lumps are found, concrete batches are to be rejected consequences of which may cause disruption of work, costly rework, and loss of valuable time. Occurrences of lumps and balls must be avoided to ensure the best interest of all the parties involved – the concrete producer, contractor and the client. This experimental study was conducted to find the causes of lumps and balls formation in high slump truck-mixed concrete and to develop procedures to avoid them. Experiments were conducted at the CSR Rinker Concrete Plant in Miami, Florida and 17 concrete batches were tested. The testing variables were the materials discharge rate, truck load, headwater content and mixing revolutions. It was found that using standard discharge rate and truck load, a headwater content of 30-40% and 80-100 initial mixing revolutions at a speed of 12 rpm could produce lumps-free concrete. On the basis of these experiments, a set of remedies is suggested and recommendations are made for further experimentation.

KEYWORDS

High slump concrete, Truck-mixed concrete, Lumps and balls, Concrete uniformity, Concrete mixing

INTRODUCTION

While ready mixed concrete can be produced in a number of ways, the most common production methods are the *central-mixed concrete* and the *truck-mixed concrete*. In the United States, the overwhelming majority of ready mixed concrete producers use truck-mixed concrete due to its low capital cost and greater coverage area. Truck mixing allows more flexibility to a concrete producer to produce customized specifications concrete (NRMCA 2002).

In the truck mixed concrete, all raw ingredients are charged directly in the truck mixer. Most or all water is usually batched at the plant. The mixer drum is turned at charging (fast) speed during the loading of the materials. There are three production techniques for truck mixed concrete:

Concrete mixed at the job site: While traveling to the job site the drum is turned at agitating speed (slow speed). After arriving at the job site, the concrete is completely mixed. The drum is then turned for 70 to 100 revolutions, or about five minutes, at mixing speed to ensure uniform mixing.

Concrete mixed in the yard: The drum is turned at high speed or 12-15 rpm for 50 revolutions. This allows a quick mixing of the batch. The concrete is then agitated slowly while driving to the job site.

Concrete mixed in transit: The drum is turned at medium speed or about 8 rpm for 70 revolutions while driving to the job site. The drum is then slowed to agitating speed till it reached the job site.

Mixing concrete in a truck is different from mixing it in a central batch plant. Mixing blades in a plant lift and drop the concrete. Width limitations on truck mixers dictate the use of spiral blades that first move the concrete down towards the head end of the drum then back up the central axis towards the discharge end. Therefore, concrete mixed in a truck mixer is normally not as uniform as concrete from a central batching plant (Gaynor 1996).

The non-uniformity of truck mixed concrete causes the development of *head packs* and *concrete balls*. Head packs refer to sand streaks that appear when discharging the final portions of the concrete. Head packs are produced when sand is loaded before coarse aggregates. The sand packs in the head of the drum and breaks loose after about half the load has been discharged. Usually the head pack goes unnoticed because the sand gets mixed into the concrete before it reaches the chute. Despite the absence of visual clues, head packs are undesirable because they cause variations in slump, air content, and strength. Head packs can be avoided by loading sand and cement together after 50-75% of the coarse aggregates and water has been charged in the truck (Gaynor 1996).

Concrete balls are round lumps of cement, sand and coarse aggregates, typically the size of a lemon to a baseball. Such lumps and balls jeopardize the structural integrity of hardened concrete by forming weakened zones and by increasing the potential for soil intrusion. Moreover, they reduce the freeze-thaw resistance of hardened concrete. Their occurrence usually indicates problems with batching and mixing, such as improper batching sequencing or not enough revolutions of the mixing drum (Suprenant 1992). Gaynor (1996) indicates that in low-slump concrete (2-3 inch slump), the concrete balls are usually grinded during mixing operation. However, in high slump concrete (over 6 inch slump), such balls doesn't breakup due to high fluidity of concrete. Hence a special loading and mixing sequence is desirable for high slump concrete to produce lumps-free concrete.

RESEARCH OBJECTIVES AND SIGNIFICANCE

Florida Department of Transportation (FDOT) Standard Specifications for Road Bridge Construction (2000) section 346-7.4.1 requires that concrete should be free from lumps and balls of cementations material. When lumps and balls are found, concrete batches are to be rejected, consequences of which may cause disruption of work, costly rework and loss of valuable time. This research was conducted with the objective of determining the reasons for the formation of lumps and balls in high slump truck mixed concrete and to find the remedies to avoid them. The production of lumps-free concrete is in the best interest of all parties involved in the project, i.e. the concrete producer, contractor and the client. There is very little published material available on this topic, hence this research paper is expected to be a valuable source of information for the concrete producers.

METHODOLOGY

The uniformity of truck-mixed concrete depends on a number of factors such as the Mixing sequence, Mixing time, Charging and mixing speed, Discharge rate (speed at which the materials are charged into the mixing truck through chute), Concrete load (total volume of concrete), Initial revolutions (number of revolutions of truck mixer after addition of materials but before slump adjustment), and Headwater content (water added to the mixer before the addition of materials). The first three factors have been extensively studied by a number of researchers (Gaynor 1996; Suprenant 1992; Gaynor and Mullarky 1975) and standardized by different organizations such as FDOT. Hence the other four factors were selected as variables for this research. All the experimentation was carried out at CSR Rinker Concrete Plant in Miami, Florida, USA.

Concrete Mix Design

FDOT standard mix design for drilled shafts (Class 4: #06-0281) was used throughout the study. The mix proportions are shown in Table 1.

TABLE 1
MIX PROPORTIONS FOR FDOT CLASS IV CONCRETE (PER CUBIC YARD)

Cement (Type I)	298 lbs	Coarse aggregates	1667 lbs	Superplasticizer	23.8 oz
Slag	447 lbs	Fine aggregates	1053 lbs	Total water content	308 lbs
Slump	7-9 inch	Compressive strength	4000 Psi	Air content	3.6%

Concrete Truck Mixer Specifications

A single 10 cubic yards truck mixer was used throughout the study. The truck had a drum volume of 473 ft^3, an agitating speed of 2-6 rpm, mixing speed of 12-16 rpm and water storage capacity of 80 gallons. The truck mixer was manufactured by McNeilus Inc. (Model KX6-414) and approved by National Ready Mix Concrete Association (NRMCA).

Mixing Sequence

The standard mixing sequence adopted by CSR Rinker concrete plant was used in this study. The details are as follows:

1. With the drum rotating at charging mode (2-4 rpm), headwater and admixtures were introduced.
2. Next the discharging of coarse and fine aggregates into the truck was initialized.
3. While this process continued, cementitious materials were ribbon fed into the aggregates stream. The aggregates continued to discharge in the truck during the whole process (even when the cementitious materials were finished).
4. Finally, the tail water was added.

The truck mixed the concrete using a certain number of initial revolutions at a mixing speed of 12 rpm. This was followed by slump water adjustment (the concrete was initially batched for a slump of 3" as the computerized batching program is designed for this slump) to increase the concrete slump up to 7-9 inch. That water is termed as *jobsite allowable.* Additional 30 revolutions were given at mixing speed (12 rpm) after each addition of jobsite allowable water until the desired slump was achieved.

Experimental Details

The experimentation was conducted in two phases. In the first phase, discharge rate, concrete load and the headwater content were taken as variables. While in the second phase, headwater content and initial revolutions were selected as the variables. The following tests were carried out in each phase: *Slump test, Air content test and the Density test.* The standard ASTM testing procedures were followed. The concrete was passed through a 2.5 inch sieve to collect concrete lump and balls and then discharged into another empty truck as illustrated in Figure 1. The size and number of lumps and balls were recorded for each batch and selected samples were sent to the laboratory for sieve analysis to determine their composition.

Figure 1: Test arrangement to collect concrete lumps and balls

TEST RESULTS

The test results are summarized in Table 2. The effect of each variable on the number of lumps formed is explained below:

Discharge Rate

Figure 2a shows the relation between the discharge rate and the number of concrete lumps. It is evident from the figure that the optimum discharge rate to minimize concrete lumps is 200 lbs/sec. Hence this discharge rate was used throughout further experimentation.

Concrete Load

The relation between the concrete load and the number of lumps (normalized to load size) came out to be linear as depicted in Figure 2b. This is very straightforward as the smaller concrete load would have less number of concrete lumps and balls. The loads of 3 yd^3 and 5 yd^3 are not commercially feasible and hence rejected. Since the normalized number of lumps between the loads of 7 yd^3 and 9 yd^3 were same, it was decided to select 9 yd^3 as the optimum concrete load.

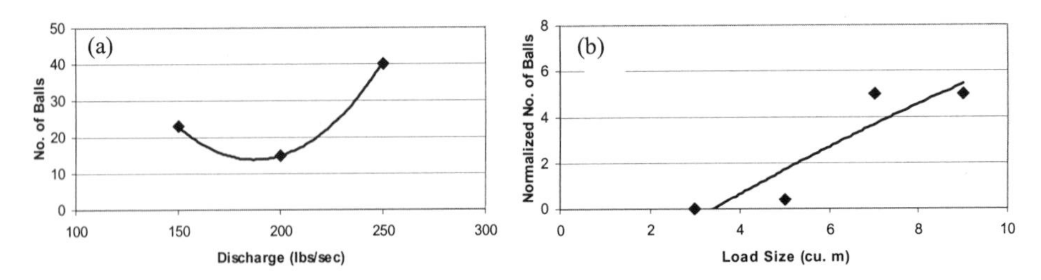

Figure 2: Effect of discharge rate and concrete load on number of concrete lumps and balls

TABLE 2
SUMMARY OF TEST RESULTS

Set	Load (yds^3)	Initial Rev.	Head Water (%)	Discharge Rate (lbs/sec)	Slump (inch)	Air Content (%)	Density (lbs/ft^2)	Number of Lumps and Balls						
								1″-2″	3″-4″	5″-6″	7″-8″	9″-10″	>10″	Total
A (Phase I)	9	40	90	150	7.8	2.00	142.0	5	6	4	7	1	0	23
	9	N.A.^	90	200	8.3	2.00	140.8	4	8	3	0	0	0	15
	9	N.A.^	90	250	8.5	2.00	142.0	6	7	8	9	5	5	40
B (Phase I)	3	48	90	200	8.0	2.25	139.6	0	0	0	0	0	0	0
	5	N.A.^	90	200	7.5	2.00	141.2	0	2	0	0	0	0	2
	7	38	90	200	7.5	2.00	140.4	7	5	12	11	0	0	35
	9	31	90	200	7.5	2.00	140.8	8	9	15	8	2	3	45
C (Phase I)	9	40	81*	200	7.8	2.25	140.4	0	2	3	3	0	1	9
	9	35	78*	200	8.5	2.50	140.0	6	7	17	6	2	6	44
	9	35	68*	200	8.0	2.00	141.2	1	2	1	1	0	0	5
	9	49	64*	200	7.5	2.50	140.0	0	0	0	0	0	0	0
D (Phase II)	9	55	50	200	8.3	3.20	140.6	9	13	11	2	3	2	40
	9	55	30	200	8.8	2.80	141.6	1	4	0	1	0	0	6
	9	55	20	200	9.0	2.80	141.0	3	3	3	0	0	1	10
E (Phase II)	9	55	30	200	8.8	2.80	141.6	1	4	0	1	0	0	6
	9	75	30	200	8.8	2.60	140.6	1	2	2	2	0	0	7
	9	100	30	200	8.5	2.50	141.6	0	0	0	0	0	0	0

* designed headwater ratios were 80%, 75%, 70% and 65%, respectively. Figures in Table show actually measured percentages.
^ N.A.: not available or not recorded.

Headwater Content and Initial Revolutions

After the first two set of experiments in Phase I, it was realized that the headwater content and initial revolutions could have a combined effect on the number of lumps. Hence in the next three sets, these factors were tested by keeping one as a constant and varying the other. The data for concrete batches with 10 or less number of lumps is plotted in Figure 3 using an average curve (approximated by eyeballing). Since 10 or less number of lumps in a 9 yd^3 load are very minimal, such concrete was considered as fairly uniform. The initial revolutions are plotted on the horizontal axis as they are usually specified in the specifications.

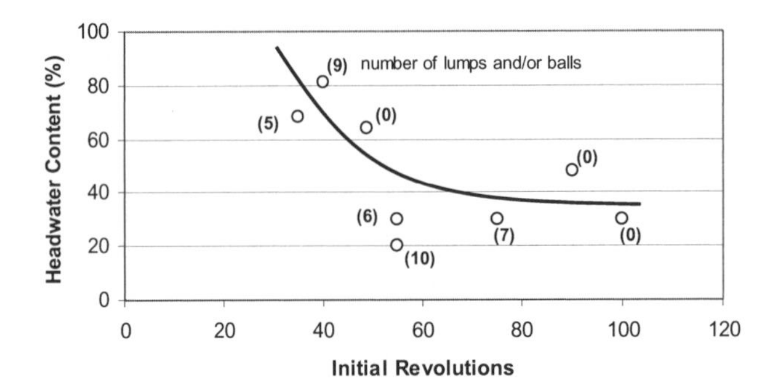

Figure 3: Combined effect of headwater content and initial revolutions on reduction of lumps and balls

It is clear from Figure 3 that with increasing numbers of initial revolutions, the headwater content can be decreased in order to produce lumps-free concrete. However, the headwater content should not be less than 25%, as the lower headwater resulted in head packs. It was further observed that the size of concrete lumps was smaller when lower headwater content was used with greater number of initial revolutions. The optimum headwater content was found to be between 30%-40% with initial revolutions ranging between 80 to 100.

Composition of Concrete Lumps

A visual inspection of lumps and balls indicated that they made up of cement, coarse and fine aggregates. The composition of most of the lumps and balls seemed to be uniform with few exceptions where dry sand and coarse aggregates were found in the center of the lump. This might be due to too low headwater content followed by low number of initial revolutions.

Figure 4: Composition of concrete lumps and balls

A representative number of samples of concrete lumps and balls were taken from different sets of batches and sent to the laboratory for sieve analysis. The results are shown in Figure 5.

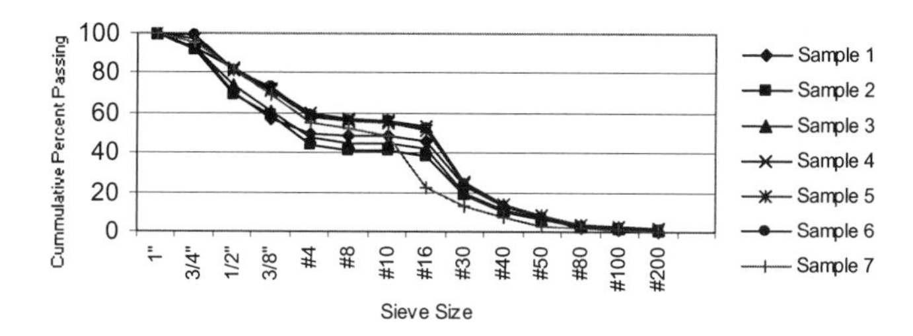

Figure 5: Sieve analysis of concrete lumps and balls

The sieve analysis indicates a gap grading with concentration of small size coarse aggregates, fine sand and cement. This is very obvious as these materials are more prone to form lumps and balls in the presence of water. The percentage of cement and water in the samples could not be determined due to the limited scope of experiments.

The size of concrete lumps varied between 1 inch to 14 inch and most bathes showed a combination of various sizes. However, smaller lumps were found in batches with low headwater and more initial revolutions. This is due to the fact that less headwater reduces the chances of cement and sand to form clutches and if formed, high initial revolutions break them. Figure 6 shows the different sizes of concrete lumps and balls found in a particular batch.

Figure 6: Lumps and balls found in a concrete batch

Concrete Batches with Lumps and Balls

Based on the interviews conducted with concrete plant managers and supervisors, the following actions could be taken to break lumps and balls in concrete, if allowed by the client and contractor.

1. If the concrete lumps are large, rotate the concrete drum at charging speed for 2-3 minutes and then at mixing speed for 3-5 minutes. This action could break the large lumps and makes concrete more uniform.

2. If small size concrete balls are found, sieve the concrete before discharging it on the chute. However, this action could not be taken if the concrete is directly pumped from the truck.

CONCLUSIONS

Based on the results of this study, the following conclusions can be drawn.

1. Headwater content and initial revolutions have a combined effect on the number of lumps and balls produced. A low headwater content followed by higher number of initial revolutions could minimize or even eliminate concrete lumps. The optimum combination was found to be 30-40% headwater with number of initial revolutions between 80-100 at a speed of 12 rpm.

2. The optimum discharge rate was found to be 200 lbs/sec.

3. The concrete load does not have a great effect on the number of lumps and balls produced. A load of 9 yd^3 was found to be commercially viable.

4. Mixing sequence has an effect on the formation of lumps and balls. A different mixing sequence could produce different results even if other factors are controlled.

5. If allowed by the client and contractor, the larger concrete lumps can be broken by rotating the drum at charging speed for 2-3 minutes and then at mixing speed for 3-5 minutes. Smaller balls could be sieved before discharging into the chute.

RECOMMENDATIONS

This research study was conduced with the aim to identify different factors that produce lumps and balls in high slump concrete. Due to the high cost of experimentation, only one set of data was collected for each factor. This may produce erroneous results hence further testing for each factor is recommended before any solid conclusions are drawn. Moreover, different concrete plants use different mixing sequences; hence more plants should be studied to support the results of this research.

ACKNOWLEDGEMENTS

The authors highly appreciate the support and help of following individuals during this research: *Dr. Sastry Putcha, Mr. Ken Blanchard, Mr. Leigh Merkert* and *Ms. Robbin Dano* from Florida Department of Transportation; *Mr. Jerry Haught* from CSR Rinker Concrete Plant, Miami, Florida; and *Dr. John Sobanjo* from Florida State University, Tallahassee, Florida.

REFERENCES

Florida Department of Transportation (FDOT). (2000). Standard Specifications for Road Bridge Construction. [online]. http://www11.myflorida.com/specificationsoffice/. Accessed on June 09, 2002.

Gaynor, R.D. (1996). Avoiding Uniformity Problems in Truck-Mixed Concrete. *Concrete Journal* **August 1996**, 570-572.

Gaynor, R.D., and Mullarky, J.I. (1975). Mixing Concrete in a Truck Mixer. *Publication No. 148. National Ready Mix Concrete Association* (NRMCA), Maryland 20910.

NRMCA (National Ready Mix Concrete Association). (2002). Ready Mix Concrete. [online] *http://www.nrmca.org*. Accessed on June 09, 2002.

Suprenant, B.A. (1992). Mixing Concrete in the Truck Takes Proper Procedures. *Publication #J921112*, The Aberdeen Group, UK.

Advances in Building Technology, Volume I
M. Anson, J.M. Ko and E.S.S. Lam (Eds.)

ADVANCES IN CONCRETE PLACEMENT METHODS AND FINISHING TECHNIQUES FOR HIGH PERFORMANCE INDUSTRIAL PAVEMENTS IN AUSTRALIA

Peter Ashford

Faculty of Architecture, Building and Planning, University of Melbourne,
Parkville, 3010, Australia

ABSTRACT

The quality and overall performance of concrete industrial floor slabs has improved dramatically as a result of technological advances in concrete supply, placement methods, the finishing process, curing and joint sealants. Placement and finishing techniques have improved substantially in recent years with larger area pours now resulting in a reduction in construction time. Client expectations for a high quality, hard wearing concrete industrial floor slab with long term performance and minimal maintenance costs has become the norm. Both designers and contractors now require specialised expertise to achieve these results. In order to improve the quality of industrial floors it has been necessary to establish working relationships between engineers, developers and the concrete contractors. This close relationship allows communication between the parties and provides a mutual understanding of all the issues associated with the design and construction methods to be employed.

KEYWORDS

High performance pavements, Vibrating screeds, Laser screeds, Scraping straight edges, Bump cutters, Ride on helicopters, Panfloats, Burnish finish, Surface tolerances, Flatness and Levellness.

INTRODUCTION

This paper reviews the most current practices in Australia for construction of high performance industrial pavements. The different floor pouring / placement methods associated with fabric reinforced, steel fibre reinforced and post-tensioned designs are highlighted together with levelling techniques and the degree of burnish finish achieved. The paper investigates through a series of six case studies, a comparison of construction methods and pour types including, long strips, double width strips with removable rails and large area pours together with floor tolerances.

The size of the pour areas and shape thereof often predetermine the choice of both the vibrating screed equipment and trowelling equipment necessary to achieve the slab finish required whilst maintaining surface tolerances. The case studies represent a variety of pour sizes and shapes and indicate the selection of equipment used by a range of specialist concrete. The paper concludes with a "Table of Characteristics and Comparisons" with an indication of future directions, developments and trends.

CONSTRUCTION TECHNIQUES & EQUIPMENT

Pouring Methods

- Pour widths, pour lengths, and the set up of reinforcement once documented and detailed, usually predetermine the concrete placement method.
- Direct placement from the truck for low slump, low shrinkage concrete.
- Reinforcement chaired up and inspected prior to the pour, but direct placement not possible.
- The alternative is to chair up reinforcement progressively but cover is suspect.
- Hence the advantage of steel fibre reinforced slabs.
- Concrete pumps with placement booms provide an efficient method of placing concrete for both post-tensioned and steel reinforced slabs when the reinforcing has been set up prior to the pour.
- However, with limited roof heights, the horizontal reach of placement booms is limited.

Established Levelling Techniques

- The plan shape and pour area size has a major impact on the levelling techniques used during construction. The traditional "Long Strip" layout requires accurate set up and levelling of the edge formwork with hand screeds or mechanical beam type vibrating screeds used to level the surface.
- Double width long strips with a removable intermediate rail are more efficient since the pour area can be doubled and construction time thereby reduced. Removable rails can also be used on large area pours.
- Large area pours with dimensions in excess of 25m (approx.) in each direction can be levelled by taking spot levels (optical or laser) on the concrete surface at regular grids (up to 3.6m) and then power floated between the spot levels. Floor tolerances expected with this method of construction are perhaps + or – 4mm in 3.0m at best. The alternative would be to use a removable intermediate rail which would require the vibrating screeds to span 12.5m (their upper limit) and floor tolerances would be unlikely to be improved. To achieve higher levels of flatness and levelness a laser guided hydraulic driven mobile cutting and vibrating screed can be used.

Slab Specifications for Finish & Tolerances

- Various trowelling techniques are used to produce either a light, medium or heavy burnish finish. Early trowelling with panfloats followed by trowelling with the standard blades and continually working the surface over an extended period which results in a hard wearing, abrasive resistant, burnish finish, (often referred to as "black concrete").
- Australian specifications typically call for a slab to be constructed within + or – 3mm in 3.0m with a cross fall not compounding over 10m and not to exceed 5mm from RL and a differential across all joints of + or – 1mm maximum. More recently however, larger buildings have been specified using the " American Standard Specifications for Tolerances for Concrete Construction and Materials, ACI #117R-90 ". This covers the specification and measurement of F_F (floor flatness) and F_L (floor levelness) values for concrete slabs.

Equipment Overview

The equipment used and the procedures to be adopted for placement, screeding and finishing varies significantly from contractor to contractor in accordance with their available equipment, crew size, quality of surface finish required, their expertise and their specialisation in constructing particular pour area types. Recent improvements in placement, finishing methods and concrete technology have allowed the pour size and joint spacing to be increased without comprising finish, flatness or overall cost. The range of equipment currently used includes:

- Optical and laser levels
- Hand screeds
- Power trowels (magic screeds)
- Scraping straightedges and bullfloats
- Mechanical vibrating screeds, solid steel sections or truss type vibrating screeds
- Laser guided hydraulic driven mobile cutting, levelling and vibrating screeds
- Pan floats (pizza pans) for walk-behind and ride-on trowels
- Single blade walk behind helicopter trowels , double and triple blade ride ons

CASE STUDIES

For comparative purposes, the project highlights were covered in the following order:
- Project description / pour areas
- Specifications / floor finish and tolerances
- Concrete placement method
- Screeding and levelling techniques / equipment
- Trowelling equipment / finishing / duration
- Difficulties experienced and comments

Each case study is complimented by two digital images which highlight the specific project features.

1. " Country Road " Clothing Distribution Warehouse

- The total floor area of 5,200 m^2 was divided into 18 long strip pours each 50m long by varying widths up to 5.85m. Pouring commenced on completion of the exterior cladding with areas up to 290m^2 per pour. This was the first burnish finish fibre reinforced slab constructed in Victoria.
- A hard wearing burnish finish was specified for both the racking and order collation areas. Tolerances were specified at + or – 3mm in 3.0m with a cross fall no compounding over 10m and not to exceed 5mm from RL with + or –1mm across joints. The constructed slab met these requirements and was checked using an "F meter " at F_F 70 and F_L 40.
- The concrete was placed direct from the truck chute, having reversed down the long strips.
- Truss type vibrating screeds were used and traversed the length of the strip. Edge formwork comprised of timber with a 50 x 4mm steel flat nailed vertically onto the timber so that the top edge was some 10mm above the timber. Optical levels to within 2mm were taken at 900mm centres along the steel edge. A 2.4m wide bump cutter was used across the strips to remove any surface irregularities.
- Twin blade ride-on power trowels commenced early with panfloats, followed by approximately 4 hrs. of helicopter blade trowelling with the blade angle progressively increasing as time passed. At this stage the concrete was quite dark in colour, characteristic of a burnish finish. The machines were equipped with two containers of aliphatic alcohol (a set retarder), which was sprayed onto the surface to adjust non-uniform setting rates. A chlorinated rubber spray on curing compound was applied immediately after the trowelling had ceased. The typical 290m^2 long strips took on average 9 hrs. to complete.
- Some steel fibres were exposed on the surface with occasional fibre tearout at the saw cut joints. The combination of a relatively dry mix design and the fibre reinforcement made the concrete difficult to work.

Figure.1 Truss vibrating screed, FRC slab

Figure.2 Ride on power trowel, long strip

2. " QP2 " A Supermarket Chain High Bay Warehouse and Distribution Centre

- The warehouse comprised a 9 storey high bay racking system with a 500mm thick ground slab to support the superstructure. This slab covered 7,730m^2 and was poured in 8 stages, covering 46m x 168m with each pour approximately 23m wide x 42m long (966m^2).The "distribution centre" slabs were 180mm thick and post-tensioned. This area covered some 15,200m^2 in 8 adjacent pours each approximately 38m x 50m ,(1,900m^2 each pour).
- The above slabs were specified to have a hard wearing burnish finish. Tolerances were specified at + or – 3mm in 3.0m with all levels not to exceed 5mm from RL. (Achieved F_F 50, F_L 40)
- Both slab types were poured using concrete pumps with placement booms. The post-tensioned slab of 1900m^2 required the use of 2 pumps and 2 vibrating screeds.
- Double and triple width pour strips were necessary to screed these large areas. Truss type vibrating screeds were used in both cases, each utilising removable rails set to the correct height / levels The rails were removed on completion of the screeding and localised rectification works done. The screeds spanned up to 11.5m to achieve the pour widths required. Optical levelling of the supporting steel edge piece to within 2mm was done at 1.2m centres. Long handle 3.0m wide scraping straightedges were used in the transverse direction to take out surface irregularities.
- The post-tensioned slab required the use of two, twin blade and one, triple blade ride-on power trowels and several walk behind machines . Early panfloat trowelling was followed by 3 hrs. of helicopter blade trowelling which achieved the light burnish finish as specified. Each slab pour required a double size crew with an average 14 hrs. to complete
- A continuous supply of concrete for the 500m^3 "high bay " slab pours, often extended the completion time. Concrete placement and vibration was difficult with closely spaced large diameter reinforcing bars in both the top and bottom of the 500mm slab. Cold and windy conditions extended the trowelling requirements.

Figure.3 Removable rails, double width strip

Figure.4 Initial trowelling with panfloats

3. " Chiffley Drive " Paper Distribution Warehouse

- A combination of reinforced and post-tensioned concrete designs with total floor area of 12,000m^2. Reinforced slabs 250mm thick poured in long strips from 9.0-13.5m wide with saw cuts on 6m x4.5m grid and maximum pour area 785m^2. Post-tensioned slabs 180mm thick with ducts at 1500mm crs. each way with both movement and construction joints. Maximum pour area 2088m^2.
- The slab was specified to be a hard burnish finish with floor tolerances of + or -4mm in 3.0m and not more than 5mm from datum. Achieved + or – 3mm in 3.0m.
- Concrete was placed using 2 placement boom concrete pumps with 2 separate concrete crews.
- The concrete was compacted with immersion vibrators, roughly shovelled level and then spot optical levels taken to suit the power float (magic screed) with a 3.6m blade utilised to screed and float the concrete surface. Permissible floor tolerances made this screeding and levelling technique more economical and faster in contrast to using long hand screeds. A long handle 3.6m wide scraping straightedge was used in the transverse direction.
- 4 walk behind power trowels with panfloats commenced at 5 hrs. and were followed by up to 5 twin blade ride-on power trowels with helicopter blades and pans. With 2 concrete crews of 8, the final trowelling ceased at 12 hrs. with a spray on curing compound applied immediately thereafter.
- The slabs was completed with a large concrete crew utilising basic equipment as the slab tolerances were not difficult to achieve. Optical levels were taken over distances up to 50m.

Figure.5 Power float, large area pour

Figure.6 Two concrete pumps, optical levels

4. " Penguin Books " Warehouse

- The slab was redesigned to a steel fibre reinforced slab 180mm thick. The total floor area was 21,000m^2 poured in 21 stages on a 3 wide x 7 long grid pattern. Pours in the 11m high racking area were 870m^2 each (28m x 31m)
- A hard wearing burnish finish was specified. Floor tolerances were specified by the supplier of the man-up turret trucks (14m lift). The specification required a 6mm maximum difference in elevation of any 2 points 2.4m apart in the direction of truck travel and a 2mm maximum difference in elevation of any 2 points across the aisle at the outside of the load wheel path. There was a 1.5mm maximum difference in elevation of any 2 points 300mm apart in the load and steer wheel path. An F_F 50 slab was considered adequate to meet these requirements. Over the 9 areas poured in the high rack area, the averages achieved were F_F 70 and F_L 60. A comparative optical level grid survey was carried out at 2m centres with < 2mm variance.
- The fibre reinforced concrete was placed directly from the chute with two trucks being able to discharge at once.
- A Somero laser screed was used to cut, screed and vibrate the concrete. A uniform head of concrete in front of the screed was no longer required and the screeding time was significantly less, resulting in substantial labour savings and the ability to level / screed much larger areas of concrete. Long handle 2.4m wide scraping straightedges were used in the transverse direction to take out surface irregularities. Edge formwork was checked with a laser level at 1.5m centres.

- Two walk behind and two ride-on power trowels were used to finish the slab. Initial trowelling with panfloats commenced 4 hrs. after the first concrete delivery. At 6 hrs. the pans were removed and helicopter blades continued with the slab starting to burnish up at 7hrs. The slab was completed at 9 hrs. and sprayed immediately with a curing compound.
- The concrete was placed with a 28m wide working face (the full width of the pour) which created some relatively cold faces when there were delays in truck deliveries. This had a direct effect on the trowelling and subsequent work required on the slab surface where fresh and older concrete occur, side by side. . Consistent concrete deliveries and uniform setting rates must be maintained. With 6 pours per week, the concreter's progress was restricted by the slower progress of the roof and wall cladding.

Figure.7 Somero laser screed, large area pour

Figure.8 Direct concrete placement

5. " Tsusho " Toyota Spare Parts Warehouse

- This fabric reinforced concrete slab 175mm thick covered an area of 25,000m^2 and was poured in 24 stages. The pours were approximately 34m long x 30m wide (1020m^2).
- Tolerances specified at + or – 3mm in 3.0m, within + or -5mm from RL and 2mm differential across joints. An optical survey on completion indicated slab tolerances were within 1.5mm.
- Placement was by concrete pump with a hydraulic boom arm and required two set up positions to cover the area. Two trucks fed the concrete pump.
- Levelling completed by laser screed set up on ladder tracks which were supported on large 4 legged stools which neatly fitted in between the reinforcing mesh grid. A long handle 2.4m wide scraping straightedge were also used. Achieved F_F 60 F_L 50 tolerances.
- Walk behind and ride-on power trowels with panfloats and blades completed the hard burnish finish within 11 hrs.
- Each slab pour area was sofcut sawed at 8 hrs. to a depth of 40mm to divide the area into 4 panels. Danley flat plate dowels 50 x10 at 600mm crs in prefabricated cages were located under these saw cuts with a 55mm high crack inducer detailed. At 24 hrs. the joint was recut to 50mm and additional cuts made to reduce the panel areas to 7.5m x 5.6m. Pour joints were constructed with Danley 6mm thick diamond plate dowels at 600mm crs.

Figure.9 Danley diamond dowel joint

Figure.10 Laser screed, ladder rails & stools

6. “ Toll Holdings ” Transportation Distribution Warehouse

- The total floor area covered 31,680m^2 and was poured in 20 stages on a 2 wide x 10 long grid pattern. Each pour measured approximately 36m x 44m (1,584m^2). The slab was 160mm thick post-tensioned with ducts at 2.1m centres each way. On completion of the stressing, the 900mm wide edge strip between the slab and external precast walls was poured.
- The specification was extremely brief and called for a standard burnish finish. Floor tolerances were specified at minimum F_F 50 and F_L 40. After each pour, an “F” meter was run over the floor in lines around the perimeter. Average readings for the slabs completed were F_F 62 and F_L 50.
- The slabs were poured using a concrete pump with a placement boom. The hydraulic arm could not reach beyond 27m, so 9-10m of flexible rubber pump hose supported on timber blocks had to be attached to the end of the pump arm in order to reach the 36m wide pour.
- Screeding and levelling was done using a Somero laser screed. Heavy duty aluminium removable ladder rail tracks were set up on stools above the tendon ducts, these extended the full width of the pour and included a ramp section to pass over the edge formwork. A long handle 3.6m wide scraping straightedge was used to smoothen out minor surface irregularities. The laser screed used two passes to achieve the tolerances required. Timber edge formwork had a 30 x 5 steel edge plate nail fixed at 450mm centres and checked for level at 1.2m centres using a laser level with circular bubble on the staff.
- Three, twin blade ride-on power trowels were used to finish the slab. Initial trowelling with panfloats commenced 4.5 hrs. after the first concrete delivery. At 6.5 hrs, helicopter blades were used with the slab starting to burnish up at 8hrs. The slab was completed at 13 hrs. with a crew of 12. “ F “ meter readings were taken prior to the slab being sprayed with a curing compound.
- Additional immersion vibrators were required to ensure adequate compaction of the concrete under the tendon ducts (normally the laser screed is sufficient). Setting up the 10m of rubber hose extension to the pump slowed down the concrete placement process.

Figure.11 Laser screed on post-tensioned slab

Figure.12 Ladder tracks & concrete pump

SUMMARY & CONCLUSIONS

The concrete contractors in Australia, have in recent years, imported a wide range of equipment best suited to meet their own specialised construction method. The variety of equipment purchased, enables the contractors to meet the surface finish and surface tolerance specifications particular to their market sector. There remains significant market segmentation within the industrial ground slab concrete industry. This is particularly noticeable with those contractors who have imported expensive laser screed equipment and now only construct large area pours to F_F and F_L requirements / specifications.

The six case studies represent a range of slab pour types, levelling and finishing techniques adopted in Australia, but it is still primarily the engineer who dictates the pour area / layout. Pour areas around or in excess of 1000 m^2 are becoming the norm with the traditional large concrete crews being replaced by the use of laser guided screed equipment. A summary Table 1. follows.

Table 1.
Characteristics & Comparison of Case Studies

CASE STUDY NO.	TOTAL FLOOR AREA SQ.M.	CONC. POUR AREA SQ.M.	POUR AREA TYPE/ METHOD	SLAB DESIGN	PLACE-MENT METHOD	TOLER-ANCES (Achieved)	SURF-ACE FINISH	LEVELS & SCREED TYPE	TROWEL METHOD & FINISHING
1	5,200	290	Long strip	Steel fibre	Direct from truck	3mm/3m 5mm RL 1mm joint (F_F 70 F_L40)	Hard burnish	Optical Truss vibrating screed	Ride-on panfloats Ride-on blades
2	7,730	966	Double strip	Steel rein.	Concrete pump	3mm/3m 5mm RL (F_F50 F_L40)	Hard burnish	Optical 2 Truss vibrating screeds	Ride-on panfloats Ride-on blades
3	12,000	2,088	Large area	Post-tension & Steel rein.	Concrete pump	4mm/3m 5mm RL (3mm/3m)	Hard burnish	Optical Power floats	Ride-on panfloats Ride-on blades
4	21,000	870	Large area	Steel fibre	Direct from truck	F_F 50 F_L 40 (F_F70 F_L60)	Hard burnish	Laser Laser vibrating screed	Ride-on panfloats Ride-on blades
5	25,000	1,020	Large area	Steel rein.	Concrete pump	3mm/3m 5mm RL 2mm joint (F_F60 F_L50)	Hard burnish	Laser Laser vibrating screed	Walk behind panfloats
6	31,680	1,584	Large area	Post-tension	Concrete pump	F_F 50 F_L 40 (F_F62 F_L50)	Standard burnish	Laser Laser vibrating screed	Ride-on panfloats Ride-on blades

COMMENTS ON TABLE 1. Characteristics and Comparison of Case Studies

Although there is no correlation between tolerances specified in (mm/3.0m and within RL) to (F_F and F_L) tolerances, measurements and verification of actual tolerances achieved are normally carried out over the slabs using an " F " meter.

ACKNOWLEDGEMENTS

The author would like to acknowledge the support and assistance from the Building Contractors, Project Managers, Engineers, Consultants and Concrete Contractors, for access to sites and documentation.

Building Contractors / Project Managers
Baulderstone Hornibrook, Vaughan Constructions, Pritchard Builders, Griffin Project Management, Hansen Yuncken

Engineers and Consultants
CSR Construction Services, Scott Wilson Irwin Johnston, The O'Neill Group, Ancon Beton P.L. John McGovern, Structural Systems

Concrete Contractors
Anglo Italian Concrete, V&G Concrete, Decco Constructions.

Advances in Building Technology, Volume 1
M. Anson, J.M. Ko and E.S.S. Lam (Eds.)

REVIEW OF DESIGN AND CONSTRUCTION INNOVATIONS IN HONG KONG PUBLIC HOUSING

Daniel WM Chan[1] and Albert PC Chan[2]

[1] Research Fellow, Department of Building & Real Estate, The Hong Kong Polytechnic University, Hung Hom, Kowloon, Hong Kong, China
[2] Associate Professor, Department of Building & Real Estate, The Hong Kong Polytechnic University, Hung Hom, Kowloon, Hong Kong, China

ABSTRACT

Public housing has exhibited a profound impact on the social and economic development of Hong Kong. During the past few decades, the Hong Kong Housing Authority (HKHA) has been committed to the construction of a substantial number of high-rise public residential blocks to meet its target housing demand. In order to cope with such a massive public housing development, the concept of 'construction process re-engineering' has been applied to these projects, especially in the significant improvements in production times and costs without compromising quality.

This paper presents a review of the current innovations in the design, construction and management of high-rise public housing towers in Hong Kong. These innovations include: (1) standardization of block designs and construction sequences; (2) large panel steel formwork systems; (3) precast concrete facade elements; (4) extensive use of prefabricated building components; and (5) submission of alternative tenders for shorter project durations. It was suggested that considerable efforts should be dedicated to the research, development and dissemination of advances in building technology (e.g. modularization and prefabrication), coupled with innovations in project delivery and management principles (e.g. project partnering and strategic alliance).

KEYWORDS

Design innovations, construction technologies, construction process re-engineering, public housing, Hong Kong.

INTRODUCTION

Public housing has brought a significant influence on the social and economic development of Hong Kong. The public housing programme is one of Hong Kong's prides over its history and has developed into one of the most cohesive and far-sighted programmes in the Asia-Pacific Region. The Hong Kong Housing Authority (HKHA) controls one of the world's largest public housing stocks, comprising 1,005,869 flats in 361 housing estates/courts as at 31 March 2001 (HKHA, 2001), amounting to house a population of 3,235,200. With its planned public housing production programmes extending into the 21st century, the Government's commitment to housing construction will continue to grow. In order to cope with such an extensive public housing programme, current strategies in relation to design principles and construction technologies applied to the public housing projects are reviewed and discussed, especially in the context of expediting the construction process, minimizing the construction cost and enhancing the quality levels of constructed facilities.

CURRENT INNOVATIONS FOR ENHANCING PROJECT PERFORMANCE

'Process Re-engineering' is a new production philosophy of proven value in the manufacturing industry (Mohamed and Yates, 1995). The initial popularity and apparent success of this new production philosophy in manufacturing has instigated researchers and practitioners to apply its concepts to the construction industry. It aims to achieve substantial improvements in production times and costs without lowering quality.

An attempt was particularly made to provide some useful insights for shortening construction durations in these public housing projects. This improvement in construction time performance can be accomplished by re-engineering the design and construction processes. It was confirmed through face-to-face interviews with the construction professional staff of the HKHA and project managers of leading building contractors, that there is considerable potential for more time savings if such major changes are introduced in the '***superstructure***' construction. Therefore, the following innovations are currently employed in pubic housing construction in Hong Kong, for accelerating the construction process which mainly focus on reducing the time for superstructure erection.

Standardization of Block Designs, Construction Sequences and Durations

Speed of construction is obviously important in achieving the expeditious rate of building and development required to meet the unprecedentedly enormous housing programme in Hong Kong. To enhance the buildability of project design, the Hong Kong Housing Department (the executive arm of HKHA) introduced a new series of standard block designs (Harmony Types 1, 2, 3 and New Cruciform Type) in the early 1990s to house some one million people by the end of the decade. Its architects and engineers carried out detailed investigations into optimizing construction programmes, leading to in-house standard norms and related guidelines for construction durations for each of the standard block designs, with specific breakdowns for piling and building contracts, as indicated in Table 1.

Moreover, face-to-face interviews with senior construction professionals of the HKHA, together with a series of case study projects under construction, reached a consensus agreement on construction progamming. It was manifested from the observations on their master construction programmes that the major construction activities of a public housing block can be classified into several fundamental primary work packages, as illustrated in Figure 1. Their detailed definitions can be sought in Chan and Kumaraswamy's (1999) publication. Standard construction sequences for these projects can lead to a shorter delivery time on the basis of an optimum resource allocation.

TABLE 1
SCHEDULE OF EXPECTED CONSTRUCTION PERIODS OF STANDARD PUBLIC HOUSING BLOCKS
(AS OF 15 OCTOBER 1996)

Type of block	Number of storeys	Piling contract		Building contract		Total construction duration
		Without caps (months)	With caps (months)	With caps (months)	Without caps (months)	(months)
Harmony 1	41	6	9	29	26	35
Harmony 2	41	6	9	29	26	35
Harmony 3	30	6	9	25	22	31
New Cruciform	38	6	9	28	25	34
Concord 1	41	6	9	29	26	35
Concord 2	31	6	9	26	23	32

To further promote the constructability, the block designs are more 'spacious' and of a standard module form. They also take account of a number of important factors, including the need to offer a flexibility in the flat mix while maintaining consistently high standards of quality in construction and detail. The standard modular flat designs have been developed which incorporate more standard prefabricated building components and work details, as well as an efficient and simple construction sequence, so as to minimize labour content on-site and speed up the construction process

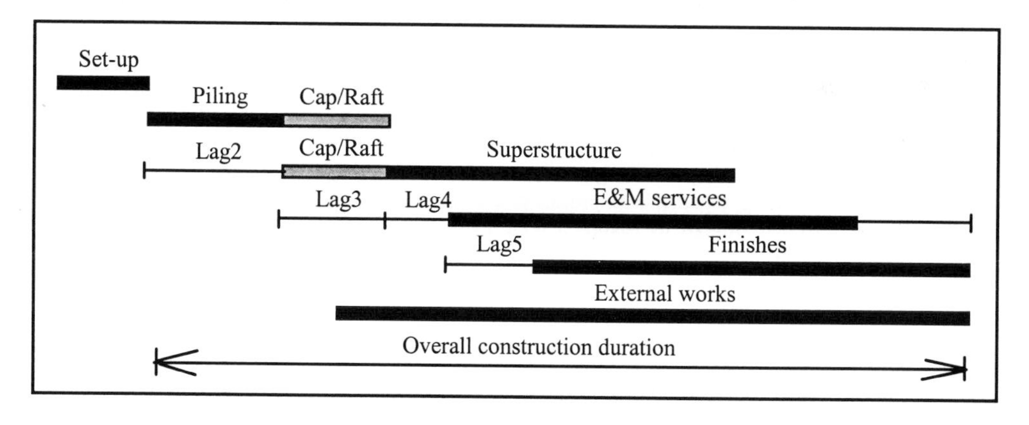

Figure 1: Typical master construction programme for standard public housing blocks

Large Panel Steel Formwork Systems

Due to the high costs of labour, market forces have driven more and more mechanization (or mechanized methods of construction) into the international construction industry (Wong and Yau, 1999). A high level of mechanization is also applied to the public housing construction industry of Hong Kong. To expedite flat production and for better quality control, the building contractors are mandatorily required to use large panel steel wall forms for wall erection and table forms for floor slab construction since the mid-1990s. This mandatory method could only be achieved by employing tower cranes to lift the heavy metal formwork. Since the reinforced concrete superstructure constitutes the most critical activity in the construction programme, the speed and efficiency provided by large panel metal formwork, as assisted by the site cranage provided by tower cranes, has significantly helped in completing projects nearer to the original schedule.

Most of the building contractors have found that the construction of each typical floor in a single Harmony Type 1 block can be accelerated from a 8-day working cycle in the early 1990s to a 6-day cycle in the mid-1990s with the wall elements lying on the critical path, specially if the design is directed towards the use of prefabricated elements and large panel steel formwork systems. Subsequent 'trial' studies indicated that casting cycles per floor of each block can be further reduced to 4 days (see Table 2), and thus compressing the original standard schedule for superstructure erection (26 months) of a standard Harmony Type 1 block of 40 domestic floors by 4 months (Chan, 1998).

TABLE 2

FOUR-DAY WORKING FLOOR CYCLE FOR A TYPICAL HARMONY DOMESTIC BLOCK

Activity	Duration
1. Install precast facade; fix reinforcement and concealed conduits to wall	1 day
2. Install large panel wall form; pour concrete to wall	1 day
3. Strike large panel wall form; install semi-precast concrete slab (for domestic areas); erect aluminum slab form (for corridor and lift lobby)	1 day
4. Fix concealed conduits and reinforcement to topping slab; pour concrete to topping slab	1 day
Typical floor cycle time :	**4 days**

In order to achieve such a 4-day construction cycle, it was also suggested to place concrete to typical floors by pump in view of relieving the expected high workload demand on tower crane for other site operations. It was recommended to provide a concrete placing boom at each floor under construction to pour concrete through distribution pipelines to any required position across the whole floor (pump rate of 60 m^3/hour to over 120m high above ground). There should be one tower crane and one concrete boom placer for each block, together with precast plank (slab) yards, precast facade yards and concrete batching plants operated on-site to ensure a continuous workflow among them. Another benefit accruing from employing metal formwork is that contractors can re-use their own sets of steel formwork for future projects. System formwork can be much more widely adopted if flat dimensions are modularized. Furthermore, high strength concrete and steel reinforcement are also recommended because of not only saving overall material costs but also the reduction in structural member sizes.

Precast Concrete Facade Elements

The use of prefabrication techniques has marked a turning point in the construction industry. Advantages include less time and reliance on site labour, easier site inspection, as well as greatly improved design details and quality control. With the labour shortages prevalent in the building industry, prefabrication together with the greater use of standardization and modular flat design are now the essential principles in the design of standard domestic blocks in Hong Kong (Chan, 1998).

As a result, another innovation in the Hong Kong public housing construction is the introduction of precast concrete facade elements for the Harmony and Concord block series, which has been highly recommended by the Housing Department since 1991. In order to achieve a finish with better quality, these elevation facades are often cast at site precasting yards complete with cast-in aluminum window frames and mosaic tiles. Though the adoption of precast facades will necessitate an increase in the initial capital investment of contractors, this can be set off by savings in in-situ formwork for the facade elements in the long run and also by the higher quality of finished concrete elements. The use of precast facades now becomes mandatory in all Harmony Block and Concord Block building contracts.

There are two kinds of methods to install the precast facades with bay windows i.e. pre-installation and post-installation. Pre-installation lies on the critical path of the construction programme and hence affects the floor cycle time, whereas post-installation is not critical but facade connections between wall and slab would have to be made very carefully in order to ensure water-tightness and overcome leakage problems. Much more use of precasting and off-site prefabrication should be encouraged for reducing construction durations and maintaining consistently high quality standards of the products.

Increasing the Extent of Prefabrication of Building Components

To expedite the construction programmes and enhance the quality levels of finished concrete products, the dependence on wet finishing trades has been significantly reduced. The HKHA has increased the large scale use of factory produced building components for standard types of housing blocks since the mid-1990s to ensure consistency in the standard of the major finishing and fitting materials. Apart from facade panels, other prefabricated products now being used widely include semi-precast concrete floor slabs, staircases, lightweight concrete partition walls, fabric reinforcements, aluminium windows, cooking bench and sink units, kitchen and bathroom doorsets, and stainless steel collapsible entrance gates (Chan, 1998). A more recent 'trial' project of using prefabricated bathroom units was carried out at Tseung Kwan O Area 73A Phase 2 Project and confirmed successful (HKHA, 2001). Modular packaging is also employed more extensively e.g. rebar in many situations is designed to be delivered pre-assembled for rapid fixing on-site.

The quality of precast elements is usually higher than that of the constructed in-situ components. However, their site assembly is more sensitive to poor quality control, for example, at the joints when connecting components. As is common to all prefabrication, most of the work is conducted in the factory, leaving little to be done on-site. This increases the likelihood of more efficient, high-quality and faster construction being achieved (Rosenfeld, 1994). Unlike the traditional methods of construction, these prefabricated components have a higher inherent quality, as well as more accurate profiles and dimensions so that fast-track construction will not be at the expense of poor quality.

Alternative Tenders for Reduced Construction Durations

In order to shorten standard construction periods, the HKHA is launching trials on different tendering methods, including (1) offering competent contractors more opportunities for proposing their own construction methods and solutions with 'shorter' construction durations (e.g. with a 4-day floor cycle) during the tender process – the Alternative Tender B which was first introduced in 1995 (HKHA, 2001); (2) negotiated fast-track contracts with prominent contractors for shorter completion periods; negotiated contracts encourage flexibility and early consideration of construction issues in the design process; and (3) selective tendering – awarding the contract to a contractor who has exhibited effective site management and supervision over the previous projects, as well as a proven track record of outstanding performance (Chan ,1998). Hence, tenders are either given financial incentives or a higher chance for winning future building contracts, to develop innovative construction techniques for faster construction schedules, but of course without compromising quality and safety.

Another new strategy proposed by the HKHA for shortening construction durations is an increased use of 'value engineering' in construction projects which would lead to more design-and-build contracts, particularly for a type of home purchase scheme (the Private Sector Participation Scheme, PSPS) where the building contractors may be responsible for both the design and construction of a new housing block (HKHA, 2001). This scheme has been introduced since 1978 and a number of measures have been implemented to encourage faster construction and building quality enhancement for new projects. For instance, the construction periods can be reduced by overlapping the design phase and construction phase, as well as by enhancing the buildability of project design (Rowlinson, 1999). Other innovative alternatives in construction procurement methods in the industry may also be considered in addition to the traditional type of contracts.

SUMMARY

This paper reviewed various current strategies for improving project performance of public housing in terms of time, cost and quality, by means of 're-engineering' a series of design and construction processes. These strategies embrace standardization of block designs and construction sequences, mandatory use of large panel steel formwork systems, precast concrete facade elements and other standardized prefabricated building components. Other processes that need to be 're-engineered' include the submission and evaluation of alternative tenders for a shorter project delivery from contractors.

To conclude, all these major technological improvements, coupled with advances in the current design and management strategies, have been observed to be effective in upgrading the present quality levels of concrete products and finishes, while variations in design work and project delays have been minimized. Chan and Kumaraswamy (2002) recently consolidated from their research study a series of technological and managerial strategies for reducing the typical floor cycle time for a standard public housing block design and hence compressing the overall project duration.

As the largest provider of housing in the Region, the HKHA is in a pioneering position of generating momentum for the introduction of a wide variety of innovative construction methods (e.g. Jump-form system formwork covering central core walls and other domestic structural walls for all wings of the building in floor cycles of 4 days), industrialized building components (e.g. modularization and prefabrication), and modern management principles (e.g. project partnering and strategic alliance) into the industry as the future core direction for building construction. Indicative site photos are given in Appendix 1 for further illustration.

REFERENCES

Chan, D.W.M. (1998). *Modelling Construction Durations for Public Housing Projects in Hong Kong.* PhD thesis, The University of Hong Kong, Hong Kong.

Chan, D.W.M. and Kumaraswamy, M.M. (1999). Modelling and predicting construction durations in Hong Kong public housing. *Construction Management and Economics*, 17(3), 351-362.

Chan, D.W.M. and Kumaraswamy, M.M. (2002). Compressing construction durations: Lessons learned from Hong Kong building projects. *International Journal of Project Management*, 20(1), 23-35.

Choi, J.C.W. (1998). *A Comparative Study on Architectural Design between the Harmony Block and Concord Block.* BSc(Hons) dissertation, The Hong Kong Polytechnic University, Hong Kong.

Construction and Contract News (2001). Public housing – Precast construction. *Construction and Contract News*, April 2001, Issue No. 2, China Trend Building Press Ltd., Hong Kong, 42-51.

HKHA (1999). *Annual Report 1998/1999*, Hong Kong Housing Authority (HKHA), Hong Kong.

HKHA (2001). *Annual Report 2000/2001*, Hong Kong Housing Authority (HKHA), Hong Kong.

Mohamed, S. and Yates, G. (1995). Re-engineering approach to construction: A case study. *Proceedings of the 5th East Asia-Pacific Conference on Structural Engineering and Construction*, 25-27 July 1995, Gold Coast, Queensland, Australia, 775-780.

Rosenfeld, Y. (1994). Innovative construction methods. *Construction Management and Economics*, 12(6), 521-541.

Rowlinson, S. (1999). "Chapter 2: A definition of procurement systems", In *Procurement Systems – A Guide to Best Practice in Construction* (edited by Steve Rowlinson and Peter McDermott), E & FN Spon, London, UK, 27-53.

Wong, A.K.D. and Yau, A.S.K. (1999). Public housing construction in Hong Kong: From building traditionally to the era of rationalization. *Group Proceedings: Group 1 – Construction*, The Hong Kong Institution of Engineers (HKIE), 2(1), 11-17.

APPENDIX 1

Standard Harmony block design of four wings (HKHA, 1999, p.33)

Large panel steel wall form (HKHA, 1999, p.36)

Tower crane and concrete boom placer (Construction and Contract News, 2001, p.42)

Concrete batching plant on-site (Construction and Contract News, 2001, p.50)

Complete precast concrete façade element (Construction and Contract News, 2001, p.45)

Semi-precast concrete floor slab panel (Construction and Contract News, 2001, p.43)

Precast concrete façade and slab yards (Construction and Contract News, 2001, p.44)

Jump-form system formwork for central core wall (Choi, 1998, p.89)

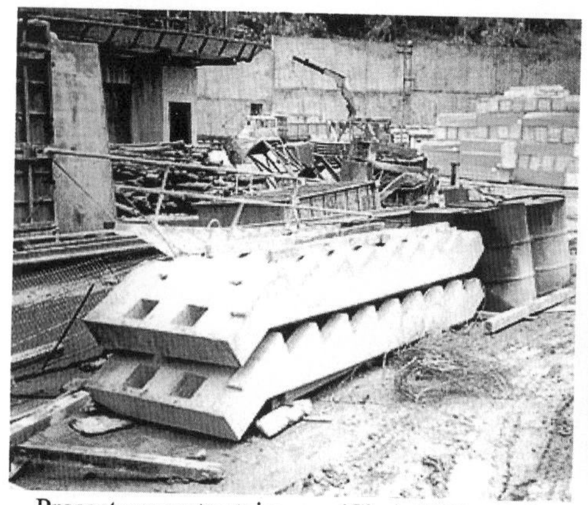

Precast concrete staircases (Choi, 1998, p.38)

Prefabricated bathrooms (HKHA, 2001, p.34)

Advances in Building Technology, Volume 1
M. Anson, J.M. Ko and E.S.S. Lam (Eds.)

STATIC LOADING TEST OF A SELF-BALANCED APPROACH TO A PILE IN THE YANGTZE RIVER

Dai Guoliang[1] Gong Weiming[1] Ji Lin[2] You Qinzhong[2]

1College of Civil Engineering, Southeast University, Nanjing, Jiangsu province, China
2Commander Department of Runyang yangtze River Highway bridge, Zhenjiang, Jiangsu province, China

ABSTRACT

Pile foundations are increasingly applied to the rapid development of long-span bridges in our country. In any kind of pile foundation a static loading test must be undertaken to determine the ultimate compressive bearing capacity of a single pile under criterion GBJ7-89. The Beicha bridge test pile of the Runyang Highway Yangtze River Bridge located in the Yangtze River between Zhenjiang and Yangzhou, whose ultimate bearing capacity is up to 40,000 kN, is hardly possible to test in conventional methods. The self-balanced method (Osterberg, Load-cell) is a new static load test method, almost not limited by testing sites and ultimate bearing capacity, which can directly determine the ultimate side resistance and tip resistance. This paper discusses the method successfully applied to the test pile, instrumented with strain gauges and the method for converting the results of the self-balanced test to the conventional loading test results is put forward, which suits the design method. The results also indicate the method will have expanding prospects for testing bridge piles under conditions of high bearing capacity and for special testing sites.

KEYWORDS: static load test, self-balanced test, side resistance, tip resistance, conversion method

1 INTRODUCTION

Pile foundations have been increasingly applied with the rapid development of long-span bridges in our country. In any kind of pile foundation a pile static loading test must be undertaken in order to determine the ultimate bearing capacity of a single pile under criterion GBJ7-89. The ultimate bearing capacity of the tested pile, located in the Yangtze River, is up to 40,000 kN. So a conventional static load test would have been prohibitively costly and, also difficult to arrange in the available time and space. Hence, the self-balanced method (Osterberg, Load cell) was selected. This is almost not limited by the test sites and ultimate bearing capacity, and can directly determine the ultimate side resistance and tip resistance. This paper describes the method, which was successfully applied in the test pile, and a method converting the results of the self-balanced test to the traditional loading test results is put forward, which suits the design process well. The results indicate the method will also have very good prospects in testing bridge piles under the conditions of high bearing capacity and for special testing sites.

2 MAIN TEST RESULTS

2.1 Principles of Self-balanced Test

The self-balanced technique incorporates a sacrificial hydraulic jack-like device (load-cell) placed at or near the base of the pile to be tested. When hydraulic pressure is increased, the load-cell expands, pushing the shaft upward and the base downward. The upward movement of the load-cell top plate is the movement of the shaft at the loading cell location and it is measured by means of displacement transducers extending from the loading cell top plate to the ground surface. The downward movement of the load-cell base plate is also measured by means of displacement transducers extending from the load-cell bottom plate to the ground surface. It is important to realize that the upward and downward load movements are not equal. The upward load movement is governed by shear resistance characteristics of the soil or rock or both along the upper part of the shaft, whereas the shear resistance of the soil or rock or both along the lower part of the shaft and the compressibility of the soil or rock or both, below the pile toe, govern the downward load movement. At the start of the test the pressure in the load-cell is 0 and the load-cell assembly carries the self-weight of the pile at the location of the load-cell structurally. The test consists of applying load increments to the pile by incrementally increasing the pressure in the load-cell and recording the results upward and downward displacements. The first pressure increments transfer the pile self-weight from the assembly to the load-cell fluid. The load-cell load determined from the hydraulic pressure reading at completed transfer is the self-weight value and it is reached at minimal movement (i.e., separation of the load-cell plates). When the full self-weight of the pile has been transferred to the pressure in the load-cell, a further increase of pressure expands the load-cell; that is, the top plate moves upward and the bottom plate moves downward.

2.2 Test procedure

The test pile diameter is 1.5 m, 89 m in length and with 8 soil-rock interfaces. Every interface is instrumented with four strain gauges. The load-cell location is 10 m away from the pile toe. Load tests were carried out generally in accordance with JTJ 024-85: Specifications for Design of ground and Foundation of Highway bridges and Culverts, in detail as follows:

(1) load increment

Each load increment was 1/15 of estimated ultimate bearing capacity; that is: 5330kN, 8000kN, 10670kN, 13300kN, 16000kN, 18670kN, 21340kN, 24000kN, 26670kN, 29340kN, 32000kN, 34670kN, 37340kN, 40000kN.

(2) displacement observation

Displacement was measured at 5, 10, 15, 30, 45, 60 minutes in the first hour, every 30 minutes after the first hour (under the control of computer).

(3) stable standard

Each load increment was maintained until the rate of movement did not exceed 0.1 mm in every one hour and lasted for two times within 1.5 hours.

(4) determination of pile ultimate bearing capacity

1) for clay, powder soil and sandy soil

$$Q_u = \frac{Q_{uT} - W}{\gamma} + Q_{uB} \quad (1)$$

Where Q_u is the ultimate compressive bearing capacity; Q_{uT} is the ultimate bearing capacity of upward pile shaft; Q_{uB} is the ultimate bearing capacity of downward pile; W is the weight of upward pile shaft; γ is the coefficient of upward pile side resistance. For clay and powder soil, γ is 0.8, while for sandy soil the value is 0.7.

2) for rock

$$Q_u = Q_{uT} + Q_{uB} \quad (2)$$

2.3 Test results

2.3.1 Pile movement versus applied load

Figure 1 presents the results of movements of top plate and bottom plate of load-cell. From the figure the upper and lower piles don't reach their ultimate bearing capacity. When the load reaches the maximum 20 000 kN, the elastic deformation of pile–soil system below the load-cell is 3.94mm(The settlement of the bottom plate when loading reached 20 000kN minus the residual settlement the bottom plate when unloading is up to 0), while its plastic deformation is 11.89mm(the residual settlement the bottom plate), and the elastic deformation of pile–soil system along the load-cell is 1.47mm(The settlement of the top plate when loading reached 20 000kN minus the residual settlement of the top plate when unloading is up to 0), while its plastic deformation is 3.09mm(the residual settlement the top plate). At the same time the compressive deformation is 3.09mm (the displacement of top plate of the load-cell minus the displacement of pile head when loading reached 20 000kN). So the ultimate compressive bearing capacity is calculated as:

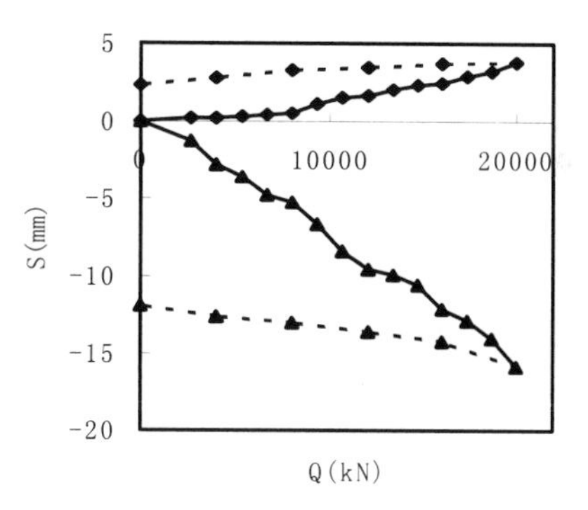

Figure 1: Pile movement versus applied load

$$Q_u \geq \frac{20000 - \pi \times 0.75^2 \times 89 \times 24.5}{0.8} + 20000 = 40180 \text{ kN} \quad (3)$$

2.3.2 Load distribution along pile

The strains measured within the pile by the strain gauges are used to estimate the load transferred along the pile to the soil, rock socket, and rock mass below the pile tip. Figure 2 is a plot of load distribution along the pile with the applied load. The distribution is the same as that of the conventional, that is, the axial force generally decreases with an increase in depth away the load-cell because of the effect of soil resistance, while axial force in the load-cell location is maximal. The load distribution along the pile when unloading is approximately the same as the loading. When unloading until 0, every section has some axial force because there still exist the compressive deformation and the soil resistance along the pile.

2.3.2 side resistance distribution along pile

Side resistance between two sections is obtained by the division of the area of surface between the two sections and the minus of axial force of the two sections, subtracting the self-weight of pile shaft between the two sections in the upper part of pile. Figure 3 represents the side resistance distribution along the pile. Judging from figure 3, the side resistance of every section increases generally with the increase of applied load. The side resistance near the load-cell first comes into effect. At the same time, it is can be concluded that the side resistance has transferred the load into other soil layer before it reaches its ultimate value.

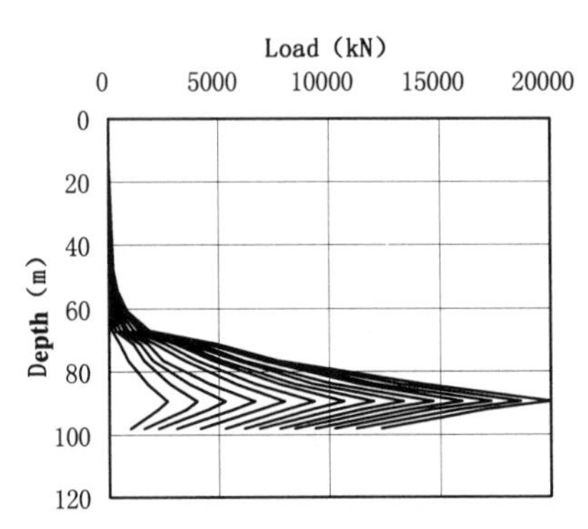

Figure 2: Load distribution

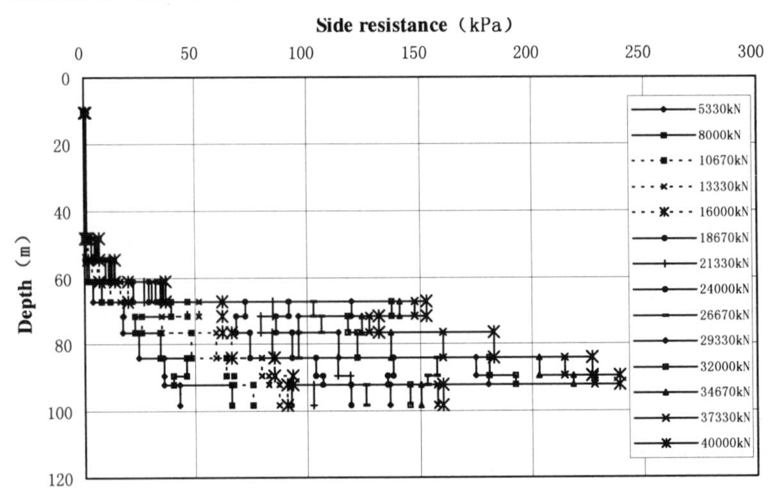

Figure 3: Side resistance distribution

3 CONVERSION TO CONVENTIONAL RESULTS

The bearing behavior of a self-balanced pile is different to that of compressive pile and tensile piles. The behaviors of load transfer of the three types of piles are shown in figure 4.

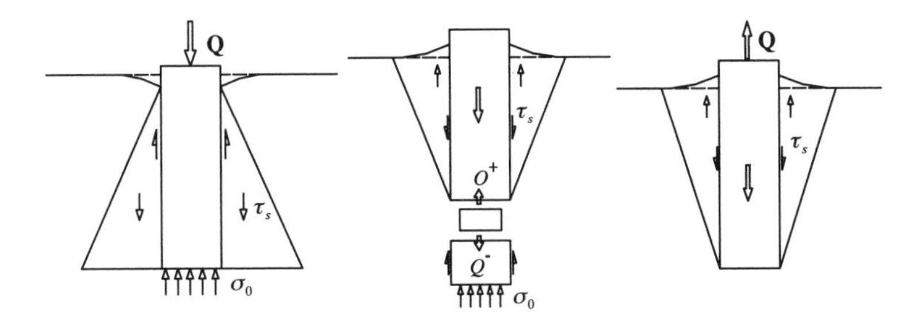

(a) compressive pile (b) self-balanced pile (c) tensile pile

Figure 4: Load transfer behavior

The displacements of the top and bottom plates of the load-cell are measured and the strains are also measured by strain gauges (see Figure 5) in the self-balanced test. The axial force distribution is obtained by calculating strains and sectional stiffness of pile. Therefore the side resistance distribution can be derived. So the equivalent pile head load-displacement can be obtained in terms of the relationship with side resistance and pile-soil displacement, and the relationship with load and displacement of a bottom plate of load-cell. In this analysis, assumptions are made as follows:

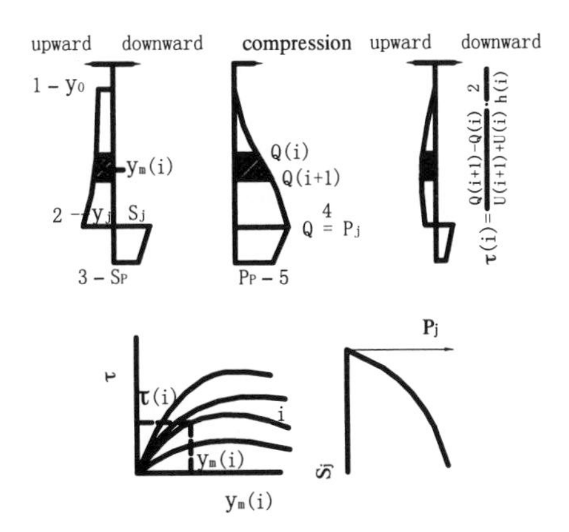

Figure 5: Relationship between axial force and side resistance and displacement

1—pile head displacement; 2—displacement of load-cell; 3—pile tip displacement; 4—load-cell force; 5—pile tip axial force.

(1) The stress-strain relationship of the pile material is elastic;
(2) Each sectional strain can be obtained from axial forces in the upper and lower sections of the unit and average sectional stiffness.
(3) The relationship between load-displacement of pile base and of side resistance and pile-soil displacement are the same as those for the conventional tests.

The upper pile shaft is divided into n units, where axial force and displacement at unit i is represented by P(i) and S(i) respectively in the self-balanced method. Their values can be expressed as follows:

$$P(i) = P_j + \sum_{m=i}^{n} f(m)\{U(m) + U(m+1)\}h(m)/2 \quad (4)$$

$$S(i) = S_j + \sum_{m=i}^{n} \frac{P(m) + P(m+1)}{A(m)E(m) + A(m+1)E(m+1)} h(m) = S(i+1) + \frac{P(i) + P(i+1)}{A(i)E(i) + A(i+1)E(i+1)} h(i) \quad (5)$$

where P_j is the axial force in the pile at unit i=n+1(load-cell force); S_j is the displacement of the pile at unit i=n+1 (the displacement of bottom plate of load-cell); f(m) is the side resistance between pile-soil interface at unit m (between units i~n) (upward is plus); U(m) is the perimeter of the pile at unit m; A(m) is the sectional area of the pile at unit m; E(m) is the elastic modulus of the pile; h(m) is the length of unit m.

$S_m(i)$, the midpoint displacement of unit i, can be expressed as follows:

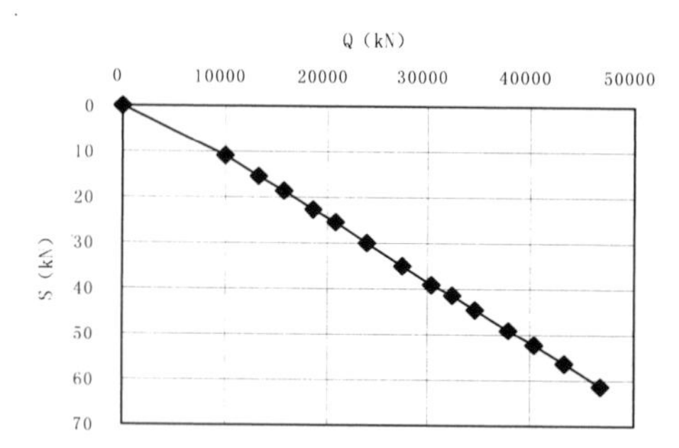

Figure 7: Equivalent load-displacement curves

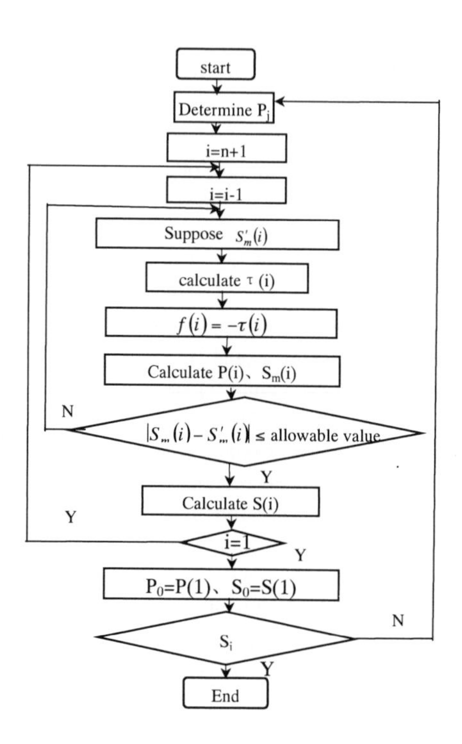

Figure 6: the flow chart of calculation

$$S_m(i) = S(i+1) + \frac{P(i) + 3P(i+1)}{A(i)E(i) + 3A(i+1)E(i+1)} \cdot \frac{h(i)}{2} \quad (6)$$

Substitute equation (4) into equation (5) and (6), can obtain:

$$S(i) = S(i+1) + \frac{h(i)}{A(i)E(i) + A(i+1)E(i+1)} \cdot \left\{2P_j + \sum_{m=i+1}^{n} f(m)[U(m) + U(m+1)]h(m) + f(i)[U(i) + U(i+1)]\frac{h(i)}{2}\right\} \quad (7)$$

$$S_m(i) = S(i+1) + \frac{h(i)}{A(i)E(i) + 3A(i+1)E(i+1)} \cdot \left\{2P_j + \sum_{m=i+1}^{n} f(m)[U(m) + U(m+1)]h(m) + f(i)[U(i) + U(i+1)]\frac{h(i)}{4}\right\} \quad (8)$$

when i=n,

$$S(n) = S_j + \frac{h(n)}{A(n)E(n) + A(n+1)E(n+1)} \left\{2P_j + f(n)[U(n) + U(n+1)]\frac{h(n)}{2}\right\} \quad (9)$$

$$S_m(n) = S_j + \frac{h(n)}{A(n)E(n) + 3A(n+1)E(n+1)}\left\{2P_j + f(n)[U(n) + U(n+1)]\frac{h(n)}{4}\right\} \quad (10)$$

Using the equations (3)～(10) and the relationship between $\tau(i)$ and $y_m(i)$, f(i) can be expressed as a function of $y_m(i)=S_m(i)$. Thus $\tau(i)$ can be obtained and f(i) can be derived through relationship $f(i) = -\tau(i)$. P_j can be obtained through the relationship of P_j and S_j. Therefore, 2n equations about S(i) and $S_m(i)$ can be set up. Figure 6 represents the flow chart.

Figure 7 is the plot of conversion of the test pile into equivalent pile head load-displacement curve.

4 CONCLUSIONS

(1) In this paper the self-balanced pile test method was successfully applied to a pile of Beicha bridge of the Yunyang Highway Yangtze River Bridge, with an ultimate compressive bearing capacity up to 40,000 kN and locating in the Yangtze river, which is hardly possible for a conventional test. This method can determine the side resistance and base resistance and the curves of load-displacement. The results indicate the method will also have excellent prospects in bridge piles tests under conditions of high bearing capacity and testing sites, which are difficult for conventional pile tests to be performed.

(2) A method converting the self-balanced test results to conventional test results is put forward based on three assumptions. The converted results are of benefits to the designer.

Reference

[1] Jori Osterberg (1989). New device for load testing driven piles and drilled shaft separates friction and end bearing. Piling and Deep Foundations.，421～427

[2] 前田良刀(1996). 第二东名东海大府高架桥工区における壁基础原位置载. 基础の工，5：60～66

[3] Ulrich Vollenweder (1997). Prufung nach dem pfahfusspresserfahren. Zurich:Journey of ETH, 6:42～53

Advances in Building Technology, Volume 1
M. Anson, J.M. Ko and E.S.S. Lam (Eds.)

THE DESIGN AND CONSTRUCTION OF TWO INTERNATIONAL FINANCE CENTRE HONG KONG STATION

Gibbons C.[1], Lee A.C.C.[1], MacArthur J.M.[2], Wan C-W.[1]

[1]Ove Arup & Partners, Hong Kong, [2] Ove Arup & Partners, London

ABSTRACT

Two International Finance Centre (IFC Tower) due for completion in mid 2003 will be the tallest building in Hong Kong and 5th tallest in the world at 420m high. It is located on the harbour front of the Central District, the business centre of Hong Kong. This paper describes the some of the geotechnical design considerations, the structural design development and construction of the tower. Specifically, the paper addresses the large diameter cofferdam solution which was adopted for the foundation; the lateral response of the structure; the issues influencing the floor system; and the detailed design of the megacolumn and outrigger lateral stability system. The paper also addresses construction-led aspects of the design adopted to reduce the construction time of the tower.

KEYWORDS

Tall Building, Composite Construction, Outriggers, Construction-led Design.

INTRODUCTION

The IFC Tower forms part of Hong Kong Station Development on the Central Reclamation Hong Kong. Developers of the site are Central Waterfront Properties (CWP) - a joint venture between Sun Hung Kai Properties Ltd, Henderson Land Development Co Ltd, Bank of China Group Investment Ltd and the Hong Kong & China Gas Co Ltd. The Mass Transit Railway Corporation (MTRC) is also a development partner of the joint venture company. Architects for the project are Cesar Pelli & Associates and Rocco Design Limited with Ove Arup & Partners (Arup) providing structural and geotechnical engineering consultancy services. Mechnical and electrical consultancy services were provided by J. Roger Preston Ltd.

The tower comprises 88 storeys with a basement level of -32.0mPD and a level to the top of the roof feature of +420.0mPD. The general footprint of the tower is 57m x 57m. Towards the top, a series of staggered step-backs reduce the plan dimensions to 39m x 39m at roof level. In all, it provides a gross floor area over 180,000m^2 of grade A office accommodation. Figure 1 shows a model of the tower on completion. One of the main requirements of the brief was to provide open plan floors and to incorporate a large degree of flexibility in order that the requirements of major tenants (such as financial institutions) could be accommodated.

The tower is just one element of a major new commercial development in central Hong Kong. The development as a whole provides office, retail, hotel and serviced apartment accommodation and accommodates Phase 2 of the MTRC Hong Kong Airport Railway Station in the 5 level basement below. It also houses major transportation facilities for the local bus companies together with several floors of car parking. The total constructed floor area of the Hong Kong Station Development amounts to 650,000 square metres.

The Main Contractor for the tower, and the remainder of the development, was a joint venture (ESJV) between E. Man and Sanfield Contractors. The steelwork for the tower was awarded in two stages - both as nominated subcontracts. The steel for the megacolumns below 6/F level (5,000 tonnes) was fabricated and installed by NKK Corporation of Japan. The steelwork for the tower above 6/F (19,000 tonnes) was awarded to a joint venture (NSJV) between NKK Corporation and Sumitomo Corporation.

The construction of the raft commenced in January 2000. At the time of writing, June 2002, the central core was constructed to level 82/F.

TOWER CONFIGURATION

The basic plan configuration of a typical floor is shown in figure 3. The core at ground level is 29m x 27m with perimeter walls 1.5m and 1.25m thick. The size of the core was essentially driven by the need to maximise the efficiency of the vertical transportation system to serve the tower. The core is of conventional reinforced concrete - studies had shown that there was a significant cost advantage compared to steel and steel/composite alternatives. Initial cost studies were also conducted on core designs comprising grade 60 (28-day cube strength in N/mm²), grade 80 and grade 100 concrete. These revealed that no great advantage was gained by changing from the grade 60 option, as savings in lettable floor space did not significantly compensate for the additional cost and construction implications of using a higher grade.

The fundamental requirement for flexible office layouts, and the desire to maximise the panoramic views, necessitated that the perimeter structure should be kept to a minimum. This led to an outrigger lateral stability solution employing eight main megacolumns (two per face) with small secondary columns in the four corners. The outriggers mobilised the columns directly without the need for transfer through a belt truss system. A less accentuated belt truss system was incorporated into these floors, however, to accept the heavy plant room floor loadings, and transfer the secondary corner column loads (as described below) into the megacolumns.

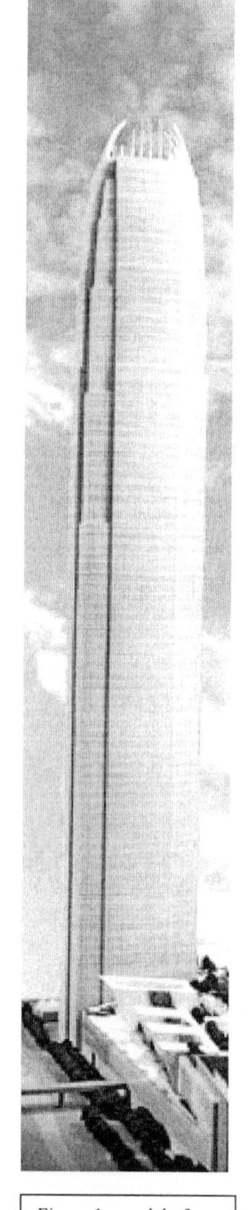

Figure 1: model of completed tower

Studies showed that three outrigger levels were required, these were located at the levels R2/F to 33/F, R3/F to 55/F and R4/F to 67/F. As the triple floor height steel outrigger trusses coincided with refuge and double mechanical floor levels, their locations were effectively a compromise between the optimum structural arrangement and constraints imposed by vertical planning and lift zoning. In any event, the system was effective in that the amount of structure occupying useable floor area was minimised. The small corner columns support gravity load only. There are three ones of such columns, each extending a maximum of 20 storeys, which are supported off transfer trusses at each of the outrigger levels. The loads for the secondary columns are effectively 'collected' by these trusses and transferred to the megacolumns.

At the higher levels, where the corners of the floor plate are stepped back, these columns are removed with the floor plate cantilevering from the megacolumns. Consideration was also given to the removal of these secondary columns throughout. However, it was concluded that the cost of the additional weight of floor steel involved, and the likely problems associated with tolerances of large cantilever floor systems, warranted that these small columns 300x 300mm should remain.

GEOLOGY AND FOUNDATION DESIGN

The geology in the vicinity of the tower is fairly typical of Hong Kong. Grade III granite, having a permissible bearing capacity of 5MPa, is found approximately 35 metres below ground level although it shelves off steeply to the west of the tower footprint. Above this are layers of decomposed granite and alluvium, and over the top 20 metres there is a layer of fill, the land only having been reclaimed three years earlier. Dynamic compaction had been carried out on certain areas of the fill to enable diaphragm walling to take place. With a total tower load of 5,200MN, foundations obviously had to be taken to rock head.

Surrounding the tower is a five level basement with a low-level commercial podium built above it. The southern most wall of the basement box also forms part of the adjacent Hong Kong Station Phase 1 structure. With trains under operation, it was essential that movements in this vicinity should be kept to a minimum. To achieve this top down methods of construction were adopted for the general basement construction outside the footprint of the tower which enabled potential movements to be minimised.

In order to optimise the construction programme, it was considered appropriate for the tower to be built using a more conventional bottom up technique. The initial foundation solution envisaged 2.5 metre diameter bored piles, belled out to 3 metres. A large diameter cofferdam, encompassing the plan form of the tower, would then be constructed using diaphragm walling techniques to enable excavation down to the pile cap level. This would enable the cap to be cast in open excavation and the whole tower constructed from the pile cap level within the cofferdam. The circular (compression ring) nature of the cofferdam eliminated the need for internal props to provide lateral support the excavation.

Figure 2: Excavation within the temporary cofferdam

Consequently excavation within the cofferdam could be carried out unhindered (figure 2). The cofferdam provided very stiff lateral support and consequently further ensured that the ground movements were kept to a minimum, particularly with respect to the operational MTRC tunnels running alongside the development.

The foundation sub-contractor Bachy Soletanche Group proposed an alternative foundation solution which was adopted in the final works. With a reasonably constant rock head level (as determined from site investigation) it was proposed to construct the 61.5 metre diameter cofferdam down to bedrock using a 1.5 metre thick diaphragm wall keyed into the rock. Using three reinforced concrete ring beams, and lowering the external water table by eight metres, excavation could then take place to rock head level and the pile cap/raft be cast bearing directly on rock, thus omitting the need for bored piling. Conditions on site proved to be slightly different from those anticipated, with a localised depression to the South East of the footprint of the rock head level down to a depth of approximately 55 metres, a depth too great to allow open excavation in this area. In order to overcome this a mixed foundation solution was adopted. Over much of the area the original raft solution was adopted. Mass concrete fill was then used locally, between rock head level and the underside of the cap, in the areas where the rock head sloped away. Locally, at the location of deepest rock, barrettes were installed from ground level to transfer the pile cap/raft loads to the bearing stratum, use also being made of the cofferdam panels to transfer vertical load.

Construction of the tower commenced in earnest in January 2000 with the first pour of the 6.5m thick reinforced concrete raft. With a total concrete volume of almost 20,000 cubic metres for the entire raft, it was considered not feasible to cast the raft in one continuous pour. The first pour was 5,000 cubic metres in volume and covered the entire area of the cofferdam. A further eight pours each covered half the area of the cofferdam, with the vertical construction joints rotated on plan through 90 degrees for each successive layer. A final capping layer comprising 3000 cubic metres of concrete, again covered the entire area of the cofferdam in a single pour.

LATERAL RESPONSE

Wind tunnel studies were performed on the tower by Rowan Williams Davies & Irwin Inc. (RWDI) in accordance with the Hong Kong Buildings Department Practice Note PNAP 150. This included topographical studies, a force balance assessment of the loads and monitoring of cladding pressures. A second confirmatory wind tunnel study was undertaken by Cermak Peterka Peterson (CPP). The two studies agreed to within 6%.

Studies highlighted the dominance of crosswind response for this particular building. The resulting global characteristic base bending moments and base shears in the orthogonal directions were 19,000MNm and 128MN. The combination factors used in the derivation of diagonal design forces were $\pm 0.79F_x \pm 0.79F_y$.

The predicted period of the building is 9.1 seconds, which accords quite well with the simple H/46 approximation. Lateral accelerations were predicted and compared against the NBCC, ISO and Davenport Criteria for occupancy comfort. In doing this a variable structural damping was built into the analysis equivalent to 0.8% percent at 1year return events, varying linearly to 2.0% for 50 year return events. Under these conditions the most critical accelerations, in terms of impact on human comfort, occurred in the 5-10 year return typhoon event range. The accelerations were however deemed acceptable for office occupancy in Hong Kong without the need for supplementary damping.

The lateral deflections under wind loading, including the second order P-delta effects of gravity load, were H/450 in the orthogonal direction and H/380 in the diagonal, where H is the height of the building above pile cap level.

FLOOR SYSTEM

Above the 6/F the tower comprises typical office floors and trading floors, with a design imposed loading of (3+1)kPa and (4+1)kPa respectively. Below 6/F level the floors are reinforced concrete - commensurate with the podium and basement construction. Initial designs were conducted to compare prestressed concrete and composite steel/concrete floor systems for the floors above 6/F. The concrete solution comprised a 275mm post-tensioned slab with 2000mm x 650mm deep perimeter reinforced concrete band beams. The composite schemes comprised 125mm thick slabs acting compositely with permanent decking supported at up to 3m intervals on a variety of steel beam options. Although studies showed that the cost of the concrete floor was slightly less, and that the anticipated cycle times for the two systems were similar, the additional costs for the columns and foundations due to the increased dead loading showed that, overall, the composite solution was preferable.

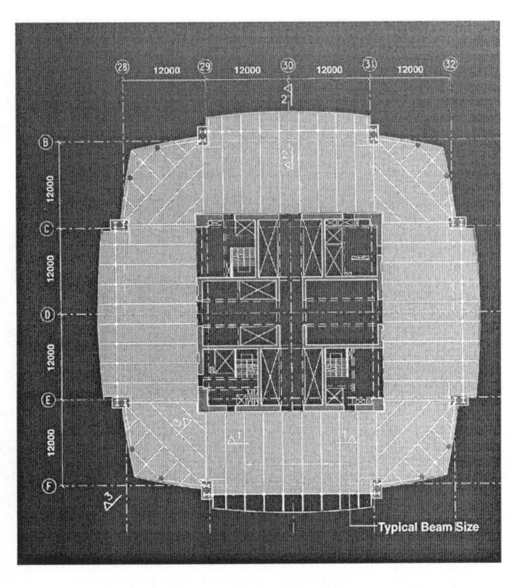

Figure 3: Typical floor framing

The typical floor-to-floor height is 4.2m with a dimension from underside of ceiling to the top of floor slab of 1.2m. A number of composite floor solutions were investigated which offered varying degrees of service/structure integration. Of these, asymmetric fabricated tapered beam and composite truss solutions presented the lightest (though not necessarily the cheapest) solutions. Such systems were, however, not favoured due to the need for maximum flexibility in the layout of main and tenant services within the floor. The solution that was adopted comprised 460mm deep steel secondary beams spanning (11.4m to 13.5m) from the core to 900mm deep primary girders spanning 24m between the main columns. One of the key features of the layout of the floor was the inclusion of a significant diagonal beam which, in conjunction with the primary girder on the main faces, provided a continuous 'tension ring' around the floor plate. This was deemed necessary to enhance robustness and provide direct buckling restraint to the columns. This peripheral primary beam arrangement provided a zone around the core where services could be installed beneath the steel beams requiring only small penetrations through the main primary girder for minimal services which needed to access the building perimeter. The 24m primary girder comprised an asymmetric fabricated section and was structurally continuous with the megacolumns.

The limiting criterion for the design of the floor system was the vertical inter-storey deflection limits at the facade of 20mm. In achieving this limit, the effects of potential differential shrinkage and long term creep effects of the primary beams on adjacent floors, were considered in addition to patterned imposed load. These effects along with the axial shortening of the core and columns were incorporated in the construction presets to ensure that the building was constructed within acceptable tolerance.

The weight of floor beam steel is 36 kg/m^2 based on gross constructed floor area.

MEGACOLUMNS

The megacolumns comprise composite steel/concrete in which the steel elements are encased with reinforced concrete. Table 1 presents a summary of the columns sizes at various heights in the building together with details of steel content. Extensive studies were carried out to establish the most appropriate form of the columns in terms of cost and buildability and to optimise the size and steel/concrete ratios.

Concrete encasement was adopted to enable the maximum use of steel in the form of reinforcement rather than the less cost-effective use of structural sections. Reference 1 describes initial studies that were performed to investigate the optimum proportions of those elements that contribute to the lateral stiffness of the tower. This included an assessment of the optimum proportions of the steel and concrete in the megacolumns (as the proportions of the core walls and outrigger elements) to minimise the initial capital structural cost of the tower. In addition, reference 1 describes subsequent optimisation studies that were performed considering the value of the useable floor space occupied by the structure.

Figure 4: Column Climbform system

One key issue was the need to maximise the buildability of the megacolumns. As a consequence, the structural steel component was split into a number of sub sections that could be lifted and connected with ease. In addition, it was considered at the outset the concrete encasement of the megacolumns would be formed using self climbing formwork (figure 4). This latter initiative effectively removed the reliance on the cranes in lifting column formwork between floor levels, thereby maximising their efficiency in lifting structural steel components and reinforcement. The formwork system effectively comprised hinged forms that wrapped around the column. Prior to jumping to the next floor, the form were unfolded to form a single plane on the façade side of the column. The formwork could then climb without interfering with the pre-installed main structural steel elements. Following the passage of the form, a few light trimming steel members could be then added to complete the floor framing immediately adjacent to the column and façade.

TABLE 1

MEGACOLUMN SCHEDULE

Level	overall dimensions	no. of sub-stanchions	Averaged weight of steel (tonne/m)	Percentage of reinforcement bar	Design effective length (m)	concrete grade
B5 to 6/F	2.3m x 3.5m	6	9.7	4.0 %	30.3	60D
6/F to 32/F	2.3m x 3.5m	3	2.7	3.5 %	24.0	60D
33/F to 52/F	1.85m x 3m	2	1.1	3.0 %	19.2	60D
53/F to 69/F	1.4m x 2.6m	2	0.9	3.0 %	14.8	45D
70/F to 77/F	1.2m x 0.9m	1	0.6	2.0 %	12.8	45D
78/F to Roof	1m x 0.75m	1	0.5	2.0 %	8.9	45D

The effect of the megacolumn buckling and restraints provided by the floor diaphragms was a key consideration given the massive nature of the columns and the relatively thin nature of the individual floor diaphragms. A second order non-linear analysis was undertaken to investigate the interaction of the spring stiffness of the floor diaphragms and the column buckling. This enabled the effective length of the column to be determined and quantified the forces to be resisted by the floor diaphragms (see table 1). The inclusion of a continuous substantial primary beam tie connecting all the megacolumns facilitated the mobilisation of the floor diaphragms in providing the necessary column restraint. It should be noted that this perimeter tie also enhanced the overall robustness of the tower.

The Contract for the structural steelwork was split into two parts in order to gain an overall programme advantage.

The steelwork in the megacolumns up to 6/F level formed one contract, with all the steel above that level forming the larger follow-on steelwork contract. In the initial contract, the steel in the megacolumns was installed by mobile cranes from ground floor level into the temporary cofferdam. The optimisation studies had shown that there were significant commercial advantages in increasing the stiffness of the columns in this first contract over and above that required for strength. It effectively minimised the stiffness requirement, and hence the impact on the tower cranes (of lower capacity than the mobile cranes), in the follow-on contract.

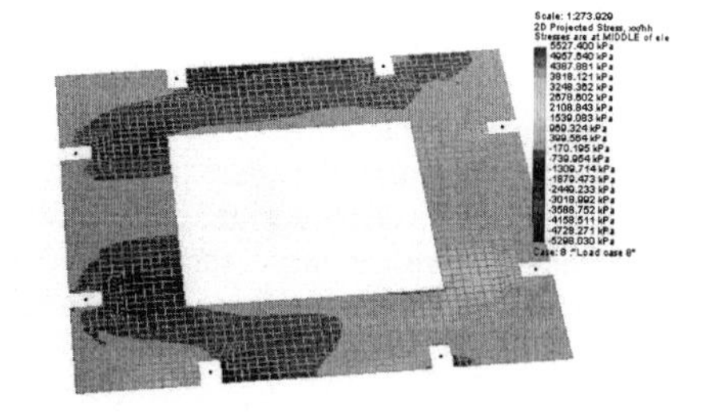

Figure 5: Finite Element analysis of floor diaphragms to assess stiffness and resulting forces in resisting column buckling.

Hence a construction-led optimisation of proportioning of structural material. The total weight of steel in the megacolumns is equivalent to 49kg/m² over the gross constructed floor area of the tower.

OUTRIGGERS

At the outset it was apparent that, in order to satisfy the access requirements in and around the mechanical floors, a steel outrigger system would be required. A punched concrete wall type outrigger was considered, however, this provided insufficient stiffness given the openings that had to be accommodated.

The steel truss outriggers pass through, and are cast within, the core walls. Optimisation studies had suggested that the axial stiffness ratios for the bottom, middle and upper outriggers should be 1 : 0.81 : 0.64 respectively (ref 1). Initially, it was considered that only the top and bottom booms of the outriggers would pass through the core - the longitudinal shears between the booms being resisted by the reinforced concrete of the core walls. However, due to the need for large openings through the core wall to accommodate the requirement for significant M&E access, it became apparent that a steel truss would be required to supplement the strength and stiffness of the perforated core.

In the analysis of the outrigger system, a detailed assessment of the local deformations at the core wall/outrigger interface was made, with the stiffness of the overall lateral stability system modified accordingly. Extensive analysis was carried out to assess the precise characteristics of the interface between the steel truss and concrete wall components to ensure strain compatibility - thereby minimising the potential for cracking in the core. The analysis took into account the flexural stiffness of the shear studs and anchorage that were used to transmit the forces from the outriggers to the concrete core. From the analysis it was found that to achieve compatibility of strain, it was necessary to restrict the stress in the shear studs to half their design capacity – essentially restricting the flexural performance of the studs to the elastic range, thereby maximising stiffness.

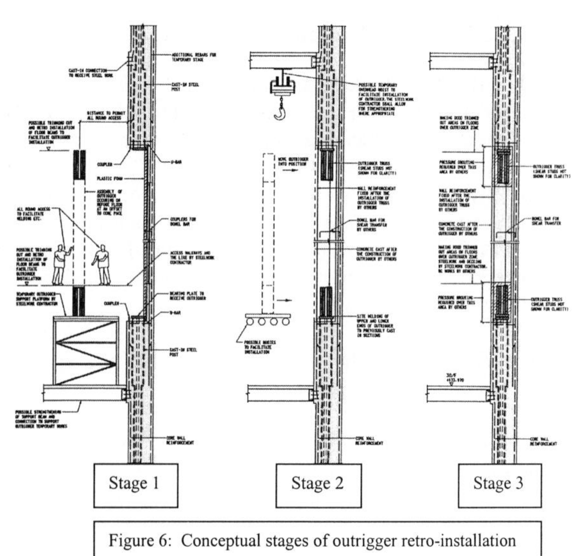

Figure 6: Conceptual stages of outrigger retro-installation

One of the key issues concerning steel outriggers in composite tall building construction, particularly where such large outriggers are required, is their potential impact on the construction programme. The core of such buildings can be constructed with relative speed and efficiency using a climbform system. Typical cycle times achieved on the core were 3-4 days. Clearly, stopping the climbform at the outrigger levels to permit the steelwork contractor to install the outrigger is not conducive to optimising continuity of labour usage, or minimising the construction programme.

Figure 7: Retro-installation of steel outrigger within the core wall

Figure 6 shows conceptually, the retro-outrigger installation approach adopted on the tower. Stage 1 involved reducing the outer core wall thickness from (typically) 1000mm to 300mm over the full height of the outrigger zone. Above the outrigger zone, the wall thickness reverts to (typically) 1000mm. Although this involved an adjustment to the climbform system at the locations of the change in wall thickness, it did permit the climbform to pass through the outrigger zone in advance of steelwork installation. In Stage 2, and with the climbform continuing to construct the upper levels, the outrigger elements were located in position and assembled. Figure 7 shows the partial installation of the outrigger within the zone corresponding to the thinned core wall. Stage 3, with the welding of the outrigger completed, the core wall was retro-concreted. This included a 100mm thick grouted layer at the top of the outrigger zone at the interface with the widened section of the wall. Figure 8 shows a view of the tower during construction with the climbform progressing beyond the outrigger zone and the partially completed steel outrigger.

Figure 8: Climbfrom construction of the core progressing prior to completion of the steel outriggers.

DIFFERENTIAL AXIAL SHORTENING

One key considerations is the differential shortening between the core and the perimeter columns during the construction of the tower and the long term effects post completion. This is particularly important in composite tall buildings in which the deformation characteristics of the steel intensive columns and the large core can be markedly different. Prior to construction a number of specific material tests were carried out to quantify the modulus and shrinkage characteristics of the proposed concrete mix. This was necessary because of the limited data available on the precise characteristics of grade 60 concrete incorporating 25% PFA. From these studies, Arup undertook a comprehensive study to quantify the differential shortening and the necessary pre-sets to be built into the construction of vertical elements.

The outrigger connections to the megacolumns incorporated a series of packing shims at the contact surfaces. This enabled the outriggers to be effective during construction to enable the tower to resist any typhoon winds that may have occurred. In addition, it allowed the packing shims to be removed, or added, as required (with the assistance of small jacks) to enable differential movements between the outrigger and columns to occur during construction. Using this approach, it effectively prevented the build-up of very large internal forces that would otherwise have been generated in the event that the outrigger and columns were rigidly, and permanently, connected together from the outset.

CONCLUSIONS

This paper has presented a brief overview of some of the design and construction-led innovations that have been adopted in the construction of the IFC Tower, the tallest building in Hong Kong.

REFERENCES

1. Chan, C.M., Gibbons, C., MacArthur, J.M., *"Structural Optimisation of the North East Tower, Hong Kong Sation"*, Fifth Int. Conf. on Tall Buildings, Hong Kong, Dec. 98.
2. Gibbons, C., Lee, A.C.C., MacArthur, J.M., "The Design of the North East Tower, Hong Kong Station", *Fifth Int. Conf. on Tall Buildings,* Hong Kong, Dec. 98.

Advances in Building Technology, Volume I
M. Anson, J.M. Ko and E.S.S. Lam (Eds.)

The construction of deep and complex basements under extremely difficult urban environment – 3 representing projects in Hong Kong

Raymond W M Wong

Division of Building Science & Technology, City University of Hong Kong

ABSTRACT

Hong Kong is famous for its congested urban environment and hilly topography, which impose enormous constraints in the construction of buildings. Cases like constructing very tall buildings, either of residential or non-residential nature, up to 50 or even 60 storeys high, and in close proximity of MTR tunnels, near sensitive slopes, or in the middle of an old town area, are very common. Such building examples are often consisting of very deep basement, with depth sometimes more than 20m.

The author has identified 3 project cases in recent Hong Kong which can best illustrate the mentioned features. These projects include the redevelopment of the Lee Gardens Hotel, the construction of the Festival Walk and the construction of the Cheung Kong Center. The author wishes to generalize these 3 projects, each has applied rather unique and innovative concepts and techniques in construction, and highlight how engineers and constructors in Hong Kong tackle such extreme situations and have the jobs neatly accomplished.

KEYWORDS

Basement construction, difficult construction environment, common construction problems and solutions

INTRODUCTION

Deep basements are becoming very popular as buildings constructed higher and higher in Hong Kong. Numerous engineering problems are likely to be encountered as construction works are going deeper and deeper down into the ground. Engineers and builders have to face a lot of problems such as the existence of complicated sub-soil, overcoming of tremendous soil pressure, the provision of complicated temporary support works, working in congested underground or sensitive nearby environment. In particular in most modern development projects, the scale of construction tends to be growing bigger and bigger, that makes planning and construction of deep basement become much more difficult and complex.

At the same time, the construction of deep basement is very expensive, time consuming, inconsistent and sensitive to the quality of planning and management of individual projects. The most important of all, such works are highly hazardous, both to human operatives working within and the life and properties of third parties that within the vicinity. Though very technical or engineering in nature it seems, to implement jobs of that level of complexity is of no doubt associated with quite a lot of managerial challenges. Such as, in the preparation of a highly efficient working programme, monitoring and rectifying the progress of works in case problems arising, or in resources planning where materials, labours and plant equipment are involved.

From the construction point of views, there are many methods to construct large-scaled and deep basement. Deep basement can be constructed using some traditional ways such as cut and fill or bottom up methods. These methods are very effective when dealing with certain jobs which is simpler in nature. On the other hand where basement is going deeper and the surrounding environment getting more complex and sensitive, bottom up method may be a more appropriate option to construct (Fig. 1).

Under typical urban environment, the situation where a basement is to be constructed can be very complex. Scenario such as working in congested site (Fig. 2), working very close to sensitive buried structure (eg. adjacent to a section of railway tunnel), or the site is located close to relatively unstable slope, are common examples that can be found everywhere in Hong Kong. Or, in certain cases, the design of a basement structure is very deep, the size of the basement is exceptionally large, sub-soil condition is very complicated, are again typical situations that make basement construction becomes very difficult.
Besides, there are situations where a new basement is required to construct simultaneously to replace an old one. Or due to the requirement of working under fast-track schedule, some basement works need to be carried out at the same time with the new foundation or even with part of the future superstructure. This will create very difficult coordination problems that involve various contractors and complicate the contractual position of the entire job. The process to make decision arriving at the best solution is of no doubt an agony journey for both engineers and builders.

Fig. 1 – Very large basement job with top-down and bottom up approach working at the same time

Fig. 2 – Example of basement constructed in very congested down-town environment

The main objective of this paper is to find a solution for the mentioned problems for there should be no absolute cure or one-step formula for construction works of such scale and complexity. Most often, the extent and scale of work for complex basement projects are usually very broad with a huge contract sum. It is difficult to compare alternatives base on previous data or project cases. Accurate cost analysis or work study is difficult to carry out either for there is almost without standard ground to make comparisons. Every project, though look relatively similar from certain indicating factors, is in fact unique in itself. A great number of random and uncontrollable variances are likely to arise during the courses of work. This makes planning and scheduling almost cannot be exact. The actual effectiveness of works is highly depended on the as-constructed site environments. Besides, the quality of the management and the executing parties, as well as the problems solving ability of the frontline personnel, also seriously affects the performance and effectiveness of works.

By the use of 3 recent basement projects, the writer tries to bring up some very representing project cases and demonstrate how works can be arranged to fit in such complicated environment and have the job accomplished in a safe and efficient manner.

LEE GARDEN REDEVELOPMENT PROJECT

The project located in a 5,750 m^2 site, which was abutted on 3 sides to small roads from 12 to 20m wide, and the remaining side adjoining a 17-storey residential building of 35 years old. In addition to the congested environment, the project also required to demolish the 22-storey Lee Garden Hotel, with a 2-level basement structure in it. In the redevelopment, a new 50-storey office building, constructed in structural steel with a RC core, together with a 4-level basement, was to be built. There were quite a number of difficulties to be overcome in the project. For example, almost all the new foundations and the required ground strengthening and permanent basement supporting works had to be carried out in the old basement before it could be demolished. As a result the old basement could only be demolished in small sections to allow for room and to cope with other associated works. At the same time when part of the basement was being demolished and cleared, temporary or sometime permanent supporting structures have to be built as soon as possible to infill the void until the old basement was completely replaced by the new. In order to gain more time, provisions were also made for the construction of the future building including the central core in RC as well as part of the new basement constructed in top-down method. Due to such constraints, it is comprehensible that limited mechanical plant could be used during the entire process.

Difficulties and Uniqueness

Though project of this kind is not exactly uncommon in Hong Kong, the following features, however, still make the project quite unique and thus imposed certain technical difficulties.

- A 2-level old basement structure covered the entire site area was to be demolished.
- A 4-level basement of the same size was to be built to replace the old, on top of which is a 5-level podium structure. This made the site extremely lack of working space.
- Due to the old basement could not support those heavy equipments (like the RCD) which were required for the construction of the foundation. Hand-dug caissons were thus used in this case.
- Because of the inevitable sectioning and phasing arrangement involved in the demolition of the old basement and to replace it with a new structure, complicated planning and construction jointing provisions were required (see attached demolition phasing plan).
- Part of the future building structure was carried out at the same time while doing the major substructure works. This included the construction of the foundation and raft for the central core of the future tower.
- The congested environment made storage and transportation arrangement within site very difficult. At a result of this, huge temporary loading platforms were provided at different locations (required to relocate from time to time) in order to store the steel stanchions, to station mobile cranes or other excavating machines etc.

Demolition of the Old Basement and Construction of the Substructure

In order to have the old basement demolished and replaced by the construction of a new, as well as to make way and facilitate the construction of the new 50-storey tower block, the following works were being carried out.

Demolition support and ground stabilization

- A series of temporary supporting system in the form of a steel strut frame was erected before various stages of basement demolition.
- A grouted wall was formed along the site perimeter down to 3m below bedrock level (av. 25m deep) as a means of ground water control.
- Demolish part of the basement slab along the perimeter wall to give way for the construction of hand-dug caissons, which were used later as cut-off wall for the new basement structure. Totally 244 caissons of 1.2m diameter were constructed for the purpose.

Construction of new foundations

- Hand-dug caissons were formed as foundation for the future building within the old basement. Totally 49 caissons with diameter ranging from 1.6m to 5.0m were constructed.
- The central part of the old basement was demolished to provide working space for the carrying out of the foundation and raft for the future building core.
- Hand-dug caissons and the caisson raft were constructed to support the central core of the new building. Excavation and construction of the 5m diameter caissons was done at the bottom of a 800m^2 pit that formed after the demolition of the center part of the old basement (Fig. 3).

Figure 3 – A pit formed further down from the partly demolished old basement structure for the construction of the foundation and raft for the future building tower.

Figure 4 – Layout showing the extremely complicated construction phasing as seen at the peak period.

Changing over and provisional works for the new building

- Part of the old basement structure along the perimeters was demolished to allow space to build the capping beam on top of the caisson wall.
- Similarly, part of the old basement structure was demolished to provide room for the erection of steel stanchions as column support to future building. These columns could enable the superstructure be constructed at the same time with the basement that was built using top-down method.
- The building core was constructed from the raft up to ground level. Since there was very limited working space within the central pit, traditional timber formwork was used in the construction of the core. The core would act later as a lateral support for the ground floor slab which served also as a separating plate to facilitate the construction of the top-down basement.
- Demolish the remaining old basement, section by section, and covered the space immediately with the new ground floor slab. Phasing and junctioning arrangement was the most difficult part of work here (Fig. 4).

Construction of the Basement

The basement was constructed using top-down method. In order to allow excavated material could be removed conveniently, two outlet points were provided on site. A grab lifter was set up on top of one of the outlets for the removal of spoil. As usual top-down construction does, excavation started from the top level downward until it reached a depth of about 5m where it would be shored and strutted. Basement slab would then be constructed with connection made to the central core, steel columns and the caisson wall using couplers. The works repeated until it reached the bottom level of the basement where the caisson caps and other ground beams were finally constructed.

THE FESTIVAL WALK

The Festival Walk, situated on a 21,000 sq m site on a stump of a small hill near the Kowloon Canton Railway (KCR) and the Mass Transit Railway (MTR) stations. The project exhibited a lot of complications during the course of its construction, basically inherited from the special nature of the job such as its unfavorable topographical and geotechnical environment, working within the railway and tunnel lines of two very busy railway networks, the requirements of constructing a very large building with exceptionally deep basement, as well as some access problems.

Topography and Geology of Site

The site is situated on a narrow terraced strip of land along the Tat Chee Avenue (TC Av.), which measured about 290m x 80m in size. The existing ground levels vary from +29mPD to +36mPD along the west boundary on TC Av. side, and from +19mPD to +26mPD along the east boundary on the KCR and Kowloon Tong (KT) Station side. In order to cope with the aerial height restriction requirements and to achieve the development potential allowed under the condition of sale of the site, there is 4 levels of basement and 3 levels of semi-basement in the development, with the deepest level being some 36m below TC Av. Completely decomposed granite was encountered at ground level, with the presence of many corestones. Rockhead on the TC Av. side varied between 6m to 65m deep. At the northern portion of site, bedrock was very close to the surface, but sloping downward to about 60m deep near the KT Station. The MTR tunnels run directly through the center of the site, thus confined the geometrical design of the sub-structure. As a result, the lowest excavation levels to the north and south of the tunnels are at -1mPD and +8mPD respectively, with the section above the MTR tunnels not exceeding +13mPD so as to ensure at least 3m cover above the tunnel structures.

Site Formation and diaphragm wall construction

The site formation contract required the contractor to excavate and remove the soil material on the terraced site from averaged +34.5mPD on TC Av. side down to +19.5mPD on the KT Station side. This involved initially a total of about 180,000 cu m of excavation.
To support the sides of the excavation, a 1.2m thick diaphragm wall was constructed around the site perimeter as well as along the MTR tunnels as a cut-off between the basement and the tunnel structure. The walls generally extended to the rockhead, which varied between 6m and 65m below ground level. Due to the shallow-laying of bedrock, the toe of the diaphragm wall in many locations were formed well above the final formation level of the basement. Underpinning works to extend the wall down to the final level were thus required, which would be carried out in parallel with the construction of the top-down basement in the following contract.

Since chiseling was not allowed within 10m of MTR tunnels, “stitch drilling” was being used to overcome underground obstructions before the forming of the diaphragm walls. This was done by drilling a series of 450mm diameter holes in row, so that any boulder or corestone could later be cut into fragments small enough to be removed by grab. Site formation works were in general phased from the north to south. The main reason was to maintain an entrance/exit point, which situated at the south-western corner of the site, for the removal of the overall 460,000 cu m excavated spoil efficiently during the course of the site formation and basement construction processes.

Phasing and Removal of Spoil

Site formation works were carried out in roughly 6 phased, unsymmetrical sections, according to the convenience of cutting and dividing of the complicated building layout with construction jointings. The main strategy in the scheduling of site formation was to have the northern portion completed as the earliest possible, so that this portion could hand over to the main contractor for the construction of the basement and superstructure. The reason is straight forward, a circular ramp leading to the basement carpark is located here. With the circular ramp completed, it could be used as the access for vehicle to enter into the basement which constructed using top-down method. Without which, the last phase of formation on the southern tip of site could not be carried out in full scale. By that time, the only entrance/exit into the site was still in the southern tip.

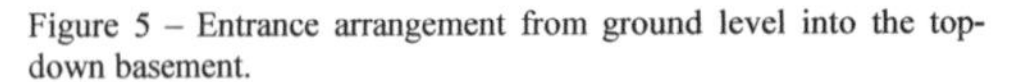

Figure 5 – Entrance arrangement from ground level into the top-down basement.

Figure 6 – A complicated junction where various major phasing sections of the superstructure and basement meet

Construction of the Basement Structure

Based on the initial formation level on +19.5mPD, there are 4 levels of full basement and 3 levels of semi-basement to be constructed. As for all top-down basement construction, the first floor plate to be constructed is important for it signifies the commencement of the basement work by providing the separating plate and lateral strut such that basement excavation can be started from there on. In this project, the first plate was located on the +19.5mPD level (Fig. 5). The basement construction also followed the 6-phased arrangement in conjunction roughly with the site formation sequences (Fig. 6). Instead of working with the basement and superstructure at the same time The superstructure was constructed in an advanced stage from the 1st and 2nd portions in staggered section onwards. This was to accommodate enough working room until the basement excavation could be started in a more efficient manner, as well as to allow the additional weight of the upper structure to balance the buoyancy effect during the excavation. With the first two to three levels of the top structure being maintained at its typical cycle, an entrance point to the basement below was then formed at the edge of position between portion 2 and 3.

Figure 7 – The portion of the basement slab located above the MTR tunnel tubes. Note the thickened slab serving as a transfer membrane spanning over the tunnel and the steel stanchion for the support of the basement structure during top-down construction

Figure 8 – The lateral support system erected during the basement excavation process to stabilize the existing MTR pedestrian entrance/ventilation shaft which is located on the right side of the photo

From this entrance point, excavation to the basement and the construction of the semi-basement structure above the +19.5mPD plate proceeded simultaneously according to the preliminary 6-phased arrangement, again, in staggered sections with the separation of carefully located construction jointing. For the lowest 4 levels of basement with headroom averaged at 3.2m that used as parking spaces for private car, part of which were excavated and constructed using a "Double Bit" method. This could produce a higher headroom such that excavation could be done using normal-sized excavating machines, as well as to allow the entrance of dumping vehicles for removal of spoil. To provide the required protection to the MTR tunnels and to prevent heaving while large volume of soil was being removed, some of the basement slabs, especially those around the MTR tunnels, were deliberately thickened for the purposes (Fig. 7).

There were some locations where the progress of the basement construction works had been significantly interrupted. At the northern edge and the adjoining corner along the TC Av. where the rockhead was laying shallowly well above the final formation level of the lowest basement, the toe of the diaphragm wall panels had to be extended further downward until it reached the final level below the lowest basement. This was done by in-situ underpinning method. Sections of the diaphragm wall panels were constructed after the removal of the bedrock, layer by layer, with vertical junctions being connected by the provision of steel couplers. This section being interrupted measured about 100m. To allow for the continual progress to the upper structure and to stabilize the effect of isolating this part of the basement structure, a row of steel strut was erected, which provided the lateral support between the diaphragm wall and the base plate at the +19.5mPD level.

Another major area of interruption came from the works around the MTR pedestrian entrance and ventilation shaft. This sensitive structure was initially protected by a row of in-situ bore piles. A 4-layered temporary steel strut and shoring system was erected to stabilize the structure while excavation proceeded (Fig. 8).

CHEUNG KONG CENTER (Redevelopment of the previous Hilton Hotel)

The project situated in a 9,650m^2 site in the down-town area of Central. The new building is a 62-storey composite structure, with a 22m x 27m reinforced concrete inner core encased in a 47 m x 47m external steel frame, together with a 6-level basement that constructed over the 2-level old basement of Hilton Hotel.

Demolition and Foundation

The previous Hilton Hotel was to be demolished at the commencement of the project. After the site was cleared, the works that followed were the construction of the diaphragm walls and the bored foundation for the new building. All the diaphragm walls employed in the project were of 1.2m thick reinforced concrete. The perimeter walls helped to support and stabilize the ground during the construction of the new basement, as well as to act as the permanent basement wall. There was a 37m-diameter shaft pit formed in the middle of the site for the construction of the core wall for the future tower. Large diameter bored piles were used as foundation for the new building. The bored piles were basically in two standard sizes. Eight of the piles were 6m in diameter and dug manually for supporting the superstructure. 20 piles for the support of the 6-level basement structure were of 1.5m diameter and dug mechanically using grabs and chiselling method.

Forming a 37m-diameter shaft pit and the construction of the core wall

Before the carrying out of the basement construction using a top-down method, the first major work below ground was to construct the central core of the main building tower, the foundation of which rested on bedrock about –28m from ground level. Instead of constructing the central core in a top-down manner, the core was built bottom up. This could be done by the forming of a pit large enough to house the core structure and its foundation (Fig. 9). A pit was thus formed with the sides supported by panels of 1.2m-thick RC diaphragm wall. When the pit was excavated down to the required formation level, a 5m-deep RC raft was constructed as foundation for the core. The core wall on the lowest basement was constructed on top of the raft (Fig. 10).

Figure 9 – A 37m dia. shaft and eight 6.5m dia. manual-dug caissons were formed to construct foundation raft for the 62-storey building

Figure 10 – The construction of the core wall using a jump-form system inside 37-m dia. shaft.

Construction of the basement

After the completion of the bored piles, steel stanchion was erected on top of each pile at its formation level as support to the basement slabs during the construction process using a top-down sequence. In order to allow the core wall and the structural frame to proceed to a safe separating distance, the ground floor slab was cast after the core wall had been completed up to the 9th level. With the temporary diaphragm wall that formed the 37m-diameter shaft gradually being demolished, the basement slab bound by the 8 super-columns was cast and connected to the core wall structure as soon as a stage of excavation completed. This made the basement structure at the center very rigid and from thereon, excavation to the sides continued, with the central part acting as a base to shore-support the newly excavated sides (Fig. 11). The floor system in the basement was of flat slab design with dropped panel around column heads. Average slab thickness was 400mm.

As a means to expedite the progress in the basement construction, the "double-bit" method was adopted (Fig. 13). The principle of this method is to have 2 levels of the basement excavated at a same time. At the bottom of the excavated, then cast the slab of the lower basement. And from this level, usual floor form would be erected and have the upper basement slab cast afterward. The basement further below would be repeated using the same principle. Temporary shores were installed in certain positions to stabilize the sides due to the depth of the excavation in the double bit. To facilitate the removal of large volume of excavated materials, several temporary openings were formed on the basement slab so that the excavated soil could be removed by lifting grabs, excavating machines or dumper truck entering into the basement through temporary ramp within the basement (Fig. 12).

Figure 11 – Center portion of the basement acting as a base to support the excavated sides during the top-down process.

Figure 12 – One of the openings formed inside the basement slab for the purpose of spoil removal. Temporary ramp would be erected in this opening to allow dumping vehicles entering into the basement for the purpose.

Figure 13 – Another openings for spoil removal. The construction of the intermediate slab using "Double Bit" method can be observed.

CONCLUSION

Besides the 3 cases as referred to in this paper, there are countless other similar basement projects being executed recently in Hong Kong. A brief summary listing some of the commonality of these projects can be drawn as a conclusion to finalize the discussion here.

- Basements of this kind are usually very big (say up to 10,000m^2) and very deep (below 20m).
- Majority of the basements are constructed in a top-down manner. Some other methods such as combining top-down and bottom up, or combining open-cut and top-down arrangement, can sometimes be seen.
- Complicated coordination problems and teething arrangement often exist between various major contracts or other major building works.
- Layout planning especially in phasing and sectioning of the job forms a very important consideration mastering the success of the project.
- Dynamic layout arrangement is usually required for the removal of the excavated spoil from the basement. This may involve the forming of temporary ramp, provision of special equipment, or the taking over of part of the completed building as temporary access in an advanced stage.
- Diaphragm walling is the most common cut-off provision being used.
- System formwork can hardly be applied for most basement jobs due to access problems as well as the confined working condition inside the excavated.
- Constructing the basement in "double bit" arrangement is becoming common.
- Protection and safety measures in particular to the life and property of third parties are highly concerned in basement jobs. Accident in this area is maintained at a relatively very low rate.
- Progress of work can hardly be predicted or monitored accurately due to the existence of numerous unforeseeable problems during the construction process.

From the investigator's point of view the suggestion to improve such situations may be quite simple. Engineers and builders in Hong Kong should have sufficient experience in handling these kinds of complicated basement jobs. Enhancement in the following areas may be simple and straight-forward solutions to improve the performance of such projects.

- Spend more resources to improve house-keeping works on site (e.g. better ventilation, lighting, access provision or tidier working place, remember it is working within basement).
- Lengthen the duration of the construction period where possible. Differences between fast track and quality should find a balance.
- Improve planning and supervision quality, other than just rely on software and log record.

As for most construction jobs, technicality is often the least of the problem. On the other hand, human factors, planning and management concerns are the most determining issues in reality. This is even more factual for basement project.

REFERENCE

RAMOND W M WONG, A Review on Common Technology Employed for the Construction of Buildings in HK, 1st International Structural Engg & Construction Conference, Hawaii, USA, 2001.

RAMOND W M WONG, Construction of a Semi-buried Building – A Super-sized Shopping Mall: The Festival Walk, Megacities 2000 Conference, Dept. of Architecture, University of HK, 2000.

DAVID SCOTT & GOMAN HO, Design and Construction of 62-Storey Cheung Kong Center, Symposium on Tall Bldg Design & Construction Technology, Beijing, 1999.

M J TOMLINSON, Lateral Support of Deep Excavations, Proceedings of The Institution of Civil Engineers, 1970.

G N GILLOTT AND JAMES C K LAM, Construction of Site Formation and Foundation Works in Building Projects, Building Construction in Hong Kong, Building Departments, HKSAR, 1998.

G J TAMARO, Deep Foundations in the Urban Environment, Habitat and the High-Rise, Council on Tall Building and Urban Habitat, Dutch Council on Tall Buildings, Amsterdam, 1995.

Advances in Building Technology, Volume 1
M. Anson, J.M. Ko and E.S.S. Lam (Eds.)

NOVEL THREE-STEP CONSOLIDATION METHOD TO ELIMINATE FOUNDATION SOIL LIQUEFYING

Min Xie, Zhengxun Xie

Nanjing Gulou Town Construction Develop Company,
Nanjing, 210016, China

ABSTRACT

This paper introduces a novel three-step method to consolidate liquescent foundation soil. The three steps are: first, drain the ground water to lower the underground water level by drilling well points; second, inject clay solutions into the soil through these well points, drain water again, and repeat this process several times; third, use machine to tamp the ground by shocking and compressing. The key to this method: before the first step is carried out, a thin waterproof wall of clay concrete is to be built around the foundation plane. Because all the materials (clay solutions and low-grade concrete) used in this technology are very cheap, the engineering cost is lower than, and the efficiency higher than, those of the usual consolidation methods, such as the heavy tamping (dynamic compaction) method, the sand or granulate stone compaction stake method, dry jet mixing method, etc.. The safety and reliability of this method has also been proved. In Haikou city (Hainan China), there is a real project example, in which the liquescent foundation soil has been consolidated by this novel method. The geological features of the project area are very bad due to the three geological faults. The geological formation there is composed of soft clay and loose sand, and high underground water level of this area makes the situation even worse. Despite all these difficulties, the consolidation method has been successfully applied.

This paper is organized in 5 parts:

(1) Background of the researching subject
(2) Basic principles of the method
(3) The detailed measures taken (consolidation scheme)
(4) A real project example
(5) Conclusions

KEYWORDS

Foundation, Liquefying, Water Flow Consolidation, Clay Solution Cementing, Compressing Consolidation

1. BACKGROUND OF THE RESEARCHING SUBJECT

There're many similarities between the Los Angeles Earthquake (Jan. 1994, USA) and the Kobe-Osaka Earthquake (Jan. 1995, Japan). The hypocenters were located directly beneath both cities, and both earthquakes caused horrible damage. What troubled the earthquake experts most was not the collapses of those modern buildings, but the extensive destruction of major expressways, rail lines, and bridges, etc. Without the availability of roads, the rescue work was hard to carry out, and thousands of lives were lost because of the delay. In the Kobe-Osaka Earthquake, the expressway system along the area of Kobe, Osaka and Nagoya was severely destroyed, and nearly half of Japan's economy was paralyzed, causing great loss. Consider the reasons why the expressways were damaged so greatly. In several places, the geological structures are really bad, the cracking and sinking of the land is hard to resist, but these areas can sometimes be avoided in the reconstruction by proper layout and path selection. What threatens the road most is the liquefying of the road foundation over large areas. In construction practice, general methods such as the stake foundation method, the deep mixing method, and the transfer soil method are used to prevent foundation from liquefying. But these methods are very expensive and work consuming, which limit them from wide application. Even in those developed countries such as USA and Japan, engineers hesitate to apply these methods. Besides, in the region where underground water level is high and loose sand layer is thick, even if stakes are used to consolidate the foundation, the effects are still not satisfactory. The initial earthquakes will gradually liquefy the soil between stakes, subsequent major earthquake and following earthquakes will put down the stakes, thus invalidating the consolidation method. Currently, many scholars and engineers are trying to find the countermeasure to eliminate the liquefying of large areas of soil foundation. The author began to pay attention to this problem after reviewing the Kobe-Osaka Earthquake, and has made a breakthrough after several years of exploring,

2. BASIC PRINCIPLES OF THE METHOD

To prevent foundation soil from liquefying, the research must be based on the preconditions of soil liquefying. As is well known, the two preconditions are high underground water level, and low clay content in the sand. And the discriminant is as follows:

$$\begin{cases} N_{63.5} < N_{cr} \\ N_{cr} = N_O[0.9 + 0.1(ds - dw)]\sqrt{3/\rho_c} \end{cases} \tag{1}$$

Where:

$N_{63.5}$: Measured value of model penetration test for liquefying soil

N_{cr}: The critical number of model penetration test for liquefying soil, variable according to earthquake magnitude

N_O: The standard number of model penetration test for liquefying soil

ds: The depth of test point of penetration for liquefying soil (m)

dw: The depth of underground water level (m)

ρ_c: The percentage of clay particles

If the underground water level can be restrained, or the clay content of sand can be increased, then the critical number of model penetration tests N_{cr} can be greatly decreased, and the possibility of foundation soil liquefying will be eliminated. This is the most simple and most ideal method. Solely depending on increasing the density of foundation soil and enhancing the measured number of model penetration tests $N_{63.5}$, it is hard to eliminate foundation liquefying, and it is also not economic. For example, in coastal area, underground water level dw is about 2.0m. The practical depth of model penetration test point of loose sand is set at 15.0m, and suppose the clay content of sand ρ_c is below 3%, then the computed critical number of model penetration test $N_{cr} = 2.2N_0 = 22$. But the surveyed $N_{63.5}$ of normal loose sand is mainly between 6 and 10, therefore, it is very difficult to acquire the ideal number of model penetration test (>22) by solely tamping and compressing, besides, the cost is very high.

If through the using of some techniques, the underground water level can be lowered, or even controlled forever, and the clay content can be increased from lower than 3% to 6%, then it is easy to meet the requirements for eliminating foundation liquefying. First, if the underground water level can be lowered or controlled forever, then the density of loose sand will be enhanced during the draining process, thus the number of model penetration test is increased. This is the procedure of water flow consolidation. Second, after the underground water level is lowered, then dw is close to ds, or even greater than ds, consequently outside the liquefying conditions, and the current N_{cr} is equal or smaller than N_o (set as 10). The number of model penetration tests can now be met with ease. Third, if ρ_c can be increased from lower than 3% to around 6%, then N_{cr} will be $N_o/\sqrt{2} = 0.7N_o$, and the number of model penetration test is easier to meet. The loose sand, which has been consolidated by the procedure of clay solution cementing, also contributes greatly to eliminating liquefying. At last, if the foundation soil is compressed by tamping and compressing, then it will be safer and liquefying is more reliably eliminated.

3. THE DETAILED MEASURES TAKEN (CONSOLIDATION SCHEME)

1) Build Waterproof Wall

To control the underwater level effectively, a waterproof wall must be built around the effective control area of the construction foundation. Theoretically speaking, a waterproof wall only need hold back water and not sustain load, therefore it can be built using cheap low-grade concrete. The thickness of the waterproof wall is approximately 600mm, depending on the convenience of construction, and its depth is determined by the desired control depth. Generally the wall should reach the waterproof layer below the loose sand layer. The construction materials are mixed with clay, coal fly ash and grits, and the intensity is about C5. If necessary, a little cement can also be added, so as to make low-grade concrete.

2) Drain Soil Water Through Well Points

Set well points on a 10.0m×10.0m grid, and drain water from the foundation soil through these well points.

3) Inject Clay Solution Into Well Points

Inject high pressure clay solution (can also be made from the drilling mud, or mixture of clay powder, coal fly ash and water) from the well points into the loose sand layer. The quantity of solution (clay content) are determined by different needs.

4) Drain Water Again

Drain the water again, so as to make the consolidation process and the cementing process simultaneous.

5) *Tamping And Compressing By Machines*

Tamping and compressing follows, in order to eliminate the holes and bumps in the crust surface.

6) *Check And Approve*

Check the project by a light penetration test and load test, obtain the survey data, compare with the design requirements, and then decide whether to accept or not.

4. A REAL PROJECT EXAMPLE

1) *Project Information*

A real application relates to the Chengli Fruit Vintage Factory, which takes up an area of 62,692.00m^2, and has a building area of 23,792.00m^2. For a light factory, the floor board load requirement is 250kPa.

2) *Geological Conditions*

The project is located in Fourth Period Alluvium. The area is geologically located in the triangle forward by the Qiongya big fault, the South Dujiang small fault, and the Qiongshan seismic belt. There are 10 layers in the soil from the surface to 15m below. The layers are thin, discontinuous, and appear interlaced. The sixth layer of silt is lentoid, and its thickness varies from 0 to 7.0m. The surveyed number of model penetration test $N_{63.5}$ of three loose sand layers was around 6.0; the underground water level is about 2.0m, and the stratum's allowable bearing strength fk varies from 70kPa to 210kPa, and there is no solid layer available to bear stakes.

3) *Measures Taken*

(1) Construct Waterproof wall:

Drill and place stakes (diameter of $\phi 600$), deep into the sixth and seventh layers (waterproof mud layer), The depth is 12.0m. Construct a waterproof wall using low-grade concrete, and its intensity is C5.

(2) Drain water through well points

Well points are set on a 10.0m×10.0m grid, the diameter of well points is $\phi 100$, and depth 8.0m. The steel bar cage can be used repeatedly. If adopting the light well points drainage method, the cost will be even less.

(3) Inject clay solution

The clay solution is a mix of clay powder, coal fly ash and water. The quantity of the clay solution should make sure that the clay content of all the three sand layers reaches 6%.

(4) Tamping and compressing

A heavy bulldozer (more than 100kw) was used three times for compression.

(5) Quality control (The author will write a special summary on details of this subject)

(a) Underground water level check

(b) Number of model penetration test check, or take light penetration test and load test.

(6) Economic benefits

Compared with the tamping method, approximately ￥350,000 was saved. Compared with the rock filling stake method,￥200,000 was saved; compared with the jet cementing method, the saving is approximately ￥2,760,000. What is the most important is that the reliability is superior to that of any other method, and the satisfactory results will be permanent.

5. CONCLUSIONS

1) Theoretical analysis and practical engineering trials prove that the three-step method, novel technique to eliminate foundation liquefying, is of the highest reliability, the easiest to carry out, and the cost is the most economic.

2) The research and improvements on the special instruments (high-pressure mud pumps, valves, squirt gun, and water pipes) will further promote the effectiveness of this method.

References:

1. Design code for Structure Foundation GBJ789
2. Seismic Design code of Building Structure GBJ11-89

CONSTRUCTION TECHNOLOGIES:

EARTHQUAKE ENGINEERING

Advances in Building Technology, Volume I
M. Anson, J.M. Ko and E.S.S. Lam (Eds.)

TESTING OF A TRANSFER PLATE BY PSEUDO-DYNAMIC TEST METHOD WITH SUBSTRUCTURE TECHNIQUE

A. Chen[1], C.S. Li[1], S. S.E. Lam[1] and Y.L. Wong[1]

[1]Department of Civil and Structural Engineering, The Hong Kong Polytechnic University

ABSTRACT

A pseudo-dynamic testing system is developed with the incorporation of the substructure technique. Newmark-α method is used to perform the numerical integration. The pseudo-dynamic system has been implemented to study the seismic response of a transfer plate in a 18-story high-rise building. Structural system comprises a transfer plate at the first floor with evenly spaced columns below, and shear walls above the transfer. A test specimen in ¼-scale was used to model the ground floor, transfer plate and the second floor. All the upper stories were simulated numerically. Pseudo-dynamic tests were conducted using three types of time-history records. When subject to an El-Centro earthquake record with maximum acceleration at 32%g, the shear wall remained undamaged whereas the transfer plate was severely damaged. The experimental study has indicated that the transfer plate system has insufficient strength to resist possible strong earthquake action.

KEYWORDS

Pseudo-dynamic test method, Substructure, Experiment, Transfer plate, Seismic, Time-history, High-rise building

INTRODUCTION

Pseudo-dynamic test method is an experimental technique for simulating the time-history response of structural members. It combines the realism of shaking table tests with the economy and convenience of quasi-static testing. Compared with the shaking table test, it has many advantages. It enables the testing of large-scale specimens. This is especially important for reinforced concrete members as we can observe the propagation of cracks and progress of failure. Concentrated mass can also be easily assumed, whereas in the shaking table tests additional loading blocks have to be used. In the last thirty years, many researches have been conducted to develop this method. Among others, Mahin and Shing (1985) and Takanashi and Nakashima (1987) have provided detail description of the method. Nakashima *et al* (1992) have also developed a system capable of performing real-time pseudo-dynamic tests. Development in recent years has already made the pseudo-dynamic testing method a reliable experimental tool. To facilitate the testing of a multi-degree-of-freedom system, a pseudo-dynamic testing system was developed and verified (Chen *et al* 2000). It is the objective of this study to extend the capability of the system to include the substructure technique. The newly developed system is also used to study the seismic response of a transfer plate system when subject to the earthquake action.

SUBSTRUCTURE TECHNIQUE

Equilibrium condition of a dynamic system is expressed while considering inertia force, damping, resistance and applied force as

$$Ma + Cv + R = f \tag{1}$$

M and C are the mass and damping matrix respectively. f is the applied load vector. a and v represent the acceleration and velocity vector respectively. Using the substructure technique, the resisting force R can be separated into two components.

$$R = r + \overline{K}d \tag{2}$$

r is the experimental component to be modeled by the test specimen. $\overline{K}d$ is the analytical component to be simulated numerically (Nakashima *et al* 1990). $\overline{K}$ is the stiffness matrix and d is the displacement vector.

In a pseudo-dynamic test, the applied load f is expressed in the form of a time-history record. As a result, Equation (1) has to be solved using an integration scheme. Newmark Method was first introduced to provide accurate and stable solution (Mahin and Shing, 1985). Shing *et al* (1991) applied implicit time integration, whereas Bursi *et al* (1994) introduced an integration scheme with adaptive time steps. In this study, a Newmark-α method (Hilber *et al* 1977) is used to perform the numerical integration. For a general multi-degree-of-freedom system and without substructure, algorithm of the pseudo-dynamic test can be expressed in the following form.

$$Ma_{i+1} + (1+\alpha)Cv_{i+1} - \alpha Cv_i + (1+\alpha)R_{i+1} - \alpha R_i = (1+\alpha)f_{i+1} - \alpha f_i \tag{3}$$

$$d_{i+1} = d_i + \Delta t v_i + \Delta t^2[(\frac{1}{2} - \beta)a_i + \beta a_{i+1}] \tag{4}$$

$$v_{i+1} = v_i + \Delta t[(1-\gamma)a_i + \gamma a_{i+1}] \tag{5}$$

The subscripts i and i+1 indicate two consecutive time increments at time t_i and t_{i+1} respectively. Δt is the time interval or $(t_{i+1}-t_i)$. α, β and γ are parameters governing numerical properties of the algorithm. To incorporate the substructure technique, Equation (3) is modified in considering Equation (2).

$$Ma_{i+1} + (1+\alpha)Cv_{i+1} - \alpha Cv_i + (1+\alpha)r_{i+1} - \alpha r_i + \overline{K}(1+\alpha)d_{i+1} - \overline{K}\alpha d_i = (1+\alpha)f_{i+1} - \alpha f_i \tag{6}$$

After some mathematical manipulation using Equations (4) to (6), we have

$$[\overline{M} + \Delta t^2 \beta \overline{K}(1+\alpha)]d_{i+1} = g_{i+1} - \Delta t^2 \beta(1+\alpha)r_{i+1} \tag{7}$$

Where

$$g_{i+1} = \overline{M}[d_i + \Delta t v_i + \Delta t^2(\frac{1}{2} - \beta)a_i] + \Delta t^2 \beta[(1+\alpha)f_{i+1} - \alpha f_i - Cv_i - (1+\alpha)(1-\gamma)C\Delta t a_i + \alpha r_i + \overline{K}\alpha d_i] \quad (8)$$

and

$$\overline{M} = M + (1+\alpha)C\Delta t\gamma \quad (9)$$

Equation (7) represents an implicit expression, since r_{i+1} depends on d_{i+1}. An iterative solution procedure is implemented in the manner described as follows.

Considering an approximate solution at iteration k,

$$[\overline{M} + \Delta t^2 \beta \overline{K}(1+\alpha)]d_{i+1}^{(k)} = g_{i+1} - \Delta t^2 \beta(1+\alpha)r_{i+1}^{(k)} + e_{i+1}^{(k)} \quad (10)$$

Here, the superscript k represents the results obtained from the kth iteration. Assuming a converged solution is obtained at iteration k+1,

$$[\overline{M} + \Delta t^2 \beta \overline{K}(1+\alpha)](d_{i+1}^{(k)} + \Delta d_{i+1}^{(k)}) = g_{i+1} - \Delta t^2 \beta(1+\alpha)(r_{i+1}^{(k)} + \Delta r_{i+1}^{(k)}) \quad (11)$$

$\Delta d^{(k)}_{i+1}$ and $\Delta r^{(k)}_{i+1}$ are the corrections at the kth iteration. Subtracting Equation (11) from Equation (10), we obtain the displacement corrections $\Delta d^{(k)}_{i+1}$.

$$K^* \Delta d_{i+1}^{(k)} = -e_{i+1}^{(k)} \quad (12)$$

with

$$K^* = \overline{M} + \Delta t^2 \beta(1+\alpha)K' + \Delta t^2 \beta(1+\alpha)\overline{K} \quad (13)$$

K' and $\overline{K}$ are stiffness matrices of the test specimen and the analytical component respectively. Improved solution at the end of the kth iteration is given by

$$d_{i+1}^{(k+1)} = d_{i+1}^{(k)} + \Delta d_{i+1}^{(k)} \quad (14)$$

$$r_{i+1}^{(k+1)} = r_{i+1}^{(k)} + \Delta r_{i+1}^{(k)} \quad (15)$$

The iterative procedure is repeated until all the corrections are less than the tolerance. The pseudo-dynamic test continues with progressive increase in the time domain, t ,via the time increment Δt.

TEST SPECIMEN OF A TRANSFER PLATE SYSTEM

Figure 1 shows a high-rise building having 17 typical floors and a single story podium. Structural system at typical floors comprises a simple shear wall system with no lintel beams connecting the shear walls.

Floor slabs are 160mm thick. The shear walls are supported by a 1500mm thick transfer plate, and below the transfer are evenly spaced columns of 900mm square. The high-rise building has been designed according to non-seismic provisions.

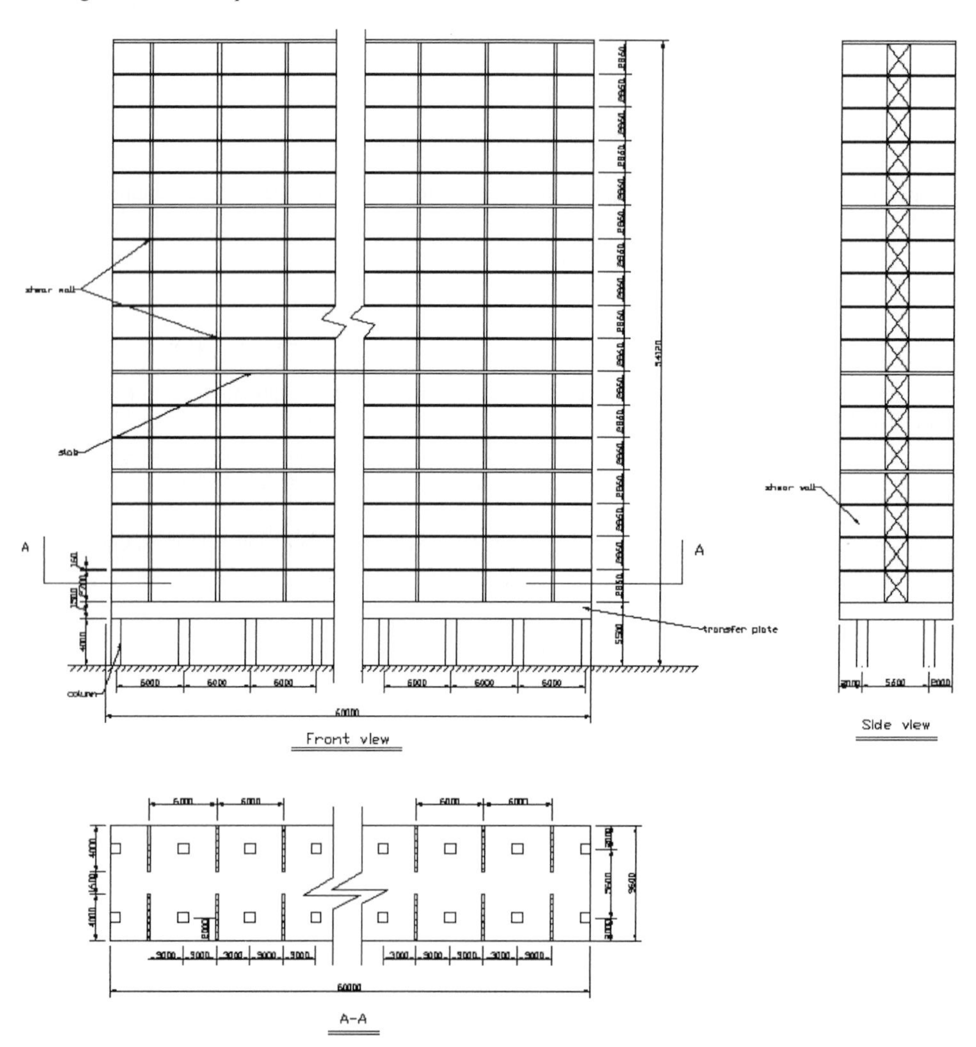

Figure 1. 18-story building with a transfer plate at the first floor.

In applying the pseudo-dynamic test method with substructure technique, upper stories of the high-rise building were simulated numerically as shown in Figure 2. A ¼-scale specimen was fabricated to represent the transfer plate and second floor of the building. Further simplifications were made in considering symmetries of the structure in both directions. Lateral displacements were applied at the first and second floors through two actuators. Normal strength concrete with maximum aggregate size at 10mm was used and the measured cube strengths f_{cu} are tabulated in Table 1. High yield deformed bars (f_y=460MPa) and mild steel bars (f_y=250MPa) were used for the main reinforcement and links respectively. To prevent the columns from failure before the transfer plate, the columns were strengthened by epoxy-grouting.

Table 1. Measured cube strength of concrete.

Member	Cube strength (MPa)
Columns	42.9
Transfer plate	41.1
Shear wall	32.9
1st floor slab	32.9

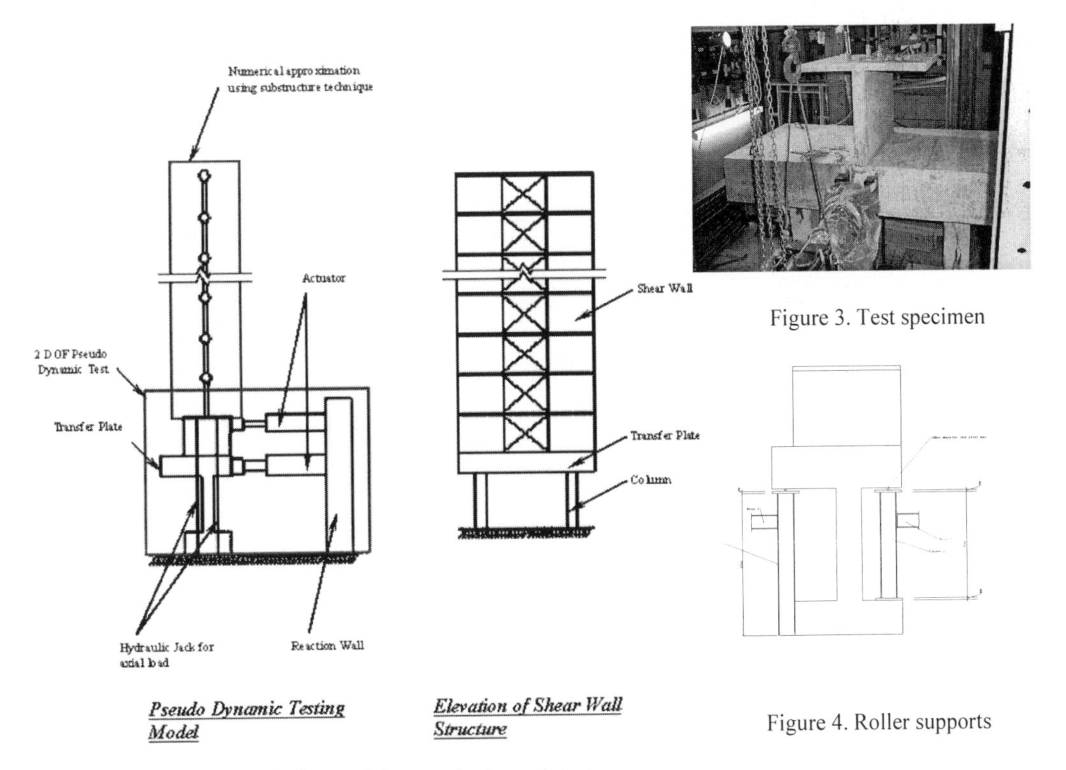

Figure 3. Test specimen

Figure 4. Roller supports

Figure 2. Setup of the pseudo-dynamic test

Figure 3 shows the test specimen. Strain gauges were installed to measure the strains in the reinforcements and concrete. Displacement transducers were installed to record the deformations. Constant axial loads acting on the transfer plate and the columns were applied through two manually controlled hydraulic jacks. To restrain rotation of the transfer plate during application of the lateral displacements, four steel columns were used to act as roller supports as shown in Figure 4. The roller supports were installed after the application of the axial loads.

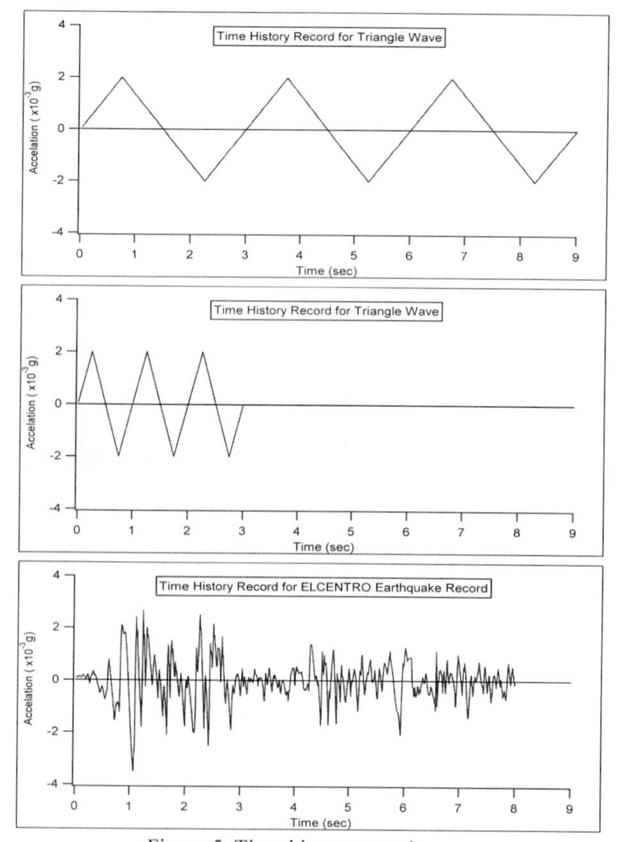

Figure 5. Time-history records

To estimate the initial stiffness at the two floors of the test specimen, lateral displacement was applied to one of the floors while the actuator at the other floor was fixed against any lateral movement. All the upper floors were assumed to have the same stiffness as the second floor, and were simulated numerically via substructure technique. Damping ratio of all the floors was assumed to be at 5%.

In simulating the seismic action, three types of loading histories were considered as shown in Figure 5, namely triangular waves and the El-Centro earthquake record. All the records were compressed according to the similitude rule. The two triangular waves were used to estimate response of the transfer plate within the elastic range. Finally, a series of El-Centro earthquake records with increasing magnitude of acceleration was applied.

EXPERIMENTAL RESULTS AND DISCUSSIONS

Figure 6 shows the lateral displacements at the transfer level ("1/F") and the second floor ("2/F") when subject to the El-Centro earthquakes with maximum acceleration at 16%g and 32%g. The two responses at 16%g and 32%g are virtually in proportional to each other. There appears to be no appreciable change in the lateral stiffness of the two floors. In the course of the experiment, no crack was observed on the shear wall. This indicates that the responses were generally elastic.

Figure 7 shows the deflection at the centre of the transfer plate when subject to the El-Centro earthquakes with maximum acceleration at 16%g and 32%g. There was substantial increase in the central deflection.

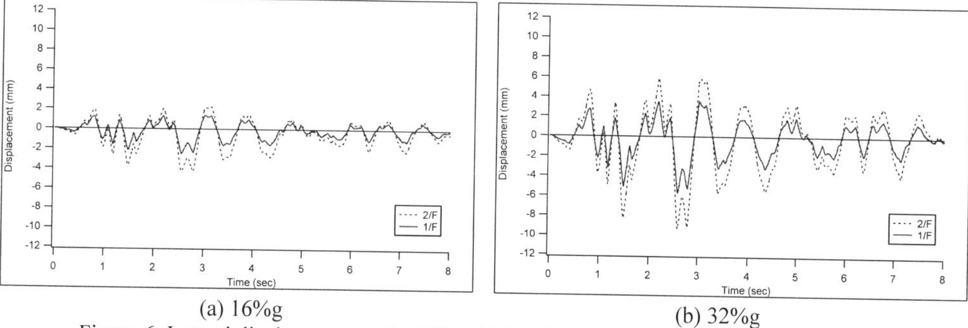

(a) 16%g (b) 32%g

Figure 6. Lateral displacement at the 1/F and 2/F when subject to El-Centro earthquakes

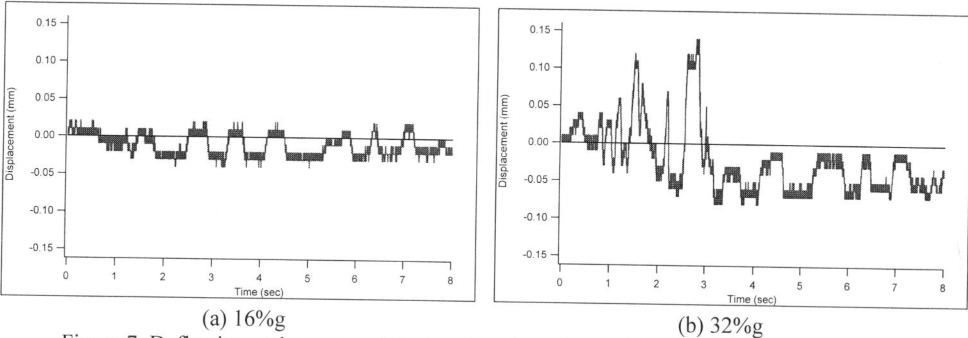

(a) 16%g (b) 32%g

Figure 7. Deflection at the center of the transfer plate when subject to El-Centro earthquakes

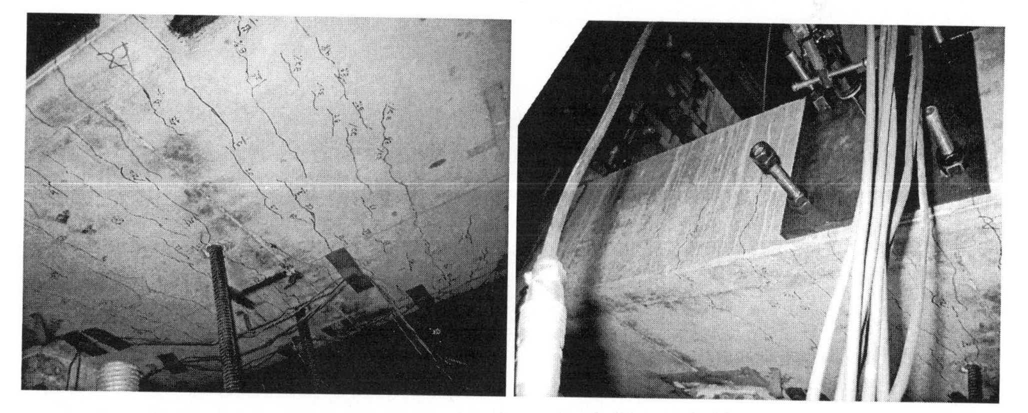

Figure 8. Cracking pattern viewed from below Figure 9. Cracking pattern viewed from one end

When subject to 32%g earthquake, there was considerable cracking at the bottom of the transfer plate at the vicinity below the shear wall. Figures 8 and 9 show the cracking patterns at the bottom of the transfer plate. All the cracks are flexural in nature and propagated along the length of the shear wall. It is worth noticed that at the initial application of the axial force through the hydraulic jacks, some cracks were also found at the bottom of the transfer plate. In general, when the maximum acceleration of the earthquake increased, new cracks were observed with more extensive coverage.

CONCLUSIONS

A pseudo-dynamic testing system is developed to examine the seismic performance of a transfer plate. Newmark-α method is used to perform the numerical integration and substructure technique is applied. A test specimen in ¼-scale was used to represent lower stories of a 18-story building. Pseudo-dynamic tests were conducted using three types of time-history records. Test results have indicated that the shear walls remained elastic throughout the loading histories, whereas the transfer plate was severely damaged when subject to an El-Centro earthquake record with a maximum acceleration at 32%g. The experimental study has indicated that the transfer plate system has insufficient strength to resisting strong earthquake action.

ACKNOWLEDGEMENT

The writers are grateful for the financial support from the Research Grant Council of Hong Kong through a RGC research grant (RGC no: PolyU 5026/98E) and The Hong Kong Polytechnic University (Project no: G-V853).

REFERENCES

Bursi, S.O., Shing P.S.B. and Zorica R.G. (1994). Pseudodynamic Testing of Strain-softening Systems with Adaptive Time Steps. *Earthquake Engineering and Structural Dynamics*, **23**, 745-760.

Chen A., Lam S.S.E. and Wong Y.L. (2000). Verification of a pseudodynamic testing system. *Proceedings of the Int. Conf. on Advances in Structural Dynamics*, Hong Kong, **2**, 851-858.

Hilber H.M., Hughes T.J.R. and Taylor R.L. (1977). Improved Numerical Dissipation for Time Integration Algorithm in Structural Dynamics. *Earthquake Engng & Structural Dynamics*, **20**, 283-292.

Mahin, S.A. and Shing P.S.B. (1985). Pseudodynamic Method for Seismic Testing. *J. St. Eng. ASCE*, **111(7)**, 1482-1503.

Nakashima et al (1990). Integration Technique for Substructure Pseudodynamic Test. *Proceedings 4th US National Conference Earthquake Engineering*, Palm Springs, CA II, 515-524.

Nakashima M., Kato H. and Takaoka E. (1992). Development of Real-time Pseudodynamic Testing. *Earthquake Engineering and Structural Dynamics*, **21**, 79-92.

Shing P.B. and Mahin S.A. (1984), *Pseudodynamic Test Method for Seismic Performance Evaluation: Theory and Implementation*, University of California, Berkeley, California, USA.

Shing, P.B., Vannan M.T. and Cater E. (1991). Implicit Time Integration for Pseudodynamic Tests. *Earthquake Engineering and Structural Dynamics*, **20**, 551-576.

Takanashi K. and Nakashima M. (1987). Japanese activities on on-line testing. *J. Eng. Mech. ASCE*, **113**, 1014-1032.

Advances in Building Technology, Volume I
M. Anson, J.M. Ko and E.S.S. Lam (Eds.)

SEISMIC DESIGN AND INELASTIC SEISMIC RESPONSE OF LUMPED MASS MDOF DUAL STRUCTURE*

Jie Jing[1], Lieping Ye [1], Jiaru Qian[1]
[1]Department of Civil Engineering, Tsinghua University, Beijing 100084, China

ABSTRACT

Based on the concept of seismic control, dual structure systems aiming at improving the shortcoming of traditional seismic structures was proposed for seismic design model by the second author. The seismic design performance objectives at different earthquake levels were previewed first in this paper. Under small earthquake and moderate earthquake levels, the elastic response spectrum method and equal cyclic energy criteria are adopted respectively to determine the parameters of the secondary structure in the dual seismic structure system. The inelastic seismic response under the severe earthquakes showed that the structure designed based on the suggested method had a suitable seismic performance.

KEYWORDS

Seismic design, dual structure system, energy concept, inelastic seismic response, performance objective levels

INTRODUCTION

In traditional seismic structure, the demand for the structure stiffness, stability, energy absorption ability is usually satisfied by the structure itself, which results in an uncertain seismic performance, especially the damage and inelastic deformation of some important structure elements are very serious under moderate and severe earthquakes and the repairing expense may be very high. With the development of seismic control method, researchers proposed the concept of Damage Controlled Structures (DCS) (J.J.Connor, et al, 1997), which makes use of the energy dissipation of secondary

* This research is supported by Chinese national natural science foundation, No. 59895410.

structure to protect primary structure (original structures) and reduces the repairing expense. After Kobe earthquake, the design using DCS became a trend (Akira Wada, et al, 1997). In the new version Chinese Building Seismic Design Code (CBSDC: GB50011-2001), the energy dissipation building structure is adopted.

Recently, the performance-based seismic design had become widely accepted. The structure seismic performance should be designed or predicted for different earthquake levels. But the current seismic design code does not suggested. In CBSDC: GB50011-2001, element design and the limit elastic story drifts are demanded at the small earthquakes level, while reinforcement detailing requirements are presented to prevent collapse of the structure.

The second author of this paper put forward the concept of dual seismic structure system, which is used as the design model for passive controlled seismic structure. In dual seismic structure system, primary structure withstands loads at service limit state; secondary structure dissipates earthquake input energy (L.P. Ye, 2000).

This paper is focused on the seismic design based on the performance-based design concept for lumped dual seismic MDOFS structures with middle-long initial period, i.e. $T_g<T_0<5T_g$, where T_0 is the initial period of the structure, T_g is the site predominant period. The elastic response spectrum method is adopted to determine the stiffness of secondary structure under small earthquake, while equal cyclic energy criteria is adopted to determine yielding strength of secondary structure under moderate earthquake. The seismic performance of the structure under severe earthquakes is calculated to check the design results.

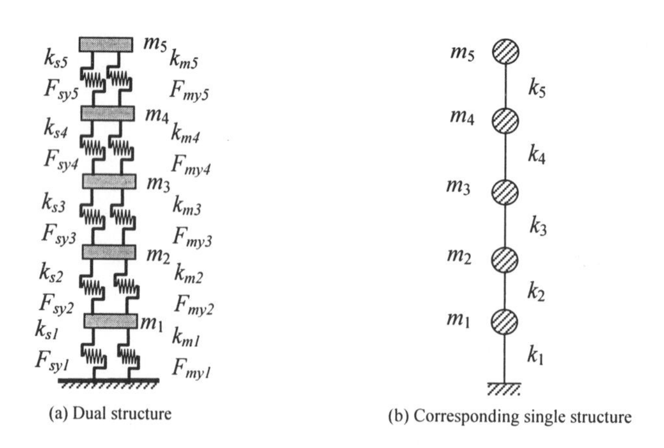

Figure 1: Analysis model

DUAL SEISMIC STRUCTURE SYSTEM AND ITS SEISMIC OBJECTIVES

Analysis Model and Parameters

The analysis model for dual seismic structure system used in this paper is shown as Figure 1(a). Figure

1(b) shows the corresponding single structure model, in which the structure parameters is same as the primary structure in the dual structure. The main parameters of the model are: (1)The mass distribution is uniform along height (m_i=m); (2)The story initial stiffness is uniform (k_{si}=k_s, k_{mi}=k_m); (3) The Clough degraded bilinear hysteresis model (Figure 2), which is fit for reinforced structures, is adopted both for primary structure and secondary structure. In the hysteresis model, the post-yielding stiffness equals to zero, and the degraded stiffness factor α equals to 0.4; (4). Mass proportional Damping is considered in analysis, the damping coefficient h=0.05m. Elcentro NS, Hachinohe NS and Kobe NS earthquake records are used for excitation input. Considering the earthquake design level 8° in CBSDC: GB50011-2001, the peak acceleration A_{max} takes 70Gal, 200Gal and 400Gal corresponding to small, moderate and severe earthquakes, respectively. The elastic and inelastic time history analysis was based on program package CANNY developed by K. N. Li.

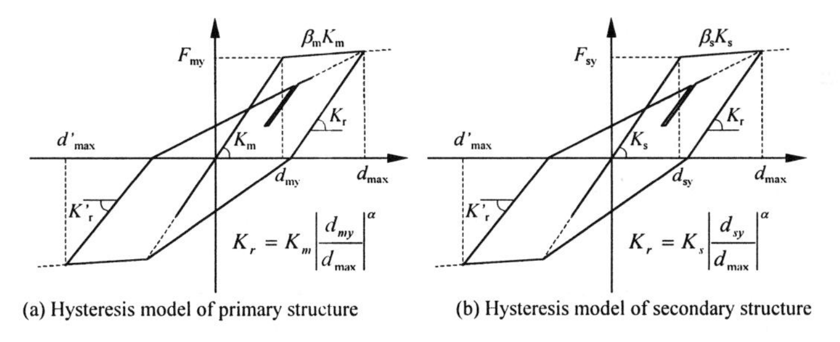

(a) Hysteresis model of primary structure (b) Hysteresis model of secondary structure

Figure 2: Hysteresis model

Seismic Objectives of DUAL STRUCTURE SYSTEM

According to the seismic demand under different earthquake levels, the seismic performance objectives of primary and secondary structure for dual structure are suggested as: (1) Primary and secondary structure remain in elastic stage at small earthquake level (Figure 3(a)). (2) Secondary structure yields but primary structure remains elastic or just reaches yielding critical state at moderate earthquakes level (Figure 3(b)). (3). Under severe earthquakes, primary and secondary structure are both in inelastic stage, but the damage of the secondary structure is some more serious than the primary structure (Figure 3(c)).

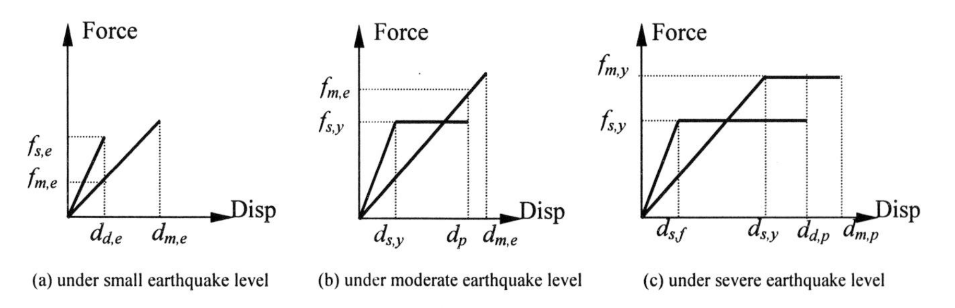

(a) under small earthquake level (b) under moderate earthquake level (c) under severe earthquake level

Figure 3: Seismic objectives of dual structure system

SEISMIC DESIGN OF DUAL STRUCTURE SYSTEM

Definition of Structural Parameters

$d_{m,e}$: maximum elastic displacement of the corresponding single structure at small and moderate earthquakes levels;

$d_{d,e}$: maximum elastic displacement of dual structure at small earthquake level. It is chosen as seismic objective for dual structure under small earthquake;

$d_{d,p}$: maximum displacement of dual structure at moderate earthquakes level. It is chosen as seismic objective for dual structure under moderate earthquake;

$F_{m,e}$: maximum elastic force response of the corresponding single structure under small and moderate earthquakes;

$F_{d,e}$: maximum elastic force response of dual structure system under small earthquake, $F_{d,e}=f_{m,e}+f_{s,e}$;

$f_{m,y}$: yielding strength of primary structure;

$f_{s,y}$: yielding strength of secondary structure;

T_m: initial period of the corresponding single structure;

T_d: initial period of the dual structure;

K_m: equivalent elastic stiffness of the corresponding single structure or primary structure in dual structure;

K_s: equivalent elastic stiffness of secondary structure in dual structure;

α_K: stiffness ratio of secondary structure to primary structure, $\alpha_K=K_s/K_m$;

α_d: maximum displacement ratio of dual structure to the corresponding single structure at small earthquake level, $\alpha_d=d_{d,e}/d_{s,e}$;

η_d: maximum displacement ratio of dual structure to the corresponding single structures at moderate earthquake level;

$\alpha_{s,f}$: ratio of yielding strength of secondary structure to the maximum elastic force response in the corresponding single structure;

μ_s: equivalent ductile factor of secondary structure, i.e. the ratio of the maximum displacement of dual structure to the yielding displacement of secondary structure;

The displacement Response of DUAL STRUCTURE SYSTEM at small earthquake level

Based on the elastic response spectrum method of CBSDC: GB50011-2001, the displacement response can be determined as:

The corresponding single structure,

$$d_{m,e}=\frac{F_{m,e}}{K_m}=\frac{\alpha_1\cdot G_{eq}}{K_m}=\frac{\left(\frac{T_g}{T_m}\right)^{0.9}\alpha_{\max}G_{eq}}{K_m} \tag{1}$$

Dual structure,

$$d_{d,e}=\frac{F_{d,e}}{K_m+K_s}=\frac{\left(\frac{T_g}{T_d}\right)^{0.9}\alpha_{\max}G_{eq}}{K_m+K_s} \tag{2}$$

The maximum displacement ratio α_d can be determined as:

$$\alpha_d = \frac{d_{d,e}}{d_{s,e}} = \left(\frac{K_m}{K_m + K_s} \right)^{0.55} = \left(\frac{1}{1+\alpha_K} \right)^{0.55} \tag{3}$$

From above Eqn.3, the stiffness K_s of the secondary structure can be obtained as,

$$K_s = (\left(\frac{1}{\alpha_d} \right)^{\frac{1}{0.55}} - 1) \cdot K_m \tag{4}$$

The relationship between the story stiffness k and structure equivalent stiffness K is shown in Eqn.5. Because the corresponding single structures and dual structure system are both lumped structures, following relation can be obtained,

$$\frac{k_s}{k_m} = \frac{K_s}{K_m} \tag{5}$$

Thus, with the known α_d, the stiffness ratio α_k can be expressed as:

$$\alpha_k = \frac{k_s}{k_m} = \frac{K_s}{K_m} = \left(\frac{1}{\alpha_d} \right)^{\frac{1}{0.55}} - 1 \tag{6}$$

TABLE 1
CALCULATION EXAMPLES UNDER SMALL EARTHQUAKE LEVEL

α_d	Stories	α_K(Eqn. 6)	$d_e \cdot \alpha_d$ (CANNY)	$d_e \cdot \alpha_d$ (Predition)	Difference (%)
0.7 (ElcentroNS T_0=1.05s)	1	0.9127	5.586mm	6.542mm	-14.6
	2		5.423mm	6.042mm	-10.2
	3		5.055mm	5.235mm	-3.4
	4		4.052mm	4.444mm	-8.8
	5		2.450mm	2.883mm	-15.0
0.8 (KobeNS T_0=1.439s)	1	0.5004	11.54mm	11.376mm	1.4
	2		9.938mm	10.16mm	2.2
	3		8.455mm	8.2mm	3.1
	4		7.418mm	6.806mm	9.0
	5		4.31mm	4.454mm	-3.2
0.9 (HachinoNS T_0=1.82s)	1	0.2111	9.626mm	8.951mm	7.5
	2		8.24mm	7.989mm	3.1
	3		8.027mm	7.328mm	9.5
	4		7.125mm	7.526mm	-5.3
	5		4.724mm	5.231mm	-9.7

Table 1 shows the comparison of prediction displacement response based on above design method to time history analyzed results for the dual structure at small earthquake level. Three initial period structures are analyzed: 1.05s, 1.439s and 1.82s. The design objective is chosen as α_d=0.7, 0.8 and 0.9, respectively. It could be seen that the difference between the prediction and analyzed results is acceptable for practical design.

The displacement Response of DUAL STRUCTURE SYSTEM at moderate earthquake level

Known from Figure 3(b), primary structure remains elastic while secondary structure yields under moderate earthquake level. Based on equal cyclic energy criteria proposed by the second author (L. P. Ye, 2001), the energy dissipation capacity of secondary structure of dual structure equals to the energy dissipation demand of the corresponding single structures. The derivation is:

E_D, the dissipated energy demand for secondary structure for displacement objective $\eta_d d_{m,e}$ of dual structure:

$$E_D = (1-\eta_d)\cdot d_{m,e}\cdot\frac{1}{2}\cdot(\eta_d F_{m,e} + F_{m,e}) = \frac{1}{2}(1-\eta_d^{\ 2})F_{m,e}d_{m,e} \tag{7}$$

E_C, energy dissipation capacity of secondary structure:

$$E_C = f_{s,y}\cdot\frac{f_{s,y}}{K_s}\cdot\frac{1}{2} + f_{s,y}(d_{d,p} - d_{s,y}) + 4\beta\cdot f_{s,y}(d_{d,p} - d_{s,y}) \tag{8}$$

Where β, the hysteresis energy dissipation factor, usually equals to 0.33 for RC structures. Let Eqn.7=Eqn.8, following result can be obtained:

$$\alpha_{s,f} = \frac{1-\eta_d^{\ 2}}{(4.64 - 3.64/\mu_s)\eta_d} \tag{9}$$

Based on above Eqn.9, the ratio of yielding strength $\alpha_{s,f}$ will be obtained when μ_s and η_d is given.

Figure 4 gives the response results comparison of dual structure system under the moderate earthquake for μ_s=2 and varying displacement objective ratio η_d. It could be seen from Figure 4 that the average analyzed results have a reasonable agreement with prediction value of the suggested method.

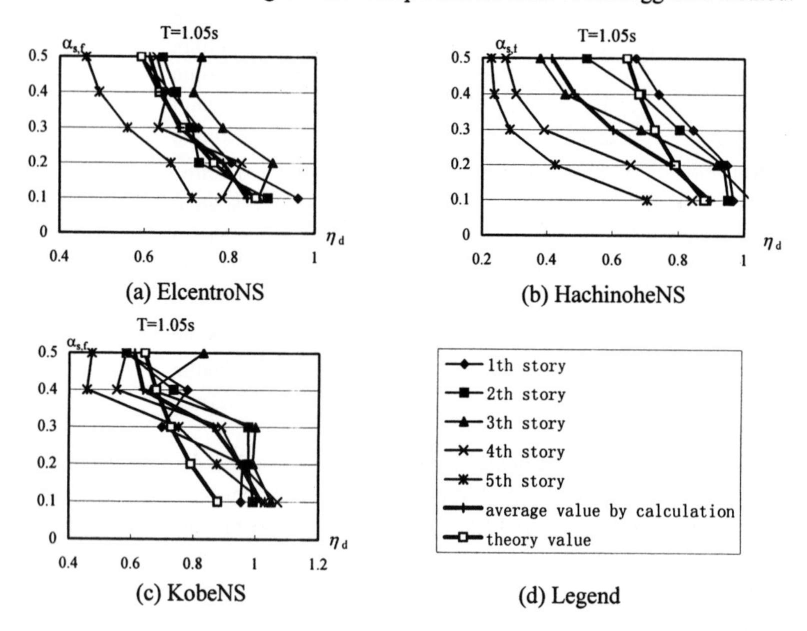

Figure 4: the effects of dual structure system under the moderate earthquakes

THE MAXIMUM INELASTIC RESPONSE OF DUAL STRUCTURE SYSTEM UNDER SEVERE EARTHQUAKE LEVEL

Based on the stiffness and yielding strength determined under the small and moderate earthquake levels previously, displacement responses of dual structure, with the design parameters α_K=1.25, $\alpha_{s,f}$=0.4 and μ_s =2, under the severe earthquake level are shown in Figure 5(a). Compared with the corresponding single structure, the story displacement response was reduced about 16~40%. Figure 5(b) shows the ductility factors. It can be seen that the ductility factor of primary structure was much decreased compared with the corresponding structure. And the ductility factor of secondary structure was within a reasonable value and also some smaller than that of the corresponding structure.

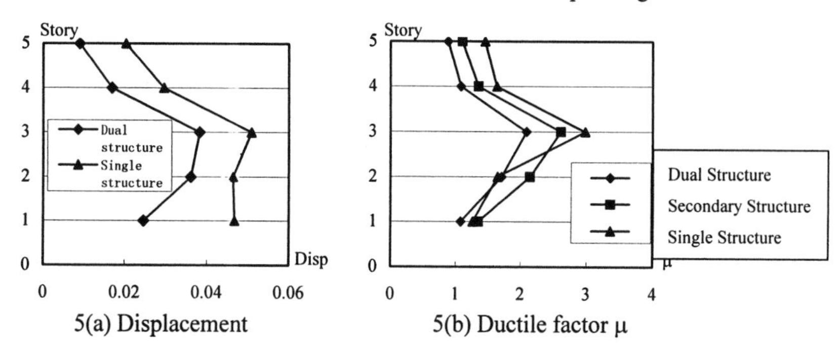

Figure 5: the effects of dual structure system under the major earthquakes

CONCLUSIONS

The following conclusions are obtained based on suggestion and analysis of this paper:

(1) It is available adopting dual structure system to resist earthquakes.

(2) The stiffness K_s of secondary structure can be determined by elastic response spectrum at the design objective of small earthquakes.

(3) The yielding strength $f_{s,y}$ of secondary structure can be determined by equal cyclic energy criteria at moderate earthquake level.

(4) The structural performance under severe earthquakes level is acceptable when the objective parameters are properly chosen.

REFERENCE

Akira Wada, et al. (1997). Seismic Design Trend of Tall Steel Buildings After the Kobe Earthquake, Japan

National Standard Building Seismic Design Code operating panel. (2001). *Building Seismic Design Code*. GB50011-2001, Beijing

Jie Jing. (2000). Applications in High-rise Buildings about Dual Seismic Structure Systems, *16th Structural Proceedings of Tall Buildings*, Shanghai, China

J.J.Connor, et al, (1997). Damage-Controlled Structures. I: Preliminary Design Methodology for

Seismically Active Regions, *Journal of Structural Engineering*, 1997.4, 423-431
L.P. Ye. (2000). Dual Seismic Structure Systems, *Building Structure*, 2000.4
L.P. Ye. and S. Otani. (1999) Maximum Seismic Displacement of Inelastic Systems Based on Energy Concept, *Earthquake Engineering and Structure Dynamics*, 1999 (6), 1483-1499
L.P. Ye. et al. (2001). Equal cyclic energy criteria and its application in the seismic and damping design. *The 5th Chinese-Japanese Structural Technique Communion*, 515-522, Xi'an, China
SEAOC Vision 2000 Committee. (1995). Performance-based Seismic Engineering, *Report Prepared by Structural Engineering Association of California,* Sacramento, California, U.S.A

Advances in Building Technology, Volume I
M. Anson, J.M. Ko and E.S.S. Lam (Eds.)

SEISMIC ASSESSMENT OF LOW-RISE BUILDING WITH TRANSFER STRUCTURE

J.H. Li[1], R.K.L. Su[1], A.M. Chandler[1] and N.T.K.Lam[2]

[1] Department of Civil Engineering, The University of Hong Kong, Pokfulam Road, Hong Kong, SAR, PRC
[2] Department of Civil & Environmental Engineering, The University of Melbourne, Parkville, Victoria 3052, Australia

ABSTRACT

An overview of the structural performance of a transfer structure in Hong Kong under potential seismic actions is presented. A hypothetical but realistic low-rise building model has been developed with 7 storey reinforced concrete frame structure and reinforced concrete transfer beams at first floor. Structural design has been based on the British Standard BS8110 and local practices. Parametric analyses of the moment-curvature relationship of each component have been conducted. By adopting the displacement-based (DB) approach, various seismic assessment methodologies, including response spectrum analysis (RSA), manual calculation and pushover analysis (POA) have been carried out. The deformations induced by the predicted seismic actions in Hong Kong are compared with those arising from POA in terms of average lateral drift ratios and maximum inter-storey drift ratios of the building. Factors influencing the performance of transfer structures are highlighted and discussed. This paper also provides a general indication of seismic vulnerability of common low-rise transfer structures in regions of low to moderate seismicity.

KEYWORDS

Earthquakes, wind, transfer, low-rise buildings, spectral, displacement, drift, soil

INTRODUCTION

Unlike other countries, transfer structures are widely used in Hong Kong, for both low-rise and high-rise buildings, in order to provide flexibility in different architectural or commercial usages above and below the transfer structures. As for low-rise transfer buildings, column type structures above the transfer structures (usually transfer deep beams) are transferred via the beam to the column type structures below. It is, however, well known by seismic engineers (Scott *et al.*, 1994, Pappin *et al.*, 1999) that under cyclic earthquake loads, concentrated stresses and large lateral displacement demands may occur at a locations where there is significant change in stiffness. Critical review of the design and construction practices of those kinds of transfer structures is urgently required. In the present study, static pushover analysis for the idealised transfer structure in Hong Kong has been carried out to identify the critical failure mechanism and to determine the pre- and post-elastic load-displacement behaviour of the structure. A displacement-based procedure has been used to assess the capacity of the structure to accommodate the strength and displacement demand generated by the soil surface response.

STRUCTURAL DETAILS OF THE 7 STOREY TRANSFER BEAM BUILDING

A hypothetical but realistic two-dimensional 7 storey reinforced concrete structure model (as shown in Figure 1) has been developed with reference to a Hong Kong school building. This model is a low-rise framed structure of 28.9m high, 23.7 width and 6 floors with a transfer beam (TB) in its first floor. Base columns are assumed to be fully fixed at ground level. Detail schedules of columns and beams are shown in Figure 1. The idealized building has been designed for the distribution of loads given in Table 1 and the design wind load derived in accordance with the Hong Kong Wind Design Code for a 50 year Return Period. It is found that the design is controlled by the loading combination case with respect to dead load and live load only. From the distribution of moments, axial forces, and shear forces generated by program ETABS Version 7.22 (Habibullah *et al.*, 1999), designs of those beams above 1st floor and between gridline ① and gridline ② are the same as B4 (see Figure 1b), those above 1st floor and between gridline ② and gridline ③ are the same as B5 (see Figure 1b).

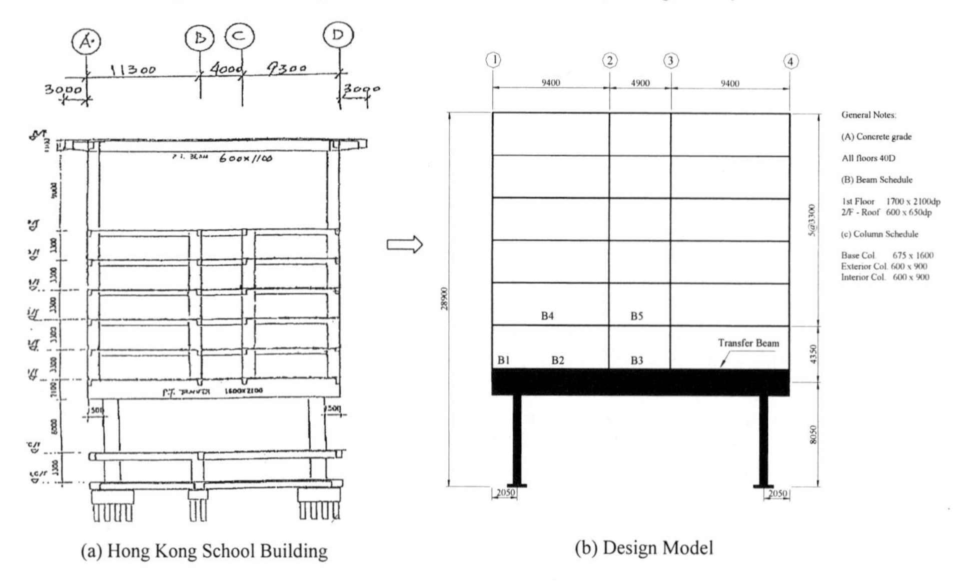

(a) Hong Kong School Building (b) Design Model

Figure 1: Example transfer building

DUCTILITY CAPACITY OF COLUMNS AND BEAMS

The moment curvature analyses for columns and beams have been conducted using the computer program RESPONSE (Collins *et al.*, 1997). The results indicate that at ultimate axial loading condition, the columns including the base columns and the frame columns have nearly reached their design ultimate inelastic strength, whilst the beams still have a large range of curvature ductility. Therefore columns are likely to be more vulnerable than beams under certain loadings. The idealized moment curvature relationship for the base columns at ultimate axial loading, utlised for the subsequent push over analysis, is shown in Figure 2. This 675 x 1600 rectangular column is reinforced along the section by 6 layers of totally 28 nos. 40mm diameter high tensile steel reinforcement. The column is confined by 10mm diameter circular hoops at a uniform spacing of 300mm centres. An independent check (based on Priestley, 1995; Priestley & Calvi, 1997; Kowalsky *et al.*, 1995) has been undertaken to assess the limiting curvatures in base columns, which gives the ultimate curvature ϕ_u of value 5 rad/1000 m, and the effective (notional) yield curvature ϕ_y of value 2 rad/1000. This leads to a curvature ductility of 2.5, which is in good agreement with the results from the RESPONSE analysis. For the limitation of the length of the paper, only the result of Moment-Curvature relationship of base column has been shown in graphical form (see Figure 2). The detailed results of all columns and beams have been given in Table 2.

TABLE 1
DISTRIBUTION OF LOADS

Load		Roof Level		Typical Floor		1st Floor	
UDL (kN/m)	DL	68		53		120	
	LL	8		11		11	
Point Load at Each Floor (kN)	Gridline	DL	LL	DL	LL	DL	LL
	①	282	35	212	53	212	53
	②	429	54	322	80	322	80
	③	429	54	322	80	322	80
	④	282	35	212	53	212	53

Note: DL = Dead Load, LL = Live Load

SEISMIC ASSESSMENT OF BUILDING

Modified lateral stiffness of the building

It is not straightforward to determine the actual stiffness of reinforced concrete as it depends on a number of factors including the dynamic stiffening, aging, steel ratio and extent of cracks in concrete under tensile loads. Typical elastic modulus of concrete can be determined either from codes of practice or from uni-axial compressive test at 40% and 10% of ultimate axial load, respectively (Lee *et al.*, and BSI, 1985). In addition, the elastic modulus of reinforced concrete may be slightly increased by 10% to 20% due to the presence of reinforcement. It was pointed out by Priestley (1998) and by Priestley and Kowalsky (1998) that section stiffness may not be considered as a fundamental section property. The stiffness of the section would be altered by changes in axial load ratio or flexural reinforcement content, due to the non-linear behaviour of the reinforced concrete and the variation of the effective sectional area for cracked sections. The stiffnesses of vertical members have therefore been modified herein, under ultimate conditions, with the factors applied to the stiffness of column and beam being 0.8 and 0.5, respectively. These factors were derived according to design values presented in the above references, modified appropriately to allow for the relatively low levels of seismic displacement demand. The natural periods are 1.40sec, 0.43sec, and 0.25sec for mode 1, mode 2, and mode 3, respectively.

Ductility and drift demand of the building

Response spectra analysis based on the modal superposition method is restricted to linear elastic analysis. This analysis can simulate the maximum response of joint displacements and member forces due to a specific ground motion loading without carrying out the lengthy integration process. There are computational advantages in using the response spectrum method of seismic analysis for prediction of displacement and member forces in structural systems. The method involves the calculation of only the maximum values of the displacement and member forces in each mode using smooth design spectra that are the average of several earthquake motions.

By using the modified stiffness model as well as assuming 5% and 3% viscous damping, a detailed response spectrum analysis (RSA) was implemented herein using the computer package ETABS (1999). Rock and soil acceleration response spectra (shown in Figure 3) as used in the RSA were generated by Sheikh (2001) based on the Shau Kei Wan reclamation site with soil depth of 52m, for different return periods, near-field and far-field events. It is found that the 2500-year return period far-field event is the critical case affecting the structures. Therefore only the corresponding drift results for the building will be presented in the following sections due to the limitation of the length of this paper.

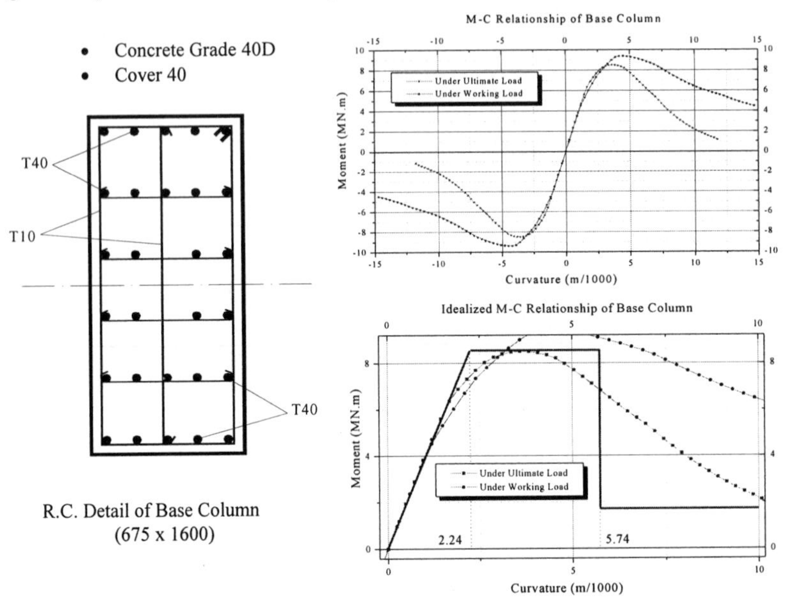

Figure 2: Idealized moment curvature relationship

TABLE 2
SUMMARY OF THE MOMENT-CURVATURE ANALYSES

Member (refer to Figure 1)	Load Condition	Effective (notional) Yield Curvature ϕ_y (rad/1000 m)		Ultimate Yield Curvature ϕ_u (rad/1000 m)		Curvature Ductility Ratio ϕ_u/ϕ_y		Ultimate Moment Capacity (MNm)	
Columns									
Base	Ultimate Axial Load (=17MN)	2.24		5.74		**2.56**		8.0	
Exterior	Ultimate Axial Load (=7MN)	4.2		9.82		**2.34**		2.0	
Interior	Ultimate Axial Load (=7MN)	3.77		9.07		**2.41**		1.7	
Beams									
	Axial Force	Positive	Negative	Positive	Negative	Positive	Negative	Positive	Negative
B1	0	1.5	2.03	19.76	22	13.17	10.84	1	12.46
B2	0	2.34	1.91	15.84	22.49	6.77	11.77	42.34	23.31
B3	0	2.68	13.76	7.07	22.51	2.64	1.64	37.33	3.61
B4	0	8.36	9.48	92.16	27.43	11.02	2.89	0.836	1.68
B5	0	6.0	6.0	76.67	76.67	12.78	12.78	0.18	0.18

Note: The R.C. arrangements of columns and beams are symmetric and asymmetric, respectively, about their principal geometric axes.

The elastic seismic displacement demand profiles determined by RSA are shown in Figure 4b, and corresponding interstorey drift ratios are shown in Figure 5. Comparisons of the results for one mode only with those for all modes indicate that higher mode effects can be ignored when analyzing the low-rise building. The maximum seismic roof displacement is 144mm and gives a 1/200 average drift ratio. At effective height, taken according to Priestley (1995) as 0.75h, the displacement demand computed as determined by RSA is 108mm giving an effective drift ratio of 1/268. The displacement at TB level is 16mm and gives the drift ratio in base column of 1/500.

The displacement spectrum at a notional 5% critical damping is normally provided from seismic hazard studies. However Wong (2000) and Su *et al.* (2002) find that damping value for typical buildings in Hong Kong is approximately 0.5-0.7%. Considering the moderate shaking intensity for a 2500-year earthquake event in Hong Kong, 3% damping ratio might be taken for the RSA of buildings in Hong Kong. Therefore for this case, the displacement demand is multiplied with a scale factor of around 1.2. The revised displacement demand and drift demand of 3% damping are shown in Figure 4b and Figure 5, respectively.

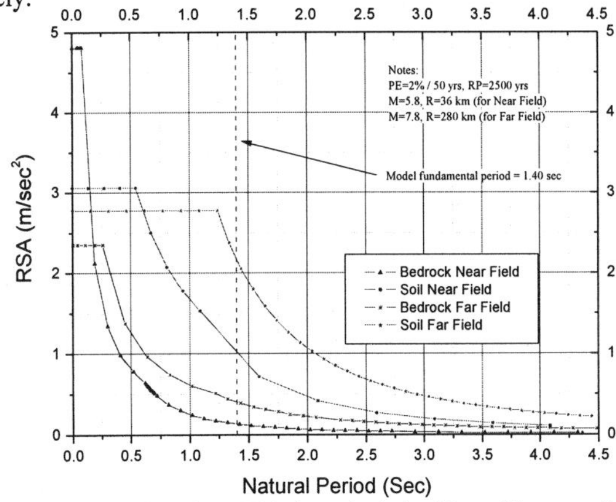

Figure 3: Response spectral acceleration (5% damping) for Hong Kong rock and soil sites

Overall ductility and drift capacity of building

As an alternative to manual calculation, the displacement capacity of building structures as a whole may be obtained by static POA (Calvi and Pavese, 1995; Park, 1996; Krawinkler and Seneviratna, 1998; Mwafy and Elnashai, 2001). POA is a simple tool for evaluating the behaviour of structures responding into the inelastic range under lateral loading. Essentially, a lumped mass single-degree-of freedom (SDF) system is created and under lateral loading the correct ratio of its mass and initial stiffness is determined such that initial period of vibration matches the fundamental natural period of vibration of the original multi-degree-of-freedom (MDOF) model. The initial natural period matching will automatically result in the response displacement time-histories at any point on the MODF structure that is proportional to that of the SDOF model, provided that the response is governed by a single fundamental mode. The linear and nonlinear behaviours are sensitive to the chosen pattern of lateral load. The load patterns adopted may not give good estimates of dynamic interstorey drift ratios over the entire height, since different load patterns may induce different localized yield mechanisms in a structure, and furthermore the demands will depend on the frequency characteristics of the ground motion and hence determine which, if any, of those localized mechanisms will activated and amplified in an earthquake. Bearing in mind the above considerations, it is herein recommended that three load patterns are selected for carrying out such POA. The first loading pattern (P1) is determined by iteration such that its shape is proportional to the deflections shape of the building throughout the elastic range of response. The second load pattern (P2) attempts to capture the soft storey effect and provides an essentially constant loading above the TB level, being linearly decreased to zero from the TB to ground level. The third load pattern (P3) attempts to capture the higher mode effects and is proportional to the SRSS combination interstorey shear obtained from RSA discussed above.

Program ETABS (Habibullah *et al*, 1999) has been used to accomplish the POA described above. The roof displacement profiles at effective (notional) yield limit and ultimate limit are shown in Figure 4b, and interstorey drift ratios at each stage are shown in Figure 5. It is found that the ultimate roof displacements, which are sensitive to different load patterns, range from 204mm to 220mm, and those at TB level range from 23mm to 34mm. Load pattern (P2) gives the smallest ultimate roof displacement capacity, but gives the largest displacement at TB level. Pattern (P3) gives the smallest value of 23mm at TB level, which is in very close agreement with the value 22mm predicted by hand calculation for ultimate axial load. The displacement ductility capacity for all load patterns is rather low and less than 2. This indicates that selecting only one load pattern as the POA is not enough to predict the critical case, and hence two patterns (P2 and P3 described above) are recommended. Displacement at effective height is 157mm at ultimate condition.

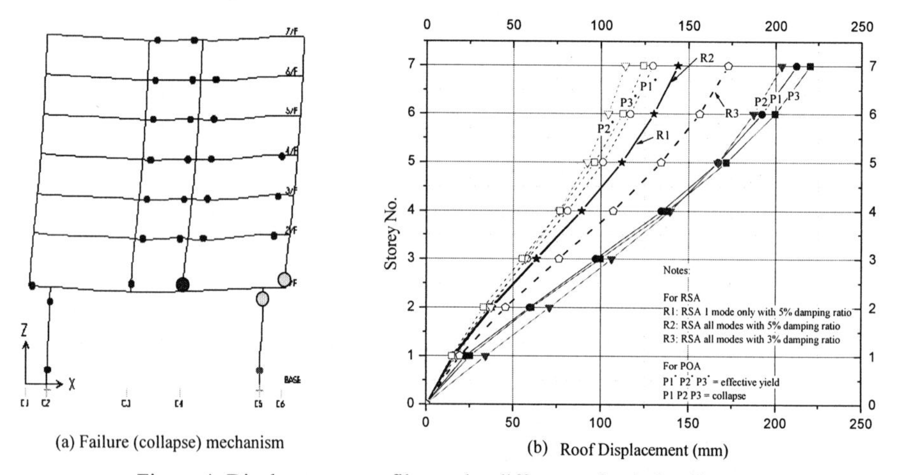

(a) Failure (collapse) mechanism (b) Roof Displacement (mm)

Figure 4: Displacement profiles under different seismic loadings

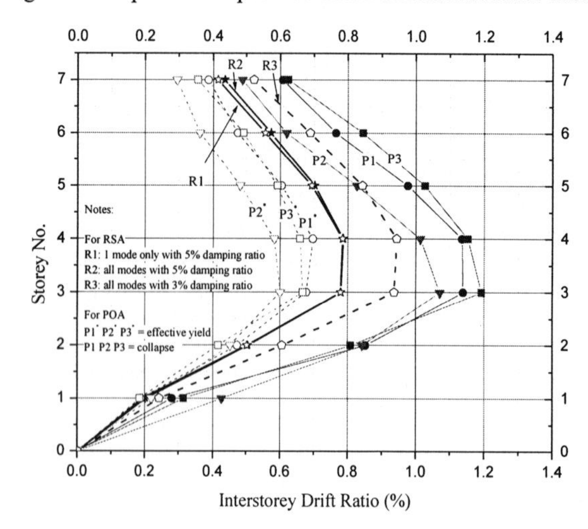

Figure 5: Interstorey drift ratios under different seismic loadings

Comparison of seismic demand with capacity

The comparison of the maximum interstorey drift demand and capacity over the entire height of the building is an effective way to identify any local distortional effect and hence any local overloading at particular floor levels. The RSA reveals that a relatively high seismic interstorey drift demand occurs at 3rd and 4th floor rather than at roof level, at effective height, or at TB level. The maximum lateral displacement demand at TB level as determined in the RSA is 16mm corresponding to a 2500-year return period, far-field earthquake at soil site. The ultimate displacement capacity from POA is 23mm, which agrees well with the value of 22mm estimated by using manual calculation. When comparing the maximum expected seismic displacement demand with capacity, a safety margin of 1.4 can be found. Also the same safety margin value of 1.4 is attained when comparing the lateral displacement demands with the ultimate displacement capacities at roof level and effective height respectively. Hence the building has a relative low probability of ultimate collapse in the form of a soft storey failure under seismic action in Hong Kong. But the drift demand will exceed the smallest drift capacity by POA under P2 at the upper part of the building (see Figure 5) if we take the 3% as the analysing damping ratio, and therefore local damages (short span beams shown in Figure 4a) of the upper part of structure may occur.

It is found that all the displacement shapes from POA are similar to those predicted by RSA. This result matches with Krawinkler and Seneviratna (1998) who highlighted the fact that the deflection profiles obtained by POA and by dynamic analysis are closely correlated for low-rise structures. The results clearly reveal that the POA described above can produce reliable results for situations where higher mode effects are minor and the soft storey effects are slight.

CONCLUSIONS

This paper has provided, for the first time in Hong Kong, a rigorous and comprehensive structural seismic assessment of a typical low-rise transfer structure designed to local codes and practices.

The major conclusions of this study are as follows:

1. Design of a low-rise building in Hong Kong is controlled by dead load rather than wind load, which is different from the high-rise building. Therefore it is not meaningful when assessing the capacity of those types of building to simply compare the seismic displacement demand with those from the design wind load.
2. Structural seismic assessment for Transfer Structures in Hong Kong requires specific account to be taken of the acute lack of ductility arising in columns supporting the transfer beam (TB) as well as the frame columns above TB level. Failure mechanism (refer to Figure 4a) shows that the short spanned beams connecting columns are prone to form plastic hinges, and therefore reach this ultimate capacity and collapse finally. Concerns on the design of those beams should be given.
3. POA procedure is sufficient to assess the displacement capacity of a low-rise building since the deflection shapes produced match well with those from the RSA due to minor higher mode effects, whilst for high-rise building POA is not recommended (Su. *et al.*, 2001). It is also found that only one load pattern to predict the critical displacement and drift capacities of building is not enough. Two loading patterns, namely uniform loading pattern (P2) and SRSS loading pattern (P3), are recommended herein.
4. The acute lack of ductility of Hong Kong transfer structures, at both member and global levels indicates that the ratio of ultimate and notional yield displacement capacities (~2) is relatively low. The possibility of sudden brittle-type failures of those columns of such structures, under excessive seismic displacements, is therefore a matter of some concern.
5. Comparisons among those inter-storey drift ratios from various analyses indicate that detailed local checks are required when assessing the performance of a low-rise transfer building rather than only focusing concern on the roof level, the effective height level and the TB level.

6. The building has a low probability of ultimate collapse in the form of a soft storey failure under seismic action in Hong Kong when taking the damping ratio as 5%, but damages may occur at upper parts of the building (especially to short span beams) if the damping value is reduced to 3%.

ACKNOWLEDGEMENTS

The methodology described in this paper forms part of the outcome of a major strategic research programme to address seismic risk in Hong Kong. The Hong Kong programme, directed by the second and third authors, has been undertaken at the University of Hong Kong since 1998, as has received continuous funding from the Research Grants Council (RGC) under grants HKU7023/99E and HKU7002/00E. The authors wish to thank the RGC for their generous support.

References

Building Development Department (1983) *Code of Practice on Wind Effects: Hong Kong – 1983*, Government of Hong Kong, Hong Kong
British Standards Institution (BSI) (1985). *Code of Practice for Design and Construction* (BS8110 Part 2), *British Standard, Structural Use of Concrete*.
Calvi, G.M. and Pavese, A. (1995). Displacement Based Design of Building Structures. *European Seismic Design Practice*, Balkema, Rotterdam, 127-132.
Collins, M.P. and Mitchell D. (1997) *Prestressed Concrete Structures*, Response Publications, Canada.
Habibullah, A. (1999*), ETABS (version 7.22) Three Dimensional Analysis of Building Systems, User's Manual*, Computers & Structures Inc.
Krawinkler, H. and Seneviratna, G.D.P.K. (1998). "Pros and cons of a pushover analysis of seismic performance evaluation", Engineering Structures, **20**(4-6), 452-464.
Lee, P.K.K., Kwan, A.K.H. and Zheng, W. (2000). "Tensile strength and elastic modulus of typical concrete made in Hong Kong", *Transactions of Hong Kong Institution of Engineers*, **7**(2), 35-40.
Mwafy, A.M. and Elnashai, A.S. (2001). "Static pushover versus dynamic collapse analysis of RC buildings", *Journal of Engineering Structures*, 23(5), 407-424.
Pappin, J.W., Kwok, M.K.Y. and Chandler, A.M. (1999). Consideration of Extreme Seismic Events in the Design of Structures in Hong Kong. *Proceedings of Conference on Construction Challenges into the Next Century*, 98-105.
Priestley, M.J.N. (1995). Displacement-Based Seismic Assessment of Existing Reinforced Concrete Buildings. *Proceedings of the Fifth Pacific Conference of Earthquake Engineering*, Melbourne, 225-244.
Priestley, M.J.N. and Calvi, G.M. (1997). "Concepts and procedures for direct displacement-based design and assessment", *Proceedings of the Workshop on Seismic Design Approaches for the 21st Century,* June 1997, Slovenia.
Priestley, M.J.N. and Kowalsky M.J.(1998) Aspects of Drift and Ductility Capacity of Rectangular Cantilever Structural Walls, *Bulletin of the NZNSEE,* **31:4,** 246-259.
Priestley, M.J.N. (1998) Brief Comments on Elastic Flexibility of Reinforced Concrete Frames and Significance of Seismic Design, *Bulletin of the NZNSEE* , **31:4,** 246-259.
Scott, D.M., Pappin, J.W. and Kwok, M.K.Y. (1994) Seismic Design of Buildings in Hong Kong, *Transactions of the Hong Kong Institution of Engineers.* **1:2,** 37-50.
Sheikh, M.N. (2001). *Simplified Analysis of Earthquake Site Response with Particular Application to Low and Moderate Seismicity Regions, MPhil Thesis*, The University of Hong Kong
Su R.K.L, Chandler A.M, Lam N.T.K. and Li J.H. (2000) Motion induced by distant earthquakes: the assessment of transfer building structures. *International conference on Advances in Structural Dynamics (ASD2000)*, the Hong Kong Polytechnic University, **1**:249-256
Su R.K.L, Chandler A.M and Li J.H. (2002) *Dynamic testing and modeling of existing buildings in Hong Kong* (submitted to Transactions of HKIE)
Wong J.C.K. (2000) *Identification of non-linear damping of tall buildings by random decrement, PhD Thesis*, City University of Hong Kong, Hong Kong, PRC

Advances in Building Technology, Volume 1
M. Anson, J.M. Ko and E.S.S. Lam (Eds.)

ULTIMATE DRIFT RATIO OF NON-DUCTILE COUPLING BEAMS

Z.Q. LIU[1], S.S.E. LAM[2], B. WU[3] and Y.L. Wong[2]

[1]School of Civil Engineering, Harbin Institute of Technology, Harbin, China
[2]Department of Civil and Structural Engineering, The Hong Kong Polytechnic University, Hong Kong
[3]Department of Civil Engineering, South China University of Technology, Guangzhou, China

ABSTRACT

Seismic performance of coupling beams designed according to non-seismic provisions is examined. Coupling beam specimens in 1/4 scale were tested under cycles of end shear. Characteristics of the specimens include high main reinforcement ratio, about 3-4%, and large spacing of links with 90-degree hooks. Based on the test data, 140mm is recommended to be the optimal spacing of links. Furthermore, coupling beams must not be heavily reinforced. Since this would lead to substantial reduction in the displacement ductility and ultimate drift ratio. Finally, an empirical relationship is proposed to estimate the ultimate drift ratio of non-ductile coupling beams.

KEYWORDS

Reinforced concrete, Coupling beams, Seismic, Non-ductile, Ultimate drift ratio, Ductility

INTRODUCTION

Coupling beams are commonly used in high-rise buildings to link up the shear walls. Seismic performance of coupling beams has been an area of strong research interest (Paulay 1971). Paulay and Binney (1974) and Barney *et al* (1980) examined the use of diagonal bars as the main reinforcements. Tassios *et al* (1996) and Harries *et al* (2000) studied the performance of different types of coupling beams. However, there exists very few research studies related to the seismic performance of coupling beams designed according to non-seismic requirements. In our earlier study (Liu *et al* 2002), cyclic load tests were carried out on 5 coupling beam specimens designed according to non-seismic design requirements. Characteristics of the specimens include high main reinforcement ratio, about 3-4%, and large spacing of links with 90-degree hooks. As a result, such coupling beams are non-ductile with limited ductility. Subsequent to the earlier study, two new specimens were prepared and tested. It is the objective of this study to combine our test data with those obtained by others to develop an empirical relationship to predict the ultimate drift ratio of non-ductile coupling beams.

EXPERIMENTAL SETUP

Two coupling beam specimens (specimens L-1 and L-2) in 1/4 scale were prepared and tested in this study. Reinforcement details of these two specimens were designed so that comparisons can be made with two other specimens (specimen L-A and L-D) reported in the previous study (Liu *et al* 2002). Figure 1 shows the reinforcement details. Furthermore, all the links used in the specimens have 90-degree hooks.

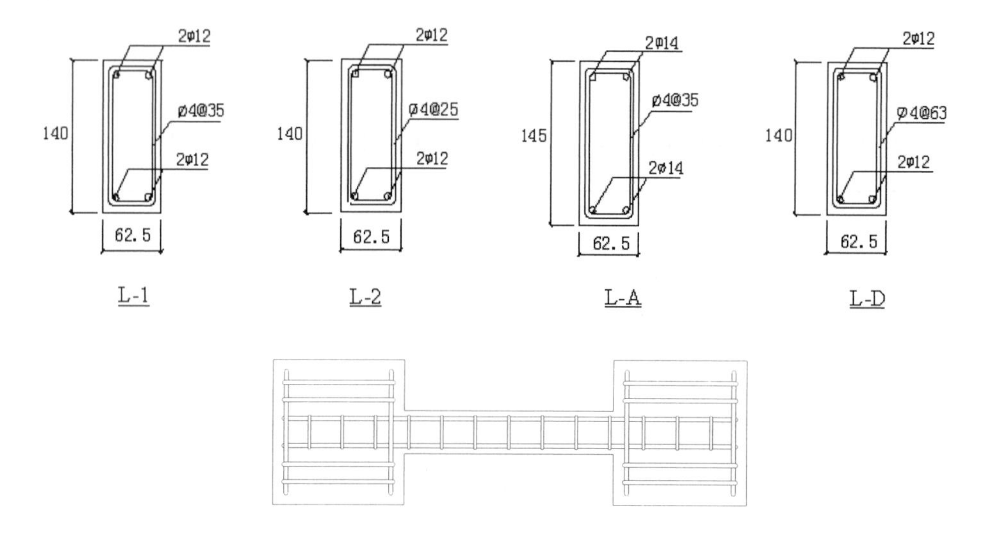

Figure 1. Reinforcement details of the specimens

Basic characteristics of the specimens are shown in Table 1. B, H, h and L are the breadth, overall depth, effective depth and clear span of a specimen. λ is the shear span ratio and is defined as half clear span divided by the effective depth of a specimen. A_{st} is the main reinforcement, and ρ is the amount of A_{st} expressed as a fraction of the cross-sectional area. Transverse reinforcement ratio is defined as A_{sh}/S_tB. A_{sh} is the total area of links in the transverse direction of bending and S_t is the spacing of links.

Table 1. Basic characteristic of the specimens

Specimen	BxH (mm)	h (mm)	L (mm)	λ	A_{st}	ρ	Links	A_{sh}/S_tB
L-1	62.5×140	124	750	3.0	2Φ12	2.92%	Φ4-2legs@35	1.15%
L-2	62.5×140	124	750	3.0	2Φ12	2.92%	Φ4-2legs@25	1.61%
L-A	62.5×145	128	750	2.9	2Φ14	3.85%	Φ4-2legs@35	1.15%
L-D	62.5×140	124	750	3.0	2Φ12	2.92%	Φ4-2legs@63	0.64%

Normal strength concrete with maximum aggregate size at 5mm was used in this study. The measured average compressive strength f_{cu} of specimens L-1 and L-2 is 38.1MPa and 37.3MPa for specimens L-A and L-D. The measured yield strength and ultimate strength of the reinforcement are tabulated in Table 2. Φ12 and Φ14 are deformed bars, whereas Φ4 is a round bar.

Table 2. Mechanical properties of reinforcements (MPa)

Specimens	Reinforcement	Yield stress	Ultimate strength
L-1 and L-2	Φ12 bars	386	582
	Φ4 steel wire	284	391
L-A and L-D	Φ12 bars	393	590
	Φ14 bars	422	630
	Φ4 steel wire	273	374

Tests were conducted using an existing loading frame installed in the Harbin Institute of Technology. Figure 2 shows the loading apparatus. The test setup represents typical loading condition in a coupling beam in that the coupling beam is subject to end shears. A load cell was installed to measure the shear load. Transducers were used to measure the relative displacements between the top and bottom stubs and the displacements within the plastic hinges.

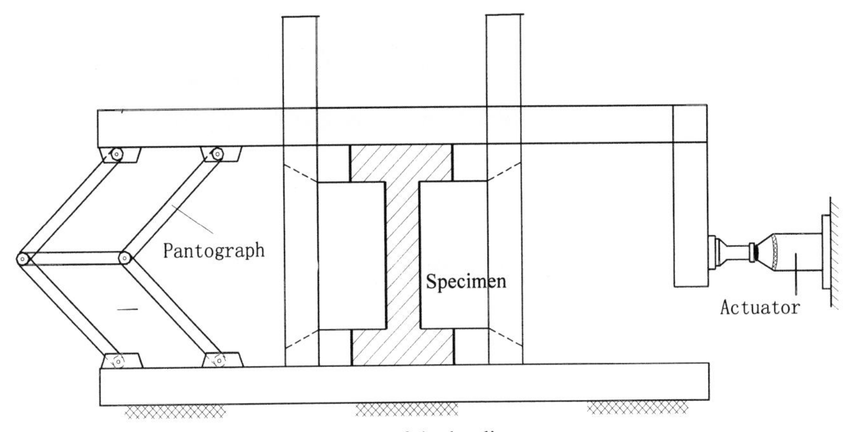

Figure 2. Setup of the loading apparatus

The specimens were subject to a series of positive and negative loading cycles. Firstly, shear load was applied (the first push cycle) until obvious reduction in stiffness was observed. The corresponding displacement was denoted as U1. Two complete cycles of push-pull loading were performed to achieve the targeted displacement δ at δ=U1. The loading cycles continued with progressive increase in the targeted displacement, i.e. δ=U2, δ=U3, and so on. The loading cycles terminated when there was more than 20% drop in the load-carrying capacity from the maximum.

TEST RESULTS

When subject to cycles of push-pull loading, failures of all the specimens were due to shear. Figure 3 shows conditions of the specimens at failure. Formation of cracks in specimens L-1 and L-2 are similar to each other. In the first few loading cycles, flexural cracks were found within the plastic hinges at the two ends. Subsequently, shear cracks developed with spalling of the concrete cover within the plastic hinges. No obvious opening of the 90-degree hooks was observed in the tests.

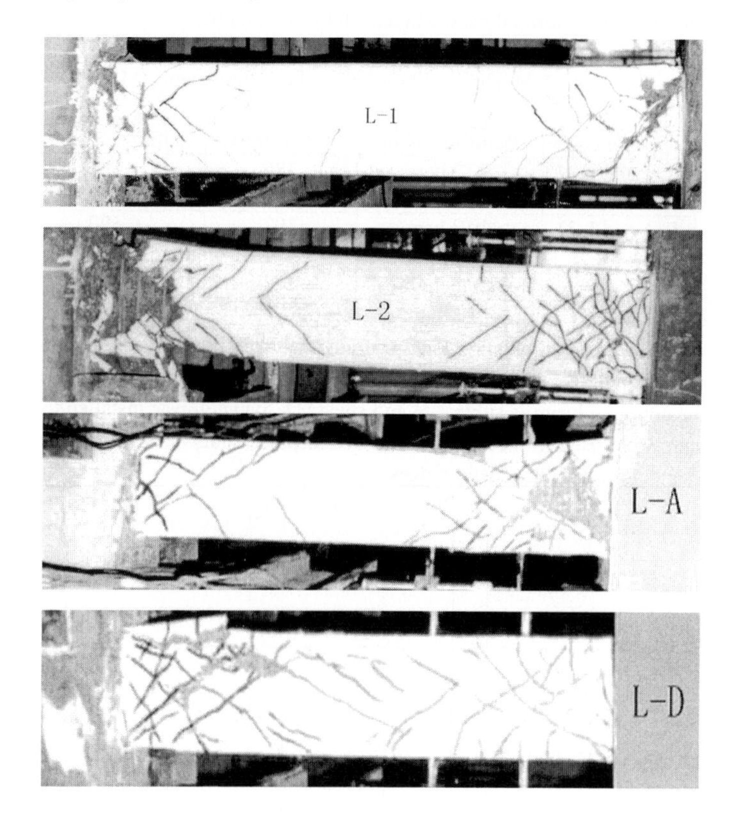

Figure 3. Modes of failure of the specimens

At the advanced stage of the loading history, deformations of all the specimens were mainly due to shear. Similarly, many inclined cracks were observed when approaching failure. However, for specimens L-1 and L-2 having smaller spacing of links, cracks were found to be concentrated at the two ends. On the contrary, the cracking pattern of specimen L-D (having the largest spacing of links) was more extensive and covers over the whole length of the beam.

HYSTERESIS BEHAVIOR

Figure 4 shows the hysteresis loops of the specimens. P is the shear load acting on a specimen. Test results are summarized in Table 3. For specimens L-1 and L-2, these are obtained from the negative envelops. For specimens L-A and L-D, these are estimated by taking the average of the values obtained from the positive and negative envelops. Yield condition is defined according to Figure 5. Failure load and displacement at failure are the corresponding values at 20% drop in the load-carrying capacity from the maximum. Displacement ductility factor is the ratio of the displacement at failure to the yield displacement. Ultimate drift ratio is the ratio of the relative displacement at the two ends of a beam at failure against the clear span of the beam.

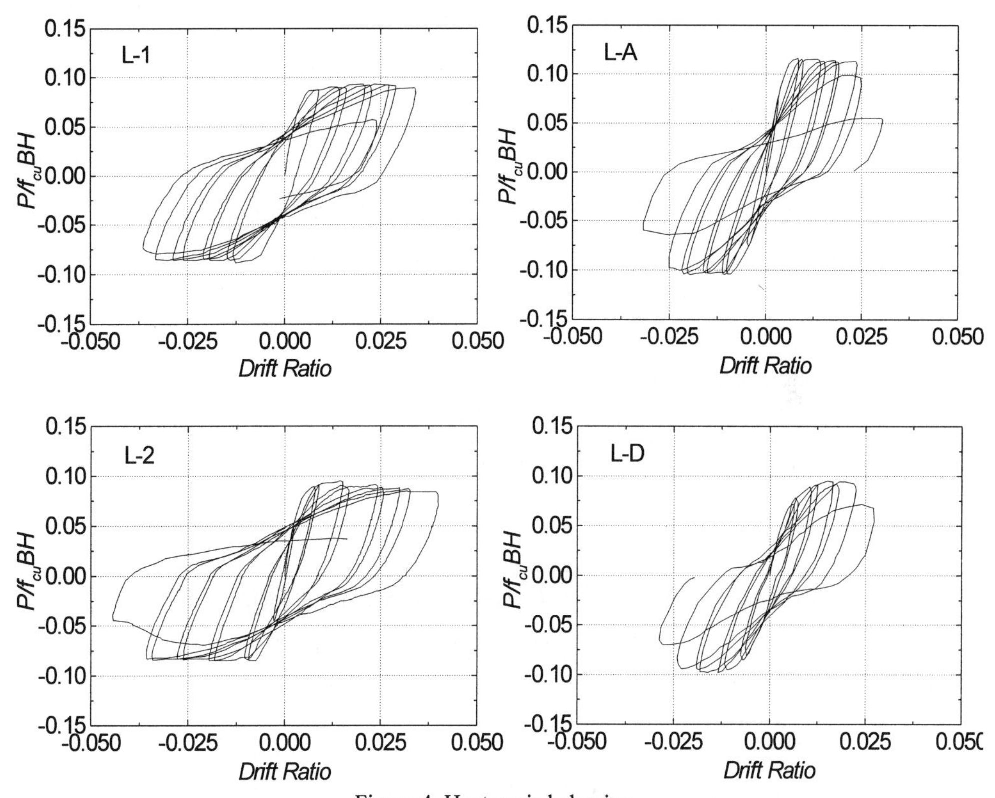

Figure 4. Hysteresis behavior

Table 3. Test results

Specimen	L-1	L-2	L-A	L-D
Yield load P_y (kN)	26.85	26.44	32.28	25.7
Yield displacement Δ_y (mm)	4.18	4.48	4.28	5.03
Ultimate load P_m (kN)	30.30	30.00	37.23	31.01
Ultimate displacement Δ_m (mm)	12.45	12.20	6.52	12.55
Failure load P_u (kN)	23.62	22.64	29.79	24.53
Displacement at failure Δ_u (mm)	27.87	28.67	20.04	19.66
Ultimate drift ratio	1/26.9	1/26.2	1/37.4	1/38.1
Displacement ductility factor	6.67	6.40	4.68	3.91

Apart from having different spacing of links, specimens L-1, L-2 and L-D have the same reinforcement details. When the spacing of links is reduced from 63mm (specimen L-D) to 35mm (specimen L-1), the ultimate drift ratio increases by 42% and the displacement ductility factor by 71%. However, when the spacing of links is further reduced to 25mm (specimen L-2), there is no obvious gain in either the ultimate drift ratio or displacement ductility factor. This indicates that the optimal spacing of links could be around 140mm (i.e. 4x35mm in full scale).

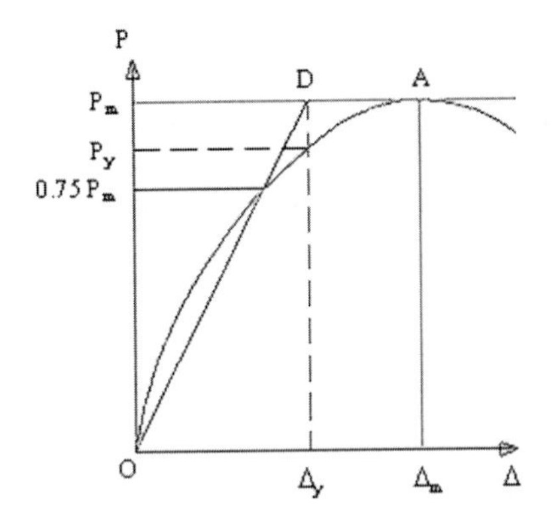

Figure 5. Definition of yielding

Main reinforcements in specimens L-A is 32% more than that specified in specimen L-1. This leads to a 23% increase in the ultimate load P_m. However, there is substantial reduction in both the ultimate drift ratio and displacement ductility factor by 28% and 30% respectively. This demonstrates the use of heavily reinforced beams may serious reduce the ductility and is therefore undesirable for seismic resistance.

EMPIRICAL RELATIONSHIP FOR ULTIMATE DRIFT RATIO

In order to develop an empirical formula to predict the ultimate drift ratio of non-ductile coupling beams, test data obtained from this study are correlated with those obtained by others. Basic properties of the selected test data are summarized in Table 4. There are 13 test data, of which 7 are obtained from this study.

Table 4. Test data

Specimen	BxH (mm)	h (mm)	L (mm)	λ	ρ	A_{sh}/S_tB	Ultimate drift ratio	Reference
L-1	62.5×140	124	750	3.0	2.92%	1.15%	1/26.9	Present study
L-2	62.5×140	124	750	3.0	2.92%	1.61%	1/26.2	Present study
L-A	62.5×145	128	750	2.9	3.85%	1.15%	1/37.4	Liu *et al* (2002)
L-D	62.5×140	124	750	3.0	2.92%	0.64%	1/38.1	Liu *et al* (2002)
L-C1	75×140	124	625	2.5	3.65%	1.15%	1/31.4	Liu *et al* (2002)
L-C2	75×140	124	625	2.5	3.65%	1.15%	1/28.6	Liu *et al* (2002)
L-E	100×150	124	875	3.5	5.47%	1.12%	1/31.7	Liu *et al* (2002)
C2	102×169	159	423	1.33	0.70%	1.10%	1/21.9	Barney *et al* (1980)
C5	102×169	159	423	1.33	0.70%	1.10%	1/22.5	Barney *et al* (1980)
C7	102×169	159	847	2.66	0.70%	1.10%	1/19.6	Barney *et al* (1980)
CB-1A	130×500	484	500	0.52	0.36%	1.03%	1/20.5	Tassios *et al* (1996)
CB-1B	130×300	284	500	0.88	0.61%	1.03%	1/19.0	Tassios *et al* (1996)
B3	182×300	267	750	1.40	1.29%	0.55%	1/39.8	Pam *et al* (2002)

In comparing the test data of specimens C2, C5 and C7, influence of the shear span ratio λ to the ultimate drift ratio appears to be of secondary importance. For lightly reinforced specimens CB-1A and CB-1B and for heavily reinforced specimens L-C1, L-C2 and L-E, the shear span ratio λ has minor effect on the ultimate drift ratio. Therefore, the shear span ratio λ is not considered as a parameter in the empirical relationship.

Based on the observations obtained from the various test results, the main reinforcement ratio ρ and the transverse reinforcement ratio A_{sh}/S_tB are identified as the controlling factors affecting the ultimate drift ratio θ. After examining several mathematical expressions, an empirical relationship is proposed to be in form of

$$\theta = \frac{A_{sh}}{s_t B}(0.0516 - 0.0095\rho + 0.0009\rho^2) \quad (1)$$

Equation (1) is obtained using the nonlinear regression technique. Figure 6 plots the variation of the ultimate drift ratio against the main reinforcement ratio ρ. Figure 7 shows the correlation between the predicted values and the experimental data. Also presented in dotted lines is the 10% error zone. The experimental data generally fall within 10% of the values predicted by the empircial relationship, Equation (1), indicating reasobably accurate predictions can be obtained by Equation (1). Lastly, applicable range of the empirical relationship is limited to normal strength concrete and with the lateral reinforcement ratio within the range of $0.5\% < A_{sh}/S_tB < 1.6\%$.

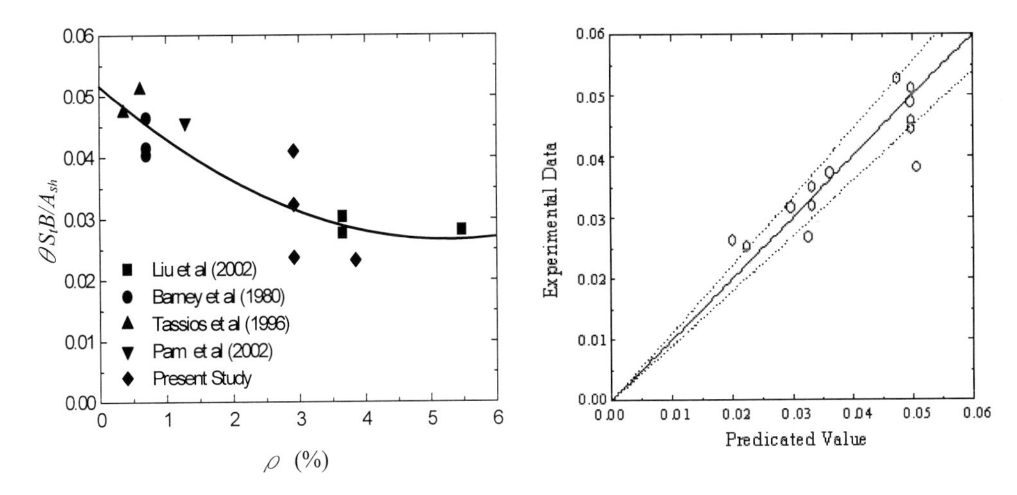

Figure 6. Variation of θ against ρ

Figure 7. Prediction of ultimate drift ratio

CONCLUSIONS

Coupling beam specimens in 1/4 scale designed according to non-seismic design provisions were tested under cycles of shear loads. Characteristics of the specimens include high main reinforcement ratio, about 3-4%, and large spacing of links with 90-degree hooks. Based on the test data, 140mm is recommended to be the optimal spacing of links. Furthermore, coupling beams must not be heavily reinforced. Since this would lead to substantial reduction in the displacement ductility and ultimate drift ratio.

Combined with the test data obtained by others, an empirical relationship is proposed to estimate the ultimate drift ratio of non-ductile coupling beams. Applicable range of the empirical relationship should be limited to normal strength concrete and with the lateral reinforcement ratio within the range of $0.5\% < A_{sh}/S_tB < 1.6\%$.

ACKNOWLEDGEMENT

The authors are grateful to the financial supports from the Research Grants Council of Hong Kong (RGC No: PolyU 5038/99E).

REFERENCES

Barney G.B., Shiu K.N., Rabba B.G., Fiorato A.E., Russell H.G. and Corle W.G. (1980). *Behavior of Coupling Beams under Load Reversals.* Research and Development Bulletin RD068.01, PCA, Research and Development/Construction Technology Laboratories.

Harries K.A., Gong B. and Shahrooz B.M. (2000). Behavior and Design of Reinforced Concrete, Steel, and Steel-Concrete Coupling Beams. *Earthquake Spectra*, **16(4)**, 775-799.

Liu Z.Q., Lam S.S.E. and Wu B. (2002). Seismic Performance of Non-ductile Coupling Beams. *Proceedings of the Structural Engineering World Congress (SEWC 2002)*, Oct 9-12, 2002, Yokohama, Japan.

Pam H.J., Su R.K.L., Lam W.Y., Li J., Au F.T.K., Kwan A.K.H. and Lee P.K.K. (2002). Designing Coupling Beams and Joints in Concrete Buildings for Improved Earthquake Resistance. *Proceedings of Recent Developments in Earthquake Engineering*, HKIE Structural Division, 17 May 2002, 91-105.

Paulay T. (1971). Coupling beams of reinforced concrete shear walls. *Journal of Structural Division, ASCE*, **97(3)**, 843-861.

Paulay T. and Binney J.R. (1974). Diagonally reinforced coupling beams of shear walls. Paper no 26, *ACI Publication SP-42*, **2**, 579-598.

Tassios T. P., Moretti M. and Bezas A. (1996). On the Behavior and Ductility of Reinforced Concrete Coupling Beams of Shear Walls. *ACI Structural Journal*, **93(6)**.

Advances in Building Technology, Volume 1
M. Anson, J.M. Ko and E.S.S. Lam (Eds.)

EFFECT OF UNCERTAIN EVOLUTIONARY PSD FUNCTION OF SEISMIC EXCITATION ON PEAK RESPONSE OF MDOF SYSTEMS

S.S.Wang and H.P. Hong

Department of Civil and Environmental Engineering, University of Western Ontario, London, Canada, N6A 5B9

ABSTRACT

Seismic excitation is inherently non-stationary with respect to time and frequency. The non-stationary excitation or process is commonly characterized by the evolutionary power spectral density (PSD) function which may be represented by a time modulating function and a PSD function of a stationary process. The present study describes an analysis procedure for evaluating probability that the peak response of a multi-degree-of-freedom system to non-stationary excitation exceeds a specified level. The analysis considers that the parameters controlling the evolutionary PSD function are uncertain. This uncertain arises from the consideration that a structural system is designed to withstand all potential earthquake threats. The model used for the PSD function of the stationary process is the Kanai-Tajimi model and the exponential modulating function is employed for the time modulating function. Parametric studies are carried out using the proposed procedure for assessing the adequacy of the so-called complete quadratic combination rule that is commonly employed in aseismic design.

KEYWORDS

Peak response, Probability, Non-stationary excitation, complete quadratic combination rule, evolutionary power spectral density function.

INTRODUCTION

The combination rules with the response spectra are often used to evaluate the peak responses of multi-degree-of-freedom (MDOF) linear systems subjected to ground motions. The response spectra are usually obtained by the mean or mean plus one standard deviation of the normalized response spectra of a suite of ground motions and the ground motion parameters such as peak ground acceleration. Such spectra may have different probability of exceedance since the peak responses at different periods are not fully correlated. If the uniform hazard spectra (UHS) which have equal probability of exceedance at different periods are taken place, the adequacy of use of the combination rules to evaluate the peak responses of MDOF should be verified. Recently, it has been found that the peak response of MDOF systems to stationary ground motions calculated by the complete quadratic combination (CQC) rule with the UHS may be conservative or unconservative depending on the modal periods and contributions of each mode (Hong and Wang 2002). However, the ground accelerations are non-stationary with respect to both time and frequency. The effect of the non-stationary excitation on the use of the CQC rule with UHS to estimate the peak response of MDOF systems was considered in this study. The analysis was carried out by modeling the non-stationary excitation as the evolutionary power spectral density (PSD) function with uncertain parameters. In particular, the Kanai-Tajimi model with exponential modulating function was used to define the evolutionary PSD function. The first-passage probability distribution developed by Vanmarcke (1975, 1976) was used to represent the probability distribution of the peak responses. A set of UHS to non-stationary ground accelerations was established. Using the obtained UHS, the probability associated with the peak responses of MDOF systems to non-stationary ground motions evaluated by the CQC rule was assessed. Details of analysis procedure and the obtained results were given in the following section.

NON-STATIONARY EXCITATION AND STRUCTURAL RESPONSES

Non-stationary Excitation

The non-stationary ground motion may be characterized with an evolutionary PSD function that consists of a time modulating function $A(t)$ and a PSD function of stationary process $G_x(\omega)$ (Priestley 1967). The evolutionary PSD function (one-sided), $G_x(\omega,t)$, can be expressed as:

$$G_x(\omega,t) = |A(t)|^2 G_x(\omega), \tag{1}$$

The Kanai-Tajimi PSD function employed in this study takes the following form (Tajimi 1960)

$$G_x(\omega) = \frac{\omega_g^{\ 4} + 4\xi_g^{\ 2}\omega_g^{\ 2}\omega^2}{\left(\omega_g^{\ 2} - \omega^2\right)^2 + 4\xi_g^{\ 2}\omega_g^{\ 2}\omega^2} G_0, \tag{2}$$

where ω (rad/sec) is the frequency, G_0 is the intensity of the ideal white noise; ω_g (rad/sec) is the filter frequency that determines the dominant range of input frequencies; the damping coefficient ξ_g is a parameter that indicates the shape of the power spectral density.

The exponential modulating function adopted in this study can be written as (Shinozuka and Sato 1967):

$$A(t) = A_0(e^{-b_1 t} - e^{-b_2 t}), \ \ b_2 > b_1, \tag{3}$$

where the b_1 and b_2 are dependent on the strong motion duration, T_0, and the fraction of rise time, ε; and the scaling factor A_0 which equals $1/\max(e^{-b_1 t} - e^{-b_2 t})$ is used in the present study.

Structural Responses to Non-stationary Ground Motion

Consider a linear MDOF system with ω_i and ξ_i denoting the frequency and damping ratio of the i-th mode subjected to a non-stationary ground acceleration with evolutionary PSD function $G_x(\omega, t)$. The response $R(t)$ of the system is also a non-stationary random process with its one-sided evolutionary PSD function, $G_R(\omega,t)$, given by:

$$G_R(\omega,t) = \sum_i^n \sum_j^n C_i C_j M_i(\omega,t) M_j^*(\omega,t) G_x(\omega), \tag{4}$$

where C_i is the effective participation factor of i-th mode; the asterisk denotes the complex conjugate; and $M_i(\omega,t) = \int_0^t h_i(\tau) A(t-\tau) e^{-i\omega\tau} d\tau$, where $h_i(\tau) = \frac{1}{\omega_i\sqrt{1-\xi_i^2}} e^{-\xi_i\omega_i\tau} \sin\left(\omega_i\sqrt{1-\xi_i^2}\,\tau\right)$.

PROBABILISTIC ANLYSIS OF PEAK RESPONSE

Statistics of Random Variables

To represent the uncertainty in seismic excitation for all potential earthquake events, the model parameters ω_g, ξ_g and G_0 in Kanai-Tajimi PSD function and the parameters T_0 and ε controlling the time modulating function were considered as random variables. The statistics of ω_g, ξ_g, G_0, and T_0 given by Vanmarcke and Lai (1980, 1982) and the statistics of ε provided by Balendra et. al (1991) were adopted. The recommended statistics of these parameters were shown in Table 1.

TABLE 1
STATISTICS OF MODEL PARAMETERS IN KANAI-TAJIMI PSD FUNCTION

Model parameter	Mean	Coefficient of variation	Standard deviation	Distribution type
ω_g (rad/sec)	19.06	0.427	8.139	Gamma
ξ_g	0.316	0.427	0.135	Lognormal
G_0 (cm^2/sec^3)	35.32	2.454	86.675	Lognormal
ε	0.159	0.58	0.092	Uniform
τ(sec)	10.09	0.90	9.081	Gumbel

Probability of Exceedance of Peak Response

Consider a linear system subjected to non-stationary ground acceleration with the evolutionary PSD function $G_x(\omega, t)$ given Eqn. 1. The probability that the peak response of the system, R, is within the prescribed barriers $\pm r$ during (0, t), $L_R(r,t)$, can be found in the literature (Vanmarcke 1975, 1976). The parameters in $L_R(r,t)$ depend on the first three moments of peak response, which are related to the evolutionary PSD function of peak response. Since the evolutionary PSD function of ground motion depends on a set of random variables **X**, $L_R(r,t)$ is also conditioned on **X**. To emphasize the conditional relation, the notation $L_{R|X}(r,t)$ is employed to replace $L_R(r,t)$.

The unconditional probability that the peak response exceeds the prescribed barriers $\pm r$ during (0, t), P_e(r), can be evaluated by FORM using (Madsen et. al 1986):

$$P_e(r) = \int_{\Omega} \left(1 - L_{R|X}(r,t)\right) f_{\mathbf{X}}(\mathbf{x}) dz d\mathbf{x} = \int_{g \le 0} \phi(z) f_{\mathbf{X}}(\mathbf{x}) dz d\mathbf{x}, \tag{5}$$

where Ω denotes the domain of $\mathbf{X}$, $f_{\mathbf{X}}(\mathbf{x})$ represents the joint probability density function of $\mathbf{X}$,

$$g = z - \Phi^{-1}\left(1 - L_{R|X}(r,t)\right), \tag{6}$$

represents an auxiliary limit state function, $\phi(\bullet)$ is the standard normal probability density function, and $\Phi^{-1}(\bullet)$ is the inverse of the standard normal distribution function $\Phi(\bullet)$.

Fractile of Peak response

Given an evolutionary PSD function and structural parameters, Eqns. 5 and 6 can be used to calculate the probability that the (absolute value of) peak response is less than a specified value. Alternatively, the peak response corresponding to a specified probability of exceedance P_{Te} or a target reliability index β_T, r_T, can be carried out iteratively. The fractiles of the peak response for a series of SDOF systems with a range of natural periods of vibration and damping ratios can be used to form the response spectra. These response spectra can be viewed as the UHS since they have equal probability of exceedance P_{Te}, and the seismic risk was reflected by considering the uncertainty in the parameters of evolutionary PSD function.

Note that for practical applications, rather than using the fractile of MDOF systems r_T, the modal combination rules with peak modal responses obtained from the response spectra are commonly employed. Since the CQC rule is most widely used in aseismic design, the study is focused on the verification of adequacy of using this rule with UHS to evaluate the peak responses of MDOF systems.

NUMERICAL ANALYSIS

Peak Response of SDOF System with Equal Probability of Exceedance

For establishing the UHS, the probability of non-exceedance levels, 1-P_{Te}, were set as 0.57 and 0.84 (i.e., β_T = 0.177 and 1). The obtained results for the non-stationary excitation represented by (N) were shown in Figure 1. Comparisons of these fractiles to the 0.57- and 0.84-fractile of peak responses to stationary excitation represented by (S) were also shown in Figure 1. This figure indicates that the peak responses obtained for stationary excitation are different from those for non-stationary excitation. This is expected because the ground accelerations are modulated by time modulating function. Note that the magnitude of the peak response to non-stationary ground acceleration depends on the scaling factor A_0 of the time modulating function. Using the results given in the figure it can be shown that the ratios of the responses and the ratios of the logarithmic of responses between any two curves at different frequencies are not constant. The logarithmic of these ratios are slightly different as well. These imply that the peak responses at different natural frequencies are not fully correlated. Therefore, the peak responses cannot be represented accurately by the product of or the logarithmic of product of a deterministic function of the natural frequency and a random variable.

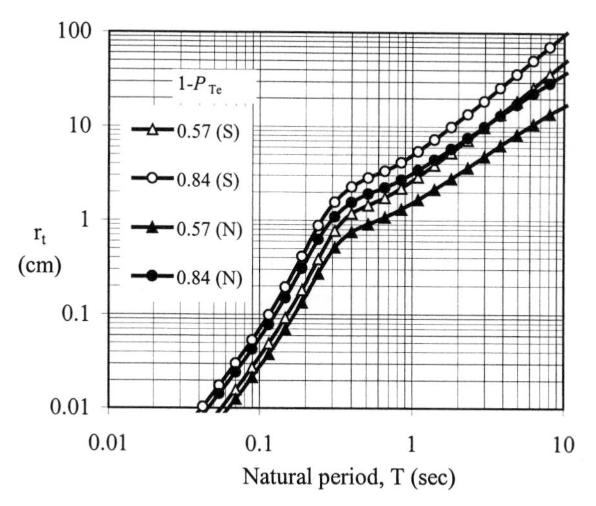

Figure 1: Peak response of SDOF systems to stationary and non-stationary excitation

Probability of Peak Response of Two-degree-of-freedom System

For a linear two-degree-of-freedom (2DOF) system, the peak response denoted by r_a can be calculated by CQC rule using the following expression (Der Kiureghian 1981, Chopra 2000):

$$r_a = \left((C_1 r_{1T})^2 + 2\rho_{12} C_1 C_2 r_{1T} r_{2T} + (C_2 r_{2T})^2\right)^{1/2}, \tag{7}$$

where r_{1T} and r_{2T} denoting the peak modal responses with the specified probability of exceedance P_{Te} can be obtained directly from the UHS shown in Figure 1 at ω_1.and ω_2; and ρ_{12} is the correlation coefficient given by Der Kiureghian (1981); C_1 and C_2 represent the effective participation factors. The Eqn 7 can also be written as:

$$r_a = \left((1-\zeta)^2 + \operatorname{sgn}(C_1 C_2) 2\rho_{12}(1-\zeta)\zeta + \zeta^2\right)^{1/2} r_{ABS}, \tag{8}$$

where sgn( ) returns the sign of the argument and $\zeta = |C_2 r_{2T}| / r_{ABS} = |C_2 r_{2T}| / (|C_2 r_{2T}| + |C_1 r_{1T}|)$ represents the contribution of the second mode.

By simple mathematical manipulation, it can be shown that given the value of ζ the probability that the peak response of the 2DOF system exceeds r_a can be evaluated using Eqn. 8 independent of the actual value of the r_{ABS}, except that r in Eqn. 5 is replaced by r_a. Note that by assigning different values of ζ from 0 to 1, different percentage of the contribution due to each mode to the system can be obtained. In particular, ζ equal to zero and one means the system is SDOF system with natural frequency equal to the ω_1 and ω_2, respectively.

Consider a 2DOF structure with T_1 = 0.2 sec and T_2 = 0.1 sec. The reliability indices β of peak responses exceeding r_a calculated using Eqn 8 were shown in Figure 2. The β for stationary ground motions were also shown in the figure. The figure indicates that β vary with ζ and β_T. The β is almost identical to β_T when ζ is close to 0 or 1; otherwise it is distinctively differ from β_T. This suggests that when the contribution due to one of modes is dominant the use of CQC rule with the UHS is adequate which is expected. However, if the contributions of both modes to the system response are significant, the use of the CQC rule with UHS could lead to different safety levels. For this 2DOF system, the CQC rule under- and over-estimates the peak response of the system for $\text{sgn}(C_1C_2) > 0$ and $\text{sgn}(C_1C_2) < 0$, respectively.

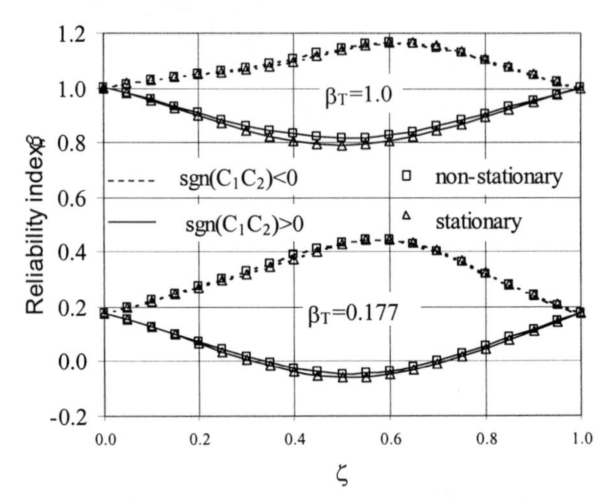

Figure 2 Reliability index of the peak response of 2DOF system using CQC rule with the UHS

Based on the above, it may be concluded that use of the CQC rule with the UHS may lead to conservative or unconservative results. To quantify such error, the ratio between the peak response r_a estimated by the CQC rule and the (1-P_{Te})-fractile peak responses, r_T, η, was calculated and shown in Figures 3a and 3b. To investigate the influence of the modal periods, other two 2DOF systems with T_1 = 0.2 sec and T_1/T_2 = 1.2, and T_1 = 1.5 sec, T_1/T_2 = 2.0 were also considered. The results for stationary ground motions were shown in the figures. These results indicate that the peak response to non-stationary ground motions obtained using CQC rule with the peak modal response from the UHS may not be adequate. It may over- or under- estimate, depending on the modal contribution and the natural periods of the structures. The conclusions for non-stationary excitations are similar to those for stationary excitations. However, the over- and under-estimation for the former seems to be larger than that for the latter.

In general, it seems that the errors (i.e., absolute value of 1-η) appear to be larger for the cases with sgn(C_1C_2)<0 than for the cases with sgn(C_1C_2)>0. The under- or over-estimation are less than about 10% and 20% if the modal contribution is less than 10% and 20% of the absolute sum of the effective modal peak response, respectively. However, CQC rule may lead to severe errors in other cases.

CONCLUSIONS

The probability analysis of the peak response was carried out using non-stationary evolutionary random vibration analysis to investigate the effect of uncertain non-stationary of seismic excitation on peak response of MDOF systems. In the analysis, the uncertainty in seismic excitation due to all potential earthquakes was considered using non-stationary PSD function with uncertain parameters.

Further, the adequacy of using CQC rule with the modal responses from the ordinates of the uniform hazard spectra was assessed.

The results show that the use of CQC rule with the modal responses obtained from the UHS is adequate if the peak responses of MDOF system are dominated by the response of one mode. However, if the contributions of different modes to the peak responses of a MDOF system are about equally significant, the use of CQC rule with the UHS leads to significant under- or over-estimation of the peak response. Such under- or overestimation depends on the modal periods of the system, and the effective modal participation factors.

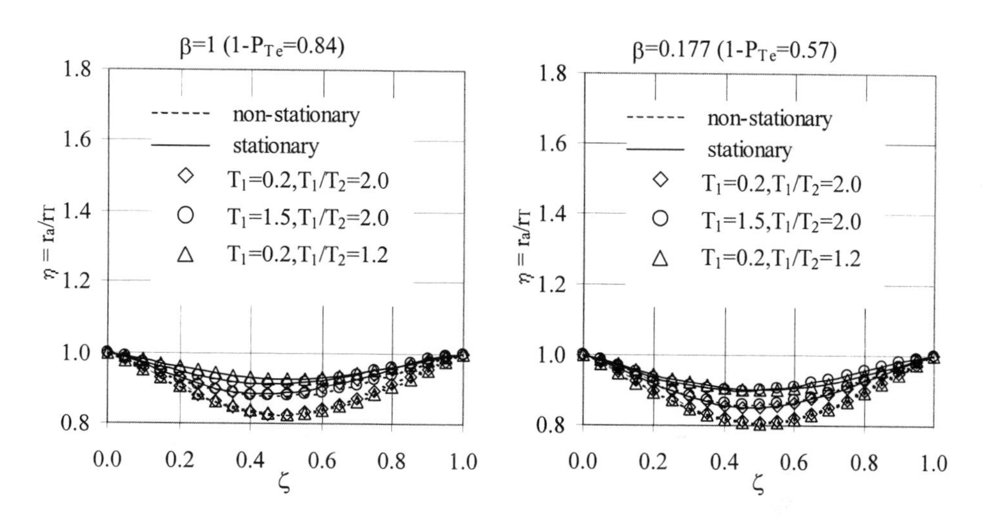

Figure 3a Ratio of the peak response of the CQC rule to the $(1-P_{Te})$ fractile response, $sgn(C_1C_2)>0$

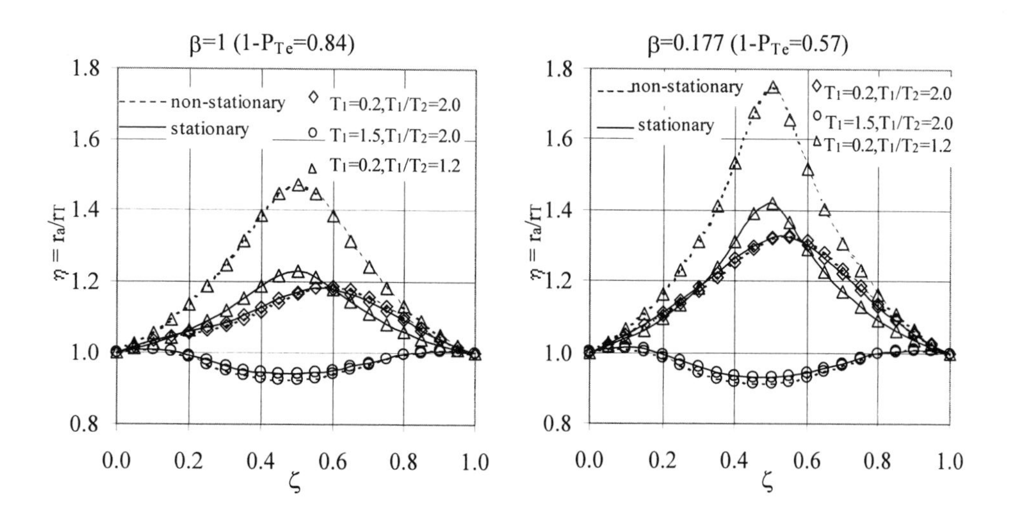

Figure 3b Ratio of the peak response of the CQC rule to the $(1-P_{Te})$ fractile response, $sgn(C_1C_2)<0$

ACKNOWLEDGEMENTS

The financial support of the Natural Science and Engineering Research Council of Canada is gratefully acknowledged.

REFERENCES

Balendra T, Quek ST and Teo YP. (1991) Time-variant reliability of linear oscillator considering uncertainties of structural and input model parameters. *Probabilistic Engineering Mechanics* **6:1**,10-17.

Chopra AK. (2000) *Dynamics of structures: Theory and application to earthquake engineering*, Prentice-Hall, Englewood Cliffs, New Jersey.

Der Kiureghian A. A response spectrum method for random vibration analysis of MDOF systems, *Earthquake Engineering and Structural Dynamics*, 1981, **9**: 419-435.

Hong HP, Wang SS (2002) Probability analysis of peak response of MDOF systems with uncertain PSD function. *Earthquake Engineering and Structural Dynamics* (in process).

Lai SP. (1982) Statistical characterization of strong motions using power spectral density function. *Bulletin of the Seismological Society of* America **72:1**, 259-274.

Madsen HO, Krenk S, Lind NC. (1986). *Methods of structural safety*, Prentice-Hall, Englewood Cliffs, N. J.

Priestley M.B. (1967). Power spectral analysis of non-stationary random process. *Journal of Sound and Vibration* **6:1**, 86-97

Shinozuka M. and Sato Y. (1967) Simulation of nonstationary random processes. *Journal of Engineering Mechanics Division ASCS* **93:EM1**, 11-40.

Tajimi H. (1960) A statistical method of determining the maximum response of a building structure during an earthquake, *Proceedings, Second Word Conference on Earthquake Engineering*, **2**, 782-796 Tokyo, Japan.

Vanmarcke EH, Lai SP. (1980) Strong-motion duration and RMS amplitude of earthquake records, *Bulletin of the Seismological Society of America*, **70:4**, 1293-1307.

Vanmarcke EH. (1976) *Structural response to earthquakes, in Seismic Risk and Engineering Decision* (Eds. Lomnitz, C. and Rosenblueth, E.), Elsevier Scientific Publishing Company, Amsterdam.

Vanmarcke EH. (1975) On the distribution of the first-passage time for normal stationary random process, *Journal of Applied Mechanics*, **42:Ser. E**, 215-220.

Advances in Building Technology, Volume 1
M. Anson, J.M. Ko and E.S.S. Lam (Eds.)

Simplified Method and Parametric Sensitivity Analysis for Soil-Pile Dynamic Interaction under Lateral Seismic Loading

Xiaochun Xiao [1] Shichun Chi[1] Gao Lin [1] John Alfano[2]

[1.] Department of Civil Engineering, Dalian University of Technology,
Dalian, 116023, P.R.China
[2.] Department of Civil and Environmental Engineering Cornell University,
Ithaca, NY, 14850, USA

ABSTRACT

Pile foundation for buildings erected on soft soil sites are often subjected to lateral loading induced by wind, sea wave and earthquakes. In many cases, the horizontal loads at the pile head usually constitute the primary constraint that affects the deformation of underlying piles or pile groups, so the behavior of lateral loaded pile has great significant for pile foundation design. In this paper, a simplified model based on Beam on Nonlinear Dynamic Winkler Foundation (BNDWF) for soil-pile dynamic interaction analysis under lateral loading is presented, and the corresponding computer program (NASPI) is developed using the *Wilson-θ* method. Through comparisons with the rigorous Finite Element analysis in time domain, it is shown that the model is simple and effective. Sensitivity analysis of the parameters including the effect of pile-soil stiffness ratio, pile slenderness ratio and earthquake excitation level has been performed. The statistical results indicate that the pile-soil stiffness ratio play the most important role in the soil-pile interaction.

KEYWORDS

Soil-pile dynamic interaction, BNDWF, lateral seismic loading, parametric sensitivity

INTRODUCTION

The response of piles and their load carrying capacity depends on the soil-pile interaction. Pile foundations of different type of structure are often subjected to lateral loading induced by the environment acting such as wind, sea wave and earthquakes. The lateral loads exerted at the pile head are basic factor that governs the response of single pile or pile groups. Since the early 1960s, many researchers have concerned themselves with piles undergoing dynamic lateral loads. Quite variety methods such as field tests, laboratory tests, numerical analyses have been performed.

Field experiments were initially conducted in the early 1960s by Matlock and Reese. As pointed out by Meymand, field experiment has the advantage that take into consideration in-site stress conditions, however seismic loads can be neither realistically nor easily reproduced. Matlock first completed his tests on laterally loaded piles in soft clays (Matlock 1962) and published a set of design *p-y* (resistance-deflection) curves at various depths. Novak was the first who attempted to analyze the dynamic response of single pile using continuum theory in frequency domain. He modeled the pile as an assembly of beam elements and their soil reactions were obtained based on a plane strain condition assumption (infinitely long pile). Nogami *et al.* developed closed-form solution for vertical and horizontal modes of vibration of an end-bearing pile by making use of rigorous three-dimensional theory in time domain. To obtain analytical closed-form solution, George Gazetas *et al.* used dynamic Winkler model with realistic frequency-dependent 'springs' and 'dashpots' to account for pile-soil-pile interaction. At same time many researchers studied pile vibration using finite element method or boundary element method in time domain and frequency domain.

In our research, a simplified model based on Beam on Nonlinear Dynamic Winkler Foundation (BNDWF) for soil-pile dynamic interaction under lateral seismic loading is presented. The validity of this simplified model is calibrated by comparing the results with those of Finite Element Method (FEM). Sensitivity analysis of the factors which affect the response of soil-pile system such as pile soil stiffness ratio, pile slenderness ratio and design earthquake level was examined. The calculation results indicate that pile soil stiffness ratio is the most sensitive one among these factors.

OUTLINE OF SIMPLIFIED MODEL AND METHOD

The single pile is modeled as a beam based on BNDWF. To develop the governing equation, the beam is discretized into a series of N-1 discrete beam elements with a total of N nodes. Element analysis, each beam element is considered individually as forces acting at the nodes, and it cause displacements and rotations of the nodes. The element model is shown in figure 1, and the relationship between nodal forces and nodal displacements given by:

$$\{F\}^e = [k_p]^e \{\Delta\}^e \tag{1}$$

where, $[k_p]^e$ is pile element stiffness, $\{F\}^e$ and $\{\Delta\}^e$ are forces and nodal displacements vectors for the same element, $\{F\}^e = [F_i \quad F_{i+1}]^T = [y_i \quad M_i \quad y_{i+1} \quad M_{i+1}]^T$, $\{\Delta\}^e = [\Delta_i \quad \Delta_{i+1}]^T = [u_i \quad \theta_i \quad u_{i+1} \quad \theta_{i+1}]^T$. The soil reaction has been modeled with a spring and dashpot element connected at each node along the pile. A graphical representation of soil element is shown in the middle part of figure 1. The terms, as pictured, are m_i, the lumped mass of the beam element at node i, k_{si}, the spring stiffness representing the soil modulus of deformation at node i, c_{si}, the damping coefficient of the soil at node i. This model gives the soil reaction shown in the following equation:

$$F_i = \begin{Bmatrix} y_i \\ M_i \end{Bmatrix} = \begin{Bmatrix} k_{si}u_i + c_{si}\dot{u}_i + m_i(\ddot{u}_i + \ddot{u}_{g_i}) \\ 0 \end{Bmatrix} \tag{2}$$

The $k_{si}u_i$ term is the spring force, the $c_{si}\dot{u}_{si}$ term is the damping force of the dashpot, the $m_i(\ddot{u}_i + \ddot{u}_g)$ term is the inertial force of the pile, the $\ddot{u}_{g_i}$ term is the acceleration due to an earthquake, but other dynamic forces such as wind or wave loads can also be used.

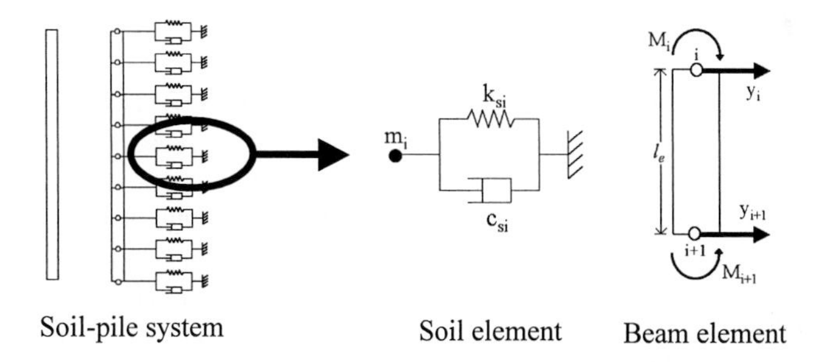

Figure 1: Discretized soil and beam elements

The k_{si}, c_{si} of every node can be calaulated as follows:

$$k_{si} = E_s A \quad \text{and} \quad c_{si} = A\sqrt{\rho E_s} \tag{3}$$

where, ρ is the soil density and A is the area of nodal control. The soil modulus of deformation E_s can be solved through the following simple equation:

$$E_s = 2(1+\mu) \times G \tag{4}$$

where, μ is definied as the soil Poisson's ratio, G is the soil dynamic shear modulus, The current shear modulus (G) is related to the initial shear modulus (G_0) through a function of the shear strain (γ). The Hardin-Drnevich model is used in our research and is represented as follows:

$$\frac{G}{G_0} = \frac{1}{1 + \dfrac{\tilde{\gamma}_i}{\gamma_r}} \tag{5}$$

where the effective shear strain across an element is $\tilde{\gamma}_i \approx 0.65(\gamma_i)_{\max}$ and (γ_r) is a reference shear strain value. After the first computational iteration, the shear modulus changes and has to be recalculated.

In order to calculate the initial shear modulus (G_0), the empirical equations will be used. The equation differs depending on the soil type. In this research project only a single, homogeneous layer of sand is considered. The empirical equation for sand is:

$$G_0 = \frac{3270(2.97-e)^2}{1-e} \times \sqrt{\sigma_0'} \tag{6}$$

where, σ_0' is the average of the principle stresses, and e is the voids ratio. G_0 and σ_0' are in units of kiloPascals (kPa). As Equation 5 shows, the shear modulus is dependent on the principle stresses. Since the stress level changes with depth, a different shear modulus will have to be computed for each node. The largest principle stress is approximated as $\sigma_1 \approx \rho g x$, x is depth from the ground surface. The smallest principle stress is approximated as $\sigma_3 \approx K_y \sigma_1$, where $K_y = \mu/(1-\mu)$ and μ is definied as the soil static Poisson's ratio. The intermediate principle stress σ_2 is taken to be equal to σ_3. All three principle stresses are then averaged to calculate σ_0'.

GOVERNING EQUATION AND SOLUTION METHODOLOGY

The simplified form of the governing equation for the whole system is represented as:

$$[K]\{\Delta\}+[C]\{\dot{\Delta}\}+[M_p]\{\ddot{\Delta}\}=-[M_p]\{\ddot{u}_g\} \tag{7}$$

where, $[K]$, $[C]$ and $[M_p]$ are global stiffness matrix, global damping matrix and mass matrix of pile respectively. $[K]=[k_s]+[k_p]$, and $[k_s]$ is soil spring stifness matrix. The $\{\Delta\}$, $\{\dot{\Delta}\}$, and $\{\ddot{\Delta}\}$, are the displacement, velocity, and acceleration vectors respectively. This research employs the *Wilson-θ* Method (Clough, R. 1996) for numerical integration of the equation (7), and θ=1.4 guarantees unconditional stable convergence. Once solutions for lateral displacements are found, they will be substituted into element equation to find the resulting shear forces and bending moments in the pile.

VERIFICATION OF SIMPLIFIED MODEL

Basing on above mentioned theory, a FORTRAN code with name of NASPI was developed. In the code, there are two main loops run through the program for calculating the outputs. The first loop begins with the dynamic calculations that constitute the *Wilson-θ* Method. The step-by-step integration method of the dynamic equation is accomplished as the acceleration, velocity, and displacement of each node are calculated for each earthquake time-step. After the final time-step, the loop completes and the soil modulus is recalculated for the dynamic condition. The dynamic calculations are then iterated until the convergence criteria are met.

In the FEM code, the system that consists of pile and surrounding soil is idealized as an assemblage of 3D finite elements. At the each node there are three translational degrees of freedom. The deformation of the pile, including bending, will be derived from these nodal translational displacements and element stresses. To accommodate the flexibility of pile under lateral loading, the piles are modeled as H_{11}, a kind of incompatible 3D element, which is improved from eight-node isoparametric hexahedral elements by adding 3 initial nodes. The incompatible element can be similarly dealt with the common eight-node isoparametric hexahedral elements after the condensation, but when to calculate the element stresses, the initial nodes' contribution will be taken into consideration. The pile will be regarded as linear elastic during the whole seismic responses, and for soil, equivalent linear method will be adopted in FEM analysis. This method for modeling nonlinear hysteretic behavior of soil was developed by Idriss and Seed (1968) *et al.* Because of its simplicity, it is the most popular method for site response analysis today. The fundamental assumption of the method is that a spring-dashpot model may approximate the dynamic response of a nonlinear hysteretic material satisfactorily if the properties of such model are chosen appropriately (i.e., strain level dependency of these material properties).

A simple sample problem is input to the program and the output is compared with FEM for test the validity of the NASPI code. A single end-bearing pile is modeled as a prismatic, cylindrical beam with a length of 15 meters and a diameter of 0.6 meters. The pile has a modulus of deformation of 25,000 MPa, a density of 2,400 $^{kg}/_{m^3}$, and a Poisson's ratio of 0.142. The pile head is flush with the ground surface and free to rotate and translate. The soil is modeled as a single, homogeneous layer of sand with a density of 2,000 $^{kg}/_{m^3}$, a voids ratio of 0.65, a static Poisson's ratio of 0.37, and a reference strain of 0.001. The input earthquake is the part of El Centro type shown in figure 2. The duration is 8.0 seconds with a time-step of 0.01. It has a design level of 0.2g for this problem. Since the response of pile head is most important in design, the displacement; shear force and bending moment time history of pile head are shown in figure 3, 4 and 5 respectively.

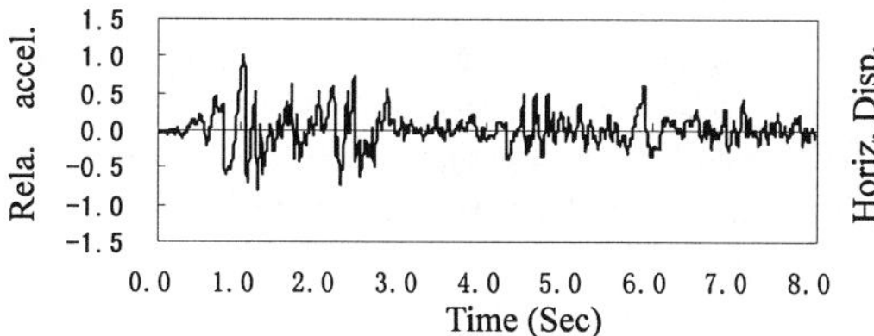

Figure 2: Input earthquake time history

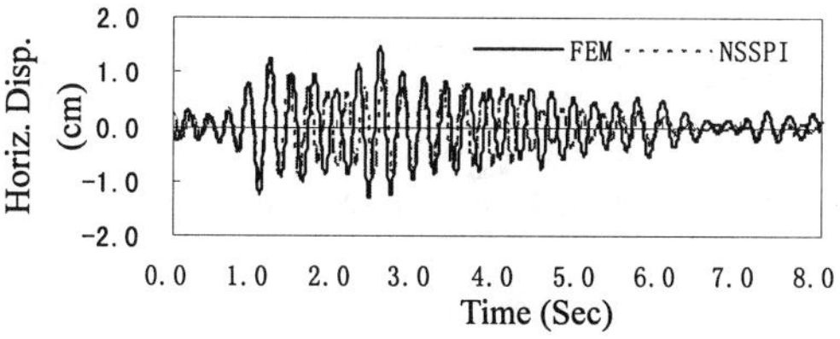

Figure 3: Comparison of pile head displacement time histories for NASPI and FEM

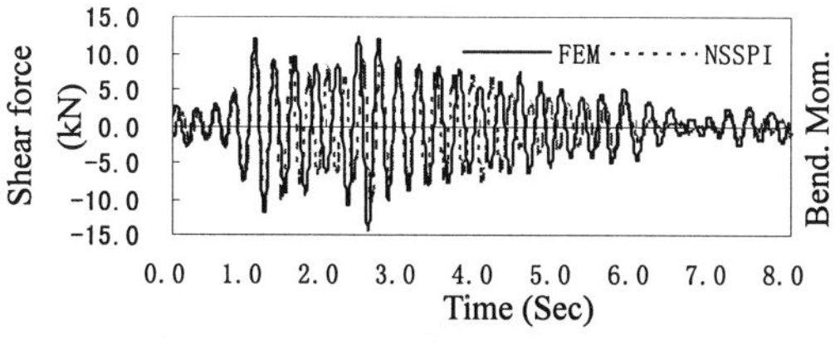

Figure 4: Comparison of pile head shear force time histories for NASPI and FEM

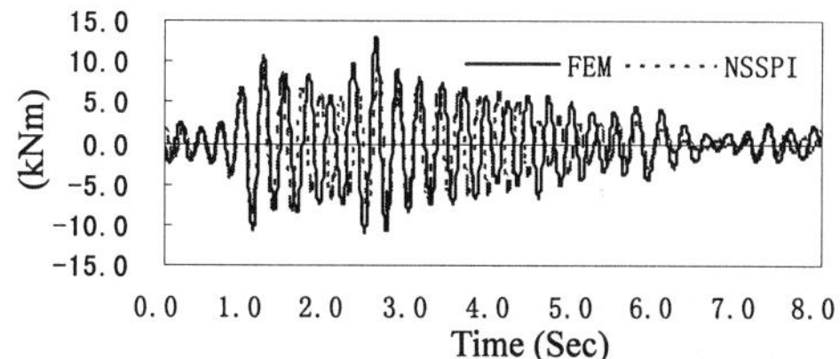

Figure 5: Comparison of pile head bending moment time histories for NASPI and FEM

Figure 3 shows the horizontal displacements of the pile head over the entire duration of the earthquake. The solid line represents the FEM analysis and the dashed line represents the output of NASPI. Initially, the two methods match up quite well for the small amplitudes of motion. As the amplitude increases, the results for NASPI fall slightly below that of the FEM analysis. This would indicate a slight overestimation of the soil stiffness. As the amplitude begins to decrease, NASPI does so more rapidly than FEM. This would indicate a slight underestimation of the damping coefficients in NASPI. These observations are not of significant importance at this point. The results for the two codes match up well enough to consider the NASPI code valid. Slight variations are to be expected due to the simplicity of the NASPI code as well as other approximations used in the solution process. For the purpose of this research, these calculations show that NASPI works properly. Similar comparisons can be made in considering the shear force and bending moment experienced by the pile head (figures 4 and 5).

PARAMETRIC SENSITIVITY ANALYSIS

Gazetas (1984) is earlier researcher to study effects of dimensionless parameters in frequency domain. He studied single piles undergoing vertical seismic loading, and believed "such results would be useful not only for developing an improved understanding of the mechanics of the problem and checking the accuracy of sophisticated solutions, but also for making preliminary design estimates in practice". The parameters of importance in this study were E_p/E_s (the pile-to-soil stiffness ratio), L/D (the slenderness ratio of the pile), and f/f_1 (the frequency ratio of excitation function to fundamental frequency of the soil). Thus, a parametric study using NASPI in time domain, would illustrate the usefulness of the code developed herein as well as the importance of various input factors. The input parameters to NASPI could be varied such that studies of E_p/E_s and L/D could be conducted. The effects of varying the design earthquake level a_i/g were also considered. Each parameter was considered at 3 different levels indicative of a high, middle, and low range for that particular parameter. Each level is shown in table 1. If a study were to consider every possible combination of levels and parameters, a total of 3×3×3=27 trials would have to be conducted. However, these 27 trials can be adequately represented by only 9 trials where each parameter is tested three times as shown in Table 2.

TABLE 1
CALCULATION TRIALS AND RESULTS

factor / level	L/D (A)	E_p/E_s (B)	a_i/g (C)
1	10	23.8	0.1
2	20	94.3	0.2
3	40	312.5	0.4

The horizontal acceleration and bending moment in the pile head were chosen as the objective functions since they are used as the inputs for what would be the base excitation of the superstructure and the constraint values for the pile design respectively. Once parameters were collected, the values for acceleration and moment were all normalized relative to the maximum value for each. The normalized values corresponding to the three trials involving each level were summed. This process was also completed for comprehensive target value that is a weighted average of the acceleration and moment for each trial and presented in Table 2. The weighted average coefficients are determined according to the cost of superstructure and pile foundation. In this research they are valued 2/3 and 1/3 roughly.

TABLE 2
CALCULATION TRIALS AND RESULTS

trial \ factor	L/D (A)	E_p/E_s (B)	a_i/g (C)	Max. Horiz. Acceleration real (m/s^2)	Max. Horiz. Acceleration Norm (%)	Max. Bending Moment Real (kN m)	Max. Bending Moment Norm (%)	Results for Comprehensive Target Values
1	1	1	1	0.379	11.21	444	0.34	7.59
2	1	2	2	1.740	51.48	25700	19.92	40.96
3	1	3	3	3.380	100.00	129000	100.00	100.00
4	2	1	2	0.844	24.97	794	0.62	16.85
5	2	2	3	2.200	65.09	12200	9.46	46.61
6	2	3	1	1.720	50.89	33000	25.58	42.45
7	3	1	3	0.525	15.53	979	0.76	10.61
8	3	2	1	1.310	38.76	9340	7.24	28.25
9	3	3	2	1.945	57.54	17900	13.88	42.99

TABLE 3
STATISTICAL ANALYSIS FOR COMPREHENSIVE VALUES

Factor	S	Dof	T	F-value	Critical values $F_{0.05}$	Critical values $F_{0.01}$	Sensitivity	Rank
A	716.27	2	358.13	3.52	19.00	99.00	*	3
B	4428.01	2	2214.01	21.77	19.00	99.00	**	1
C	1467.49	2	733.75	7.21	19.00	99.00	*	2
Tolerance	203.42	2	101.71					
Sum	6815.20	8						

Note: The F-value is defined in reference (Chen, K. 1996).

The symbols *, **, *** represent insensitive, sensitive and very sensitive respectively.

A statistical analysis for comprehensive target values was conducted and presented in table 3. The F-values for each factor are computed as shown in the above table. The F-value is calculated using the different levels as mere data points. The magnitude of the level is not important, and thus the sensitivity is independent of the differences in level sizes. Depending on F-value relative to the critical values, the sensitivity of comprehensive values is determined. For $F<F_{0.05}$, the factor is said to be insensitive. For $F_{0.05}<F<F_{0.01}$, the factor is sensitive. For $F>F_{0.01}$, the factor is very sensitive. From this table, it has been determined that comprehensive values are sensitive to changes in stiffness ratio. These results correspond well with the qualitative analysis.

CONCLUSIONS

NASPI presents an alternative method for evaluating soil-pile interactions, which is an important step to properly evaluating the soil-pile-superstructure interactions in design practice. NASPI meets the goals of being both simple and efficient, thus it is a valuable tool in practical preliminary design. The use of NASPI in the parametric study of three important dimensionless parameters: slenderness ratio, stiffness ratio, and earthquake design level, shows that output results are most sensitive to the stiffness ratio.

REFERENCES

Chen, K. (1996) *Experiments Design and Analysis*, Tsinghua Univ. Publishing House, Beijing.

Clough, R., Penzien, J. (1996). *Dynamics of Structures*, 2nd Edition, McGraw-Hill, New York

Fan, K., *et al.* (1991) "Kinematic Seismic Response of Single Piles and Pile Groups," *Journal of Geotechnical Engineering, ASCE*, **(117)12,** 1860-1879.

Gazetas, G. (1984). "Seismic Response of End-bearing Single Piles," *Soil Dynamics and Earthquake Engineering*, **3(2),** 82-93.

Gazetas, G., *et al.* (1991) "Dynamic Interaction Factors for Floating Pile Groups," *Journal of Geotechnical Engineering, ASCE,* **117(10),** 1531-1548.

Gazetas, G., et al. (1995) "Simple Methods for the Seismic Response of Piles Applied to Soil-Pile-Bridge Interaction," *Proceedings, 3rd International Conference on Recent Advances in Geotechincal Earthquake Engineering and Soil Dynamics*; Vol. III, pp. 1547-1556.

Lok, M.H. (1999) "Numerical Modeling of Seismic Soil-Pile-Superstructure Interaction in Soft Clay," *Ph.D. Thesis*, Univ. of California, Berkeley.

Markis N., Gazetas, G. (1992) "Dynamic Pile-Soil-Pile Action. Part II: Lateral and Seismic Response", *Earthquake Engineering and Structure Dynamic*, **21(2),** 145-162.

Matlock, H. & Reese, L. C. (1960). "Generalized solutions for laterally loaded piles", *Journal of Soil Mechanics and Foundation Division, ASCE,* **86:SM5**, 63-91

Matlock, H. (1962) "Correlations for Design of Laterally Loaded Piles in Soft Clay", A Report to Shell Development Company, Houston, Texas, Engineering Science Consultants.

Meymand, P.J. (1998) "Shaking Table Scale Model Tests of Nonlinear Soil-Pile-Superstructure Interaction in Soft Clay," *Ph.D. Thesis*, Univ. of California, Berkeley.

Nogami, T., *et al.* (1988) "Time domain flexural response of dynamically loaded piles," *Journal of Engineering Mechanics Division, ASCE*, **114(9),** 1512-1525.

Nogami, T., *et al.* (1992) "Nonlinear Soil-Pile Interaction Model for Dynamic Lateral Motion," *Journal of Geotechnical Engineering, ASCE,* **118(1),** 89-106.

Novak, M. (1974). "Dynamic Stiffness and Damping of Piles", *Can. Geotech. J.*, **11(4)**, 574-598.

Novak, M. (1977). "Soil-Pile Interaction", *Proc. 6th World Conf. Earthquake Eng.*, New Delhi, **IV**, 97-102.

Prakash, S. (1981). *Soil Dynamics*, McGraw-Hill, Inc. US, 220-270.

Winkler, E. (1867). "Die Lehre von der Elastizitat und Kestigkeit," Verlag, 182.

CONSTRUCTION TECHNOLOGIES:

FRP COMPOSITE, RETROFIT AND REPAIR TECHNOLOGIES

Advances in Building Technology, Volume 1
M. Anson, J.M. Ko and E.S.S. Lam (Eds.)

FRP REINFORCED ECC STRUCTURAL MEMBERS UNDER REVERSED CYCLIC LOADING CONDITIONS

Gregor Fischer, Victor C. Li
ACE-MRL, Department of Civil and Environmental Engineering, University of Michigan

Abstract

This paper reports on an investigation of FRP reinforced ECC flexural members with respect to their load-deformation behavior, residual deflection, damage evolution and failure mode. The paper briefly reviews particular aspects of conventional FRP reinforced concrete members, such as interfacial bond strength, flexural crack formation, composite deformation behavior, and brittle failure mode and presents a comparison to FRP reinforced ECC under reversed cyclic loading conditions. The combination of structural FRP reinforcement and ECC matrix in flexural members is characterized by a non-linear elastic response with stable hysteretic behavior, small residual deflections and ultimately gradual compression failure. Compatible deformations of FRP reinforcement and ECC lead to low interfacial bond stress and prevent composite disintegration by bond splitting and cover spalling. Flexural stiffness and strength as well as crack formation and widths in FRP reinforced ECC members are found effectively independent of interfacial bond properties due to the tensile deformation characteristics of the cementitious matrix.

Keywords: ECC, FRP reinforcement, composite, interfacial bond, deformation compatibility, damage tolerance

Introduction

Conventional FRP reinforced concrete members are typically over-reinforced and designed to fail in compression of concrete rather than by tensile rupture of the FRP reinforcement. Consequently, FRP reinforced concrete members possess relatively low ductility due to the brittle failure mode of concrete in compression as compared to tensile yielding of reinforcement in properly designed steel reinforced concrete members. Concepts to overcome this deficiency include ductile compression failure of concrete by providing confinement reinforcement or utilizing fiber reinforced concrete (Naaman and Jeong, 1995; Alsayed and Alhozaimy, 1999) as well as hybrid FRP reinforcement with inherent ductility (Harris et al., 1998). These concepts may provide a more gradual failure mode under monotonic loading conditions as compared to tensile failure of FRP reinforcement, however, under reverse cyclic loading conditions are unable to maintain their energy dissipation capabilities due to the unrepeatable nature of their inelastic deformation mechanism, i.e. concrete crushing, fiber pullout, or partial tendon rupture.

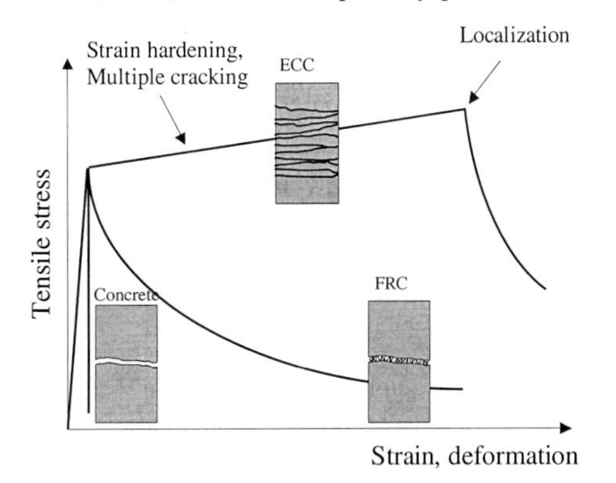

Figure 1 Schematic stress-strain behavior of cementitious matrices in tension

This paper presents results of an investigation on the effect substituting brittle concrete with a ductile, fiber reinforced cementitious composite in combination with structural FRP reinforcement under reverse cyclic loading conditions. This engineered cementitious composite (ECC) represents one type of high performance fiber-reinforced cement composites (HPFRCC), which are designed with the intent of obtaining a high toughness composite material with pseudo strain-hardening and multiple cracking properties. While concrete in tension fails in a brittle manner upon reaching its cracking strength, ECC undergoes a strain-hardening phase analogous to that of metals (Fig.1). Beyond formation of first cracking in the cementitious matrix, ECC is designed to increase its composite tensile stress up to strain levels on the order of several percent. In contrast to localized deformation in concrete and conventional FRC (Fig.1), ECC accommodates imposed tensile deformations by formation of uniformly distributed multiple cracking with small individual crack widths (<200μm).

The interaction of elastic reinforcement and ductile cementitious matrix with respect to interfacial bond mechanisms, composite damage evolution, member deformation capacity, and failure mode are experimentally verified and contrasted to FRP reinforced concrete. The combination of elastic FRP reinforcement and ductile ECC takes advantage of the material properties of the constituent materials and more importantly of their synergistic interaction. In particular, the interfacial bond mechanism in FRP reinforced ECC is affected by this interaction, where compatible tensile deformations of reinforcement and matrix on a macro-scale prevent relative slip and therefore reduce activation of interfacial bond stress (Fischer and Li, 2002).

Material composition and properties

The ECC matrix used in this particular study utilized 1.5%-Vol. of Polyethylene fibers, which are added during the mixing procedure and randomly oriented in the hardened composite. The cementitious matrix in ECC consists of cement, sand (maximum particle size 0.3mm), water, superplasticizer, and common admixtures to enhance the fresh properties of the composite.

Material properties obtained from this composition are a tensile first cracking strength of 4.5MPa at 0.01% strain and an ultimate tensile strength of 6.5MPa at approximately 4% strain (Fig.2a). In compression, ECC has a lower elastic stiffness compared to concrete as well as larger strain at reaching its compressive strength due to the lack of large aggregates. The compressive strength of this particular version of ECC was 80MPa. Beyond ultimate, the compressive stress drops to approximately $0.5f'_c$ with subsequently descending stress at further increasing deformation (Fig.2b).

Concrete utilized coarse aggregates (maximum particle size 10mm), cement, water, and superplasticizer to enhance the fresh properties of the mix. The compressive strength of concrete used in this study was 50MPa at a strain of 0.2% (Fig.2b).

The longitudinal reinforcement of the specimens investigated in this study was provided by Aramid-FRP with a ribbed surface geometry similar to that of conventional steel reinforcement. The material properties according to the specifications of the manufacturer are a tensile elastic modulus of 54GPa, average tensile strength of 1800MPa, and tensile strain capacity of 3.8%.

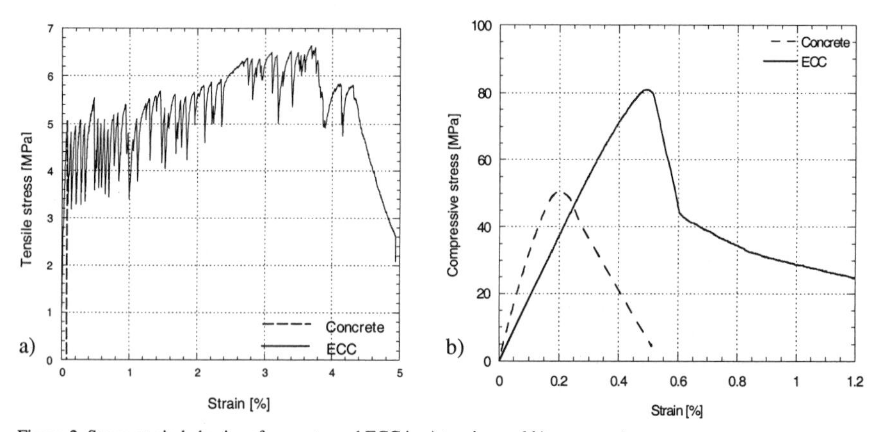

Figure 2 Stress-strain behavior of concrete and ECC in a) tension and b) compression

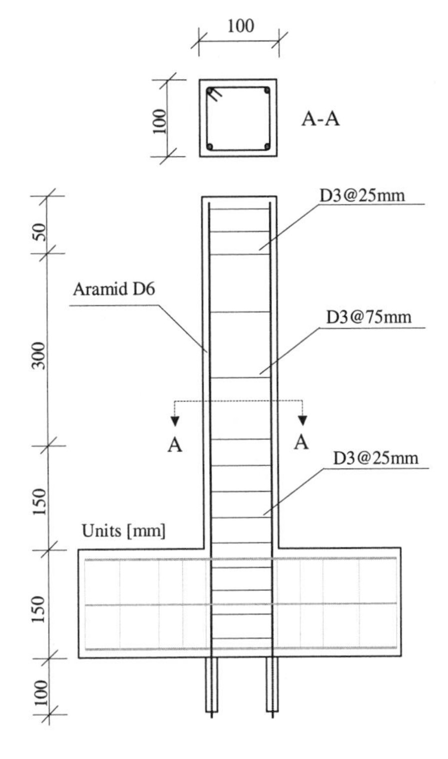

Figure 3 Specimen configuration

Specimen configuration and testing procedure

In this paper, results from tests of four different specimen configurations are presented, which were longitudinally reinforced with four FRP bars (∅6mm), arranged symmetrically relative to both axes (Fig.4). These specimens were designed to exceed the compressive strength of the cementitious matrix prior to rupture of the FRP reinforcement. The experimental tests presented herein contain an R/C control specimen with transverse reinforcement (S-1), an R/ECC specimen with transverse reinforcement (S-2), an R/ECC specimen without transverse reinforcement (S-3), and an R/ECC specimen without transverse reinforcement and additional sand coating (S-4). The specimen configurations are summarized in Table 1.

The load-deformation response of FRP reinforced ECC flexural members is experimentally investigated and compared to FRP reinforced concrete using small-scale cantilever beams with 500mm height and square cross-sectional dimensions of 100mm (Fig.3). In order to provide cantilever type loading conditions, a rigid transverse beam was integrally cast with the cantilever base. Lateral loading was applied at the top of the cantilever through a loading frame equipped with a 100kN capacity actuator according to a displacement controlled loading sequence.

Load-deformation response

The load-deformation response of the tested specimens, FRP reinforced concrete as well as FRP reinforced ECC, is predominantly characterized by non-linear elastic behavior with relatively small residual deflections. As intended, failure is initiated by inelastic deformations of concrete in specimen S-1 and ECC in specimens S-2, S-3, and S-4. Ultimate failure occurred in all considered cases by reinforcement rupture due to reverse cyclic loading conditions and resulting damage on the longitudinal reinforcement under compression. The apparent differences between FRP reinforced concrete and FRP reinforced ECC in terms of load-deformation response, damage evolution, and deflection capacity are established in detail by comparing specimens S-1 and S-2. Furthermore, the behavior of specimens S-3 and S-4 is compared to specimen S-2 in order to investigate the influence of reinforcement detailing and interfacial bond properties on FRP reinforced ECC members.

Flexural cracking in specimens S-1 (R/C, with transverse reinforcement) and S-2 (R/ECC, with transverse reinforcement) occurred at relatively small flexural loads and deflections, which are for all practical purposes insignificant compared to service and ultimate deflections.

Beyond initiation of flexural cracking and at increasing drift levels, the formation of further flexural cracks is limited in specimen S-1 due to interfacial bond properties and stress transfer length between FRP reinforcement and concrete, resulting in a relatively large crack spacing of approximately 100mm (Fig.5). Therefore, imposed specimen deflections are accommodated by relatively large crack widths. The crack spacing in specimen S-2 is significantly smaller at all drift levels (Fig.5) especially at the cantilever base and similar to the crack spacing observed in this particular version of ECC in direct tension (10mm), suggesting that flexural crack formation is effectively independent of the interaction with the FRP reinforcement. The formation of bond splitting cracks in specimen S-1 and the lack thereof in specimen S-2 further indicate differences in the respective composite deformation mechanism. While the strain lag between reinforcement and concrete at a crack location causes relative slip and radial stresses exceeding the matrix tensile strength as indicated by significant formation of bond splitting cracks, compatible deformations between reinforcement and ECC in the multiple cracking stage prevent significant bond stress and relative slip.

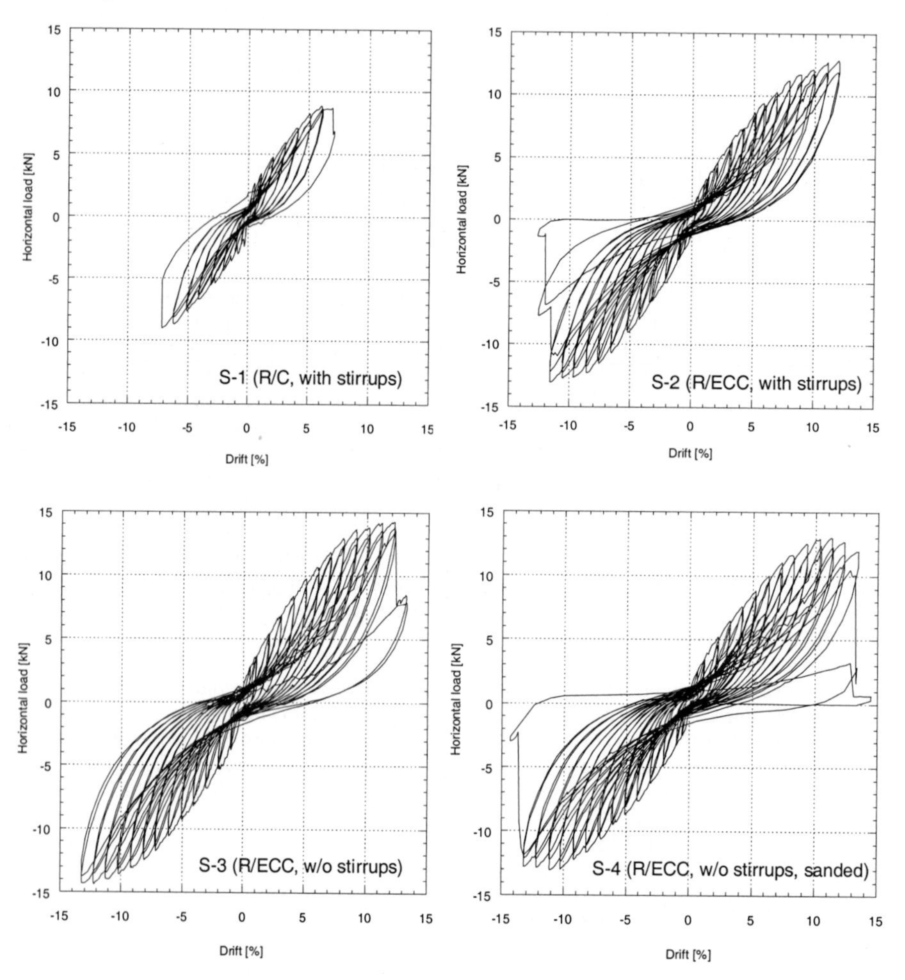

Figure 4 Load-deformation response of tested specimens

At further increasing deflection, the load-deformation behavior of specimens S-1 and S-2 is characterized by a progressive reduction of flexural stiffness at increasing applied load. A linear portion of the response cannot be clearly identified, which is due to inelastic deformations in the cementitious matrices in tension (flexural crack formation) and compression (gradual crushing). Furthermore, progressing interfacial bond failure and bond splitting cracks in specimen S-1 contribute to stiffness reduction and increasing non-linearity of the load-deformation response. Disintegration of the FRP reinforced concrete composite by crushing and bond splitting induces severe damage to the FRP reinforcement and partial rupture occurs at 7% drift and ultimate load of approximately 9kN (Fig.4).

In specimen S-2, flexural cracking continuously initiates up to relatively large deflections (7% drift), beyond which flexural stiffness reduction is influenced by ECC inelastic deformations under compression. The transition between these phases is indicated by noticeable plateau in the load-deformation curve beyond 7% drift (Fig.4). Ultimately, excessive compressive strain in ECC lead to damage in the FRP reinforcement and subsequent rupture at 12% drift and ultimate shear force of approximately 13kN. Similar to specimen S-1, the nominal tensile strength of the reinforcement is not exceeded at this point, however, damage induced by compressive and transverse deformations considerably reduces the tensile capacity.

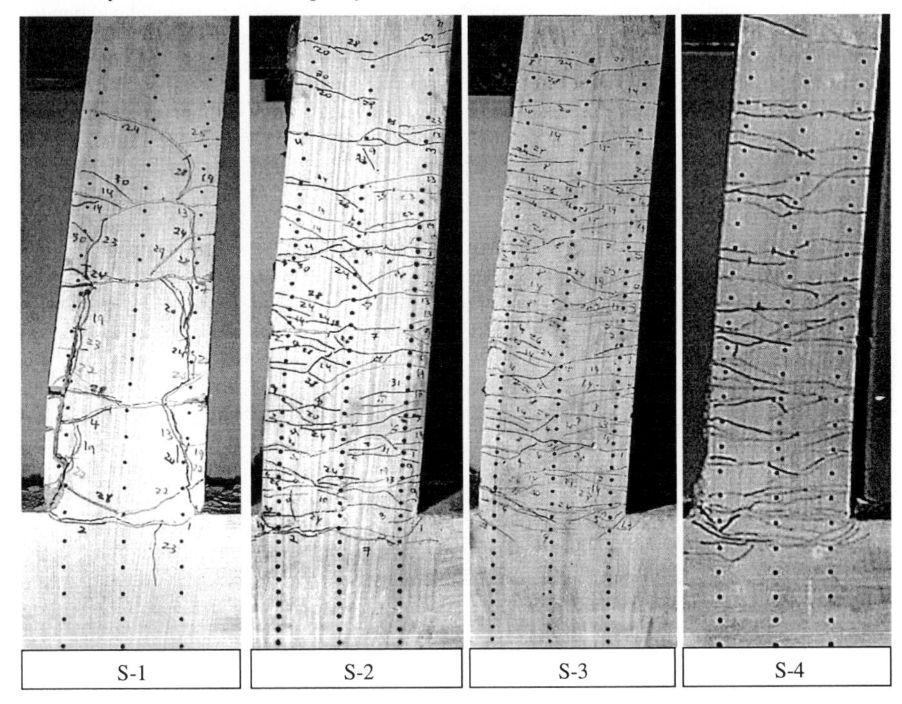

Figure 5 Deflected shape and damage pattern of specimens at 7% drift

Residual deflection and energy dissipation

The behavior of specimens S-1 and S-2 in terms of residual deflection and energy dissipation (Fig.6) is compared at increasing drift levels. In this paper, residual deflection is characterized by the ratio of permanent deflection after unloading and maximum deflection experienced at a given drift level. Energy dissipation is represented by the equivalent damping ratio, defined as the energy dissipated during a complete loading cycle normalized by the elastic energy stored in the member at a given drift level (Chopra, 1995).

Flexural crack formation primarily affects the specimen behavior at small drift levels in terms of residual deflection and energy dissipation. Incomplete crack closure after unloading at small drift levels significantly contributes to the residual deflection of specimens S-1 and S-2, which show relatively large residual deflection ratios at 0.5% drift. At increasing drift levels, this ratio declines in both specimens and reaches a minimum of approximately 0.2, however, at different drift levels. Minimum residual deflection in S-1 is observed at 3% drift and in S-2 at 6% drift, which roughly coincides with flexural crack saturation in both specimens, respectively. In specimen S-1, additional inelastic deformation mechanisms, such as bond splitting, spalling and compression failure, succeed this phase particularly beyond 5% drift, thus the residual deflection ratio increases rapidly. In specimen S-2, flexural crack saturation is achieved at approximately 7% drift, beyond which the residual deflection ratio remains constant at 0.2 due to a stable flexural deformation mechanism. At deflections above 10% drift, crushing of ECC and rotational sliding constitute a major change in the deformation mechanism, resulting in a considerable increase in residual deflection ratio.

The energy dissipation characteristics of both specimens show trends very similar to those described for the residual deflection ratio. In both specimens, the equivalent damping ratio at small drift levels is primarily affected by flexural crack formation and reaches their respective minimum at flexural crack saturation, beyond which it remains constant due to internal friction along the crack planes and increases slightly at further damage by matrix crushing, spalling or rotational sliding occurs.

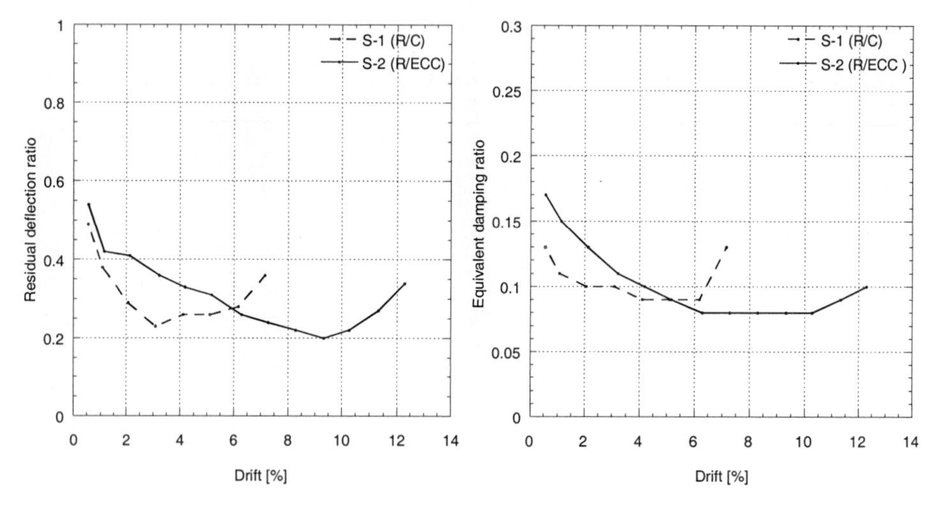

Figure 6 Residual deflection and equivalent damping in specimens S-1 and S-2

These observations suggest that flexural crack formation and saturation as well as damage induced by compression have major influence on residual deflection and equivalent damping characteristics of FRP reinforced members. While at drift levels prior to flexural crack saturation FRP reinforced ECC (S-2) shows larger residual deflection due to the larger number of flexural cracks compared to FRP reinforced concrete (S-1), it is able to maintain its flexural deformation mode beyond saturation with small residual deflections up to relatively large drift levels.

Transverse reinforcement

The comparison of specimen S-2 (R/ECC, with transverse reinforcement) (Fig.4) and S-3 (R/ECC, w/o transverse reinforcement) (Fig.4) indicates higher ultimate strength and deflection of specimen S-3 without transverse reinforcement. In view of identical longitudinal reinforcement in both specimens, this suggests that the presence of stirrups in specimen S-2 contributes to tensile strength reduction in the FRP reinforcement leading to premature failure. Inspection of specimen S-2 after completion of the test gives evidence supporting this assumption. Rupture of FRP exclusively occurred at the intersection with stirrups, which locally constrained and punctured the longitudinal reinforcement. The experimentally obtained flexural strength of specimen S-3 is nearly identical to that predicted and therefore, its ultimate strength is considered optimal at given material properties of reinforcement and matrix. Lateral support and confinement of FRP reinforcement in S-3 is entirely provided by ECC. In contrast to a discrete lateral support in members with conventional stirrups, the longitudinal reinforcement is continuously supported, confined, and protected in the ECC matrix without intermittent restraint. The comparison of specimens S-2 and S-3 suggests that transverse reinforcement should not be used in FRP reinforced ECC flexural members if shear strength requirements permit. Although transverse reinforcement is usually found to enhance effective compressive strength and strain capacity of cementitious materials (concrete, ECC), which is advantageous in compression controlled members, its damaging effect on FRP reinforcement is detrimental. From the experimental results obtained in this study it is found that the self-confining effect of ECC is sufficient to prevent reinforcement buckling and assure ductile compressive failure, such that additional confinement reinforcement is not necessary.

Interfacial bond strength

The modification of the FRP reinforcement surface by additional sand coating does not positively affect the load-deformation behavior of specimen S-4 (R/ECC, w/o transverse reinforcement, sand coating) (Fig.4) compared to S-3 (R/ECC, w/o transverse reinforcement) (Fig.4). Maximum crack widths are identical prior to localization of cracking and ultimately, flexural crack spacing and distribution are similar in both specimens, however, in specimen S-4 local crack spacing at the cantilever base is found slightly smaller and localization of cracking is to some extent delayed. The increased crack density at the cantilever base weakens the compressive strength of the matrix and promotes the formation of a rotational sliding plane, which negatively affects flexural stiffness and strength of specimen S-4.

The comparison of specimens S-3 and S-4 with respect to load-deformation behavior and flexural crack formation confirms the assumption that the composite deformation characteristics of reinforced ECC members are effectively independent of the interfacial bond properties, provided that sufficient development length is available or external anchorage is supplied.

Conclusions

The response of FRP reinforced flexural members to reversed cyclic loading conditions indicates non-linear elastic load-deformation behavior with relatively small residual deflections. While the overall load-deformation behavior of FRP reinforced concrete and ECC show similarities, detailed differences have been established in terms of their composite deformation mechanism, damage evolution and ultimate deflection capacity.

Deformation compatibility between FRP reinforcement and ECC is found to effectively eliminate interfacial bond stress and relative slip in the multiple cracking deformation regime, preventing bond splitting and spalling of ECC cover. In contrast, incompatible deformations between reinforcement and concrete cause loss of interfacial bond and composite action resulting in damage to the reinforcement and limited deflection capacity of the FRP reinforced concrete member.

While the increased flexural strength of FRP reinforced ECC compared to reinforced concrete is mainly attributed to the compressive strength of ECC, the deflection capacity is fundamentally affected by improved composite interaction. The load-deformation response of FRP reinforced ECC is dominated by flexural deformation up to relatively large drift levels and crack formation is found effectively independent of interfacial bond properties. Inelastic deformation of ECC in compression lead to flexural stiffness reduction and ultimately gradual mode of failure, however, also induced compressive strain and tensile strength reduction in the FRP reinforcement.

Despite the ductile deformation behavior of ECC in direct tension, FRP reinforced ECC members do not have significant energy absorption capacity compared to conventional steel reinforced members. Flexural crack formation is the primary dissipating mechanism and therefore, cannot be repeatedly utilized.

Acknowledgements
The research described in this paper has been supported by a grant from the National Science Foundation (CMS-0070035) to the ACE-MRL at the University of Michigan. This support is gratefully acknowledged.

References

Alsayed, S.H., Alhozaimy, A.M. (1999), "Ductility of concrete beams reinforced with FRP bars and steel fibers", Journal of Composite Materials, Vol.33, No.19, pp.1792-1806

Chopra, A.K. (1995), "Dynamics of Structures – Theory and applications to Earthquake Engineering", Prentice-Hall, Inc.

Fischer, G., Li, V.C. (2002), "Influence of matrix ductility on the tension-stiffening behavior of steel reinforced Engineered Cementitious Composites (ECC)", ACI Structural Journal Vol.99, No.1, January-February 2002, pp104-111

Harris, H.G., Samboonsong, W, Ko, F.K. (1998), "New ductile FRP reinforcing bar for concrete structures", ASCE Journal for Composites for Construction, Vol.2, No.1, February 1998, pp.28-37

Naaman, A.E., Jeong, S.M. (1995) "Structural ductility of concrete beams prestressed with FRP tendons", *Non-metallic (FRP) Reinforcement for Concrete Structures*, Proceedings of the Second International RILEM Symposium (FRPRCS-2), E&FN Spon, London

Tables

Specimen	Composite	Surface	Reinforcement ratio		Ultimate strength	
			$\rho_{longitudinal}$	$\rho_{transverse}$	Predicted	Experimental
S-1	R/C	ribbed	1.13	0.57/ 0.19	9.5	9.0
S-2	R/ECC	ribbed	1.13	0.57/ 0.19	14.2	12.5
S-3	R/ECC	ribbed	1.13	-	14.2	14.3
S-4	R/ECC	ribbed, sanded	1.13	-	14.2	13.0

Table 1 Summary of specimen configurations

Advances in Building Technology, Volume 1
M. Anson, J.M. Ko and E.S.S. Lam (Eds.)

ULTIMATE AXIAL STRAIN OF FRP-CONFINED CONCRETE

L. Lam and J.G. Teng

Department of Civil and Structural Engineering
The Hong Kong Polytechnic University, Hong Kong, China

ABSTRACT

External confinement by the wrapping of FRP sheets (or FRP jacketing) provides a very effective method for the retrofit of reinforced concrete (RC) columns subject to either static or seismic loads. For the reliable and cost-effective design of FRP jackets, an accurate stress-strain model is required for FRP-confined concrete. Central to such a model is the determination of the ultimate condition of FRP-confined concrete which is reached when the FRP ruptures. This ultimate condition is characterized by two parameters: the ultimate axial strain and the corresponding stress level which may or may not be the ultimate strength of the FRP-confined concrete (i.e. the axial compressive strength). This paper presents a new equation for the ultimate axial strain of FRP-confined concrete on which there has been much uncertainty and considerable controversy. Both the stiffness and the actual ultimate condition of the FRP jacket are explicitly accounted for in this new equation. The theoretical basis and performance of the equation are discussed in detail. The paper also suggests confined cylinder tests as a standard type of tests to determine the efficiency factor of a particular FRP product in confinement applications, defined as the ratio between the actual hoop rupture strain of FRP in FRP-confined concrete and the ultimate tensile strain from material tests generally conducted using flat coupons.

KEYWORDS

Concrete, FRP, Confinement, Stress-Strain Model, Ultimate Strain, Ultimate Condition

INTRODUCTION

One important application of FRP composites in civil engineering is as a confining material to concrete in the retrofit of existing reinforced concrete (RC) columns by the provision of an FRP jacket in the form of FRP wraps or prefabricated FRP shells. As the main function of the FRP jacket is to provide confinement to the concrete, the reinforcing fibres are mainly or only present in the hoop direction, so the jacket has little longitudinal stiffness. That is, the jacket can be simplified as a unidirectional material providing only hoop resistance to any expansion of the concrete. The ultimate condition of such FRP-confined concrete is reached when the FRP ruptures due to hoop tensile stresses; failure of insufficient vertical lap joints is excluded from consideration here. It is with this type of FRP-confined concrete that this paper is concerned. Concrete-filled FRP tubes for new construction also make use of FRP confinement but they differ somewhat in behaviour as a result of the substantial longitudinal stiffness possessed by the tube.

For the reliable and cost-effective design of FRP jackets, an accurate stress-strain model is required for FRP-confined concrete. Over the last few years, a number of stress-strain models have been proposed. These models can be classified into two categories: (a) design-oriented models, and (b) analysis-oriented models. Stress-strain models of the first category are in closed-form expressions, with key parameters directly based on test results. They are suitable for direct application in design calculations by hand and/or spreadsheets. On the other hand, stress-strain models of the second category are aimed at accurate predictions using an incremental numerical procedure. These models are particularly suitable for incorporation in computer-based numerical analysis such as a nonlinear finite element analysis. A detailed discussion of these models can be found in Monti (2001), Teng et al. (2002) and Lam and Teng (2002a).

Central to any stress-strain model for FRP-confined concrete is the determination of the ultimate condition of FRP-confined concrete which is reached when the FRP ruptures. This ultimate condition is characterized by two parameters: the ultimate axial strain and the corresponding stress level which may or may not be the ultimate strength of the FRP-confined concrete (i.e. the axial compressive strength). There are two major deficiencies in existing stress-strain models including the authors' recent design-oriented model (Lam and Teng 2001, 2002b, Teng et al. 2002) and both are closely associated with the definition of the ultimate condition of FRP-confined concrete. First, the stiffness of the FRP jacket has not been properly accounted for by existing design-oriented models in predicting the ultimate condition of confined concrete, although analysis oriented models do not suffer from this problem. The stiffness of the FRP jacket in fact controls the stress-strain response of FRP-confined concrete, particularly the ultimate axial strain (referred to simply as the ultimate strain hereafter), as shown in this paper. If the confinement effect of the FRP jacket on the ultimate strain is accounted for only through the confinement ratio (i.e., the ratio between the maximum confining pressure and the compressive strength of unconfined concrete) as in a number of existing design-oriented models (e.g. Karbhari and Gao 1997, Miyauchi et al. 1999, Lam and Teng 2002b), different expressions are needed for the ultimate strain for different types of FRP due to their different stiffnesses (Lam and Teng 2002b). Second, when calculating the maximum confining pressure in both design- and analysis-oriented models, it has been commonly assumed that FRP rupture occurs when the hoop stress in the FRP jacket reaches the tensile strength determined from material tests, with the only exception being Xiao and Wu's (2000) model. This assumption is however not valid according to observations made in tests on FRP-wrapped concrete cylinders. The hoop strain reached in FRP at rupture can be much lower than the ultimate tensile strain of FRP from material tests (Lam and Teng 2002a). The omission of this important aspect of behaviour in existing models results in uncertainty in defining the ultimate condition of FRP-confined concrete. A more detailed discussion of the limitations of existing stress-strain models for FRP-confined concrete can be found in Lam and Teng (2002a).

This paper is concerned with the ultimate axial strain of FRP-confined concrete which is one of the two parameters defining the ultimate condition of FRP-confined concrete. By explicitly accounting for the stiffness of the FRP jacket and the actual condition of FRP rupture in FRP-confined concrete, a unified equation is presented for concrete confined by different types of FRP materials. The theoretical basis and performance of the equation are also discussed in detail.

THEORETICAL BASIS

In studies on actively confined and steel-confined concrete, the axial strain at the compressive strength of confined concrete ε_{cc} has been shown to relate linearly to the confinement ratio (Richart et al. 1929, Candappa et al. 2001). Based on this approach, in several design-oriented models for FRP-confined concrete (e.g. Karbhari and Gao 1997, Miyauchi et al. 1999, Lam and Teng 2002b), the confinement effect of the FRP jacket on the ultimate strain of FRP-confined concrete ε_{cu}, which in most cases is equal to ε_{cc}, is accounted for only through the confinement ratio. A deficiency of this approach is that the stiffness of the FRP jacket is not properly accounted for. This deficiency can be explained by examining the predictions of an analysis-based model in which the stiffness of the jacket is an inherent factor in determining the interaction between the FRP jacket and the concrete core.

TABLE 1
PROPERTIES OF CONFINING MATERIALS FOR RESULTS SHOWN IN FIGURE 1

Confining material	Elastic modulus (MPa)	Rupture or yield stress (MPa)	Rupture or yield strain (%)	Thickness of confining jacket (mm)
Steel	2×10^5	300	0.15	4
CFRP	2.35×10^5	3530	1.5	0.34
GFRP	23100	462	2.0	2.6

Figure 1 shows three stress-strain curves predicted by the analysis-oriented model developed by Speolstra and Monti (1999) for concrete cylinders confined by three different confining materials: steel, CFRP and GFRP. The compressive strength of the unconfined concrete is 35 MPa while the diameter of the cylinder is 152 mm. The three materials have typical properties as given in Table 1; the thicknesses were assumed values which lead to the same ultimate tensile capacity in the hoop direction and thus the same maximum confining pressure but different jacket stiffnesses. Figure 1 shows clearly that the predicted stress-strain response, particularly the ultimate strain, depends strongly on the stiffness of the FRP jacket. The compressive strength also depends on the stiffness of the jacket, but this dependence is much weaker within the practical range of FRP material properties, which supports the stiffness-independent approach widely-adopted for compressive strength (e.g. Lam and Teng 2002c). These conclusions from the analysis-oriented model are believed to be valid also for real FRP-confined concrete, because any significant inaccuracy of the model is in the concrete but the results were obtained for the same concrete and the behaviour of concrete is reasonably well approximated in the model from limited previous comparisons (Yuan et al. 2001).

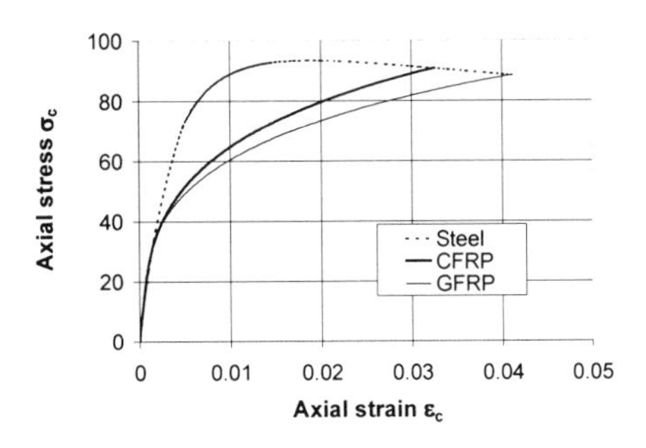

Figure 1 Stress–strain curves predicted by Spoelstra and Monti's (1999) model

The dependence of the ultimate strain of FRP-confined concrete on the stiffness of the confining jacket can also be shown by examining the constitutive model for concrete under a triaxial state of stress proposed by Ottosen (1979). This model is based on non-linear elasticity, with the properties of concrete being represented by the secant values of elastic modulus and Poisson's ratio. It has been adopted by the CEB-FIP Model Code 1990 (CEB-FIP 1993). The basic equations of the model are

$$\varepsilon_1 = \frac{1}{E_{sec}}\left[\sigma_1 - \nu_{sec}(\sigma_2 + \sigma_3)\right] \tag{1a}$$

$$\varepsilon_2 = \frac{1}{E_{sec}}\left[\sigma_2 - \nu_{sec}(\sigma_1 + \sigma_3)\right] \tag{1b}$$

$$\varepsilon_3 = \frac{1}{E_{\sec}}\left[\sigma_3 - \nu_{\sec}(\sigma_1 + \sigma_2)\right] \tag{1c}$$

where $E_{\sec}$ = secant modulus of elasticity and $\nu_{\sec}$ = secant Poisson's ratio. For confined concrete, ε_1 = ε_c = axial strain of concrete, ε_2 = ε_r = radial (lateral) strain of concrete = ε_3 = ε_θ = circumferential strain of concrete, σ_1 = σ_c = compressive stress of concrete, and σ_2= σ_3 = σ_r = lateral confining pressure. In this paper, compressive stresses and strains in concrete are taken as positive, while for the FRP, tensile stresses and strains are taken to be positive. The following equation can be obtained from Eqn. 1:

$$\varepsilon_c = -\frac{\varepsilon_r}{\nu_{\sec}} + \frac{(1 - \nu_{\sec} - 2\nu_{\sec}^2)\sigma_r}{\nu_{\sec} E_{\sec}} \tag{2}$$

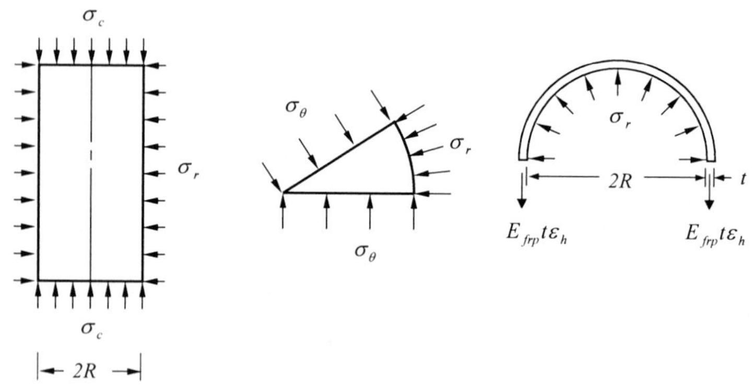

Figure 2 Confinement of concrete by an FRP jacket

The confining action of FRP to concrete is illustrated in Figure 2 where all the stresses are shown in their positive directions. The lateral confining pressure provided by FRP at any stage of loading is given by

$$\sigma_r = \frac{E_{frp} t \varepsilon_h}{R} \tag{3}$$

where E_{frp} and t = elastic modulus in the hoop direction and total thickness of the FRP jacket respectively, R = radius of the concrete core, and ε_h = hoop tensile strain in the FRP and = -ε_r due to the difference in sign convention between the FRP jacket and the concrete core. Under the ultimate condition of FRP rupture, ε_r= ε_{ru} = -$\varepsilon_{h,rup}$, where ε_{ru} is the ultimate lateral strain of the confined concrete and $\varepsilon_{h,rup}$ is the hoop rupture strain of the FRP. The secant modulus of elasticity of concrete under the ultimate condition is given by (Ottosen 1979)

$$E_{\sec u} = \frac{E_{\sec o}}{1 + 4(A-1)x} \tag{4}$$

where $E_{\sec o}$ = secant modulus of elasticity at the compressive strength of unconfined concrete and = $f'_{co} / \varepsilon_{co}$ with f'_{co} being the compressive strength of unconfined concrete and ε_{co} being the axial strain at f'_{co}; $A = 2$ for a parabolic stress-strain curve; and x is given by $x = \sqrt{J_2} / f'_{co} - 1/\sqrt{3}$ with J_2 being the second invariant of the deviatoric stress tensor under the ultimate condition. Substituting Eqns. 3 and 4 into Eqn. 2 for the ultimate condition, the following expression (Lam and Teng 2002d) can be obtained:

$$\frac{\varepsilon_{cu}}{\varepsilon_{co}} = \frac{\varepsilon_{h,rup}}{\nu_{\sec u}\varepsilon_{co}} + \frac{(1 - \nu_{\sec u} - 2\nu_{\sec u}^2)(1 + 4x)}{\nu_{\sec u}}\left(\frac{E_{frp}t}{E_{\sec o}R}\right)\left(\frac{\varepsilon_{h,rup}}{\varepsilon_{co}}\right) \tag{5}$$

where ε_{cu} = ultimate strain of the confined concrete, and $\nu_{\sec u}$ = secant Poisson's ratio of the confined concrete under the ultimate condition. The term $E_{frp}t/E_{\sec o}R$ is the confinement stiffness ratio, representing the stiffness ratio between the FRP jacket and the concrete core (Lam and Teng 2002d).

PROPOSED EQUATION

Equation 5 shows clearly the dependence of the ultimate strain of FRP-confined concrete ε_{cu} on the confinement stiffness ratio and the strain capacity of FRP at hoop rupture. The secant Poisson's ratio of FRP-confined concrete under the ultimate condition, $\nu_{\sec u}$, also depends strongly on the confinement stiffness ratio (Lam and Teng 2002d). Thus, the ultimate strain of FRP-confined concrete ε_{cu} can be predicted using the following equation:

$$\varepsilon_{cu}/\varepsilon_{co} = c + k_2\left(\frac{E_{frp}t}{E_{\sec o}R}\right)^{\alpha}\left(\frac{\varepsilon_{h,rup}}{\varepsilon_{co}}\right)^{\beta} \tag{6}$$

where c = normalised ultimate strain of un-confined concrete, k_2 = strain enhancement coefficient, and α and β are exponents to be determined. The effects of the secant Poisson's ratio $\nu_{\sec u}$ and the parameter x are reflected by the choice of appropriate values for α, β and k_2. This equation explicitly accounts for the stiffness and the actual ultimate condition of the jacket, and may thus be referred to as a stiffness-based equation. If both α and β are taken as unity, then Eqn. 6 reduces to

$$\varepsilon_{cu}/\varepsilon_{co} = c + k_2\frac{f_{l,a}}{f'_{co}} \tag{7}$$

where $f_{l,a}$ is the actual maximum confining pressure given by

$$f_{l,a} = \frac{E_{frp}t\varepsilon_{h,rup}}{R} \tag{8}$$

so that the ultimate strain is directly and only related to the actual confinement ratio $f_{l,a}/f'_{co}$ as ε_{co} is generally taken as a constant.

DETERMINATION OF α AND β

The two exponents, α and β, can be determined using test results from the open literature (Lam and Teng 2002d). In Lam and Teng (2002d), a database containing the test results of 81 plain concrete circular specimens wrapped with FRP is presented. Different types of FRP, namely, carbon FRP (CFRP), aramid FRP (AFRP) and glass FRP (GFRP) were used in these tests. The CFRP used included normal CFRP with high strength carbon fibres and HM CFRP with high modulus carbon fibres. The database includes results of the ultimate lateral strain ε_{ru}, ultimate axial strain ε_{cu} and compressive strength of confined concrete f'_{cc}. Most of the specimens were reported with corresponding FRP material properties from flat coupon tests (e.g. ASTM D3039 1995). For the rest, FRP properties from the manufacturers were reported.

Figure 3 shows plots of the strain enhancement ratio against the actual confinement ratio from the test database (Lam and Teng 2002d) for CFRP wraps and AFRP wraps respectively. In these diagrams and elsewhere in the paper, ε_{co}, which is the strain at the compressive strength of unconfined concrete, is taken as 0.002. A linear relationship clearly exists in both cases but the two trend lines are very different (Figure 3). These diagrams indicate that the strain enhancement ratio can be related to the confinement ratio for a given type of FRP, but separate expressions are needed for different types of FRP due to their difference in stiffness (Lam and Teng 2002b, Teng et al. 2002). To achieve a unified expression for the ultimate strain of FRP-confined concrete, the confinement stiffness ratio needs to be reflected, which means that the exponents, α and β, cannot both be unity.

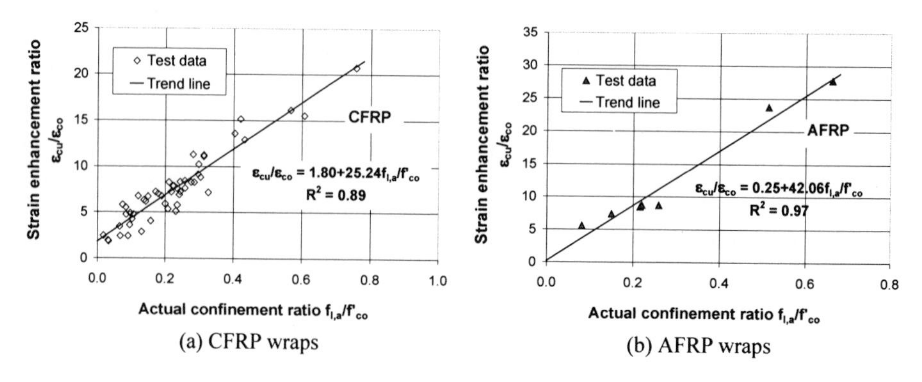

(a) CFRP wraps (b) AFRP wraps

Figure 3 Strain enhancement ratio versus actual confinement ratio

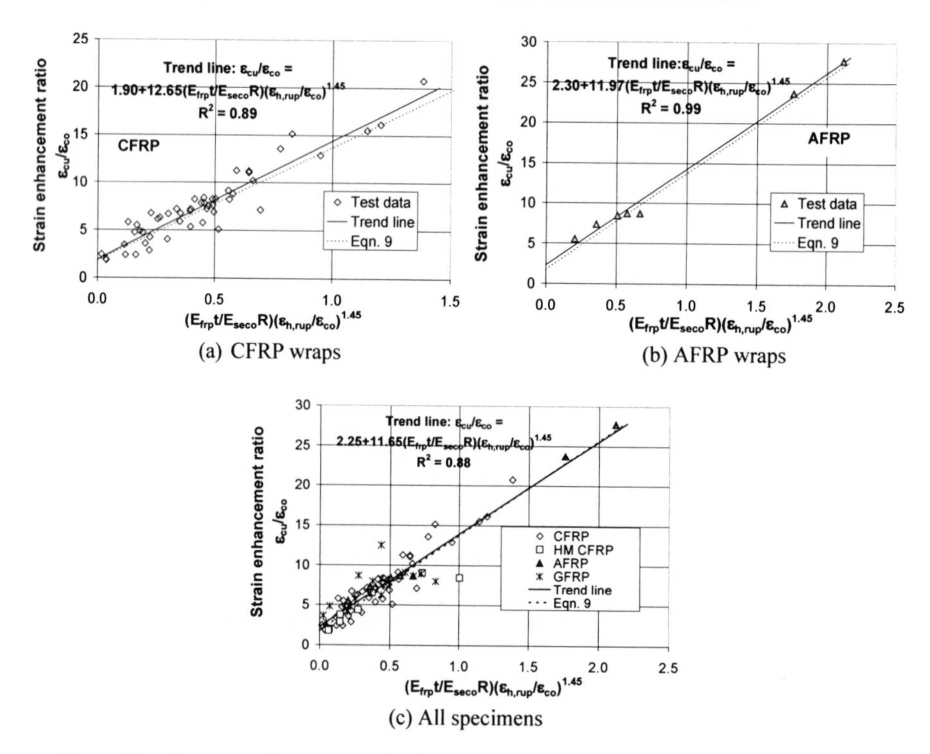

(a) CFRP wraps (b) AFRP wraps

(c) All specimens

Figure 4 Performance of proposed equation for the ultimate strain of FRP-confined concrete

Figure 4 shows comparisons of the test data with the following equation with the various coefficients of Eqn. 6 carefully calibrated from the test data (Lam and Teng 2002d):

$$\varepsilon_{cu} / \varepsilon_{co} = 1.75 + 12\left(\frac{E_{frp}t}{E_{\sec o}R}\right)\left(\frac{\varepsilon_{h,rup}}{\varepsilon_{co}}\right)^{1.45} \tag{9}$$

Figures 4a and 4b show that when plotted separately, similar trends are observed for CFRP and AFRP wraps. When all test data are plotted together (Figure 4c), the data points of HM CFRP wraps and GFRP wraps also fall near the trend line and the predictions of Eqn. 9, although the data points of GFRP wraps are rather scattered. Eqn. 9 therefore provides a unified expression for the ultimate strain of FRP-confined concrete that is accurate for different types of FRP.

FRP EFFICIENCY FACTOR

The application of Eqn. 9 requires the FRP hoop strain at rupture which cannot be taken to be the tensile strain from material tests, generally conducted using flat coupons following an acceptable standard (e.g. ASTM D-3039 1995). For most of the test specimens in the database, the FRP material properties were obtained by the researchers from flat coupon tensile tests. For the 51 CFRP-wrapped specimens, the ratio of the actual FRP hoop rupture strain ($\varepsilon_{h,rup}$) in FRP-confined concrete to the FRP rupture strain from flat coupon tests (ε_{frp}), defined as the FRP efficiency factor, is 0.583 on average (Lam and Teng 2002d). Making use of this efficiency factor, the ultimate strain of FRP-confined concrete can be expressed as

$$\varepsilon_{cu} / \varepsilon_{co} = 1.75 + 5.5\left(\frac{E_{frp}t}{E_{\sec o}R}\right)\left(\frac{\varepsilon_{frp}}{\varepsilon_{co}}\right)^{1.45} \tag{10}$$

This equation requires the user to know only the tensile strain ε_{frp} from flat coupon tests. It is worth noting that with this average efficiency factor of 0.583 for CFRP-wrapped concrete, the compressive strength equation proposed in Lam and Teng (2002c) for FRP-wrapped concrete derived mainly from test data of CFRP-wrapped specimens based on the flat coupon tensile strength and the alternative equation proposed in Lam and Teng (2002d) based on the actual hoop rupture strain of FRP provide almost identical predictions. The former equation is given by

$$\frac{f'_{cc}}{f'_{co}} = 1 + 2\frac{f_l}{f'_{co}} \tag{11}$$

where f_l is the nominal maximum confining pressure which can be obtained using Eq. 8 with $\varepsilon_{h,rup}$ replaced by ε_{frp}, while the latter equation is given by

$$\frac{f'_{cc}}{f'_{co}} = 1 + 3.3\frac{f_{l,a}}{f'_{co}} \tag{12}$$

For other types of FRP, insufficient information exists to define this efficiency factor with confidence. Even for CFRP-confined concrete, there is a considerable scatter in the efficiency factor deuced from test results (Lam and Teng 2002d). The present authors therefore recommend that for cost-effective and safe designs, a small number of FRP-confined cylinder tests, analogous to plain cylinder tests, be conducted to determine the efficiency factor of a particular FRP product in confinement applications. For this purpose, a standard confined cylinder test method should be formulated. FRP manufacturers can then supply the results of such tests as part of their product information. If this information is not available from the manufacturer, the user should conduct these tests instead.

CONCLUSIONS

This paper has been concerned with the ultimate axial strain of FRP-confined concrete. A stiffness-based ultimate strain equation has been presented for concrete confined by wrapped FRP with fibres only or mainly in the hoop direction. The theoretical basis and performance of the equation have been described in detail. The stiffness and the actual ultimate condition of the FRP jacket are explicitly accounted for in the equation so that it is accurate for different types of FRP as shown through comparisons with existing test data. For the application of this equation in design, the FRP efficiency factor (ratio between the actual hoop rupture strain of FRP in FRP-confined concrete to the tensile rupture strain from material tests) needs to be established. For this purpose, confined cylinder tests have been suggested as a standard type of tests to supply this information.

ACKNOWLEDGEMENTS

Both authors wish to thank the Research Grants Council of Hong Kong SAR (Project No: PolyU 5064/01E) and The Hong Kong Polytechnic University for their financial support.

REFERENCES

ASTM D3039/D3039M –95 (1995). Standard Test Method for Tensile Properties of Polymer Matrix Composites Materials, *Annual Book of ASTM Standards*, Vol. 14.02.

CEB-FIP (1993) *CEB-FIP Model Code 1990: Design Code*, Thomas Telford, London, UK.

Candappa, D.C., Sanjayan, J.G., and Setunge, S. (2001). Complete Triaxial Stress-Strain Curves of High-Strength Concrete. *Journal of Materials in Civil Engineering, ASCE*, **13:3**, 209-215.

Karbhari, V.M., and Gao, Y. (1997). Composite Jacketed Concrete under Uniaxial Compression–Verification of Simple Design Equations. *Journal of Materials in Civil Engineering, ASCE*, **9:4**, 185-193.

Lam, L. and Teng, J.G. (2001). A New Stress-Strain Model for FRP-Confined Concrete. *Proceedings, International Conference on FRP Composites in Civil Engineering*, Hong Kong, China, pp. 283-292.

Lam, L. and Teng, J.G. (2002a). Stress-Strain Models for Concrete Confined by Fiber-Reinforced Polymer. *Proceedings, 17th Australasian Conference on the Mechanics of structures and Materials*, Gold Coast, Queensland, Australia, pp. 39-44.

Lam, L. and Teng, J.G. (2002b). *Stress-Strain Models for FRP-Confined Concrete*, Research Centre for Advanced Technology in Structural Engineering, The Hong Kong Polytechnic University.

Lam, L. and Teng, J.G. (2002c). Strength Models for FRP-Confined Concrete. *Journal of Structural Engineering, ASCE*, **128:5**, 612-623.

Lam, L. and Teng, J.G. (2002d). Design-Oriented Stress-Strain Model for FRP-Confined Concrete. To be published.

Miyauchi, K., Inoue, S., Kuroda, T., and Kobayashi, A. (1999). Strengthening Effects of Concrete Columns with Carbon Fiber Sheet. *Transactions of The Japan Concrete Institute*, **21**, 143-150.

Monti, G. (2001). Confining Reinforced Concrete with FRP: Behavior and Modeling. *Proceedings, Workshop on Composites in Construction: a Reality*, Capri, Italy, July 2001.

Ottosen, N.S. (1979). Constitutive Model for Short-Time Loading of Concrete. *Journal of the Engineering Mechanics Division, ASCE*, **105**, 127-141.

Richart, F.E., Brandtzaeg, A. and Brown, R.L. (1929). *The Failure of Plain and Spirally Reinforced Concrete in Compression*, University of Illinois Engineering Experimental Station, Bulletin No. 185, Urbana, Illinois, USA.

Spoelstra, M.R. and Monti, G. (1999). FRP-Confined Concrete Model. *Journal of Composites for Construction, ASCE*, **3:3**, 143-150.

Teng, J.G., Chen, J.F., Smith, S.T. and Lam, L. (2002). *FRP-Strengthened RC structures*, John Wiley & Sons, Ltd., UK.

Xiao, Y. and Wu, H. (2000). Compressive Behavior of Concrete Confined by Carbon Fiber Composite Jackets. *Journal of Materials in Civil Engineering, ASCE*, **12:2**, 139-146.

Yuan, X.F., Lam, L., Teng, J.G. and Smith, S.T. (2001). FRP-Confined RC Columns under Combined Bending and Compression: a Comparative Study of Concrete Stress-Strain Models. *Proceedings, International Conference on FRP Composites in Civil Engineering*, Hong Kong, China, pp. 749-758.

Advances in Building Technology, Volume I
M. Anson, J.M. Ko and E.S.S. Lam (Eds.)

EFFECT OF SIZE ON THE FAILURE OF FRP STRENGTHENED REINFORCED CONCRETE BEAMS

Christopher K.Y. LEUNG[1], Zhongfan CHEN[2], Stephen K.L. LEE[3], Jianmao TANG[3]

[1]Department of Civil Engineering, and [3]Advanced Engineering Materials Facility
Hong Kong University of Science and Technology,
Clear Water Bay, Kowloon, HONG KONG
[2]Department of Civil Engineering, Southeast University, Nanjing, CHINA

ABSTRACT

The bonding of fiber reinforced plastic (FRP) plates is an effective and efficient method to improve the bending capacity of concrete beams. In the literature, various design methodologies have been proposed and several of them have been found to compare well with test data or to provide reasonable lower bounds. However, almost all the experimental data are obtained from laboratory-size specimens that are several times smaller than members in real structures. On close examination, different design methodologies exhibit significantly different trends of load capacity vs member size. With no test results on larger specimens, it is not possible to determine which approach is more appropriate for designing real size members. In this investigation, geometrically similar reinforced concrete beams with steel ratio of 0.01, and depth ranging from 0.2 m to 0.8 m were prepared. Some RC beams were tested as control while others were retrofitted with 2 to 8 layers of Carbon fiber reinforced plastic (CFRP) sheets to achieve the same CFRP/concrete area ratio. All the beams were tested to failure to investigate the variation of failure load with member size. The results of the present investigation are expected to provide useful information for (i) the extrapolation of small specimen results to large members of practical size, and (ii) the checking of theoretical models for design.

KEYWORDS

Reinforced Concrete, Fiber Reinforced Plastics, Structural Retrofitting, Experimental Testing

INTRODUCTION

After years in service, many concrete structures have deteriorated and are in need of repair. In some cases, due to change of usage, members need to be strengthened through retrofitting to provide a load carrying capacity higher than that of the original design. For the flexural strengthening of concrete beams, the bonding of fiber reinforced plastic (FRP) plate to the tensile side has been shown to be an effective means (Saadatmanesh and Ehsani, 1991, Triantafillou and Plevris, 1992, Meier et al, 1993,

Arduini et al, 1997). Besides extensive laboratory studies, the application of FRP to strengthen beams in real structures has also been reported by Meier (1995) and Tan (1997).

For the widespread use of FRP strengthening in practice, a comprehensive design guideline is required. Any design guideline should include a method to accurately predict the failure load of the strengthened member. In the literature, various approaches have been proposed for failure prediction of FRP strengthened members. Some approaches (Teng et al, 2000) are based on the direct fitting of experimental data. Others (El-Mihilmy and Tedesco, 2001 , Raoof and Hassanen, 2000) are based on analytical models, and test data are employed to verify the theoretical prediction. In any case, the full verification of design models relies on the availability of experimental data covering a wide range of design variables. Most available data in the literature, however, are obtained from laboratory-size specimens, that are much smaller than members in real structures. For example, out of the 127 test results on flexural strengthening from 23 studies collected by Bonacci and Maalaj (2001), over 80 of them were obtained from specimens with depth of 300 mm or less. There are only 15 results with beam depth of 400mm or above, and the largest beam was 574mm in depth (with only data from one test). In real structures, beam members are rarely below 400 mm in depth, and very often deeper than 600 mm. Therefore, very few available existing data are representative of real members.

To see the effect of size on member strength, the best way is to calculate the normalized ultimate moment M/bd^2 for sections with similar reinforcement ratio but different size. In Fig.1 below, theoretical prediction of the size effect are carried out with 5 different models (El-Mihilmy and Tedesco, 2001, Oehlers, 1992, Raoof and Hassanen, 2000, Teng et al, 2001, Ziraba et al. 1994). In the analysis, geometrically similar beams 220, 440, 660 and 880 mm in depth are considered. The beam length (L), is 12 times the depth. Loading is applied at one-third points along the span. The steel reinforcement ratio is 1%. The width of the FRP plate (E=235 GPa) is the same as the beam, and the plate thickness is 0.2% the beam depth. The cut-off point of the plate is at a distance L/12 from the support. For each curve, the normalized failure moment (M/bd^2) (Here, d is taken to be the total beam depth, as the FRP is glued to the bottom) is further normalized with its value for the 220 mm depth beam. The predicted trend of failure load vs size can then be easily compared. From Fig.1, it can be seen that the different models show significantly different effects of member size.

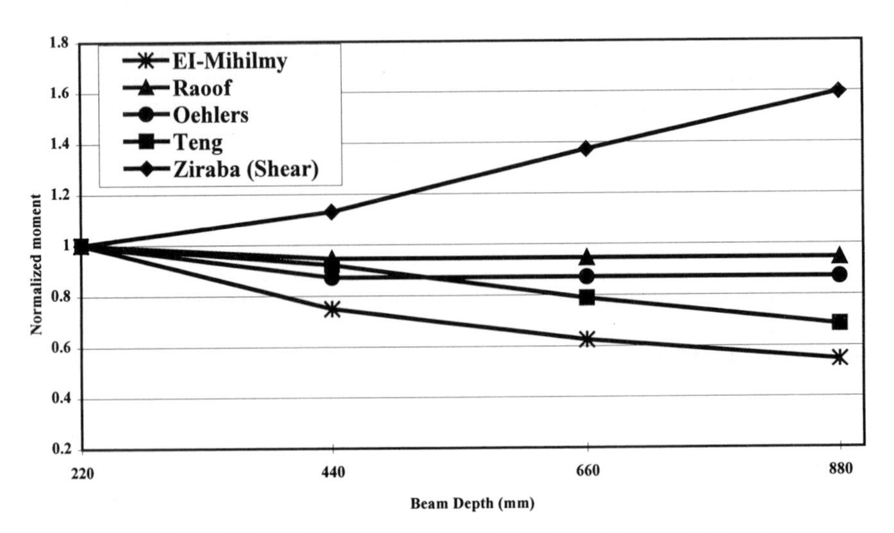

Figure1 Effect of Size on the Normalized Failure Moment of FRP retrofitted Beams

Note that each of these models has been compared with a significant number of experimental data for small specimens, with reasonable agreement achieved. However, even if we empirically 'calibrate' each model (e.g. by multiplying the predicted value with an empirical constant) to give similar results for laboratory size specimens (at 220 mm depth), the predicted failure moment for large members can differ a lot. We suspect that the increasing trend predicted by Ziraba et al.'s model is an artifact of the model, but even if this curve is taken out, the other models still show significant differences. The implication is very clear: experimental results on large members are necessary for the proper validation of design models for flexural failure of strengthened beams.

TESTING PROGRAM

Specimen Design, Preparation and Testing

To study the effect of size on the bending behaviour of FRP strengthened beams, nine reinforced concrete beams were prepared in three different sizes: 75mm (width) x 200mm (depth) x 2m (length), 150 mm x 400 mm x 4 m and 300 mm x 800 mm x 7.5 m. For all the beams, high yield steel reinforcement (σ_y=460 MPa) was employed and the thickness of cover (c) to the tensile reinforcement was 0.075d. The tensile steel reinforcement ratios in the beams (A_s/bh, where h = d-c) were respectively 1.13%, 1.09% and 1.13%. Also, all beams were over-reinforced in shear to ensure that failure would occur under bending. During testing, the loading spans for the three types of beams were respectively 1.8m, 3.6m and 7.2 m. 4 point loading was carried out, with loading applied at one-third points along the loading span. All the specimens were prepared at the same time, with concrete from a ready mix supplier. The design strength of the concrete (f_{cm}) was 30 MPa. Considering standard deviation of concrete, the mean strength should be close to 40 MPa. When the beams were tested, compression strength was also determined with concrete cubes. The average cube strength was found to be 47 MPa.

After the reinforced concrete beams were prepared and cured for over a month, 6 of the beams (2 for each size) were strengthened with bonded carbon fiber reinforced plastic (CFRP) strips. The Young's modulus of the CFRP is 235 GPa while its strength is 4.2 GPa. The CFRP is supplied in the form of a flexible sheet (0.11mm in thickness) with impregnated resin. In the retrofitting process, the bottom surface of the concrete beam was first roughened with a needle gun. A primer was then applied onto the concrete surface. Alternate layers of resin and CFRP were applied under the required thickness was attained. Note that the applied resin contains a hardener and will cause the CFRP sheet to harden and bond strongly to the concrete. For the three beam sizes, 2, 4 and 8 layers of CFRP were applied, and they were cut off at 0.2m, 0.4m and 0.8m from each support respectively. The width of CFRP was the same as the width of the beam. At least 7 days was allowed for the resin to attain maximum strength before testing of the beams were carried out.

All tests were performed with a Dartec loading system of 250 tonne capacity. In all specimens, strain gauges have been installed on both the steel reinforcements and CFRP sheet to monitor the failure process during the test. Besides the applied load, the displacement at mid-span was also measured with a LVDT. For some of the beams, the complete testing was videotaped to capture the damage processes leading to final failure.

Test Results

As mentioned above, a lot of information has been obtained from each individual test. Complete analysis of all the experimental data is still in progress and will be reported in an upcoming publication. In this short paper, we will only focus on the normalized moment for beams of various sizes, the maximum strain measured on the FRP and the percentage increase in load capacity due to strengthening. The experimental values are summarized in Table 1 below.

TABLE 1
EXPERIMENTAL RESULTS FOR BEAMS OF VARIOUS SIZES

Member	Type	Section Dimensions	Load Capacity (kN)	Max. FRP Strain (x 10^{-6})	M/bd^2 (N/mm^2)	% Increase in Load Capacity
SB1	Control	200 mm (d) x 75 mm (b)	61.02	-	6.102	-
SB2	Strengthened		69.56*	6634*	6.956*	14.0*
SB3	Strengthened		75.74	9737	7.574	24.1
MB1	Control	400 mm (d) x 150 mm (b)	212.02	-	5.301	-
MB2	Strengthened		274.14	7752	6.854	29.3
MB3	Strengthened		272.54	6976	6.813	28.5
LB1	Control	800 mm (d) x 300 mm (b)	822.78	-	5.142	-
LB2	Strengthened		1017.56	6400	6.360	23.7
LB3	Strengthened		1033.00	6673	6.456	25.5

* For specimen SB2, there was a problem with the surface preparation, so the CFRP debonded prematurely, with no concrete sticking to its surface. The result is hence an underestimate for the load capacity of a properly retrofitted beam.

From the above table, the effect of member size on the normalized moment can be observed. When going from the smallest to the largest member, (M/bd^2) decreases by roughly 15%. This size effect is considered quite moderate, considering the fact that the largest specimen is four times the size of the smallest one. In all members, failure was associated with debonding of the FRP from the bottom of the beam. The debonded region was not confined to one side of the beam but extended through the middle span. After testing, the failure surface of the FRP was inspected, and only a thin layer of stuck concrete was found. From our experience, this type of failure was induced by the opening of flexural cracks in the middle of the beam, rather than by stress concentrations near the plate cut-off point. Indeed, in the test on the largest beam, we were able to capture the failure sequence with a video camera. On replaying, it was clear that interfacial failure first started below a flexural crack near one of the loading points, and then propagated towards the end of the bonded plate.

The maximum strain measured on the FRP also suggested the presence of a size effect. Since the magnitude of measured strain also depends on the location of the gauge relative to the flexural cracks, the comparison of maximum strain between various members cannot be considered exact. However, a clear qualitative trend can be observed from the results. With a smaller specimen, it was possible for the FRP to reach a higher strain before failure. This observation is consistent with the size effect on normalized moment of the beam.

A very interesting observation from our result is that the size effect also exists for normal reinforced concrete members that are not strengthened. Indeed, if we consider the percentage increase of load capacity through FRP strengthening, the intermediate member exhibits the highest improvement, while the large and small members show about the same increase. Note that the tensile reinforcement ratio in the intermediate member was a bit smaller than the other two beams. The amount of added FRP reinforcement was hence relatively higher. This may partially explain the higher strengthening ratio. Overall, if one considers the percentage of load increase for members of all three sizes, the size effect is not too significant.

CONCLUSION

In this paper, the effect of member size on the load carrying capacity of FRP strengthened reinforced concrete beams is studied. The result shows that the size effects on the normalized moment of strengthened beams and percentage improvement in load capacity are quite moderate. To our knowledge, this investigation is the first ever study of its kind. It provides useful results for (i) extrapolation of laboratory results to real structural members, and (ii) checking of theoretical models. Based on the test data, any correct theoretical model for FRP debonding failure should not exhibit a large effect of size on the failure load.

REFERENCES

M. Arduini, A. Di Tommaso and A. Nanni (1997) Brittle Failure in FRP Plate and Sheet Bonded Beams. *ACI Structural Journal*, **94(4)**, 363-370.

J.F. Bonacci and M. Maalaj (2001). Behavioral Trends of RC Beams Strengthened with Externally Bonded FRP. ASCE Journal of Composites in Construction, **5**(2), 102-113.

M.T. El-Mihilmy and J.W. Tedesco, J.W. (2001). Prediction of Anchorage Failure for Reinforced Concrete Beams Strengthened with Fiber-Reinforced Polymer Plates. *ACI Structural Journal*, **98**(3), 301-314.

U. Meier, M. Deuring, H. Meier, and G. Schwegler, G. (1993). CFRP Bonded Sheets. *In*: *Fibre-Reinforced Plastic (FRP) Reinforcement for Concrete Structures: Properties and Applications*, edited by A. Nanni, Elsevier Science, Amsterdam, The Netherlands.

U. Meier (1995) Strengthening of Structures using Carbon Fibre/Epoxy Composites. *Construction & Building Materials*, **9**, 341-353.

Oehlers, D.J. (1992). Reinforced Concrete Beams with Plates glued to their Soffits. ASCE Journal of Structural Engineering, **118**(8), 2023-2038.

M. Raoof and M.A.H. Hassanen (2000). Peeling Failure of Reinforced Concrete Beams with Fibre-Reinforced Plastic or Steel Plates Glued to their Soffits. *Proceedings of the Institution of Civil Engineers: Structures and Buildings*, **140**, 291-305.

H. Saadatmanesh and A.M. Malek (1998). Design Guidelines for Flexural Strengthening of RC Beams with FRP Plates. *ASCE Journal of Composites for Construction*, **2**(4), 158-164.

K.H. Tan (1999). Towards a Cost-Effective Application of FRP Reinforcement in Structural Rehabilitation. *In: Proceedings of the Seventh East Asia-Pacific Conference on Structural Engineering and Construction*, Kochi, Japan, 65-73.

J.G. Teng, J.F. Chen, S.T. Smith and L. Lam. (2001). FRP Strengthened RC Structures. John Wiley & Sons, West Sussex, U.K.

T. C. Triantafillou and N. Plevris (1992). Strengthening of RC Beams with Epoxy Bonded Fibre Composite Materials. *Materials and Structures* **25**, 201-211.

Y.N. Ziraba, M.H. Baluch, I.A. Basunbul, A.M. Sharif, A.K. Azad and G.J. Al-Sulaimani (1994). Guidelines towards the Design of Reinforced Concrete Beams with External Plates. *ACI Structural Journal*, **91**(6), 439-646.

Advances in Building Technology, Volume I
M. Anson, J.M. Ko and E.S.S. Lam (Eds.)

ANALYTICAL BEHAVIOUR OF CONCRETE BEAMS REINFORCED WITH STEEL REBARS AND FRP RODS

H. Y. Leung, S. Kitipornchai, R. V. Balendran and A. Y. T. Leung

Department of Building and Construction, City University of Hong Kong,
Tat Chee Avenue, Kowloon, Hong Kong

ABSTRACT

Reinforced concrete (RC) beam embedded with steel reinforcement usually gives a ductile behaviour but low strength whereas use of fibre-reinforced-plastic (FRP) rods provides a higher ultimate strength but limited ductility. It is thus thought that a combined use of steel rebars and FRP rods in concrete beam can offer a ductile and strong performance. To verify this, an analytical model is generated. The model computes the flexural behaviour of concrete members reinforced with conventional steel reinforcing bars and novel FRP rods. The model implementation is based on the use of the true stress-strain relationships for plain concrete, steel rebar and FRP rod. The main variable considered in this study is arrangement/depth of steel rebars and FRP rods. This paper presents some particular findings from the model. Most results are also compared with that of concrete beam reinforced only with steel rebars.

KEYWORDS

Concrete, RC beam, reinforcement, steel rebar, FRP rod, flexural strength

INTRODUCTION

Reinforced concrete (RC) beam embedded with conventional steel reinforcement usually gives a ductile behaviour but low strength whereas use of novel fibre-reinforced-plastic (FRP) rods provides a higher ultimate strength but limited ductility. Instead of exploring new materials as reinforcement so as to enhance the performance of reinforced concrete beam, it is thought that a mix use of steel rebars and FRP rods in concrete beam can offer a ductile and strong performance. Tentatively, concrete beams reinforced with both steel rebars and FRP rods exhibit one of the following flexural failure modes:

1. Concrete crushing occurs before steel yields and FRP breaks.
2. Steel yields followed by concrete crushing while FRP remains intact.
3. Steel yields followed by steel breaks while FRP remains intact.
4. Steel yields followed by FRP snaps.

These failure modes depend largely on the material properties of the ingredients and the configuration of the beam section. It is apparent that the first failure mode occurs only in over-reinforced sections. Since the aim of this study is to investigate the effect of addition of FRP rods into the steel-reinforced section, the latter three failure modes are desired. In this paper, an analytical model is generated to compute the flexural behaviour of concrete section reinforced with conventional steel reinforcing bars and novel FRP rods. Some results, including the strains, neutral axis depth, moments and curvatures, are discussed.

ANALYSIS & ASSUMPTIONS

A simple analytical procedure is developed to calculate the strains and stresses in concrete and reinforcements, moment and curvature in a rectangular concrete cross section reinforced with steel rebars and FRP rods. The procedure, which is derived from force equilibrium and strain compatibility, is applicable to conventional steel reinforced sections, FRP reinforced sections, as well as concrete sections reinforced with a combination of steel rebars and FRP rods. Fig. 1 shows the rectangular concrete beam section used in this study. The basic assumptions made include (1) linear strain distribution across the depth of section, (2) no bond slip of reinforcement, (3) reinforcement stress can be obtained from its stress-strain relationship, (4) tension concrete ignored, (5) shear deformation neglected and (6) failure of section is defined by either concrete crushing or reinforcement snapping.

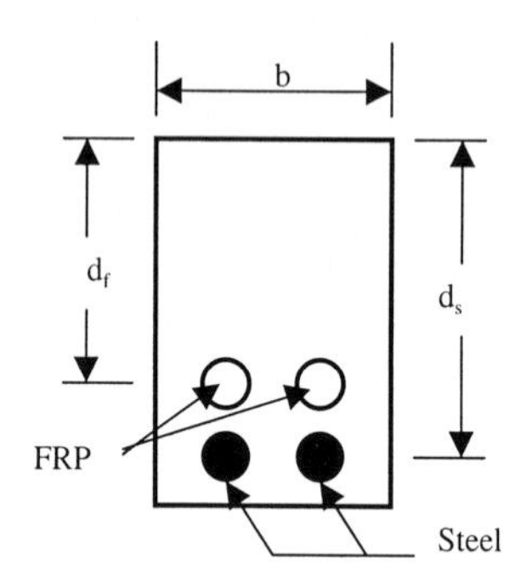

Figure 1: Beam cross section

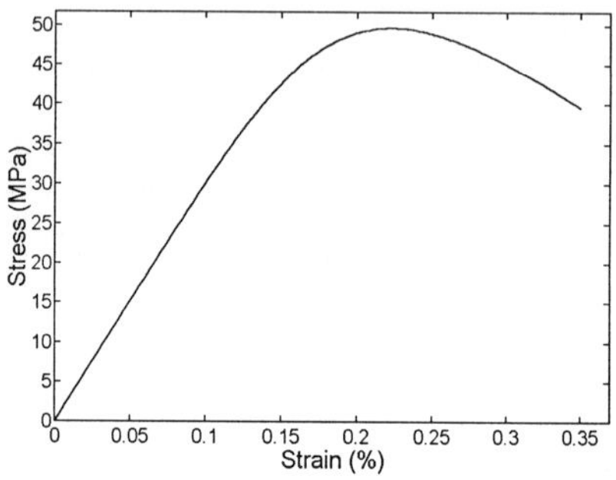

Figure 2: Stress-strain curve for concrete in compression

STRESS-STRAIN RELATIONSHIPS

Concrete in compression

The nonlinear stress-strain model of concrete in compression developed by Almusallam & Alsayed (1995) is used. This model was found to fit well the experimental results for normal strength and high strength concretes. In the model, the relationship between concrete stress (σ_c) and concrete strain (ε_c) is expressed by Eqn. 1.

$$\sigma_c = \frac{(K - K_p)\varepsilon_c}{\left[1 + \left(\frac{(K - K_p)\varepsilon_c}{\sigma_0}\right)^n\right]^{\frac{1}{n}}} + K_p\varepsilon_c \tag{1}$$

where K = initial slope of the curve; K_p = final slope of the curve; σ_0 = reference stress; and n = curve-shape parameter. The equations of all these model parameters are listed in Appendix. Fig. 2 shows a stress-strain curve of compression concrete where the ultimate concrete strength $\sigma_c^{'} = 50\,\text{MPa}$. As can be seen in Fig. 2, the stress-strain curve terminates at an assumed strain value of 0.0035.

Tensile steel embedded in concrete

The stress-strain curve of steel reinforcing bars embedded in concrete is assumed to be bi-linear and was modelled by Wang & Hsu (2001). The relationship between steel stress (σ_s) and steel strain (ε_s) can be expressed by the following two equations (Eqn. 2 and Eqn. 3):

$$\text{for } \varepsilon_s \leq \varepsilon_n \qquad \sigma_s = E_s \varepsilon_s \tag{2}$$

$$\text{for } \varepsilon_s > \varepsilon_n \qquad \sigma_s = \sigma_y \left[(0.91 - 2B) + (0.02 + 0.25B) \frac{\varepsilon_s}{\varepsilon_y} \right] \tag{3}$$

where E_s = elastic modulus of steel; σ_y = yield stress of steel; B = stress parameter; ε_y = strain of steel at yield; and ε_n = strain of steel at average stress when initial yielding occurs. All the model parameters are again given in the Appendix. This model, unlike others, accounts for the fact that the stresses in the steel bars between concrete cracks will be less than the yield stress at the cracks. An averaged value of steel stress across the cracks is thus adopted.

To facilitate generating a curve, some parameters are assumed and they include the breadth of concrete section (b = 200 mm); depth of steel rebars (d_s = 300 mm); area of steel (A_s = 157 mm^2); yield stress (σ_y = 420 MPa); and elastic modulus (E_s = 200 GPa). The resulting curve is shown in Fig. 3.

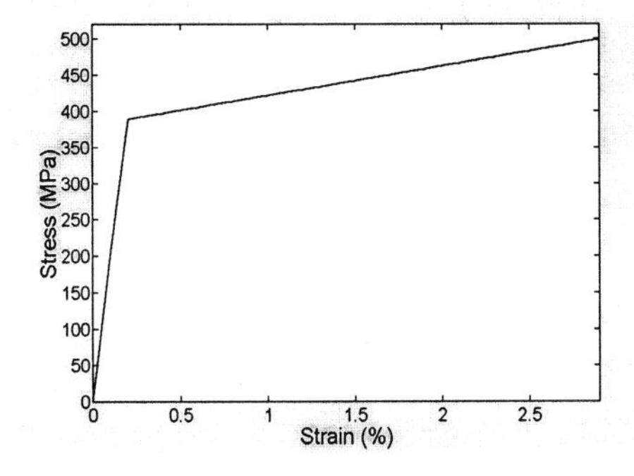

Figure 3: Stress-strain curve for steel

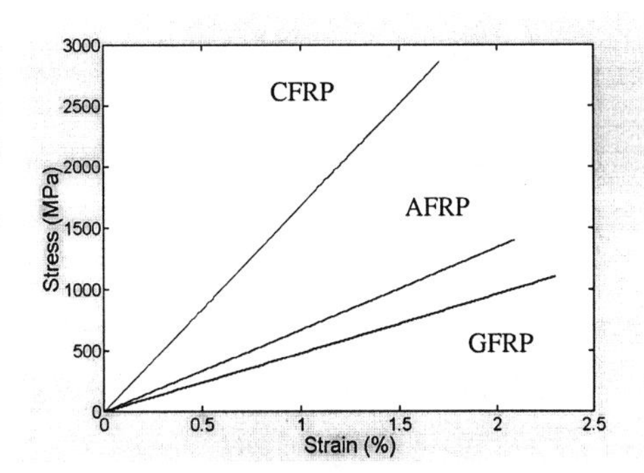

Figure 4: Stress-strain curve for FRP

Tensile FRP rod

The relationship of FRP stress (σ_f) and FRP strain (ε_f) under tension is assumed to be linear and is defined by Eqn. 4 and Eqn. 5.

$$\text{for } \varepsilon_f \leq \varepsilon_f^u \qquad \sigma_f = E_f \varepsilon_f \tag{4}$$

$$\text{for } \varepsilon_f > \varepsilon_f^u \qquad \sigma_f = 0 \tag{5}$$

where E_f denotes the elastic modulus of FRP.

The three common FRPs used in construction industry are C(carbon)FRP, A(aramid)FRP and G(glass)FRP, and some of their properties (values obtained from Leung & Burgoyne, 2001) are given in Table 1. Using these values, the stress-strain curves for the three FRPs can then be generated and they are illutrated in Fig. 4.

TABLE 1
PROPERTIES OF FRPS

Materials	CFRP	AFRP	GFRP
Elastic modulus E_f	168 GPa	67 GPa	48 GPa
Ultimate strain ε_f^u	0.017	0.021	0.023

DERIVATIONS

A strain incremental technique is used in the derivation of different parameters. The strain in the extreme fibre of concrete (ε_c^{top}) is increased in a fixed increment, from zero to the ultimate (failure) strain value, to generate the moment-curvature relationship. Fig. 5 shows the distributions of strain and stress across the cross section.

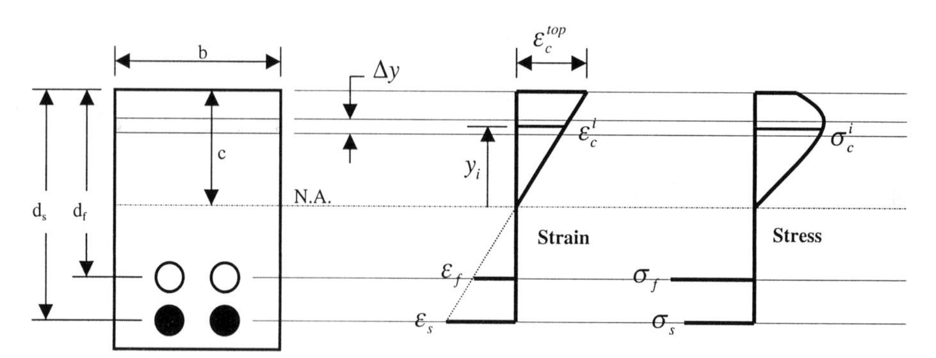

Figure 5: Strain and stress distributions

Once the value of ε_c^{top} is determined, a value of c is assumed. The compression concrete is divided into slices. Strains in the ith compression concrete slice (ε_c^i), FRP rods (ε_f) and steel rebars (ε_s) are then calculated, in terms of ε_c^{top}, by the following three equations (Eqns. 6 to 8):

$$\varepsilon_c^i = \varepsilon_c^{top} \times \left(\frac{y_i}{c} \right) \tag{6}$$

$$\varepsilon_f = \varepsilon_c^{top} \times \left(\frac{d_f - c}{c} \right) \tag{7}$$

$$\varepsilon_s = \varepsilon_c^{top} \times \left(\frac{d_s - c}{c} \right) \tag{8}$$

where y_i is the distance of the ith concrete slice measured from the neutral axis; c is the depth of neutral axis; d_f and d_s are the depth of FRP rods and steel rebars respectively. The stresses in the ith concrete slice (σ_c^i), FRP rods (σ_f) and steel rebars (σ_s) are found from the corresponding stress-strain curves.

The concrete compressive force (F_c) is determined by summation of the product of elemental stress and the corresponding elemental area ($b \times \Delta y$) over the entire compression concrete portion. It is expressed by Eqn. 9.

$$F_c = \sum_i \left(\sigma_c^i \times b \times \Delta y \right) \tag{9}$$

The tensile FRP force (F_f) and steel force (F_s) are found by multiplying the FRP stress by the FRP area (A_f) and the steel stress by the steel area (A_s) respectively (Eqns. 10 and 11).

$$F_f = \sigma_f \times A_f \tag{10}$$

$$F_s = \sigma_s \times A_s \tag{11}$$

Equilibrium of internal forces is imposed by using Eqn. 12:

$$F_c = F_f + F_s \tag{12}$$

If the above equation is not satisfied, the location of the neutral axis (c) is adjusted until force equilibrium is achieved.

The internal resisting moment (M) is then calculated by summing the moments of all internal forces about the neutral axis, and it can be expressed, by Eqn. 13, as follows:

$$M = \sum_i \left(\sigma_c^i \times b \times \Delta y \right) y_i + F_f \times \left(d_f - c \right) + F_s \times \left(d_s - c \right) \tag{13}$$

As indicated by Eqn. 14, the curvature (κ) is also found by dividing the extreme concrete fibre strain by the neutral axis depth.

$$\kappa = \frac{\varepsilon_c^{top}}{c} \tag{14}$$

A small increment on ε_c^{top} is applied and the procedure repeats. The analysis terminates when ε_c^{top} goes beyond 0.0035.

RESULTS & DISCUSSION

To generate analytical results from the above procedure, some material and geometric parameters are assumed. To do this, the model parameters used in generating the curves in Fig. 2 and Fig. 3 are adopted, GFRP is chosen as the internal reinforcement and its properties are obtained in Table 1. It should be noted that a breaking strain of 10% is imposed for tensile steel rebars embedded in concrete. Assumed that $A_f = A_s$, the single variable left is the depth of GFRP rods (d_f) and its value is altered so as to investigate its effect on the flexural behaviour of the beam section (as shown in Fig. 1). Some analytical results are generated and are shown in Figs 6 to 10.

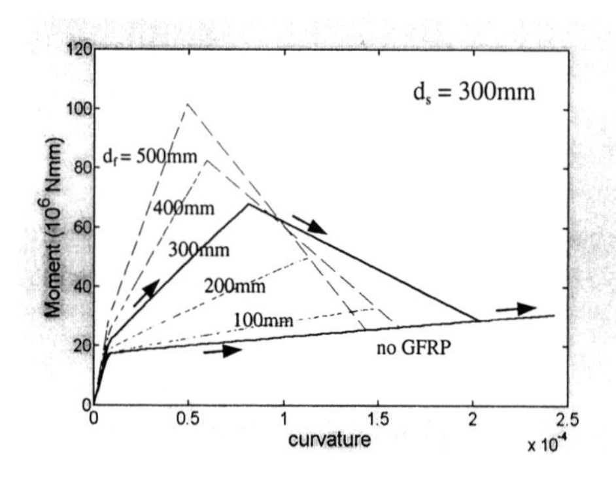

Figure 6: Moment against curvature

Figure 7: Concrete strain against moment

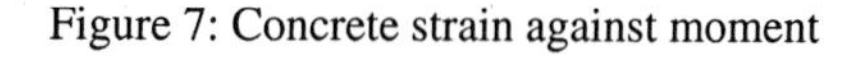

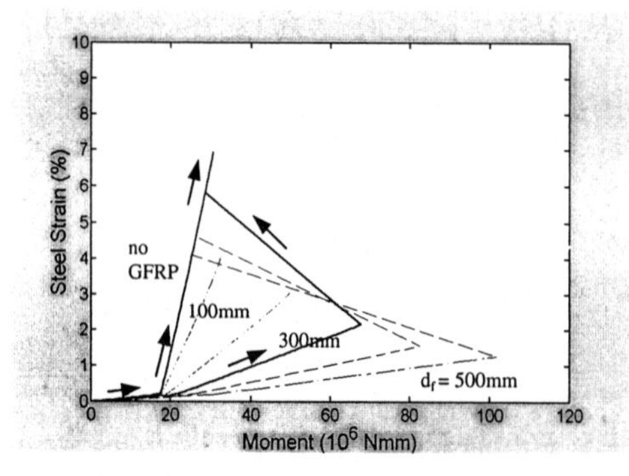

Figure 8: Steel strain against moment

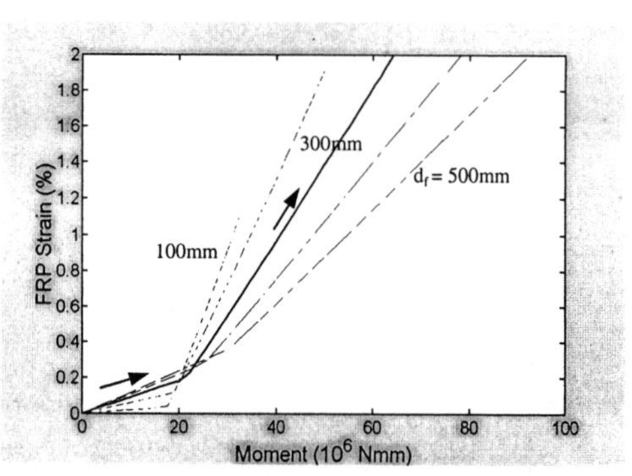

Figure 9: GFRP strain against moment

Fig. 6 shows the variation of moment against curvature for different values of d_f, the curve for no GFRP is also plotted. Since under-reinforced section is adopted, all different sections show occurrence of steel yielding. From the figure, it is clear that the beam section, with no GFRP, becomes a conventional steel-reinforced section and its moment-curvature relationship is bi-linear. Notice that when GFRP rods are added, the initial stiffness and the yield moment seem to increase slightly. When larger value of d_f is adopted, the resisting moment increases in a faster way and higher peak value is resulted. When the two curves with $d_f = 100mm$ and $d_f = 200mm$ are considered, they show an increase up to a peak and then stop. The two curves generally show a similar shape of the steel-reinforced section, except the increase in moment capacity and reduction in ductility. From Figs 7 to 9, it can be seen that the strains of steel bars and GFRP rods for

$d_f = 100mm$ and $d_f = 200mm$ are smaller than their corresponding ultimate strain values, while the concrete strain increases rapidly to 0.0035. This demonstrates that concrete crushing occurs before steel or FRP snaps. For the remaining three curves, a reverse situation is found; reinforcement breaks before concrete crushing occurs. As can be seen in Figs 8,9 and 10, the variation of steel strain remains in its plastic range while GFRP rods have reached its breaking point at failure. This explains the abrupt drop in moment-curvature relationship for $d_f = 300mm$, $d_f = 400mm$ and $d_f = 500mm$, as indicated in Fig. 6. It should also be noted that when larger value of d_f is used, the decrease in moment is more rapid. After GFRP rods have broken, the beam section contains only steel rebars and the moment-curvature curve for steel-reinforced section is followed.

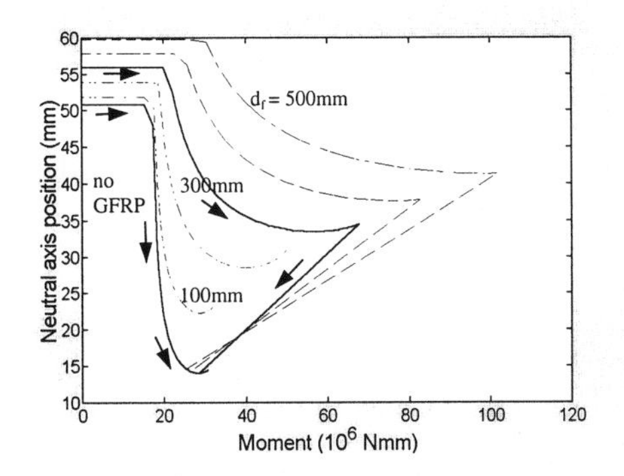

Figure 10: Neutral axis depth against moment

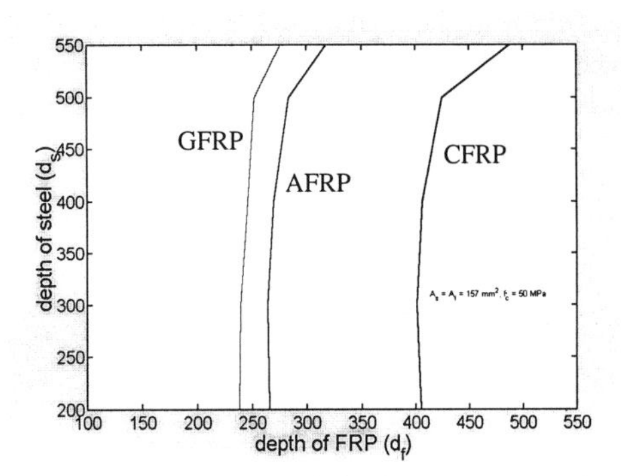

Figure 11: d_s versus d_f

It is interesting to study the neutral axis depth for section containing both kinds of reinforcements. As can be seen in Fig. 10, when no GFRP is used, the neutral axis remains fairly constant. Upon loading, it shifts upward when yielding of steel occurs. The neutral axis slightly goes downward when the section is about reaching its failure, this is explained by the fact that a reduction in concrete stress after its peak value and an increase in stresses in tensile reinforcement give rise to an increase in neutral axis depth. A similar situation is found when the depth of GFRP rods is not large ($d_f = 100mm$ and $d_f = 200mm$). When d_f is further increased, an abrupt upward shift in neutral axis is expected since the failure is governed by snapping of GFRP. Once the GFRP loses its action, the neutral axis depth must be reduced to balance the force between compression concrete and remaining steel rebars.

A similar analysis can be carried out for AFRP and CFRP using values in Table 1. Instead of plotting the strains, neutral axis depths, moments and curvatures, an interaction diagram for different failure modes is produced in Fig. 11. Three different curves are plotted for the three FRPs. Each curve represents the condition that concrete crushing and breaking of FRP rods break simultaneously. The area to the left of the curve indicates that concrete crushing occurs prior to snapping of FRP while the area to the right suggests FRP breaks followed by concrete crushing.

The current investigation provides some results on concrete flexural members reinforced with a mix of tension reinforcements, but there are limitations from a practical point of view. Firstly, it is rare to have steel rebar snapped at ultimate state as the breaking strain of steel is very much in excess of that for FRP. Besides, d_f greater than 300 mm is not common in daily practice. However, these two conditions are considered for the completeness of study.

CONCLUSIONS

From the analytical results, it is possible to conclude that:

1. Failure of steel-reinforced section with adddition of FRP rods is usually governed by concrete crushing or FRP snapping. The steel breaking failure mode is rare.
2. When steel rebars and FRP rods are employed, concrete crushing is usually accompanied by small value of FRP depth whereas breaking of FRP is normally the case for large value of FRP depth.
3. When FRP snaps, the beam section returns to the conventional steel-reinforced section.
4. Interaction diagram, which shows different failure modes, is produced.

REFERENCES

Almusallam T.H. and Alsayed S.H. (1995). Stress-strain relationship of normal, high-strength and lightweight concrete. *Magazine of concrete research* **47:170,** 39-44.

Wang T. and Hsu T.T.C. (2001). Nonlinear finite element analysis of concrete structures using new constitutive models. *Computers and Structures* **79**, 2781-2791.

Leung H.Y. and Burgoyne C.J. (2001). Analysis of FRP-reinforced concrete beam with aramid spirals as compression confinement. *Proceedings of the International Conference on Structural Engineering, Mechanics and Computation Cape Town, South Africa, edited by A. Zingoni,* Elsevier. **1,** 335-342.

APPENDIX

1. Model parameters for concrete in compression

$$n = -\frac{\ln 2}{\ln\left(\frac{\sigma_1}{\sigma_0} - \frac{K_p}{K - K_p}\right)}$$

$$\sigma_1 = \sigma_c'\left[\left(\frac{\varepsilon_1}{\varepsilon_0}\right) - \left(\frac{\varepsilon_1}{\varepsilon_0}\right)^2\right]$$

$$\varepsilon_1 = \frac{\sigma_0}{K - K_p}$$

$$\sigma_0 = 5.6 + 1.02\sigma_c' - K_p\varepsilon_0$$

for $\sigma_c' \leq 55$ MPa

$$K_p = 5470 - 375\sigma_c'$$

for $\sigma_c' > 55$ MPa

$$K_p = 16398.23 - 676.82\sigma_c'$$

$$K = 3320\sqrt{\sigma_c'} + 6900$$

$$\varepsilon_0 = \left(0.2\sigma_c' + 13.06\right)\times 10^{-4}$$

where the ultimate compressive strength of concrete (σ_c') is in MPa.

2. Model parameters for tensile steel rebars embedded in concrete

$$\varepsilon_n = \varepsilon_y(0.93 - 2B)$$

$$B = \frac{\left(\sigma_{cr}/\sigma_y\right)^{1.5}}{\rho}$$

$$\rho = \frac{A_s}{bd_s} \geq 0.5\%$$

$$\sigma_{cr} = 0.31\sqrt{\sigma_c'}$$

where σ_c' and the cracking strength (σ_{cr}) are in MPa.

Advances in Building Technology, Volume 1
M. Anson, J.M. Ko and E.S.S. Lam (Eds.)

APPLICATION OF GFRP REBAR IN CIVIL STRUCTURES

Z.J. Li[1] and Z.J. Zhang[2]

[1]Assoc. Prof., Dept. of Civil Engineering
[2]Grad. Student, Dept. of Civil Engineering
The Hong Kong University of Science & Technology
Clear Water Bay, Hong Kong

ABSTRACT:

The bonding properties between GFRP rebar and concrete are tested and analyzed to ensure feasible application of GFRP rebar in concrete structures. It is found that the bonding strength between GFRP bar and concrete is comparable with that between steel rebar and concrete. Also when there is special surface treatment, the bonding can be significantly improved. Then concrete beams with GFRP rebar embedded as reinforcement are tested and analyzed. It is found that the bonding is acceptable, and the bending capacity is larger using GFRP rebar, except that serviceability is difficult to satisfy.

KEYWORDS

GFRP, bond property, concrete structure, beam, bending

INTRODUCTION

Composites have been introduced into civil engineering structures mainly in the last decade. The application of composites in structural concrete mainly has three forms, FRP bars (tendons), sheets and composite tubes. Compared with conventional steel reinforcement, composite or fiber reinforced plastic (FRP) offers some excellent advantages. They are non-corrosion, non-magnetic, lightweight , and they have excellent handling property, high specific strength, availability in long lengths avoiding the need for lapping, and good fatigue and creep characteristics.

Nanni (1993) analyzed behavior and a design method for concrete beams with different concrete strengths, different kinds of FRP and different reinforcement ratios. Nanni et al (1996) systematically introduced components, and studied the behavior and applications of anchorage systems for FRP tendons. Dolan and Burke (1996) studied the behavior of concrete flexural members prestressed with pretensioned FRP tendons.

GFRP rebar has the advantages of high strength/weight ratio, corrosion resistance, and is also economical comparing with CFRP. Furthermore, when used as a prestressing tendon, its low Young's Modulus brings the unique benefit of lower prestress loss. Its disadvantages include non-ductility, instability at high temperature, etc. However, by carefully selecting the application situation and methodology, GFRP materials can be very useful in civil engineering structures.

In this study, the push-out method is adopted for studying the bonding property between GFRP rebar and concrete, and the results are found to be satisfactory. Then, the four-point bending test is carried out to study the effect of GFRP rebar as conventional reinforcement. It is found that bonding between GFRP rebar and concrete is acceptable but the serviceability problem prevents the full utilization of GFRP's high ultimate tensile strength.

BOND PROPERTY STUDY

The bond between the composite rod and concrete was studied using the push-out test method, developed by Li et al (1998). The material used is GFRP rod with a glass fiber volume ratio of 67% and Young's Modulus of 51 GPa, and one the diameter is 8 mm. As shown in Table 1, six different formulas are used, with 2 kinds of GFRP rod: one has a smooth surface and the other is sand blasted on the surface. On addition, specimens of formula 1 with smooth surface steel bar were tested as control specimens.

TABLE 1
FORMULAS USED FOR SPECIMEN PREPARATION

		Water	cement	sand	PVA	CEP	Metakaolin	Silicon fume	Slag
1	Plain mortar matrix (cement+sand, w/c=35%, c/s=1:1	0.35	1	1					
2	PVA powder modified matrix (1%) (cement+sand+PVA, w/c=35%, c/s=1:1	0.35	1	1	0.01				
3	Cellulose Ether Powder modified (0.2% of methocel added)	0.35	1	1		0.002			
4	cement+sand+silica fume (replace 5% of cement with s.f.)	0.35	0.95	1				0.05	
5	cement+sand+metakaolin (replace 5% of cement with metakaolin)	0.35	1	1			0.05		
6	cement+sand+slag (replace 50% of cement with slag)	0.35	0.5	1					0.5

A typical load push-out curve as a function of displacement is shown in Figure 1. The load curve shows a linear-elastic stage, a nonlinear ascending stage, then a sudden drop followed by a stable constant force stage. This indicates that the push-out process includes interfacial crack propagation phenomenon.

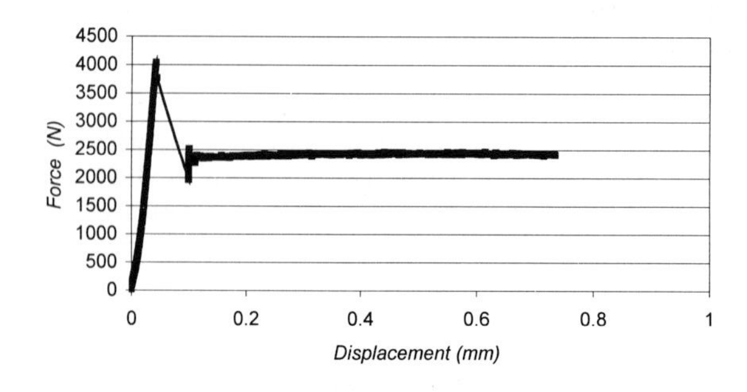

Figure 1: A typical push-out load curve as a function of displacement

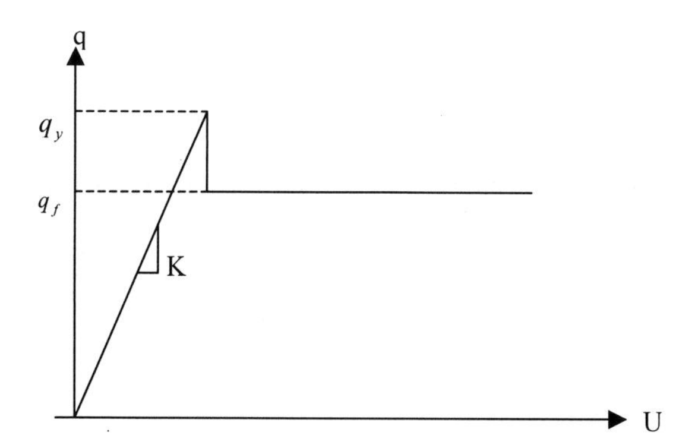

Figure 2: Constitutive model for interfacial shear force

Figure 2 shows the simplified constitutive model for the interfacial bonding stress between GFRP rod and cement mortar.

Calculation of the interfacial bonding stress was carried out using equations 1 to 5

$$
\begin{cases}
U^* = U(L) = \dfrac{P^* - q_f a}{E_s A \omega} \coth[\omega(L-a)] + \dfrac{P^* - 0.5 q_f a}{E_s A} a & (1) \\
q_f = \dfrac{\omega P^*_{\max}}{a\omega + \sinh[\omega(L-a)]\cosh[\omega(L-a)]} & (2) \\
q_y = q_f \cosh^2[\omega(L-a)] & (3) \\
\tau_y = \dfrac{q_y}{2\pi R} & (4) \\
\tau_f = \dfrac{q_f}{2\pi R} & (5)
\end{cases}
$$

where,
U: rebar displacement;
P: push force at x = L;
q: shear force per unit length acting on the rebar;
q_f: frictional shear force per unit length;
q_y: interfacial yield parameter;
a: debonding length;
L: rebar embedment length;
A: rebar cross-sectional area;
R: rebar radius;

$$\omega = \sqrt{\frac{k}{E_s A}}$$

Since the only unknown is the debonding length, a, its solution will result in the length of the debonded zone at the peak load, $P^*_{\max}$. After obtaining the value of the debonding length, a, the frictional shear bond strength q_f and the value of q_y were determined. Then bonding strength measures τ_y and τ_f were calculated, as shown in Table 2 and Table 3.

TABLE 2
RESULTS FOR THE SMOOTH GFRP ROD AND SMOOTH STEEL ROD

Mix Formula	*W*	τ_y *(MPa)*	τ_f *(MPa)*
Formula 1	14.44	6.63	6.1
Formula 2	26.85	5.32	4.07
Formula 3	26.72	7.00	5.29
Formula 4	33.91	8.47	5.48
Formula 5	27.60	8.06	6.04
Formula 6	29.55	8.30	5.98
Formula 1 with Steel rod	7.98	7.33	6.85

TABLE 3
RESULTS FOR THE SAND BLASTED GFRP ROD

Mix Formula	*w*	τ_y*(MPa)*	τ_f*(MPa)*
Formula 1	49.89	17.50	13.64
Formula 2	44.78	10.63	7.14
Formula 3	20.75	10.02	8.84
Formula 4	30.35	19.69	16.26
Formula 5	27.51	16.25	14.07
Formula 6	42.38	18.41	15.34

As shown in Table 2, the silica fume modified matrix had the maximum bond strength, which was comparable to that of the control specimen, with a smooth-surface steel rod in plain cement mortar.

It can be seen in Table 3 that GFRP rod with the sand blasting surface treatment has improved bonding with cement mortar, increasing from 67% to 196%. It can be concluded that GFRP rod with proper surface treatment can significantly improve the bonding property and hence can be utilized as normal reinforcement in concrete structures.

GFRP as normal reinforcement

Beam specimens with dimensions of 100 x 150 x 2000 mm were prepared for the 4-point bending test with an effective cross section depth d = 120 mm. Two Φ8mm sand-blasted GFRP rebars were embedded in the concrete beam as reinforcement, with a reinforcement area A_{frp} of 100.5 mm^2.

The four-point bending test arrangement at Material Testing System (MTS) is shown in Figure 3. Strain gauges are attached to the surface of the concrete beam, and two Linear Variable Differential Transducers (LVDTs) are used to measure the deflection at the bottom of the beams in the pure bending region.

Figure 4 shows the schematic layout of the test specimen, with S = 600 mm, and L = 1800 mm.

Figure 3: Set up of 4-point bending test

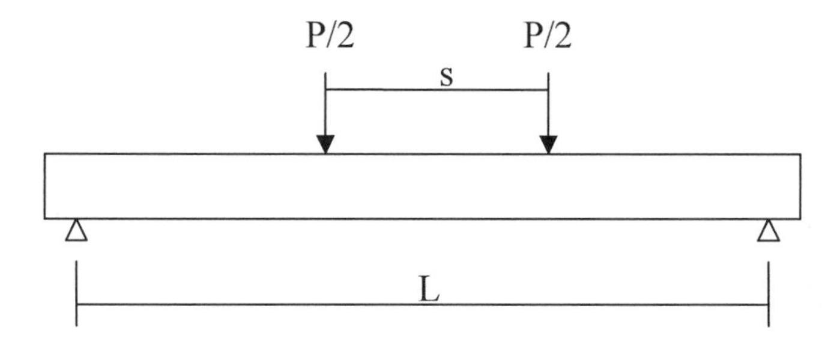

Figure 4: Schematic layout of test specimen

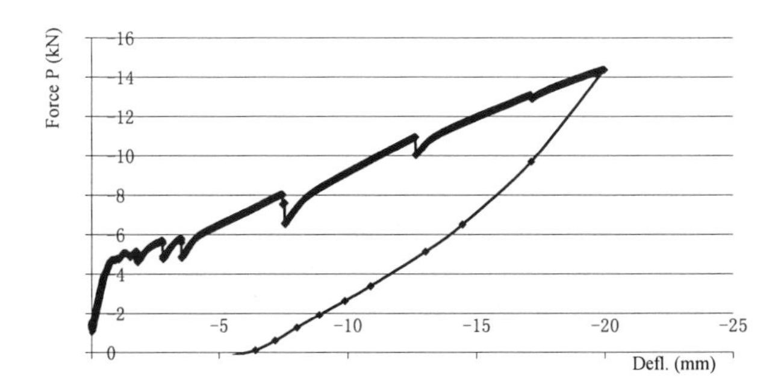

Figure 5: Force vs deflection curve for beam specimen 1

The force vs deflection relationship of beam specimen 1 is shown in Figure 5. It is loaded in deflection control mode until its deflection reaches 20 mm, and then unloaded. The maximum deflection to load span ratio is 1/82, which is far beyond the serviceability requirement. It is observed that several main cracks occurred in the pure bending region, with their depths extending to over ¾ of the beam depth. This implies that the bonding between GFRP rebar and concrete is acceptable, otherwise the beam would have failed. The turning point where the slope changes rapidly indicates the occurrence of a crack at the bottom surface of the beam where tensile stress exceeds the ultimate tensile strength of the concrete. This turning point has a value of force P as about 5 kN.

The several sudden drops in the force vs deflection curve may be caused by some partial failure in the bonding between GFRP rebar and concrete. Since the sand blasted on the GFRP rebar surface is not quite uniform, at some weak position the sand can separate from the rebar causing partial debonding, which is also observed in the push-out test.

The strain vs Time curve in Figure 6 shows the same characteristic as that of the force vs deflection curve.

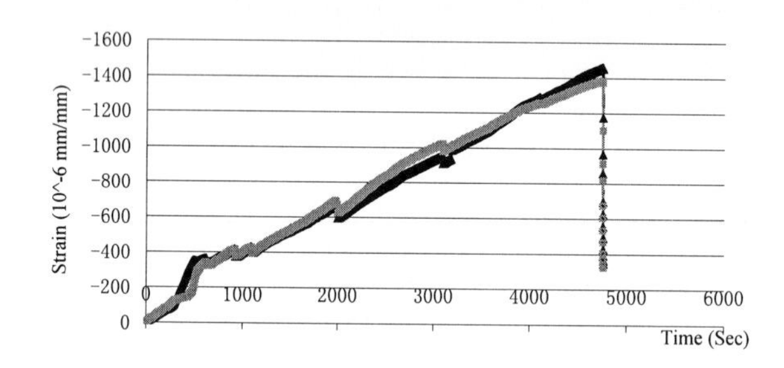

Figure 6: Strain on bottom of beam vs time for beam specimen 1

Another two specimens were tested using deflection control mode until failure occurred. Figure 7 shows the force vs deflection curve for beam specimen 2, which is quite similar to that of beam specimen 1. The turning point is at a force of about 5 kN, as is the case for beam specimen 1. The maximum deflection to load span ratio is 1/61, which is even larger than beam specimen 1. The failure mode is in shear, where the concrete cover burst out due to partial debonding between GFRP rebar and concrete, as can be seen in Figure 8.

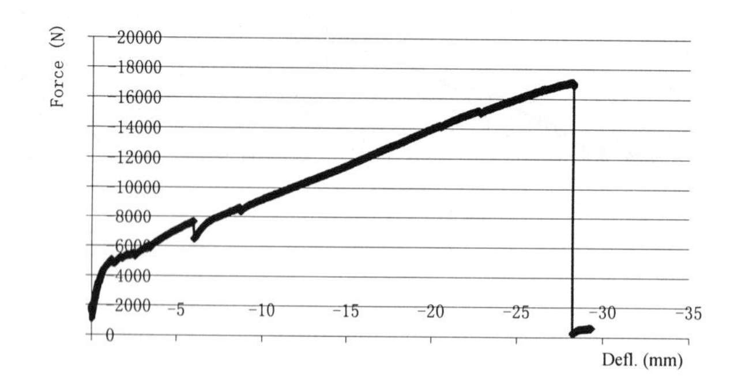

Figure 7: Force vs deflection curve of beam specimen 2

Beam specimen 3 has a behavior similar to that of beam specimens 1 and 2. The maximum deflection to load span ratio is 1/49, which is the largest among the three. Shear failure occurs and the ultimate load P_u is 19.9 kN.

All three specimens show a similar crack propagation manner. The first crack appears in the pure bending region, and then in this region several cracks subsequently occur. The bending cracks continue to extend upward, as long as the beam is still capable of carrying loading. The occurrence and propagation of shear crack leads to shear failure, and part of the concrete below the shear crack at the bottom side of the beam debonds from the GFRP bar and bursts out.

Bending Capacity Calculation

If we assume bonding between GFRP and concrete is satisfactory, and fcu is 40 MPa, and f_{frp} is 1100 MPa, hence Mu is calculated as 11kN•m, and the corresponding Pu is 36.7 kN

Shear Capacity Calculation

Similarly, we can get the shear capacity Vu = 10.7 kN, corresponding Pu = 21.4 kN

With the assumptions, shear capacity is shown to be less than bending capacity, so is aspect controls and the failure mode is a shear failure, and this agrees with the test result. The prediction of shear failure load Pu = 21 kN is close to the test results, and the difference in values may be caused by non-uniformity of specimens, inaccurate assumptions, inaccuracy in calculation method, etc.

Figure 8: Shear failure of beam specimen 2 (shear crack)

CONCLUSION

The bonding between GFRP rebar and concrete is comparable to that between steel bar and concrete. And the bond property can be significantly improved using proper surface treatment of the GFRP rebar. This, in a sense, ensures the possible application of GFRP rebar as reinforcement for concrete structures.

Due to the low elastic modulus of GFRP rebar, concrete beams with GFRP as reinforcement usually have lower flexural rigidity, so the serviceability requirement usually controls the design. Some possible ways to overcome this limitation on the usefulness of GFRP, are to carefully control the application situation to avoid the serviceability problem. Combination of GFRP reinforcement and steel reinforcement, adopting prestressing system to solve the serviceability problem are two possibilities.

References:

Chaallal, O. and Benmokrane, B. (1996). Fiber-Reinforced Plastic Rebars for Concrete Applications. *Composites*: Part B, 27B, 245-252.

Dolan C.W. and Burke C.R. (1996). Flexural Strength and Design of FRP Prestressed Beams. *Advanced Composite Materials in Bridges & Structures – 2 nd International Symposium*, Canadian Society of Civil Engineers, Montreal, Quebec, Canada, 383-390.

Larralde, J. and Silva-Rodriguez, R. (1993). Bond and Slip of FRP Rebars in Concrete. *Journal of Materials in Civil Engineering*, **5:1**, 30-40.

Li, Z.J., Xu, M.G. and Chui, N.C. (1998). Enhancement of rebar (smooth surface)-concrete bond properties by matrix modification and surface coatings. *Magazine of Concrete Research*, **50:1,** 49-57.

Maheri, M.R. (1995). An improved method for testing unidirectional FRP composites in Tension. *Composite Structures*, v33, 27-34.

Mirmiran, A. (2001) Innovative combinations of FRP and traditional materials. *FRP Composites in Civil Engineering, CICE 2001*, v2, 1289-1298.

Mulu Muruts and Ludovit Nade, Laminated Galss Fiber Reinforced Plastic (GFRP) Bars in Concrete Structures. [online].
Available: http://www.vbt.bme.hu/phdsymp/2ndphd/proceedings/muruts.pdf

Nanni A. (1993). *Fiber-reinforced-plastic (FRP) reinforcement for concrete structures : properties and applications,* Elsevier , Amsterdam ; New York, USA

Nanni A. Bakis C.E., O'Neil E., and Dixon T.O. (1996). Performance of FRP Tendon-Anchor Systems for Prestressed Concrete Structures. *PCI Journal*, **41:1**, 34-44

Uomoto, T. (2001). Durability of FRP reinforcement as concrete reinforcement. *FRP Composites in Civil Engineering, CICE 2001*, v1, 85-96.

Advances in Building Technology, Volume I
M. Anson, J.M. Ko and E.S.S. Lam (Eds.)

A GENERIC STRUCTURAL ENGINEERING APPROACH TO RETROFITTING RC BEAMS IN BUILDINGS BY ADHESIVE BONDING LONGITUDINAL PLATES

Deric J. Oehlers and Rudi Seracino
Department of Civil and Environmental Engineering, The University of Adelaide
Adelaide, South Australia, SA5005, Australia

ABSTRACT

Retrofitting reinforced concrete structures by adhesive bonding steel or FRP plates to their surfaces is an efficient and inexpensive form of rehabilitation that is rapidly gaining in importance. The main problem with adhesive bonding plates to reinforced concrete is the very weak tensile strength of the concrete that allows premature debonding of the plate that can occur even before the serviceability loads are reached. There has been published much research on premature debonding, particularly for tension face plates, from which have been developed: design rules of thumb; analyses based on stress concentrations; and those based on fracture mechanics. Much of this research may appear to be confusing for structural engineers as they are often not based on a structural engineer's understanding of the familiar behaviour of reinforced concrete. In this paper, a new, simple and generic structural engineering approach is described for designing plated structures that are based on the fundamental principles and mechanisms of behaviour familiar to structural engineers who design reinforced concrete structures. It will be shown that this new approach can be applied to longitudinal plates: of any material type such as FRP or steel; of any size; of any cross-section such as flat plates, angle sections and U-sections; and that are adhesively bonded to any position such as the tension face, compression face and sides of the RC beams or slabs.

KEYWORDS

Retrofitting, rehabilitation, plating, reinforced concrete, beams, slabs, FRP plating, steel plating.

INTRODUCTION

Many non-structural engineering researchers appear to make the problem of retrofitting reinforced concrete buildings by adhesive bonding longitudinal plates sound extremely complex and outside the scope of understanding of structural engineers. This is wrong. It is shown in this paper that the fundamental understanding that structural engineers have of the behaviour of reinforced concrete structures is essential to the understanding of the plated RC structure, and allows a simple generic design approach to be developed and used. In this paper, the retrofitting of an internal continuous beam in an RC building is used as an illustration. At first, the variety of materials that can be used is discussed, this is followed by the structural engineering reasons for choosing the positions of the longitudinal plates, which is then followed by a description of the structural engineering debonding mechanisms and where they occur. The paper is concluded with an application of a design approach.

MATERIAL AND SECTIONAL DUCTILITY

The material behaviour of the plate affects the structural behaviour of the plated section. Structural engineers are familiar with the behaviour of steel as shown in Fig.1 which in its elastic range is relatively stiff (E_s) after which there is a large ductile plateau before the onset of strain hardening at very high strains. Hence, if a steel plate is designed to yield prior to debonding then the plated section will behave in a ductile fashion and moment redistribution can occur. In contrast, fibre reinforced polymer (FRP) is a linear elastic material that is relatively flexible, then fractures without any material ductility. Hence in an FRP plated beam, the force in the FRP plate keeps increasing until crushing of the concrete or debonding, so that the curvature at failure is relatively low allowing little if any moment redistribution.

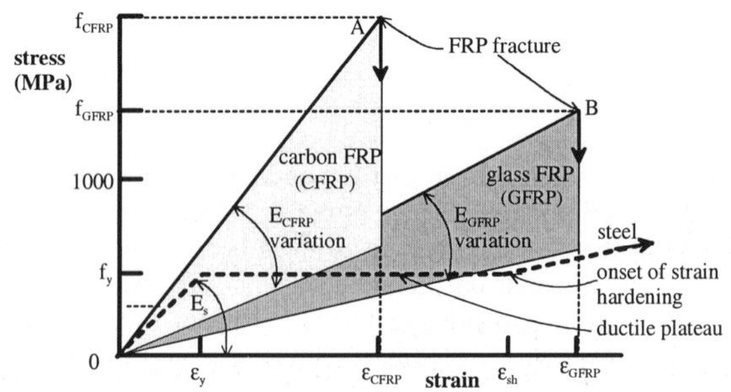

Figure 1: Plate material properties

LONGITUDINAL PLATE POSITIONS

Probably the most common form of longitudinal plating is to adhesively bond plates to the tension faces of continuous beams as shown in Fig.2, so that the internal lever arm of the axial plate force is maximised. However as structural engineers well know, the addition of tension reinforcement reduces the sectional ductility which imposes a limit to the amount of strengthening that can occur. Furthermore, tension face plates are prone to debonding due to the interaction of the plate stress fields with those of the tension reinforcing bars, such that debonding occurs at the level of the tension reinforcement.

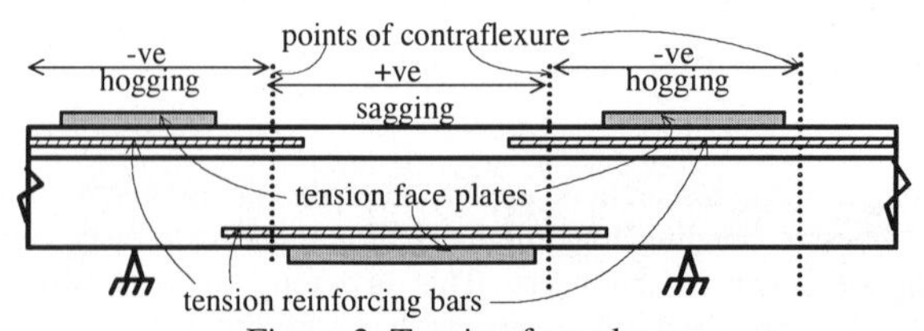

Figure 2: Tension face plates

An alternative approach is to adhesively bond plates to the sides as in Fig.3. These plates are less prone to debonding as they are less likely to interact with the tension reinforcing bar stress fields, so that the plane of debonding occurs in the concrete adjacent to the adhesive layer and not at the level of the tension reinforcing bars. Furthermore, side plates have less of an effect on the sectional ductility particularly when there are deep side plates as in A in Fig.3, in which case the sectional ductility may increase. It may also be worth noting that sides plates can also increase the shear capacity.

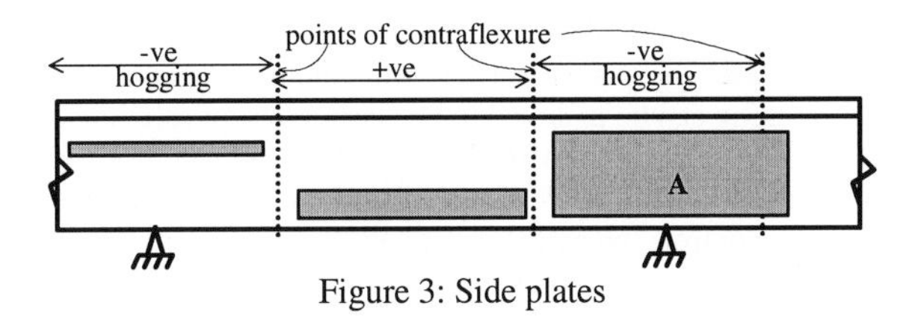

Figure 3: Side plates

PREMATURE DEBONDING MECHANISMS

An example of a retrofitted beam is shown in Fig.4. This beam reached its moment capacity at large rotations even though some debonding cracks were visible. The aim of the structural engineer is to ensure that premature debonding of the plate does not occur to such an extent as to weaken the structure or reduce its ductility. The following three major debonding mechanisms have been identified.

Figure 4: Retrofitted beam

Plate End (PE) Debonding

An example of plate end (PE) debonding is shown in Fig.5. The stress concentrations due to the termination of the plate induce cracks that start at the plate end and propagate inwards, gradually at first and then rapidly until complete debonding occurs at which the strains in the plate reduce to zero so that the beam now acts as unplated.

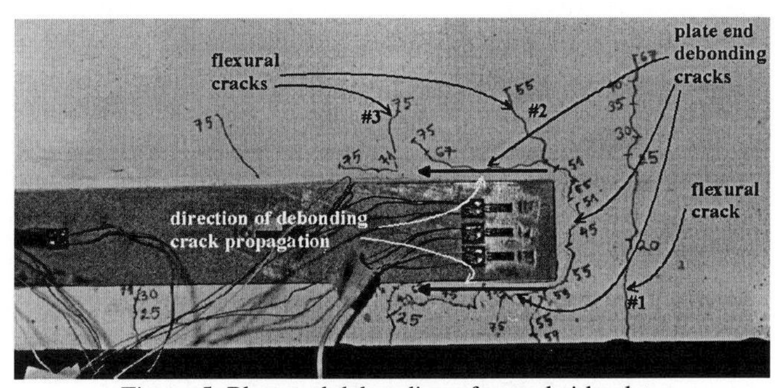

Figure 5: Plate end debonding of a steel side plate

The mechanism of PE debonding is illustrated in Fig.6 for a tension face plate, where it can be seen in Fig.6(a) that the curvature of the plated beam induces a moment M_c and axial force P_a in the plate. These stress resultants M_c and P_a, caused by the curvature of the beam induce the plate/concrete interface forces in Fig.6(a) and stresses in Fig.6(b) and (c) that result in debonding. Hence a simple solution to preventing PE debonding is to terminate the plate at a point of contraflexure where the curvature is zero. Sometimes it may not be possible to terminate the plate at a point of contraflexure due to envelopes induced by variations in load. In which case, the plates can be terminated in regions of low moment, where the section apart from tensile cracking can be assumed to be elastic; design rules are available for tension face plates (Oehlers and Moran 1990), side plates (Oehlers 2000), angle sections (Nguyen et al 1998) and compression face plates (Mohamed Ali et al 2002) to ensure that premature debonding does not occur. It can be seen that PE debonding controls the extent of plating, that is the length of plate required.

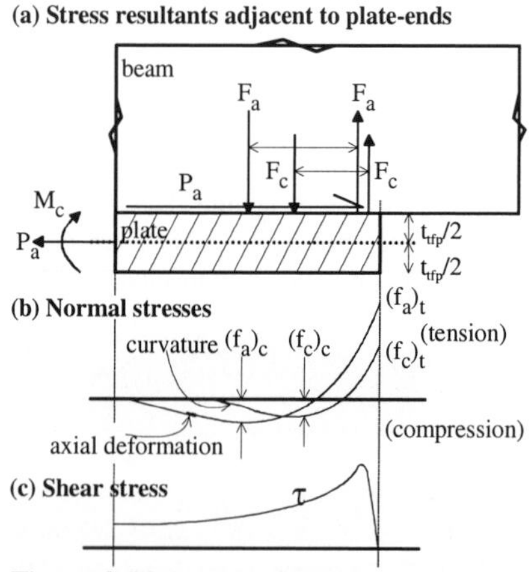

Figure 6: Plate end debonding mechanism

Intermediate Crack (IC) Debonding

Intermediate crack (IC) debonding of a tension face plate is shown in Fig.7. The interface IC debonding cracks propagate from the intercept of the intermediate cracks, induced by combinations of applied flexural or shear forces, with the longitudinal plate. The IC debonding cracks propagate towards the plate ends gradually joining up until a large length of plate between the plate ends has separated from the beam. Hence, to all intents and purposes the plate has debonded even though the plate ends may still be attached, as the plate contribution is now negligible.

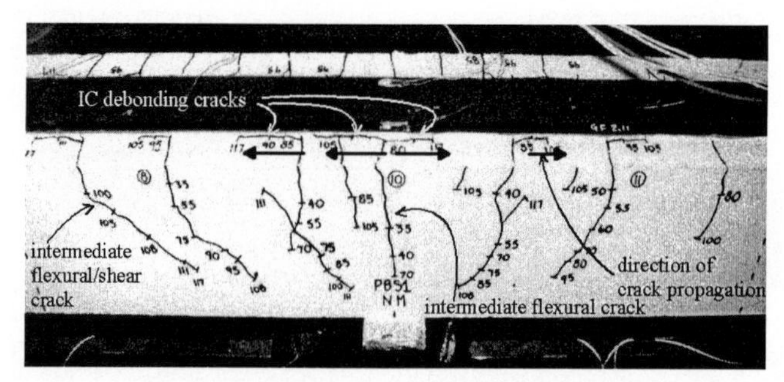

Figure 7: IC debonding of tension face plate

The mechanism of IC debonding is illustrated in Fig.8. To accommodate the formation of the flexural intermediate crack adjacent to the plate requires infinite strains in the plate which of course cannot occur, so that the strains are relieved by the formation of the interface IC debonding cracks. Much research has been done on IC debonding (Teng et al 2002). For example it has been shown the stress at which debonding occurs is proportional to $\sqrt{\left(E_p\sqrt{f_c}\right)/t_p}$ where E_p and t_p are the plate stiffness and thickness respectively; so that thin stiff plates are less likely to debond and IC debonding will usually start in regions of maximum flexure. Therefore, the thickness of a steel plate can be adjusted so that yield will occur before IC debonding, so that the cross section fails in a ductile mode that allows moment redistribution. In contrast, FRP plates will debond whilst still elastic, often at a stress of about 30% of the ultimate FRP strength, so that failure tends to be brittle and little moment redistribution can occur. It can be seen that IC debonding controls the flexural strength and ductility.

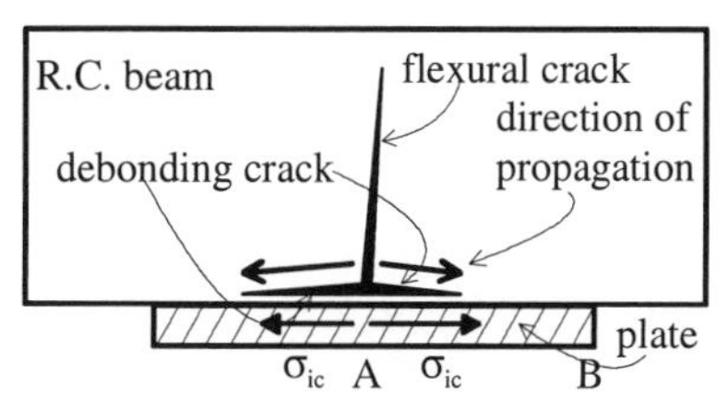

Figure 8: Mechanism of intermediate crack debonding

Critical Diagonal Crack (CDC) Debonding

Critical diagonal crack (CDC) debonding is due to the vertical shear displacement across a critical diagonal crack as shown in Fig.9(a) for a beam without stirrups and in Fig.9(b) for a beam with stirrups. The debonding crack propagates catastrophically from the intercept of the diagonal crack with the plate to the plate end. Tests have clearly shown that the longitudinal plates debond before the stirrups can be stretched so that debonding occurs at the strength of the concrete beam without stirrups. However these tests have also shown that the longitudinal plates inhibit the formation of the critical diagonal crack. Hence a lower bound to the shear load at which debonding occurs V_{cdc} is the shear strength of the beam without stirrups V_{uc} (Zhang 1997) and the contribution from the stirrups V_{us} is negligible or zero.

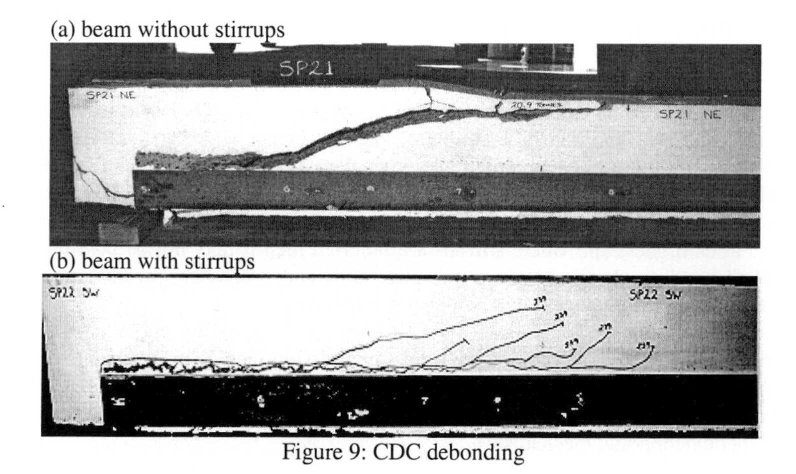

Figure 9: CDC debonding

If a shear capacity greater than V_{uc} is required for the longitudinally plated region, then methods of analysis are available for determining the vertical shear load to cause a longitudinal plate to debond V_{cdc}, which depends on whether the plate is attached to the tension, side or compression face (Jansze 1997; Mohamed Ali et al 2001, 2000, 2002). The method of analysis is illustrated in Fig.10 for a side plated beam. For each possible position of a diagonal crack, it is necessary to determine the shear load to cause the diagonal crack to develop V_{cr} and the shear resistance against sliding across this diagonal crack V_u. The intercept as in Fig.10 defines both the position of the critical diagonal crack and the vertical shear to cause debonding V_{cdc}. It is worth noting in Fig.10, that for each position of the diagonal crack, the resistance to sliding depends on the IC debonding resistance across the diagonal crack P_{ic} in the top half of the figure. It can be seen that CDC debonding relies on the IC debonding strength but IC debonding is not the same as CDC debonding as IC debonding depends on axial deformations in the plate whilst CDC debonding depend on shear deformations in the RC beam.

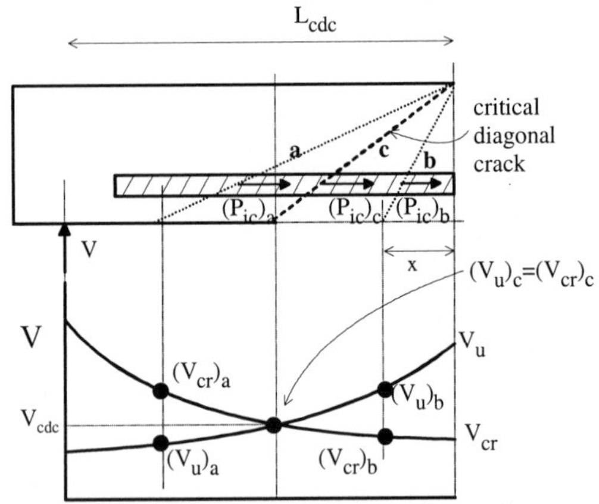

Figure 10: Analysis of CDC debonding resistance

OCCURRENCE OF DEBONDING MECHANISMS

The distribution of the major debonding mechanisms are illustrated in Fig.11(a) for tension face plates. It can be seen that IC debonding initiates in the regions of maximum moment, where the strains in the plate are at their largest, and propagate towards the plate ends. IC debonding controls the flexural capacity and flexural ductility and hence the ability of the beam to redistribute moment. CDC debonding depends on the vertical shear capacity, ignoring the contribution due to stirrups, and hence is often more likely to occur in the negative or hogging regions where the vertical shear is at its highest and where often stirrups are required. Hence in slabs where stirrups are rarely used, CDC debonding is unlikely to affect design as most slabs are designed for V_{uc} only. In contrast in the negative regions of continuous beams where the vertical shear is at its highest and where stirrups are required then it may not be possible to adhesively bond plates. PE debonding always occurs at the plate ends and propagate inwards and can be simply prevented by terminating the plate near to a point of contraflexure.

Sometimes the plates are placed or terminated in the compression face as in Fig.11(b). In which case in the compression face zones, IC debonding cannot occur but CDC and PE debonding can occur as shown. The same debonding mechanisms also apply to side plates beams as in Fig.11(c). Hence it can be seen that these are generic debonding mechanisms.

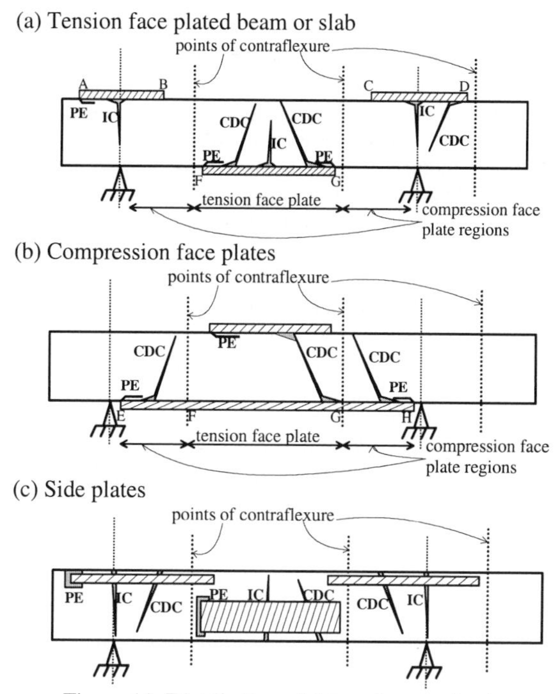

Figure 11: Distribution of debonding mechanisms

APPLICATION OF DESIGN APPROACH

The continuous member in Fig.12 is used to illustrated the generic plating approach.

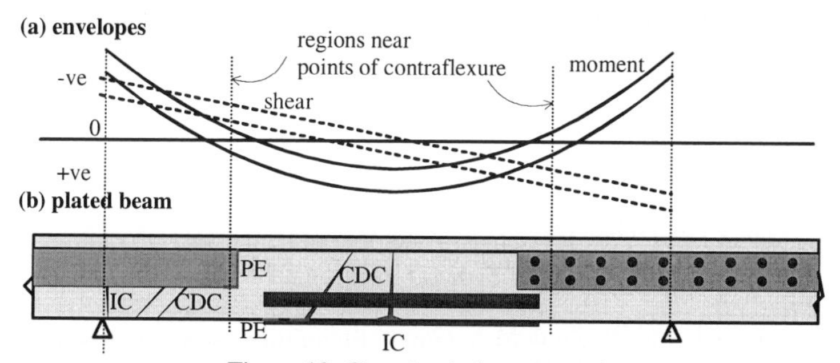

Figure 12: Generic plating approach

Let us first assume that the continuous member in Fig.12 is a slab and hence does not have stirrups, so that the vertical shear capacity is V_{uc} which is the shear resistance of the critical diagonal crack of the unplated section. The addition of plates will inhibit the formation of the critical diagonal crack which will increase the vertical shear capacity to V_{cdc}, the shear load to cause CDC debonding. V_{cdc} is now both the shear load to cause debonding as well as the shear capacity of the plated slab. Hence CDC debonding is unlikely to occur anywhere along the slab, just as long as the new design shear load due to retrofitting does not exceed V_{cdc} which is the likely scenario. PE debonding can be prevented by terminating the plates in regions of low moment so this should also not be a problem. Hence this leaves IC debonding which can occur at the positions of maximum moment in both the -ve and +ve regions. FRP plates are usually the preferred option due to their ease of application and handling and durability. These can be used in both the –ve and +ve regions if little or no moment redistribution is required. If, however, the -ve regions are required to redistribute moment to the positive region then steel plates may be the preferred option in the –ve regions as the steel plates can be designed to yield before IC debonding. Hence there is a possibility that the final design will use both steel and FRP plates.

Let us now assume that the continuous member in Fig.12 is a beam and hence does have stirrups so that the vertical shear capacity is now $V_{uc} + V_{us}$ which is the shear resistance of the unplated section. As with the slab, plate end debonding can be prevented by terminating the plate at the point of contraflexure. In the +ve region the vertical shears here are relatively small, compared with those in the –ve region, so that V_{us} may be relatively small. Just as long as $V_{cdc} < V_{uc} + V_{us}$ of the +ve region, then CDC debonding may not occur particularly as tests have shown that plating can increase V_{cdc} to twice V_{uc}. However, in the –ve region it may not be possible to prevent CDC debonding due to the high value of V_{us} in which case it may be necessary to bolt the plates as shown. As with the slabs, IC debonding and its effect on ductility may control where FRP and steel plates are used.

REFERENCES

Jansze W. (1997). *Strengthening of RC members in bending by externally bonded steel plates.* PhD Thesis, Delft University of Technology.

Mohamed Ali M.S. Oehlers, D.J. and Bradford, M.A. (2000). Shear peeling of steel plates adhesively bonded to the sides of reinforced concrete beam. *Proc. Instn Civ. Engrs Structs & Bldgs*, **140,** 249-259.

Mohamed Ali M.S. Oehlers D.J. and Bradford M.A. (2001). Shear peeling of steel plates bonded to the tension faces of RC beams. *ASCE Journal of Structural Engineering.* **Vol.127 No.12,** 1453-1460.

Mohamed Ali M.S., Oehlers D.J. and Bradford M.A. (2002). Interaction between flexure and shear on the debonding of RC beams retrofitted with compression face plates. Accepted by *Journal of Advances in Structural Engineering.*

Nguyen N.T., Oehlers, D.J., and Bradford, M.A. (1998). Models for the flexural peeling of angle plates glued to R.C. beams. *Journal of Advances in Structural Engineering.* **Vol. 1 No.4,** 285-298.

Oehlers D. J. and Moran J. P. (1990). Premature failure of externally plated reinforced concrete beams. *Journal of the Structural Division of the ASCE,* **Vol. 116, No. 4**, 978-995.

Oehlers D.J., Nguyen, N.T. and Bradford, M.A. (2000). Retrofitting by adhesive bonding steel plates to the sides of R.C. beams. Part 1: Debonding of plates due to flexure. *Journal of Structural Engineering and Mechanics, An International Journal,* **Vol.9, No.5**, 491-504.

Teng J.G. Chen J.F. Smith S.T. and Lam, L. (2002). *FRP strengthened RC structures.* Wiley, Oxford, UK.

Zhang J.P. (1997). Diagonal Cracking And Shear Strength Of Reinforced Concrete Beams. *Magazine of Concrete Research.* **Vol.49, No.178**, 55-65.

Advances in Building Technology, Volume 1
M. Anson, J.M. Ko and E.S.S. Lam (Eds.)

EFFICIENCY OF STRENGTHENED R. C. COLUMNS

Prof. Dr. M. Akram Tahir

University of Engineering and Technology, Lahore-Pakistan.

ABSTRACT

An experimental study was conducted, in which a large number of full-scale models of columns were cast and later strengthened by preparing different types of surface conditions between the existing and added concrete. The core was encased in the jacket cast from fresh concrete by placing properly tied additional reinforcement in the jacket. Five types of surface condition were studied during the strengthening program. The other parameters of the study were the strengths of concrete and steel both in the core and jacket. It was interesting to note that surface conditions of existing concrete had little effect on the strength of strengthened columns in case load was applied concentrically. However, in case of eccentrically loaded columns, the surface conditions resulting in poor bond showed drastic decrease in strength as compared to those providing better bond between old and new concrete. Another interesting fact was noted that for the efficient strengthening of the columns the strength of the two materials in the jacket should be at least equal to or greater than those of the respective core materials.

KEYWORDS:

Strengthening, eccentrically loaded columns, full-scale models, concrete jacket, surface conditions.

INTRODUCTION

During renovation of buildings often it becomes necessary to strengthen the existing columns in order to increase their load carrying capacity. The strengthening becomes inevitable due to increase in the service loads or loss of member strength. The strengthening is provided by adding an additional section, termed as jacket, around the existing column section, named as core. Additional longitudinal bars that are properly tied are provided in the jacket. The bond between the surfaces of core and jacket-concretes has sharp influence on the integrity of the strengthened column. The core is subjected to the stresses prior the strengthening is done where as the jacket has to take the share of load by transition. The strength of the materials in core and jacket is usually different. These factors make the problem of strengthening a bit more complex. In this study a number of columns were cast and strengthened. Two concretes with nominal strengths of 21 MPa and 28 MPa were used during casting and strengthening operations with two grades of steel i.e., 280 MPa and 420 MPa. The details of the test program are presented in the following section.

EXPERIMENTAL PROGRAM

A number of full-scale models were cast, strengthened, and tested to destruction. The main variables of the study were the strength of concrete and steel in the core and jacket, the eccentricity of loading, and surface condition of the core concrete. The test program was completed in three series. First of all a pilot study was conducted on 75-mm square sections strengthened to 150-mm square sections. The concrete was made using coarse river sand as fine aggregate; 12-mm maximum size crushed lime stone aggregate, and ordinary Portland cement. The concrete mixtures were designed for two grades with nominal strengths of 21 MPa and 28 MPa. Two grades of deformed mild steel with nominal yield strength of 280 MPa and 420 MPa were used as reinforcement. The columns were cast using molds in vertical direction. The columns were cured for twenty-eight days and then left for weathering for quite a some time. The test program was completed in three different series. The details are presented.

The Pilot Program

The concrete mixtures were designed for nominal strengths of 21 and 28 MPa at 28 day. Five surface conditions of the old concrete, as listed in the following, were prepared before strengthening during the pilot test program:

S1. The same old dried surface without cleaning was used for strengthening.
S2. The cover was removed of the old steel reinforcement before strengthening.
S3. The old surface was cleaned thoroughly before strengthening the section.
S4. The surface was painted and aged before strengthening.
S5. The surface was prepared wet by using cement water.

The columns were cast, cured, and then left for aging in weather. These columns were later strengthened with the five surface conditions of old concrete. The strengthened columns were cured and then tested to destruction by applying both concentric and eccentric loads. The characteristics of the columns cast and tested during the pilot study are reported in Table 1.

Table 1
THE COLUMNS CAST IN PILOT STUDY

S,No	Old Section (Core)				New Section (Jacket)			
	b, h mm	f'_c MPa	f_y MPa	A_s	*b, h* mm	f'_c MPa	f_y MPa	A_s
1	75	21	280	4#4	150	21	280	6#4
2	75	21	280	4#4	150	21	420	6#4
3	75	21	280	4#4	150	28	420	6#4
4	75	21	420	4#4	150	28	280	6#4
5	75	28	280	4#4	150	21	420	6#4
6	75	28	420	4#4	150	21	280	6#4

Note: The length of the columns was 2 m. Twenty-two columns were cast in this part of the program. The main reinforcement in core and jacket sections were tied properly using #3 bar-square ties spaced at 75 mm and 150 mm respectively.

Results of The Pilot Test Program

Table 2

EXPERIMENTAL STRENGTHS OF THE COLUMNS LISTED IN TABLE 1.

Sr. No	P, kN (e=0 mm)					P, kN (e= 50mm)				
	S1	S2	S3	S4	S5	S1	S2	S3	S4	S5
1	774	776	774	774	764	321	331	311	264	324
2				843					321	
3				950					367	
4	936			929		386	415		321	
5				914					364	
6				860					307	

The strengthened columns were cured and later tested to destruction in a 100-Ton Buckton universal testing machine. The machine may accommodate 4m long columns. It can be observed from the tabulated results that surface conditions of old concrete have little influence whatsoever on the load carrying capacity of the strengthened columns where the load was applied concentrically. The strength results of the columns listed at serial no. 1 confirm this fact when loaded concentrically. However, in case of eccentric loading, severe segregation occurred at the interface of the old and new concretes for the surface condition S4, where the old surface was painted and put for aging before strengthening. The most effective surface condition for strengthening was S2, where the old concrete cover was removed prior the strengthening was performed.

FULL SCALE MODELS

The tests on full-scale models were carried out in two series. In the first of the series, twenty-two 100x100 mm columns were cast and later strengthened to 200 x 200 mm. The height of the columns

TABLE 3

DETAILS OF THE COLUMNS CAST AND TESTED DURING 2ND SERIES.

S. No	Old Section		New Section		P, kN, e= 50mm		P, kN, e= 100mm	
	f'_c MPa	f_y MPa	f'_c MPa	f_y MPa	S1	S4	S1	S4
11	21	280	21	280	587	507		
12	21	280	21	420	653	570		
13	21	420	21	420	667	590		
14	21	420	21	280	604	534		
15	21	280	28	280	713		397	280
16	21	280	28	420	780		456	336
17	28	280	21	280	626		355	265
18	28	280	21	420	674		415	305
19	28	420	21	420	680			
20	28	420	21	280	607			

was fixed at 2500 mm. The materials used in this series were same as those employed in the pilot study. The combinations of two concrete mixtures with nominal strengths of 21 and 28 MPa were used in core and jacket. The details of the columns and experimental strengths are reported in Table 3

The comparison of axial strengths of the columns tested in series 2 is depicted in Fig.1. It is interesting to note that strength reduces appreciably with the reduction in strengths of jacket steel and concrete. The concrete strength is less effective than steel strength in enhancing the load carrying capacity of the strengthened columns

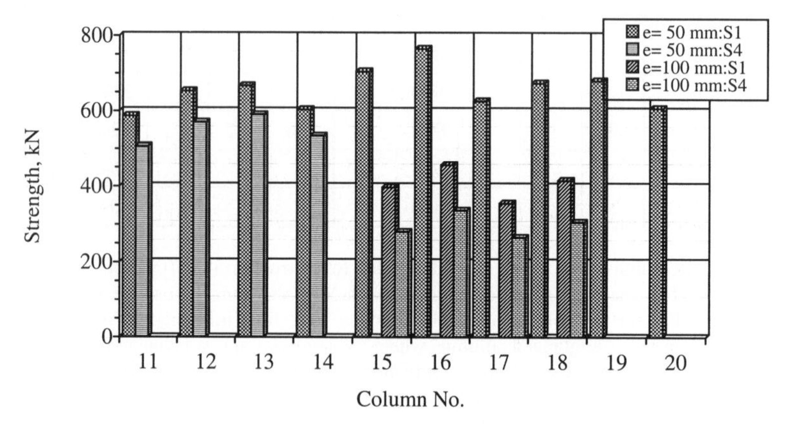

Figure 1. Ultimate strength of the columns tested in series 2.

Another set of ten columns was cast and strengthened in the third series. The length of these members was 3000 mm and core size was 150 mm square and jacket size was 300 mm square. The reinforcement in core was 4#6 bars and that in jacket was 6#4 bars. The concrete of the same two nominal strengths, 21 and 28 MPa was used in core as well as in jacket. The same reinforcement of the two nominal grades, 280 MPa and 420 MPa, was used. The details of the columns are given in Table 4 and the test results are reported in Figure 2.

TABLE 4

THE DETAILS OF THE COLUMNS CAST AND TESTED DURING SERIES 3.

Sr. No	Old Section 150x150 mm			Strengthened Section 300x300 mm		
	f'_c MPa	f_y, MPa	Reinforcement	f'_c MPa	f_y, MPa	Reinforcement
21	21	280	4#6	21	280	6#4
22	21	280		21	420	
23	21	420		21	420	
24	21	420		21	280	
25	21	280	4#6	28	280	6#4
26	21	280		28	420	
27	28	280	4#6	21	280	6#4
28	28	280		21	420	
29	28	420		21	420	
30	28	420		21	280	

The length of the columns was 3000 mm and they were strengthened with surface condition S5. These columns were properly tied to avoid bulging of longitudinal steel. The tabulated concrete and steel strengths are nominal. The loading eccentricity was kept constant at 150 mm.

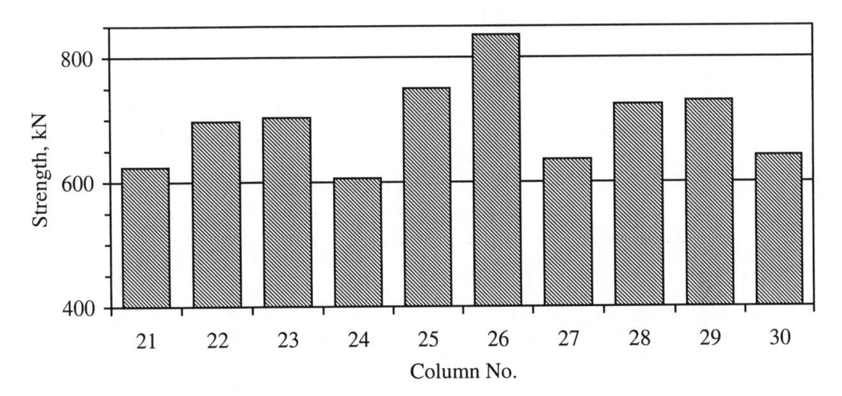

Figure 2. The axial strengths of the series 3 columns loaded with eccentricity=150 mm

DISCUSSION OF THE RESULTS

A very interesting picture emerges as the axial strengths of various columns are compared. The results of the two series are compared with respect to the change in strengths of steel and concrete both in core and jacket. The comparisons for series 2 tests are presented in Figure 3. It can be noted from Figure 3a and 3b that change in strength of steel or concrete in core does not affect the strength of strengthened column appreciably. For example, compare the strengths of the two columns, in each of the pairs (11,14), (12,13), (17, 20) and (18,19). The jackets of the two columns in each pair are cast from the same set of materials. The core of the two is also cast from the same concrete. The steel reinforcement, however, in the core of the two columns in each pair is different. It can be seen that increasing yield strength of core steel from 280 MPa to 420 MPa does not enhance the strength of the columns appreciably. The change is merely in the range of 1 to 3%. A similar picture is emerged from the comparison of the two columns in each of the pairs; (11,17), (12,18), (13,19) and (14,20). The two columns in each pair are cast from identical materials except that the concrete in the cores of the two is of different strength. It can be confirmed from Figure 3b, that increasing strength of core concrete from 21 MPa to 28 MPa does not affect the strength of column significantly. The change in strength is in the range of 1 to 6%, with an average of 3%. Similar trends can be observed from Figure 4, which confirm that core strength is not that important for eccentrically loaded strengthened columns.

However, an increase in either concrete- or steel-strength of jacket appreciably enhances the strength of strengthened column. These comparisons are based on the steel strength in jacket and are depicted in Figure 3c and those on concrete strength of jacket are shown in Figure 3d. It can be observed from Figure 3c, that strengths of the columns increase in the range of 7-12% when strength of reinforcement is changed from 280 MPa to 420 MPa. Similarly the increase in strength of jacket concrete from 21 MPa to 28 MPa enhances the strength of column by 20%. Similar observations can be made from Figure 4c and d. It may be noted that increase in the strength of jacket concrete is more effective than that of steel in increasing the load carrying capacity of the strengthened columns.

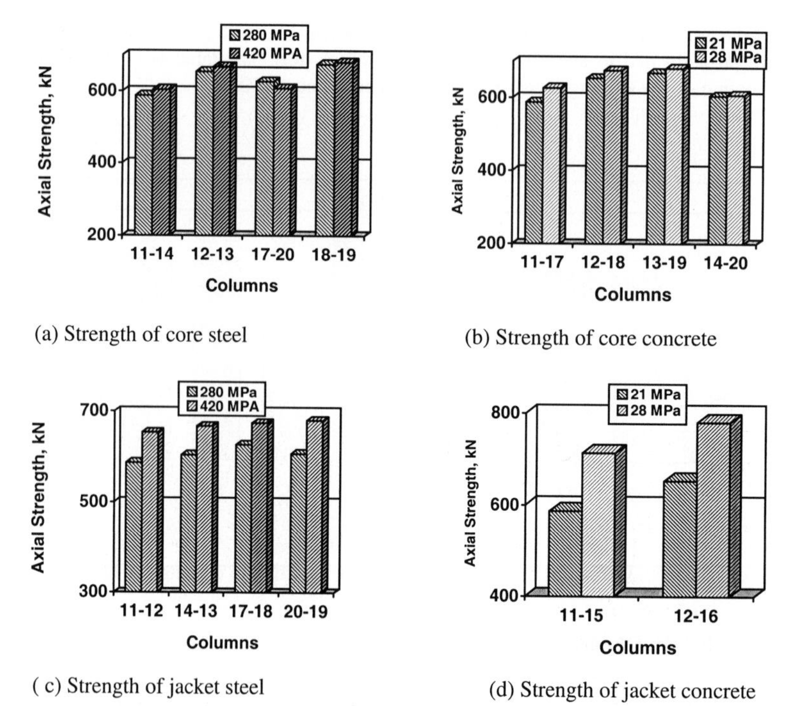

(a) Strength of core steel

(b) Strength of core concrete

(c) Strength of jacket steel

(d) Strength of jacket concrete

Figure 3. Comparison of strengths based on concrete and steel strengths in core and jacket. (Results from series 2.)

Another important observation made from the comparison of series 2 test results is that surface condition S4 causes a sever reduction in the strength of strengthened columns when loaded eccentrically. This reduction is more pronounced when eccentricity is increased from 50 mm to 100 mm. Compare the drop in strength of the columns with surface conditions S1 and S4, at serial numbers 11 to 14 and 15 to 18 as shown in Figure 1. The surface condition of old concrete, i.e.,, interface between core and jacket is an important parameter of strengthening. If a perfect bond is not established between the old and new concretes, a sever segregation is observed at testing time when load is applied eccentrically. This effect becomes more and more pronounced with increase in loading

CONCLUSIONS

Two important conclusions can be drawn from the results observed in this study. Firstly the ultimate strength of concrete and yield strength of steel in jacket should be kept at-least equal to the strengths of the two materials used in core for an efficient strengthening of existing columns. It means that use of stronger materials in jacket makes the strengthening more efficient. Secondly, there should be a perfect bond between old and new concretes at the interface of core and jacket to avoid segregation of concrete during load application. The surface of old concrete should be prepared thoroughly before the strengthening operations are carried out. It may be noted, however, that there is quite a sharp difference between the test conditions and the actual situation in buildings where the columns would have be strengthened. In these tests, the core and jacket were simultaneously stressed whereas in real

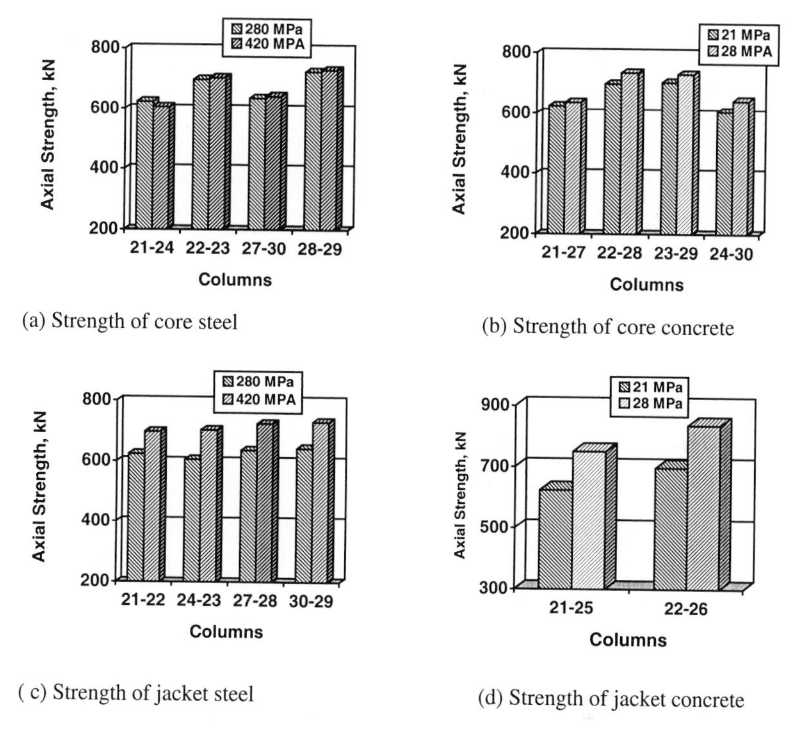

(a) Strength of core steel

(b) Strength of core concrete

(c) Strength of jacket steel

(d) Strength of jacket concrete

Figure 4. Comparison of strengths based on concrete and steel strengths in core and jacket (Results from series 3.)

buildings the core would be stressed prior the strengthening process would have been carried. Therefore it would be necessary to provide some stress transfer mechanism between old and new concrete sections. It may be provided in the form of the shear bolts, by surface roughening or by providing epoxy coating on the surface of existing section.

ACKNOWLEDGEMENTS

The author is highly indebted to the Directorate of Research, Extension, and Advisory Services of the University of Engineering and Technology, Lahore, for funding this piece of research work.

Advances in Building Technology, Volume I
M. Anson, J.M. Ko and E.S.S. Lam (Eds.)

REHABILITATION OF REINFORCED CONCRETE COLUMNS BY EXTERNAL PLATE BONDING

B. Uy[1]

[1]School of Civil and Environmental Engineering, The University of New South Wales, Sydney, NSW 2052, AUSTRALIA

ABSTRACT

Rehabilitating reinforced concrete elements using external steel plates has been applied to both beams and slabs in bridge and building construction over the last two decades. The effect of applying external steel plates increases the strength of beams and slabs. This technique can also be applied to reinforced concrete columns to increase their strength with a minimal increase in the cross-sectional dimensions of the member. This paper presents an experimental programme used to ascertain the behaviour of both short and slender columns. These columns had steel plates bonded to them by the use of both bolting or gluing and bolting. A simplified theoretical model based on composite construction techniques is then presented and calibrated with the experiments to determine the strength increase and recommendations for further research are then made. Furthermore, recent research on the compressive behaviour of carbon fibre reinforced plastic (CFRP) plates and their effects on rehabilitating reinforced concrete elements is also discussed.

KEYWORDS

buckling, columns, composite structures, concrete structures, rehabilitation

INTRODUCTION

Rehabilitation or retrofitting of reinforced concrete structures often provides a more logical and economical alternative than demolition or replacement. This has been made evident with the many applications where steel plating has been used to increase both the stiffness and strength of existing reinforced concrete slabs and beams. To further increase the range of applicability of steel plate bonding, research into the retrofitting of reinforced concrete columns needs to be carried out as illustrated in Figure 1. The advantages of this include increasing both the strength and stability of a column without a significant increase in cross-sectional dimensions. Reinforced concrete columns may need to be strengthened for a number of reasons, which may be due to increasing the gravity load on a structure, or upgrading a structure for lateral loading such as wind or earthquake.

This paper presents an experimental and analytical study to consider the effects of steel plate bonding on the strength and stability of reinforced concrete columns. The experiments involved testing both short and slender columns. A series of theoretical models is then presented and calibrated with the experiments for determining the axial and flexural strengths. The paper concludes with suggestions for further research into the rehabilitation and retrofitting of concrete structures in particular in relation to the use of carbon fibre reinforced plastic plating.

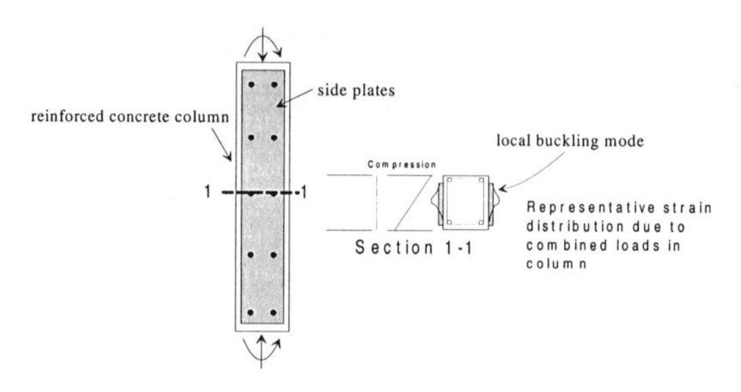

Figure 1: Application of CFRP plates to reinforced concrete columns

EXPERIMENTS

This paper will present the results of two series of experiments. Two series of experiments were conducted on short reinforced concrete columns to assess the increase in both axial stiffness and strength obtained by plate bonding. The concrete in these two series was exactly the same. A model to assess the axial stiffness and strength is then presented and calibrated with the experimental results. The following section outlines the two series of tests conducted. All of the plain reinforced concrete specimens were designed according to the Australian Standard for Concrete Structures, AS3600-1994, (Standards Australia, 1994).

Test Series 1

This set of tests consisted of constructing six identical reinforced concrete columns of 150 x 150 mm in cross-section, which were all of 450 mm in length. Each of these columns contained minimal longitudinal reinforcement consisting of four deformed Y12 bars and was detailed with sufficient lateral ties. Two plain specimens were tested and denoted as Plain 1 and Plain 2. Two specimens were tested with plates bolted to their sides and were denoted as Bolted 1 and Bolted 2. A final set of columns was tested with plates glued and bolted to their sides and these specimens were denoted Glued & Bolted 1 and Glued & Bolted 2. For this series of tests, plate coupons were tested in tension to establish their yield strength and elastic modulus, which are summarised in Table 1. Furthermore, a series of concrete cylinders were tested in compression to establish their mean compressive strength as summarised in Table 2.

Test Series 2

This set of tests also considered six identical reinforced concrete columns of 150 x 3000 mm of 815 mm in length. Again each of these columns was reinforced with minimal longitudinal reinforcement and detailed with sufficient lateral ties. This diagram also shows the location and extent of stiffeners used to initiate failure of the columns away from the ends of the specimen. The difference in these tests was that they were bonded with steel plates on only two opposite sides. Again two plain specimens were tested and denoted as Plain 1 and Plain 2, two specimens were tested with plates bolted to their sides and denoted as Bolted 1 and Bolted 2 and two specimens with plates glued and bolted to their sides were denoted Glued & Bolted 1 and Glued & Bolted 2. Also concrete cylinders were tested in compression and steel plate coupons were tested in tension to assess their stress-strain characteristics, which are again summarised in Tables 1 and 2 respectively.

TABLE 1
STEEL COUPON EXPERIMENTS

Series	Nominal Plate Thickness (mm)	Nominal Plate Width (mm)	Elastic Modulus ($x10^3$ N/mm^2)	Yield Stress (N/mm^2)
Mean	3	24	200	285

TABLE 2
CONCRETE CYLINDER TESTS

Series	Cylinder Diameter (mm)	Age at Testing (days)	Compressive Strength (N/mm^2)
Series 1	150	106	54
Series 2	150	106	54

TABLE 3
REINFORCING TESTS

	Nominal Diameter (mm)	Nominal Area (mm^2)	Yield Stress (N/mm^2)	Ultimate Stress (N/mm^2)
Mean	12	113	478	594

Short Column Tests

The short column tests, which included both Test Series 1 and 2, were conducted under pure compression in a Denison 5,000 kN capacity servo-controlled testing machine. The column ends were set in large steel plates and cast in high strength mortar to ensure uniform loading conditions and thus uniform compression loading. The short column tests were designed so that the ends were fully fixed, which virtually eliminated any second order effects from developing.

RESULTS

The following section outlines the results of each of the test series as well as summarising the results of the material property tests.

Test Series 1

The results of these tests are summarised in Table 4. The comparisons of axial strength show one that by the application of bolting a 16% increase in axial strength is achieved whereas gluing and bolting achieves a 32 % increase in axial strength. There is very little difference in axial stiffness created by steel plate bonding, although it seems to lie between 0 and 20 %. Failure modes of these columns are illustrated in Figure 2 which highlights local buckling of the steel plates and local crushing of the concrete.

TABLE 4
SERIES 1 COMPARISONS

Type	N_u Strength (kN)	N_u Average (kN)	Relative Value
Plain 1	1031		
Plain 2	1125	1078	1.00
Bolted 1	1251		
Bolted 2	1255	1253	1.16
Glued and Bolted 1	1359		
Glued and Bolted 2	1476	1418	1.32

Figure 2: Failure mode of short steel plated reinforced concrete column

Test Series 2
Results from the second series of tests were slightly different from those of Test Series 1 and the comparisons of axial strength are summarised in Table 5. Firstly, two plain samples were tested and these gave a base strength for the reinforced concrete member. Now the glued and bolted specimens achieved a lower strength than the specimens, which were bolted only, with an 11% and 20% increase respectively.

TABLE 5
SERIES 2 COMPARISONS

Type	N_u Strength (kN)	N_u Average (kN)	Relative Value
Plain 1	2214		
Plain 2	1828	2021	1.00
Bolted 1	2374		
Bolted 2	2476	2425	1.20
Glued and Bolted 1	2252		
Glued and Bolted 2	2229	2240	1.11

THEORETICAL MODEL

In order to be able to predict the strength of externally plated reinforced concrete columns, a theoretical model is proposed. The axial strength of the steel plated reinforced concrete columns is determined by employing the principle of superposition, which yields the following equation

$$N_u = N_{uc} + N_{ur} + N_{us} \tag{1}$$

where the concrete contribution to the axial strength is

$$N_{uc} = \alpha f_c A_c \tag{2}$$

where α is the factor associated with compression of the columns. The factor based on back analysis of the reinforced concrete columns was determined to be approximately 0.70 for Series 1 and Series 2. The reasons for this supposedly lower concrete strength can be explained by the fact that the cross-sections had a significantly large cover are compared with core area and thus this concrete suffered from premature spalling. The reinforcing steel contribution is given by

$$N_{ur} = f_{yr} A_r \tag{3}$$

The contribution of the steel plate was calculated as

$$N_{us} = f_{ys} A_s \tag{4}$$

when local buckling did not occur, or a reduced strength was used which was dependent on the local buckling strength, so that

$$N_{ol} = f_{ol} A_s \tag{5}$$

where the critical stress is given by

$$f_{ol} = \frac{\pi^2 E_s I_s}{A_s L_{bb}^2} \tag{6}$$

This buckling load may be higher if the adhesive restricts the steel plate from buckling out of plane. However in Series 2 the glue may have promoted buckling at a much earlier stage due to creating some initial imperfections. The steel plate undergoes a restrained buckling mode, which essentially reduces the effective length. An adequate buckling model assumes the plates are clamped at the bolt locations, however the unloaded edges are assumed to be free, which is essentially a column buckling approach with fixed ends.

COMPARISONS

The model developed for axial strength is compared with the experimental results of test Series 1 and 2. For both Series the concrete strength used a factor of 0.70 for the concrete cylinder strength as discussed previously. Tables 6 and 7 show the comparisons between the experiment and theory for the axial strengths. One can see that the model is very accurate in predicting the axial strength. The ratio of experimental to predicted strength is shown to be greater than 1.0 in all cases, thus showing that the model is conservative. The mean value for this ratio was 1.06 and 1.09 for Series 1 and 2 respectively, with a standard deviation of 0.08 in both cases. In both Series 1 and Series 2 plate buckling governed the strength of the steel plating, and this was substantiated by the experiments where plate buckling was noted to be one of the failure modes.

In addition to comparing the test results with the theoretical model proposed, it was also compared with the suggested ultimate strength and design strengths from the Australian Standards for Concrete Structures, AS3600. Tables 6 and 7 shows the ratio of the experimental strength to predicted ultimate strength, as well as the ratio of experimental to assumed design strength.

The AS3600 model was used with nominal strengths, so that the nominal concrete compressive strength was 40 MPa and the nominal reinforcing steel strength was 400 MPa as ordered. The nominal yield strength for the steel plate was taken as 250MPa. This comparison therefore resulted in some interesting observations. If one was to use the AS3600 strength model with nominal strengths, a mean value of 1.02 and 1.06 was established with a standard deviation of 0.11 and 0.08 respectively for Test Series 1 and 2.

If one compares the experimental strengths with the design strengths as suggested by AS3600, a standard deviation of 1.70 and 1.78 with a corresponding standard deviation of 0.32 for both test series. This shows that for design, the AS3600 model will certainly provide a factor of safety of approximately 2.5, by the time load factors are introduced.

TABLE 6
AXIAL STRENGTH COMPARISONS (SERIES 1)

Specimen	$N_{u.Experiment}$ (kN)	N_{uc}	N_{ur}	N_{us}	$N_{u.Predicted}$ (kN)	Ratio= ($N_{u.Experiment}$/ $N_{u.Predicted}$)	Strength Ratio (AS3600)	Design Ratio (AS3600)
Plain #1	1031	815	216	0	1031	1.00	1.09	1.82
Plain #2	1125	909	216	0	1125	1.00	1.19	1.98
Bolted #1	1251	849	216	163	1228	1.02	0.90	1.49
Bolted #2	1255	849	216	163	1228	1.02	0.90	1.50
Glued and Bolted #1	1359	849	216	163	1228	1.11	0.97	1.62
Glued and Bolted #2	1476	849	216	163	1228	1.20	1.06	1.76
					Mean	1.06	1.02	1.70
					Standard Deviation	0.08	0.11	0.32

TABLE 7
AXIAL STRENGTH COMPARISONS (SERIES 2)

Specimen	$N_{u.Experiment}$ (kN)	N_{uc}	N_{ur}	N_{us}	$N_{u.Predicted}$ (kN)	Ratio= ($N_{u.Experiment}$/ $N_{u.Predicted}$)	Strength Ratio (AS3600)	Design Ratio (AS3600)
Plain #1	2214	1890	324	0	2214	1.00	1.22	2.05
Plain #2	1828	1504	324	0	1828	1.00	1.01	1.69
Bolted #1	2374	1697	324	52	2073	1.15	1.05	1.76
Bolted #2	2476	1697	324	52	2073	1.19	1.10	1.83
Glued and Bolted #1	2252	1697	324	52	2073	1.09	1.00	1.67
Glued and Bolted #2	2229	1697	324	52	2073	1.08	0.99	1.65
					Mean	1.09	1.06	1.78
					Standard Deviation	0.08	0.08	0.32

CARBON FIBRE PLATING IMPLICATIONS

Recent studies on the compressive behaviour of carbon fibre plates restrained by a rigid medium have shown that these plates are more susceptible to local instability than steel plates, due to the high level of anisotropy which is existent in unidirectional plates, (Uy and Aw, 2001 and Uy and Yang, 2001). Further experiments have also shown that due to the very low transverse stiffness and strength, small out of plane deformations of the plate in compression can lead to transverse splitting which is a highly undesirable failure mode. Further experiments of these plates when used in conjunction with reinforced concrete columns should be conducted to assess the suitability of this material for rehabilitation purposes.

Figure 3: Failure mode of carbon fibre reinforced plastic plating under compression

CONCLUSIONS

This paper has presented a benchmark set of tests for reinforced concrete columns bonded with steel plates under axial flexural loading. The paper has illustrated that bonding steel plates to reinforced concrete columns can increase both the stiffness and strength of the member. Both bolting and gluing and bolting were investigated. Bolting steel plates was shown to produce a higher strength and stiffness than the control specimens. Gluing and bolting also produced an increase in strength and stiffness and in the case of the slender columns, this was totally effective in eliminating local buckling and slip and thus allowing full composite action to be achieved. A theoretical model has also been presented to determine the axial stiffness and strength. This model was calibrated with sufficient accuracy with the test results and in all cases the model is shown to be conservative. However further research is required before this approach could be adopted in design as only a limited amount of specimens were tested. Furthermore, given the growing use of fibre reinforced plastic plates in retrofitting concrete structures, further tests should be conducted using these forms of plates.

ACKNOWLEDGEMENTS

The author would like to thank Ms. Caroline Raleigh and Messrs Ian Bridge, Bob Rowlan, Richard Webb and Philip Yeadon all formerly of the Faculty of Engineering, University of Wollongong for assisting in undertaking the experimental programme presented in this paper. This project was sponsored by the Australian Research Council Small Grants Scheme.

REFERENCES

Standards Association of Australia, *SAA Concrete Structures Code*, SAA, Sydney, Australia, 1994, AS3600-1994.

Uy, B. and Aw, K. (2001) Local buckling of carbon fibre plastic (CFRP) plates restrained by a rigid medium, *Proceedings of the First International Conference of FRP in Civil Engineering, Hong Kong, December,* pp. 723-728.

Uy, B. and Yang, T. (2001) Local Buckling of anisotropic plates restrained by concrete, *First Asian Pacific Congress on Computational Mechanics, Sydney, November,* pp. 279-284.

CONSTRUCTION TECHNOLOGIES:

HIGH PERFORMANCE CONCRETE/COMPOSITES

Advances in Building Technology, Volume I
M. Anson, J.M. Ko and E.S.S. Lam (Eds.)

NUMERICAL SIMULATION OF THERMAL CRACKING OF CEMENT-BASED MULTI-INCLUSION COMPOSITE

Y.F. Fu[1], Y.L. Wong[1], C.A. Tang[2] and C.S. Poon[1]
[1]Department of Civil & Structural Engineering, Hong Kong Polytechnic University
[2]Lab of Numerical Test of Material Failure, Northeastern University, Shenyang, China

ABSTRACT

A newly developed 2-*D* mesoscopic thermoelastic damage (MTED) model is used to study the thermal stress field and associated fracture in a cement-based composite with multiple circular inclusions at high temperatures. It is found that the thermal stress field and the associated cracking are dominated by the interaction of inclusions, and the heterogeneity. The macrocracks occur firstly between the two inclusions which have the shortest distance apart if coefficient of thermal expansion (CTE) of the inclusion is greater than that of matrix. The resistance of inclusion with higher strength against crack is also found. The MTED model yields a good agreement between the proposed numerical simulations and the available experimental results.

KEYWORDS

Thermal stress; Thermal induced cracking; Irregular inclusion; Numerical simulation; Heterogeneity

INTRODUCTION

When a cement-based composite material is subjected to elevated temperatures, the difference in thermal deformations of the cementitious matrix and the aggregate inclusions induces thermal stresses and possibly cracking in the material. Previous investigations [Venecanin, 1990; Khoury,1992] have shown that the greater the mismatch in the mechanical and thermal properties between the matrix and the inclusion, the more significant the reduction of strength, elasticity modulus, and durability of concrete would be. The deteriorations of these properties were mainly due to the thermal induced cracking. Venecanin (1990) experimentally studied the effects of thermal incompatibility of concrete components (TICC) by measuring their CTE, and the CTE heterogeneity of aggregate inclusions. The results showed that TICC was one of the reasons for thermal cracking. Kristensen and Hansen (1994) lent a support to the Venecanin's view in their investigation of the thermal cracking of cement paste and concrete using ultrasonic pulse velocity techniques. Their experimental results indicated that cement paste and concrete could be damaged by internal cracking due to differential stresses at differential temperature between core and surface of specimen. Later, based on the fracture mechanics and micromechanics approaches, an analytical model (Sumarac et al., 1998) was proposed to study the reduction of elastic modulus with temperatures due to the mismatch in the CTE between mortar and aggregate, assuming a uniform distribution of cracks. Recently, both analytical modeling and numerical simulation (Hsueh, et al., 2001) were used to analyze residual thermal stresses and CTE in two-phase composites. It was found that the predicted CTE of a two-phase material was dependent on the Young's modulus ratio and the difference between the CTE of the two phase-materials. If the intergranular phase had a greater CTE than that of the grain, the intergranular phase and the grain would be stressed in tension and hydrostatic compression, respectively. Although the studies on the thermal stress and the thermal crack have been conducted numerically or experimentally, the studies of the thermal cracking processes – the initiation, propagation and linkage of crack in concrete at elevated temperatures is still limited.

As observed in experiments (Tijissens et al., 2001; Mier et al., 1997), concrete is a highly heterogeneous material that can be quantified in three levels: macroscale, mesoscale and microscale. The heterogeneity is an indispensable reason to cause the concentration stress and local failure (Brandt, 1996; Hsu, 1963). At room temperature, the effects of heterogeneity on the fracture processes in concrete have been reported in literatures. However, the previous studies do not consider the effects of heterogeneity among the phase materials of concrete on the stress development and cracking process under high temperatures. The complex structure of concrete is unable to be described by an analytical model. The process of thermal cracking of heated concrete has not been visually demonstrated till now.

This paper presents a series of numerical studies to quantify/qualify the effects of the interaction and CTE of multi inclusions on the thermal stresses and associated cracking in a cement-based composite. Consequently, a more realistic thermal stress field and the thermal cracking in meso-structure can be observed visually and dynamically by the newly developed mesoscopic thermoelastic damage (MTED) model (Fu et al., 2002).

NUMERICAL METHOD

Brief Outline Of Mesoscopic Thermoelastic Damage (MTED) Model

The cement-based composite is made of two phase materials (matrix and inclusion), each of which at mesoscale is heterogeneous. The interface bonding between the matrix and the inclusion is assumed to be perfect. A statistical Weibull Distribution is used to describe the heterogeneity of the phase material using a homogeneity index. The mechanical properties (such as strength, elastic modulus and Poisson ratio) are randomly allocated to each representative volume element (RVE) to account for the inherent variability in the phase material, using the Monte-Carlo method.

In this study, we implicitly assume that the macroscopic mechanical behavior of the composite is a result of the growth of microcracks, and is captured in the damage evolution law for RVE. A smeared crack concept is adopted to construct the constitutive law of RVE. The crack is not considered as a discrete displacement jump, but rather a variation in the mechanical properties according to a continuum law, such as damage mechanics. A separation function of damage is deduced to describe the softening process of the material. A full description of the MTED model is available in the paper (Fu et al., 2002). The schematic of the thermoelastic behavior law of the material is shown in Fig.1. It is assumed that the stress-strain relationship of a RVE is linearly elastic till its peak-strength is reached, and thereafter follows an abrupt drop to its residual strength. Under compression, RVE remains a residual strength after peak-strength; whereas, under tension, the residual strength of RVE is equal to zero. Parameters S and S_r are used to denote the uniaxial compressive/tensile strength and residual strength; ε^0, ε_0 are the initial thermal strain and the strain corresponding to the peak strength of RVE, respectively. This model is able to capture the complete thermal cracking processes including the initiation, propagation and linkage of cracks. In the present study, the CTE and mechanical properties of the matrix and inclusions are considered to be independent of temperature variations.

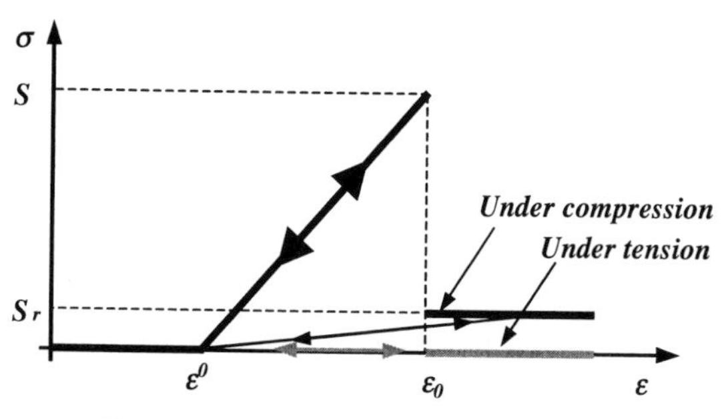

Fig.1 Constitutive relation of RVE

Numerical Specimens

Four numerical square specimens with different thermal and geometrical properties (see Table.1) are presented here to simulate the thermal induced cracking process in the composite with multi-inclusions. Specimen no.1 and no.2 with multi circular inclusions of same CTE (see Fig.2) are set to investigate the influence of the interaction between inclusions on the thermal stress field, and the only variable is the inclusion diameter. The proportion of the inclusions in Specimens no.1 and no.2 are 28.3% and 50.2%, respectively. For the former, the diameter of inclusion is 15mm. For the latter, the diameter of inclusion is 20mm. Specimens no.3 and no.4 (see Fig.2) are similar to those of Specimens no.1 and no.2 respectively, except the former ones have a higher degree of heterogeneity (h=3 in matrix and inclusions). All specimens are subjected to uniform temperature field changes (steady state) ranging from 20°C till 620°C, at an increment of 10°C.

TABLE.1
MATERIAL PROPERTIES OF SPECIMENS NO.1 TO 4

Parameter		Value	
		Matrix	Inclusion
Homogeneity index (h)	Specimen no.1 and no.2	300	300
	Specimen no.3 and no.4	3	3
Mean elastic modulus (MPa)		60000	100000
Mean compressive strength (MPa)		200	300
Poisson Ratio		0.25	0.20
Coefficient of thermal expansion (/°C)	Specimens no.1 to no.4	1.0E-5	1.1E-5
Temperature increment (°C)		10	10
Tension cutoff		0.1	0.1
Frictional angle (°)		30	30
Dimension (mm)	Specimen no.1	100×100	Φ15
	Specimen no.2		Φ20
	Specimen no.3		Φ15
	Specimen no.4		Φ20

(a) Specimen no.1

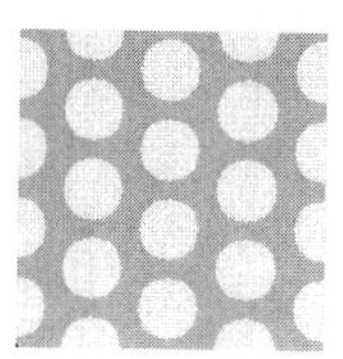

(b)Specimen no.2

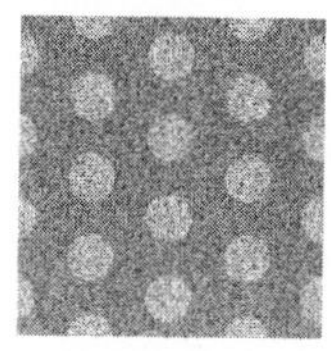

(c) Specimen no.3

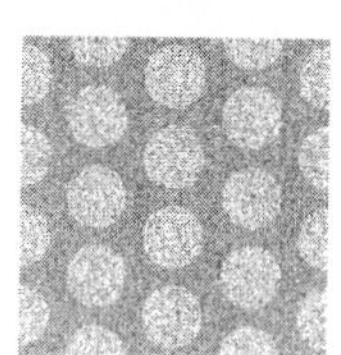

(d) Specimen no.4

Fig.2 Numerical models

NUMERICAL RESULTS AND DISCUSSION

Thermal Stress Field

Effects of interaction of inclusions

In case of Specimen no.1 with multi inclusions of 15mm diameter, a diamond-shaped region surrounded by the inclusions (see Fig.3a) is chosen to explain the influence of the arrangement of inclusions. In this region, the distance between the inclusions ***A*** and ***B*** is larger than that between the inclusions ***C*** and ***D***. Consequently, in the sub-region Ψ_1 the stress in the *X*-direction is smaller than 0, $\sigma_x<0$, and in the *Y*-direction is larger than 0, $\sigma_y>0$; whereas in the sub-region Ψ_2 the stress in the *X*-direction is larger than 0, $\sigma_x>0$, and in the *Y*-direction is smaller than 0, $\sigma_y<0$ (as shown in Fig.3c). In the locally enlarged pictures of the principal stress contours of the sub-regions Ψ_1 and Ψ_2

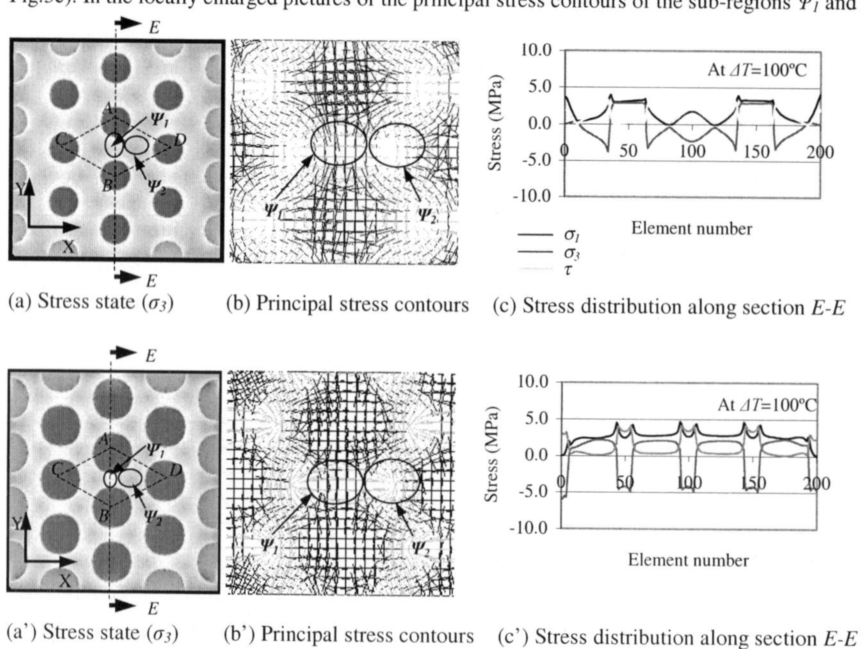

(a) Stress state (σ_3) (b) Principal stress contours (c) Stress distribution along section *E-E*

(a') Stress state (σ_3) (b') Principal stress contours (c') Stress distribution along section *E-E*

Fig.3 Thermal stress field in Specimens no.1 and no.2

(see Fig.3b), a dark line is a trajectory of a maximum principal stress and a light is the trajectory of the minimum principal stress. It is evident that the values of all principal stresses approach zero in the sub-region Ψ_2, but the principal stresses in the sub-region Ψ_1 have the highest value in the whole matrix (see Fig.3c). There are six sub-regions of low stress like Ψ_2 around each inclusion in the composite. Hence, the elements in Ψ_1 have higher probability of failure than those in Ψ_2. Similar stress states in Ψ_1 and Ψ_2 are observed in Specimen no.2 with larger multi inclusions of 20mm diameter (see Fig.3a', 3b'). The calculated results have a good agreement with those reported in the literature (Hsueh et al., 2001; Hsu, 1963). The results from Specimen no.2 further prove that the stress induced by the cross-effect of the inclusions is one of the reasons to cause the stress concentration. From Fig.3c and Fig.3c', it is found that the principal stresses between the inclusions ***A*** and ***B*** are increased by 20%. Such also occurs between the inclusions ***A*** and ***C***, or ***C*** and ***B***, or ***B*** and ***D***, or ***D*** and ***A***.

Consequently, the stress concentration occurred in the matrix due to the cross-effect of multi inclusions makes the elements in the matrix more prone to fail.

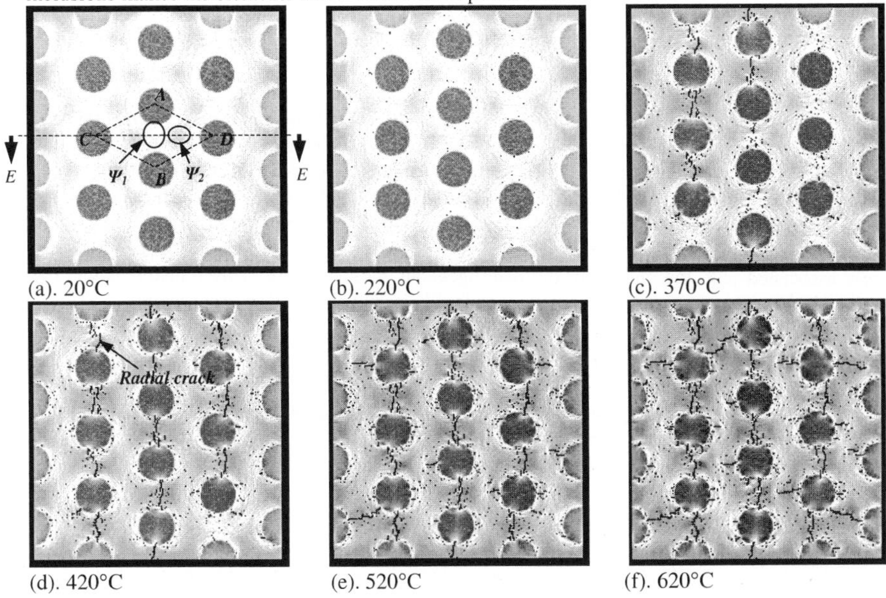

(a). 20°C (b). 220°C (c). 370°C

(d). 420°C (e). 520°C (f). 620°C

Fig.4 Thermal cracking process of Specimen no.3

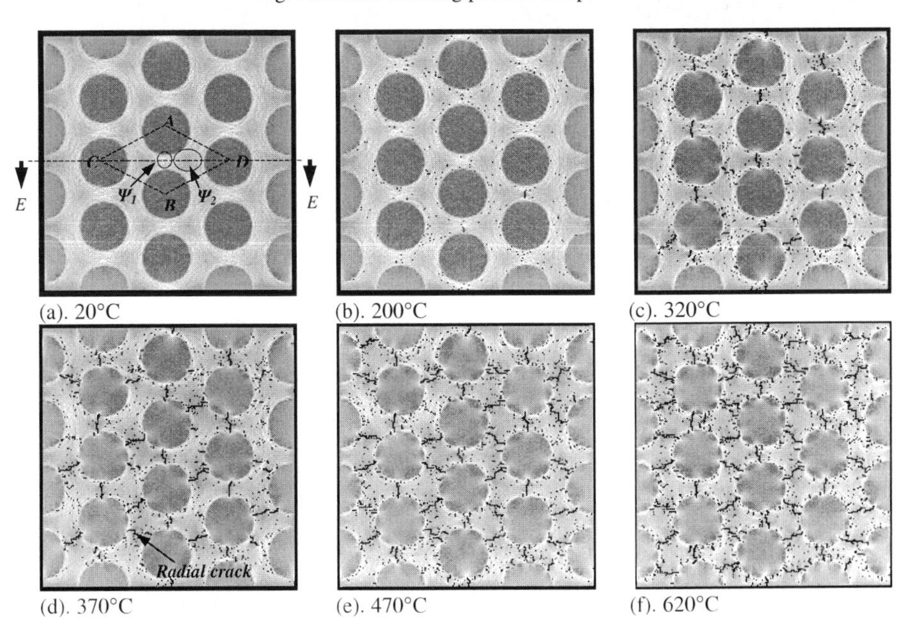

(a). 20°C (b). 200°C (c). 320°C

(d). 370°C (e). 470°C (f). 620°C

Fig.5 Thermal cracking process of Specimen no.4

Effects of heterogeneity

Fig.4a shows the thermal stress field of Specimen no.3 that has heterogeneous matrix and inclusions. Gray scale indicates minimum principal stress values. Comparing the stress distribution along the mid-section *E-E* of Specimen no.3 (Fig.6a) with that of Specimen no.1 with homogeneous phase materials (Fig.3c), it is very clear that heterogeneity causes local fluctuation of stresses in matrix and inclusion even before crack initiates (see Fig.6a), even though their macroscopic thermal stress fields are same from the view point of statistics. With increasing the temperatures, the number of the broken elements is also increased. Due to this reason, the degree of the stress fluctuation becomes severe (see Fig.6a). With the increase of number of the damaged elements, the original stress distribution is changed. For example, at temperature increment of 620°C (see Fig.4f), the high stresses have been transferred to the tips of the irregular cracks, whose distributions are closely associated with the change in the heterogeneous mesostructure. This transfer of thermal stress field with the variation of mesostructure cannot be easily observed in experiments.

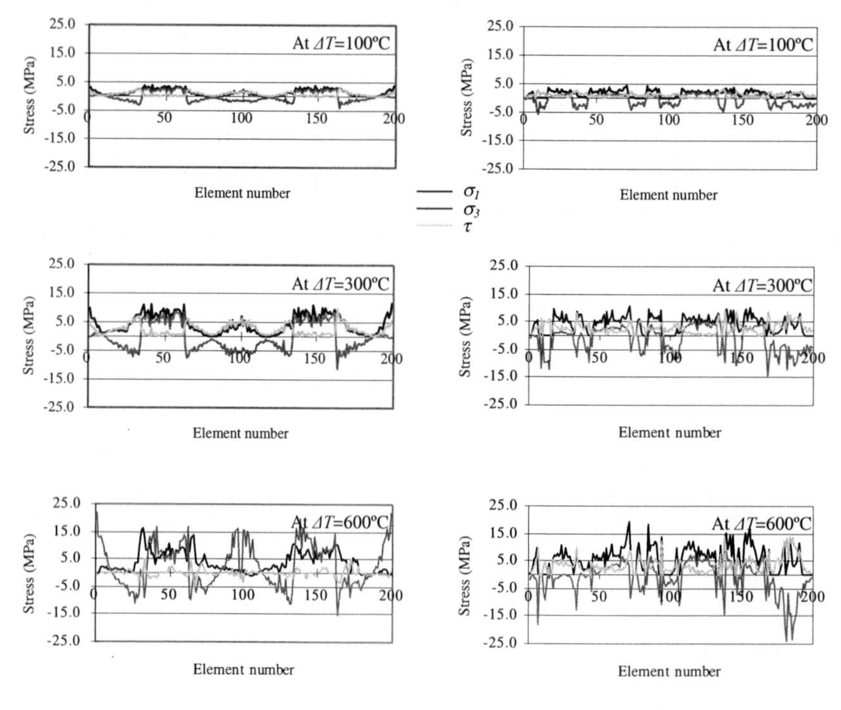

(a) Specimen no.3 (b) Specimen no.4

Fig.6 Thermal stress distribution along mid-section *E-E* of Specimens no.3 and 4

Thermal Fracture

Effects of interaction of multi inclusions

The cross-effect on stresses between the two neighborhood inclusions makes the elements in sub-region Ψ_1 more prone to fail. The commence of crack development is largely dependent on the distance between any given two inclusions. For example, for Specimen no.3, when the elements around the inclusions firstly reach their strength, cracks begin to develop at the interface of the matrix and the inclusion, and then propagate towards to the middle of the two neighborhood inclusions (see Fig.4). For Specimen no.4 with larger inclusions, cracks are firstly formed in the middle of the two neighborhood inclusions which have the shortest distance apart, rather than at the interface of the matrix and the inclusion, and then propagate towards the two neighborhood inclusions (see Fig.5b). As the diameter of inclusions increases, the distance between ***A*** and ***C***, or ***C*** and ***B***, or ***B*** and ***D***, or ***D*** and ***A*** in Specimen no.4 reduces (see Fig.5). As a result, inclined cracks making an angle of about 30° to the horizontal line are developed (see Fig.5e-f) instead of the horizontal cracks occurred in Specimen no.3 with smaller circular inclusions (as shown in Fig.4e-f).

Effects of heterogeneity

In case of Specimen no.3, a diffused damage initially appears randomly in the whole region of the matrix at temperature of 220ºC (see Fig.4b). It means that the failure of an element is dependent both on the thermal induced local stress and on the strength of the element. The strength distribution of the elements is mainly dependent on the heterogeneity at meso-scale. Under the same stress field, the elements with a higher ratio of the stress to the strength have higher probability to fail than those with a lower ratio. Consequently, the shape of the crack path is irregular, rough and bifurcate.

Another interesting phenomenon can be observed from Fig.4 and Fig.5. The crack development is resisted by an inclusion with high elastic modulus and strength. When the tip of a radial crack touches an inclusion with higher elastic modulus and strength, the inclusion obstructs the radial crack propagation. For example, in case of Specimen no.3, after the radial cracks coalesce between two-neighborhood inclusions linking up in the *Y*-direction, they do not propagate when the tips of the cracks approach such an inclusion (see Fig.4c-f). Instead, the radial cracks in the *X*-direction begin to initiate and propagate, since the stresses have been redistributed and transferred into the tips of these horizontal cracks. Thus the inclusion can play an important role to (i) resist further propagation of crack, (ii) consume more energy required by failure, and (iii) dictate the redistribution of the stresses. These results are consistent with the viewpoints in the literature (Kristensen et al., 1994) that thermal cracking probably travel in all directions in mortar due to the presence of aggregate in concrete.

The numerical results also reveal that discontinuous cracks grow simultaneously at the poles of the different inclusions. Increasing number of such cracks eventually result in the irregular radial cracks, length of which is greater than the direct distance between the two corresponding inclusions because of the irregular path of crack. It lastly impairs the cohesive properties of the bond phase material (matrix). From the observation of the thermal fracture processes, it appears that heterogeneity can be used to interpret the randomization of crack origin and crack shape. It can also be used to explain the reason of local thermal fracture.

CONCLUSIONS

In this paper, a newly developed mesoscopic thermoelastic damage model (MTED) has been implemented to numerically simulate the thermal stress field and the crack formation, extension and coalescence in a cement-based multi-inclusions composite at elevated temperatures. From the results presented in this study, some main conclusions are summarized as follows.

1. The thermal stress field being studied is focused on the effects of thermal mismatch and the arrangement of inclusions. When the CTE of inclusion is larger than that of the matrix, the inclusion and the matrix are subjected to statistically hydrostatic compression and a combination of compression and tension, respectively. A high stress region is induced at the interface between the matrix and the inclusion by the thermal mismatch, and in the area between any two-neighborhood inclusions with the shortest distance apart.

2. The initiation and path of cracks depend on the heterogeneity at meso-scale. When the CTE of inclusion is larger than that of matrix, the interaction of inclusions is another factor affecting the thermal cracking. The cracks form, propagate and coalesce firstly along direction between the two inclusions which have the shortest distance apart. When the inclusion diameter increases, the cross-effect of stress concentration around the two neighborhood inclusions increases the probability of failure of the elements locating between these two inclusions. However, when the strength of the inclusion is higher than that of matrix, the inclusion can prevent the further propagation of the cracks to a certain extent.

In this paper the temperature-dependent properties (strength, elastic modulus and CTE) of each component material of the composite at high temperatures, which can substantially affect the thermal cracking, have not been quantified. Further work on this direction will be conducted. Nevertheless, the numerical study by using the proposed MTED model offers an understanding of the thermal stress redistribution and cracking process of a cement-based composite, from which various mechanisms of thermal induced stresses and associated cracking have been identified.

REFERENCE

Venecanin SD. (1990). Thermal incompatibility of concrete components and thermal properties of carbonate rocks *ACI Materials Journal* **87:6**, 602-607.

Khoury A. (1992). Compressive strength of concrete at high temperatures: a reassessment. *Magazine of Concrete Research* **44:161**, 291-309.

Kristensen L, Hansen TC. (1994). Cracks in concrete core due to fire or thermal heating shock. *ACI Materials Journal* **91:5**, 453-459.

Sumarac D, Krasulja M. (1998). Damage of plain concrete due to thermal incompatibility of its phases. *International Journal of Damage Mechanics* **7**, 129-142.

Hsueh CH, Becher PF, Sun EY. (2001). Analysis of thermal expansion behavior of intergranular two-phase composites. *Journal of Materials Science* **36**, 255-261.

Tijissens MGA, Sluys LJ, Giessen EVD. (2001). Simulation of fracture of cementitious composites with explicit modeling of microstructural features, *Engineering Fracture Mechanics* **68**, 1245-1263.

Van Mier, Jan GM. (1997). *Fracture Processes of concrete: assessment of material parameters for fracture models*. CRC Press, U.S.A

Brandt AM. (1995). *Cement-based composites: Materials, Mechanical Properties and performance*. E& FN Spon, UK

Bažant ZP, Kaplan MF. (1996). *Concrete at high temperatures: Material properties and mathematical models*, Longman House, UK

Hsu Thomas TC. (1963). Mathematical analysis of shrinkage stresses in a model of hardened concrete, *ACI Journal*, 371-390.

Fu YF, Wong YL, Tang CA, Poon CS. (2002). Thermal induced stress and associated cracking in cement-based composite at elevated temperatures – Part I: Thermal cracking around single inclusion. *Submitted to Cement and Concrete Composite.*

Neville AM. (1993). *Properties of concrete*, 3rd Ed., Longman House, UK

Goltermann P. (1995). Mechanical predictions on concrete deterioration. Part 2: Classification of crack partterns, *ACI Materials Journal* **92:1**, 58-63.

ACKNOWLEDGEMENTS

The materials presented in this paper are some of the findings of the G-V848 research project entitled "Thermal Stress and Associated Damage in Concrete at Elevated Temperatures", of The Hong Kong Polytechnic University funded by the research Grants Council. The project is also partly supported by the NNSF of China (No.50174013).

Advances in Building Technology, Volume I
M. Anson, J.M. Ko and E.S.S. Lam (Eds.)

APPLICATIONS OF COPPER GANGUE WITH HIGH CONTENT OF Fe_2SiO_4 AS CEMENT ADMIXTURE AND FOR CEMENT PRODUCTION

Lin Zongshou   Wan Huiwen   Zhao Qian   Huang Yun

School of Materials and Engineering, Wuhan University of Technology ,
Wuhan,Hubei,P.R.China,430070

Abstract:

Copper gangue (CG) is an inactive industrial waste generated from a copper refinery. Making good use of this waste will be beneficial to the sustainable development of society. In this study, the copper gangue was firstly used as a cement admixture. The results showed that with the replacement of 5~25% CG, 3-day and 28-day strengths decreased by 5~32%. However the long-term strength of the specimens with 10% CG, especially after the samples cured for 3 months, approached that of plain mortar. Its mechanism was studied by electron probe microanalyzer (EPM), and the results indicated that a small quantity of $Fe(OH)_3 \cdot nH_2O$ slowly formed from Fe_2O_3 in the presence of $Ca(OH)_2$, and accordingly the hardened cement paste became more dense. Secondly, the CG as iron material for cement production was also researched. Optical microscope analyses showed that CG containing Cu, Zn, Ti, Mn, Ag etc microelements, can decrease clinker burning temperature and liquid viscosity, and enhance migration of the CaO and SiO_2. A blended admixture with low content of CaF_2-$CaSO_4 \cdot 2H_2O$ could take a partial mineralization effect and further benefits the clinker burning.

Key words:

Industrial waste, Copper gangue, Cement admixture, Microelement

1. Introdution

Copper residue is an industrial waste that is discharged by a copper refinery. In order to extract some useful metals, a cooling technology for removing residue is introduced by slowly cooling (48h) rather than quickly cooling so that some metals ions grow. Such copper residue contains 4~5% copper, 1.01g/t gold, 24g/t silver and 42% magnetite. After one more flotation for copper and magnetic by extraction of iron, the discharged residue is called copper gangue (CG). But CG grains mostly are in the range 0.037~0.10mm and contain a large amount of water. CG still contains approximately 48% fayalite (mostly Fe_2SiO_4) and 28% fine crystal magnetite (mostly Fe_2O_3 and less FeO), while copper content is only 0.3~0.4%.

It is well known that raw material containing high iron oxide can be used as an alterative material for cement production. Because iron oxide can form liquid phase in clinker burning, the addition of

CG can enhance migration of the CaO and SiO_2, and benefit the clinker sintering forming calcium aluminoferrite [1]. J.O.Odigure's researches [2-5] showed that abrasive slurry containing high metallic particles (Fe^{3+}, Fe^{2+}, Fe) content, could significantly improve the clinker mineral chemical composition, morphology and phase ratio. The kinetics of chemical reactions in hardening cement mortar produced from the raw mix containing metallic particles, differ from those of ordinary Portland cement using raw materials of natural origin.

This research is aimed at investigating the influence of CG used as a cement admixture on cement mortar strength and the influence of CG used as a raw material on cement clinker mineral morphology. It will be helpful to the recycling and reuse of this industrial waste.

2. Experiment

2.1 Chemical composition and microanalysis of CG

The chemical composition and mineral analysis of CG are presented respectively in Tables 1 and 2. The X-ray diffraction pattern of is shown in Fig.1, which indicates that the mineral components of CG are mainly composed of Fe_2SiO_4, Fe_2O_3, FeO and a small quantity of amorphous SiO_2.

SiO_2	Al_2O_3	Fe_2O_3	CaO	MgO	SO_3	LOI*
34.84	2.99	52.42	4.27	4.14	/	-2.36

LOI*—Loss on ignition

Table 1 Chemical composition of CG (mass %)

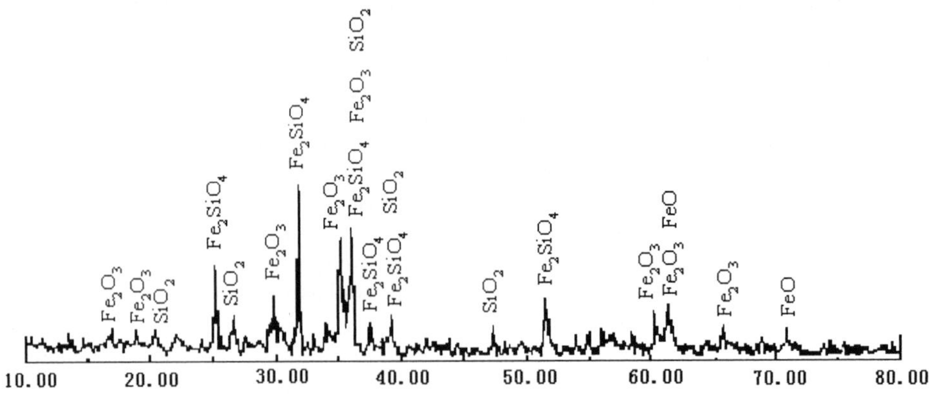

Fig. 1 X-ray diffraction pattern of the CG

Magnetite	Fayalite	Amorphous silicate	Hematite&limonite	Feldspar&kaolinite	Brass&matte
30.2	47.8	14.7	2.8	2.5	0.7

Table 2 Mineral composition of CG (mass %)

2.2 CG used as cement admixture

The CG was dried and manually dispersed. Portland cement (surface area 300m²/kg) was provided by Jinyue cement plant, the chemical composition of the clinker is shown in Table 3. CG substitution percentages by weight relative to cement were 0%, 5%, 10%, 15%, 20% and 25%, respectively. All samples were mixed for 6h in an automatic blender and then normal consistency and setting time were measured according to GB 1346-92 (Chinese standard). Specimens for testing mechanical strength, after demoulding, were cured in water 28 days at 20±2℃and then cured to prescribed ages in the ambient conditions. Compressive strengths were measured according to GB 175-92. CG substitution ratios and physical properties of cement are presented in Table 4.

SiO_2	Al_2O_3	Fe_2O_3	CaO	MgO	f_{CaO}	LOI*
19.52	5.80	4.81	62.66	2.11	2.27	0.44

LOI*—Loss on ignition

Table 3 Chemical composition of clinker (mass %)

N	CG substitution ratio (mass %)	Setting times (min)		Normal consistency (%)	Compressive strength (MPa. After days)			
		Initial	Final		3	28	90	360
1	0	104	195	29.0	25.6	52.3	68.4	69.3
2	5	125	210	28.6	24.2(0.95)	50.2(0.96)	68.0(0.99)	68.7(0.99)
3	10	138	220	28.0	22.1(0.86)	46.8(0.89)	68.7(1.01)	70.6(1.02)
4	15	155	245	27.5	20.3(0.79)	42.5(0.81)	65.3(0.95)	66.5(0.96)
5	20	168	250	27.0	18.7(0.73)	38.6(0.74)	63.0(0.92)	61.8(0.89)
6	25	176	270	26.4	17.3(0.68)	35.5(0.68)	58.9(0.86)	58.1(0.84)

*( )—strength ratio, which is the ratio of the strength of various sample to that of sample 1 at the same age.

Table 4 CG substitution ratios (mass %) and physical properties of cement

In order to study the influence of the Fe_2SiO_4 and Fe_2O_3 into CG on mechanical strength of cement and avoiding interfere with the iron phase contained in Portland cement, the following experiments were carried out. (1) Incorporating 25% CG in white Portland cement to make paste sample A. (2) At the same condition, 25% CG in Portland cement to make paste sample B (curing 3 months, curing system as before). (3) Their microstructure and composition were studied using EPM (Figs. 2 and 3).

2.3 CG used as raw material for cement production

The raw materials used for the experiments were limestone, clay, coal, gypsum, fluorite and CG. For the making of coal ash, coal was put in an earthen box and burnt in an electric furnace using silicon-heating elements. The burning temperature was 1200℃, with a duration of 6h at the maximum temperature. Coal ash content was 25.8%. The compositions of the raw materials are presented in Table 5. The compositions of the raw mixes were determined (shown in Table 6). The dried raw materials mixed in the required proportion were ground to a fineness of 6 to 8 mass % residue on an 80-μm size mesh using a laboratory ball-mill. They were then watered and manually pressed into a Φ10×30mm column. Those columns were dried and burnt in an electric furnace using silicon-heating elements. The burning temperature was gradually raised to 1450℃, the rate of rise of temperature was controlled to 200~250℃/h, with a duration of 45 min at the maximum temperature. The clinkers were cooled in the air. The clinker samples were analyzed with an optical microscope and the microstructural images are shown in Figs. 4 and 5. The result determined of the free CaO is shown in Table 6. The produced clinker and 4% gypsum were ground approaching a surface area of 310m^2/kg. The mechanical strengths of the clinkers, determined by the above method, are presented in Table 6.

Table 5 Chemical composition of raw materials

Material	Oxides contents, mass %							
	SiO_2	Al_2O_3	Fe_2O_3	CaO	MgO	SO_3	CaF_2	LOI*
Limestone	5.64	1.65	0.76	48.50	1.77			40.38
Clay	63.48	13.76	5.93	2.25	1.62			9.60
CG	34.84	2.99	52.52	4.27	4.14			-2.36
Gypsum	0.26	0.25	0.06	32.56	0.18	45.88		19.80
Fluorite	25.22	1.08	0.09	1.71	0.50		65.73	5.41
Coal ash	54.20	29.14	7.84	3.45	0.47			

LOI*—Loss on ignition

Table 6 Composition of raw mix and compressive strength of cement

N	Composition of raw mix, mass %							f_{CaO} (mass %)	Compressive strength (MPa)	
	Limestone	Clay	Iron powder*	CG	Gypsum	Fluorite	Coal*		After 3days	After 28days
7	79.41	7.84	1.92	0	0	0	10.83	2.52	28.0	56.1
8	78.34	7.37	0	3.46	0	0	10.83	2.08	27.8	56.7
9	75.27	7.03	0	3.49	2.64	0.74	10.83	1.75	29.4	60.2

Iron powder*— chemical agent containing Fe_2O_3 98.5%.
Coal*—laboratory should use coal ash, 10.83×25.8%.

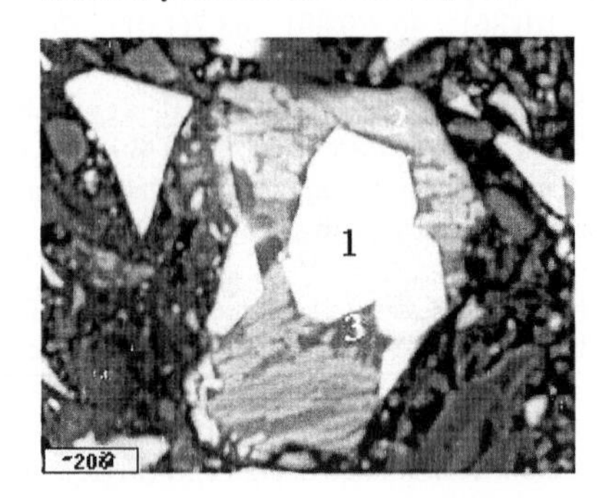

Fig. 2 EPM photograph of the sample A
1-Fe_2O_3 2,3-Fe_2SiO_4

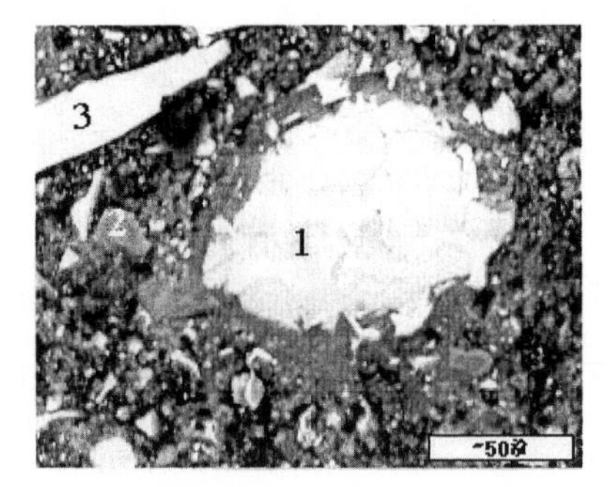

Fig. 3 EPM photograph of the sample B
1-C_2S 2-$Ca(OH)_2$ 3-Fe_2SiO_4

3. Results and discussion

3.1 Petrographic features of the CG

The mineral analyses of Table 2 show that CG comprises mainly fayalite, magnetite and amorphous silicate minerals. From Fig. 1, it is clear that the diffraction peaks of Fe_2SiO_4, Fe_2O_3, FeO and SiO_2. EPM (Fig. 2) show that Fe_2O_3 particle is partially wrapped in Fe_2SiO_4. The grains of CG are distributed mostly in the range 0.037~0.10mm with tough surface and dense structure. Its density is 5.20g/cm^3.

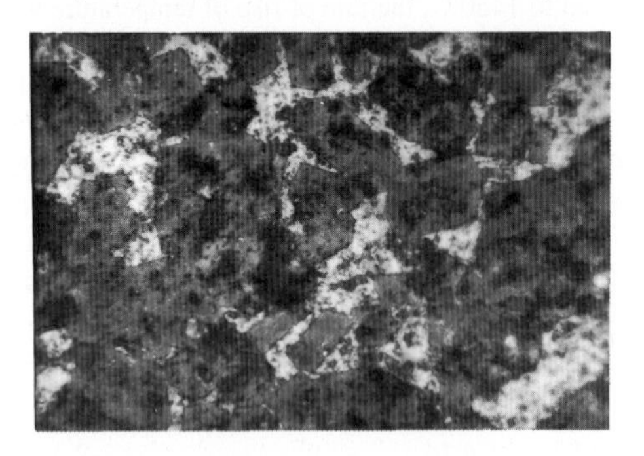

Fig. 4 Optical photograph of sample 7

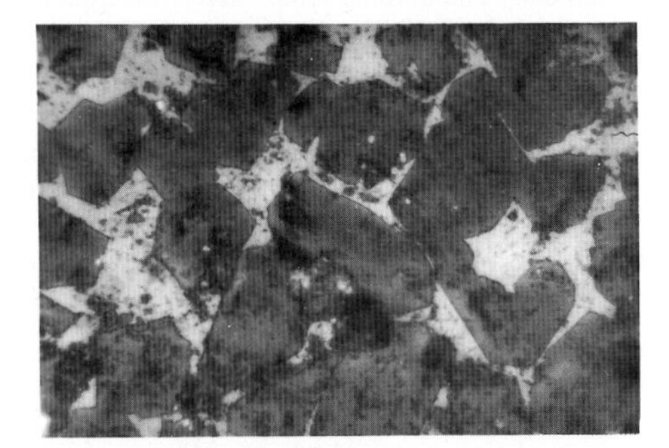

Fig. 5 Optical photograph of sample 8

3.2 The influence of CG used as admixture on strength of the cement

Table 4 shows that if the setting time of the cement is prolonged normal consistency of the cement decreased gradually with the increase of CG substitution. The effect of CG on the mechanical properties was remarkable. An investigation was made over a period of one year. It is found that the CG substitution led to a decrease of strength, and the higher the replacement, the lower the compressive strength. To increase understanding of the influence of the CG substitution on strength, the strengths at various ages are expressed in term of the strength ratio, which is presented in bracket of Table 4. It can be seen that the strength ratios of 3-day and 28-day were approximately equal at the same substitution ratio, and both reduce from 0.95(or 0.96) with 5% CG to 0.68 with 25% CG. However the strength ratios at 3 months and 1 year increased, even exceeding the control group and reached 1.02 within the substitution ratio was 10% CG. This implies that the CG has a positive effect on the long-term strength.

It is easy to understand that cement strength decreased with the increase of CG due to its inactivity. The long-term strength increase was related probably to Fe_2O_3 and FeO, which is contained in CG. The interactions between the oxides and the cement matrix could be that $Fe(OH)_3 \cdot nH_2O$ gradually form through the reaction of Fe_2O_3 in the presence of $Ca(OH)_2$. Even in the atmosphere, FeO can also slowly form $Fe(OH)_3 \cdot nH_2O$ in a solution of $Ca(OH)_2$ [6]. The volume of $Fe(OH)_3 \cdot nH_2O$ is much bigger than that of Fe_2O_3 and FeO, and thereby hardened cement paste becomes more compact. Fig. 3 shows that there is still a small quantity of unhydrated C_2S after curing for 3 months and the strength hence further increased after 3 months. Fig. 2 shows that a part of Fe_2O_3 was wrapped in Fe_2SiO_4. By composition analyses of bare Fe_2O_3 particles using EPM, the results showed that the chemical composition around the Fe_2O_3 particles differed obviously from those of Fe_2O_3 particles in the centre. This implies that a small quantity of $Fe(OH)_3 \cdot nH_2O$ probably formed on the surface of the Fe_2O_3 particles. When the samples were kept in the atmosphere for 1 year, a little $Fe(OH)_3 \cdot nH_2O$ was continuously formed increasing local stress and $Fe(OH)_3 \cdot nH_2O$ lost nH_2O increasing pore ratio. These factors resulted in only a slight decrease of the one-year compressive strengths with high content of Fe_2SiO_4 and Fe_2O_3. This phenomenon will be continuously observed.

3.3 The influence of CG on mineral composition and microstructure of clinker

Table 6 shows that 2.52% free CaO existing in sample 7 was higher than in samples 8 and 9. It indicated that CG helped clinker burning, because CG contains Cu, Zn, Ti, Mn, Ag etc microelements, which can decrease burning temperature of clinker and liquid viscosity [7]. In sample 9, free CaO was of its lowest content, indicating that CaF_2-$CaSO_4 \cdot 2H_2O$ partially shared mineralization effect, greatly accelerating sintering reaction [8]. The results showed that CG used as raw material for cement production is not disadvantageous to cement strengths. An Optical photograph of sample 7 (Fig.4) shows the predominance of hexagonal alite crystals 30~40 μ m in length, and minerals partially eroded and containing about 50% alite. Sample 8 (Fig.5), which was burnt in the same condition, has larger uniform size alite crystals predominantly 50~60 μ m in length, profile of alite clarity and containing over 55% alite. This could be attributed to early formation of a large quantity of liquid phase and consequently enhanced migration of the CaO and SiO_2.

4. Conclusion

When the CG was used as a cement admixture, the early strengths of the cement decreased with the increase of CG, however the long-term strengths of the cement, especially after 3 months, clearly increased. Analyses of EPM show that there was probably $Fe(OH)_3 \cdot nH_2O$ slowly forming through the reaction between Fe_2O_3 and water in the hardened cement paste.

CG substitution percentage by weight of cement should be controlled below 20%, as the compressive strengths at 1 year incorporating more than 20% CG decreased lightly.

CG used as a raw material, due to the Cu, Zn, Ti, Mn, Ag etc microelements contained can decrease clinker burning temperature and liquid viscosity, with advantage to clinker burning. A little CaF_2-$CaSO_4 \cdot 2H_2O$ added to raw meal, is of further benefit to clinker burning.

Reference

[1] F.M.Lea. (1984). The Chemistry of Cement and Concrete, third ed., China construction industry publishing company
[2] J.O.Odigure. (1996). Mineral composition and microstructure of clinker from raw mix containing metallic particles. Cem. Concr. Res. **26:8,** 1171-1178.
[3] J.O.Odigure. (1996). Kinetic modelling of cement raw mix containing iron particles and clinker microstructure. Cem. Concr. Res. **26:9,** 1435-1442.
[4] J.O.Odigure. (1997) Preparation of cement raw mix containing metallic particles. Cem. Concr. Res. **27:11,** 1641-1648.
[5] J.O.Odigure. (2001) Optimization of cement mortar strength from raw mix containing metallic particles. Cem. Concr. Res. **31:1,** 51-56.
[6] Cai Shaohua. (1999) Inorganic Chemistry of Elements, second ed., Zhongshan university publishing company
[7] Wang yanmou et al. (1985) Beijin International Congress on the Cement and Concrete, beijin.
[8] W.A.Kleem and I.Jawed. (1980) Seventh International Congress on the Chemistry of Cements, Paris.

Advances in Building Technology, Volume I
M. Anson, J.M. Ko and E.S.S. Lam (Eds.)

PROPERTIES OF LIGHTWEIGHT AGGREGATE FOR LIGHTWEIGHT CONCRETE PRODUCTION

Tommy Y. Lo, H.Z. Cui

Department of Building & Construction
City University of Hong Kong

ABSTRACT

The increasing demands for the speed of construction and the greater use of prefabrication units in the building industry have urged the need for exploring new innovative construction materials. The adoption of lightweight aggregate concrete in prefabrication of structural panels for building works can minimize the foundation load, reducing the size of structural members, speeds up the construction program and has better control on material quality. All lightweight aggregates are porous in nature that water absorption is much greater than normal weight aggregate. The great absorption rate of aggregate will lowers the water content at the aggregate/cement paste interface and affects critically the mix design, workability of fresh lightweight concrete. Since characteristic of lightweight aggregate varies with the production plants, operation and the raw material used in the production process, the surface characteristic, size and pore structure of the lightweight aggregate differs from one type to the others. Surface topography and pore distribution of the aggregate surface dominate the water absorption characteristic of the aggregate, affect critically the properties of the resulting lightweight aggregate concrete and the bonding with the cement matrix. This paper presented a detail study on the physical properties of a locally available lightweight aggregate made from expanded clay. Density and strength of the lightweight aggregate of different density ranges and their relationship with water absorption characteristic and internal microscopic were examined.

KEYWORDS

Lightweight aggregate, porosity; strength, water absorption, microstructure

1. INTRODRUCTION

The adoption of lightweight aggregate concrete in prefabrication of structural panels for building works can minimize the foundation load, reducing the size of structural members, speeds up the construction program, achieving better quality control and provides a green building construction environment. In other part of the world, high-rise buildings have been built with structural lightweight aggregate concrete [LWAC]. The 42 storeys Prudential Life Building in Chicago and the famous "Nordhordland" bridge are good examples of the applications of structural LWAC (1). A cost model of an eight-storey height office building compared the construction cost of using LWC and NWC reinforced concrete finding that a LWC building could save 1½ week in the construction period. Although the cost of LWAC is 5% higher than traditional normal weight concrete, the overall cost saved 4% (2).

However, the lightweight concrete currently adopted in Hong Kong is limited to non-structural applications (3,4). The available type of ready mixed lightweight concrete is limited to foam concrete or polystyrene bean concrete. Although research finding indicated that structural lightweight concrete with density below 1600kg/m^3 can be made with a locally available expanded clay lightweight aggregate (5), the development on the use of structural LWAC is hindered by the lack of research data on concrete properties, in particular the long-term durability characteristics. Moreover, overseas experience may not apply to Hong Kong because properties of the LWAC vary with the type and characteristic of the lightweight aggregate use.

2. MATERICALS

The LWA used in this experimental work is a lightweight expanded clay aggregate [LECA] produced by bloating clay under high temperature firing of over 1200°C in the rotary kiln [Figure 1 & 2]. Because of the porous nature of the aggregate core, it is extremely light in weight and the insulation property is good. The aggregate is stable, inert to acid corrosion, not attacked by fungi, non-absorbent and frost resistant.

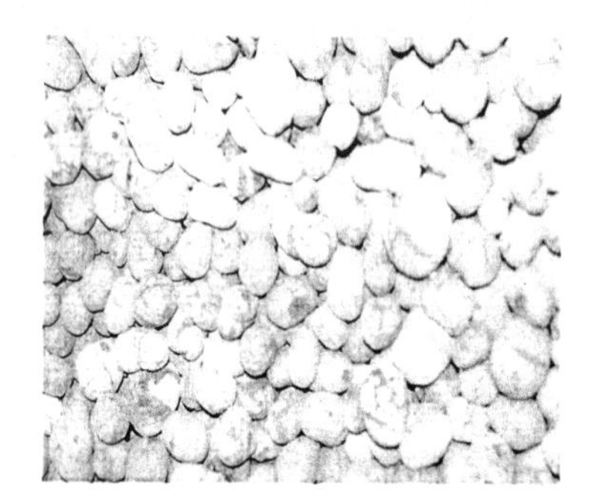

Figure 1: General view of coarse LECA (10mm size)

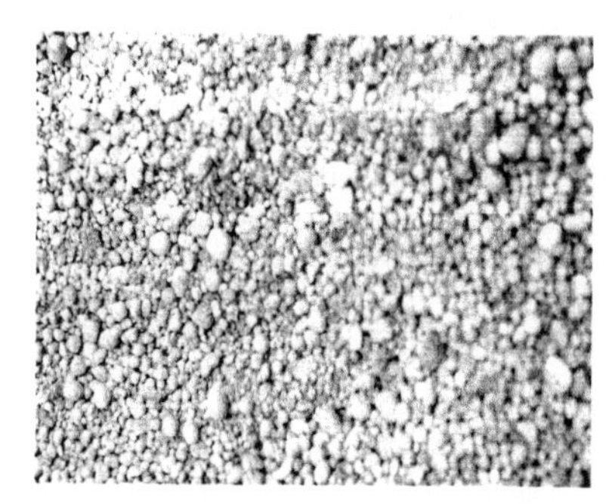

Figure 2: General view of fine LECA

3. RESEARCH SIGNIFICANCE

This paper presented the results on the physical properties, internal microstructure and water absorption of a locally available LECA aggregate. The finding is crucial in understanding the characteristic of LECA aggregate for the development of prefabricating structural lightweight members.

4. RESULTS AND DISCUSSIONS

Bulk density and strength

Most of the lightweight aggregates [LWA] are cellular in structure; the bulk densities are considerably less than those of normal weight aggregate [NWA]. As seen in table 1, 5mm, 14mm and 25mm sizes LECA aggregate were compared. The measured bulk density ranging between 450-850kg/m^3 was only ¼ to ½ the density of NWA. The oven-dry density also fall within the standard ranges of 400~1200kg/m^3 for fine lightweight aggregates and 250~1000kg/m^3 for coarse aggregate (6).

Strength and density of lightweight aggregate varies with the type of aggregate, source of raw materials and production method. Aggregates can be strong and hard or weak and fragile depending on the shape and size of aggregate. Microscopically speaking, it is related to the volume and distribution of the micro pores and whether the pores are inter-connected within the aggregate. Shell thickness is also one of the controlling factors. Simple tube compressive strength was used to examine the average strength of lightweight aggregate. Detail of the tube strength equipment was given in figure 3 (7). Results of the tube strengths of the LECA aggregate are presented in figure 4 and co-related with the bulk densities. Aggregate with a density of 450kg/m^3, 650kg/m^3 and 850kg/m^3 achieved tube strength of 35kN, 50kN and 65kN respectively. In general, aggregate with low bulk densities achieves low tube strengths.

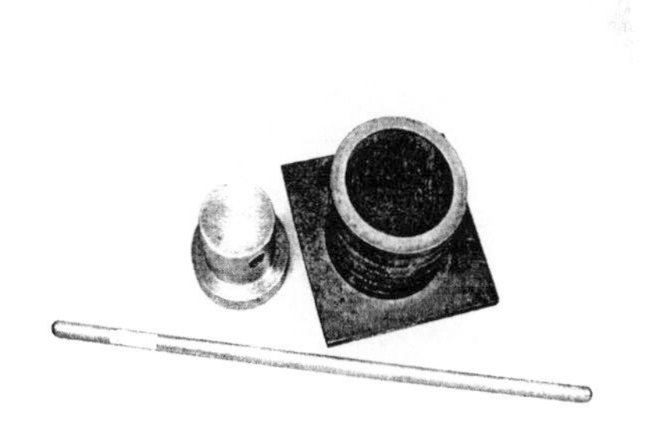

Figure 3 : Equipment for Tube strength determination

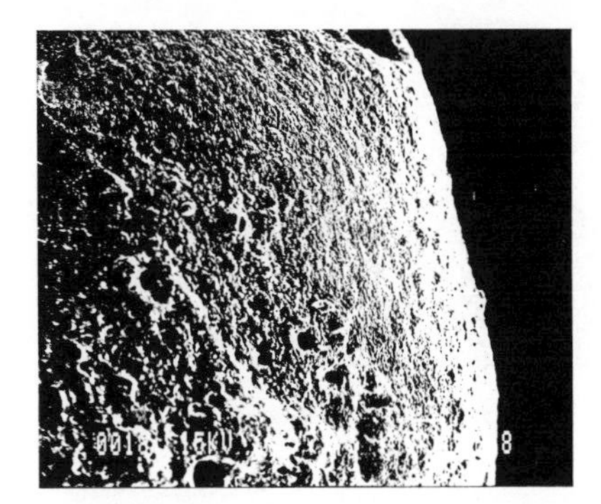

Figure 4 : Surface topography of a typical lightweight aggregate

Microstructure and water absorption of aggregate

Figure 4 shows the surface topography of a typical LECA aggregate. It is seen the aggregate is round with a rough surface texture. The porous surface of the aggregate is good in forming a mechanical bond with the surrounding mortar enhancing the aggregate/cement pastes interface, and reduces the formation of micro-cracks inside the concrete matrix (8).

Water absorption of the porous lightweight aggregates is generally higher than the normal weight aggregate. Finding from past research (9) suggested that seventy percent of the 23-h water absorption took place within the first 30s of absorption. Water absorption rate also differ significantly between cut and whole aggregates. The 1-h water absorption rate of the cut sample was 58% higher than that of whole aggregate. The great variation in water absorption rate was attributed to the presence of an outer shell that controls the ingress of water and the number of voids and fissures that appeared in the aggregate core, pore size and its distribution within the aggregate core affects critically the density, water absorption and strength of the resulting concrete (7).

The cross-sections of the aggregate microstructure were shown in Figs 5,6 and 7. The microstructure of the 3 grades of aggregate with densities 450kg/m^3, 600kg/m^3 and 850kg/m^3 were compared. For aggregate with density 850kg/m^3, it is seen that micro pores were evenly distributed throughout the aggregate particles and were not inter-connected with each other. Comparing to the aggregate of density 450kg/m^3, the internal microstructure composed with large inter-connected voids and fissures. In general, the number of inter-connected pores increases with the decrease in bulk density of aggregate.

Water absorptions of the 3 grades of aggregate were compared and results were shown in Figure 8. It indicated that the initial absorption of 6mm aggregate was greater than that of 25mm. The absorption rate of 14mm aggregate was the lowest. Around 50-65% of the 24 hours absorption took place at the first few minutes. The absorption increases with time but the absorption ranking was the same. The finding confirmed with past research result by Swamy (9). Moreover, the absorption rate was irrelevant to the size of aggregate.

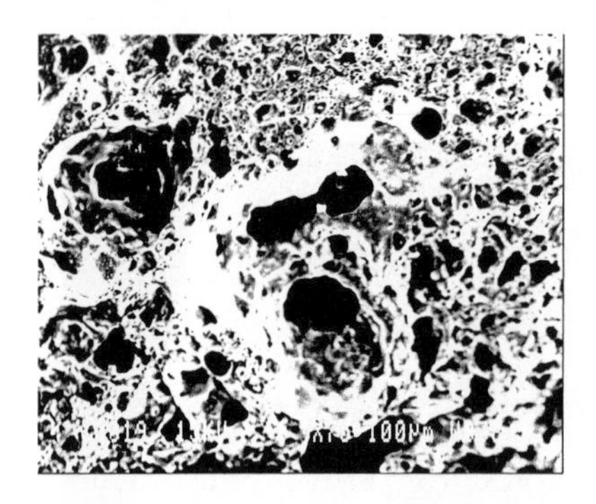

Figure 5: Lightweight Aggregate (Density 450 kg/m^3)

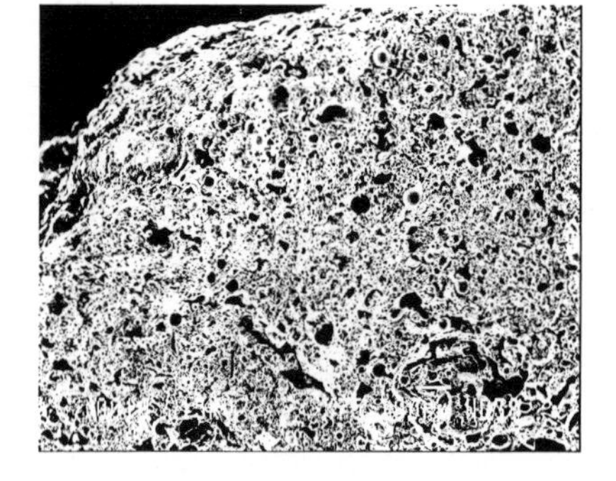

Figure 6: Lightweight Aggregate (Density 600kg/m^3)

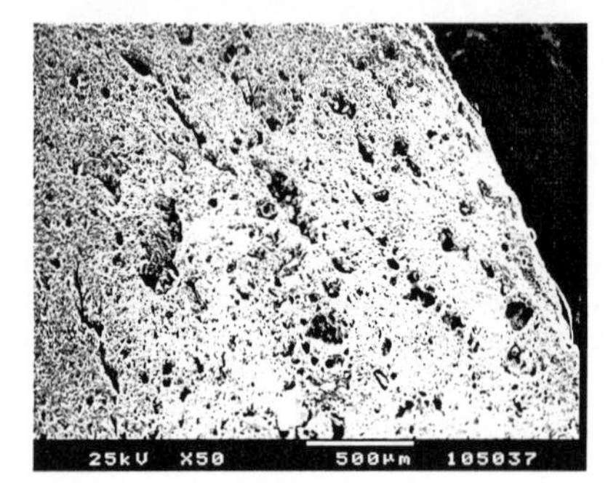

Figure 7: Lightweight Aggregate (density 850 kg/m^3)

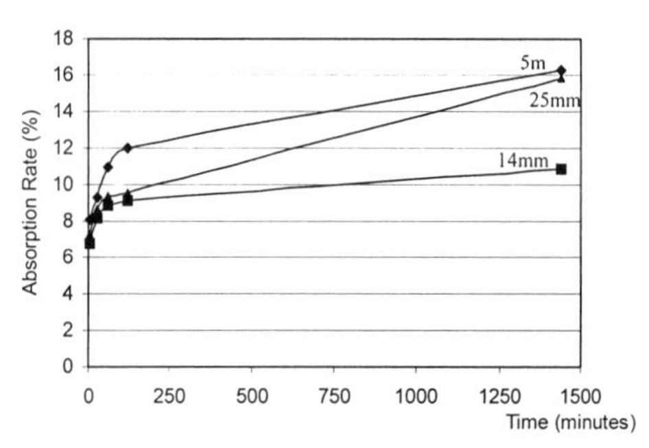

Figure 8: Water absorptions of the lightweight aggregates

Table 1 summarized the pore distribution of the 3 aggregate samples with different density. It is seen that the majority of the pore are sized between 1µm to 100µm for all three density graded aggregate. When the total volume of micro pore increases, the aggregate density decrease. This also explains why water absorption of cut (or fragmented) aggregate was greater than that of whole aggregate. When the lightweight aggregate broken, the number of open pores increases leading to a larger water absorption rate of the absorption. Therefore, a thick and dense aggregate shell is necessary in ensuring the blockage of water path passing into the aggregate core.

TABLE 1
PORE SIZE AND THE VOLUME DISTRIBUTION WITHIN CORE OF LIGHTWEIGHT AGGREGATE

Aggregate Density (kg/m^3)	Aggregate size	Pore Distribution				
		<0.01 µm	0.01-1µm	1-10 µm	10-100 µm	>100 µm
400	25 mm	0.3	87.4	728.6	171.8	7.9
600	5 mm	0.4	36.8	652.3	156.5	3.9
800	14 mm	0.2	53.5	459.6	102.1	4.5

* Above data is the average of ten samples

It is also interesting to note that the density of aggregate does not necessarily directly related to the size. For example, the density of the 5mm aggregate was greater than that of the 25mm aggregate but was lower than the density of a 14mm aggregate. This phenomenon can be explained by the presence of substantial amount of large pore (>100µm) within the 25mm sized aggregate.

5. CONCLUSION

Based on the findings, the following conclusions can be drawn:

1) Pores size of the locally available LECA aggregates mainly ranged between 1 µm to 100 µm.
2) When the total volume of micro pore decreases, the aggregate density increase.
3) The initial water absorption is 50-65% of the 24 hours absorption
4) Tube strength of aggregates is related to the pores volume and bulk density. The aggregate strength decreases with the increase of total pore volume within the aggregate core.
5) The number of inter-connected pores increases with the decrease in bulk density of aggregate.

6. ACKNOWLEDGEMENT

The work described in this paper was fully supported by a grant from the Research Grants Council of the Hong Kong Special Administrative Region, China [Project No. CityU 1037/01E]

7. REFERENCES

1. CEB-FIP. (2000). Lightweight Aggregate Concrete, part 3, International Federation for Structural Concrete..
2. Institution of Structural Engineers. (1987). "LWAC Guide: lightweight aggregate", *Institution of Structural Engineers/Concrete Society,* p.11-31. London..

3. Lo, T. Y. (1992). "The Application of Lightweight Precast Wall Systems to Housing Block of Hong Kong." *The Third International Symposium on Noteworthy Applications in Concrete Prefabrication.* Singapore, 13-15 July, (p95-101).
4. Lo, T. Y. (1995), "Lightweight Concrete - A Challenge in Housing Construction," Proceedings of XXIII IAHS World Housing Congress, Excellence in Housing : Prospects and Challenges in the 'Pacific' Century, Vol. 2, Singapore. 25-29 Sept.
5. Lo, Y, Gao, X. F. and Jeary, A. P. (1999), Microstructure of Pre-wetted Aggregate on Lightweight Concrete," Building and Environment. Vol. 34. No. 6.
6. BS3797. (1990) *Specification for lightweight aggregates for concrete.* Part 2, BSI, London. U.K.
7. GB 2839-81. (1981) *National Standard for Production of Lightweight Expanded Clay Aggregate,* People Republic of China.
8. Mays G C & Barnes R A. (1991) Performance of Lightweight Aggregate Concrete Structures in Services Royal Military Coll of Science, *Structural Engineers* Vol 69, No. 20, 351-361, 15 Oct.
9. Swamy, R.N. and Lambert, G.H. (1981). "The microstructure of Lytag aggregate", *Journal of Cement Composite and Lightweight Concrete,* Vol. 3, No. 4, Nov. pp.273-282.

Advances in Building Technology, Volume I
M. Anson, J.M. Ko and E.S.S. Lam (Eds.)

PRELIMINARY RESULTS ON THE USE OF GFRP AS AN EXTERNAL REINFORCEMENT FOR AUTOCLAVED AERATED CONCRETE

Ali M. Memari* and Ayca Eminaga*

*Department of Architectural Engineering, The Pennsylvania State University
104 Engineering A, University Park, Pennsylvania, USA

ABSTRACT

The bond strength between Glass Fiber Reinforced Polymer (GFRP) sheets and Autoclaved Aerated Concrete (AAC) surface is investigated. Specimens made of two small AAC blocks attached together with two strips of GFRP are pulled apart by using a hydrolic jack. Bond length is a variable in the study. The results show that the bond strength between AAC and GFRP is comparable to that between Carbon FRP and normal concrete surface.

KEYWORDS

Fiber reinforced polymer, autoclaved aerated concrete, bond strength, external reinforcement.

INTRODUCTION

Although Autoclaved Aerated Concrete (AAC) products have been used in Europe and several other countries in Asia and the Middle East for a few decades, this lightweight concrete has been introduced in the U.S. market primarily since 1995. According to Woodard *et al.* (1999), "This new building material combined the advantage of wood and cement products – it could be cut, drilled, and nailed like wood while maintaining high strength, excellent durability, thermal insulation, sound insulation, and fire resistance. Subsequently, it gained rapid acceptance in Europe and throughout the world." There is good optimism for the growth of AAC in the U.S. market. Chusid (1999) predicts that if the announced plans for a few more factories realize, "there will be three million cubic yards of annual capacity in North America within the next two to three years, equivalent to over 150 million square of wall, floor and roofs and enough for an estimated 10,000 to 15,000 buildings per year." Currently, the primary use of AAC in the U.S. is in the form of masonry blocks, wall and floor panels, and lintels. Panels and lintels are reinforced to carry superimposed loads. Three of the key references on AAC are Wittmann (1983), Wittmann (1992), and Aroni *et al.* (1993). The book by Wittmann (1983) includes a bibliography chapter on AAC that lists 492 documents on AAC. Even though many publications exist on AAC, only a fraction of the studies have addressed structural aspects.

Unlike Normal Weight Concrete (NWC), one of the deficiencies associated with the use of steel bars to reinforce AAC elements is the weakness of the bond between the steel surface and the surrounding AAC. According to Aroni *et al.* (1993), because of the low alkali reserve and high porosity of AAC, reinforcement should be covered with anti-corrosion material (such as protective bituminous coating). According to Hanecka (1992), because of corrosion protection coating, the bond properties and their effects on the response of AAC members under load are not sufficiently understood. Aroni *et al.* (1993) clearly mentions the deficiency: "The bond strength between the reinforcement and AAC is much lower than that in normal concrete." Because of such a problem, the anchorage of reinforcement cannot be provided by bond, and therefore, cross bars welded to longitudinal reinforcement must be used. In such cases, the anchorage is provided by the bearing of the cross bars against the AAC. The pullout strength of longitudinal bars is limited by the smaller of the spot weld capacity of cross bar-longitudinal reinforcement connections and bearing capacity of the cross bar-AAC interface (Boutros and Saverimutto 1997). Therefore, practically the bond strength between steel and AAC is ignored. However, according to Aroni *et al.* (1993), if any bond strength is to be allowed in design, the capacity has to be verified by a pullout test based on RILEM Recommendation AAC 8.1.

In an effort to find an alternative reinforcement system or an approach for supplemental reinforcement (in addition to what is usually provided in precast AAC members), this paper explores the possibility of using Fiber Reinforced Polymers (FRP) as an externally applied reinforcement. Before the experimental program is described, a brief background on the use of FRP on NWC and the basis for the experimental study reported here is mentioned.

BACKGROUND ON THE USE OF FRP ON NWC AND THE BASIS FOR THE EXPERIMENTAL STUDY

Over the past decade, FRP composites have evolved as reliable alternative reinforcing products for repair and rehabilitation of aging concrete and masonry structures. Currently, three types of fibers, namely, carbon, aramid, and glass are widely manufactured, which when the fibers are embedded in a polymeric resin (matrix), the composites are referred to, respectively, as CFRP, AFRP, and GFRP. From the three types of composites, CFRP provides the largest strength and is commonly used to strengthen existing reinforced concrete members. Given that AAC is a weaker material then NWC, the use of high strength CFRP is not justified (Vadovic 2000) and therefore, GFRP was determined to be suitable for use with AAC in this study.

The effectiveness of using externally bonded FRP composites to strengthen concrete beams has been well established. The primary objective of this study is to investigate the suitability of externally bonded GFRP composites to be used for AAC lintels and floor and wall panels. However, before experiments on beams and wall panels are carried out, the characteristics of the bond between the externally applied GFRP sheets and AAC surface should be determined. Several investigators have performed such studies for bond characteristics between CFRP and NWC. Maeda *et al.* (1997) used two small concrete blocks (100 mm x 100 mm in cross section) that were attached together with 50 mm wide strips of CFRP on two sides of the blocks, but with a gap between the blocks. The tension force to the CFRP was applied by pulling two steel rebars that were inside the concrete prisms. In that study, bond length and concrete strength was varied and the resulting data provided information on the variation of bond strength with bond length and concrete strength in addition to strain distribution. Brosens and Van Gemert (1997) have also performed tension tests on two concrete prisms (150 mm x 150 mm x 300 mm) attached with CFRP laminates on two opposite sides. Tensile force is applied on steel plates glued on the other sides (opposite to CFRP sheets) of the prisms. During the displacement controlled tests, information on the applied load, relative movement of the prisms and the strains of the

CFRP were taken. In that study, too, the bond length and concrete strength were varied. Representative results from these studies will be compared with the results of AAC tests reported in this paper.

EXPERIMENTAL PROGRAM

The diagram of the test setup for this study is shown in Figure 1. The test specimen consists of two AAC prisms (100 mm x 100 mm x 275 mm) connected together by two GFRP strips, which are attached to two opposite faces of the AAC prisms. In order to apply tension to the GFRP strips, a loading jack is placed between the two AAC blocks such that it can push them apart. The loading jack, which is operated by a hydraulic pump, has a 88 kN capacity.The test specimen is loaded in a load controlled manner. The load on the specimen is obtained from reading the pressure recorded. Displacements of the prisms are recorded using potentiometer and also an oscilloscope connected to the potentiometer. Figure 2 shows a photograph of the test setup. In the first part of the study reported here, only bond length was varied. The GFRP strips used had a width of 50 mm, but the bond length was varied from 50 mm to 200 mm at 50 mm increments.

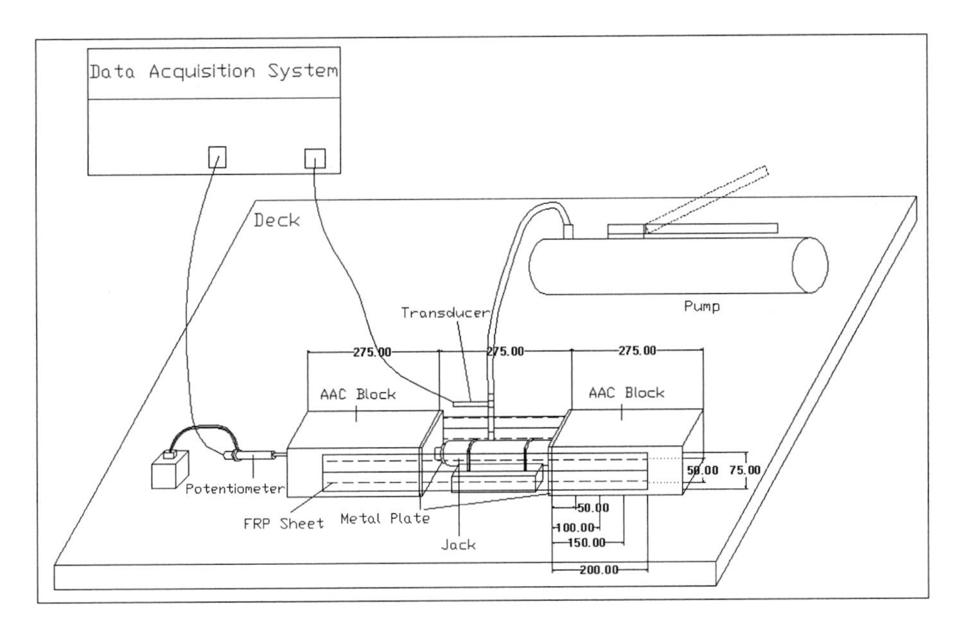

Figure 1: Illustration of test set up (dimensions are in mm).

The AAC used, which was produced by a HEBEL plant in the state of Georgia in US and now owned by Babb International, Inc., Matrix PAAC Division, had a nominal compressive strength of 5 MPa and a unit weight of 720 kg/m^3. The GFRP used is manufactured by Master Builders, inc. and is designated MBrace EG 900 E-glass fibers, which according to Tysl (2000), is more economical compared to carbon or aramid FRP for use with AAC. The matrix (resin) used consists of MBrace Primer, MBrace Putty, and MBrace Saturant.The EG 900 material used is unidirectional, weighs 900 grams/m^2, and has a strength of 525 N/mm of width. Table 1 shows the matrix of specimen configurations used in this part of the study. For each bond length, three tests were performed. Further tests are ongoing (but not reported here) to investigate the effects of GFRP width and the AAC compression strength.

Figure 2: A photograph of the test set up of specimen d1.

TABLE 1
Specimen Configurations and Experimental Results

Specimen	Bond Width	Bond Length	Bond Area	Failure Load	Failure	Bond Strength
	mm	mm	mm2	(Fu)	Displacement	(Fu/BA)
			(2*BW*BL)	KN	mm	Mpa
a1	50.00	50.00	5000.00	11.14	2.44	2.16
a2	50.00	50.00	5000.00	10.70	2.41	2.07
a3	50.00	50.00	5000.00	10.93	1.80	2.12
b1	50.00	100.00	10000.00	13.795*	2.48	1.335*
b2	50.00	100.00	10000.00	14.40	1.97	1.39
b3	50.00	100.00	10000.00	12.77	3.84	1.24
c1	50.00	150.00	15000.00	17.53	2.08	1.13
c2	50.00	150.00	15000.00	14.76	3.36	0.95
c3	50.00	150.00	15000.00	16.36	3*	1.06
d1	50.00	200.00	20000.00	17.69	3.79	0.86
d2	50.00	200.00	20000.00	18.00	2.19	0.87
d3	50.00	200.00	20000.00	17.13	3.27	0.83

*; Data could not be received. So assumptions are made regarding to other similar specimens to plot result graphs.

TEST RESULTS AND DISCUSSION

The results of the tests are listed in Table 1 and also plotted in Figures 3 to 5. Moreover, photographs of typical failure modes are shown in Figures 6 to 9 (one for each bond length). As can be seen in Figure 3, an increase in bond length does not necessarily lead to an increase in bond strength. This result has also been reached in the Maeda *et al.* (1997) study. Looking at the photographs of failed specimens helps the understanding of the behavior. Figure 6 shows the mode of failure of specimen a3 with a bond length of 50 mm. As can be seen, the failure is not a bond failure; rather, it is a shear failure in the AAC prism. Figure 7 shows the typical failure mode for specimens b1 to b3 with a bond length of 100 mm. The failure mode here seems to be a combination of bond failure and shear failure. The same type of failure can be seen in Figure 8 for specimens cl to c3. However, Figure 9 shows a pure bond failure for the longest (200 mm) bond length used.

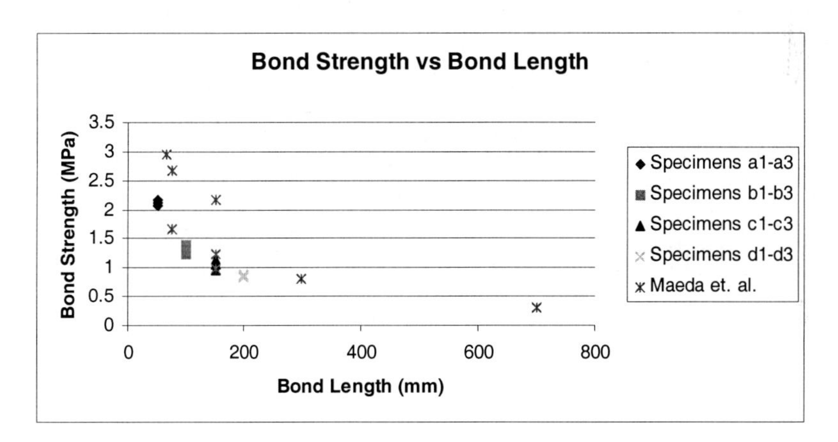

Figure 3: Graph of bond strength versus bond length.

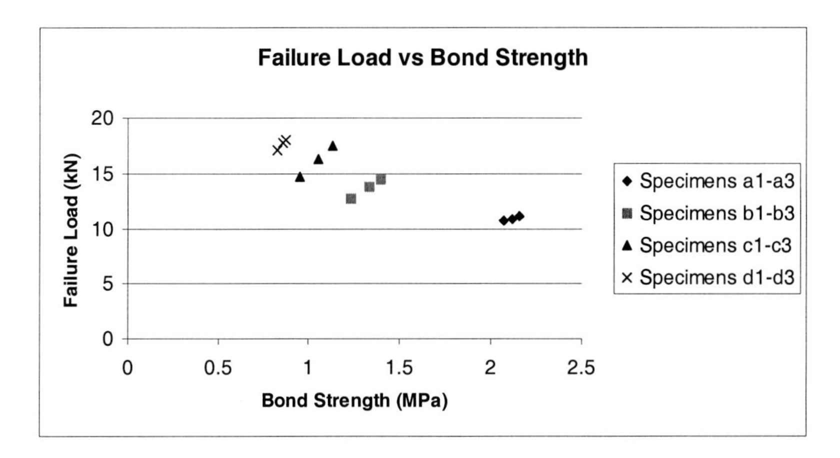

Figure 4: Graph of failure load versus bond strength.

The plotted results in Figure 3 show that bond strength between GFRP and AAC surface for longer contact surface approaches 0.83 MPa (for bond length of 200 mm) from a high value of 2.16 MPa for the short bond length of 50 mm. It should be noted that the bond strength values are only nominal values obtained by diving the failure load by the bond area (on two faces) regardless of the mode of failure. For practical situations, where the FRP sheet is applied rather continuously at the bottom of a beam, the bond strength to be expected is then the value close to that for the longest bond length, i.e., approximately 0.83 MPa. According to the study by Maeda *et al.* (1997) as shown in Table 2, bond strength between CFRP and NWC varies from 0.31 to 2.94 MPa with bond strength decreasing with increasing bond length. A significant observation here is that this range of bond strength is very close to that obtained in this study. Of course, tests on the use of GFRP on NWC will be conducted as a follow-up in this study to confirm this conclusion. Yet another observation is that while the ratio of an average bond strength for long bond lengths (0.75 MPa) from Maeda *et al.* (1997) study to the NWC compressive strength (say 43.3 MPa) is 0.017, the ratio from this study is 0.83/5.0 or 0.166, or 10 times larger. This is another reason for the suitability of the use of GFRP as an external reinforcement of AAC flexural members.

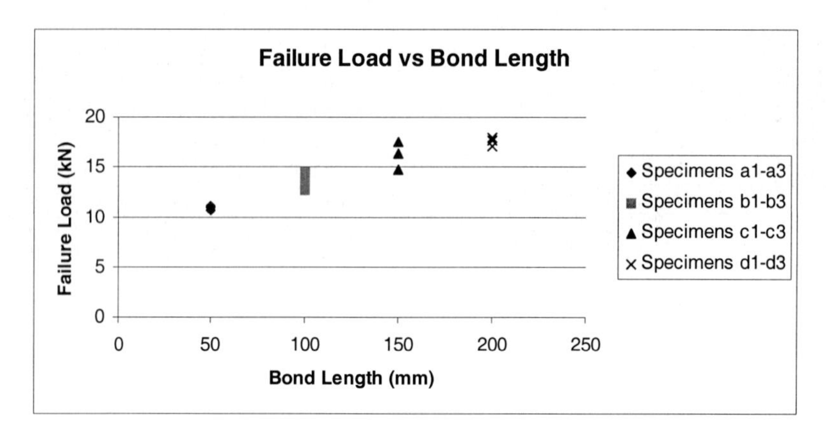

Figure 5: Graph of failure load versus bond length.

TABLE 2.
Experimental Results of *Maeda et. al. (1997)*

Bond Length(mm)	Ultimate Load (kN)	Average Bond Strength (MPa)
75	11.6	1.67
150	18.4	1.23
300	23.9	0.8
75	20	2.67
150	14.6	0.97
65	19.1	2.94
150	32.5	2.17
700	20	0.31

Figure 6: Failure of specimen a3.

Figure 7: Failure of specimen b2.

Figure 8: Failure of specimen c3.

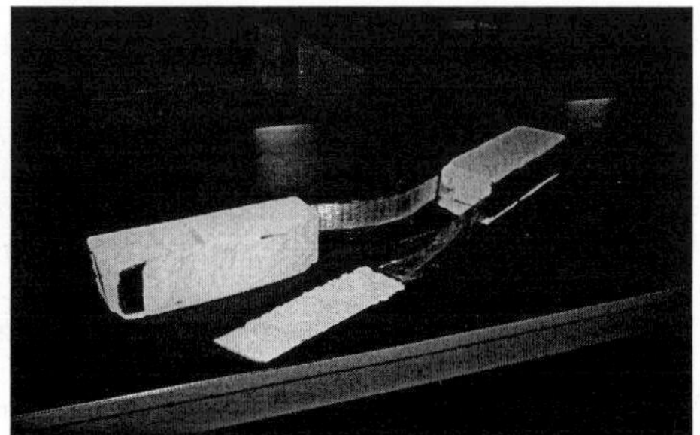

Figure 9: Failure of specimen d2.

CONCLUSION

The study has shown that the bond strength between AAC and GFRP is comparable to that between NWC and CFRP. This shows the stability of GFRP sheet to be used as an external reinforcement on NWC. Further tests are needed, however to establish the characteristics of bond and strength enhancement when GFRP sheets are applied to beams and panels for use under flexural conditions.

ACKNOWLEDGEMENT

The materials for this study were donated by Babb International, Inc. and Master Builders, Inc. and Cyprus Fulbright Commision has provided a fellowship for the second author for an M.S. Degree in Architectural Engineering at Penn State University. All contributions are gratefully acknowledged.

REFERENCES

Aroni, S., de Groot, G. J., Robinson, M. J., Svanholm, G. and Wittmann, F. H., (1993). *Autoclaved Aerated Concrete Properties, Testing, and Design – RILEM Recommended Practice*, RILEM Technical Committees 78-MCA and 51-ALC, E & FN SPON.

Boutros, M. and Saverimutto, L., (1997). Anchorage Capacity of Reinforcing bars in Autoclaved Aerated Concrete Lintels, *Materials and Structures*, **Vol. 30, No. 203**, pp. 552-555.

Brosens, K. and Van Gemert, D., (1997). Anchoring Stresses Between Concrete and Carbon Fibre Reinforced Laminates, *Proceedings of the 3rd International Symposium on Non-metallic (FRP) Reinforcement for Concrete Structures*, **Vol. 1 Oct. 1997**, pp. 271-278.

Chusid, M., (1999). Building with Autoclaved Aerated Concrete, *Masonry Construction*, **January 1999**, pp. 24-27.

Hanecka, K. (1992). The Bond Stresses of AAC and Slips of Reinforcement Bars, *Advances in Autoclaved Aerated Concrete*, Wittmann (Ed.), Balkema, Rotterdam, pp. 181-186.

Maeda, T. Asano, Y., Sato, Y., Ueda, T. and Kakuta, Y. (1997). A Study of Bond Mechanism of Carbon Fiber Sheet, *Proceedings of the 3rd International Symposium on Non-metallic (FRP) Reinforcement for Concrete Structures*, **Vol. 1 Oct. 1997**, pp. 279-286.

Tysl, S. (2000). Master Builders, Inc., Personal Communications.

Vadovic, M. (2000). Gannett Fleming, Inc., Personal Communications.

Woodard, J, Barnett, R., and Schmidt, S., (1999). Design and Utilization of Autoclaved Aerated Concrete Masonry, *Proceedings, 8th North American Masonry Conference*, **June 6-9, 1999**, Austin, Texas, pp. 335-345.

Wittmann, F. H., (1983). *Autoclaved Aerated Concrete, Moisture and Properties*, (Ed.), Proceedings of the RILEM International Symposium on Autoclaved Aerated Concrete, Lausanne, Switzerland, March 1982, Elsevier,

Wittmann, F. H., (1992). *Advances in Autoclaved Aerated Concrete*, (Ed.), Proceedings of the 3rd RILEM International Symposium on Autoclaved Aerated Concrete, Zurich, Switzerland, 14-16 October 1992, A.A. Balkema, Rotterdam.

Advances in Building Technology, Volume I
M. Anson, J.M. Ko and E.S.S. Lam (Eds.)

HIGH PERFORMANCE FIBRE REINFORCED CONCRETE

S. Mindess[1] and A.J. Boyd[2]

[1] Department of Civil Engineering, University of British Columbia, Vancouver, British Columbia, V6T 1Z4, Canada
[2] Department of Civil and Coastal Engineering, University of Florida, Gainesville, Florida, 32611, USA

ABSTRACT

Since its introduction in the late 1960s, fibre reinforced concrete (FRC) has become an increasingly common building material. It is estimated that, in North America, over 15% of the concrete produced now contains fibres of some kind. At present, almost all of the FRC is used in relatively "low level" applications: slabs-on-grade, control of plastic shrinkage, fibre shotcrete for repairs and tunnel linings, and so on. However, there are still very few truly *structural* applications of FRC; it is rarely used in conjunction with conventional steel reinforcement to improve the mechanical behaviour of concrete in civil engineering structures. As well, fibres are still most commonly used with relatively low strength concretes. In the present work, the use of fibres with both high strength concrete (HSC) and steel reinforcement is described. It is shown that with conventionally reinforced HSC, the fibres improve the post-peak behaviour of the system. Under impact loading, fibres greatly improve the toughness of the composite, particularly when the concrete is subjected to uniaxial or biaxial confinement; fibres may also change the mode of failure, from brittle shear fracture to more ductile flexural failure in both beams and plates.

KEYWORDS

Fibre reinforced concrete, plates, beams, impact loading, uniaxial confinement, biaxial confinement, steel reinforcement.

INTRODUCTION

Plain concrete is a brittle material, strong in compression, but with very low tensile stress and strain capacities. In traditional concrete construction, continuous steel reinforcing bars are used to carry the tensile and shear forces in concrete structures, and to provide some degree of ductility (or energy absorption capacity). However, since about 1970, *fibre reinforced concrete* (FRC) has become an increasingly common building material. The short, discrete, randomly distributed fibres bridge across the microcracks that develop as the concrete is subjected to stress (due to loading, drying shrinkage, thermal changes, and so on), and may impart considerable *"pseudo-ductility"* to the concrete. Thus, the FRC may continue to carry significant stresses at deflections far beyond those required to crack the plain matrix; that is, it acquires *post-peak load carrying capacity.*

Currently in North America, it is estimated that about 15% of the concrete that is placed contains fibres of some type. The majority of the FRC is used in what might be termed "low-level" (i.e., non-structural) applications: for control of plastic shrinkage cracking, in slabs-on-grade, for fibre shotcrete tunnel linings, and for repair of damaged concrete. There are still very few truly structural applications of FRC, in which it is used in conjunction with conventional reinforcing bars to help carry the design loads and deformations. Moreover, fibres are still most commonly used with relatively low strength (ordinary) concretes.

In the present work, the use of fibres with both high strength concrete (HSC) and steel reinforcement is described, with particular reference to recent experimental work carried out at the University of British Columbia. It is shown that, with conventionally reinforced HSC, fibres greatly improve the post-peak behaviour of the system. Under impact loading, fibres greatly improve the toughness of the composite materials, particularly when the specimens are subjected to lateral confinement; in these cases, the fibres may also lead to a change in the mode of failure.

COMPRESSION

Plain concrete fails in a relatively brittle manner. For HSC ($f'_c > 70$ MPa), it exhibits nearly elastic behaviour up to a stress of about 0.7 f'_c; at higher stress levels its behaviour becomes increasingly nonlinear as the microcracks which have developed begin to coalesce, leading eventually to failure. The addition of fibres to a high strength matrix has little effect on f'_c, though it tends to increase somewhat the stress level at which the σ-ε curve becomes noticeably nonlinear. However, by bridging across the cracks, the fibres lead to multiple cracking and hence the ability of the FRC to exhibit much greater strains at failure. For fibre volumes less than about 1.5%, the fibres simply lead to the formation of more (and finer) cracks, and consequently to a greater toughness. For higher fibre volumes, however, a ductile mode of failure occurs, characterized by the development of a few inclined shear cracks. The effects of high volumes of various fibres in a HSC matrix ($f'_c = 70$ MPa) are shown in Fig. 1.

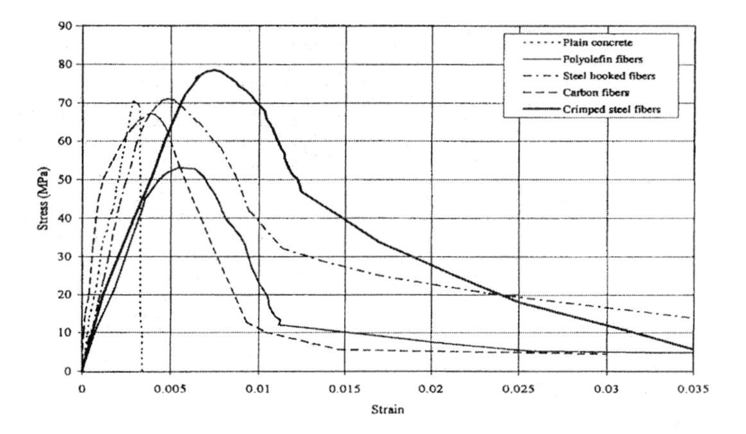

Figure 1(a): Stress-strain curves for HSC at 2.0% by volume of fibres (adapted from Campione and Mindess (1999)).

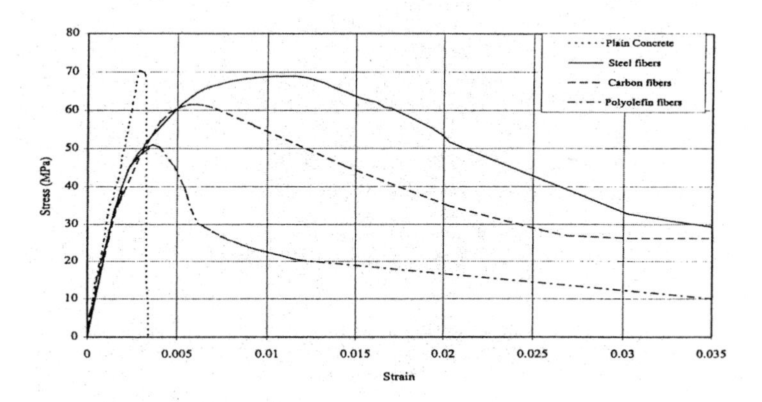

Figure 1(b): Stress-strain curves for HSC at 3.0 percent by volume of fibers (adapted from Campione, Mindess and Zingone, (1997)).

The slight reduction in strength at the highest fibre volumes is due to the difficulty of compacting the concrete fully at such fibre volumes. For similar volumes of carbon fibres, greater reductions in strength were observed, again due to difficulties in compaction, though the increases in toughness were about the same as those found with the steel fibres (Campione *et al.*, 1999b).

The real advantage of using high volume FRC, however, comes when the fibres are combined with conventional steel reinforcement. (Campione et al., 1997; Campione and Mindess, 1999; Campione et al, 1999b; Campione and Mindess, 2000; Campione, et al. 1999). When FRC cylinders containing more than 2% by volume of fibres were reinforced with spiral steel reinforcement, a number of effects were noted:

- Both the ultimate strength and the corresponding strain were increased;
- Beyond the peak load, the failure of the specimen was characterized by successive yielding of the steel spirals at different sections. This occurred because the fibres were able to carry the load transferred to the matrix when a spiral ruptured, as may be seen in Fig.2.

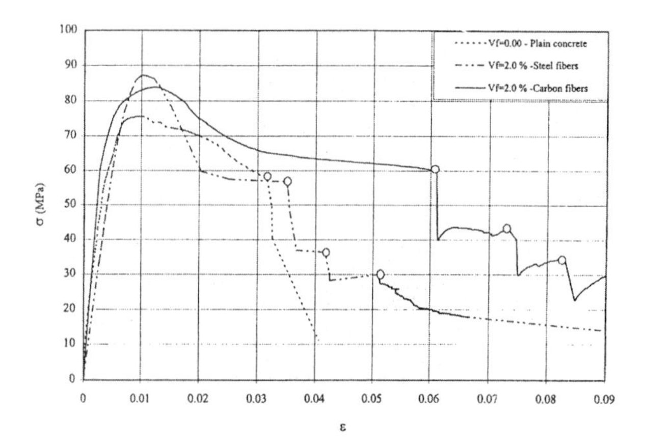

Figure 2: Stress-strain curves for confined HSC at different fiber volumes spiral reinforcement ρ=0.02592 -o failure of steel spirals. (adapted from Campione and Mindess, 1999).

- Thus, the combination of FRC and steel spirals may lead to a considerable increase in the specimen ductility (or ability to absorb energy).
- The maximum strain corresponding to the failure of the steel spirals increased.
- The concrete cover over the spiral steel exhibited less spalling; primarily, spalling was confined to the regions of rupture of the spiral steel.

Fibres combined with spiral steel reinforcement were also found to be very effective under *cyclic* compressive loading. In these tests, specimens similar to those described above were tested under cyclic loading, between 90% of the ultimate stress and 10% of the ultimate stress, to obtain the complete envelope of the cyclic σ–ε behaviour, as shown in Fig.3.

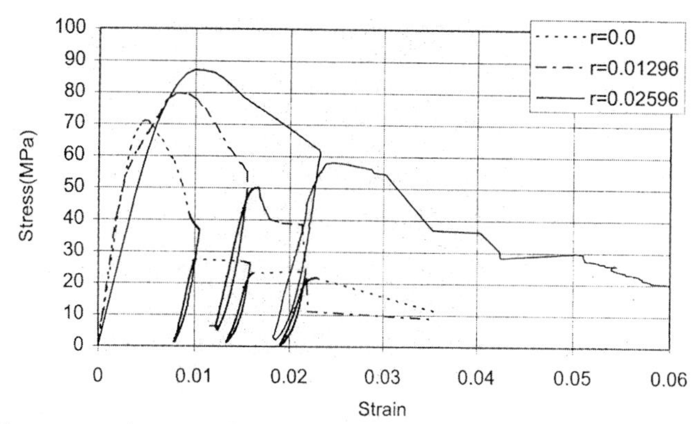

Figure 3: Cyclic stress-strain curves for confined HSC with 2.0% by volume of hooked steel fibers and different steel spiral volumes (Campione et al., 1999a)

The envelope of the cyclic loading curves corresponds closely to that of the monotonic σ–ε curve (measured on a companion specimen). In addition,

- The failure strain under cyclic loading between two fixed stress levels depends only on the maximum stress level in the matrix;
- The toughness of the material under cyclic loading is higher than that under monotonic loading.

FLEXURE

As in the case of compression, the addition of high fibre volumes also increases the residual (post-cracking) strength and energy absorption capacity (toughness) of high strength concrete beams. Under cyclic loading (Campione *et al.*, 2000) showed that FRC with a fibre volume of 3% could withstand very high strains without significant losses in strength. At even higher fibre volumes (≈3.8%), Balaguru and Franklin (1999) showed that steel FRC could withstand approximately 75% of the ultimate load at large deflections, even under reverse cyclic loading. This has obvious implications for concrete structures in severe seismic areas.

High fibre volumes are particularly effective in increasing the shear capacity of high strength, conventionally reinforced beams (Campione and Mindess, 1999; Campione *et al.*, 2000a; Campione *et al.*, 2000b). Although there was no particular increase in the load carrying capacity of the FRC beams compared to those made with a plain concrete matrix, the brittle compressive failure of plain concrete over-reinforced beams was transformed to a "ductile" compressive failure. However, the current provisions for calculating ductility factors appear not to be suitable for FRC reinforced beams.

IMPACT

Both plain concrete and FRC are highly strain rate sensitive materials, with their apparent strengths increasing with increasing strain rates. However, we are still unable to predict their behaviour under impact or dynamic loading on the basis of static tests. Also, because of the inherent difficulties in carrying out properly instrumented impact tests, the available experimental data are often *artifacts* of the particular testing system and specimen geometry employed. Nonetheless, there are enough data reported to permit the following general observations to be drawn:

- The addition of fibres to concrete increases the measured impact strength. The increase obtained in a particular test depends upon the fibre volume, fibre geometry , and matrix strength.
- As with plain concrete, the fracture energy (toughness) of FRC is increased under impact loading (Banthia *et al.*, 1994).
- The strain rate sensitivity of FRC increases both with increasing fibre content and fibre aspect ratio. This is generally attributed to the strain rate sensitivity of the fibre-matrix bond strength (Naaman and Gopalaratnam, 1983).
- Adding fibres to the concrete matrix will significantly improve the bond between the concrete and conventional steel reinforcing bars (Yan and Mindess, 1991; Yan and Mindess, 1994). The composite becomes more ductile, absorbs more energy, and exhibits a higher ultimate bond stress.

The effects of fibres can be particularly dramatic when the FRC is subjected to multiaxial loading. As shown in detail by Sukontasukkul (2001) and Sukontasukkul *et al.* (2001).

- When compression specimens are confined laterally, the mode of failure changes from the usual shear cone appearance to a columnar or vertical splitting type. Both the strength and strain at peak load are increased.
- For beams and plates, an increased lateral confining stress gradually changes the mode of failure from flexure to shear. Both the strength and toughness increase with increasing confinement.
- Higher confining stresses and higher fibre contents permit more impact energy to dissipate through the specimens, and at the same time permit the specimens to absorb more energy.

CONCLUSIONS

The combination of high strength concrete, high fibre volumes and conventional reinforcement can lead to dramatic improvements in the mechanical behaviour of concrete members, for both static and dynamic loading conditions. Despite the higher *material* costs of such systems, both the structural properties and durability of concrete structures can be greatly improved. As yet, the available building codes do not make it easy to exploit these improved properties, since most codes are based entirely on *strength* considerations, and do not take into account the *toughness* of the material. It can only be hoped that the codes will soon reflect the advances made in enhancing the properties of cement-based composite materials, so that this new generation of construction materials can be exploited to their full extent.

REFERENCES

Balaguru P. and Franklin H.S. (1999). 'High Performance User-Friendly Fiber Reinforced Composite Under Cyclic Loading', in: Reinhardt H.W and Naaman A.E. (eds.), *High Performance Fiber Reinforced Cement Composites (HPFRCC3)*, RILEM Publications, S.A.R.L.: 225-238.

Banthia N., Chokri C., Ohama Y. and Mindess S. and (1994). Fiber Reinforced Cement-Based Composites Under Tensile Impact. *Advanced Cement-Based Materials* **1:3**, 131-141.

Campione G., La Mendola L. and Zingone G. (2000). 'Flexural-Shear Interaction in High Strength Fibre Reinforced Concrete Beams' in: Rossi P. and Chanvillard (eds.), *Fibre-Reinforced Concretes (FRC) – BEFIB' 2000*, Proceedings of the Fifth International RILEM Symposium, RILEM Publications, S.A.R.L.: 451-460.

Campione G. and Mindess S. (1999a). 'Compressive Toughness Characterization of Normal and High-Strength Fiber Concrete Reinforced with Steel Spirals', in: Banthia N. MacDonald C. and Tatnall P. (eds), *Structural Applications of Fiber Reinforced Concrete*, ACI SP-182, American Concrete Institute, Farmington Hills, Michingan: 141-161.

Campione G. and Mindess S. (1999b). 'Fibers as Shear Reinforcement for High Strength Reinforced Concrete Beams Containing Stirrups', in: Reinhardt H.W and Naaman A.E. (eds.), *High Performance Fiber Reinforced Cement Composites (HPFRCC3)*, RILEM Publications, S.A.R.L.: 519-529.

Campione G. and Mindess S. (1999c). 'The Flexural Behavior of Over-Reinforced High Strength Concrete Beams Containing Fibers', in: Reinhardt H.W and Naaman A.E. (eds.), *High Performance Fiber Reinforced Cement Composites (HPFRCC3)*, RILEM Publications, S.A.R.L.: 509-518.

Campione G. and Mindess S. (2000). 'Size Effect in Compression of High Strength Fibre Reinforced Concrete Cylinders Subjected to Concentric and Eccentric Loads', in: Wittmann F.H. (ed.), *Materials for Building and Structures, EUROMAT99*. Wiley-VCH, 6, 86-91.

Campione G. and Mindess S. (2001). Behaviour of Normal Weight and Lightweight Fibre Reinforced Concrete in Compression', in: Banthia, N. Sakai K. and Gjaru O.E. (eds.), *Proceedings, Third International Conference on Concrete Under Severe Conditions (CONSEC '01)*. University of British Columbia, Vancouver, B.C.: **2:1274-1286**.

Campione G., Mindess S. and Banthia N. (2000). 'Monotonic and Cyclic Flexural Behaviour of Medium and High Strength Fibre Reinforced Concrete', in: Rossi P. and Chanvillard (eds.), *Fibre-Reinforced Concretes (FRC) – BEFIB' 2000*, Proceedings of the Fifth International RILEM Symposium, RILEM Publications, S.A.R.L.: 461-470.

Campione G., Mindess S. and Zingone G. (1997). 'Failure Mode in Compression of Fibre Reinforced Concrete Cylinders with Spiral Steel Reinforcement', in: Brandt A.M., Li V.C. and

Marshall I.H. (eds.), *Brittle Matrix Composites 5*, Woodbend Publishing Limited, Cambridge and Bigraf, Warsaw: 123-132.

Campione G., Mindess S. and Zingone G. (1999a). 'Behavior of Fiber Concrete Reinforced with Steel Spirals under Cyclic Compressive Loading', in: Azizinamini A. Darwin D. and French C. (eds.), *High Strength Concrete*, First International Conference sponsored by United Engineering Foundation, Inc., American Society of Civil Engineers: 136-148.

Campione G., Mindess S. and Zingone G. (1999b). Compressive Stress-Strain Behaviour of Normal and High-Strength Carbon-Fiber Concrete Reinforced with Steel Spirals. *ACI Materials Journal* **96:1**, 27-34.

Naaman A.E. and Gopalaratnam V.S. (1983). Impact Properties of Steel Fibre Reinforced Concrete in Bending. *International Journal of Cement Composites and Lightweight Concrete* **5:40**, 225-233.

Sukontasukkul P. (2001). *Impact Behaviour of Concrete Under Multiaxial Loading*. Ph.D. Thesis, University of British Columbia, Vancouver, Canada.

Sukontasukkul P., Mindess S., Banthia N. and Mikami T. (2001). Impact Resistance of Laterally Confined Fibre Reinforce Concrete Plates. *Materials and Structures (RILEM)* **34**, 612-618.

Yan C. and Mindess S. (1991). Bond Between Concrete and Steel Reinforcing Bars Under Impact Loading. *Brittle Matrix Composites 3*, 184-192.

Yan C. and Mindess S. (1994). Bond Between Epoxy Coated Reinforcing Bars and Concrete Under Impact Loading. *Canadian Journal of Civil Eng.* **21:1**, 89-100.

Advances in Building Technology, Volume I
M. Anson, J.M. Ko and E.S.S. Lam (Eds.)

DIFFERENTIAL SCANNING CALORIMETRY STUDY OF THE HYDRATION PRODUCTS IN PORTLAND CEMENT PASTES WITH METAKAOLIN REPLACEMENT

W. Sha

School of Civil Engineering, The Queen's University of Belfast, Belfast
BT7 1NN, UK

ABSTRACT

The hydration products in Portland cement pastes mixed with metakaolin (MK) replacement have been examined using differential scanning calorimetry (DSC). The amounts of calcium hydroxide remaining in the pastes after long ages were evaluated in particular by measuring the enthalpy of its decomposition. Reasonable agreements are demonstrated between the results from the present work and previous studies by other investigators.

KEYWORDS

Hydration products, thermal analysis, fly ash, metakaolin, Portland cement, calorimetry.

INTRODUCTION

Recent years have seen a significant amount of research on the use and properties of cement and concrete mixed with cement replacement materials. These include pulverised fuel ash (pfa, or fly ash), condensed silica fume (csf, or microsilica), ground granulated blast furnace slag (ggbs) and metakaolin (MK). The large quantity of literature published in this area has been accompanied by an increased use of these replacement materials in construction [Sindu (2000)], mostly because of environmental and economical benefits. Increased use of cement replacement materials has also been driven by the requirement for enhanced durability.

In 1999, the author and his colleagues used differential scanning calorimetry (DSC) to study the hydration products in Portland cement (PC) paste [Sha et al. (1999)]. This was extended to PC with metakaolin [Sha & Pereira (2001a)], pulverised fuel ash [Sha & Pereira (2001b)] and slag [Sha & Pereira (2001c)] replacements. The present work is a continuation of the research by Sha & Pereira (2001a), but concentrates on hydration after longer times. Specifically, the longer times are for ages beyond 230 and 190 days for the PC with 20% and 30% metakaolin, respectively. Results are also compared with previous data for samples at shorter ages.

The work in Sha et al. (1999) involves using DSC in an investigation of the thermal behaviour of hydration products in Portland cement as a function of age. The two-step loss of water from calcium silicate hydrate, dehydroxylation of calcium hydroxide, and decarbonation of calcium carbonate contribute respectively to the three major endothermic peaks in the DSC curves. Peaks due to the formation of ettringite and iron-substituted ettringite, C_4AH_{13} and Fe_2O_3 solid solution were also found.

The work in Sha & Pereira (2001a) has demonstrated that DSC analysis can be used to trace the hydration process of cement with metakaolin replacement. The pozzolanic reaction of metakaolin and its effect on cement hydration can be monitored. The technique, combined with theoretical analysis based on reaction stoichiometry, can contribute to the understanding of these processes.

In Sha & Pereira (2001b), hydration products in normal Portland cement mixed with 30% fly ash were analysed. The phases detected include calcium silicate hydrate, iron-substituted ettringite, C_4AH_{13}, calcium hydroxide, calcium carbonate and amorphous phases in fly ash. In addition, ggbs powder was investigated, revealing peaks due to the crystallisation of the amorphous phases. Due to their relatively preliminary nature and the limit of article space, the results of the powders and pastes were not closely correlated, and some of the experimental results remained unexplained with regard to their hydration/pozzolanic processes, or the resulting engineering implications. At the then stage of the research development in the area, there was a lack of proper understanding of the implications of the DSC testing results on engineering properties particularly at early ages. The result presented could not possibly give a full meaningful interpretation of the role of fly ash or slag in concrete, although there was a catalogue of data that may prove crucial in the course of further research. Chemical analyses of quenched samples either side of peak temperatures was expected to help in understanding the mechanisms involved. Unfortunately, there was some wrong assignments of the DSC peaks in Sha & Pereira (2001a,b).

The work reported in Sha & Pereira (2001c) involves an investigation of the thermal behaviour of hydration products in ggbs. The two-step loss of water from calcium silicate hydrates and dehydroxylation of calcium hydroxide ($Ca(OH)_2$) contribute, respectively, to the two major peaks in the DSC curves. Peaks due to the formation of ettringite and Fe_2O_3 solid solution were also found. The crystallisation peaks from amorphous phases in the ggbs are also significant in the DSC thermograms.

The DSC used in these studies has high temperature sensitivity, but the accuracy of the enthalpy measurement is limited to approximately 20% relative error. Compared to other analytical methods, DSC measurement of phases is fast, but the identification of phases relies on previous knowledge of their decomposition or transformation temperatures and enthalpies.

There has been other work, by de Silva & Glasser (1990, 1992, 1993), Frías & Cabrera (2001), Frías et al. (2001), Kinuthia et al. (2001), Murat (1983), Pera & Ambroise (1998), Péra et al. (1998, 2001), and Serry et al. (1984).

MATERIALS

The work reported in this paper was carried out on the same pastes as used in Sha & Pereira (2001a), using the same procedure and apparatus. Therefore, the experimental details will not be repeated here, for which Sha et al. (1999) and Sha & Pereira (2001a,b,c) can be consulted. The chemical composition, phase analysis, fineness and some other physical properties of metakaolin and cement are given in Table 1. The water/solid ratio is 0.45.

TABLE 1

CHEMICAL COMPOSITION AND SOME PHYSICAL PROPERTIES OF CEMENT AND METAKAOLIN

Compound	Cement	Metakaolin
SiO_2	21.8	54.8
Al_2O_3	4.2	41.2
Fe_2O_3	2.5	0.57
CaO	65.1	0.02
MgO	-	0.31
Na_2O	0.13	0.04
K_2O	0.72	2.27
TiO_2	-	0.01
SO_3	2.4	2.4
Loss on ignition	-	0.8
Specific gravity	3.15	2.52

RESULTS AND DISCUSSION

20% Metakaolin

Samples were analysed at two ages, 230 and 420 days, after storing continuously under water at 20°C from one day after mixing (Figure 1). A number of peaks are shown in the curve.

It has been established by a range of different authors, that the reaction of MK with CH, either as added CH or from hydrating PC, produces C-S-H gel and C_4AH_{13} and, often at a rather later stage, C_2ASH_8. At long curing periods (beyond 28 days) hydrogarnet (C_3AH_6) may develop particularly at elevated temperatures. At high MK:CH, or MK:PC, C_2ASH_8 is favoured relative to C_4AH_{13}. If carbonation occurs or carbonate is available, which is often the case, carboaluminates, $C_3A.0.5C\underline{C}.0.5CH.H_{11.5}$ and/or $C_3A.C\underline{C}.H_{11}$, form together with C_4AH_{13}. In addition, if sulphate is available in the system, this is normally the case, ettringite forms rapidly, within the first few hours. Gehlenite hydrate, C_2ASH_8, is often not crystalline in its initial stages of formation. It normally shows a weight loss band on thermogravimetry analysis (TGA) and an exotherm on differential thermal analysis (DTA) at 160–200°C, C_4AH_{13} and related carboaluminate phases at 220–280°C and hydrogarnet at 350–400°C. Ettringite usually loses its water at 80–130°C and C-S-H gel, although it loses its water over a wide temperature range, loses the greatest proportion of its water between 120 and 160°C. The DSC measures the energy change of these dehydration or dehydroxylation reactions rather than the weight loss (TGA) or differential temperature (DTA). The transformations, however, occur in the same temperature regions, in the same way that those for $Ca(OH)_2$ and $CaCO_3$ do.

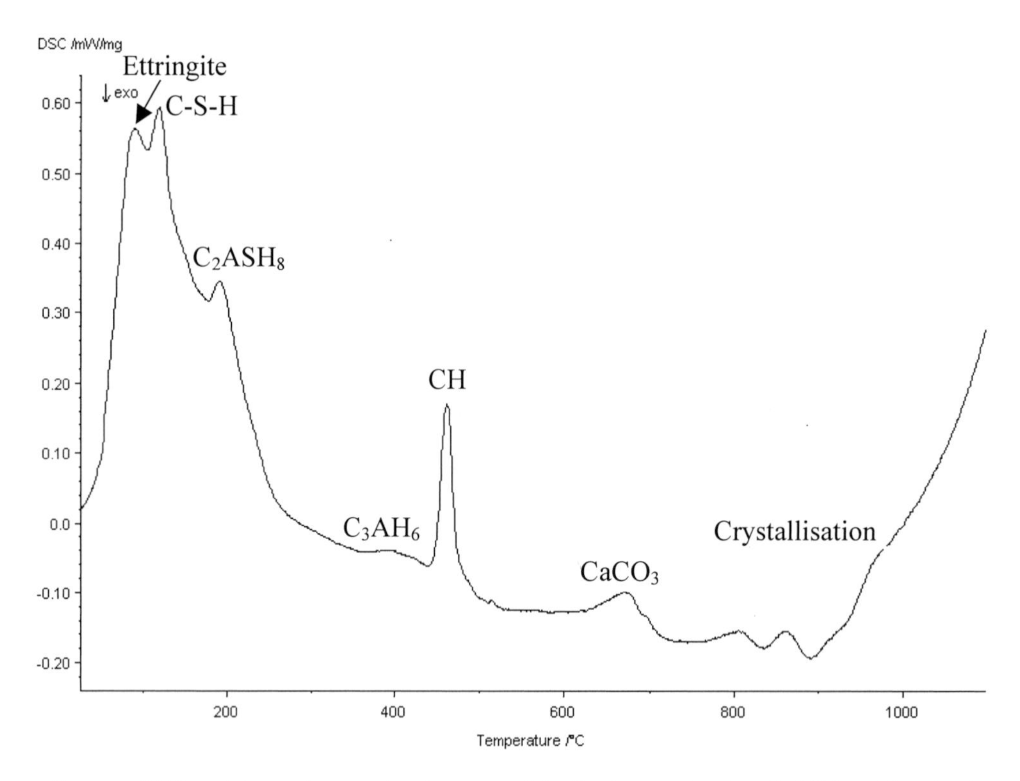

Figure 1: Differential scanning calorimetry curve of Portland cement paste with 20% metakaolin at 230 days

It was found that there was no significant change between the two ages studied, indicating that no significant reaction had occurred, including in particular the pozzolanic reaction. Calculation showed that with 15% or higher metakaolin replacement, there should be no calcium hydroxide left in the system after hydration and pozzolanic reactions are complete [Sha & Pereira (2001a)]. The large amount of calcium hydroxide ($Ca(OH)_2$, or CH) remaining in the samples was not surprising, however. Reaction products are formed in such a way that they inhibit further reaction. For example in concrete with only PC as binder, after many years of PC hydration there is still unhydrated PC remaining. Wild & Khatib (1997) conclude that replacement levels of PC by MK considerably in excess of 15% would be required to fully consume all the CH in MK-PC pastes and mortars. This is based on their hypothesis that an inhibiting layer of reaction product is formed on the MK particles. This is also discussed in Kinuthia et al. (2001). Considering that there is no calcium hydroxide after similar age in cement with 30% metakaolin replacement (see next section), it appears that the theory used in the calculations is essentially valid but the coefficients need to be adjusted. The amount of calcium hydroxide, 24 J/g, is equivalent to 2.4% of the paste, or 4.2% of the original amount of cement. This is lower than the values reported by Frías & Cabrera (2000), but the overall trend for the change of CH amount with age is in agreement with Sha & Pereira (2001a). The same qualitative comparison can be made with work by Wild & Khatib (1997), although the DSC work did not reveal the drop of CH content at around 14 days [Sha & Pereira (2001a)]. The lower CH level found in this work compared to the literature [Wild & Khatib (1997), Frías & Cabrera (2000)] should be related with the different materials and mixes used. It is difficult however to separate the influence of different metakaolin and cement compositions, different physical properties of the materials, and different water/solid ratio.

Calculating enthalpy values for combined adjacent peaks was applied to the peaks previously identified as from crystallisation of the amorphous materials [Sha & Pereira (2001a)]. The double crystallisation peak starts to appear in samples at 52 days, but the total enthalpy remains the same until 420 days, averaged at 43±5 J/g.

30% Metakaolin

For PC paste with 30% metakaolin replacement, samples at two ages were analysed, 190 and 380 days (Figure 2). As for 20% metakaolin, data for earlier ages were given in Sha & Pereira (2001a). The paste with 30% metakaolin exhibited significant acceleration of the speed and degrees of the pozzolanic reaction. Figure 3 shows DSC thermograms for pastes at the ages of 2, 24, 38 and 73 days, respectively. The separation of the crystallisation peak is complete after 38 days (Figure 3), half of the age for this in the paste with 20% metakaolin. In addition, the peak corresponding to the decarbonation of $CaCO_3$ starts to grow large at an earlier age. There is an overlap of peaks at this region, with possible additional contribution from solid-solid transformations [Bhatty (1991)]. Therefore, the obvious dual peak phenomenon in this temperature range may be attributed to this transformation. It is also interesting to note the significant peaks corresponding to gehlenite hydrate and C-S-H, at 24 days (Figure 3). These continue to increase with longer age, overlapping to cause the large peak in the curve for 73 days (Figure 3). As gehlenite hydrate and C-S-H are both pozzolanic reaction products from MK, this difference from the behaviour of 20% metakaolin paste is as expected. Figure 4 shows the variation of peak temperature as a function of age.

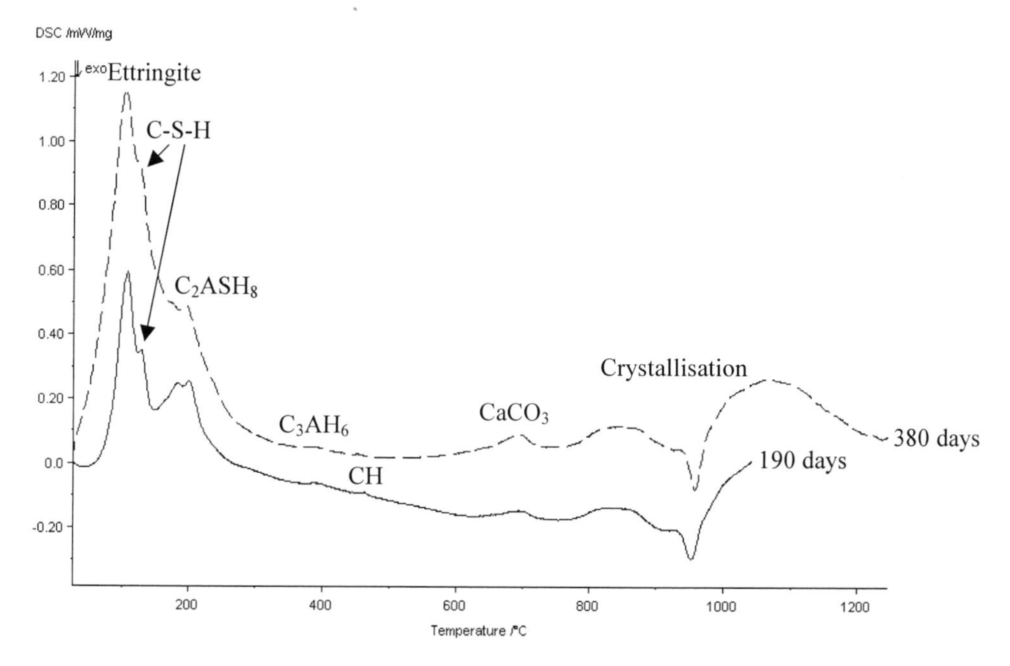

Figure 2: Differential scanning calorimetry curves of Portland cement paste with 30% metakaolin at 190 days (solid line) and 380 days (dashed line)

There is no significant change apart from the complete disappearance of the calcium hydroxide peak after 380 days, from the absence of CH peak in Figure 2. There is a peak for the sample after 190 days, but it is very small and almost invisible in the printed DSC curve.

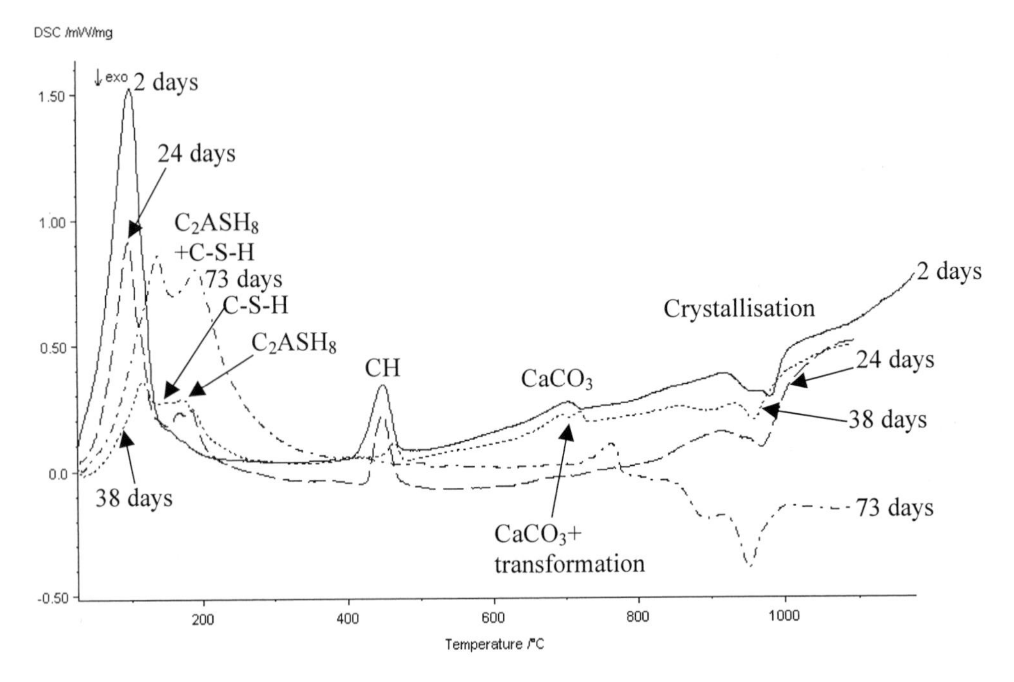

Figure 3: DSC thermograms from hardened paste containing 30% metakaolin replacement at 2 (solid line), 24 (long dashed line), 38 (short dashed line) and 73 days (dash dot line)

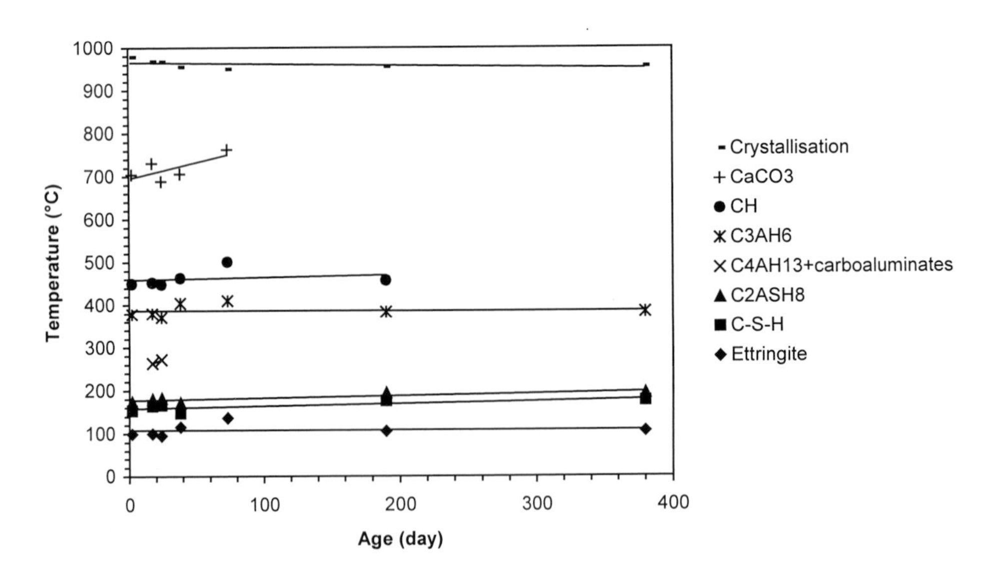

Figure 4: DSC peak temperatures from the mix with 30% metakaolin replacement

The increase of metakaolin level from 20% to 30% replacements pushes up the crystallisation temperatures. The average total enthalpy from the crystallisation peaks for ages of 73, 190 and 380 days is 106±8 Joule per gram of sample.

CONCLUSIONS

Differential scanning calorimetry analysis showed that in cement paste with 20% metakaolin replacement, the amount of calcium hydroxide after 230 and 420 days hydration remains the same, 2.4% of the paste. With 30% metakaolin replacement, there is only a minute quantity of calcium hydroxide after 190 days, and this completely disappears after 380 days.

ACKNOWLEDGEMENT

The author thanks Professor P.A.M. Basheer for discussion and the provision of concrete laboratory facilities.

REFERENCES

Bhatty J.I. (1991). A Review of the Application of Thermal Analysis to Cement-admixture Systems. *Thermochimica Acta* **189:2**, 313–350.

de Silva P.S. and Glasser F.P. (1990). Hydration of Cement Based on Metakaolin Thermochemistry. *Advances in Cement Research* **3:2**, 167–177.

De Silva P.S and Glasser F.P. (1992). Pozzolanic Activation of Metakaolin. *Advances in Cement Research* **4:16**, 167–178.

Desilva P.S and Glasser F.P. (1993). Phase Relations in the System CaO-Al_2O_3-SiO_2-H_2O Relevant to Metakaolin-calcium Hydroxide Hydration. *Cement and Concrete Research* **23:3**, 627–639.

Frías M. and Cabrera J. (2000). Pore Size Distribution and Degree of Hydration of Metakaolin-cement Pastes. *Cement and Concrete Research* **30:4**, 561–569.

Frías M. and Cabrera J. (2001). Influence of MK on the Reaction Kinetics in MK/lime and MK-blended Cement Systems at 20°C. *Cement and Concrete Research* **31:4**, 519–527.

Frías M., Sánchez de Rojas M.I. and Rivera J. (2001). The Effects of Temperature on the Formation of Hydrogarnet in Matrices Made with Metakaolin/lime Mixtures. *Seventh CANMET/ACI International Conference on Fly Ash, Silica Fume, Slag and Natural Pozzolans in Concrete* (SP 199-43), vol. 2, 22–27 July 2001, Chennai (Madras), India, ed: V.M. Malhotra, pp. 757–768.

Kinuthia J.M., Bai J., Wild S. and Sabir B.B. (2001). Thermogravimetric Analysis of Portland Cement-metakaolin-PFA Blends. *Modern Building Materials, Structures and Techniques*, The Seventh International Conference, 16–18 May 2001, Lithuania, Abstract Proceedings, p. 55.

Murat M. (1983). Hydration Reaction and Hardening of Calcined Clays and Related Minerals, I-Preliminary Invstigation on Metakaolinite. *Cement and Concrete Research* **13:2**, 259–266.

Pera J. and Ambroise J. (1998). Pozzolanic Properties of Metakaolin Obtained from Paper Sludge. *Sixth CANMET/ACI International Conference on Fly Ash, Silica Fume, Slag and Natural Pozzolans in Concrete* (SP178-52), vol. 2, Bangkok, ed: V.M. Malhotra, pp. 1007–1020.

Pera J., Bonnin E. and Chabannet M. (1998). Immobilization of Wastes by Metakaolin-blended Cements. *Sixth CANMET/ACI International Conference on Fly Ash, Silica Fume, Slag and Natural Pozzolans in Concrete* (SP178-52), vol. 2, Bangkok, ed: V.M. Malhotra, pp. 997–1006.

Péra J., Ambroise J. and Chabannet M. (2001). Transformation of Wastes into Complementary Cementing Materials. *Seventh CANMET/ACI International Conference on Fly Ash, Silica Fume, Slag and Natural Pozzolans in Concrete* (SP 199-26), vol. 2, 22–27 July 2001, Chennai (Madras), India, ed: V.M. Malhotra, pp. 459–475.

Serry M.A, Taha A.S, El-Hemaly S.A.S. and El-Didamony H. (1984). Metakaolin-lime Hydration Products. *Thermochimica Acta* **79**, 103–110.

Sha W., O'Neill E.A. and Guo Z. (1999). Differential Scanning Calorimetry Study of Ordinary Portland Cement. *Cement and Concrete Research* **29:9**, 1487–1489.

Sha W. and Pereira G.B. (2001a). Differential Scanning Calorimetry Study of Ordinary Portland Cement Paste Containing Metakaolin and Theoretical Approach of Metakaolin Activity. *Cement and Concrete Composites* **23:6**, 455–461.

Sha W. and Pereira G.B. (2001b). Differential Scanning Calorimetry Study of Normal Portland Cement Paste with 30% Fly Ash Replacement and of the Separate Fly Ash and Ground Granulated Blast-furnace Slag Powders. *Seventh CANMET/ACI International Conference on Fly Ash, Silica Fume, Slag and Natural Pozzolans in Concrete*, Supplementary Papers, 22–27 July 2001, Chennai (Madras), India, compiled: M. Venturino, pp. 295–309.

Sha W. and Pereira G.B. (2001c). Differential Scanning Calorimetry Study of Hydrated Ground Granulated Blast-furnace Slag. *Cement and Concrete Research* **31:2**, 327–329.

Sindu J. (2000). Pfa in Runways. *Concrete* **34:8**, 46–47.

Wild S. and Khatib J.M. (1997). Portlandite Comsumption in Metakaolin Cement Pastes and Mortars. *Cement and Concrete Research* **27:1**, 137–146.

Advances in Building Technology, Volume 1
M. Anson, J.M. Ko and E.S.S. Lam (Eds.)

ULTIMATE MOMENT EQUATION OF CIRCULAR HSC MEMBERS

Y. Sun and K. Sakino

Department of Architecture, Graduate School of Human Environment Studies,
Kyushu University, Hakozaki 6-10-1, Fukuoka, 812-8581, JAPAN

ABSTRACT

It has been widely accepted that confinement of concrete by transverse reinforcement is effective in upgrading deformation capacity of high strength concrete (HSC). In order for structural engineers to conduct reasonable design of confined HSC members, they are of fundament importance 1) to make clear correlation between the amount of transverse steels and the mechanical properties (e.g. ultimate moment capacity, etc) of the confined member, and 2) to develop a sound design method that can take into account effects of primary structural factors, particularly of confinement by transverse steels.

Purpose of this paper is to propose a simple design equation for computing ultimate flexural strength of circular confined HSC or NSC (normal-strength concrete) members. The emphasis is placed on HSC members. The proposed design equation has a very simple mathematical expression so that structural engineer can conduct the complicated calculation of the ultimate moment of confined concrete members by hand. Another significant feature of the proposed design equation is that it can take into account effects of circular transverse steels on the ultimate moment of confined circular members. Experimental results of confined concrete columns available are used to verify validity of the proposed equation. The predicted ultimate moments agreed very well with the measured ones, which implies that the proposed design equation could provide a powerful tool for structural engineer to appropriately and directly calculate the ultimate moment of a circular HSC member.

KEYWORDS

Circular concrete member, high-strength concrete, confined concrete, confinement effect, circular steel tube, ultimate moment, confinement design

INTRODUCTION

Due to its high load-carrying capacity and durability, HSC has recently gained increasing use in high-rise building structures and highway constructions. To promote use of HSC in structures located on seismic regions, prevention of brittle failure mode of the HSC becomes an important issue.

Confinement of concrete by transverse reinforcement is effective in enhancing ductility of the HSC.

Traditionally, the transverse spirals or hoops have been widely used to confine concrete members, and their beneficent effects on the mechanical behavior of concrete members have been studied by many researches as reviewed in the ACI state of art report [ACI, 1997]. Nevertheless, there has not yet been a reliable and comprehensive design method enabling structural engineers to simply compute the ultimate capacities of confined HSC columns. Particularly true is this for the circularly confined HSC columns due to lack of information on the ultimate flexural behavior of circular columns.

The authors have conducted an integrated experimental and analytical studies of mechanical behavior of circularly confined HSC columns, aiming at setting up a rational design method for the ultimate capacities of the circular columns. In the authors' previous experimental works [Sun and Sakino, 1999, 2000], the test columns were confined by circular steel tubes with various thickness in lieu of common circular hoops, to quantitatively study confinement effect of circular transverse reinforcement on the mechanical properties of the "pure" confined concrete. Based on their experimental data, the authors have developed an unified stress-strain model for the circularly confined concrete [Sun and Sakino, 2001]. This confinement model has been proved applicable to the concrete confined by conventional circular hoops, the concrete confined by circular steel tubes, and the concrete in circular CFT column. Furthermore, as the first step toward simplification of ultimate capacity design for the circular HSC columns, Sun and Sakino [2001] have also proposed a equivalent rectangular stress block for the compressed concrete in a circular column section and a design procedure to predict ultimate capacities of circularly confined HSC columns. However, as can be seen later, that method involves the use of complicated expressions and still needs help of a computer.

To overcome these shortcomings in the previous design method, this paper propose a much simpler design equation for computing the ultimate flexural strength of circularly confined HSC members. The mathematical expression of the proposed design equation, which defines the correlation between the axial load and the ultimate moment for a given circular column, is developed by utilizing the rectangular equivalent stress block proposed by the authors [Sun and Sakino, 2001]. Study of this paper will focus on circular columns confined by steel tube. Development of the design equation for calculation of ultimate moment of the columns confined by common circular hoops is in progress, and will be reported in near future.

OUTLINE OF EQUIVALENT RECTANGULAR STRESS BLOCK

Interaction diagram between the axial load N and the ultimate moment (referred to as N-M interaction diagram hereafter) is usually adopted to compute the flexural strength of concrete columns. Generally, for a given column, the N-M interaction diagram can be obtained by conducting flexural analysis, i.e., by computing the moment-curvature response of the column section under various levels of axial load. If a sound stress-strain model for the confined concrete is assumed, the flexural analysis can give fairly accurate prediction of the flexural strength, in which confinement effect of transverse steels can be taken into account. This general approach, however, is very tedious and needs help of a computer program, since it involves an iterative procedure to find the depth of the neutral axis for the internal forces to balance the external applied load.

On the other hand, if the stress state of the compressed concrete at the peak moment can be simplified, it is possible to simplify the calculation procedure of the N-M interaction diagram. By conducting parametric study of flexural response for circularly confined columns, the authors have proposed an equivalent rectangular stress block to replace the actual stress state for the compressed concrete in a circular column section as the first step toward simplification of flexural strength calculation [Sun and Sakino, 2001]. Figure 1 shows outline of the proposed stress block for the circularly confined concrete. As shown in Figure 1, the equivalent stress block has width of αD_c and depth of βX with uniform stress f_{cc}, where D_c is the diameter of core concrete, X is the depth of the neutral axis, and f_{cc} ($=K f_p$) is the strength of confined concrete. Utilizing the stress block, the axial compressive force N_B and the first moment M_B sustained by the concrete at ultimate state can be calculated as follows:

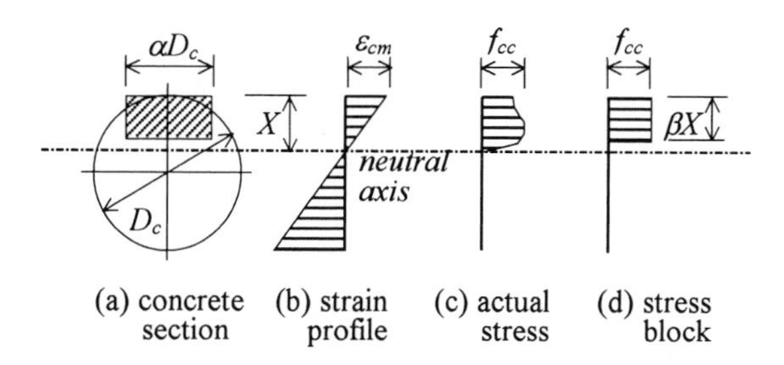

Figure 1: Equivalent rectangular stress block for the compressed concrete in a circular column

$$N_B = \alpha\beta K f_p D_c X, \qquad M_B = \alpha\beta K f_p D_c X\left(\frac{D_c}{2} - \frac{\beta}{2}X\right) \tag{1}$$

where α and β are the stress block parameters, K is the strength enhancement ratio of confined concrete, f_p is the unconfined concrete strength, D_c is the diameter of concrete core, and X is the depth of the neutral axis. Based on the theoretical results concerning with the flexural behavior of circularly confined concrete columns, the authors have developed mathmatical expressions to compute the stress block parameters α and β in forms of

$$\alpha\beta = A(K, X_n) - B(K, X_n)\frac{f_p}{42}, \quad \frac{\beta}{2} = C(K, X_n) - D(K, X_n)\frac{f_p}{42} \tag{2}$$

where

$$A(K, X_n) = \frac{0.723 + 0.061K}{0.112 + X_n}X_n, \quad B(K, X_n) = \frac{0.048K^{-2}}{0.072K^{-1.5} + X_n}X_n \tag{3}$$

$$C(K, X_n) = (0.476 + 0.051K)\left(1 - 0.132{X_n}^2\right), \quad D(K, X_n) = 0.017\left[1 - (0.024 + 0.187K){X_n}^2\right] \tag{4}$$

in which X_n is the normalized depth of the neutral axis. The strength enhancement ratio K of concrete confined by circular steel tube can be obtained by applying famous Richart formula as follow

$$K = \frac{f_{cc}}{f_p} = 1 + 4.1\frac{2}{D/t - 2}\frac{f_{yt}}{f_p} \tag{5}$$

where f_{cc} is the confined concrete strength, f_{yt}, t, and D are the yield strength, thickness, and outside diameter of the steel tube, respectively.

FEATURES OF THE ULTIMATE *N-M* INTERACTIVE DIAGRAM

To investigate features of the ultimate *n-m* interactive curve for circularly confined concrete columns and get fundamental information for development of design equation, parametric flexural study was conducted. Figure 2 shows details of the sample section as well as varying ranges of the main variables. As obvious in Figure 2, parametric study covered a wide range of concrete strength and steel amount, which implies the wide applicability of the to be proposed design equation in terms of the concrete strength and the amount of longitudinal bars.

To compute the ultimate *n-m* interactive curve using the equivalent stress block, they are assumed that

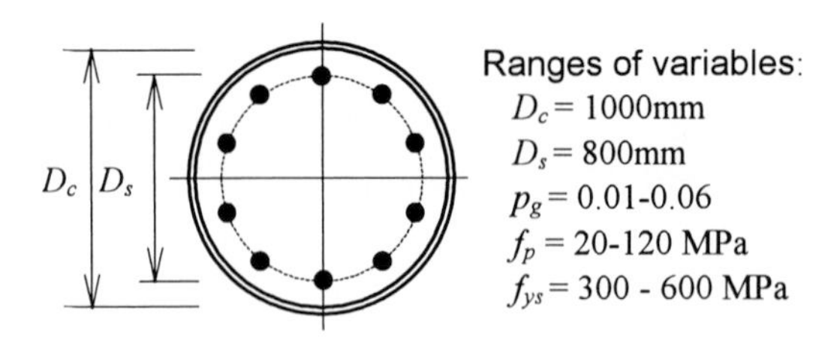

Figure 2: Details of the sample column section

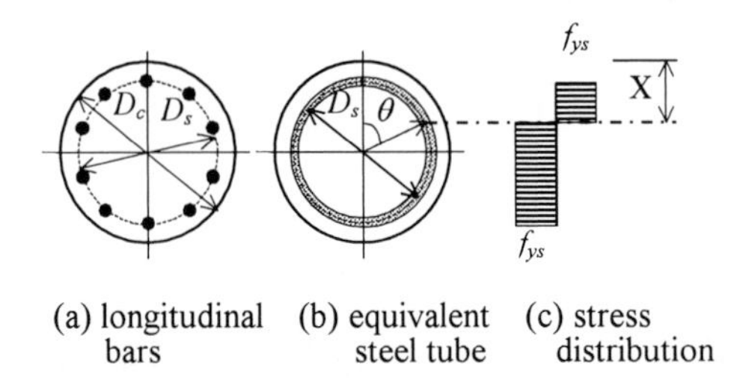

Figure 3: Concept of the equivalent axial steel tube of the longitudinal bars

1) the tensile strength of concrete can be neglected, 2)the concrete sustained axial force as well as moment can be replaced by those of the stress block shown in Figure 1, 3) the longitudinal bar is a rigid-plastic material, and 4) the longitudinal bars can be replaced with an equivalent axial steel tube having identical reinforcement area (see Figure 3).

Based on these assumption, the calculation procedure for ultimate *N-M* interaction diagram doesn't involve iterative calculation, and can be summarized as follows.

1) Calculate the strength enhancement ratio K of confined concrete by substituting D/t ratio and yield strength f_{yt} of the steel tube and the unconfined concrete strength f_p, into Eq. 5.
2) Give an initial value (e.g. 0.02π) to θ, the radial angle of the section and calculate the normalized depth of the neutral axis X_n by Eq. 6.

$$X_n = X/D_c = 0.5(1 - D_s \cos\theta / D_c) \tag{6}$$

   where D_c is the diameter of confined core concrete, and D_s is the distance between the centroids of the longitudinal bars. (see Figure 3)
3) Calculate the stress block parameters $\alpha\beta$ and $\beta/2$ from Eq. 2 through Eq. 4 and compute the axial force N_B and moment M_B sustained by the concrete by using. 1
4) Based on the last two assumptions above mentioned, the axial force N_s and moment M_s sustained by longitudinal bars can be calculated as

$$N_s = \frac{f_{ys} p_g D_c^2 (2\theta - \pi)}{4}, \quad M_s = \frac{f_{ys} p_g D_c^2 D_s \sin\theta}{4} \tag{7}$$

   where f_{ys} and p_g are the yield strength and the steel ratio of the longitudinal bars, respectively.
5) The normalized axial force n as well as the normalized moment m corresponding to the given radial angle θ are then given by

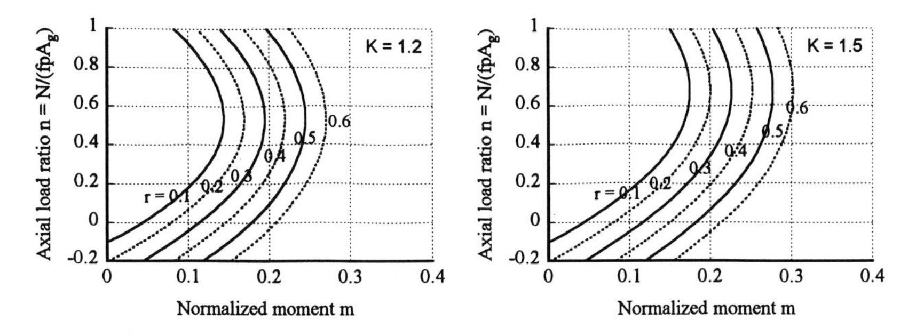

Figure 4: Examples of the calculated *n-m* interactive diagrams of the sample columns

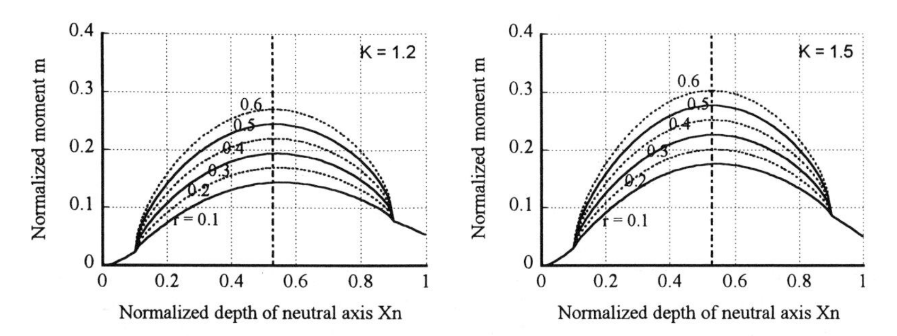

Figure 5: Relationships between the normalized neutral axis depth *Xn* and the moment m

$$n = \frac{N_B + N_S}{A_g f_p} = \frac{4}{\pi}\alpha\beta \bullet K \bullet X_n + \frac{1}{\pi}\frac{f_{ys}p_g}{f_p}(2\theta - \pi) \tag{8}$$

$$m = \frac{M_B + M_S}{A_g D_c f_p} = \frac{4}{\pi}\alpha\beta \bullet K \bullet X_n\left(\frac{1}{2} - \frac{\beta}{2}X_n\right) + \frac{1}{\pi}\frac{f_{ys}p_g}{f_p}\frac{D_S}{D_c}\sin\theta \tag{9}$$

6) The normalized ultimate *n-m* interaction diagram can be completely determined by incrementing θ till $\theta = \pi$, and repeating the calculation step 2) through step 5) above.

Figure 4 plots several ultimate *n-m* interactive curves of the sample column section obtained utilizing the calculation procedure described above, while Figure 5 shows relationships between the normalized ultimate moment *m* and the neutral axis depth. In Figures 4 and 5, the confinement effect of circular steel tube is represented by the strength enhancement ratio K of confined concrete as defined by Eq. 5, and the factor *r* expresses the normalized steel ratio of the longitudinal bars and is given by $r = f_{ys}\rho_g/f_p$.

From Figures 4 and 5, one can observe the following features for the ultimate *n-m* interactive curves of circularly confined concrete columns:

1) For a circular column confined by the steel tube with a specific K value, the *n-m* interactive curves expand outwards as increment of the amount of longitudinal bars. On the other hand, the normalized moment *m*, as a function of the axial load ratio *n*, tends to reach its maximum at a axial load level that is nearly independent of the amount of longitudinal bars.
2) The axial load ratio n_o at the peak point of each *n-m* interactive curve increases with the K value, while the peak moment m_o at each *n-m* interactive curve increases as increment of both the K value

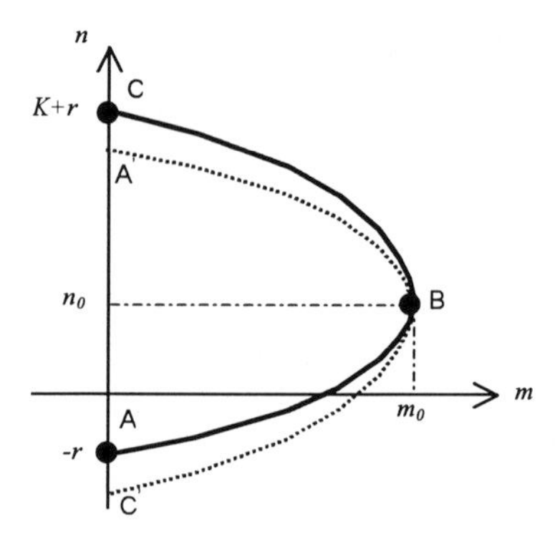

Figure 6: Idealization of the ultimate *n-m* interactive diagram

and the amount of longitudinal bars.

3) Regardless of the confinement degree (i.e. the *K* value) and the amount of longitudinal bars, each ultimate *n-m* curve reaches its peak moment as the normalized neutral axis depth X_n is between 0.52 through 0.55 with a mean value of 0.53.

PROPOSAL OF A DESIGN EQUATION

The calculation method using the equivalent stress block does really simplify the computation of flexural strength for the circularly confined members as compared with the method using the general flexural analysis. However, this method is still complicated as can be seen from Eq. 2 through Eq. 4. To provide a simpler and more powerful tool enabling structural engineers to more quickly and reasonably conduct the confinement design, a design equation is developed in this section.

Based on the results of parametric study described in the previous section, the ultimate *n-m* interactive curve can be idealized as the curve ABC shown in Figure 6. Obviously, to represent the features observed in the ultimate *n-m* interactive diagrams, the design equation, which defines the relationship between the normalized moment *m* and the axial load ratio *n*, should satisfy the following conditions:

$$m\big|_{n=-r} = 0, \qquad m\big|_{n=K+r} = 0$$
$$m\big|_{n=n_0} = m_0, \qquad \left.\frac{\partial m}{\partial n}\right|_{n=n_0} = 0 \tag{10}$$

where m_0 and n_0 are the peak moment of each *n-m* curve and the corresponding axial load ratio, respectively.

The four conditions given by Eq. 10 can determine a single cubic curve, but a cubic curve containes four constants, which might results in new complexity. By carefully analyzing the ABC curve shown in Figure 6, one can find that dividing the *n-m* curve into two parts, the ascending portion AB and the descending part BC, will result in a much simpler expression for the *n-m* curve. In fact, the AB curve can be taken as the ascending half of a parabola ABA', while the BC curve as the descending half of a parabola C'BC.

Considering that the both parabolas, ABA' and C'BC, have the line defined by $n=n_0$ as their common symmetric axis, the mathematical expression for the ultimate *n-m* curve can be simply written as follow:

$$m = \begin{cases} m_0\left[1-\left(\dfrac{n-n_0}{n_0+r}\right)^2\right], & n \le n_0 \\ m_0\left[1-\left(\dfrac{n-n_0}{n_0+K+r}\right)^2\right], & n > n_0 \end{cases} \tag{11}$$

The two parabolas defined by Eq. 11 are continuous at the peak point. Apparently, the proposed design equation is a two-parameter model. Only if the peak moment m_0 and the corresponding axial load ratio n_0 are given, one can completely determine the ultimate *n-m* curve.

As described in the previous section, each *n-m* curve reaches its peak as the normalized depth of neutral axis is between 0.52 and 0.55 with a mean value of 0.53 regardless of the values of both K and r. Hence, substituting X_n=0.53 into Eq. 3 and Eq. 4, and then Eq. 8 and Eq. 9, one can obtain close approximations for m_0 and n_0. Noting that as X_n=0.53,

$$\theta \approx \frac{\pi}{2} \Rightarrow \sin\theta \approx 1.0$$

exists, and by conducting regression analysis on the approximations of m_0 and n_0, the following expressions were derived to compute m_0 and n_0:

$$\begin{aligned} n_0 &= \frac{1}{\pi}\left(0.1K^2 + 1.3K - 2.2K^{-1} f_p \times 10^{-3}\right) \\ m_0 &= \frac{1}{\pi}\left[0.31K + (0.61K - 0.85) f_p \times 10^{-3} + r\frac{D_s}{D_c}\right] \end{aligned} \tag{12}$$

It can be seen from Eq. 11 and Eq. 12 that the proposed design equation enables structural engineers to calculate the ultimate flexural strength by hand.

VERIFICATION OF THE DESIGN EQUATION

The experimental results of bending tests by Sun and Sakino [2000] are used to validate the proposed design equation. The previous tests comprised of eleven columns made of HSC. These columns were confined in circular steel tubes having inner diameter of 250mm and five kinds of wall thickness varying between 1.2mm and 4.5mm. The concrete strengths were 55.4MPa and 89.2MPa. In each column, the longitudinal steel was ten D13 deformed bars (12.7mm diameter) placed uniformly along the perimeter with 25mm of cover to the steel centroids to give a steel ratio of 2.54%.

The circular steel tubes were fabricated by welding folded steel plate having the targeted thickness at laboratory. The steel tubes were used only to confine concrete rather than to directly sustain the axial stress due to external loads.

Table 1 lists the experimental outlines of these test column along with the measured ultimate moments as well as the calculated results. For comparison, the theoretical predictions based on the stress block are also given in Table 1. It can be seen that the theoretical moments obtained by the proposed design equation agree very well with the experimental results, having the same accuracy as the calculated results by the equivalent stress block.

TABLE 1: OUTLINES OF THE PREVIOUS TESTS AND PRIMARY RESULTS

Column	fp (MPa)	t (mm)	fyt (MPa)	K	n	Mexp (kNm)	Mc1 (kNm)	ratio	Mc2 (kNm)	ratio
No.1	55.4	1.15	297	1.20	0.2	87.0	83.9	1.037	84.3	1.032
No.2		1.15	297	1.20	0.4	110.9	103.9	1.067	105.0	1.056
No.3		2.2	308	1.40	0.4	126.1	112.3	1.122	113.7	1.110
No.4		2.2	308	1.40	0.7	145.0	120.7	1.201	120.6	1.202
No.5	89.2	1.15	297	1.13	0.2	111.2	114.1	0.974	110.5	1.006
No.6		1.55	290	1.17	0.2	110.0	115.3	0.954	110.6	0.995
No.7		1.55	290	1.17	0.3	131.7	135.6	0.971	130.6	1.008
No.8		3.1	304	1.35	0.5	173.0	170.3	1.016	171.2	1.011
No.9		3.1	304	1.35	0.7	175.0	168.3	1.040	169.4	1.033
No.10		4.5	375	1.58	0.5	198.3	189.7	1.046	189.8	1.045
No.11		4.5	375	1.58	0.7	211.0	201.7	1.046	202.0	1.045
							Mean	1.043	Mean	1.049
							St. Dev.	0.071	St. Dev.	0.061

Note: n = axial load ratio
Mexp = the measured ultimate moments
Mc1 = the calculated moment based on the proposed design equation
Mc2 = the calculated moment by the equivalent stress block
Ratio = the measured moment / the calculated moment

CONCLUSIONS

A simple design equation has been proposed in this paper to calculate the ultimate flexural strength of concrete columns retrofitted by circular steel tubes. The Proposed design equation has a very simple expression and can take into account confinement effect of steel tubes on the flexural strength. Because of its simplicity and comprehensiveness, the proposed design equation can provide a powerful tool enabling structural engineers to conduct rational retrofitting design for the existing concrete columns.

REFERENCES

ACI 441R-96. (1997). *High-Strength Concrete Columns: State of the Art*, ACI Structural Journal, May-June, pp. 323-335.

Sun Y. and Sakino K. (1999). *Axial Behavior of the HSC Confined by Circular Steel Tubes*, Procs. of the JCI, Vol. 21, No. 3, pp. 589-594. (in Japanese)

Sun Y., et al. (1999). *Axial and Flexural Behavior of Confined HSC Columns*, Procs. of the Annual Convention of the AIJ, Vol. 3, pp. 763-770.(in Japanese)

Sun Y. and Sakino K. (2000). *A Comprehensive Stress-Strain Model for High-Strength Concrete Confined by Circular Transverse Reinforcement*," Procs. of the 6th ASSCS-2000 International Conference on Composite Structures, Los Angeles.

Sun Y. and Sakino K. (2001). *Simplified Design Method for Ultimate Capacities of Circularly Confined High-Strength Concrete Columns*. The ACI special Publication SP193, pp.571-585.

Advances in Building Technology, Volume 1
M. Anson, J.M. Ko and E.S.S. Lam (Eds.)

PERFORMANCE TEST OF A MODEL FOR PREDICTING STRENGTH DEVELOPMENT OF FLY ASH-CONCRETE

Tahir M. A.[1] and Nimityongskul P.[2]

[1]Professor, Department of Civil Engineering, University of Engineering and Technology, Lahore, Pakistan
[2]Program Coordinator, Structures and Construction Program, School of Civil Engineering, Asian Institute of Technology, Bangkok, Thailand.

ABSTRACT

The main features of a model that takes into account the chemical composition and physical properties of the binder materials for predicting strength development of fly ash-concrete are discussed. The model is based on the Feret's Law enriched by the concept of maximum paste thickness. The performance of the model is tested using the strength data from an experimental program in addition to the data adopted from several researchers. The model predictions agree very well with the experimental data. In an extensive theoretical study the sensitivity of the model to the changes in chemical composition of cement and fly ash is also evaluated. It is found that the model performance is just in line with the accepted principles of the hydration and pozzolanic chemistry.

KEYWORDS:

Strength, chemical composition, fineness, fly ash, hydration, maximum paste thickness, pozzolana.

INTRODUCTION

Tahir and Nimityongskul (1998) presented a model for predicting strength development of fly ash-concrete. The model bases its predictions on the chemical composition and fineness of both the Portland cement and fly ash incorporated in the concrete. Further, it considers the effect of aggregate content and grading on the strength development of concrete. The model was calibrated and validated using data from a huge experimental program conducted at AIT (Tahir, 1998) in addition to the data adopted from several researchers around the globe. The nature of the data involved in calibration and validation made the model site independent and it could be used for design of fly ash concrete mixtures quite confidently. In this study the performance of the model is tested in view of the established facts from the chemistry of hydration and pozzolanic reactions. The chemical compositions of the cement and fly ash are varied within the practical limits and influence is evaluated on the strength development of concrete. The selected material compositions are adopted from published data.

THE MODEL

The model is described briefly; the detailed derivation of the model has been presented elsewhere (Tahir, 1998). The model predicts the strength on the basis of the four Bogue Compounds in Portland cement and four major oxides in the fly ash, namely CaO, SiO_2, Fe_2O_3, Al_2O_3, and the alkali contents of the binder, in terms of the equivalent Na_2O. In the following a brief outline of the model is presented, the detailed derivation of the model are presented elsewhere. The strength development of a fly ash concrete is based on the modified Feret's Law (De Larrard and Tondat, 1993) as given below:

$$f_c^{'}(t) = A\ \Lambda^{B}\left[\frac{V_h(t)}{V_c + V_f + V_w + V_v}\right]^2 \qquad (1)$$

where

$$V_h(t) = V_h(28) + \Delta V_h(t) \qquad (2)$$

In which $V_h(28)$ denotes the solid volume of hydrated pasted at the age of 28-day and $\Delta V_h(t)$ denotes the change in the former with time. The solid volume at any age t, $V_h(t)$), is expressed in terms of Hyd_i and Poz_i, the hydraulic and pozzolanic parameters, r_W and r_F, the physical parameters and NaEq, the sodium equivalent of the mixture, as

$$\begin{aligned} V_h(t) = {} & \alpha \log\frac{t}{28} + \sum_{i=1}^{6}(\beta_i + \beta_i^{'}\log\frac{t}{28})\,Hyd_i + \sum_{i=1}^{4}(\gamma_i + \gamma_i^{'}\log\frac{t}{28})\,Poz_i \\ & + (a + a'\log\frac{t}{28})\,r_W + \sum_{i=1}^{2} b_i + b_i^{'}\log\frac{t}{28})\,[r_F]^{\frac{1}{i}} + (c + c'\log\frac{t}{28})\,NaEq \end{aligned} \qquad (3)$$

$$V_h(t) = \sum_{i=1}^{6}\beta_i\,Hyd_i + \sum_{i=1}^{4}\gamma_i\,Poz_i + a\,r_W + \sum_{i=1}^{2} b_i\,[r_F]^{\frac{1}{i}} + c\,NaEq \qquad (4)$$

$$\Delta V_h(t) = (\alpha + V_h(28))\log\frac{t}{28} \qquad (5)$$

The 6-hydration parameters have been defined as follows:

$$Hyd_1 = \Omega_C\frac{C_2S}{C_2S + C_3S}\,x\,\frac{C_3A}{G_C} \quad ; \quad Hyd_2 = \Omega_C\frac{C_2S}{C_2S + C_3S}\,x\,\frac{C_4AF}{G_C}$$

$$Hyd_3 = \Omega_C\frac{0.994\,C_2S}{G_C} \quad ; \quad Hyd_4 = \Omega_C\frac{C_3S}{C_2S + C_3S}\,x\,\frac{C_3A}{G_C}$$

$$Hyd_5 = \Omega_C\frac{C_3S}{C_2S + C_3S}\,x\,\frac{C_4AF}{G_C} \quad \text{and} \quad Hyd_6 = \Omega_C\frac{0.75\,C_3S}{G_C} \qquad (6)$$

in which C_3A, C_2S, C_3S and C_4AF denote the weights of the four Bogue Compounds respectively from the cement per unit volume of the concrete mixture, G_C denotes the density of cement and Ω_C denotes the fineness index of the Portland cement which can be obtained from the Blaine surface areas of the Portland cement used and a reference cement, F_c and F_{co}. F_{co} is taken equal to300 m^2/kg.

$$\Omega_C = \sqrt{\frac{F_C}{F_{CO}}} \qquad (7)$$

The four pozzolanic parameters are defined in the following; when $SiO_2 < S_d$:

$$Poz_1 = \Omega_F\frac{2.85\,SiO_2}{G_F}\,x\,\frac{1.321\,CaO}{CH_p}$$

$$Poz_2 = \Omega_F\frac{2.85\,SiO_2}{G_{CF}}\,x\,\frac{0.215\,C_2S + 0.487\,C_3S - 0.305\,C_4AF}{CH_p}$$

$$Poz_3 = \Omega_F\frac{Al_2O_3}{G_{CF}} \quad \text{and} \quad Poz_4 = \Omega_F\frac{Fe_2O_3}{G_{CF}} \qquad (8)$$

in which G_F and Ω_F respectively denote the density and fineness index of fly ash. The latter is defined in terms of the fineness of fly ash used, F_F, and that of a reference fly ash, F_{Fo}, as follows:

$$\Omega_F = \left[\frac{F_F}{F_{FO}}\right]^{1/3} \tag{9}$$

where F_{F0} is taken equal to 300 m^2/kg. The symbols CaO, SiO_2, Al_2O_3 and Fe_2O_3 denote the weights per unit volume of concrete of the calcium oxide, silica, alumina and ferric oxide respectively from the fly ash. The symbols CH_P and S_d appearing in the above expressions respectively denote the net calcium hydrate available for taking part in the pozzolanic reaction and the silica demand of the mixture. The two are calculated as follows:

$$CH_p = 1.321\ CaO + 0.215\ C_2S + 0.487\ C_3S - 0.305\ C_4AF \tag{10}$$

$$S_d = 0.541\ CH_p \tag{11}$$

The four parameters Poz_1 to Poz_4 are defined in the following; when $SiO_2 \geq S_d$:

$$Poz_1 = 1.541\ \Omega_F\ S_{co}\ \frac{1.321\,CaO}{G_F}$$

$$Poz_2 = 1.541\,\Omega_F\ S_{co}\ \frac{0.215\,C_2S + 0.487\,C_3S - 0.305\,C_4AF}{G_{CF}}$$

$$Poz_3 = \Omega_F\ S_{co}\ \frac{Al_2O_3}{G_{CF}} \quad \text{and} \quad Poz_4 = \Omega_F\ S_{co}\ \frac{Fe_2O_3}{G_{CF}} \tag{12}$$

in which, S_{co} denoted the silica coefficient and is defined as follows:

$$S_{co} = \left[\frac{SiO_2}{S_d}\right]^{0.2} \tag{13}$$

The symbol G_{CF} appearing in the above equations denotes the density of that material, which takes part in the hydration reaction from the fly ash and Portland cement, to produce pozzolanic $C_3S_2H_3$:

$$G_{CF} = 0.649\ G_C + 0.351\ G_F \tag{14}$$

The only chemical parameter is equivalent sodium content of the mixture NaEq, expressed as:

$$NaEq = NaEq(OPC) + NaEq(FA) \tag{15}$$

Where NaEq(OPC) and NaEq(FA) denote the weights of the alkali oxides from Portland cement and ash. The two alkali oxides (Na_2O and K_2O) are expressed as equivalent of Na_2O. The two physical parameters water-binder ratio, r_W and fly ash-cement ratio, r_F are defined as:

$$r_W = \frac{W}{C+F} \tag{16}$$

$$r_F = \frac{F}{C} \tag{17}$$

Where W, C and F denote the weights of water, cement and fly ash respectively per unit volume of concrete. The values of the 29 constants in Eqn. 2 are given in the following table. These values were determined by calibrating the model from a versatile data.

Constant	β_1	β_2	β_3	β_4	β_5	β_6	γ_1	γ_2	γ_3	γ_4
Value	-26.2	22.9	2.49	6.66	-5.76	1.06	.169	.025	.594	.279
Constant	β'_1	β'_2	β'_3	β'_4	β'_5	β'_6	γ'_1	γ'_2	γ'_3	γ'_4
Value	-24	12.2	1.49	15.57	-6.42	-.90	.0022	.0435	.436	-.087
Constant	a	a'	b_1	b_2	b'_1	b'_2	c	c'	α	
Value	-.0351	.937	-.321	.0656	-.0138	.0344	3.27e-4	-6.7e-5	-.0171	

PERFORMANCE TEST

The model was calibrated and validated using data from a large number of researchers from several countries around the globe. The details of the data have been published by Tahir (1998). The scatter of the predicted and test data is plotted in Figure 1. It can be seen that model predictions agree very closely to the experimental observations. In addition to this a performance test was run on the model using Portland cements and fly ashes of various chemical compositions. The solid volume resulting from the hydration and pozzolanic reactions was computed from the strength data of the mixtures. It was estimated by the model. The trends were tested and compared with the standard norms of the chemistry of hydration and pozzolanic reactions.

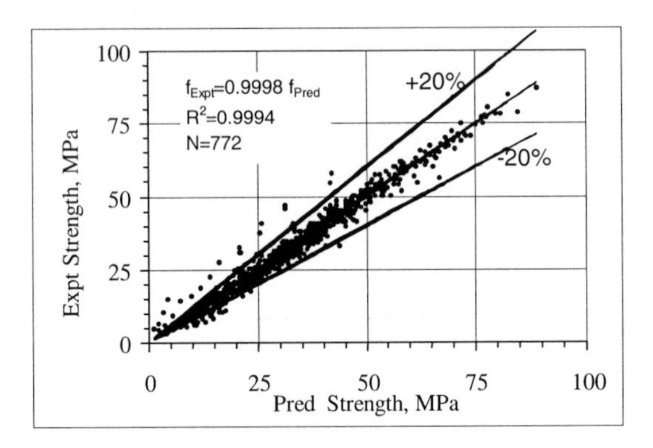

Figure 1. Scatter of model predictions and experimental data

Once the calibration and validation of the model was performed, the performance of the model was tested using Portland cements and fly ashes of different chemical compositions and fineness. In the first part of the performance test a Portland cement of constant composition was used with fly ashes of different compositions. The properties of a partial list of the binders are given in Tables 1 and 2.

TABLE 1

VARIATION OF OXIDE CONTENTS OF FLY ASH CONSIDERED IN THE PERFORMANCE TEST

ID	%SiO_2	%Al_2O_3	%Fe_2O_3	%CaO	%Na_2O	%K_2O	Remarks
S-20%	20						Variation
S-40%	40	15	10	5	2.0	1.9	of
S-60%	60						SiO_2
C-05%				5			Variation
C-15%	30	15	10	15	2.0	1.9	OF
C-25%				25			CaO
F-10%			10				Variation
F-25%	30	15	25	5	2.0	1.9	of
F-40%			40				Fe_2O_3
A-15%		15		5			Variation
A-30%	30	30	10	5	2.0	1.9	of
A-45%		45		5			Al_2O_3

TABLE 2

PROPERTIES OF THE CEMENT CONSIDERED IN THE PERFORMANCE TEST

$\%C_2S$	$\%C_3S$	$\%C_3A$	$\%C_4AF$	$\%Na_2O$	$\%K_2O$	Surface Area
20	55	10	10	0.17	0.50	300 m^2/kg

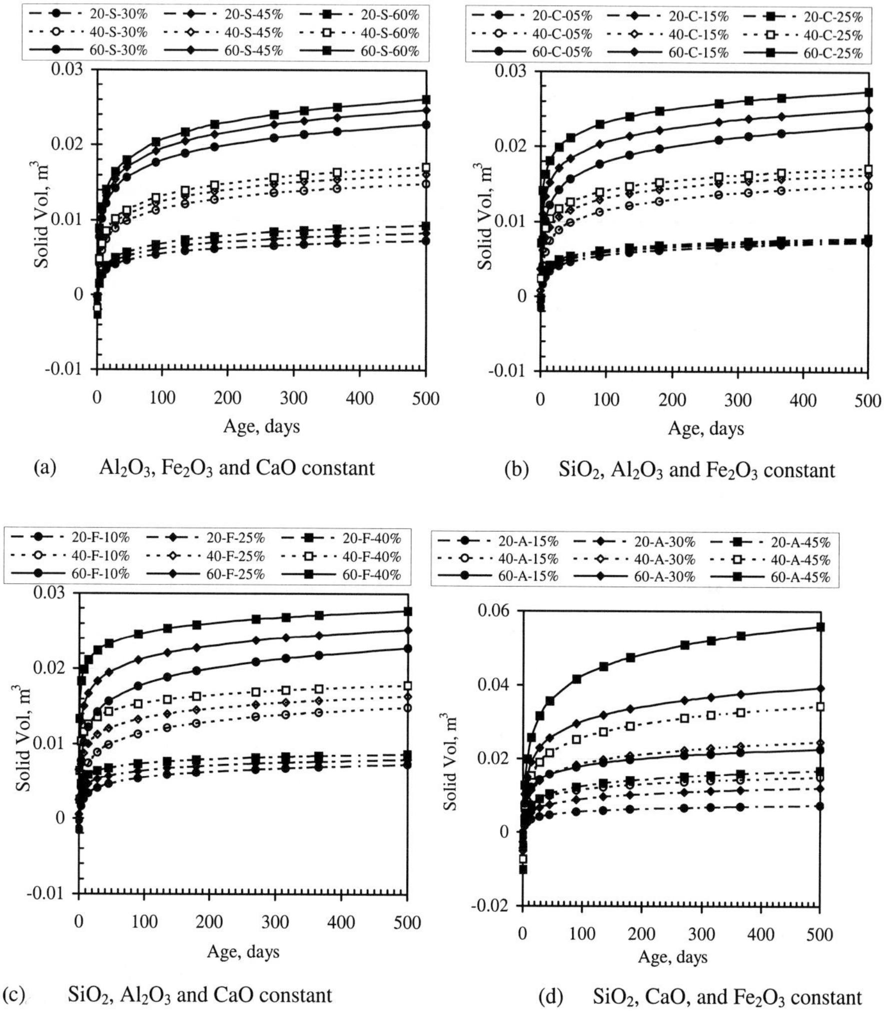

(a) Al_2O_3, Fe_2O_3 and CaO constant

(b) SiO_2, Al_2O_3 and Fe_2O_3 constant

(c) SiO_2, Al_2O_3 and CaO constant

(d) SiO_2, CaO, and Fe_2O_3 constant

Figure 2. Change in pozzolanic solid volume due to variation in oxide composition of fly ash.

However, the rate of change of the solid volume with age is much slower compared to that caused by the same variation in the silica content. It can be observed from the figure that the higher the calcium content of the ash, the earlier the pozzolanic reaction starts contributing to the solid volume. The change in the solid volume is nonlinear with the change in calcium content of ash. For all replacement ratios, an increase in Fe_2O_3 content of ash causes an increase in the solid volume. The influence of Al_2O_3 on the pozzolanic reaction is shown in Fig 2d. The solid volume increases appreciably with an increase in alumina content of the ash. It can be noted that for the same increase in the two oxide contents, Al_2O_3 produced the solid volume about five times more than that produced by Fe_2O_3. This phenomenon is just in line with the observed facts from the hydraulic and pozzolanic behavior of the low and high calcium fly ashes

In the second part of the performance test the Bogue's compounds are varied as shown in Table 3. The properties of ash used with these cements are listed in Table 4. The results of this study are shown in Fig. 3.

TABLE 3

VARIATION IN THE MAIN COMPOUNDS OF THE CEMENT CONSIDERED IN THE PERFORMANCE ANALYSIS

ID	$\%C_2S$	$\%C_3S$	$\%C_3A$	$\%C_4AF$	$\%Na_2O$	$\%K_2O$	Remarks
C3S=50%	30	50	8	9	0.17	0.5	Variation
C3S=45%	30	45	8	9	0.17	0.5	of
C3S=40%	30	40	8	9	0.17	0.5	C_3S
C2S=30%	30	50	8	9	0.17	0.5	Variation
C2S=25%	25	50	8	9	0.17	0.5	of
C2S=20%	20	50	8	9	0.17	0.5	C_2S
C3A= 8%	30	50	8	7	0.17	0.5	Variation
C3A=10%	30	50	10	7	0.17	0.5	of
C3A=12%	30	50	12	7	0.17	0.5	C_3A
C4AF=5%	30	50	8	5	0.17	0.5	Variation
C4AF=7%	30	50	8	7	0.17	0.5	of
C4AF=9%	30	50	8	9	0.17	0.5	C_4AF

TABLE 4

PROPERTIES OF THE FLY ASH CONSIDERED IN THE PERFORMANCE ANALYSIS

$\%SiO_2$	$\%Al_2O_3$	$\%F_2O_3$	$\%CaO$	$\%Na_2O$	$\%K_2O$	Surface Area
29.2	19.2	13.2	23.4	2.00	1.90	300 m^2/kg

It can be observed from Figure 3 that an increase in C_4AF causes increase in solid volume and strength at all ages. On the other hand an increase in C_3A reduces the solid volume of hydration, which means that an increase in C_3A causes reduction in the strength of hydrated paste at all ages. This observation is in line with the established facts of the hydration chemistry. The increase in C_3S or C_2S causes increase in solid volume of hydration at all ages. However, the change caused by C_2S variation is more pronounced at later ages as compared to that caused by C_3S. This observation is again in line with the established facts of hydration chemistry.

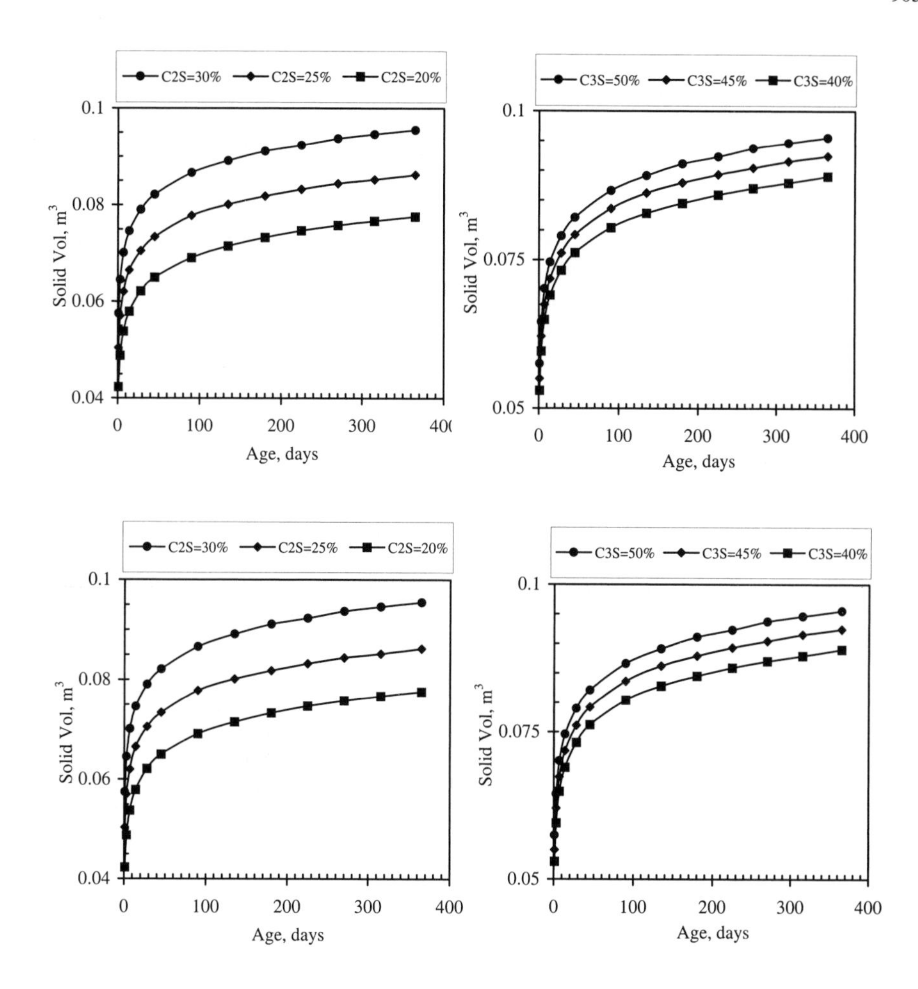

Figure 3. Change in solid volume due to variation in compound composition of cement.

CONCLUSIONS

The quality of calibration and validation of the presented model is found quite satisfactory and the predicted strengths agree closely to the experimental strengths. It has been observed from the performance tests on the model that an increase in SiO_2, CaO and Fe_2O_3 results in increase of strength at all ages and the relationship is quite similar for the three oxides. However, the change in Al_2O_3 content affects the strength in quite a different way. It is further seen that an increase in C_3A content of cement results in decrease of strength at al ages, which is a confirmation of the basic fact of the chemistry of hydration and pozzolanic reactions. The model predictions further confirm that an increase in C_4AF causes increase in the solid volume of hydration and hence in strength of hardened paste. All the tests and observations confirm the site independence of the model and hence it can be used confidently in designing mixtures incorporating fly ash of variable characteristics.

REFERENCES

Tahir, M. A. and Nimityongskul, P., (1998). "Model for Predicting 28-day Strength of Fly ash-Concrete", *Proceeding of Japan Concrete Institute*, Vol. 20, No. 2, pp. 139-144.

Tahir, M. A., (1998). "Model for Predicting Strength Development of Concrete Incorporating Fly ash of Variable Chemical Composition and Fineness", *Doctoral Dissertation,* AIT, Bangkok.

De Larrard, F. and Tondat, P., (1993). "Sur la contribution de la topologie du squelette granulaire a la resistance en compression du beton.", *Materials and Structures*, Vol. 26, pp. 505-516, (In French).

Advances in Building Technology, Volume 1
M. Anson, J.M. Ko and E.S.S. Lam (Eds.)

MIXTURE DESIGN FOR HIGH-PERFORMANCE CONCRETE FOR SLIP-FORM PAVEMENT IN JIANGXI PROVINCIAL NATIONAL HIGHWAY 323

Wu Shaopeng[1,2] , Zhou Mingkai[1] , Zeng Ming[1] and Ouyang zhengxiang[1]

[1]Key Laboratory for Silicate Materials Science and Engineering Ministry of Education, Wuhan University of Technology, Wuhan 430070,China
[2]College of Transportation, Southeast University, Nanjing 210096,China

ABSTRACT

Raw materials such as Ninguo ordinary Portland cement (produced by Anhui province), crushed lime stone, Fengcheng fly ash, mixture additive, etc were used in optimizing the design of mix proportions for slip-form paving cement concrete pavement in Jiangxi provincial national highway 323. Experimental results indicated that these raw materials could fully meet the engineering requirements and the slip-form paving cement concrete had good workability and mechanical properties. And, it was economically reasonable and feasible. This was proved with the achievement of more than 90 kilometers of highway and 20,000m^3 cement concrete. The proposed methods for controlling and adjusting mix proportions are helpful for practical highway construction projects.

KEYWORDS: Slip-form paving, Optimizing design, Mixture proportion, Workability

1. INTRODUCTION

Slip-form paving cement concrete is a special pavement material. Its mix proportions are central to one of the pivotal constructional techniques in ensuring high-performance slip-form paving cement concrete pavements in a high-grade highway. The technical requirements for mix proportions of cement concrete pavement for National Highway 323 are as follows: Firstly, it has suitable workability. The slump of fresh cement concrete is less than 5 cm, and it should match the speed of paving and meet the capacity of mechanical vibration. Secondly, it should have excellent mechanical properties. Bending strength after 28 days is 5.2MPa, and the structural bending strength should not be less than 4.5MPa. Thirdly, it should have good durability. The service life of the slip-form paving cement concrete pavement is required to be 10 years longer than that of an ordinary concrete pavement. Lastly, it should be economically reasonable and feasible. In this study, many attempts have been tried to fulfill the above requirements for highway construction, and a satisfactory result was achieved on a practical project.

2.RAW MATERIAL

The ordinary Portland cement(32.5 degree) was produced by the Ningguo cement works, Anhui province. A middle coarse river sand was employed as fine aggregate while a 531.5mm crushed limestone as coarse aggregate. Fengcheng Class II dry fly ash was used as mineral admixture. The chemical additives were used Type FDN-R water-reducer and Type JS-2 water reducer and retarder, which were produced Hubei and Hunan, respectively.

3. MIXTURE PROPORTION DESIGN

3.1MIXTURE PROPORTIONS

To achieve the mixture proportions for slip-form paving cement concrete it was first necessary to determine its water/cement ratio and sand ratio. There is an empirical formula to calculate water/cement ratio, which designs mixture proportions mainly in terms of bending strength. But the mixture proportion should be adjusted by the accumulated experience. In fact, water to cement ratio has a greater influence on compressive strength on than bending strength to a certain extent. Except for cement type and aggregate factor, cement content is the most significant factor affecting the bending strength of cement concrete. Sand rate of slip-form paving cement concrete is higher than that in ordinary cement concrete, so as to ensure the adhesion and compactness of cement concrete. There is an optimum sand ratio for a specific aggregate and water to cement ratio, with which the best properties of cement concrete can be achieved. According to our experience, when total volume of cement mortar, including cement, water, sand and air void, was 0.55~0.59m^3 in each cubic meter at concrete, the comprehensive properties were much better. Of course, the experimental result was decisive. Based on the engineering demands, the primary mix proportions requirements are shown in table 1.

TABLE 1
TECHNICAL REQIREMENT OF MODIFIED SLIPFORM PAVEING CEMENT CONCRETE NATIONAL HIGHWAY 323

Strength requirement	Setting time	Slump	Workability
Bending strength after 18 days is more than 5.2 MPa	Initial setting time is 6-7 hours and final setting time is 8-10 hours at 20.	The slump of fresh cement concrete was less than 6cm. Constructional slump is more than 2cm.	Excellent adhesion and workability, no weeping

The design of mixture proportions is based on experience based parameter and formulae. In order to agree with the requirements for bending strength and relevant technical properties, experiments were carried out to test and adjust mixture proportions in the lab. The results are showed in table 2.

TABLE 2

TEST-EXPERIMENTAL MIXTURE PROPORTION RESULTS

NO.	Water cement ratio	Sand ratio (%)	Mixture additive (%)	Material content of per cubic meter (kg/m^3)				
				Cement	Fly ash	Water	Sand	Stone
1	0.41	36	JS-2 1.2	300	80	156	687	1220
2	0.42	36	FDN-R 1.5	300	80	159	687	1220
3	0.445	36	FDN-R_1 1.0	300	80	169	666	1245
4	0.43	36	FDN-R 1.4	320	50	159	692	1229
5	0.43	36	JS-2 1.2	320	50	159	692	1229
6	0.44	36	FDN-R_1 1.0	320	50	163	692	1229
7	0.41	36	FDN-R 1.4	280	100	156	689	1225
8	0.46	36	FDN-R_1 1.0	280	100	175	688	1222
9	0.426	36	JS-2 1.2	280	100	162	688	1222
10	0.49	38	JS-2 1.2	280	60	167	747	1218
11	0.466	38	FDN-R 1.4	280	60	158	747	1218
12	0.51	38	FDN-R 1.4	280	60	174	747	1218
13	0.45	36	FDN-R 1.4	300	80	169	685	1218
14	0.48	36	FDN-R 1.4	280	60	163	706	1254
15	0.46	36	FDN-R 1.4	260	100	167	697	1238
16	0.44	36	FDN-R 1.4	390	0	172	691	1212
17	0.43	36	FDN-R 1.4	350	60	176	671	1194
18	0.42	36	FDN-R 1.4	350	60	171	671	1194

3.2 EXPERIMENTAL RESULTS

Experimental results are showed in Table 3.

TABLE 3

EXPERIMENTAL PROPORTIONS OF THE MIX OF TABLE 2

NO.	Slump mm			Bending strength MPa			Workability
	Initial	1/2h	1h	3d	7d	28d	
1	140	20		5.18	4.81	5.70	Good adhesion, little bleeding
2	60	30		5.51	5.29	5.62	Good adhesion, little bleeding
3	75	30	10	5.00	4.88	6.40	Good adhesion, little bleeding
4	160	35	15	5.78	6.87	6.92	Good adhesion, little bleeding

5	150	40	10	5.44	5.91	5.61	Excellent adhesion, little bleeding
6	75	20		5.53	6.22	6.40	Good adhesion, little bleeding
7	165	105	60	5.33	6.64	6.83	Excellent adhesion, no bleeding
8	50	25		4.42	5.53	7.20	Excellent adhesion, no bleeding
9	65	25		5.00	5.73	5.80	Excellent adhesion, no bleeding
10	35	15			5.27	4.90	Good adhesion, little bleeding
11	55	25		4.91	5.24	6.07	Good adhesion, little bleeding
12	65	35		4.13	4.78	6.00	Good adhesion, little bleeding
13	65	30	25	3.01	4.53	5.50	Good adhesion, little bleeding
14	55	30	30	3.35	4.48	5.10	Common adhesion, a little bleeding
15	75	40	20	3.00	3.71	5.40	Good adhesion, little bleeding
16	60	20		3.65	4.18	5.80	Common adhesion, bleeding
17	55	25		2.61	4.06	5.00	Excellent adhesion, little bleeding
18	90	45	35	3.00	4.34	5.20	Excellent adhesion, little bleeding

Firstly, workability: the orthogonal test analysis showed that primary factors influencing the slump was the additive and cement content next, and then water to cement ratio.
Secondly, bending strength: the most important factor influencing strength was cement content, water to cement ratio next, and then cement type. Cement content had the greatest influence among them. During the construction of slip-form paving cement concrete, water cement ratio, water content and cement content were strictly controlled. It is important to use a water-reducing agent and air-entraining agent properly. Based on the research results, the constructional mixture proportions were as follows.

TABLE 4
RECOMMENDED MIXTURE PROPORTION

Cement	Fly ash	Water	Sand	Broken stone	Mixture additive.
300	60	151	704	1225	JS-2 4.32

3.3 ADJUST AND CONTROL MIXTURE PROPORTION

Mixture proportion test and adjusting, including workability of fresh mixing cement concrete, air void content, density and strength was done to make these relevant standards meet the requirement.

3.3.1 ADJUST MIXTURE PROPORTION

Firstly, according to the recommended mixture proportion, the feasibility test was conducted, and raw materials, adhesion of the fresh concrete, bleeding and surface status of the concrete were taken into account. The result in table 5 showed that the fresh concrete had good workability.

TABLE 5
TESTING RESULT OF FRESH-MIXING CEMENT CONCRETE

Item	Slump (mm)	Density (kg/m^3)	Air void content (%)	Workability
Initial	60	2390	5.0	Excellent adhesion, little bleeding
1/2h	35	2410	4.3	Excellent adhesion, little bleeding
1h	25	2420	3.9	Excellent adhesion, no bleeding

Secondly, the air void content was determined. The air void content of freshly mixed cement concrete was measured based on Test JTJ 053-94. It agreed with the requirement for freeze-resisting properties as to the air content and met the pavement engineering need. Thirdly, the mixture proportion of freshly mixed cement concrete was adjusted. The mixture proportion test assumed the density method in this study. The density of fresh-mixing cement concrete was 2390 kg/m^3 after vibrating, which was less than designing density. Because of the bleeding, a conversion coefficient was used as follows.

$$\beta = R_1/R_2 2390/2444 = 0.978 \quad (1)$$

$$ScSd \times \beta \quad (2)$$

Mixture proportions after adjusting are as shown table 6.

TABLE 6
MIXTURE PROPORTION AFTER ADJUSTING

Cement	Fly ash	Water	Sand	Broken stone	Mixture additive
293	59	148	688	1198	JS-2 4.3

Lastly, the bending strength was determined. The beam samples were made based on the relevant standards. In this test, the water content was kept constant, but the water to cement ratio was increased or decreased 0.05, respectively. Two groups of samples were tested for determining bending strength. The results are shown in table 7. Test result agreed to the requirement.

TABLE 7
TEST RESULT OF BENDING STRENGTH

Type	Bending strength after 7 days (MPa)	Bending strength after 28 days (MPa)
W/C = 0.42	5.28	6.40
W/C = 0.47	5.15	6.20
W/C = 0.37	5.50	6.50

3.3.2CONTROL CONSTRUCTIONAL MIXTURE PROPORTION

It is necessary to adjust and control mixture proportion according to ambient temperature, transport distance, raw materials, especially water content of the aggregates, proportion of sand and crushed stone after mixing. The purpose is to ensure mixture proportion parameters fulfill the design requirements for concrete workability, bending strength and durability.

Firstly, raw materials on site must be tested when they are received. Slip-form paving is mechanized construction as a large scale. During construction, cement, sand, crushed stone and mixture additives are tested regularly. Only qualified material can enter the site. It was necessary to adjust mixture to meet constructional requirements when the quality of qualified material fluctuated. The regular test result of Hailuo P42.5 Portland cement showed that its quality fluctuated only slightly and the properties remained stable. Properties of fly ash were up to standard except for color changes. Water-reducer content varied a little with single mixture additive. The calculation system broke down easily with higher injection of mixture additive. Screening tests indicated that modulus of fine sand and the grading of 5-16mm of crushed stone varied greatly, but other relevant standards qualified.

Secondly, the precision of the mixing system was calibrated regularly. The precision of slip-form paving cement concrete is required to the higher than that of ordinary cement concrete and be calibrated before construction. Measuring error requirements were as follows: cement±1%, fly ash±2%, crushed stone ±2%, water±1% mixture additive±2%. If the measuring error was out of the requirement, the measuring system had to be re-calibrated during construction. If the automatic measuring system broke down, the pavement construction had to be stopped until the breakdown was fixed. When a trial batch was used on the 323-highway construction site, bleeding was found in the pavement. The reason was that the error of the binding material system beyond the requirement limits. The measuring system for chemical additives easily broke down or went out of control for low contents and complex composition. So it was important to test and calibrate that measuring system.

Thirdly, the quantity of the sand, crushed limestone and water were adjusted according to the moisture contents of sand and crushed stone, fineness modulus of sand and gradation of aggregate. Moisture contents of the raw materials varied with weather conditions, locations and exposure time etc., hence, the moisture contents of these materials had to be kept monitored. Sand ratio and fineness modulus differs for different batches. Sand ratio was adjusted when fineness modulus varied obviously. Increasing sand ratio was based on pavement requirement. Sand ratio for slip-form paving cement concrete, fineness modulus, water content of raw material and stone proportions were tested every day before mixing in 323 Engineering. The tests were performed by professional quality inspectors. On the construction site on Highway 323, the mix proportions were adjusted according to slump changes. With these measures, the quality of the pavement was further improved.

Lastly, the mixture proportions were adjusted according to paving conditions on site. The quality of the slip-form paving depends on weather conditions, pavement conditions, concrete delivering distance. So it is necessary to adjust mixing parameters according to paving condition to ensure high-quality of pavement. The 323 Engineering experienced all sorts of strict tests, including high-temperature, long-distance transportation and bad constructional conditions. When the transporting distance was about 18km and the temperature in summer was 47, the slump of cement concrete reduced quickly. In order to improve the slump and to meet the slump requirement of more than 20mm on site, a water-reducing agent was accordingly added. Dense mortar layer on the pavement surface was 2 to 4mm thick. In order to ensure the quality of cement concrete, a finer aggregate was chosen, and the reducing-water agent content was increased and the water content reduced accordingly. Transporting assembly line had to be organized well, avoiding traffic jam and breakdown.

Freshly mixed cement concrete conformed to the requirement for slip-form paving by testing and controlling the constructional mixture proportion. Acceptance quality of the pavement engineering approached a good level and achieved complete success.

4. CONCLUSIONS

4.1 The concrete prepared with Niangua ordinary Portland cement, crushed limestone, Fengcheng class II fly ash and water-reducers was able to meet the requirements for slip-form paving construction in Jiangxi Provincial National Highway 323.

4.2 The cement concrete prepared based on the mixture proportions in table 6 exhibited suitable workability, excellent mechanical properties and economy.

4.3 The mixture proportion designed for the slip-form paving concrete in Highway 323 engineering had been firstly used in slip-form paving construction of Jiangxi Province. These satisfactory results were achieved is proved by the completion of the qualified construction projects including more than 90 kilometers of highway and 20 thousands cubic meters cement concrete.

References

[1].Rongqing Yan (1997). Paving Machine And Constructional Technology Of Slip-form Paving Cement Concrete. People's Communications Press Of China(Beijing)

[2]Daoyuan Zhu(1999). Multi-statistical Analysis And Software, Southeast University's Press Of China(Nanjing)

[3]Chuhang Wu(2000), Paving Construction And New Technology Of Cement Concrete, People's Communications Press Of China(Beijing)

Advances in Building Technology, Volume 1
M. Anson, J.M. Ko and E.S.S. Lam (Eds.)

ADVANCES IN THE STUDY OF FATIGUE BEHAVIOR OF HIGH PERFORMANCE CONCRETE IN CHINA

XIAO Jian-Zhuang and CHEN De-Yin

Department of Building Engineering, Tongji University, Shanghai 200092 P.R. China

ABSTRACT

With the wide application of high performance concrete in civil engineering, in such as high-speed railroads, maglev transrapid, offshore platforms and super high-rise buildings, the fatigue behavior of high performance concrete has had more and more attention by civil engineers. Based on the investigations reported in relevant home references, using the comparative analysis method, the fatigue behavior of high performance concrete is reviewed and the factors affecting fatigue behavior are studied, including fatigue damage mechanism analysis, fatigue failure characteristic and fatigue concept design. Finally some suggestions on the fatigue-resistance of high performance concrete are given.

KEY WORDS

High performance concrete, fatigue test, fatigue behavior, fatigue design, number of cycles, cumulative damage

HIGH PERFORMANCE CONCRETE

Concrete, since the 19th century, has developed rapidly from plain concrete to reinforced concrete, pre-stressed concrete, and special concrete into which additives are added. Although the new members in the family of concrete----high strength concrete and high performance concrete [1], came into use only a few years ago, as they meet the demands of suitable strength and excellent durability in structures, they will be used widely in civil engineering in the future.

FATIGUE PROBLEMS ON STRUCTURES OF HIGH PERFORMANCE CONCRETE

Under repeated load, many structures in civil engineering have collapsed suddenly, in which the stress is less than the design value. This is called fatigue failure [2]. Since World War II, many such accidents have occurred. Before the 1970's, the fatigue problems of concrete structures were not paid so much attention as steel structures. This is because fatigue failure occurred infrequently when the permissible stress method was adopted in structural design. But with the development and wider application of concrete structures, fatigue damage can not be ignored in some structures under repeated load, especially for some important structures in a condition of high stress, such as high-speed railroads, maglev transrapid, offshore platforms and super high-rise buildings. Therefore, several scholars in China have carried out many tests and conducted research on the fatigue behavior of concrete. It should be pointed out that although high performance concrete or high strength concrete is not mentioned in much literature, as the relevant performance of the concrete used in the tests approaches or reaches the basic behavior of high performance concrete such as high strength, good fluidity, or excellent durability, they can be effectively considered as high performance concrete.

FATIGUE TESTS ON HIGH PERFORMANCE CONCRETE

Different methods are used in classifying fatigue tests. For example, fatigue tests can be classified as low-cycle fatigue tests and high-cycle fatigue tests according to number of cycles. Because the stress level in low-cycle fatigue tests is high, the tests are called fatigue tests under high stress, while high-cycle fatigue tests are called fatigue tests under low stress. Fatigue tests may also be classified as constant-amplitude loading tests and variable-amplitude loading tests [3] which can be sub-classified as one stage constant-amplitude, multi-stage constant-amplitude, incremental stages loading and so on. For this aspect of loading status, the fatigue tests are classified as under compression, tension, flexure, flexure and shear. The main fatigue tests in China in recent years are listed and described in Table 1. Based on those advances in the study of the problems of fatigue performance of high performance concrete nowadays is put forward in this paper.

FACTORS AFFECTING FATIGUE BEHAVIOR OF HIGH PERFORMANCE CONCRETE

Effect of Admixtures

The common mechanism of fatigue failure of concrete is described as follows. There are subtle defects such as subtle cracks, holes, low-strength interfaces or impurities. When the defects are subjected to loading, a stress concentration is produced. Under repeated load, damage emerges and accumulates so much so that structures fail. Fatigue failure is a process of cumulative damage. But adding some special admixtures such as silicon ash, and fly ash and others can reduce initial defects in concrete. The admixtures can prevent cracks from emerging and developing, and the fatigue behavior of the concrete is raised. Another kinds of admixture are fibers, including steel fiber, carbon fiber and polypropylene fiber. These fibers can be included singly or mingled. The fatigue behavior of high strength concrete is improved markedly by mixing proper contents of steel fiber [4]. The reason is obtained easily. Under repeated load, the initial defects in concrete develop into tiny cracks and stress concentrations appear at the tip of tiny cracks. However the cracks are stopped by the intersections with steel fibers. At the same time stress concentration is released by steel fibers. So fatigue behavior is improved. But when the number of cycles increases further, the ability of steel fibers holding cracks is limited so that cracks expand across steel fibers and finally the concrete fails from cumulative damage exceeding the limit value. It is suggested in the test [5] that the compressive strength of steel

TABLE 1

MAIN TESTS ON FATIGUE PERFORMANCE IN CHINA

Investigator	Strength (MPa)	Mixture of concrete	Size (mm^3)	Feature	Type
Wu Peigang Zhao Guangyi Bai Liming	76.4 84.7 77.6	Water cement ratio:0.330 Sand cement ratio: 1.204 Stone cement ratio:2.458 Water reducer:0.010	Prism $70.7^2 \times 212.1$ $100^2 \times 300$	S_{max}=0.80,0.75,0.70 0.65 S_{min}=0.10 f=0.10Hz Constant-amplitude	Axial compres sion
Ju Yang Fan Chengmou	30	Water cement ratio: 0.5 Sand cement ratio: 2.213 Sand cement ratio: 2.213 Fiber ratio: 0.012	Cylinder specimens Effective size: 100×250	S_{max}=0.75,0.65 S_{min} = 0.05 f=7.5Hz Variable-amplitude	Axial compres sion
Wu Peigang Zhao Guangyi Zhan Weiwei	73.7 65.7 77.9 83.9 69.9	Water cement ratio: 0.33 Sand cement ratio: 1.204 Stone cement ratio: 2.458 Water reducer: 0.010	Prism: $100^2 \times 100$ $100^2 \times 400$ $100^2 \times 400$	S_{max}=0.90,0.80,0.75 0.70,0.65,0.60 S_{min} =0.10 f=5, 10, 15Hz	Tension (Split axial, bending tension)
Li Yongqiang Chen Huimin	51.74	Water cement ratio:0.40 Sand cement ratio:1.16 Stone cement ratio:.47	Prism: $100^2 \times 515$	S_{max}=0.9,0.8,0.75, 0.675,0.6,0.5 S_{min}=0.09,0.08, 0.075,0.0675, 0.06,0.05, f=10Hz	Flexure
Yan Yun Sun Wei	80 100 120	Water cement ratio: 0.28 Silicon / cement: 0.20 Fiber ratio: 0.015	Prism: $100^2 \times 400$	S_{max}>0.8 or<0.8 S_{min} =0.1 S_{max} f=5-7Hz,or15-20Hz	Flexure
Zhao Shunbo Zhao Guofan Huang Chengkui	38.3-48.5	Mixed with shear steel-typed steel fiber, the maximum grain size:10mm, NF-B water reducer, the shear steel fiber, steel fiber reinforced pre-stressed concrete beams	Beam: $150 \times 300 \times 2300$ Shear-span ratio: 1.75	S_{max}=0.6 S_{min} =0.1 S_{max} f=8-10Hz	Flexure and shear
Yu Maohong Zhan Junhai Don Wenshan Zhang etc.	25.0	Water cement ratio: 0.5 Sand cement ratio: 4	Thin-walled cylinder tube, $\Phi 115 \times 330$ δ =20	τ / σ =0.45,0.25, σ =-10MPa f=4Hz	Compre ssion and torsion

Note: S_{max}, S_{min}—the maximum stress σ_{max}/f_c and the minimum stress σ_{min}/f_c under fatigue load
σ_{max}, σ_{min}—the stress of upper limit and lower limit, MPa

fiber concrete under static loading is closed to that of normal strength concrete, but there are obvious differences in fatigue behavior, such as fatigue life. The fatigue life of steel fiber concrete is 7.9-13.7 times that of concrete without steel fibers. It is also proved by the test that steel fiber is effective in aspects of fatigue behavior in reinforced pre-stressed concrete beams [6].

When part of the cement is replaced with silicon ash in high strength concrete [4], variation happens in structures due to the characteristics of the interfacial zone between aggregate and matrix. The effect of silicon ash and fly ash eliminates tropism and enrichment of the crystal of $Ca(OH)_2$. The boundary between interface and matrix blurs and this means the weakness of the interface is alleviated or disappears. So the interfacial bond between aggregate and cement is strengthened and interface failure will be avoided when loading is applied. Thus, the fact that silicon ash has an overall effect on fatigue behavior can be explained by the reduction in amount and size of the initial cracks.

The following conclusions can be obtained from the comprehensive analysis of test results described in Table 1. There have different effects improving fatigue behavior depending on the different admixtures used. Steel fiber is effective for confining the prolongation of cracks especially, while silicon ash is good at reducing the source of cracks. Carbon fiber has good rigidity with poor toughness and polypropylene fiber is the other way round. Under the same conditions, carbon fiber improves the fatigue behavior of concrete more effectively than polypropylene fiber. But if fiber and silicon ash are intermixed, fatigue behavior is improved greatly because they make full use of their respective advantages and make up for each other's shortcomings. In the test [4], it is suggested that when the number of cycles is the same, high strength concrete in which the same ratio of silicon and steel fiber are mixed can increase 2 times in strength, and when the level of stress is the same, the number of cycles is increased over 10 times. If carbon fiber and polypropylene fiber are mixed together, a superimposed effect exists and fatigue behavior is improved greatly because the polypropylene fiber increases the tenacity of concrete and carbon fiber improves the ability to resist load. When the volume content of carbon fiber and polypropylene fiber reach 0.015 and 0.009 respectively, fatigue life [7] can be increased by up to 10 times that of pure concrete.

Effect of Loading

The effect of loading is referred to as the effect of stress level and sequence of loading. Generally, when its stress level is high, fatigue life is short; and when stress level is low, fatigue life is long. However, there are different opinions about the exact relationship between them. For example, one researcher thinks that single logarithm relationship exists [8], and others think that double logarithm relationship exists [9].

The varieties of fatigue behavior affected by stress level mostly lie in the way of cracks develop and the speed of cumulative damage. When the stress level is low, at the beginning of stage, cracks are few and short. With the cycles increasing, these cracks develop steadily or increase slowly. With a number of cycles increasing further yet, new cracks emerge so much so that structures fail due to new cracks and old cracks intersect and run through. When its stress level is high, at the early stage there are more cracks. With number of cycles increasing further, old cracks develop continuously and new cracks also emerge rapidly. Eventually the cracks intersect and become so wide that the structures fail. As for high strength concrete, it is suggested by the test [10] that under high stress, the fatigue life of high strength concrete is greater than that of normal strength concrete under the same conditions, while under a low stress level, normal strength concrete has a longer life.

The effect of the sequence of variable-amplitude loading on fatigue behavior is different. The loading history in which the early stress level is lower (or higher) and the late stress level is higher (or lower) is often referred to. Different load sequences are mainly attributed to transformation of cumulative damage. According to this idea, controlling the number of early stress level cycles is very important. When the cycles are less than a critical number, the subtle cracks in concrete tend to close back; otherwise the subtle cracks tend to develop. It is inferred from the tests [11, 12] that when the number of cycles at low (or high) stress levels at the early stage is less than the critical number, the surplus fatigue life improves much more than the corresponding life when it only remains in high (or low) stress level.

Relationships Between Admixtures Effects and Loading

The effect of admixtures on fatigue behavior is related to the sequence of loading. This conclusion is obtained from the relationships between the varieties and content of admixtures and the sequences of loading. Fatigue behavior is obviously increased when the proper content is controlled according to various stress levels. Also a critical number for the early stress level exists depending on the various loading regimes. The critical number is related to the behavior of concrete. It is concluded from the test [7] that when fibers are intermixed, corresponding to a certain stress level, the flexure fatigue behavior is improved as the content increases. In the study [6] on the shear fatigue behavior of concrete with fiber, it is clear that steel fibers are effective in controlling the development of cracks while the function of steel fibers is also related to the cyclic characteristics. In the tests under variable-amplitude loading [11, 12], the low-high loading sequence, fatigue life improves as long as under low stress, the number of cycles do not exceed a critical number. The critical number is different for various concrete. The detailed relationship will need to be studied further. From the above descriptions, it can be deduced that fatigue behavior is affected by the inner characteristics of the concrete and the exterior factors of loading.

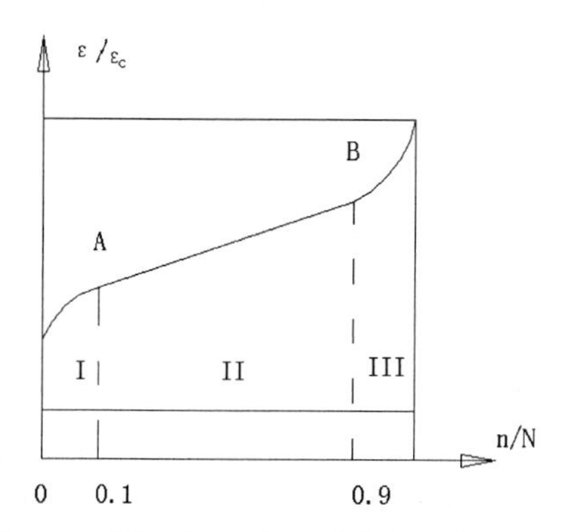

Figure 1: The ideal curve of the three-phrased fatigue deformation development

FATIGUE DEFORMATION OF HIGH PERFORMANCE CONCRETE

Fatigue Deformation Characteristics

First, under repeated load, fatigue failure takes a long time, but at the moment of failure, it is brittle. Moreover the internal force is complex and it is more appropriate to relate fatigue failure with deformation. As for high performance concrete and high strength concrete, when failure happens, the brittleness is obvious and ductility is weak. Second, during the course of cumulative damage, damage is usually measured by deformation and elastic modulus. Finally many tests suggest that the characteristics of fatigue deformation of concrete under repeated load are similar [13]. With stress level and number of cycles increasing, the longitudinal fatigue deformation increases. A three-phased law described by the ideal curve line in Figure 1 has been proposed. In the figure, the phase of deformation develops rapidly before point A and longitudinal strain increases slowly after that. The second phase is the relatively steady phase. This phase also fits the normal working phase of concrete structures under repeated load. The longitudinal strain varies as a linear function as the number of cycles increases. This phase ends of point B. After B, the failure phase is entered. The longitudinal strain increases sharply, develops rapidly and failure occurs.

Criterion of Fatigue Failure

Deformation of concrete under repeated load demonstrates general characteristics and its durability behavior of a concrete structure declines as a result of cumulative damage. Although the final failure has not emerged, the durability has reduced so much that structures are not in a safe condition any longer. Based on this, the deformation due to cumulative damage is looked as the criterion to judge fatigue failure. This indicates [14] that when the residual strain of high strength concrete reaches 35% longitudinal total strain while the number of cycles is up to 85% of the total number of cycles, high strength concrete has lost the ability to resist repeated load. Under repeated load [6], the failure characteristics for oblique section in beams is that when oblique cracks have appeared under static loading, the strain in the shear-compression zone at the tip of the individual oblique cracks mostly comes up to the ultimate compressive strain.

FATIGUE DESIGN OF HIGH PERFORMANCE CONCRETE

Fatigue design of concrete can be divided into conventional fatigue design and fatigue design based on probability.

Conventional Fatigue Design

The size of the section of members is designed according to static loading and the fatigue check is carried out at the critical points in the section. Based on the requirement for structure design, it is necessary for the structural resistance to be greater than the internal force. The internal force is a function of repeated load, while the resistance of the structure is a function based on the fatigue strength. So in the study of conventional fatigue design, the first thing is to determine fatigue strength. Fatigue strength can be acquired by fatigue tests on the concrete under the basic constant-amplitude repeated load.

The relationship between stress and number of cycles at failure (S-N) needs to be obtained and the fatigue limit diagram needs to be considered to obtain the relationship between the effect of constant-amplitude and variable-amplitude. A series of studies on fatigue behavior of concrete under conditions of tension, compression, flexure, shear, and torsion have been carried out, it is rarely reported that relationships among them are found. The fatigue limit diagram is such a chart [3] that describes the relationship between cyclic loading regimes. From the chart, variable-amplitude loading can be transformed to constant-amplitude loading in order to acquire fatigue strength. However in this field, the studies so far are insufficient and are affected by the randomness of the date. So further study needs to be carried out to obtain the correct fatigue limit curve.

Fatigue Design Based on Probability

Nowadays, although reliability is considered in fatigue design, the design still actually depends on combination of experience and probability. Two problems that need to be solved are for given reliability how to calculate the critical number and for given critical number how the reliability can be checked. Most of the research abroad and at home is mainly focused on the distribution of random variables and the relationship of S-N under certain distributions.

CONCLUSIONS

1) Both similarities and differences are found in the fatigue performance of normal concrete and high performance concrete.
2) The factors affecting fatigue behavior of high performance concrete, including the different varieties and contents of admixtures, are related to certain stress level and the sequence of loading.
3) In the study of fatigue behavior of normal strength concrete, fatigue strength is mostly focused upon. While for high performance concrete, the deformation characteristics can be seen as important and should be investigated under repeated load.
4) Study of the fatigue behavior of high performance concrete should be combined with the research

on its durability.

5) Admixtures, including silicon ash, fly ash, steel fiber, carbon fiber and polypropylene fiber, can enhance the fatigue behavior of high performance concrete. However further study is needed.

REFERENCES

[1]Yves Malier. (1992). *High performance concrete from material to structure*. E and FN SPON

[2]Guo Zhenhai. (1999). *The theory of reinforced concrete* . Press of Tinghua University

[3]Xu Hao. (1988). *Fatigue Strength*. Press of High Education

[4]Sun Wei and Yan Yun.(1994). Study on the impact and fatigue performance and mechanism of high strength concrete and steel fiber reinforced high strength concrete. *Journal of Chinese Civil Engineering* 27:5, 20~27.

[5]Deng Zongcai.(2000). Study on compressive behavior of steel fiber reinforced concrete. *Building Structures* 30:9,53~55.

[6]Zhao Shunbo, Zhao Guofan and Huang Chengkui.(2000). Research on the shear fatigue behavior for steel fiber reinforced prestressed concrete beams. *Journal of Chinese Civil Engineering* 33:5, 35~39.

[7]Zhang Shaobo, Hua Yuan and Jiang Zhiqing.(1998). Experimental study on flexural fatigue behavior of hybrid fiber reinforced concrete. *Journal of Mechanics in Engineering*, 59~63(supplement)

[8]ACI Committee 215. (1974). Considerations for design of concrete structures subjected to fatigue loading. *Jounal of ACI* 30:2, 97~121.

[9]Shi Xiaoping, Yao Zukang, Li Hua and Gu Minhao.(1990). Study on flexural fatigue behavior of cement concrete. *Journal of Chinese Civil Engineering* 23:3, 11~22.

[10]Zhao Guangyi , Wu Peigang and Zhan Weiwei.(1993). The fatigue behavior high-strength concrete under tension cyclic loading. *Journal of Chinese Civil Engineering* 26:6, 13~19.

[11]Ju Yang and Fan Chengmou.(1995). Enhancement action in fatigue of steel fiber reinforced concrete. *Journal of Chinese Civil Engineering* 28:3, 66~77.

[12]Ju Yang and Fang Chengmou.(1995). Fatigue action of steel fiber concrete under variable-amplitude axial compressive loading. *Journal of Harbin University of Architecture and Engineering* 28:3, 77~82.

[13]Ou Jinping and Lin Yanqing.(1999). Experimental study on performance degradation of plain concrete due to high-cycle fatigue damage. *Journal of Chinese Civil Engineering* 32:5,15~22.

[14]Wu Peigang, Zhao Guangyi and Bai Liming.(1994). Fatigue behavior of high strength concrete under compressive cyclic loading. *Journal of Chinese Civil Engineering* 27:3, 33~40.

[15]Li Yongqiang and Che Huimin.(1998). A study on the cumulative damage to plain concrete due to flexural fatigue. *Journal of the Science of Railway in China* 19:2, 52~59.

[16]Yu Maohong, Zhao Junhai, Don Wenshan, Zhang Chunyan and Zhang Ning.(1998). Experimental studies on the fatigue strength of concrete under the complex stress state. *Journal of Xi'an Jiaotong University* 32:3, 89~92.

Advances in Building Technology, Volume 1
M. Anson, J.M. Ko and E.S.S. Lam (Eds.)

HIGH PERFORMANCE POLYMER CONCRETE

R Y Xiao and C S Chin

Department of Civil Engineering, University of Wales Swansea, UK SA2 8PP

ABSTRACT

This paper presents experimental investigation and analysis of high performance polymer concrete (HPP) in which volume fraction of fibres was varied. Test results indicate that both compressive, tensile strength of concrete could be increased with the inclusion of fibre. Moreover, the use of fibres to improve the shear resistance has also been shown to be very effective for beams. HPP fibres have significantly increased the ultimate shear strength of reinforced concrete beam.

KEYWORDS: high performance polymer (HPP); superplasticizer; compressive strength; flexural strength; cylinder splitting strength; shear strength; design formulae.

INTRODUCTION

Cementitious materials in the form of concrete are attractive for use as construction materials since they can be readily molded and complex shapes may be fabricated. Concrete lends itself to a variety of innovative designs as a result of its many desirable properties. Concrete is brittle and weak in tension; these characteristics have limited its use. The development of fibre reinforced concrete has provided a technical basis for improving these deficiencies (1).

The main reason for the growing interest in the contribution of fibres in cement-based materials is the desire to increase the tensile properties of the basic matrix. However recent developments made polypropylene fibres the most widely used fibre in concrete, with the UK usage for 1995 estimated at 1000 tonnes. Polypropylene fibre provides a simple effective and economical means, even in small quantities, have dramatic effect on suppression of early cracking and thereby eliminate the main causes of weakness (2).

Goldfein (3) firstly suggested polypropylene fibres as an admixture to concrete in 1965 for the construction of blast-resistant buildings for the U.S. Corps of Engineers. His work comprised the incorporation of fibres in mortar and neat cement. This research gave the incentive for the early trials on polypropylene film fibre in concrete by Shell International Chemical Co. Ltd. who gave the material the name Caricate (4). For centuries, man has attempted to reinforce concretes with various types of fibres. There are biblical references to the use of straw fibers in the molding of bricks. Later, the Roman Empire used hair fibres in structural mortars. Today, fibres are used in concrete to minimize cracking. Cracking happens in concrete when external stress exceeds strain capacity. By increasing concrete's strain capacity, stresses from intrinsic and external forces can be better managed. Fibre reinforcement redistributes stress in concrete and therefore cracking can be minimized (5).

The concept of using traditional steel and polypropylene fibres as reinforcement is not new, it is well known that both flexural and ductility of concrete could be increased with the inclusion of fibre. Moreover, the use of fibres to improve the shear resistance has also been shown to be very effective for beams. Handling steel fibres for concrete mixing may cause construction hazardous while traditional polypropylene fibres tend to form bundles in the concrete mixing if the volume ratio is high which will affect the strength contribution.

These are the reasons why that the new generation of HPP has been designed and produced. This type of fibre has the similar to that of steel fibres and could also be bent and has dents on its surface. Clearly, this will increase its workability in the concrete also improve its bond strength with the mortar of concrete. HPP fibre is extruded from 100% virgin polypropylene and mechanically deformed in a proprietary shape to maximize anchorage for reinforcement in cementitious composites. S-152 High Performance Polymer fibre is manufactured as a coarse filament with an engineered contoured profile. HPP fibre, because of the low modulus, provides good toughness response for large crack openings. Typical relevant properties of HPP fibres are shown in Table 1 [5].

TABLE 1

HPP FIBRE PHYSICAL PROPERTIES

PROPERTIES	VALUE
Specific Gravity	0.91
Nominal Filament Diameter	0.9mm
Fibre Length	50 mm
Aspect ratio	55.6
Deformation Spacing	3 /cm
Elastic Modulus	3.5 kN / mm^2
Ultimate Elongation	15 %
Water Absorption	0

RESEARCH SIGNIFICANCE

The research reported in this paper shows the effect of HPP fibres in improving the compressive strength, tensile strength and shear capacity of concrete. The test results obtained prove that HPP fibre could improve material properties and clearly increase the shear strength of the reinforced concrete beams Combination of minimum links and fibre reinforcement forms an effective system of shear reinforcement in a structural member. Experimental results suggest a simple means of predicting the flexure, cylinder splitting strength, and ultimate shear strength of fibre reinforced concrete beam. The paper presents useful data on material properties of fibre reinforced concrete, particularly for test data separating fibre contents.

TEST PROGRAM

Ordinary Portland cement, HPP fibres, crushed gravel with maximum size of 6mm coarse aggregate, beach sand and tap water were mixed to make the concrete. The specimens were cast from two different mixes. Mix A had cement: water: sand: coarse aggregate (6mm) ratio of 1: 0.52: 1.83: 3.25 and Mix B had a cement: water: sand: coarse (6mm) aggregate ratio of 1:0.51: 1.61: 2.57. A variety of mixers have been used in practice, some requiring an adjustment to the existing equipment, some none at all. Previous research has shown that he tumble form of mixer is a good way of mixing the fibres. The absence of blades in the tumble mixer helped to ensure good fibre distribution in the mixture. The aggregates, sand, and cement were added with water and mixed for three minutes. The polypropylene fibres were then evenly sprinkled over the surface, and mixing resumed for three minutes. Finally, superplaticizers were poured to the mix without pre-mixed with other admixtures.

6 non-reinforced beams (100 x 100 x 500mm), 48 cubes (6 x 6 x 6in), 10 cylinders (6 x 12in) and 12 reinforced beams (120 x 150 x 1000mm) were cast from each batch of mix. All specimens

were cast perpendicular to the testing direction were then removed from the molds and stored in air at room temperature after 24 hours until testing at 30 days of age.

PRESENTATION OF TEST RESULTS

TABLE 2
RESULTS OF CUBE, CYLINDER AND BEAM TESTS

Beams	Fibres content	Web steel	Concrete strength			First shear crack strength	Ultimate shear strength
			Compressive strength	Cylinder splitting tensile strength	Modulus of rupture		
	%		N/mm^2	N/mm^2	N/mm^2	N/mm^2	N/mm^2
Mix Type A							
N1	-	No	40.95	2.74	4.38*	1.47	1.51
NF1	0.8	No	46.79	3.02	4.28*	1.78	1.97
W1	-	Yes	40.95	2.74	4.38*	1.88	1.91
WF1	0.8	Yes	46.79	3.02	4.28*	1.97	2.60
Mix Type B							
N2	-	No	40.86	2.47	4.34†	1.54	1.66
NF2	0.8	No	45.79	2.98	3.86†	1.54	1.72
NF3	1.0	No	45.53	2.26	5.01†	1.60	1.75
NF4	1.2	NO	43.46	2.40	5.66†	1.91	2.02
W2	-	Yes	40.86	2.47	4.34†	2.40	2.58
WF2	0.8	Yes	45.79	2.98	3.86†	2.58	2.71
WF3	1.0	Yes	45.53	2.26	5.01†	2.58	2.71
WF4	1.2	Yes	43.46	2.40	5.66†	2.58	2.72

Compressive Strength

Nine cubes had no fibres and twenty-seven cubes had HPP fibres of varying amount (0.8%, 1.0%, 1.2%) were cast to make more effective comparison between plain concrete and fibrous concrete reinforced by HPP fibres. Three cubes were tested for each batch of mixes to obtained the average value. The compressive strength increase with the percentage of fibres but there is a drop in compressive strength as percentage of fibres is over 1%; however the figure shows that the compressive strength is still higher than that of plain concrete. This can be concluded that HPP fibres have increased the compressive strength of concrete. The results from the cube crushing tests are shown in Table 2.

Flexural Strength

The flexural strength for fibrous concrete that contained 0.8 per cent of HPP fibres is slightly lower than that plain concrete. This applied to flexural test for both Mix Type A and B. However the flexural strength is increased when 1.0 and 1.2 per cent of HPP fibres were added. The flexural strength increased to a maximum value of 30% for beam that contained 1.2 per cent of HPP fibres. The failure pattern of the samples highlighted the benefit of inclusion of fibres. The plain sample failed with a sudden brittle where as the fibrous samples failed in more ductile manner.

Cylinder Splitting Strength

The cylinder splitting strength is increased by 20% when 0.8 per cent of HPP fibres were added to the mix. But the splitting strength is decrease when more than 0.8 per cent of HPP fibres were added. The splitting strength for cylinder that contained 1.0 and 1.2 per cent of HPP fibres became close to the plain cylinder. This may be explained that 0.8 per cent of HPP fibres had increased the splitting strength to the maximum value of 20%. The failure pattern of the samples that were similar to the samples used in flexural beam tests again highlighted the properties of fibrous concrete. The test results can be seen in Table 2.

RELATIONSHIP BETWEEN COMPRESSIVE AND TENSILE STRENGTH

The compressive strength of concrete is its property commonly considered in structural design but for some purposes the tensile strength is of interest [(6)]. It is possible to form an empirical relationship between the compressive strength and the tensile strength of the concrete. The relationship however is not directly proportional because as the compressive strength increases the tensile strength increases at different rate. A number of ratios have been suggested but they are all subjected to variation by a number of factors. As the tensile strength is so small it is much more sensitive to variation in constituents and production. The amount of fine and coarse aggregate and the overall grading of the aggregate can also affect the ratio.

Eight cylinders and four beams of different fibre contents have been tested. The test results obtained from the flexure and cylinder splitting tension test suggest two empirical relationship between compressive and indirect tensile strength of the fibrous concrete as follows:

$$f_t = 0.21 * (f_{cu})^{2/3} * \left(\frac{1}{V_f}\right)^{0.45} \quad (1) \qquad\qquad f_f = 0.398 * (f_{cu})^{2/3} * V_f^{\,0.85} \quad (2)$$

where f_f = Flexure strength of fibrous concrete, N/mm^2
f_t = Cylinder splitting strength of fibrous concrete, N/mm^2
f_{cu} = Characteristic strength of fibrous concrete, N/mm^2
V_f = Percentage of fibres added into the mix, %

Figure 1: Relation between tensile strength and compressive strength and percentage of fibres

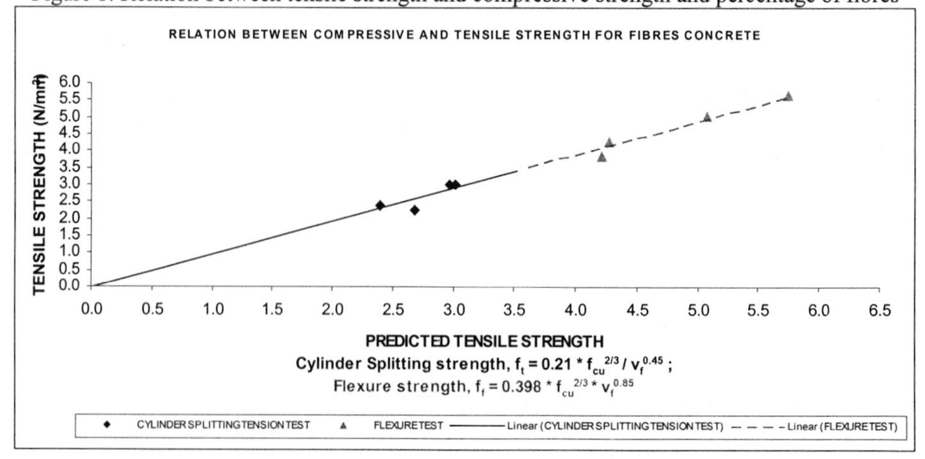

Figure 1 shows good straight-line correlation between the tensile strength obtained from both flexure and cylinder splitting tension test and the suggested equation stated above. The comparison between test results and predicted strength using equation (1) and (2) are shown in Table 3.

TABLE 3
COMPARISON OF TEST RESULTS FOR PREDICTED AND EXPERIMENTAL TENSILE STRENGTH

Specimen	Fibres percentage	Tensile strength	Predicted tensile strength	$\left(\frac{Experimental}{Predicted}\right)$ (Tensile strength)	
	%	N/mm^2	N/mm^2		
		Mix Type A			
AC1 [C]	-	2.74	-	-	
AC2 [C]	0.8	3.02	3.01	1.00	
AF1 [F]	-	4.38	-	-	
AF2 [F]	0.8	4.28	4.27	1.00	
		Mix Type B			
BC1 [C]	-	2.47	-	-	
BC2 [C]	0.8	2.98	2.97	1.00	
BC3 [C]	1.0	2.26	2.68	0.84	
BC4 [C]	1.2	2.40	2.39	1.00	
BF 3 [F]	-	4.34	-		
BF 4 [F]	0.8	3.86	4.21	0.92	
BF 5 [F]	1.0	5.01	5.07	0.99	
BF 6 [F]	1.2	5.66	5.74	0.99	
[C] Cylinder [F] Flexure Beam		Mean		0.96[C]	0.98[F]
		Standard Deviation		0.08[C]	0.04[F]

PREDICTION OF ULTIMATE SHEAR STRENGTH

As Ref. [9], the development of a general simple formula to predict the shear strength of fibres reinforced concrete beams is critical to the successful application of fibres as shear reinforcement in practice. However, from the test results reported here, it appears that the splitting tensile and the flexural strength are the important material parameters that reflect the ultimate shear strength.

From the previous report [9], parameters that influence the ultimate shear strength other than flexure and splitting strength are the shear span-effective depth ratio and the reinforcement ratio. It is not possible to include the effect of beam size in this formula because all tests were on a single beam size. However, computational studies seem to indicate similar shear strength predicted when beams of different size scale were entered. But, ultimate shear strength reduces with the beam size particularly the beam depth; that is, larger beams are proportionately weaker than smaller beam [8]. If so, an additional factor indicating dependence on effective depth of the beam must be included. Thus, the shear strength of a fibre reinforced concrete beam are expressed as:

$$f_v = (0.368) * (f_f * f_t)^{5/6} * \left(\frac{1000 * A_s}{b_v * a_v * d^{0.6}} \right)^{1/3} \quad (3)$$

where f_v = Predicted ultimate shear strength of fibre reinforced concrete beam;
f_f = Flexure strength of fibrous concrete, N/mm^2;
f_t = Cylinder splitting tensile strength, N/mm^2;
A_s = Area of tension steel provided, mm^2;
b_v = Width of the beam, mm;

a_v = Shear span, mm;

As proposed by Sharma [10], the ultimate shear strength of a fibres reinforced concrete beam with web reinforcement can be expressed as

$$v_{nf} = f_v + f_s \tag{4}$$

where v_{nf} = Nominal shear strength of fibres reinforced concrete section in combined bending and shear, N/mm^2;

f_v = Ultimate shear strength of the fibrous concrete, N/mm^2;

f_s = Shear strength due to web reinforcement, N/mm^2.

Shear strength due to web reinforcement f_s can be evaluated by using equation (5) as proposed in BS 8110 [11],

$$f_s = \left(\frac{0.95 * f_{yv}}{b_v}\right) * \left(\frac{A_{sv}}{S_v}\right) \tag{5}$$

where f_{yv} = Characteristic strength of links

b_v = Breadth of the section, mm;

A_{sv} = Total cross-section of links at the neutral axis, at a section, mm^2;

S_v = Spacing of links along the member, mm.

Figure 2 shows good straight-line correlation between the ultimate shear strength and the predicted ultimate shear strength through the origin. To show that equation (3) give a reasonable estimate of the ultimate shear strength of fibres reinforced concrete beams, results from this investigation are compared in Table 4. All the beams with fibres reinforcement are used in the correlation. The ratio of experimental to predicted shear strength is calculated. There is a definite trend, as shown in Table 4, that fibres reinforcement in beams increase the ultimate shear strength of beams

Figure 2: Relationship between ultimate shear strength and material and geometrical properties. (Unit for horizontal axis are in N$^{5/3}$ / mm$^{2/3}$)

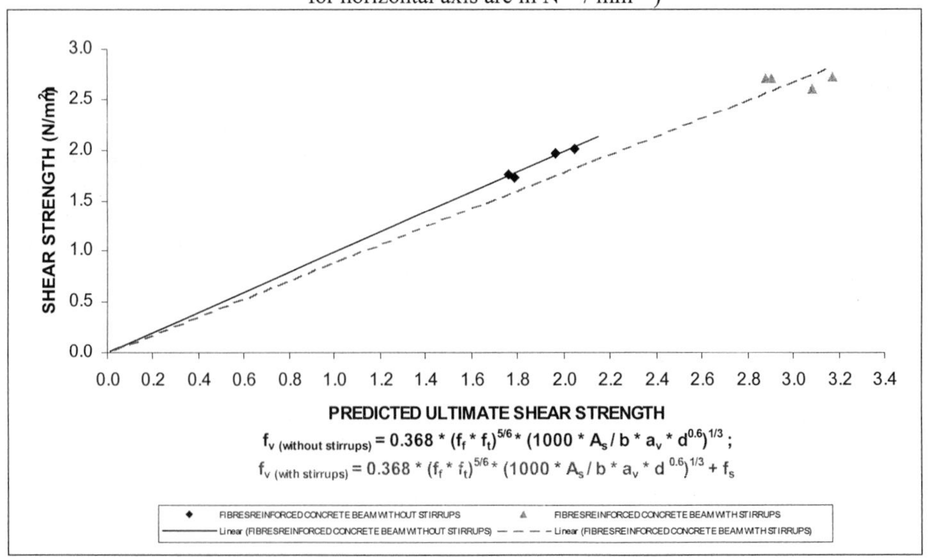

TABLE 4
COMPARISON OF TEST RESULTS OF BEAMS FAILING IN SHEAR

Beam No.	$\left(\frac{a}{d}\right)$	Fibre content	Ultimate shear strength (USS)		Percentage increase of USS	Predicted USS	$\left(\frac{Experimental}{Predicted}\right)$ (USS)
			With fibres	Without fibres			
		%	N/mm^2	N/mm^2	%	N/mm^2	
Mix Type A							
N1	2.57	-		1.51	-		
NF1	2.57	0.8	1.97		30.5	1.96	1.00
W1	2.57	-		1.91	-		
WF1	2.57	0.8	2.60		36.1	3.08	0.84
Mix Type B							
N2	2.57	-		1.66	-		
NF2	2.57	0.8	1.72		3.6	1.78	0.97
NF3	2.57	1.0	1.75		6.0	1.76	1.00
NF4	2.57	1.2	2.02		21.7	2.05	0.98
W2	2.57	-		2.58	-		
WF2	2.57	0.8	2.71		13.0	2.90	0.93
WF3	2.57	1.0	2.71		13.0	2.88	0.94
WF4	2.57	1.2	2.72		14.0	3.17	0.86
					Mean		0.94
					Standard Deviation		0.06

CONCLUSIONS

From the experimental observations reported in this paper, the following conclusions can be drawn:

1) Polypropylene fibres (S-152) are effectively in increasing the compressive, tensile, and shear capacity of concrete.
2) Flexural strength increased with fibre percentage, as there was a 30 per cent increase in tensile strength for the 1.2 per cent fibres beam while the splitting tension test has a 20 per cent increase in tensile strength with the 0.8 per cent HPP fibres cylinder but decrease in tensile strength was observed with the 1.0 and 1.2 per cent fibres cylinder. The increase in splitting strength with fibre percentage however was less well defined. Anyway, the failure pattern of the all the samples highlighted the benefit of the inclusion of HPP fibres. The plain sample failed with a sudden brittle failure where as the fibrous samples failed in more ductile manner. This shows that HPP fibres can reduce crack widths and prevent brittle failure of concrete. The results from both tests were used to generate empirical formulae, relating the compressive and tensile strengths of the fibrous concrete.
3) However, HPP fibres have significantly increased the ultimate shear strength of reinforced concrete beam. As reported by Sharma [11], for a practical solution, tests have shown that a combination of minimum links and HPP fibre forms an effective system of shear reinforcement in a structural member. With these combinations, HPP fibres have changed the failure mode of the specimen beams from shear to flexure failure. Fibres reinforced concrete beams have more ductility and significant amount of energy absorption than normally reinforced concrete beams. The presences of polypropylene fibres (S-152) in concrete restrict propagation of cracks and allow more uniform cracking. Reasonable estimation shear strength of HPP fibres reinforced concrete beams can be made by using the semi empirical equation derived.

References

(1) J. J. Beaudoin, "Fibre-Reinforced Concrete", *Canadian Building Digest*, CBD-223, originally published April 1982.

(2) Ir. Trevory Gregory, One-Day Seminar on Fibre Reinforced Concrete, Concrete and Steel Structures Into the 21st Century, 5th May 1995.

(3) Goldfein, S., "Fibrous Reinforcement for Portland Cement", Modern Plastics, 1965, 156-159 (April).

(4) D. J. Hannant, "Fibre Cements and Fibre Concretes", John Wiley & Sons.

(5) Synthetic Industries, Fibre Reinforced Concrete Division, Copyright 1999.

(6) A. M. Neville, "Properties of Concrete", Fourth edition, Longman.

(7) P. J. F. Wright, " Comments on an Indirect Tensile Test on Concrete Cylinders ", Magazine of Concrete Research (London), V.7, No. 20, July 1955, pp. 87-96.

(8) Kong & Evans, "Reinforced and Prestressed Concrete", Third edition.

(9) Victor C. Li, Robert Ward, and Ali M. Hamza, "Steel and Synthetic Fibers as Shear Reinforcement" *ACI Structural Journal* / September – October 1992, 89-M54, pg. 499-508.

(10) A. K. Sharma, "Shear Strength of Steel Fiber Reinforced Concrete Beams", *ACI Structural Journal* / July – August 1986, 83-56, pg. 624-628.

(11) British Standards Institution, BS 8110-1:1997:Concrete Design, Section 2 & 3, Code of Practice for Design and Construction.

Advances in Building Technology, Volume 1
M. Anson, J.M. Ko and E.S.S. Lam (Eds.)

STUDY ON PHOSPHOGYPSUM FLY ASH LIME SOLIDIFIED MATERIAL

Zhou Mingkai[1] Shen Weiguo[1] Wu Shaopeng[1,2] Zhao Qinglin[1]

1. Key Laboratory for Silicate Materials Science and Engineering Ministry of Education, Wuhan University of Technology, Wuhan 430070,
2. College of Transportation, Southeast University, Nanjing 210096

ABSTRACT

A new type of solidified material mainly composed of phosphogypsum, fly ash and lime was prepared in this study. Its mix proportions were optimized. The test results suggested that the ratio of phosphogypsum and fly ash at 1 : 1 and the content of the lime solidified material 6~8% was the optimum mix proportion. This kind of solidified material could be used as road base course material when its compacting ratio is 93~96%. It could also be used as sub-base material when its compacting ratio is around 93%. Crushed stone stabilized with it could be used as road base material. The microstructure of this kind of materials was observed with SEM and the mechanism of strength formation is discussed.

KEYWORDS: Phosphogypsum, Fly ash, Solidified material, Road base

1. INTRODUCTION

Recycling of solid waste is significant to sustainable development. Phosphogypsum and fly ash are the vital solid waste sources in China. Large amounts of them are left without treatment. The lime-fly ash road base material has been widely used in China. But its early age strength was very low [1.2], and this limited its extensive usage. The development of a new type of solidified material is described in this study. The solidified material is mainly composed of phosphogypsum, fly ash and lime. Its strength at early age was much higher than that of the ordinary lime-fly ash road base materials. It can be used as road base material and sub-base material. It can also be used as road base filling material. When stabilized with crushed rock it can be used as road base material.

2 MATERIALS AND TEST METHODS

2.1 MATERIALS

Phosphogypsum was obtained from Xianfan Liming Chemistry Ltd and the fly-ash from Xiangfan lignite power station. The stabilizer was made of lime and 10% of additive. The crushed rock was obtained from Xianshan stone plant, Xangfan, Hubei province.

TABLE 1:CHEMICAL AND MINERALOGICAL COMPOSITIONS OF THE RAW MATERIALS (w%)

materials	chemical composition								
	SiO_2	Fe_2O_3	Al_2O_3	CaO	MgO	TiO_2	SO_3	P_2O_5	IL
PG	3.62	0.05	0.08	31.57	0.17	0.06	42.48	1.05	19.88
FA	6.20	5.98	6.70	8.50	1.80	0.49	—	—	3.50
ST	1.42	0.13	0.38	87.79	1.64	1.80	4.05	—	4.05

Notes: PG-phosphogypsum; FA-fly ash; ST-stabilizer

2.2 TEST METHODS

Tamping test of phosphogypsum-fly ash-lime solidified material was carried out conforming to the specifications of test method T0804-94, which was published by the Ministry of Communication of China. The unconfined compressive strength was carried out conforming to test method T0804-94. The phosphogypsum-fly ash-lime solidified material was shaped within a $\Phi 5.0 \times 5.0 cm^3$ mould while the phosphogypsum-fly ash-lime solidified material stabilized crushed rock was shaped within a $\Phi 10 \times 10 cm^3$ mould.

3 RESULT AND DISCUSSION

The effect of the composition of phosphogypsum-fly ash-lime solidified material on the dry density and strength of the solidified material was studied. And the suitable mixture proportions of phosphogypsum-fly ash-Lime solidified material were optimized by means of the 7 days strength. By providing silica and calcium oxide respectively, fly ash and lime also play an important role in the solidified material. In this paper it described how the crushed rock stabilized with phosphogypsum-fly ash-lime solidified material was prepared as road base material.

3.1 THE PROPORTION OF THE PHOSPHOGYPSUM-FLY ASH-LIME SOLIDIFIED MATERIAL

In this test the effect of mix proportions and the influence of stabilizer on the strength of phosphogypsum-fly ash-lime solidified material was studied. The content of stabilizer varied from 4:96 to 10:90. The testing cylinder was made with compactness ratio of 96%. The phosphogypsum-fly ash-lime solidified material was stabilized with ashed lime stabilizer, slaked lime on quick lime. The strength and maximum dry density of phosphogypsum-fly ash-lime solidified material with different proportions are listed in table 2.

TABLE 2: MIX PROPORTION TEST OF PG-FA-QL SOLIDIFIED MATERIAL

No	ST : PG : FA	γ_{dm}(g/cm^3)	W_{pt}(%)	R_{7d}(MPa)
1-1	8 : 20 : 72	1.401	18.34	2.56
1-2	8 : 30 : 62	1.446	17.62	2.76
1-3	8 : 40 : 52	1.464	16.35	3.13
1-4	8 : 50 : 42	1.487	16.21	3.24
1-5	8 : 60 : 32	1.491	16.21	3.19
1-6	8 : 70 : 22	1.498	15.76	3.10
1-7	8 : 80 : 12	1.493	14.91	3.04
1-8	4 : 50 : 46	1.487	16.21	2.01
1-9	6 : 50 : 44	1.487	16.21	2.76
1-10	10 : 50 : 40	1.487	16.21	2.65
*1-11	8 : 0 : 92	1.312	19.30	0.90

Notes: ST-stabilizer; *: ST is slaked lime; +: ST is quicklime; γ_{dm}-Maximum dry density Notes; W_{pt}-Optimum water content; R_{7d} is compressive strength

Table 2 show that the maximum dry density of the phosphogypsum-fly ash-lime solidified material increased with the increase of the phosphogypsum content since the phosphogypsum has a density higher than that of the fly ash. When the ratio of fly ash and phosphogypsum was about 1:1, the phosphogypsum-fly ash-lime solidified material had the highest strength. When the stabilizer content of phosphogypsum-fly ash-lime solidified material was about 6% the solidified material obtained the highest strength. From the results we can preliminarily concluded that the suitable proportion is the ratio of fly ash and phosphogypsum of about 1:1 and the stabilizer content about 8~9%.

3.2 COMPACTNESS RATIO OF THE PHOSPHOGYPSUM-FLY ASH-LIME SOLIDIFIED MATERIAL

The phosphogypsum-fly ash-lime solidified material is a material with high porosity. The porosity was dependent on the forming pressure and compacting work while the products (specimens) were made. Porosity can govern the properties of the materials, especially affecting material strength. High porosity will result in low material strength, whereas decreasing porosity can enhance material strength. The cylindrical specimens of this study were compacted to the compactness ratio of 93% and 96%. The strengths at 7 and 28 days are listed in table 3.

TABLE 3: PG-FA-QL SOLIDIED MATERIAL'S COMPRESSIVE STRENGTH AT DIFFERDNT COMPACTION RATIO

No	ST : PG : FA	93% compaction ratio	96% compaction ratio
		R_{7d}/R_{28d}(Mpa)	R_{7d}/R_{28d}(MPa)
2-1	8 : 30 : 62	1.57/2.21	2.76/3.75
2-2	8 : 40 : 52	1.89/2.34	3.13/4.97
2-3	8 : 50 : 42	2.04/2.85	3.24/4.86
2-4	8 : 60 : 32	1.97/2.67	3.19/4.51

Table 3 shows that the strength of the phosphogypsum-fly ash-lime solidified material with the compactness ratio of 96% had a much higher strength than that with the compactness ratio of 93%. This strength is satisfactory when the material is used as road base material or sub-base material. If the compactness ratio of the phosphogypsum-fly ash-lime solidified material is below 93%, it can only be used as road base filling material.

3.3 PHOSPHOGYPSUM-FLY ASH-LIME SOLIDIFIED MATERIAL STABILIZED CRUSHED ROCK

The base course material of high grade road must have a great deal of aggregate as framework. we use the phosphogypsum-fly ash-lime solidified material to stabilize crushed rock. The compactness ratio was 98%. The test result of the strength is listed in table 4.

TABLE 4: COMPRESSIVE STRENGTH OF CRUSHED ROCK STABILIZED BY ST-PG-FA MATERIAL AND LIME FLY ASH

No	Mix proportion ST : FA : PG : CS	γ_{md}/(g/cm^3)	W_{pt}/ (%)	R_{7d}/ (MPa)	Cv/ (%)
3-1*	5 : 30 : 65	1.880	9.90	0.823	5.81
3-2+	5 : 30 : 65	1.885	9.90	0.995	3.88
3-3	5 : 30 : 65	1.885	9.90	1.100	7.82
3-4	5 : 10 : 20 : 65	1.887	9.50	1.200	5.38
3-5	5 : 15 : 15 : 65	1.890	9.40	1.480	8.84
3-6	5 : 20 : 10 : 65	1.891	9.30	1.680	6.85

Notes: ST-stabilizer; *: ST is slaked lime; +: ST is quicklime;
γ_{dm}-Maximum dry density Notes; W_{pt}-Optimum water content

The strength of the phosphogypsum-fly ash-lime solidified material stabilized crushed rock was about 40~100% higher than that of the lime-fly ash stabilized crushed rock. The strength of this material is about the range of the standard of "Lime-Fly ash Solidified Materials" of Ministry of Communication of China. Hence, it can be used as road base material for various grades of highway.

3.4 THE MECHANISM OF STENGTH FORMATION

When the lime was added into fly ash, the pozzolana reaction took place. The reaction can be qualitatively described as fellow [3].

$$xCa(OH)_2+SiO_2+nH_2O \rightarrow xCaO \cdot SiO_2 \cdot (n+x)H_2O$$

$$yCa(OH)_2+Al_2O_3+nH_2O \rightarrow yCaO \cdot Al_2O_3 \cdot (n+y)H_2O$$

When there is plenty of phosphogypsum ($Ca_2SO_4.2H_2O$), the product $yCaO \cdot Al_2O_3 \cdot (n+y)H_2O$ was substituted by AFt. The phosphogypsum hastened the pozzolana reaction also. So the strength of the phosphogypsum-fly ash-lime solidified material is much higher than that of the lime-fly ash road base material[4].

$$3Ca(OH)_2+Al_2O_3+3CaSO_4\cdot 2H_2O+26H_2O \rightarrow 3CaO\cdot Al_2O_3\cdot 3CaSO_4\cdot 32H_2O$$

The SEM images of the phosphogypsum-fly ash-lime solidified material show the microstructure of material and the crystal of phosphogypsum.

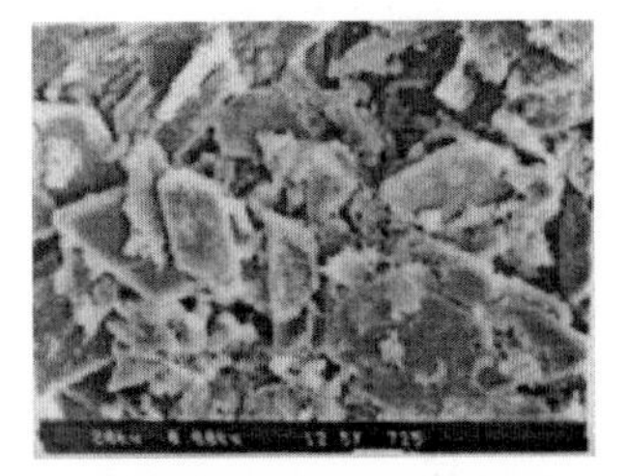

a. MICROSTRUCTURE OF MATERIAL

b. FLAKY CRYSTAL OF PHOSPHOGYPSUM

Fig 1: SEM PICTUREOF 28d PG-FA-QL BINDER

The phosphogypsum-fly ash-lime solidified material is a material with a large number of pores. The morphology of the phosphogypsum is flake-like crystal[1] while AFt is needle-like crystal[5], according to the SEM images. The fractional force between phosphogypsum grains is larger than that between fly ash grains. This also contributed to the strength formation also.

4 SUMMARY AND CONCLUSIONS

A suitable proportion for the phosphogypsum-fly ash-lime solidified material was a ratio of fly ash to phosphogypsum at 1:1 and a stabilizer content was about 8~9%.

The strength of the phosphogypsum-Fly Ash-Lime Solidified Material with a compactness ratio of 96% had a much higher strength than that with a compactness ratio of 93%. A 96% one can be used as road base material or sub-base material. Materials with a compactness ratio of 93% and below can only be used as road base filling material.

The strength of the phosphogypsum-fly ash-lime solidified material stabilized crushed rock was about 40~100% higher than that of lime-fly ash stabilized crushed rock. This strength reached the required range of lime-fly ash solidified materials of the standard of the Ministry Communication of China. Hence, it can be used as road base material for various grades of highway.

When there is sufficient plenty phosphogypsum($Ca_2SO_4.2H_2O$), the product of $yCaO\cdot Al_2O_3\cdot (n+y)H_2O$ was substituted by AFt. The phosphogypsum hastens the pozzolanic reaction also. So the strength of the phosphogypsum-fly ash-lime solidified material was much higher than the lime fly-ash road base material. The fractional force between phosphogypsum grains is also larger than that between the fly ash grains. This also contributed to the strength formation.

References

[1]Hou Changjun, Huo Danqun, ect. Performance and application of Phosphogypsum [J]. Chemistry industry and mine technology, 1997, 26 (2): 50-52

[1] Fu Buojie, Chen Liding, Yu Xiubo, ect. The new trait and countermesure of china environment[J]. Environment Science, 2000, 21 (5): 104-106

[2] Caijun Shi, Robert L. Day, etc. Chemical activation of Blended cement made with lime and natural pozzolans[J]. Cement and Concrete Research, 1993, 23(6):1389-1395

[3]Weiping Ma, Paul W. Brown, etc. Hydrothermal Reactions of fly ash with $Ca(OH)_2$ and $CaSO_4$ $2H_2O$[J]. Cement and Concrete Research, 1997, 27(8):1237-1248

[4]P.K.Mehta. Scanning elector micrographic studies of effringife formation[J]. Cement and Concrete Research, 1976, 6(2):169-181

INDEX OF CONTRIBUTORS

Volumes I and II

KEYWORD INDEX

Volumes I and II

RECEIVED
AUG 2 9 2003